International Conference on Power Electronics and Power Transmission (ICPEPT 2021)

Journal of Physics: Conference Series Volume 2108

Xi'an, China
15 - 17 October 2021

ISBN: 978-1-7138-5079-3
ISSN: 1742-6588

Printed from e-media with permission by:

Curran Associates, Inc.
57 Morehouse Lane
Red Hook, NY 12571

Some format issues inherent in the e-media version may also appear in this print version.

Institute of Physics
Dirac House, Temple Back
Bristol BS1 6BE UK

Phone: 44 1 17 929 7481
Fax: 44 1 17 920 0979

techtracking@iop.org

Additional copies of this publication are available from:

Curran Associates, Inc.
57 Morehouse Lane
Red Hook, NY 12571 USA
Phone: 845-758-0400
Fax: 845-758-2633
Email: curran@proceedings.com
Web: www.proceedings.com

TABLE OF CONTENTS

Preface

The 2021 International Conference on Power Electronics and Power Transmission (ICPEPT 2021) was held on October 15-17, 2021 in Xi'an, China. ICPEPT 2021 is to bring together innovative academics and industrial experts in the field of Power Electronics and Power Transmission to a common forum. The primary goal of the conference is to promote research and developmental activities in Power Electronics and Power Transmission and another goal is to promote scientific information interchange between researchers, developers, engineers, students, and practitioners working all around the world. The conference will be held every year to make it an ideal platform for people to share views and experiences in Power Electronics and Power Transmission and related areas.

The conference model was divided into two sessions, including oral presentations and keynote speeches. In the first part, some scholars, whose submissions were selected as the excellent papers, were given 15 minutes to perform their oral presentations one by one. Then in the second part, keynote speakers were each allocated 30-45 minutes to hold their speeches. More than 100 participants attended the meeting. There were over 20 experts and scholars in the area of Power Electronics and Power Transmission representing different famous universities and institutes around the globe to form Conference Committees.

In the keynote presentation part, we invited three professors as our keynote speakers. The first keynote speakers, Assoc. Prof.Jinsong Tao, from School of Electrical Engineering and Automation, Wuhan University, China was invited to present his talk *Operation Mode Analysis and Coordinated Control Strategy Research of Multi-terminal Network with DC Microgrid.* Assoc. Prof. Sohrab MIRSAEIDI, from Beijing Jiaotong University, China was our second keynote speakers. He presented a talk: *Improvement of Capacitor-Commutated-Converter-Based and Fault-Current-Limiting-Based Commutation Failure Prevention Approaches in HVDC Transmission Networks.* In this talk, the structure of an improved Controllable Commutation Failure Inhibitor (CCFI) is presented which obviates the main drawbacks of the existing capacitor-commutated-converter-based and fault-current-limiting-based strategies. Assoc. Prof. ANWAR ALI, from Zhejiang Sci-Tech University, China as our finale keynote speakers. He delivered a speech: *Design and Development of a Small Satellite Subsystems-AraMiS-C1.*

We are glad to share with you that we received lots of submissions from the conference and we selected a bunch of high-quality papers and compiled them into the proceedings after rigorously reviewed them. These papers feature following topics but are not limited to: 1. Power Electronic Technology. 2. Electric Power System. All the papers have been through rigorous review and process to meet the requirements of International publication standard.

We are really grateful to the International/ National advisory committee, keynote speakers, session chairs, organizing committee members, student volunteers and administrative assistance of the management section of University, including accounts section, digital media and publication house. Also, we are thankful to all the authors for contributing a large number of papers in the conference, because of which the conference became a story of success. It was the quality of their presentations and their passion to communicate with the other participants that really make this conference series a great success.

The Committee of ICPEPT 2021

Journal of Physics: Conference Series **2108** (2021) 011001 doi:10.1088/1742-6596/2108/1/011001

Committee member

Conference Chairman
Assoc. Prof. Jinsong Tao, Wuhan University, China
Assoc. Prof. ANWAR ALI, Zhejiang Sci-Tech University, China
Assoc. Prof. Sohrab MIRSAEIDI, Beijing Jiaotong University, China

Program Committees
Dr. TASHER ALI SHEIKH, Mie University, India
Dr. Arockia Selvakumar. A, VIT University, India
Dr. P. PANDIYAN, National Institute of Technology, Trichy
Dr. KULWANT SINGH, Kurukshetra University, Kurukshetra
Dr. BADRUL HISHAM, UNIVERSITI TEKNIKAL MALAYSIA MELAKA, India
Assoc. Prof. Parneeta Dhaliwal, I.K. Gujral Punjab Technical University, Punjab
Assoc. Prof. Saman Abdanan Mehdizadeh, Agricultural Sciences and Natural Resources University, Iran

International Technical Program Committees
Dr. Vineet Dahiya, K R Mangalam University, India
Dr. UDIT MAMODIYA, Poornima University, Jaipur, India
Dr. Khalid Hamid, National Chiao Tung University, China
Dr. V. VIJAYA KISHORE, Ravindra College of Engineering for Women, Kurnool, India

Peer review declaration

All conference organisers/editors are required to declare details about their peer review. Therefore, please provide the following information:

- **Type of peer review: Double-blind**

- **Conference submission management system: AI Scholar System**

- **Number of submissions received: 189**

- **Number of submissions sent for review: 181**

- **Number of submissions accepted: 103**

- **Acceptance Rate (Number of Submissions Accepted / Number of Submissions Received X 100): 54.5%**

- **Average number of reviews per paper: 2**

- **Total number of reviewers involved: 60**

- **Any additional info on review process:**
- **Step 1. Each of selected paper will be reviewed by two/three professional experts in the related subject area.**
- **Step 2. Review Reports received from the experts will be judged by one of the editors either Review Reports are logical or not?**
- **Step 3. If not logical, then editor can assign new reviewer or can also judge at his/her own.**
- **Step 4. If logical, then Review Reports will be sent to authors to modify the manuscript accordingly.**
- **Step 5. Authors will be required to revise their papers according to the points raised.**
- **Step 6. Revised version will then be evaluated by the editor for the incorporation of the points raised by the reviewers.**
- **Step 7. Then the editor will send the revised manuscript to the reviewers again for re-evaluation.**
- **Step 8. If the reviewers approve the revise version of the manuscript, then will be accepted for publication.**

- **Contact person for queries:**
Xuexia Ye
publication@keoaeic.org
AEIC Academic Exchange Information Centre

Please submit this form along with the rest of your files on the submission date written in your publishing agreement.

The information you provide will be published as part of your proceedings.

Study on Key Technologies of Power Dispatching Control System under the Background of Cloud Computing

Siming Zhou [1], Guanqiang Liu [2], Hai Qin [1], Tian Xia [1]

[1]Power dispatching control center of Guizhou Power Grid Co., Ltd., Guiyang 550007, Guizhou

[2]NARI-TECH Nanjing Control Systems Ltd., Nanjing, Jiangsu 211106

Abstract. In order to adapt to the development and application of power dispatch control systems in the future cloud computing environment, this study proposes to make full use of the elastic characteristics of cloud computing to construct a new power dispatch control system architecture. First, we analyze the shortcomings of the traditional power dispatch control system architecture, from the perspective of architecture complexity, operation and maintenance, universality and disaster tolerance. Secondly, combining the characteristics of the existing dispatch control system, we propose a new power dispatch control system architecture that conforms to the cloud computing platform. Finally, we emphatically analyzed the key technologies for key breakthroughs needed to realize the power dispatching control system architecture. Through the comparison of traditional and new power dispatch control system architectures, it can be considered that the new power dispatching control system architecture has greatly improved reliability, flexibility, safety and economy, and has certain advantages.

1. Introduction

With the development of cloud computing in recent years, domestic Internet companies and hardware equipment manufacturers have begun to get involved in the field of cloud computing. At present, the application of cloud computing in the power system is mainly concentrated in equipment virtualization and distributed storage for the purpose of big data analysis. There are only a few theoretical articles on the application of cloud computing in power dispatch control systems, and there are no practical applications in China. Wang D et al. proposed a new power grid power dispatching control system architecture based on cloud computing technology, and gave the actual application plan [1]. Yang SC et al. proposed a new architecture and key technical support for automatic intelligent dispatching of power grids based on situational awareness [2]. Through an in-depth analysis of the basic concepts and core features of cloud computing, Cao Yang pointed out that virtualization and distribution are the core technologies of cloud computing, and studied the advantages and disadvantages of cloud computing technology in the application of power dispatching systems [3]. Most foreign research on the application of cloud computing technology focuses on big data analysis, and the application of cloud computing in power dispatch control systems has not been reported [4].

Under the background of the development trend of cloud computing technology, this paper focuses on how the existing power dispatching control system can adapt to the future cloud computing environment, and tries to establish an application support framework and data processing mode suitable for cloud computing technology in the system by changing the mismatched architecture and design mode in the existing system, in order to give full play to the advantages of cloud platform elastic computing and storage, eliminate the real-time processing performance bottleneck in the system, and improve the system's data access, storage, analysis, integration and expansion capabilities accordingly.

2. Drawbacks of Traditional Power Dispatching Control System

In the architecture of traditional power dispatching control system, a fixed topology is usually used, that is, each server runs a fixed module to achieve specific functions, and by the combination of multiple servers and multiple modules to realize the complete function of the power dispatching control system [5].

The architecture of traditional power dispatching control system is shown in Figure 1.

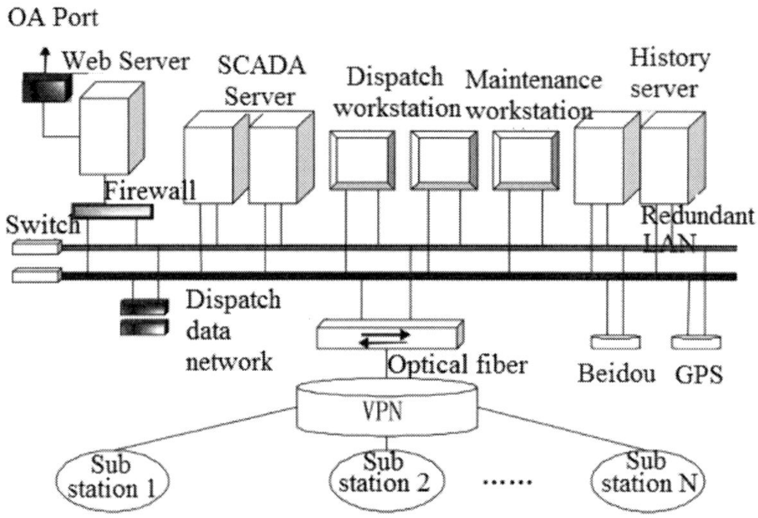

Figure 1. Architecture of traditional power dispatching control system

Such architectural model has shown multiple drawbacks as follow:

a) Complex architecture and high coupling. In order to ensure the reliability of the dispatching control system, the traditional architecture provides dual-unit backup guarantee for some core modules, so that when one of the nodes fails, the backup node will take over the original work [6]. The application of this technology makes the system architecture extremely large, and it is difficult to increase redundant nodes. Moreover, the equipment installation and configuration process would be very complicated, and the components need to be highly coupled and compatible.

b) Difficult to maintain. Although the main structure has a dual-unit backup guarantee, there are still a large number of single points of failure in the remaining parts, and once a single point of failure is encountered, the related investigation work will inevitably affect the complete operation of the system. In addition, in the traditional case, the deployment of dispatching control system operation and maintenance personnel is determined according to the number of system equipment, so with the development of smart grid, information resources will increase rapidly, and the corresponding management workload will also increase sharply, which means more manpower consumption and greater maintenance costs [7].

c) Poor universality. Different manufacturers produce their own functional systems, and different functional systems use different hardware structures, which requires special maintenance means or special technicians, so it is difficult to achieve unified specification.

d) Poor disaster tolerance. As the nerve center of smart grid, once the dispatching control system fails, local power grid events may quickly expand to the whole power grid, and even cause power grid collapse [8]. Therefore, it requires the dispatching control system to have better disaster tolerance. However, under the traditional architecture, due to the non-uniform application standards and strict requirements for the operating environment, it is difficult to achieve large-scale and most application disaster backup. This means that each server needs the same server as a disaster backup. Obviously, the cost is too high. Therefore, it is difficult to have a general disaster recovery scheme under the traditional architecture.

3. New Power Dispatching Control System Based on Cloud Computing

3.1 Cloud Computing Model

Power dispatching control problems in multiprocessors and distributed systems, have always been the research focus of computer science. At this stage, with the development of cloud computing, task scheduling under the cloud platform has also been favoured by many scholars, and more and more researchers have invested in the task scheduling research under cloud computing. Cloud computing power dispatching control can be understood as a mapping process. Through virtualization technology, the software and hardware facilities of the cloud platform are abstracted as resources, and the user's request is regarded as a task. The role of power dispatching control is to allocate necessary and appropriate cloud resources to perform tasks for users' submitted tasks based on the status and prediction information of both resources and tasks under certain constraints [9]. After the task is completed, the processing result is returned and resources are released. Therefore, a good power dispatching control mechanism can coordinate and allocate resources in the cloud environment, effectively improve system performance, shorten task running time, and reduce resource consumption.

There are two basic starting points for cloud computing power dispatching control: on the one hand, transparently allocate resources for tasks submitted by users; on the other hand, through a certain algorithm scheduling, the resources in the cloud and the tasks submitted by the user are matched with the maximum utilization rate, the utilization efficiency of the resources in the cloud is improved, and the user is provided with better and more efficient services. Cloud computing is the inheritance and development of distributed computing and grid computing, so it has a larger scale and complexity, and the resources in the cloud environment are heterogeneous, dynamic, distributed, and autonomous. Therefore, for cloud platforms, a good and efficient scheduling strategy is essential.

The cloud computing power dispatching control is very similar to the dispatching control process of grid computing. Cloud computing power dispatching control model is shown in Figure 2.

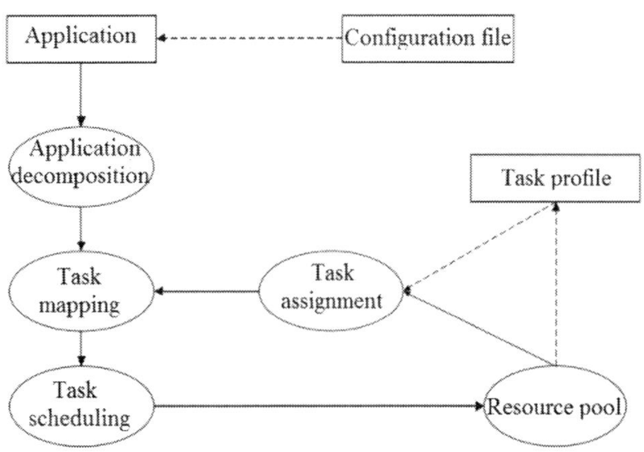

Figure 2. Power Dispatching Control Model Based on Cloud Computing

In general, the execution process of cloud computing power dispatching control is as follows [10]:

a) Decompose the application according to the application configuration file;

b) Resource selection based on resource pool and resource pool configuration file;

c) Perform task mapping and calculate the resources required by the task;

d) Assign resources to the tasks submitted by users based on resource configuration files and resource pool usage.

Task scheduling problem is an NP-hard problem, because resources and tasks are dynamically changing, which is not conducive to task scheduling [11]. Therefore, the scheduling algorithm can only find a relatively optimal solution to the task scheduling problem, and the scheduling service quality QoS is used as the evaluation standard.

The objectives of cloud computing power dispatching control include the following aspects

a) Shortest span. The span refers to the completion time of the task, that is, the time it takes from the cloud computing system to execute the first task submitted by the user to the end of the last task. The length of the span directly affects the user experience. The shorter the span, the less time it takes, and the less the user needs to pay, and this is what the user expects.

b) Quality of service (QoS). Service quality refers to the pros and cons of the performance of the system as reflected by the cloud computing platform when providing services. The level of QoS directly determines whether users will continue to use the cloud computing platform. QoS in the cloud platform is mainly reflected in the performance of task scheduling, that is, the allocation of virtual resources [12]. In a heterogeneous cloud computing platform, how to efficiently schedule virtual resources is the key to improving service quality.

c) Load balancing. Load balancing is a key indicator for measuring system resource scheduling in cloud computing, which refers to the balanced use of virtual resources under the cloud computing platform to avoid the situation where some nodes are overloaded and some nodes have no tasks to do. In order to achieve this effect, it is necessary to use a reasonable task scheduling strategy to balance the relationship between resources and tasks while making full and reasonable use of virtual resources.

d) Economic benefit. Cloud computing is a pay-as-you-go service, so for service providers and users, cost is a factor that needs to be balanced [13]. For users, reducing service fees is what they want; while for service providers, it is desirable to reduce costs while ensuring overall revenue.

3.2 System Architecture

In order to solve many problems existing in the traditional dispatching control system, this study applies cloud computing technology to the architecture design of dispatching control system. The whole system adopts network architecture, as shown in Figure 3.

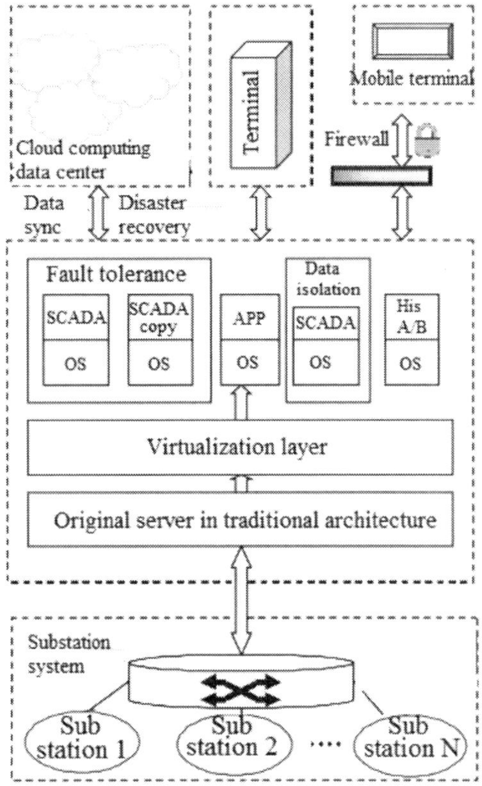

Figure 3. Achiteture of new power dispatching control system

Compare the traditional and new architectures from the overall architecture of the dispatching control system, information processing subsystem, information transmission subsystem, human-computer interaction subsystem, etc., as shown in Table 1.

Table 1 Comparison of new and traditional power dispatching control system

Item	Traditional power dispatching control system	New power dispatching control system
Overall architecture	Localized resource	Networked resource
Information processing subsystem	Core server dual unit standby; physical switching after failure; can withstand the abnormality of a unit; no remote disaster recovery	All servers are standby for each other; synchronous memory switching after failure; able to withstand n-1 server exceptions; multiple sites are disaster recovery sites for each other
Information transmission subsystem	Traditional special telecontrol channel; point to point	Dispatching virtual private data network; "point to multipoint" network structure
Human-computer interaction subsystem	Web services; dispatching workstation; management workstation	Web services; virtual workbench; authorized mobile terminal

The new dispatching control system uses virtualization technology to transform the traditional servers into cloud computing data center for unified management

a) Realize the decoupling of hardware and software is realized. Each module of the dispatching control system no longer depends on a specific server, but the operation resources are specified by the cluster control center in the form of virtual machine [14]. Similarly, any changes to the hardware will not affect the operation of virtual machines in the resource pool.

b) Realize the virtual machine system managed in the form of files. After the modules of the system become files, it means that they can be managed in the form of files, move, modify and copy on the resource pool, and achieve high-level reliability guarantee with the cooperation of certain monitoring means.

c) Realize the centralized management of multiple servers. At this time, the management unit is the resource pool [15]. Through the management center, the operation status of all servers and virtual machines can be monitored, and operation instructions can be issued to any link, greatly simplifying the maintenance difficulty of the system.

4. Key Technologies of Cloud Computing Power Dispatching Control System

4.1 Repackaging of Dispatching Master Station System

The dispatching center and each sub plant station use the data dedicated line service provided by the communication operator to access the communication transmission network through leased optical channel, and build VPN on this basis to establish the power dispatching data network. The SCADA server of the dispatching center and the communication integration server of each substation are a node in the network, and are connected according to "point to multipoint" to realize dispatching communication through VPN.

The dispatching master station system is repackaged, the server and network equipment in the original dispatching control system architecture are fully utilized, and it is integrated into a resource pool by virtualization technology. Each original server becomes an equal member of the pool, has the same topology connection mode and the same management configuration mode, and has a comprehensive dual network redundancy structure at the physical level. Each module of the master station system is encapsulated in the form of virtual machine files, and its specific operation resources are allocated by the resource pool control center. The communication between each module of the system is completed by the virtual logic network.

4.2 Construction of Dispatching Control Data Center

At the physical level, the power dispatching control system communicates with the outside world through a unified outlet to form a local cloud computing data center. At the same time, the construction method of cloud computing data center can be copied in different environments, integrate dispatching automation resources in multiple places, and build multiple remote cloud computing data centers. On this basis, the interconnection of dispatching control systems in different regions is constructed to make the dispatching resources call across regions, realize the application level disaster recovery mechanism, and form a strong reliability guarantee. Multiple cloud computing data local dispatching master station systems reach a cooperative relationship with remote cloud computing data centers to realize mutual disaster recovery support. Under normal conditions, the local master station system completes the synchronization of virtual machine files with the remote cloud computing data center regularly, and establishes a resource mapping mechanism to ensure normal network communication after system migration. When an exception occurs, because the remote end has completed the synchronization of virtual machine files, it can quickly organize relevant resources according to needs, execute virtual machine files, and finally realize the overall remote restart of the system. The two sides that backup each other shall establish a detection mechanism with both automatic and manual modes, prompt when the other party's fault is detected, and carry out disaster recovery with the consent of the administrator, so as to avoid the failure of heartbeat detection caused by long-distance network transmission.

4.3 Construction of Dispatching Terminal System

The dispatching terminal system runs multiple scheduling applications on the hosted virtual machine in the form of creating programs running sandbox, and sends them to different terminals through the network. There is only one source program in the terminal system. In case of multiple users, multiple memory copies will be copied to ensure the consistency of the scheduling operation platform used by different users. In order to ensure system reliability, the new dispatching control system architecture replaces the standby mode of key modules in the traditional architecture by running memory redundancy. In the process of function upgrading, all users can upgrade synchronously only by changing the source program file. In addition, the man-machine interface of network transmission can be transmitted not only to fixed workstations, but also to on-site mobile devices through 4G network, making the dispatching work more intuitive and efficient.

In order to ensure system reliability, the new dispatching control system architecture replaces the standby mode of key modules in the traditional architecture by running memory redundancy. Under normal operation, the system assigns two different servers to run the same virtual machine file, generates two copies of the same memory data, and performs the same input and output work to the two copies of memory data at the same time. Under abnormal operation, if the main virtual machine or host server fails, the standby server will take over the original work within a few milliseconds to ensure the continuous operation of the virtual machine. At the same time, the data center will select another standby server and generate standby memory data again.

4.4 Data Transmission and Secure Storage

Through the analysis of the storage technology of power dispatch control data, including data separation, data recovery, data storage, data backup and other aspects. In the cloud computing environment, electric power companies can select different storage locations for corresponding data information through important coefficients and basic performance through the highly centralized storage technology of private cloud, so that different types of data can be isolated from each other, so as to effectively avoid data loss and leakage problems. The use of cloud computing can back up the information of power companies in real time, enabling power companies to restore data in a short period of time when emergencies occur.

5. Conclusion

The widespread adoption of cloud computing technology will bring a brand new revolution to the power dispatch control system. In order to adapt to the development and application of power dispatch control systems in the future cloud computing environment, this study proposes to make full use of the elastic characteristics of cloud computing to construct a new power dispatch control system architecture. First, we analyze the shortcomings of the traditional power dispatch control system architecture, from the perspective of architecture complexity, operation and maintenance, universality and disaster tolerance. Secondly, combining the characteristics of the existing dispatch control system, we propose a new power dispatch control system architecture that conforms to the cloud computing platform. Finally, we emphatically analyzed the key technologies for key breakthroughs needed to realize the power dispatching control system architecture.

References

[1] Wang Ding, Qian Kejun, Gao Yidan, et al. Cloud computing platform technology and its application in power grid dispatching[J].Power System and Clean Energy, 2015, 31(4): 72.
[2] Yang Shengchun, Tang Biqiang, Yao Jianguo, et al. Automatic intelligent dispatching architecture and key technologies of power grid based on situation awareness[J].Power System Technology, 2014, 38(1): 33-39.
[3] Cao Yang, Gao Zhiyuan, Yang Shengchun, et al. Application of Cloud Computing Mode in Power Dispatching System[J].China Electric Power, 2012, 45(6): 19-22.

[4] Ren-Bo W U , Quan-Xi Y U , Wang J , et al. Research on Key Technology of Power Dispatching Control System Based on Cloud Computing[J]. Techniques of Automation and Applications, 2019, 23(2): 56.
[5] Xiao X . Research on Integrated Construction Model of Main Distribution Network of Power Dispatching Control System[J]. Telecom Power Technology, 2019, 18(3): 101-105.
[6] Chen Jiaqian, Qian Haifeng, Lu Bing, et al. Design of Human-Machine Interface of City-County Electric Power Dispatching and Control Integration System[J]. Zhejiang electric power, 2013, 032(005):65-68.
[7] Chen A , Xie D , Yu S , et al. Comprehensive evaluation index based on droop control for DC distribution system dispatching[J]. International Journal of Electrical Power & Energy Systems, 2019, 106:528-537.
[8] Xing Y , Lang Y , Li Q , et al. Discussion on pattern of centralized operation and maintenance for multilevel smart power dispatching control system[J]. Power System Protection and Control, 2018, 46(15):142-148.
[9] Lin J . Cloud Computing Based System Design for Power Grid Dispatching and Control Training Simulation[J]. Automation of Electric Power Systems, 2017, 41(14):164-170.
[10] Wang J , Wang L , Dong G , et al. Research and Implementation of The Fault Information System FOR Relay Protection Based on Cloud Computing Technology[J]. Journal of Physics Conference Series, 2020, 1650:032003.
[11] Huang Y, Wang X, Liu X. Research on the Information Security Status Quo of the Application of Cloud Computing in Power Dispatching Automation System[J]. Information and computer, 2017, 000(021):188-189.
[12] Yuan B, Xu Y. Construction analysis of cloud computing technology in new power dispatching automation system [J]. Automation application, 2019 (01): 120-121.
[13] Sun Ke. Application of cloud computing model in power dispatching system [J]. Low carbon world, 2017 (16): 57-58.
[14] Chen Yong. Application of cloud computing mode in power dispatching system [J]. Introduction to scientific and technological innovation, 2016,13 (31): 75-76.
[15] Pan Yuanli, Li Quan. Architecture analysis of power dispatching automation system based on cloud computing technology [J]. Hebei electric power technology, 2016,35 (01): 4-7.

Electromechanical Transient Modeling and LVRT Parameter Identification of Large Capacity Full Power Converter Wind Turbines Based on PSASP Program

Guoqiang Lu [1a*], Xiangyu Tao[2b], Chunmeng Chen [1c], Jiatian Gan[1d], Xufeng Zhao[2e]

[1] Electric Power Research Institute of State Grid Qinghai Electric Power Company, Xining, China

[2] China electric power research institute, Beijing, China

[a*] lgqiang2015@163. com , [b] txyepri@163. com , [c] chenchunmeng@163. com,

[d] dianaellena@126. com, [e] zhaoxufeng@epri.sgcc.com.cn

Abstract—Full power converter wind turbine is the main type of wind power, so the simulation calculation needs to establish accurate model parameters. This paper analyzes the model structure of PSASP program according to its low voltage ride through control and physical characteristics, and puts forward the parameter identification method of LVRT characteristics of full power converter wind turbine, and to use the LVRT data of 5. 5MW unit for parameter identification and simulation verification. This paper proposes that the electromechanical transient simulation can ignore the part of the generator model of the full power converter wind turbine, and simulates the grid side converter with the controlled current source. The main characteristics of LVRT are determined by the control system of the converter. In order to do the parameter identification, it is necessary to calculate and analyze the control characteristics of multiple measured data. First, determine the control mode, then determine the control parameters to complete the parameter identification. In this paper, the modeling conditions and model structure of the full power converter wind turbine are confirmed. The correlation between the parameters during the LVRT fault and the parameters during the LVRT recovery period and the LVRT characteristics is analyzed. In this paper, a parameter identification method is proposed to analyze the active current and reactive current during the LVRT fault, which has strong physical significance and operability. Based on the actual LVRT characteristics of 5. 5MW wind turbine, the parameter identification and simulation are carried out to verify the correctness of the method.

1. Introduction

Wind power in China has developed rapidly in recent years. At present, it is mainly concentrated in the three north areas. A large number of wind turbines connected to the power grid will have an all-round impact on the power grid, including overvoltage, low voltage, power angle stability, dynamic stability and subsynchronous oscillation[1-10]. Electromechanical transient analysis is still the main technical means of power grid simulation. Large scale full power wind power access brings stability problems. Firstly, electromechanical transient analysis needs to be carried out. After finding the problems, electromagnetic transient analysis can be carried out according to the situation. Therefore, the electromechanical model parameters of wind turbine, especially the low voltage ride through model parameters, are the basic conditions for accurate simulation. According to the current development trend, full power converter wind turbine is mainly used for full power wind power. For accurate simulation, it is necessary to identify the model parameters of typical high-capacity full power converter, especially the low-voltage ride through parameters.

Over the years, a lot of research has been carried out on the modeling and parameter identification of full power converter wind turbine and full power wind power at home and abroad[11-19], many test and identification methods are proposed. Based on the general model structure of direct drive permanent magnet wind turbine, document [14] proposes a method using particle swarm optimization algorithm for parameter identification, and obtains parameters with good adaptability, but it does not analyze multiple faults and is not related to the actual physical process. Reference [15] proposed a method to identify the model parameters of doubly fed induction generator based on short-circuit current, which can better identify some doubly fed machine parameters, but did not identify the LVRT model parameters. Reference [16] proposed a new method for decoupling identification of grid side control parameters of doubly fed wind turbine based on frequency domain method. Reference [17] proposed a frequency domain identification method for rotor side controller of doubly fed wind turbine by adding pseudo-random signal. However, in general, these documents mainly model and identify the parameters of the wind turbine, and do not measure and model the most important low-voltage ride through characteristics of the wind turbine. Therefore, their results can not solve the most urgent needs of simulation calculation.

Starting with the physical structure and general control logic of full power converter wind turbine, this paper analyzes the basic structure of full power converter wind turbine model in PSASP program. Finally, combined with the actual characteristics of LVRT, the simulation model parameters are identified, and the simulation measurement is checked, and more accurate results are obtained.

2. Control model of wind turbine with full power converter

The most considered faults in electromechanical transient simulation are various types of short-circuit faults, which are low-voltage ride through faults of various types and amplitudes for wind turbines. Therefore, in this sense, the accurate simulation of low-voltage ride through control is the most concerned in electromechanical transient simulation.

According to the above and PSASP instructions, the control logic of the grid side converter in electromechanical transient simulation is shown in Figure 5.

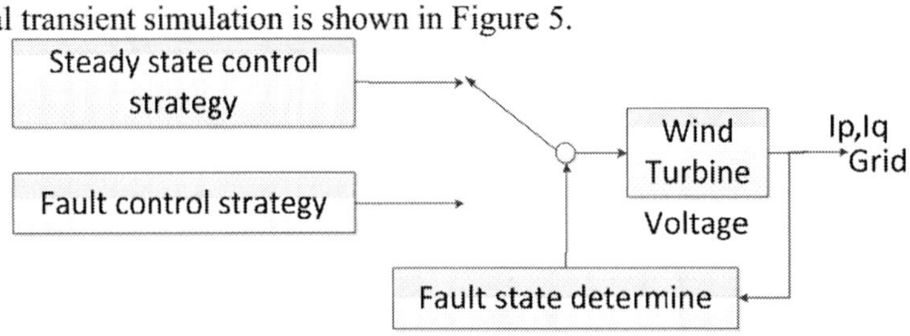

Fig. 1 Overall diagram of converter control model

1) The active power control model is shown in Figure 2. In the figure, $U_{DC,ref}$ is the DC voltage reference value, U_{DC} is the DC voltage feedback value, T_1 is the DC voltage measurement time constant, K_{PUDC} , K_{IUDC} are the PI control parameter of the voltage outer loop, $I_{d,\max}$, $I_{d,\min}$ are the maximum and minimum limit amplitudes of the d-axis current, $I_{d,\mathrm{grid,ref}}$ is the d-axis current command of the grid side current source. It can be seen from Figure 6 that the active power control of the grid side converter is not directly based on the active power command, but by maintaining the stability of the DC bus voltage. At the same time, it also controls the DC bus voltage to avoid deviation under normal operation;

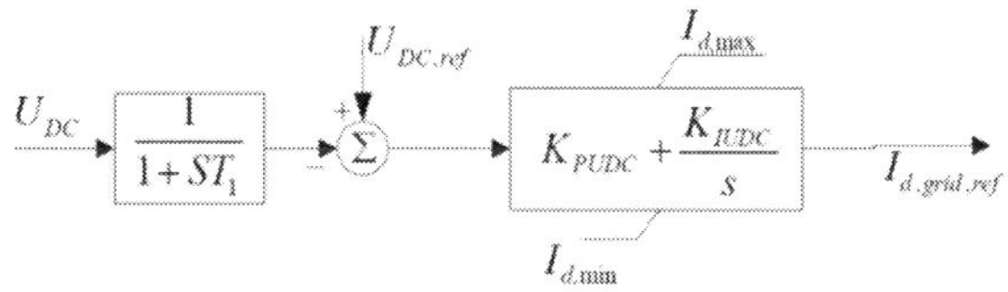

Fig. 2 Active Power Control under Normal Conditions

2) The reactive power control model is shown in Figure 3, which is divided into two control modes. One is the terminal voltage control mode in the upper part, which calculates the reactive power setting value according to the deviation between the terminal voltage setting V_{ref} and the terminal voltage feedback V_{reg} ; There is also a more commonly used power factor control mode, as shown in the following half. According to the active power P_{gen} and power factor PF_{ref} , the given value of reactive power under the current power is calculated. Further, the given value of reactive power is converted into reactive current value to control the q-axis current of grid side converter.

It is noted that the main steel beams of three specimens are supported at the support beam as shown in Fig. 1. Each secondary beam (steel channel)

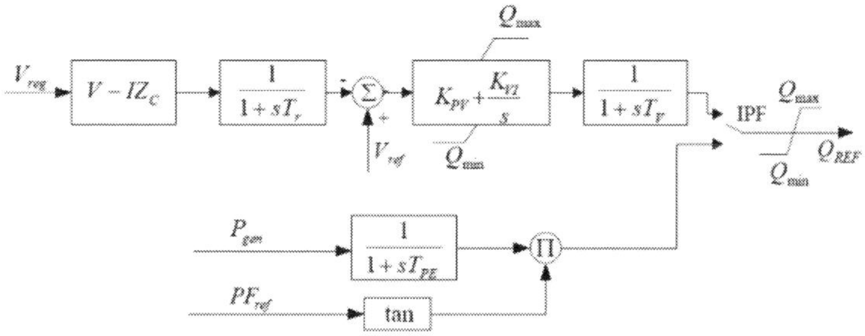

Fig. 3 Reactive Power Control under Normal Conditions

The control model under LVRT state is generally controlled according to two or three stages: fault period, fault removal time and fault recovery stage. The specific time period is divided as follows:

1) During failure. The active current control is controlled according to the formula $I_{P\_LVRT} = K_{1\_LP\_LV}V_t + K_{2\_LP\_LV}I_{P0} + I_{PSET\_LV}$, where I_{P_LVRT} is the given value of active current during LVRT fault, $K_{1\_LP\_LV}, K_{2\_LP\_LV}, I_{PSET\_LV}$ are relevant constants, V_t is the terminal voltage during fault, and I_{P0} is the initial active current value before fault; The reactive current is controlled according to the $I_{Q\_LVRT} = K_{1\_LQ\_LV}(VL_{IN} - V_t) + K_{2\_LQ\_LV}I_{Q0} + I_{QSET\_LV}$, where I_{Q_LVRT} is the given value of the active current during the LVRT fault, $K_{1\_LQ\_LV}, K_{2\_LQ\_LV}$, I_{QSET_LV} are the relevant constant, V_t is the terminal voltage during the fault, V_{IN} is the threshold voltage of the LVRT, and I_{Q0} is the initial active current value before the fault;

2) Failure recovery phase. During this period, the general active current recovers according to a certain slope; The reactive current is recovered according to the immediate recovery or exponential curve.

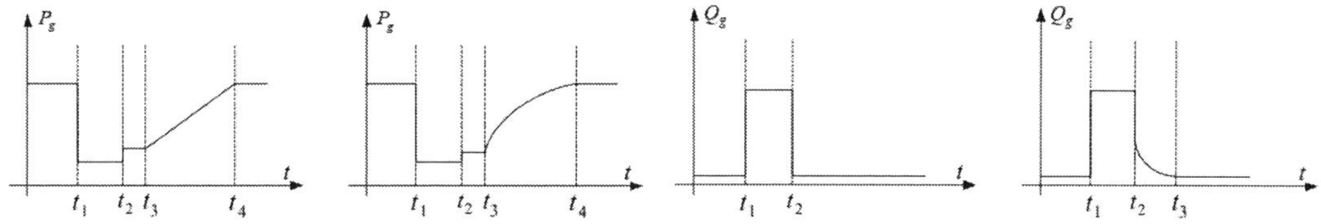

(a) Active power control under LVRT (b) Reactive power control under LVRT

Fig. 4 Active and reactive power control under low voltage ride through

3. Parameter identification of large capacity full power converter wind turbine

According to the above, the model parameters of wind turbine with full power converter mainly include two aspects:

1) Control parameters of active and reactive power in normal state; Normal state mainly refers to the operation state of voltage and current disturbance corresponding to the terminal voltage between 0. 9-1. 1pu during non LVRT and HVRT. Due to the principle of PQ decoupling control characteristics of wind turbine, this part of characteristics is generally difficult to be obtained by field test.

2) Control parameters of active and reactive power during LVRT. Because the most common faults in power grid are various short-circuit faults, the low-voltage ride through characteristics of wind turbine are the focus of simulation model.

According to the above analysis, the basic principles of parameter identification for large capacity full power converter wind turbine are as follows:

(1) Identification basis and method of control parameters of active and reactive power in normal state: it is difficult to carry out the measurement of these parameters because the wind turbine is normal. Therefore, it mainly refers to the design parameters provided by the manufacturer or the typical parameters provided by the simulation program;

(2) Identification basis of active and reactive power control parameters during LVRT: at present, the LVRT characteristics of wind turbine are mainly carried out by means of voltage drop on the primary side of the generator end with special equipment. This method has high requirements on equipment and high test cost, which is difficult to carry out in a large area. Another method is to connect the relevant control board to the digital analog hybrid simulation platform from the field, test the LVRT control characteristics of the corresponding control board through the test on the digital analog hybrid simulation platform, and carry out parameter identification on this basis. Therefore, the active and reactive power characteristics during LVRT are mainly identified according to the actual characteristics obtained by these two ways.

(3) Identification method of active and reactive power control parameters during LVRT: combined with the active power control structure introduced in the first part of this paper, the measured LVRT process is divided into two parts: fault period and fault recovery. The power and voltage in the two processes are used to convert the current, calculate the unit value of the corresponding active current and reactive current and the corresponding current limit value, and take values based on the calculation results of multiple faults to form stable calculation parameters.

Compared with the doubly fed wind turbine, the large-capacity full power converter wind turbine has been simplified more, but there are still many model parameters, and the parameters are interrelated, so the identification work must be carried out carefully. Combined with the measured characteristics, even if only the LVRT characteristics are considered, there are still up to 16 or even 20 tests to be compared according to the requirements of relevant standards. Each test includes two groups of active and reactive curves, and a total of 32 groups of curves are required to be compared. Finally, up to 32 groups of curves need to be accurately simulated with a set of parameters.

Therefore, the basic flow of wind turbine parameters for large capacity full power converter is as follows:

1) Determine the basic control mode based on the manufacturer's design data and the communication with the manufacturer on the normal / LVRT control strategy;

2) According to the measured characteristics of LVRT, the key control parameters are calculated and analyzed. If the parameters under different working conditions are inconsistent, it is necessary to further verify with the manufacturer and consider the consideration of parameter value;

3) Combined with the measured characteristics of LVRT, the simulation results of model parameters are checked and improved to ensure the accuracy of simulation results.

The standard LVRT characteristics under 16 working conditions are shown in Table 1.

Table1 16 Typical fault conditions

No	Voltage dip pu.	Fault time ms	Fault type	Power level
1	0. 20	625	3phase/2phase	Full power
2	0. 35	920	3phase/2phase	Full power
3	0. 50	1214	3phase/2phase	Full power
4	0. 75	1705	3phase/2phase	Full power
5	0. 20	625	3phase/2phase	Small power
6	0. 35	920	3phase/2phase	Small power
7	0. 50	1214	3phase/2phase	Small power
8	0. 75	1705	3phase/2phase	Small power

The measured active power characteristics of the three-phase symmetrical fault of the wind turbine are shown in Fig.5 and Fig.6.

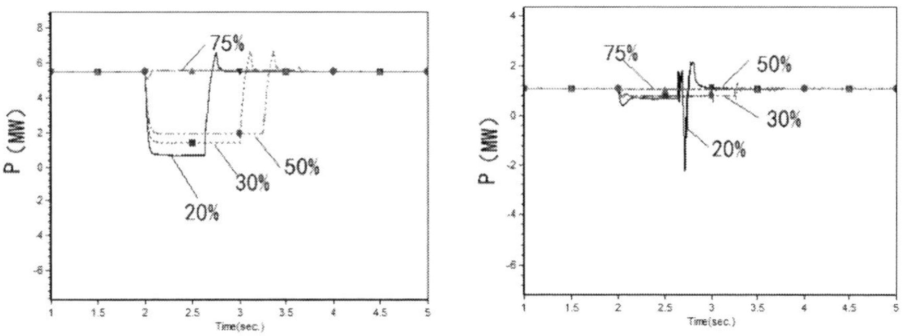

(a) active power

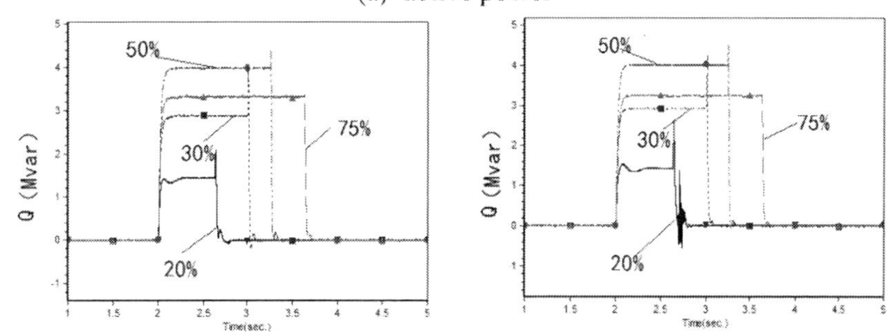

(b) reactive power

Fig. 5 100% and 20% power three-phase fault, LVRT characteristics

Identify the model parameters of the process. According to the identification principles in the third part of this paper, first determine the active control strategy, and calculate the active and reactive current during the LVRT process, as shown in Table 3.

Table2 Analysis of active and reactive power characteristics of symmetrical fault

No	Voltage dip pu.	IP. pu.	IQ. pu.	I. pu.	Power level
1	0. 20	0. 7	1. 43	1. 59	Full power
2	0. 35	0. 7	1. 45	1. 61	Full power
3	0. 50	0. 7	1. 45	1. 61	Full power
4	0. 75	1. 34	0. 79	1. 55	Full power
5	0. 20	0. 7	1. 43	1. 59	Small power
6	0. 35	0. 4	1. 45	1. 51	Small power
7	0. 50	0. 29	1. 44	1. 47	Small power
8	0. 75	0. 26	0. 77	0. 81	Small power

The data in Table 2 are analyzed as follows:

1) The active current is not a fixed value in the process of LVRT, so the power control mode should be selected for the active current in the process of LVRT;

2) The maximum unit value of reactive current is about 1.45, and the maximum value of total current is about 1.61. It is determined that the active and reactive current control in the process of LVRT is reactive priority, that is, the current capacity of the converter is allocated according to the reactive demand first, and the remaining capacity is allocated to the active power control;

3) To sum up, it is determined that reactive power is preferred in the process of LVRT, active power is constant power control, reactive current limit value is 1. 45, total current limit value is 1.61, and the corresponding calculated active current limit value is 0.7.

Determine the reactive power control strategy, and calculate the reactive power characteristics in the process of symmetrical fault, as shown in Table 3:

Table3 Analysis of reactive power characteristics of symmetrical fault

No	Voltage dip pu.	0. 9-V_fault pu.	IQ. pu.	Kq pu.	Power level
1	0. 20	0. 7	1. 43	1. 99	Full power
2	0. 35	0. 55	1. 45	2. 69	Full power
3	0. 50	0. 4	1. 45	3. 59	Full power
4	0. 75	0. 15	0. 79	5. 28	Full power
5	0. 20	0. 7	1. 43	1. 99	Small power
6	0. 35	0. 55	1. 45	2. 7	Small power
7	0. 50	0. 4	1. 44	3. 6	Small power
8	0. 75	0. 15	0. 77	5. 31	Small power

The data in Table 4 are analyzed as follows:

1) The maximum limit value of reactive current is 1. 45;

2) The control coefficient of reactive current for different faults is inconsistent, and the maximum value is 5. 31. Considering that for other faults, 5.31 multiplied by 0.9-voltage in column 3 in the table will exceed the reactive current limit value of 1.45, which is still consistent with the actual situation, so the reactive current control coefficient can be taken as 5.31.

To sum up, the identification results of active and reactive power control under symmetrical fault are summarized as follows:

1) The active power is constant power control, and the limit value of active current needs to be further determined in combination with asymmetric fault;

2) The reactive current control coefficient is 5.31 and the current limit value is 1.45.

Using the above parameters for simulation verification, the comparison curve between the three-phase fault simulation results and the measured results is shown in Figure 6, which is in good agreement.

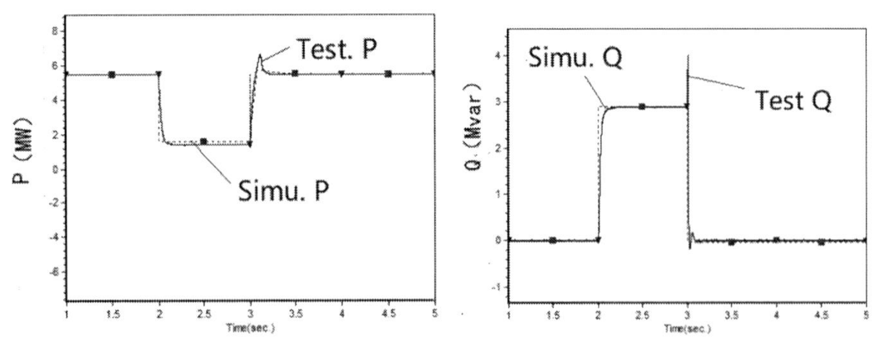

Fig. 6 Active and reactive power simulation vs. measurement

The identification results of active and reactive power control under symmetrical fault and asymmetric fault are summarized as follows:

1) The active power is controlled by constant power, the limit value of active current is 0.7 for symmetrical fault and 1.26 for asymmetric fault, and the value is 1.26 after comprehensive consideration;

2) The reactive current control coefficient is 5.31 and the current limit value is 1.45.

3) During LVRT, the control is reactive power priority.

From the above analysis, simulation and measured comparison, according to the comparative analysis of measured characteristics of LVRT, the control mode and parameters of large capacity full power converter wind turbine can be determined, and the actual characteristics of wind turbine can be accurately simulated with a set of parameters.

4. Conclusion

Based on the results and discussions presented above, the conclusions are obtained as below:

(1) The electromechanical transient simulation model of large capacity full power converter wind turbine generator ignores the generator, considers the converter as the controlled current source, and takes the active and reactive power control commands as the current directly injected into the power grid, which is more simplified than the physical device;

(2) Large capacity full power converter wind turbines generally use generator double windings and two converters in parallel to achieve high power. In the actual modeling process, the two converters need to be combined and equivalent. This paper deduces the relevant equivalent method;

(3) Based on the actual characteristics of wind turbine LVRT, combined with the manufacturer's LVRT design strategy, the key model parameters of LVRT in the simulation model are identified by analyzing the active current and reactive current in various faults. The method has strong physical significance and small amount of calculation. The identification results can accurately simulate the active and reactive power characteristics of multiple working conditions, and solve the problem of LVRT characteristic identification and accurate simulation, which is the most key problem in power system simulation.

Acknowledgments

This work was financially supported by the State Grid Qinghai electric power company science and technology project (52280719003N:Research on power system simulation and measurement modeling of SVG, SVC and other reactive power compensation devices).

References

[1] Cai Xu, Chen Gen, Zhou Dangsheng, et al. . Review and prospect on key technologies for offshore wind power converters[J]. Journal of Global Energy Interconnection, 2019,2(2): 102-115(in Chinese) .

[2] Fan Mingtian, Zhang Zuping, Su Aoxue, et al. Enabling technologies for active distribution systems [J]. Proceedings of the CSEE, 2013, 33(22): 12-18(in Chinese).

[3] Wang Weisheng,Zhang Chong,He Guoqing, et al. Overview of research on subsynchronous oscillations in large-Scale wind farm integrated system[J]. Power System Technology,2017,41(4):1050-1060(in Chinese)

[4] Deng Wangbo,Wang Haiyun,Chang Xiqiang, et al . Analysis of influencing factors of subsynchronous oscillation in large-scale direct-drivern wind farm[J],High Voltage Apparatus,2019,55(6) :218-224(in Chinese) .

[5] Huang Lingling,Cao Jialin,Zhang Kaihua,et al . Status and prospects on operation and maintenance of off wind turbines[J] . Proceedings of the CSEE,2016,36(3):729-738(in Chinese) .

[6] Nian Heng,Cheng Peng,He Yikang,Review on operation techniques for DFIG-based wind energy conversion systemms under network faults[J] . Proceeings of the CSEE,2015,35(16): :4184-4197.

[7] Yao J,Li H,Chen Z,et al. Enhanced control of a DFIG-based wind-power generation system with series grid-side converter under unbalanced grid voltage conditions[J]. IEEE Transction on Power Electronics,2013,28(7):3167-3181
[8] Nian H,Cheng P,Zhu Z Q. Coordinated direct power control of DFIG system without phase-locked loop under unbalanced grid voltage conditions[J]. IEEE Transactions on Power Electronics,2016,31(4):2905-2918.
[9] Carroll J,McDonald A,McMillan D. Reliability comparison of wind turbines with DFIG and PMG drive trains[J]. IEEE Transactions on Energy Conversion,2015,30(2):663-670.
[10] Li Hui,Liu Shengquan,Ran Li,et al. Overview of condition monitoring technologies of power converter for high power grid-connected wind turbine generator system[J] . Transactions of China Electrotechnical Society,2016, 31(8)： 1-10(in Chinese)
[11] Liu Tao,Ye Xiaohui,Wu Guoyang,et,al. An activer power control model of wind power generating unit suitable for medium-and long-term dynamic simulation of power grid,Power System Technology,2014,38(5): :1210-1215(in Chinese)
[12] Xu Yin,Chen Ying,Mei Shengwei,Review on wind turbine models for power system transient simulations[J]. Automation of Electric Power Systems,2011,35(9): 100-107(in Chinese)
[13] Clark K， Miller N W， Sanchez-Gasca J J. Modeling of GE windturbine-generators for grid studies ［R］. 2009.
[14] Huang Hua,Pan Xueping,Li Jiawei,et,al. Parameter setting method and measurement verification of LVRT model for direct drive permanent magnet wind turbine generator[J]. 2019, ,39 （4）， 155-160(in Chinese)
[15] Pan Xueping,Yin Ziyin,Ju ping,et,al. Model parameter identification of DFIG based on short circuit current[J]. Electric Power Automation Equipment,2017,37(11): 27-31 (in Chinese)
[16] Pan Xueping,Wen Chaorong,Ju Ping,et,al. A frequency-domain based method to identify parameters of grid side converter controller for doubly fed induction generators[J],2015,39(3): 634-638 (in Chinese)
[17] Pan Xueping， Ju Ping， Wen Rongchao， et al. Decoupling estimation of parameters in rotor side controller of DFIG-based wind turbine by frequency domain method [J]. Automation of Electric Power System， 2015， 39(20)： 19-25(in Chinese).
[18] Wu Tao,Yu Xiao,Guo Jiayang,et,al. Empirical analysis on model and parameters of grid-connected doubly fed induction generators in transient stability computation[J]. Power System Technology,2011,35(1) : 100-105(in Chinese)
[19] Fu Rong,Sun Wanpeng. Parameter identification of DFIG based on trajectory characteristic analysis[J]. Electric Power,2013,46(7): 100-104(in Chinese)

Application of space vector pulse width modulation algorithm based on programmable logic array in power system

Zhihai Cheng[1a*], **Hui Liu**[1b*]

[1]School of Energy and Mechanical Engineering, Shanghai Electric Power University, Shanghai 200120, China

[a*]czh.cn@163.com, [b*]2652956071@qq.com

Abstract—In the traditional power system, the voltage regulation and frequency modulation control method used for boiler induced draft fan is limited in control accuracy. With the improvement of furnace pressure change accuracy, higher requirements are put forward for motor control algorithm. This paper studies permanent magnet synchronous motor (PMSM) as the executive part of induced draft fan servo control system,and adopts FOC control method based on space vector pulse width modulation (SVPWM).Firstly, the simulation model of SVPWM pulse width modulation algorithm is established by MATLAB to provide theoretical support for the subsequent debugging algorithm; When using FPGA to realize SVPWM modulation algorithm, FPGA has the advantages of high reliability and high real-time processing, which provides theoretical support and application guidance for the realization of high-precision motor control of boiler induced draft fan in power system.

1. Introduction

The boiler air flow classifier is mainly composed of secondary air, tertiary air,impeller, transmission system, etc. During the operation of the production line, the materials enter through the feed pipe of the classifier under the attraction of the induced draft fan. At the same time, the air enters the scouring area through the secondary air inlet evenly distributed on the cone under the attraction of the induced draft fan, and the coarse and fine mixed powder is used for air screen scouring. In order to improve the air classification accuracy of the boiler, it is necessary to improve the permanent magnet synchronous motor of the induced draft fan control accuracy and control stability of the drive system.

There are three main PMSM control strategies: voltage regulation and frequency modulation control (VF), direct torque control (DTC) and vector control (FOC). The VF control method is open-loop control, which has simple control and low precision. Motor out of step will occur under heavy load. This is the traditional motor control method of boiler induced draft fan[1]. DTC control realizes high-performance dynamic control through direct tracking of flux linkage amplitude and torque. It does not need complex calculation formulas such as complex spatial coordinate changes, and is insensitive to motor parameters. It has simple structure, fast torque response and good robustness. However, it is protected by intellectual property rights. At present, only ABB company applies this control method[2].FOC control is to add phase control on the basis of controlling the voltage and current of three-phase AC, detect the magnetic pole position of motor rotor in real time to determine the voltage phase of three-phase stator winding, and realize current control torque. FOC control is a closed-loop control with high control accuracy and flexible control modes (including id = 0 control;

maximum torque current ratio control, constant flux control, flux weakening control, etc.), but the calculation of FOC control is complex and requires a variety of spatial coordinate transformations [3].

In this paper, the simulation model of SVPWM is established by establishing simulink of MATLAB, and the "three loop" control simulation of PMSM is completed, which provides a theoretical basis for circuit debugging. According to the simulation results, the motor selection and drive circuit design are completed. The SVPWM control algorithm is realized through anti fuse FPGA to complete the bottom drive control of permanent magnet synchronous motor.

2. Principle of space vector pulse width modulation

The essence of SVPWM pulse width modulation is the modulation mode of ideal sine wave and ideal third harmonic. On the one hand, this modulation mode ensures the maximum voltage utilization; On the other hand, most motors are three-phase windings with a winding subspace angle of 120° , and the third harmonic of phase voltage does not affect the line voltage, and such modulation mode will not affect the synthetic magnetomotive force of the motor[4]. Figure 1 shows a voltage waveform (approximately trapezoidal) consisting only of fundamental wave and third harmonic:

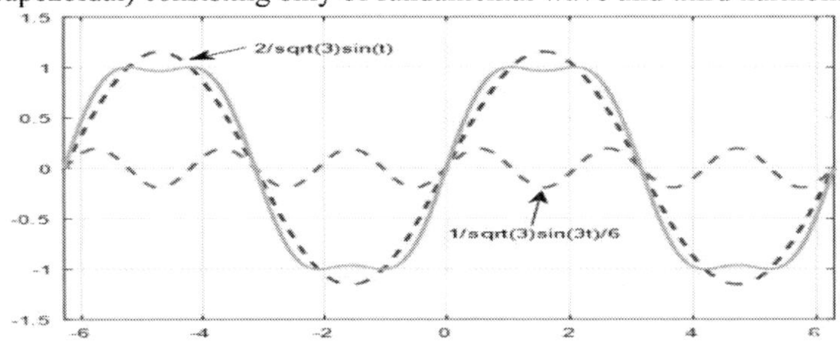

Figure 1. Ideal SVPWM modulation waveform

The SVPWM method adopts the average equivalent principle, that is, the average value is equal to the given voltage vector by combining the basic voltage vector in one switching cycle. At a certain time, the voltage vector is obtained by different combinations of two adjacent non-zero vectors and zero vectors in time. The action time of each voltage vector is controlled in a sampling period to make the voltage space vector close to rotating according to the circular track, and the actual flux generated by different switching states (Sa, Sb, Sc) of the inverter approaches the ideal flux circle.

3.Simulation Implementation of space vector pulse width modulation algorithm and analysis of vector control simulation results

The implementation of SVPWM algorithm is mainly divided into several steps: sector judgment calculation, voltage vector selection and action time calculation, sector vector switching point calculation and three-phase PWM pulse trigger.

3.1sector judgment of reference voltage vector

According to figure 2, U_a and U_b are the components of the reference voltage U_{OUT} on the α axis and β axis. Further simplify the analysis of the above conditions. The sector where the reference voltage vector U_{OUT} is located can be judged by the three formulas Uβ, $\sqrt{3}U\alpha - U\beta$, $-\sqrt{3}U\alpha - U\beta$.

$$\begin{cases} U_1 = U_\beta \\ U_2 = \frac{\sqrt{3}}{2}U_\alpha - \frac{U_\beta}{2} \\ U_3 = -\frac{\sqrt{3}}{2}U_\alpha - \frac{U_\beta}{2} \end{cases} \tag{1}$$

If U1 > 0, a = 1, otherwise a = 0;

If U2 > 0, then B = 1, otherwise B = 0;

If U3 > 0, then C = 1, otherwise C = 0;

Define the number of sectors N=4C+2B+A to obtain the corresponding relationship between the number of sectors N and the sector where the actual U_{OUT} is located, as shown in Table 1[5]:

Table 1 correspondence between sector number N and actual sector

N	3	1	5	4	6	2
Fan area code	I	II	III	IV	V	VI

3.2 calculation of sector voltage vector action time

According to the principle of space voltage vector action equivalence, taking U_{OUT} in sector I as an example, the action time of adjacent non-zero voltage vector T_4、T_6, zero voltage vector T_0、T_7 and in sector I is sorted out, as shown in equation (2)[6]:

$$\begin{cases} T_4 = \dfrac{3U_\alpha T}{2U_{dc}} - \dfrac{1}{2}T_6 = \dfrac{3U_\alpha T_s}{2U_{dc}} - \dfrac{1}{2}\dfrac{\sqrt{3}U_\beta T_s}{U_{dc}} = \dfrac{\sqrt{3}T_s}{U_{dc}}\left(\dfrac{\sqrt{3}U_\alpha}{2} - \dfrac{U_\beta}{2}\right) = \dfrac{\sqrt{3}T_s}{U_{dc}}U_2 \\ T_6 = \dfrac{\sqrt{3}U_\beta T_s}{U_{dc}} = \dfrac{\sqrt{3}T_s}{U_{dc}}U_1 \\ T_7 = T_0 = \dfrac{T_s - T_4 - T_6}{2} \end{cases} \tag{2}$$

Where is the U_{dc} bus voltage and T_s is the time period, U_1, U_2, U_3 which is determined by equation (3).

$$\begin{cases} U_1 = U_\beta \\ U_2 = \dfrac{\sqrt{3}}{2}U_\alpha - \dfrac{U_\beta}{2} \\ U_3 = -\dfrac{\sqrt{3}}{2}U_\alpha - \dfrac{U_\beta}{2} \end{cases} \tag{3}$$

Similarly, the action time of different voltage vectors in other sectors can be calculated.

3.3 determination of sector vector switching point

According to the average value equivalence principle in sector I, the switching time of different voltage vectors in sector I is shown in equation (4).

$$\begin{cases} T_a = \dfrac{T_s - T_4 - T_6}{4} \\ T_b = \dfrac{T_4}{2} + T_a \\ T_c = \dfrac{T_6}{2} + T_b \end{cases} \tag{4}$$

The relationship between three-phase voltage switching time switching points Tcm1, Tcm2 and Tcm3 and each sector number N is shown in the table below:

Table 2 time switching points of each sector T_{cm1}, T_{cm2} and T_{cm3}

N	1	2	3	4	5	6
T_{cm1}	T_b	T_a	T_a	T_c	T_c	T_b
T_{cm2}	T_a	T_c	T_b	T_b	T_a	T_c
T_{cm3}	T_c	T_b	T_c	T_a	T_b	T_a

3.4. Simulation modeling and results of space vector pulse width modulation

The overall model of space vector control SVPWM is established in Simulink of MATLAB, as shown in Figure 2, which includes four parts: sector judgment, calculation of action time of basic voltage vector, determination of sector vector switching point and three-phase voltage generation module.

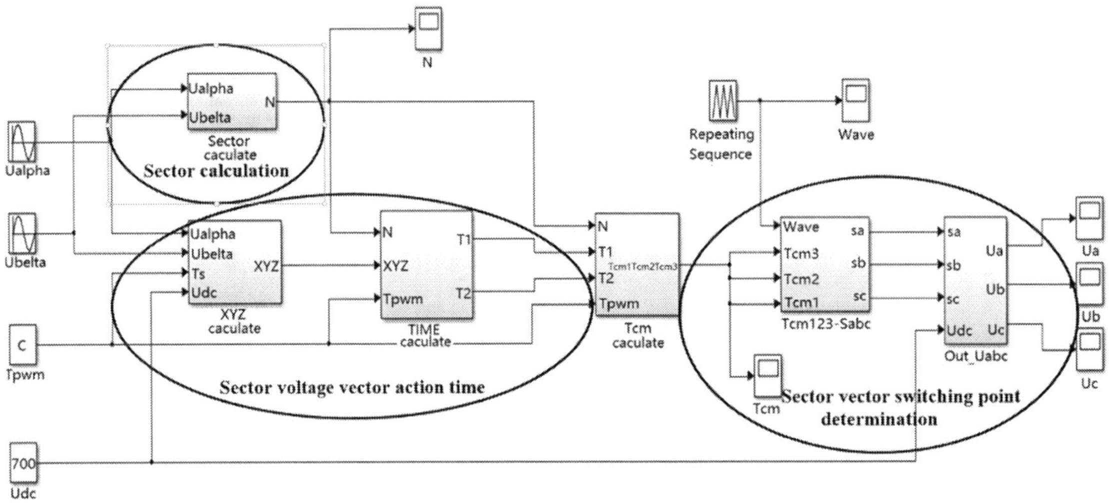

Figure 2. Simulink modeling of SVPWM

The simulation results are shown on left, where Fig. 2 is the sector number N on right is the voltage vector switching time point.

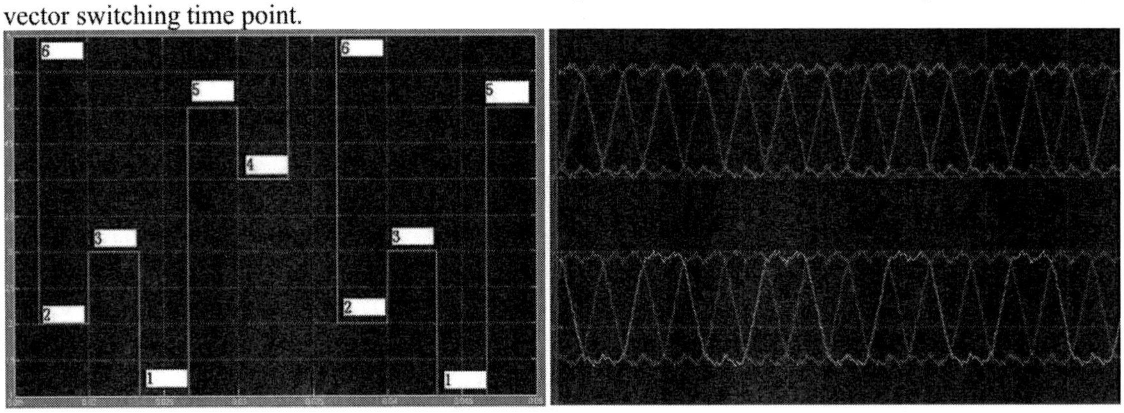

Figure 3. Simulation results of SVPWM module

4. Implementation of programmable logic array for space vector pulse width modulation

According to the Top-Down design idea, each functional module of SVPWM in quartus environment is designed. Combined with the in-depth analysis of SVPWM principle, the top-level design file of SVPWM algorithm will call each functional module. The system structure is shown in Figure 4.

FPGA top-level file uses MATLAB to generate simulated three-phase sinusoidal periodic angle data of motor winding to make DDS file, which is realized by clarke transform module, sector selection module, vector action time calculation module, switch selection module and so on.

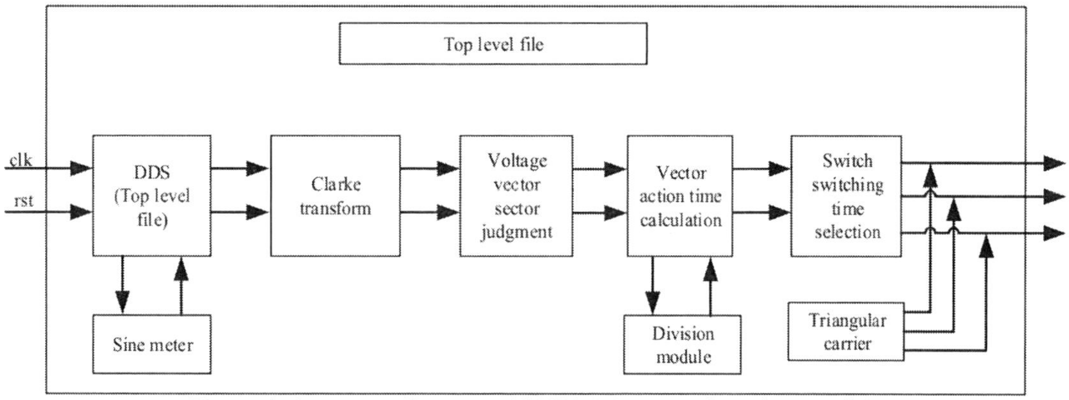

Figure 4. Structure diagram of SVPWM system based on FPGA

The low-speed start of the motor is realized through the acceleration and deceleration module in FPGA. In the design process, the carrier frequency is set to 500Hz, the data period of three-phase winding angle is 0.1024s, and the motor speed is calculated to be 150r / min combined with the time sampling point. The simulation results of the SVPWM control algorithm implemented by FPGA are shown in Figure 5, including sinusoidal two-phase voltage inputs U_a and U_b, three-way switch switching action time and triangular carrier setting.

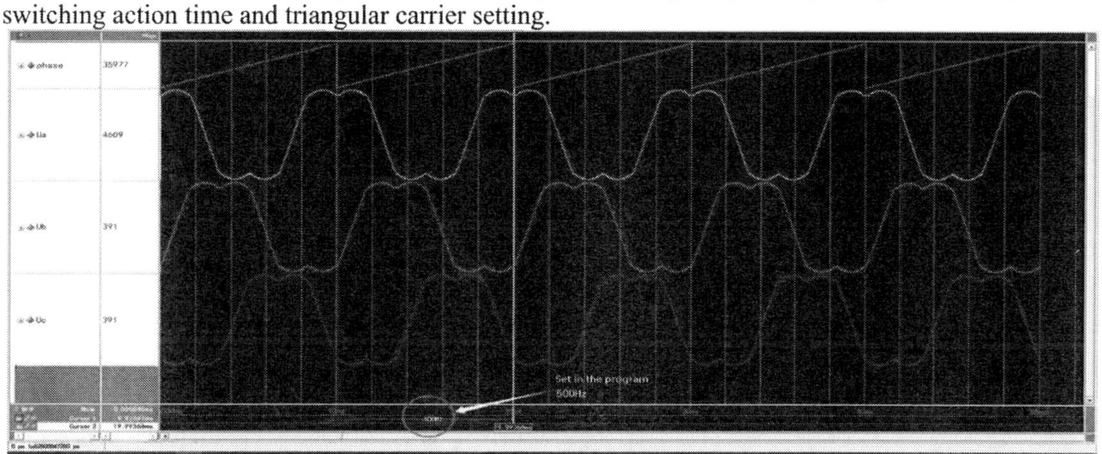

Figure 5. FPGA implementation results of SVPWM

It can be seen from Figure 5 that in FPGA simulation, the cycle alternation of sector n is 3 → 1 → 5 → 4 → 6 → 2, and the saddle shaped modulation waveform is realized through SVPWM algorithm to realize the accurate control of permanent magnet synchronous motor.

5. Conclusion

In this paper, the space voltage vector SVPWM model is built by Matlab / Simulink. At the same time, the SVPWM control algorithm is realized by software program and hardware circuit design on FPGA to complete the bottom drive control of permanent magnet synchronous motor. The motor control of induced draft fan in power system adopts FPGA to realize the bottom control algorithm, which has the advantages of high real-time and high reliability. In the future, the motor closed-loop control algorithm can be integrated in FPGA. At the same time, the device adopts anti fuse FPGA to improve the control accuracy, control reliability and real-time of the whole control system and meet the control requirements of various induced draft fans in the future.

Reference

[1] Wang Yang. Design of AC voltage regulation and frequency modulation control elevator system based on PLC [D]. Jiangsu: Suzhou University, 2011.

[2] Pan Feng, Yan Geng long, Yuan Wei Hua, et al. Direct torque control of permanent magnet synchronous motor based on double sliding mode [J]. Journal of electrotechnics, December 2018, Vol. 33, supplement 2:427-433.

[3] Cai Gang. Design and implementation of motor drive system for four rotor UAV Based on FOC [D]. Shaanxi: Chang'an University, 2016.

[4] Li Jinying, Fu Chengyu, Chen Xinglong, et al. Permanent magnet synchronous motor drive controller and modulation mode [J]. Motor and control applications, 2010, Vol. 37, No. 7: 28-31.

[5] Peng Bo, Xiao Dengming, long Bo. Simulation implementation of three-phase inverter SVPWM based on Simulink [J]. Industrial control computer, 2011, Vol. 24, No. 7: 90-92.

[6] Xiao Chunyan. Research and implementation of voltage space vector pulse width modulation technology [M]. Jiangxi: School of information engineering, Nanchang University, 2005.

Research on Operation Reliability Evaluation of Power-generating Unit

Shuai Di[1,a]

[1]Thermal Control Technology Institute, Datang Northeast Electric Power Test & Research Institute Co., Ltd, Changchun 130000, China

[a]dtdby0431@163.com

Abstract—Objective and reasonable evaluation is the basis of power-generating unit optimal operation, but evaluation method based on single weight can't adapt to the high accuracy requirement of comprehensive evaluation. In order to achieve the accurate evaluation of power-generating unit operating reliability, this paper firstly calculates subjective weight and objective weight of target attribute by analytic hierarchy process method and entropy weight method. It calculates the associated weight of target attribute by grey relational degree method, and it obtains the linear combined weight by relaxation factor. Then, the power-generating unit operating reliability is evaluated by improved TOPSIS (Technique for Order Preference by Similarity to an Ideal Solution), and the evaluation result is obtained. Finally, the relaxation factor is continuously evaluated from 0 to 1 to perform sensitivity analysis. This paper evaluates the operation reliability of nine wind turbines in a wind farm, and the evaluation result is closer to actual operation state of wind turbines. This method has high reliability and practical value, and it provides a new technical means for making a reasonable maintenance plan of wind turbines.

1. Introduction

In recent years, the wind power has developed rapidly. At the same time, many shortcomings and problems have been exposed, which directly lead to the low operation level and the low maintenance level of wind farms. They have a negative impact on the operation hours improvement and power generation efficiency. The operation state and deterioration trend of wind turbines are hidden in SCADA data. In order to realize the intelligent operation and maintenance of large wind farms, it is necessary to analyze SCADA data reasonably and obtain several indexes for evaluating wind turbine reliability. It can achieve the reasonable evaluation and wind turbine reliability ranking. Based on evaluation result, it can provide a basis for wind turbine maintenance and classify the maintenance priority of wind turbines in the same health state. It can realize the intelligent operation and maintenance of large wind farms and improve the economic benefits.

Through statistical analysis of power curve, wind energy utilization curve and power generation efficiency, the operating state and power generation performance of units are evaluated in literature [1-2]. Literature [3] evaluated the performance of different wind turbines types according to the reliability of actual power curve. Based on the correlation of SCADA data, literature [4] studies unit health status and proposes quantitative evaluation method. In literature [5-7], methods such as neural network, clustering, fuzzy, K-nearest neighbor method and principal component analysis are used to evaluate the power generation performance of units. In literature [8], Extreme Learning Machine was used to train the neural network model, and which was used to evaluate the working condition of the gear box. In Literature [9], Online Sequential Extreme Learning Machine was used to evaluate the state of transmission system. Literature [10-12] realize the abnormal identification of the whole wind turbine based on data-driven method.

Based on SCADA data of nine wind turbines, this paper obtains the combined weight by AHP, entropy weight method and grey correlation method. It uses improved TOPSIS method to comprehensively evaluate the wind turbines reliability, and it provides a basis for intelligent operation and maintenance.

2. Reliability Comprehensive Evaluation Index of Wind Turbine

The equipment reliability evaluation indexes are usually used to evaluate the operating state of power-generating unit. MTTR is the mean repair time of trouble shoot. T_{MTBR} is the mean interval time of trouble shoot. FN is the failure number. MRTBF is the average running time without failure. A_i is the average availability.

MRTBF refers to the average working time between two adjacent failures in a certain period of time. A_i is the percentage between available hour number and available running time. MRTBF and A_i are the key indicators to evaluate equipment reliability, and they are the main basis to distinguish reliability of power-generating units. T_{MTBR} is the average time between equipment failures. The larger these indicators are, the healthier the unit is, and they are called benefit indicators.

FN refers to the failure number of power-generating unit. MTTR refers to the average equipment failure time, which is the measure of reliability. The smaller these indicators are, the less healthy the unit is, and they are called cost indicators.

The traditional reliability evaluation method of wind turbine is setting a threshold value for a single evaluation index, and the turbines are divided into three grades: normal, sub-health and serious.

The wind turbines which are in same health grade don't have a proper order, and the maintenance priority can' t be determined. Wind turbine operating conditions are complex, and they have gray scale to a certain degree. There are many indexes which affect the wind turbine reliability. In order to obtain accurate ranking results and make reasonable maintenance plan of wind turbine, it is necessary to carry out Multi-attribute comprehensive evaluation.

3. Determine the Evaluation Model of Combination Weight

3.1 Technique for order preference by similarity to an ideal solution

Topsis method can evaluate and sort by calculating Euclidean distance between target attributes and ideal solutions. This method makes full use of original data information, and it solves the multi-objective decision problem of limited solutions. It has the advantages of real, intuitive and reliable. The decision matrix Y is $m \times n$ data matrix. Each attribute is a column, and a data sample is a row. The decision matrix is processed by dimensionless data index, and the evaluation matrix Z is obtained.

Multiplying each attribute of evaluation matrix by corresponding weight to obtain the weighted normalized evaluation matrix X:

$$X_{ij}=w_j \cdot z_{ij} \tag{1}$$

The ideal solution is the set of each attribute optimal values, which is an imaginary optimal scheme. The negative ideal solution is the set of each attribute worst value, which is an imaginary worst solution.

For the benefit index, ideal solution is $\max_i x_{ij}$, and negative ideal solution is $\min_i x_{ij}$. For the cost index, ideal solution is $\min_i x_{ij}$, and negative ideal solution is $\max_i x_{ij}$.

Calculating relative proximity L_i:

$$L_i = \frac{d_i^-}{d_i^+ + d_i^-} \tag{2}$$

In formula (2), d_i^+ and d_i^- are the Euclidean distance. Li is between 0 and 1, and it is the basis for evaluating ranking. The closer is to 1, the stronger the unit reliability is.

The calculation process amplifies w_j influence on the evaluation result. The sum of w_j^2 is not equal to 1, and the evaluation result will also be affected. The standard Euclidean distance can effectively weaken the influence and obtain more accurate evaluation results.

$$d_i^+ = \sqrt{\sum_{j=1}^{n} w_j \left(z_{ij} - z_j^+\right)^2} \tag{3}$$

$$d_i^- = \sqrt{\sum_{j=1}^{n} w_j \left(z_{ij} - z_j^-\right)^2} \tag{4}$$

It can be seen from Equations (3) and (4) that the weight is more reasonable.

3.2 Subjective and objective combination weight

The entropy weight method calculates objective weight by data difference information.

If the information entropy is larger, the information provided is smaller. The less it plays in comprehensive evaluation, the less its objective weight is.

The objective weight of each evaluation index is $\{\beta_j\}$. Entropy weight method relies too much on data samples and objective information, and the objective weight obtained can not be accurately and scientifically used for comprehensive evaluation.

Analytic hierarchy process calculates the subjective weight of object attributes by establishing comparison matrix. It can be used for qualitative and quantitative analysis of multi-attribute decision problems.

Subjective and objective weights are combined for scientific evaluation. $\{\alpha_j\}$ is Calculated by AHP method, and $\{\beta_j\}$ is calculated by entropy weight method.

According to the principle of minimum discrimination information, the combined weight of entropy AHP is:

$$\psi_j = \frac{\sqrt{\alpha_j \beta_j}}{\sum_{l=1}^{n} \sqrt{\alpha_j \beta_j}} \quad (j=1,2, \ldots, n) \tag{5}$$

3.3 Modification of grey weight

Wind turbine operating conditions are complex, and operation datas in a certain degree have grayscale properties.

In order to build a more scientific reliability evaluation model, the grey relational weight is multiplied by a relaxation factor d to adjust weight linearly.

Calculating correlation coefficient $P_i(k)$:

$$P_i(k) = \frac{\min_i \min_k \left|C_k^{\psi} - C_k^i\right| + \rho \max_i \max_k \left|C_k^{\psi} - C_k^i\right|}{\left|C_k^{\psi} - C_k^i\right| + \rho \max_i \max_k \left|C_k^{\psi} - C_k^i\right|} \tag{6}$$

In the formula, $\rho \in [0,1]$, and ρ take 0.5 generally.

The grey weight of each evaluation index is $\{w_j\}(j=1,2,\ldots,m)$.

The linear combination weight is:

$$\phi_j = \frac{w_j + d\varphi_j}{\sum_{j=1}^{n} w_j + d\varphi_j} \tag{7}$$

In the formula, d is the relaxation factor, which changes from 0 to 1 and adapts to the established model in this paper. It can adjust the correction range of grey weight on entropy AHP weight.

3.4 Evaluation model based on grey entropy AHP combined weight

This paper uses TOPSIS method to evaluate the reliability of wind turbines. It replaces the Euclidean distance with standard Euclidean distance. It replaces the individual weights with linear weights. The evaluation process is shown in Figure 1.

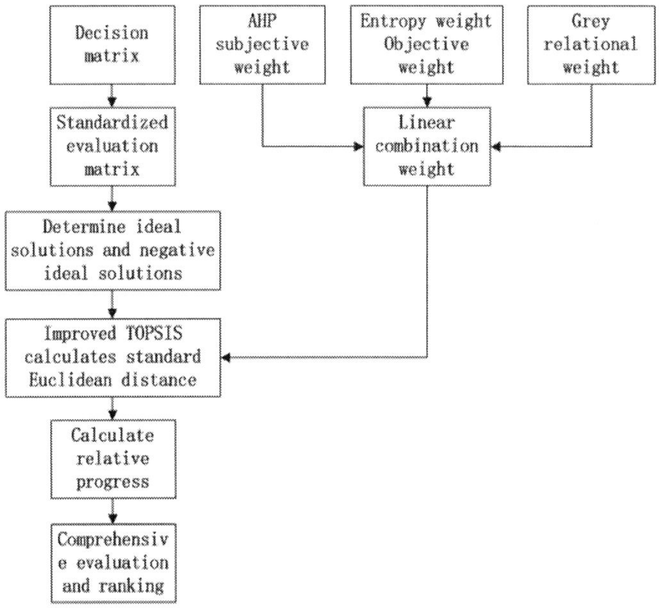

Fig.1 Grey entropy AHP comprehensive evaluation model

4. Application Cases

4.1 Application of grey entropy AHP combined weight evaluation model

In this paper, reliability comprehensive evaluation and ranking of nine wind turbines in a wind farm are carried out. Now they are setting as No.1 to No.9.

It conducts dimensionless processing on SCADA data of nine wind turbines and obtains the normalized decision matrix, as shown in Table 1.

Table.1 Normalized decision matrix

No.	MRTBF	A_i	T_{MTBR}	MTTR	FN
1	1.00	0.99	1.00	0.00	1.00
2	0.47	0.85	0.54	0.34	0.97
			...		
9	0.02	0.75	0.02	0.91	0.28

Objective weight can be obtained by entropy weight method: $\beta=$ (0.351, 0.068,0.359, 0.095, 0.127).

Subjective weight can be obtained by AHP method: $\alpha=$ (0.386, 0.386, 0.092, 0.092, 0.043).

After calculation, CI=0.024, RI=1.12, CR=0.0213, and CR<0.1. It passes the consistency test.

Combined weight is obtained: $\varphi=$ (0.419, 0.184, 0.207, 0.107, 0.084).

Grey correlation weight is obtained: $w=$ (0.162, 0.325, 0.167, 0.179, 0.167).

Taking the linear correction factor d as 1. Grey entropy AHP linear weight is obtained: $\phi=$ (0.290, 0.254, 0.187, 0.143, 0.125).

The standard Euclidean distance D^+ between each scheme and ideal solution is calculated by TOPSIS method. The standard Euclidean distance D^- between each scheme and negative ideal solution is also calculated by TOPSIS method. The relative proximity L_i is obtained.

D^+ = (0.378, 0.483, 0.665, 0.747, 0.373, 0.412, 0.656, 0.926, 0.737).

D^- = (0.922, 0.662, 0.644, 0.546, 0.737, 0.713, 0.631, 0.339, 0.522).

L_i = (0.709, 0.604, 0.492, 0.422, 0.663, 0.633, 0.489, 0.268, 0.414).

In a comprehensive order from largest to smallest: No.1> No.5 > No.6 > No.2 > No.3 > No.7 > No.4 > No.9 > No.8.

4.2 Comparison of different assessment models

The following are the ranking results of different evaluation models:

The comprehensive order of entropy weight method is: No.1> No.5> No.2> No.6> No.7> No.3> No.4> No.9> No.8.

The comprehensive order of AHP method is: No.5> No.1> No.6> No.2> No.7> No.3> No.4> No.9> No.8.

The comprehensive order of grey correlation method is: No.5> No.6> No.1> No.2> No.3> No.7> No.4> No.9> No.8.

The comprehensive order of entropy AHP method is: No.1> No.5> No.6> No.2> No.3> No.7> No.4> No.9> No.8.

The comprehensive ordering of grey entropy AHP method is: No.1> No.5> No.6> No.2> No.3> No.7> No.4> No.9 > No.8.

4.2.1 Comparison of single threshold assessment model

It sets thresholds for a single evaluation index to conduct comprehensive evaluation, and the reliability of each wind turbine is shown in Table 2. The sorting result accords with actual situation. Four healthy wind turbines are lined up in front. Three sub-health wind turbines are lined up in the middle. Two serious wind turbines are lined up in the behind. It proves the validity and correctness of five sorting results.

Table.2 Reliability status

healthy state	health	sub-health	serious
number	No.1 No.2 No.5 No.6	No.3 No.4 No.7	No.8 No.9

Compared with the above results, it is not accurate to use a single weight to calculate. The ranking results of entropy AHP weight and grey entropy AHP linear weight are excellent, and the ranking results are exactly the same. The sorting results of No. 1 and No. 5 wind turbine are the same as entropy weight method. The sorting results of No. 6 and No. 2 wind turbine are the same as AHP method. The sorting results of No. 3 and No. 7 wind turbine are the same as grey correlation method. The sorting results of No.4, No.9 and No.8 wind turbine are the same as AHP, entropy weight and grey correlation method.

It shows that the combination weight and linear combination weight are more reasonable and practical. It reflects subjective factors, objective factors, and gray factors. It can represent the true state of wind turbines.

4.2.2 Comparison of combined algorithm evaluation models

The ranking results of entropy AHP weight and grey entropy AHP linear combination weight are consistent. This shows that the grey weight has little influence on reliability comprehensive ranking of wind turbines. The relaxation factor is introduced to adjust grey weight, and the sensitivity analysis results are shown in Fig. 5. With the increase of relaxation factor, the distance of each relative closeness degree is obvious and uniform. This paper takes the relaxation factor as 1 to establish evaluation model.

The grey entropy AHP combined weight evaluation model can effectively divide the maintenance priority.

No.8 and No.9 wind turbine are in serious condition, but it can be seen from Figure 5 that No.8 is obviously more serious than No.9.

The No.8 wind turbine needs priority maintenance to adjust fan state and eliminate hidden trouble.

No.1 is in much better state than No.2, No.5 and No.6.

No.3, No.4 and No.7 have little difference in operating state. This method can effectively divide the maintenance priority. The higher the priority is, the sooner maintenance is required.

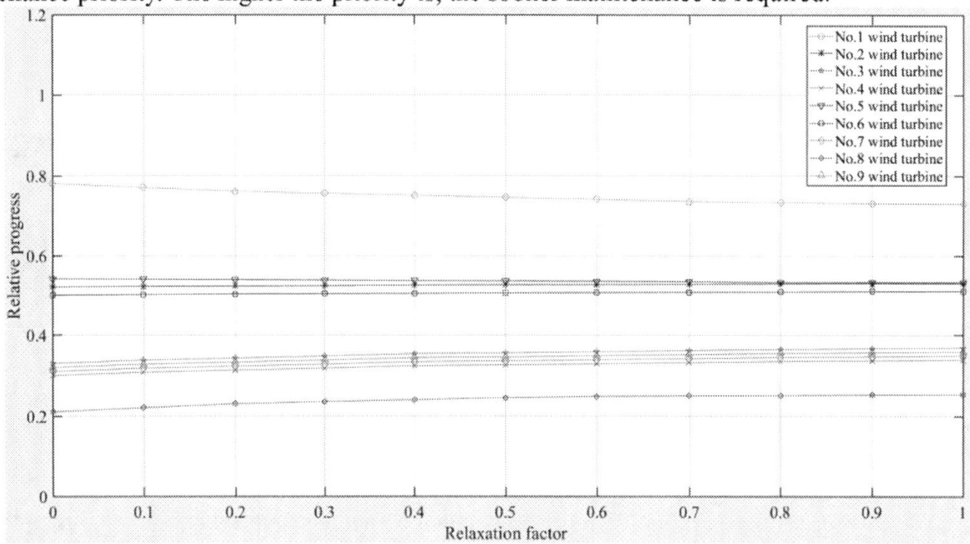

Fig.2 Sensitivity analysis

5. Conclusion

In this paper, the subjective weight and objective weight are calculated by AHP method and entropy weight method. The grey weight is calculated by grey correlation method, and linear combination weight is obtained. The linear combination weight is applied to the improved TOPSIS method, and the improved method can comprehensively evaluate the wind turbine reliability. The result shows that this method has higher accuracy.

When it comes to indicators that can't be defined by specific thermodynamic model in the follow-up study, it should complement information with expert knowledge and experience, and it shouldn't seek absolute objectification results. The comprehensive evaluation model and weight determination method presented in this paper can provide useful comparison and reference for similar problems.

References

[1] KUSIAK A, ANDRE W, ANOOP V. Monitoring wind farms with performance curves [J]. IEEE Transactions on Sustainable Energy, 2013, 4(1): 192-199.

[2] DAI Shuoming. Analysis of power generation performance evaluation methods for wind turbines [J]. Technological Development of Enterprise, 2016, 35(9): 95-96.

[3] LAPIRA E, BRISSET D, ARDAKANI H D, et al. Wind turbine performance assessment using multi-regime modeling approach [J]. Renewable Energy, 2012(45): 86-95.

[4] YANG Wenxian, COURT R, JIANG Jiesheng. Wind turbine condition monitoring by the approach of SCADA data analysis [J]. Renewable Energy, 2013, 53: 365-376

[5] LI S, WUNSCH D C, HAIR E O', et al. Compararive analysis of regression and artificial neural network models for wind turbine power curve estimation [J]. Journal of Solar Energy Engineering, 2001, 123 (4): 327-331.

[6] USTUNTAS T, SAHIN A D. Wind turbine power curve estimation based on cluster center fuzzy logic modeling [J]. Journal of Wind Engineering and Industrial Aerodynamics, 2008, 96(5): 611-620.
[7] JIA Xiaodong, JIN Chao, MATT B, et al. Wind turbine performance degradation assessment based on a novel similarity metric for machine performance curves [J]. Renewable Energy, 2016, 99: 1191-1201
[8] Qian P, Ma X, Cross P. Integrated data-driven model-based approach to condition onitoring of the wind turbine gearbox[J]. IET Renewable Power Generation, 2017, 11(9): 1177-1185.
[9] Qian P, Ma X, Zhang D. Estimating Health Condition of the Wind Turbine Drivetrain System[J]. Energies, 2017, 10(10).
[10] Yang H, Huang M., Lai C, et al. An approach combining data mining and control charts-based model for fault detection in wind turbines[J]. Renewable Energy, 2018, 115: 808-816.
[11] Sun P, Li J, Wang C, et al. Condition assessment for wind turbines with doubly fed induction generators based on SCADA data[J]. Journal of Electrical Engineering & Technology, 2017, 12(2): 689-700.
[12] Sun P , Li J , Wang C , et al. A generalized model for wind turbine anomaly identification based on SCADA data[J]. Applied Energy, 2016, 168: 550-567.

Journal of Physics: Conference Series **2108** (2021) 012005 doi:10.1088/1742-6596/2108/1/012005

Starting System and Safety Protection of Explosion-proof Diesel Engine for Coal Mine Based on Compressed Air

Kun Zhu[3], JiquanLiu[2,4*], Zhiwei Yan[1,2] and Lei Zhang[1,2]

[1] National Engineering Laboratory for Coal Ming Machinery, Taiyuan, Shanxi, 030006, China

[2] Tai Yuan Research Institute of China Coal Technology&Engineering Group, Taiyuan, Shanxi, 030006, China

[3]China Coal Technology&Engineering Group, Beijing, 100013, China

[4]College of Mechatronics Engineering, NUC, Taiyuan, Shanxi, 030051, China

*Corresponding author's e-mail: mkyljq@126.com

Abstract: The starting and safety protection system is an important part ofthe explosion-proof diesel engine, and there are malfunctions in the explosion-proof diesel engine, such as difficulty in starting and failure in protection.As a gas working medium, compressed air is safe, reliable, and advantageous in acquisition and storage.Therefore, analyzing the starting principle and safety protection requirements of explosion-proof diesel engines and studying the correlation between the two systems, a set of starting and safety protection systemsusing compressed air as the working medium is designed in this paper. Moreover, the performance requirements of the main components in the system are studied, analyzing and calculating the operating parameters in the system, so that the design plan can be optimized. The problems appearing instarting and safety protection of explosion-proof diesel engines are solved through this system.

1. Introduction

The application of a large number of modern coal mining machinery not only improves the efficiency of coal mining work, but also reduces the work intensity of workers, improving workplace safety.As a special power system device, explosion-proof diesel engines such as monorail cranes, rail locomotives, explosion-proof rubber tire vehicles, special power stations, and mobile pumping stations are currently widely used in equipment in explosive environments such as coal mines. Moreover, owing to a series of advantages such as strong power, flexible fuel installation and convenient maintenance, explosion-proof diesel engines have been recognized by the majority of users. Additionally, as a branch of the diesel engine group, explosion-proof diesel engines have all the necessary system components as ordinary diesel engines. Meanwhile, due to the particularity that explosion-proof diesel engines work inthe environment with explosive gases, explosion-proof diesel engines must have explosion-proof performance avoiding ignition and detonation and blocking the transmission of detonation. In addition, starting system should provide sufficient speed and torque to overcome the frictional resistance when the internal mechanisms of the explosion-proof diesel engine are stationary as well as the compression resistance when the piston compresses the air. Moreover, the protection system must realize self-alarm and shutdown when the operating parameters of the explosion-proof diesel engine are abnormal and the airexceeds the standard. Nowadays, there are mechanical, pneumatic, and electrical systems in the international and domestic explosion-proof diesel engine starting systems, and for the protection systems,there areair protection and electrical protection, which have not unified the two, separating the internal logical connection of the two systems and resulting in a series of problems such as system mismatch and poor performance. Therefore, in this paper, based on the integrity of the explosion-proof diesel engine system and comprehensively considering the internal logical connection between starting and safety protection, compressed air is used as a unified working medium to design a set of economical, practical, and highly reliablestartingand safety protection system of explosion-proof diesel engine, which has explosion-proof performance and can smoothly start the explosion-proof diesel engine.

2.Principles of Starting and Safety Protection ofExplosion-proof Diesel Engines

2.1Principle of Compressed Air Starting

Starting the explosion-proof diesel engine must rely on starting system. When the explosion-proof diesel engine is stationary, external force must be used to drive the crankshaft to rotate, forcing the connecting rod piston to compress air.At the end of the compression stroke, the diesel is injected into the combustion chamber to form gas with high temperature andpressure to start normally. After the normal start, the explosion-proof diesel engine can operate normally. Besides, there must be sufficient starting torque and starting speed during starting. The starting speed refers to the minimum speed of the crankshaft during the normal operation of the explosion-proof diesel engine, which ensures that the explosion-proof diesel engine can inhale a sufficient amount of fresh air and has a certain temperature to meet the necessary conditions for diesel combustion. Starting torque refers to the minimum system torque required to overcome the internal frictional resistance and air compression resistance of the explosion-proof diesel engine during starting.Only when the speed provided by starting system is greater than the starting speed and the torque provided by starting system is greater than the minimum system torque can the explosion-proof diesel engine start smoothly[1].

In the compressed air starting, the potential energy of compressed air is converted into the rotational kinetic energy of air motor, driving the crankshaft to rotate, so that the explosion-proof diesel engine can start up normally[2].Fig.1 is a flow diagram of compressed air. When compressed air is started, it should meet the following requirements:

$$n_s = n_m \geq n_{min} \tag{1}$$

$$T_s = T_m \geq T_r \tag{2}$$

n_m is the output speed of the air motor (rp/m), n_s is the Starting speed (rp/m), n_{min} is the minimum speed of the crankshaft (rp/m), T_s is the Starting torque (Nm), T_m is the output torque of the air motor (Nm), T_r is the Resistance distance of explosion-proof diesel engine (rp/m).

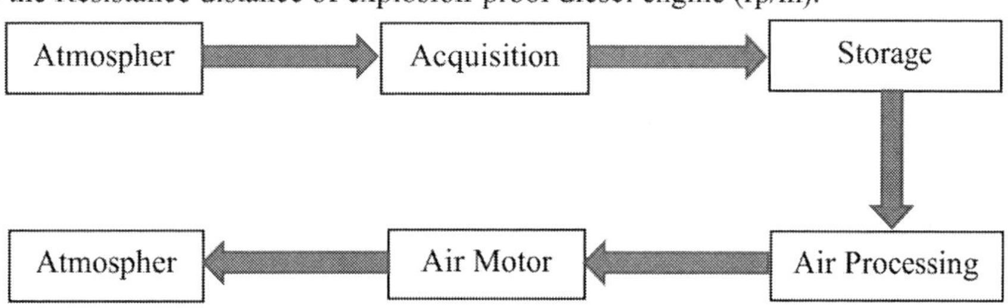

Fig.1 Airflow route map

Control and starting system is the core subsystem of starting system, which must be given priority at the beginning when designing starting system.It ensures a smooth start, and prevents gear collision as well as secondary start-up. In addition,it can also realize air acquisition and compression, filter processing, shunt control, power output, etc[2]. Fig.2 is a control and starting system.

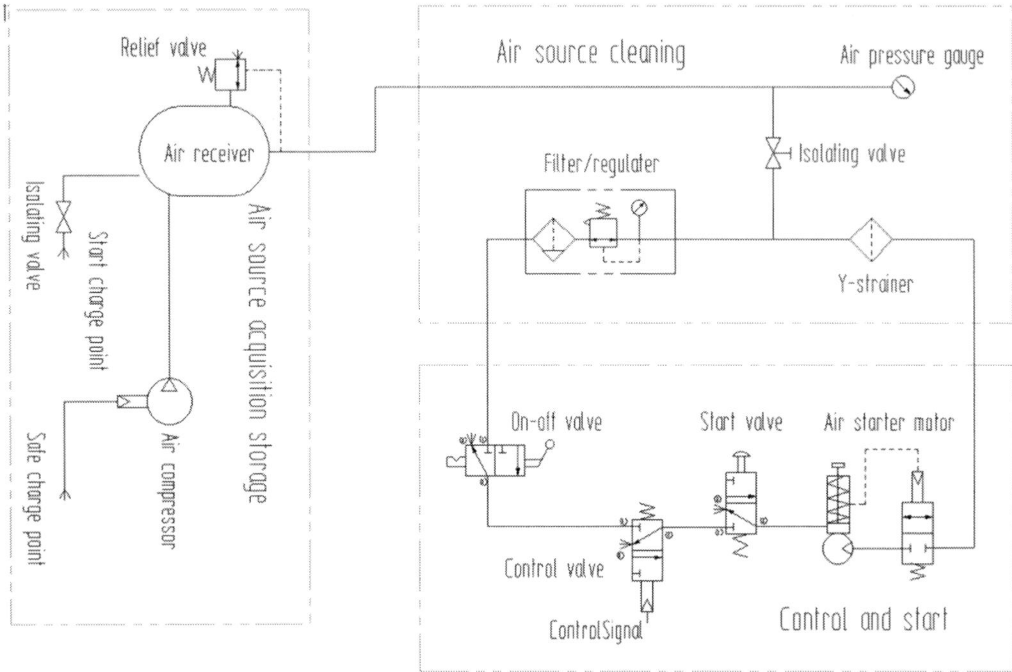

Fig.2 Control and starting system

Compressed air flows out of the storage device. After processing, the water and other impurities are discharged, which is divided into two paths, namely power source and control air source. Then, the air is intervened to flow to various pneumatic control valves, controlling the meshing state of the pinion in the air motor and the ring gear of the explosion-proof diesel engine. Therefore, the engagement before starting and the separation after starting can be realized,so that the normal starting can be controlled. Moreover, the power air source flows to the air motor, which drives the motor to rotate to start the explosion-proof diesel engine.

2.2 Safety Protection Principle of Explosion-proof Diesel Engine

The Gas explosion is one of the most serious disasters in coal mines, which is destructive and sudden.It must meet three conditions at the same time. In other words, for gas concentration, 5%<concentrationgas<16%; for ignition source, ignition temperature>650; for combustion-supporting agent (oxygen) concentration, oxygen>12%. In addition, explosion-proof diesel engines are prone to backfire in the intake air during operation, and there are residual sparks in the exhaust gas. The exhaust temperature reaches 600 degrees. The surface of the body is prone to form high temperatures, etc.Therefore, the basic principle of explosion-proof diesel engine protection is to block the ignition source and reduce the surrounding temperature changes.Table 1 is a list of factors that cause the gas explosion and the normal operation of explosion-proof diesel engines.

Table 1 Factor table

FactorM ethod	Air inlet	Air vent	The machine surface	Between the parts	Exhaust gas temperature	Engine oil	Coolant
Manifestation method	Tempering ignite	Sparks ignite	High-temperature ignition	Fire in the gap	High-temperature ignition	Abrasion	High temperature
Processing method	Flame arrester	Flame arrester	Cooling and insulation	Structural transformation	Purification device	Pressure control	Temperature control

According to Table 1, when dealing with explosive factors, structural modifications to the body can be used. For example, in the design, for the component gap factor, the surface roughness of the part can be increased, and the edge thickness can be greater than 13mm, so that the explosion can be avoided. In addition, special devices can also be added for direct explosion protection, such as factors including the air inletandthe air vent. Usually,an air intake flame arrester or exhaust fence is added. Moreover, it is also feasible to monitor the operation of the explosion-proof diesel engine in real time through monitoring means.Usually, the monitored information is converted into a pneumatic signal,hydraulic signal, or voltage signal, controlling the actuator to stop the explosion-proof diesel engine when an abnormal state such as oil pressure monitoring and water temperature monitoring occurs[4].

3.CompressedAir Starting and Protection System Composition of Explosion-proof diesel engine

Analyzing principle, the system composition is distinguished by functional form, which is mainly composed of air source acquisition and storage, air source cleaning, status sensing, logic valve group, actuator and supporting joint pipelines, as is shown in Fig.3.

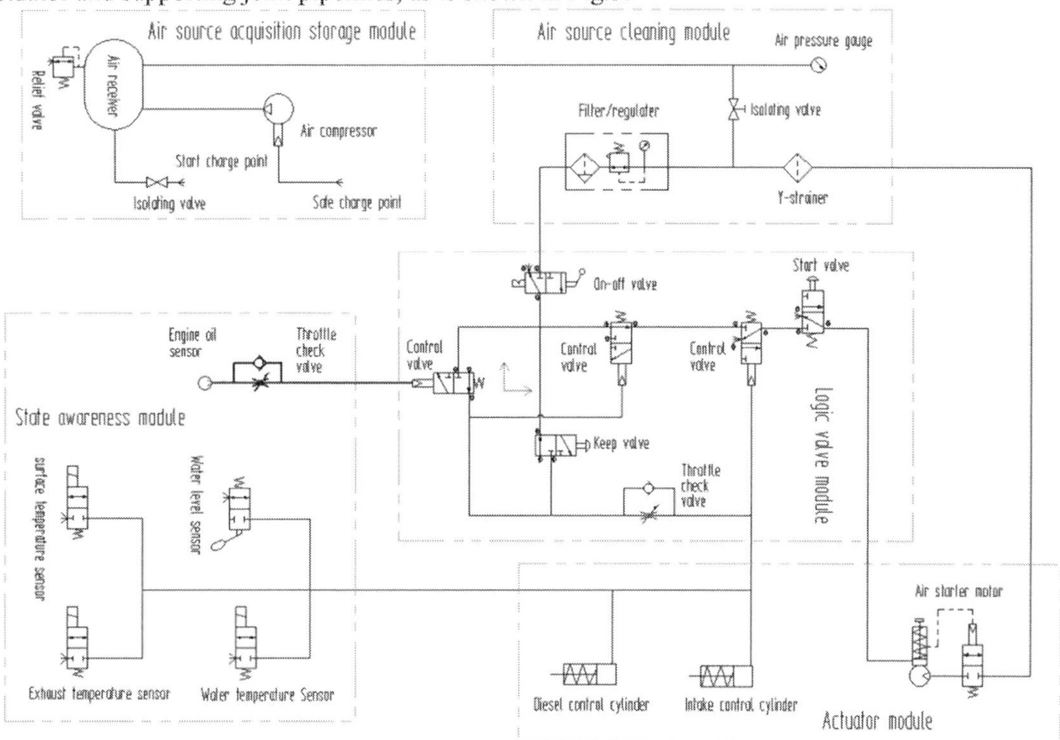

Fig.3 Compressed air starting and protection system diagram of explosion-proof diesel engine

Air source acquisition and storage are mainly composed of air pumps, air storage tanks, one-way valves, safety valves, ball valves, and drain valves.The function of the air pump is to compress the external air that will be stored in the air storage tank to supply the system air source when the explosion-proof diesel engine starts and works normally, which can be driven by gearboxes, hydraulic motors, air motors, and explosion-proof motors. In addition, due to the need for explosion-proof, the air pump casts water channels in the pump body and uses circulating water for forced cooling. The maximum temperature of the shell is less than 150˚ .

One-way valves and safety valves restrict the direction of air flow and control the maximum pressure of the air storage tank. Air can only enter the air storage tank from the air pump inlet without reverse flow. What is more, when the pressure at the outlet of the air pump reaches 0.8MPa which is set by the unloading valve, the pressure will be relieved, and the pressure in the air storage tank cannot exceed the maximum pressure of 1MPaso that the two together protect air source acquisition and storage module.

The air storage tank is a pressure vessel that stores high-pressure air compressed by the air pump, which must have a certain degree of corrosion resistance and rust resistance, and the pressure level should meet the design requirements. Additionally, the drain valve is installed at the bottom of the air storage tank to discharge the water formed by air cooling and condensation, since the condensed water will squeeze the effective volume of the air storage tank. Meanwhile, low temperatureswill freeze and block the air path,which is harmful to the entire system.

Air source cleaning is mainly composed of different filters impurities such as the water and dust that enter the system, cleaning the air source, so that the entire system can work safely and effectively.

The state perception is mainly completed by four necessary sensors, namely water temperature sensor, water level sensor, exhaust temperature sensor, and surface temperature sensor.

The logic valve group is mainly composed of pneumatic valves with different functions. According to the control mode, it can be divided into manual button valves, pneumatic control valves, hydraulic control valves, etc. Besides, the start and protection of explosion-proof diesel engines are controlled by combining different valves.

The actuator is composed of the motor body and the delay valve, which converts the potential energy of the compressed air in the air storage device into kinetic energy to start the explosion-proof diesel engine. In addition, the selection of the air motor is controlled by the starting requirements of the explosion-proof diesel engine.

4. Explosion-proof Diesel Engine Starting and the Selection Calculation of Safety Protection System Component

When designing the system, due to safety and performance requirements, the compressed air pressure of the air storage tank is usually set to 0.8MPa, the control air pressure is generally lower than 0.7MPa, and the pressure safety factor of each component is 2. Besides, the opening temperature of the water temperature sensor is 100˙ , the opening temperature of the exhaust temperature sensor is 70˙ , and the opening temperature of the surface temperature sensor is 150˙ . In addition, the water level of the water level sensor is determined according to the design requirements of the exhaust air treatment box. Meanwhile, in the selection process, the working pressure of the pneumatic valve is generally 0-8MPa, the ambient temperature is 10-80 degrees, and the flow rate is 10L/min. The flow rate of the safety valve is greater than the flow rate of the air pump.

4.1 Motor Starting Torque

The output torque, displacement, and working pressure of the air motor are necessary parameters when selecting the appropriate model from the product catalog. In theory, the necessary and sufficient condition that the explosion-proof diesel engine can smoothly start is that the starting speed and torque of the air motor are greater than the minimum starting speed and torque of the explosion-proof diesel engine[5]. The minimum torque of the explosion-proof diesel engine is calculated by the formula (3):

$$\mathrm{T}_{\min} = 9550 P_{s\min} n_{\min} \tag{3}$$

P_{smin} isthe minimum starting power of explosion-proof diesel engine.

Considering the transmission friction and internal leakage loss of the air motor, the output torque of the air motor is:

$$\mathrm{T}_{\mathrm{m}} = \mathrm{T}_{\min} / \eta_{\mathrm{t}} = 9550 P_{s\min} n_{s\min} / \eta_{\mathrm{t}} \tag{4}$$

T_{m} is the output torque of air motor, η_{t} isStarting transmission efficiency.

When the high-pressure pneumatic motor, the pressure will be reduced, and the air potential energy will be transformed into the rotational kinetic energy of the air motor so that the starting is completed. The displacement calculation formula is as follows:

$$V_{\mathrm{m}} = 2\pi \mathrm{T}_{\mathrm{m}} / p_{\mathrm{w}} \tag{5}$$

V_{m} is thedisplacement of air motor, p_{w} is therated working pressure of air motor, which is less than 0.8MPa.

4.2 Design of the Air Storage Tank

As the core component of air acquisition and storage, the air storage tank belongs to the category of pressure vessels, whichare usually welded by stainless steel. Moreover, the volume is determined by the number of starts. The amount of air consumed at one start is calculated by the formula (6):

$$m_k = \rho_k n_s t V_m = 2\pi\rho_k n_s t T_m / p_w \tag{6}$$

m_k is the air consumption at k times of starting, ρ_k is theair density at k times of starting t isstarting time, which is usually 5 seconds;

After starting k times, the remaining amount of the compressed airin the air storage tank is:

$$m_0 - \sum_{k=1}^{n} m_k = p_k V_L / RT \geq 0 \tag{7}$$

m_0 is the initial air volume in air storage tank, V_L is theair tank volume, R -is the air constant, p_k is the air pressure at k times of starting, T is the temperature of the air in the air storage tank, which remains unchanged in a steady status.

It is assumed that the compressed airin the air tank container changes with constant volume and temperature, namely

$$p_k / p_0 = m_k / m_0 \tag{8}$$

From formulas (6), (7), and (8) it is obtained that

$$V_L = \left(m_0 - \sum_{k=1}^{n\leq 4} m_k = \right) m_0 RT / m_k p_0 \tag{9}$$

5.Conclusions

As one of the important systems in explosion-proof diesel engines, the starting and safety protection system is difficult to be designed. In this paper,considering the requirements of the explosion-proof diesel engine in starting and operation protection, the starting and safety protection system of the explosion-proof diesel engine are jointly analyzed to study the principle of compressed airstarting and safety protection and calculate the main component parameters.

Based on the results and discussions presented above, the conclusions are obtained as below:

(1) It is revealed that the start-up process of explosion-proof diesel engines must be carried out under the premise of meeting safety.

(2) It isdesigned a set of explosion-proof diesel engine compressed airstarting and protection systems that meets the use of coal mines

The compressed airstarting and protection systems can be applied to equipment that uses explosion-proof diesel engines as power sources in coal mines, such as electric locomotives, monorail cranes, trackless rubber-tired vehicles, etc. In addition, when designinga specific system, the differences between different products should be considered to unify and harmonize the generality and differencesand intelligence is the direction of its development.

Acknowledgement

This work is supported by Scientific Research Project of Shanxi Tiandi Coal Machinery Equipment Co., Ltd. [grant numbers, M2021-MS11]

References

[1] Kevin Hoag, Brian Dondlinger.Vehicular Engine Design Springer Nature Book,2016
[2] An Siyuan, Study on the starting mode of the explosion-proof diesel engine of the support truck[J]. Machine Tool & Hydraulics, 2017.4
[3] Rao Junliang, Chen Yipei, Hu Xiaohui.Design of the protection circuit for the gas-starting motor of the explosion-proof trackless rubber-tired vehicle[J].Mining Machinery,2012.11

[4] Li Ruijian, Diagnosis and treatment of starting failure of Detroit diesel engine on the offshore platform[J]. Diesel Engine, 2019.9.
[5] Tang Mingxiang, Tang Yufeng, Ding Xuemei et al. Development and application of an experimental device for the starting system of a diesel engine[J]. Journal of Chongqing University of Technology (Natural Science), 2020. 5

Simulation analysis of underwater working condition of ship rudder blade based on Fluent

Haowei Zhao[1a*], Guifeng Liu[1b*], Jianggui Han[1c]

Mechanical Engineering, Naval University of Engineering, Wuhan 430032, China

[a*]1130441457@qq.com, [b*]2838739726@qq.com, [c]921174821@qq.com

Abstract—The rudder blade is an important power plant of the ship. Removing damage and failure is the premise of the normal operation of the ship。In this paper, the numerical calculation method is used, combined with K- ε (k-epsilon model) and S-A (spalart allmaras) turbulence model, the underwater working conditions of ship rudder blade are analyzed, and the influence of water flow impact at different angles of attack on the damage failure of rudder blade components is discussed. The simulation results show that when the ship is sailing at constant speed, with the change of water attack angle, the main distribution range of rudder blade stress deviates, and the lift and resistance of rudder blade are also gradually increasing, which is easy to damage the parts with concentrated stress. It has theoretical guiding significance for the protection and repair of rudder blade damage.

1. Introduction

During the navigation operation of the ship, the rudder blade plays an important role as a power mechanism, which can not only ensure the navigation stability and operability of the ship, but also improve the economy and rapidity of the ship through the optimization of structural functions[1]. The rudder blade system works in high humidity and high salinity waters, and bears high-power loads such as torque, impact and vibration, resulting in complex and diverse failure forms and affecting the normal operation of the ship[2].When the ship is sent to the dock for overhaul after long-term operation, the main maintenance problems of the rudder blade are steel plate fracture, impact loss and cleaning of marine attachments, and the damaged parts of the rudder blade are mainly concentrated on the upstream side of the rudder blade and the upstream surface of the steering. Therefore, it is of great significance for rudder blade protection and repair to simulate the working condition of rudder blade and explore the relationship between its component failure and the transient load caused by water flow impact.

Common rudder blades can be divided into single plate rudder and streamlined rudder according to different section shapes. The streamlined rudder is generally selected for long-distance navigation, so the representative streamlined rudder is selected for analysis.

2. Overview of Hydrodynamic Performance of Rudder Blade

The hydrodynamic performance analysis of the rudder blade is to obtain the dimensionless working coefficient of the rudder blade during underwater steering by calculating the overall working condition of the rudder blade at different angles of attack. The equivalent loads on the rudder blade can be obtained by converting the specified formula, so as to evaluate the possible damage of each part of the rudder blade.

When a ship is sailing on the water surface, the rudder blade will be affected by the force of water flow on it, which is hydrodynamics. As shown in Fig. 1, when the ship is working, water flows to the rudder blade at speed v and angle of attack φ, and the rudder blade is affected by the rudder pressure at this time. The rudder pressure on the rudder blade is equivalent to the resultant force P, and the operating point of the resultant force P is the intersection of the resultant force P and the chord of the rudder blade, which is related to the motion direction of the hull or rudder blade. The resultant force P can be decomposed into P_d in the direction of water flow and P_l in the direction perpendicular to the direction of water flow. It can also be decomposed into tangential force P_t and normal force P_n perpendicular to the string direction according to the string direction. The torques produced by the normal force N on the leading edge of the rudder blade and the rudder rod are M_0 and M[3].

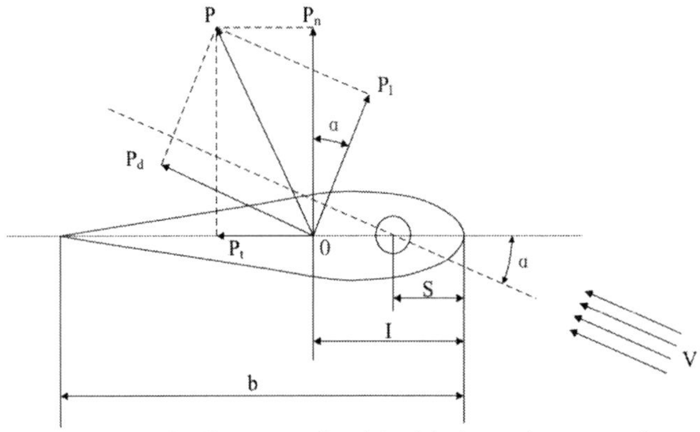

Fig. 1 Schematic diagram of rudder blade under water force

The hydrodynamic force and torque generated by the rudder blade under the action of water flow can be expressed by dimensionless coefficients. That is, the resultant force coefficient C, the lift coefficient C_l, the drag coefficient C_d, the tangential force coefficient C_t, the normal force coefficient C_n, the moment coefficient C_m and the rudder rod moment coefficient C_{m0}. The hydrodynamic performance analysis of the rudder blade is to obtain the variation of the hydrodynamic and dimensionless hydrodynamic coefficients of the rudder blade through the calculation of the rudder blade working at different angles of attack. Among them:

$$\begin{cases} C_n = \dfrac{P_n}{\frac{1}{2}\rho v^2 A}; \quad C_t = \dfrac{P_t}{\frac{1}{2}\rho v^2 A} \\ C_l = \dfrac{P_l}{\frac{1}{2}\rho v^2 A}; \quad C_d = \dfrac{P_d}{\frac{1}{2}\rho v^2 A} \\ C_m = \dfrac{M}{\frac{1}{2}\rho v^2 A b} \end{cases} \tag{1}$$

In the type, ρ is the density of water, kg/m^3; υ is flow velocity, m/s; A is the lateral projection area of the rudder blade, m^2; B is the profile width of the side of the rudder blade, m.

The correlation of the above dimensionless coefficients is as follows:

$$\begin{cases} C = \sqrt{C_n^2 + C_t^2} = \sqrt{C_d^2 + C_l^2} \\ C_n = C_l \cos\alpha + C_d \sin\alpha \\ C_t = C_d \cos\alpha - C_l \sin\alpha \\ C_l = C_n \cos\alpha - C_t \sin\alpha \\ C_d = C_t \cos\alpha + C_n \sin\alpha \end{cases} \tag{2}$$

At present, there are generally three methods for hydrodynamic performance analysis of rudder blades: hydrodynamic empirical formula, reference to the data of rudder blade model and simulation by using CFD (Computational Fluid Dynamics) software. The advantages and disadvantages of the three methods are shown in Tab. 1. Therefore, using CFD software to simulate the working condition of rudder blade has a high degree of simulation.

Tab. 1 Common analysis methods of hydrodynamic performance

Method for analysis of hydrodynamic performance	advantages	disadvantages
The hydrodynamic empirical formula is adopted	Simple calculation	The simulation is poor, and the error is large with the actual working situation
Refer to rudder model data	The model is similar to the actual rudder blade in geometry, motion and dynamics, and the experimental results are relatively accurate	The model experiment process is long, the experiment cost is relatively high, and the experimental condition is different from the actual condition
Applying CFD software	The experimental working condition is close to the actual working condition, and the experimental cost is relatively low	There are requirements for computer computing power

3. Calculation Process of Hydrodynamic Performance of Rudder Blade

3.1. Modeling and meshing

In this study, the hydrodynamic performance of streamlined rudder blades was analyzed, and the model of the rudder blade was semi-suspended. The NACA0015 airfoil profile was selected as the experimental object, and the hydrodynamic performance of the rudder blade under different water attack angles was calculated respectively, so as to make theoretical preparation for the underwater fault analysis of the rudder blade.

NACA0015 airfoil profile, chord length is 1.5m, aspect ratio $\lambda=1$, the modeling environment is set as Reynolds number $Re=0.81\times10^{-6}$, the rudder blade water attack Angle is 0°, 10°, 20°, 30° and 40° respectively, and the water velocity is set as 10m/s.

The boundary conditions of simulated flow field in the moving environment of rudder blade should be combined with the governing equation to make the simulation close to the actual working situation. Environmental boundaries include inlet, outlet, wall and far field boundary condition. The inlet was given the initial velocity of the flow field, and the underwater operation of the rudder blade was simulated. The outlet was set as the pressure outlet, the wall was set as the wall with no slip boundary, and the far field boundary of the fluid was set as the flow field was stable and undisturbed.

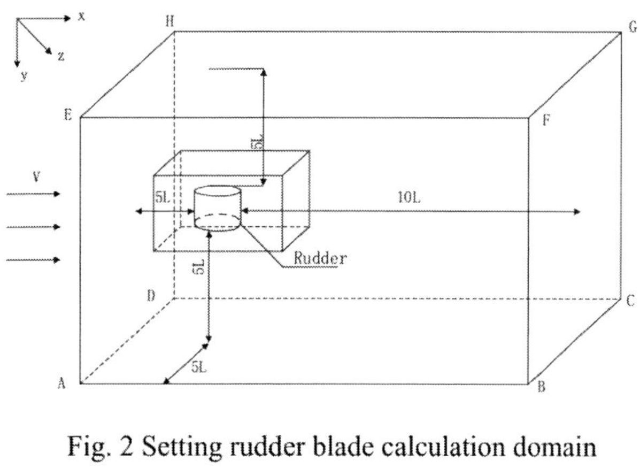

Fig. 2 Setting rudder blade calculation domain

The setting of rudder blade calculation domain is as follows: domain surface ADHE is set as inlet surface, and external calculation domain surface BCGF is set as outlet surface; the external domain surfaces ABCD, EFGH, ABFE and DCGH are set as two symmetrical planes. The surface along the - X direction and the surface along the X direction in the inner calculation domain are the boundary interfaces for rudder blade hydrodynamic calculation. The rudder blade modeling is shown in Fig.3 and Fig.4.

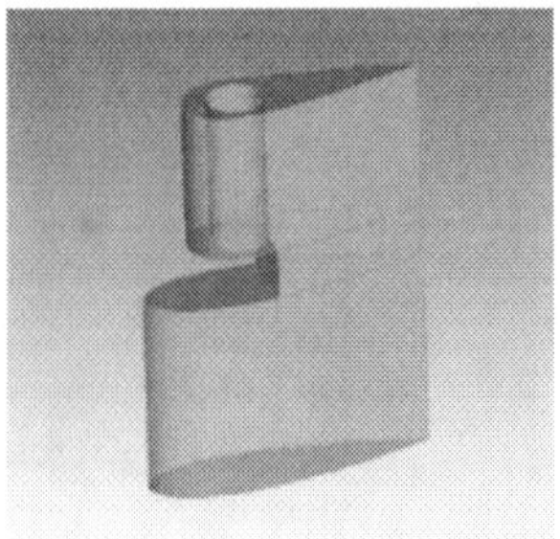

Fig. 3 Rudder blade modeling

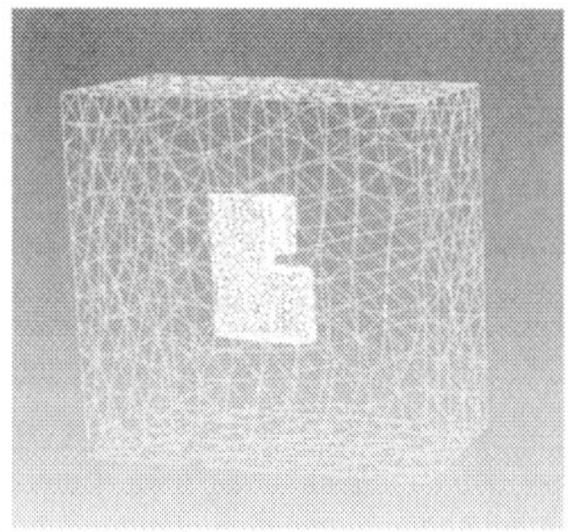

Fig. 4 Grid division of rudder blade and computing domain

3.2. Calculation results and analysis of hydrodynamic performance of rudder blade

In this chapter, five flow directions are used in the numerical simulation (0 °, 10 °, 20 °, 30 °, 40 °). The hydrodynamic performance of the rudder blade is simulated and calculated, and the pressure distribution cloud diagram of the rudder blade and the lift and drag coefficients of the rudder blade are obtained, and compared with the theoretical calculation results of the rudder blade, the influence of the angle of attack on the working performance of the rudder blade is analyzed.

3.2.1. Pressure distribution cloud diagram

The pressure distribution clouds of 0° and 20° are shown in Fig.5. In the diagram of each rudder blade, the right side is the upflow side, and the left side is the upflow side in the diagram of the back of the blade.

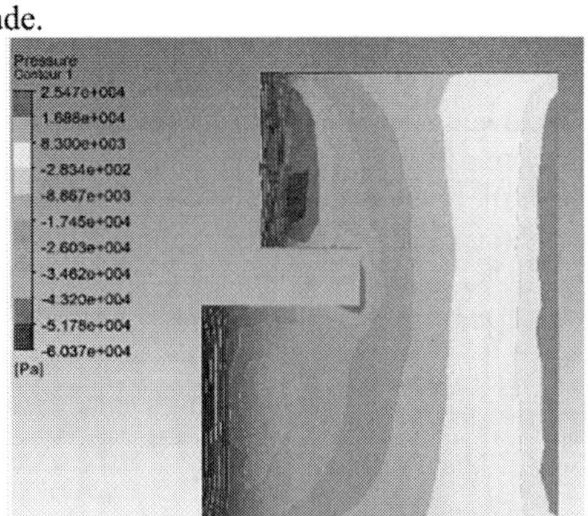

(a) Rudder blade positive

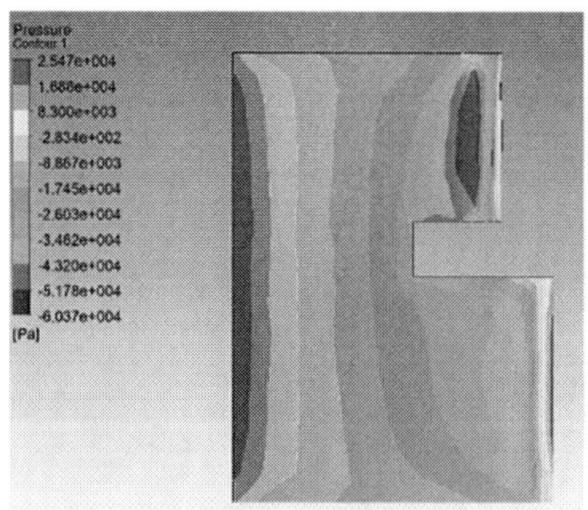

(b) Rudder blade back

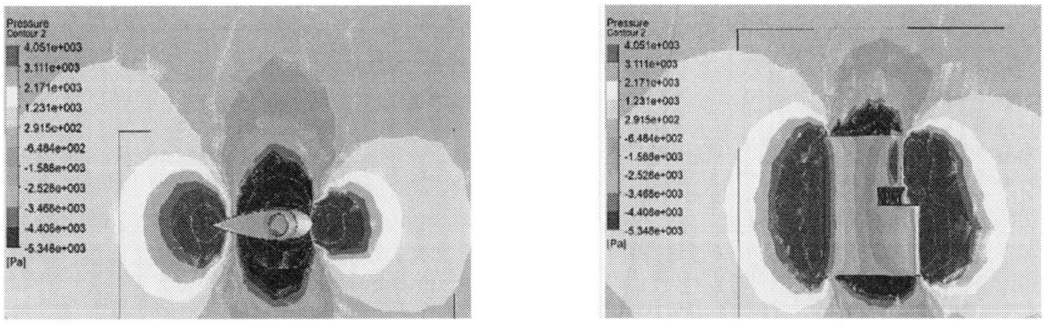

(c) x-y Cross section stress cloud (d) z-x Cross section stress cloud

Fig. 5 Cloud diagram of rudder blade pressure distribution at 0 ° angle of attack

According to the above pressure cloud diagram of the rudder blade, it can be concluded that under the condition of flow attack Angle, the average velocity of water flow on the rudder blade surface is smaller than that on the back of the rudder blade, so it is easy to produce pressure difference, namely suction.

3.2.2. Curve of rudder blade lift and drag coefficient

The lift force and drag coefficient curves of the semi-suspended streamlined rudder with section NACA0015 at different Angle of attack are as follows:

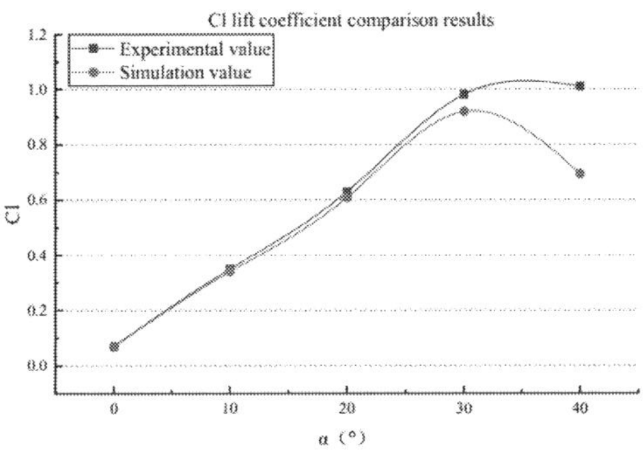

Fig. 6 Lift coefficient curve of rudder blade at different angles of attack

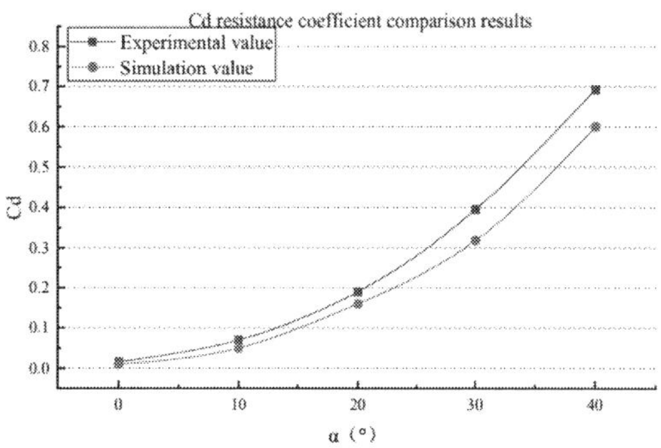

Fig. 7 Drag coefficient curve of rudder blade at different angles of attack

In the simulation, it can be seen that the stress on the upstream side of the rudder blade is large when the rudder blade is working, and the stress on the upstream surface of the rudder blade is large when steering, which is easy to impact and damage the rudder blade surface.

4. Conclusion

In this paper, s-A model and standard k -ε model were used to analyze the viscous flow field of the rudder blade under different working conditions, to study the stress distribution, lift and drag curves of the rudder blade under different flow attack angles, and to explore the relationship between the failure of the rudder blade component and the transient load caused by water impact. The main conclusions are as follows:

(1) Under the transient impact of water flow, the stress of rudder blade is concentrated during navigation, which is easy to cause damage or failure of rudder blade.

(2) When the ship is running, the main stress of the rudder blade is distributed on the inflow side and inflow surface, which is consistent with the fault problems encountered in practical maintenance. It can be verified that the transient impact of water flow is an important factor of the mechanical failure of the rudder blade.

In this experiment, the underwater working condition of rudder blade is simulated and analyzed, which is verified with the actual maintenance working condition, and the ship damage is analyzed in principle, which provides a guiding idea for ship maintenance and fault protection.

References

[1]Feng Song, Donghua Jiang, Tianling Dai. (2019) Rudder Blade Selection and Adaptive Parameter Optimization .J. Ship Standardization Engineer, 52(04) : 76-81.
[2]Deming Wu, Shouren Li, Shan Wang. (1995) Reliability Design of Ship Product Reliability Engineering .M. Harbin: Harbin Engineering University Press.
[3]Qinsheng Mei. (1981) Rudder for ship. People's Communications Press. Beijing.
[4]Jinqi Shi, Qingeng Zhang, Liwen Wang. (2016) Dynamic Simulation of Torpedo's Electric Servo Actuator .J. Torpedo Technology, 24(06): 463-468.
[5]Fei Peng, Yong Huang, Zhun Li. (2020) Finite Element Analysis of Strength of Vibrating Screen Vibrator Shaft Based on ANSYS Workben ch .J. Coal Mine Machinery, 41(3): 173-176.
[6] Xiongke Zhang, Zhong Wang, Hui Fan, et. (2021) Simulation analysis of impact load on actuating mechanism of double degree of freedom. Ship Science and Technology, 43(3): 19-23
[7]Yang Yong, Wang Zaigao, Huang Chao. (2021) Analysis of Hydrodynamic Characteristics of Balanced Rudder Based on Fluent .J. Ship Materials & Market,29(05): 27-31.
[8]Yuan Pei-yin, Zhao Yu, Lei Lin, Qin Huai-tao, Zhou Chuang, Li Dong-ying. (2021) Simulation analysis of dynamic characteristics of collision between ship and dock based on LS-DYNA. Ship Science and Technology, 43(3): 29-32

Comparing and Analysis of Test and Evaluation Methods for Connected Vehicle Communication Antenna System

Tong Zhu[1a], Guotian Ji[1b*], Guokai Jiang[2c], Yifu Ding[3d], Jiaxu Feng[2e], Changyuan Wang[3f]

[1] China Automotive Technology & Research Center Co., Ltd., No. 68, East Xianfeng Road, Dongli District, Tianjin, China.

[2] CATARC Software Testing (Tianjin) Co., Ltd. No. 68, East Xianfeng Road, Dongli District, Tianjin, China.

[3] China Automotive Research New Energy Vehicle Inspection Center (Tianjin) Co., Ltd., No. 68, East Xianfeng Road, Dongli District, Tianjin, China.

[a] zhuzhu75@126.com, [b*] jiguotian@catarc.ac.cn, [c] jiangguokai@catarc.ac.cn, [d] dingyifu@catarc.ac.cn, [e] fengjiaxu @catarc.ac.cn, [f] wangchangyuan@catarc.ac.cn

Abstract—With the development of Connected Vehicle, various of on-board applications and functions rely on stable, reliable and fast-response wireless connections furtherly, making it more urgent and widespread for vehicles to be equipped with relevant antenna systems such as positioning and communication. There are many types of vehicle antenna systems, together with a wide range of installation locations and a complex electronic environment which brings a huge challenges to the development of antenna manufacturers and the matching of automobile OEMs. In order to carry out the R&D, matching and testing of the vehicle antenna system better, various parties have proposed a multi-dimensional test and evaluation methods, but there is still a lack of horizontal evaluation between them. This paper analyses the requirements of automotive antenna system evaluation, sorts out and compares the technical paths of various common solutions, summarizes the characteristics and differences of various solutions, and provides a research foundation for further optimizing the vehicle-level antenna system test and evaluation program.

1. Introduction

With the in-depth integration of the automotive industry, electronics industry and communications industry, consumers' demand for automobiles has evolved from a simple transportation tool to more safety, comfort, and intelligence requests. Therefore, the on-board wireless communication function has been greatly developed. As an indispensable and important conversion device for wireless communication systems, the antenna is responsible for the mutual conversion function between the guided wave of the transmission line and the electromagnetic wave in space. The various applications of the vehicle make the number of vehicle antennas reach more than a dozen, and its wireless communication function highly depends on the underlying RF performance of the vehicle antenna system. Therefore, OEMs and antenna suppliers are paying more attention to the antenna system. For R&D and testing, various parties have proposed a variety of vehicle-mounted antenna test programs from different dimensions. This article will research the technical characteristics and differences of each solution which start from the needs of the automotive industry for antenna systems, and promote the further development of vehicle antenna system testing and evaluation programs.

2. On-board Antenna Type and Distribution

With the further deepening of vehicle intelligence and networking, the realization of its sensing and communication functions is closely related to the antenna. The on-board antennas are abundant in variety and increasing in number. The frequency distribution is shown in Figure 1.

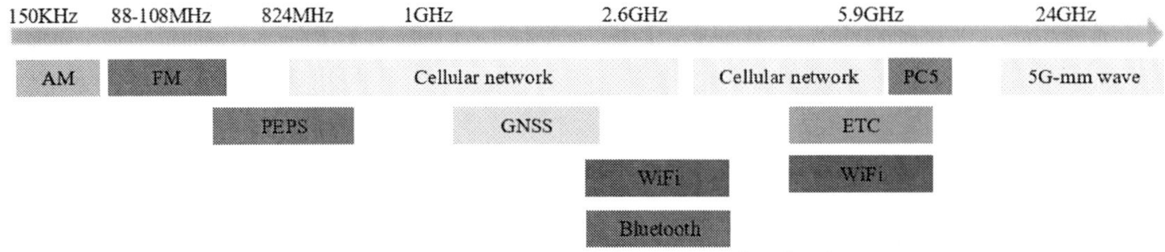

Fig.1 Antenna type and frequency distribution

Generally, it is installed in different positions of the vehicle according to the characteristics of the antenna function and application [1]. The installation position of the common on-board antenna is shown in Figure 2.

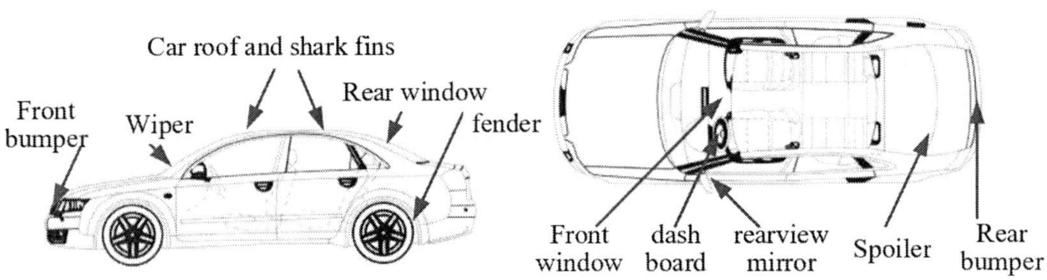

Fig. 2 Common vehicle antenna installation location

According to the functional requirements realized by the vehicle antenna, the different types of vehicle antenna installation positions are:

- Broadcast Radio Antenna

This type of antenna is used in AM, FM, DAB and other broadcast radio communications, mainly installed in: car window glass, roof short pole or shark fin, trunk or mudguard long mast, rear-view mirror, bumper, etc.[2] .

- Cellular Network Antenna

This type of antenna is used in cellular mobile communications such as GSM, GPRS, EDGE, CDMA (EV-DO), WCDMA, HSPA, LTE, 5G, etc. It is mainly installed on: roof short poles or shark fins, roof smart antennas (integrated), Dashboard, centre console, rear-view mirror, bumper, etc.

- Direct Communication Antenna

This type of antenna is used in PC5, DSRC direct communication, and is mainly installed in: roof shark fins, rear-view mirrors, roof smart antennas (integrated), etc. [3].

- Global Navigation Satellite System(GNSS) Antenna

This type of antenna is used in GPS, Beidou, Glonass, Galileo and other satellite positioning and positioning communications, and is mainly installed in: roof shark fins, roof smart antennas (integrated), dashboards, centre consoles, etc.

- Infotainment and Communication antenna

This type of antenna is used in Bluetooth, WLAN and other infotainment communications, mainly installed in: dashboard, centre console, roof shark fin, rear-view mirror, etc.

- Smart Entry System Antenna

This type of antenna is used in PEPS, PKE, RKE and other automotive intelligent entry systems, and is mainly installed on the roof, the rear of the car, the side of the car, etc.

Among the above types of antennas, cellular wireless communication antennas, direct communication antennas, and satellite navigation antennas are closely related to vehicle safety and are the focus of this paper.

3. Analysis of Requirements for Testing and Evaluation of Vehicle Antenna Systems

The radio frequency performance of the vehicle antenna system directly affects the performance of the vehicle's cellular wireless communication, direct communication and global satellite navigation. Therefore, the antenna system, as the necessary underlying support for vehicle perception and communication functions, has received more and more attention from OEMs and Tier 1 for its radio frequency performance.

The performance of the vehicle antenna system is affected by many factors, including:

- The configuration of the installation location, the material of the car body, and the installation angle.
- Vehicle occlusion and reflection.
- Layout of wiring harness.
- Harassment from other electronic systems of the vehicle.

The radio frequency performance of the vehicle antenna is not affected by a single component or a single factor [4]. Therefore, there is a significant difference between antenna component-level testing and vehicle-level testing. Its performance is a system problem. R&D and testing need to be comprehensively designed and investigated from the vehicle level. With the deepening of industry research, all parties have derived a variety of different technical solutions based on different foundations and ideas. However, the automotive industry does not have a unified standard to regulate the evaluation system and test methods, resulting in differences in the actual loading performance of the vehicle antenna system, which has a certain restrictive effect on the development of the industry. This article will further research the principles and characteristics of each technical programs, and explore the direction of future technical routes.

4. Research of the Test Program Technical Path

4.1. Single Antenna Passive Radio Frequency Performance Test (Test Program 1)

Due to the large size of the vehicle-level full-wave anechoic chamber and the high construction cost, some OEMs still test the circuit characteristics and the passive radio frequency performance of the single antenna(see Figure 3) supplemented by the vehicle field function verification to research and development.

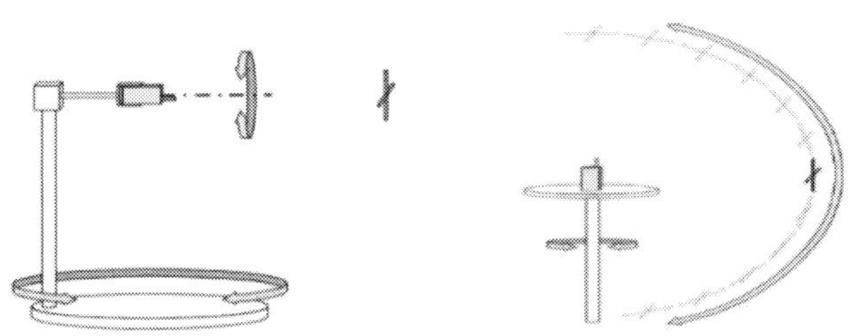

Fig.3 test of single antenna

Due to the small size of the single antenna, the solution has low requirements on the size of the full-wave anechoic chamber, and the test of the antenna unit is more convenient. However, the program also has greater risks, which are mainly manifested in:

- Only use the CW signal to test the passive radiation characteristics such as the pattern, gain, polarization, etc., without considering the OBU and actual communication signal characteristics, it is difficult to evaluate the receiving sensitivity of the radiation power meter.
- Without considering factors such as actual vehicle installation and matching scenarios, vehicle electronic and electrical architecture configuration, electromagnetic compatibility, etc., only the radio frequency characteristics of the antenna can be obtained.
- Once the vehicle function verification effect is not good, it is difficult to locate the problem and analysis quantitatively, and the rectification is difficult.

Therefore, the test plan is changed to more applications and the R&D and testing of radio and infotainment antennas. As more and more antennas are involved in driving safety issues, this solution is not enough to support R&D. OEMs, testing institutions and scientific research institutions have gradually shifted the research direction to vehicle-level antenna testing (see test program 2, 3, and 4).

4.2. Passive RF Performance Test for Vehicle-level Antenna (Test Program 2)
Different from Test Plan 1, the passive radio frequency performance test program of the whole vehicle antenna considers the actual installation position of the antenna, the material of the vehicle body, the installation angle, the layout of the wiring harness, and the vehicle blocking reflection. An example of the test layout is shown in Figure 4.

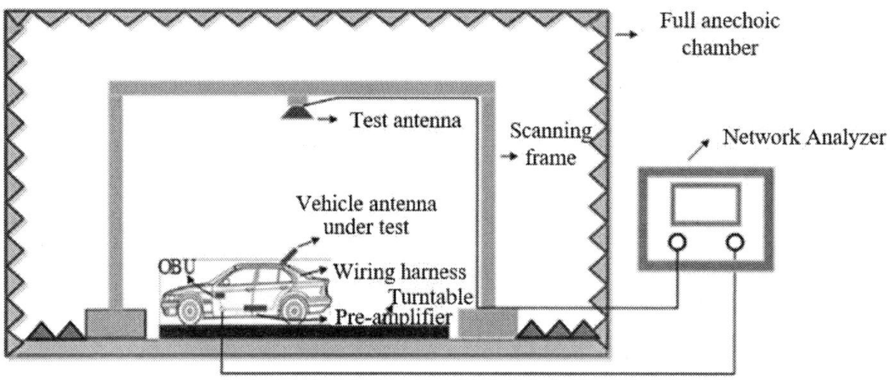

Fig.4 Example of layout of passive radio frequency performance test for vehicle antenna

4.3. Integrated Test for Antenna System Active Radio Frequency Performance on the basis of Vehicle Radiation Aperture (Test Program 3)
The function of the vehicle system depends on the underlying radio frequency performance of the antenna system, so it is very necessary to test and evaluate the antenna system under the actual vehicle installation state. Due to the configuration of the antenna installation position, the vehicle body material, the installation angle, the layout of the wiring harness, the vehicle occlusion and reflection, and the disturbance of other vehicle electronic systems, there is a significant difference between the antenna component level test and the vehicle level test. So it is the most direct and effective to test the vehicle antenna system composed of OBU, wiring harness and antenna [5].

The ideal solution for the active test of the vehicle-level antenna system (OBU-Harness-Antenna) is to use the far-field or compact range solution for testing. The test items generally include the transmitting power and receiving sensitivity corresponding to the upper hemispherical surface of the car (as shown in Figure 5) or a specific angle in the upper hemispherical surface (as shown in Figure 6).

For the far-field test program, the distance between the test antenna and the vehicle under test satisfies the following relationship:

$$r = 2D^2/\lambda \tag{1}$$

where r is the test antenna and the vehicle under test;D is the radiation size (the size of the metal shell of the vehicle connected to the antenna is the default if unknown);λ is the wavelength.

Taking the C-V2X antenna of a passenger car as an example, the far-field distance between the test antenna and the vehicle under test will exceed 1,000 meters.

There are strict requirements for the outdoor far field. In addition to the size of the site, in order to ensure the consistency and reproducibility of the test, it is also necessary to ensure a relatively clean electromagnetic field environment around the site to avoid co-frequency and adjacent-frequency interference. And considering the further expansion of the on-board frequency, the size of the outdoor far field still has the possibility of further expansion. Therefore, the field size factor and electromagnetic environment factor are a big challenge for the outdoor far field.

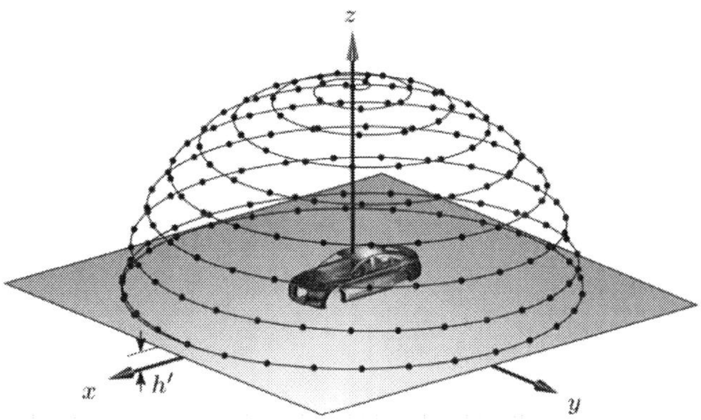

Fig.5 Example of vehicle active test range (upper hemisphere).

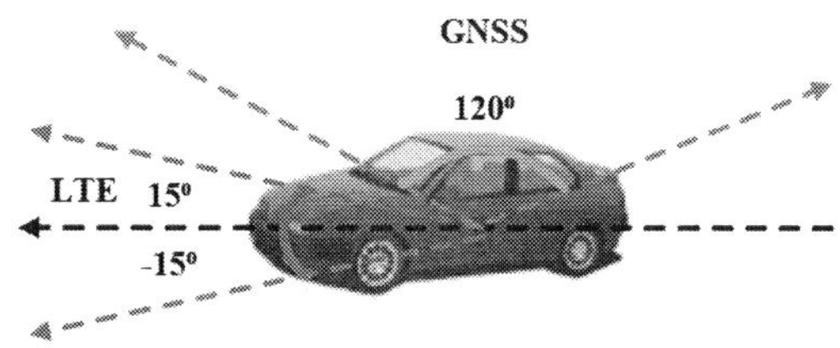

Fig.6 Example of vehicle active test range (a specific angle on the upper hemisphere).

In order to avoid electromagnetic environment interference and compress the test distance at the same time, a compact range solution with mature technology can be adopted, as shown in Figure 7. The compact range program converts spherical waves into plane waves through the reflecting surface, and reduces the size of the vehicle's darkroom to tens of meters on the premise of achieving the far-field purpose. At the same time compact range program can form a certain size of quiet zone space. After measuring the quiet zone emission level, amplitude taper, amplitude fluctuation, phase error, and cross polarization, the far-field requirement can be achieved.

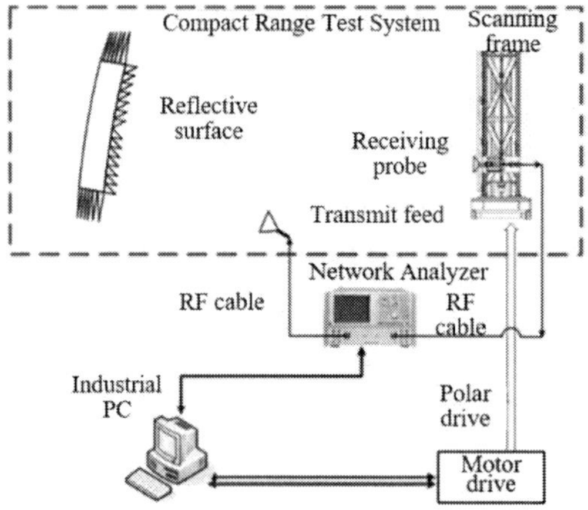

Fig.7 Example of compact range test program.

4.4. *Simplified test for vehicle-level active RF performance of antenna system (Test Program 4 and 5)*

The tight field test program has mature applications in the communications industry, but has not yet been applied in the automotive industry. The main reasons are:

- The connected vehicles has developed rapidly in recent years, and its realized functions involve safety. The testing of communication antennas has received more and more attention from OEMs, but it is still in the initial stage of development;
- Due to the large quiet space corresponding to the size of the car, the load-bearing capacity of the turntable used by the car is relatively high, which makes the cost of the compact range higher.

Taking into account the necessity of active testing of vehicle-level antenna systems, some research institutions and program suppliers have turned to research simplified schemes for vehicle-level antenna systems.

4.4.1. *Simplified plan for step-by-step testing (Test Program 4)*

A simplified test program is designed based on the assumption that the electronic disturbance of the vehicle will not affect the antenna system. Firstly, the program tests the passive radio frequency performance of the vehicle antenna (Program 2), obtains the vehicle-level pattern of the antenna, and tests the receiving sensitivity of the OBU's transmit power meter by conduction. Through the mathematical derivation and combination of the directional pattern and the OBU transmit/receive characteristics, the vehicle-level active radio frequency performance is calculated.

This solution is simple to operate, but will have certain differences due to the different electromagnetic characteristics generated by the electrical and electronic components of different vehicles.

4.4.2. *Simplified plan for equivalent radiation aperture (Test Program 5)*

The equivalent radiation aperture program is based on the assumption that the scattered current of the antenna decays rapidly on the metal surface of the vehicle body, and the antenna radiation area is limited to a certain size around the antenna. According to the formula (1), when the radiation size D is further reduced, the test distance between the test antenna and the vehicle under test will be effectively reduced.

The program ensures the integrity of various key factors by directly testing the active radio frequency indicators of the vehicle antenna. However, since the radiation size is affected by many factors such as signal characteristics, body material, installation angle, etc., its range is difficult to determine.

5. Test Program Comparison

Through analysing the research of the above 5 kinds of antenna test program, the advantages and disadvantages can be analysed which is shown in Table 1.

Table.1 Antenna test program comparison.

program	Test method description	Advantages	Disadvantages
1	Single antenna passive test	Quickly obtain antenna single circuit characteristics and passive radio frequency characteristics	Without considering factors such as actual vehicle installation and matching scenarios, vehicle electronic and electrical architecture configuration, EMC, etc., it is difficult to troubleshoot antenna matching problems
2	Passive test of vehicle antenna (Near to far field conversion)	The test distance is short, the test is easy to implement, and the passive radio frequency indicators such as the antenna pattern of the whole vehicle can be obtained	Cannot directly perform active RF index test
3	Active test of vehicle antenna (Far field, compact range)	The test distance is shortened by reflecting surface or phased array technology, and the active radio frequency index test is realized	The cost is high, and the project is difficult to realize
4	Active test of vehicle antenna (Simplified step by step)	Easy to implement	Unable to obtain the electromagnetic influence of the electronic and electrical components on the vehicle
5	Active test of vehicle antenna (Simplified equivalent radiation aperture)	The test system is simple, the test is easy to implement, and it can verify the influence of the electromagnetic characteristics of electronic and electrical components on the antenna system	Difficult to obtain equivalent radiation size

6. Conclusion

The rapid development of vehicle intelligence and networking has made the types and numbers of antennas on vehicles increasingly abundant. The wireless communication function is highly dependent on the underlying radio frequency performance of the vehicle antenna system. It has become an industry consensus that the test and evaluation should be carried out in conjunction with the entire vehicle. In order to solve this problem, various test schemes have emerged. How to comprehensively consider test evaluation schemes from different dimensions such as test accuracy and convenience has become a key issue of industry research. This article analysed the technical characteristics and advantages and disadvantages of each program to help promote the further development and improvement of the vehicle-mounted antenna system test and evaluation program.

References

[1] Rabinovich V, Alexandrov N, Alkhateeb B. Automotive antenna design and applications[J]. 2010.
[2] GM Worldwide Engineering Standard, "Vehicle AM/FM, T-DAB/T-DMB, RFR, Gain, Directionality, and Impedance Test", [S], 2018.

[3] 3GPP, "Study on enhancement of 3GPP Support for 5G V2X Services (Release 16)," Tech. Report TR 22.886, Revision 16.2.0, Dec. 2018.
[4] Saporetti M.A., Foged L., Casta M.S., et al. Description and results: Antenna measurement facility comparisons [J]. IEEE Antennas and Propagation Magazine, 2017, 59(3) : 108-116.
[5] 3GPP, "User Equipment (UE) Over The Air (OTA) performance; Conformance testing," Tech.Specif. TS 37.544 v14.5.0, 2018.

Multistage noise reduction processing for vibration signal of hydropower units

Yousong Shi [1a*], Jianzhong Zhou [1b*]

[1] School of civil and hydraulic engineering, Huazhong university of science and technology Wuhan 430074, China

[a*] d201880942@hust.edu.cn, [b*] jz.zhou@hust.edu.cn

Abstract—In actual field testing environments of hydropower units, unit vibration signals are often contaminated with noise. In order to obtain the real vibration signal, a multi-stage vibration signal denoise method SG-SVD-VMD is proposed for the guide bearing nonlinear and non-stationary vibration signals. And the root mean square error (RMSE) and signal to noise ratio (SNR) are used to evaluate the noise reduction ability of eight methods. The results show that the noise-canceling ability of this proposed method has improved to some extent. It can effectively suppress the noise of the hydropower units vibration signals. This method can effectively identify the shaft track and the running state of hydropower units.

1. Introduction

The operating environment of the hydroelectric unit is harsh, especially the frequent change of working conditions of pumped-storage generating units, which leads to strong nonlinear and non-stationarity vibration of the unit, meanwhile, there is strong noise in the process of shafting vibration data collection. How to quickly and effectively extract the characteristics of such signals has always been a key and difficult point in the field of signal processing[1].

In recent years, many experts and scholars have done a lot of research on the noise reduction of vibration signals of hydropower units and achieved remarkable results. Wavelet transform[2,3] essentially is a band-pass filter. It is more suitable for the decomposition of linear and steady-state signals. For the non-linear and non-stationary vibration signals of hydropower units, the effect of noise reduction by using wavelets is not particularly desirable. Empirical mode decomposition (EMD) [4,5] has high-frequency resolution and adaptive decomposition characteristics, but there are some defects of model mixing and end effect. In order to overcome the problem of different mode aliasing in EMD, Dragomiretskiy et al. [6] proposed variational mode decomposition(VMD) method based on optimal to solve the variational problem, the given signal is decomposed into finite bandwidth component sets. Singular value decomposition [7] (SVD) is different from the above method in principle. It reconstructs the signal and achieves nonlinear filtering through orthogonal decomposition of high dimensional Hankle matrix and effective singular value screening. The core idea of Savitzky Golay(SG) Filter is also to perform weighted filtering on the data in the window, but its weight is obtained by least square fitting of a given high-order polynomial. The advantage of this method is that it can keep the change information of signal more effectively while filtering and smoothing. To improve the analysis accuracy of the pendulum signal of hydropower units, the characteristic frequency components are extracted. Thus, a multi-stage noise reduction method SG-SVD-VMD is proposed for the vibration monitoring data of the upper, lower, and water guide bearings, and verify its effectiveness.

2. SG-SVD-VMD fusion signal denoising principle

2.1. SG smoothing filtering

SG smoothing filter[8] is widely used in data stream smoothing and noise reduction, it is using the least square method to fit the signal sequence by moving the window in the time domain, which can effectively remove the noise without changing the shape and width of the original signal. First, a window array with $2M+1$ sampling points centered on x_i is constructed, and then a p-order polynomial $q(n)$ is constructed to fit the array:

$$q(n)=\sum_{k=0}^{p}a_k n^k,\quad -M\le n\le M,\ p\le 2M+1 \tag{1}$$

Defined function:

$$C=\sum_{n=-M}^{M}(q(n)-x(n))=\sum_{n=-M}^{M}\left(\sum_{k=0}^{p}a_k n^k-x(n)\right) \tag{2}$$

When Eq (1) gets the minimum value, the filtering effect is the best, and then all the fitting points of the original data can be obtained by moving the window array. In the process of filtering, the noise part that deviates too much from the normal trend curve is eliminated, which plays the role of smooth filtering, to reduce the signal to deal with the noise. The signal obtained after smooth filtering is as follows:

$$f(i)=\sum_{k=-M}^{M}h(k)y(x-k) \tag{3}$$

where $f(i)$ is the filtered signal, $h(k)$ is the sampling response of the smoothing filter, $y(x)$ is a group of data in the original signal, $x=-N,\cdots,0,N$.

2.2. Singular Value Decomposition (SVD)

As an orthogonalization decomposition method, the key to SVD noise reduction lies in the construction of the Hankel matrix and the selection of effective singular values[9].

Given a signal sequence of $\{v_i\}$, Hankel matrix is constructed as follows:

$$\boldsymbol{H}=\begin{bmatrix} v(1) & v(2) & \cdots & v(q) \\ v(2) & v(3) & \cdots & v(q+1) \\ \vdots & \vdots & \ddots & \vdots \\ v(d) & v(d+1) & \cdots & v(N) \end{bmatrix} \tag{4}$$

where $N=d+q-1, d>q$, N is the length of the acquired signal.

By SVD of the matrix $\boldsymbol{H}$, we can get:

$$\boldsymbol{H}=\boldsymbol{U}\boldsymbol{\Delta}\boldsymbol{V}^{\mathrm{T}}=\sum_{i=1}^{d}\theta_i u_i v_i^{\mathrm{T}} \tag{5}$$

in Eq(5), u_i and v_i are orthogonal vectors $\boldsymbol{U}\in\mathbf{R}^{d\times d}$ and $\boldsymbol{V}\in\mathbf{R}^{q\times q}$, respectively. θ is the singular value of $\boldsymbol{H}$; The expression of the diagonal matrix $\boldsymbol{\Delta}$ is as follows:

$$\boldsymbol{\Delta}=diag(\theta_1,\theta_2,\cdots,\theta_d) \tag{6}$$

in Eq(6): θ_i satisfies $\theta_1\ge\theta_2\ge\cdots\theta_d\ge 0$.

For noise-free signals, the diagonal matrix Δ is full rank. For a noisy signal, the effective singular values are usually concentrated in a larger part. After the Hankel matrix is obtained, the noise reduction effect of SVD is only affected by the effective singular value screening. Commonly used effective singular value screening methods include median filtering, mean filtering, differential spectral filtering, etc[10-11]. In this paper, we choose to mean filtering for noise reduction.

2.3. Variational mode decomposition (VMD)

VMD is an adaptive, non-recursive variational mode decomposition method proposed by Dragomiretskiy et al. in 2014, which can decompose the signal into the sum of finite IMF components. The main steps of VMD are as follows [12] :

For each mode function $u_k(t)$, the Hilbert transform is used to calculate the corresponding analytic signal, and then its one-sided spectrum is obtained.

$$f_s = \left[\delta(t) + \frac{\mathrm{j}}{\pi t} \right] * u_k(t) \tag{7}$$

For each mode function $u_k(t)$, the spectrum of each mode is modulated to the corresponding baseband by aliasing the exponential term of its corresponding center frequency

$$B = \left\{ \left(\delta(t) + \frac{\mathrm{j}}{\pi t} \right) * u_k(t) \right\} \mathrm{e}^{-\mathrm{j} w_k t} \tag{8}$$

The Gaussian smoothing method of demodulation signals is used to estimate the bandwidth of each modal signal, i.e. the gradient square norm, and then solve the variational problem with constraints. The constrained variational expression is

$$\begin{cases} \min \left\{ \sum_k \left\| \partial(t) \left[\left(\delta(t) + \frac{\mathrm{j}}{\pi t} \right) * u_k(t) \right] \mathrm{e}^{-\mathrm{j} w_k t^2} \right\|_2^2 \right\} \\ \text{s.t.} \sum_k u_k = f(t) \end{cases} \tag{9}$$

in Eq(9), $\{u_k\} = \{u_1 \cdots u_k\}$ is K IMF components obtained by decomposition, $\{\omega_k\} = \{\omega_1 \cdots \omega_k\}$ is the frequency center of each component, * is the convolution, $\partial(t)$ is the derivative of the function at time t, $\delta(t)$ is the unit impulse function.

The quadratic penalty factor α and the Lagrange multiplication operator $\lambda(t)$ are used to solve the solution of Eq (9), and the constrained variational problem is turned into an unconstrained problem, i.e:

$$L(\{u_k\}, \{\omega_k\}, \lambda) = \alpha \sum_k \left\| \partial_t \left[\left(\delta(t) + \frac{\mathrm{j}}{\pi t} \right) * u(t)_k \right] \mathrm{e}^{-j\omega_k t} \right\|_2^2 + \left\| f(t) - \sum_k u_k(t) \right\|_2^2 + \left\langle \lambda(t), f(t) - \sum_k u_k(t) \right\rangle \tag{10}$$

The alternative direction multiplier method is adopted to solve the above variational problem, and the extended Lagrangian saddle point is sought by alternatively updating u_k, ω_k and λ_{n+1}. The solution of the variational problem is as follows:

$$\hat{u}_k^{n+1}(\omega) = \frac{\hat{f}(\omega) - \sum_{i-1, i \neq k}^{k} \hat{u}(\omega) + \frac{\hat{\lambda}(\omega)}{2}}{1 + 2\alpha(\omega - \omega_k)^2} \tag{11}$$

Similarly, the updating method of the center frequency is

$$\hat{\omega}_k^{n+1} = \frac{\int_0^\infty \omega |\hat{u}(\omega)|^2 \, \mathrm{d}\omega}{\int_0^\infty |u(\hat{\omega})|^2 \, \mathrm{d}\omega} \tag{12}$$

The iteration stop criterion is as follows:

$$\sum_k \left\| \hat{v}_k^{n+1} - \hat{v}_k^n \right\|_2^2 \Big/ \left\| \hat{v}_k^n \right\|_2^2 < \varepsilon \tag{13}$$

Finally, K IMF components are obtained according to the frequency domain characteristics of the actual signal. When the signal is denoised by VMD, the presetting of mode number K value and the selection of effective modes after VMD decomposition affect the denoising effect of the final signal. In this paper, the minimum Babbitt distance(MBD) method is used to determine the optimal K value by calculating the MBD between the modal components after VMD decomposition under different K values. Further, the effective component is determined according to the mutual correlation between each component and the original signal, and the effective component is fused to obtain the denoised signal.

2.4. SG-SVD-VMD fusion signal denoising

In order to filter out the noise components of shafting vibration data and improve the signal-to-noise ratio of the data. We proposed a multi-level noise reduction method combining SG, SVD, and VMD, and the specific process is shown in Fig. 1.

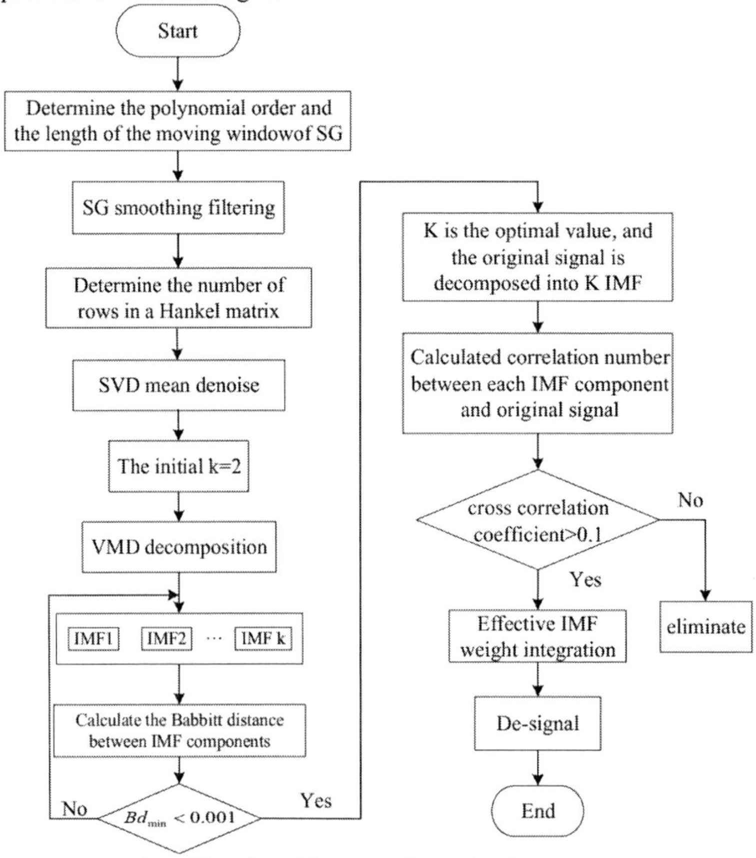

Fig.1 Signal multistage noise reduction process

In this paper, we set the polynomial order of the SG smoothing filter as 5 and the length of the moving window as 41. Meanwhile, set the number of rows for the reconstructed Hankel matrix to 8.

In order to reveal the performance of the proposed method in noise reduction, we selected the SNR and RMSE as evaluation criteria. The measured data of shafting vibration of a power station were de-noised at multiple levels. As can be seen from Table 1, After SG-SVD-VMD denoising, the shafting vibration data has higher SNR and smaller RMSE, and the data before and after de-noising were shown in Fig. 2, and the spectrum diagram before and after shafting vibration denoising are shown in Fig. 3. Meanwhile, As can be seen from Fig.2 and Fig.3, The Shafting vibration condition monitoring data by

SG-SVD-VMD multi-stage noise reduction, some noise components have been filtered out. In conclusion, it can be seen that the SG-SVD-VMD fusion multi-stage noise reduction method has a good noise reduction effect and universality.

Table 1 The SNR and RMSE of each signal denoise method

		X_3	Y_3	X_4	Y_4	X_5	Y_5
VMD	SNR	17.6353	16.0648	19.7347	16.3642	5.9473	4.8077
	RMSE	0.0818	0.0810	0.0570	0.0863	0.1828	0.2108
SG-VMD	SNR	16.0114	14.1497	15.2229	14.1068	5.5645	4.8988
	RMSE	0.0987	0.1010	0.0956	0.1118	0.1911	0.2086
SVD-VMD	SNR	17.5892	16.0299	16.8659	16.3438	8.5060	4.8702
	RMSE	0.0823	0.0813	0.0791	0.0864	0.1362	0.2093
SVD-SG-VMD	SNR	16.0286	13.7879	13.2089	14.0245	5.5583	4.9030
	RMSE	0.0985	0.1053	0.1205	0.1129	0.1912	0.2087
VMD-SG	SNR	17.6333	16.0305	19.7115	16.3774	5.8921	4.7716
	RMSE	0.0819	0.0813	0.0571	0.0861	0.1840	0.2117
VMD-SVD	SNR	17.5842	16.0022	19.6979	16.3266	5.8383	4.7849
	RMSE	0.0823	0.0816	0.0572	0.0866	0.1851	0. 2114
VMD-SVD-SG	SNR	17.6131	15.9934	19.7435	16.3497	5.8492	4.7694
	RMSE	0.0821	0.0817	0.0569	0.0864	0.1849	0.2118
VMD-SG-SVD	SNR	17.5872	15.9790	19.6888	16.3201	5.8118	4.7769
	RMSE	0.0823	0.0818	0.0572	0.0867	0.1857	0.2116
SG-SVD-VMD	**SNR**	**18.7629**	**16.7913**	**19.7437**	**16.3663**	**6.2178**	**4.9064**
	RMSE	**0.0719**	**0.0745**	**0.0568**	**0.0862**	**0.1772**	**0.2085**

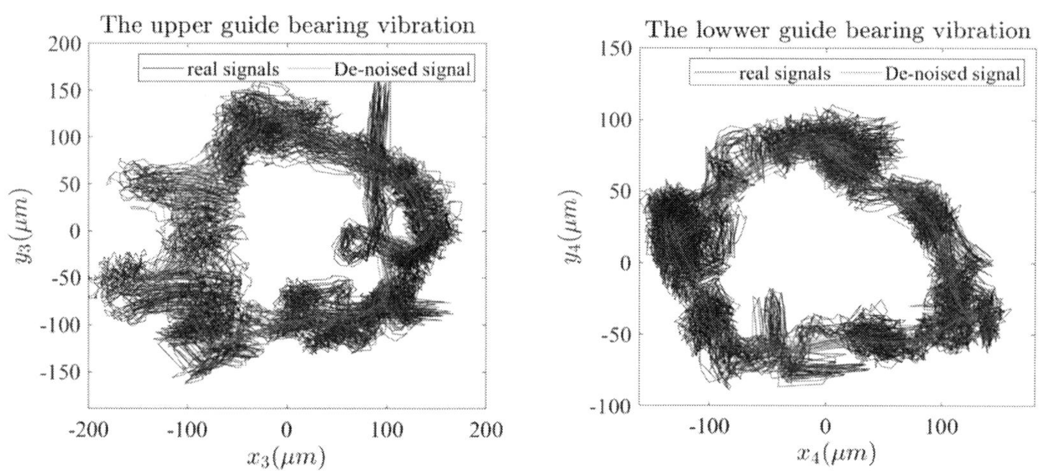

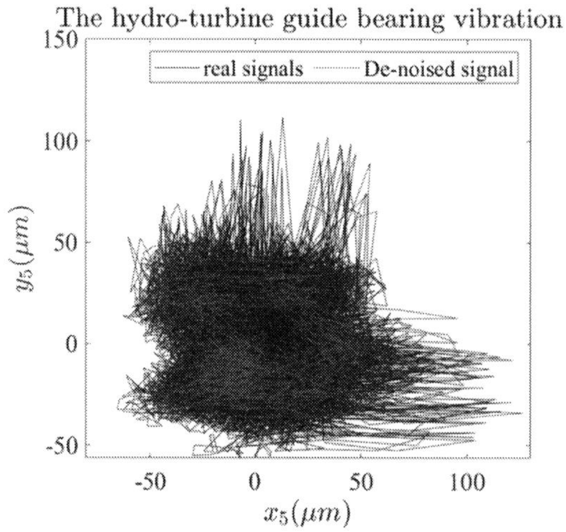

Fig.2 shafting vibration before and after de-noising

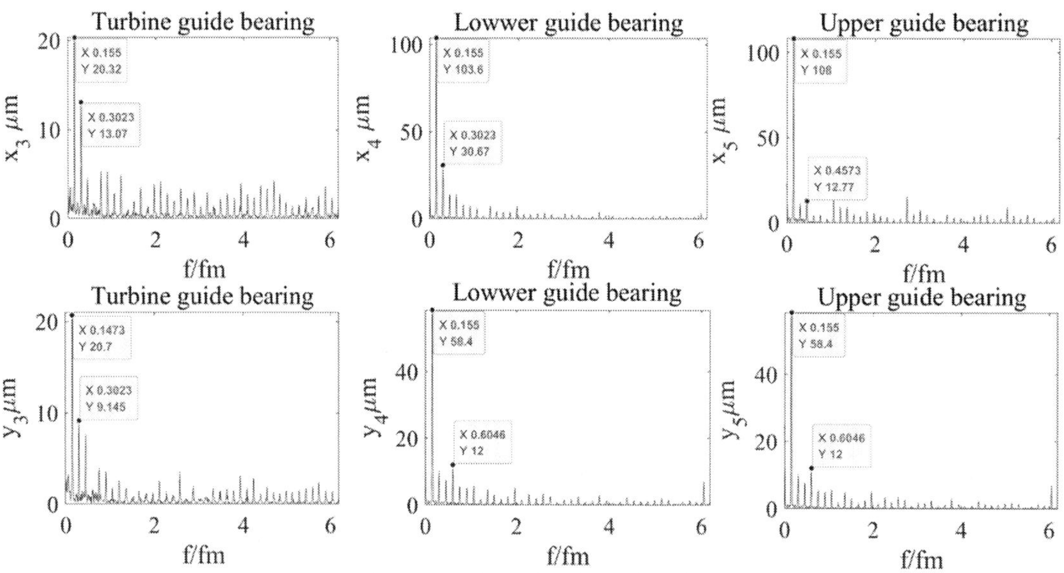

(a) Spectrum diagram before shafting vibration denoising

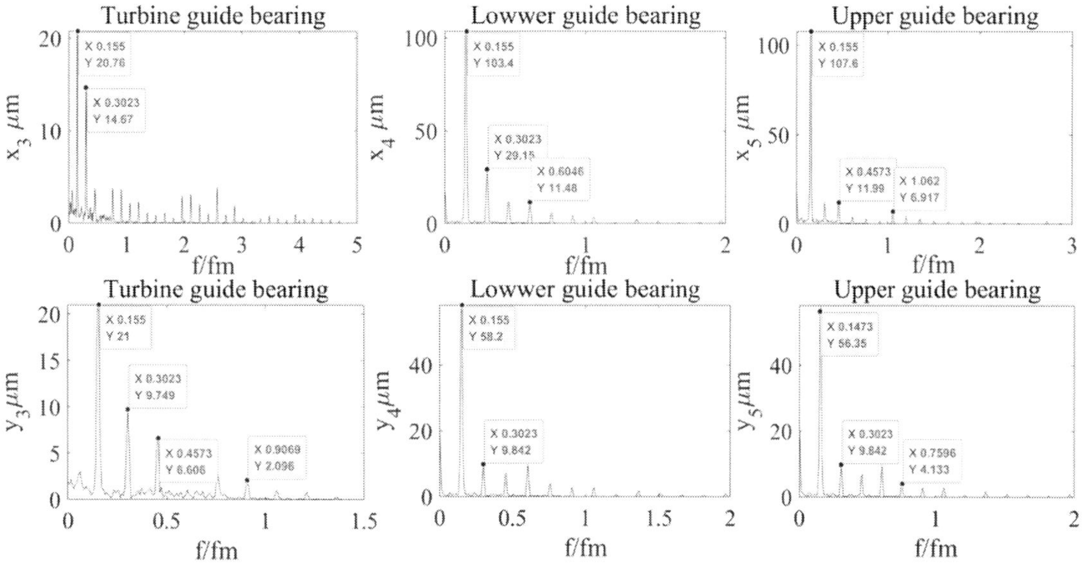

(b) Spectrum diagram after shafting vibration denoising
Fig.3 Spectrum diagram of the shafting vibration

3. Conclusion

In this paper, a multi-stage noise reduction method SG-SVD-VMD for vibration signal of hydropower units is proposed. Firstly, the SG filter was used to smooth the filtering, and then SVD was used to de-noise and reconstruct the filtered signals. Finally, the reconstructed signal is decomposed by VMD, the number of decomposition is determined by the MBD and the effective components are determined by the coefficient of correlations between the decomposed signal and the original signal to obtain the real vibration signal of the unit. The results show that the parameters of SNR and RMSE for the SG-SVD-VMD method have an enhancement to some extent compared with the other denoising methods. Real vibration signals of hydropower units demonstrate the proposed method's performance of noise reduction. It can effectively remove the noise to timely and accurately obtain the true state of the unit. The proposed method provides a new tool for rotor system fault diagnosis, has good application prospects.

Acknowledgments

This work was supported by the National Key R&D Program of China (GrantNo.2016YFC0402205), National Natural Science Foundation of China(U1865202), National Natural Science Foundation of China(52039004) Fundamental Research Funds for the Central Universities , HUST : 2021JYCXJJ042

References

[1] Shi Y , Zhou J , Lai X , et al. Stability and sensitivity analysis of the bending-torsional coupled vibration with the arcuate whirl of hydro-turbine generator unit[J]. Mechanical Systems and Signal Processing, 2021, 149:107306.

[2] An Xueli, Jiang Dongxiang, Liu Chao, et al. Wind farm power prediction based on wavelet decomposition and chaotic time series. Expert Systems with Applications, Vol. 38, Issue 9, 2011,p. 11280-11285.

[3] Zhang Xin, Feng Naizhang, Wang Yan, et al. Acoustic emission detection of rail defect based on wavelet transform and Shannon entropy. Journal of Sound and Vibration, Vol. 339, 2015, p. 419-432.

[4] Luo Xingqi, Xue Yangang, Wang Han Singular data restoration for hydraulic generator fault based

on EMD-wavelet analysis. Journal of Hydroelectric Engineering, Vol. 31, Issue 2, 2012, p. 245-250.

[5] Wang Han, Luo Xingqi, Xui Yangang, et al. EMD and index energy-based extraction of the draft tube dynamic characteristic of hydraulic turbine. Journal of Hydroelectric Engineering, Vol. 31, Issue 5, 2012, p. 286-291.

[6] Dragomiretskiy K, Zosso D. Variational mode decomposition [J]. IEEE Transactions on Signal Processing, 2014, 62(3): 531-544.

[7] Stewart G W. On the early history of the singular value decomposition [J]. SIAM Review, 1993, 35(4): 551-566.

[8] Li Y , Li K , Lu Q . Applying segmentation and classification to improve performance of smoothing[J]. Digital Signal Processing, 2021, 109(11):102913.

[9] Zhong J , X Bi, Shu Q , et al. Partial Discharge Signal Denoising Based on Singular Value Decomposition and Empirical Wavelet Transform[J]. IEEE Transactions on Instrumentation and Measurement, 2020, PP(99):1-1.

[10] Jia Y , Li G , X Dong, et al. A novel denoising method for vibration signal of hob spindle based on EEMD and grey theory[J]. Measurement, 2021, 169:108490.

[11 Das I , Arif M T , Maung A , et al. An Improved Hilbert-Huang Transform for Vibration-Based Damage Detection of Utility Timber Poles[J]. Applied Sciences, 2021, 11(7).

[12] Li J , Yao X , Wang H , et al. Periodic impulses extraction based on improved adaptive VMD and sparse code shrinkage denoising and its application in rotating machinery fault diagnosis[J]. Mechanical Systems and Signal Processing, 2019, 126(JUL.1):568-589.

Current Status and Prospects of Gas Turbine Technology Application

Jiangpeng Li[1, †], Ziti Liu[2, *, †] and Ruoxuan Ye[3, †]

[1]Northeastern University, Shenyang, Liaoning, 110819, China

[2]School of Aeronautics and Astronautics, Huazhong University of Science and Technology, Wuhan, Hubei, 430074, China

[3]Leeds College, Southwest Jiaotong University, Chengdu, Sichuan, 610097, China

[*]Corresponding author's e-mail: U201811599@hust.edu.cn

[†]These authors contributed equally.

Abstract. The gas turbine is widely used in various fields, including powering aircraft, ships, trains, and electrical generators. This paper reviews multiple researches about two usages of gas turbines, including power generation and propulsion in aerospace. To be specific, two types of gas turbines have been considered in the power generation section. The first one is the micro-scale turbine, and its working principle has been introduced in section 2.1.1. In addition, six diverse kinds of gas turbines, sorted by a different manufacturer, are introduced in 2.1.2, and it has been found out that, compared to its counterpart, EnerTwin is obviously more sustainable. At the same time, both of them generally cost the same. The second type of gas turbine is used in a combined cycle power plant (CCPP), a popular power station. The working principle of CCPP is introduced in 2.2.1, while several optimization methods are illustrated in 2.2.2, including solar thermal power methods and other novel methods. The result indicates that the most popular method of optimizing the combined cycle gas turbine is integrating an additional unit. One of those outstanding technics is the integrated solar-combined cycle, contributing to 64% of fuel saving with 2.8% of output reduction.

1. Introduction

The gas turbine is widely used in various fields, including powering aircraft, ships, trains, and electrical generators [1, 2]. There are a lot of advantages it has. For example, the small size makes it more convenient than most reciprocating engines [3], low lubricating reduces maintenance costs, and the exhaust's waste heat can be used to improve efficiency [4]. Generally, the gas turbine is a continuous and internal combustion engine whose working cycle is assumed to be the Brayton cycle. Four thermodynamic processes are isentropic compression, isobaric combustion, isentropic expansion, and heat rejection [5]. And gas turbine has three main components, compressor, combustor, and a turbine on the same shaft as compressor [6].

From various types of gas turbines, the micro gas turbine is a new kind of energy-saving and environment-friendly power supply [7], which is suitable to apply in urban, rural, and remote areas. It has been one of the most promising technologies globally for power generation, and the requirements of installation gas grow every day [8]. As for the large-scale power plants, installing renewable power plants is ascending extremely fast, and relevant research is also publishing unprecedentedly fast. However, the instability of renewable power generation is still an unconquerable difficulty under the modern technological level [9]. It has been widely acknowledged that a steady electrical power supply is the foundation of modern industrial manufacture [10]. Although the power plant, which is based on unrenewable fuels, has many drawbacks in the environmental aspect, the climatic independence, which is a sign of stability and geographical flexibility, is a very outstanding advantage. Hence, the continuous research and optimization which focus on the unrenewable power plant are vital for industrial maintenance and development [11]. Compared to the conventional power station, which only uses the steam turbine, CCPP has many outstanding merits, including high efficiency.

This article reviews the current development of gas turbines from two parts: power generation and aircraft propulsion. The working principle and status quo of micro gas turbines and combined cycle power plants are analysed in detail, and the future development of gas is forecasted to lay the foundation for the further development of gas turbines.

2. Microturbine

2.1. Principles of Micro Gas Turbine

Looking at the components of a micro gas turbine, all kinds of gas turbines generally have the same components, like compressor, combustor, and turbine. Compared to conventional design [12], the micro gas turbine system has several special designs to improve its ability. For example, most gas turbines use axial compressors and turbines, but micro gas turbines use the centrifugal compressor and centripetal turbines [11]. Although the axial compressor is compact, safe, and efficient, it is more suitable for large flow. In contrast, the centrifugal compressor has such advantages as small volume and high single-stage pressurization. Moreover, the centripetal turbine is also more suitable for the micro turbine, as the small flow loss and residual velocity could improve the efficiency, especially for the flow in a small volume. According to the working principles, there are two kinds of gas turbine, simple gas turbine, and generative gas turbine. They can be operated on both open and closed cycles, while the open ones are more common [13].

For the open simple gas turbine, the air is continuously drawn into the compressor, where it is compressed to high pressure. Then, the compressed air flows into a combustion chamber, which will be burnt, resulting in combustion products at an elevated temperature. The combustion products expand through the turbine and are discharged to the surroundings. Part of the turbine work developed is used to drive the compressor, and another is used to generate power, drive the vehicle, or supply electricity. The workflow is shown in Figure 1.

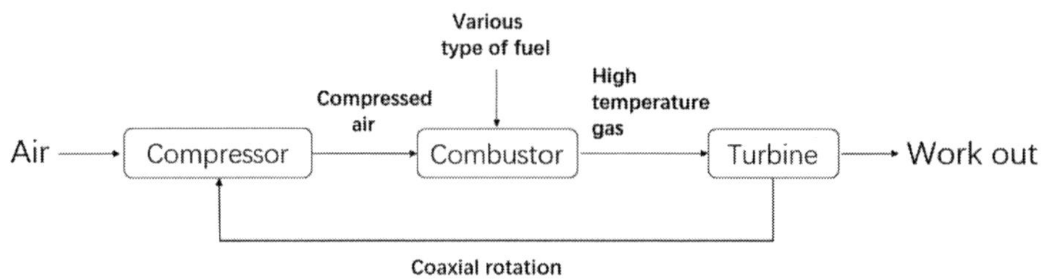

Figure 1. Workflow of the simple gas turbine

The open, generative gas turbine has an additional part when working, the generator. As the temperature of exhaust from the turbine is normally well above the ambient temperature, it always causes loss irrevocably when the gas is discarded directly to the surroundings. However, the generator avoids this potential loss, which makes the exhaust preheat the air in the compressor before entering the combustor, reducing the amount of fuel required. From the workflow of the generative gas turbine in Figure 2, it is clear that there are basically the same steps as the previous simple one. Still, the exhaust will not be expelled directly but flow into the regenerator to preheat air in the compressor. To a large extent, this design increases fuel use efficiency and becomes more sustainable [14].

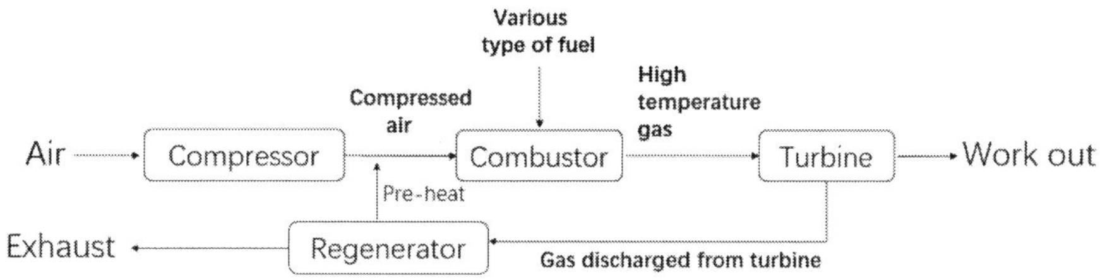

Figure 2. Workflow of the generative gas turbine

If ignoring irreversibility when air passes each component, in other words, there is no pressure drop by friction, and the air is at constant pressure through the heat exchangers. And if the stray heat transfers to the surroundings are zero, the processes through the turbine and compressor could be isentropic. The whole working cycle is assumed as the ideal Brayton cycle, which is drawn in Figure 3. There are totally four stages in the Brayton cycle, 1-2 is the compression process at constant entropy in the compressor, whose pressure ratio is p_2/p_1, 2-3 is the endothermic process at constant pressure in the combustor, 3-4 is expanding at constant entropy in the turbine, whose temperature ratio is T_3/T_2, and 4-1 is exothermic process at constant pressure.

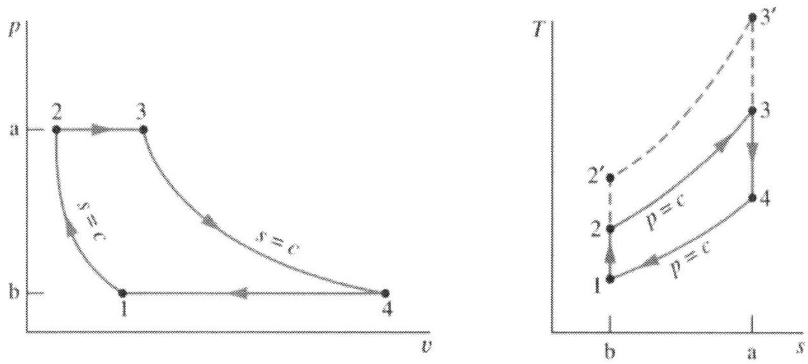

Figure 3. p-v and T-s diagrams of Brayton cycle

On the T-s diagram, area 2-3-a-b-2 represents the heat added per unit of mass, and area 1–4–a–b–1 is the heat rejected per unit of mass. On the p–v diagram, area 1–2–a–b–1 represents the compressor work input per unit of mass, and area 3–4–b–a–3 is the turbine work output per unit of mass. The enclosed area for the former is the network, and the latter is the heat added per unit of mass.

2.2. Current Status of Micro Gas Turbine

Distributed generation and combined heat and power generation (CHP) are two popular aspects of micro gas turbines, that they have gradually substituted some traditional machines. There are three microgrid benefits summarized, energy security, economic benefit, and clean energy integration. The most popular example of developing microgrids is the 'New York Prize', a $ 40M competition initiated by New York State to push the microgrid from feasibility studies to implementation [15]. For CHP, it is an environmentally friendly generation way that mainly uses renewable energy, such as solar, biofuel, and biomass. For example, biomass is a decomposable matter from plants or animals, avoiding NOx and Sox emissions in the combustor [16].

The micro gas turbine is applied in lots of CHP and microgrids. On the one hand, although the micro gas turbine is slightly inefficient at generating electricity, the energy utilization rate of the cogeneration system could exceed a lot of large units, which makes it is sustainable and suitable for applying in the city, countryside, and even remote area. On the other hand, it is lighter and has fewer rotating parts compared to internal combustion engines, so avoid expensive operation and maintenance costs. Hence, combined heat and power generation is a powerful competitor for an internal heat engine.

With the development of technology and the sustainable consciousness of the human, there are different kinds of micro gas turbine products invented and sold in several companies. Nowadays, the leader in the micro gas turbine market is Capstone, followed by Flexenergy, with a market share totally more than 95%. And from the forecast by Forecast International to 2029, the market share of both companies will increase more and more [17].

Table 1. Market Share in 2029.

Company	Total Sales volume	Production Value (Million dollars)	Market Share (%)
Capstone	3480	5.2057	72.8
Flexenergy	407	1.239	17.3

Moreover, Forecast International projects a jump from the range of 600 to 750 per year to between 900 and 1,050 from 2021 to 2029. This leap is partly attributed to two European companies, Bladon and Micro Turbine Technologies (MTT), entering the market. Therefore, in this report, several representative companies and their products will be chosen for analysis, includes the famous companies, Capstone and Flexenergy, a newly developed company, MTT, and a Chinese company, ENN Power.

3. Combined Cycle Power System

3.1. working principle of CCPP

The combined cycle is shown on the T-s diagram in Figure. 4. The upper cycle (1-4) is a Brayton cycle, and the lower cycle (a-f) is a Rankine Cycle. For the Brayton cycle part, in processes 1-2, the injected air is compressed and gains relatively high pressure. In processes 2-3, intensive combustion of syngas (a combination of injected air and fuel) occurs in the combustor. In processes 3-4, combustion, which has ultra-high temperature and pressure, expanded in the gas turbine and produced work. In progress 4-1, the exhaust gas pours out of the gas turbine and enters the HRSG. Thus, processes 4-1 are the "connection" of the Brayton and Rankine cycles in the CCPP.

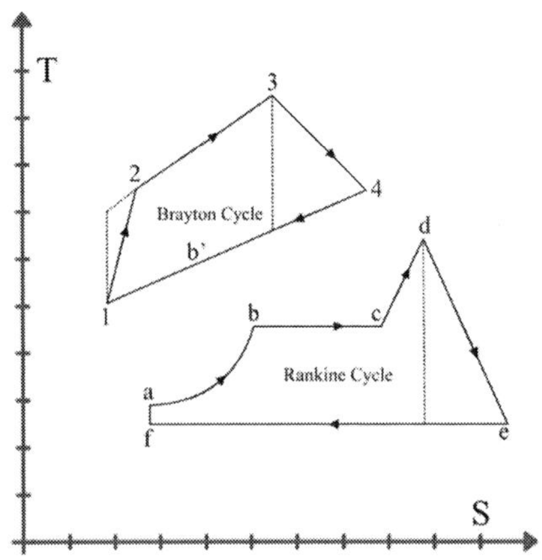

Figure 4. T-s diagram of the combined cycle

For the Rankine cycle part, process a-d occurs in the HRSG. Specifically, a-b mostly occurs in the economizer and water-cooling wall where the supercooled water becomes saturated water and stored in the steam drum. Process b-c mostly occurs in the water-cooling wall where evaporation happens. The steam and liquid phases are separated in the steam drum, and the steam goes to the superheater where process c-d occurs. The saturated steam becomes superheated in the superheater, and it will drive the steam turbine to rotate in process d-e. The process e-f means the exhaust steam becomes liquid phase in the condenser (cooling tower), and it will be pumped back to the economizer.

 2108 (2021) 012009 doi:10.1088/1742-6596/2108/1/012009

Figure. 5 is a comprehensive diagram of CCPP. From the knowledge of combustion, it has been proved that the theoretical temperature of combustion will increase if the injected air is preheated. Thus, an air preheater is built at point 1, which is the inlet of the compressor. After that, the injected air gains a higher pressure at point 2, which is the compressor. Then, the intensive combustion of syngas, which combines air and fuel, occurs in the combustor (Point 3). The combustion produces exhaust gas with ultra-high pressure and temperature to drive the turbine at point 4. The shaft on the turbine, which is connected to the generator, enables the generator to rotate with the turbine simultaneously to produce electricity. The pressure of exhaust gas decrease because it has already driven the turbine. However, the temperature is still high enough to heat the steam.

The blue tube (Point 5) in the HRSG is the economizer, and part of the supercooled water becomes saturated in these tubes. The saturated water is stored in the steam drum and evaporates in the water-cooling wall (Point 6). The steam phase and liquid phase get separated from each other in the steam drum. After that, the steam acquires a very high temperature and pressure in the superheater (Point 7). The superheated steam drives the steam turbine in point 7, producing electricity like the gas turbine. The exhaust steam gets cooled and becomes a liquid phase in the condenser. Finally, the liquid water is pumped back to the economizer.

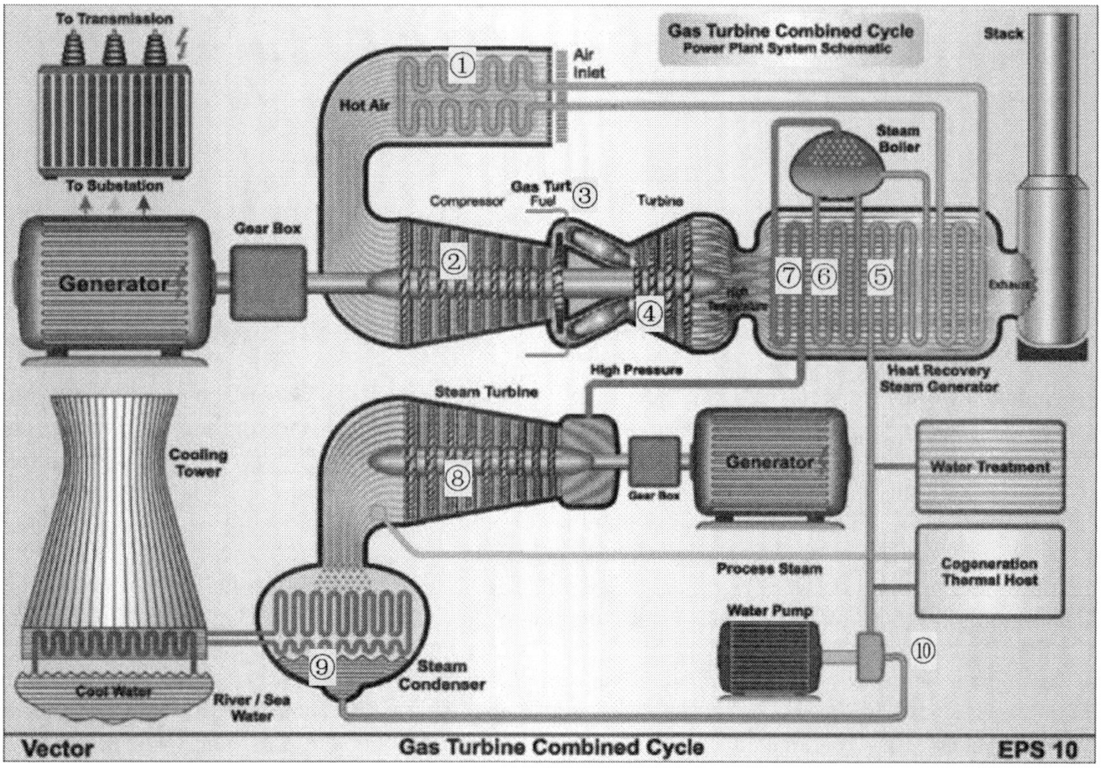

Figure 5. A comprehensive diagram of the CCPP [18]

3.2. Current status and relevant research of CCGTs

The efficiency of the gas turbine is relatively high. Specifically, General Electric (GE) manufactured a 605 MW turbine (9HA) with an efficiency of 62.22% when its temperature is as high as 1540°C [19]. In 2018, GE founded an 826MW turbine (9HA.02), which has over 64% in the combined cycle. And this progress was due to the manufacturing and combustion breakthrough [20]. In addition, the 7HA turbine from GE achieved a gross efficiency of 63.08% in March 2018 [21].

3.2.1. Optimizing CCGTs through solar thermal energy. As the prices of fossil fuels and natural gas are increasing, there is an ascending interest in improving the efficiency via renewable energy, including concentrating solar power (CSP) [22]. The novel concept of integrating solar thermal energy in CCPP is called the ISCC (Integrated Solar-Combined Cycle). Compared to a single CSP plant or a CCPP plant, ISCCs have some additional advantages. Including a higher thermo-dynamical efficiency [22] and a lower cost of integrating a heliostat field in an existing CCPP than building the entire CSP plant [23]. It is worth noting that upgrading the CCPP via CSP is not contradictory to what is mentioned in section 1 because the CCPP is an independent system that can function steadily with or without the additional CSP system. Figure. 6 is a typical ISCC with parabolic troughs with thermal oil as the heat transfer fluid (PT-HTF).

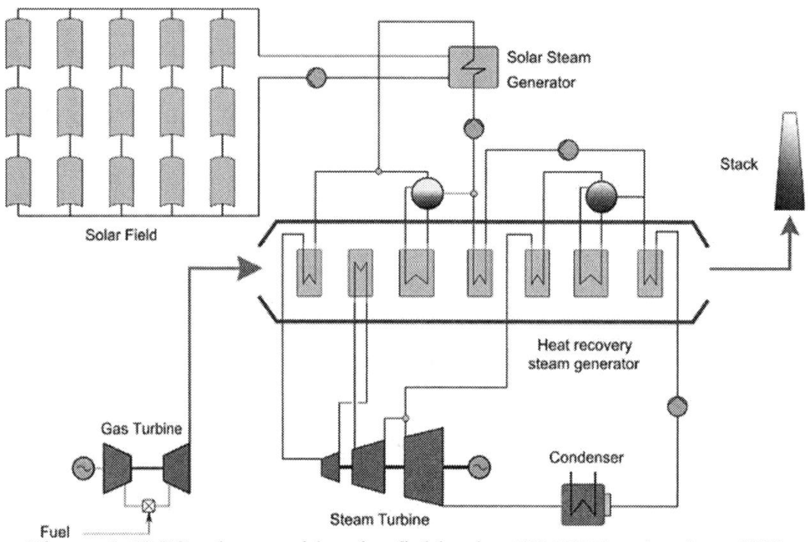

Figure 6. ISCC scheme with solar field using PT-HTF technology [22]

Ahmad M. Abubaker et al. [24] integrated a CSP system into a CCPP system. Specifically, parabolic trough collectors (PTCs) were installed to preheat the air at the inlet of the combustor chamber. The PTCs were then used to control the temperature of the compressor's inlet by driving an absorption cooling cycle (a single effect LiBr/H2O system). Their result indicated that a 7.2% increase in thermal efficiency and a 27.7 MW ascent of power output were achieved. They also established a highly accurate equation, which is based on linear regression, to predict the system's performance. They claimed that the optimum thermal efficiency was in the range of 53.1%-61.11% with a capital cost of 86.5-158.3 Mil$. Finally, they defined a favorable optimum point at an efficiency of 57.86% with a capital cost of 131.8 Mil$, and the payback period equals 2.8 years, which means it is a feasible novel cycle.

Evangelos Bellos et al. [25] established analysis and optimization on a solar-assisted gas turbine. PTCs are included in the model to achieve less consumption of fuels and a more environmentally friendly system. Their research aims to find the minimum size of the heliostats field while achieving a relatively appropriate fuel consumption. Their final result indicated that a gas turbine that integrated a CSP system which consists of 1050 heliostats, consumes 0.3389kg/s of natural gas and produces about 14.81 MW of electricity. When comparing the upper data to the initial gas turbine, which produces 15.27 MW of electricity while consuming 0.9404 kg/s of natural gas, the use of the CSP system contributes 64% of fuel saving with 2.8% of output reduction. Finally, they claimed that integrating an additional CSP on the CCGTs leads to considerable fuel conservation with slight power reduction.

3.2.2. Optimizing CCGTs through other methods. In the academic field, scholars have many novel concepts of optimizing the CCGTs. Roberto Carapellucci et al. [26] investigated a new concept for optimizing the existing CCGTs by integrating an additional unit on the current gas turbine. This additional unit enables the superheated steam to be used in three different ways. As shown in Figure. 7, in option A, the superheated steam is directly injected into the combustor of the current turbine. In option B, the superheated steam is used to promote methane reforming. And then, the syngas of reformed methane and steam is used as the fuel of the existing combustor. In option C, the syngas are also used as the fuel of the combustor, which is in the supplemental gas turbine. In option B&C, the heat recovery is greater than the conventional method since the produced syngas absorbs the heat from superheated steam thermally and chemically. Their result shows that option A has the best economic performance, with the highest operational simplicity when the power augmentation is lower than 50%. Differently, when the power augmentation is larger than 80%, option C is more competitive than option A.

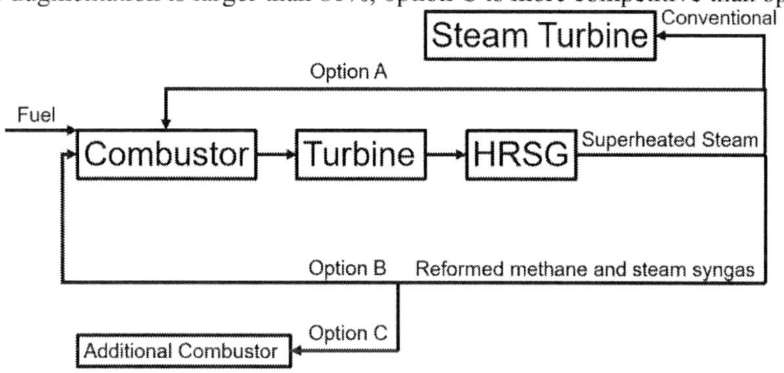

Figure 7. Schematic diagram of Carapellucci, R's research

Hany A.Al-Ansary et al. [27] pointed out that a proven method of upgrading the efficiency of CCGTs is declining the inlet temperature of the compressor, especially during the hottest time in the summer. Hence, they investigated the prospect of using a TIAC (hybrid Turbine Inlet Air Cooling). The TIAC is made up of a mechanical chilling system which is followed by an evaporative cooling system. Figure. 8 and 9 are comprehensive diagrams of the mechanical chilling system, including dry and wet cooling condensers. In this research, Al-Ansary defines 4 different scenarios, which are shown in Table 2. Their result showed that the TIAC systems have the ability to boost the output of CCGTs by 10% or more (in ISO condition). In addition, the comparison of four different scenarios showed that scenario 4 is the most attractive option.

Table 2. Four scenarios defined in Hany A.Al-Ansary et al's study

Case I	Mechanical chilling only with dry cooling of the condenser
Case II	Mechanical chilling only with wet cooling of the condenser
Case III	Hybrid TIAC system with dry cooling of the condenser
Case IV	Hybrid TIAC system with wet cooling of the condenser

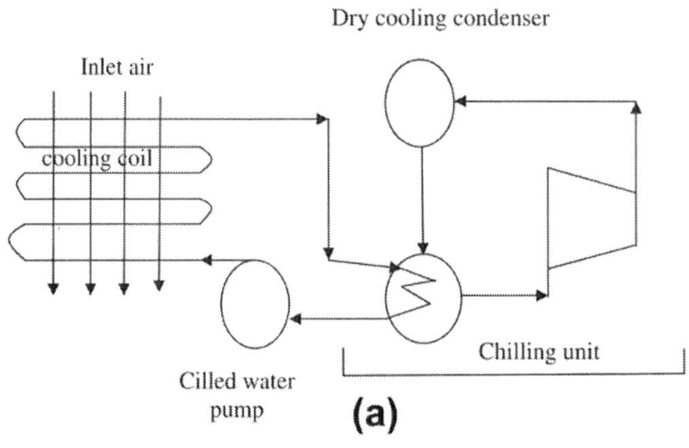

Figure 8. Dry cooling condenser [27]

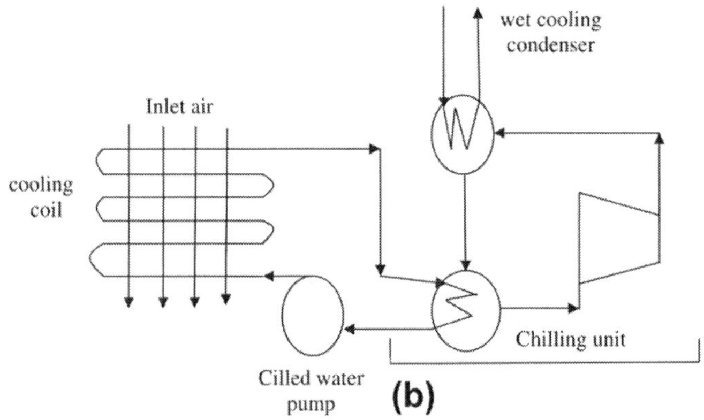

Figure 9. Wet cooling condenser [27]

Andrea Giugno et al. [28] integrated an additional high-efficient heat pump in a CCPP as a flexibility enhancement to improve the global efficiency and operational flexibility in part-load operation. Specifically, the heat pump allows crews to modify the intake temperature of the compressor, which is a positive way of increasing power production. They also pointed out that the combined cycle's efficiency is more influenced by the auxiliary consumption than the ambient temperature. Thus, selecting the appropriate size of the heat pump is important. They analyzed the climatic data and its correlations with energy market conditions and then established a model to calculate the thermoeconomic potential. Their result shows that the ratio of electrical consumption of an air-cooled heat pump to the maximum electrical production increase is 1:100. Based on their result and the characteristic of the heat pump, Giugno points out that the electrical market conditions could harm the profits of installation even under favorable climatic potential. Hence, they came up with a concept that "using the off-design curves and optimization algorithm in performing coupling analysis appears to be more effective."

4. Conclusion

(1) Micro gas turbine has been widely applied in various areas, like providing electricity and heat to residential buildings, making business buildings more sustainable, and producing more stable and efficient energy for remote countryside and island. However, it has not been applied to hybrid cars, as several difficulties make efficiency low, frequent engines stop, and lack of space for big heat exchangers. Hence, in the future, to realize the application, it is essential to ensure the efficiency of the micro turbine

with a small volume to reduce the energy cost, cut the production cost to be affordable, and reduce the noise level to ensure the passenger comfort.

Moreover, methods for prospective health monitoring are usually used for large scale gas turbines, rarely for micro gas turbine. However, it can largely increase the availability and reduce the cost for operating and maintenance. The future research could focus on estimating and predicting performance degradation of gas turbine to achieve condition-based maintenance. The reason why is sudden failure mechanisms, such as bearing failure or object damage, are hard to be predicted. These may lead to fuel consumption increasing, uneconomical operation and greenhouse gases developing.

(2) One of the most popular optimization methods of CCGT is to integrate another additional unit instead of directly modifying the specific characteristics of the gas turbine. Among these integrated technologies, ISCC may be one of the most developed applications. Many countries have installed concentrated solar power plants, which means that the application of solar thermal energy is a relatively mature technology. Therefore, integrating mature technology into the existing CCGT may be a viable option. Also, the reference (2) indicates that one of the investigated ISCCs model can contribute to 64% of fuel saving with 2.8% of output reduction, which means that this technic has considerable expectation.

References

[1] Antonanzas, J., Jimenez, E., Blanco, J. (2014) Potential solar thermal integration in Spanish combined cycle gas turbines. Renew Sust Energ Rev., 37: 36–46.
[2] Dersch, J., Geyer, M., Herrmann, U. (2004) Trough integration into power plants—a study on the performance and economy of integrated solar combined cycle systems. Energy., 29: 947–959.
[3] Bellos, E., Tzivanidis, C., Antonopoulos, K.A. (2017) Parametric analysis and optimization of a solar assisted gas turbine. Energy Conversion and Management. Energy Convers. Manag., 139: 151–165.
[4] Carapellucci, R., Giordano, L. (2019) Upgrading existing gas-steam combined cycle power plants through steam injection and methane steam reforming. Energy., 173: 229–243.
[5] Al-Ansary, H.A., Orfi, J.A., Ali, M.E. (2013) Impact of the use of a hybrid turbine inlet air cooling system in arid climates. Energy Convers. Manag., 75: 214–243.
[6] Giugno, A., Sorce, A., Cuneo, A. (2021) Effects of market and climatic conditions over a gas turbine combined cycle integrated with a Heat Pump for inlet cooling. Appl. Energy., 290:116724.
[7] Omair, A. (2016) Current Status and Prospects of Supercomputing Used for Gas Turbine Engines Design. J Energ Eng., 2:82–92.
[8] Kobayashi, Y., Gengo, T., Ando, Y. (2009) Current status and future prospects for large-scale SOFC combined cycle system. J of H Tem Society., 35: 256–262.
[9] Clarke, D.R., Oechsner, M., Padture, N.P. (2012) Thermal-barrier coatings for more efficient gas-turbine engines. Mrs Bulletin., 37: 891–899.
[10] Waitz, I.A., Gauba, G., Tzeng, Y.S. (1998) Combustors for micro-gas turbine engines. J Fluid Eng-t ASME., 120: 109.
[11] Tayarani-Bathaie, S.S., Vanini, Z., Khorasani, K. (2014) Dynamic neural network-based fault diagnosis of gas turbine engines. Neurocomputing., 125:153–165.
[12] Chang, I., Bae, J., Park, J. (2016) A thermally self-sustaining solid oxide fuel cell system at ultra-low operating temperature (319 C). Energy., 104: 107–113.
[13] Kvamsdal, H.M., Jordal, K., Bolland, O.A. (2007) Quantitative comparison of gas turbine cycles with CO2 capture. Energy., 32: 10–24.
[14] Blakey, S., Rye, L., Wilson, C.W. (2011) Aviation gas turbine alternative fuels: A review. P combust Inst, 33: 2863-2885.
[15] Hirsch, A., Parag, Y., Guerrero, J. (2018) Microgrids: A review of technologies, key drivers, and outstanding issues. Renewable and Sustainable Energy Reviews., 90: 402-411.

[16] Martinez, S., Michaux, G., Salagnac, P. (2017) Micro-combined heat and power systems (micro-CHP) based on renewable energy sources. Energy Conversion and Management., 154: 262-285.
[17] Rory, P. (2020) Turbomachnery International: Microturbines are find their niche. https://www.turbomachinerymag.com/view/microturbines-finding-their-niche.
[18] Saad F. A. (2021) Gas Turbine Combined Cycle Power Plant. https://www.shutterstock.com/image-vector/gas-turbine-combined-cycle-power-plant-142631011.
[19] Kellner, T. (2016) Here's Why The Latest Guinness World Record Will Keep France Lit Up Long After Soccer Fans Leave | GE News. https://www.ge.com/news/reports/bouchain.
[20] Millas, T. (2017) HA technology now available at industry-first 64 percent efficiency | GE News. https://www.ge.com/news/press-releases/ha-technology-now-available-industry-first-64-percent-efficiency.
[21] Quesnel, M. (2018) GE's HA Gas Turbine Delivers Second World Record for Efficiency | GE News. https://www.ge.com/news/press-releases/ges-ha-gas-turbine-delivers-second-world-record-efficiency.
[22] Antonanzas, J., Jimenez, E., Blanco, J., Antonanzas-Torres, F. (2014) Potential solar thermal integration in Spanish combined cycle gas turbines. Renewable and Sustainable Energy Reviews., 37: 36–46.
[23] Dersch, J., Geyer, M., Herrmann, U., Jones, S. A., Kelly, B., Kistner, R., Ortmanns, W., Pitz-Paal, R., Price, H. (2004) Trough integration into power plants—a study on the performance and economy of integrated solar combined cycle systems. Energy., 29: 947–959.
[24] Abubaker, A. M., Ahmad, A. D., Singh, B. B., Akafuah, N. K., Saito, K. (2021) Multi-objective linear-regression-based optimization of a hybrid solar-gas turbine combined cycle with absorption inlet-air cooling unit. Energy Conversion and Management., 240: 114266.
[25] Bellos, E., Tzivanidis, C., Antonopoulos, K. A. (2017) Parametric analysis and optimization of a solar assisted gas turbine. Energy Conversion and Management., 139: 151–165.
[26] Carapellucci, R., Giordano, L. (2019) Upgrading existing gas-steam combined cycle power plants through steam injection and methane steam reforming. Energy., 173: 229–243.
[27] Al-Ansary, H. A., Orfi, J. A., Ali, M. E. (2013) Impact of the use of a hybrid turbine inlet air cooling system in arid climates. Energy Conversion and Management., 75: 214–223.
[28] Giugno, A., Sorce, A., Cuneo, A., Barberis, S. (2021) Effects of market and climatic conditions over a gas turbine combined cycle integrated with a Heat Pump for inlet cooling. Applied Energy., 290: 116724.

Experimental PIV investigation of the PZT fans array coupling effect at high Reynolds numbers

Yiyang Chen[1a], Hao Ding[1b], Ziwen Wang[1c], Jianxin Li[2d], Jiahuan Cui[1,3e*]

[1]ZJUI Institute, Zhejiang University, Haining, 314400, China

[2]China Automotive Technology and Research Center Co, Ltd, Tianjin, 300300, China

[3]School of Aeronautics and Astronautics, Zhejiang University, Hangzhou, 310007, China

[a]yiyangc.18@intl.zju.edu.cn, [b]haod.19@intl.zju.edu.cn, [c]ziwen.19@intl.zju.edu.cn, [d]lijianxin@catarc.ac.cn, [e*]jiahuancui@intl.zju.edu.cn

Abstract—Piezoelectric fan arrays are being increasingly emphasized for heat dissipation in small-sized electronic devices. In this study, PIV experiments were conducted to investigate the flow fields induced by piezoelectric fan arrays with different vibration modes and pitches at high Reynolds numbers ($324< Re <509$) in a stationary air environment. As a result, when the PZT fan array is vibrating in-phase, the saddle points in the time averaged flow field are formed and separated gradually as the pitch increases, the remnant vortex and the induced vortex interact to form a jet with a periodic oscillation in the direction. Jet velocity reaches a maximum at $P = 3A$. In counter-phase vibration, saddle points are separated from one region under large pitches, the interaction of counter-rotating induced vortices forms a vertical upward jet. The morphology of induced and remnant vortices with different vibration modes and array pitches are responsible for the jet formation and flow field pattern. The interaction of counter-rotating vortices in counter-phase vibration leads the jet intensity higher than in-phase vibration induced jet, the optimal setting of the PZT fan under this study is determined as $P = 2.5A$ with counter-phase vibration. The experimental results provide validation for the simulation study and give guidance to the application

1. Introduction

Rotary type fans have been widely applied in industrial cooling systems due to high performance and efficiency. However, with the development of electric components which tend to shrink in size, rotary type fans are limited by narrow space because the performance and efficiency of rotary type fans degrade rapidly when fans' size is too small. The manufacturing dimensions of rotors, bearings and other components also make rotary type fans difficult to popularized used in electronic components cooling. Nowadays, piezoelectric (PZT) fans with the advantages of low power consumption, high reliability, and the small geometric deformation to generate a significate airflow are considered as the better option for tiny electronic components heat dissipating [1, 2].

Toda was the first to investigate the fluid field generated by a single PZT fan in 1978 and made a preliminary study of PZT fan application about cooling [3, 4]. Subsequently, computational simulations and Particle Image Velocimetry (PIV) experimental validation were applied to study the flow field characteristics induced by the PZT fan. Kim et al. used PIV to experimentally investigate the flow field around a vibrating cantilever plate and explained the air flow induced by the cantilever plate is a pseudo-jet flow dominated by vorticial structures. Vortex size, strength and other information is also obtained by wavelet analyze [5]. Ebrahimi et al. conducted the experimental and numerical simulation studied on 3D mean jet downstream and vortex structures produced by the oscillation PZT fan with different oscillating frequency, amplitude and geometric parameter. The results revealed that vibration amplitude and width of PZT fan blade are the main factors that affect the shape of jet [6]. Akshat Agarwal et al. presented a three-dimensional analysis of a single unconfined PZT fan. Phase locked PIV technique is applied to experimental research the formation and shedding of vortices. The structures of vortices appear as a horseshoe shape and then evolve into a hairpin vortex [7]. The aerodynamic performance of PZT fans has also been studied both theoretically and experimentally. Eastman et al. measured the thrust of vibrating oscillating cantilever in the range of Reynolds number from 500 to 5000 and gave out a simple concept sidewall design to maximum the performance based on their experiment data [8]. Stafford and Jeffers studied the aerodynamic performance of a vibrating piezoelectric blade based on experiment under different operating conditions and confinements. The measurement of pressure-flow rate characteristics shows that stall characteristics exists in time-average velocity fields between maximum efficiency and maximum static pressure. Flow reversal dominates at maximum static pressure. A recommended operating range was given between maximum efficiency and maximum flow rate [9].

The flow field induced by a single PZT fan has been well studied. PZT fans array will be more widely used in thermal management than individual fans. However, only a few studies have focused on the flow field details of the PZT fans array. Kimber et al., experimentally investigated the aerodynamic coupling effect of PZT fan arrays. Both in-phase and counter phase vibration mode are explored for both face-to-face and edge-to-edge arrangement. The result showed that for the face-to-face arrangement, the fluid damping is promoted when operates in in-phase mode, and the coupling effect became stronger as the pitch got smaller. The oscillation for the face-to-face configuration behaved in an uncoupled manner when the pitch is above 6 times of the amplitude [10]. Choi et al. simulated the unsteady flow around a vibrating cantilever PZT fan pair operating at different phase angle using a commercial software. They found the interaction between the counter-rotating vortex generated by each cantilever is depended on the phase angle difference. And they also concluded that the cantilever pair performed most effectively in generating airflow when vibrating in counter-phase mode and performed worst when vibrating in in-phase mode [11]. In another research done by Choi et al., by evaluating time-averaged axial velocity and the mass flow rate downstream. They found the best distance between two PZT fans of counter-phase vibrating mode and explained that the performance of in-phase vibrating mode is inferior to the single cantilever regardless of the distance between the beams [12]. Hales and Jiang studies the effect of pitch on the performance of the PZT fan arrays in a confined environment numerically. They also concluded that the performance of the counter-phase oscillation is superior to the in-phase oscillation across the tested pitch range. They highlighted that the formation of vortices is a determining factor for the optimal pitch [13]. In their following publication, the effect of phase variation between the two PZT fan blades in a confined environment is studied numerically. An asymmetrical flow domain is observed when the blade pair is not operating in in-phase or counterphase mode. The skewness of the downstream velocity profile is depended on the interaction of vortex generated between two bladed during an oscillation cycle [14].

Most of the above literature used numerical simulations to investigate the effects of phase and distance on the velocity and flow downstream of the PZT fans array, giving optimal placement distance intervals. There are some simulations to investigate the coupling effect of distance and phase on adjacent fans by measuring PZT fan amplitudes. However, very few experiments have been conducted to verify the coupling effects and performance of PZT fan arrays for validation. In this study, the flow field pattern of the PZT fan array at high Reynolds number was experimentally determined by the PIV technique to investigate the variation of the coupling effect. The Reynolds number range taken in this study is $324 < Re < 509$, and the coupled flow field caused by the larger amplitude shows different characteristics from previous studies. The large amplitude is more consistent with the application of PZT fans to heat dissipation. Meanwhile, the spacing of PZT fans as a variable was taken more finely. The velocity and vorticity fields induced by two different phase difference vibrations is measured in an unconfined stationary air environment. The time-averaged velocity field helps visualize the coupling pattern of the flow field and the air-driven performance of the PZT fans array. The ensemble-averaged vortex field explains the formation of the velocity field and shows the instantaneous coupling effects at different pitches.

The structure of the article is as follows. Section 2 introduces the setup of the PIV experiment and the data processing method. Section 3 shows the flow field characteristics of different degrees of coupling effect captured by PIV and gives explanations. And section 4 summarizes the conclusions obtained from the above contents.

2. Experiment Methodology

2.1. PZT fan characteristic parameters and fan array set up

This paragraph gives the settings and specific parameters of the PZT fans array. All PZT fan blades in the research have the same geometrical parameters length l_b = 65mm, width w_b = 15 mm, thickness t_b = 0.1 mm and resonance frequency f_r = 58 Hz. The visualization parameters are labelled in Fig. 1. The PZT Fan array consisted by two PZT fan blades is mounted on two *x-y* movable manual adjustment translation stages at the bottom of a container box in the central position. Distance between two PZT fan blades can be regulated by the two translation stages. The range of distance between two blades can be adjusted from 3mm to 28mm. Manual adjustment translation stages reading by micrometer caliper has accuracy 0.01mm, distance between two blades with a reading error of ± 0.001 mm. The different vibrating amplitudes of single PZT fan reached by a series of voltages in the range of 5.52 mm < A < 8.67 mm with an uncertainty of ± 0.09mm, corresponding to 1 pixel in the image. The range of deflection in this study was determined as 0.085 < A/l_b < 0.133. The characteristic trailing edge velocity in this study is defined as

$$U_t = 4fA \tag{1}$$

which is the average velocity of the trailing edge in a cycle. The PZT fan amplitude varies for different degrees of coupling. Due to the piezoelectric effect, even if the signal generator sends the same voltage signal to the PZT fan array, they may consume different amounts of power. In this study, the power difference due to different coupling levels was measured to be less than 1%. Therefore, PZT fans arrays driven by the same voltage signal at different pitches are considered to have the same power dissipation together for comparison. The same Reynolds number is used to represent the single PZT fan and PZT fans array driven by the same amplitude voltage signal. The Reynolds number is defined as follows:

$$Re = \frac{U_t w_b}{4v} = \frac{fAw_b}{v} \tag{2}$$

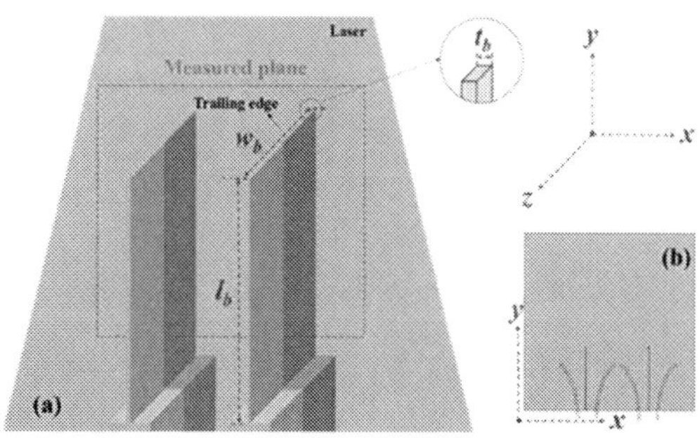

Fig. 1 Geometric parameters of the PZT fan array and the illuminated plane for the PIV experiment, with frame (b) PIV experimental view of computing plane (*x-y* plane)

2.2. Particle Image Velocimetry experiment set up

The experiment system for Particle Image Velocimetry (PIV) is shown in Fig. 2. The size of the transparent acrylic glass container box is 300mm×300mm×300mm, which is large enough to reduce the sidewall effect. Transparent outer walls facilitate observation and laser penetration. The closed container box avoids the interference of the outside flow field. The bottom of the container box is fixed on an X-Y translational platform. The translational platform is driven by a Daheng Optical GCD-202002M stepper motor with a resolution of 0.001mm.

A 527 nm double-pulse Nd:YLF laser (VLite-Hi-527-40, BeamTech) is used to illuminate the experimental measurement area. The time interval Δt between the two image frames used for the PIV computing is set as 60 μs. A high-speed camera (FASTCAM Nova S12 type 200KS-M-32G, Photron) is used to record the images with a capture rate of 5800Hz, which is 100 times of the frequency f of the oscillating PZT fans array, so that 100 pairs of PIV images can be obtained in one oscillation period, equivalent to 5800 pairs of PIV images per second. The image resolution is set to 1024×1024 pixels, 12 bits per pixel. The pulsed laser and high-speed camera are controlled by a laser pulse synchronizer (TSI model 6130036 LaserPulse). Before the PIV experiment, the air-olive oil droplets mixture is deposited for 5 minutes and then ejected from a Laskin nozzle to form stationary aerosols.

The transient example images obtained from the shooting are processed by the commercial software TSI Insight 4G. The entire image processing is divided into three parts. In pre-process, all pixels are subtracted to the average brightness of the corresponding pixels of all images to reduce the effect of the background color on the image processing results. Then multiplication operation is used for the pixel points to increase the brightness.

In the second step, the pre-processed images are further processed. The particle velocity was computing using a recursive Nyquist grid with an overlapping grid spacing of 50%. The size of the interrogation window was set to 32 × 32 pixels, ensuring that at least 15 particles exist in the interrogation area. An FFT correlator is chosen as the correlation engine, and a Gaussian Peak with the signal to noise ratio of 1.5 is used for the peak engine. Finally, the computed images are post-processed, and the data are validated globally by a standard deviation speed range filter with a standard deviation factor of 7, followed by local validation using a median test method with a 5 × 5 pixels neighborhood size. Empty spaces in the data are filled recursively with a 5 × 5 pixels neighborhood size local averaging method. Finally, all data were smoothed with a filter size of 5 × 5 pixels and a σ value of 0.8

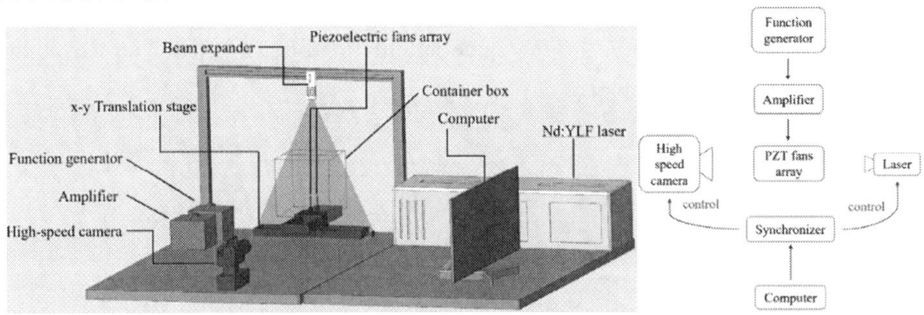

Fig. 2 Schematic of PIV experiment setup

3. Results and discussion

In this study, the amplitudes of the PZT fans array at different pitches at multiple Reynolds numbers are determined. Fig. 3(a) and Fig. 3(b) define the in-phase (IP) vibration and counter-phase (CP) vibration, which correspond to PZT fans pair phase differences of 0° and 180°, respectively. Due to the coupling effect, the amplitude of single PZT fan when vibrating in-phase and counter-phase will change as the coupling effect increases. Fig. 3(c) shows the blade vibration amplitude varies with blade pitch at five different Reynolds numbers. The vibration amplitude of PZT fans array is divided by the single blade amplitude of its corresponding Reynolds numbers. Under in-phase vibration mode, the amplitude is larger than single blade vibration amplitude. The coupling effects at different Reynolds numbers follow a similar trend. Starting from $P = 1A$, the amplitude gradually decays from $1.2A$ to near $1A$. When the blade pitch is greater than $3.5A$, the coupling effect is weak. The coupling under counter-phase vibration weakens the vibration amplitude of the blades, the coupling effect decays as the blade pitch increases and the coupling strength of the two vibration modes is similar at different blade pitches. The same Reynolds number and pitch intervals that are of interest for the following study, coupling effects persist under these conditions.

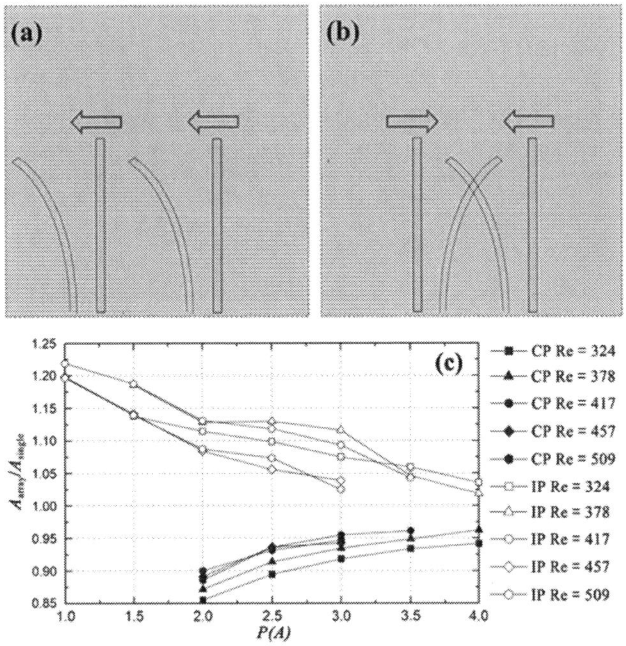

Fig. 3 Amplitude variation due to coupling effect of PZT fans array: (a) schematic of in-phase vibration; (b) schematic of in-phase vibration. (c) Amplitude ratio profile under different Reynolds numbers and pitches

3.1. Time-averaged flow field

In the current study, the time-averaged flow field of a single PZT fan is obtained from the PIV measurements as a contrast to study the effect of PZT fans array coupling effect. 2000 pairs PIV images are captured at 100x the PZT fan intrinsic frequency, which equals to 5800hz, then time averaging computing is applied to the 20 periods images to obtain the flow field. Fig. 4 shows the contour of time-averaged velocity magnitude (U_{mag}) in *x-z* planes in the case of $Re = 417$. The point $x = 0$mm, $y = 0$mm corresponds to the position of the tip of the leaf of PZT fan at rest. The previous study indicated that the flow patterns are similar with different Reynolds numbers. In general, the flow pattern in the x-y plane shows a highly symmetrical "Y" shape. [5]

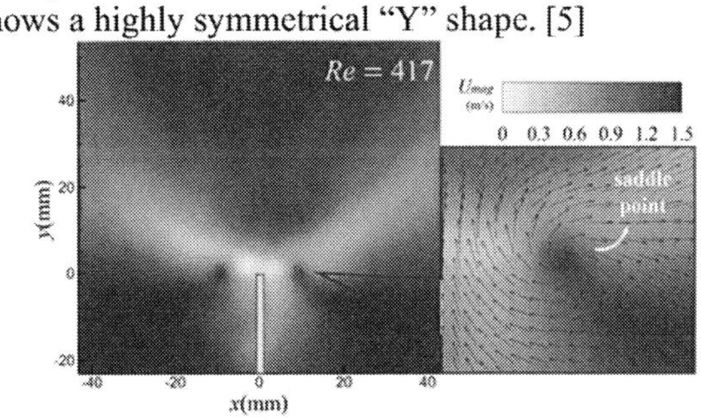

Fig. 4 Time-averaged velocity magnitude (U_{mag}) field of single PZT fan at $Re = 417$.

The high-speed area at the tip of the blade gradually evolves into two oblique upwards jets. The velocity magnitude between two jets is relatively low. The two saddle points are on both sides of the leaf tip, below the jet. The formation of two saddle points is related to the fan-induced formation of two vortices. The vortex generated at phase $\tau = 0.25$ by a single PZT fan is shown in the Fig. 5. The definition of τ and the phase-locked PIV measurement for generating images will be mentioned in the next section. The leftward motion of the leaf tip induces a CW vortex. Above the CW vortex is the remnant vortex generated by the last vibration cycle. In addition, due to the motion of the blade body, the pressure difference causes the air near the blade body to move in the direction of its motion and upwards. A blade body-induced vortex in the opposite direction of the blade tip-induced vortex is generated. Due to the high strength of the Tip-induced vortex and the continuous presence for a while during the entire periodic vibration of the PZT fan to keep the position nearly fixed. Therefore, two saddle points appear in the time-averaged flow field at the locations of the CW and CCW tip induced vortex cores, respectively.

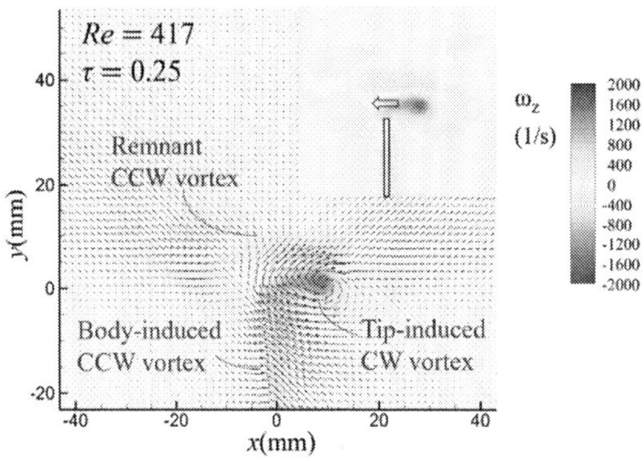

Fig. 5 Vortex induced by the periodic motion of the PZT fan.

Fig. 6 shows the time-averaged flow field pattern of PZT fans array with different pitches P between two blades. Symmetrical "Y" shape flow field similar to the single blade is presented when $P = 1.5A$. With the increase of the interval, the flow field exhibits different characteristics. When $P = 2.5A$, a new high-speed region appears between the two jets, and a low-speed region appears in the region encircled by the two blades. At $P = 2.5A$, the area of the low velocity region continues to expand, and the saddle points are starting to show at the top of the region. As the pitch continues to increase, the intermediate high velocity region evolves into a new jet. When $P = 3.5A$, in the middle low speed region with the same height with blade tips, three saddle points appears. The two saddle points below are in the position of the CW and CCW vortices cores. The interaction of the vortices forms a new saddle point and leads to the creation of the jet. The inverse flow observed by Choi et al. [11] can be found in the Fig. 6. The periodic motion of the blades creates two vortices between the blades. Although these two counter-rotating vortices are generated at different phases, the time averaging within the cycle causes the horizontal component of the velocity to be eliminated and the two counter-rotating vortices act to generate the reversed flow. The formation of vortices will be discussed in detail in the next section. The velocity field induced by the low-deflection PZT fans array is presented as two jets starting from the blade tip vertically upward and gradually combining to form a low velocity region with one saddle point [11]. The high Reynolds number flow field at the corresponding pitch retains two pseudo-jets to the side and has two additional saddle points at the position of vortex core in the low velocity region.

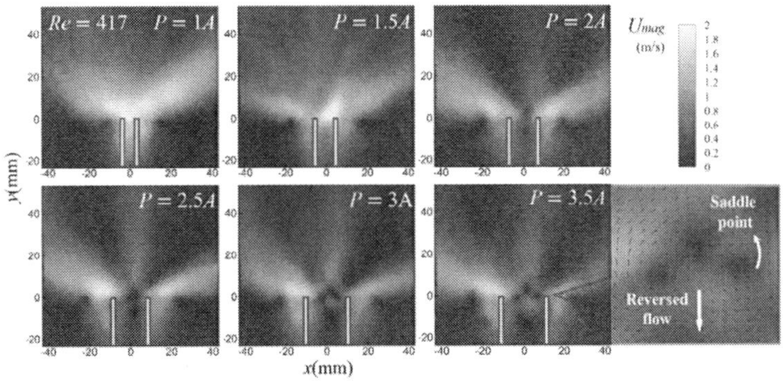

Fig. 6 Time-averaged velocity magnitude (U_{mag}) field of IP PZT fans array with different pitches at $Re = 417$.

Fig. 7 (b) shows the time-averaged velocity profile of PZT fans array alone the line y = 0mm and 0mm < x < 40mm. In the case $P = 1A$ and $P = 1.5A$, velocity magnitude increases monotonically along the x-axis, with a high velocity magnitude near the tip of the blade and decreasing to zero at 40mm. The velocity magnitude peak appears when the blade spacing is greater than $P = 2A$, and the horizontal coordinate of the velocity magnitude peak becomes lager with increasing blade spacing. The case $P = 2.5A$ has the highest velocity magnitude peak along the x-axis. Because of the vortices interaction in blade spacing, the velocity magnitude near the midspan line is small when $P > 2A$. Due to the blade displacement, the dropping tendency mitigates and causes the velocity magnitude residual when approaching the image edge. The velocity profile alone the midspan line is also investigated as shown in Fig. 7 (c). Alone the line x = 0mm and 0mm < y < 40mm, contrary to the monotonic decreasing of $P = 1A$ and $P = 1.5A$, when $P > 2A$, there is an upward trend in velocity for a period near the tip of the blade, a turning point appears at y = 10mm. Before y = 10mm, the velocity magnitude of the case $P > 2A$ is much smaller than the case $P = 1A$ and $P = 1.5A$, but relatively far away the final velocity is almost the same at different pitches. The change in velocity magnitude profile monotonicity is related to the generation of saddle points.

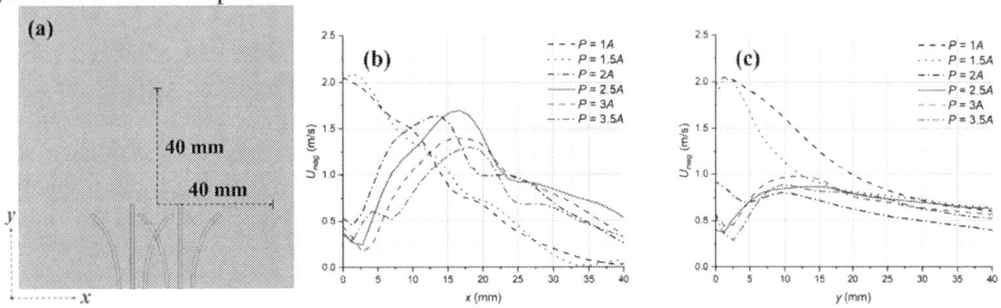

Fig. 7 Velocity magnitude at streamwise and spanwise lines of IP PZT fans array with different pitches: (a) schematic of the lines for data extraction. (b) Velocity magnitude (U_{mag}) profile of the flow field along the line $y = 0$, $0 < x < 40$, (c) velocity magnitude (U_{mag}) profile of the flow field along the line $x = 0$, $0 < y < 40$.

Time-averaged flow field of PZT fans array with counter-phase vibration is also obtained from 2000 images over 20 cycles, which is shown below in Fig. 8. The flow field is highly symmetrical. The flow field pattern does not change significantly as the blade pitch increases, a vertical upward jet and two lateral jets are captured at different blade pitches. The reversed flow is also present in the flow field pattern. The intensity of the lateral jet is weaker compared to the in-phase vibration-induced jet. The strength of the lateral jet increases as the blade pitch increases. The coupling effect of the PZT fans array weakens the strength of the lateral jet by reducing the maximum tip velocity in a vibration period. When $P > 3A$, three saddle points in the center of the blade spacing appear. Unlike the in-phase vibration-induced flow field, the low velocity region that becomes saddle points exists at low pitches. This is due to the fact that counter-phase vibrations can induce vortices in the spacing of the array at low pitches. Reverse rotation of the vortex core is too close to make saddle points difficult to distinguish and connect into a low velocity region. The causes of jet and reversed flow formation are different from those of in-phase vibration-induced flow, it will be claimed more clearly in the next section. Two counter-rotating vortices in the same phase are beneficially coupled, the symmetry of the vibration allows the flow velocity on the midspan line to have no component in the x-axis. The high Reynolds number flow field maintains the similar characteristics as we mentioned before, with two lateral jets and two saddle points located below that can be identified in the velocity magnitude flow field.

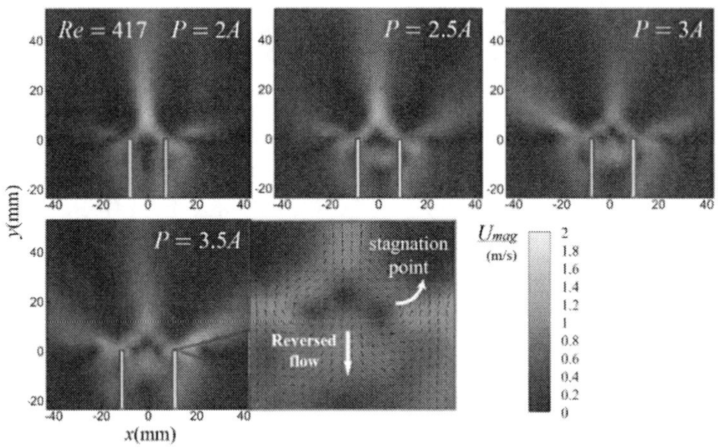

Fig. 8 Time-averaged velocity magnitude (U_{mag}) field of CP PZT fans array with different pitches at $Re = 417$.

The velocity magnitude of counter-phase vibration in the axial direction was also sampled and velocity profiles were plotted for different blade pitches. Fig. 9 (b) and Fig. 9 (c) is the axial velocity magnitude extracted from the horizontal line and the midspan line as analyzed before. Along the $+x$ direction, the velocity magnitude shows a similar changing mode of variation at different blade pitches. Velocity magnitude appeared the maximum peak from $x = 7$mm to $x = 17$mm, and with the increase of blade pitch, the curve with peak velocity point gradually moved forward. The change in spacing also allows the high-speed area to be expanded. In the line alone x = 0mm, the velocity of the counter-phase induced flow field is smaller and the reduction of blade amplitude due to vibration is the cause of this phenomenon. Along the midspan line, the velocities at different blade pitches all reach a maximum at y = 10 mm, followed by a slow decrease. In addition, at high pitch, the velocity shows a period of decreasing trend in the first few millimeters in the direction of $+y$-axis at the tip of the leaf due to the gradual formation of three stagnation points. Compared to the velocity characteristics of the flow induced by in-phase vibration mode along the midspan line, the velocity under counter-phase has a higher closing velocity at the end due to a slower decrease and has higher peak.

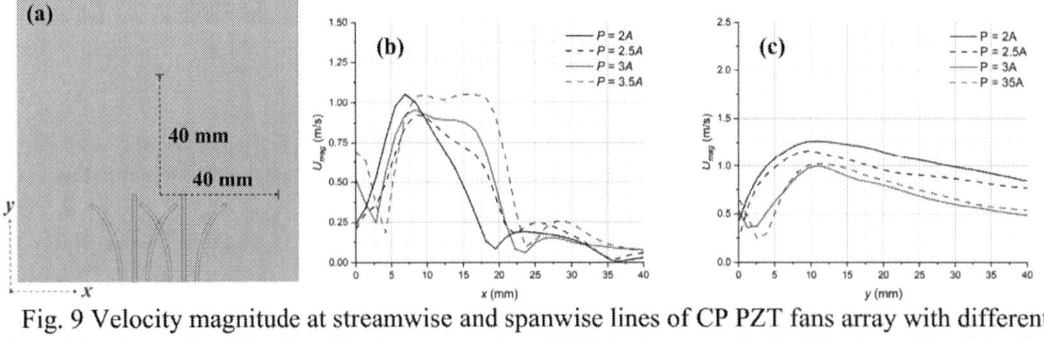

Fig. 9 Velocity magnitude at streamwise and spanwise lines of CP PZT fans array with different pitches: (a) schematic of the lines for data extraction. (b) Velocity magnitude (U_{mag}) profile of the flow

field along the line $y = 0$, $0 < x < 40$, (c) velocity magnitude (U_{mag}) profile of the flow field along the line $x = 0$, $0 < y < 40$.

3.2. Vortex evolution

In this paper, the vibration period of a PZT fan was divided into different phases. The formation process of velocity field and vortex is obtained by the ensemble-averaged method under different phases. the phase tau is defined as $\tau = t/T$, where t represents the time and T represents the period of vibration of a single PZT fan. The evolution of the vortex is obtained from the ensemble-averaged images of 8 different phases. The time when the PZT fan blade swings to the far right is defined as 0. For an array of two PZT fans, the reference blade is set to the right blade and the time corresponding to the phase is 0 when the right PZT fan swings to the far right. The definition of the specific position at phase τ is shown in Fig. 10.

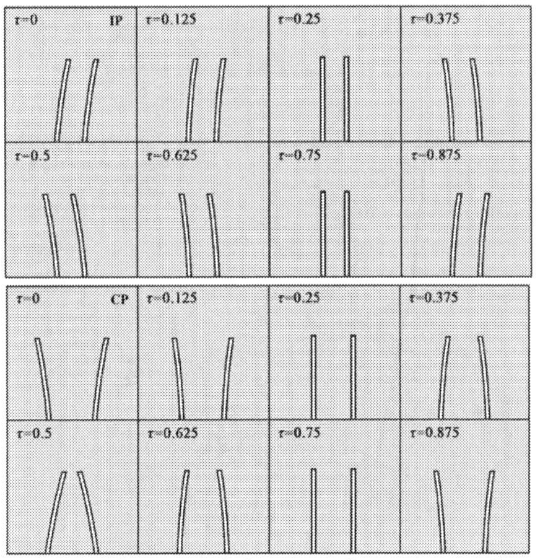

Fig. 10 Definition of the phase τ of the PZT fans array of periodic vibrations.

As the blade pitch increases, the intermediate jet induced by PZT fans array with in-phase vibration gradually strengthens, and stagnation points are formed. In this section, the case of $P = 1A$, $P = 2A$ and $P = 3.5A$ is analyzed separately. At $P = 1A$, the velocity field is shown as a Y-shaped jet. $P = 2A$, the intermediate jet is generated, and stagnation points appear. $P = 3.5A$, three stagnation points are clearly visible. The images in Fig. 11 were subjected to the ensemble-averaging method. 200 pairs of images were used to generate location-specific ensemble-averaged images, and the results of the ensemble-averaged processing at 8 different locations were presented. Fig. 11(a) shows the vorticity image of one cycle of vibration from the rightmost position under in-phase vibration when $P = 1A$ at $Re = 417$. When $P = 1A$, a CW vortex and a CCW vortex are generated in one vibration cycle. For the phase $\tau = 0$ and $\tau = 0.5$, the PZT fans blade is at right and left ends, the blade tip speed is zero and the vortex generated is the weakest. Take $\tau = 0$ as an example, with the blade to swing, a CW vortex is generated, when $\tau = 0.25$, the blade moves to the midspan line, the strongest vortex is generated by the blade tip. When $P = 1A$, due to the small spacing, in-phase vibration of the two blades produced only one vortex, the overall characteristics are similar to the single blade vibration. When $P = 2A$, as shown in Fig. 11(b), the maximum vorticity is minimum at $\tau = 0$. As the blade moves to the left, two CW vortices are generated from the tips of the two blades, and the two vortices can be clearly distinguished because the blade spacing is large enough to vortex generation and existence. The vortex intensity reaches its maximum at $\tau = 0.25$, and as the blade oscillates in reverse after $\tau = 0.5$, two new CCW vortices are generated, the remnant CW vortices gradually merge at the periphery. In addition, a vortex core with relatively large vortex intensity between the two blades can be clearly observed at $\tau = 0.25$ and $\tau = 0.75$. Since the blade positions are still relatively close, the vortex core induced by the leading blade cannot be effectively distinguished, which leads to the low velocity region in the velocity magnitude field between the two blades in the flow field pattern. When the blade pitch P reaches $3.5A$,

Journal of Physics: Conference Series **2108** (2021) 012010 doi:10.1088/1742-6596/2108/1/012010

which has shown in Fig. 11(c). The evolution of the vortex is close to the flow field evolution process induced by the vibration of two independent PZT fans. During the vibration of the PZT fans to the left, two CW vortices are generated from the lobe tips, respectively, while the remnant CCW vortices generated in the previous oscillation cycle continue to propagate upward in evolution. Unlike the case when the $P = 2A$, the remnant vortex generated by the two blades do not merge, they propagate separately because a part of the newly generated vortices occupies the areas between two PZT fan blades. The newly generated vortex interacts with the remnant vortex produced in the previous cycle of the other blade to produce an upward jet. At $\tau = 0.25$ and $\tau = 0.75$, the two blades produce two vortices of higher intensity each because the blade tip velocity reaches its maximum. However, the vortex generated by the leading blade is obviously compressed, the compressed vortices in $\tau = 0.25$ and $\tau = 0.75$ resulting in two compressed vortices on both sides of the midspan line, which are corresponding to the velocity field calculated by the time-averaged method as two saddle points below. And above these two points, as the velocities in different directions are cancelled out, a new saddle point is created above these two points because the velocities in different directions are cancelled by the algorithm. Unlike the case where the vortices generated by the two blade tips are not separated, in the case of separation, the vortex core strength generated by the leading blade in vibration is stronger than that generated by the following blade when it is compressed. In the whole vibration cycle, due to the asymmetry of the blade motion, although the jets in the time-averaged flow field are symmetric, the ensemble-averaged flow field corresponding to each phase is highly asymmetric, and the upward jets are periodically biased to the left and right sides with the movement of the blade.

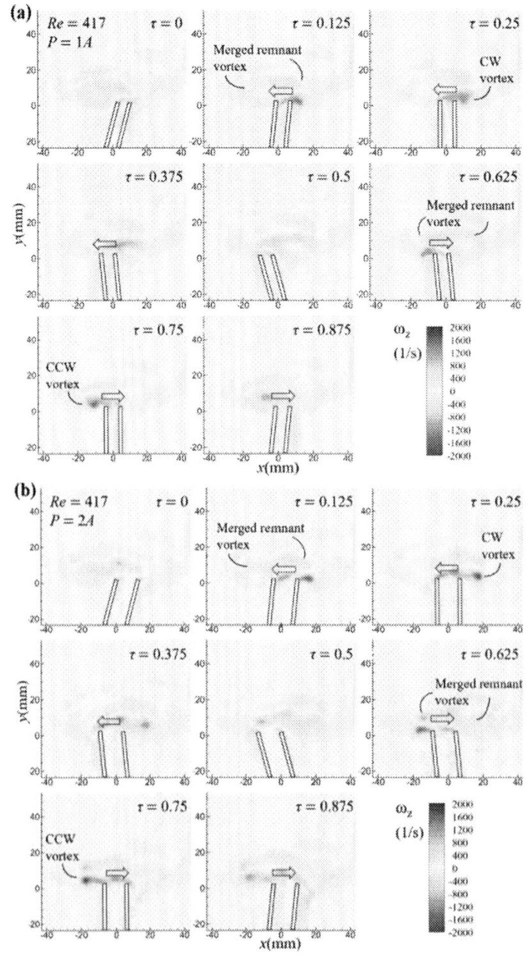

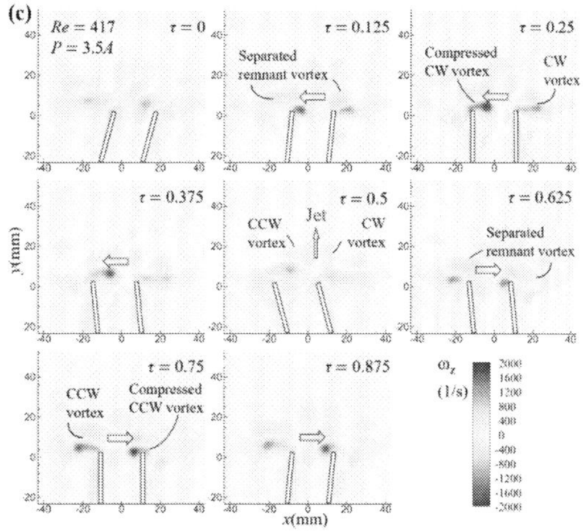

Fig 11. Evolution of vorticity (ω_z) in one cycle of IP PZT fans array at Reynolds number $Re = 417$, with pitches (a) $P = 1A$, (b) $P = 2.5A$ and (c) $P = 3.5A$

To better reveals the details of the flow field at different pitches, the ensemble-averaged images with $\tau = 0.25$ for different blade spacing at Reynolds number $Re = 417$ are shown in Fig. 12. At $\tau = 0.25$, the tip-induced vortex has the maximum intensity, because the blade has the maximum tip velocity. At $P = 1A$, the vortices generated by the vibration of the two PZT fans merge due to the lack of space, as the remnant vortex of the previous cycle is located above the newly generated vortex. At $P = 1.5A$, the tip-induced vortex tends to separate, and when the blade pitch is expanded to $2A$, the tip-induced vortex is almost separated, and it can be observed that the vortex generated by the leading blade already has a more obvious vortex core. When $P = 2.5A$, the tip-induced vortex is completely separated, and the vortex generated by the leading blade is severely compressed, which becomes stronger than the vortex induced by the backing blade. The remnant vortex generated during the last cycle of vibration is separated. A part of the CW vortex appears between the two remnant vortices and interacts with them to produce an upward jet. As the blade pitch continues to increase, the compression of the vortex induced by the leading blade is relieved.

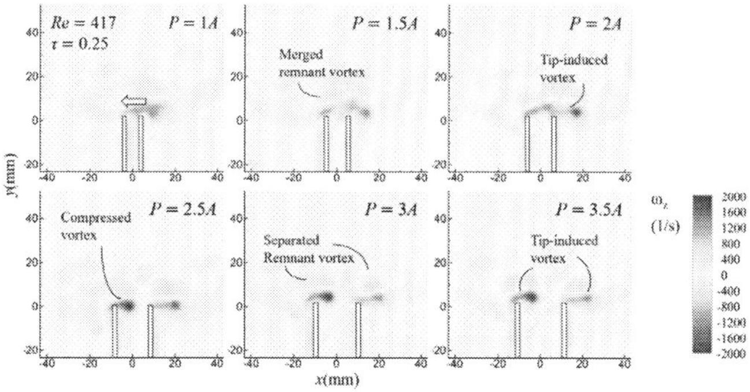

Fig. 12 Phase-locked vorticity field induced by IP PZT fans array with different pitches at Reynolds number $Re = 417$ and $\tau = 0.25$.

The coupled flow fields of the PZT fans array at different Reynolds numbers were also investigated, and the results are shown following. Fig. 13 shows the coupled flow field for different Reynolds numbers at $P = 2A$. From the above conclusion, $P = 2A$ is a critical distance for the coupled flow field, and at this distance, the coupled flow field for all three different Reynolds numbers maintains the similar characteristics, the tip-induced vortex is not completely separated, and the remnant vortex surrounds the whole tip-induced vortex at the periphery. However, with the increment of the Reynolds numbers, the induced vortex also produces a slight separation, the center strength of the remnant vortex becomes weaker, the evidence supports the idea that larger PZT fans amplitude leads to the weakening of the coupling effect of PZT fans. So, the conclusions drawn in the previous section should only apply to the vibrating PZT fans with a specific range of Reynolds numbers between 378 and 457. More studies are needed on the change in the degree of coupling due to Reynolds number variation.

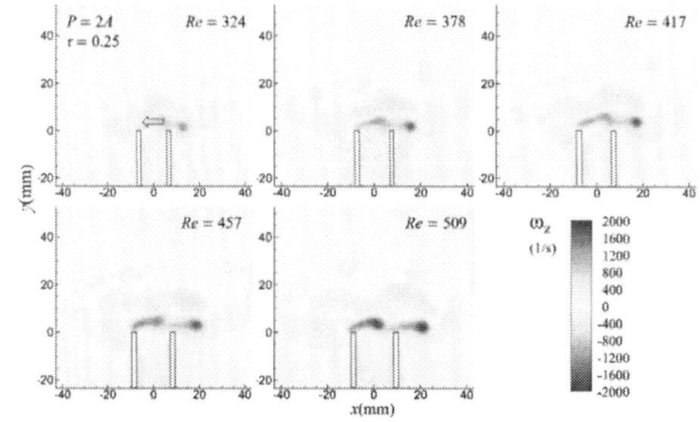

Fig. 13 Phase-locked vorticity field induced by IP PZT fans array with pitch $P = 2A$ at different Reynolds numbers in the range of 324< *Re* <509.

The vortex evolution under counter-phase vibration shows different characteristics with in-phase vibration. The counter-phase vibrational ensemble-averaged images at two different pitches were investigated as shown in Fig. 14. Fig. 14. (a) display the case of $P = 2A$, this is the shortest distance to ensure that fan blades do not touch each other. The PZT fan blade vibrating in the opposite direction produces a vortex also in the opposite direction. At the time of $\tau = 0$, the tips of the double blades of counter-phase vibration are biased to the sides and are in the open state. With the inward contraction of both sides of the blade, the left and right blades produced CCW vortex and CW vortex respectively, and the size and intensity of the vortex reached the maximum at $\tau = 0.25$. After that, the two induced generated vortices converge toward the midspan line and propagate upward. The remnant vortex structure exists throughout the vibration cycle and does not dissipate, and this pair of interacting vortices works together to produce an upward jet. Therefore, unlike the unstable jet under in-phase vibration of the PZT fans, the upward jet in counter-phase vibration mode is continuous and stable in the direction. After $\tau = 0.5$, the two PZT fans blades start to move to both sides, two CW vortex and CCW vortex are generated from the left and right blades respectively, the intensity of the vortex reaches the maximum at $\tau = 0.75$ but due to the lack of space between the two blades, the vortex is compressed. After that, the intensity of both vortices decreases, and each propagates to both sides and dissipates. The details of the vortex's evolution in the case of $P = 3.5A$ is shown in Fig. 14 (b). There is no significant difference between the different pitches. Compared to $P = 2A$, the degree of compression of the vortices is reduced at $\tau = 0.75$, and the cores of both vortices are more pronounced and farther apart. Therefore, two lower saddle points, corresponding to the two vortex cores, and one higher saddle point, corresponding to the zero-velocity point generated by the two vortices coupling, can be observed in the time-averaged image with a large pitch.

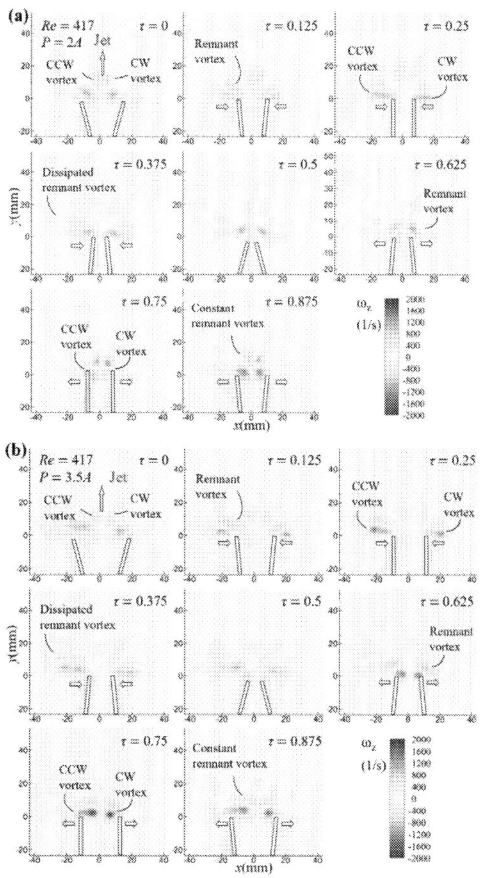

Fig 14. Evolution of vorticity (ω_z) in one cycle of CP PZT fans array at Reynolds number $Re = 417$, with pitches (a) $P = 2.5A$ and (b) $P = 3.5A$

At $\tau = 0.75$, the blades of PZT fans are in vertical position. The two blades are moving to the two sides, and two vortices in opposite directions are generated in the space between the PZT fans array, and the compression of the vortices varies at different distances. Fig. 15 investigates the coupling of the flow field by comparing the compression of the vortices at different Reynolds numbers. At $P = 2A$, the intensity of the induced vortex is low, but the intensity of the remnant vortex is relatively high. As the pitch increases, the degree of vortex extrusion decreases, and the intensity of the induced vortex increases while the intensity of the remnant vortex decreases. The core of the induced vortex has started to become clear at $P = 2.5A$, and the time averaged velocity flow field image also indicates that the stationing point is formed near $P = 2.5A$. As pitch continues to increase after reaching $3A$, the coupling of the flow field weakens significantly and the remnant vortex changes from a compressed circle to a bar shape. Half side of the image is similar to the flow field image induced by a single PZT fan.

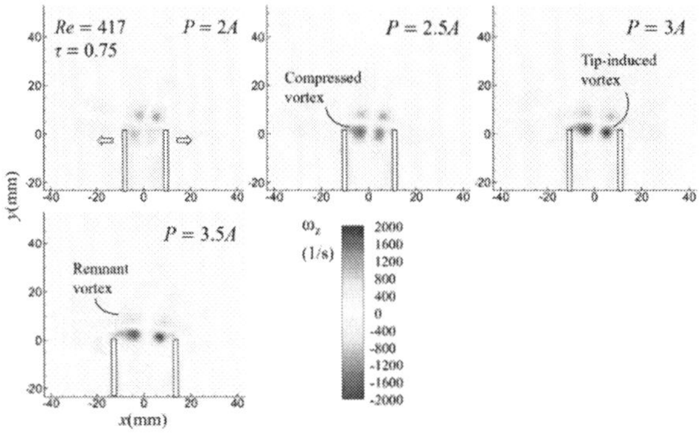

Fig. 15 Phase-locked vorticity field induced by IP PZT fans array with different pitches at Reynolds number $Re = 417$ and $\tau = 0.25$.

The increase of Reynolds number makes the intensity of the vortex increase, but the change of Reynolds number also changes the flow field pattern. Like the in-phase vibration, Increasing the Reynolds number brings a change in the flow field pattern similar to increasing the pitch. Fig. 16 shows the flow pattern changes various Reynolds numbers. At low Reynolds numbers, the cores of the two reversed vortices are not obvious and the whole vortex is squeezed in the horizontal direction and the vortex is in an elliptical shape. And at $Re = 509$, it can be observed that the vortex becomes a more uncompressed circle. The core of the vortex becomes apparent at Re = 417, and at $Re = 457$ or greater, focusing on the left part of the pattern, the remnant vortex begins to combine with the CW vortex produced by the blade body to wrap the CCW vortex produced by the blade tip.

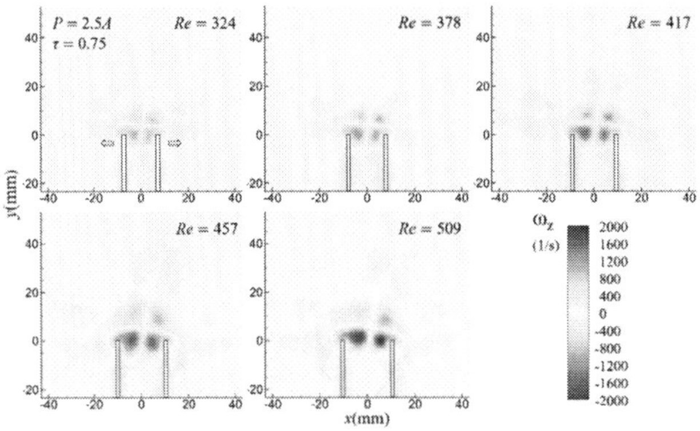

Fig. 16 Phase-locked vorticity field induced by IP PZT fans array with pitch $P = 2A$ at different Reynolds numbers in the range of $324 < Re < 509$.

The downstream flow velocity of the PZT fan determines the effectiveness of the heat dissipation. In a coupled array of PZT fans, the jet along the midspan line will provide the main air flow to the target components. In the in-phase and counter-phase vibration modes, in the range of Reynolds numbers of interest in this study, the jet appears when $P > 2A$. Fig. 17 (a) illustrates the maximum velocity magnitude of the jet at $y = 20$ mm when $Re = 417$. The counter-phase vibration has a greater downstream velocity than the in-phase because the two vortices generated are beneficially coupled. As the distance increases, the beneficial coupling effect decreases, resulting in a decrease in jet strength, especially at pitch between $2.5A$ and $3A$. The jet generated by the in-phase vibration reaches its maximum at $P = 3A$. The lack of space for vortex formation in the PZT fan array when the distance is too close, and the vortex generated in the space is the condition for jet generation. When the distance is too far, the coupling effect also weaken. The facts lead to the optimal pitch for jet generation in in-phase vibration, which was determined as $P = 3A$ for the conditions of $Re = 417$ in this study.

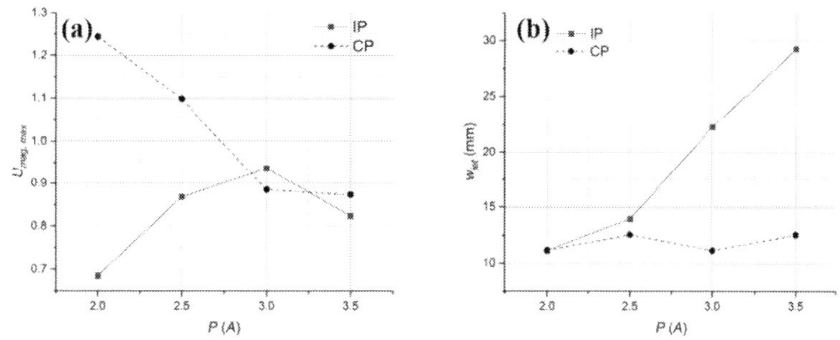

Fig. 17 maximum jet velocity ($U_{mag,\ max}$) and width (w_{jet}) of IP and CP vibrating PZT fans array.

The area affected by the jet is also a parameter that affects the cooling effect. The width of the jet at $y = 20$ mm was determined and shown in Fig. 17 (b). The jet width of counter-phase vibration remains stable and distinct smaller than the in-phase vibration. The in-phase jet width increases significantly with increasing pitch. This difference in performance stems from the different reasons for jet generation. At high Reynolds number, the large amplitude causes the two vortices generated by a single PZT fan to propagate to both sides. As the blade gap widens, the propagation restriction of the vortex on one side of the PZT fan array to the other side is diminished, allowing the deflection of airflow more severe. This also allows PZT fans with large pitches of in-phase vibration to have a larger area for high speed. While the direction of the jet generated by the counter-phase vibration of the symmetrical flow field remains stable, the intensity of the jet varies with distance but remains constant in the high-speed area. Due to the difference in stability and speed of the jets, counter-phase vibration is the more beneficial vibration method for heat dissipation. The optimal Pitch was determined to be around $P = 2.5A$. More experiments on the flow rate need to be done in the follow-up studies.

4. Conclusion

In this study, the time-averaged and assembled-averaged flow patterns of PZT fans array with different spacing at high Reynolds number are investigated to research the coupling of PZT fans array at different spacing and Reynolds number. The performance of a piezoelectric fan is measured by the maximum speed and width of the jet. The following are the conclusions drawn from this study.

1. As the space in the middle of the PZT fans array increases, the in-phase vibration generates jets and three saddle points appear and separate in time-averaged flow field. The jet originates from the averaging of the deflected flow which generated by the periodic interaction of an induced vortex and a counter-rotating remnant vortex. The velocity of jet reaches the highest at $P = 3A$, and the jet width increases with the increase of pitch.
2. In counter-phase vibration, the spacing allow the persistence of induced vortex, causes the jet to persist and produce three saddle points. The maximum velocity of the jet is greater than the in-phase induced jet velocity due to the beneficial coupling of counter-rotating induced vortices. The jet intensity decreases significantly after $P = 2.5A$, considering the energy consumption per unit length, the optimal location is determined at $P = 2.5A$.
3. The presence of the vortex core leads to the formation of the saddle points which start generated at $P = 2A$, and the induced vortex and the remnant vortex begin to separate when PZT fans vibrate in-phase with $P > 2.5A$. Saddle points separated at under the same pitch. The induced vortex of the PZT fan coupled with the remnant vortex from the last cycle of the adjacent fan generates the jet. While the flow field pattern of the counter-phase vibrating PZT fan is similar at different pitches, saddle points and the position of induced vortices core begin to separate with the increase pitches.
4. The time-averaged PZT fans array flow fields at high Reynolds numbers have two lateral jets and a more pronounced upper lateral saddle point. This is not consistent with the previous simulation results using low deflected PZT fan blades. The flow field with the same pitch at different Reynolds numbers shows a weaker coupling effect with increasing Reynolds numbers. Therefore, the effective

range of Reynolds numbers in this experimental study was determined to be 324< *Re* <509.

Acknowledgments

This work is supported by the Nature and Science Foundation of Zhejiang Providence: LQ19A020003. This work was also supported in part by the Zhejiang University/University of Il-linois at Urbana-Champaign Institute, and was led by Principal Supervisor Jiahuan Cui.

References

[1] Maaspuro M (2016) Piezoelectric oscillating cantilever fan for thermal management of electronics and LEDs — A review. Microelectron Reliab 63: 342-353. https://doi.org/10.1016/j.microrel.2016.06.008
[2] Hales A Jiang X (2018) A review of piezoelectric fans for low energy cooling of power electronics. Appl Energ 215: 321-337. https://doi.org/10.1016/j.apenergy.2018.02.014
[3] M. Toda (1978) Theory of air flow generation by a resonant type PVF2 bimorph cantilever vibrator, Ferroelectrics, 22:1, 911-918. https://doi.org/10.1080/00150197908239445
[4] M. Toda (1981) Voltage-Induced large amplitude bending device-PVF2 Bimorph-its properties and applications, Ferroelectrics, 32:1, 127-133. https://doi.org/10.1080/00150198108238683
[5] Kim YH, Cierpka C, Wereley ST (2011) Flow field around a vibrating cantilever: coherent structure eduction by continuous wavelet transform and proper orthogonal decomposition. J Fluid Mech 669: 584-606. https://doi.org/10.1017/S0022112010005318
[6] Ebrahimi ND, Eldredge JD, Ju YS (2020) Three-dimensional characteristics of the jet flows induced by a pitching plate in a quiescent fluid. J Fluid Mech 887: A25. https://doi.org/10.1017/jfm.2020.2
[7] Agarwal A, Nolan KP, Stafford J, Jeffers N (2017) Visualization of three-dimensional structures shed by an oscillating beam. J Fluid Struct 70: 450-463. https://doi.org/10.1016/j.jfluidstructs.2017.02.013
[8] Eastman A, Kiefer J, Kimber M (2012) Thrust measurements and flow field analysis of a piezoelectrically actuated oscillating cantilever. Exp Fluids 53: 1533-1543. https://doi.org/10.1007/s00348-012-1373-6
[9] Stafford J, Jeffers N (2017) Aerodynamic performance of a vibrating piezoelectric blade under varied operational and confinement states. IEEE T Comp Pack Man 7(5): 751-761. https://doi.org/10.1109/TCPMT.2017.2666879.
[10] M.L. Kimber, R. Lonergan, S.V. Garimella (2009) Experimental Study of Aeordynamic Damping in Arrays of Vibrating Cantilevers. Journal of Fluids and Structures 2(8):1334-1347. https://doi.org/10.1016/j.jfluidstructs.2009.07.003
[11] Choi, Minsuk & Lee, Sang-Youp & Kim, Yong-Hwan (2012) On the flow around a vibrating cantilever pair with different phase angles. European Journal of Mechanics - B/Fluids 34:146–157. https://doi.org/10.1016/j.euromechflu.2012.01.021.
[12] Choi, Minsuk & Cierpka, Christian & Kim, Yong-Hwan (2014) Effects of the distance between a vibrating cantilever pair. European Journal of Mechanics - B/Fluids 43: 154–165. https://doi.org/10.1016/j.euromechflu.2013.08.006.
[13] Hales, Alastair & Jiang, Xi (2019) Geometric optimisation of piezoelectric fan arrays for low energy cooling. International Journal of Heat and Mass Transfer 137: 52-63. https://doi.org/10.1016/j.ijheatmasstransfer.2019.03.086.
[14] Hales, A. and Xi Jiang (2019) Optimisation of low energy cooling through phase variation between adjacent piezoelectric fan blades. International Journal of Heat and Mass Transfer 139: 362-372. https://doi.org/10.1016/j.ijheatmasstransfer.2019.05.004

Journal of Physics: Conference Series **2108** (2021) 012011 doi:10.1088/1742-6596/2108/1/012011

Research on small lightweight electromagnetic interference measurement system based on patrol unmanned aerial vehicle

Pingyuan Liu[1a*], Nian Wu[2a*], Junwen Yao[1b*], Xianbin Ke[2b*], Shiguang Bie[2c*], Feng Liu[2d*]

[1]Guangdong Power Grid Co. , Ltd, No. 757 East Dongfeng Road, Yuexiu District, Guangzhou City, Guangzhou and 510000, China

[2]Electric Power Research Institute Wuhan Nanrui Co. , Ltd, No. 143 Luoyu Road, Hongshan district, Wuhan City, Wuhan and 430074, China

[1a*]plainliu@qq.com, [2a*]1175791947@qq.com, [1b*]867271037@qq.com, [2b*]1169271991@qq.com, [2c*]414790841@qq.com, [2d*]215273428@qq.com

Abstract—Unmanned aerial vehicle (UAV) has been widely used in power system inspection. Most of the existing UAVs are equipped with visible and infrared cameras for fast full spectrum photography and fault monitoring of transmission lines. When the transmission line has defects such as aging, broken strand and serious corrosion, the traditional image recognition, infrared and other observation methods can not identify these defects well. In order to realize the rapid detection and judgment of broken strand, this paper analyzes the characteristics of electromagnetic interference of transmission line, and constructs the electromagnetic interference measurement system based on patrol UAV. The research results of this paper can realize the real-time detection and alarm of broken strand, and provide technical support for line inspection.

1. Introduction

Transmission line inspection UAV has the ability of high-altitude, long-distance, fast and self-operation. It can cross mountains and rivers to quickly inspect the transmission line, greatly improving the speed and efficiency of transmission maintenance and repair[1]. As an important technical means of intelligent machine inspection, UAV inspection has been widely used in the system. The application of UAV inspection technology in power industry provides a new method and means for power line inspection and large-scale disaster prevention and mitigation inspection. When the transmission line has defects such as aging, broken strand and serious corrosion, the traditional image recognition, infrared and other observation methods can not identify these defects well, but the change of electromagnetic interference level is an important characterization of the defects[2].

Therefore, combined with the large-scale application of existing UAV and the research on corona characteristics of transmission line wire, this paper studies the correlation between broken strand defect of transmission line wire and high-frequency electromagnetic spectrum distribution, establishes a lightweight integrated communication system for receiving and processing electromagnetic interference information applied to small UAV, and realizes the positioning of serious corona discharge parts of wire, and real-time detection of aging, broken strand and serious corrosion defects can not be distinguished by image.

2. Measuring Principle

2.1. Electromagnetic interference characteristic

Within a certain distance, the height of the line wire nearest to the measuring point has an impact on the speed of the electromagnetic interference level decreasing with the distance. The lower the height of the line wire, the faster the attenuation[3]. The transverse attenuation rate of electromagnetic interference is a complex function, which is related to the frequency and the distance from the interference source. For the extremely low frequency (3-30 kHz) and low frequency (30-300 kHz) bands, the measurement antenna will be in the near field or induction field, and the attenuation rate of electromagnetic interference is $\frac{1}{R^2}$, R is the distance from the wire to the antenna. For the intermediate frequency (300-3000khz) band, with the antenna gradually away from the line, the electric field formed by corona pulse gradually changes from the induced field to the surface wave field at the radial distance $\frac{\lambda}{2\pi}$ from the wire, where λ is the wavelength, and the unit is m. The attenuation rate of the surface wave field is $\frac{1}{R}$. When the frequency is from 3 MHz to 30 MHz, the electric field is the combination of surface wave field and radiation field[4].

The transverse attenuation near the line can be expressed by the following formula:

$$E = f(R) = E_0 + 20k\lg(\frac{R_0}{R}) \tag{1}$$

Where, E is the electric field component level of electromagnetic interference field strength at the linear distance R , expressed by $dB(\mu V / m))$. k is the attenuation coefficient, which is 1.65 in the frequency band of 0.5mhz-1.6mhz ; E_0 is the electric field component level of electromagnetic interference field strength at the reference distance $R_0 = 20m$, expressed by $dB(\mu V / m))$.

2.2. Detection principle

Based on the analysis of electromagnetic interference characteristics of transmission line, when the transmission line is in normal operation, the magnetic field signal detected around the line is also normal. When the transmission line fails, the fault point will produce the relative difference of fault signal, which will lead to the change of the current of the transmission line in the transmission process[5]. When the EMI antenna is close to the high-voltage transmission line, the broken line will produce the alternating magnetic field, and then produce the interference signal, and the signal will fluctuate in a large relative difference, Therefore, the wire surface state can be judged by detecting the radio signal.

The working principle and basic structure of the inspection UAV for detecting broken strand defect of transmission line by electromagnetic interference are shown in the block diagram in Fig.1:

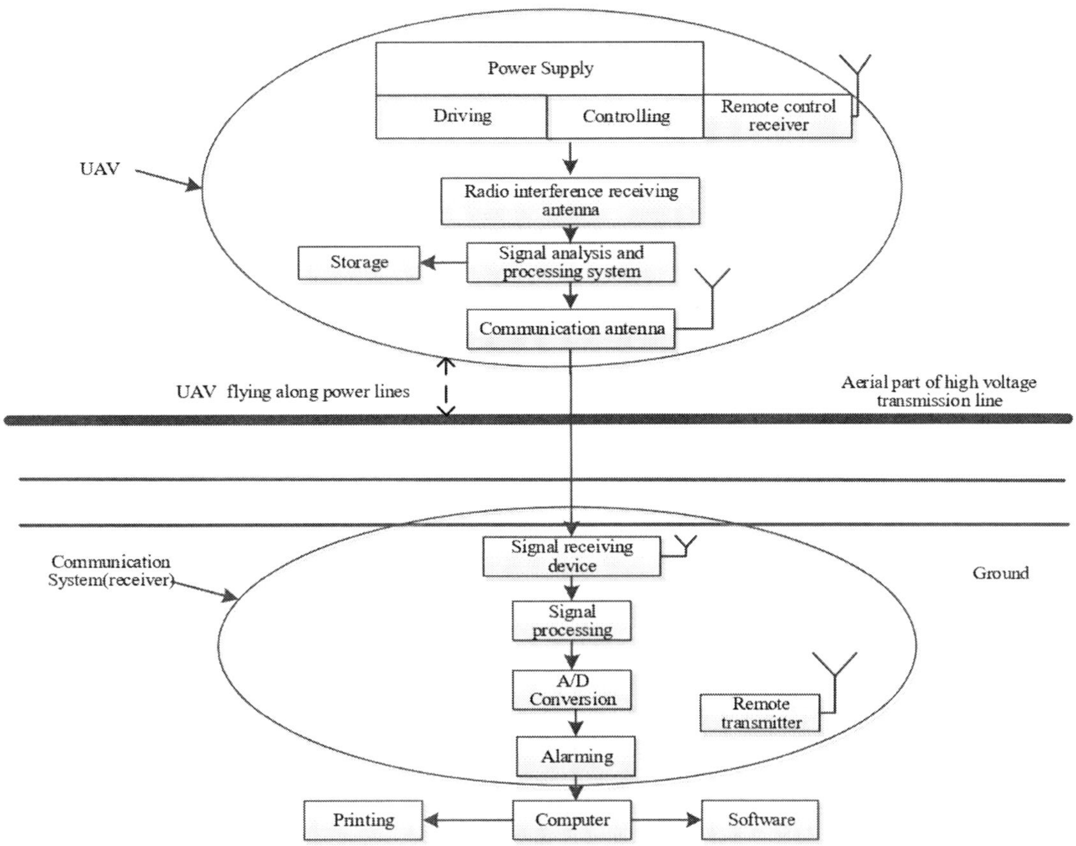

Fig.1 Working principle diagram

When the transmission line is running normally, the magnetic signal detected around the line is also normal. When the transmission line is in fault, the relative difference of the fault signal will be produced at the fault point, and the current of the transmission line will change during transmission. The detected radio around the transmission line will reflect the change of the current in the line in real time, therefore, by detecting the radio signal, we can get all kinds of characteristic information of the line signal, and determine the position of the fault by the GPS positioning system, at the same time, image the fault point.

After the ground operator starts the UAV through the remote control device, along with the patrol UAV's movement in the high-voltage transmission line, the radio interference receives the transmission line status signal which the antenna detects, the signal is processed by the amplification denoising analysis system, with the threshold value for decision, the problem line is automatically executed beyond the threshold interval photo storage, and at the same time the signal is sent to the ground control end to remind[6]. With the flight of the patrol UAV, the detection can be completed on the high-voltage transmission line abnormal fault detection points for a detailed Quantitative analysis, diagnosis of high-voltage transmission line broken, damage fault.

3. System Design

As a fault detection and diagnosis system for high-voltage transmission lines, it has the functions of real-time monitoring, collecting, processing, transmitting and storing the radio interference signals generated by transmission line faults, combined with the recording information of the radio test data, it can be used to detect and diagnose the faults such as broken and scattered strands, which is convenient for the staff to find and analyze the fault points. The communication function of the system makes the detection information more comprehensive and the diagnosis position more accurate. In addition, the system has the storage function of the time information and the location information, it provides a great help to the follow-up data processing and analysis.

The small lightweight EMI measurement system based on patrol UAV is composed of data acquisition unit, central processing unit and data transmission unit. The overall structure of the system is shown in Fig.2:

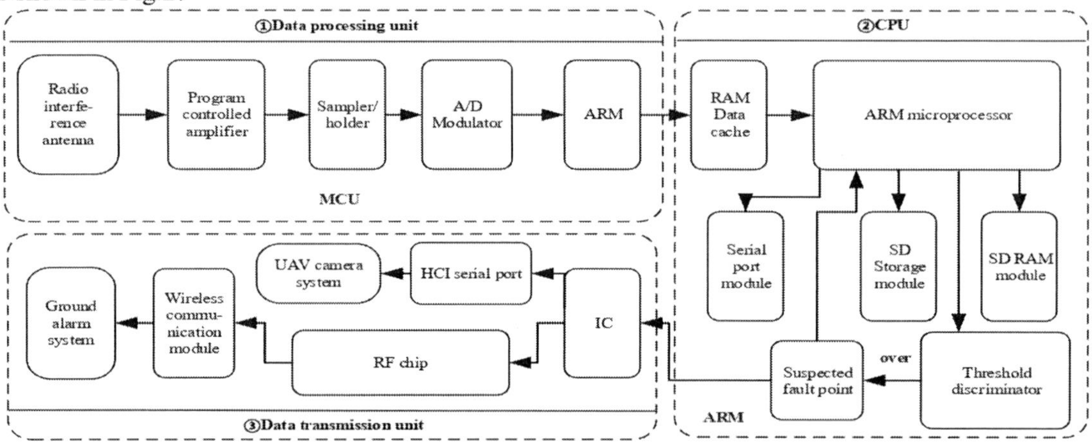

Fig.2 Overall system structure diagram

(1) Data acquisition unit: MCU is used as the main controller in the front-end acquisition part, which mainly completes the front-end processing of electromagnetic interference signal input, A/ D analog-to-digital conversion, data storage and transmission and other functions.

The hardware is divided into three parts, the first part is the antenna, the second part is the front-end signal processing circuit, including signal conditioning and trigger acquisition circuit, etc. .

(2) Central processing unit: the central processing platform based on ARM mainly uses the ARM microprocessor as the main control unit, designs the corresponding peripheral circuit modules of the chip, transplants the embedded Linux operating system, and realizes the control of the front-end acquisition circuit of the MCU, as well as the reading, analysis and processing of the collected data, photo execution, fault location, fault detection, fault detection and so on Alarm, memory and other functions[7].

Hardware design ARM the most simple system, including ARM processor, SDRAM circuit, as well as the system involved in the expansion of applications, mainly SD card memory module, serial port module, Positioning Module, warning module. The software part includes porting the embedded Linux operating system platform, controlling the acquisition path of the front-end acquisition part, the functions include real-time storage of data with time information and location information, serial transmission, integrated fault diagnosis and analysis of fault data record information combined with radio interference, taking photos of fault points and warning.

(3) Data transmission unit: composed of RF chip, related circuit and antenna. Taking the RF chip as the main control unit, the corresponding peripheral circuit module of the chip is designed. The modulated digital signal is converted to digital analog, converted and filtered, and then sent to the ground flying hand through the antenna to remind the failure point.

The hardware is divided into two parts, the first part is the RF chip and signal processing circuit, including signal conditioning and trigger transmission circuit, analog-to-digital Conversion Circuit, etc. . The software part includes the trigger transmission program, the analog-to-digital Conversion Program, the Signal Processing Code, the fault warning program and so on.

4. Conclusions

Conclusion based on the analysis of electromagnetic interference characteristics of transmission line, combined with the large-scale application of existing UAV and the research on corona characteristics of transmission line wire, this paper studies the correlation between broken strand defect of transmission line wire and high-frequency electromagnetic spectrum distribution, establishes a lightweight integrated communication system for receiving and processing electromagnetic interference information applied to small UAV, and realizes the positioning of serious corona discharge parts of wire, and real-time detection of aging, broken strand and serious corrosion defects can not be distinguished by image.and provides strong technical support for the condition assessment of transmission line conductor. In the future, this technology can be used in line maintenance and repair to improve the level of intelligent power grid inspection.

Acknowledgments
This work was financially supported by China Southern Power Grid Corporation(item number: GDKJXM20190010(031000KK52190001)).

References
[1] Transmission and Distribution Committee of the IEEE Power Engineering Society. ANSI/IEEE Std 430-1986, "IEEE standard procedures for the measurement of radio noise from overhead transmission lines and substations." New York, USA, 1986.
[2] J.J.LAFOREST, M.BARETSKY, JR.D.D, (1966). "MACCARTHY. Radio Noise Levels of EHV Transmission Lines Based on Project EHV Research." IEEE Transactions on Power Apparatus and System 1966,85(12):1213－1230,1966.
[3] TANG Mingwen, DAI Lihao, LIN Zhaohui, et al, (2013) "Application of unmanned aerial vehicle in inspecting transmission lines." Electric Power, 2013, 46(3): 35-38.
[4] HU Yi, (2007). "Analysis on operation faults of transmission line and countermeasures." High Voltage Engineering, 2007, 33(3):1 -8.
[5] ZHANG Wenfeng, PENG Xiangyang, CHEN Ruiming, et al, (2014). "Intelligent diagnostic techniques of abnormal heat defect in transmission lines based on unmanned helicopter infrared video." Power System Technology, 2014, 38(5): 1334-1338.
[6] WANG Rui, WANG Ke, XU Xiaogang, et al, (2014). "Design of ground measurement controlstation and transportation support system for unmanned aerial vehicle on power transmission line inspection." Guangdong Electric Power, 2014, 27(2): 87-92.
[7] Olsen R G, Schennum S D and Chartier V L, (1992). "Comparison of several methods for calculating power line electromagnetic interference levels and calibration with long term data." IEEE Transactions on Power Delivery,1992,7(2):903－913.
[8] PENG Xiangyang, ZHONG Qing, CHEN Ruimin, et al, (2014). "Intelligent diagnostic techniques of transmission lines corona discharge defect based on UAV ultraviolet detection." High Voltage Engineering, 2014, 40(8): 2292-2298.

Research on residual current warning technology based on big data analysis

You Li[1a*], Chuanfu Xia [1b], Yuan Zhuang[1c], Baoquan Liao[1d], Shuai Huang[1e]

[1] State Grid Sijishenwang Location Based Service(Beijing) Co.,Ltd., Beijing, 102209, China

[a*]liyou@sgitg.sgcc.com.cn, [b]xiachuanfu2897@163.com, [c]zhuangyu1@sgitg.sgcc.com.cn, [d]317685243@qq.com, [e]17710187156@163.com

Abstract—Residual current is the main cause of electrical fire. The traditional residual current monitoring methods often give delayed alarm, false alarm or even missed alarm, which cannot be applied in practice. This paper put forward the residual current detection solution based on big data analysis. The big data analysis based on residual current, temperature and humidity could detect residual current anomaly information as early as possible, realize fault monitoring and early warning. This method was applied to important facilities such as substation, data center. It could reduce or eliminate fire hazards, and provide decision basis for electrical fire warning.

1. Introduction

With the development of economy, electrical fire accidents are increasing year by year in China. From 2012 to 2016, there were 1,595,000 fires in China, among which electrical fires accounted for 29.5% of the total and reached 471,000, causing 14,440 casualties and direct economic losses of 19.8 billion YUAN [1]. From January to October in 2020, a total of 196,000 fire accidents occurred, among which electrical fires accounted for 32.1% of the total, far exceeding fires caused by other factors [2]. According to the causes of electrical fire, electrical fire accidents are mostly caused by leakage current generated by grounding fault short circuit [3-5].

When the insulation of the power lines decreases, leakage currents can be monitored and cause grounding fault. It is an important factor for electrical fire. When the leakage current is small, there may not be an immediate fire. However, with the increase of leakage current, the temperature of fault point will rise rapidly, and eventually lead to the occurrence of fire [6]. Therefore, residual current fire monitoring system is a very critical equipment, which has been widely used in electrical fire monitoring [7-11].

However, there are shortcomings for the traditional residual current monitoring systems, that processing method is relatively simple and that the intelligence degree is low. These monitoring systems often give delayed alarm, false alarm and even missed alarm, that causes great economic losses to the people. The prevention and control of electrical fire safety should focus on pre-prevention and follow the principle of the early detection, prevention and control.

This paper proposed the residual current monitoring and early warning scheme based on big data analysis, which is applicable residual current monitoring of important facilities such as substations and data centers. The residual current, temperature and humidity are collected in real-time using newly designed and developed senor terminal and management terminal. The new algorithm based on the processing and analysis of big data could find out the abnormal residual current as early as possible, accurately predict the risk, reduce or eliminate fire hazard, and provide decision basis for the electrical fire warning.

2. System architecture

The residual current monitoring and early warning system is mainly composed of three devices, namely sensing terminal, management terminal and monitoring terminal. The overall architecture of the system is shown in Fig. 1.

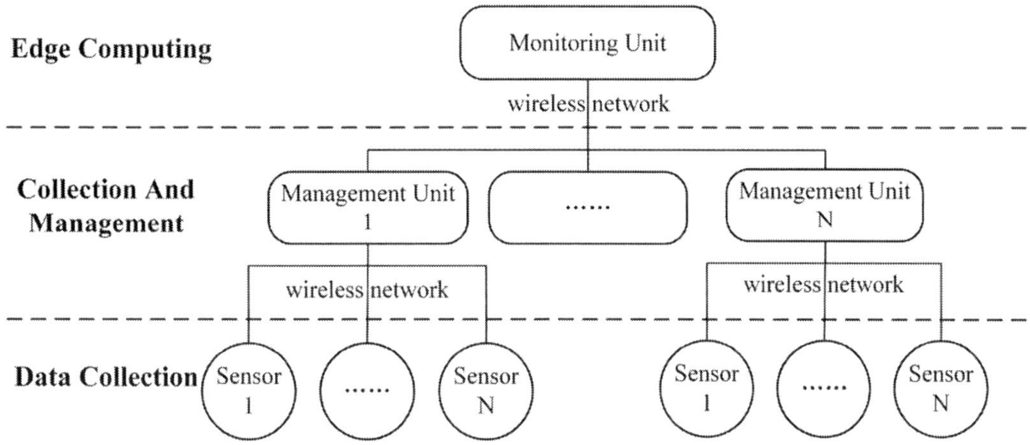

Fig. 1 Overall architecture design of the system

2.1. Sensing terminal

The sensing terminal collects the residual current, ambient temperature and humidity of each feeder branch of the AC power system. Ambient temperature and humidity are one of the important parameters for residual current analysis. The integration of ambient temperature, humidity and residual current monitoring can provide effective data support for the early warning analysis of residual current. The key technical indicators are shown in the Table 1.

Table 1. Key technical indicators of sensing terminal

Item	Residual current	Ambient temperature	Relative humidity
Measurement range	10mA~10000mA	-20°C ~ +85°C	0-100%
Measurement accuracy	0.5 level	±1°C	±1%

2.2. Management terminal

The management terminal can collect the residual current and other data from the sensing terminal, and to store and record steady-state recording and transient recording of fault alarm using wireless communication.

The fault wave-recording is for management terminal to record steady-recording of residual current and dynamic recording of the fault data in a specific time when the fault occurs. The function is used mainly to achieve data traceability and data backup, and to provide accurate and reliable data support for fault analysis. The leakage current, temperature and humidity. The leakage current, temperature and humidity are recorded for wave-recording.

The key technical indicators are as follows:

(1) Steady-state recording. The recording period is 1s, and the recording time is 6 months;

(2) Transient recording. The sampling frequency of recording is 4K Hz. It can record the transient waveform from 1s before leakage to 10s after leakage. A maximum of 500 transient recording waves can be recorded

2.3. *Monitoring terminal*

The monitoring terminal is the core of data collection, calculation, and management for residual current monitoring system. There are the functions including data collection and data analysis.

2.3.1. *Data collection*

Real-time data collection. The monitoring terminal can collect the leakage current, temperature and humidity and device status. The monitoring component can collect and display all the data and status information.

History data collection. It includes historical alarm data and wave recording data. The history data is stored in the form of files in SD cards through the file system. User can view a maximum of 1000 pieces of historical data through the device interaction and copy data through to USB storage.

2.3.2. *Data analysis*

Fault location. The monitoring terminal saves the sensor ID configured on the connected sensing terminal and identifies the specific branch ID. When the leakage current alarm exceeds the alarm threshold, the monitoring terminal determines the branch alarm to locate the specific branch.

Fault warning. The thresholds are set for the monitoring terminal according to the real-time and history monitoring data values. Based the early-warning algorithm of big data analysis, it can analyse the trend change of leakage current, temperature, and humidity. When fault risks are found, the early warning is sent for user.

3. Methods and data analysis

3.1. *Comprehensive analysis algorithm*

Early warning analysis of residual current, is a comprehensive consideration to the occurrence of residual current principle, reason, and environmental factors such as temperature, humidity condition. Based on the residual current, temperature, humidity data in a period, the algorithm can analyse the tendency of residual current. When lope-change trend or absolute value of residual current exceed the threshold, the algorithm will give a early warning.

The principle of the algorithm is shown as follow. When residual current reaches a certain value, the ambient temperature will accumulate and increase faster than residual current value. In addition, in some cases, due to the poor insulation of the cable, the leakage may be large or small, and the occurrence time is short. Therefore, the fault cannot easily be detected just using the temperature value. However, the accumulated temperature can be determined to a certain extent, and then the comprehensive analysis of environmental humidity value can be forecasted.

The main flow of the algorithm is shown in Fig. 2.

(1) Data collection and monitoring. Collect and monitor the ambient temperature, humidity and residual current value regularly.

(2) Calculation of difference.

When temperature value is collected, the difference between the current temperature value and the last temperature value is calculated and stored as a signed number.

$$\Delta T = T_n - T_p \tag{1}$$

Where, ΔT is the temperature difference, T_n is the current temperature value, T_p is the last temperature value.

(3) Difference analysis

When the temperature exceeds the threshold, the change of slope rate for the storage temperature differences is calculated and analysed. If the slope rises gradually, it indicates that the temperature has begun to rise and is close to the threshold value. When that happens, residual current values begin to monitored.

(4) Residual current monitoring

When the current residual current exceeds the threshold, start fast sampling. that is increasing the frequency of residual current sampling originally and starting humidity data analysis.

(5) Humidity data analysis

The purpose of humidity data analysis is to detect transient faults. The principle is that the increase of humidity will have a greater impact on leakage. If the cable insulation skin is broken, the increase of humidity will lead to an increase of residual current. That may lead to a large leakage current in an instant state. The slope algorithm can be used to analyze the trend of humidity change over a period.

If the humidity increases constantly and reaches the maximum value in a certain period, a preliminary warning can be given at this time to warn the environment of possible leakage caused by insulation damage, and double sampling can be carried out at the same time. If the residual current value exceeds the threshold, upgrade the alarm or direct alarm. If it is not, the conventional algorithm is used to analyze the trend of residual current and carry out early warning.

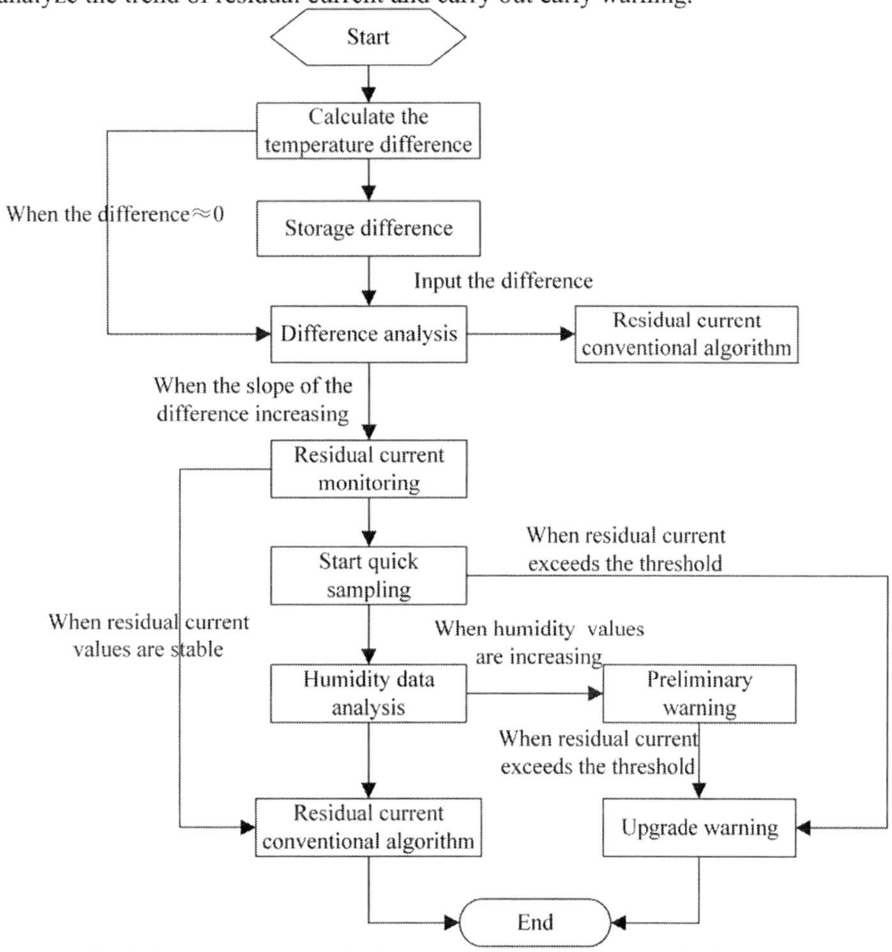

Fig. 2 Flow chart of residual current comprehensive analysis algorithm

3.2. Residual current conventional algorithm

Conventional early warning algorithm is based on the variation rule of residual current value of different fire accidents: when the residual current value has been abnormal, it will show a certain trend characteristic, so we can use the variation trend of residual current value to realize the monitoring and early warning of electrical fire accidents.

Viewing the change of accident fire, there are generally three categories: one is ongoing fault, one is gradual fault, the other is mutant fault. If the gradual fault is not handled in time, it will gradually develop into a mutant fault, and the change situation is irreversible. Therefore, conventional data analysis will focus on three scenarios:

(1) Ongoing type

If the residual current value suddenly reaches the threshold of warning, it indicates that the fault is occurring and an alarm will be generated immediately. The thresholds for warning can be set based on application scenarios.

(2) Gradual type

Warning conditions are that the slope of residual current value increases constantly to a certain a threshold or that the residual current values increase to the threshold of warning.

The sample dataset A of monitoring is shown in Figure 3. Within 24 hours, the residual current values continued to increase slowly. When the value exceeded to the threshold at point a which was set to 3.5A, the warning was given out.

(3) Mutant type

When the leakage fault happens, residual current value continues to increase. The residual current values not exceed the threshold; however, the slopes of residual current exceed the threshold. Then the warning will be given out.

The sample dataset B of monitoring is shown in Fig.3. The residual current values continued to increase, but no exceeded the threshold of warning that was set to 3.5A. However, the slope of residual current values also increased and exceeded the threshold at point b and c which set as 0.5. Then the warning was given out.

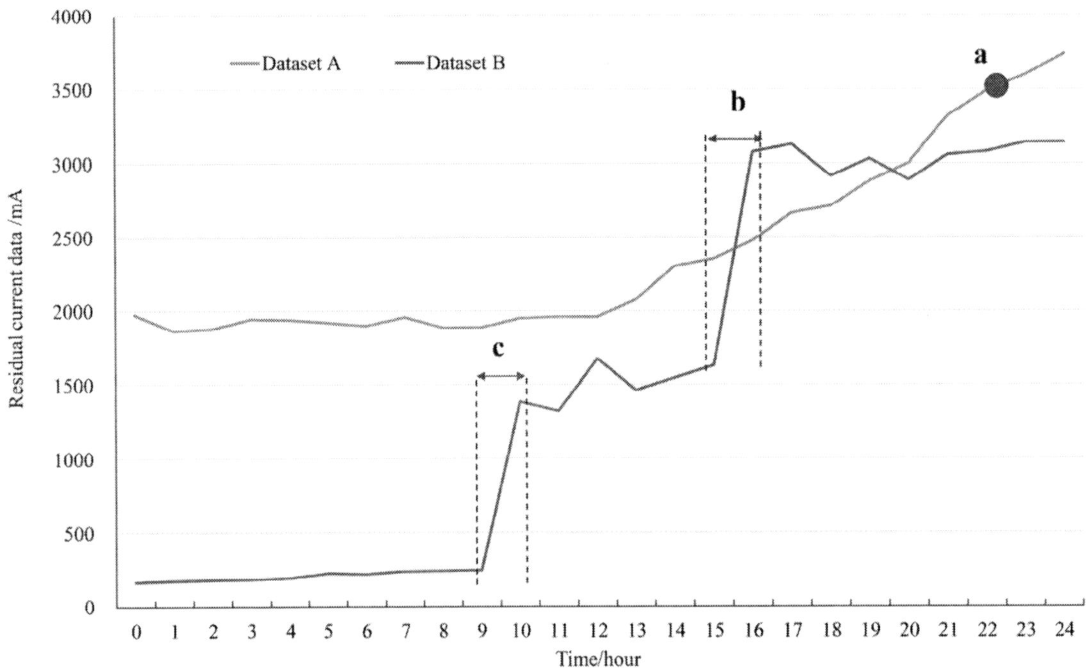

Fig. 3 The Sample data of residual current conventional analysis algorithm

4. Conclusion

The research work has successfully proposed the residual current monitoring and early warning scheme based on the big data analysis. The residual current, temperature and humidity were collected in real-time using newly designed and developed senor terminal and management terminal. The new algorithm based on the processing and analysis of big data could find out the abnormal change of residual current as early as possible and realize the fault monitoring and early warning. It is useful for important facilities such as substations and data centers to reduce or eliminate the fire hazard, and provide decision basis for the electrical fire warning. In the future, it is considered that neural network algorithm will be used for residual current monitoring and early warning to avoid threshold configuration.

Acknowledgments
This work is supported by State Grid Information & Telecommunication Group CO.,LTD (52680020002J).

References
[1] Zhao J. (2017) Statistical analysis of electrical fire. Fire Services, 12 :71-72.
[2] He S. L., Zhu J. (2020) Risk assessment of terminal substation fire accident: taking a substation in Ulanqab as an example. Industrial Safety and Environmental Protection, 46:24-27.
[3] Wu Z. Q., Wang D. K., Wu L., Zhao H. L. (2020) Analysis on the cause and Prevention technology of electrical fire. China Fire, 7:63-66.
[4] Gao K. (2013) Alarm and monitoring system for electric fire protection and electric fire prevention. Building Electricity, 32:54-56
[5] Liu J. (2011) Discussion on residual Action Current alarm System for fire prevention. China New Technologies and Products, 3:255-256.
[6] Li D. J. (2020) Electric fire monitoring system and electric fire prevention. Fire Protection Today, 5:52-53.
[7] Huang B., Han J., Zhou T. L. (2019) Research on residual current electric fire monitoring system. China Computer & Communication, 31:86-87.
[8] Xie Z. R., Shan L. H., Shi Q. B., Zhang H. (2019) Discussion on advance technology of early warning and monitoring for electrical fire. Building Electricity, 38:39-44.
[9] Li X. B., Zhong C., Zhang X. J. (2017) The analysis of residual current based on fast Fourier transforms. Fire Science and Technology, 36:359-362.
[10] Zhang C. P., Feng Y. (2016) Design of multi-sensor combined detector for electric fire. Fire Science and Technology, 35:1726-1728.
[11] Li H. W., Shen J. B., Tang Y., Li H., Yu X. L. (2011) Application status and development trend of residual current electric fire monitoring system. Fire Technique and Products Information, 5:23-25.

Research on the Operation Model and Strategy of Lithium Battery to Provide Frequency Regulation in the Power System

Yu Shi[1a*], Zhong Zhang[2b*], Jing Wang[1], Hao Li[1], Xiang Zhao[1], Xintong Yang[1]

[1] Power Economic Research Institute of Jilin Electric Power Co., Ltd., Changchun, 130000, China

[2] School of Electrical Engineering, Dalian University of Technology, Dalian 116024, China;

[a*]jyyshiyu@126.com, [b*]zzhang2017@hotmail.com

Abstract—With the development of smart grid and energy internet, energy storage will play an important role in maintaining the power balance and providing frequency regulation in future power system. Consequently, energy storage faces a trade-off problem between the energy market and frequency regulation market. In this work, an operation model and strategy of energy storage to provide frequency regulation is proposed. Taking lithium battery as an example of energy storage, its technical feasibility to provide frequency regulation is firstly discussed. Then, a day-ahead energy market model in the power system is constructed to calculate the revenue of lithium battery. Finally, the operation strategy of lithium battery to provide secondary frequency regulation (namely AGC) is proposed based on the model of opportunity cost. Case study is conducted on a modified IEEE 6-bus system. It shows that the reduced marginal revenue of lithium battery in the energy market increases with the growth of its declared capacity for frequency regulation. Consequently, the optimal declared frequency regulation capacity for lithium battery is the amount, at which the reduced marginal revenue in the energy market is equal to the compensation price of frequency regulation.

1. Introduction

Under the development of smart grid and energy internet, the penetration level of renewable energy generation (REG) will keep increasing [1]. Currently, the proportion of coal-fired power generation has decreased to 49.1% in China. The proportion of non-fossil energy generation has reached 44.8%. REG represented by wind power and PV has greater uncertainty and volatility, which calls for more and more ancillary services such as frequency regulation and reserve in the power system.

To reduce the adverse impact of REG on the operation of the power system and ensure its safe operation, ancillary services market will further be developed. It is in favor to exploit and stimulate more ancillary service sources. Energy storage (ES), as a new type of flexible resource, has great potential in providing ancillary services. However, energy storage takes the roles of both load and generation due to its charging and discharging behavior. Thus, the analysis of cost and benefit of ES in ancillary services market becomes an urgent issue. In this regard, this paper takes the AGC as an example to investigate the cost and benefit of ES in the ancillary services market.

ES brings multiple benefits to the power system. From the perspective of generation side, it can promote the consumption of REG and reduce the curtailment of wind and PV power. From the perspective of the grid, ES contributes to peak-shaving through charging and discharging process.

From the perspective of demand side, ES can smooth the power fluctuations and provide emergency reserve to improve the quality of power supply [2].

The ancillary services from the generation side can hardly satisfy the requirements of peak shaving and reserve in modern power system. Increasing attention has been paid to the ancillary service potential of ES and flexible resources in the demand-side. The national grid in UK has clearly defined frequency response products provided by ES and load response. Relevant literature has carried out preliminary research on the operation and benefit analysis of ES participating in the ancillary service market [3]-[6].

Lit. [7] proposes a transaction model for aggregators of distributed ES participating in the ancillary services market of peak shaving. It investigates the operation mode of Data System Automation Program (DSAP) to provide peak shaving. Lit. [8] addresses the day-ahead and real-time market clearing issues considering the participation of ES and demand-side response. Lit. [9] investigates the operation strategy of ES and electricity vehicle (EV) in the frequency regulation market under the market rules for California independent system operators (ISO). Lit. [10] proposes a realization scheme for battery storage used for frequency regulation based on a distributed autonomous control. The scheme optimizes the frequency control strategy for a combined ES and REG system. Lit. [11] proposed a cooperative control framework for a wind energy conversion system (WECS) and compressed air energy storage (CAES). CAES can provide frequency regulation to release the impact of wind power on the system. Lit. [12] established a dual-objective complementary control model for ES providing frequency regulation service. The objective function consists of minimizing power deviation and frequency regulation cost.

Lit. [13] constructs a frequency regulation market considering the participation of wind, PV, hydropower, thermal power and ES. The market is cleared based on sequential bidding mode and marginal price principle. It demonstrates that the frequency regulation performance of the system is improved greatly when considering multiple types of participants. Lit. [14] proposed a control method for ES to provide frequency regulation service. The goal is to maximize the benefits of frequency regulation considering the degradation cost of ES. The frequency regulation performance of ES is compared with thermal power generators in Lit. [15]. It is concluded that ES has the advantages of fast response speed and high stability in frequency regulation. With the deepening research on ES providing ancillary services, its advantages of flexibility and rapidity have gradually become prominent. Consequently, increasing attention has been paid to ES in providing ancillary services.

In summary, technical issues of ES providing ancillary services have been studied in relevant literature from perspectives of control methods and performance analysis. However, the operation strategy of ES to provide frequency regulation service need to be further studied. Lithium battery, as a typical ES, has become an important ancillary service resource in the power system. This study takes lithium battery as an example of ES to studies its operation strategy to provide AGC frequency regulation to the power system. The modelling method and optimal strategy is proposed for lithium battery to participate in the energy market and the AGC market.

The work of this paper is organized as follows. Section 2 analyses the frequency regulation performance of ES. Section 3 constructed the operation model of ES. The optimal operation strategy of lithium battery to provide AGC service is proposed in Section 4. Case studies are conducted in Section 5, followed by conclusions in Section 6.

2. Technical Feasibility of Lithium Battery to Provide Frequency Regulation to Power System

2.1. Frequency regulation principle of lithium battery

Lithium battery realizes frequency response control through the power control of its grid-connected converter. The frequency control is realized through two steps of control process [15].

2.1.1.DC/AC converter control
DC/AC converter control consists of two parts: inner loop control and outer loop control. The inner loop control realizes the decoupling control of inner loop currents. The outer loop control realizes integrated inertia control and droop control.

In the control methods for lithium battery when responding to system frequency deviation, active power-frequency droop control is a commonly used control method. The active power increment and frequency deviation satisfy the following relationship:

$$\Delta P_{s,t} = -\Delta f_t / \delta_s \tag{1}$$

where $\Delta P_{s,t}$ is the active power increment of lithium battery at time t; Δf_t is the system frequency deviation at time t; δ_s is frequency regulation coefficient of lithium battery.

The input or output active power of lithium battery is determined by system frequency deviation and the frequency regulation coefficient. When the frequency deviation of the system exceeds the set limit, the corresponding active power input or output of the lithium battery will change. In the operation the state of charge (SOC) of lithium battery is required to be restricted within a safety range, for example [10%, 90%].

2.1.2.Operation status control of lithium battery
For conventional generators, the power regulation range is within their minimum and maximum technical outputs. For lithium battery, the theoretical adjustment range in frequency regulation process is the full power range, namely $[-P_{\max}, P_{\max}]$, where $P_{\max}$ is the maximum charging and discharging power.

2.2. Frequency regulation performance of lithium battery
Lit [15] tested the frequency regulation response speed of a thermal power unit and an ES with the same installed capacity. The test found that under the same disturbance, the time required for the ES to respond to frequency deviation and reach a stable output is about 1 s. While the time required for the thermal power unit is about 6 s. It indicated that the frequency regulation response speed of ES is faster than that of the thermal power unit.

Due to the fast response capability, ES can provide a certain amount of active power compensation for the power system in the early stage of frequency fluctuations. Thus, the power system can respond quickly to frequency deviations and reduces the maximum system frequency deviation.

3. Operation Model of and Profit Analysis for Lithium battery in Day-ahead Energy Market

3.1. Operation model of lithium battery
The operation of lithium battery is restricted by its maximum charging/discharging power and the boundary of SOC. Since lithium batteries has a relatively fast adjustment rate, the ramping constraint is generally not considered. Power and SOC constraints of lithium battery are constructed as follows:

$$0 \le p_{ch,t}^{es} \le p_{\mathrm{ch,max}}^{es} \tag{2}$$

$$0 \le p_{dch,t}^{es} \le p_{\mathrm{dch,max}}^{es} \tag{3}$$

$$Soc_t^{es} = Soc_{t-1}^{es} + p_{ch,t}^{es}\eta_{ch}\mathrm{T} - p_{dch,t}^{es}/\eta_{dch}\mathrm{T} \tag{4}$$

$$Soc_{min}^{es} \le Soc_t^{es} \le Soc_{max}^{es} \tag{5}$$

where $p_{ch,t}^{es}$, $p_{dch,t}^{es}$ is the charging and discharging power of lithium battery, respectively; $p_{\mathrm{ch,max}}^{\mathrm{es}}$, $p_{\mathrm{dch,max}}^{\mathrm{es}}$ is the maximum charging and discharging power respectively; Soc_t^{es} represents the SOC of lithium battery, $Soc_{\mathrm{min}}^{\mathrm{es}}$, $Soc_{\mathrm{max}}^{\mathrm{es}}$ is the lower and upper limit of SOC; η_{ch}, η_{dch} is the charging and discharging efficiency, respectively; T if the length of a dispatch interval.

Since the battery cannot be charged and discharged simultaneously, two "0-1" variables and their mutual exclusion constraint are introduced. Eq. (2~3) are then replaced by Eq. (6~7).

$$0 \le p_{ch,t}^{es} \le b_{ch,t} p_{ch,max}^{es} \tag{6}$$

$$0 \le p_{dch,t}^{es} \le b_{dch,t} p_{dch,max}^{es} \tag{7}$$

$$b_{ch,t} + b_{dch,t} \le 1 \tag{8}$$

where $b_{\mathrm{ch},t}$, $b_{\mathrm{dch},t}$ represents the charging and discharging state of the battery.

3.2. Profit modelling of lithium battery in the energy market

Lithium battery has charging and discharging states in the energy market. In order to maximize its profits, it will charge when energy price is low and discharge when energy price is high. The net income of lithium battery is the discharge income minus charge cost and its operation costs.

3.2.1Operation cost of lithium battery

The usage life of a lithium battery depends on the depth of discharge in each cycle. Lit. [17] proposes an operation degradation cost model for lithium battery as shown in equation (9)

$$C_{\mathrm{D},t} = \frac{C_I}{2N_{life}Q_0}(p_{\mathrm{ch},t}^{\mathrm{es}} + p_{\mathrm{dch},t}^{\mathrm{es}})\mathrm{T} \tag{9}$$

where $C_{\mathrm{D},t}$ is the degradation cost of lithium battery; C_I is the initial investment cost of lithium battery system; N_{life} is the average cycle life of lithium battery, Q_0 is the rated storage capacity of lithium battery.

The operation and maintenance of lithium battery system also incurs costs. The operation and maintenance cost per unit time is related to the charging and discharging power:

$$C_{\mathrm{M},t} = K_{\mathrm{M},t}(p_{\mathrm{ch},t}^{\mathrm{es}} + p_{\mathrm{dch},t}^{\mathrm{es}})\mathrm{T} \tag{10}$$

where $C_{\mathrm{M},t}$ is the operation and maintenance cost of lithium battery; $K_{\mathrm{M},t}$ is the operation and maintenance cost coefficient.

3.2.2.Benefit of lithium battery

The net profit of lithium battery is the discharge income minus the charge cost and its operation costs. Taking one day as a dispatch unit with N periods, the profit in a day is expressed as:

$$f^{\mathrm{es}} = \sum_{t=1}^{\mathrm{N}}(p_{\mathrm{dch},t}^{\mathrm{es}}\lambda_t - p_{\mathrm{ch},t}^{\mathrm{es}}\lambda_t - C_{\mathrm{D},t} - C_{\mathrm{M},t})\mathrm{T} \tag{11}$$

where λ_t is the electricity price at time t.

4. Operation Strategy of Lithium Battery to Provide AGC Frequency Regulation

4.1. Energy market modelling considering the participation of lithium battery

The opportunity cost of lithium battery to provide ancillary services comes from the loss of revenue in the energy market. In this reason, a market clearing model of energy market is established considering market participants including thermal power units, wind power units, lithium battery and loads.

The objective function of the energy market is to minimize the total operation cost from thermal power units and lithium battery:

$$(\mathrm{P}) \min f = \sum_{t=1}^{\mathrm{N}}\sum_{i=1}^{n_g} Cost_{i,t}^{\mathrm{g}} + \sum_{t=1}^{\mathrm{N}}(C_{\mathrm{D},t} - C_{\mathrm{M},t}) \tag{12}$$

$$Cost_{i,t}^{\mathrm{g}} = a_i^{\mathrm{g}} + b_i^{\mathrm{g}} p_{i,t}^{\mathrm{g}} + c_i^{\mathrm{g}} {p_{i,t}^{\mathrm{g}}}^2 + u_i^{\mathrm{g}} x_{i,t}^{\mathrm{g}} \tag{13}$$

where, the first term in Eq. (12) represents operation cost of conventional generators, which consist of generation cost and start-up cost as expressed in Eq. (13); $p_{i,t}^{\mathrm{g}}$ is the power of thermal unit i; u_i^{g} is the start-up cost of thermal unit i; $x_{i,t}^{\mathrm{g}}$ represents the state of start-up; a_i^{g}, b_i^{g}, c_i^{g} is the coefficient of the quadratic cost function of a thermal unit; $C_{\mathrm{D},t}$ and $C_{\mathrm{M},t}$ is the degradation cost and maintenance cost of lithium battery, which is given by Eq. (9) and (10), respectively.

The objective function of the market clearing problem is subjective to the following constraints.
Power balance constraint:

$$\sum_{i=1}^{n_g} p_{i,t}^{\mathrm{g}} + p_t^{\mathrm{re}} + p_{\mathrm{dch},t}^{\mathrm{es}} - p_{\mathrm{ch},t}^{\mathrm{es}} = L_t \tag{14}$$

where L_t represents system load; p_t^{re} is the power of renewable energy generation.

Power boundary constraint of thermal units:

$$b_{i,t}^{\mathrm{g}} P_{i,\mathrm{min}}^{\mathrm{g}} \leq p_{i,t}^{\mathrm{g}} \leq b_{i,t}^{\mathrm{g}} P_{i,\mathrm{max}}^{\mathrm{g}} \tag{15}$$

Ramping constraint of thermal units:

$$p_{i,t}^{\mathrm{g}} - p_{i,t-1}^{\mathrm{g}} \leq \Delta P_{i,\mathrm{max}}^{\mathrm{g,u}}, \forall i, t \tag{16}$$

$$p_{i,t-1}^{\mathrm{g}} - p_{i,t}^{\mathrm{g}} \leq \Delta P_{i,\mathrm{max}}^{\mathrm{g,d}}, \forall i, t \tag{17}$$

Start-up and shut-down constraint of thermal units:

$$b_{i,t}^{\mathrm{g}} - b_{i,t-1}^{\mathrm{g}} = x_{i,t}^{\mathrm{g}} - y_{i,t}^{\mathrm{g}}, \forall t \tag{18}$$

$$\sum_{\tau=\max\{1,t-\mathrm{U}_i+1\}}^{t} x_{i,\tau}^{\mathrm{g}} \leq b_{i,t}^{\mathrm{g}}, \forall t \tag{19}$$

$$\sum_{\tau=\max\{1,t-\mathrm{D}_i+1\}}^{t} y_{i,\tau}^{\mathrm{g}} \leq 1 - b_{i,t}^{\mathrm{g}}, \forall t \tag{20}$$

where $b_{i,t}^{\mathrm{g}}$ represents the operation state of thermal unit i; $P_{i,\mathrm{min}}^{\mathrm{g}}$, $P_{i,\mathrm{max}}^{\mathrm{g}}$ the minimum and maximum power output of a thermal unit, respectively; $\Delta P_{i,\mathrm{max}}^{\mathrm{g,u}}$, $\Delta P_{i,\mathrm{max}}^{\mathrm{g,d}}$ is the maximum upward and downward ramping rate, respectively; $y_{i,t}^{\mathrm{g}}$ represents the state of shut-down of a thermal unit; U_i, D_i is the minimum start-up and shut-down time, respectively.

Power constraints on the renewable energy generation:

$$0 \leq p_t^{\mathrm{re}} \leq P_t^{\mathrm{re}} \tag{21}$$

where P_t^{re} is the total forecast power of renewable energy generation.

4.2 AGC opportunity cost of lithium battery to provide AGC

Opportunity cost refers to the maximum benefit that an investment or a behaviour can obtain from alternative schemes.

The opportunity cost of ES to provide AGC frequency regulation refers to the reduced revenue in the energy market. As shown in Eq. (22), the opportunity cost is equal to the difference between the revenues in the energy market before and after providing AGC frequency regulation capacity.

$$C_{\mathrm{OP}} = Pr_{\mathrm{e}} - Pr_{\mathrm{e}}' \tag{22}$$

where C_{OP} is AGC opportunity cost of the lithium battery; Pr_{e} and Pr_{e}' is the revenue of lithium battery in the energy market before and after providing AGC frequency regulation capacity.

In this work, it is assumed that the bidding strategy of lithium battery in the AGC market is symmetrical, namely provide the same upward and downward AGC frequency regulation capacity.

Under the energy market environment presented in Section 4.1, AGC opportunity cost based on simulation calculation can be obtained through the following steps:

4.2.1.Read in the parameters of generators, lithium battery, and load.

4.2.2. Construct the energy market model considering of the participation of lithium battery, namely Eq. (4~10, 12~21).

4.2.3. Solve the energy market clearing problem and calculated the revenue of lithium battery in the energy market, denoted by Pr_e.

4.2.4.Assume lithium battery provides certain amount of AGC capacity to the system. Then, the operation boundary of lithium battery in the energy market are updated as shown by Eq. (23~26).

4.2.5.Resolve the market clearing problem after updating the operation boundary of lithium battery and obtain the revenue in the energy market, namely Pr_e'.

4.2.6. Calculate AGC opportunity cost, namely $Pr_e - Pr_e'$.

$$p_{\mathrm{ch,max}}^{es\prime} = p_{\mathrm{ch,max}}^{es} - R_{\mathrm{D}} \tag{23}$$

$$p_{\mathrm{dch,max}}^{es\prime} = p_{\mathrm{dch,max}}^{es} - R_{\mathrm{U}} \tag{24}$$

$$Soc_{\mathrm{min}}^{\mathrm{es}\prime} = Soc_{\mathrm{min}}^{\mathrm{es}} + R_{\mathrm{U}} t_{\mathrm{AGC}} \tag{25}$$

$$Soc_{\mathrm{max}}^{\mathrm{es}\prime} = Soc_{\mathrm{max}}^{\mathrm{es}} - R_{\mathrm{D}} t_{\mathrm{AGC}} \tag{26}$$

where R_{U} and R_{D} is the upward and downward AGC biding capacity of lithium battery.

4.3 Operation strategy of lithium battery to provide AGC to the power system

Lithium battery can participate in the energy and AGC market simultaneously. For each unit of capacity, it has two options, namely i) participating in the energy market to make profits by charging and discharging at different periods, ii) participating in the AGC market to obtain compensation benefits. To maximize the profits, the decision can be made by comparing the opportunity cost of unit battery capacity and the AGC compensation price. For example, when AGC compensation price is 0.16 yuan/kW in the power system, and the calculated opportunity cost of per unit battery capacity is 0.2 yuan/kW, it indicates that the profit of lithium battery through providing AGC service is lower than the reduced cost in the energy market. Thus, it prefers to participate in the energy market.

In this study, AGC opportunity cost of lithium battery is calculated successively. Then, decisions are made according to the opportunity cost and the AGC compensation price. As long as the opportunity cost of per unit AGC capacity is less than AGC compensation price, lithium battery can make more profits by providing AGC service. Therefore, the optimal AGC bidding capacity can be obtained through the simulation process shown in Fig. 1.

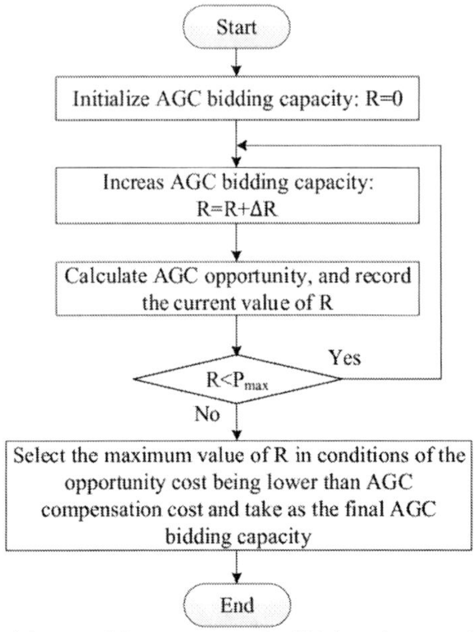

Fig. 1. Decision-making process for lithium battery in the market

5. Case Studies

5.1 Introduction to the test case

Simulation is conducted on a modified IEEE 6-bus system, as shown in Fig. 2. The system contains three conventional units: G1, G2 and G3, connected to bus 1, 2 and 6, respectively; one wind power connected to bus 4; and one lithium battery with a capacity of 20MW/100MWh connected to bus 5. The parameters of the three conventional units are shown in Table 1, including the maximum and minimum technical output, the quadratic function coefficient of the generation cost, the fuel price.

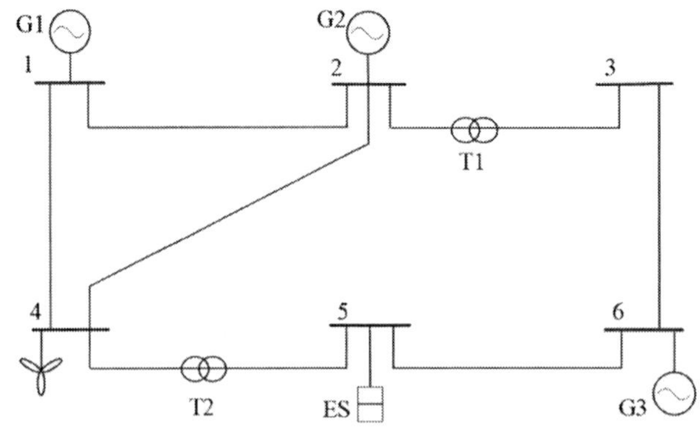

Fig. 2. Single-line diagram of IEEE 6-bus system

TABLE I. GENERATOR DATA

UNIT	Min output MW	Max output MW	Quadratic cost function coefficient			Fuel price $/MBtu
			a Mbtu	*b* MBtu/MWh	*c* MBtu/MWh^2	
G1	90	220	176.9	13.5	0.0004	5
G2	10	100	129.9	32.6	0.001	5.5
G3	10	20	137.4	17.6	0.005	5.2

5.2 Energy market clearing results

The energy market results and clearing price is shown in Fig. 3 and Fig. 4. The clearing prices in the energy market are determined by the marginal generation cost of the conventional units. The higher is the output of the conventional units, the higher is the market clearing price. The lithium battery is charged during the low price hours and discharged during the peak price hours. It arbitrages the price spread between the charging and discharging period in the energy market to make profits.

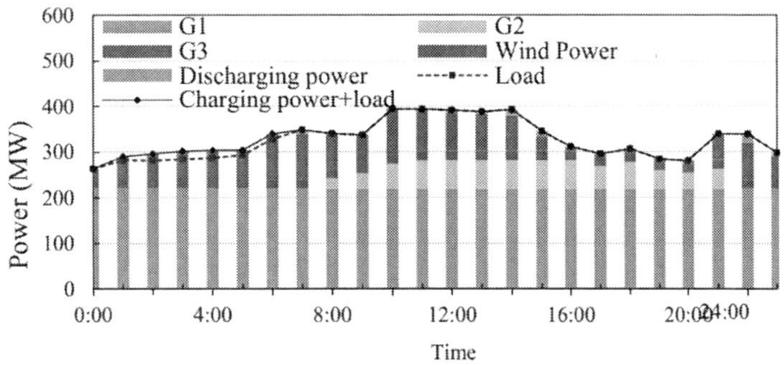

Fig. 3. Energy market results

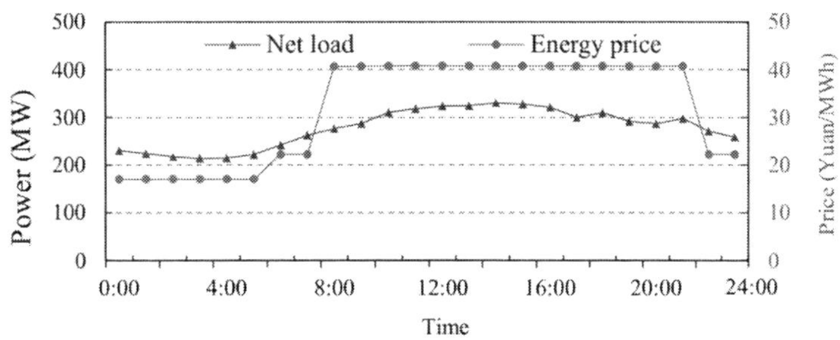

Fig. 4. Load curve and clearing price in the energy market

5.3 Operation results and revenue of lithium battery after providing AGC service

The opportunity cost of providing AGC for Lithium battery is obtained through a simulation method. The trend of the SOC of lithium battery when providing different AGC capacities is shown in Fig. 5. The trend of SOC can also reflect its revenue in the energy market to certain extend. When the AGC capacity is 0 MW, i.e. when lithium battery only participates in the energy market, it has the highest revenue in the energy market. The utilisation rate of its capacity reaches maximum, with the maximum SOC close to 100 MWh and the minimum SOC close to 22 MWh. When the AGC capacity increases to 10 MW, the maximum amount of electricity stored in lithium battery is less than 80 MWh. When AGC capacity increases to 12 MW, the highest SOC of lithium battery is less than 70 MWh, indicating that the utilization rate of lithium battery in the energy market is further reduced. When the AGC capacity increases to 18 MW, the SOC of lithium battery changes between 20~40 MW. The corresponding profits in the energy market become even low.

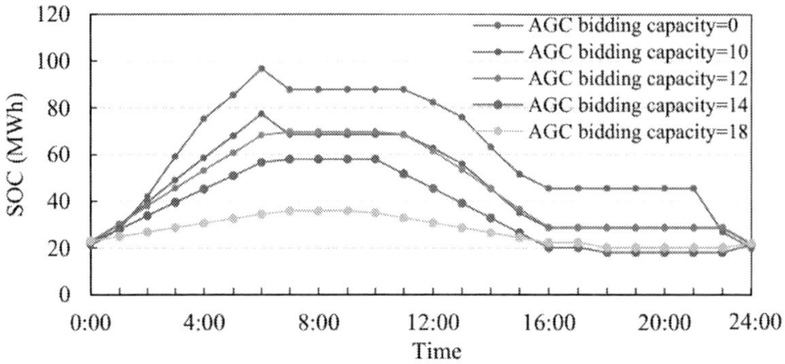

Fig. 5. SOC of energy storage under different AGC bidding price

Fig. 6 illustrates the profit changes of lithium battery in the energy market against different AGC capacities. The reduced revenue is the AGC opportunity cost of lithium battery. As the AGC bidding capacity increases, the revenue of lithium battery in the energy market decreases. Its opportunity cost increases faster and faster. The slope of the opportunity cost curve can be regarded as the marginal opportunity cost, which also increases as the bidding capacity getting larger, indicating that the AGC resources become more and more scarce.

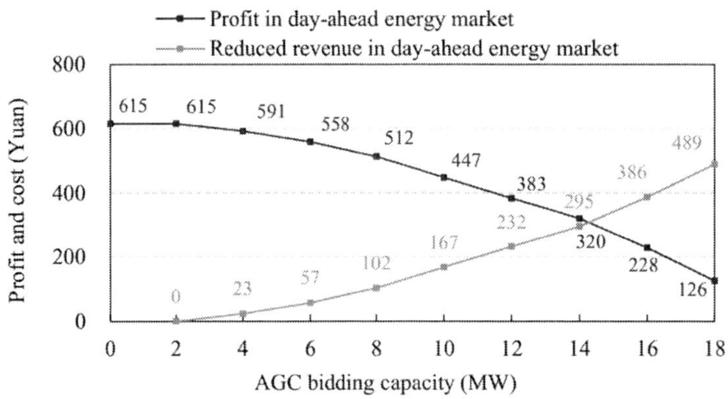

Fig. 6 Profit and reduced revenue of lithim battery in the energy market after providing AGC frequency regulation capacity

The market strategy for lithium battery is to reasonably allocate the its capacity in the energy market and AGC market to maximize the total profits. The opportunity cost per unit of AGC capacity is calculated and shown in Fig. 7. It shows that the larger AGC bidding capacity is, the higher is the opportunity cost per unit of AGC capacity. When the value is higher than the revenue per unit of capacity in the AGC market, the total revenue obtained by lithium battery begins to decrease.

As shown in Fig.7, if the AGC compensation price is at 0.8 yuan/MW and AGC bidding capacity reaches 12 MW, the opportunity cost per unit of AGC capacity is equal to the AGC compensation price. In this condition, the total revenue from lithium battery in the energy market and AGC market reaches the maximum. If the AGC bidding capacity is higher than 12WM, the total revenue of the lithium battery will decrease.

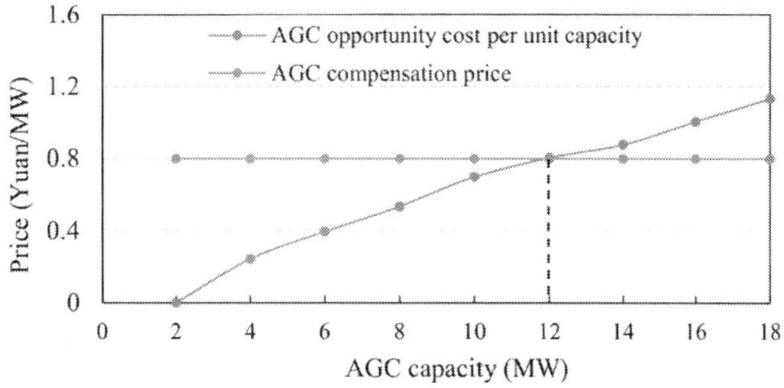

Fig. 7. Marginal opportunity cost per unit AGC capacity

6. Conclusions

In this paper, an operation model and strategy of lithium battery to provide frequency regulation is proposed. The technical feasibility lithium battery to provide frequency regulation is analysed. An day-ahead energy market clearing model with the participation of lithium battery storage is established. The reduced revenue in the energy market, namely the opportunity cost of lithium battery to provide frequency regulation is obtained through successive simulations. Through comparing the opportunity cost with the AGC compensation price, an operation strategy is proposed for lithium battery to allocate its charging and discharging capacity in day-ahead energy market and the AGC frequency regulation market.

After lithium battery provides AGC frequency regulation capacity, its upper power limit in day-ahead energy market will decrease. While the available range of its SOC in the energy market also becomes smaller. As a result, its revenue in the energy market decreases.

The marginal reduced revenue, namely the opportunity cost of lithium battery in day-ahead energy market increases as it provides more and more AGC frequency regulation capacity. Consequently, optimal frequency regulation capacity for lithium battery is the amount, at which the marginal reduced marginal revenue in the energy market is equal to the compensation price of frequency regulation.

Acknowledgments

This work was financially supported by the Top Ten Strategic Projects of Jilin Electric Power Co., Ltd of State Grid in 2020.

References

[1] China Electric Power Statistics and Data Center. Analysis and Forecast of National Power Supply and Demand Situation from 2020 to 2021[J]. China Power Enterprise Management, 2021(04): 91-94.

[2] Xie M, Wei W, Li J, et.al. Research on Combined Dispatching Mode of Energy Storage in Energy Market and Regulation Market[J]. Electric Power Construction, 2019, 40(02): 20-28.

[3] Xue Y.S, Luo Y.H., Li B.J., et.al. A Review of Interruptible Load Participating in System Reserve[J]. Automation of Electric Power, 2007, 31(010): 1-6.

[4] Han B, Lu S, Xue F, et al. Day-Ahead Electric Vehicle Aggregator Bidding Strategy using Stochastic Programming in an Uncertain Reserve Market[J]. IET Generation, Transmission & Distribution, 2019, 13(12): 2517-2525.

[5] Konda S R, Panwar L K, Panigrahi B K, et al. Optimal Offering of Demand Response Aggregation Company in Price-Based Energy and Reserve Market Participation[J]. IEEE Transactions on Industrial Informatics, 2018, 14(2): 578-587.

[6] Tang Z, Liu Y, Wu L, et al. Reserve Model of Energy Storage in Day-ahead Joint Energy and Reserve Markets: A Stochastic UC Solution[J]. IEEE Transactions on Smart Grid, 2020. (Early Access)

[7] Liu D N, Gao Y, Zhang T T, et al. Research on strategy of distributed energy storage aggregators participating in peak load regulation auxiliary service[J]. IOP Conference Series: Earth and Environmental Science, 2021, 687(1): 012140.

[8] Khazali A, Kalantar M. Optimal dispatch of responsive loads and generic energy storage units in a co-optimized energy and reserve market for smart power systems[J]. Taylor & Francis, 2016, 44(20):

2285-2297.
[9] Vatandoust B, Ahmadian A, Golka M, et al. Risk-averse optimal bidding of electric vehicles and energy storage aggregator in day-ahead frequency regulation market[J]. IEEE Transactions on Power Systems, 2019, 34(3): 2036-2046.
[10] HyungSeung K, Hong J H, Choi I S. Implementation of Distributed Autonomous Control Based Battery Energy Storage System for Frequency Regulation[J]. Energies, 2021, 14(9): 2672-2672.
[11] Abouzeid S I, Guo Y F, Zhang H C. Cooperative control framework of the wind turbine generators and the compressed air energy storage system for efficient frequency regulation support[J]. International Journal of Electrical Power and Energy Systems, 2021, 130. (Early Access)
[12] He T Y, Li S N, Wu S J, et al. Biobjective Optimization-Based Frequency Regulation of Power Grids with High-Participated Renewable Energy and Energy Storage Systems[J]. Mathematical Problems in Engineering, 2021. (Early Access)
[13] Sun Z.H., Zhao J.Q., Yang Y.H., et.al. Clearing Model of Regulation Market with Participation of Multi-Type Source of Wind-PV-Hydro-Thermal-Storage[C]. Proceedings of the 2019 Annual Conference of the Electric Power Market Committee of the Chinese Society for Electrical Engineering.
[14] Liu Q K, Liu M b, Lu W T. Control Method for Battery Energy Storage Participating in Frequency Regulation Market Considering Degradation Cost [J]. Power System Technology, 2021: 1-9. (Early Access)
[15] Wang F K, Xie L R, Qiao Y. Analysis on Frequency Modulation Performance of Power System Improved by Battery Energy Storage[J]. Automation of Electric Power Systems, 2021， 1-13.
[16] Gonzalez-Garrido A, Saez-De-Ibarra A, Gaztanaga H, et al. Annual optimized bidding and operation strategy in energy and secondary reserve markets for solar plants with storage systems[J]. IEEE Transactions on Power Systems, 2019, 34(6): 5115-5124.
[17] Zhang H T, Qin W P, Han X Q, et. Multi-time Scale Optimization Scheduling Scheme of Microgrid Energy Management. Power system technology, 2017,41(05):1533-1542.

Journal of Physics: Conference Series **2108** (2021) 012014 doi:10.1088/1742-6596/2108/1/012014

Research on Security Access of Power Internet of Things Terminals Based on Edge Computing Technology

Yong Tang[1,a], Linghao Zhang[1,b], Juling Zhang[1,c], Siyu Xiang[1,d], He Cai[2,e*]

[1]State Grid Sichuan Electric Power Research Institute, Chengdu High-tech Zone, Sichuan Province, China

[2]School of Electrical and Electronic Engineering, North China Electric Power University, Changping District, Beijing, China

[a]email: 18415820@qq.com, [b]email: 16100178@qq.com, [c]email: zjl725@qq.com, [d]email: 280552598@qq.com, [e*]email: 120212201260@ncepu.edu.cn

Abstract—In view of the current lack of unified security authentication and control for the power Internet of Things terminal equipment, at the perception level of the power Internet of Things, the perception layer terminal access control, front-end authentication technology realization and terminal equipment abnormal behavior detection methods are proposed. This method enhances the communication security between power equipment and edge nodes, and ensures the safe and stable operation of the power Internet of Things.

1. Introduction

In recent years, driven by the rapid growth in global energy and power demand[1], the power grid has gradually evolved from a traditional power grid to a smart grid, from a single regional power grid to a large-scale power grid across provinces and regions. At the same time, with the rapid development of the Internet, the power grid has been transformed into an energy Internet in which energy and information are deeply coupled. In terms of equipment status monitoring, it realizes real-time monitoring of equipment operating status and environmental status on the power generation side, grid side, and customer side, and intelligent analysis is performed[2]. Power generation forecasting, grid dispatching, equipment operation and maintenance, and grid smart operation are all inseparable from the power Internet of Things built by various power smart terminal equipment and transmission networks. The power Internet of Things plays an important role in promoting the realization of all-round real-time perception, fault recovery, and security control at the energy internet level[3]. The traditional power Internet of Things has been developed for many years, but the power Internet of Things, especially the security research at the terminal level, is still in its infancy. Due to the lack of unified security certification and control for terminal equipment at the perception layer, the situation of network security at home and abroad has become increasingly severe in recent years, which has caused many criminals to take advantage of the virtual intrusion and use illegal equipment and viruses to invade the power network, resulting in a blackout in Ukraine, 2019 Latin American countries blackouts, etc. Therefore, how to promote the evolution of the traditional power Internet of Things to a new form of full-service perception and ubiquitous power Internet of Things, and how to fully cope with the new international situation, new risks and new challenges of the new form of the Internet of Things are the questions that the State Grid Corporation of China needs to answer[4].

2. Basic introduction to edge computing

Edge computing is a new type of computing paradigm that performs computing tasks at the edge of the network. Data computing tasks can be divided into two parts: the uplink Internet of Everything service and the downlink cloud service. The edge in edge computing refers to any computing, storage, and network resources between the path from the source of data to the cloud computing center. According to the specific requirements of different application services and actual scenarios, the edge can be one or more resource nodes on this path. The edge concept of edge computing is relative to the cloud computing center. Edge resources include a variety of devices. In the power Internet of Things, such as smart meters, non-intrusive user power load collection terminals, and line-side data collection terminals Etc.; Edge nodes are scattered near users, line protection and measurement equipment, have many different types of resources, and are independent of each other[5]. The purpose of edge computing is to integrate these independently distributed resources. These resources are separated in the spatial dimension, but are close to the power Internet of Things terminal in the network dimension, and can be executed at the appropriate edge nodes.

2.1 Features of edge computing

(1) Low latency

Low latency is a recognized feature in edge computing academia[6]. Because the deployment of edge computing is close to the terminal equipment of the power Internet of Things, massive data is directly processed locally, avoiding the delay and congestion caused by local uploading to the cloud, which is of great significance for businesses that require timely feedback. For example, for the relay protection equipment on the line side, when the sensor detects abnormal data, after receiving the terminal data and judging the fault, the edge node located nearby can issue the protection instruction in time, switch the line quickly, and avoid the power grid cascading failure[7].

(2) Distributed deployment

Compared with traditional centralized servers, decentralized computing nodes are distributed near terminal devices, which can process data more safely, and avoid terminal node failures caused by large-scale data tampering and attacks on cloud servers[8]. In the power Internet of Things, some terminal devices continuously upload monitoring information to the cloud center. Massive terminal devices cause a large amount of data congestion, increase the burden on the network, and affect the upload of other important services of the power grid[9]. Therefore, by deploying edge computing nodes, relying on its distributed technical advantages, the business data can be temporarily stored to the edge nodes, and when the cloud needs to detect data, the required data can be sent to the cloud.

(3) High bandwidth

Due to the distributed deployment characteristics of edge computing, simple and real-time data processing and analysis can be performed locally without uploading all data or information to the cloud, reducing network transmission pressure and avoiding network congestion caused by business uploads. Therefore, the network transmission rate can be greatly increased.

(4) High security

Since the computing node is at the edge of the network, the edge node encrypts the data after receiving the data, and then uploads it to the central node through the transmission network. Compared with traditional terminal equipment, encryption is performed close to the equipment to ensure data security from the source of the data as much as possible[10]. In the power Internet of Things, user load data and power protection service data can be encrypted at edge nodes, which is of great significance for some power services with high real-time requirements.

3. Ubiquitous Power Internet of Things Model

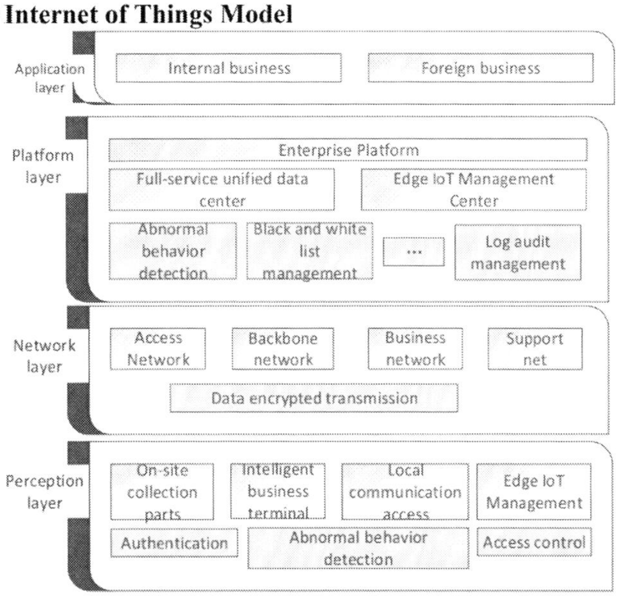

Fig.1 Structure of Ubiquitous Power Internet of Things

The ubiquitous power Internet of Things system structure is shown in Figure 1, including the perception layer, network layer, platform layer and application layer. The application layer includes the internal business and external business of the grid company, the internal business includes production operation and operation management, and the external business mainly includes integrated energy services. The platform layer mainly includes a unified data center for the entire business of the power grid and an edge management center of the power Internet of Things. The main function of the platform layer is to perform abnormal behavior detection, black and white list management, and log management for power equipment. The network layer mainly includes the access network, the backbone network, the business network and the support network. The network layer guarantees the normal data transmission and data security of each device node to the cloud center, edge node to cloud node. The perception layer includes on-site data collection components, intelligent business terminals, local communication access modules, and edge computing IoT management modules. The main functions of the perception layer include device identity authentication, abnormal behavior detection, access control, access audit, and log playback.

4. Terminal security mechanism for power Internet of Things

4.1 Power terminal equipment identity authentication

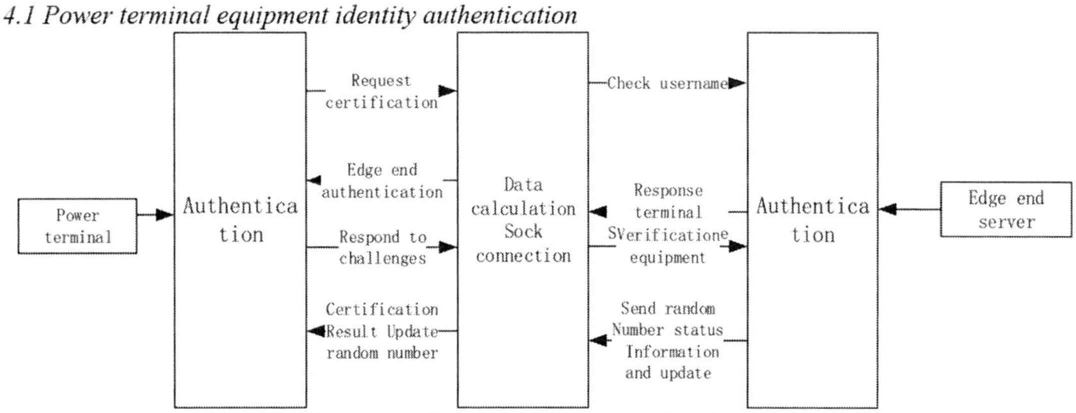

Fig.2 Schematic diagram of identity authentication process

When power terminal equipment is connected to the power Internet of Things, nodes will periodically send identity verification requests. Because the edge computing nodes in the power Internet of Things generally have small data storage space, their computing capabilities can only meet small-scale task processing. Therefore, the authentication method of challenge and response is adopted, and the joint authentication method of HASH encryption and pseudo-random number is used. The schematic diagram of the identity authentication process of the power edge equipment is shown in Figure 2.

The power terminal requests authentication, connects to the edge node by sock connection, the edge terminal checks its user name, the edge terminal responds to the terminal and sends a challenge, the power terminal responds to the challenge sent, and sends the response back to the edge server, and the edge server verifies that the device is successful. The edge terminal sends a random number to update the status information of the device, and the power terminal receives the authentication result, updates the random number, and completes identity authentication.

4.2 Access audit and log playback

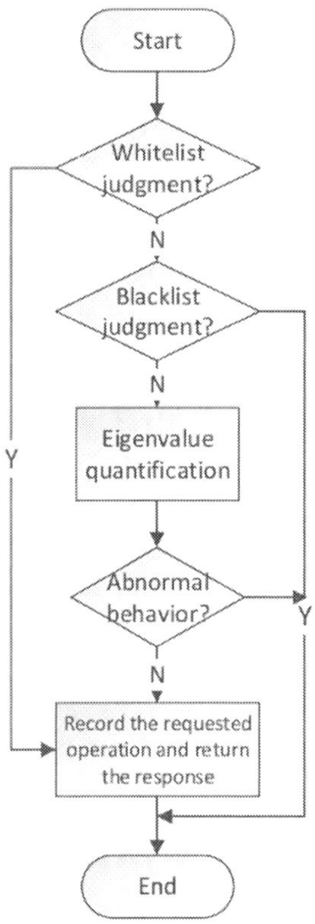

Fig.3 Access audit flow chart

Access review is mainly to review the hardware physical address of the terminal equipment connected to the power Internet of Things, the data content request sent, and whether it contains illegal characters and other content, and save important logs uploaded by the terminal. First, determine whether the access device belongs to the white list. If the device belongs to the white list, record the requested operation and return a response. If the device is not in the whitelist, it is judged whether it is in the blacklist. If it is not in the blacklist, its characteristic value is quantified and its behavior is compared with the library. If it is a normal behavior, its request operation is recorded and the response is returned.

4.3 Abnormal behavior detection

By extracting the characteristics of the attack behavior of the power Internet of Things, the specific threshold is obtained, and a detection result is judged by comparing with the empirical threshold. The rule base matching method is adopted for detection, that is, the rule base inspection is performed within an interval of the threshold. The flow chart of abnormal behavior detection is shown in Figure 4.

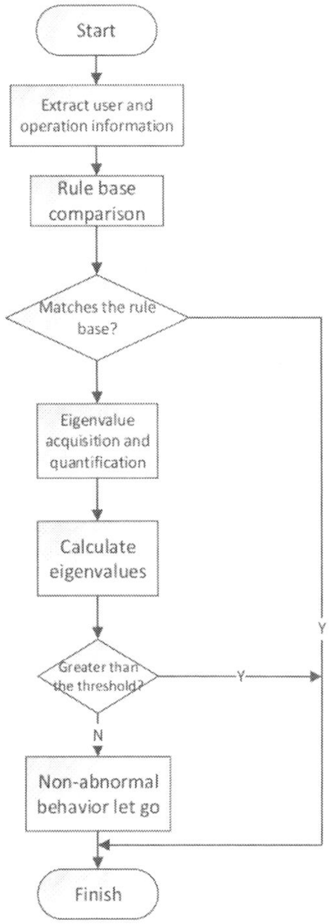

Fig.4 Flow chart of abnormal behavior detection

The specific process of abnormal behavior detection: First, the edge node extracts device information and operation information, and compares it with the rule base in the node. If the behavior belongs to the rule base, the process ends; otherwise, the feature value is extracted and quantified, and calculate its characteristic value; if the characteristic value is less than the threshold, it is judged to be a non-abnormal behavior, and if it is greater than the threshold, the process ends.

5. Conclusion

Based on edge computing, this paper adopts the edge security access process of the power Internet of Things that combines terminal device security authentication, access audit, log playback, and abnormal behavior detection. Compared with the original traditional power Internet of Things device access process, at the edge node, relying on the edge node for device identity authentication and abnormal behavior detection, the method proposed in this paper is suitable for implementation in resource-constrained terminal devices, and can improve the equipment Certification efficiency.

Acknowledgments

We would like to express our heartfelt thanks to all those who have helped this research work. This project is supported by the company's project "Ubiquitous Power Internet of Things User Behavior Intelligent Perception and Prediction Technology Research Based on Big Data and AI Technology"(SGSCDK00XXJS2100178).

References

[1] Lu S.L., Ding J, Zhang H.B., (2021) Research and Thinking on the Standard System of Intelligent IOT Sensing Technology for New Power System.J . Electric Power Information and Communication Technology,19(08):39-46.

[2] Liu W.Q., Tan Y.H., Deng X., (2021) End-side cloud and fog collaboration model for power cyber-physical system.J . Journal of Electric Power System and Automation,1-7.

[3] Gong G.J., Luo A.Q., Chen Z.M., (2018) Cyber-physical system of active distribution network based on edge computing.J . Power grid technology,42(10):3128-3135.

[4] Cai M.Y., Fen S.Y., Du H.W., (2019) Edge node perception adaptive data processing method for ubiquitous power Internet of Things.J . High voltage technology,45(06):1715-1722.

[5] Chen L., Chen H.Z., Deng S., (2014) Research on Security Access Control Method of Electric Power Intranet Terminal.J. Electric Power Information and Communication Technology,12(06):1-5.

[6] Hou B., (2021) Research on Surveillance Video Acquisition and Adaptive Playback System Based on Edge Computing .D . Inner Mongolia University.

[7] Lu S.L., Ding J., Zhang H.B. (2021) Research and Thinking on the Standard System of Intelligent IOT Sensing Technology for New Power System.J . Electric Power Information and Communication Technology,19(08):39-46.

[8] Wu Y.C., Li M.L., Peng W., (2020) Security Protection Design for Edge Computing of Power Internet of Things.J . China Science and Technology Information,Z1:84-85.

[9] Zhao Z.X., Li H.F., Hou W., (2021) Research and Practice of Power Internet of Things in State Grid Jilin Electric Power Infrastructure.J . Journal of Electrical Engineering,16(02):166-173.

[10] Li F.R., Yi Y.P., Shi W., (2021) Distribution Big Data Technology of Active Distribution Network Based on Edge Computing.J. Software Guide,20(08):160-165.

Simulation Analysis of Current in Metal Sheath of Power Cable Considering Temperature Characteristics

Jian Fang[1], Min Zhang[1], Wenxiong Mo[1], Bo Huang[1], Jiaxing He[1] and Xiang Lin[1*]

[1]Electric Power Test & Research Institute of Guangzhou Power Supply Bureau of Guangdong Power Grid Co., Ltd., Guangzhou, 510410, China)

[*]shore2021@foxmail.com

Abstract—This paper proposes a cable broadband circuit model considering the temperature characteristics. This paper establishes the equivalent circuit model of the cross-bonded power cable and proposes the broadband parameter calculation method considering the temperature. The cable parameters are calculated by the vector fitting method. Based on the operating current and cable temperature data of the cable on a certain day, the simulation analysis of the cable sheath current considering the temperature is carried out using Simulink. Finally, the simulation results show that the cable sheath current is significantly affected by temperature.

1. Introduction

The widespread use of power cables has brought severe challenges to cable operation, maintenance and overhaul. As the service life increases, power cables continue to age, and cable defects will occur during operation, which will eventually cause operation failures [1]. The cable metal sheath current is one of the important indicators to judge the operating status of cables. Therefore, simulating and analyzing the characteristics of the cable sheath current is important.

To limit the sheath-induced voltage and sheath current, the cable generally adopts the grounding method of metal sheath cross-bonding. Literature [2] proposed a cross-bonded cable model under normal operation but did not consider the leakage current. Literature [3,4] further considered the induced current part and the leakage current part and proposed a power frequency lumped parameter model for calculating the induced current and the leakage current respectively. Based on the commonly used cable frequency domain simulation model, the cable joints and along the cable impedance change are strengthened, which can be used to accurately calculate the transmission line parameters and cable resonance analysis.

Cable parameters are also affected by the temperature of the cable. The temperature of the cable will change during operation, which will affect the resistivity of the cable conductor and the dielectric constant of the insulating material [5]. However, few cable modeling studies consider both the broadband characteristics and temperature characteristics of the cable. Therefore, based on the study of the broadband circuit model of the cross-bonded cable and the cable temperature characteristics, this paper establishes a cable broadband equivalent model and analyzes the cable sheath current by simulation.

2. Cable Broadband Circuit Model Considering Temperature Characteristics

2.1. Equivalent circuit model of the cross-bonded cable

To suppress the current in the metal sheath of the cable, the metal sheath of the single-core cable generally adopts a three-section cross-bonded grounding method. The major section of a typical cross-bonded cable is shown in Fig 1.

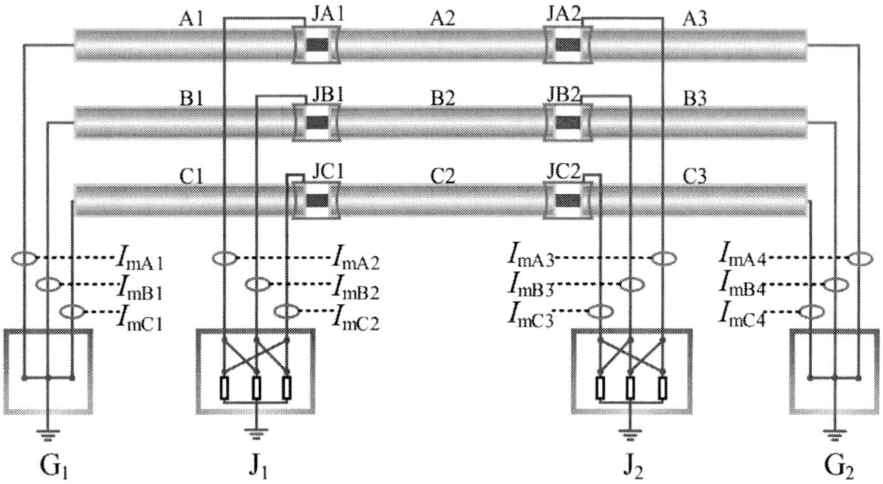

Fig 1 Major section of the cross-bonded cable

The connection mode of the metal sheath inside the cross-bonded box J1 is: the cable section A1 is connected with section B2, the cable section B1 is connected with section C2, and the cable section C1 is connected with the cable A2 section. In the same way, in the cross-bonded box J2, the metal sheath connection mode is: the cable section A2 is connected with section B3, the cable section B2 is connected with section C3, and the cable section C2 is connected with the cable section A3. Therefore, the sheath current of the cross-bonded cable flows through three loops, A1-B2-C3, B1-C2-A3, and C1-A2-B3.

The I-shaped lumped parameter model is used to establish the equivalent circuit of the cross-bonded cable, as shown in Figure 2.

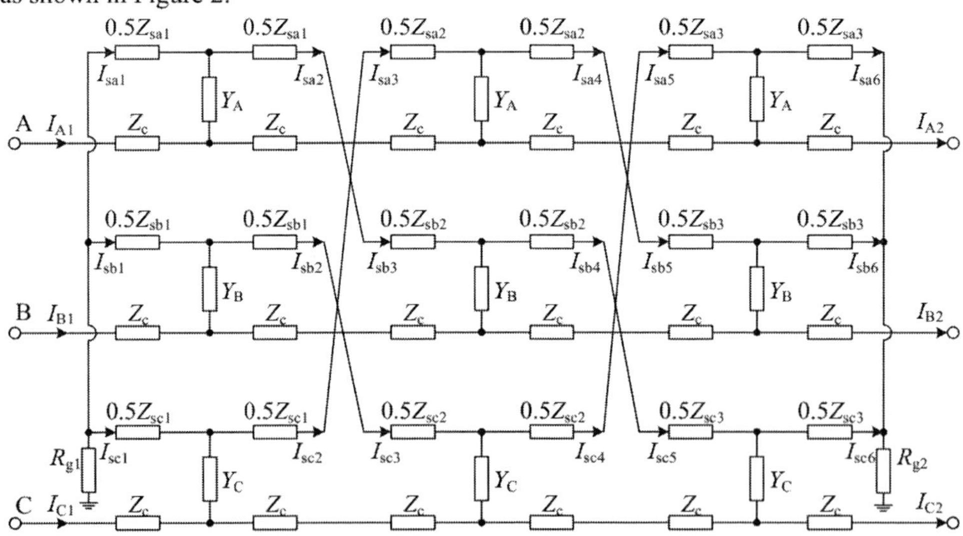

Fig 2 Equivalent circuit model of cross-bonded cable

In Fig 2, the length of each cable section is l, and the unit impedance of the cable core is $Z_0(j\omega)$, then $Z_c(j\omega)=0.5lZ_0(j\omega)$; $Y_A(j\omega)$, $Y_B(j\omega)$, $Y_C(j\omega)$ is the insulation admittance of each section of the three-phase cable, $Z_{sa1}(j\omega)$~$Z_{sa3}(j\omega)$, $Z_{sb1}(j\omega)$~$Z_{sb3}(j\omega)$, $Z_{sc1}(j\omega)$~$Z_{sc3}(j\omega)$ are the metal sheath impedances of each phase in each section; $I_{sa1}(j\omega)$~$I_{sa6}(j\omega)$, $I_{sb1}(j\omega)$~$I_{sb6}(j\omega)$, $I_{sc1}(j\omega)$~ $I_{sc6}(j\omega)$ are the currents flowing through the metal sheath of each phase, $I_{A1}(j\omega)$~$I_{C1}(j\omega)$ are the currents flowing into each phase core of the cross-bonded cable, and $I_{A2}(j\omega)$~$I_{C2}(j\omega)$ are the currents flowing out of each phase core of the cross-bonded cable.

2.2. Calculation of cable broadband equivalent circuit parameters considering temperature characteristics

To improve the accuracy of the metal sheath current simulation analysis, this paper considers the broadband characteristics of power cables. The calculation formulas of cable unit conductor frequency-dependent impedance $z_c(j\omega)$, unit metal sheath frequency-dependent impedance $z_s(j\omega)$ and unit insulation frequency-dependent admittance $y_d(j\omega)$ are as follows:

$$z_c(j\omega)=\frac{1}{2\pi r_c}\sqrt{j\omega\mu_0\rho_c}+\frac{j\omega\mu_0}{2\pi}\ln\frac{r_s}{r_c}+\frac{1}{2\pi r_s}\sqrt{j\omega\mu_0\rho_s} \tag{1}$$

$$z_s(j\omega)=\frac{\rho_s}{\pi\left(r_s^2-r_d^2\right)}+j\omega\left\{\frac{\mu_0}{8\pi}-\frac{\mu_0}{2\pi}\left[\frac{r_s^2\left(r_s^2-2r_d^2\right)}{\left(r_s^2-r_d^2\right)^2}\ln r_s+\frac{r_d^4}{\left(r_s^2-r_d^2\right)^2}\ln r_d+\frac{r_d^2}{2\left(r_s^2-r_d^2\right)}\right]\right\} \tag{2}$$

$$y_d(j\omega)=j\omega\frac{2\pi}{\ln\frac{r_s}{r_c}}\varepsilon_0\varepsilon_r \tag{3}$$

Where μ_0 is the vacuum permeability, ε_0 is the vacuum dielectric constant, r_c is the radius of the cable conductor, r_d is the outer diameter of the cable insulation layer, r_s is the outer diameter of the cable metal sheath, and ε_r is the relative permittivity of the cable insulation material, which is expressed by plural $\varepsilon_r=\varepsilon_r'-j\varepsilon_r''$.

For the XLPE cable, the influence of temperature on the dielectric constant can be ignored, only the influences on the resistivity of the conductor and the metal sheath are considered. The calculation formulas of cable unit conductor frequency-dependent impedance $z_c(j\omega, \theta_c, \theta_s)$) and unit metal sheath frequency-dependent impedance $z_s(j\omega, \theta_s)$ considered the temperature characteristics are shown in (4) and (5).

$$z_c(j\omega,\theta_c,\theta_s)=\frac{1}{2\pi r_c}\sqrt{j\omega\mu_0\rho_c(\theta_c)}+\frac{j\omega\mu_0}{2\pi}\ln\frac{r_s}{r_c}+\frac{1}{2\pi r_s}\sqrt{j\omega\mu_0\rho_s(\theta_s)} \tag{4}$$

$$z_s(j\omega,\theta_s)=\frac{\rho_s(\theta_s)}{\pi\left(r_s^2-r_d^2\right)}+j\omega\left\{\frac{\mu_0}{8\pi}-\frac{\mu_0}{2\pi}\left[\frac{r_s^2\left(r_s^2-2r_d^2\right)}{\left(r_s^2-r_d^2\right)^2}\ln r_s+\frac{r_d^4}{\left(r_s^2-r_d^2\right)^2}\ln r_d+\frac{r_d^2}{2\left(r_s^2-r_d^2\right)}\right]\right\} \tag{5}$$

Where θ_c is the cable conductor temperature, θ_s is the cable sheath temperature, $\rho_c(\theta_c)$ is the temperature-corrected cable conductor resistivity, and $\rho_s(\theta_s)$ is the temperature-corrected cable metal sheath resistivity. The calculation formula of the temperature-corrected resistivity $\rho(\theta)$ is shown as:

$$\rho(\theta)=\rho_{20}\left[1+\alpha_{20}\left(\theta-20\right)\right] \tag{6}$$

Where ρ_{20} the resistivity of the material at 20°C, α_{20} is the temperature coefficient of resistance of the material, and θ is the temperature of the material.

3. Parameter Fitting of Power Cable Broadband Circuit Model

In this paper, the vector fitting method is used to calculate the cable frequency variation model parameters. The principle of the vector fitting method is: suppose the frequency domain response function of a certain system is $F(j\omega)$, the frequency domain response function after fitting is $F_{\text{fit}}(j\omega)$, which is expressed as:

$$F_{\text{fit}}(j\omega) = j\omega\left(\sum_{i=1}^{n}\frac{c_i}{j\omega - a_i} + h\right) + d \tag{7}$$

Where c_i is the residue, a_i is the pole, h and d are real numbers, and n is the number of poles.

The low-order rational function expression of the approximate cable frequency parameters is obtained by the vector fitting method, and the vector fitting result can be equivalent to the series-parallel circuit of passive components. The cable frequency-dependent impedance can be equivalent with an *R-L* series circuit, and the cable frequency-dependent admittance can be equivalent with a *G-C* parallel circuit, as shown in Fig 3.

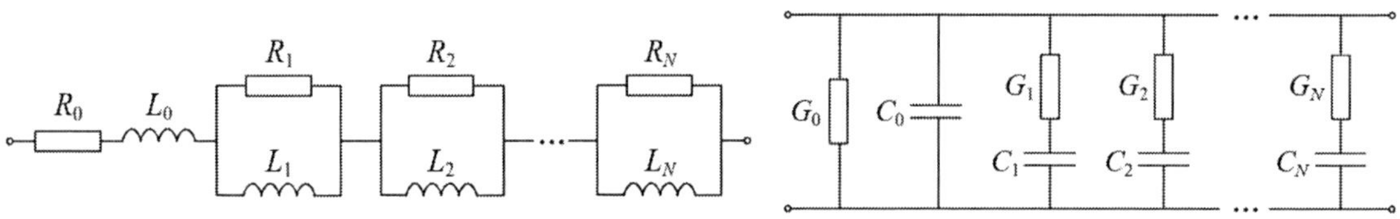

(a) *R-L* series circuit (b) *G-C* parallel circuit

Fig 3 Equivalent circuits of cable frequency-dependent parameters

The corresponding relationship between the vector fitting results and the parameter values of the components in the equivalent circuits of Figure 3 is as follows: for the frequency-dependent impedance, DC resistance $R_0=d$, inductance $L_0=h$, resistance $R_i=c_i$, inductance $L_i= -c_i/a_i$, $i=1, \cdots, N$; for the frequency-dependent admittance, DC conductance $G_0=d$, capacitance $C_0=h$, conductance $G_i=c_i$, capacitance $C_i=-c_i/a_i$, $i=1, \cdots, N$.

The cable parameters of a certain 110 kV XLPE cable per unit length are shown in Table 1. In the process of vector fitting, it is considered that the cable conductor temperature θ_c and the sheath temperature θ_s are both 20°C.

Tab 1. Main parameters of the 110kV XLPE cable

Parameter	Value
Voltage rating (kV)	110
Conductor cross-sectional area (mm^2)	1000
Length of each cable section (m)	480
Diameter of cable conductor d_c (mm)	38.4
Outer diameter of insulation layer d_d (mm)	74.4
Outer diameter of metal sheath d_s mm)	80.4
Outer diameter of outer sheath d_e (mm)	87.6
Resistivity of cable core at 20°C ρ_{c20} (Ω·m)	1.75×10^{-8}
Resistivity of metal sheath at 20°C ρ_{s20} (Ω·m)	21.4×10^{-8}
two-phase space between AB and BC s (m)	0.27
Distance between cable and temperature measuring point L (m)	0.5

According to the results obtained by vector fitting *d*, *h*, c_i and a_i, the resistance, the inductance, conductance, and the capacitance of each level in Fig 3 are calculated. The fitting parameters are shown in Tab 2-Tab 4.

Tab 2 Fitting parameters of the *R-L* equivalent circuit for conductor frequency-dependent impedance

Resistance	Value (Ω/m)	Inductance	Value (H/m)
R_0	3.3842×10^{-5}	R_0	1.5129×10^{-7}
R_1	1.8112×10^{3}	R_1	2.0119×10^{-3}
R_2	1.3598×10^{3}	R_2	6.9649×10^{-4}
R_3	1.6632×10^{2}	R_3	3.8737×10^{-4}
R_4	1.2127×10^{1}	R_4	1.9587×10^{-4}
R_5	3.5548×10^{-3}	R_5	6.8873×10^{-5}
R_6	4.3656×10^{-4}	R_6	2.0681×10^{-5}

Tab 3 Fitting parameters of the *R-L* equivalent circuit for sheath frequency-dependent impedance

Resistance	Value (Ω/m)	Inductance	Value (H/m)
R_0	2.5727×10^{-4}	R_0	5.2695×10^{-7}
R_1	1.2655×10^{-5}	R_1	1.7925×10^{-7}
R_2	1.8530×10^{-5}	R_2	7.0997×10^{-7}
R_3	1.0937×10^{-6}	R_3	1.4781×10^{-6}
R_4	5.5284×10^{-7}	R_4	3.07138×10^{-5}
R_5	6.8012×10^{-8}	R_5	2.2521×10^{-5}
R_6	2.9605×10^{-8}	R_6	5.5859×10^{-5}

Tab 4 Fitting parameters of the *G-C* equivalent circuit for insulation frequency-dependent admittance

Conductance	Value (S/m)	capacitance	Value (F/m)
G_0	1.2719×10^{-7}	C_0	1.8128×10^{-10}
G_1	2.6302×10^{1}	C_1	4.6655×10^{-4}
G_2	2.6285×10^{1}	C_2	3.6654×10^{-4}
G_3	1.4252×10^{-3}	C_3	3.6978×10^{-5}
G_4	1.4178×10^{-3}	C_4	3.6989×10^{-5}
G_5	3.7055×10^{-8}	C_5	2.0586×10^{-7}
G_6	1.7480×10^{-9}	C_6	3.2981×10^{-7}

Substituting the data in Tab 2-
Tab 4 to the equivalent circuit in Fig 3, the equivalent circuit of the conductor frequency-dependent impedance, the sheath frequency-dependent impedance, and the insulation frequency-dependent admittance can be obtained.

4. Case Study

Simulink is used to simulate the cable model, and the main parameters of the model are shown in Tab 1. Based on the cable broadband equivalent model, the current of the metal sheath is simulated. The simulation results are as shown in Fig 4.

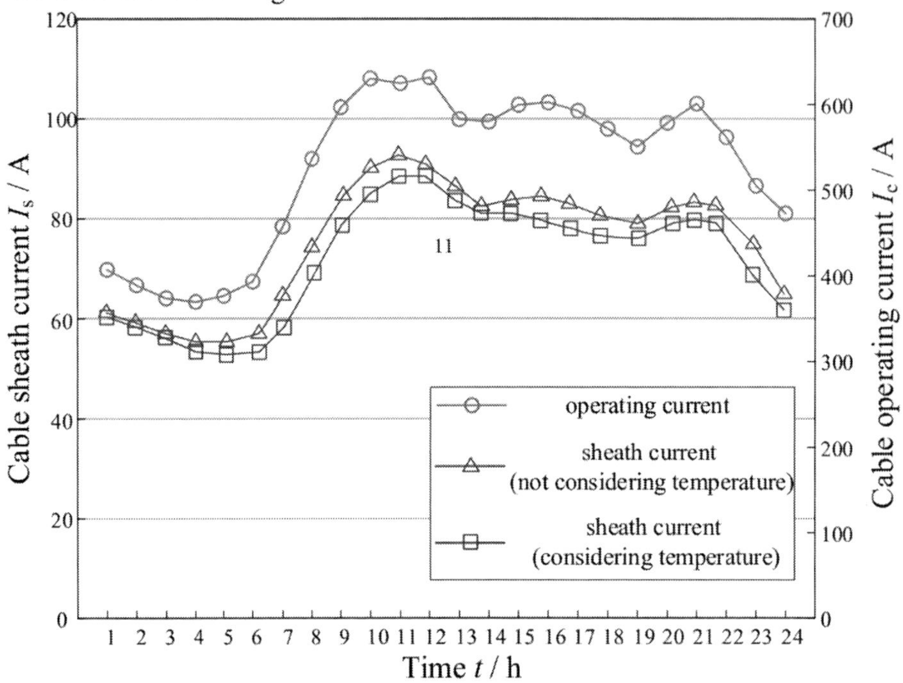

Fig 4 Simulation results of cable sheath current

As can be seen from **Fig 4**, the magnitude of the sheath current simulation value is related to the cable operating current, and the changing trend of the sheath current is consistent with the changing trend of the cable operating current. The cable temperature has a significant impact on the sheath current, but it has little effect on the changing trend of the sheath current, and the simulation value of the sheath current when the temperature is not considered is higher than cable sheath current considering the temperature. Comparing the two cable sheath current curves, we can see that the maximum difference between the two current curves appeared at the 9th hour, with an absolute value of 3.67A and a relative error of 4.2%.

The requirement for the live test results of the cable sheath ground current is: the ground current should be less than or equal to 100A, and the ratio of the ground current to the load current should be less than 20%. When the cable operating current increases, the temperature of the cable will increase accordingly and will affect the current value of the cable sheath. According to the current simulation results of the cable metal sheath considering the influence of temperature, it can be seen that the influence of temperature will increase the resistance of the cable metal sheath and cause the current value of the metal sheath to decrease. Therefore, when judging the operating state of the metal sheath, the influence of temperature should be considered. This article suggests improving the criterion of the ground current of the metal sheath and reduces the threshold of the normal value of the metal sheath current by 5% to exclude the impact of temperature on the judgment of the operating state of the cable metal sheath.

5. Conclusion

This paper proposes a cable broadband circuit model that takes into account the temperature characteristics. Taking the major section of the cross-bonded cable as the research object, the equivalent circuit model is established, and the cable broadband parameter calculation method considering the temperature characteristics is proposed. The cable parameters are calculated by the vector fitting method. Based on the operating current and cable temperature data on a certain day, a simulation analysis of the cable sheath current is carried out. The simulation results show that the cable sheath current is significantly affected by temperature. The simulation value of the cable sheath current considering the temperature is lower than the simulation value of the cable sheath current not considering the temperature, and the maximum difference reaches 4.2%. In summary, the cable broadband circuit model considering the temperature characteristics proposed can improve the accuracy of the cable metal sheath current calculation, and then provide guidance for the monitoring and identification of the cable operating state.

Acknowledgments
This work is supported by the Science and Technology Project of China Southern Power Grid: the Research and Development Project of Intelligent Flexible Detection Integrated Device (GZHKJXM20200083), completed in the Key Laboratory of Medium-voltage and Low-voltage Electric Equipment Inspection and Testing of China Southern Power Grid, China.

References
[1] Orton, H. (2015) Power Cable Technology Review. High Voltage Engineering, 41: 1057-1067.
[2] Marzinotto, M., Mazzanti, G. (2015) The Feasibility of Cable Sheath Fault Detection by Monitoring Sheath-to-Ground Currents at the Ends of Cross-Bonding Sections. IEEE Transactions on Industry Applications, 51: 5376-5384
[3] Yan, L., Peng, F.,Chen, X., et al. (2008) Study on Sheath Circulating Current of Cross-linked Power Cables. In: 2008 International Conference on High Voltage Engineering and Application. Chongqing. pp: 645-648.
[4] Yang, Y., Hepburn, D. M., Zhou, C., et al. (2015) On-line Monitoring and Trending of Dielectric Loss in a Cross-bonded HV Cable System. In: 2015 IEEE 11th International Conference on the Properties and Applications of Dielectric Materials. Sydney, pp: 301-304.
[5] Gao, Q., Wei, Z., Sun, G., et al. (2015) Temperature-dependent Power System Optimal Power Flow. Automation of Electric Power System, 39: 824-827. (in Chinese)

Design and electromagnetic analysis of switched reluctance linear generator in electric vehicle

Jieyu Liu[1*], Ye Tang[1], Wenyu Cheng[1] and Changpeng Li[2]

[1]State Grid Liaoning Marketing Service Center, Shenyang, Liaoning 110006, China

[2]State Grid Benxi Electric Power Supply Company, Benxi, Liaoning 117000, China

*1142101056@ncepu.edu.cn

Abstract. The structural characteristics and basic design requirements of design of switched reluctance linear generator are introduced. According to the general principles and design experience of electromechanical design, the dimension of design of switched reluctance linear generator is determined, and the two-dimensional and three-dimensional static finite element models of design of switched reluctance linear generator are established in Flux 2D and Flux 3D respectively. By comparing the simulation results of the two-dimensional model and the three-dimensional model, it is found that the difference between the real simulation results of the 2D model and the 3D model is small, and the calculation time cost of the 2D finite element simulation is much lower than that of the 3D model. Therefore, the subsequent work of this paper adopts the 2D finite element model. Finally, the static electromagnetic field and electromagnetic characteristics of the design of switched reluctance linear generator are analyzed by finite element analysis.

1. Introduction

The dimensions of the main structural parameters of DSRLG(Design of Switched Reluctance Linear Generator) are determined firstly. Then, according to the dimension of the motor, the finite element method was used for modeling. Considering the end effects of the two sections of the linear motor, the 2D and 3D static finite element models of DSRLG were established respectively, and the electromagnetic analysis of DSRLG was carried out. By comparing the results of 2D and 3D electromagnetic analysis, the finite element modeling method in the subsequent optimization process was selected.

2. Dimension design of DSRLG

Due to the current literature for bilateral linear motor design is less, also did not form a standardized design criteria[1-4]. In this paper, the author studies on bilateral linear motor is made by rotating the Switched Reluctance Machine(SRM) development change, such as smooth and symmetry, the determination of the dimension of DSRLG will refer to traditional rotary SRM design principles and design experience [5]. Combined with the structural particularity of linear motor and the consideration of bilateral stator structure, the specific values of each structural parameter in the design process were adjusted, and the structural parameters of DSRLG were obtained [6]. In order to make reasonable use of the existing universal SRM design criteria and experience, the design process thought of DSRLG in this paper adopts: firstly, the design criteria of traditional rotary SRM and the design experience of linear motor are referred to determine the size of unilateral Switched Reluctance Linear

Machine(SRLM) generated by rotating SRM to be paved in the circular direction [7]. Then, the structure size of bilateral DSRLG is obtained by adjusting the structural parameters and making symmetrical changes. Considering the overall size of the generator suitable for reciprocating wave power generation environment and the existing manufacturing process and cost of the motor, the stator and actuator stack thickness L_s,L_m,total width L and air gap g of single-side SRLM are determined to be 90mm, 90mm, 174mm and 0.5mm respectively. The dimension of the width L_a of the actuator is calculated as follows:

$$L_a = k_1 \cdot L \tag{1}$$

where k_1 is specified as the ratio of the width of the actuator to the total width. According to the general design rules of rotary SRM, the value range of k_1 is generally between 0.4 and 0.7. However, in the linear motor, since it is paved along the radial direction from SRM, there is no corresponding rotor shaft in SRM. Therefore, the $k_1$value of the linear motor can be slightly smaller. In this case, k_1 is valued at 0.29.

The dimensions of the yoke thickness H_m are calculated as follows:

$$H_m = \frac{(L_a - L_b)}{2} \tag{2}$$

In formula (2), L_b is the width of the actuator shaft.As can be seen from the foregoing, there is no moving sub-shaft in the linear motor, so L_b is 0.

According to the design experience of rotating SRM, the thickness of stator yoke H_s is generally slightly smaller than that of actuator yoke. In this example, the dimension of the stator yoke thickness H_s is tentatively set at 18.5mm.

The stator slot depth H_{sslot} can be obtained by the following formula (3) :

$$H_{sslot} = L - L_a - H_s - g \tag{3}$$

The dimension of the actuator slot depth H_{mslot} is calculated as follows:

$$H_{mslot} = L_a - H_m \tag{4}$$

According to the general design rules of rotating SRM, the relationship between stator extreme width B_s and stator yoke thickness H_s is generally:

$$H_s = k_2 \cdot \frac{B_S}{2} \tag{5}$$

Among them, k_2 is the scale coefficient, which is generally between 1.2 and 1.4 in rotary SRM design. In the design of the linear motor in this example, considering the motion and structure characteristics of the linear motor, its output pulsation may have a greater impact on the structural strength of the motor. In order to reduce the output pulsation and noise of the motor, the proportional coefficient k_2 in this example should be large, which is temporarily set as 1.85.

Similarly, according to the design criterion of rotational SRM, the relation between the pole width of the actuator B_m and the yoke thickness of the actuator H_m is generally:

$$H_m = k_3 \cdot \frac{B_m}{2} \tag{6}$$

In the formula, k_3 is the proportional coefficient, which is usually between 1.2 and 1.4 according to design experience. In this example, the proportional coefficient k_3 of the linear motor is larger than that of the rotating motor, which is tentatively set as 1.9.

Through the above design steps, a reasonable dimension of the single-side SRLM can be obtained. In order to eliminate the serious normal electromagnetic force of unilateral SRLM and improve the power density of the motor, a symmetric stator can be placed on the other side of the actuator, and the actuator is changed from a single salient pole structure to a symmetric double salient pole structure, so as to obtain the corresponding bilateral stator type DSRLG. The dimensions of DSRLG are shown in Table 1.

Table 1. Geometric dimensions of the DSRLG

Geometric dimension	Dimension /mm
air gap g	0.5
The stator pole pitch T_s	40
Mover pole pitch T_m	60
The stator yoke thickness H_s	18.5
Mover yoke thickness H_m	25
Stator stack thickness L_s	90
Mover stack thickness L_m	90
Stator pole width B_s	20
Mover pole width B_m	26
Stator slot depth H_{sslot}	105
Mover slot depth H_{mslot}	12.5

3. Finite element modeling and simulation

The finite element method is adopted in this paper to establish the two-dimensional finite element model and three-dimensional finite element model of DSRLG respectively, and the calculation results of the two are compared. Based on the accuracy of the calculation results and the calculation time, the appropriate finite element model was selected and applied to the electromagnetic analysis of DSRLG, as well as the sensitivity analysis.

The process of using Flux 2D and Flux 3D to establish the finite element model of DSRLG is similar, as shown in the Figure 1.

The relationship of flux and magnetic common energy with respect to position and current calculated by Flux 2D and Flux 3D DSRLG models are shown in Fig 2 respectively. By comparing the calculation results of the two models, it is found that the main error of the two models is that the flux value of the two models is smaller than the flux value of the three models when the actuator is in the misaligned position. Similarly, the magnetic common energy value obtained by two-dimensional finite element model simulation is smaller than that obtained by three-dimensional finite element model simulation when the actuator is in the misaligned position. This is because the three-dimensional finite element model can take into account the longitudinal MFL of the motor, while the two-dimensional finite element model does not take into account the longitudinal MFL and the MFL effect at the misaligned position of the motor are more serious, so the flux value obtained by 3D simulation is slightly larger than that obtained by 2D simulation. Although there are some errors in the simulation results of the two-dimensional model and the three-dimensional model,the accuracy of DSRLG model established by Flux 2D can be guaranteed. The calculation time of the two-dimensional finite element model is only about one-tenth of that of the three-dimensional model of the same size. Therefore, after considering the accuracy of the model and the cost of calculation time, Flux 2D model is selected as the electromagnetic calculation method in the subsequent analysis.

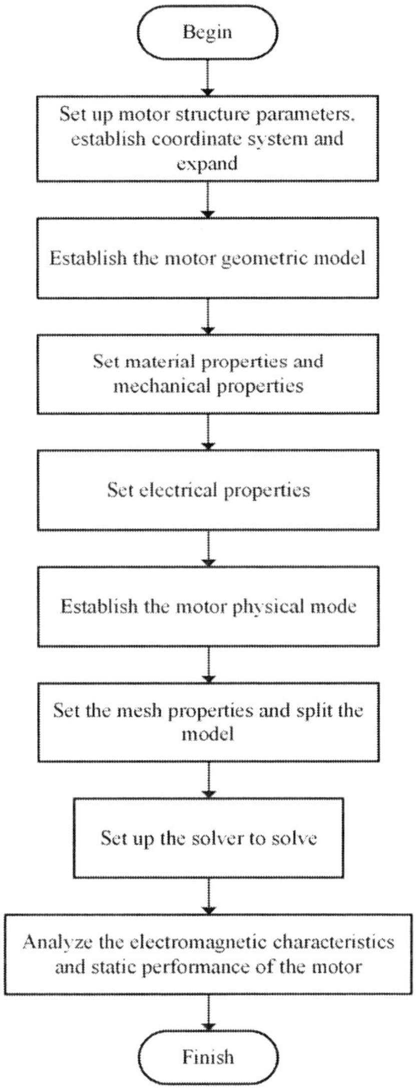

Figure 1. Workflow of DSRLG finite element modelling

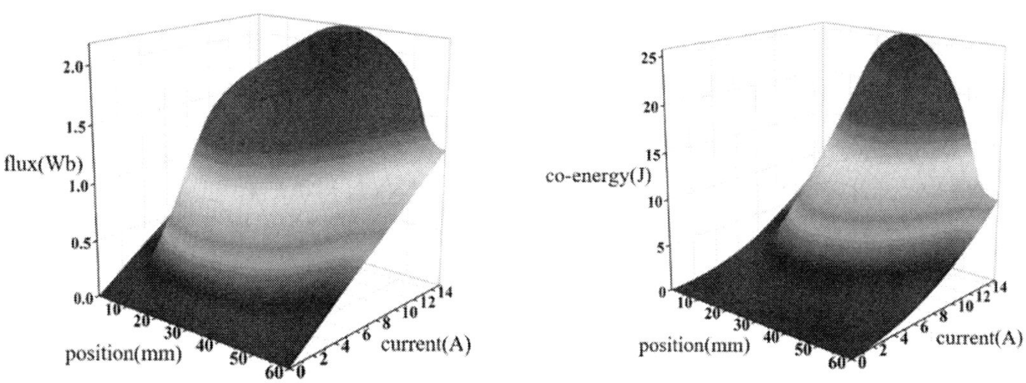

Figure 2. Variation trend of flux and co-energy with position and current

4. Electromagnetic analysis of the DSRLG

A complete actuator period of DSRLG studied in this paper is 60mm. According to the common design custom, the alignment position of the stator salient pole and the axis of the actuator salient pole is defined as 30mm, and the position of the reverse movement along the direction of the actuator movement is defined as 0mm. The position 30mm along the direction.

Because in the solver setting of the DSRLG static finite element model, the position is set from 0mm to 60mm, increase by 2mm as an interval step, increase by 1A as a step from 0A to 15A, and obtain the data of the relationship between the static flux flux and magnetic common energy of DSRLG and the change of current and the position of the actuator. Fig 3 shows the three-dimensional curve of flux flux and magnetic common energy, in which the Z axis respectively represents flux flux and magnetic common energy. The X-axis and Y-axis in the Figure 3 represent the current and the position of the actuator, respectively.

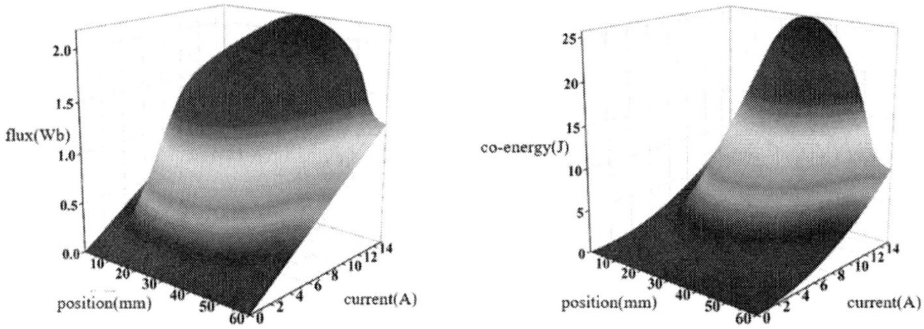

Figure 3. Variation trend of flux and co-energy of DSRLG with current and position

As can be seen from the figures above, the trend of the flux - position curve is the same for any current value. When the actuator is at 0mm of the asymmetrical position, the flux value is low. When the actuator moves from 0mm to 30mm, the flux reaches its maximum value. If the current value at this point is larger, the variation trend of the flux chain gradually tends to be nonlinear. The larger the current is, the more serious the saturation phenomenon of the magnetic circuit will be, and the more obvious the nonlinear variation trend will be. When the actuator continues to move from the position of 30mm to the position of the next asymmetrical position of 60mm, the flux decreases in a nonlinear to linear trend, and finally reaches the lowest value. It is obvious from the figure that the value of the flux linkage is symmetric about the 30mm point where the position of the actuator is perfectly aligned. For any position of the actuator, the flux - current curve is gradually increasing. It should be noted that at the misaligned position of the actuator, the flux flux increases linearly with respect to the current; However, the increase of the current of the flux linkage increases non-linearly with the increase of the current value as the alignment position of the actuator increases. This is because, when the actuator is in the misaligned position, the magnetic circuit reluctance is large, and the magnetic circuit is always in the unsaturated state during the process of the gradual increase of current. When the actuator is in alignment position, the magnetic circuit reluctance is the least, and the magnetic circuit is gradually saturated with the increasing current. The variation trend of the three - dimensional magnetic co-energy - position - current curve is roughly similar to that of the magnetic linkage curve.

In addition to the three-dimensional curves of flux linkage and magnetic co-energy, the static finite element model can also obtain the curves of the static electromagnetic force of DSRLG at different current currents changing with its position, as shown in Fig 4 below. It can be seen from the figure that the static electromagnetic force increases first and then decreases when the electromaneuvers move from the unaligned symmetrical position of 0mm to the aligned position of 30mm. This is because in the process from 0mm to 10mm, the edge of the actuator salient pole has not coincided with the stator salient pole, and the inductance change rate has not reached the maximum. When the position of the actuator reaches around 10mm, the edge of the salient pole of the actuator just begins to coincide with the salient pole of the stator. At this time, the inductance change rate reaches the maximum, so the static electromagnetic force reaches the maximum. After that, as the position of the actuator gets closer to the alignment position, the inductance change rate decreases gradually, and finally the static electromagnetic force reaches the lowest value at 30mm. In addition, it can be seen from the figure that the static electromagnetic force tends to increase with the increase of current at the fixed position of the actuator. With the increase of current value, the static electromagnetic force gradually tends to be non-linear. This phenomenon is more obvious the closer to the aligned position. This is because, near the alignment position of the actuator, the magnetic circuit reluctance is small, and with the increase of the excitation current, the ferromagnetic material gradually becomes saturated, so the growth of the static electromagnetic force tends to be gentle.

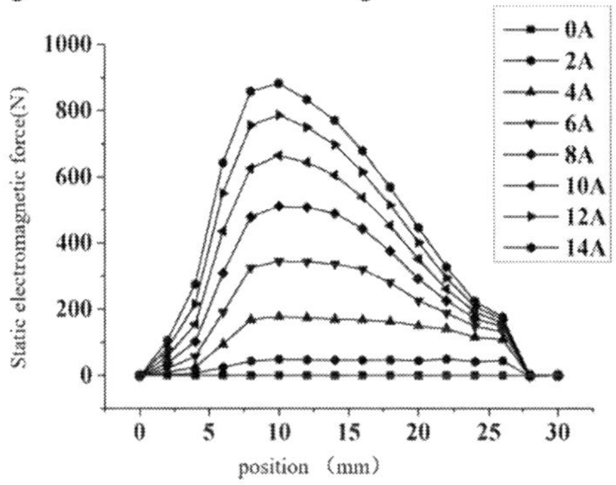

Figure 4. Variation trend of static electromagnetic force with position

5. Conclusion

According to the application requirements of wave linear power generation and the design criteria and experience of the traditional rotary SRM, the structure parameters of the corresponding unilateral SRLM are determined, and the unique structural characteristics of the bilateral DSRLG are combined to improve the dimension of the DSRLG studied in this paper, which lays a foundation for the follow-up research.

Based on the finite element analysis, the finite element modeling method applicable to this study was determined by comparing the simulation calculation results and calculation time cost of 2D finite element simulation and 3D finite element simulation.

Through the data processing of the DSRLG two-dimensional static finite element simulation results, the three dimensional curves of the flux flux, the magnetic co-energy and the static electromagnetic force with respect to the position and the current are obtained respectively, and the electromagnetic characteristics of DSRLG are analyzed accordingly.

References

[1] Torrey D.Switched reluctance generators and their control. IEEE Transactions on Industrial Electronics,2002,49(1):3-14

[2] Chang Y C, Liaw C M. On the Design of Power Circuit and Control Scheme for Switched Reluctance Generator. IEEE Transactions on Power Electronics, 2008, 23(1): 445-454

[3] Chang Y C, Liaw C M. Establishment of a Switched-Reluctance Generator-Based Common DC Microgrid System. IEEE Transactions on Power Electronics, 2011, 26(9): 2512-2527

[4] Ferreira C A, Jones S R, Heglund W S, et al. Detailed design of a 30-kW switched reluctance starter/generator system for a gas turbine engine application. IEEE Transactions on Industry Applications , 1995, 31(3): 553-561

[5] Kioskeridis I, Mademlis C. Optimal efficiency control of switched reluctance generators. IEEE Transactions on Power Electronics , 2006, 21(4): 1062-1072

[6] Chen H, Wang X, Gu J J. Sliding mode control of Switched Reluctance linear generator system. International Conference on Networking, Sensing and Control. IEEE, 2009: 779-782

[7] Chen H, Zhan Y, et al. Multiobjective Optimization Design of Tubular Permanent Magnet Linear Launcher . IEEE Transactions on Plasma Science, 2019, 47(5): 2486-2492

Development of the Line Protection Measurement and Control Device based on Domestic MCU

Chengxuan ZHAO[1a*], Yuantao WANG [2b], Chuankun NI [1c], Chen HU [2d], Yunlong XU [1e], Wei Ma [2f], Yaxin Su[1g]

[1] XJ Group Corporation, Xuchang 461000, Henan, China;

[2] State Grid Hangzhou Power Supply Company, Hangzhou 310000, Zhejiang, China

[a*]new2020@163.com, [b]809995399@qq.com, [c]xjtc_nichuankun@126.com, [d]qzhuchen@126.com,[e]xjtc_xuyunlong@126.com, [f]2jnyz-mawei@163.com, [g]xjtc_suyaxin@126.com

Abstract—In this paper, the dual MCU hardware architecture of protection measurement and control device was presented by analyzing the current situation of domestic MCU and chip selection. The problems existing in ADC sampling accuracy and ADC integral non-linearity were studied. The design scheme of high-precision ADC acquisition system based on dual MCU architecture was proposed, which could meet the requirements of technical specifications by using oversampling and adaptive processing of sampling nonlinear area. In terms of reliability, the EMC performance of the device was promoted by the improvement of outlet and ADC acquisition circuit.

1. Introduction

The shortage of global semiconductor industries and the weakness of supply chain problems are concerned. With the development of domestic chip industry, the domestic chip technology in the field of relay protection is gradually improved[1].

The development of the low-voltage protection and control device based on domestic chip is restricted by existing production and manufacturing process of domestic chips. The main control chip is selected from other industrial sectors[2]. Based on the technical performance of existing domestic chip, through the optimized design of existing relay protection algorithm, function, and hardware and software architecture, overcoming the insufficient of peripherals, main control ADC chip accuracy, domestic component reliability, etc.[3] to achieve the design goal which is the domestic chip protection performance is not less than existing relay protection performance. Based on the current status of domestic main control chip without memory ECC (Error Checking and Correction or Error Correction Coding) verification mechanism [4]-[5], it is necessary to study mistake proofing under existing architecture. In the meanwhile, working on the per-fection of self-check logic, improve the robustness of hardware architecture and reliability of software pro-gram, in order to avoid the incorrect action of protection.

This paper analyzes the current status of MCU technology, and proposes the hardware and software design scheme and performance improvement measurement of the line protection and control device. The influence of the non-linear error precision of the ADC was overcome, and proposes the improvement measurement of electromagnetic compatibility for domestic chip components. The development of domestic low-voltage line protection and control device has great practical significance of the autonomous control of the production and manufacturing of the power system secondary protection control equipment.

2. Domestic MCU Technology Status

At this stage, domestic MCU cannot be compared with foreign brands in the market share or technical advancement. 8 MCU still occupied the mainstream market, accounting for about 50%, and 16/32-bit MCU accounts for about 20%. The applications of domestic MCU are concentrated in low-end electronic products. Compared the three domestic 32-bit general MCU chip parameters, Zhaoyi innovative GD32F450ZI is excellent in performance, storage capacity, ADC external channel, peripheral interface, etc., which could be used for low-voltage line protection and control device.

Table 1 Comparison of parameters of three domestic 32-bit MCU

Manufacture	Zhaoyi Innovation	Zhixin microelectronic	Yatley Technology
Type	GD32F450ZI	SCM610	AT32F413RBT7
Kernel	CORTXE-M4	CORTXE-M4	CORTXE-M4
Basic frequency	200M	150M	200M
FLASH	2048KB	1024KB	256K
SRAM	512KB	256KB	64K
ADC	12bit 24 external channel	12bit 4 external channel	12bit 2 external channel
GPIO	114	100	55
UART	8	8	5
Ethernet	1	2	1
CAN2.0B	2	0	2

The GD32F450ZI MCU of Zhaoyi Innovation is a 32-bit universal micro-controller based on the ARM® CortexTM-M4 processor with low interruption delay time, low cost debugging features. High integration and enhanced features make the controller suitable for high performance and low power applications. The controller supports powerful and scalable instruction sets, including universal data processing I/O control tasks, enhanced data processing bit domain operations, DSP (digital signal processing), and floating point arithmetic instructions. Software is running in built-in SRAM, FLASH, which has the advantages of low power and excellent electromagnetic compatibility performance.

3. The Design of Line Protection and Control Device

3.1. Hardware architecture

Reliability is the basic requirement of the relay protection device. Factors affecting the reliability of the device hardware system are mainly included: the quality of the components, the manufacturing process level, the rationality of the circuit design, the integration of components [6]. Circuit design still occupies an extremely important location in the design of the hardware system. In order to achieve high reliability, the hardware design based on domestic MCU uses a single board with dual MCU redundant, and two systems work parallel [7].

For analog sampling link, the sampling data for start MCU and protect MCU are used for logic data calculation of protect start and action respectively. Dual ADC uses mutual independent analog sampling loops and analog domain power supplies to maximize guarantee the independence of dual ADC and reduce the interference between dual ADC. For logical judgment link, start MCU and protect MCU program run in independent memory respectively, logic judgment duplication, which is overcoming domestic master chips does not have ECC check function and avoiding the protection mis-operation or rejection caused by memory errors [8]. For export execution link, start MCU control outlet relay power supply, protect MCU control relay outlet, and serial export control improves the reliability of the device.

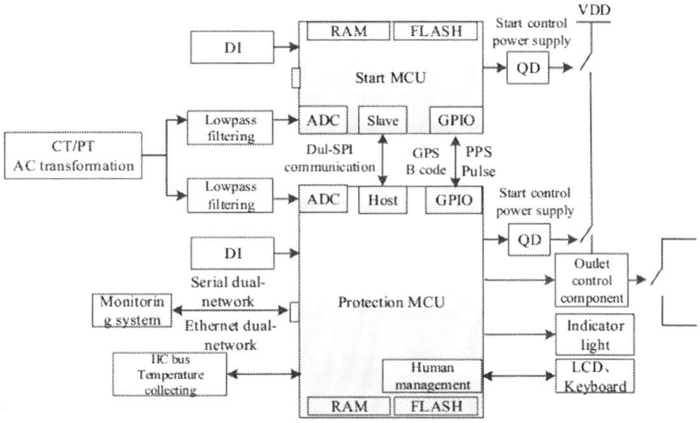

Fig.1 Hardware architecture of protection measurement and control device based on domestic MCU

The interface management functions such as LCD interaction, keyboard and indicator control were deployed in protect MCU. The start MCU and protection MCU collect the external DI simultaneously, and complete the protective value synchronization of two MCU through inside communication. IIC bus is equipped with RTC chip clock input, temperature monitoring, 3-way 485/232 communication multiplexing with print function. Completing the time synchronization function of B code and PPS pulse for start and protection MCU by GPIO communication [9].

3.2. Oversampling design

The analog sampling accuracy is the core indicator of the relay protection device. The ADC module which was discussed in this paper is integrated in the MCU chip. The analog-digital converter which is using successive approximation method, with a theoretical sampling resolution is 12 bits, and its acquisition voltage range is 0～3V. The SNR (signal-to-noise ratio) of the ADC in domestic MCU is 66dB, and the actual effective bit is 10.7 bits. According to the effective bits of the AD and the resistance parameter of the acquisition loop, the accuracy of the analog sampling which is contained the detection of the protection current, protective voltage, and measuring current could be calculated [10]. As shown in Table 2, the resolution of the protective current and measuring current is 1LSB0.184A and 1LSB0.0184A, respectively, which does not satisfy the accuracy requirements of protection current which is ± 2.5% or ± 0.01In, and the measured current which is ± 0.2%.

Over-sampling techniques can not only improve the chip sampling resolution, but also can reduce the effect of sampling noise[11]. According to the derived result of the oversampling rate, the relationship between the quantized SNR and the oversampling rate M may satisfy the following formula. The higher the oversampling rate M, the higher the SNR.

$$SNR_{\text{oversample}}(dB)=6.02N+1.76+10\lg(M)$$

The relationship between the ADC sampling digital encoding (ENOB) and the quantized signal-to-noise ratio (SNR) satisfies the following formula.

$$ENOB=\frac{SNR-1.76}{6.02}$$

As can be seen from the formula, when the sampling frequency is doubled, the in-band noise will decrease 3dB, and the measured resolution will increase by 1/2 bits. Therefore, sampling at 4 times the sampling frequency to increase the resolution by 1 bit. M = 64, SNR increases 18 dB, ENOB adds 3 digits[12]. As the number of samples increases, the number of data valid bits also increases the calculation load of the CPU, which is not conducive to the overall performance of the device. According to Table 2 test data analysis, when the oversampling set to 64 times, it is possible to meet the accuracy requirements and ensure the small CPU load. Therefore, this paper is set in accordance with 64 times.

Journal of Physics: Conference Series **2108** (2021) 012017 doi:10.1088/1742-6596/2108/1/012017

Table 2 Comparison of analog sampling accuracy after oversampling

Analog Channel	Applied Value (V/A)	Displayed Value (V/A)	Accuracy (%)
Phase A protection voltage	57.740	57.767	0.047%
Phase B protection voltage	57.740	57.714	-0.045%
Phase C protection voltage	57.740	57.833	0.161%
Phase A protection current	1.0	1.006	0.600%
Phase B protection current	1.0	1.007	0.700%
Phase C protection current	1.0	1.007	0.700%
Phase A measurement current	1.0	1.000	-0.100%
Phase B measurement current	1.0	0.999	-0.100%
Phase C measurement current	1.0	0.998	-0.200%

3.3. *Software Architecture*

Based on the principle of functional hierarchical design of platform module and flexible deployment of service function, the device software architecture design is divided into task scheduling modules, interrupt modules, multiple real-time service task modules, and hardware drive modules.

The real-time multi-task scheduling module is the core of the entire software system, all interrupts and tasks are reasonably ordered under the policy control of the scheduling module [13]. Interrupt mainly includes sampling timer interrupts, timing scheduling interrupts, frequency capture interrupts, and keyboard interrupts. Real-time service tasks include: system main task, AD sampling computing tasks, frequency computing tasks, protect analog computing tasks, DI acquisition tasks, protect logic discriminating tasks, key value processing tasks, and display tasks. Each task has a different priority, and the task is communicated and synchronized by the message and mail [14]. The hardware drive layer completes the communication and data interaction between the MCU and the underlying hardware, including the ADC conversion, and DI acquisition, buttons and LCD, LED, time synchronization, Ethernet communication and other hardware communication interfaces. [15]

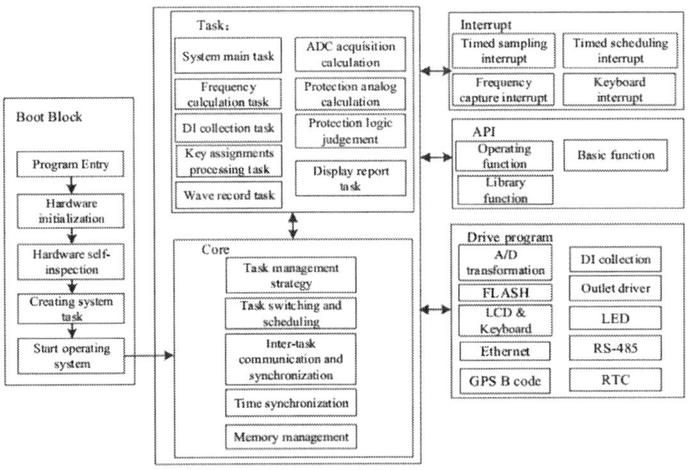

Fig.2 The software system framework diagram of protection measurement and control device based on domestic MCU

The MCU chip integrates only one Ethernet network port communication interface, and the dual ethernet communication function of the device is completed by two MCUs. Double MCU is used by dual SPI channel to communication [16], the protect MCU is used as the SPI communication host and the start MCU is used as the SPI communication slave, and the dual ethernet communication function is implemented by virtual network card technology. The data of station control layer A received by protection MCU is analysed and processed by the communication module of protection MCU. The data of station control layer B received by start MCU is sending to the virtual network by SPI communication bus for analysing and processing.

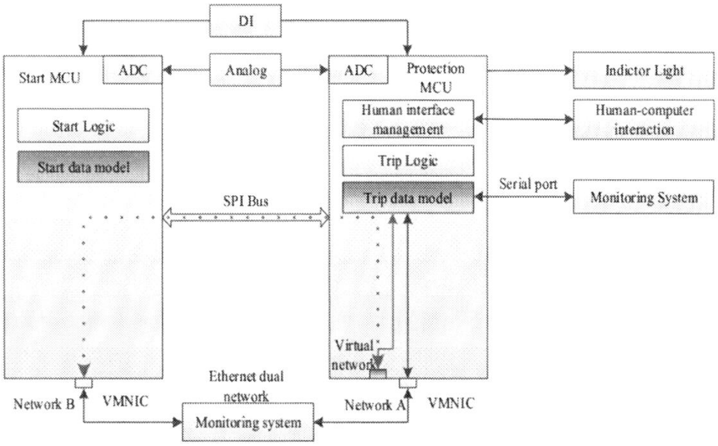

Fig.3 Dual Ethernet communication function

4. Overcoming an impact of integral nonlinear error of ADC

In the ADC static precision indicator, INL (Integral Nonlinear) is the maximum deviation of analog-to-digital transmission curve and the ideal condition of the analog-to-digital curve. INL represents the degree of the actual transfer curve offset the ideal transfer curve, in a percentage or LSB unit [17]. After removing static disorders and gain errors, INL usually represents the following publicity:

$$\Delta INL = \left[(V_D\text{-}V_{zero})\,/\,V_{zero}\text{-}D\right], 0 < D < 2^n - 1$$

Where n is the resolution of the ADC, V_D indicating the analog voltage corresponding to the digital code D, V_{zero} indicating the quantized voltage corresponding to the digital encoding output is 0.

In the proto test and manufacturing stage, the error is large when current channel is below 2A which is not met the accuracy requirements. By testing the ADC module integral nonlinear curve inside the M4 [18], the most severe integral nonlinear area of domestic ADC is 2000.

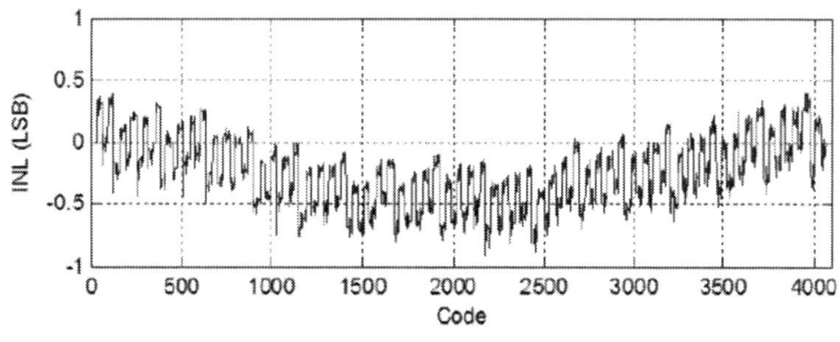

Fig.4 Integral nonlinear characteristic curve of domestic ADC

M4 internal integration ADC is single pole, in order to achieve a bipolar electrical amount A/D transform, the AC input loop needs to increase the bias circuit [19]. Design the bias voltage URef in the front end of the AC input voltage, the maximum range of ADC input voltage is 3V,and the bias voltage set to $U_{Ref} = 1/2U_N = 1.5V$.

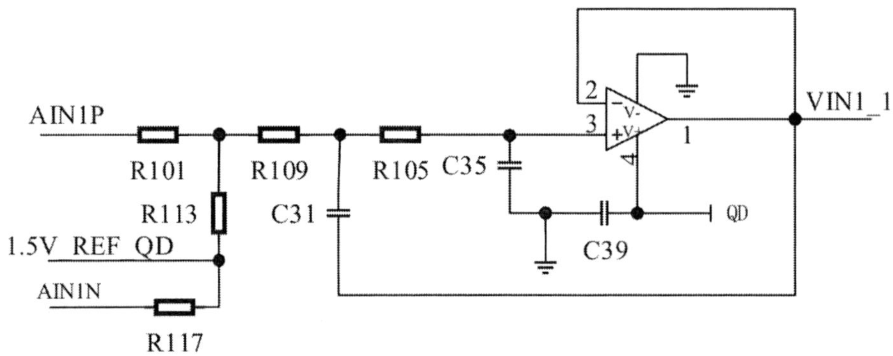

Fig.5 DC bias voltage circuit

Since the full scale original sample value of 3V is 4096, applying DC bias voltage of the original sample value of 1.5V is 2048, and the original sample value is floating up and down at 2048 when the analog magnitude is small. According to the integral nonlinear characteristic curve of the simulated ADC module, as shown in Fig. 4, its integral non-linearity is the worst in the intermediate area, that is, when the ADC reference range of 0 ~ 3V is 1.5V, the analog acquisition error is the largest.

In this paper, in an analog acquisition design, it is mainly to escape the derived area with large non-linear points by adjusting the reference voltage. The original 1.5V_REF_BH / QD raised voltage is formed by the 3V reference voltage through the high-precision resistance division of the two resistance value of 10 kΩ. Adjusting the voltage resistance resistance to 11kΩ and 10kΩ and the raised voltage is adjusted to 1.429V (3V / 21KΩ * 10kΩ), in order to avoid the worst nonlinear area of domestic ADC points, and improve the collection accuracy of domestic ADC.

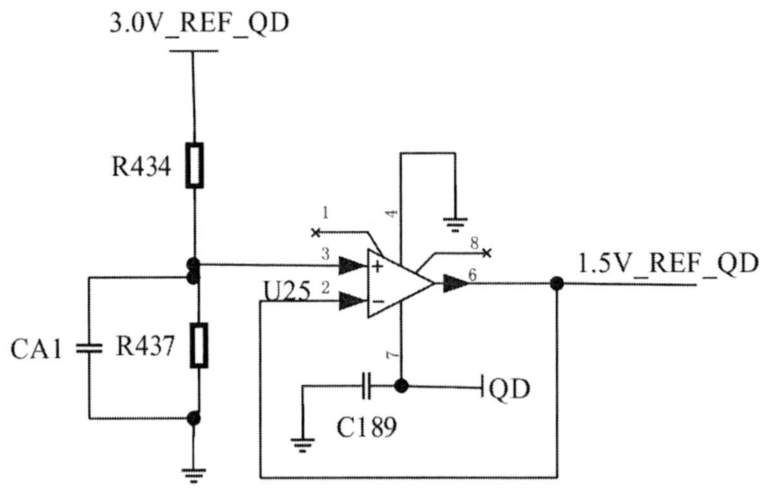

Fig.6 Bias voltage adjustment circuit

Take protection current A as an example, before and after the raising voltage modification, the analog acquisition value is contrasted as shown in Table 3. According to the test results, adjust the raised voltage by avoiding the nonlinear area of the ADC points, improves the analog sampling accuracy of domestic ADC effectively, especially the 0.2A to 2A small magnitude range analog sampling accuracy is significantly improved.

Table 3 Improved analog sampling accuracy comparison

In/A	Before improvement	Absolute error	After improvement	Absolute error
5	5.01	0.20%	5.006	0.12%
4.5	4.517	0.34%	4.499	-0.02%
4	4.023	0.46%	3.999	-0.02%
3.5	3.511	0.22%	3.508	0.16%
3	3.015	0.50%	3.007	0.14%
2.5	2.52	0.40%	2.506	0.12%
2	2.031	0.62%	1.998	-0.04%
1.5	1.548	0.96%	1.497	-0.06%
1	1.086	1.72%	0.994	-0.12%
0.5	0.532	0.64%	0.491	-0.18%
0.2	0.234	0.68%	0.197	-0.06%

5. Device electromagnetic compatibility capability improvement research

5.1. Electromagnetic compatibility improvement of the DO loop

Complex electromagnetic environments provide high requirements for electromagnetic compatibility performance based on MCU domestic chip relay protection hardware platform. For the domestic low-voltage relay protection hardware platform designs, components that need to consider electromagnetic compatibility performance include optical coupling, digital isolation, RS232, RS485, 245, 4245, LDO, DC/DC, TVS tube, connector, etc. [20] In selection and testing, the isolation voltage, breakdown voltage, electrostatic isolation level, etc. should be focused to ensure that domestic components meet electromagnetic compatibility requirements.

The outlet circuit which is controlled by protection MCU outputs the control signal through the expanded bus, after the Dalinkton optocoupler, the outlet relay is driven. In a fast transient immunity test, under the condition of no outlet, the optical coupler breakdown due to the fast transient pulse interference, the device sends false report of outlet loop breakdown. Co-capacitance between the collector and the emitter and filtering the electrical transient pulse interference to enhance the stability of the outlet driving optocoupler. [21] It is verfired that all related circuits of the device perform IV stage electricity transient pulse anti-disturbance test results meet the requirements of the test standards.

5.2. EMC improvement of ADC acquisition loop

In analog acquisition function module, the reference voltage (3.0V_Ref_BH / QD) and the lift voltage (1.5V_REF_BH / QD) determine the accuracy of analog sampling and dual ADC self-test. Since the large area of the analog acquisition module and the use of the domestic MCU, the long rise of the reference voltage is extremely susceptible to electromagnetic interference. In the actual electrical fast transient pulse immunity test and surge immunity test, the reference voltage and the lift voltage would occur level fluctuations due to the electromagnetic pulse interference.

Add the pair of capacitance on the reference voltage output and the raised voltage output and each analog sampling loop to enhance the anti-interference ability of the reference voltage source and the lifting voltage[22]. The actual test verifies that the ground capacitance has a good filtering and stabilizing effect, and the results of the IV level electrical fast transient impulse immunity test and IV surge immunity test of all related circuits of the device meet the requirements of the test standards.

6. Test verification

The low-voltage line protection measurement and control device based on domestic MCU developed in this paper has passed the type test of the National Electrical Instrument Quality Supervision and Inspection Center (Report No.: WJ19570), and the device has passed 42 testing indexes. The device has been put into trial operation on the 10kV Chahua I line of 220kV Chapeng Substation in Anyang, Henan. So far, the trial operation of the device is in good condition.

7. Conclusion

The line protection measurement and control device based on domestic MCU developed by this paper takes into account the terminal compatibility and functional compatibility with existing mainstream supply devices, and achieves rapid replacement with existing devices. At the same time, the device performance indicators are equivalent to existing devices. The independent level of the relay protection industry has effectively guaranteed the safe and stable operation of state grid.

With the new opportunities for the development of the domestic chip industry, the field of power grid relay protection has gradually begun to use domestic chips. On the one hand, it is conducive to create a good domestic chip ecology, which really promote the maturity of domestic chips. On the other hand, by gradually increasing the engineering application of fully autonomous controllable devices, the long-term operating data is monitored in real-time. The long-term operating data is verifying the stability and reliability of its long-term operation, which is providing a strong support for the gradual replacement of autonomous controllable devices.

Acknowledgments

This work was financially supported by science and technology project of State Grid Corporation of China (5500-202046258A-0-0-00).

References

[1] HU Zhiqiang. Review on domestic production based on the self-controlled technology [J]. Cyberspace Security, 2018, 8(9): 90-97.
[2] WANG Nong,Lei Haijun,Yuan Yuan, et al. Accuracy Analysis and Design of Multi--Channel Analog Acquisition Systems Constructed by Domestic Key Components[J].ComputerMeasurement&Control,2020,28(10):236-240.
[3] GUO Changrong. Research on reliability environment testing technology of domestic processor based on coordination of hardware and software ［D］. Guangzhou： South China University of Technology， 2018.1-69.
[4] YUE Feng,SHI Zhiwei,Dong Jincai,MOU Jinli. Reliability design and test for relay protection devices applied in smart grid ［J］.Huadian Technology,2020,42(02):50-57.
[5] GUO Wenhui. Design and implementation for the hardware system of high performance communication main control board [D]. University of Chinese Academy of Science, 2020.1-91.
[6] YE Bei,HE Qiang. Development of Current Signal Output Device With Homemade MCU [J]. JOURNAL OF SHANGHAI SHIO AND SHIPPING RESEARCH INSTITUTE, 2020,43(02):15-20.
[7] XIE Li, ZHOU Hualiang, YU Tongwei, et al. Research and development on the ultra-high-speed transmission line protection based on the concept of the equal transfer[J].Power System Protection and Control, 2019,47(20):178-187.
[8] ZHOU Hao, SHI Lei, PENG Tao, LI Songze, et al. Analysis and countermeasures of single event upset soft errors in a relay protection device[J]. Power System Protection and Control, 2021, 49(7): 144-149.
[9] DING Ming,CHENG Qing,LI Lin,et al. Research and Development of a Local Module Applied to the Main Equipment of the Third Generation Intelligent Substation [J]. High Voltage Apparatus, 2020,56(12):226-234.
[10] GUO Shuhong.Design and Research of Digital Decimation Filter in Σ-ΔA/D Converter [D]. Harbin Engineering University,2020.1-75.
[11] HUANG Sheng-Wen,Li-Jen Chang,Soon-Jyh Chang. A 14-BIT LOW POWER 2-MS/S SAR ADC WITH RESIDUE OVERSAMPLING[J]. 2020,27(3):
[12] CHEN Xing,Lin Jianting,Mao Yue,Han Dong. Design of a high-precision ADC for the application of monitoring electric power in power grid[J].Dianzi Jishu Yingyong,2018,44(12):

[13] SUN Haiyong. Research on Formal Verification Method of Embedded Operating System [D]. UNIVERSITY OF ELECTRONIC SCIENCEAND TECHNOLOGY OF CHINA, 2020.1-98.
[14] TAN Weibin,ZHANG Wei,QUAN Li,et al. Development of a new on-line monitoring device for distribution network lines[J]. Power System Protection and Control, 2019,47(01):158-165.
[15] GUO Yong， WANG Fuhe， PAN Chaoqiong,et al. Design of the Intelligent Monitor Terminal for the Wind PowerHeating Station Based on ARM [J].Power System and Clean Energy,2020,36(03):82-88.
[16] WANG Jun, Xu Nan, Hui Liang. The Implementation of SPI Interface Communication Based on DSP+MCU [J]. INFORMATION&COMMUNICATIONS,2020,(1):35-36.
[17] LIU Jie,WANG Xuan,GONG Ke,et al.Sub-pocosecond level clock jitter measurement technique based on ADC noise distribution[J].MICROELECTRONICS＆COMPUTER, 2020,37(03):71-75+82.
[18] PENG Wei,ZHANG Jun,CHENG Jie, et al.A Novel Nonlinearity Testing Method of ADC[J].Microeletronics, 2019,49(03):441-446.
[19] JIAO Zhihao,Zhang Ruizhi,Jin Kai, et al. A new network redundancy device for smart substation and its implementation[J].JOURNAL OF XI'AN JIAOTONG UNIVERSITY,2020,54(05):109-116.
[20] LI Youjun， ZHOU Hualiang， ZHENG Yuping, et al. Error-tolerant Design and Application of Relay Protection Device Against Unexpected Memory Bit Change[J]. Automation of Electric Power Systems, 2019,19(04):1-7.
[21] XU Ji-yang,WEN Hao. EMC design of engine ECU input signal ［J］.Electronic Design Engineering, 2019,27(20):143-146.
[22] MENG Lingzhi， WANG Yan， LI Haifeng.Reliability design and test for relay protection devices applied in smart grid ［J］.ELECT R OMECHANICAL COMPONENTS, 2019,39(05):8-13.

Transmission line condition assessment technology based on multi-source Parameter Fusion

Wentao Dai[1], Minzhen Wang[1b*], Jiankai Guo[2], Zheng Li[3], Dongliang Qiao[1], Ziwen Shang[1]

[1] Changchun Institute of Technology, Changchun, 130012, China

[2] State Grid Changchun Electric Power Supply Company, Changchun, 130012, China

[3] State Grid Dandong Electric Power Supply Company, Dandong, 118000, China

[b*]wangminzhen@ccit.edu.cn

Abstract—Aiming at the shortcomings of current transmission line state evaluation, which refers to the isolation of scalars and the low accuracy of evaluation methods, a transmission line state evaluation technology based on multi-source parameter fusion is proposed. This paper collects and selects characteristic data such as line test inspection information, operation and maintenance records, and power grid operating parameters as indicator quantities. The membership function is used to express the distribution state of the index quantity, and the distribution state is used as the data to establish a comprehensive evaluation model of the line operation state based on the long- and short-term memory network. The data tag is used to train the deep network to establish the mapping relationship between index parameters and operation state. Experiments show that the evaluation accuracy of the proposed method is significantly higher than that of the traditional shallow method, which can provide theoretical guidance for the operation and maintenance of overhead transmission lines.

1. Introduction

Transmission line is an important part of power system. Correctly judging the operation state of transmission line is of great significance to improve the security of power system. Transmission line condition assessment technology is a hot issue in current research[1]. At present, there are many theoretical methods of transmission line condition assessment, but there is no unified standard, and there is a subjective influence in determining the weight of each index. To solve this problem, this paper proposes a transmission line state evaluation method based on multi-source parameter fusion evaluation method, establishes the transmission line state evaluation index system, uses the 3-scale multi-source parameter fusion method to determine the weight of each index, and then obtains the comprehensive evaluation result of transmission line state through fuzzy synthesis[2]. The idea and model of deep learning are introduced into the condition evaluation of overhead transmission lines in the main network, and a condition evaluation system of 110 and 220 kV overhead transmission lines based on multi-source parameter fusion is proposed. In this study, the multi-source data is selected as the typical index quantity, the membership function is used as the data marker, and the multi-source parameter fusion model is input as the training sample[3]. The correctness and effectiveness of the system are verified by an example. The system can output operation state level with high accuracy, and can guide manual condition based maintenance and repair.

2. Multi-source parameter fusion evaluation of transmission line state

2.1. Transmission line status multi-source parameter feature acquisition

Import the collected multi-source data into the processing tool, first check the field interpretation, data source, code table and other description information of each data, and then preprocess the data. The main process is as follows: data cleaning mainly includes the scrubbing of default values, format content and logic errors[5]. Aiming at the problems of independent evaluation indexes, various types and complex evaluation system structure of overhead transmission lines, a layered idea is proposed, and a three-layer evaluation system is constructed, as shown in the figure 1.

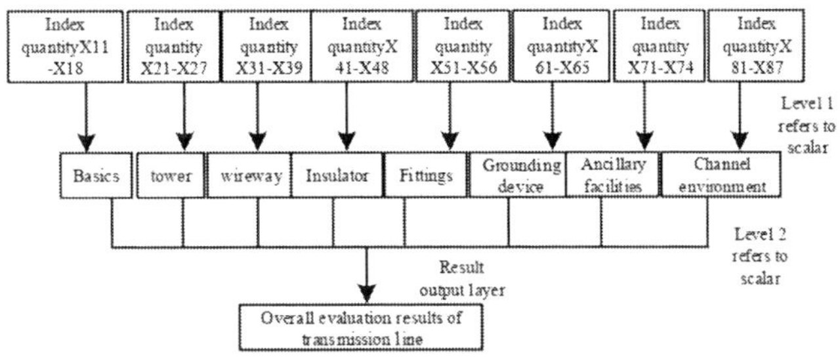

Fig. 1 Comprehensive evaluation model of transmission line

In the figure, the first layer refers to a scalar, which includes the state quantity of the device unit; The second layer is composed of eight 2-level scalars composed of equipment units; The third layer is the output layer, which uses deep learning algorithm for data mining and data fusion of level 1 and 2 indicators. The key evaluation indexes are extracted by correlation analysis and principal component analysis. The membership function establishes the parameters contained in the function distribution according to expert experience, and represents the output results in the form of set. This paper adopts the function form of the combination of triangle and semi trapezoid, as shown in the figure 2.

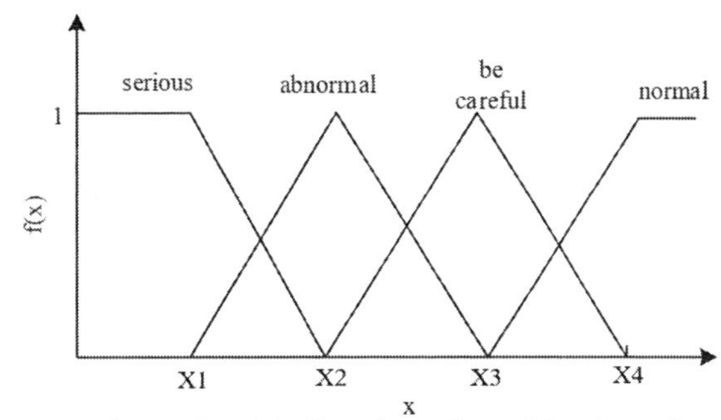

Fig. 2 Distribution diagram of membership function of combination of semi trapezoid and triangle

At the same time, the membership functions of four evaluation levels are given.

$$f_1(x)=\begin{cases}1, & x_{,,}\ x_1 \\ \dfrac{x_2-x}{x_2-x_1}, & x_{1,,}\ x<x_2 \\ 0, & x..x_2\end{cases} \quad (1) \qquad f_2(x)=\begin{cases}\dfrac{x-x_1}{x_2-x_1}, & x_{1,,}\ x<x_2 \\ \dfrac{x_3-x}{x_3-x_2}, & x_{2,,}\ x<x_3 \\ 0, & x..x_3 \text{ or } x<x_1\end{cases} \quad (2)$$

Where: $F_i(x)$ is the membership value corresponding to different Lj $(j = 1,2,3,4)$ evaluation grades; $X1 \sim X3$ are the boundary values of membership functions corresponding to different operating states. In most cases, its value is general, which is taken in this paper $x_1 = 10, x_2 = 40, x_3 = 70$. Through the in-depth data mining of multi-source parameter fusion, the transmission line state reliability interval $[P(L_1), P(L_2), P(L_3), P(L_4)]$ can be output. The result is a fuzzy set. In order to obtain more accurate state evaluation results, the percentile weighting method is used to score the operation state of the transmission line, and the formula is the calculation formula of the operation state grade of the transmission line.

$$H=\sum_{j=1}^{1} m_j p(l_j) \quad (3) \qquad m_j=\begin{cases}10, & j=1\\40, & j=2\\70, & j=3\\90, & j=4\end{cases} \quad (4)$$

Set the data vector set as $X=\{\mathbf{X}_i, i=1,2,\ldots,n\}$ the number of data points is n, the sample dimension is D, and the multi-source parameter is n. x is divided into n disjoint classes $C_1, C_2, \ldots, C_{N_c}$ various multi-source parameter centers are $\overline{\mathbf{M}}=\{\overline{\mathbf{m}}_r, r=1,2,\ldots,N_c\}$. The scoring values corresponding to the status level are shown in the table 1 below.

Table 1 Scoring values corresponding to status level

Status level	Score value	State description
serious	0≤H<20	Some important values exceed the warning value, and power outage maintenance is required
abnormal	20≤H<60	Some important values are close to the warning value, and maintenance shall be arranged first
be careful	60≤H<80	Some important quantities are missing close to the warning value, and the operation needs to be monitored
ordinary	80≤H<100	Each state quantity is within the normal value, and the transmission line operates normally

In the lightning risk assessment of overhead transmission lines, due to the different influence of each characteristic factor, the weight must be reasonably allocated according to the importance of each factor. When the comprehensive evaluation result of transmission line state is determined, when any level of index is good, general, abnormal or serious state, the comprehensive evaluation result of transmission line state shall be the most serious state.

2.2. Evaluation algorithm of transmission line state defects based on multi-source Parameter Fusion

When evaluating the units of the line, it is necessary to study not only the deduction of individual state quantity, but also the total deduction of components. The condition evaluation criteria for components are shown in the table. Under the condition of normal state, the score reduction of each state quantity unit and the total score reduction of components must meet the standards of the table 5. When only one of them meets the criteria in Table 2, it is specified as the state of attention; When the score reduction of each state quantity unit reaches the standard in Table 2, it is specified as abnormal or serious state.

Table 2 Line unit evaluation criteria

Line unit \ state	normal		be careful		abnormal	serious
Basics	total	individual	total	individual	individual	individual
tower	<15	≤10	≥15	12~25	30~33	45
Conductor and ground wire	/	≤10	/	12~25	30~33	45
Insulator string	<17	≤10	≥17	12~25	30~33	45
Fittings	<15	≤10	≥15	12~25	30~33	45
Grounding device	<25	≤10	≥25	12~25	30~33	45
Ancillary facilities	/	≤10	/	12~25	30~33	45

Channel environment	<25	≤10	≥25	12~25	30~33	45
Basics	/	≤10	/	12~25	30~33	45

The evaluation objects include things, strategies, methods, etc. The important feature of extenics evaluation method is that it can not only obtain the level degree of the evaluated object's own state in quantity, but also reflect the dynamic process of the evaluated object from quantitative change to qualitative change to a certain extent. The expression of classical domain is as follows:

$$M_{0j} = \left(O_{0j},\ C, V_{0j}\right) = \begin{bmatrix} O_{0j} & c_1 & v_{0j1} \\ & c_2 & v_{0j2} \\ & \cdots & \cdots \\ & c_n & v_{0jn} \end{bmatrix} = \begin{bmatrix} O_{0j} & c_1 & \langle a_{0j1}, b_{0j1} \rangle \\ & c_2 & \langle a_{0j2}, b_{0j2} \rangle \\ & \cdots & \cdots \\ & c_n & \langle a_{0jn}, b_{0jn} \rangle \end{bmatrix} \quad (5)$$

Where, O_{0j} represents j levels divided: $c_n(n = 1, 2, \ldots n)$ represents the characteristics of state level O_{0j}; v_{0jn} (j = 1,2,... N) is the category of quantity value formulated by c_n for feature c_n, that is, the category of data obtained by each state level for the corresponding characteristic The result of condition based maintenance evaluation of transmission line is more in line with the actual requirements.

2.3. Implementation of transmission line condition assessment

Short term condition evaluation of overhead transmission line, that is, real-time condition evaluation. According to different meteorological conditions, it can be divided into non disaster weather, disaster weather and extreme disaster weather. Transmission line status level can directly or indirectly represent various information of transmission line status. According to the degree of deterioration, it is divided into grade I, grade II, grade III and grade IV. Based on the characteristics of transmission line condition evaluation and combined with the suggestions of relevant experts of transmission line, the transmission line condition equivalence is divided into "normal", "attention", "abnormal" and "serious", and the specific meanings of deterioration I and II are explained respectively.

Table 3 Meaning of transmission line status type

Status type	Status score	Deterioration level	meaning
normal	70~100	I	The value is within the specified range, and the line operates normally
be careful	50~75	II	If the value is within the specified range and tends to exceed the range, it shall be repaired according to the original plan
abnormal	25~50	III	A small part of the state quantity slightly exceeds the specified range, so it is necessary to monitor it and arrange maintenance
serious	0~25	IV	Some important state quantities exceed the specified range and may deteriorate, so maintenance shall be arranged as soon as possible

The multi-source parameter fusion method mainly studies complex and uncertain systems or chaotic laws, takes data-driven as the core, collects data for specific problems or businesses, and realizes the effective fusion of multi-source data, so as to construct the mathematical analysis and mining model of data and obtain the problem interpretable results. Big data research does not conflict with the traditional theoretical system, and it is also an important part of the scientific research system.

3. Analysis of experimental results

Analysis of experimental results in order to verify the effectiveness of this method, 2000 transmission line cases confirmed to have abnormal defects / faults recorded by the power grid production management system are selected as the training set, and 500 transmission line cases tracked and observed after early warning are selected as the test set to jointly form the sample database of this paper. The learning rate is set to 0.03 and the number of iterations is set to 100. Programming language: Python, programming platform: pycharn, computer CPU: 23ghz, Intel Core i5, memory: 16GB, operating system: osx10.13.6. The index for evaluating the performance of the model is accuracy, which is defined as the number of samples n whose predicted operation state level is consistent with the actual operation state level. The ratio to the total number of samples n is shown in the formula.

$$1 \quad A=\frac{N_c}{N}\times 100\% \qquad (6)$$

Take the current and historical transmission line test and patrol information, operation and maintenance records, and the membership of power grid operation parameters as inputs to predict and evaluate the transmission line status after one week. Further, the Tomek links method is used to randomly select 70% of the samples from the fused data as the training set and the rest as the test set data. The "average temperature", "precipitation", "relative humidity", "average air pressure", "average wind speed" and "lightning recording times" are selected as the input characteristics. The real-time fault status (status) of a certain type of equipment in the same meteorological area "1" indicates that the equipment fails, and "0" indicates that the equipment fails to report) is used as the output to establish a random forest model. The results are as follows:

Table 4 Evaluation of fault state prediction results

Estimate/ True value	fault	Non fault
fault	209	25
Non fault	5	1168

The figure 3 shows the variation of training set accuracy of three models (LSTM / SVM / BPNN) with the number of iterations.

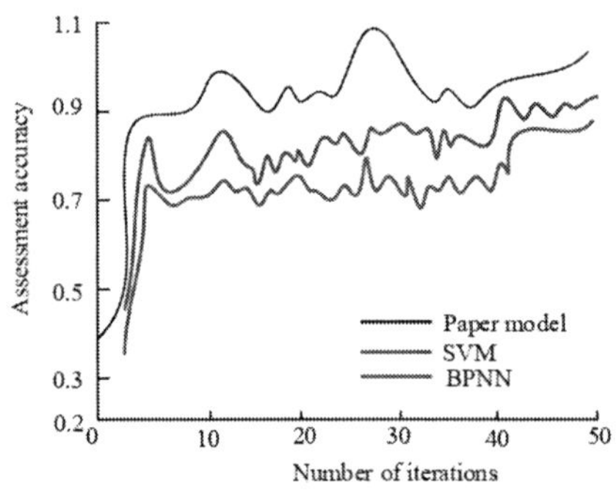

Fig. 3 Change trend of state evaluation accuracy

The figure 4 shows the accuracy of the three models (LSTM / SVM / BPNN) on the training set and the test set.

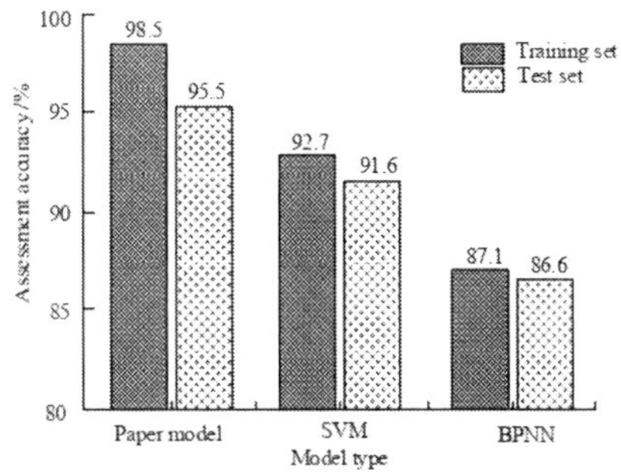

Fig. 4 Evaluation results of transmission line operation status

It can be seen from the figure that the data sequence of cross entropy loss function is as follows: paper model < SVM < BPNN, that is, the difference between the transmission line state predicted by paper model and the actual state is the smallest, so the weights w and B of paper model reach the optimal value. The evaluation accuracy of the training set is sorted as follows: paper model > SVM > BPNN. The accuracy increases with the number of iterations, and gradually reaches a stable state. The order of evaluation accuracy of training set and test set is: paper model > SVM > BPNN. In terms of the accuracy of the training set, the paper model improves the accuracy of SVM and BPNN by 6.2% and 8.7% respectively, while the paper model improves the accuracy of SVM and BPNN by 3.9% and 8.9% respectively.

4. Conclusion

The established transmission line condition evaluation model integrating multi-source data such as line test and patrol information, operation and maintenance records and power grid operation parameters is conducive to formulating timely and effective manual patrol strategy. Through the example analysis and result comparison, the evaluation accuracy based on multi-source parameter fusion network is higher, which is higher than other common evaluation methods, and the gap between the evaluated state reliability is more obvious. In the next step, the architecture improvement and parameter optimization strategy of multi-source parameter fusion model can be studied to further improve the evaluation accuracy and realize the overall state evaluation of power system in a wide range.

References

[1] H Zhang, X Li, L Zhang, W Wang, B Shi. 2020, Discovering Station Patterns of Urban Transit Network with Multisource Data: Empirical Evidence in Jinan, China. KSCE Journal of Civil Engineering, 25(2):1-12.
[2] S Fang, X Pan, S Xiang, C Pan. 2020, Meta-MSNet: Meta-Learning based Multi-Source Data Fusion for Traffic Flow Prediction. IEEE Signal Processing Letters, PP(99):1-1.
[3] J Li, X Jia, X Lv, Z Han, J Liu, J Hao. 2019, Opportunistic routing with data fusion for multi-source wireless sensor networks. Wireless Networks, 25(6):3103-3113.
[4] Z Zhang, K Fu, X Sun, W Ren. 2019, Multiple Target Tracking Based on Multiple Hypotheses Tracking and Modified Ensemble Kalman Filter in Multi-Sensor Fusion. Sensors, 19(14):3118-3119.
[5] ED Hoegg, S Godin, J Szpunar, R Lobinski, DW Koppenaal, RK Marcus. 2019, Coupling of an atmospheric pressure microplasma ionization source with an Orbitrap Fusion Lumos Tribrid 1M mass analyzer for ultra-high resolution isotopic analysis of uranium. Journal of Analytical Atomic Spectrometry, 34(7):1387-1395.
[6] E Anastasiou, A Castrignano, K Arvanitis, S Fountas. 2019, A multi-source data fusion approach to assess spatial-temporal variability and delineate homogeneous zones: A use case in a table grape vineyard in Greece. Science of The Total Environment, 684(SEP.20):155-163.

Journal of Physics: Conference Series **2108** (2021) 012019 doi:10.1088/1742-6596/2108/1/012019

Approach for Provincial distribution automation system in Outage analysis

Zhihao Zhang [1,2a*], Yulin Qian[1,2b*], Diandian Zhou[1,2c], Kexin Zhang[1,2d]

[1] NARI GROUP CORPORATION/STATE GRID ELECTRIC POWER RESEARCH INSTITUTE, NanJing 211102, China,

[2] NARI TECHNOLOGY CO., LTD., NanJing 211102, China

[a*] 18913322736@163.com, [b*]qianyulin@sgepri.sgcc.com.cn, [c]zhoudiandian@sgepri.sgcc.com.cn, [d]zhangkexin@sgepri.sgcc.com.cn

Abstract—People's lives are becoming more and more inseparable from electricity, and the requirements for the quality of power supply are getting higher and higher. Because of most of the faults in the power system originate from the distribution network, the need to standardize the management of power outages and improve the efficiency of research and judgment has become more and more significant. However, in China, the differences in the development of automation construction in various places and the outage information of different departments require that power outage business applications should try to achieve accurate analysis of outage information under different scenarios. This paper designs a power outage analysis approach, which comprehensively analyzes distribution network data such as substations, lines, and meters, and establishes a signal credibility image, which flexibly mines and analyzes outage events from multiple dimensions and effectively improves the fault tolerance of fault location.

1. Introduction

The distribution network is the final link of power transmission in the power system which directly affects the power quality of users. Because there are many barriers between sharing data and integration of business systems, there is still a lot of room for improvement in the efficiency of fault repairs and the implementation quality of planned power outages in the related business of power outages in the distribution network. Up to now, the coverage of distribution network terminal equipment in most areas is limited, and many lines have not yet been automated in China. Even in the same area, automated construction will also receive differential investment. These caused a lot of difficulties and problems such as spending a lot of time on site surveys, repeating maintenance of information in different systems and difficulty in aggregating information.

In the field of distribution network fault location technology, most researches focus on fault location methods based on terminal signals such as monitoring Feeder Terminal Unit (FTU) and Distribution Terminal Unit (DTU). The common methods are artificial intelligence algorithms such as genetic algorithm and ant colony algorithm. Due to the complexity of the distribution network that it is not only necessary to realize the discrimination of wrong information, but also to make probability inferences on the basis of incomplete information, the algorithm selection needs to be fully considered[1-3]. DS evidence theory is very effective for probabilistic reasoning. Compared with traditional probability theory methods, it can better grasp the uncertainty of the problem by fusing evidence from multiple sources of evidence, and has achieved good results in many current fields. Compared with traditional power outage analysis applications, the power outage analysis approach in this paper adopts a hierarchical and hierarchical research and judgment mechanism, and the configuration of power outage sensing conditions is more flexible. On this basis, an improved strategy is proposed, which combines signal credibility profile and d-s theory algorithm Perform fusion judgments, while studying and judging logic to take into account multiple types of scene conditions, it also solves the problem of inaccurate fault location when the signal is abnormal or conflicts[4].

2. Overall Design

Even in more developed areas in China, the construction of distribution automation is still in the large-scale pilot stage. The core point is to better complete fault repairs under existing conditions, and to transform the weak points of the grid to achieve Reasonable deployment of resources[5-8]. This requires that the design of power outage analysis applications not only ensure the timely accuracy of real-time fault perception to improve the efficiency of emergency repairs, but also need to rely on data analysis and verification to improve the reasonable standardization of power outage management from the side, and to explore and analyze weak links to support accurate investment in the project.

Provincial power grid companies already have relatively stable business systems in the fields of dispatching, operation inspection and marketing. The main problem is that the same type of data comes from different systems. Too many repetitive and conflicting invalid data increase data processing risks and screening costs. On the other hand, part of the data is only used at a single professional level that the data sharing is insufficient. The provincial deployment architecture of the distribution automation master station application has advantages in terms of construction costs and data fusion and other sharing aspects. It is easier to achieve standardized management of data aggregation, and it is more convenient to carry out benchmarking in the same industry.

The information of the distribution network can be divided into static information and dynamic information. The static information mainly includes the topology information of the distribution network line and the relationship between primary and secondary equipment. The dynamic information covers the distribution network operation data, including the substation signal, the line signal and the courts signal.

3. Hierarchical Grading Judgment

The fault location of the distribution network and analysis of the suspected fault zone are the most important part of the power outage management function. Power supply companies with a high level of automation will install a large number of intelligent detection terminals, which can obtain a large number singals of teleindication and telemetering. Even if some signals are mishandled or missed, the automation master station can also combine other signals and fault record information, or use DTU, FTU to send signals for offline fault analysis. Almost all fault location methods that rely only on a certain type of distribution network line signal have their limitations, resulting in a greater possibility of misjudgment, and it is not easy to analyze the nature and scope of the outage. In this case, if the fusion analysis of the signals in the EMS station or the distribution transformer signals is combined, not only the research and judgment rules can be enriched, but the reliability of the analysis results can be ensured to a large extent. This chapter mainly introduces the resource integration methods between the various systems of the distribution network, and elaborates key technologies such as main distribution network model splicing and fault research and judgment rules[9-12].

3.1. Rules of Hierarchical Grading Judgment

The fault signal source includes multi-level alarm information of substation-line-transformer-user. The primary problem of the entire fault analysis process is how to effectively and accurately analyze the nature of the power outage based on the multi-source fault alarm information, locate the fault zone, and select the most suitable topological search starting point.

This paper divides power outage events into deterministic and non-deterministic events, and analyzes and judges power outage events by hierarchical priority. Deterministic power outage events refer to faulty power outages, planned power outages, temporary power outages, etc., and deterministic events are placed at the top of the research and judgment logic. This logic is not actively associated with the results of the underlying research and judgment, but is only positioned for passive association. Non-deterministic power outage events include the power outage events associated with the real-time power outage signal in the courts and the work order reported by the user. The system can quickly match according to incomplete information, automatically find the most relevant power outage event, suspected event from the power outage event pool, and determine whether the signal is a related sub-signal in a certain deterministic event. The signal priority is designed such that the signal alarm information in the substation has priority over the line, and the line alarm information has priority over the courts. The distribution transformer signal with higher reliability at the station area level has priority over the user-level fault information. For example, when the system can complete the fault diagnosis and judgment by combining the substation-level and feeder-level alarm information, there is no need to consider the accuracy of the court-level information. When the qualitative event is completed, the theoretical station area power loss signal set and the actual station power loss signal set are checked to support the courts signal remediation and credibility portrait modeling. The processing logic of hierarchical research and judgment runs relatively independently. According to the level and priority, the lower-layer research and judgment analysis needs to consider the upper-layer research and judgment results, and when the signal is relatively complete, the upper-layer research and judgment does not need to consider the lower-layer signals or the judgment results. Cooperate with each level to check, and analyze the nature of the power outage and the fault interval according to the priority of the criterion.

3.2. Process Of Power Outage Research And Judgment

When the graphic model data is complete and reliable, the fault location uses the fault time point alarm information uploaded from information sources at all levels to confirm the approximate location of the device and the starting point of the topology search, and then complete the device location based on the topology analysis. This method usually requires the combination of better quality graphics and model data. If the data signal quality is guaranteed only, the research and judgment logic can be simplified to first query whether there are related signals at the level according to the discrete power failure signal, and then group the signals according to the feeder and branch lines, and first determine whether the entire line is out of power or the branch line is out of power, then analyze the most likely trip position, and then locate the fault area.

When the signal is missing in the substation or the feeder signal is missing, or the line has not yet realized the automatic terminal coverage, it will directly lead to the failure to be associated with the deterministic power outage event, and the complete logic power outage event research and judgment cannot be completed. At this time, the most likely trip position can only be inferred through the court-level signal, combined with the topological analysis of the static correlation relationship between the branch line and the distribution transformer. In some extreme cases, it is necessary to manually verify the authenticity of non-deterministic events and the nature of power outages. If a branch line fails and there is no circuit breaker upstream, it will cause the entire line to fail. This event will be classified as a power failure of the entire line. In addition, in order to ensure that the accuracy of the station-level signal reverse analysis is not affected by signal misreporting, it is necessary to combine real-time telemetry data to assist in research and judgment[13-19].

Due to the different construction conditions of multiple systems in different regions, there are differences in signal delay and accuracy. The fusion of multi-level data brings more logical directions for research and judgment, and may also have a negative impact on the results of research and judgment. For example, when logic conflicts occur at different levels, additional algorithms are needed to reduce the coupling of signals. As an uncertain reasoning method, d-s evidence theory is suitable for fusing multiple evidences and comprehensively considering multiple evidences to make decisions to improve the reliability of failure analysis.

4. Closed-loop checking method of fault signal based on d-s theory

4.1. Signal quality analysis

In areas with a high degree of distribution network automation, a large number of fault signals can be obtained when the distribution network fails. According to the rules of hierarchical grading judgment described in the previous chapter, the automation system can accurately and quickly locate the fault zone. However, currently in many areas with low automation or rural distribution networks, intelligent monitoring terminals such as DTU and FTU are usually not installed, and their automated construction and operation and maintenance schemes mostly use complete sets of switches as the mainstay, supplemented by fault indicators. The fault indicator will cause signal false alarm or signal loss due to the diversity of the environment and the operation mode of the distribution network. These poor signal quality will cause erroneous analysis of fault research and judgment.

In addition to DMS systems, systems in other fields such as EMS systems, PMS systems, SG186 systems, and electricity consumption information collection systems also provide data related to the distribution network. Each system data has real-time mismatch and accuracy problems. For example, the sampling frequency and real-time performance of the courts data of the electricity consumption information collection system cannot accurately match the real-time performance of the EMS system, and there are also a large number of signal quality problems such as false alarms, false alarms and non-reports in the power outages of the courts. In this case, the traditional fault location method is very likely to be misjudged, causing repair personnel to be unable to accurately locate the fault section, reducing the efficiency of fault repair and the reliability of power supply.

Some of these signal quality problems can be identified through anti-calibration analysis. For example, the signal in the substation has a protection signal, but there is no circuit breaker action signal. At the same time, the DTU/FTU has a switch action signal and a short-circuit accident total signal. Or if the fault indicator has an over-current flop action, it is inferred that the circuit breaker action signal in the station is missed, which is a clearly identifiable signal missed problem. In another part of the situation, both false positives or false negatives may occur. For example, the fault indicator on the branch line has an overcurrent flop action, but there is no power failure signal sent in the downstream courts. It may be the fault indicator malfunctioning, or the distribution transformer signal may be lost. These signal quality problems cause that it is not easy to analyze according to logical rules.

4.2. Evidence Theory Algorithm Based on Probability Secondary Distribution

When the distribution network fails, multi-source fault information can be obtained. If each type of fault alarm information is independent and correct, they can determine the fault area. However, the actual alarm information has a chance of false positives and missed reports, which makes the fault area identified by each type of fault information have the possibility of error. As an uncertain reasoning method, d-s evidence theory is suitable for fusing multiple evidences and comprehensively considering multiple evidences to make decisions to improve the reliability of failure analysis. Its main characteristics can be summarized as follows: it satisfies weaker conditions than Bayesian probability theory, does not need to know the prior probability, can well represent "uncertainty", and is widely used to deal with uncertainties such as information fusion and intelligence cases. In d-s evidence theory, the recognition framework U is a complete set composed of mutually exclusive and limited research objects.m is the basic probability assignment function of U, $m: 2^U \rightarrow [0,1]$, and satisfies:

$$m(\emptyset) = 0 \qquad (1)$$

$$\sum_{A \subseteq U} m(A) = 1 \qquad (2)$$

m(A) is the basic credibility of A, formula (1) indicates that the credibility of the empty set is always 0, and formula (2) indicates that the sum of the credibility of all propositions is 1.

Assuming that n pieces of evidence are obtained from n information sources, the basic probability distribution is $m_1, m_2, m_3, \cdots\cdots, m_n$. If they are independent of each other, then formula (3) can be used to express the synthesis rule of multiple evidence bodies.

$$\mathrm{m}(A) = \begin{cases} 0\,, & A = \emptyset \\ \frac{1}{1-k} \sum_{A_i = A} \prod_{1 \le j \le n} m_j(A_i)\,, & A \ne \emptyset \end{cases} \quad (3)$$

The evidence conflict coefficient is formula (4).

$$k = \sum_{A_i = \emptyset} \prod_{1 \le j \le n} m_j(A_i) \quad (4)$$

k reflects the degree of conflict between the various evidences, between 0 and 1. The larger the k, the higher the degree of conflict.

When there is no conflict between the evidences, after the fusion of evidence theories, the trustworthiness of high-trust targets can be further strengthened. But if the fault information is falsely reported, the evidence generated by the falsely reported signal may conflict with the evidence obtained from other correct information. When the degree of conflict of evidence is large or even completely conflicted, there will be results that are contrary to the actual situation. At this time, the possibility of misjudgment is very high, so further improvement is needed to adapt to the actual operating conditions of the distribution network.

In the description of signal quality mentioned earlier in this article, it is very likely that one piece of evidence will completely negate another piece of evidence in such situations as refusal to move, false positive, and underreport. Therefore, this article adopts a method which pre-processes each branch evidence and then assigns weight values according to the degree of relevance between the evidences, and analyzes and calculate according to the new distribution probability. What needs to be explained here is that the results after the redistribution probability and the results before the allocation will change in value, but the relative probability of each element of the fusion result remains the same.

4.3. Closed Loop Check of Differentiated Failure Analysis

When counting the actual false signal data on site, the analysis shows that signal false alarms and omissions are generally regular and repeatable. Based on the massive historical data of power equipment, including false alarms, false alarms, refusal to move, and other error information, this paper establishes a data warehouse for the credibility of fault signals. Take court signals as an example. At present, most of the on-site station area power outage signals and power recovery signals in China are temporarily forwarded by the use and acquisition system. Its data quality is characterized by low real-time signal accuracy, but its historical data accuracy rate is over 90%. By checking real-time data and historical data, doing probability statistics from dimensions such as time and data defects, and establishing a multi-level mapping of equipment-time-signal type-defect type-event probability, the logical conflicts between multi-source data can be reduced.

After a fault occurs, the system preprocesses the fault signal according to the hierarchical and hierarchical judgment logic. If the priority judgment rule is met, the decision result can be directly given through the hierarchical and hierarchical judgment rule. If the priority judgment rule is not met, the system needs to generate multiple judgment evidences based on the fault information and the network topology to construct a set of suspicious fault areas. Then, the suspected failure area obtained from various types of failure information is combined with equipment credibility data, and a new probability value is calculated according to certain rules to give evidence, and the fusion calculation is carried out through the improved d-s evidence theory, and finally the decision result is obtained. According to the decision result and the static topological relationship of the line, the theoretical signal set can be analyzed. Comparing with the actual signal set, the whole link can realize closed-loop processing, and the processing flow is shown in Fig.1.

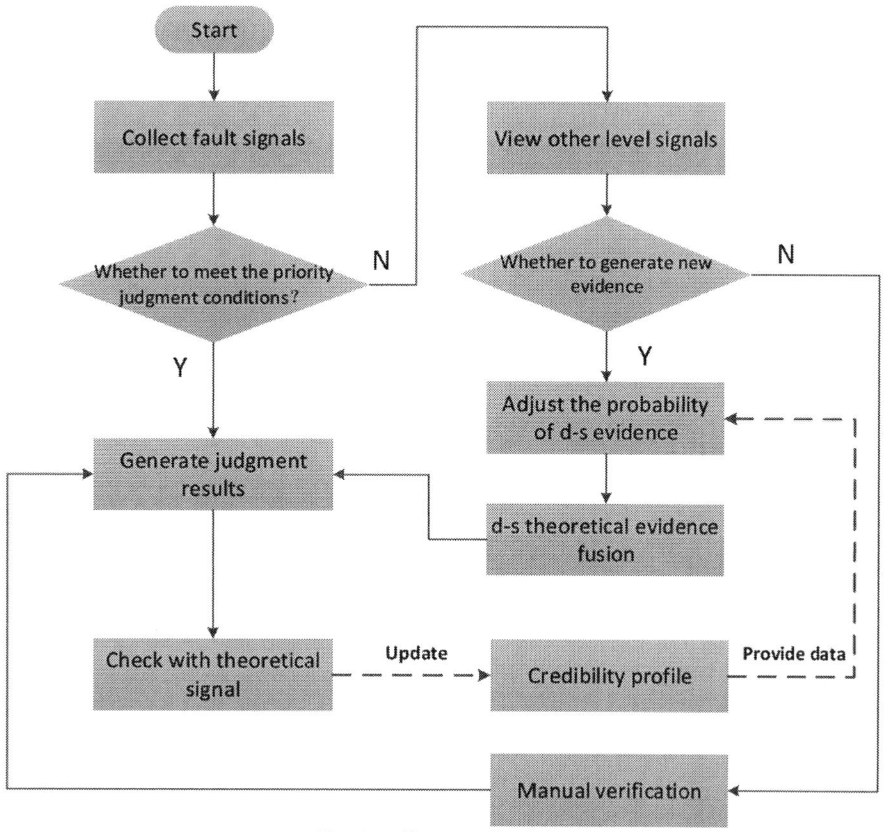

Fig.1 Failure analysis process

5. Case Analysis

The example given in this article is a feeder automation distribution network based on the cooperation of circuit breakers and section switches. This kind of distribution network line is generally relatively simple, mainly divided into a radial network structure and a ring network structure. In addition to the main components of the normal distribution network, in order to facilitate the location of the fault area, a fault indicator with remote signaling function is generally installed at an important location of the feeder. Take a radial network structure line as an example. As shown in Fig.2, CB represents a circuit breaker, F represents a fault indicator, S represents a sectional switch, and T represents a distribution transformer station area.

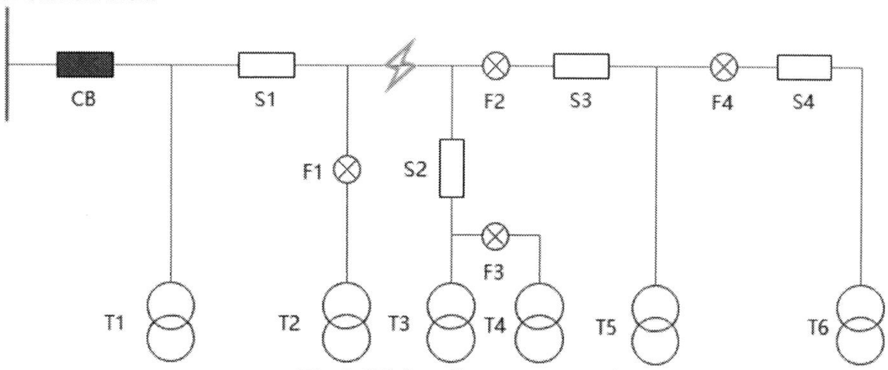

Fig.2 Wiring diagram example

For example, short-circuit fault occurred in the S1-S2-S3 trunk line area. According to the hierarchical classification rules, if the signal is complete, the fault area (S1, S2, S3) can be analyzed by combining the signal in the substation and no teleindications of F2. When a malfunction of F2 happens, the system receives fault-related signals such as circuit breaker CB opening and closing, overcurrent protection signal, F2 overcurrent flop action, and power outage signals in T1, T2, T3, T4, T5 and T6 courts. Afterwards, we successively received the repetition signals from all courts except T4 and T5, and users in the T4 reported failure signals. The fault area inferred from the outgoing circuit breaker and the fault indicator signal is (S3, S4), the inferred range is (S1, S2, S3) based on the power loss signal of T4, and the inferred range is (S3) based on the P5 power loss signal. ,S4), the range inferred based on the user's failure signal is (T4), (S1, S2, S3). According to the traditional fault research and judgment method, when a false alarm occurs in F4, the research and judgment result will be a fault in the downstream area of F4, namely (S3, S4), which will cause the application to capture the fault, but the fault area is not correctly located.

According to historical statistics, the ratio of line faults to courts faults on this line is about 6 to 1, so:

$$\begin{cases} M_1\{(S_3,S_4),(P_4)\} = \left\{\frac{6}{7},\frac{1}{7}\right\} \\ M_2\{(S_1,S_2,S_3)\} = 1 \\ M_3\{(S_3,S_4)\} = 1 \\ M_4\{(S_1,S_2,S_3),(P_4)\} = \left\{\frac{6}{7},\frac{1}{7}\right\} \end{cases} \tag{5}$$

The substation signal is generally very reliable. The false alarm rate of the T5 signal is 70%, the false alarm rate of the T4 signal is 20%, the false operation probability of F1 is 10%, the false operation probability of F2 is 40%, and the accuracy probability of the fault indicator signal is 54%, the T5 signal is accurately sent with a probability of 30%, and the user reporting a failure is a non-deterministic event. The correction factor can refer to the signal accuracy rate of the station area to be 0.8. After the second treatment, the improved evidence is shown in formula 6:

$$\begin{cases} M_1'\{(S_3,S_4),(T_4),\emptyset\} = \{0.46,0.08,0.46\} \\ M_2'\{(S_1,S_2,S_3),\emptyset\} = \{0.8,0.2\} \\ M_3'\{(S_3,S_4),\emptyset\} = \{0.4,0.6\} \\ M_4'\{(S_1,S_2,S_3),(T_4),\emptyset\} = \{0.69,0.11,0.2\} \end{cases} \tag{6}$$

The four pieces of evidence are fused by formula (3). According to formula (4), the calculation result is 0.81, which is substituted into formula 7:

$$\begin{cases} \mathrm{M}\{(S_3,S_4)\} = 0.029 \\ \mathrm{M}\{(S_1,S_2,S_3)\} = 0.937 \\ \mathrm{M}\{(P_4)\} = 0.0012 \\ \mathrm{M}\{\emptyset\} = 0.067 \end{cases} \tag{7}$$

According to the fusion result, it can be judged that in the case of signal false alarm or incomplete terminal coverage, the decision result of the fault area of the evidence theory algorithm combined with the secondary distribution of probability is (S1, S2, S3). The result is consistent with the original setting, and the influence of signal quality on the result is well avoided.

6. Conclusion

This paper closely combines the concerns of various professionals on power outages, and conducts detailed analysis and research on the fault location methods of the distribution network, and realizes a wider monitoring range and more precise description of power outage information under differentiated conditions. The method of constructing a structured power outage information resource pool provides ideas for further accumulating the value of enterprise data and enhancing the effectiveness of data sharing. At the same time, it also provides a foundation for building a data center with open data and shared ecology. With the automatic transformation of medium-voltage lines and the vigorous advancement of TTU, the monitoring range of the system will be extended to low-voltage branch circuits and meter boxes, and the accuracy and completeness of signals at all levels will be qualitatively improved. This will provide a strong data guarantee for the further improvement of the quality of power supply services.

Acknowledgments
We would like to express sincere gratitude for the support provided by Science and Technology Project of State Grid Corporation "Research and Application of Key Technologies of Distribution Network Topology Self-identification Based on Cloud-Edge Collaboration"(5400-202028111A-0-0-00).

References

[1] Rahman Dashti,Mohsen Ghasemi, Mohammad Daisy. (2018) Fault location in power distribution network with presence of distributed generation resources using impedance based method and applying π line model. Energy(), doi:10.1016/j.energy.2018.06.111.
[2] LIU Zhuan, GUO Quanhui, LIU Juan, WANG Chang. (2017) Analysis on distribution network fault repair time base on statistical data[J]. Electric Power, 2017, 50(5):84-87.
[3] Wenbo Li, Jianjun Su, Xin Wang, Qian Ai. (2020) Fault location of distribution networks based on multi-source information. Global Energy Interconnection(1), doi:10.1016/j.gloei.2020.03.005.
[4] Alireza Bahmanyar, Sadegh Jamali. (2017) Fault location in active distribution networks using non-synchronized measurements. International Journal of Electrical Power and Energy Systems(), doi:10.1016/j.ijepes.2017.06.018.
[5] CUI Lizhong, ZHANG Ruixue, LIU Tao, et al. (2017) Reliablity calculation and equipment layout planning for complex distribution automation system[J]. Automation of Electric Power Systems, 2017, 41(20):84-91. DOI: 10.7500/AEPS20170302005.
[6] GUO Ningming, YANG Fei, QIN Jian, CHEN Xiangxun. (2015) Grid fault location method based on genetic algorithm and signal spectrum analysis[J]. Automation of Electric Power Systems, 2016, 40(15):79-85. DOI: 10.7500/AEPS20150907003.
[7] A.Borghetti, S.Corsi, C.A.Nucci, M.Paolone , et al. (2006) On the use of continuous-wavelet transform for fault location in distribution power systems. Electrical Power and Energy Systems, 2006(28):608-617.
[8] K.Ramar, E.E.Ngu. (2010) A new impedance-based fault location method for radial distribution systems[A], Power and Energy Society General Meeting[C],2010IEEE:1-9.
[9] WANG Wei. LIU Chang, LIU Chaofeng. (2018) Study on evaluation system of urban water supply safety based on multi-factor information fusion[J]. Journal of Safety Science and Technology, 2018, 14(11): 180-185.
[10] Georgilakis P S , Hatziargyriou N D. (2015) A review of power distribution planning in the modern power systems era: Models, methods and future research[J]. Electric Power Systems Research, 2015, 121:89-100.
[11] WANG Li, BAI Jing, ZHANG Yiran. (2013) A new method for conflict evidence processing in D-S theory[J]. Journal of Electronic Measurement and Instrumentation , 2013.13(28):8497-8501.
[12] SUN Rui. Research on D-S evidence theory based information fusion and its application in reliability data processing [D]. (2011) Hanzhou: University of Electronic Science and Technology of China, 2011.
[13] WANG Yansong, ZONG Xueying, YI Jingcheng. (2018) Fault-tolerant algorithm for fault location in distribution network[J]. Electric Power Automation Equipment, 2018, 38(4):9-15.
[14] HUAN Jiajia, WANG Gang, LI Haifeng, WANG Longjun. (2013) A comprehensive outage management model based on multi-objective decision analysis[J]. Power System Technology, 2013, 37(10):2802-2807.
[15] ZHENG Wenjie, HUANG Jiajian, SUN Chuan, WANG Longjun. (2016) Reliablity evaluation based on outage character classification for multi-level power grid[J]. Electric Power

Automation Equipment, 2016, 36(3):129-135.
[16] HAN Wei. (2018) Research and implementation of DS evidence theory in mobile medical platform[D]. Beijing: Beijing University of Posts and Telecommunications, 2018.
[17] WANG Yongwei, LIU Yunan, YANG Yingjie, et al. (2013) New approach for conflict process in D-S evidence theory[J]. Computer Engineering and Design, 2013.34(12):4316-4320.
[18] WANG Qiujie, JIN Tao, LIU Jun. (2019) Hierarchical analytic model for fault location of distribution network considering FTU missing and false positives[J]. Electric Power Automation Equipment, 2019, 39(1):147-153.
[19] ZHU Yuanpei, XING Ningzhe, Ji Yutong, et al. Fault location algorithm of integrated data network for power system based on interactive detection[J]. Automation of Electric Power Systems, 2017, 41(4):150-155. DOI: 10.7500/AEPS20160614002.

Design and implementation of active frequency control algorithm for virtual power plant based on flexible resource rapid response

Bin Li[1a*], **Junjie Yu**[1b*],

[1,2] Zhongshan Power Supply Bureau of Guangdong Power Grid Co., Ltd., Zhongshan, Guangdong 528400, China

[a*]libin@gdzs.csg.cn, [b*]yujunjie@gdzs.csg.cn

Abstract: At present, the power generation resources in the power system show a clean and decentralized development trend. These distributed generation resources are small in scale and lack of special regulation system. Aiming at the uncertain and intermittent output characteristics of photovoltaic distributed generation virtual power plant(VPP), this paper puts forward the scheme of active frequency response intelligent control, completes the hardware and communication design, algorithm design and software implementation. Finally, the actual engineering practice is carried out in three photovoltaic VPP in Zhongshan City, Guangdong Province, which verifies the feasibility of relevant schemes and algorithms, and provides a certain basis for the next research.

1. Introduction

Under the current situation, the power generation resources in the power system show a clean and decentralized development trend. These distributed generation resources are small in scale and scattered in layout. At present, there is no special regulation system, which can not fully meet the requirements of the rapid promotion of power marketization.

Virtual power plant (VPP), as an independent controllable system containing flexible loads and a variety of distributed generators, organically integrates distributed generators, loads, energy storage devices, converters and monitoring and protection devices, which can flexibly switch between grid connected operation and isolated island operation modes, so as to greatly improve the reliability of power supply security and main network friendliness. [1]There are a lot of flexible resources existed in the VPP, such as various energy storage equipment, new energy power generation equipment, etc. This kind of equipment responds quickly. By adjusting its power output, the active power of the VPP can quickly respond to the frequency fluctuation of the external network, so that the VPP can provide active frequency support to the outside world.[2]

In the photovoltaic distributed generation VPP, its power output is random and intermittent, which is not conducive to maintaining the active power balance of the photovoltaic distributed generation VPP, and even serious frequency oscillation.[3] In this case, how to adjust effectively and control the frequency of photovoltaic distributed generation VPP intelligently needs further exploration. The purpose is to quickly and effectively realize the frequency adjustment to ensure the stable and reliable operation of the system.[4]

2. Design of active frequency control algorithm for photovoltaic distributed generation VPP

The project adopts the cluster primary frequency regulation control algorithm in the VPP based on feedback optimization. The structure of feedback optimization is shown in Figure 1.

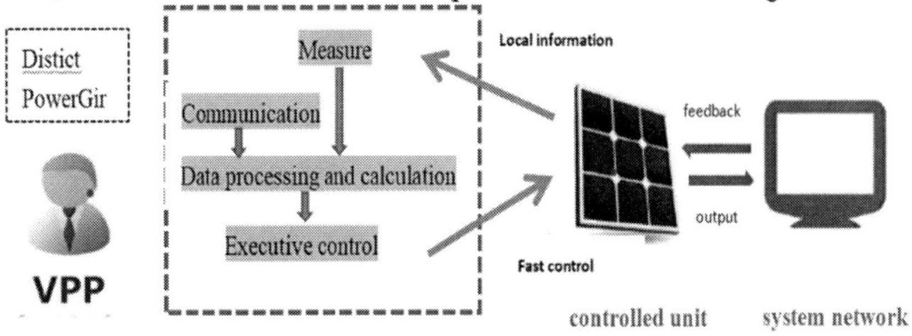

Fig. 1 Feedback optimization control structure

Specifically, after the VPP collects the frequency deviation of the parallel node, the active power gap in the current state is calculated according to the deviation and the reported regulation characteristics of the VPP, and the optimal power flow solution algorithm based on second-order cone relaxation is used to dynamically solve the active power allocation problem. After the active power is distributed, each controlled equipment is controlled in place. At this time, the power gap is calculated again according to the measurement information, the above optimization problem is corrected, and the iteration converges to the optimal solution.

2.1. *Active frequency response control process of VPP*

When the power grid frequency changes beyond the manually defined active frequency response dead zone, the fast control device of the VPP forms the control target command according to the active frequency response active frequency droop characteristics (realized by setting the frequency and active power broken line function), and can distribute the control command of each unit according to the control constraint conditions and the operating conditions of the unit. It shall be forwarded to the monitoring system and the unit for execution through fast Ethernet.[5]

The schematic diagram of active frequency response control process (Fig. 2) is as follows:

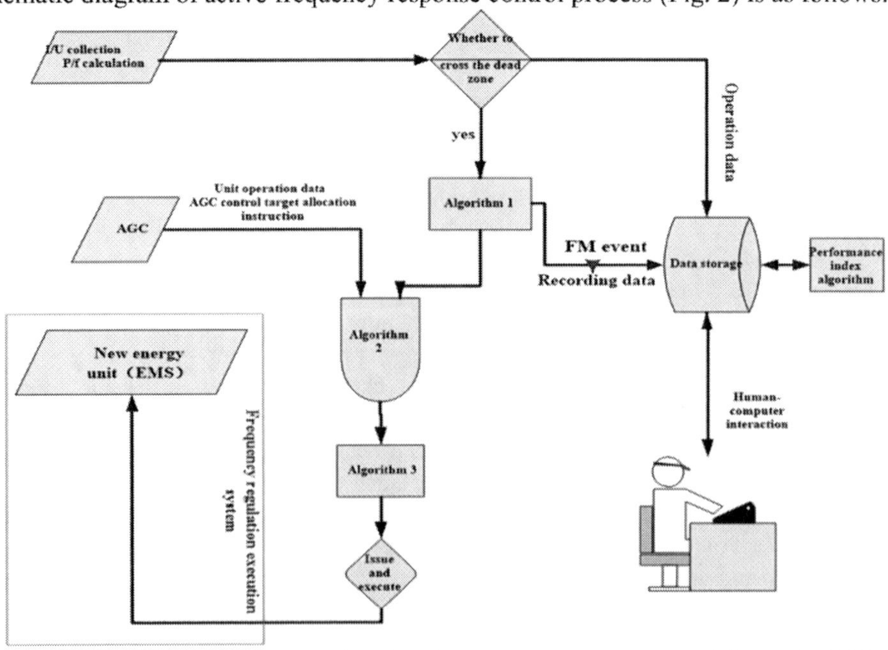

Fig. 2 Schematic diagram of active frequency response control flow of VPP

1) Algorithm 1 - Initial value of FM target command: when the power grid frequency changes beyond the manually defined dead zone of active frequency response, the system forms the initial value of control target command according to the active frequency droop characteristics of active frequency response (realized by setting the broken line function of frequency and active power).

2) Algorithm 2 - FM target command execution value: coordinate control logic judgment according to control constraints and AGC control target value to form control target command execution value.

3) Algorithm 3 - Active control allocation command for each unit: combined with the operation information of the unit, the active control command of each unit is formed through the definition of control command allocation strategy, and transmitted to the monitoring system through fast Ethernet or directly distributed to each unit for execution.

4)Performance index algorithm: It mainly includes theoretical contribution power, actual contribution power, primary frequency modulation contribution rate, correct frequency modulation action rate, and real-time frequency modulation response.

2.2. *Active frequency control algorithm of photovoltaic distributed generation VPP*

2.2.1. *Initial value algorithm for FM target command*

The active power variation is realized by the given active power frequency droop characteristic curve function (Fig. 3):

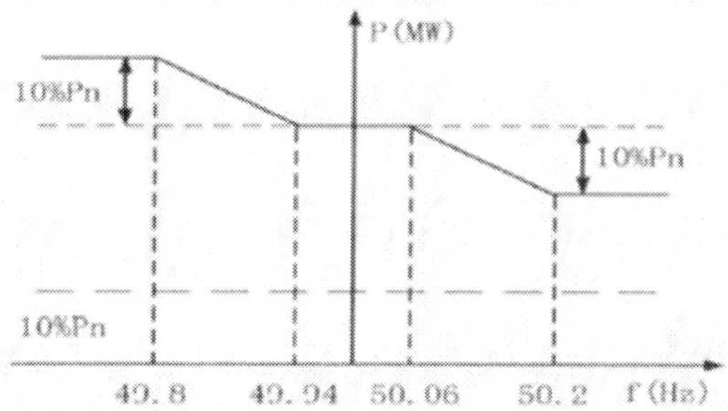

Fig. 3 Schematic diagram of active frequency response active power frequency droop characteristics of photovoltaic power station

$$P_{target} = P_0 - P_N * \frac{(f - f_d)}{f_N} * \frac{1}{\delta\%} \quad (1)$$

Among $P_N * \frac{(f - f_d)}{f_N} * \frac{1}{\delta\%} = \Delta P \quad (2)$

Parameter definition:

- FN, system rated frequency: 50Hz
- f_d,Action threshold for active frequency response(for Photovoltaic power station):dead band is set to ± 0.06hz; At f>50.06Hz, f_d=50.06Hz; At f<49.94Hz, f_d=49.94Hz
- PN, rated power: installed capacity, unit MW or kW, pay attention to unit consistency
- $\delta\%$,Adjustment rate: 3% for photovoltaic power station

Input:

- f,Real time frequency value of parallel node
- P0,Initial value of parallel power

Output:

- Ptarget,FM control target initial value

Constraints:

- When ｜ΔP｜>PN*δ%， then ΔP= PN*δ%
- When high frequency disturbance （ f>50.06Hz for Photovoltaic power station），P0<10%PN ， It can no longer be adjusted downward， as Ptarget = P0
- When low frequency disturbance （f<49.94Hz for Photovoltaic power station），P0 is Maximum generating capacity（There is no up regulation space, and there is no need to reserve spare FM capacity.），then Ptarget = P0

2.2.2. Coordinated control algorithm with AGC

The FM target command execution value is algebraic sum of AGC command value (PAGC) and active frequency response adjustment ΔP.

Among ΔP= Ptarget - P0

That is, the execution value of FM target command:

$$P=PAGC+\Delta P= PAGC+ Ptarget - P0 \quad (3)$$

Constraints:

- When grid frequency f>50.1Hz ， and AGC Instruction value（PAGC）>P0， then P=Ptarget
- When grid frequency f<50.1Hz ， and AGC Instruction value（PAGC）<P0， then P=Ptarget

2.2.3. Instruction allocation algorithm for target value unit

The optimal control algorithm for the characteristics of VPP, on the premise of meeting the control target value of active frequency response coordinated with AGC and various safety constraints of power grid and equipment, and combined with the operation status of the unit, allocates the overall control target command, which can provide the following distribution strategies (consistent adjustment direction, smooth adjustment and optimization strategy):

Proportional distribution method:

P1=P2= …=Pn= P/n(n is the number of controllable units) (4)

Constraints: single unit control difference ΔPn meets the climbing rate requirements.

Similarity margin distribution method:

P1=P10+（P-P0）/n (5)

…

Pn=Pn0+（P-P0）/n (6)

Constraints: single unit control difference ΔPn meets the climbing rate requirements.

2.2.4. Real time index algorithm of active frequency response

The adjustment of active frequency response frequency step disturbance process of photovoltaic power station is shown in the following figure (Figure 4).

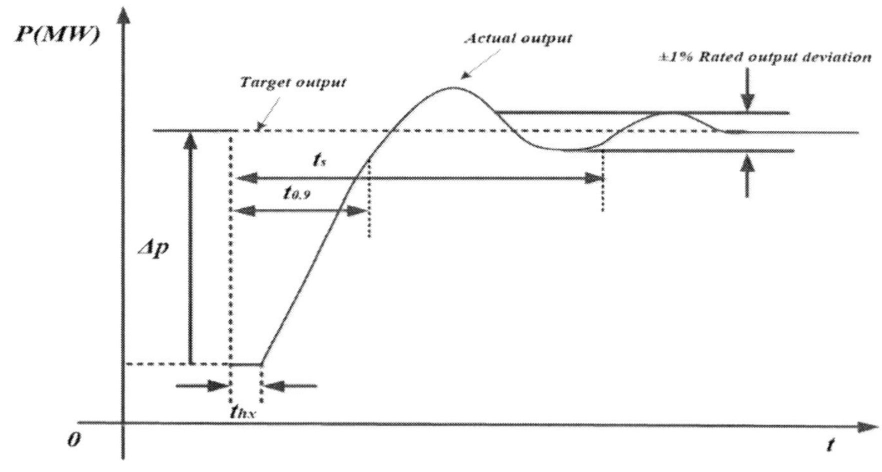

Fig. 4 Adjustment diagram of active frequency response frequency step disturbance process of photovoltaic power station

1) Load response lag time of active frequency response

The load response lag time of active frequency response refers to the time required for the VPP from the grid frequency crossing the dead zone of active frequency response to the load change of the VPP.

2) Adjustment and stabilization time of active frequency response

t0.9: When the power grid frequency changes beyond the dead zone of the active frequency response of the VPP, the load adjustment amplitude of the active frequency response of the VPP shall reach 90% of the maximum load adjustment amplitude of the theoretical calculated active frequency response within the required time;

ts:When the power grid frequency change exceeds the active frequency response dead band of the VPP, within the required time, the average deviation between the actual output of the VPP and the response target of the VPP shall be within ± 1% of the rated active output of the VPP.

3)Active frequency response performance index of photovoltaic power station[5]

Load response lag time of active frequency response： thx≤2s

Adjustment and stabilization time of active frequency response： t0.9≤5s、 ts≤15s

3. Engineering implementation of algorithm

At present, photovoltaic power generation accounts for the largest proportion of distributed power generation in Zhongshan, Guangdong. The real-time monitoring networking schemes adopted by photovoltaic power stations are different according to the type of grid connected inverter and the power response time of inverter.

The intelligent control sub station of the VPP is deployed in Changhong Photovoltaic Station, Galanz Photovoltaic Station and Midea Photovoltaic Station to complete self-discipline control.

According to the actual situation of field equipment of Changhong Photovoltaic Station, Galanz Photovoltaic Station and Midea Photovoltaic Station, two equipment transformation schemes in Photovoltaic Station are designed as follows:

1) After fast control communication transformation, the inverter is directly connected to the intelligent control substation of the VPP to realize the fast power control of the photovoltaic inverter. Support steady-state AGC / AVC control of photovoltaic power station and primary frequency regulation of photovoltaic power station.

2) The inverter is connected to the intelligent control substation of the VPP through the existing station monitoring system to realize the conventional power control of the photovoltaic inverter. Support steady-state AGC / AVC control of photovoltaic power station.

The layout transformation scheme of demonstration application in Changhong Photovoltaic Station, Galanz Photovoltaic Station and Midea Photovoltaic Station is as follows (Table 1):

Tab. 1 Demonstration application scheme of photovoltaic power plant

number	Point distribution equipment	Total capacity (MW)	inverter		Transformation scheme
			Single capacity (kW)	Total number	
1	Changhong Photovoltaic Station	13.16	500	18	Option 1
			630	2	
			30	44	Option 2
			20	52	
2	Galanz Photovoltaic	52.38	40	859	
			28	560	
3	Midea Photovoltaic Station	23.87			
Summary statistics		Total controllable capacity (MW)	89.41	Fast controllable total capacity (MW)	10.26

After the transformation of demonstration application, all demonstration application photovoltaic power stations have the function of coordinating and controlling with the intelligent control master station of VPP. The total capacity of controllable photovoltaic power stations is 89.41MW, of which the power generation output capacity of photovoltaic power stations supporting rapid power control is 10.26MW.

The active frequency step test is conducted at the substation of Changhong VPP. The disturbance test parameters are shown in the following table (Table 2):

Tab. 2 Table of test parameters

number	Parameter name	Parameter value
1	Adjustment coefficient (%)	2
2	Installed capacity (MW)	10.26
3	Sampling period (MS)	100
4	Step upper limit frequency (Hz)	50.1
5	Step lower limit frequency (Hz)	49.9

The test results of active frequency step disturbance are shown in the following table (Table 3):

Tab. 3 Table of test results

number	Frequency disturbance type	Step experiment disturbance value (Hz)	Response time (s)	Adjustment time (s)	Target adjustment	Active power before step (MW)	Active power after step (MW)	Control deviation (%)
1	Step up disturbance	50.15	0.201	0.201	-0.513	3.00456	2.41456	2.60
2	Step down disturbance	49.8	0.199	0.199	1.026	1.09745	2.1135	1.00

According to the statistical analysis in the above table, the average active power regulation time of the active frequency step test of the substation of Changhong VPP (that is, when the power grid frequency change exceeds the active frequency response dead band of the VPP, the average deviation between the actual output of the VPP and the response target of the VPP shall be within ± 1% of the rated active output of the VPP within the required time) is about 200ms, It shows good fast response performance.

4. Conclusion

By effectively adjusting and intelligently controlling the frequency of the photovoltaic distributed power VPP, maintaining its active power balance and eliminating serious frequency oscillation, the frequency adjustment can be realized quickly and effectively to ensure the stable and reliable operation of the system.

The active frequency control algorithm design of VPP based on flexible resource and rapid response is based on the active frequency intelligent control project of photovoltaic distributed generation VPP in Zhongshan City, Guangdong Province. The results of this practice meet the predetermined verification objectives, and reflect the good fast response performance of the algorithm under certain constraints. This engineering practice, aiming at the intermittent characteristics of new energy after grid connection, provides a certain research method and engineering demonstration for its intelligent frequency control.

However, for the frequency control of photovoltaic VPP and other flexible resources, there are still many aspects that need to be further studied, such as the participation of more constraints in model optimization, the improvement of algorithm convergence performance and computational efficiency, etc. These are the topics to be studied in the next step.

Acknowledgments

Supported by science and technology project of Guangdong Power Grid Co., Ltd (Item No: GDKJXM20185069(032000KK52180069))

References

[1] Zhinong Wei;Shuang Yu;Guoqiang Sun . Concept and development of virtual power plant [J]. Power system automation, 2013, 37 (13): 1-9

[2] Defu Cai; Peng Wang; Bingke Yan . Analysis of voltage and frequency withstand capacity of distributed photovoltaic in Hubei Province [J]. Hubei electric power, 2020, 44 (3): 31-36

[3] Junjie Yu ;Guomin Liu ;Yongxin Yang . Research on distributed collaborative control of virtual power plant with source- network-load interaction [J]. China's new technologies and products, 2020,8 (bottom): 12-13

[4] The project team of intelligent operation optimization management integrated platform development and demonstration application for "Source -Network" interactive virtual power plant.Research report on intelligent control subsystem of distributed generation virtual power plant [R]. Zhongshan, Guangdong: Zhongshan Power Supply Bureau, 2021

[5] Northwest energy regulatory market [2018] No. 41 document. Work plan for promotion and application of fast frequency response function of new energy stations of northwest power grid [R]. 2018

Complete characteristic curve processing method based on improved backpropagation neural network and Logarithmic curve projection

Baonan Liu[1a], **Jianzhong Zhou**[1b*]

[1]School of civil and hydraulic engineering, Huazhong university of science and technology, Wuhan, 430074, China

[a]D201981043@hust.edu.cn, [b*]jz.zhou@hust.edu.cn

Abstract—The calculation of transient process is the basis of the design and construction of pumped storage power plant, which directly affects operation stability of pumped storage units. However, for satisfying the design of uniform and smooth flow in both directions, the complete characteristic curves of pumped turbine show a significant S feature, which brings difficulties to the interpolation calculation in the transient process as there are crossover and overlap phenomenon in pump and turbine working conditions. In this paper, a transformation method for complete characteristic curves based on logarithmic curve projection and improved backpropagation neural network is presented to solve the above problem. Furthermore, the feasibility and accuracy of the method are verified through the comparison of load rejection condition with on-site measurement, the results show that the proposed method overcomes the multi-value problem that exists in the original curve, in especial makes the small opening region well expressed. The simulation based on the transformed curve reaches a high similarity of transient process with the field test, both in the trend and extreme values, which provides great convenience for the calculation of transient process.

1. Introduction

Pumped storage power plant, as an effective solution to incorporate renewable energy and ensure the stability of power grid, has been vigorously developed for the past few years [1]. In the design of power plants, the primary thing is to conduct the calculation of transient processes for the purpose that the power plants can smoothly and safely convert between multiple working conditions. However, the analytical model of pump-turbine has not been obtained till now because of strong nonlinearity and time variation characteristics. Traditionally, the calculation of turbine adopts the complete characteristic curves given by the manufacturers, while there exist serious phenomena of crossover, aggregation, bending, and multi-values at both ends of the curves [2]. Specifically speaking, the pump-turbine has remarkable S characteristics where small changes in unit speed will cause sharp changes in the unit flow and unit torque. Hence, accurate calculation of the transient process of pumped storage power plants has become a great challenge to the economical and efficient operation of power plants.

Scholars have conducted fruitful research on the processing of complete characteristic curves. According to the dimensionless similar parameters of the pump, Suter et al. [3] proposed Suter transformation to deal with the complete characteristic curves of the pump-turbine, which effectively eliminated the multi-value phenomenon and brought convenience for interpolation calculation. Liu et al. [4] put forward an improved transformation method on the basis of Suter transformation, which effectively avoided the occurrence of serious crossover and overlap situations. Yang et al. [5]

constructed a B-spline space curved surface for the representation of the complete characteristics of pump-turbine, which effectively eliminated the multi-valued and non-corresponding problems existing in the plane curve representation. Chen et al. [6] presented an equal opening line method to better express the small opening characteristic and make a good connection between the zero opening characteristic and the finite opening characteristic. Nevertheless, there are still some defects in the existing treatment methods. The curves processed by the Suter transformation show uneven distribution under different opening degrees, and there are still serious aggregation and torsion at both ends of the curves. Several parameters have been added with the improved Suter transformation method, while there is no guidance for the selection of accurate parameters. The appropriate B-spline function is difficult to construct and the equal opening line method raises high requirements on the quantity of complete characteristic curves.

Based on the above discussion, a complete characteristic curve processing method based on improved backpropagation neural network and logarithmic curves projection (IBP-LCP) is proposed in this paper. The complete characteristic curves are first transformed by projection along the logarithmic curve, then the nonlinear relation of transformed curves is learned by introducing the improved BP neural network. The proposed method overcomes the multi-value of the original characteristic curves and makes the small opening region well expressed, which improved the calculation speed and precision of the transient process for pumped storage power plants.

The rest of this paper is organized as follows. In section 2, the proposed IBP-LCP processing method is elaborated in detail. The feasibility and accuracy of the proposed method are verified by the case of load rejection in a real pumped storage power plant in Section 3. Section 4 presents the conclusion of this study.

2. Processing method for complete characteristic curve

An accurate model of the pump-turbine is of great significance for the calculation of transient process. Nevertheless, the complete characteristic curves of pump-turbine have a significant S region, as shown in Fig.1, which got into trouble when conduct regulation guarantee calculation. In this paper, a novel processing method is proposed. The logarithmic curve projection is adopted to transform the complete characteristic curves in the first place, which overcome the multi-value problem without changing the continuity and smoothness of the original curves. Furthermore, the improved BP neural network is introduced to excavate the internal correlation of the transformed curves, thus the curves can be expanded reasonably.

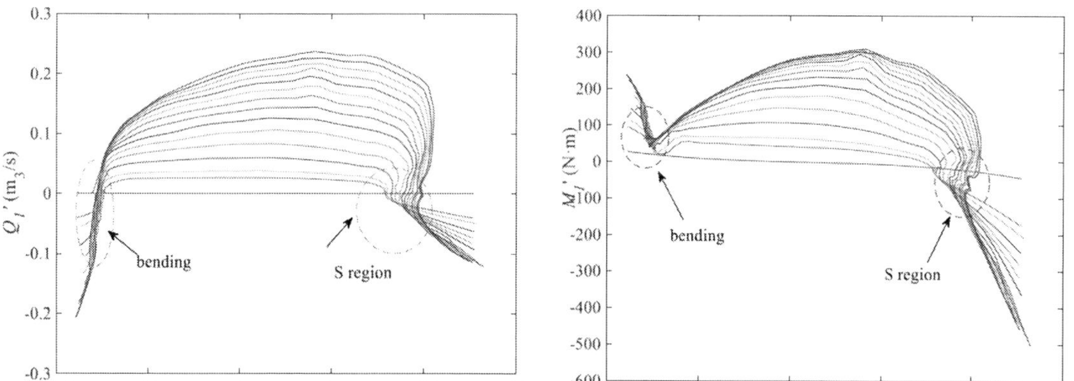

(a) Flow characteristic curve (b) Torque characteristic curve

Fig.1 The complete characteristic curves of pump-turbine.

The Logarithmic curve projection takes $x = \alpha_1 / e^{y1}$ as the abscissa to describe the characteristics of pump-turbine, the transformation formula is shown as follows:

$$\alpha_1 = x_{11} / x_{11r} \tag{1}$$

$$v_1 = Q_{11} / Q_{11r} \tag{2}$$

$$x = \alpha_1 / e^{v1} \tag{3}$$

Where, x_{11} Q_{11} are unit rotational speed and unit flow, respectively. x_{11r} Q_{11r} are the corresponding rated values. The transformed curves after LCP are shown in Fig.2. It is obvious that the transformed curve maintains the continuity and smoothness of the original curves and the curves are relatively uniform under different opening degrees. Compared with large opening degrees, there is no special point from small opening to zero opening degree, and there is no multi-value problem. Nevertheless, the transformed curves still have some shortcomings, as follows: (1) the abscissa of the curves is not aligned and the distribution of the data points in each curve is uneven. (2) the measured curves from the manufacturer are too few to meet the requirements of interpolation.

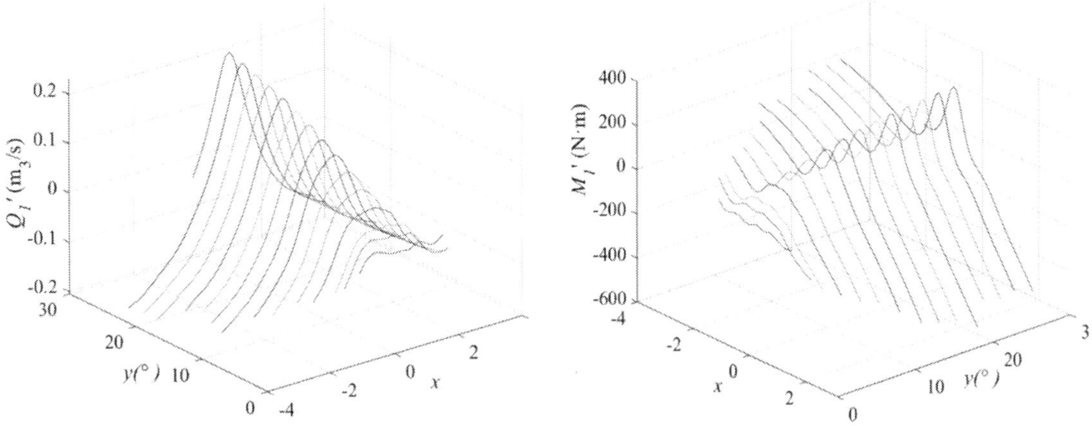

(a) Flow characteristic curve (b) Torque characteristic curve

Fig.2 The complete characteristic curves of pump-turbine after logarithmic curve projection.

The BP neural network has strong nonlinear mapping ability and flexible network structure, which can be used to describe the nonlinear relationship of characteristic curves overcoming the above shortcomings [7]. For the problems that exist in traditional BP neural network, i.e. easy to form a local minimum but not the global optimal, many times of training, low learning efficiency, slow convergence, we adopt improvement strategies by adding dynamic parameter method and optimization of learning factors to solve. The dynamic parameters are expressed as follows:

$$\Delta W(k+1) - \delta \Delta W(k) = a(1-\delta)E(k)/W(k) \tag{4}$$

Where, δ is an additional parameter, the range is $0 \le \delta \le 1$. The essence of dynamic parameters is that, when the amount of correction is insufficient, the next amount of correction will be increased and vice versa. By adjusting the dynamic parameters, the oscillation trend in the learning process of the neural network can be reduced and the local minimum value can be avoided or reduced.

In order to reduce the number of iterations and speed up convergence, it is necessary to adjust the learning factor reasonably. Accordingly, the learning factor is modified by changing the step size according to the output error. The formula is as follows:

$$a = a + \frac{\lambda \times \left[E_p(n) - E_p(n-1)\right]}{E_p(n)} \tag{5}$$

Where, a is the learning rate, and λ is adjustment step size, which is usually between 0 and 1.

3. Model validation

To verify the performance and accuracy of the proposed processing method, a case study is conducted by implementing it in a real pumped storage power plant. When large electrical equipment failure or the power supply network outlet circuit breaker suddenly trip, the unit had to dump the load to maintain torque balance, while the huge residual energy causes the speed of the unit to rise quickly and the pressure of the water system to changes notably. Based on that, the load rejection condition, as the running tracks of unit span most of the characteristic curves back and forth, especially the hump domains and the S-shaped region, is implemented as the typical extreme transient process for comparison.

The IBP model, as shown in Fig.3, consists of double input (x, y), double output (Q, M), and double hidden layers taking account of the coupling relation between Q and M. In the training process of the BP neural network, the maximum number of iterations is 300, the number of each hidden layer node is 15 which is selected by orthogonal test, the initial learning rate is 0.1, and the training objective error is 10^{-5}. The hidden layer adopts the activation function of *tansig*, and the output layer adopts the linear activation function *purelin*. After the procedure of encryption and extension, the treated curves are shown in Fig.4, the results show that the inherent law of characteristic curves can be generalized from the existing data sequence which can be attributed to the strong self-learning and predictive ability of the IBP neural network.

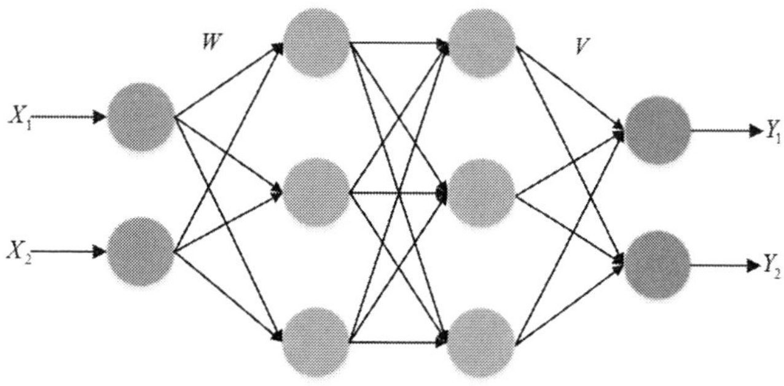

Fig.3 The structure diagram of BP neural network.

Subsequently, the simulation of load rejection is performed in MATLAB. The default working condition is set to a complete load rejection, 733m water level in the upstream reservoir, and 181m water level in the downstream reservoir. The guide vane closes in 20s. other boundary conditions can be found in Ref. [8]. The transient process of key nodes, i.e. the process of rotational speed change, the pressure of volute, and the pressure of the draft tube are extracted for comparison with field test.

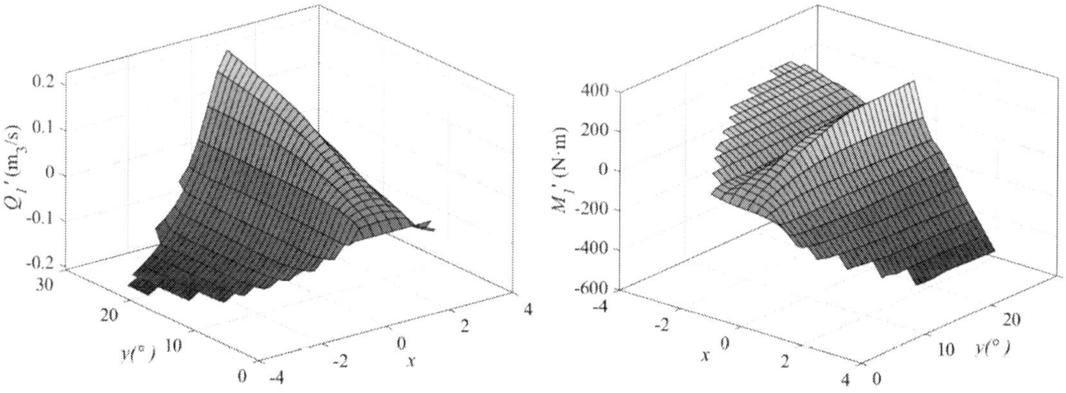

(a) Flow characteristic curve (b) Torque characteristic curve

Fig.4 The complete characteristic curves after IBP-LCP transformation

As can be seen from Fig.5, the simulation using the transformed complete curves shows an analogical transient process with field test. To be more specific, the occurrence time of the maximum relative rotational speed of the simulation model and the measured data are identical, same thing happens with the maximum water pressure of volute. Though for the rotational speed, volute, and draft tube, some deviation appears after 20 seconds of load rejection, the development trend of transient process achieves high similarity. In addition, IBP-LCP fits the nonlinear relationship of characteristic curves, the flow and torque characteristics could be obtained by just inputting the guide vane opening and unit speed, which greatly improves the calculation speed and accuracy of the transient process.

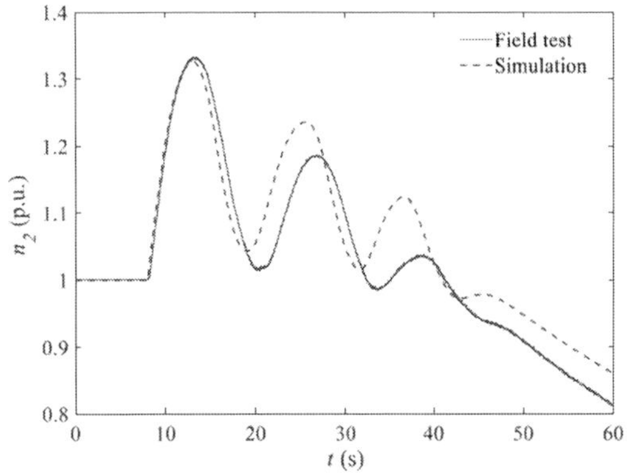

(a) Rotational speed

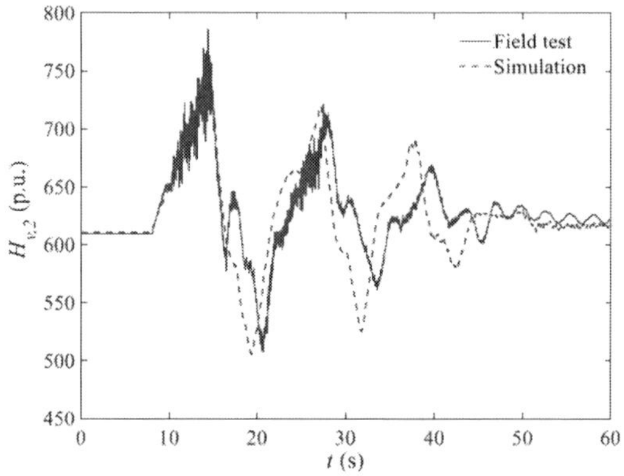

(b) Pressure of volute

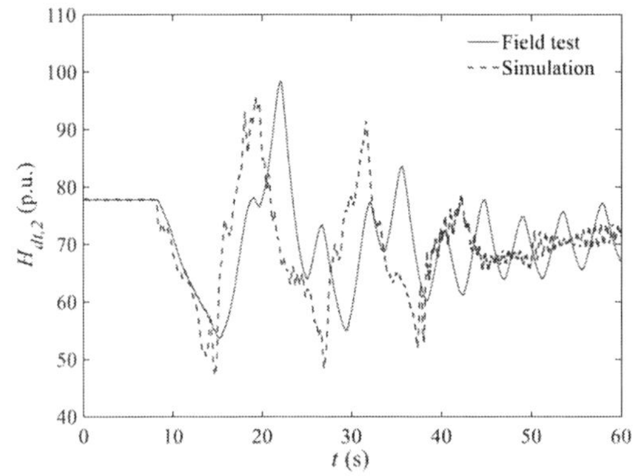

(c) Pressure of draft tube

Fig.5 Comparison of transient process between the simulation model and the field test

4. Conclusion

In this paper, a complete characteristic curve processing method based on IBP-LCP is proposed. By using the treatment method, multi-value problem in complete characteristic curves of pump-turbine has been eliminated and reasonable expansion of the characteristic curve will not cause the omission of the original data with help of IBP. In addition, there is a good connecton between the finite opening characteristic and the zero opening characteristic. Furthermore, without changing the continuity and smoothness of the original image, interpolation can be carried out, or the computer can be used for calculation. Compared with the measured data, the simulation based on the treated curves processed by the proposed method has a consistent transient process trend and extreme values with on-site measurement, which can serve as the basis for the calculation of transient processes.

Acknowledgment

This paper was supported by the National Key R&D Program of China (Nos.2016YFC0402205).

References

[1] Mwt A , Sry A, Xxt A , et al. Risk and profit-based bidding and offering strategies for pumped hydro storage in the energy market[J]. Journal of Cleaner Production, 256.
[2] Zhang, Tian P, Zhou, et al. An Improved Autoencoder and Partial Least Squares Regression-Based Extreme Learning Machine Model for Pump Turbine Characteristics[J]. Applied Sciences, 2019, 9(19):3987.
[3] Suter P. Representation of pump characteristics for calculation of water hammer[J]. Sulzer Technical Review, 1966, 4(66):45-48.
[4] Liu Q, Liu D. Some Problem on Hydraulic Transient Calculation for Pumped Storage Power Station[C]. Proceedings of the International Symposium on Pumped Storage Development, Beijing: Water Resources and Electric Power Press, 1990: 30-50. (in Chinese).
[5] Yang J , Yang J . B-Spline Surface Construction for the Complete Characteristics of Pump-Turbine[C]// Power and Energy Engineering Conference (APPEEC), 2012 Asia-Pacific. IEEE, 2012.
[6] Chen Nai-xiang, Zhang Yang-jun, Zhu Bao-shan, el al. New Presentation of the Complete Characteristics of Pump-turbine and the Calculation of Complex Operating Conditions[J]. Journal of Tsinghua University(Science and Technology), 1996, 36(7):62-66. (in Chinese).
[7] Ding, S., Su, C. & Yu, J. An optimizing BP neural network algorithm based on genetic algorithm. Artif Intell Rev 36, 153–162 (2011).
[8] Lai X , Li C , Zhou J , et al. A multi-objective optimization strategy for the optimal control scheme of pumped hydropower systems under successive load rejections[J]. Applied Energy, 2020, 261:114474-.

The Heat Transfer Coefficient Predictions in Engineering Applications

Junchi Wan[1a*]

[1]Shandong University, Jinan, China, 250002

[a*]201800180151@mail.sdu.edu.cn

Abstract—Most engineering applications have boundary layers; the convective transport of mass, momentum and heat normally occurs through a thin boundary layer close to the wall. It is essential to predict the boundary layer heat transfer phenomenon on the surface of various engineering machines through calculations. The experimental, analogy and numerical methods are the three main methods used to obtain convective heat transfer coefficient. The Reynolds analogy provides a useful method to estimate the heat transfer rate with known surface friction. In the Reynolds analogy, the heat transfer coefficient is independent of the temperature ratio between the wall and the fluid. Other methods also ignore the effect of the temperature ratio. This paper summarizes the methods of predicting heat transfer coefficients in engineering applications. The effects of the temperature ratio between the wall and the fluid on the heat transfer coefficient predictions are studied by summarizing the researches. Through the summary, it can be found that the heat transfer coefficients do show a dependence on the temperature ratio. And these effects are more obvious in turbulent flow and pointing out that the inaccuracy in the determination of the heat transfer coefficient and proposing that the conjugate heat transfer analysis is the future direction of development.

1. Introduction

A flow boundary layer is created near the surface when viscous fluid flows over a solid surface. Usually, there is a temperature difference between the main flow and the wall. Not only does the boundary layer have a drastically changing velocity gradient, but also the temperature changes drastically. Therefore, convective heat transfer occurs intensively in the boundary layer. This phenomenon occurs in many engineering applications, such as the blades of gas turbines, the walls of heat exchangers, and the heat dissipation surfaces of integrated circuits [1]. Common convective heat transfer can be divided into forced convection and natural convection. But in simplify, both types of convective heat transfer can divide the boundary layer flow into the laminar flow and turbulent flow. The upstream flow has a great influence on the heat transfer coefficient in a laminar boundary layer, including leading-edge conditions and wall temperature changes. In the turbulent boundary layer, the heat transfer coefficient is not sensitive to the slow flow direction change of the wall temperature [2].

This paper mainly predicts the heat transfer coefficient, including analogy, experimental, and numerical methods. This paper discusses these methods in detail to analyze the existing shortcomings. As the basis of convective heat transfer, these methods are used and studied in many research. However, many papers do not consider their accuracy. Through comparison, the influence of the temperature ratio on the heat transfer coefficient and the future development trend of the heat transfer coefficient prediction techniques are shown clearly.

2. Methods for predicting the heat transfer coefficient

Newton's cooling equation for calculating the rate of convective heat transfer Q is expressed as

$$Q = hA(T_w - T_\infty) \quad (1)$$

Where h is the heat transfer coefficient, A is the area of heat transfer, T_w is the wall temperature of solid, and T_∞ is the temperature of the main fluid. In engineering calculations, the temperature difference $(T_w - T_\infty)$ is always positive.

Newton's cooling equation is only a definition formula of the heat transfer coefficient h. It does not reveal the connection between the heat transfer coefficient and the related factor that affect it. The prediction method of heat transfer coefficient is obtained based on Eq. (1) and these connections.

2.1. Analogy method

The analogy method refers to the method of obtaining the basic relationship between two different physical phenomena due to the similarity in the governing equations by measuring one of the phenomena. In boundary layer heat transfer, the most basic and most commonly used analogy is the Reynolds analogy. It is an analogy between heat and momentum transfer due to molecular motion. It provides an available tool to use the shear stress to calculate the heat transfer coefficient for both laminar and turbulent flows. The general expression of the Reynolds analogy is

$$Nu = \frac{C_f}{2} RePr \quad (2)$$

C_f is the skin friction coefficient; it can be obtained from experiments, Nu is the Nusselt number, Re is the Reynolds number, Pr is the Prandtl number. These three dimensionless numbers are defined as

$$Nu = \frac{hl}{k} \quad Re = \frac{U_\infty l}{v} \quad Pr = \frac{uc_p}{k} \quad (3)$$

Here k is the thermal conductivity, v is the kinematic viscosity, u is the dynamic viscosity, U_∞ is the fluid velocity, c_p is the specific heat at constant pressure, l is the characteristic length. Through combining Eq. (2) and Eq. (3), the expression of the heat transfer coefficient h can be determined

$$h = \rho U_\infty c_p \frac{C_f}{2} \quad (4)$$

So h can be determined by Eq. (4), all the fluid properties are obtained at the reference temperature $T_m = (T_w + T_\infty)/2$ [3]. However, the Reynolds analogy has a limitation that the Prandtl number must be assumed as one. Although for most gas and water which the Prandtl number is close to one, Eq (2) is still valid, the error can be ignored. Later, Chilton and Colburn [4,5] modified the Reynolds analogy, known as the Reynolds-Colburn Analogy

$$Nu = \frac{C_f}{2} RePr^{\frac{1}{3}} \; (0.6 < Pr < 60) \quad (5)$$

Zeng et al. [6] found that the experimentally measured air-side heat transfer coefficients were consistent with the modified Reynolds analogy, verifying its applicability.

The modified Reynolds analogy and some other analogies have a large number of applications in engineering. Wang et al. used the heat-momentum transfer analogy to correlate the single-phase heat transfer data in microfin and showed good accuracy [7]. A semi-empirical model based on an analogy between heat and momentum transfer in turbulent flow was proposed by Arsenyeva et al. to predict the heat transfer for flows in channels of a plate heat exchanger [8]. Similarly, KWON et al. developed a model using a heat-momentum transfer analogy to estimate the evaporative heat transfer coefficient in complicated cooling systems [9]. Conversely, the measured heat transfer coefficient can be used to calculate the skin friction coefficient through the Reynolds analogy to study the effect of roughness, like Abuaf et al. did on turbine airfoils [10]. Also, in the turbine airfoil, the naphthalene sublimation technique based on a heat and mass transfer analogy was applied by Häring et al. to measure the heat transfer coefficient [11].

2.2. Experimental method

The calculation formulas of the heat transfer coefficient are obtained through the experiments, and they are the basis of engineering design. These formulas are also called correlations, in which the heat transfer coefficient to be determined included in the Nusselt number. Usually, dimensional analysis is used to derive the non-dimensional number that reflects the physical parameters' inner relationship. For example, dimensional analysis is used on the forced convection heat transfer. It shows that the Nusselt number can be expressed as a function of Reynolds number and Prandtl number [3].

$$Nu = f(Re, Pr) \tag{6}$$

But the determination of the specific function form is empirical. Usually, people determine the constants in Eq. (6) through experimental data of different Reynolds numbers and Prandtl numbers, as Lienhard concluded in flat-plate boundary layers. Researchers are constantly conducting researches on correlations, and some new and more accurate correlations are proposed for the same type of problem [2].

However, correlations are only applicable to the convective heat transfer problems on the surface of some objects with simple and regular shapes, such as the outer surfaces of plates, cylinders and spheres, and the inner surfaces of circular tubes non-circular channels. For objects with more complicated shapes, like the turbine blades, it is so challenging to use correlations for calculation. In this case, it is necessary to back to the basic Newton's cooling equation (1). That the heat transfer coefficient h is expressed as

$$h = \frac{q}{T_w - T_f} \tag{7}$$

q is the heat flux. Andreini et al. experimented on an open-loop wind tunnel with a cooling system of a high-pressure gas turbine leading-edge model to determine the heat transfer coefficient distribution on the internal surface, using the Eq. (7) [12]. An electrically heated sheet applied on the internal surface was used to imposed a constant heat flux. The wall temperature was measured by wide band thermochromic liquid crystals, and the fluid temperature was measured by T-type thermocouples. Similarly, the heat transfer coefficient distribution behind the rib is determined by Davletshin et al., using constant-temperature heat flux generated from a printed circuit board and wall temperature measured by resistance thermometers [13]. Will et al. conducted similar experiments on spheres of various diameters [3]. It can be seen that the temperature obtained by experimental measurement is the basis. In addition to using Eq. (7), the energy conservation equations and the lumped parameter approximation were also used [6,10].

2.3. Numerical method

The numerical solution of the convective heat transfer problem refers to using various numerical methods to solve the governing equation, including the Navier-Stokes equation and energy conservation equation, etc. under fixed conditions. The fixed conditions include the velocity, pressure, temperature and other relevant conditions at the initial moment and on the boundary. Usually, the solution process needs to use computational fluid dynamics (CFD) software to obtain the temperature distribution of the fluid and the heat flux at the wall. Then the heat transfer coefficient predictions are obtained by Eq. (7) [14].

2.4. Discussion

Back to Eq. (4) derived from the Reynolds analogy, it can be seen that the heat transfer coefficient depends on the fluid properties obtained at the reference temperature $T_m = (T_w + T_\infty)/2$. If the temperature differences between the wall and free stream are small, the variations in these properties would be small. But when the temperature differences are relatively large, the accuracy of the predictions will decrease, so the effect of the temperature ratio between the wall and free stream should be considered. Eq. (7) is the key basis when using experiments or numerical methods for the heat transfer coefficient predictions. The fixed thermal boundary conditions are imposed, like constant heat flux, and the constant heat transfer coefficient is assumed. As shown in Fig. 1, the heat flux q is linear with the wall temperature; the slope is the result. But it should be pointed out that such an assumption is not accurate because this hypothesis set that the thermal boundary condition does not affect the heat transfer coefficient. In fact, the h would change with the changing wall temperature. For predicting the heat transfer coefficient more accurately in engineering applications, the dependence of the heat transfer coefficient on the wall-to-fluid temperature ratio is worth research.

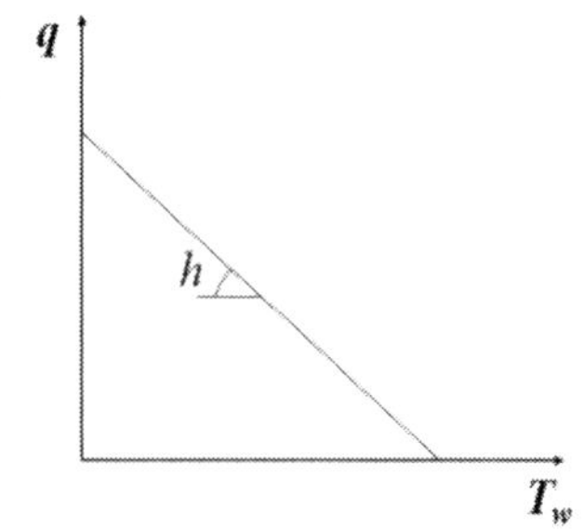

Fig. 1. The line under the constant heat transfer coefficient

3. Effects of the temperature ratio on the heat transfer coefficients

To figure out the temperature ratio effects, Fitt et al. proposed an equation $Nu=Nu_i(T_w/T_\infty)^n$ under constant Reynolds number [15]. They conducted experiments of airflow on a flat plate with zero pressure gradient, using the transient technique to make sure the properties of tested airflow were constant. The Reynolds number per meter was 2.7 x 10^7 and the Mach number was 0.55. With fixed air temperature and changing wall temperature by forced convection, the range of the wall-to-fluid temperature ratio was 0.6 to 1.3. Within this range, with the help of a power law, the measured value of n is between -0.28 and -0.21. Some numerical calculations were also done by them, shown similar results of turbulent flow compared to the experiments, but the effect of the temperature ratio in laminar flow is very small. It can be shown that the heat transfer coefficients in the turbulent boundary layer exist a dependency on the wall-to-fluid temperature ratio. The effect is proved; the next step is to quantify it. This work was done by Maffulli and He [16]. They used CFD with the Spalart-Allmaras turbulence model to analyze the external heat transfer coefficients on a 2D Nozzle Guide Vane profile under three different temperature ratios (0.95, 0.75 and 0.6). Both laminar and turbulent conditions were considered. The result was shown in Fig. 2, which shows that the values of h at the trailing edge in 0.6 temperature ratio are about 25% higher than those in 0.95 temperature ratio.

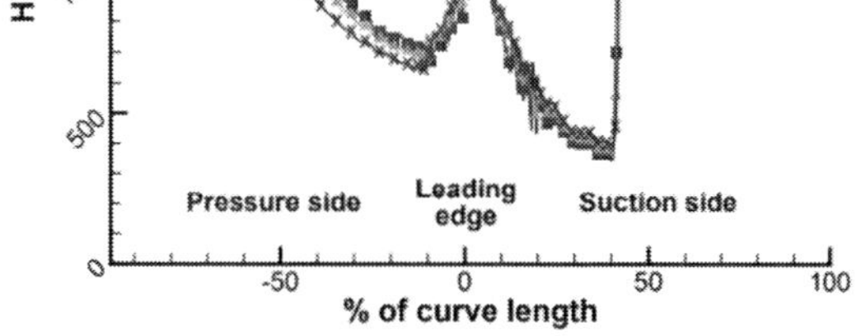

Fig. 2. Dependence of external heat transfer coefficient distributions on the temperature ratio [16]

Obviously, the temperature ratios have fewer effects on the laminar flow area, the 0% to 40% of curve length. This result is consistent with the conclusions of Fitt et al. It can be confirmed that the effects of the temperature ratio on the heat transfer coefficient are much smaller in the laminar flow state than in the turbulent flow state.

4. Conclusion

Whether it is a flat plate with a simple shape or a steam turbine blade with a complex structure, the heat transfer coefficient of the surface should be accurately predicted to obtain the heat transfer characteristics. It plays a very important role in the design of various mechanical parts in engineering applications. Three different methods for determining the heat transfer coefficient and their applications have been summarized in this article. The factors that affect the accuracy of the results have been discovered from the basic principles, and the effects of the wall-to-fluid temperature ratio are proposed. By summarizing the research of other authors, it is concluded that the temperature ratio has a certain effect on the heat transfer coefficient. This influence is almost absent in laminar flow but more obvious in a turbulent flow. The difference in heat transfer coefficients can even reach 25% on the suction side of the blades at two temperature ratios with a relatively large difference. Fundamentally, the effect of the temperature ratio is caused by the fixed thermal boundary conditions in the calculation process. Because in the process of convective heat transfer, the thermal boundary conditions are affected by the interaction between the fluid and the wall, so it will change as the heat transfer process progresses. This problem can be solved if applying the coupled solid-fluid conjugate heat transfer analysis because the boundary conditions are taken as the calculation results. The rationality and accuracy of the simulation results of the heat transfer problem would be much improved. But for now, the conjugate heat transfer analysis is still in development, and the use cost is relatively expensive. So the non-conjugate methods are still most frequently used. In the future, the continuously developing conjugate heat transfer analysis techniques would better solve the heat transfer problem in engineering applications.

Acknowledgements
I am very grateful to professor Rosic, who gave me the ideas for my paper in his busy schedule and provided related materials to help me complete this paper.

References

[1] Forooghi, P., Stripf, M., & Frohnapfel, B. (2018) A systematic study of turbulent heat transfer over rough walls [J] International Journal of Heat and Mass Transfer, 127, 1157–1168.
[2] Lienhard, J. H. (2020) Heat Transfer in Flat-Plate Boundary Layers: A Correlation for Laminar, Transitional, and Turbulent Flow [J] Journal of Heat Transfer, 142 (6), 061805.
[3] Will, J. B., Kruyt, N. P., & Venner, C. H. (2017) An experimental study of forced convective heat transfer from smooth, solid spheres [J] International Journal of Heat and Mass Transfer, 109, 1059–1067.
[4] Chilton, T. H., & Colburn, A. P. (1934) Mass transfer (absorption) coefficients prediction from data on heat transfer and fluid friction [J] Industrial & Engineering Chemistry, 26 (11), 1183–1187.
[5] Colburn, A. P. (1964) A method of correlating forced convection heat-transfer data and a comparison with fluid friction [J] International Journal of Heat and Mass Transfer, 7 (12), 1359–1384.
[6] Zeng, Z., Sadeghpour, A., & Ju, Y. S. (2018) Thermohydraulic characteristics of a multi-string direct-contact heat exchanger [J] International Journal of Heat and Mass Transfer, 126, 536–544.
[7] Wang, C. C., Chiou, C. B., & Lu, D. C. (1996) Single-phase heat transfer and flow friction correlations for microfin tubes [J] International Journal of Heat and Fluid Flow, 17 (5), 500–508.
[8] Arsenyeva, O. P., Tovazhnyanskyy, L. L., Kapustenko, P. O., & Demirskiy, O. V. (2014) [J] Generalized semi-empirical correlation for heat transfer in channels of plate heat exchanger. Applied Thermal Engineering, 70 (2), 1208–1215.
[9] KWON, H., PARK, G. C., & MIN, B. J. (2008) Experimental study of interfacial shear stress for an analogy model of evaporative heat transfer [J] Journal of Nuclear Science and Technology, 45 (6), 487–497.

[10] Abuaf, N., Bunker, R. S., & Lee, C. P. (1997) Effects of surface roughness on heat transfer and aerodynamic performance of turbine airfoils [J] Journal of Turbomachinery-Transactions of the ASME, 120 (3), 522–529.
[11] Häring, M., Bölcs, A., Harasgama, S. P., & Richter, J. (1994) Heat transfer measurements on turbine airfoils using the naphthalene sublimation technique. [J] Journal of Turbomachinery-Transactions of the ASME, 117 (3), 432–429.
[12] Andreini, A., Burberi, E., Cocchi, L., Facchini, B., Massini, D., & Pievaroli, M. (2015) Heat transfer investigation on an internal cooling system of a gas turbine leading edge model. [J] Energy Procedia, 82, 222–229.
[13] Davletshin, I., Mikheev, A., Mikheev, N., & Shakirov, R. (2020) Data on distribution of heat transfer coefficient and profiles of velocity and turbulent characteristics behind a rib in pulsating flows [J] Data in Brief, 33, 106485.
[14] Defraeye, T., Blocken, B., & Carmeliet, J. (2012) CFD simulation of heat transfer at surfaces of bluff bodies in turbulent boundary layers: Evaluation of A forced-convective Temperature wall function for mixed convection [J] Journal of Wind Engineering and Industrial Aerodynamics, 104-106, 439–446.
[15] Fitt, A. D., Forth, C. J. P., Robertson, B. A., & Jones, T. V. (1986) Temperature ratio effects in compressible turbulent boundary layers [J] International Journal of Heat and Mass Transfer, 29 (1), 159–164.
[16] Maffulli, R., & He, L. (2013) Wall temperature effects on heat transfer coefficient [J] Journal of Propulsion and Power.

Journal of Physics: Conference Series **2108** (2021) 012023 doi:10.1088/1742-6596/2108/1/012023

Frequency tracking and tuning control of wireless power transfer system

Ziyi Liu[1], Hao Zhan[1]

[1] China-EU Institute for Clean and Renewable Energy, Huazhong University of Science and Technology, Wuhan, Hubei, 430074, China

*Corresponding author's e-mail: m201971287@hust.edu.cn

Abstract. A tuning control method with frequency tracking function is proposed to improve the degradation of efficiency caused by coil detuning in magnetically-coupled resonant wireless power transfer.Firstly,the detuning mechanism is studied by combining the AC impedance characteristics of the series resonant circuit,and the feedback control circuit is constructed by using a modified phase-locked loop, which outputs a variable frequency PWM wave to regulate the operating frequency of the high frequency inverter and maintains the phase difference between the inverter output voltage and the original side current within the error range.The Matlab/Simulink simulation results show that the design can successfully transfer the system to a new resonant state with short regulation period and high control accuracy,which can effectively improve the transmission efficiency and load power of the system.

1. Introduction

In recent years magnetically-coupled resonance wireless power transfer(MCR-WPT)systems are gradually replacing the traditional power supply[1] in cell phones, robots,and electric vehicles by virtue of their advantages such as freedom from physical connection and flexibility of power supply.To ensure the stability and efficiency of the MCR-WPT system[2],its operating conditions need to be maintained in the resonant state[3].However,changes in transmission distance and system parameters in the actual environment will cause the resonant frequency of the system to shift[4]-that is, detuning will occur,thus affecting the transmission performance of the system.Therefore,the adaptation of the resonant frequency is a key technology for the MCR-WPT system[5].

The frequency tracking and tuning control proposed in this paper avoids the reduction of efficacy due to long-term operation of the system by controlling the on-off of the switch tubes of the high-frequency inverter link so that its output frequency always adapts to the changing system parameters with the same inherent resonant frequency in the transmitter.The system parameters are determined through simulation,and the modified phase-locked loop(PLL)maintains the inverted square wave voltage in the same frequency and phase as the transmitter current so that the input impedance of the system is purely resistive during resonance, which can provide a theoretical basis and solution for studying the MCR-WPT technology with frequency regulation.

2. Magnetically-coupled resonant wireless power transfer system model

2.1. MCR-WPT system structure

Figure 1 shows the MCR-WPT system using a series-series (S-S) topology.

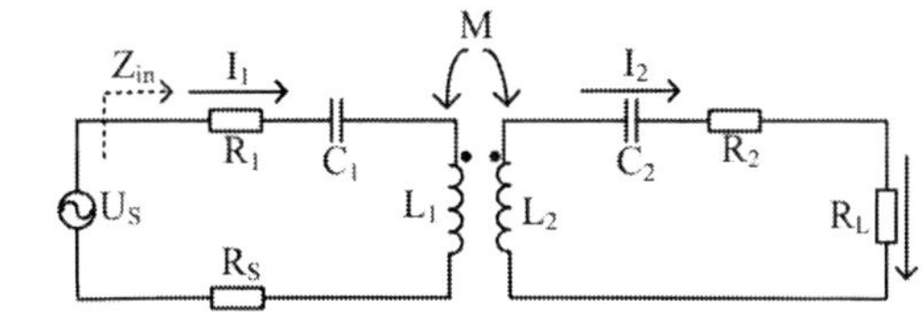

Figure 1. S-S type mutual inductance model.

U_s is the fundamental component of the high-frequency power supply,and R_s,R_1,R_2 are internal resistances.L_1,L_2 are the inductance of the coil at the transmitting end and the receiving end respectively; M is the mutual inductance;C_1, C_2 are the compensation capacitors;R_L is the load.

When the high frequency power supply operates at ω,the primary equivalent impedance Z_1,the secondary equivalent impedance Z_2 can be expressed as:

$$\begin{cases} Z_1 = R_1 + j\omega L_1 + \dfrac{1}{j\omega C_1} + R_s \\ Z_2 = R_2 + j\omega L_2 + \dfrac{1}{j\omega C_2} + R_L \end{cases} \tag{1}$$

where $M = k\sqrt{L_1 L_2}$ (k is the coupling coefficient).

The Kirchhoff's voltage law (KVL) equation is written for the mutual inductance model:

$$\begin{cases} Z_1\dot{I}_1 - j\omega M\dot{I}_2 - \dot{U}_s = 0 \\ \dot{I}_2 Z_2 - j\omega M\dot{I}_1 = 0 \end{cases} \tag{2}$$

This leads to the system output power and efficiency through equation:

$$\eta = \frac{P_{out}}{P_{in}} = \frac{I_2{}^2 R_L}{U_s{}^2 \,\mathrm{Re}(I_1)} = \frac{\omega^2 M^2 R_L}{Z_2\left[Z_1 Z_2 + (\omega M)^2\right]} \tag{3}$$

The optimal efficiency point corresponding to the frequency ω_0 can be found by making $\dfrac{d\eta}{d\omega} = 0$:

$$\omega_0 = \frac{1}{\sqrt{L_1 C_1}} = \frac{1}{\sqrt{L_2 C_2}} \tag{4}$$

At this point η takes the maximum value η_{max}:

$$\eta_{\max} = \frac{(\omega_0 M)^2 R_L}{(R_L + R_2)^2 (R_s + R_1) + (R_L + R_2)(\omega_0 M)^2} \tag{5}$$

Looking in from the power side of Figure 1, introduce the reflected impedance $Z_r = \dfrac{(\omega M)^2}{Z_2}$,the MCR-WPT system can be equated to a passive one-port network with an input impedance of:

$$Z_{in} = Z_1 + Z_r = \mathrm{Re}(Z_{in}) + j\,\mathrm{Im}(Z_{in}) \tag{6}$$

It is known that $Im(Z_{in})=0$,so impedance angle $\varphi = \arctan\dfrac{\mathrm{Im}(Z_{in})}{\mathrm{Re}(Z_{in})} = 0$ in (6), which means that Z_{in} presents a purely resistive form[6], the system resonates at frequency ω_0, and the most significant feature is that the voltage and current at the transmitter are in the same frequency and phase[7].

2.2. Influence of detuning on transmission characteristics

In actual operation, due to the error of the coil inductance value and other factors, the resonant frequency will gradually shift, if ω_0 is maintained as the original value, the complete resonant state is destroyed, that is, detuning occurs, the formula (6) is no longer a pure real number.The efficiency and power curves under detuning condition are shown in Fig.2.When the primary side is detuning, η_{max} occurs at a frequency significantly shifted away from the fully resonant frequency of 300KHz; relatively, the detuning phenomenon has a less obvious effect on the power.

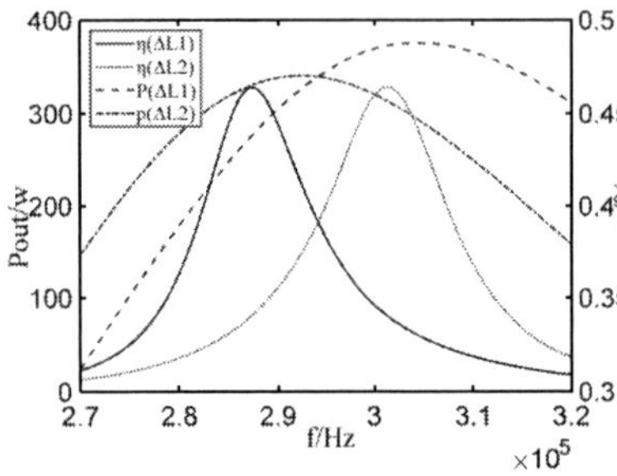

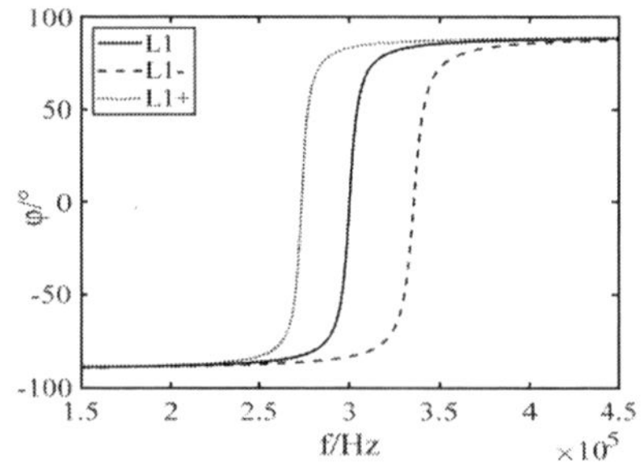

Figure 2. η and P_{out} at detuning operating conditions. Figure 3. The offset curve of φ with f.

As can be seen,the change of inductance at the transmitter ΔL_1 is the main factor causing the frequency shift.Therefore,ω_0 needs to satisfy the following condition when the system resonates:

$$\omega_0 = \omega_1 = \frac{1}{\sqrt{L_1 C_1}} \tag{7}$$

ω_1 is the intrinsic resonant frequency of L_1.The curve of φ shifted with the operating frequency is shown in Fig.3.When L_1 occurs incremental or decremental detuning,$\varphi(\omega_0)\neq 0$,ω_1 changes to ω_1' and no longer satisfies the formula (7).

3. Frequency track control method

In order to solve the detuning problem,this paper designs a frequency tracking control method,when ω_1 is shifted to ω_1',the operating frequency of HF inverter is changed online to ω_0', so that the adjusted system conforms to (8) and ensures a new resonant working state:

$$\omega_0' = \omega_1' \tag{8}$$

In this paper,we propose a frequency tracking and tuning control method based on an improved PLL.The control block diagram is shown in Figure 4.The primary current sampling signal $u_{in}(\theta_{in})$ and the feedback control signal are processed by the phase and frequency detector(PFD) composed of second-order generalized integrator (SOGI) to obtain the impedance angle information u_q,u_q is calculated by the loop controller(ARC) to obtain the control input u_c, according to which the voltage controlled oscillator(VCO) generates the feedback control signal $u_{out}(\theta_{out})$ whose frequency varies linearly with u_c,and the pulse frequency modulation(PFM)signal generator generates the frequency adjustable PWM waves V_1-V_4 under the action of u_{out},which completes the frequency tracking control of the high frequency inverter through the driving circuit.The control system is shown in Figure 5.The implementation method of each main part is described below.

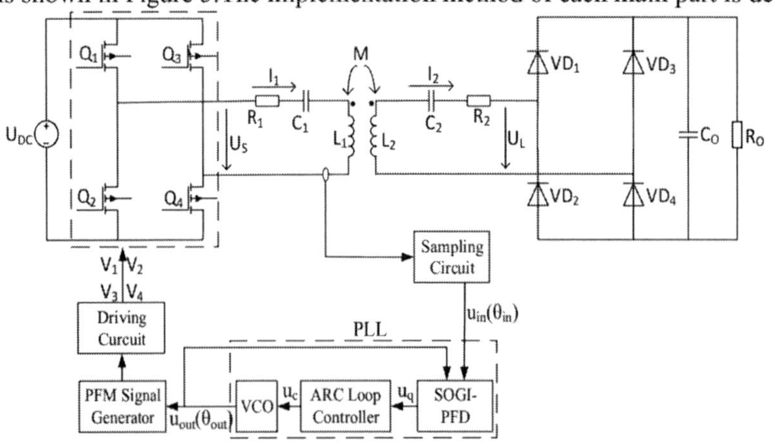

Figure 4. The proposed frequency tracking tuning control.

3.1. phase and frequency detector with second-order generalized integrator

The model of SOGI-PFD is shown in Figure 5.

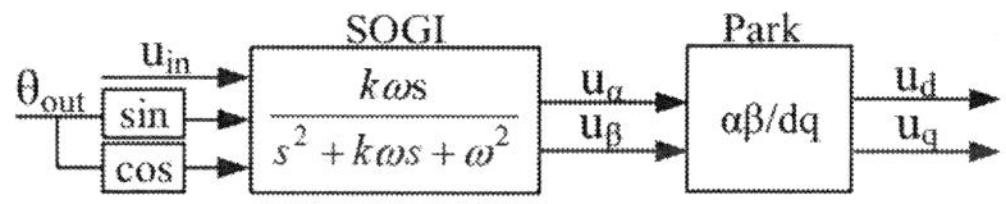

Figure 5. The SOGI-PFD.

When the primary signal is $u_{in}=U_{in}\sin\theta$, set actual phase as $\hat{\theta}$,SOGI extracts two orthogonal sinusoidal signals from it as:

$$\begin{cases} u_\alpha = U_{in}\sin\hat{\theta} \\ u_\beta = U_{in}\cos\hat{\theta} \end{cases} \tag{9}$$

After the Park coordinate transformation,we get $u_q=U_{in}\sin(\hat{\theta}-\theta)$.When the system is stable, the difference between $\hat{\theta}$ and θ is very small,according to the equivalence infinitesimal principle $\lim\limits_{\hat{\theta}\to\theta} U_{in}\sin(\hat{\theta}-\theta)=U_{in}\sin(\hat{\theta}-\theta)$,u_q can be approximated as the phase difference value.

The SOGI structure introduced in this paper is not affected by the original edge,and high dynamic response can be achieved by designing reasonable parameters.

3.2. loop controller and voltage controlled oscillator

In this paper, a second-order active proportional integral filter is used for loop control.Set τ_1,τ_2 as the parameters to be designed.Let the transfer function of ARC be:

$$F(s)=\frac{1+s\tau_2}{s\tau_1} \tag{10}$$

VCO is an integral section in the S domain and the transfer function is:

$$N(s)=\frac{K_{VCO}}{s} \tag{11}$$

K_{VCO} is the gain, and to ensure the sensitivity of VCO, K_{VCO}= 20000Hz/v.

Introduce the intrinsic frequency and damping coefficient:

$$\begin{cases} \omega_n = \sqrt{\dfrac{K}{\tau_1}} \\ \xi = \dfrac{\tau_2}{2}\sqrt{\dfrac{K}{\tau_1}} \end{cases} \tag{12}$$

The PLL closed-loop transfer function is:

$$H(s)=\frac{K_d F(s)N(s)}{1+K_d F(s)N(s)}=\frac{2\xi\omega_n s+\omega_n^2}{s^2+2\xi\omega_n s+\omega_n^2} \tag{13}$$

The loop gain $K=K_{VCO}K_d$.In order to obtain a better transient response,the damping factor ξ= 0.707 is taken; 0.5 times the fundamental frequency is chosen as the bandwidth to effectively filter out high frequency noise.

$$B_L=\int_0^{+\infty}|H(j2\pi F)|\,dF\approx\frac{\omega_n}{8\xi}\left(1+4\xi^2\right) \tag{14}$$

Let B_L=300000π rad/s,so $\omega_n\approx 592384.4$ which substituted into (14) yields the parameters of the ARC of τ_1=2.39e-6,τ_2=1.181e-8.

Figure 6 shows the second-order loop bode diagram of the PLL. Whenξ=0.707, the phase marginγ=65°, the loop can work stably and maintain a strong immunity to interference. In addition to this, when there is a frequency difference in the input signal, the phase difference at steady state can be guaranteed to be 0 due to the presence of a near-ideal product link in its transfer function (10).

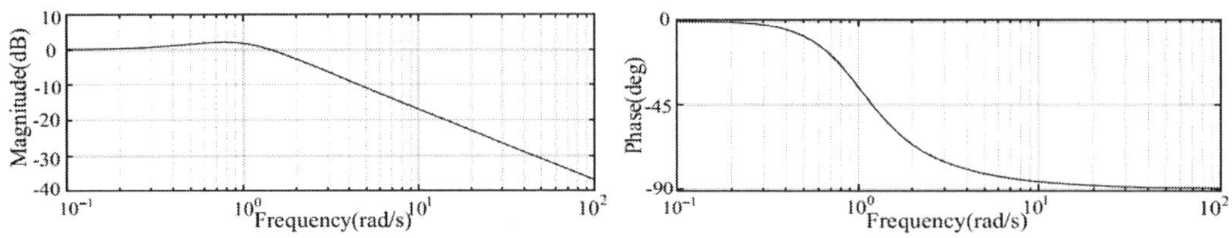

(a)Magnitude (b)Phase

Figure 6. The Bode diagram of PLL system.

4. Simulation experiments and results

4.1. Space considerations

In this paper, the system is tested using Matlab/Simulink, and the circuit parameters are shown in Table 1.The system is detuned twice in the operation.The input impedance angle is shown in Figure 7.At 0.002s,φ first undergoes a short overshoot phase and then frequency tracking control can quickly achieve tuning; at 0.004s, the PLL can still precisely control the fluctuation range of φ.

Table 1. The main parameters.

Symbol	Valuc
L_1 (L_2)	24.7e-6
C_1 (C_2)	11.4e-9
R_1 (R_2)	1.1
R_L	10
U_{DC}	40
M	3e-6

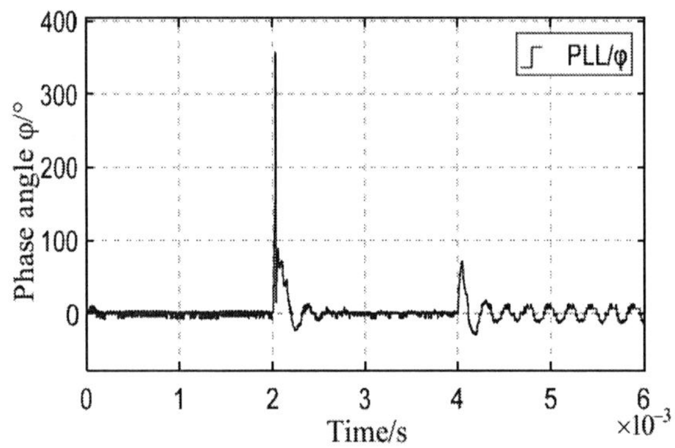

Figure 7. The impedance angle.

The system voltage and current waveform are shown in Fig8.In 8(a),At 0.002s,the resonant frequency drifts, and the figure shows that U_1 and I_1 are no longer in phase; then the PLL circuit starts frequency tracking and the phase difference gradually decreases; finally the system shifts to the new resonant frequency *f*=273540Hz.

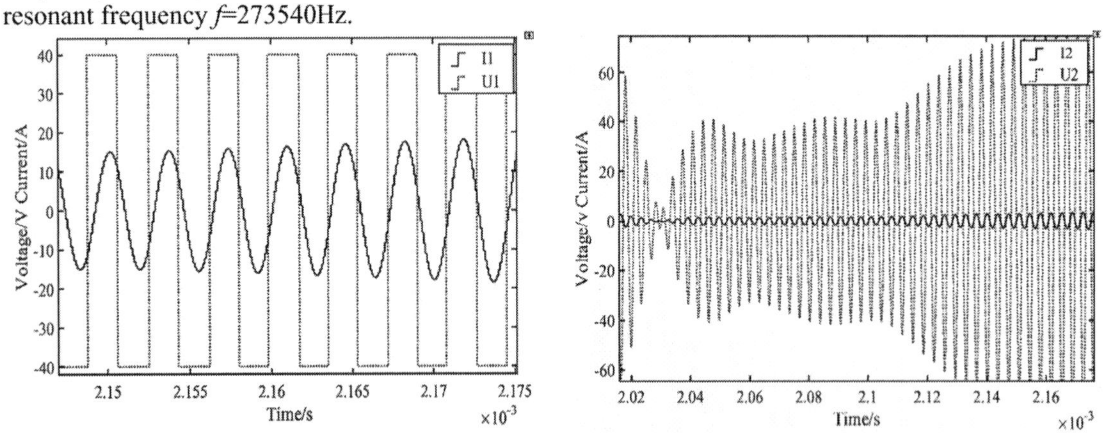

(a)Transmitter voltage U_1 and current I_1. (b)Receiver voltage U_2 and current I_2.

Figure 8. The system voltage and current waveform.

In 8(b),when the system is adjusted, the U_2 waveform is more stable and the amplitude increases compared with that of before detuning, indicating that the frequency tracking effect of the PLL circuit can improve the quality of the load voltage waveform while increasing the transmission efficiency.

4.2. Simulation Data Analysis

Based on the simulation results, the output power P_{out} and the system efficiency η are plotted with the degree of detuning (time),as shown in Figure 9.

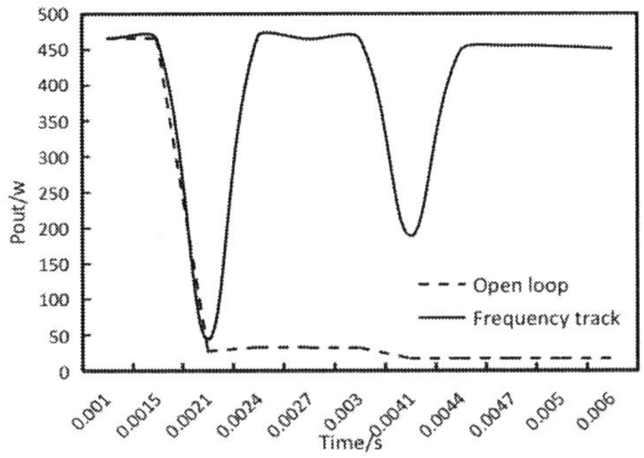

(a)P_{out} varies with the degree of detuning

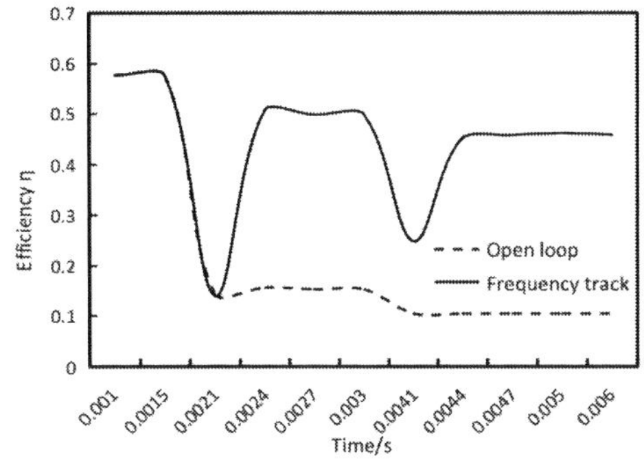

(b)η varies with the degree of detuning

Figure 9. Transmission characteristics.

Taking the open-loop system as the reference, the occurrence of detuning has an irreversible influence on the system - the frequency is shifted from the global optimal resonance point and leads to power and efficiency degradation, while the tuning method proposed in this paper cannot completely eliminate the adverse effect, but the local optimal resonance point can be found after 3e-4s of frequency tracking, thus improving the system characteristics and substantially recovering P_{out} and η.

5. Conclusion

The MCR-WPT system needs to change the driving frequency to improve the transmission performance due to the detuning caused by the change of coil parameters in actual operation.To address this problem, this paper proposes a frequency tracking control tuning method through mathematical modeling of the detuning mechanism, which quickly reverts the phase difference between the inverter output voltage and the transmitting coil current within the error range after the system detuning, and actively adjusts the system to a new resonant state with high control accuracy.It can be verified by simulation that the transmission efficiency and load power of the MCR-WPT system can be effectively improved after adopting frequency track.

Acknowledgments

This work was financially supported by China-EU Institute for Clean and Renewable Energy, Huazhong University of Science and Technology.

References

[1] Zhu, Y., Zhang, H., Wang, Z., Cao, X., & Zhang, R. (2021). Design of a high-power high efficiency multi-receiver wireless power transfer system. Electronics, 10(11), 1308.

[2] Gao P, Tian Z, Tian X. Analysis of the Frequency Characteristics of MCR-WPT System with Asymmetric Coils Structure. IEEJ transactions on electrical and electronic engineering. 2021.

[3] Yang J, Shi Y, Wei WY, Shen H. A wireless power transfer system based on impedance matching network. International journal of RF and microwave computer-aided engineering. 2020;30(12):n/a.

[4] Lin Y, Zhao Z. Topology and control strategy on transformerless wireless power station for future electric transportation systems. International transactions on electrical energy systems. 2021.

[5] Moon J. High-frequency capacitive wireless power transfer technologies. Journal of Power Electronics. 2021.

[6] Tang J, Zhang Q, Cui C, Na T, Hu T. An Improved Hybrid Frequency Pacing Modulation for Wireless Power Transfer Systems. IEEE transactions on power electronics. 2021;36(11):12365-74.
[7] Tang L, Jeng S, Chang E, Chieng W. Variable-Frequency Pulse Width Modulation Circuits for Resonant Wireless Power Transfer. Energies (Basel). 2021;14(12):3656.

Simulation Research on Electric Field Distribution of Insulator String in ULTRA High Voltage Transmission Line

Weihao Shen[1]

[1] Institute of electrical from Xi'an JongTong University, Xi'an, Shaanxi, China, 710049

*Corresponding author's e-mail: shenweihao@stu.xjtu.edu.cn

Abstract. Insulators are one of the important components of overhead transmission lines. Because of their excellent characteristics, insulators have obvious advantages in the selection of external insulation in UHV AC transmission projects. The potential distribution of insulator string is not even due to the shape characteristics of insulator string, the structure of metal tool, the low conductivity of silicon rubber material and the stray capacitance between insulator string and wire, iron tower and metal tool. When the surface field intensity of insulators and fittings exceeds the initial field intensity, corona discharge will occur, which will affect the electromagnetic environment and the operation characteristics of insulating materials. Therefore, it is of great significance to determine the potential distribution of insulator strings of UHV transmission lines for detecting low and zero value insulators, implementing voltage equalization measures effectively, and ensuring the safe and stable operation of transmission lines. Based on Ansys software of 500kv transmission line insulator string, this paper establishes a three-dimensional model, the method of using finite element numerical simulation respectively with no equalizing ring and equalizing ring in both cases the umbrella skirt on simulation, the distribution of electric field along the roots, and further studies the equalizing ring the effect of different parameters on the electric field distribution. Finally, an optimal solution is established.

1. Introduction

As the insulation support of overhead transmission line, insulator plays an important role in EHV transmission project. Its performance directly affects the safe operation of the whole line. However, due to the very high voltage level, the length of insulator string increases sharply. Affected by the distributed capacitance of tower and high-voltage conductor, the electric field distortion of high-voltage conductor side and tower grounding side of insulator string is very serious. Usually, a grading ring is installed on the insulator string to increase the distributed capacitance between the insulator and the conductor, so as to effectively improve the voltage distribution of the insulator string. The grading ring can improve the potential distribution of the composite insulator and control the field strength in the high electric field area, prevent corona discharge, leakage trace and electric erosion, reduce audible noise, reduce the influence of ion current under DC field, and eliminate the degradation of non ceramic materials caused by corona [1]. Therefore, the optimal design of grading ring is an indispensable part in engineering application [2].

For decades, people have proposed various methods to measure the field strength distribution of insulator string, such as capacitance equivalent circuit method, analog charge method and mutually related field-circuit method [3, 4]. However, there is an inherent problem to be overcome in the measurement process, that is, the arrangement of measuring probe changes the electric field distribution of insulator string, which brings some systematic errors, and the test method has a long cycle and high cost, difficult to implement. In recent years, due to the great improvement of computer software and hardware, finite element method is widely used to solve electromagnetic field problems. It is not limited by environment and conditions, and can change the size and position of the calculation model at will, with good flexibility. The difficulty of optimizing the design of pressure equalizing ring is greatly simplified [5, 6, 7, 8].

In this paper, ANSYS finite element simulation software is used to simulate and analyze the potential and electric field without and with voltage equalizing ring, and further study the influence of various parameters of voltage equalizing ring on the electric field intensity distribution with voltage equalizing ring. It provides a good theoretical basis for installing grading ring on UHV transmission line in engineering.

2. The establishment of 3-D model

According to the actual situation of 500kV AC transmission line in China, a three-dimensional model of 500kV AC tower insulator string is established. The main parameters of the model in the modeling process are shown in Table 1. The tower type is wine glass tower, and the influence of the hollow network structure of the tower on the model simulation results is ignored. The insulator on the cross arm adopts vertical type I suspension mode, and the model of 4-bundle conductor is LGJ400 / 35.

Table 1. Main parameters of simulation model.

Essential information	Main simulation parameters
Length of insulator structure/cm	445
Mandrel diameter/cm	2.4
Sheath thickness/cm	0.6
Large and small umbrella diameter/cm	17, 8.4
Big umbrella spacing/cm	8
Diameter of pressure equalizing ring/cm	40
Diameter of equalizing ring pipe/cm	6
Shielding depth/cm	5
Height, width and thickness of tower/cm	3300, 400, 50
Length, width and thickness of cross-arm/cm	1000, 400, 50
Relative dielectric constant of FRP	7.2
Relative dielectric constant of silicone rubber	4.3

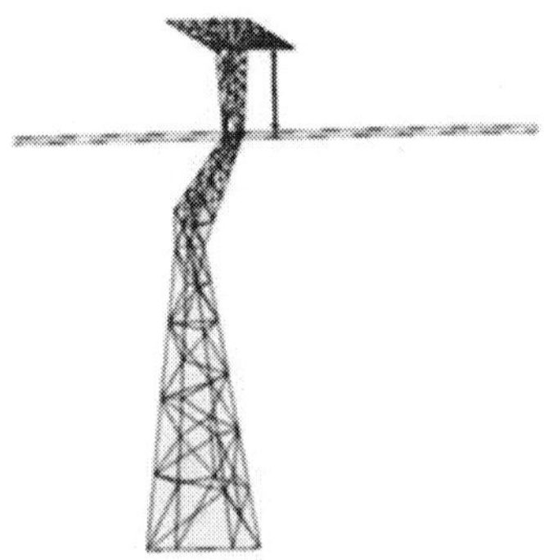

Figure 1. Tower concept drawing [5].

Although the electric field distribution of insulator string is not a strictly axisymmetric electric field, from the perspective of Engineering approximation, this paper considers that its electric field distribution is axisymmetric to the tower, so in the subsequent research, 1/2 tower model is taken for Simulation Research (as shown in Figure 1).

At the same time, some parameters of the model are also appropriately simplified during the research process:

(1) In the process of modeling, the influence of tower on the electric field distribution of insulator string can not be ignored, but its shape can be appropriately simplified, and the tower and cross arm can be simplified into two cuboids, which are made of steel;

(2) In the process of modeling, the influence of conductor and connecting fittings on the electric field distribution of insulator string can not be ignored, but the appropriate length can be selected to establish the three-phase conductor model. In this paper, the length of conductor is 400cm, and the material of conductor and connecting fittings is aluminum;

(3) In the process of modeling, it is generally considered that the phase to phase distance of three-phase AC is large, and the grounding tower has a certain shielding effect on the phase to phase electric field, so the phase to phase influence can be ignored, and only the simplified model of medium phase V-string can be established;

(4) The size of the umbrella skirt on the protective sleeve and insulator string is made of silicone rubber, and the insulator core rod in the protection is made of glass fiber reinforced plastic;

(5) Move the boundary at infinity to an appropriate distance close to the insulator, that is, replace the infinity boundary with a finite boundary. This treatment will enhance the electric field near the insulator, but because the overall size of the composite insulator is very small and the power line is very sparse not far away, the error to the calculation result is also small.

Finally, the excitation and boundary conditions are added to the established 1/2 tower simulation model: during the simulation, the lower end of the sub is connected with hardware and wires to add high-voltage excitation:$U_0 = 500/\sqrt{3} \times \sqrt{2} \times 1.1$; add 0 excitation to the upper side of tower, cross arm, insulator and the outer surface of air domain.

The final ANSYS modeling results are shown in Figure 2 and Figure 3 below:

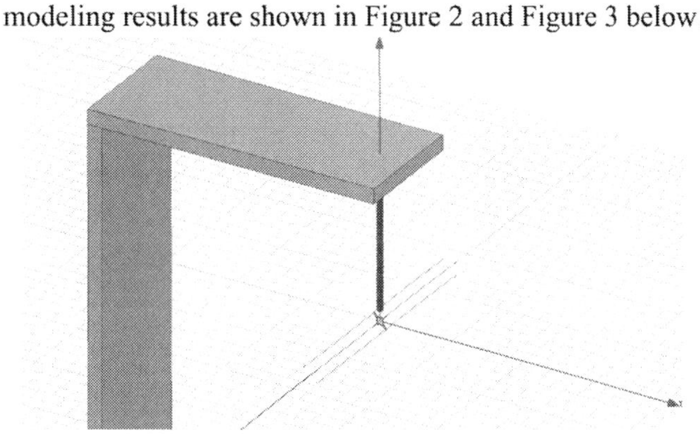

Figure 2. Model of 1/2 tower insulator string without grading ring.

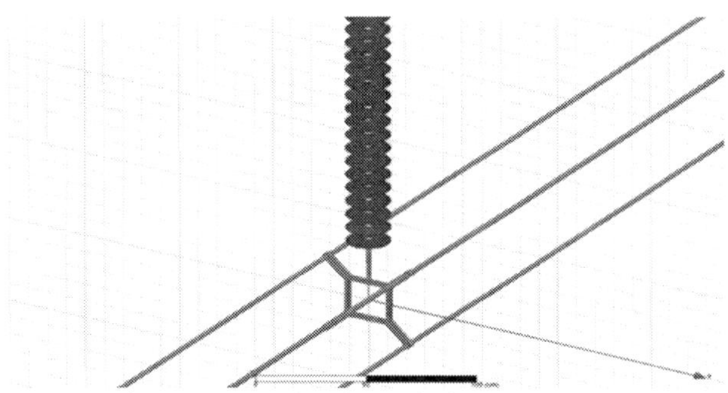

Figure 3. Local enlarged view of model.

3. Analysis of simulation results

3.1. Effect of pressure equalizing ring with or without on electric field and potential distribution

The 500kV AC-tower-insulator string 3D model established above can analyze the electric field intensity distribution along the connection between the root of the umbrella skirt and the protective sleeve (Figure 4) :

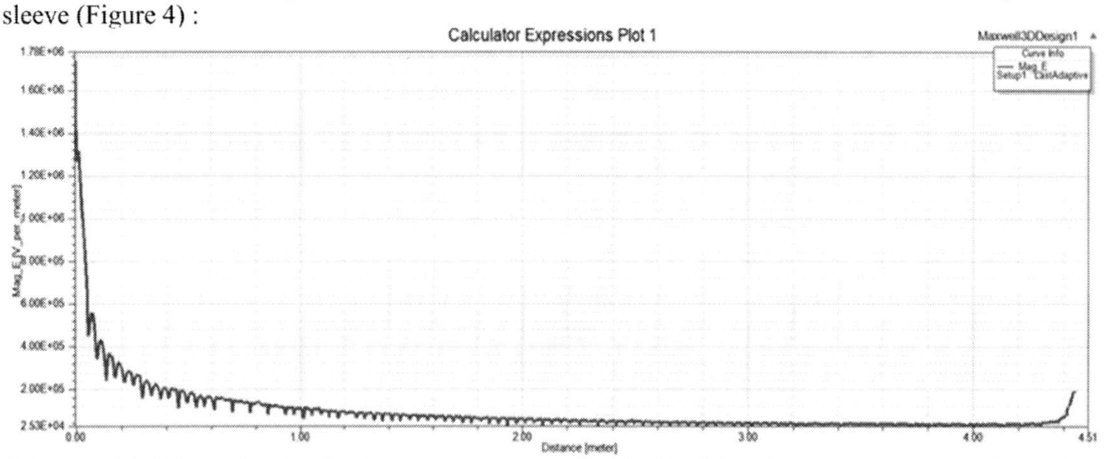

Figure 4. Field intensity distribution along the root of umbrella skirt without pressure equalizing ring.

As can be seen from the image, the electric field distribution along the root of the umbrella skirt is not uniform when there is no pressure equalizing ring, and the electric field intensity and potential reach the maximum at the point close to the connection point of the fittings and protective sleeve. Due to its shape characteristics, metal structure and low conductivity of silicone rubber material, the potential distribution of insulators attenuates rapidly from the high voltage end. Such potential distribution generates a high electric field at the adjacent high voltage end and the ground end. If the electric field intensity on the insulator surface exceeds the corona starting field, a corona discharge will occur. Obviously, in the absence of voltage equalizing ring, the electric field intensity on the high voltage side of insulator string far exceeds the corona field intensity.

Firstly, it is considered to install a pressure equalizing ring at the high voltage end of the insulator string on the basis of the above simulation model. The tube radius of the pressure equalizing ring is R= 6cm, the ring radius is R=40cm and the shielding depth is D =5cm, as shown in Figure 5.

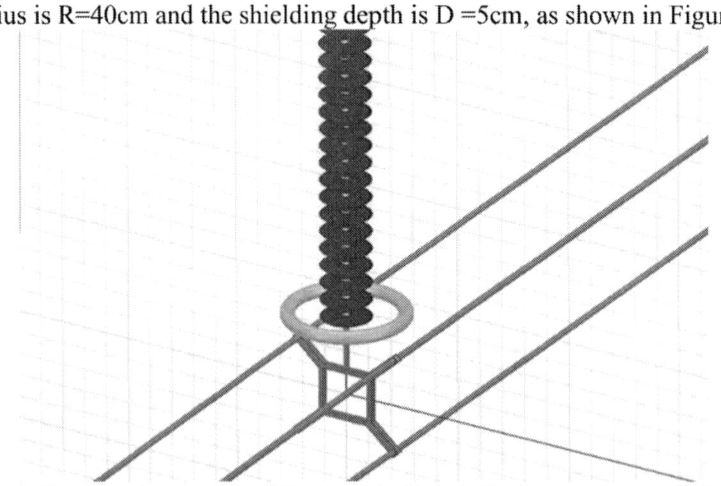

Figure 5. Partial enlargement of the model after adding pressure equalizing ring.

The simulation results show that the electric field intensity along the root of the umbrella skirt is significantly inhibited by the installation of a pressure equalizing ring at the high voltage end, namely at the bottom of the insulator string. At this time, the electric field intensity of the insulator string reaches its maximum value at the distance of about 40cm (39.58cm) from the connection between the protective sleeve and the harness.

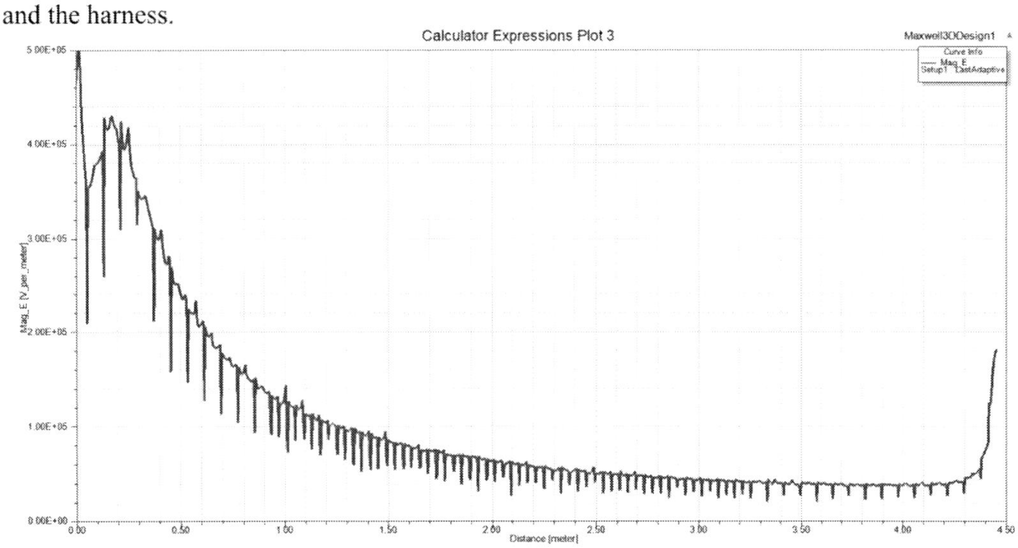

Figure 6. Field intensity distribution along the root of umbrella skirt with pressure equalizing ring.

After the voltage equalizing ring is installed at the high voltage end of the insulator string, the electric field intensity of the high voltage side along the root of the umbrella skirt changes obviously (as is shown in Figure 6), but the position where the maximum field intensity appears is almost unchanged. Meanwhile, the electric field intensity of the low voltage side does not change significantly. The potential distribution along the line has no significant change. Therefore, we focus on the influence of several different parameters of the pressure equalizing ring on the distribution of electric field intensity.

3.2. Effect of shielding depth on electric field intensity

The installation of pressure equalizing rings with different shielding depths is studied. This paper keeps the diameter of the pressure equalizing ring D=40cm and the diameter of the pressure equalizing ring D= 6cm unchanged, then changes the shielding depth H to increase gradually from 5cm, and obtains the electric field intensity distribution curve along the root of the insulator string umbrella skirt when H =10cm, H =15cm, H =20cm and h=25cm respectively, and takes out the maximum group of electric field intensity of each group separately. Excel charts were drawn for comparison (Figure 7, Table 2).

Table 2. Maximum electric field intensity at different shielding depths.

Shielding depth/cm	Maximum electric field intensity/ (V/m)
5	499086
10	467158
15	447526
20	523203
25	685307

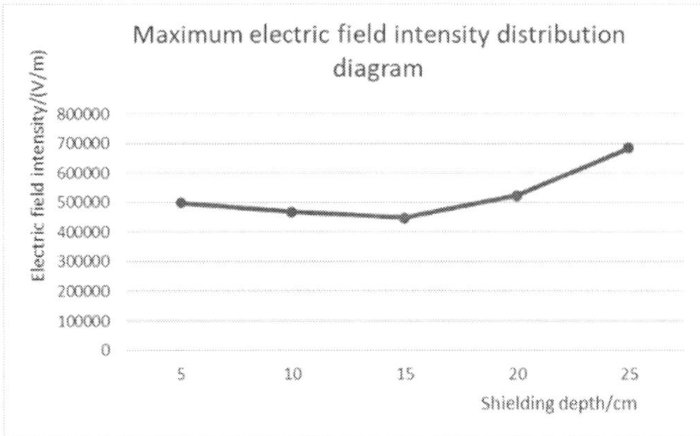

Figure 7. Maximum electric field distribution diagram at root of umbrella skirt with different shielding depth.

As can be seen from Figure 8, with the increase of shielding depth, the maximum electric field intensity E_max on the surface along the root of insulator string skirt decreases first and then increases. This is because the main position of electric field distortion of pressure sharing ring changes with the change of shielding depth. In addition, too deep shielding depth will lead to a substantial decrease in the lightning resistance level of insulators. Therefore, the field intensity distribution at the high voltage side of composite insulators can be improved by adjusting the shielding depth appropriately. The minimum value is 447526V/cm when the shielding depth is 15cm.

3.3. Effect of pressure equalizing ring diameter on electric field intensity

The pressure equalizing rings with different diameters were studied. This article keeps the shielding depth of pressure equalizing ring h=15cm, the diameter of pressure equalizing ring tube D= 6cm, then changes the diameter of pressure equalizing ring from 30cm, takes a set of data every 10cm, and gets d= 30cm, D=40cm, D=50cm, D=60cm, D=70cm, respectively. The electric field intensity distribution curve along the root of insulator string skirt when D=80cm, and the maximum group of electric field intensity of each group was taken out separately to draw excel charts for comparison (Figure 8, Table 3).

Table 3. Maximum electric field intensity at root of umbrella skirt with different diameters of pressure equalizing ring.

Diameter of pressure equalizing ring/cm	Maximum electric field intensity/ (V/cm)
30	512564
40	447526
50	496325
60	709674
70	648777
80	635806

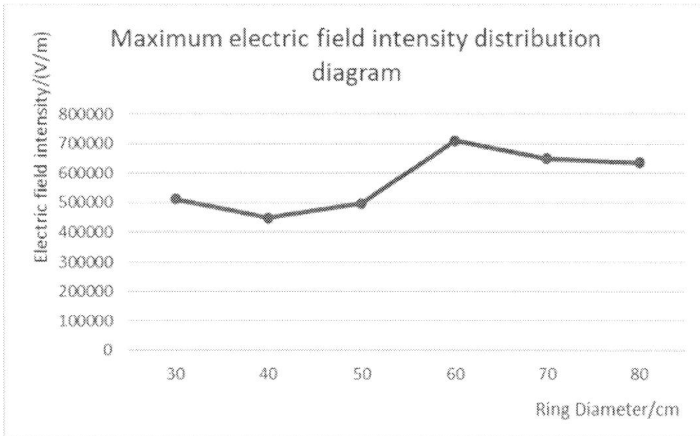

Figure 8. Distribution diagram of maximum electric field intensity at root of umbrella skirt with different pressure equalizing ring diameters.

It can be seen from Figure 8 that as the diameter of pressure equalizing ring increases, the maximum electric field intensity E_max on the surface along the root of insulator string skirt decreases first and then increases. The minimum value is 447526V/cm when the diameter of the pressure equalizing ring is 40cm, so it is more appropriate when the diameter of the pressure equalizing ring is about 40cm.

3.4. Effect of pressure-equalizing ring diameter on electric field intensity

The pressure equalizing rings with different diameters of pressure equalizing rings were studied. This article keeps the shielding depth of pressure equalizing ring h=15cm and the diameter of pressure equalizing ring D =40cm, then changes the diameter of pressure equalizing ring tube from 4cm, takes a group of data every 1cm, and obtain the electric field intensity distribution curve along the root of insulator string umbrella skirt when D =4cm, D =5cm, D =6cm, D =7cm and D =8cm, respectively. The maximum group of electric field intensity of each group was taken out separately and an Excel chart was drawn for comparison (Figure 9, Table 4).

Table 4. Maximum electric field intensity at root of umbrella skirt with different ring diameters.

Loop diameter/cm	Maximum electric field intensity/ (V/cm)
4	539495
5	504197
6	447526
7	443634
8	448346

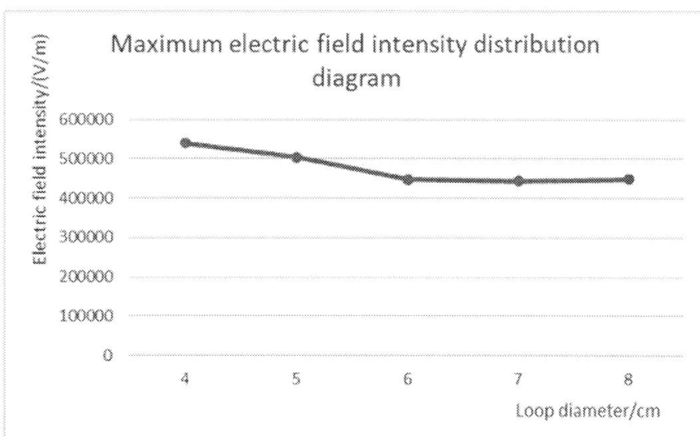

Figure 9. Distribution diagram of maximum electric field intensity at root of umbrella skirt with different pipe diameters.

It can be seen from Figure 9 that the maximum electric field intensity E_max on the surface along the root of insulator string skirt decreases first and then increases with the increase of the diameter of pressure equalizing ring tube. The minimum value is 443634V/cm when the diameter of pressure equalizing ring pipe is 7cm, so it is more appropriate when the diameter of pressure equalizing ring pipe is about 7cm.

4. Conclusion

The distribution of electric field intensity at the root of umbrella skirt of 500kV UHV insulator string is approximately u-shaped, and the maximum electric field intensity of insulation part appears near the connection of high voltage end mandlet, protective sleeve and fittings. After installing a voltage equalizing ring at the high voltage end of insulator string, the electric field intensity at the root of insulator string parachute skirt can be greatly reduced. Meanwhile, changing the structural parameters of voltage equalizing ring will also affect the distribution of electric field intensity at the root of insulator string parachute skirt. The optimal solution can be obtained when the shielding depth is about 15cm, the diameter of pressure equalizing ring is about 40cm, and the diameter of pressure equalizing ring is 7-8cm. The conclusion can provide technical support for the design of new pressure equalizing ring of 500kV transmission line in the future.

Acknowledgements

It took the author nearly two months to finish writing this paper. In this process full of struggle, I have encountered countless difficulties and many warm-hearted help. Here, I especially wanted to thank my instructor, Ms. Shen Yao, for her tireless guidance and help and selflessly revising the paper for me, There will be no final completion of this paper.

In addition, I would also like to thank my dear classmates and friends such as Jia Yu, fan Sixue and Qian Xingyou for their help in the collection of data sources and standardization of format of my thesis; when I was busy, they reminded me of the paper matters to pay attention to and the tasks to be completed in time, which was conducive to my smooth completion and submission.

References

[1] Lemiao Wang, Electrical performance analysis and fault diagnosis of uHV line insulators [D] Shandong University of Technology, 06 (08), 2020.

[2] Dengyun Li, Kai Zhu, Zhicheng Li, Yang Liu, Maowu Liu, Kaipei Liu, Optimal Design of pressure Sharing ring of 1100 kV DC High Voltage Generator based on orthogonal test [J] Journal of Wuhan University, Vol. 54, No. 7, Jul. 2021.

[3] Qinghui Zhang, Hanyang Zan, Minghua Zhang, Surface field intensity calculation method of UHVDC transmission line based on simulated charge method [J] Journal of Qingyuan Polytechnic, Vol.8, No.1, JAN, 2015.

[4] Jiajing Li, Zhen Dong, Kai Qu, Calculation of distribution voltage of insulator string of 500 kV line based on field circuit connection method [J] High Voltage Apparatus, Vol.54, No.8, Aug, 16, 2018, 0021-0026.

[5] Guiyang Chen, Li Cheng, Limming Wang, Huikun Pei, Xing Yang. Optimization design of electric Field Distribution for 500KV AC composite insulator. SHAANXI ELECTRIC POWER, Vol.43, No.3, 2015.

[6] Yan Tang, Shangang MA, Zhaonian LI, Yan Yang, Simulation analysis of 750 kV transmission line insulator string voltage distribution based on ANSYS [J] Journal of Qinghai University(Natural Science Edition), Vol.30, No.1, Feb, 2012.

[7] Mengting Wei, Feng Huo, Jiangjun Ruan, Congpeng Huang, Xiaodong Wan, Electric field distribution calculation and parameter optimization design of voltage equalizing ring for inverted V insulator string in UHV AC transmission line [J] Advanced Technology of Electrical Engineering and Energy, Vol.35, No.10, Oct, 2016.

[8] Li Zhou, The optimization design of pressure equalizing ring for 750kV transmission line is discussed [J] Decision Exploration (Middle), Nov. 2020.

Application of Automatic Code Generation Based on PSIM in Digital Control BUCK Circuit

Yang LU[1] and Puqiong YANG[1*]

[1] School of Electrical Engineering, University of South China, Hengyang, Hunan ,421001, China

[*]Corresponding author's e-mail: yangpuqiong@usc.edu.cn

Abstract. In recent years, digital control has been widely used in the field of power electronics. Automatic code generation, as a graphical software writing method, has brought great convenience to the writing of electronic control software due to a series of advantages such as easy to write and easy to simulate and verify. Simcoder in PSIM is an automatic code generation software that enables the automatic generation of complete DSP control program projects. In this article, the current and voltage loops of a typical average current controlled BUCK circuit are designed as an example, and the use of PSIM and Simcoder is described in detail. Based on the main circuit and Matlab parameter calculations, the model is simulated and the self-generated code is downloaded into the TMS320F28335 to test the output voltage and current of the hardware circuit. The circuit output has the advantage of stable output voltage with small steady-state error and low voltage overshoot, demonstrating the practicality and efficiency of automatic code generation on this project.

1. Introduction

As a graphical software writing method, automatic code generation has been widely used in the field of power electronic digital control in recent years due to a series of advantages such as easy writing and easy simulation verification, which had brought great benefits to the writing of electronic control software. Conveniently, Simcoder in PSIM is a software that can automatically generate a complete set of DSP control program engineering, and can generate product-level readable and fast in execution embedded code.

Reference[1] used graphical programming to establish a 160km/h standard EMU air-conditioning inverter power control strategy model under the environment of automatic code generation, and executed the automatic code generation process; Reference [2] proposed a method based on Simulink The CAN underlying module library design method of automatic code generation technology used S-Function to design and configure CAN channel information modules and CAN message transceiver modules that can import DBC files, edit TLC files, and used RTW (real-time workshop) to implement Simulink Module automatic code generation. Reference [3] is based on the use of DSP's quadrature encoded pulse (QEP) module and M/T algorithm to design the motor speed detection system, and the detection results were transmitted to the upper computer through the DSP serial port, which realized the real-time display of the motor speed. Reference [4] used automatic code technology to realize the design platform of automobile electronic control system, realized the development of the drive module required by the control system through the Matlab platform, and designed the hardware module of the automobile control system, created the software model of the automobile electronic control system, and used the fuzzy controller to Adjust the PI parameters in the control system. Finally, the designed automobile control system is tested. The test results show that the designed software has certain advantages in efficiency, and it has good resistance and stability in the actual system control process.

This paper first introduces the principle and work flow of automatic code generation based on PSIM, then builds a digital model of average current- mode control, introduces the design of the circuit parameters, and introduces the use of PSIM and Simcoder. After software simulation, the Simcoder The generated code is downloaded to TMS320F28335 and run. Through experiments on the hardware circuit, the validity of the automatically generated code on the circuit is verified.

2.Principle and process of automatic code generation

2.1 Principle

SimCoder is a tool for automatic code generation with the help of PSIM software. The control circuit can be converted into a digital control program by this tool, and the circuit can be actually replaced by DSP for another simulation, and finally the control program that has been verified by simulation can be simulated. Burn into the DSP chip, and do control and communication through the DSP to verify the correctness of the design circuit and controller.

2.2 Work process

As shown in Figure 1, when using automatic code generation, first build a PSIM model according to the design requirements of the control software, and convert the control loop in the model to a digital control loop. The SimCoder tool in PSIM can convert the digital control loop into a complete DSP project. Complete DSP project, compile the project and burn it into the DSP chip, control the actual hardware circuit, and receive the data from the hardware circuit through the interface of the DSP board.

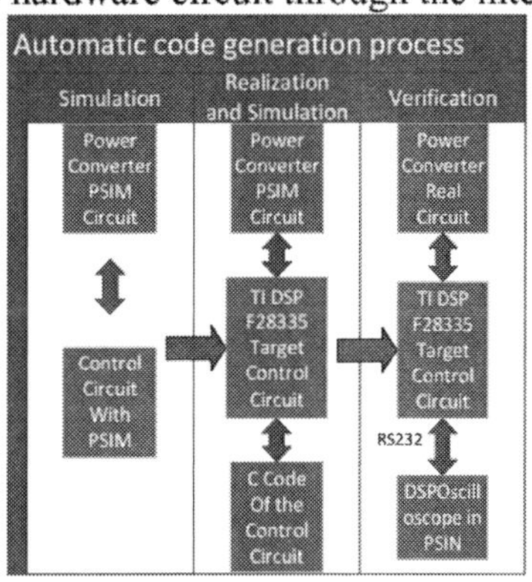

Figure 1. Automatic code generation process

2.3 Features

Power electronics analysis, design, simulation and physical verification can be performed Simultaneously. It is highly efficient. It is mainly reflected in the graphical software design. There is no need to consider the grammatical problems of complex programming languages. Only need to focus on the research and development of algorithms, which can be easily controlled complex control logic and algorithms; efficiency is also reflected in the ability to simulate, through simulation, design vulnerabilities can be found and resolved before the software is deployed, which greatly reduces the time cost of debugging. Under the PSIM, the programming of the program is completed by the way of establishing the hardware circuit and the programming is carried out, so that the developer who does not know the programming of DSP can easily complete the programming of the program, and quickly enter the field of digital control.

3.Build a digital control BUCK converter circuit model

As shown in Figure 2, the BUCK circuit adopts average current- mode control, adopts double closed loop control of voltage loop and current loop. The outer loop is a voltage loop to adjust the voltage error and generate the current command of the current loop, which is different from the inductor current to generate PWM generation. The main parameters of the BUCK circuit are: the input voltage is 50V, the output voltage is 24V, the input capacitance is two 100uF/250V electrolytic capacitors in parallel, the inductance is 365uH, and the output capacitance is three 100uF/250V electrolytic capacitors. The capacitors are connected in parallel, and the switching frequency is 40kHz.

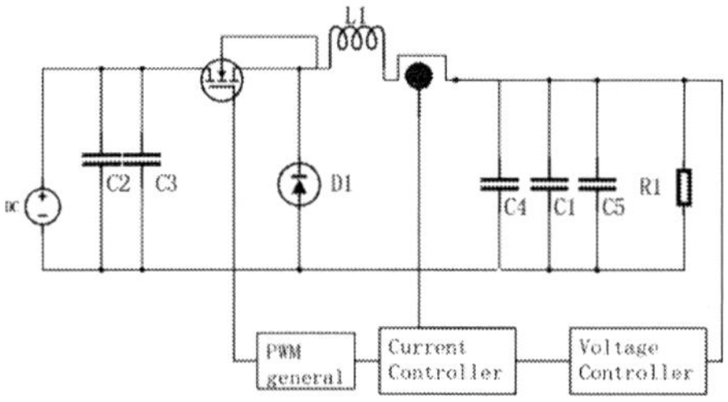

Figure 2. Digital controls the basic topology of the BUCK converter

According to the control requirements of the converter, use PSIM tools to build a control model, including a control loop based on average current and a soft start circuit.

3.1 Average current- mode control loop parameter calculation

3.1.1 Current loop design

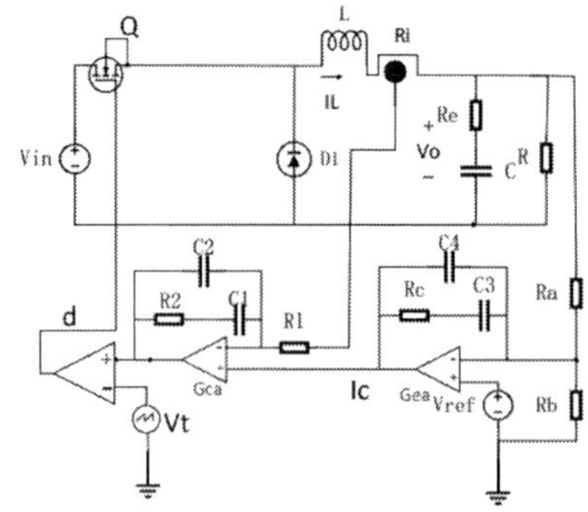

Figure 3. Average current control chart

The average current loop design is shown in the figure. Using the state average method, as shown in figure 3 above, we can get

$$L\frac{dIL}{dt} = dVin - Vo \tag{1}$$

Ignoring the changes of Vo and Vin, we can get from

$$\frac{\tilde{I}}{\tilde{d}} = \frac{Vin}{sL} \tag{2}$$

Considering the current sensing ratio and the gain of PWM, we can get

$$H_i(s) = \frac{I\tilde{L}Ri}{\tilde{d}Vt} = \frac{RiVin}{sLVt} \tag{3}$$

Aiming at the first-order system, the current error amplifier (GCA) can be designed using the second-class error amplifier method as shown in Figure 4.

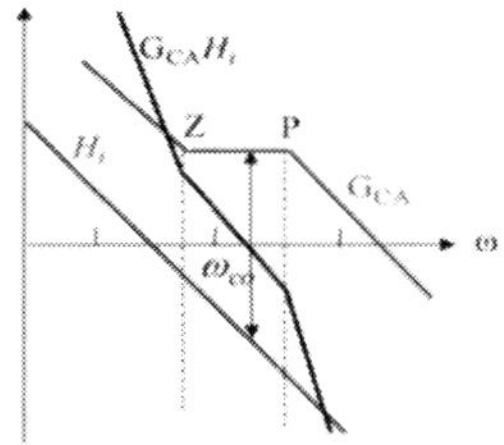

Figure 4. Two error amplifier design pattern

Since the PWM control voltage can only intersect with its sawtooth wave signal once in a cycle, the maximum bandwidth (w_co) of the current loop is limited by the rising slope of Vcon less than the rising slope of the PWM sawtooth wave (Vt). The rising slope of Vcon can be sensed The measured inductance current falling slope is determined by GCA amplification, and can be obtained from the above limitation

$$(\frac{Vo}{L})RiG_{CA,\max}(w_{co}) = Vtf_s \tag{4}$$

After finishing formula (4), we can get

$$G_{CA,\max}(w_{co}) = \frac{\tilde{V}_{con}}{Ri\tilde{IL}} = \frac{Vtf_sL}{VoRi} \tag{5}$$

From formulas (3) and (5) and use $G_{CA,\max}(w_{co})H_i(w_{co}) = 1$ we can get

$$\frac{Vtf_sL}{VoRi}\frac{RiVi}{w_{co}LVt} = 1 \tag{6}$$

Formula (6) can be rearranged to get

$$w_{co,\max} = \frac{Vinf_s}{Vo}(rad/s) \tag{7}$$

$$f_{co,\max} = \frac{f_s}{2\pi D} \tag{8}$$

From equation (8), it can be seen that if the design is designed with the rising slope of the control voltage Vcon, the theoretical maximum current loop bandwidth may be higher or too close to the switching frequency, so it is impossible to set this value, generally limited to the noise ratio, bandwidth The selection of (w_{co}) can be set at 1/4~1/8 of the switching frequency. Once the bandwidth (w_{co}) is selected, the K-factor method can be used to make the second type of error amplifier $z = w_{co}/K$, $p = w_{co}/K$.

3.1.2 Voltage loop design

Generally, the response speed of the voltage loop is much lower than that of the current loop. Therefore, when the voltage loop is modeled, the current loop can be regarded as ideal, that is, the response of the sensed inductor current and its command is regarded as 1, namely

$$\frac{Ri\tilde{IL}}{Ic(s)} = 1 \tag{9}$$

$$\frac{\tilde{V_O}}{\tilde{IL}} = R\frac{1+\frac{s}{w_z}}{1+\frac{s}{w_0}}, w_z = \frac{1}{C\,\mathrm{Re}}, w_p = \frac{1}{CR}s \tag{10}$$

Based on this assumption, the equivalent circuit of the voltage loop of the average current- mode control can be simplified as shown in Figure 5. From Figure 5, we can get

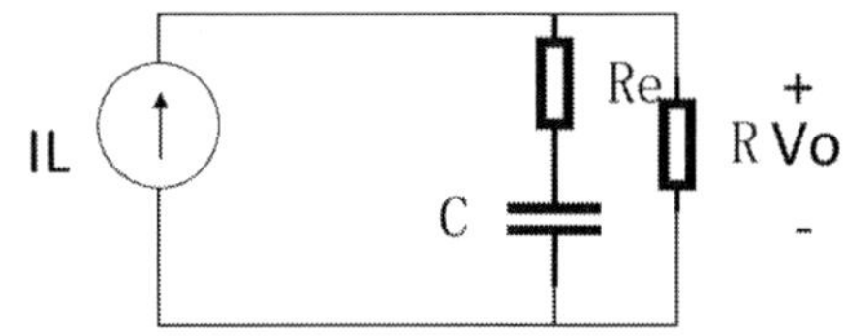

Figure 5. Voltage loop equivalent

Using equations (9) and (10), the control flow chart of the voltage loop can be drawn as shown in Figure 6, and the voltage error difference amplifier (GEA) can also be designed using the aforementioned two types of error amplifiers as shown in Figure 7. The bandwidth of the voltage loop can be designed to be 1/3~1/5 of the bandwidth of the current loop.

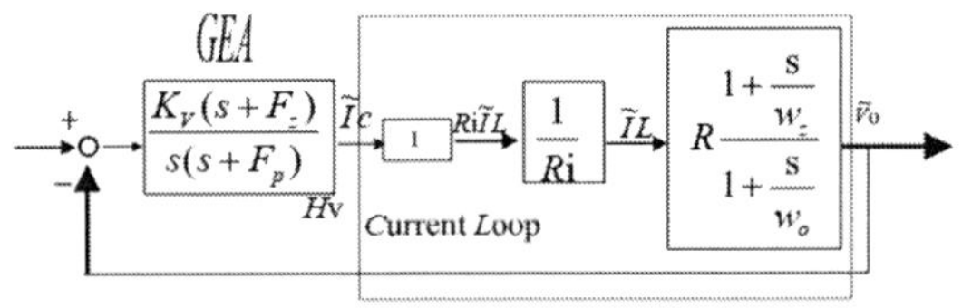

Figure 6. Voltage loop control block diagram

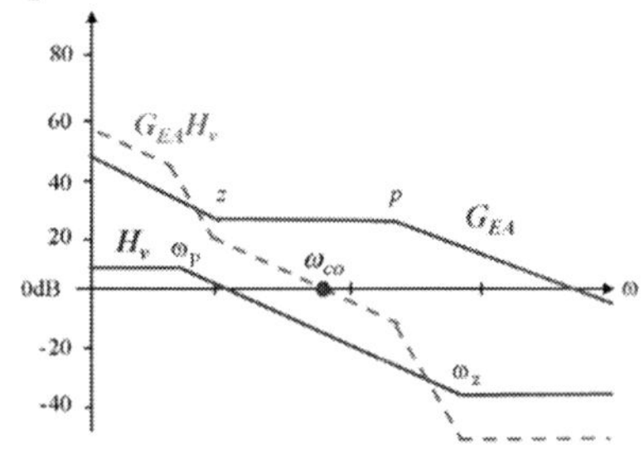

Figure7.Voltage Circuit GEA Bode Diagram

According to the above-mentioned current loop voltage loop model and error amplifier design method, using MATLAB to write a program, the Bode plot of average current control is obtained, as shown in Figure 8 is the Bode plot of the current loop, Figure 9 is the Bode plot of the voltage loop, and Figure 8 is the bode plot of current loop.

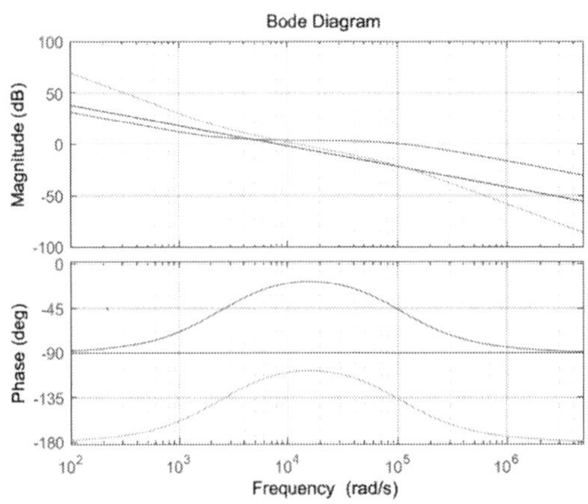

Figure 8. Bode plot of current loop

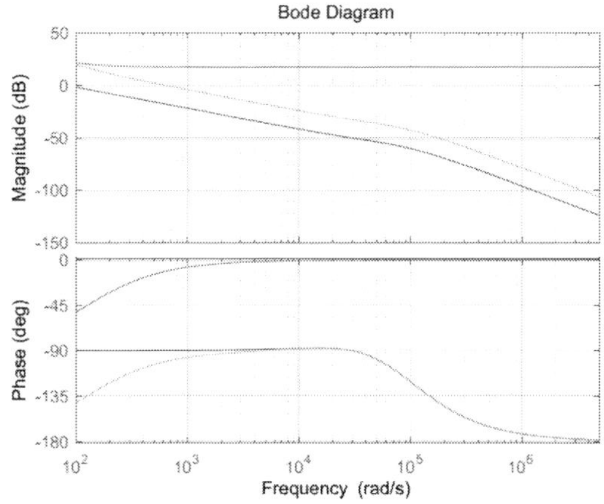

Figure 9. Bode plot of voltage loop

3.2 PSIM average current- mode control circuit design and simulation

Based on the above design, draw the PSIM analog circuit, as shown in Figure 10. The voltage and current waveforms are shown in Figure 11. The load has a sudden change at 0.04s. As shown in the figure, the output voltage also changes when the load jumps, but it quickly responds to the set value and remains stable.

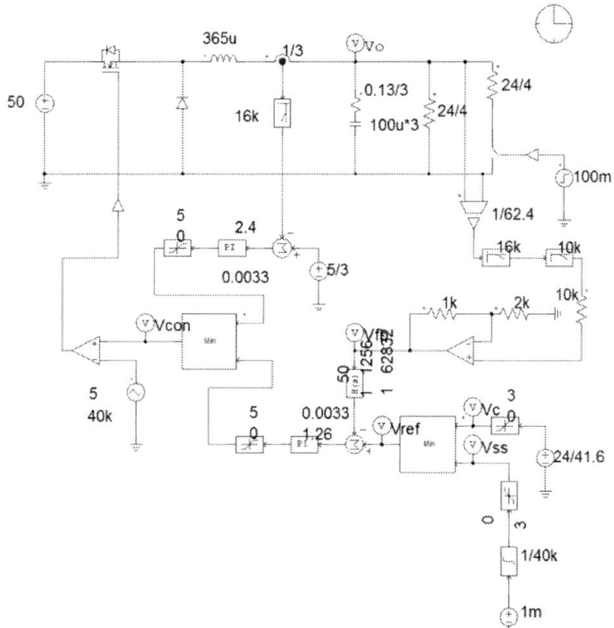

Figure 10. PSIM Average Current Control Circuit

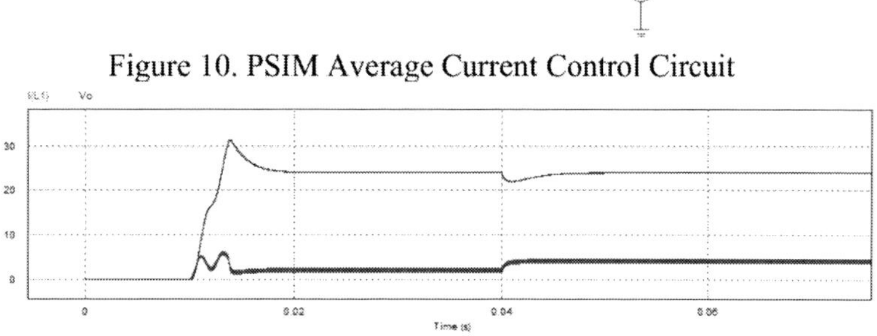

Figure 11. Voltage and current output

4.Automatic code generation and computer testing

In order to verify the function of the generated code, test on the hardware BUCK circuit, and connect the BUCK circuit hardware structure to the DSP c2000 board card series. Figures 12 and 13 show the BUCK hardware topology and the DSP digital control hardware circuit.

Figure 12. BUCK hardware topology

Figure 13. DSP digital control hardware circuit

Download the automatically generated code to the DSP control board. Since the resistance load is used at this time, the load jump is not set, and the voltage and current waveforms of the oscilloscope output corresponding to the simulation are obtained, as shown in Figure 14.the output voltage and current value, as shown in Figure 15.

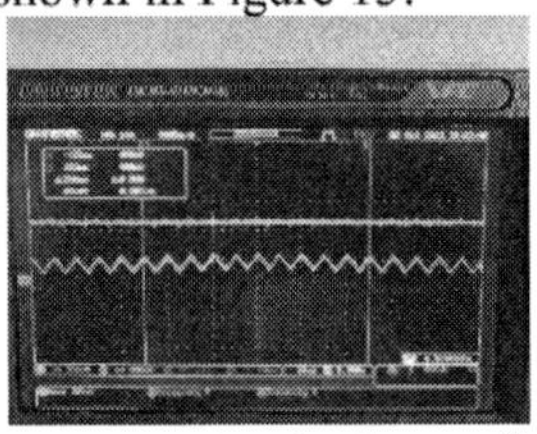

Figure 14. Voltage current output waveform

Figure 15. output voltage and current value

Using the automatic code generation method, software functions can be debugged in the software simulation link, and most software problems can be corrected, which greatly shortens the time for prototype debugging. In addition, graphical programming improves the operability and controllability of the realization of software functions. While more effective reducing the loopholes in the function design, the software writing time is greatly reduced. For general electronic control software with functions such as working sequence, fault diagnosis, closed-loop modulation, communication, data storage, etc., from building the software according to design requirements to starting prototype testing after software deployment, it usually takes 1-2 weeks, which is more efficient than handwritten software Significantly improved.

5.Conclusion

This paper explains the basic principle of using PSIM to realize automatic code generation. Taking the BUCK circuit with average current- mode control as an example, it introduces the construction of PSIM software model, model simulation and software deployment, and shows how automatic code generation is used in power control software development. In the application process. After hardware physical testing, the control software function realized by automatic code generation is the same as the simulation result, which can greatly shorten the software development time and improve the development quality.

Acknowledgments

This work was supported by the National Magnetic Confinement Fusion Energy Development Research Project (Grant No.2019YFE03070000), the National Natural Science Foundation of China (Grant No.11675074), and Hengyang Science and Technology Plan Project (Grant No.202002042159).

References

[1] Hou Y.C.,Yang D.J.,Zhen Y.W.,Zhao X.Q.,Shen L.,Shen Z.Z.(2021)Application of automatic code generation in train air conditioning power supply software[J].Electrical Drive, 51(12):41-46(in Chinese).

[2] Wang W.,Mo G.X.,Shen J.,Xie Y.B.,Wang W.M.(2020)The design of CAN underlying module library based on Simulink automatic code generation technology[J].Control and Information Technology, (03):93-96(in Chinese).
[3] Cui Y.X.,Chang Y.J.,Zhang Q.,Zheng J.(2019)Realization of speed measurement algorithm based on automatic generation of DSP code[J].Journal of Guilin Institute of Aeronautics and Astronautics,24(04):3-7(in Chinese).
[4] Gang J.Z.,Li Y.P.,Wen G.R.(2019)Automobile control system design based on automatic code generation technology[J].Electronic Design Engineering, 27(18):166-169.(in Chinese).

Design of vision system for high voltage transmission line obstacle Removal Robot

Yijiang Wu*, **Zhicong Dong, Huibin Zhou**

Zhongshan Power Supply Bureau of Guangdong Power Grid Corporation, Zhongshan, Guangdong, 528400, China

*wuyijiang@zs.gd.csg.cn

Abstract—Aiming at the problem of low visual co direction rate of high-voltage transmission line obstacle Removal Robot in traditional system, a new vision system of high-voltage transmission line obstacle removal robot is designed. In the hardware part of the system, the storage structure of DSP, TMS32DC541Q and image sensor are designed; at the same time, through the visual foreign object recognition of high-voltage transmission line obstacle removal robot, the visual image threshold of high-voltage transmission line obstacle removal robot is calculated, the visual judgment of high-voltage transmission line obstacle removal robot is realized, and the system design is completed. The experimental results show that compared with the traditional vision system, the designed vision system of high-voltage transmission line obstacle Removal Robot effectively improves the degree of the same direction with the original image and meets the practical application requirements.

1. Introduction

Power system plays an important role in all aspects of life, production and entertainment, and power transmission system is an indispensable part of power system. In recent years, foreign matters, kites, aluminum foil, plastic paper, plastic bags, etc. are often blown onto high-voltage transmission lines under the influence of bad weather such as strong wind, rain and snow. Foreign matters contacting rain and snow can easily lead to line short circuit, and kites with metal wires will directly lead to short circuit fault. Short circuit and discharge faults of power lines caused by foreign matters in high-voltage lines occur from time to time. The vision system of high voltage transmission line obstacle removal robot uses computer to realize human vision function and recognize the objective three-dimensional world system. At present, the hot research topic at home and abroad is the third generation machine vision system. The system adopts high-speed image processing chip and parallel algorithm, which has high intelligence and universal adaptability. The goal is to simulate people's high vision. The research on obstacle Removal Robot usually transmits the image to the host computer through camera and wireless transmission equipment, and determines the damage of transmission line through the image processing of the host computer. According to the working nature of the obstacle removal robot, the automatic obstacle Removal Robot combining inspection robot technology and machine vision technology can independently identify and remove foreign objects by using machine vision and intelligent control. At the same time, the monitoring system composed of camera, wireless transmission equipment and host computer can monitor the action of the robot, If necessary, adjust and control its action to achieve the purpose of accurate action.

2. Hardware design of vision system for high voltage transmission line obstacle Removal Robot

2.1. Digital Signal Processing(DSP)

In the hardware design of vision system of high voltage transmission line obstacle removal robot, this paper selects DSP as the core hardware of digital signal processing to process the images collected by high voltage transmission line obstacle removal robot [1]. DSP brings great opportunities to the development of digital signal processing because of its unique stability, large-scale integration, repeatability, programmability and easy to realize adaptive processing. DSP processors are applied and optimized in the field of digital signals. Their special hardware structure is more suitable for processing digital signals than ordinary microprocessors, making the products easier to realize and maintain. Real time digital signal processing is an important technical field of digital signal processing [2]. The system can complete the specified processing function in a limited time, that is, before the external signal changes, and the signal processing speed must be greater than the signal update speed[3]. Digital signal processing generally converts analog signal into digital signal, and then converts it into analog signal output after digital signal processing. In fact, some digital signal processing systems only need to output digital signals without D / A converter, while some systems input digital signals without A / D converter [4-5]. For pure digital systems, only the core part of digital signal processor is required. In this vision system, the image pixels to be processed are integers, so the fixed-point processing DSP chip is considered. In order to simplify the later image processing algorithm, the black-and-white image sensor is selected to collect the gray image. The space occupied by the pixels of the gray image is small, so the required data memory does not need to be too large [6]. To sum up, the system selects TMS320C5410 (later replaced by C5410) as the main chip of image processing template.

2.2. Storage structure of TMS32DC541Q

For the storage part of the system, C5410 memory is set to three independent spaces: program storage space, data storage space and I / O storage space [7-8]. The data storage space of 64K words includes dual access RAM (C DARAM) with two access operations per instruction cycle of 8K words and single access RAM (SARAM) with only one access operation per instruction cycle of 56K words.

The C5410 chip provides a ROM with 16K word capacity. The contents of the high 2K word ROM in the 16K word ROM are defined by TI company. The contents of the 2K word program space are shown in Table 1.

Tab. 1 C5410 list of on-chip ROM high 2K address range

Serial number	Program accessor address range	Memory content
1	F000h~FBFFh	Bootstrap loader
2	FC00h~FCFFh	256 word u-law companding table
3	FD00h~FDFFh	256 word A-Law companding table
4	FE00h~FEFFh	256 word sine function value lookup table
5	FF00h~FF7Fh	Internal self test procedure
6	FF80h~FFFFh	Interrupt vector table

As shown in Table 1, the program storage space of C5410 can address 64K program storage space without expansion. Their on-chip ROM, dual addressing ram and single addressing RAM can be mapped to the program space through software, and when the storage unit is mapped to the program space, the processor can automatically address their address range [9]. C5410 has two universal Vo. In order to access more universal Iro, the host communication parallel interface and synchronous serial interface can be configured to be used as universal I / O. External I / O can also be extended to allow C5410 to access 64K of external I / O space.

2.3. Image sensor

Image sensor chip is the core of vision system. At present, the popular image sensors mainly include CCD (electric device) image sensor and CMOS image sensor. The function of the system is to identify the foreign matters on the high-voltage line. There is no requirement for the clarity and color of the image. This disadvantage basically has no impact on the system. According to the system requirements, the CMOS black-and-white image sensor OV5017 produced by omnivision company is selected as the image acquisition chip of the system, The 8-bit A / D converter in the chip can convert video signals at an adjustable rate of 0-50FPS, which is synchronized with the actual pixel rate. Digital image signals with 256 gray levels can be directly output. It is convenient for the main chip to directly carry out the task of digital image processing. The chip has the following characteristics: the photosensitive area is 1 / 4 inch; progressive scanning; Internal 8-bit AID converter; Programmable control function; The frame rate can reach sofps, 384×288 pixels; SV or 3.3V power supply.

3. Software design of vision system for high voltage transmission line obstacle Removal Robot

After the hardware platform is built, the software design of DSP and image sensor, and the software design of memory. The software flow chart of the whole system is shown in Figure 1.

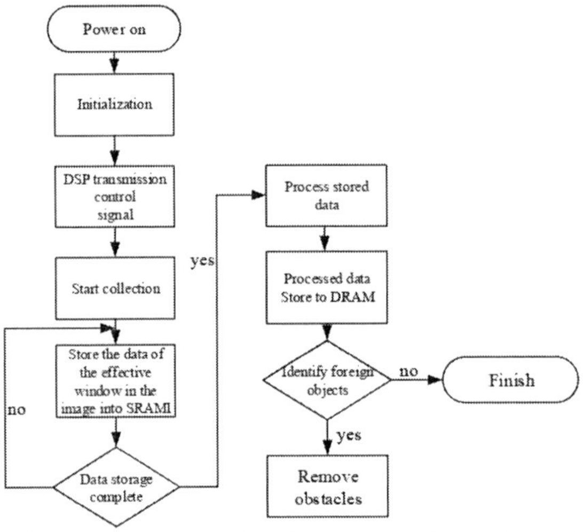

Fig. 1 Software flow of vision system of high voltage transmission line obstacle Removal Robot

As shown in Figure 1, the main processes of the three steps in the figure will be described in detail below, and the specific contents are as follows.

3.1. Visual foreign object recognition of high voltage transmission line obstacle Removal Robot

In the design of vision system software for high-voltage transmission line obstacle removal robot, it is difficult to find an algorithm to identify all kinds of foreign objects because the properties of foreign objects on high-voltage transmission line are uncertain and their shape, color and texture are unknown. Image binarization can convert the image into black-and-white gray-scale images, and separate the transmission line from the background. An image includes target, background and noise. The most common method to extract the target from a multi value gray image is to set a threshold value of θ, and use θ to divide the image data into two parts: the pixel group greater than θ and the pixel group less than θ. For example, if the input image is $f(x, y)$ and the output image is $g(x, y)$, there is formula (1).

$$g(x, y) = \begin{cases} 1, f(x, y) \geq \theta \\ 0, f(x, y) < \theta \end{cases} \quad (1)$$

This processing method of using θ to divide the image $f(x, y)$ into target and background regions is binarization, also known as threshold segmentation.

3.2. Calculating the visual image threshold of high voltage transmission line obstacle Removal Robot

The contrast between all objects and the background is almost the same, the target can be separated from the background as long as the correct closed value is selected. In fact, in most cases, the gray value of the image background is not constant, and the contrast between the object and the background also changes in different cases. In this paper, Otsu algorithm is used to process the collected images. Assuming that the gray level of an image is 0 ~ m-1 and the number of pixels of gray value i is n_i, the total number of pixels N is:

$$N = \sum_{i=0}^{m-1} n_i \quad (2)$$

The probability p_i of each value is:

$$p_i = \frac{n_i}{N} \quad (3)$$

Then it is divided into two groups $C_o = \{0 - T - 1\}$ and $C_1 = \{T - m - 1\}$ with T. the probability of each group is as follows: assuming that the probability of C_o is ω_o, there is formula (4).

$$\omega_o = \sum_{i=0}^{T-1} p_i \quad (4)$$

Assuming that the probability generated by C_1 is ω_1, there is formula (5).

$$\omega_1 = \sum_{i=0}^{m-1} p_i \quad (5)$$

Assuming that the average value of C_o is μ_o, there is formula (6).

$$\mu_o = \sum_{i=0}^{T-1} \frac{ip_i}{\omega_o} \quad (6)$$

Assuming that the average value of C_1 is μ_1, there is formula (7).

$$\mu_1 = \sum_{i=0}^{m-1} \frac{ip_i}{\omega_1} \quad (7)$$

Therefore, the average gray value μ of all samples is:

$$\mu = \omega_o \mu_o + \omega_1 \mu_1 \quad (8)$$

The variance σ^2 between the two groups is calculated by the following formula:

$$\sigma^2 = \omega_o(\mu_o - \mu)^2 + \omega_1(\mu_1 - \mu)^2 \quad (9)$$

This method is not only simple, but also can get satisfactory results regardless of whether the histogram of the image has obvious bimodal or not. Therefore, this method is the best method for automatic threshold selection for the images collected by the system.

3.3. Visual judgment of high voltage transmission line obstacle Removal Robot

(1) For the simulated transmission line images under the collected test conditions, through the reduction of redundancy and binarization processing, the image contour of one transmission line is clear and has less interference. At the same time, it can be seen that the projection position and inclination of the transmission line change in a small range. Therefore, the position of the line can be correctly located by restricting the linear projection position and inclination, Eliminate the interference of other linear segments. After correctly positioning the line position, it can be seen that the transmission line without obstacles is basically completely separated from the background after binary

processing, and the straight lines at the left and right edges of the transmission line are clearly visible. For the transmission line with obstacles, its projection will interrupt the transmission line, and the lines at the left and right ends of the line or the color of the transmission line will change due to the obstacles. Therefore, whether there are obstacles on the transmission line can be judged by the change of transmission line pattern. The specific steps are as follows: Establishing a two-dimensional array, calculate the p value in the parameter expression according to the Hough transform formula and the range of angle B, and accumulate the two-dimensional array with the same parameters. When the whole image is scanned, the points with the maximum value and the second largest value in the accumulated array correspond to the parameters of the straight lines at the left and right edges of the transmission line. According to the equation of the line where the two edges of the transmission line are located, the image pixels on the two lines are found, and the value of the pixels is used to judge whether there are obstacles on the transmission line. There are two cases:

A: The color of the obstacle is very different from that of the transmission line. When the image is binary processed, the obstacle and the transmission line are two different colors. The color of the part of the transmission line blocked by the obstacle is different from the color of the transmission line itself. In fact, most of the foreign objects hanging on the transmission line are kites, and the color of kites is relatively bright, which is in sharp contrast to the color of the transmission line, so this situation is common in obstacle recognition.

B: There is little difference between the color of the obstacle and the color of the transmission line. When the image is binarized, the obstacle and the transmission line are the same color. Even if the transmission line is blocked by the obstacle, the color of the blocked part and other parts is the same. In this way, the location of the obstacle can not be determined only by judging the color of the transmission line. After binarization, there is an obvious boundary between the transmission line and the surrounding background. Within the specified detection range, if the detected image pixels are the same as the image pixels of the transmission line, this section is the location of the obstacle.

In fact, the background of four split high-voltage transmission lines is also very simple, basically only blue sky and towers. At the same time, there are few auxiliary components of the four split high-voltage line, and there is only a cross shaped isolation rod between the two towers, which has no impact on the image processing and recognition of a single line. Therefore, the algorithm is also feasible in practical application.

4. Experiment

In the following, a comparative experiment is designed in this paper. Firstly, a high-voltage transmission line is randomly selected and other external factors affecting the identification results are ignored. Secondly, the traditional system is used for visual recognition of high-voltage transmission line obstacle removal robot, and the big data processing technology is used to analyze the same direction rate between the collected image and the original image. The vision system of high-voltage transmission line obstacle removal robot designed in this paper is used to operate in the same steps. Finally, six sets of experimental data were collected, and the collated data is shown in Figure 2.

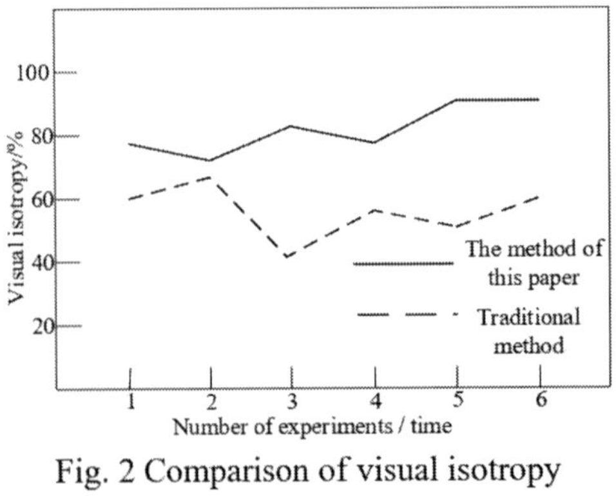

Fig. 2 Comparison of visual isotropy

Figure 2 shows the comparison curve of Visual Co orientation rate of the two systems. According to the experimental results in Figure 2 and the data generated in the experimental process, it is concluded that the vision system of high-voltage transmission line obstacle removal robot designed in this paper effectively improves the accuracy of obstacle recognition of high-voltage transmission line and the degree of CO orientation with the original image. Thus, it brings a broader development space for the development of high-voltage transmission line obstacle removal.

5. Conclusion

The obstacle clearing robot has been preliminarily applied to some high-voltage transmission lines, and some problems have also appeared in its practical application. For example, it can only run between two towers without obstacle crossing function and can not patrol the whole length of the line; When foreign matters on the line are detected, because the camera is on the PTZ on the top of the robot, the algorithm for judging the relatively accurate position of foreign matters and the robot is very cumbersome, which is likely to cause that the robot can not achieve the expected effect when removing foreign matters independently. For these shortcomings, the following improvements can be made in the follow-up research: in view of the defect that the robot can only move between two towers, the mechanical walking mechanism of the robot is improved, and its driving wheel is changed into the structure of two half wheels. When walking between two towers, the two half wheels are combined into one driving wheel. When the tower obstacle needs to be crossed, The front driving wheel is divided into two half wheels, which are combined into one driving wheel after crossing the obstacle. At this time, the rear driving wheel plays a supporting role. Similarly, the rear driving wheel is divided into two half wheels. When crossing obstacles, the front driving wheel plays a supporting role. In this way, the front and rear driving wheels cooperate with each other to successfully cross the tower obstacles, so as to patrol the whole process of the transmission line.

Acknowledgment

This work was supported by Research and application of key technology and equipment of intelligent tree obstacle clearing machine (robot) (No.GDKJXM20201944).

References

[1] Juang L H , Zhang S. (2019) Intelligent Service Robot Vision Control Using Embedded System[J]. Intelligent automation and soft computing, 25(3):451-458.
[2] Wang Y , Shan M , Yue Y , et al. (2019) Vision-Based Flexible Leader-Follower Formation Tracking of Multiple Nonholonomic Mobile Robots in Unknown Obstacle Environments[J]. IEEE Transactions on Control Systems Technology, 1-9.
[3] Deshpande S , Walambe R . (2020) Differential Gaming approach with safety parameter for Mobile Robot to circumvent a Dynamic Obstacle - ScienceDirect[J]. IFAC-PapersOnLine, 53(1):477-482.
[4] Ramesh D , Kathail A , Koditschek D E , et al. (2020) Modulation of Robot Orientation Via Leg-Obstacle Contact Positions[J]. IEEE Robotics and Automation Letters, PP(99):1.
[5] Huang L , Wu G , Tang W , et al. (2021) Obstacle Distance Measurement Under Varying Illumination Conditions Based on Monocular Vision Using a Cable Inspection Robot[J]. IEEE Access, PP(99):1.
[6] Khalilpour S A , Khorrambakht R , Damirchi H , et al. (2020) Tip-trajectory tracking control of a deployable cable-driven robot via output redefinition[J]. Multibody System Dynamics, (4):1-28.
[7] Eak A , At B , Di A , et al. (2020) Audio-visual based non-line-of-sight sound source localization: A feasibility study[J]. Applied Acoustics, 171.
[8] Aladem M , Rawashdeh S A . (2019) A Combined Vision-Based Multiple Object Tracking and Visual Odometry System[J]. IEEE sensors journal, 19(23):11714-11720.
[9] Vega J , Caas J M . (2019) Open Vision System for Low-Cost Robotics Education[J]. Electronics,

8(11):1295.

Demonstration and Optimization of Floor Height of 220kV GIS Room in the Indoor Substation

Yan Li[1*], Peng Wu [1], Lin Xing [1], Ning Wang [1], Rong Cheng [1], Yuan Hu [1]

[1]State Grid Hebei Economic Research Institute, Shijiazhuang, Hebei, 050000, China

[*]Corresponding author's e-mail: ly_0306@163.com

Abstract. With the development of cities, the construction of indoor substations and the use of gas insulated switchgear (GIS equipment) are becoming more and more widespread. In this paper, the height of 220kVGIS room in the indoor substation is demonstrated and optimized. On the basis of the collection and analysis of the basic data of each major equipment manufacturer, the height of GIS room is analysed from the expansion, maintenance needs and pressure test, which are two limiting factors. Finally, the optimal height of 220kVGIS room is determined.

1. Introduction

Indoor substations are more and more widely used in urban construction and development because of their advantages of land saving and a building's appearance in harmony with its surroundings[1-2].

At present, in the design scheme of indoor substation with 220kV and below voltage level, the height of GIS room is usually about 9m, which generally needs to occupy two floors of the main factory building. If the height of the GIS room is optimized, it can be effectively lowered to make full use of the space above the GIS room, which is of great significance to reduce the construction area and floor area of the substation and reduce the project investment.

The floor height of GIS room should consider the following two factors: (1) GIS expansion and maintenance needs[3-5]; (2) GIS pressure test. The following is an optimization of the height of GIS room based on the collection and analysis of basic data of each major equipment manufacturer.

2. The height of GIS equipment

In order to determine the height of GIS room, first of all, it is necessary to make clear the height of GIS equipment. In this paper, the equipment data of 220kV GIS major manufacturers are statistically analysed, as shown in Table 1.

Table 1. 220kV GIS equipment data statistics of main manufacturers.

The serial number	The manufacturer name	Interval height/m
1	PG Toshiba	3.2
2	Shanghai Siemens	3.3
3	Xindongbei Electric	3.2
4	Shandong Taikai	3.3
5	Beijing Beikai Electric	3.3

6	Hv Hipot Electric	3.3
7	Shanghai Electric	3.3

According to the analysis of the equipment specifications of the above manufacturers, the following data is determined as the basic information for layout optimization: The height of the interval equipment is not greater than 3.3 meters.

3. Study and optimization of floor height of 220kV GIS room

The following two constraints of GIS height are considered respectively.

3.1. Height requirements for GIS expansion and maintenance

According to the investigation of GIS manufacturers and operation units, the fault probability of GIS equipment is very small, and the general fault can be solved by on-site debugging or local replacement of faulty parts. If there is a need to return to the plant for overhaul, the SF6 gas can be evacuated completely, and then the GIS can be divided into parts for shipment intervals using the same transportation method as the expansion[6].

Expansion and maintenance can be carried out in three ways:

- The external wall of the equipment room adopts detachable lightweight composite wall panel, and the whole wall can be detachable to realize GIS integral interval transportation. Lifting and maintenance can be achieved through the detachable wallboard as a whole, and lifting in the GIS room is no longer considered. However, the late expansion construction of this way is complex, which affects the bearing capacity and durability of the wall, and its reliability needs to be investigated.
- Increase the longitudinal size of the GIS room, widen the transportation channel, ensure the maximum unit transportation at the expansion or maintenance interval, and realize the installation of the expansion equipment in place. The height of the GIS room can be reduced, but the GIS area needs to be increased.
- Use the hook of GIS room to lift the expansion or maintenance interval equipment to the expansion position across the interval.

In order to achieve long-term development, it is necessary to reduce the footprint and make full use of space. It is recommended to use cross-interval lifting to meet the height requirements of expansion and maintenance.

Considering cross-interval lifting, the height requirement of GIS room is:

The height requirement of GIS room =3.3m + 3.3m + 0.4m (construction margin) = 7m.

Generally speaking, the spacing equipment is dismantled into small components for hoisting. Therefore, the floor height of GIS room is set at 7 meters to meet the requirements of expansion and maintenance.

3.2. Height requirements of GIS pressure test

For GIS equipment, the following voltage test methods for connecting cables are generally adopted:

- Add test air casing (220kV casing height is 2.5m). In this way, due to the need to consider the distance between the air casing and the surrounding walls and the top plate is not less than 1.8 meters[7-8], the height of the pressure test is:

  The height of the pressure test=3.3m+2.5m +1.8m =7.6m.
- If the GIS room space is insufficient, the GIL branch bus can also lead out the outdoor to test, without increasing the height of the GIS room because of the test. However, the GIL branch bus must be provided by GIS manufacturers in this way, which increases the equipment cost, is not convenient for bidding and procurement, and the utilization rate is low.
- GIS equipment manufacturers can also provide a fully closed test casing, using this equipment can not only achieve the local pressure test, but also to solve the problem of small space caused by the optimization of equipment plant. However, the test device has some problems,

such as the small capacity of the equipment, and only 1 ~ 2 phases in a single interval can carry out the withstand voltage test at the same time. At present, it is still in the research and development stage of related technologies and products, and it is not mature.

- If there is still a large space margin in the GIS room (for example, the construction scale of this period is only part of the equipment interval), the test casing should be placed horizontally or tilted during the pressure test, as shown in Figure 1. During the expansion, if there are too many intervals or insufficient space in this period and the safety distance requirements for horizontal or inclined placement of the test casing are not met, the voltage withstand test in special cases can be carried out by renting GIS electromagnetic voltage transformer with power output (PTV equipment).

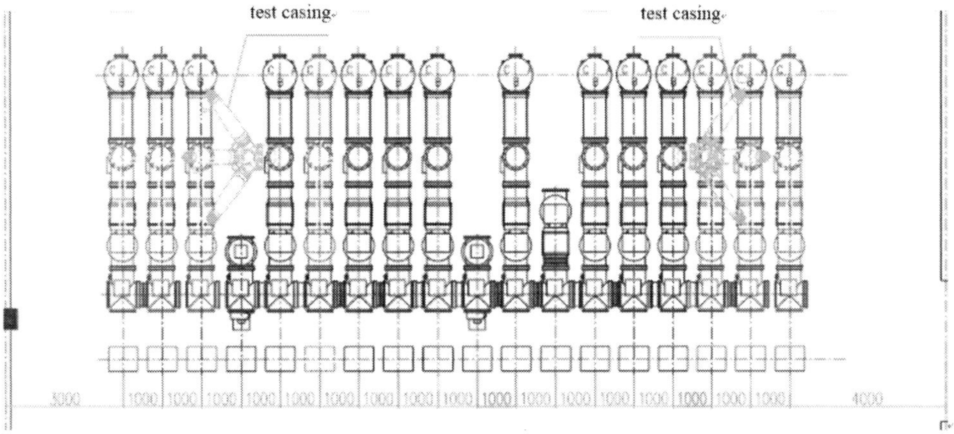

Figure 1. Schematic diagram of horizontal or inclined placement of test casing.

The principle of PTV method is to apply a low voltage on the secondary side of PT, and a high voltage is induced on the primary side of PT and applied to the GIS switch, so as to provide high voltage for the field GIS switch withstand voltage test. In this way, the demand for pressure test equipment can be greatly reduced in the field, and the occupation of the test casing to the height can be avoided, so as to achieve a significant reduction in the height.

In addition, the PVT can be directly used as a voltage transformer in the interval of GIS equipment after the completion of the voltage withstand test, so it has high practicability. The real picture is as follows:

Figure 2. PTV real picture.

The height of the 220kV GIS room is limited by the pressure test, which is 0.6m higher than the height required for expansion and maintenance. When space is sufficient, the test casing can be tilted or placed horizontally. And because the PVT technology has been relatively mature, it has been widely used in foreign countries, and the rental price is low. Through technical and economic comparison, if the space is not enough to meet the safety distance requirements of tilted or horizontal placement of the test casing, the pressure test can be carried out by leasing PTV equipment.

Therefore, it is recommended to use this pressure test method to reduce the height requirement to less than 7m.

Therefore, considering GIS maintenance and expansion, pressure test and other factors, it is recommended that the floor height of 220kV GIS room be no less than 7 meters.

4. conclusion

In this paper, based on the analysis of the data of major GIS equipment manufacturers, comprehensive consideration of GIS maintenance and expansion and pressure test and other factors, through the use of spare space, horizontal or tilted placement of test casing and the use of new PVT equipment, the 220kV GIS room height requirement is optimized to no less than 7 meters. This is of great significance to reduce the building area and floor area of the indoor substation and reduce the project investment.

The suggestion of floor height of 220kV GIS room in this paper has guiding significance to the construction of indoor substation.

Acknowledgments

Here, first of all, I would like to thank my master Lin Xing. In the topic selection of the article, the determination of the research route and the whole process of writing the article, she gave careful guidance and provided me with a lot of data and suggestions. Her professional research in the field of high voltage and profound insights on the subject have benefited me a lot. At the completion of this article, I would like to extend my heartfelt thanks and high respect to my teacher's hard work.

References

[1] Li, X.F., Zhen, L., Zhou, X.W., et al. (2010) Breakdown failure analysis of 220kV gas insulated switchgear. China Power, 43(1): 43-45.
[2] Liu, J.H., Yao, M., Huang, C.J., Experimental research on partial discharge localization in GIS using ultrasonic associated with electromagnetic wave method [J]. High Voltage Engineering, 2009, 35(10): 2458-2463.
[3] DL/T5352-2006 Guide for withstand voltage and insulated test of gas-insulated metal-enclosure switchgear on site[S]. China Electric Power Press, Beijing.
[4] DL/T5352-2006 Technical specification for design of high voltage distribution units[S]. China Electric Power Press, Beijing.
[5] DL/T618-2018 Guide for hand-over test of gas-insulated metal-enclosed switchgear on site[S]. China Electric Power Press, Beijing.
[6] Li, R.M., Wang, L.F., Discussion on high voltage test of electric power equipment [J]. Science and Technology Innovation Guide, 2014(34): 75.
[7] He, Z.G., Safety protection measures for high pressure test. Value Engineering, 2012, 31(4): 38
[8] Liu, Y.Z., Yuan, L., Discussion on safety protection measures of high pressure test [J]. The Scientific Herald, 2013(5): 215.

Pulse interference identification and suppression for OFDM signal

Wei Wei[1*] , Tao Peng [1] and Li Ye [1]

[1] State Grid Hubei Marketing Service Center (Measurement Center),Wuhan,Hubei,430000,China

*Corresponding author's e-mail: 1036243440@qq.com

Abstract. The new generation smart meter is the core node of information acquisition and flows in the power Internet of Things (IoT). Because it is powered by the low-voltage power line, various short-time pulse interference signals in the line can be easily coupled to the received signal of the communication module embedded in the meter, resulting in the degradation of communication performance. In view of this situation, an identification-suppression algorithm of the short-time pulse-interference signal is proposed. The core idea is to comprehensively judge the existence of the interference signal and its specific position in the time domain based on the energy ratio of adjacent OFDM symbols and the change of mean square error value between frequency domain amplitude signals. Based on this, an interference energy elimination scheme with an adaptive threshold is realized. Theoretical analysis and simulation results show that the scheme has low computational complexity, high accuracy of pulse interference identification and position judgment, and can effectively reduce the energy of pulse interference signal, and thus improving the decoding performance of OFDM while it receives signals.

1.Introduction

Smart meter, as the basic and key asset of smart grid and Ubiquitous Electric Internet of Things (UEIOT), has functioned as a comprehensive sensor in low-voltage power grid from the initial device to count the amount of the power consumption. It is not only the source to obtain the users' power consumption information, but an important source to receive the operation data in the power grid [1-2]. The electricity meter is powered by the power lines, which connect with a large number of various types of power loads. The startup and operation of those power loads will cause pulse interference signals in a short time arising from the lines, and will decrease the decoding performance of the electricity meter, due to the presence of residual interfere energy coupled into the received signals of the communication module embedded in the electricity meter, despite the preprocessing of those pulse interferes.

Orthogonal frequency-division multiplexing (OFDM) technology [3] has been the standard for numerous types of communications [4-5], thanks to its advantages in terms of spectrum efficiency and channel equalization complexity, including various types of communication standards that have been widely used in electricity meters [6-8]. Pulse interference has the characteristics of random burst, short duration and high interference amplitude [9-11], and its interference energy in the time domain will diffuse in the time-frequency conversion procedure when OFDM receives signals, which leads to the disturbance exerted to a large amount of frequency-domain signals, resulting in the dramatically decreasing decoding performance for the received signals.

Since the pulse interference will partially cause the abnormal amplitude of the received signals, the commonly used method to eliminate the noise is to detect the amplitude of the input signal and then to set the signal over the prescribed threshold to zero, in order to suppress the interference signal energy. On the other hand, in the OFDM communication system, the characteristics of multi-carrier signals make its emission signal encounter Peak-to-Average Power Ratio (PAPR) in the time domain. Besides, the signal transmission is also influenced by the channel time-varying, multi-path transmission and channel noise, resulting in sharp fluctuations of the amplitude envelope of received signals that are not affected by pulse interference, thus, the threshold with constant value cannot be set in turn. For the OFDM system with Phase-Shift Keying (PSK) modulation, this paper introduces a short-time pulse interference signal recognition and energy suppression algorithm in the procedure of receiving signals, for which the focus is to calculate the energy ratio of the adjacent OFDM symbols and the mean squared error between amplitude signals in the frequency domain. According to the changes of the previous values, it is to make sure that whether there is a short-time pulse interference signal in the received signal or not, and to ensure the concrete position of the interference signal in the time domain, in order to realize the suppression of pulse interference energy based on an adaptive dynamic soft threshold. The presented theoretical analysis and simulation results show that the proposed method has low computational complexity, high accuracy of pulse interference identification and position judgment, which can effectively suppress the energy of pulse interference signal and improve the decoding performance of receiving signal.

2.Identification and Suppression Algorithm of Short-Time Pulse Interference Signal

2.1 System Interference Model

Assuming that the communication module in the electricity meter adopts OFDM technology, with N subcarriers and PSK modulation technique, the calculation procedure to transform the frequency domain signal into the time domain one by the sender, i.e. inverse discrete Fourier transform (IDFT), is expressed as below.

$$x(k) = IDFT(X) = \sum_{n=0}^{N-1} X(n)e^{j\frac{2\pi k}{N}n}, k = 0,..., N-1 \tag{1}$$

where $X = [X(0), X(1),..., X(N-1)]$ represents frequency domain signal, and $x = [x(0), x(1),..., x(N-1)]$ stands for the corresponding time-domain signal, which is rewritten as $\bar{x} = (x_{N-M},..., x_{N-1}, x_0, x_1,..., x_{N-1})$ after inserting the cyclic prefix signal with length M.

On the other hand, the received signal can be expressed as

$$r = H\bar{x} + W + I \tag{2}$$

where H stands for channel matrix in the transmission, and W stands for the channel white noise that can meet the characteristics of complex Gaussian distribution, and I stands for the short-time pulse interfere signal of the received signals by the external influence factors.

After removing the cyclic prefix at receiving end, the frequency domain signal can be obtained by means of discrete Fourier transform (DFT). According to the principle of OFDM communication, the frequency domain signal can be expressed as:

$$R = \Lambda X + F_N I + F_N W \tag{3}$$

where Λ is a diagonal matrix, and F_N is an N-order Fourier transform matrix.

It is seen from Eq. (3) that the signal corresponding to the channel white noise after DFT can still meet the characteristics of the white noise distribution. Moreover, for the short-time pulse signal interference, only a small proportion of elements in the time domain have large absolute values with the remaining equal to 0, and the interference energy will be extended to all the frequency domain signals of the influenced OFDM symbols after DFT. This phenomenon will reduce the signal-to-noise ratio for a higher proportion of elements in the corresponding coding block, which will sharply decrease the signal decoding performance.

Interference is a perpetual problem in the field of communications. So far, there are many mature technical solutions for the identification and suppression of short-time pulse interference. However, in OFDM systems, these traditional solutions face the following difficulties:

(1) The multi-carrier characteristics of OFMD signal leads to high PAPR;

(2) Channel multipath transmission, channel noise and other related factors will cause more serious envelope fluctuation of the received signal;

(3) Communication channels generally have a high time variance, and the generation and numerical distribution of short-time pulse signals have heavy randomness;

(4) In the transmission procedure, the communication signal may have huge fading changes due to the presence of obstructions and other factors, resulting in significant energy fluctuation of OFDM symbols at different positions even without pulse interference.

The previous factors may cause large fluctuations in the envelope and energy distribution of the received signal by OFDM, introducing great trouble in the determination of pulse interference, interference location and denoising threshold setting, which makes some traditional interference identification and suppression algorithms not applicable to the technical field of communications.

2.2 Interference Identification and Localization Algorithm

For the OFDM communication system with phase modulation (PSK) and N subcarriers, if there is short-time pulse interference in its communication environment, the following interference identification scheme will be added to its signal processing, which was shown below:

Step 1: If the Cyclic Redundancy Check (CRC) detection result of a received packet signal after decoding is wrong, it is considered that it may be subject to short-time pulse interference, and the identification of short-time pulse interference is started for the received signal;

Step 2: Assuming that the received signal contains K OFDM symbols, the received signal in the time domain after removing the cyclic prefix will be noted as:

$$y = (\underbrace{y_{1,0},\ldots,y_{1,N-1}}_{1^{st}\ symbol}, \underbrace{y_{2,0},\ldots,y_{2,N-1}}_{2^{st}\ symbol},\ldots, \underbrace{y_{K,0},\ldots,y_{K,N-1}}_{K^{st}\ symbol}) \tag{4}$$

Step 3: Applying discrete Fourier transform (DFT) to these time-domain symbols sequentially, the frequency domain signal is obtained as

$$Y = (\underbrace{Y_{1,0},\ldots,Y_{1,N-1}}_{1^{st}\ symbol}, \underbrace{Y_{2,0},\ldots,Y_{2,N-1}}_{2^{st}\ symbol},\ldots, \underbrace{Y_{K,0},\ldots,Y_{K,N-1}}_{K^{st}\ symbol}) \tag{5}$$

Step 4: Calculating the energy of the K frequency-domain symbols in Eq. (5) and note as $P=[P_1,P_2,\ldots,P_K]$, where

$$P_k = \sum_{n=0}^{N-1} \left|Y_{k,n}\right|^2 \tag{6}$$

Step 5: Calculating the amplitude signal of the K frequency-domain symbols in Eq. (5) and the mean squared errors (MSE) between the adjacent amplitude signals, expressing as

$$MSE_k = \frac{\sum_{n=0}^{N-1} (\left|Y_{k+1,n}\right| - \left|Y_{k,n}\right|)^2}{N}, k=1,\ldots,K-1 \tag{7}$$

Step 6: Based on Eq. (6), the energy ratio between the adjacent symbols can be calculated, namely,

$$R_k = \frac{P_{k+1}}{P_k}, k = 1, \ldots, K-1 \tag{8}$$

It is noteworthy that since the PSK modulation mode with constant amplitude is adopted, the OFMD signal has equal amplitude in the frequency domain. The parameter MSE_k in Eq. (7) represents the similarity between the frequency domain amplitude signals of the $k+1$ th and k th OFDM symbols, and a larger value means weaker similarity. On the other hand, the parameter R_k in Eq. (8) represents the similarity between the energy of the $k+1$ th and k th OFDM symbols, and the similarity is higher if the value is close to 1. Considering that the two adjacent OFDM signals with convergent channel fading, the following conclusions listed in Table 1 are drawn with respect to the four cases that the two adjacent OFDM symbols encounter short-time pulse interference or not.

Table 1. Parameters with Different Interferes

	(none, none)	(with, none)	(none, with)	(with, with)
MSE_k	≈0	Relatively large	Relatively large	Relatively large
R_k	≈1	>>1	<<1	≈1

Step 7: Sorting the frequency domain symbols in Eq. (5) in descending order according to the symbol energy size. Based on the sparse characteristics of the short-time pulse interference signal in the time domain, the symbols smaller than the middle default energy value are considered free of pulse interference;

Step 8: Based on these default symbols that are not interfered with by the pulse, it is to check whether there are symbols interfered by the short-time pulse signal among the remaining ones, for which the procedure is structured as follows:

8.1) Set threshold $C = 1.5 \times \min(MSE_1, MSE_2, \ldots, MSE_{K-1})$;

8.2) Assuming that the kth symbol is supposed to have not interfered, and the adjacent one on the left side (i.e., the $(k\text{-}1)^{th}$ symbol) has not been checked, the decision algorithm is depicted as

$$D_{k-1} = \begin{cases} without\ \text{int}erference, if\ R_k \geq 2/3, and\ MSE_k \leq C \\ with\ \ \text{int}erference, else \end{cases} \tag{9}$$

8.3) Assuming that the kth symbol is supposed to have not interfered, and the adjacent one on the right side (i.e., the $(k+1)^{th}$ symbol) has not been checked, the decision algorithm is depicted as:

$$D_{k+1} = \begin{cases} without\,\text{int}erference, if\ R_k \leq 1.5, and\ MSE_k \leq C \\ with\,\text{int}erference, else \end{cases} \tag{10}$$

8.4) By making use of the previous algorithms in subsection 8.2 or 8.3, it is to check whether the undecided symbols are disturbed or not sequentially, until the adjacent left and right symbols of all the symbols undisturbed are checked;

8.5) If there are still some symbols in undecided status after the completion of Step 8.4), they will behave like the followings:

2108 (2021) 012028 doi:10.1088/1742-6596/2108/1/012028

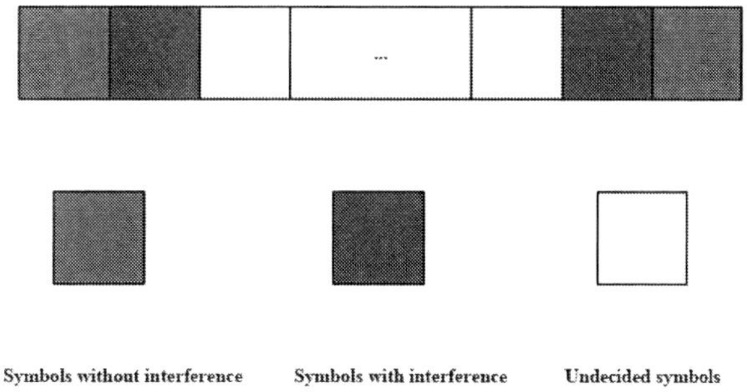

Figure 1. Schematic diagram of the scene when undecided symbols appear

In view of this, let suppose an undecided symbol with number a, and find a symbol with the closest time to itself that does not interfere. If the symbol is set as b, the decision algorithm of the symbol with number a is given by:

$$D_a = \begin{cases} without\ \ interference, if\ \frac{P_a}{P_b} \le 1.5, and\ \sum_{n=0}^{N-1}(|S_{a,n}| - |S_{b,n}|)^2 / N \le C \\ with\ \ interference, else \end{cases} \quad (11)$$

Upon the accomplishment of Steps 8.1-8.5, the judgment of all the symbols can be completed.

Considering the limited ability of error correction of channel coding, if a high proportion of the received signals encounter short-time pulse interference, the result of decoding error cannot be changed, even if the interference can be correctly identified. Therefore, the starting point of the design of this paper is that only a small percentage of the received signal may have interference signals in the signal, so the received signals are ranked by the energy magnitude, and by default half of the symbols with less energy are not interfered with; then, by utilizing the characteristics of the fading coefficients of adjacent symbols experiencing convergence channel. The mean square error and energy ratio of the nearest non-interference symbol and the symbol to be determined on the time axis. They are used to comprehensively evaluate whether the undecided symbol is affected by pulse interference, so as to ensure the short-term high-precision pulse interference identification and localization algorithm.

2.3 Adaptive Soft Threshold Based Interference Suppression Algorithm

After completing the identification and localization algorithm of the previous short-time pulse interference signal, if all symbols in the received signal are judged as non-interference symbols, it is considered that the signal-to-noise ratio of the signal is lower than the decoding threshold, leading to the failed decoding, which is reported to the high-level protocol for sequential processing. If it is determined that there are interfered symbols in the received signal, the following interference suppression will be applied to these symbols:

Step 1: Assuming that the k_1 th symbol is judged as undisturbed, and based on this symbol, it is to judge the k_2 th symbol as disturbed.

Step 2: According to the interference decision principle, it is known that $P_{k_2} / P_{k_1} > 1.5$, therefore, the energy of the pulse interference signal causes the abnormal energy value of the disturbed symbol. For the k_2 th time domain signal $y_{k_2} = [y_{k_2,0}, \ldots, y_{k_2,N-1}]$ in Eq.(4), it is to set an adaptive threshold Δ_{k_2} to carry out the following peak clipping processing on the signal amplitude, in order to eliminate the energy of pulse interference signal:

$$\bar{y}_{k_2,n} = \begin{cases} y_{c,n}, if \left|y_{c,n}\right| \le \Delta_{k_2} \\ \Delta_{k_2} \times y_{c,n} / \left|y_{c,n}\right|, if \left|y_{c,n}\right| > \Delta_{k_2} \end{cases}, n = 0, \ldots, N-1 \tag{12}$$

where the setting principle of the threshold Δ_c is to make the energy of the k_2 th time domain symbol after peak clipping approximately equal to the energy of the k_1 th time domain symbol, namely, $\sum_{n=1}^{N} \left|\bar{s}_{c,n}\right|^2 / P_{k_1} \approx 1$;

Step 3: Replace the new time domain symbol obtained after peak clipping with the original symbol, and then re-perform the reception processing process for the new signal, including channel equalization (if any pilot signal falls in the interfered symbol) and channel decoding;

Step 4: Whether or not the new signal is successfully decoded, the results are reported to the top level for processing.

The energy suppression algorithm of short-time pulse interference signal is based on the energy level determined as non-interference symbol, and it is to carry out peak clipping for suppression processing on the disturbed time-domain symbol through the adaptive soft threshold method of Eq. (12), so that the processed signal energy tends to be consistent with the energy of non-interference symbol. This method can achieve a good balance between noise energy suppression and energy preservation of the original communication signal, in order to improve the signal-to-noise ratio of the frequency domain signal obtained by the discrete Fourier transform of the interfering time-domain symbols, which provides some help for the improvement of decoding performance.

2.4Simulation Results

In this section, simulation tool by MATLAB is used to comprehensively verify the interference identification accuracy of the previous algorithm and the improvement of the decoding performance after noise suppression.

Here is an example of the QPSK modulation. The number of subcarriers is $N = 512$ and the cycle lasts with length of 1/4 symbol. The transmitted signal includes 32 OFDM symbols, and the transmitting channel is AWGN channel, of which the signal-to-noise ratio of the signal in the non-interference area is 0dB. Moreover, in the two interference areas, short-time pulse interference satisfying Gauss distribution occurs, and the signal-to-noise ratio of the interference area is equal to -6dB and -8dB, respectively. The amplitude of the received time domain signal, the location of the interference area and the soft threshold peak clipping are shown in the figure below:

 2108 (2021) 012028 doi:10.1088/1742-6596/2108/1/012028

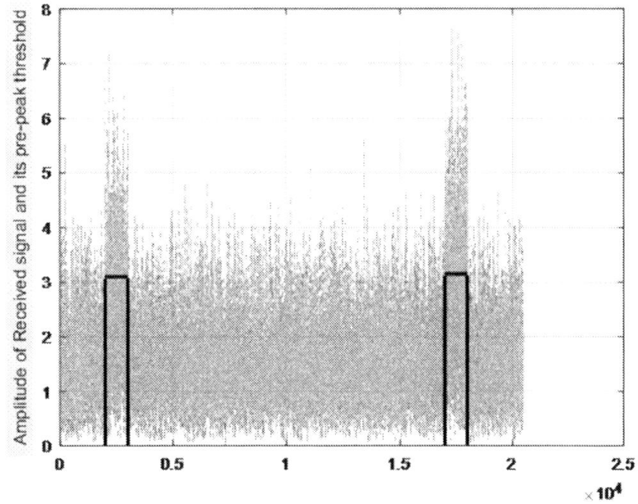

Figure 2. Received signal and its peak clipping of threshold in the disturbed area

The MSE values between adjacent symbols of the previous 32 OFDM symbols are depicted in the following figure:

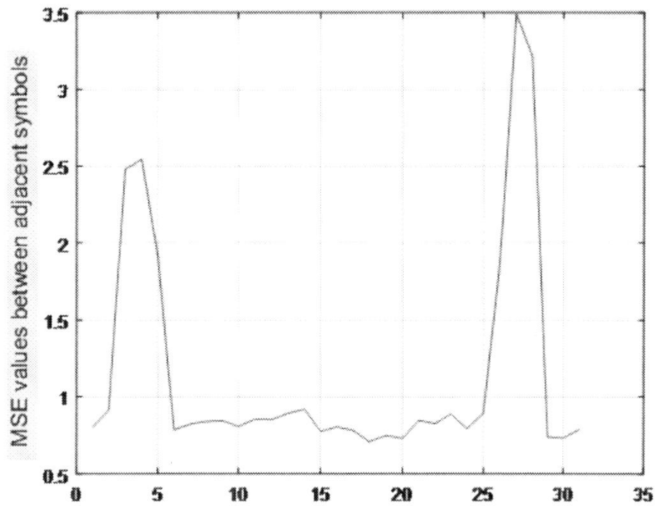

Figure 3. The MSE values between two adjacent OFDM symbols

The energy ratio between the adjacent symbols of the previous 32 OFDM symbols is shown in the figure below:

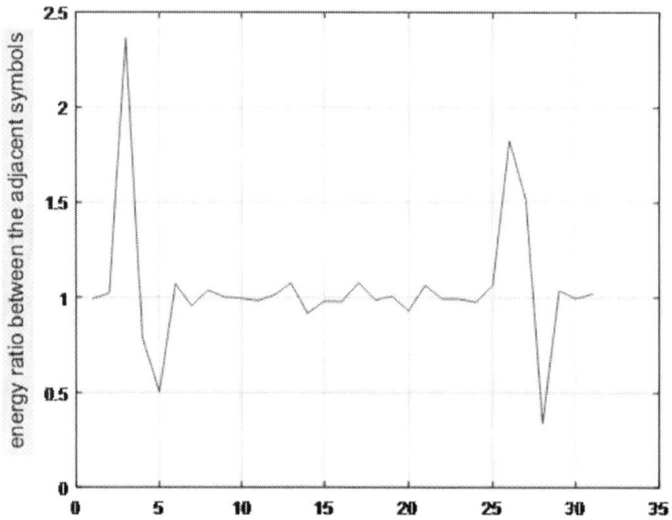

Figure 4. The energy ratio between the two adjacent OFDM symbols

It can be observed from Figs. 2 to 4 that the pulse interference signal will lead to the local extreme value of MSE and energy ratio between the adjacent OFDM symbols, which can provide a reliable reference for interference identification, positioning and threshold peak clipping. The interference identification algorithm based on MSE and energy ratio can be adaptive to different communication environments and ensure high-accuracy decision results. Moreover, the noise energy suppression based on adaptive soft threshold can effectively remove the interference energy in the pulse interference area, and it will not introduce negative impact on the signal in the non-interference area.

In order to verify the improvement of decoding performance after noise suppression, modulation coding schemes of BPSK +1/2 turbo and QPSK +1/2 were taken as examples. Assuming that the channel is AWGN signal, and 1/32 of former and 1/16 of the received signals in time domain are interfered by short-time pulses with an interference signal ratio 10dB, the two following figures show the varying curves of packet error rate of the signals without pulse interference signal, with pulse interference signal and with noise suppression, in terms of the different signal-to-noise ratios of the two previous different modulation coding schemes, respectively.

It can be seen from the simulation results in Fig. 5 that when BPSK +1/2 turbo scheme is adopted, although only 1/32 of the time domain signal is interfered by short-time pulses, the noise energy will be extended to more frequency domain signals without denoising, resulting in a heavy loss of decoding performance, where the decoding performance decreases by about 5 dB at 10-3 packet error rate. On the other hand, after introducing the denoising algorithm, the noise energy is significantly suppressed, and about 4.2 dB signal processing can be gained.

As can be seen from the simulation results in Fig. 6, when the QPSK +1/2 turbo scheme is adopted, due to that 1/16 time-domain signal are interfered by the short-time pulse interference, the proportion of frequency-domain signal affected by its energy exceeds the upper limit of error correction ability of signal channel coding. In light of this, even if the signal-to-noise ratio in the undisturbed area can increase continuously, it still cannot be decoded correctly, which means that the receiver will not be able to correctly receive the communication packet data. After introducing the noise energy suppression principle, all decoding errors can be avoided when the signal-to-noise ratio of the undisturbed area is high. Although the error rate remains at a high level, the receiver can receive the correct data packet again.

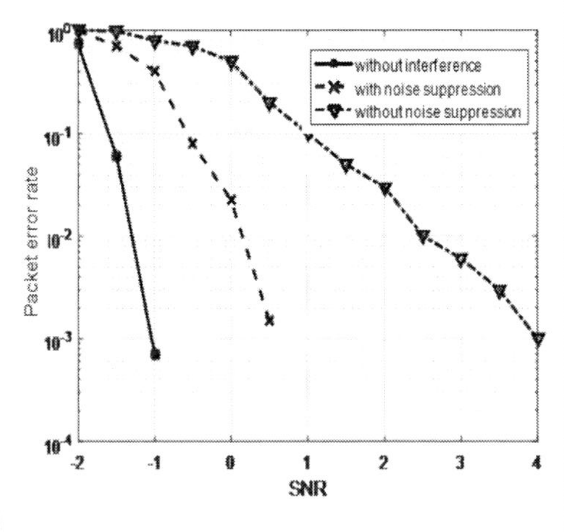

Figure 5. Packet error rate of1/2Turbo+ BPSK signal.

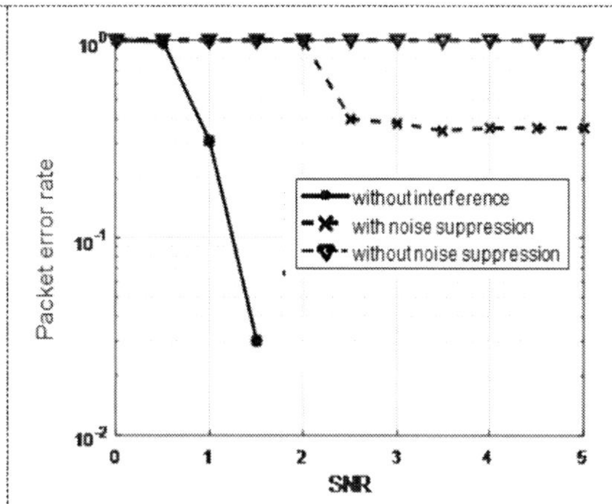

Figure 6. Packet error rate of1/2Turbo+ BPSK signal.

3. Conclusions

Taking the influence of short-time pulse interference signals onto the communication module embedded in the smart meter of the OFDM system into consideration, this paper presents a simple but efficient algorithm for noise interference identification, localization and suppression algorithm. The algorithm can handle the problem of pulse signal recognition caused by the fluctuating envelope of OFDM signals, and realize accurate judgment and accurate localization of interference. Its adaptive soft threshold principle can significantly suppress noise energy, thus, it can help to improve significant signal processing gain to the decoding process of the received signal affected by short-time pulse interference.

Acknowledgements

The authors would like to thank the State Grid Corporation of China for its great support of this research. This paper is funded by the key technology research of smart IoT meter for energy service (52153220001Y) of State Grid Hubei Electric Power CO., LTD.

References

[1] F Vallee F, Klonari V, Lisiecki T, et al. (2013) Development of a probabilistic tool using Monte Carlo simulation and smart meters measurements for the long term analysis of low voltage distribution grids with photovoltaic generation. J. International Journal of Electrical Power & Energy Systems., 53(dec.):468-477.
[2] Tasdighi M, Ghasemi H, Rahimi-Kian A. (2014) Residential Microgrid Scheduling Based on Smart Meters Data and Temperature Dependent Thermal Load Modeling. J. IEEE Transactions on Smart Grid., 5(1):349-357.
[3] Cimini L. J. (1985) Analysis and simulation of a digital mobile channel using orthogonal frequency division multiplexing. J. IEEE Trans. Commun., vol. 33, no. 7, pp. 665–675.
[4] https://wi-sun.org/
[5] https://g3-plc.com/home/
[6] IEEE Standard for Local and metropolitan area networks—Part 15.4: Low-Rate Wireless Personal Area Networks (LR-WPANs) Amendment 3: Physical Layer (PHY) Specifications for Low-Data-Rate, Wireless, Smart Metering Utility Networks[S].

[7] Pateromichelakis E, Shariat M, Quddus A. Ui. (2013) On the evolution of multi-cell scheduling in 3GPP LTE / LTE-A. J. in IEEE Communications Surveys & Tutorials, vol. 15, no. 2, pp. 701-717.
[8] WANG D M, ZHANG Y, WEI H. (2016) An overview of transmission theory and techniques of large-scale antenna systems for 5G wireless communications. J. Science China Information Sciences., 12:77-84.
[9] Arkhipov R M, Arkhipov M V, Babushkin I, et al. (2021) Envelope Area and Electric Pulse Area Interference in Excitation of Quantum Systems by Few-cycle Attosecond Light Pulses. J. JETP Letters, 22:1-5.
[10] Happe S , Kranz H G , Krause W. (2002) Real-time suppression of stochastic pulse shaped noise for on-site PD measurements. In: International Symposium on Electrical Insulating Materials. IEEE, New York. pp.65-72.
[11] Morishima Y, Oka I, Ata S . (2010) Pulse Interference Suppression Effects by Modified Viterbi Decoding. J. Proceedings of the. ITE annual convention, 51:9-14.

Design of Synaptic Neuronal Circuit Based on Memristors

Lin Ma, Yi Tong and Lin He*

College of Electronic and Optical Engineering, College of Microelectronics, Nanjing University of Posts and Telecommunications, Nanjing, Jiangsu, 210023, China

Corresponding author's e-mail: helin@njupt.edu.cn

Abstract: To solve the problems of poor learning efficiency and low accuracy caused by the single fixed synaptic weight in the traditional artificial neural network. On the foundation of the improved memristor model, this paper designs a synaptic neuronal circuit based on the natural memory characteristics of the memristor. This synapse is composed of six memristors. The resistance of the memristor is changed by adding a periodic square wave to update the synaptic weight. This circuit can realize signed synaptic weighting， which has certain linear characteristics. Finally, two synaptic weight update methods are proposed based on this circuit, and the validity of the design is verified through Spice simulation experiments.

1.Introduction

The memristor [1] is considered to be the best device for designing synaptic circuits due to its non-volatility, memory, nanometer size, and high integration. In recent years, scholars have studied more and more synaptic circuits. Literature [2] proposed a synaptic circuit composed of a single memristor, which can simulate the relative plasticity of synapses, but the controllability is poor. It can achieve "positive" synaptic weights. The synaptic circuit proposed in [3] consists of two memristors, but it can only achieve "positive" synaptic weights, which has limitations. Literature [4] introduced a transistor as a switch to realize the linear operation of the synaptic weights, which improved the accuracy of the weights, but the linearity characteristics were poor. Literature [5] proposed a synaptic circuit composed of four memristors, which can realize the "negative", "zero", and "positive" of the synaptic weights, but the weight processing process still does not have linear characteristics and is accurate Degree is low. Literature [6] is composed of two memristors and two resistors, which can realize the "negative", "zero", and "positive" of the weight, but it increases the power consumption and the circuit size, and the weight processing process is not continuous. Affect the speed of neural network training. Based on the above, this paper proposes a synaptic circuit composed of six memristors and proposes two-weight processing methods, which not only realizes the "negative", "zero", and "positive" of the synaptic weight but also has a good Circuit compatibility and linearity characteristics.

2 Memristor model

2.1 Proposal of the memristor

As shown in Figure 1, the existing basic components have always lacked a component that can express the relationship between the charge and magnetic flux of the charge memristor. In 1971, Chua defined the relationship between the two based on the completeness principle of circuit theory [7]. The defined circuit element [8] is called a memristor.

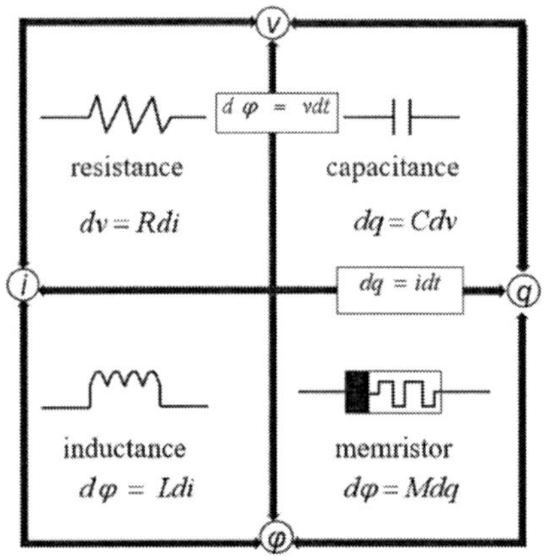

Figure 1. Relation diagram of four basic elements

According to the literature [9], the resistance of the HP memristor is:

$$M(t) = R_L \cdot x(t) + R_H[1 - x(t)] \tag{1}$$

$$x(t) = \frac{w(t)}{D} \in [0,1] \tag{2}$$

x represents the internal state of the memristor, which is the ratio of the doping thickness(w) to the film thickness(D).R_L and R_H are the minimum memristive value and the maximum memristive value of the memristor, respectively. The state equation corresponding to the state is:

$$\frac{dx}{dt} = ki(t) \tag{3}$$

$$k = \frac{u_v R_L}{D^2} \tag{4}$$

u_v represents the average mobility of ions, i represents the current flowing through the memristor, and k is the ratio of the changing speed of the state variable to the current flowing through the memristor[10].

2.2 Threshold Memristor Modeling

To better simulate the nonlinear migration phenomenon of ions inside the memristor, equation (3) can be improved as:

$$\frac{dx}{dt} = ki(t) \cdot f(x) \tag{5}$$

$f(x)$ is the window function, and *joglekar* functions are used in this article:

$$f(x) = 1 - (2x - 1)^{2p} \tag{6}$$

The above model has a simple structure and is easy to implement, but it does not have threshold characteristics. Based on this, it is necessary to add a threshold function to this model as follows:

$$g(x) = \begin{cases} 0, v \le v_T \\ 1, v > v_T \end{cases} \tag{7}$$

So the state equation can be further rewritten as:

$$\frac{dx}{dt} = ki(t) \cdot f(x) \cdot g(x) \tag{8}$$

When the input voltage is greater than the memristor threshold, the memristor resistance changes. The resistance of the memristors do not change if it does not reach the memristor threshold voltage. The memristors used below are all of this model.

3 Synapse circuit design

Figure 2 shows the specific neuron circuit. All synaptic circuits share an active load. During the operation of the circuit, the output voltage of each synaptic circuit is converted into a corresponding current signal by a differential amplifier and finally converted by a load circuit. For the corresponding voltage signal output.

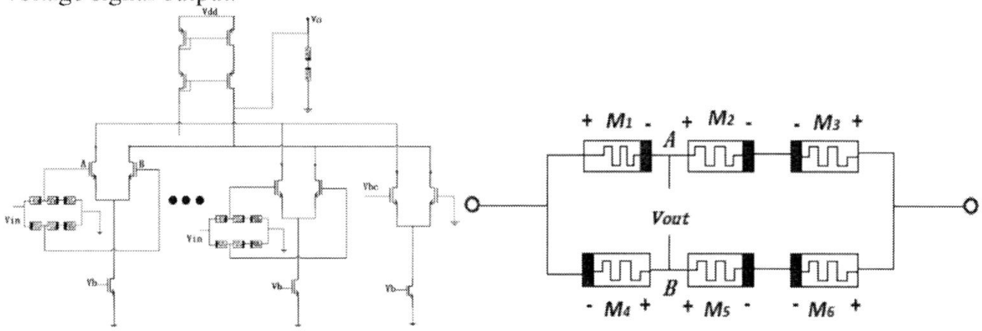

Figure 2. Neuron circuit

The synapse circuit consists of six memristors. When a forward or reverse voltage is applied to the input, the resistance of the memristor increases or decreases depending on its polarity. Each circuit contains three memristors. when a strong forward voltage is applied, the resistance of the forward-biased memristor becomes smaller, the voltage at A point increases, and the resistance of the reverse-biased memristor increases, and the voltage at B point decreases.

The memristor components of this synapse circuit are M_1, M_2, M_3, M_4, M_5, M_6. Define the input voltage as v_{in},, then the voltage at point A is:

$$v_A = \frac{M_2+M_3}{M_1+M_2+M_3} v_{in} \tag{9}$$

The voltage at point B is:

$$v_B = \frac{M_5+M_6}{M_4+M_5+M_6} v_{in} \tag{10}$$

The voltage difference between point A and point B is (9)-(10), which can be obtained:

$$v_A - v_B = \frac{M_2+M_3}{M_1+M_2+M_3} v_{in} - \frac{M_5+M_6}{M_4+M_5+M_6} v_{in} \tag{11}$$

And because the memristors M_2 and M_3 are connected in reverse series, and the total resistance is $M(t)$, we can get:

$$M(t) = M_2(t) + M_3(t) \tag{12}$$

Select memristor M_2, M_3, M_5, M_6 as the same type of memristor, which can be obtained by formula (1), the initial values of M_2 and M_3 at zero time:

$$M_2(0) = R_L \cdot x(0) + R_H[1 - x(0)] \tag{13}$$

$$M_3(0) = R_H \cdot x(0) + R_L[1 - x(0)] \tag{14}$$

$x(0)$ is the state of the memristor at zero time, the resistance value of M_2, M_3[11]can be written as:

$$M_2(t) = R_L \cdot [x(t) + x(0)] + R_H[1 - x(t) - x(0)] \tag{15}$$

$$M_3(t) = R_H \cdot [x(t) + x(0)] + R_L[1 - x(t) - x(0)] \tag{16}$$

Substituting formulas (15) and (16) into (12), we can get:

$$M(t) = M_2(0)+M_3(0) \tag{17}$$

From (17), the memristor is connected in reverse series, the total resistance value is the initial value and the sum of the resistance values is a constant, denoted as M, and then into equation (11):

$$v_A - v_B = \frac{M}{M_1+M} v_{in} - \frac{M}{M_4+M} v_{in} = \left(\frac{(M_4-M_1)M}{(M_1+M)(M_4+M)}\right) v_{in} \tag{18}$$

Formula (18) can be expressed as:

$$v_A - v_B = k v_{in} \tag{19}$$

The changing trend of synapse weight k is:

$$\begin{cases} k > 0, M_4 > M_1 \\ k = 0, M_4 = M_1 \\ k < 0, M_4 < M_1 \end{cases} \quad (20)$$

4 Experimental results and analysis

4.1 HP Memristor with Threshold

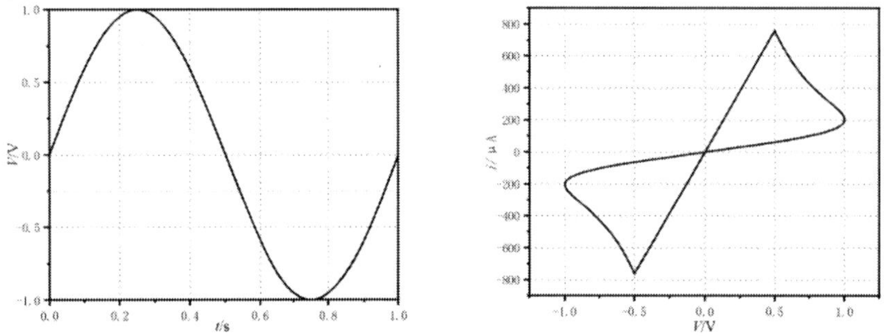

Figure 3 Simulation diagram of threshold memristor characteristics

Set the memristor threshold voltage to 0.5V, Figure 3 shows the V-I curve of the memristor when the input voltage is a sin function.

4.2 Update of Synaptic Circuit Weights

4.2.1 Forward voltage updates synaptic weights

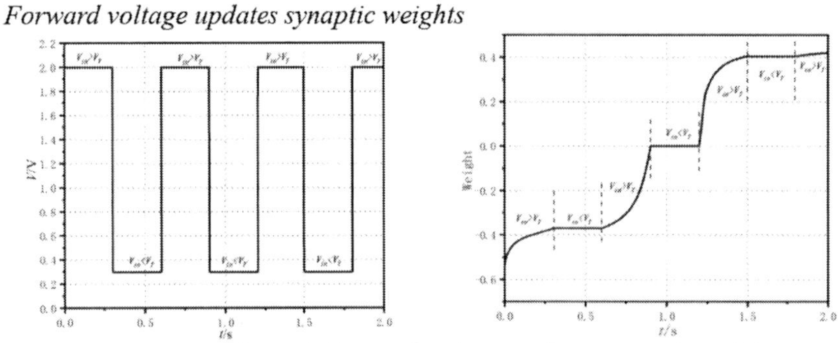

Figure 4. Forward voltage updates synaptic weights

Define parameters R_L =100Ω, R_H =15kΩ, D =10nm, $u_v = 10^{-14}m^2s^{-1}v^{-1}$, P =1, v_T =0.5V, $M_1(0)$=14kΩ, $M_4(0)$=1kΩ, Other memristor resistance values are 1kΩ.

As shown in Figure 4, input a two-second periodic voltage, and the synapse weight changes. Based on the different initial resistance settings of the memristors M_1 and M_4, the synaptic weights can be updated and maintained only with the forward voltage, which greatly improves the learning efficiency of the neural network.

4.2.2 Bidirectional voltage update synaptic weight

Define parameters R_L=100Ω,R_H=15kΩ,D=10nm,u_v=$10^{-14}m^2s^{-1}v^{-1}$, P=1, v_T=0.5V, $M_1(0)$=8kΩ, $M_4(0)$=8kΩ, Other memristor resistance values are 3kΩ.

As shown in Figure 5, input the forward and reverse voltages for 1.5 seconds respectively, and the changes in the synaptic weight. Based on the same initial resistance settings of the memristors M_1 and M_4. To realize the update of synaptic weight. The continuity of the input voltage can be achieved through an external circuit, and the synapse weight has a linear characteristic.

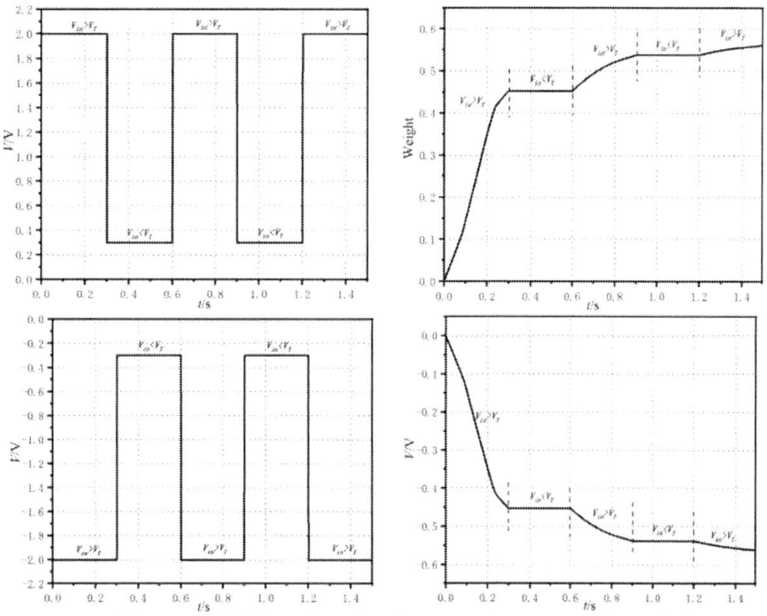

Figure 5. Bidirectional voltage update synaptic weight

From the above simulation results, it can be seen that the above two methods can realize the update of synapse weights, and the methods are stable, indicating that this synapse is beneficial to the smooth progress of the neural network and the test. It should be noted that the operation and operation of the entire neural network circuit are based on impulse signals. This is mainly because of the stability and ease of implementation of the pulse signal. At the same time, the memristive synaptic circuit is easier to control the weight under the pulse signal, such as the change time of the memristor resistance value, the change time of the synapse weight, so Need a duty cycle and so on, further thinking, within the range of the pulse period, the power consumption of the circuit can be better reduced, to avoid the waste of energy.

5.Conclusions

This paper proposes a synaptic neuron circuit. Both of the two synaptic weight adjustment methods used can achieve signed synaptic weights, which can maintain a certain linear characteristic in the process, and the synaptic weights are stable and controllable. The synapse circuit is only composed of memristors. Compared with the traditional neuron circuit, the circuit compatibility is higher and the synapse weight can be stored in time. The effectiveness of this synaptic neuron circuit function is verified by Spice simulation experiments. Therefore, this design is of great significance to the research of neural network circuits and has a certain reference value.

References

[1]L. Chen, W. Zhou, C. Li, and J. Huang, "Forgetting Memristors and Memristor Bridge Synapses with Long- and Short-term Memories," *Neurocomputing,* vol. 456, no. 9, 2021.

[2]S. Shin, K. Kim, and S. M. Kang, "Memristor Applications for Programmable Analog ICs," *IEEE Transactions on Nanotechnology,* vol. 10, no. 2, pp. 266-274, 2011.

[3]M. H. Vo, "Multilayer Neural Network with Synapse Based on Two Successive Memristors," *The Open Electrical & Electronic Engineering Journal,* vol. 12, no. 1, pp. 132-147, 2018.

[4]Jiu Yang, Lidan Wang, and Duan Shukai Duan, "Design and Application of a Reverse Series Memristive Synapse Circuit," Science China: Information Science, vol. 000, no. 003, pp. P.391-403, 2016.

[5]Hyongsuk *et al.*, "Memristor Bridge Synapses," *Proceedings of the IEEE,* 2012.

[6]L. Wang, X. Wang, S. Duan, and H. Li, "A spintronic memristor bridge synapse circuit and the application in memristive cellular automata," *Neurocomputing,* vol. 167, no. Nov. 1, pp. 346-351, 2015.
[7]G. Lecerf, J. Tomas, and S. Saïghi, "Excitatory and Inhibitory Memristive Synapses for Spiking Neural Networks," in *IEEE International Symposium on Circuits & Systems*, 2013.
[8]Duan *et al.*, "Composite Characteristics of Memristor Series and Parallel Circuits," *International Journal of bifurcation and chaos in applied sciences and engineering,* 2015.
[9]Thomas and Andy, "Memristor-based neural networks," *Journal of Physics D Applied Physics,* vol. 46, no. 9, p. 093001, 2013.
[10]S. Wen, X. Xie, Z. Yan, T. Huang, and Z. Zeng, "General memristor with applications in multilayer neural networks," *Neural Networks,* p. S0893608018301084, 2018.
[11]H. Kim, M. P. Sah, C. Yang, T. Roska, and L. O. Chua, "Neural Synaptic Weighting With a Pulse-Based Memristor Circuit," *Circuits and Systems I: Regular Papers, IEEE Transactions on,* vol. 59, no. 1, pp. p.148-158, 2012.

Research on real time simulation modeling method of large scale MMC electromagnetic transient

Lei Wang[1], Hongjun Zhang[1], Hui Hu[1], Liping Hao[1], and Wei Xu[2*]

[1]Liupanshui Power Supply Bureau of Guizhou Power Grid Co., Ltd., Liupanshui, Guizhou, 553000, China

[2]Shanghai YISU Information Technologies Co.,Ltd., Minhang, Shanghai, 201100, China

[*]Corresponding author's e-mail: wei.xu@yisuworld.com

Abstract. Modular multilevel converter (MMC) contains a large number of power electronic switching devices. The modeling method based on switching circuit model needs a lot of resources and the simulation speed is slow, so it is difficult to realize large-scale real-time simulation of electromagnetic transient. A MMC electromagnetic transient numerical modeling method based on ideal transformer model (ITM) is presented. Firstly, the MMC system is divided into the main circuit network and the sub module group network by ITM method, and the error caused by decoupling delay in serial and parallel real time simulation is compensated respectively by interpolation prediction and advanced interpolation prediction. Secondly, the capacitor in sub module is discreted respectively by trapezoidal integration method, backward Euler method and Gear-2 method. Based on the above numerical integration, the difference equations of capacitance voltage, capacitance current and output voltage of half bridge and full bridge sub modules are derived. Then, in order to improve the calculation speed, a simplified numerical model of half bridge and full bridge sub module based on switching function is proposed. Finally, the MMC based on switching circuit model runs off-line simulation in the simulation software, and the above MMC numerical modeling method runs real-time simulation in Speedgoat real-time simulator. The off-line and real-time simulation results of the MMC numerical modeling method and the switching circuit model are compared. And the simulation results verify the feasibility and effectiveness of the above MMC numerical modeling method in real time simulation.

1. Introduction

Modular multilevel converter (MMC) has been applied in the field of flexible DC transmission due to its advantages of unbalanced operation, fault ride through ability and low switching loss. But in practical engineering application, it is usually necessary to test and verify the feasibility and effectiveness of its hardware equipment and control system. Among them, hardware in loop test is an efficient test method to verify the control protection strategy, which requires that the electromagnetic transient simulation of the main circuit including MMC can achieve synchronization with the real clock. However, MMC has a large number of power electronic switch devices. If traditional modeling method based on switching circuit is adopted, it will be difficult to simulate in real-time mode. At present, most literatures have proposed modeling methods for MMC high efficiency electromagnetic transient simulation, such as Thevenin equivalent model of MMC bridge arm unit [6-8], MMC average model [9-11], L\RC model of MMC sub module [12] and electrical order reduction decoupling model [13]. These methods can improve the real-time simulation modeling of MMC. For example, literature [14] based on literature [8] adopts backward Euler method to equivalent the arm unit of MMC bridge with Thevenin, and realizes the real-time simulation of 601 levels MMC system based on RT-LAB platform.

At present, the commercial simulation platform which can realize the real-time simulation of MMC electromagnetic transient is mainly RT-LAB and RTDS. Among them, RT-LAB has launched MMC HVDC hardware in kind simulation solution earlier than RTDS. Both of them are real-time simulation modeling of valve body by FPGA, but for the same number of modules, RTDS needs more FPGA resources than RT_LAB. The state space node algorithm used by RT-LAB needs to calculate the equivalent circuits of Thevenin or Norton of each partition network in advance. The general program of equivalent circuit pre calculation and overall circuit solution is difficult to realize. The interface algorithm of transmission line model adopted by RTDS needs to adjust the parameter value of additional circuit according to different circuit parameters [16]. And the valve control circuit based on L/RC model needs a small simulation step, so it is not suitable for multi CPU core parallel simulation. On the other hand, RT-LAB and RTDS are commercial real-time simulation platforms, and their hardware configuration costs are more expensive, and the cost of later expansion and upgrading is also high.

In this paper, a real-time simulation numerical modeling method based on ideal transformer model method (ITM) is proposed. The detailed description and derivation of the method are given in detail from the circuit segmentation principle, error compensation of circuit decoupling delay, detailed and simplified numerical modeling principle of half bridge and full bridge sub module. In the aspect of simulation analysis, firstly, taking 21 level MMC inverter as an example, the accuracy of serial simulation is compared with the MMC segmentation model based on several interpolation prediction; Then, taking the full bridge sub module as an example, the simulation accuracy and acceleration effect of the sub module based on several discrete methods are compared; Finally, taking 21 level half bridge MMC flexible distribution system as an example, the simulation of MMC multi-core parallel simulation based on the above modeling principle is compared with the original overall circuit model in power flow regulation and uninterruptible power supply.

2. Equivalent model of MMC based on ITM

2.1. Circuit model segmentation of MMC based on ITM

Circuit model segmentation is based on the idea of grouping, which can divide a higher order power system network into several lower order independent sub networks. The interface algorithms between the associated networks mainly include: node splitting method, state space nodal (SSN), damping impedance method (DIM) and ideal transformer model method (ITM) [17]. The node splitting method and SSN algorithm need to pre-process each sub network with Thevenin equivalent or Norton equivalent. The dim method and ITM method both use controlled source to replace the associated sub network to realize the simultaneous operation simulation of the whole network. The dim method has higher stability region and simulation accuracy than the ITM method, but when the circuit parameters change, the dim method needs to reset the impedance parameters. For the convenience of implementation, ITM method is used to segment the circuit model of cascaded MMC, as shown in Figure 1.

In fact, because of circuit decoupling, in order to avoid the algebraic loop between the main network and the sub network, it is necessary to insert the delay unit between the voltage and current signal in series simulation mode. The decoupling delay unit will cause the capacitor voltage of MMC segmentation model to lose control under some circuit parameters, and then lead to serious instability of MMC output waveform.

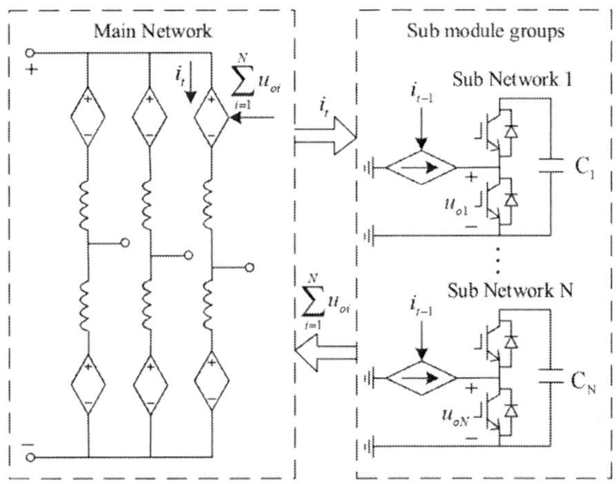

Fig. 1 MMC segmentation model based on TIM method

2.2. Error compensation method of MMC segmentation model

In this paper, the interpolation prediction method is used to compensate the error caused by the delay unit, and the next step electric quantity is interpolated and extrapolated according to the historical value. The common interpolation extrapolation methods include flat wave interpolation, least square method, linear interpolation, and parabolic interpolation. The basic principle is shown in figure 2. Among them, the flat wave prediction interpolation has the same slope as the least square method, which is mainly used for armature current prediction in BPA version of EMTP. Linear interpolation is used to predict angular velocity in BPA of EMTP. Parabolic interpolation can be regarded as the prediction based on the slope change rate (second order derivative) of two lines formed by adjacent historical points. Because the above interpolation extrapolation methods use the first few step history values for prediction, the prediction effect of discrete variables with mutation characteristics is not ideal, so the last step history value can be directly used for such variables or mutation points.

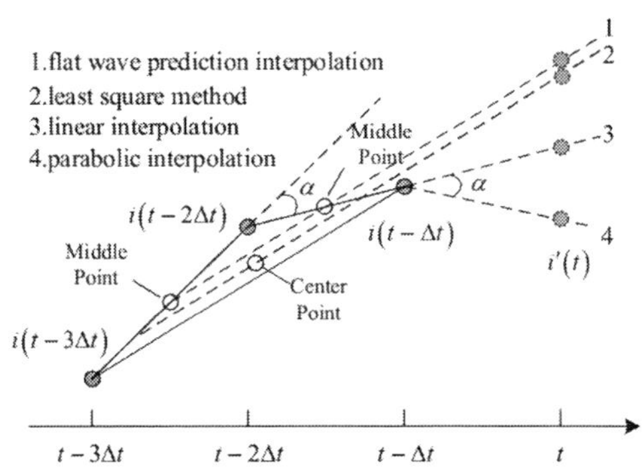

Fig. 2 Several interpolation extrapolation methods

The corresponding interpolation formulas in the figure above are as follows:

$$i'(t)=\frac{5}{4}i(t-\Delta t)+\frac{1}{2}i(t-2\Delta t)-\frac{3}{4}i(t-3\Delta t) \tag{1}$$

$$i'(t)=\frac{4}{3}i(t-\Delta t)+\frac{1}{3}i(t-2\Delta t)-\frac{2}{3}i(t-3\Delta t) \tag{2}$$

$$i'(t)=2i(t-\Delta t)-i(t-2\Delta t) \tag{3}$$

$$i'(t)=3i(t-\Delta t)-3i(t-2\Delta t)+i(t-3\Delta t) \tag{4}$$

When the single core serial simulation is used, the output voltage of the sub module is not suitable for interpolation prediction because of the sudden change of the output voltage, and the waveform of the bridge arm current is relatively smooth due to the existence of the bridge arm reactor. In order to solve the algebraic loop problem between the main network and the sub network, the delay unit is added to the current channel of the bridge arm to make interpolation prediction, and the output voltage of the sub module directly uses the present value.

When multi-core parallel simulation is used, if the main network and sub module groups are respectively configured in different CPU cores, there is a fixed delay between CPU cores, which makes the bridge arm current and sub module output voltage channel have one delay. Therefore, the bridge arm current can be compensated by the advanced interpolation prediction method, while the output voltage of the sub module directly uses the historical value of the previous step, and the voltage delay is indirectly compensated [18]. For example, in the calculation results of formula (1) and formula (2), the leading interpolation prediction formulas obtained by using the corresponding interpolation prediction are

$$i''(t)=\frac{33}{16}i(t-\Delta t)-\frac{1}{8}i(t-2\Delta t)-\frac{15}{16}i(t-3\Delta t) \tag{5}$$

$$i''(t)=\frac{19}{9}i(t-\Delta t)-\frac{2}{9}i(t-2\Delta t)-\frac{8}{9}i(t-3\Delta t) \tag{6}$$

Electromagnetic transient numerical model of sub modular network

After the MMC model is segmented by ITM method, the electrical decoupling between each sub module is realized. In addition, the output voltage and capacitor voltage signals of each sub module are only needed when the main network and sub module groups operate simultaneously. Therefore, the electromagnetic transient value of each sub module network can be directly established, and the difference equations of the above related variables of each sub module can be directly solved in order to facilitate the efficient operation of MMC real-time simulation..

2.3. Discretization of sub module capacitor

For the electromagnetic transient simulation of MMC sub module network, it is necessary to discretize the capacitor elements in the sub module. The commonly used discretization methods are trapezoidal integral method (A-stability), backward Euler method (A-stability), gear-2 method (A-stability), Simpson method (conditional stability) and Runge Kutta method (conditional stability). In this paper, the trapezoidal integral method, backward Euler method and gear-2 method are selected to discretize the capacitor elements in the sub module. The volt ampere characteristics of the capacitor are as follows

$$u_C(t)=u_C(t-\Delta t)+\frac{1}{C}\int_{t-\Delta t}^{t} i_C(t)dt \tag{7}$$

If trapezoidal integral method is adopted, the expression of capacitive current can be obtained from equation (7)

$$i_C(t)=\frac{2C}{\Delta t}u_C(t)-\frac{2C}{\Delta t}u_C(t-\Delta t)-i_C(t-\Delta t) \tag{8}$$

If backward Euler method is adopted, the expression of capacitive current can be obtained from equation (7)

$$i_C(t)=\frac{C}{\Delta t}u_C(t)-\frac{C}{\Delta t}u_C(t-\Delta t) \tag{9}$$

If gear-2 method is adopted, the expression of capacitive current can be obtained from equation (7)

$$i_C(t)=\frac{3C}{2\Delta t}u_C(t)-\frac{2C}{\Delta t}u_C(t-\Delta t)+\frac{C}{2\Delta t}u_C(t-2\Delta t) \tag{10}$$

In order to facilitate the derivation and expression of the sub module numerical model based on the above three numerical integration methods, equation (8), equation (9) and equation (10) can be written into the following unified Norton equivalent form:

$$i_C(t) = u_C(t)/R_C + I_{Ch} \tag{11}$$

Where, R_C is the equivalent resistance of the capacitor; I_{Ch} is current history terms. The specific value of R_C and I_{Ch} is determined by the discretization method

$$R_C = \begin{cases} \frac{\Delta t}{2C} & trapezoidal \\ \frac{\Delta t}{C} & backward\ Euler \\ \frac{2\Delta t}{3C} & \text{Gear-2} \end{cases} \tag{12}$$

$$I_{Ch} = \begin{cases} -i_C(t-\Delta t) - \frac{2C}{\Delta t}u_C(t-\Delta t) & trapezoidal \\ -\frac{C}{\Delta t}u_C(t-\Delta t) & backward\ Euler \\ \frac{C}{2\Delta t}u_C(t-2\Delta t) - \frac{2C}{\Delta t}u_C(t-\Delta t) & \text{Gear-2} \end{cases} \tag{13}$$

Among the above discretization methods, trapezoidal integration method is widely used in electromagnetic transient simulation of power system because of its good stability and high simulation accuracy, but it may produce numerical oscillation. Although the simulation accuracy of backward Euler method is not as good as trapezoidal integration method, it does not produce numerical oscillation and has a more simplified calculation. When the simulation step is small, the simulation accuracy is acceptable. Gear-2 method does not produce numerical oscillation, and its simulation accuracy is between trapezoidal integration method and backward Euler method.

2.4. Electromagnetic transient numerical model of half bridge sub module

According to equation (11), after the half bridge sub module network is discretized by trapezoidal integration method, backward Euler method or gear-2 method, the electromagnetic transient discrete adjoint circuit has the same form, which can be unified into the circuit structure as shown in Figure 3. The only difference is that capacitor equivalent resistance R_C and current history terms I_{Ch} have different equation, which is equation (12) and equation (13). Therefore, for the half bridge sub module, the electromagnetic transient numerical model based on the above discretization method can be derived and expressed in a unified form.

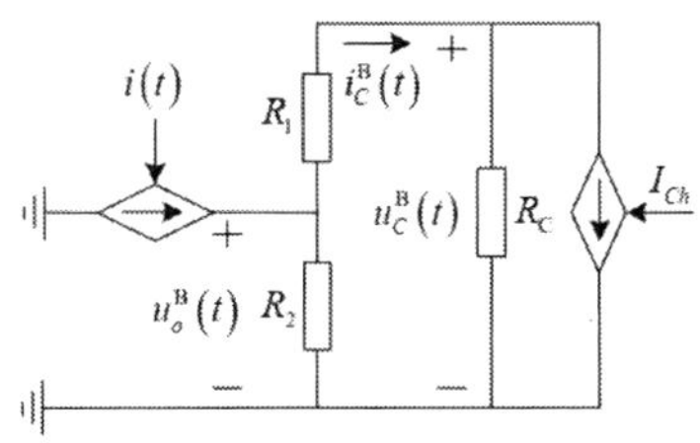

Fig. 3 Discrete companion circuit of half bridge module

According to the circuit model in Figure 3, the difference equations of half bridge sub module capacitor voltage $u_C^B(t)$, capacitor current $i_C^B(t)$ and sub module output voltage $u_o^B(t)$ can be derived from KCL equation and VCR equation of each component

$$u_C^B(t) = \frac{R_2 R_C}{R_1 + R_2 + R_C} i(t) - \frac{(R_1 + R_2) R_C}{R_1 + R_2 + R_C} I_{Ch} \tag{14}$$

$$i_C^{\mathrm{B}}(t)=\frac{R_2}{R_1+R_2+R_C}i(t)+\frac{R_C}{R_1+R_2+R_C}I_{Ch} \tag{15}$$

$$u_o^{\mathrm{B}}(t)=R_2\left(i(t)-i_C(t)\right) \tag{16}$$

Where, the superscript B represents the half bridge MMC sub module, and the capacitance equivalent resistance R_C and current history term I_{Ch} are determined by the discretization method, and formula (12) and formula (13).

The values R_1, R_2 are determined by their respective trigger pulse States,

$$R_i=\begin{cases}R_{\mathrm{on}} & S_i=1\\ R_{\mathrm{off}} & S_i=0\end{cases}\quad (i=1,2) \tag{17}$$

Equations (14) - (17) constitute the electromagnetic transient numerical model of half bridge MMC sub module network, and each sub module numerical model is operated independently. The overall equivalent of MMC proposed in reference [6-7] is also based on the ITM segmentation method, which is different from that in this paper. It only splits the main circuit and the overall unit of bridge arm, and then makes the overall equivalent of Thevenin for the unit of bridge arm. In this paper, each sub module is modeled separately, and then the output voltage is summed. But the two are equivalent, because the main circuit only needs the total output voltage of the bridge arm sub module to calculate when it is divided based on the ITM method. However, if Thevenin's integral equivalence is applied to the bridge arm element, the unnecessary summation of concentrated resistance parameters will be increased.

2.5. Electromagnetic transient numerical model of full bridge sub module

Similar to the numerical modeling of half bridge sub module, the electromagnetic transient adjoint circuit of full bridge sub module also has a unified structure after discretization by trapezoidal integral method, backward Euler method or gear-2 method, as shown in Figure 5. The difference is that the values of capacitance equivalent resistance R_C and current history term I_{Ch} are different.

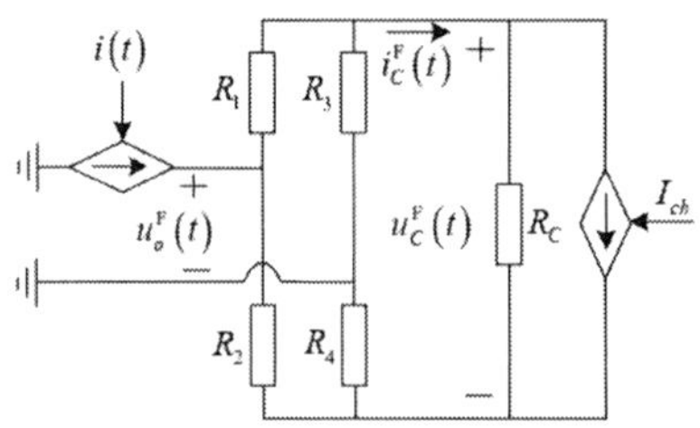

Fig. 4 Discrete companion circuit of full bridge module

According to the circuit model in Figure 4, the difference equations of full bridge sub module capacitor voltage $u_C^{\mathrm{F}}(t)$, capacitor current $i_C^{\mathrm{F}}(t)$ and sub module output voltage $u_o^{\mathrm{F}}(t)$ can be derived from KCL equation and VCR equation of each component

$$u_C^{\mathrm{F}}(t)=\frac{AR_C}{B+CR_C}i(t)-\frac{BR_C}{B+CR_C}I_{Ch} \tag{18}$$

$$i_C^{\mathrm{F}}(t)=\frac{A}{B+CR_C}i(t)+\frac{CR_C}{B+CR_C}I_{Ch} \tag{19}$$

$$u_o^{\mathrm{F}}(t)=\frac{E+DR_C}{B+CR_C}i(t)-\frac{AR_C}{B+CR_C}I_{Ch} \tag{20}$$

Where:

$$\begin{cases} A = R_2R_3 - R_1R_4 \\ B = R_1R_3 + R_1R_4 + R_2R_3 + R_2R_4 \\ C = R_1 + R_2 + R_3 + R_4 \\ D = R_1R_2 + R_1R_4 + R_2R_3 + R_3R_4 \\ E = R_1R_2R_3 + R_1R_2R_4 + R_1R_3R_4 + R_2R_3R_4 \end{cases}$$

Where, the superscript F represents the full bridge MMC sub module, and the capacitance equivalent resistance R_C and current history term I_{Ch} are determined by the discretization method, and formula (12) and formula (13).

The values R_1, R_2, R_3, R_4 are determined by their respective trigger pulse States,

$$R_i = \begin{cases} R_{\text{on}} & S_i = 1 \\ R_{\text{off}} & S_i = 0 \end{cases} \quad (i = 1,2,3,4) \tag{21}$$

It is worth noting that: unlike trapezoidal integration method, after discretization by backward Euler method and gear-2 method, the historical term in the difference equation does not include the capacitive current term, and the difference equation of capacitive voltage and sub module output voltage does not need the capacitive current term for iterative calculation. Therefore, when we do not pay attention to the capacitive current of sub module, the difference equation of capacitive current using backward Euler method or gear-2 method can be omitted to reduce the amount of calculation.

3. Simplified numerical model of submodular network

Comparing the numerical models of half bridge and full bridge sub module, it can be seen that the difference equation of full bridge sub module is much more complex than that of half bridge sub module, and its computation is mainly focused on the calculation of the coefficient of difference equation, which will increase the computational burden of processor and limit the scale of real-time simulation of MMC. In this paper, a relatively simplified numerical model of half bridge and full bridge sub modules is proposed, which is suitable for the large-scale real-time simulation of half bridge or full bridge MMC.

3.1. Simplified numerical model of half bridge sub module

According to the control and operation characteristics of the half bridge sub module, the output characteristics of the sub module can be divided into two states: capacitor voltage on and capacitor voltage off. If the conventional switching function method is used for modeling, the stability will be poor due to the decoupling delay between AC and DC networks. Therefore, the sub module DC network can be directly discretized, and the excitation current source can be taken as the switch function of the bridge arm current. The simplified equivalent circuit is shown in Fig. 5.

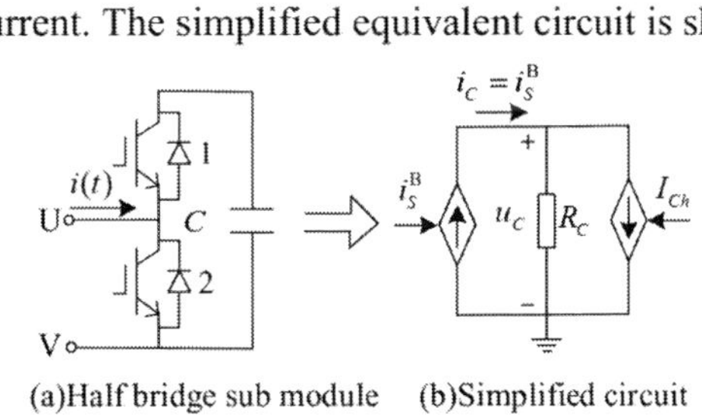

(a)Half bridge sub module (b)Simplified circuit

Fig. 5 Simplified circuit of half bridge sub module

In Fig. 5 (b), the current source of the injected capacitor is determined by the bridge arm current and the half bridge sub module switching function

$$i_S^B(t) = S^B i(t) \tag{22}$$

$$S^B = \begin{cases} 1 & S_1 = 1, S_2 = 0 \\ 0 & S_1 = 0, S_2 = 1 \end{cases} \tag{23}$$

If the trapezoidal integral method is adopted, the difference equation of capacitor voltage in Figure 5 (b) can be deduced as

$$u_C(t)=u_C(t-\Delta t)+S^{\mathrm{B}}\frac{\Delta t}{2C}\left(i(t)+i(t-\Delta t)\right) \tag{24}$$

If the backward Euler method is adopted, the difference equation of capacitor voltage in Figure 5 (b) can be deduced as

$$u_C(t)=u_C(t-\Delta t)+S^{\mathrm{B}}\frac{\Delta t}{C}i(t) \tag{25}$$

If the Gear-2 method is adopted, the difference equation of capacitor voltage in Figure 5 (b) can be deduced as

$$u_C(t)=\frac{4}{3}u_C(t-\Delta t)-\frac{1}{3}u_C(t-2\Delta t)+S^{\mathrm{B}}\frac{2\Delta t}{3C}i(t) \tag{26}$$

The output voltage of the half bridge sub module is determined by its switching function and capacitor voltage

$$u_o(t)=S^{\mathrm{B}}u_C(t) \tag{27}$$

The basic principle of the simplified half bridge sub module numerical method is to transform the time-varying characteristics of the switching resistance in the sub module circuit model into the time-varying characteristics of the current injected into the capacitor. From the difference equation of the simplified model, it can be seen that the difference of calculation burden between the above discretization methods is small, and the calculation burden based on gear-2 method is the largest.

3.2. Simplified numerical model of full bridge sub module

According to the control and operation characteristics of the full bridge sub module, the simplified equivalent circuit of the similar full bridge sub module is shown in Figure 6. The capacitor switching state has three states: positive input, cut-off and negative input, so the current mode of capacitor injection is different from that of half bridge sub module.

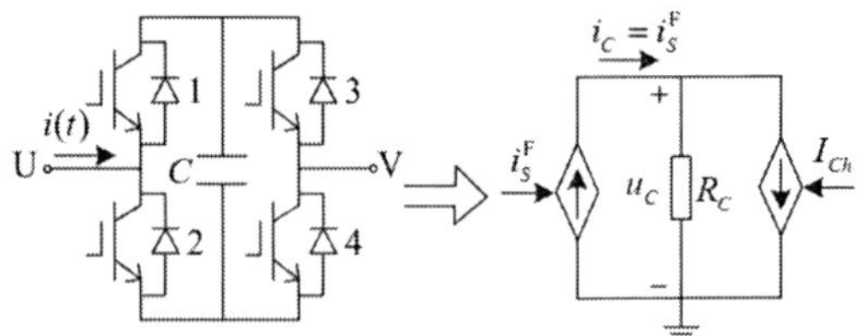

(a)Full bridge sub module (b)Simplified circuit

Fig. 6 Simplified circuit of full bridge sub module

In Fig. 6 (b), the current source of the injected capacitor is determined by the bridge arm current and the switching function of the full bridge sub module

$$i_S^{\mathrm{F}}(t)=S^{\mathrm{F}}i(t) \tag{28}$$

$$S^{\mathrm{F}}=\begin{cases}1 & S_1=S_4=1\wedge S_2=S_3=0\\ 0 & S_1=S_3=1\vee S_2=S_4=1\\ -1 & S_2=S_3=1\wedge S_1=S_4=0\end{cases} \tag{29}$$

The capacitor voltage difference equation of full bridge sub module is similar to that of half bridge sub module. For the convenience of comparison with the detailed numerical model, the capacitance voltage difference equation of the simplified model based on trapezoidal integral method, backward Euler method or gear-2 method is unified as the following form

$$u_C(t)=S^{\mathrm{F}}R_Ci(t)+R_CI_{Ch}^{*} \tag{30}$$

Where, R_C of is the same as that of (12), and I_{Ch}^{*} is

$$I_{Ch}^{*}=\begin{cases}u_C(t-\Delta t)+S^F\frac{\Delta t}{2C}i(t-\Delta t) & \text{trapezoidal}\\ u_C(t-\Delta t) & \text{backward Euler}\\ \frac{4}{3}u_C(t-\Delta t)-\frac{1}{3}u_C(t-2\Delta t) & \text{gear}-2\end{cases} \tag{31}$$

The output voltage of the full bridge sub module is determined by its switching function and capacitor voltage

$$u_o(t)=S^F u_C(t) \tag{32}$$

Equations (28) - (32) can form a simplified numerical model of full bridge sub module network. Compared with the detailed numerical model, it can be seen that the difference equation coefficients of capacitor voltage and output voltage are much simpler than the detailed numerical model.

However, the simplified full bridge sub module numerical method can also simulate the dynamic characteristics of the capacitors of each sub module, and the amount of calculation is greatly reduced. It is suitable for the real-time simulation of more large-scale full bridge MMC and can realize the real-time simulation verification of capacitor voltage balance.

4. Analysis of simulation example

The hardware platform used for example analysis is Speedgoat real time target machine: the processor of the real-time simulator is Intel Core i7-7700k, the number of cores is 4, the main frequency is 4.20GHz, and the memory is 4GB;

4.1. Comparison of simulation accuracy between detailed and simplified numerical models

Take the numerical model of full bridge sub module as an example. In the simulation software, the detailed and simplified numerical models of full bridge sub module based on trapezoidal integration method, backward Euler method and gear-2 method are modeled and compared with the IGBT circuit model in the simulation software. The test adopts the circuit shown in Figure 4, in which the excitation source adopts the sinusoidal current source with amplitude of 10 A and frequency of 50 Hz, the initial value of capacitor voltage is set to 1 kV, the capacitor is set to 0.005 F, the pulse signal takes a fixed period of 0.0001 s, and the simulation step is set to 50μs. Set 0s-0.0005s interval to adopt forward operation; 0.0005s-0.015s section adopts negative operation; The current source is cut off between 0.015s and 0.002s, but the trigger pulse is given continuously. The simulation results of capacitor voltage, capacitor current and output voltage of sub module based on the above model are shown in Fig. 7 and Fig. 8.

The simulation results of the detailed numerical model show that the detailed numerical model of the full bridge sub module based on trapezoidal integration, backward Euler method or gear-2 method is highly consistent with the simulation results of the circuit model. The simulation results of capacitor voltage and output voltage of sub module are almost the same as that of the circuit model based on trapezoidal integral method, because the circuit model in simulation software is also discretized by trapezoidal method; The simulation accuracy of backward Euler method is low, especially in the simulation results of capacitance voltage, which is mainly shown that the simulation results are less than the other two discrete methods; The simulation accuracy of gear-2 method is between trapezoid integral method and backward Euler method. In a word, the above simulation results are basically in accordance with the truncation error law of the numerical integration method.

The simulation results of the simplified numerical model show that the simplified numerical model of the full bridge sub module based on trapezoidal integral method, backward Euler method or gear-2 method is basically consistent with the simulation results of the circuit model. Compared with the simulation results of detailed numerical model, it can be seen that the capacitor voltage does not charge or discharge slightly, as shown in Fig. 8 (a) amplification part, and the output voltage of its sub module does not drop on the switching element, as shown in Fig. 8 (b) amplification part, which is consistent with the modeling characteristics. In addition, the simplified numerical model based on trapezoidal integral method is closer to the simulation results of the circuit model.

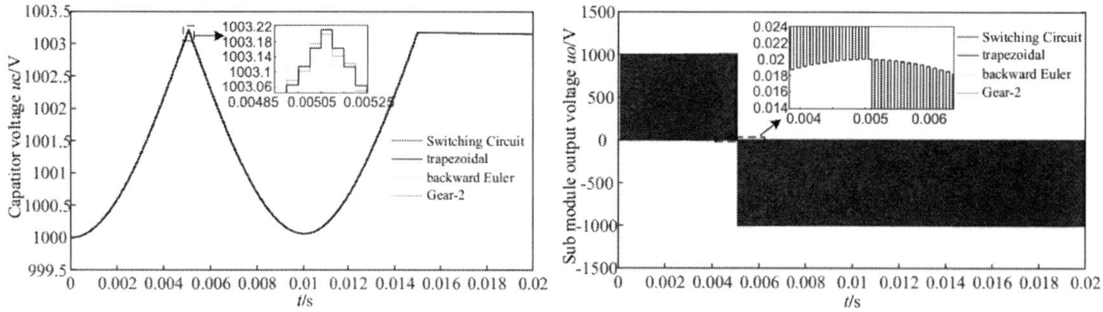

Fig. 7 Simulation results of detailed numerical model of full bridge sub module

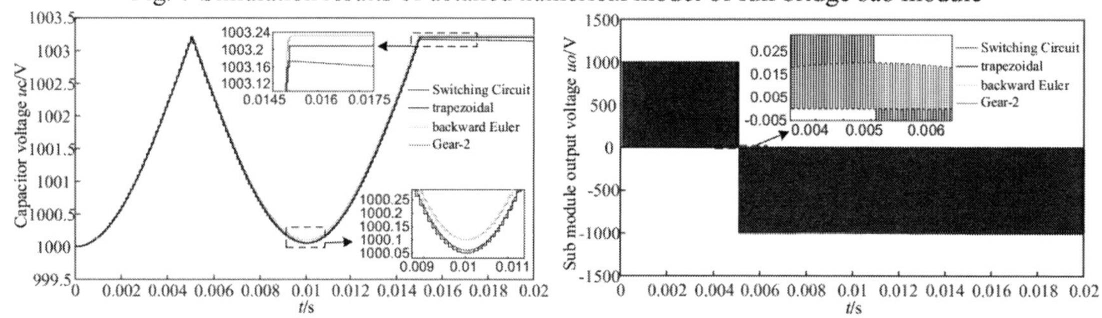

Fig. 8 Simulation results of Simplified numerical model of full bridge sub module

4.2. Comparison of computation efficiency between detailed and simplified values

Taking the numerical model of full bridge sub module as an example, according to section 2.1, using ITM method to segment the MMC model, the electrical decoupling between each sub module is realized, the operation of each sub module is independent of each other. Therefore, the real-time simulation here still uses the test circuit in 4.2. The detailed numerical model simplified numerical model and circuit model are real-time and run in Speedgoat real-time simulator. The real-time simulation step is set to 50μs. For the detailed numerical models of backward Euler method and gear-2 method, the calculation term of capacitive current is omitted here; In view of the small difference in the calculation burden of the simplified numerical model among the above discretization methods, only the acceleration effect based on trapezoidal integral method is listed here. The real-time simulation acceleration effects of the above models are shown in Table 1.

It can be seen from the real-time simulation results that when the number of sub modules is small, the acceleration effect of the detailed numerical models based on several discretization methods has little difference, for example, when simulating 60 sub modules, they only occupy nearly 1/6 of the CPU resources. The simulation of 60 sub module switch circuit model has occupied nearly 4/5 of the CPU resources. This is because the basic framework of state space method is used in circuit model calculation. In real-time operation, the system state matrix and inverse matrix need to be adjusted continuously according to different switch states, so the calculation amount is large. For the numerical model, because it is more direct than the circuit model, it has a good acceleration effect.

When the number of sub modules is large, the switch circuit model cannot run in real-time mode, so the scale of MMC real-time simulation is limited; For the detailed numerical model, the trapezoidal integral method needs to calculate the capacitor current for the difference iteration, so its acceleration effect is slightly lower than that of the backward Euler method and gear-2 method. With the increase of the number of sub modules, the computation efficiency of the backward Euler method and gear-2 method are better; For the simplified numerical model, the calculation of the difference equation is much less than that of the detailed model, so its computation efficiency is the best and it can realize the real-time simulation of large-scale MMC based on the computation efficiency trend of real-time simulation.

Table 1. Computation efficiency comparison of full bridge sub module groups.

Types of models	Execution Time (μs)		
	60 sub modules	120 sub modules	240 sub modules
Switch circuit model	40	—	—
Detailed model (trapezoid)	8	12	20
Detailed model (backward Euler)	8	11	17
Detailed model (Gear-2)	8	11	18
Simplified model (trapezoid)	6	7	9

4.3. Comparison of parallel real time simulation based on half bridge MMC

Parallel simulation is one of the development trends of MMC real-time simulation, such as PC cluster, CPU-CPU and CPU-FPGA. In this paper, taking the multi-core parallel simulation as an example, the parallel simulation of MMC segmentation model is compared with the switch circuit model. The simulation object is a 21 level MMC flexible distribution network, and its main circuit structure is shown in Figure 9.

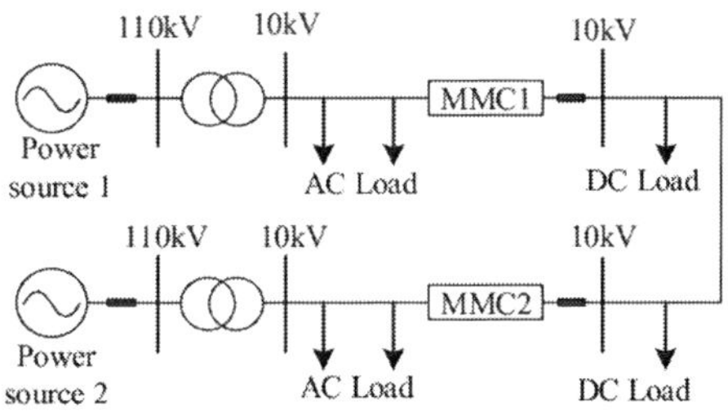

Fig. 9 AC/DC flexible distribution network based on MMC

Because of the large number of sub modules based on the switch circuit model, it is difficult to be real-time, so it runs in the offline simulation mode. The MMC segmentation model based on numerical model adopts three core parallel simulation mode. The main circuit and control system are configured in one core, and the sub module group and voltage modulation are evenly configured in the other two CPU cores. Equation (6) is used as the prediction unit, while the MMC sub module adopts the detailed and simplified numerical models based on trapezoidal integral method, and the simulation step is set to 50μs.

In case 1, the active power command of MMC2 changes from 0 to -600kW at 2.5s, and the reactive power command changes from 0 to - 300kvar at 3.5s. The simulation results are shown in Figure 10; In case 2, a branch of MMC2 is disconnected due to fault at 3.5s, and the control of MMC2 is changed from PQ control to V/F control in order of uninterruptible supply power for the load. The simulation results are shown in Figure 11.

It can be seen that the simulation results of MMC segmentation model based on advanced prediction and sub module numerical model are basically consistent with the MMC switching circuit model. Different from the serial simulation, the parallel simulation needs to predict the bridge arm current in advance. And the output voltage of the sub module directly uses the value of the previous step, so there is still one delay at the voltage mutation point of the sub module which causes deviation of the simulation results. In addition, compared with the simplified numerical model, the detailed numerical model of the sub module is closer to the switch circuit model in most frequency band. However, the simulation results in Section 5.2 show that the simplified numerical model takes less CPU resources and is more suitable for large-scale MMC real-time simulation.

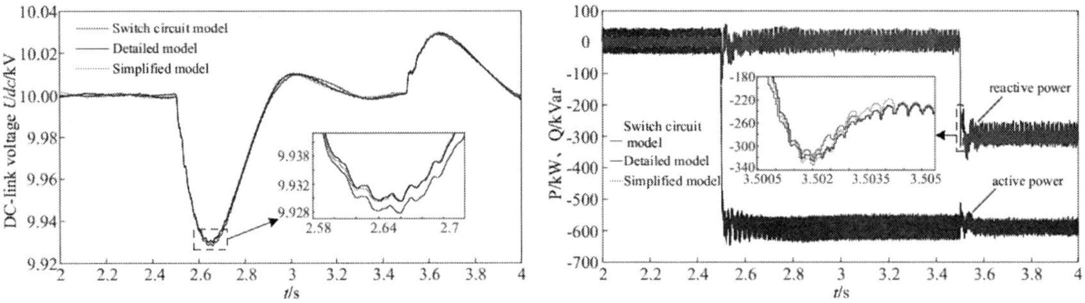

Fig. 10 Comparison of simulation results based on power flow regulation

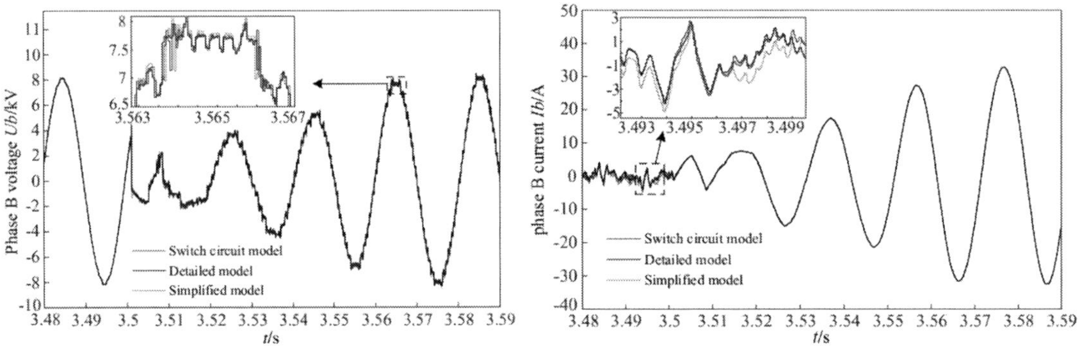

Fig. 11 Comparison of simulation results based on uninterruptible power supply

5. Conclusion

If the scale of MMC real-time simulation is small, the detailed numerical model based on trapezoidal integral method is recommended; if the scale of real-time MMC simulation increase, the detailed numerical model based on gear-2 method or backward Euler method are recommended to enlarge the scale of MMC model; When the above detailed numerical models cannot run in real-time mode, the simplified numerical model based on trapezoidal integral method can be recommended to realize the real-time simulation of larger scale MMC. On the other hand, if the scale of MMC simulation is small, serial simulation based on interpolation prediction is recommended. When the scale of MMC simulation is large, parallel simulation based on advanced interpolation prediction can be used.

Subsequently, the research will focus that the simplified numerical model of MMC run in FPGA, and the parallel real-time simulation step size is no more than 1 μs, which can improve the simulation accuracy and be suitable for large-scale real-time simulation of MMC.

References

[1] Yang X, Lin Z, Zheng Q, et al. (2013) A review of modular multilevel converters [J]. Proceedings of the CSEE, 33(6) : 1-14

[2] Li C,Xie Z,Lin W,et al. (2015) Accurate valve loss calculation method and analyzing platform for medium and high-frequency MMC[J].Proceedings of the CSEE,35(17) :4361-4370(in Chinese) .

[3] Wang H,Tang G,He Z, et al. (2015) Power losses calculation of modular multilevel converter[J].Automation of Electric Power Systems,39(2) :112-118(in Chinese) .

[4] Kurt F. (2010) Modern HVDC PLUS application of VSC in modular multilevel converter topology[C]//IEEE International Symposium on Industrial Electronics (ISIE). Bari, Italy,2010:3807-3810.

[5] Huang H. (2009) Multilevel voltage-sourced converters for HVDC and FACTS applications[C], Cigre, SC B4 2009 Bergen Colloquium. Bergen, Norway,1-8.

[6] Gnanarathna U N, Gole A M, Jayasinghe Rohitha P. (2011) Efficient modeling of modular multilevel HVDC converters (MMC)on electromagnetic transient simulation programs[J]. IEEE Transactions on Power Delivery, 26(1):316-324.
[7] Wang P, Cui X. (2013) A Time-Domain Equivalent Model of Modular Multilevel Converter and Its Fast Algorithm [J]. Power System Technology, 37(8):2180-2186(in Chinese).
[8] Saad H, Dufour C, Mahseredjian J, et al. (2013) Real time simulation of MMCs using the state-space nodal approach[C]//The International Conference on Power Systems Transients (IPST2013) in Vancouver,Canada,June 18-20.
[9] Rohner S,Weber J,Bernet S. (2011) Continuous model of modular multilevel converter with experimental verification[C]//Energy Conversion Congress and Exposition. Phoenix, AZ: IEEE:4021-4028.
[10] Guo G, Hu X, Wen J, et al. (2015) A Large-Scale Sub Module Group Based Algorithm for Modeling and High-Speed Simulation of Modular Multilevel Converter[J]. Power System Technology, 39(5):1226-1232(in Chinese).
[11] Zhou S, Lin W, Yao L, et al. (2015) Generic Averaged Value Models for Two-level VSC and MMC[J]. Automation of Electric Power Systems, 39(12):138-144(in Chinese).
[12] Kaijian Ou,Trevor M,Bruce Warkentin, et al. (2014) Research and Application of Small Time-step Simulation for MMC VSC-HVDC in RTDS[C], International Conference on Power System Technology, Oct 20-22,2014,Chengdu,China:877-822.
[13] Xu J, Zhao C, Liu W. (2013) Accelerated model of ultra-large scale MMC in Electromagnetic transient simulations[J]. Proceedings of the CSEE,33(10):114-120(in Chinese).
[14] Xiong Y, Zhao C, Liu Q, et al. (2016) Modeling of Real-time Simulation and Hardware-in-the-loop Experiments for Modular Multilevel Converters[J]. Automation of Electric Power Systems,40(21):84-89(in Chinese).
[15] Wang W, Zhu J, Li W, et al. (2015) SSN-Based RT-LAB Simulation of MMC-HVDC System[J]. Southern Power System Technology, 9(6):22-27(in Chinese).
[16] Sumek E, Rudi W. (2015) Real-Time Performance of FPGA-based MMC Valves for HVDC Systems [C]//IECON 2015-41st Annual Conference of the IEEE Industrial Electronics Society,Nov 9-12, Yokohama,Japan:4506 - 4512.
[17] Yue C, Zhou X, Li R. (2004) Study of parallel approaches to power system electromagnetic transient real-time simulation[J]. Proceedings of the CSEE, 24(12):1-7(in Chinese).
[18] Zhang H, Hao Z, Chen Z, et al. (2017) Modeling Method for Modular Multilevel Converter in Real Time Simulation [J]. Automation of Electric Power Systems, 41(7):120-126(in Chinese).

Journal of Physics: Conference Series **2108** (2021) 012031 doi:10.1088/1742-6596/2108/1/012031

Remote monitoring system design of power substation based on GPRS

Yicheng Xiong[1], Qian Liu[2], Maojun Deng[3] and Zhilei Niu [3]

[1] Elect & Computer Engineering, Oregon State University , Corvallis, USA

[2] Computer Information and Technology Faculty, The National University of Malaysia, Beijing, 100017, China

[3]XJ Electric Co.,Ltd. , Xuchang , Henan, 461000, China

[*]Corresponding author's e-mail: xycphil@gmail.com

Abstract. This paper studies the design scheme of a substation remote monitoring system based on GPRS, and gives the solution about poor real-time and imperfect function of centralized substation monitoring system. The working principle diagram of the measurement and control device is introduced in detail, and the modules of analog acquisition and switching input are designed. The substation remote monitoring software based on GPRS communication is developed to realize the remote real-time monitoring of electrical and switching parameters and other functions, such as data query, fault early warning and energy efficiency analysis. The experimental results show that the system has the characteristics of real-time, multi-function and high reliability. It is of great significance to implement defensive maintenance, improve power supply quality, save energy and reduce consumption.

1.Introduction

In 1979, Japan developed the world's first CNC system SDC-1 for distribution substations. The United States, Germany, France, Italy and other countries have successively developed their own products, such as the IMPac modular protection and control system developed by the United States and the substation integrated automation system LSA678 in xikouzi, Germany. Countries have made continuous breakthroughs in this field, and substation monitoring system has been widely used[1].

The research on Substation Monitoring Technology in China began in the 1980s. In recent years, with the rapid development of power system related technologies in China, ensuring the safe operation of power system has become the focus of relevant research. In case of sudden failure or temporary power failure in power system, it will bring huge economic losses and material impact to life and production. Therefore, at this stage, there is an urgent need for a technology to realize remote real-time monitoring of power system equipment[2].

Under the long-term sustainable development strategy, the related applications of power system restrict the rapid economic development. Accelerating the optimization of power system and improving the monitoring ability of power system equipment is the primary content of current research. The power system has many distribution transformer nodes, wide geographical coverage and extremely complex power grid lines. The power energy is transmitted from the power plant to the transmission line through the transformer and accepted by each power user through the step-down transformer. In this process, there are many links that need the remote real-time monitoring system to monitor each node of the power equipment[3-4] . In the process of transmitting monitoring data, if wired communication is used, it will bring great working pressure to technicians and hardware laying personnel, so the wireless network technology for power equipment monitoring system is particularly important. At present, SDMs communication network and communication platform are widely used, but due to the limitation of data transmission mode, the system can not meet the use requirements when running[5-7]. Based on GPRS wireless communication network, this paper will design substation remote real-time monitoring system and deeply study the detection and maintenance of power system.

2.Overall design of substation remote monitoring system

Substation is an important node for power system control, regulation, protection and monitoring. The substation remote monitoring system uses advanced computer technology, modern electronic technology, communication technology and information processing technology to recombine and optimize the functions of substation secondary equipment (including relay protection, control, measurement, signal, fault recording, automatic device and telecontrol device), and monitor the operation of all substation equipment. Through the mutual exchange of information and data sharing among various equipment in the substation integrated automation system, the tasks of substation operation monitoring and control are completed.

The substation remote monitoring system replaces the conventional secondary equipment of the substation, simplifies the secondary wiring of the substation, and improves the safe and stable operation level of the substation. The integration of functions is the biggest feature of substation remote monitoring system. It is based on computer technology, data communication and information sharing.

The common substation remote monitoring system adopts the centralized control structure composed of front-end machine and background machine (general control unit). The front-end machine completes the functions of data input and output, protection, control and monitoring, and the background machine (general control unit) completes the functions of data processing and remote communication. The remote monitoring system in this paper adopts a hierarchical and distributed structure, as shown in Figure 1.

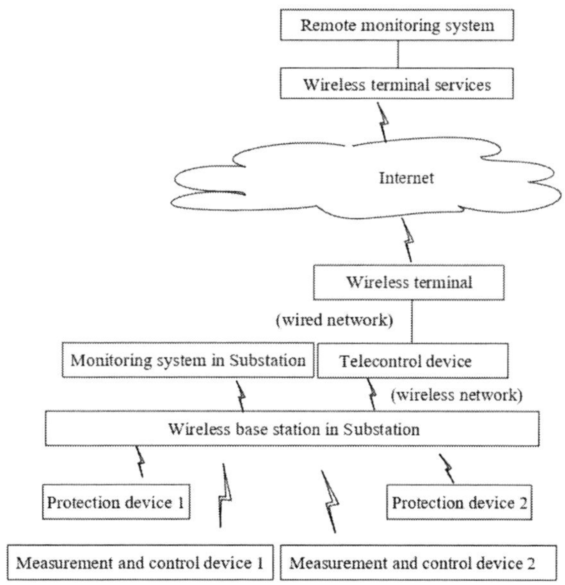

Fig. 1 The dual configuration structure diagram

The two-layer distributed control system structure of whole station control level (station control level) and local unit control level (bay level) is set according to the control level and object of the substation. The equipment layer is the protection, measurement and control, safety automatic device and other secondary equipment with monitoring and acquisition function in the station. These equipment are equipped with wireless network card. The secondary equipment of the equipment layer collects, stores and processes the primary and secondary information, sends it to the station control layer through the wireless base station in the substation, and can accept the control information of the station control layer. The station control layer generally includes integrated monitoring system, wave recording device, data network shutdown, synchronous clock, etc. the equipment in the station control layer is physically independent. Collect, process, store, display and forward the equipment information in the station through the wireless network, and operate and control the relevant equipment through the wireless network.

Dispatching centers at all levels collect substation operation information through telecontrol devices (data network shutdown), analyze and make decisions, and determine the economic and stable operation mode of power grid. As the external communication controller of the substation, the telecontrol device is an important part of the substation remote monitoring system. It creates a real-time database as the center of data processing to meet the real-time requirements of the dispatching master station. The telecontrol device communicates with the wireless terminal through the wired network and connects to the Internet network through the wireless terminal. A corresponding wireless terminal service device at the remote end receives the Internet network information and is connected to the remote monitoring system through the wired network.

3.Design of measurement and control device in equipment layer

In the substation, the status of primary equipment or other secondary equipment is collected through the secondary circuit, transmitted to the monitoring system in the substation through wireless communication, and received the control command of the monitoring system in the substation to control the primary equipment or other secondary equipment. As shown in Figure 2.

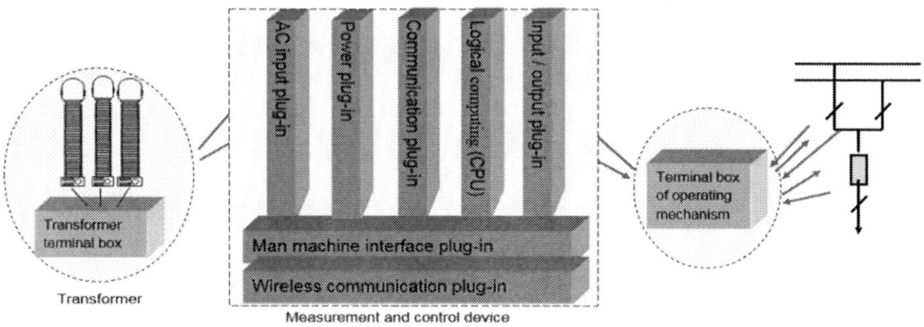

Fig. 2 The dual configuration structure diagram

See Table 1 for specific function configuration of measurement and control device.

Tab.1 List of functional configuration of measurement and control device

NO.	Function	Explain
1	Telemetry	Including voltage, current, power, frequency and direct flow processing
2	Remote signaling	Including the processing of single point remote signaling and double point remote signaling
3	remote control	Processing including selective control and direct control
4	Remote adjustment	realizing the acquisition and control of three groups of gears
5	Communication	Including communication with monitoring system and debugging software
6	Timing function	Time synchronization modes such as B code, pulse or SNTP can be selected

3.1.Design of analog *acquisition module*

The measurement and control device mainly completes the functions of analog quantity acquisition and switching value input and output. Analog quantity acquisition includes DC sampling and AC sampling. The schematic diagram of analog acquisition is shown in Figure 3.

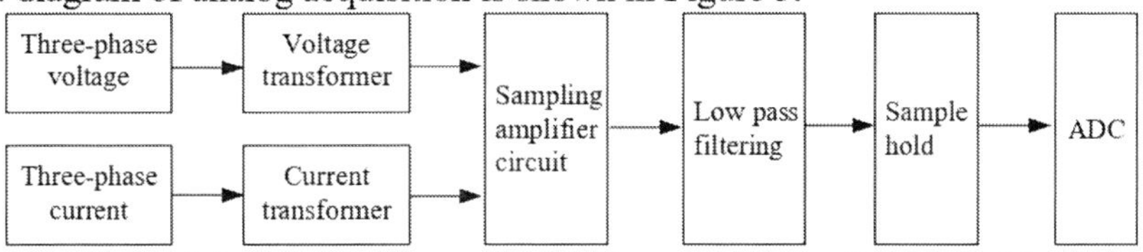

Fig. 3 Schematic diagram of analog acquisition Sample hold

The voltage and current signals obtained from voltage transformer Pt and current transformer CT cannot be directly input to CPU for data processing, but must go through signal conditioning circuit and A/D conversion circuit. Signal conditioning is to convert the U and I analog signals obtained from Pt and CT to the voltage range required by ADC through amplification circuit, low-pass filtering, sampling and holding. The A/D converter converts the analog quantity into digital quantity, and then transmits it to the CPU for calculation and processing of electrical parameters, so as to obtain various electrical parameters such as the effective value of u and I, active power, reactive power, power factor and harmonic.

3.2.Switching value I / O module

The function of the switching value input module is to realize the remote signal collection of the monitoring system in the substation, that is, by monitoring the auxiliary normally open and normally closed contacts of the switchgear in the substation, so as to identify the switching status of power devices such as load switch, circuit breaker, capacitor and transformer. Remote signal acquisition is vulnerable to electromagnetic interference, so low-pass filter circuit and photoelectric isolator are added to the switching value input module of measurement and control device. The switching value input through low-pass filter circuit and photoelectric isolator is shown in Figure 4.

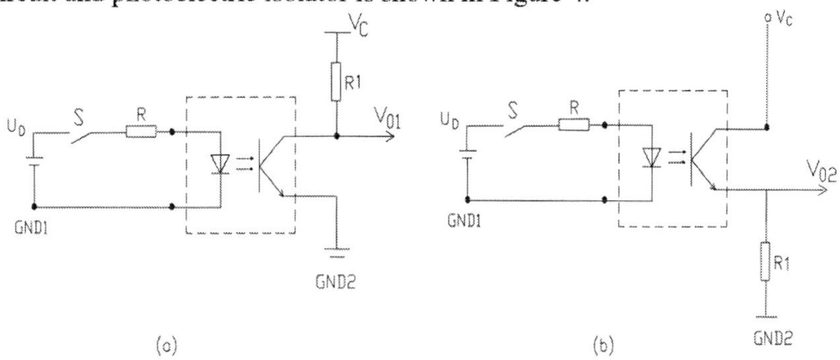

Fig. 4 Input diagram of switching value

Chattering elimination shall be considered during switching value input. The software shall capture the first jump and start timing. When the delay time expires, it is still in the high-level state, and the displacement is effective. This event shall be officially recorded. Delay is generally taken as 5s.

The function of switching value output module is to realize the remote signal control of monitoring system in substation, that is, to control the opening and closing of power device switches in substation. In order to effectively resist interference, the switching value output signal adopts the export mode of "start + export" double command to further improve the reliability of switching value output. The device startup relay control circuit and outlet control circuit are shown in Figure 5 and Figure 6.

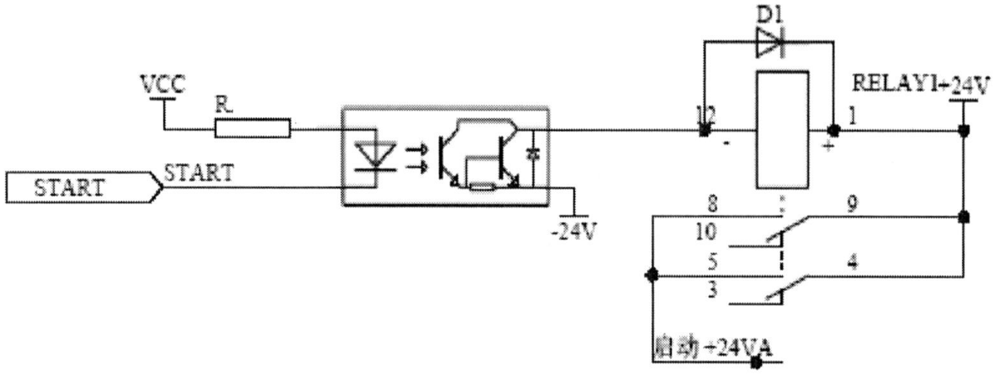

Fig. 5 Schematic diagram of starting control circuit

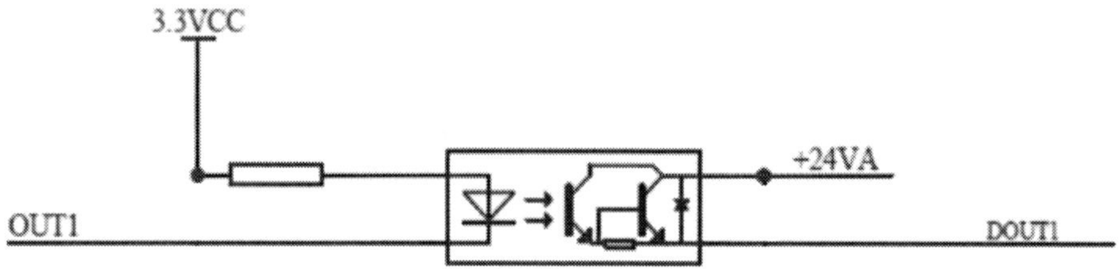

Fig. 6 Schematic diagram of outlet control circuit

The VCC in the figure is 3.3V. In the starting circuit, start is the control input of the starting relay from the CPLD (complex programmable logic device) chip of the CPU board. When "start" is low, the starting relay relay1 acts, the normally open contact is closed, and the starting + 24va voltage rises to + 24V. In the outlet circuit, out1 is the control input of 1 outlet relay, which is also from the CPLD chip of CPU board. When out1 is low level and + 24va voltage is + 24V, dout1 output drives the outlet relay corresponding to the outlet board to act, and the corresponding normally open contact is closed to control the outlet output.

3.3. GPRS communication module

GPRS wireless communication technology has the characteristics of high transmission rate, always on-line, low communication cost, convenient power supply and maintenance, wide coverage and so on[8-9]. The measurement and control device is connected with the main station of the monitoring system in the substation through an independent GPRS link . The GPRS module of the measurement and control device adopts the square M590 GPRS module. After the measurement and control device establishes a point-to-point protocol (PPP) link with the operator through the M590 GPRS module, it uses the at command at + tcpsetup = < n >, < IP >, < port > to establish a GPRS link with the master station of the monitoring system in the substation, where n is the link number, which can only be 0, 1, 2, 3, and IP is the address of the master station of the monitoring system in the substation to be linked, Port is the port number provided by the main station of the monitoring system in the substation. After the GPRS link between the measurement and control device and the main station of the monitoring system in the substation is established successfully, the measurement and control device can communicate with the main station of the monitoring system in the substation. The GPRS link establishment process is shown in Figure 7.

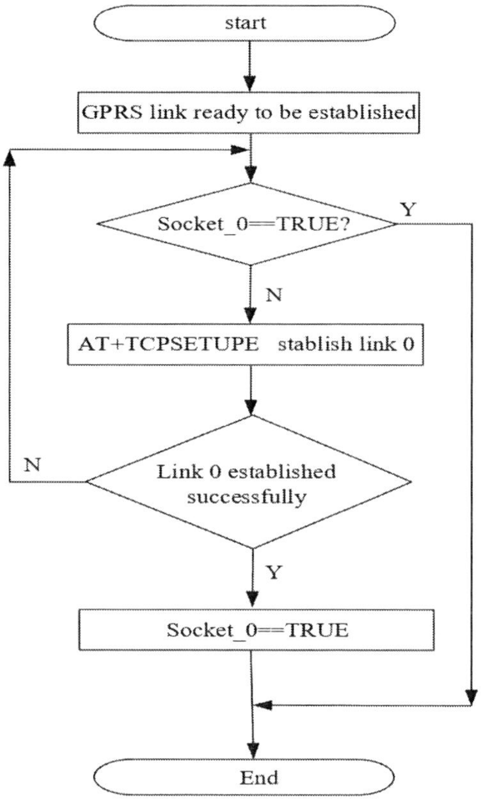

Fig.7 GPRS link establishment flow chart

4.Software and hardware design of integrated monitoring system

The networking mode of the monitoring system is distributed, and all servers can be deployed. Even if the equipment manufacturers are different, as long as they are in the same region and on the same platform, the system can realize unified management, and the platforms do not affect each other. Each platform manages itself, and the overall operation of the platform will not be affected by each single point of failure. In any platform, there is its own management system. Like the control protocol, this system is unified, which can authorize users in this range, and the resources between them can be shared, which is also to ensure the interoperability of resources in the whole network. For the operation status and performance status of the primary and secondary equipment in the substation, the monitoring system can make the personnel on duty clear at a glance, which has been optimized in terms of equipment maintenance, labor intensity and man hour cost, so as to ensure the safe and reliable normal operation of the power system.

The hardware of the monitoring host adopts the server, which runs the integrated monitoring system software, can communicate with the Bay layer and other station control layer equipment, and realize the functions of analyzing, displaying and storing real-time communication data. The basic functions of the integrated monitoring system are as follows:

4.1.Data acquisition, processing and storage

*4.1.1.*Analog quantity includes electric quantity such as current, voltage, active power, reactive power, frequency and power factor, and non electric quantity such as temperature.

*4.1.2.*Switching value acquisition includes position signals of circuit breaker, disconnector and grounding knife switch, protection action signals, operation monitoring signals and tap position signals of on load voltage regulating transformer, and realizes functions such as timing acquisition, equipment abnormal alarm, event sequence record (SOE), operation record, logic operation, etc.

4.2.Alarm processing

4.2.1. In the stage of the first accident alarm, if the second alarm occurs, it can be handled in the same way, and the first alarm will not be overwritten.

*4.2.2.*In case of advance alarm, except that the processing method is the same as that of accident alarm, the sound and information color can be different from accident alarm, and can selectively send information to the distance.

4.3.Control operation

*4.3.1.*With local / bay level / station control level / remote, multi-level control, with necessary safety inspection and error proof locking, to complete the control of switch and knife switch; Adjustment of tap of main transformer; On / off of protection function pressing plate; Control functions such as signal reset and equipment start and stop.

*4.3.2.*The monitoring system provides two anti error locking functions at the station control layer. One is to cooperate with the special microcomputer anti error system to complete the anti error locking function of the whole station; The other is the software anti error locking function embedded in the system. The user can complete the design of station control layer locking logic through a friendly software locking logic definition tool. The two functions can be used together to complete the anti error locking of station control layer.

*4.3.3.*The monitoring system has the operation monitoring function, allowing the operator to work in one workstation, and the supervisor can monitor on another workstation; When one workstation fails, the operator and supervisor can operate and monitor on the same workstation. In order to prevent misoperation, step-by-step operation is adopted under any control mode, i.e. selection, calibration and execution, and operator and supervisor passwords and line codes are set at the station level to ensure the safety and correctness of operation.

4.4.Interface

It can monitor the operation parameters and equipment status of main electrical equipment through the display. The screen supports double screen display, and the screen operation supports stepless zoom. It can roam smoothly and has the function of guide map. The main display screen includes:

*4.4.1.*Main electrical wiring diagram, including real-time values showing equipment operation status, power flow direction, main electrical quantities (current, voltage, frequency, active power, reactive power), etc.

*4.4.2.*Operation diagram of computer monitoring system: display the equipment configuration and connection status of computer monitoring system in graphic mode and color change.

*4.4.3.*Configuration diagram of secondary protection, reflecting the switching condition and setting value of each set of protection.

5.Conclusion

In general, the GPRS based remote monitoring system developed in this paper has high real-time performance, comprehensive functions, superior performance and stable operation. It is of great significance to implement defensive maintenance, improve power supply quality and save energy and consumption. This design idea of applying wireless communication mode to substation remote monitoring system is the innovation of this paper.

Acknowledgments

Under the careful guidance of Deng Maojun and Niu Zhilei, I successfully completed the compilation, revision and finalization of this paper. Thank you two mentors, my parents and my friends. You are my strong backing.

References

[1] Ding Shuwen, Huang Xuncheng,Hu Qizhou. Principle and application of substation integrated automation [M]. First edition. Beijing: China power for publishing, 2003: 27-28.
[2] Peng Zhiqiang, Zhang Qibing,Su Dawei,etc. Remote operation and maintenance technology of substation monitoring system based on GSP [J]. Power automation equipment, 2019, 39 (4): 210-216.
[3] Xiang Chunzhi, Lv Shulin. Remote monitoring system for ship operation status under wireless network [J]. Ship science and technology, 2017, 39 (18): 73-75.
[4] Cuiyu,Wuyi,Zhangzhi,etc.Design and implementation of relay protection intelligent mobile operation and maintenance system based on power wireless virtual private network [J]. Power system protection and control, 2018, 46 (23): 175-181.
[5] Xu Xinqian, Xulai, Chenjie,etc. Cable channel emergency monitoring system based on mobile wireless sensor network[J]. China power, 2020, 53 (1): 66-71 + 99.
[6] Xiong Haijun, Zhanghai. Research on wireless network structure design strategy of power equipment monitoring system[J].Electrical measurement and instrumentation, 2020, 57 (21): 24-31.
[7] Yu Jinhui, Zheng Xiaoyi, Zhao Shiyan,etc. Development of remote monitoring system for port mechanical equipment based on WLAN [J]. Mechanical design and manufacturing, 2018 (4): 269-272.
[8] Guo Yaohua. Wireless temperature monitoring system of intelligent substation equipment based on ZigBee and GPRS network [J]. Electrical technology, 2016, 11(A): 44-45. Instrument technology and sensors, 2014 (1): 79-82
[9] He Biyi, Zhou Guoping, Ding Jianqiang,etc. Design of a monitoring system for box type substation based on WSN and GPRS [J]. Electrical technology, 2016, 11(A): 44-45.

Research on Secondary Cable Identification Based on the Principle of Electromagnetic Induction

Li Zhu Bao[1a*], Jia Yan Yu[1b*], Yu Wu[1c*], Hao Wu[1d*], Zhi Jun Tang[1e*], Jun Cheng Wang [1f*]

State Grid Jiangsu Electric Power Co., Ltd. Maintenance Branch

[a*]20201249047@nuist.edu.cn, [a*]10385681@qq.com, [b*]saltfishmail@sina.com, [c*]19105037@qq.com, [d*]29789531@qq.com, [e*]422816108@qq.com, [f*]304869298@qq.com

Abstract: When the operation and maintenance station performs cable maintenance, installation measures, etc., due to the lack of direct and effective tools, and the complicated wiring of the operation and maintenance station equipment, the wiring diagram is inconsistent with the actual situation, which brings risks to the removal of waste cables. As an indispensable safety tool in the power industry, the cable identification device is an important means to ensure the safety of maintenance personnel. In view of the current situation, this paper analyzes the electric field around the cable line of the operation and maintenance station, and simplifies the simulation of the line, and proposes a secondary cable identification device based on the principle of electromagnetic induction, which can quickly identify the target cable, which greatly improves the cable identification performance. Efficiency reduces the time for cable identification and improves safety.

1.Introduction

As China's economy shifts from a stage of rapid development to a stage of high-quality and stable development, power cables have been widely used in power transmission and distribution in power systems, especially in operation and maintenance stations and urban distribution networks[1]. Power cables are often laid in cable tunnels or cable trenches, and many cables with different voltage levels are often laid in parallel in a cable tunnel or cable trench[2]. When the substation is carrying out protection technology transformation, when the cable in the cable trench needs capacity increase, transformation and line path change, it is necessary to correctly identify the target cable and find out the direction and function of the original cable before the demolition and transformation of waste cable can be carried out. At present, when dismantling and transforming used cables in substations, most of them use comparison cable signs, or choose to refer to drawings to identify secondary cables[3]. When the old substation is carrying out the transformation of waste cables, due to the long age of the substation, the secondary cable signs are not standardized or the handwriting is blurred, the signs are missing, and a few cable signs are marked incorrectly, which makes the success rate of cable identification through cable signs low and serious waste of time. The cable laying situation in the substation is shown in Figure 1:

At present, the commonly used methods for identifying cables are the audio signal method and the pulse signal method, which are mainly aimed at the path finding method of adding signals under power outage conditions[4]. However, in actual operations, many cables to be identified are running under power, which cannot fully meet the identification requirements for live cables. Therefore, how to identify live cables quickly and accurately has always been a difficult problem for cable frontline workers to solve urgently.

Figure 1 Diagram of cable laying in cable trench

In response to the above problems, this paper proposes a secondary cable identification device based on the principle of electromagnetic induction[5]. Based on the principle of electromagnetic induction, a conductor in an electric field will generate induced charges in a changing electric field. By measuring the relationship between the potential of the induction electrode relative to the ground and the actual voltage on the power line, the actual voltage value of the power line can be obtained. The voltage level of the power line can realize the identification of the cable line.

2. Principle and basis of cable identification

2.1 Principle analysis

There is a special substance around the charged object. This invisible and intangible substance can transmit energy, which makes people begin to explore the magnetic field. Faraday gave the expression of electromagnetic induction phenomenon, the magnitude of induced electromotive force e is determined by the rate of change of magnetic flux through the coil:

$$e = -\frac{d\phi}{dt} = -\frac{d}{dt}\int B \cdot dS$$

There are three main reasons for the change of magnetic flux:

(1) In a constant magnetic field, the relative movement of the conductor causes the magnetic flux of the closed loop to change. As shown in the figure, in a static magnetic field, the magnetic induction intensity B is constant, and the conductor ab moves to the right along the horizontal rail. A current that changes with time will be generated in the closed conductor [6]. The electromotive force e generated at both ends of the conductor ab is called motion electromotive force.

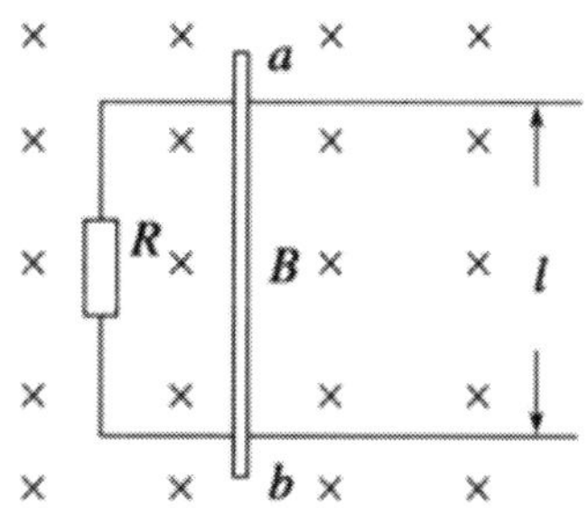

Figure 2 The induced electromotive force generated by a moving conductor in a magnetic field

(2) The closed conductor loop remains unchanged, and the magnetic field changes. At this time, an induced current will be generated in the closed conductor, and the resulting electromotive force is called induced electromotive force. Since the loop does not change, the magnitude of the induced electromotive force is only related to the time t. At this time, the first-order partial conductance of the magnetic induction intensity time is shown in the following formula:

$$e = -\frac{d}{dt}\int B \cdot dS = -\frac{d}{dt}\int \frac{d}{dt}(B \cdot dS)$$

$$= -\int \left[\frac{\partial B}{\partial t} \cdot dS + B \cdot \frac{\partial (dS)}{\partial t}\right]$$
$$= -\int \frac{\partial B}{\partial t} \cdot dS$$

(3) The magnetic induction intensity changes with time, and the loop also changes. At this time, the induced electromotive force e_1 generated when the two circuits are unchanged and the motional electromotive force e_2 when the magnetic induction intensity is constant are as follows:

$$e = e_1 - e_2$$
$$= -\int \frac{\partial B}{\partial t} \cdot dS - \int B \cdot \frac{\partial (dS)}{\partial t}$$

At this time, the second term of the above equation e_2 can be rewritten as:

$$-\int B \cdot \frac{\partial (dS)}{\partial t} = -\int_l B(v * dl) = \int_l (v * B) \cdot dl$$

It can be concluded from the above formula that the induced electromotive force can be superimposed and extended to any closed loop in space. As long as the magnetic flux in the loop changes, the electromotive force will be induced in the closed loop. A changing electric field will induce a changing magnetic field, and a changing magnetic field will induce a changing electric field. This interdependent and restrictive relationship between electric and magnetic fields forms the phenomenon of electromagnetic induction.

2.2 Transmission model of cable line

The analog charge method uses a set of discrete virtual charge points to equate the uneven current flowing in the actual conductor, so that the electric field generated by the analog charge and the electric field generated by the current in the actual conductor have the same boundary conditions, and then all the points are solved. Match point potential.

The cable line uses alternating current for electric energy transmission. the magnetic fields generated by the alternating current rotate and superimpose or cancel each other, and adjacent lines will also generate alternating magnetic fields to superimpose, which makes the electromagnetic environment of the cable line more complicated [7]. If three-dimensional it is very difficult to calculate the space magnetic field. The wavelength λ of the cable line is 6*10^6m, which is much longer than the length of the cable line itself. Therefore, the induced current generated by the alternating magnetic field is ignored, and the eddy current effect caused by the magnetic field on the surface of the conductor is ignored.

When applying the analog charge method for analysis and calculation, certain conditions need to be met. The following simplifications are made here:

(1) The cable line is regarded as infinitely long, and the cable lines are arranged at equal intervals;

(2) Ignore the distortion effects of lightning conductors, ground wires, insulators and fittings on the electric field;

(3) The influence of magnetic field is not considered;

(4) The electric field is regarded as a two-dimensional field;

3.Device design

3.1 Hardware design

3.1.1 Overall design

The cable identification device designed in this paper uses a single-chip microcomputer as the control chip, which mainly includes signal acquisition module, filtering and amplification module, A/D conversion module, single-chip module, alarm module, wireless communication and human-computer interaction module, etc. The overall structure of the system as shown in Figure 2:

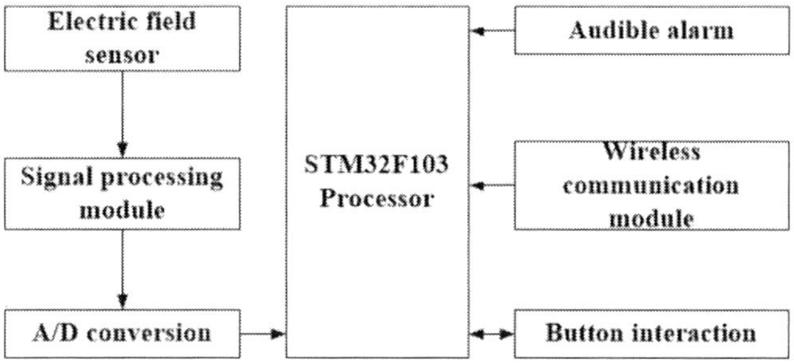

Figure 2 Overall structure

The signal collected by the sensor is transformed and processed, and the harmonic is eliminated through the filter circuit. The output voltage signal is mV level. It needs to be amplified before it can be sampled by the A/D module of the single chip microcomputer. The sampled results are sent to the host through the wireless module, and the host will judge after receiving them. The identified results are output through acoustic and optical signals, and display the collected electric field value.

3.1.2 Filter circuit

Due to the large amount of electromagnetic interference in the measurement environment, the signal output by the sensor will contain a lot of noise, which will cause signal distortion in severe cases and affect the accuracy of the measurement. This requires a filter circuit to filter out the useless signals as much as possible. Composition, leaving a useful signal. The cable identification device designed in this paper is mainly used to measure the power frequency voltage of the cable line. The cut-off range of the frequency is set to 1000 *Hz*, so a low-pass filter is selected. Among the common low-pass filters, the frequency amplitude characteristics of the Butterworth filter are relatively flat, especially in the low frequency band, which is approximately a straight line, which meets the requirements of the designed cable identification device.

The simplest filter is an *RC* low-pass filter, which consists of a resistor and a capacitor. After the *RC* filter, a voltage follower circuit is added to form a first-order active low-pass filter as shown in Fig. 3.

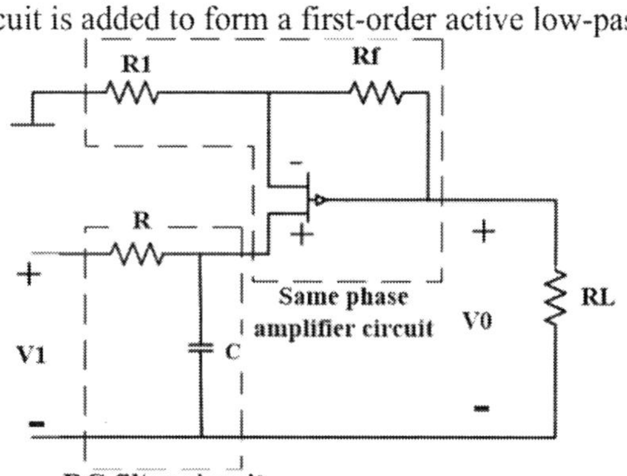

Figure 3 First-order low-pass filter circuit

3.1.3 A/D conversion

The output signal of the sensor has analog signals such as voltage, current, pressure, temperature, and the processor (such as a single-chip microcomputer) can only recognize the digital signal logic 0 and 1. These analog signals cannot be directly collected by the single-chip microcomputer, and need to be converted into what the single-chip microcomputer can do. The calculation can be performed after the recognized digital signal, which requires an analog-to-digital conversion circuit (Analog to Digital Convert) to achieve.

The A/D module used in this article is the built-in A/D module of the STM32F103 single-chip microcomputer, with a resolution of 12bit, a reference voltage of 3.3V, and an input analog voltage of

0-5V. The working principle of this A/D converter is shown as in Fig. 4.

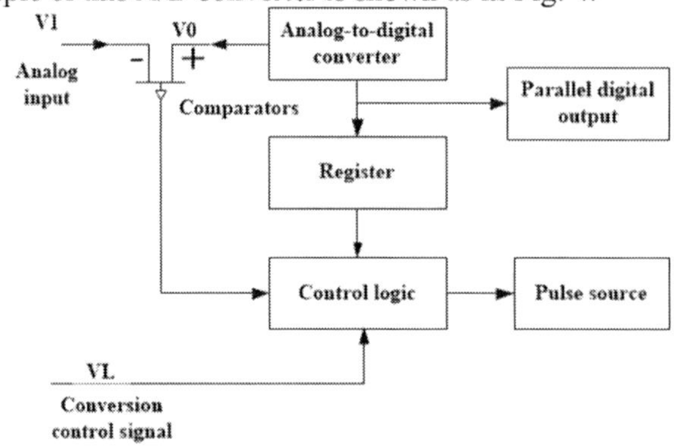

Figure 4 A/D converter working principle

The STM32F103 has a built-in 16-channel 12-bit A/D converter, providing 4 modes (single, continuous, voltage comparison, timer automatic), and the conversion rate is 1MHz, 500kHz, 125kHz, 31.25kHz. The external address of the A/D conversion control register ADC_CTL is 0xD0. Considering the real-time change of the measurement position, the continuous conversion mode is used, and the control bit ADC_CNTNU_CVT is set to 1.

3.1.4 Button design

The button is the input function that realizes the human-computer interaction. When the single-chip microcomputer is required to produce certain specific actions, you can press the preset button, and the single-chip computer detects that the level of the circuit connected to the button has changed and sends out the corresponding instruction. To achieve specific functions.

There are 4 buttons in the device design, respectively K1, K2, K3 and K4. Among them, K1 is the button reset, which is used when the system needs to be initialized; K2 is used to start measurement. When K2 is pressed, the single-chip microcomputer sends the data sent by the sensor Carry out A/D conversion; K3 is used to test the alarm circuit, which will drive the buzzer and light up the diode after pressing; K4 is the switch button, which can be turned on/off after long pressing.

3.2 Software design

The functions to be realized by this device are as follows:

(1) It can display the sensor data after processing;

(2) According to the results of the data, judge whether it is the target cable, and if the conditions are met, an alarm will be issued;

(3) The key can realize the corresponding human-computer interaction function.

3.2.1 Main program design

Long press the switch button, the single-chip microcomputer is powered on, enters the initialization program, and performs self-checking on each module. When the start measurement button is pressed, the A/D conversion module starts to work, converts the voltage signal of the sensor into a digital signal and sends it to the single-chip microcomputer. The microcontroller analyzes the data and sends the data to the host for processing. The driver generates the corresponding waveform and generates the data. If the data is displayed as a target cable, the microcontroller will execute the alarm program, drive the buzzer, and make the LED lights flash alternately.

3.2.2 Data processing

When the measurement button is pressed (set ADC_PD=0, the control register ADC_SINGLE_CVT in the analog-to-digital conversion is set to 1, and the A/D conversion is started. After the conversion is completed, an interrupt time is generated. Because there are 2 sensors, it is necessary to turn on the continuous conversion mode (ADC_CNTNU_CVT=1), and turn on the analog/digital conversion comparison function (EN_ADC_CMP=1). After the input is converted, it is compared with the preset threshold in the comparison buffer (ADC_CMP_V). When the input value is greater than (When ADC_BIG=0) or less than (ADC_BIG=1) setting value, an interrupt signal will be generated to the MCU to execute the corresponding function.

3.2.3 Judgment process

When the voltage values of the two sensors are collected, the single-chip microcomputer first performs the difference calculation on the sensor values, and compares the calculated difference with the preset threshold. If the difference is less than the preset value, it is judged that there is no target cable. Terminate the calculation; when the difference is greater than the preset value, go to the next step. Compare the value of sensor 1 with the threshold of the target cable. If it is greater than the threshold, the electrification condition is met, execute the electrified program, and terminate the operation; if it is less than the threshold, enter the next step. Compare whether the value of sensor 2 meets the electrified threshold, if so, execute the electrified procedure; otherwise, execute the no electrified procedure, and the procedure ends.

4. Device test

Through the construction of the system hardware circuit and the debugging of the software program, in order to demonstrate the reliability of the device, measurements will be carried out in the substation environment to verify the performance of the device. The main test content is to select the cable line of the substation, measure the surrounding environment without interference, check the reliability of the device and determine the reliable range of identification, and combine the experimental data to summarize the reliability of the device.

The measurement area is located directly below the cable line, and the distance from the cable is in the range of 0 cm to 20 cm. Take the ground directly below the middle phase as the origin, and set 1 measuring point on the left and right, a total of 3 measuring points. Each measuring point measures the value of the distance of 0 cm, 5 cm, 10 cm and 15 cm at the same time. Due to the measurement error of the sensor, the unit data is rounded. The measurement data is shown in Table 1.

Table 1 Measurement data

Measuring point coordinates (cm)	Measuring value 1 (V/cm)	Measuring value 2 (V/cm)	Difference (V/cm)
(-1, 0)	3.04	2.61	0.43
(-1, 5)	2.61	2.16	0.45
(-1, 10)	2.17	1.72	0.45
(-1, 20)	1.73	1.30	0.43
(0, 0)	2.97	2.51	0.46
(0, 5)	2.52	2.07	0.45
(0, 10)	2.09	1.66	0.43
(0, 20)	1.68	1.24	0.44
(1, 0)	3.07	2.62	0.45
(1, 5)	2.63	2.21	0.42
(1, 10)	2.21	1.75	0.46
(1, 20)	1.78	1.33	0.45

According to the data in the table, the device is reliable and the best measuring distance is in the range of 5 cm.

5.Conclusion

The cable identification device designed in this paper is easy to operate, has a high safety factor, and has broad application prospects and market demands. The feasibility and practicability of the equipment are verified through the trial application of the actual cable operation on site. The equipment has extremely high accuracy and adaptability, which can further improve the safety and efficiency of power engineering construction and cable general survey, and provide great help to avoid misjudgment of identifying cables.

This article mainly completes the following tasks:

(1) The electromagnetic environment around the transmission line was studied, and the cable line was simplified according to the analog charge method and a mathematical model was established; according to the field strength distribution characteristics of the transmission line space, a non-contact cable identification method was explored, and verified that the method is indeed feasible in the experiment.

(2) Filter and amplify the voltage signal from the sensor and send it to the single chip microcomputer for A/D conversion and analysis. Determine whether it is the target cable according to a certain algorithm process. If so, send an audible and visual alarm to improve the maintenance personnel.

References

[1] Gao Guihua. Design of non-contact ultra-high voltage electroscope based on field strength distribution [D]. Xidian University, 2014.

[2] Yang Pengfei, Wen Xiaolong, Ni Xiaoming, Peng Chunrong. Research on the spherical shell-type non-contact AC test device[J]. Journal of Electronics and Information, 2021, 43(06): 1637-1643.

[3] Tan Yumiao. Research and realization of non-contact high voltage DC electroscope[D]. Chongqing University of Technology, 2018.

[4] Chen Jiandong, Wang Dongpeng, Fan Daoqing, Chen Tao, Dong Hualiang, Li Changxin, Yu Bing. Design of a non-contact high-voltage detection device [J]. Electronic Technology, 2017, 46(05): 33-35+29.

[5] Huang Qin. Research on the Algorithm of Intelligent Video Early Warning System for Transmission Line Breakdown Prevention [D]. South China University of Technology, 2020.

[6] Gu Yijun, Mo Haibo, Liang Tao, Wang Gonghan. Distributed optical fiber sensing technology-based monitoring system for preventing external damage to distribution lines[J]. Rural Electrification, 2020(10): 38-43.

[7] Hong Bo, Zhang Chenxi, Wang Xu. Monitoring and early warning system for high-voltage cable protection against external damage based on Michelson interference and image recognition[J]. Zhejiang Electric Power, 2021, 40(03): 79-84.Qi Dongchun,Chen Xinghua,Zhu Yiwen,Zhang Qi. A new type of wind-resistance cable net for narrow suspension bridges and wind-resistance cable element for its calculation[J]. Structures,2021,33:

[8] Hu Yujing,Hu Wei,Yang Wen,Wu Junjie,Chang Yan,Yin Fucheng,Xie Yi. The corrosion analysis and prevention of secondary cable joints used in outdoor terminal boxes of substation in humid environment[J]. IOP Conference Series: Earth and Environmental Science,2021,621(1):

[9] Sun Zhongyu, Xu Bingyin, Wang Wei, Chen Heng, Liu Yang, Wei Xinchi. Modeling and Simulation of Cable Fault Pulse Current Ranging System[J]. Automation of Electric Power Systems, 2021, 45(04): 142-147.

[10] Jiang Yixin, Chen Feng. Development of a live secondary cable identification instrument[J]. Information and Communication, 2017(02):282-283.

Journal of Physics: Conference Series **2108** (2021) 012033 doi:10.1088/1742-6596/2108/1/012033

NB-IoT based method for monitoring the tilt status of transmission towers

Yan Zheng*, Yanfeng Wu, Jianchuan Gao and Shu Cui

State Grid Dalian Electric Power Supply Company, Dalian, 116000, China

*Corresponding author's e-mail: zhengy3_dl@ln.sgcc.com.cn

Abstract. The current monitoring of transmission tower tilt status mainly uses a combination of manual and sensor data collection, which not only has low monitoring efficiency and high cost, but also makes it difficult to guarantee the monitoring accuracy. In response to the problems of the above research, the NB-IoT-based transmission pole tower tilt state monitoring method will be studied. Six-axis attitude sensor GY-521, wind sensor and tension sensor are used to collect the pole tower status data. The NB-IoT network is designed to transmit the noise reduction processed data to the base station, which is transmitted to the upper computer. The data is analyzed in the upper computer using LSTM network to get the monitoring results of the tilt status of the pole tower. The experimental results show that the maximum root mean square error of the proposed monitoring method is only 0.02701, which is much smaller than the comparison method, and the monitoring efficiency is high and the performance is more suitable for practical applications.

1. Introduction

In the natural environment and the role of external conditions, tower foundation from time to time will occur slip, tilt, settlement, cracking and other phenomena, thus causing the deformation of the tower, tower tilt or even inverted tower break line tower tilt will cause the tower guide ground line unbalanced force, causing the tower force changes, resulting in electrical safety distance is not enough, affecting the normal operation of the line. Inverted tower line will make the power supply line in paralysis, seriously affect people's production life, causing huge losses [1]. Therefore, in the daily maintenance of power system, we need to focus on the monitoring and management of transmission equipment, especially the transmission tower state. In the early days, the monitoring of the status of transmission towers by relying on manual patrol inspection not only required high technical level of inspectors, but also had cumbersome inspection steps and extremely low monitoring efficiency[2]. Non-destructive monitoring of transmission towers using ultrasonic pulse velocity in transmission towers in different states, although it reduces the cost investment of manual monitoring, but it cannot evaluate the tower tilt state from the perspective of data, and there are limitations in its use [3]. Nan Yinggang et al. used fiber-optic grating stress sensors to collect the stress changes in towers under wind and ice loads, so as to build a finite element model of the tower to achieve monitoring of the tilt state of the tower. This method only takes into account the influence of a limited number of factors on the tower tilt, which is only applicable to the monitoring of tower tilt in some areas and is not universally applicable [4]..

In recent years, many experts and scholars at home and abroad have studied the characteristics of NB-IoT that are different from other wireless communication technologies and optimized its shortcomings in the current stage [5]. The NB-IoT data can be directly connected to the official platform of telecom operators, which is more convenient for development and practical management. In this paper, NB-IoT technology is applied to monitoring the tilt status of transmission towers, and a set of sensor-based transmission tower tilt detection method is designed to obtain and analyze and upload transmission tower tilt status information.

2. NB-IoT based method for monitoring the tilt status of transmission towers

2.1. Transmission tower tilt status data collection

Tilt angle of transmission tower is an important parameter to measure the state of the tower, use at the top of the column of angle sensor installed in the transmission tower, tower angle and the mathematical relationship between the sensor output, the specific size of the transmission tower angle can be calculated. In this study, a 6-axis attitude sensor GY-521 is selected for the transmission tower tilt angle measurement [6], which uses an MPU6050 acceleration sensing chip to digitally output the 6-axis rotation matrix data of the fusion algorithm, which solves the problem of inter-axis difference compared to other multi-component combination sensors. Table 1 below shows the performance parameter table of MPU6050.

Tab. 1 Table of performance parameters of MPU6050

Number	Performance Parameters	Specific values
1	currents	3-6V
2	current	<10m A
3	range (of scales or measuring equipment)	Acceleration: ±16g Angular velocity: ±2000°/S
4	measurement dimension	Acceleration: 3 Angular velocity: 3 dimensions Stance angle: 3 dimensions
5	Data output frequency	100HZ (baud rate 115200). 20HZ (baud rate 9600)
6	baud rate	115200 9600
7	stability	Acceleration: 0.01g Angular velocity: 0.05°/S
8	Stability of attitude measurement	0.01°
9	resolution (of a photo)	Acceleration: 6.1e-5g Angular velocity: 7.6e-3°/S

When the transmission tower is in a tilted state, the gravitational acceleration of the tower has components on all three axes, and these components are combined to be the gravitational acceleration [7], using trigonometric functions to calculate the angle at which an axis deviates from its original position (start by putting the tilt sensor as close to the original position as possible for reference). FIG. 1 below shows the relationship between acceleration and inclination angle for the three axes.

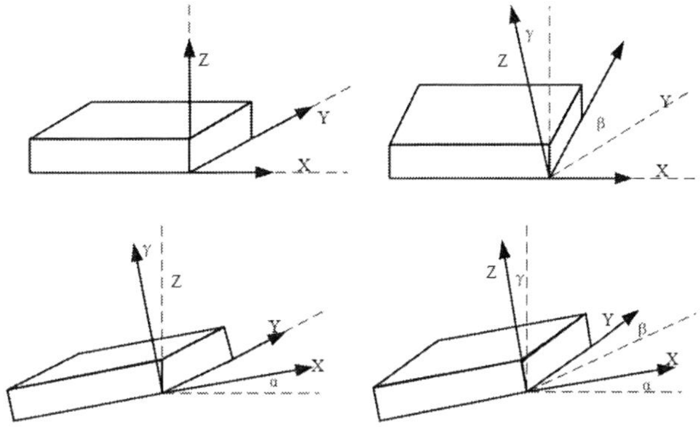

Fig. 1 Relationship between triaxial acceleration and inclination angle

From FIG.1, the calculation formula of inclination angle can be deduced:

$$\alpha = \tan^{-1}\left\{\frac{A_X}{\sqrt{A_Y^2 + A_Z^2}}\right\} \quad (1) \quad \beta = \tan^{-1}\left\{\frac{A_Y}{\sqrt{A_X^2 + A_Z^2}}\right\} \quad (2) \quad \gamma = \tan^{-1}\left\{\frac{\sqrt{A_X^2 + A_Y^2}}{A_Z}\right\} \quad (3)$$

In the above formula, α , β and γ are the deviation angles corresponding to the X, y and Z axes of the triaxial acceleration respectively. Since the sensor is installed at the cross arm of the tower top and the waist of the tower, ignoring the influence of the tower material deformation of the tower itself, in the actual measurement[8], only the gravity acceleration changes in the x-axis and Y-axis directions need to be collected, that is, the along-line tilt angle and the transverse tilt angle of the tower. To improve the efficiency of monitoring of transmission tower tilted state, this study uses NB-IoT network transmission sensor acquisition of transmission tower tilted state data.

2.2. NB-IoT Network Design

In the network architecture of NB-IoT, it includes: NB-IoT terminal, E-UTRAN base station, home subscriber server (,HSS), mobility management entity (MME), service gateway (S-GW) and PDN gateway (P-GW); service capability exposure function (SCEF) [9]. The Service Capability Open Unit SCEF and the corresponding interfaces T6 and S6t are used for non-IP data transfer. Among them, T6 interface is used to connect SCEF and MME; S6t interface, to connect SCEF and HSS. the NB-IoT network architecture for pole tower tilt status monitoring is shown in FIG. 2.

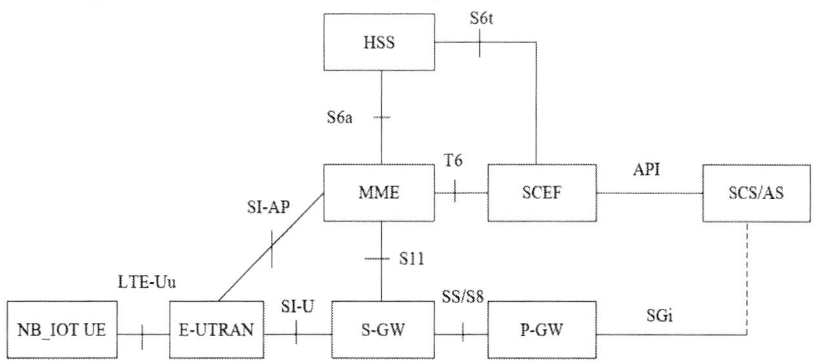

Fig. 2 NB-IoT network architecture diagram

The NB-IoT communication network acts as a bridge in the whole monitoring process, connecting measurement units such as gyroscopes, accelerometers and attitude sensors at the bottom layer and data processing center at the top layer. The bottom sensor network is connected to the NB-IoT network to provide monitoring data for the tilt status of transmission towers. The NB-IoT network has two power saving modes to reduce the operating power consumption of hardware devices. In this study, the NB-IoT network is selected to operate in eDRX mode in order to meet the real-time monitoring of the tilt status of transmission poles. In eDRX mode, a TAU cycle consists of two state cycles, Connect eDRX and Idle eDRX. In Connect eDRX, it is necessary to listen to the PDCCH and receive downlink data with a maximum receive cycle of 10.24s . The communication protocol of each terminal in NB-IoT network is COAP communication protocol. CoAP restricted application protocol is mainly used in IoT scenario. CoAP protocol is a one-to-one protocol to achieve communication between terminals and servers.

2.3. Transmission tower tilt state data processing

The two filters give an all-pass estimate of the tilt angle, and the complementary filter model is simple enough to provide accurate tilt angle data over a wide frequency band. The complementary filter based tilt angle sensor data fusion method can have the advantages of both accelerometer and gyroscope. The mathematical model of the complementary filter is described as follows.

$$\theta_a = \eta \times (\theta_a + \omega \times \Delta t) + (1-\eta) \times \theta_{acc} \quad (4)$$

Where, θ_a is the inclination angle; η is the complementary filter factor; ω is the angular velocity measured by the gyroscope.The tilt angle of the transmission tower measured by the accelerometer is θ_{acc} .The magnitude of the time constant τ ($\tau \in [0,1]$). Suppose the sampling interval of the sensor is Δt when the sensor collects transmission towers, then the time constant for the filtering process by the complementary filter is defined as:

$$\tau = \frac{\eta \Delta t}{1-\eta} \leftrightarrow \eta = \frac{\tau}{\tau + \Delta t} \quad (5)$$

After processing the transmission tower tilt status monitoring data collected by the sensors according to the above, the data is transmitted via the NB-IoT network base station to the upper computer for analysis, and the LSTM neural network is used to realize transmission tower tilt status monitoring.

2.4. Implementing transmission tower tilt condition monitoring

The repetitive hidden layer module of the LSTM neural network has three gate controllers and a memory unit. The three gate controllers are the input gate, the forgetting gate and the output gate. The gate controllers mainly include the Sigmoid function and the dot product operation. The input vector of the input gate at the time t is :

$$a_i^t = U_i \times s_{t-1} + W_i \times x_t + b_i \quad (6)$$

Where, at time t ,when the LSTM network is processing the sensor data, the input vector of the input gate is a_i^t ; s_{t-1} is the output vector of the hidden layer at $t-1$; when the LSTM network is processing sensor data at time t ,the input vector of the network as a whole is x_t . W_i is the weight matrix between the input gate and the neural network input layer. U_i is the weight matrix between the input gate and the LSTM hidden layer at the previous moment; b_i is the offset of the input gate. The input vector of the output gate at the time t is:

$$a_o^t = U_o \times s_{t-1} + W_o \times x_t + b_o \quad (7)$$

Where, at time t ,when the LSTM network is processing the sensor data, the output t vector of the input gate is a_o^t ; W_o is the weight matrix between the output gate and the neural network input layer; U_o is the weight matrix between the output gate and the LSTM hidden layer at the previous moment; b_o is the offset of the output gate. The output vector of the output gate at time t is

$$g_t = f_o\left(a_o^t\right) \quad (8)$$

Where, g_t is the output vector of the output gate at time t; $f_o(\ )$ is the sigmoid function. The input of the implicit layer of the LSTM consists of the input vector and the output of the implicit layer at the previous moment, and the function is hyperbolic tangent function. The tower tilt data collected by each sensor when stationary and different external forces are applied to the transmission towers are selected as the training data of the LSTM network.

3. Experimental studies

3.1. Experimental data and preparation

In this paper, a wineglass tower in a 110 kV transmission line of a regional power grid is selected for study and analysis. the transmission line pole tower meets the normal environmental requirements for line ice cover load, construction conditions, and wind load. The towers have a tower height of 34.8 m, a call height of 29.8 m, and a root opening of 5.77 m. The towers at different locations in the line were selected as the study objects, and the corresponding sensors were installed and sensor communication networks were formed according to the design requirements of the two monitoring methods. According to the data of meteorological department, the tilt state of the pole tower is monitored under different winds with the influence factor of wind on the tilt of the pole tower. The modeling analysis results of the simulation software are used as standard values to calculate the monitoring errors of the two monitoring methods. The monitoring time of the two monitoring methods during the detection process is also counted.

3.2. Experimental results and analysis

Table 2 below shows a comparison of the monitoring results of the two transmission tower tilt monitoring methods.

Tab. 2 Experimental data

Wind Power Rating	Distance to data base station/m	NB-IoT monitoring methods		RTK monitoring methods	
		Monitoring mean squared error MSE	Average monitoring time/ms	Monitoring mean squared error MSE	Average monitoring time/ms
4	100	0.02654	14.1	0.03564	18.2
4	300	0.02698	15.9	0.03789	20.6
5	100	0.02661	14.6	0.03691	19.4
5	150	0.02634	15.2	0.04135	21.9
6	100	0.02667	15.7	0.04277	22..5
6	200	0.02701	15.3	0.04563	26.7

From the data in table 2 above, it can be seen that the maximum root mean square error of monitoring the tilt state of transmission towers by the monitoring method in this paper is only 0.02701, which is smaller than the root mean square error of the RTK monitoring method. Moreover, the average monitoring time of the method in this paper fluctuates very little under the influence of different levels of wind, while the average monitoring time of the RTK monitoring method has more obvious fluctuations. In addition, from the data in table 2 above, it can be seen that the monitoring efficiency and accuracy of the RTK monitoring method are more obviously affected by the distance of the monitoring object.

4. Conclusion

Transmission tower tilt can lead to abnormal electricity network system, and severe can lead to a crash, it will eventually result in serious economic loss and serious safety accidents. By monitoring the basic condition of the transmission tower, it is of certain significance to reduce the occurrence of accidents and economic losses. In order to improve the accuracy of transmission tower state monitoring and to avoid the transmission tower lean phenomenon, affecting the safety of transmission and personnel work in this paper, we study the transmission tower tilt condition monitoring method based on NB-IoT. Using NB-IoT technology of high-speed communication performance, this method improves the transmission tower tilted condition monitoring accuracy and efficiency. In the future, in-depth research will be carried out on tower condition monitoring in different environments, with the help of a variety of sensors and sensor data fusion algorithms, so as to improve the universal applicability of transmission tower inclination monitoring methods.

Acknowledgment

This work was supported by the Science and Technology Project (Research on the Online Monitoring Device of Pole Tower Inclination Based on NB-IoT Technology) of State Grid Liaoning Electric Power Co., Ltd. (Grant No.2021YF-07) .

References

[1] Lei Wen, Gangqiang Kong, Hossam Abuel-Naga, Qingsong Li,Zhendong Zhang. (2020). Rectification of Tilted Transmission Tower Using Micropile Underpinning Method. Journal of Performance of Constructed Facilities, 34(1):04019110-04019110.

[2] Jin Yi , Wang Wei,Pei Ling, Chen Xin,Song Bao. (2021). The Research and Implementation Transmission Line Tower Rod and Monitoring System Using Reverse Network RTK Technology. Journal of Physics: Conference Series, 1894(1):012044.

[3] Takalkar S.M.. (2020). Non-Destructive Condition Monitoring of Transmission Line Towers & Foundation. Power Engineer Journal, 22(1): 16-26.

[4] Nan Yinggang , Xie Wenping, Min Li,Cai Shunshuo, Ni Jiasheng, Yi Jun, Luo Xiaoyu, Wang Ke, Nie Ming,Wang Chang, Peng Gang Ding, Guo Tuan. (2020). Real-Time Monitoring of Wind-Induced Vibration of High-Voltage Transmission Tower Using an Optical Fiber Sensing System. IEEE Transactions on Instrumentation and Measurement, 69(1): 268-274.

[5] Xinbo Huang , Long Zhao, Ziliang Chen, Cheng Liu. (2019). An online monitoring technology of tower foundation deformation of transmission lines. Structural Health Monitoring, 18(3):949-962.

[6] Sunita Malhara ,Vijay Vittal. (2010). Mechanical State Estimation of Overhead Transmission Lines Using Tilt Sensors. IEEE Transactions on Power Systems, 25(3):1282-1290.

[7] Norikazu Fuse, Yoshiharu Shumuta, Atsushi Naganuma, Jun-ichi Tani, Yasuhiko Hori. (2019). Corrosion Rate Estimation Based on Sensor Monitoring: Field Test Validation in Transmission Towers. Corrosion, 75(7): 839-847.

[8] R. Karami-Mohammadi, M. Mirtaheri, M. Salkhordeh, M. A. Hariri-Ardebili. (2020). Vibration Anatomy and Damage Detection in Power Transmission Towers with Limited Sensors. Sensors, 20(6): 1731.

[9] Parrino Stefano , Peruzzi Giacomo , Pozzebon Alessandro. (2021). LoPATraN: Low Power Asset Tracking by Means of Narrow Band IoT (NB-IoT) Technology.. Sensors (Basel, Switzerland), 21(11):3772.

Design and Characteristics of CMOS Inverter based on Multisim and Cadence

Haoran Xu[1*], Jianghua Ding[1] and Jian Dang[2]

[1]Changan University, China, Xi'an,710000

[1]Henan University of science and technology, China, Luoyang, 471000

[2]The Gunston school, Centreville, the United States, MD 21617

[*]Corresponding author's e-mail: 181404110203@stu.haust.edu.cn

Abstract. Known as complementary symmetrical metal oxide semiconductor (cos-mos), complementary metal oxide semiconductor is a metal oxide semiconductor field effect transistor (MOSFET) manufacturing process, which uses complementary and symmetrical pairs of p-type and n-type MOSFETs to realize logic functions. CMOS technology is used to build integrated circuit (IC) chips, including microprocessors, microcontrollers, memory chips (including CMOS BIOS) and other digital logic circuits. CMOS technology is also used in analog circuits, such as image sensors (CMOS sensors), data converters, RF circuits (RF CMOS), and highly integrated transceivers for various types of communications. Based on multisim14.0 and cadence, the characteristics and performance of CMOS inverter are studied by simulation.

1. Introduction

Multisim is a windows-based simulation tool launched by National Instruments (NI) Co., Ltd. It is suitable for board-level analog/digital circuit board design [3] and includes graphical input of circuit schematics, circuit hardware description language input. It has a large simulation analysis capabilities. With Multisim, engineers can interactively build circuit schematics and simulate the circuit. Multisim refines the complex content of SPICE simulation, so that engineers can quickly capture, simulate and analyze new designs without knowing in-depth SPICE technology, which also makes it more suitable for electronics education. PCB design engineers and electronics educators can complete a complete integrated design process from theory to schematic capture and simulation to prototype design and testing through Multisim and virtual instrument technology.

Candence is a large, professional EDA (Electronic Design Automation) software, which can realize almost all the contents of electronic design, layout design, experimental simulation, etc. Software with similar functions include mentor graphics, Synopsys and so on. But compared with other EDA software cadence, cadence has more powerful functions in circuit simulation, circuit diagram design, layout design and wiring. In addition, cadence also shares data with many semiconductor companies and establishes a process library for simulation, which is convenient for users to carry out simulation. The process used in this paper is smic18mm.

The static complementary CMOS combinational logic gate is actually formed by expanding the static CMOS inverter into a logic gate with multiple inputs, so the inverter is discussed here first. The inverter is used in digital integrated circuits to invert the phase of the input signal by 180 degrees, that is, the gate circuit that performs the inversion operation. At this time, the MOS tube is working in the on-off state. But when the MOS tube works in the transition zone, it can play an amplifying role and can be used in audio amplifiers or clock oscillators. The circuit structure is composed of a PMOS tube and an NMOS tube. Here, the pull-down network and the pull-up network are composed of a single MOS tube to form a static complementary circuit [5].

Noise margin is a measure of design margins to ensure circuits functioning properly within specified conditions [9]. When an inverter is transitioning from a logic high to a logic low, there is an indistinct region in which we cannot consider the voltage either low or high. It is at this precise moment that we consider it to be our noise margin. the CMOS inverter has good logic buffer characteristics, in that, its noise margins in both low and high states are large. Noise margin happened when a inverter connect to another inverter, there could be a noise, which means that the Vout could be higher or lower, and if the VOH is lower than VIH, the signal will not be treated as a logic high. What we do is we need some margin that to make sure that we have enough space for the noise margin. Since The VOH in CMOS inverter is close to the supply voltage, and the minimum of the voltage is also close to zero. The CMOS inverter has the great noise margin. To ensure that transistors switch properly under specified noisy conditions, circuits must be designed with specified noise margins. Figure 1 illustrates the noise margin and the terms [9].

Noise margin high : NMH = VOH – VIH

Noise margin low : NML = VIL – VOL

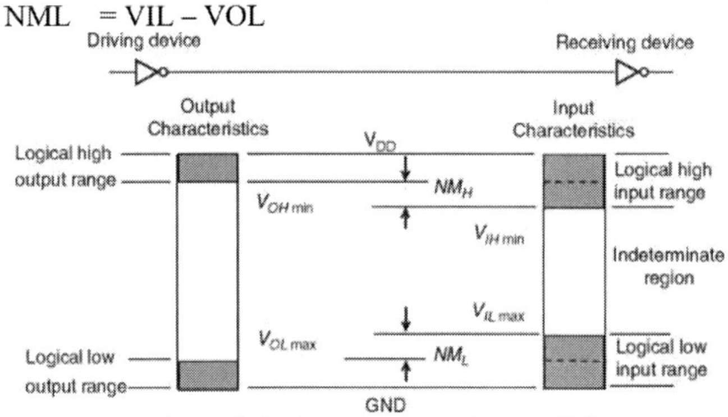

Figure 1. Noise margin and terms [9]

This paper aims to study the properties and characteristics of COMS inverter through two different platforms: Multisim and cadence, so as to provide more reference for other inverter research later.

2. The design of inverter base on Multisim

2.1. Structure of CMOS inverter

Static complementary CMOS gate-level circuits are the most widely used type of logic gates. Because of its good stability, good performance and low power consumption, it is widely used in the design of integrated circuits. The static complementary CMOS gate-level circuit is a combination of a pull-up network (PUN) and a pull-down network (PDN), and its structure is shown in the Figure 2 in the static complementary CMOS gate circuit design, a pull-down network is generally first designed according to the logic solution calculation formula to be implemented, and then a pull-up network is made according to the logic structure of the pull-down network. The specific implementation is to exchange the serial-parallel mode of the MOS transistors in the pull-down network and the N/P attributes of the MOS transistors to obtain a new network that is the pull-up network. In terms of operation, for NMOS tubes, series connection means to do the method operation, parallel means do the addition operation, and vice versa [5].

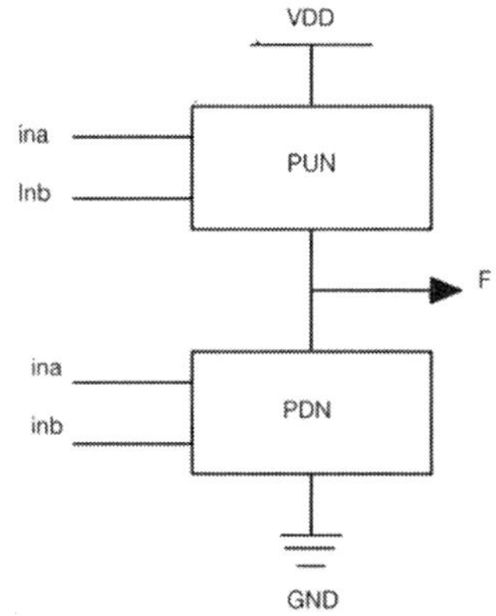

Figure 2. Two-input complementary logic gate structure

2.2. The design principle of inverter based on Multisim

What the NOT gate realizes is that the logic function is out = -in, which means that the input signal is reversed, so it is also called an inverter. The gate level diagram is as in Figure 3.

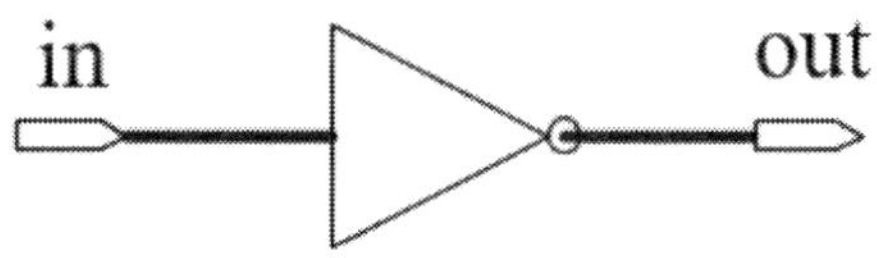

Figure 3. The gate level diagram [2]

The corresponding CMOS structure is as follows in Figure 4: it is composed of a PMOS and NMOS, PMOS is pulled up and connected to VDD; NMOS is pulled down and connected to GND.

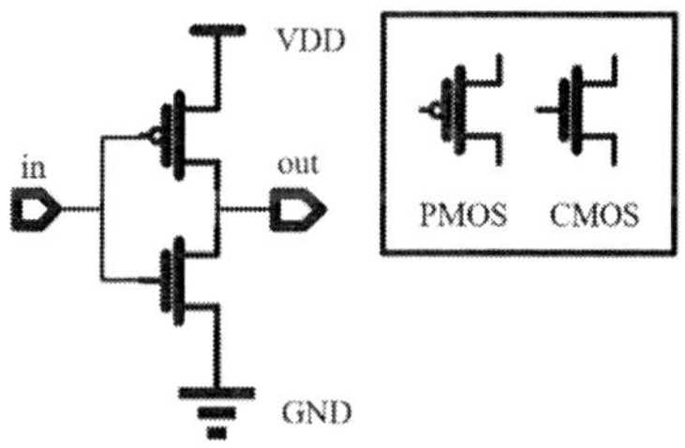

Figure 4. Corresponding CMOS structure [2]

When the input is 0 (low level), the output is 1 (high level), that is, the output is connected to the power supply VDD; when the input is 1 (high level), the output is 0 (low level), that is, the output must be connected to GND (ground). Therefore, the equivalent circuit of the above NOT gate CMOS circuit is as follows:

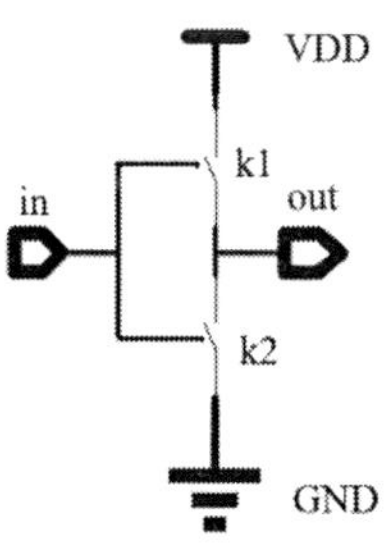

Figure 5. NOT gate CMOS circuit [2]

As in Figure 5, when the input is 0, switch k1 is closed and k2 is opened, that is, the output is connected to VDD, and the output is 1. When in is 1, the switch k1 is opened and k2 is closed, that is, the output is connected to GND, and the output is 0.

PMOS is equivalent to a switch, as follows in Figure 6, when the G terminal is low, the D terminal and the S terminal are connected.

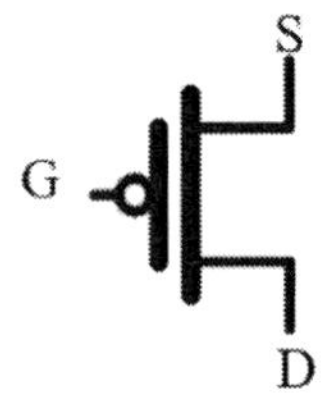

Figure 6. PMOS [2]

NMOS is equivalent to a switch, as follows in Figure 7, when the G terminal is high, the D terminal and the S terminal are connected.

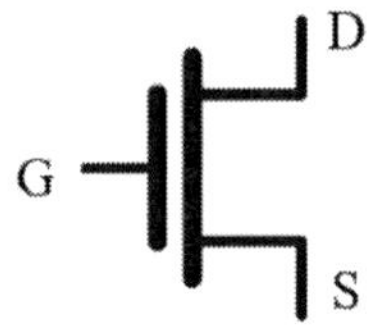

Figure 7. NMOS [2]

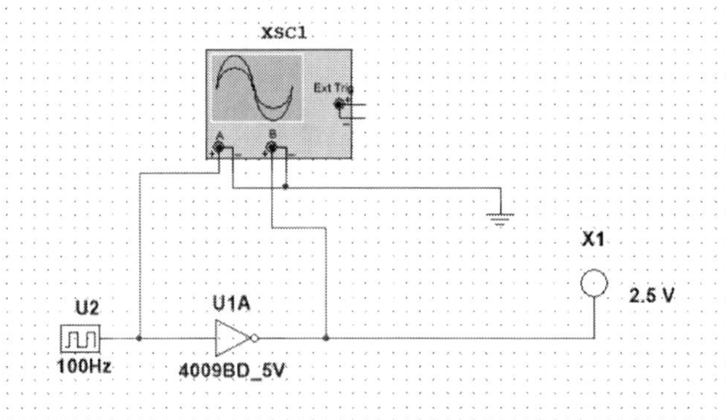

Figure 8. Inverter

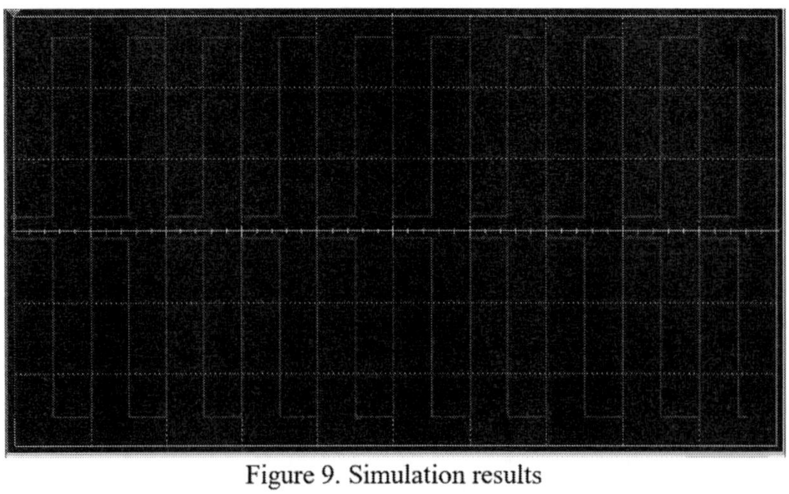

Figure 9. Simulation results

Through the simulation of the circuit in Figure 8, during the test time, the input waveform (Figure 9) is ultimately opposite to the output waveform displayed on the oscilloscope. When the input is high, the output is exactly low, and the error is within the allowable range.

3. Transient characteristics of CMOS inverter based on Multisim

The transient characteristics of the inverter can be expressed by the rise time and fall time of the inverter. The rise time$t_r$is the time required to make the output level of the inverter from 0.1vdd to 0.9vdd, and the fall time$t_f$is defined as the time required for the output level to fall from 0.9vdd to 0.1vdd. Set the input high level inverter to 1.8V and the input low level to 0V [4].

Similarly, the output high level and low level are 1.8V and 0V respectively. It can also be found that the potential of input and output is just the opposite, which satisfies$V_{out} = \overline{V_{in}}$. According to the design requirements, select the appropriate aspect ratio, so that the inverter input flip level is exactly half of the supply voltage, $V_{it} = 0.5V_{DD}$. Through transient simulation, the rise time and fall time of inverter are obtained. It can be found from Figure 10 that the threshold voltage vit is approximately equal to 0.9V, and it is not difficult to find that the rise time is about 1US and the fall time is about 0.7us.

When NMOS transistors transmit at high level, threshold loss occurs, while PMOS transistors transmit at low level, threshold loss occurs. As a result, the CMOS process combines two different MOS transistors. Moreover, when the output is high or low, only one MOS transistor in the CMOS inverter is on. Theoretically, there is no current between the power supply and the ground, and there is no static power consumption. The power consumption of the CMOS inverter is very low, which is also the biggest advantage of the CMOS inverter circuit [1].

Figure 10. DC voltage transmission characteristic curve of CMOS inverter

4. Design principle of inverter based on Cadence

The not gate (inverter) has an input and an output. The input is in and the output is out. When the input is high level, the output is low level, when the input is low level, the output is high level. Output $OUT=\overline{IN}$, The truth table is shown in Table 1.

Table 1. Not gate (inverter) truth table

IN	OUT
0	1
1	0

Inverter used in digital integrated circuits can reverse the phase of the input signal by 180 degrees, which is the gate circuit for reverse operation. At this time, the MOS transistor works in the on-off state. But when the MOS transistor works in the transition region, it can play the role of amplification, and can be used in audio amplification or clock oscillator [6].

The inverter circuit is composed of a NMOS transistor (N transistor) and a PMOS transistor (P transistor). The source of P transistor is connected with high level V_{DD}, and the source of N transistor is connected with low level V_{SS}. Input in controls the conduction and cutoff of P transistor and N transistor. When the input in level is low, the P transistor is on, the N transistor is off, and the output out level is high; When the input in level is high, the N transistor is on, the P transistor is off, and the output out level is low. In this way, the logic operation of the inverter (not gate) can be achieved. The inverter (not gate) circuit and transient simulation results are shown in the Figure 11, Figure 12 and Figure 13.

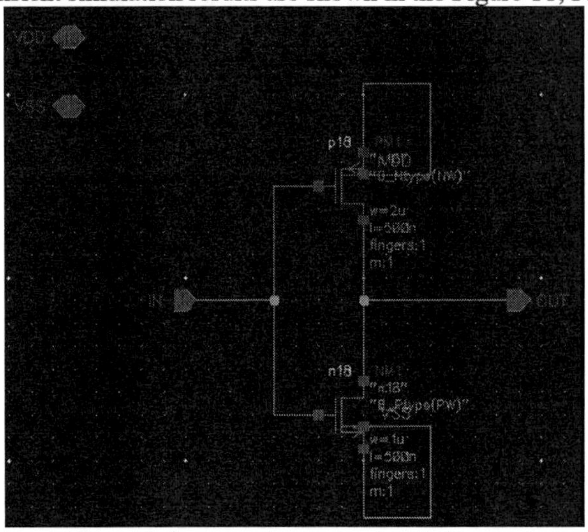

Figure 11. Inverter (not gate circuit diagram)

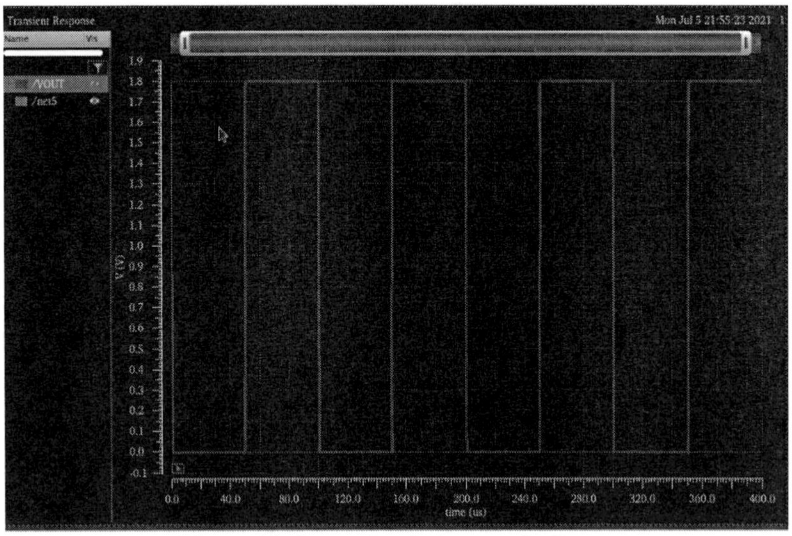

Figure 12. Input simulation results Input

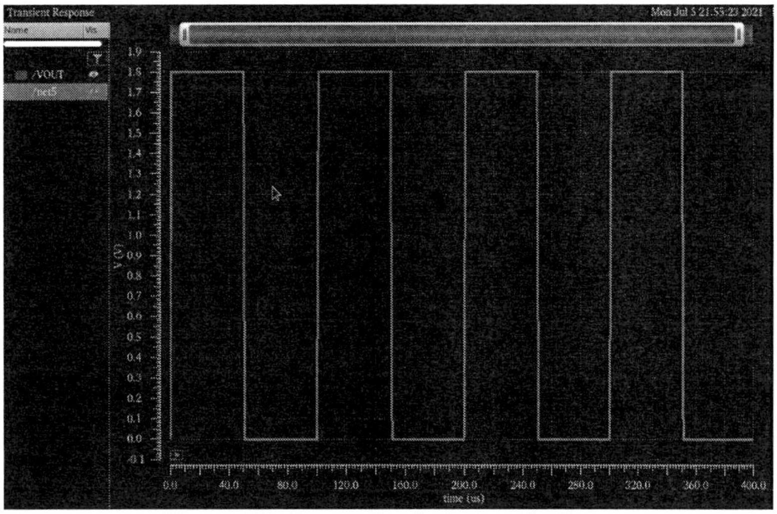

Figure 13. Output simulation results Input

5. Transient characteristics of CMOS inverter based on Cadence

The transient characteristics of the inverter can be expressed by the rise time and fall time of the inverter. The rise time $t_r$is the time required to make the output level of the inverter from 0.1vdd to 0.9vdd, and the fall time t_f is defined as the time required for the output level to fall from 0.9vdd to 0.1vdd. Set the input high level of Figure 11 inverter to 1.8V and the input low level to 0V [8].

Similarly, the output high level and low level are 1.8V and 0V respectively. It can also be found that the potential of input and output is just the opposite, which satisfies$V_{out}= \overline{V_{in}}$. According to the design requirements, select the appropriate aspect ratio, so that the inverter input flip level is exactly half of the supply voltage, $V_{it} = 0.5V_{DD}$. Through transient simulation, the rise time and fall time of inverter are obtained. It can be found from Figure 14 that the threshold voltage vit is approximately equal to 0.9V, and it is not difficult to find that the rise time is about 1US and the fall time is about 0.7us.

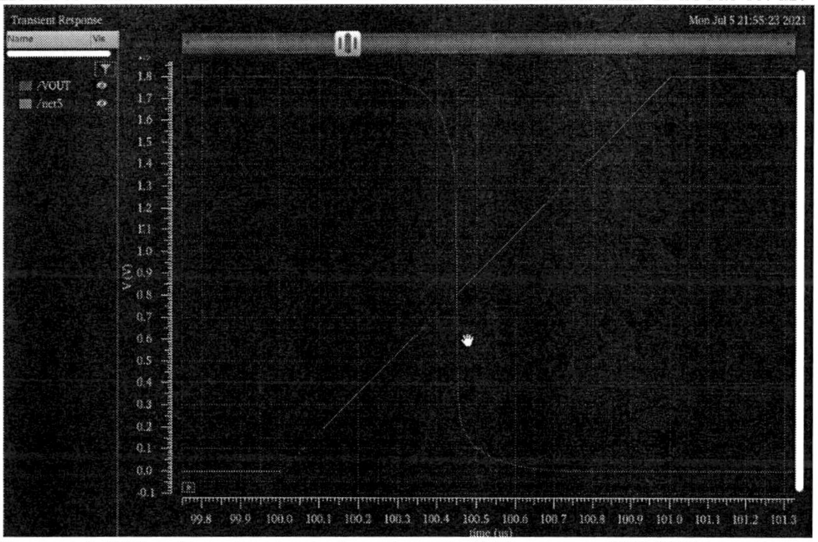

Figure 14. DC voltage transmission characteristic curve of CMOS inverter

When NMOS transistors transmit at high level, threshold loss occurs, while PMOS transistors transmit at low level, threshold loss occurs. As a result, the CMOS process combines two different MOS transistors. Moreover, when the output is high or low, only one MOS transistor in the CMOS inverter is on. Theoretically, there is no current between the power supply and the ground, and there is no static power consumption. The power consumption of the CMOS inverter is very low, which is also the biggest advantage of the CMOS inverter circuit [7].

6. Conclusion

In this paper, the classic static CMOS logic gate circuit of the inverter is simulated by Multisim and cadence simulation software, and the correct simulation results are obtained. The rise time and fall time of the inverter and the relationship between the input and output level and the power supply voltage are verified. On this basis, the NAND gate is optimized and simulated by using pseudo NMOS structure.

The structure has the following excellent characteristics: (1) the number of quality tubes is reduced by nearly half and the power consumption is greatly reduced. (2) The load terminal is connected with smaller loads, which increases the effective utilization area of the layout. (3) The processing difficulty is small and the processing cost is reduced (4) The CMOS noise immunity is high.

Because CMOS inverter is one of the basic units of static CMOS logic circuit and an important part of integrated circuit design, it is very important to verify its working principle and switching characteristics.

References

[1] Qian Xiang, Lu Yaqing. Characteristic analysis and Simulation of CMOS inverter based on cadence [J] Communication power technology, 2021.
[2] Yan Shi Fundamentals of Digital Electronic Technology (Sixth Edition) Higher Education Press.
[3] Chen Tong. Application of Multisim virtual simulation experiment in circuit teaching [J] Electronic Test, 2021(15): 127-128+80.
[4] Zhang Lu, Zhang Yadong, Yin Huaxiang. Research status of CMOS inverter circuits based on TMD materials [J] Micro-Nanoelectronic Technology, 2021, 58(03): 185-195.
[5] Wang Wenbin, Yang Dingkun, Luo Kun, Ouyang Lin. Cadence-based static CMOS gate circuit simulation design [J] Science and Technology Innovation, 2019 (24): 64-65.
[6] Wang Wenbin, Yang Dingkun, Luo Kun, Ou Yanglin. Static CMOS gate circuit simulation design based on cadence [J] Scientific and technological innovation, 2019, 24.
[7] Qian Xiang, Lu Yaqing. Characteristic analysis and Simulation of CMOS inverter based on cadence [J] Communication power technology, 2021.
[8] Chen Zhongjian, Gan Xuewen, Jin Haiyan, Zhao BAOYING. Integrated circuit principle and design [M] Beijing: Peking University Press, February 2006.
[9] Xinghao Chen, Nur A. Touba, Fundamentals of CMOS design, 2009.

Wireless power transfer system based on frequency and impedance matching hybrid adjustment against system detuning

Zhengnan Yuan[1,2*], Pengcheng Fu[2], Guangshuai Lu[2], Pengfei Cao[2]

[1] Lanzhou Institute of Physics, Lanzhou, 730000, China

[2] School of Information Science and Engineering, Lanzhou University, Lanzhou, 730000, China

[*]Corresponding author's e-mail: yuanzhn19@lzu.edu.cn

Abstract. System detuning caused by a variation in the distance between the transmitting and receiving terminals can greatly reduce the transmission power and efficiency of a magnetic resonance-coupled wireless power transmission (WPT) system, which limits the WPT application scope. This paper proposes a magnetic resonance coupling wireless power transmission system, which is based on jointly and continuously adjustable frequency compensation (CAFC) and two-transistor-controlled variable capacitor circuits (TCVCs). Therefore, this system can reach the resonant state by using CAFC and two-TCVCs when the transmission distance is changed. The proposed system can adaptively adjust combinations of the operating frequency and equivalent compensation capacitor's capacitance to achieve impedance matching avoiding the phase difference caused by the imaginary part of the impedance, thus maintaining stable transmission efficiency under the condition of transmission distance variation. Compared to the traditional magnetic coupled resonant circuit based on impedance matching or variable resonant frequency, the proposed system achieves higher efficiency and stability and dynamic distance adaptation.

1.Introduction

Wireless power transmission (WPT) is an emerging technology that has a broad application prospect in the fields of daily consumer electronic products, biomedical implants, electric vehicle charging, and other consumer electronic products [1]. Magnetic coupling resonance is an important technique in WPT. The basic principle of magnetic coupling resonance is that exchanging the energy at resonance is more efficient than at non-resonance [2]. Due to the usage of strong coupling resonance technology, the magnetic coupling resonance wireless power transmission (MCR-WPT) can achieve high power and efficiency [3]. However, the application of the MCR-WPT has been limited by its poor stability due to its sensitivity to the changes in the external conditions, such as distance and receiver load [4]. Particularly, the change in the operational distance between the transmitter and receiver can result in impedance mismatch, thus decreasing the transmission efficiency [5].

Generally, there are two common methods for transmission efficiency improvement under different distances or detuning conditions. The first method is based on impedance matching, which includes using DC-DC converters for tuning the load impedance and passive circuits for impedance matching. Since the power converter can be flexibly controlled by altering the equivalent impedance of the load, the DC-DC converters for tuning load impedance have been widely discussed [6-8]. Since the optimal energy efficiency conditions change with the coupling coefficient k, the greatest challenge is how to feedback control to locate the maximum energy transmission efficiency conditions. Besides DC-DC converters for altering the load impedance method, another common approach is using the passive circuits for impedance matching. According to the maximum power transfer theorem, this method requires impedance matching with the power source. However, this approach is suitable only for very low power applications (below 1 W) because it has a relatively high power loss in the power source resistance. The second method is to use variable resonant frequency methods, which include the operating frequency changing and using reconfigurable resonant circuits and coils. The former can achieve higher efficiency than the latter by adjusting the resonant frequency with the load resistance [9]. However, only when the load resistance is much larger than the optimum load resistance, the transmission efficiency can be obviously improved [6, 10]. Moreover, even when the frequency is adjusted to a suitable value, the equivalent input impedance of capacitance and inductance parameters can also change due to the distance change. Thus, by adjusting only the frequency, the circuit cannot fully restore the resonance state. The latter usually involves a switchable capacitor array [11], reconfigurable coils [10, 12], or their combinations [13, 14] for adjusting the resonant frequency in order to achieve the optimal energy transmission efficiency conditions when the load changes. Although using a large number of switchable capacitors or inductors can provide wide discrete changes in the resonant frequency, the complexity and size of a circuit are greatly increased since numerous switches and the corresponding driving circuits and control are required. Moreover, the impedance can be adjusted only in a few coil points, which makes the circuit adjustment has a certain error and can fail to achieve the optimal transmission efficiency.

In order to address the above-mentioned problems, this paper proposes a continuously frequency-and-impedance matching hybrid adjustment WPT system (CFIMHA-WPT) based on jointly and continuously adjustable frequency compensation and multiple transistor-controlled variable capacitors (TCVCs) to obtain the maximum transmission efficiency under detuning caused by variations in transmission distances. The proposed architecture is inspired by one of the previous works [15], where the equivalent capacitance of the TCVC was simply controlled with a DC voltage similar to varactors so that the zero voltage switching frequency of the system could be adjusted smoothly. Furthermore, in order to overcome the problem of limited adjustment accuracy and the conflict between a large adjustable range and a long diode conduction period of a single TCVC, there are two main improvements in the proposed system: (1) the accuracy of the impedance matching adjustment can be effectively improved by connecting one large TCVC and one small TCVC in series, thus obtaining the maximum transmission efficiency under different distance conditions; (2) in addition to the effect of frequency adjustment, jointly and continuously adjustable frequency compensation and two TCVCs are used together to adjust the circuit to a new resonance state. Compared to the traditional tuning circuits, the proposed system with simple circuits can achieve higher adjustment accuracy and easily extend the adjustment range and dynamic distance adaptation because the capacitance circuits are continuously adjustable.

2.Proposed continuously frequency-impedance matching hybrid adjustment WPT system

Based on the previous analysis[12,16], when $\omega = 1/\sqrt{LC_T} = 1/\sqrt{L_R C_R}$, where C_T and C_R denote the equivalent capacitances, and L and L_R are the equivalent inductances of the transmitter and receiver, respectively, the condition of resonance can be satisfied. According to the two-port network theory [17], the input and output impedance of a system can be respectively expressed as:

$$Z_{in} = R_1 + \frac{(\omega M)^2 (R_1 + R_R)}{(R_1 + R_R)^2 + (\omega L_R - 1/\omega C_R)^2} - j\left(\frac{(\omega M)^2 (\omega L_R - 1/\omega C_R)}{(R_1 + R_R)^2 + (\omega L_R - 1/\omega C_R)^2} - (\omega L - 1/\omega C_u)\right), \tag{1}$$

$$Z_{out} = -R_R + \frac{(\omega M)^2 (R_R - R_1)}{(R_R - R_1)^2 + (\omega L - 1/\omega C_u)^2} + j\left(\frac{(\omega M)^2 (\omega L - 1/\omega C_u)}{(R_R - R_1)^2 + (\omega L - 1/\omega C_u)^2} - (\omega L_R - 1/\omega C_R)\right); \tag{2}$$

where R_1 and R_R are equivalent resistances, and L and L_R are equivalent inductances of the transmitter and receiver, respectively; C_R is the equivalent capacitance of receiver; M is mutual inductance; C_u is equivalent compensation capacitance. Therefore, when the transmission distance changes, the imaginary parts of the input and output impedances can be set to zero by adjusting the frequency ω or C_u or combinations of ω and C_u, which means that the system can obtain the resonant state again.

The basic idea, according to (1) and (2), is to tune operating frequency or the equivalent compensation capacitor or inductance or their combinations to achieve impedance matching avoiding the phase difference caused by the imaginary part of impedance. If the inductance is used as phase compensation, circuit complexity will increase, which may cause stability problems. In addition, increasing the compensation inductance can also produce a certain energy loss [18]. Therefore, besides frequency adjustment, the capacitance is adjusted by using TCVC as a circuit whose capacitance value can be adjusted by controlling the transistor via feedback voltage [15]. When the control voltage changes, the on and off states of the capacitor are controlled by the diode turning on or off, respectively, and the equivalent capacitance can be calculated using the on-time t_{on} [15]. In order to resolve conflict between a large adjustable range and a longer diode conduction period, two or more TCVCs connections is proposed in this manuscript. Take two TCVCs connections as an example, the structural diagram of the CFIMHA-WPT is presented in Figure 1 (C_{u1} and C_{u2} represent capacitances of the first and second TCVCs, and $C_{u1} > C_{u2}$). The two-TCVCs circuit structure is shown in Figure 2. The basic working principle of the CFIMHA-WPT based on two-TCVCs is as follows. When a longitudinal distance between the transmitter and receiver coils changes, the current and voltage phases are collected by the current and voltage detection modules and then transmitted to the phase discriminator to determine the phase difference. Then, the phase difference signal is transmitted in the form of voltage to the controller, and the output frequency of the inverter circuit and the value of the TCVCs are adjusted by the controller based on the received signal. Considering that the adjustment accuracy of a TCVC is inversely proportional to its adjustable range, the adjustable range of the TCVC is defined by the voltage-controlled large-variable-capacitance circuit (C_{u1}), and the adjustable precision of the TCVC is defined by the voltage-controlled small-variable-capacitance circuit (C_{u2}). The system can reach a more accurate resonance state by adjusting C_{u1} and C_{u2} simultaneously. Therefore, the accuracy of the TCVC adjustment can be effectively improved by connecting one C_{u1} and one C_{u2} in series. Coupled with the effect of frequency adjustment, the two methods are used together to adjust the circuit to a new resonance state better.

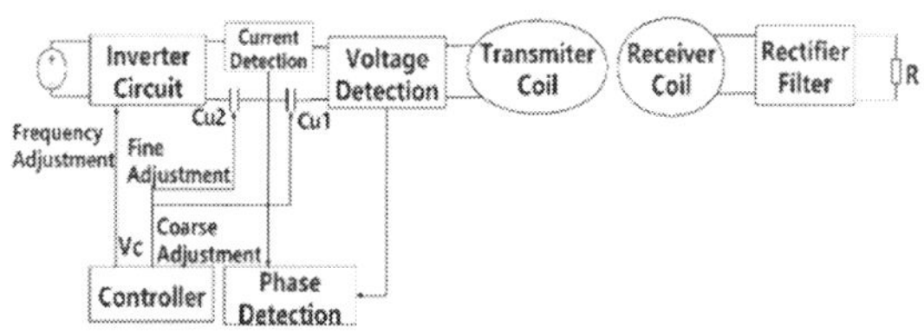

Figure 1. The structural diagram of the CFIMHA-WPT based on the TCVC.

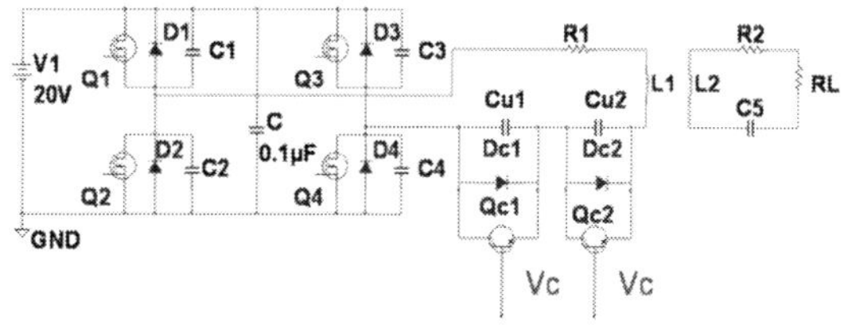

Figure 2. The structure diagram of two-TCVCs.

3.Experimental prototype analysis

According to the previous theoretical and simulation analyses, the experimental circuit was built. The hardware consisted of a power module, an inverter module, a transmission module, and two-TCVCs, as shown in Figure 3. When the system was under the resonance conditions (i.e., the distance between the two coils was 10 mm), the phase difference between the voltage and current was equal to zero, as shown in Fig.4a. When the transmission distance between the two coils was longer than 10 mm, a large phase difference between the voltage and current occurs, which means that the system was in the detuned state [7, 19]. In this case, the phase difference between the voltage and current was very smaller by using the two-TCVCs than that using one TCVC, as shown in Figure 4(b) and (c). It can be speculated that a further increase in the number of TCVCs could improve the system's transmission efficiency, but it will also increase the complexity and reduce the stability of the system. Therefore,

Journal of Physics: Conference Series **2108** (2021) 012035 doi:10.1088/1742-6596/2108/1/012035

using two-TCVCs is considered to be the optimal solution from the aspect of the overall system performance. Especially, the voltage and current of a transmitter can reach the same phase when the frequency adjustment and two-TCVCs are used simultaneously, which can make the system work at the resonance state, as shown in Figure 4(d). The experimental results are in agreement with the above-mentioned theoretical results.

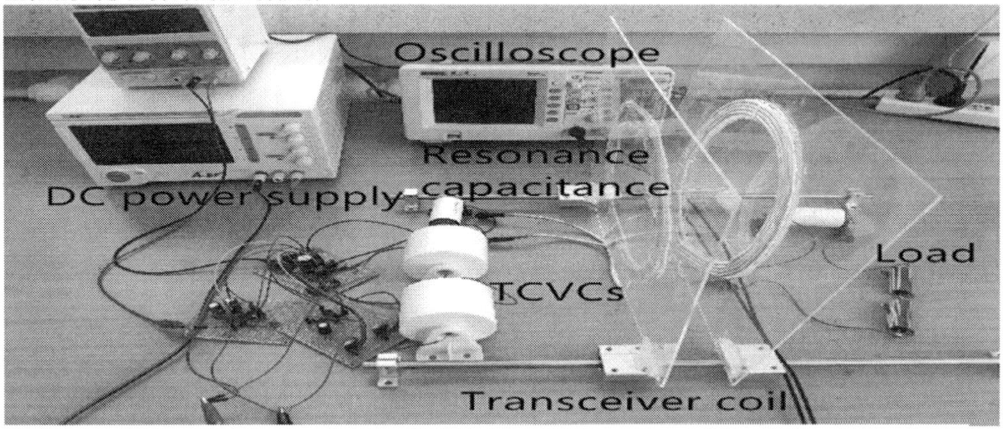

Figure. 3. The photo of the CFIMHA-WPT based on the TCVCs.

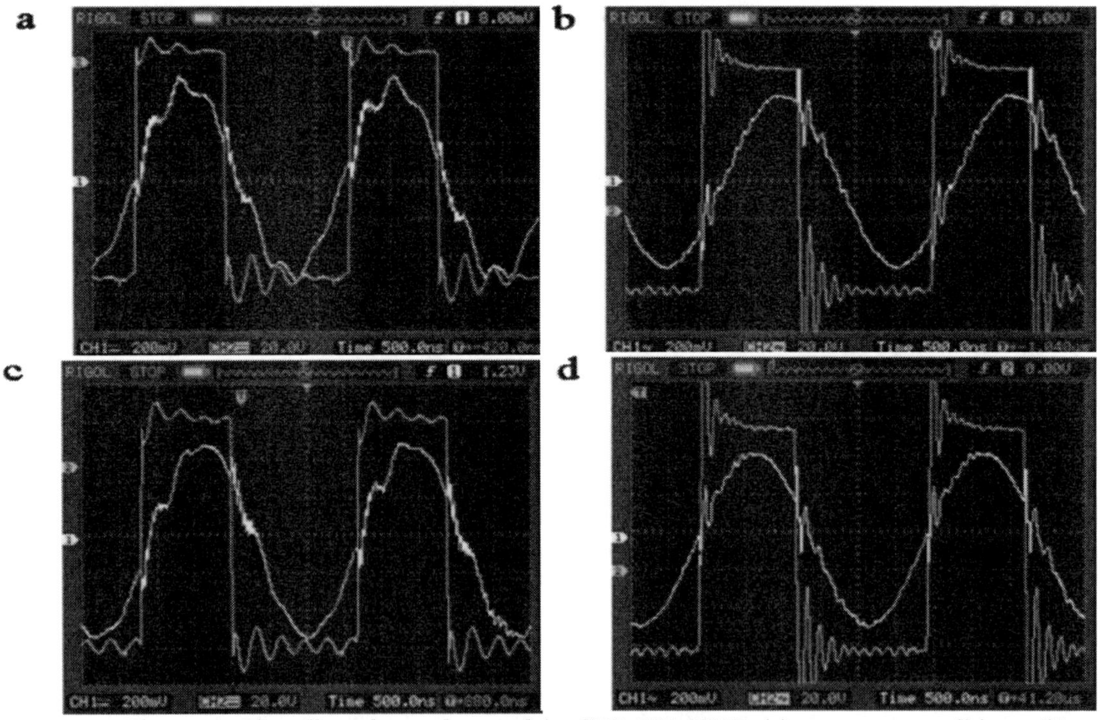

Figure. 4. The adjusted waveforms of the CFIMHA-WPT: (a) resonance conditions; (b) one TCVC; (c) two-TCVCs; (d) both frequency adjustment and the two-TCVCs.

The measured transmission efficiency and the corresponding distance change curves under the joint effect of two-TCVCs and frequency adjustment are shown in Figure 5. The results show that the adjustment effect of two-TCVCs is better than that of single TCVC. Furthermore, the best transmission efficiency of the system is achieved by jointly using frequency adjustment and two-TCVCs; also, the transmission efficiency achieved in this way decreases much more slowly than that of the other adjustment methods at a long distance between the two coils. The transmission distance was limited to 50 mm in the experiments due to the low-power source and low-frequency source in our laboratory. It should be noted that for the tested distances, the experimental results were in agreement with the theoretical results; thus, all the drawn conclusions are reliable.

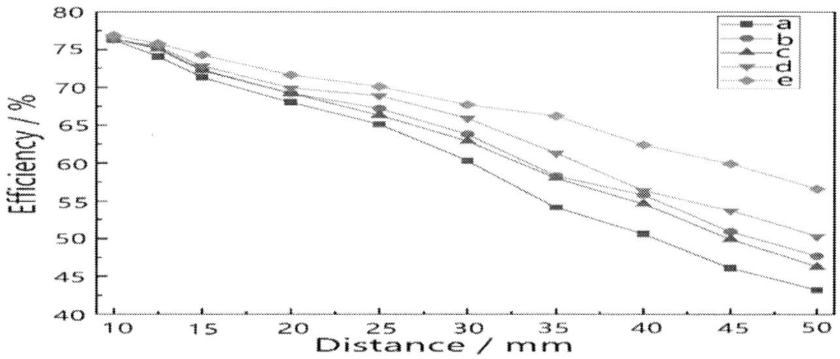

Figure 5. The efficiency curve of the CFIMHA-WPT for different transmission distances: (a) without any adjustment; (b) with frequency adjustment; (c) with one TCVC; (d) with two-TCVCs; (e) with joint usage of frequency adjustment and the two-TCVCs

4.Conclusion

In this paper, a method for mitigating the detuning in a dynamic wireless power transfer system based on the joint effect of two-TCVCs and frequency adjustment is proposed. The proposed CFIMHA-WPT can realize a continuous, adjustable capacitance to achieve the best impedance matching easily, and it is simpler and has a smaller volume than that used the capacitor matrix structure. Both simulation and experimental results show that the proposed CFIMHA-WPT with joint usage of the frequency adjustment and two-TCVCs has better transmission efficiency at variable transmission distance than the system with single TCVC adjustment type. The analysis results also speculate that when a larger number of TCVCs are used in the system, higher transmission efficiency can be achieved under the conditions of variable long transmission distance. Consequently, the proposed system can significantly improve the transmission efficiency of a system for different transmission distances. The proposed method can be used in low-power against system detuning applications design, such as consumer electronics, as well as in higher-power applications design.

Acknowledgments

The authors would like to acknowledge the financial support by the National Natural Science Foundation of China under Grant 61804071, the Natural Science Foundation of Gansu Province (No. 20JR10RA614), the Open Fund of Key Laboratory of Time and Frequency Primary Standards, CAS, Fundamental Research Funds for the Central Universities (lzuxxxy-2019-tm04).

References

[1] Buja,G., Bertoluzzo, M., and Mude, K. N. (2015) Design and Experimentation of WPT Charger for Electric City Car. IEEE Trans. Ind. Electron., vol. 62, no. 12, pp. 7436-7447.
[2] Hou, S. L., Wang, J., Fu, W. N., and Sun, M. (2011) A Comparative Study Between Novel Witricity and Traditional Inductive Magnetic Coupling in Wireless Charging. IEEE Trans Magn, vol. 47, no. 5, pp. 1522-1525.
[3] Liu, F., Yang, Y., Ding, Z., Chen, X., and Kennel, R. M. (2018) A Multifrequency Superposition Methodology to Achieve High Efficiency and Targeted Power Distribution for a Multiload MCR WPT System. IEEE Trans. Power Electron., vol. 33, no. 10, pp. 9005-9016.

[4] Abou Houran, M., Yang, X., and Chen, W. (2018) Magnetically Coupled Resonance WPT: Review of Compensation Topologies, Resonator Structures with Misalignment, and EMI Diagnostics. Electronics, vol. 7, no. 11.
[5] Gao, P., Tian, Z., Pan, T., Wu, J., and Gui, W. (2018) Transmission efficiency analysis and optimization of magnetically coupled resonant wireless power transfer system with misalignments. Aip Adv., vol. 8, no. 8.
[6] Zhong, W. X., and Hui, S. Y. R. (2015) Maximum Energy Efficiency Tracking for Wireless Power Transfer Systems. IEEE Trans. Power Electron., vol. 30, no. 7, pp. 4025-4034.
[7] Mai, R., Liu, Y., Li, Y., Yue, P., Cao, G., & He, Z. (2017) An active-rectifier-based maximum efficiency tracking method using an additional measurement coil for wireless power transfer. IEEE Trans. Power Electron., vol. PP, no. 1, pp. 716-728.
[8] Zhong, W., and Hui, S. (2016) Charging time control of wireless power transfer systems without using mutual coupling information and wireless communication system. IEEE Trans. Ind. Electron., vol. 64, no. 1, pp. 1-1.
[9] Patil D., Sirico, M., Gu, L., and Fahimi, B. (2016) Maximum efficiency tracking in wireless power transfer for battery charger: Phase shift and frequency control. 2016 IEEE Energy Conversion Congress and Exposition (ECCE), pp.1-8.
[10] Zhong, W., and Hui, S. (2017) Reconfigurable wireless power transfer systems with high energy efficiency over wide load range. IEEE Trans. Power Electron., vol. 33, no. 7, pp. 6379-6390.
[11] Kennedy, H., Bodnar, R., Lee, T., and Redman-White, W. (2018) A self-tuning resonant-inductive-link transmit driver using quadrature symmetric delay trimmable phase-switched fractional capacitance. IEEE J Solid-St. Circ., pp. 1-13.
[12] Hu, J., Lee, C. K., and Lu, H. C. (2019). Reconfigurable Wireless Power Transfer Systems for Distance Adaptation. 2019 IEEE International Symposium on Circuits and Systems (ISCAS). IEEE, pp. 1-4.
[13] Beh, T. C., Kato, M., Imura, T., Oh, S., and Hori, Y. (2013) Automated impedance matching system for robust wireless power transfer via magnetic resonance coupling. EEE Trans. Ind. Electron., vol. 60, no. 9, pp. 3689-3698.
[14] Yu, T. C., Huang, W. H., and Yang, C. L. (2019) Design of dual frequency mixed coupling coils of wireless power and data transfer to enhance lateral and angular misalignment tolerance. IEEE J. Electromagn., RF, Microw. Med., vol. 3, no. 3, pp. 216-223.
[15] Tian, J., and Hu, A. P. (2017) A dc-voltage-controlled variable capacitor for stabilizing the zvs frequency of a resonant converter for wireless power transfer. IEEE Trans. Power Electron., vol. 32, no. 3, pp. 2312-2318.
[16] Yang, G., and Liang, H. (2018) Adaptive frequency measurement in magnetic resonance coupling based WPT system. Measurement, vol. 130, pp. 318-326.
[17] Jayathurathnage, P., Vilathgamuwa, M., and Simovski, C. (2018) Revisiting Two-Port Network Analysis for Wireless Power Transfer (WPT) Systems." 2018 IEEE 4th Southern Power Electronics Conference (SPEC), pp. 1-5.
[18] Yao, Y., Wang, Y., Liu, X., and D Xu. (2018) Analysis, design and optimization of lc/s compensation topology with excellent load-independent voltage output for inductive power transfer. IEEE Trans. Transp. Electrification, vol. 4, no. 3, pp. 767-777.
[19] Shui, H., Yu, D., Yu, S. S., Iu, H.,and Cheng, H. (2020) An autonomous impedance adaptation strategy for wireless power transfer system using phase- controlled switched capacitors. IEEE J. EM. Sel. Top. P., pp. 1-1.

Research and classification of wireless power transfer with relative application

Muchen Qiu[1*]

[1]Electronic and electric college of TianGong University, TianGong University Tianjin, China, 300387 *Corresponding author's e-mail:acad_qiumc@163.com

Abstract. In this paper, wireless power transfer (WPT) will be discussed in its research and application. The basic information like the reason why devices choose to use wireless power and the concept of typical WPT types is introduced. The wireless power transfer system can be divided into 5 diverse types by the theory of power transmitting, and the working principle will be introduced separately. Then, the detail of WPT via magnetically-coupled resonance is mainly discussed, including the analyzing methods, optimization scheme and the current problems. Finally, this paper give two examples on medical use and mobile phone battery charging.Using the technology of WPT, the problem of the medical tiny robots battery charging is solved, and it provides a more convenient method of charging the mobile phones without the cables. The wireless power transfer technique has potential advantages and broad prospects, which is of vital significance in power transfer development.

1. Introduction

With the consumption of the traditional fossil energy, we are facing the emergency of energy exhaustion, so that scientists began to search for reliable new energy to alleviate the current energy crisis. At present, electricity is the most ideal final energy form, which can be converted into the ultimate energy through clean energy such as wind energy, water energy, nuclear energy. Supported by various government policies and development of new technologies, electricity has been widely used in factory production, daily life, transportation energy and other aspects. [1-2]In recent years, the development of electric vehicles and the universal use of mobile phones have played a crucial role in promoting the development of wireless power transfer（WPT） technology.

The research on wireless power transfer (WPT) technology began in the 1980s. Nikola Tesla, an American scientist, first carried out WPT experiment. In 1899, he successfully lit a distant incandescent lamp with 50 kHz AC [3]. By the early 1960s, American Raytheon company William C. brown and others had done a lot of research on radio energy transmission, laid the experimental foundation of WPT and made this concept a reality [4]. In 1968, an American aeronautical engineer called Peter E. Glaser, proposed the idea of using microwave to transmit electric energy from solar satellites to the ground, that is, to establish a space solar power station (SPS) in geosynchronous orbit [5].DOE and NASA jointly organized a study to demonstrate the concept of SPS program and affirmed its feasibility from 1977 to 1980,. In response to the energy crisis, the United States, Japan and other major developed countries have successively carried out research on space solar power plants, which has greatly promoted the development of WPT technology [6-7]. And by the 1990s, the university of auckland conducted deeper research on WPT[8].

This paper will focus on the research and application of current wireless power transfer technology, especially WPT with magnetically-coupled resonance and give a brief introduce of the WPT technique.

2. Analysis

2.1. the background of developing WPT

WPT technology opens a broader research field for the development of modern energy. Compared with traditional wired power transmission, the wireless power transmission can solve a series of economic and security problems caused by transmission cable failure, and become more convenient for individuals. And with the growing consumer group, to a certain extent, the related industries have promoted the development of WPT technology. Wireless power transfer is a way to complete energy transmission without cable transmission for both transmitters and receivers. In comparison with traditional energy transmission methods, it has a lower integrated cost, effectively reduces environmental constraints, and a more prominent advantage is that it can accurately control the power outflow distance and also achieves the ability to transfer energy to a moving target, which is essentials to the fields of Aerospace.

For illustrations, its design of intelligent network, system structure to improve transmission efficiency, and the expansion and application of receivers are the current research direction. Along with the demand for electrical vehicles and mobile phones, wireless power transfer has a large percentage in the charging methods. In comparison with transmitting power through a cable, one noticeable merit of the WPT is able to charge the battery in relative shorter time with higher amount of energy. Until now it has been one century and the appearance of electric vehicles and mobile phones boosts the applications of wireless power transfer, which indeed refers to transmit energy or charge the battery [9]. Charging with wireless equipment usually refers to wireless power transfer. The battery of daily electronic devices, such as mobile phones, laptops, have requirements about close-distance WPT [10].

2.2. the classification of common WPT

The wireless power transfer technique can be divided into 5 different types, These will be discussed below [11].

2.2.1. WPT via inductive coupled

The fundamental frequency AC power is rectified, filtered, and converted into high-frequency AC power, which is input to the transmitter through the resonant compensation network, and finally supplied to the load with inductive coupling. The power is relatively high and it has a high efficiency in a short distance. The basic construction is shown below.

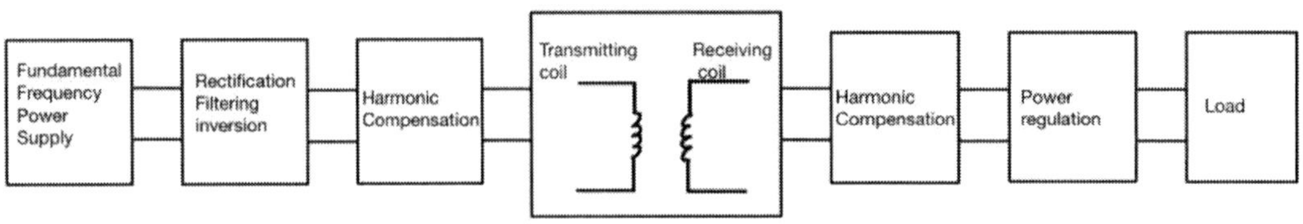

Fig1. the basic structure of WPT via inductive coupled

2.2.2. WPT via magnetically-coupled resonance

This is similar to the resonance of the sound, when it comes to this method, the transmitting winding will generate a resonance magnetic field at a specific frequency when the electricity is provided.

The working principle of the magnetically coupled resonance type WPT is that the high frequency power supply outputs high frequency alternating current to the transmitting antenna, and the receiving antenna and the transmitting antenna are coupled by the magnetically coupled resonance, which enables efficient wireless transmission of energy from the transmitting end to the receiving end. The received energy can be directly supplied to the load after rectifying and filtering by the load drive circuit[12-13].

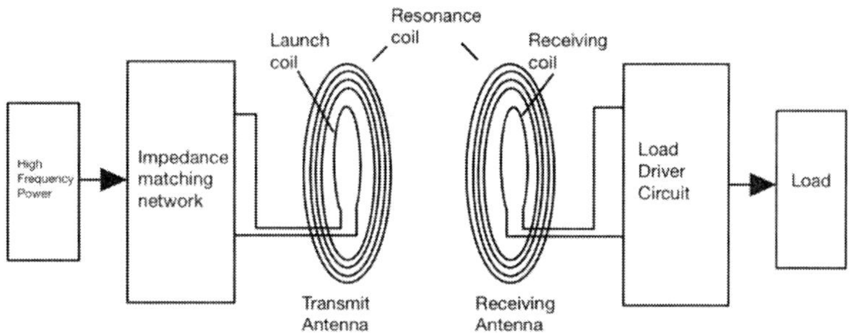

Fig2. Basic structure of WPT via magnetically-coupled resonance
(Source: Research Status and Application of Wireless Power Transmission Technology)

2.2.3. Laser wireless power transfer

The transmitter of the whole system convert the electricity into laser with Stimulated radiation. The laser with modulation will transmit to the receiver via the air. When the receiver gets the laser, the electricity will be used with photoelectric conversion. The laser can transmit the power at long distance, however, the transmitting efficiency is easily influenced by the quality of the atmosphere.

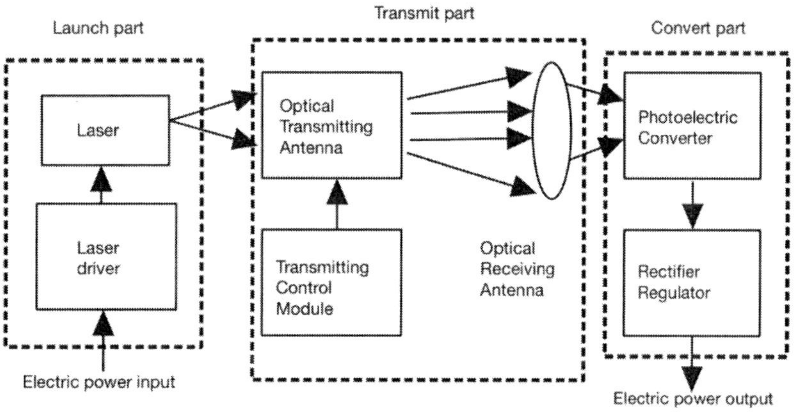

Fig3. Basic structure of WPT via laser
(Source: Research Status and Application of Wireless Power Transmission Technology)

2.2.4. Microwave wireless power transfer

It is using transmit antenna to convert the electricity into microwave energy and transmitting it directly through the atmosphere to the receiver. The energy will eventually be used as DC power. It is able to achieve long distance transmission and has less energy loss.

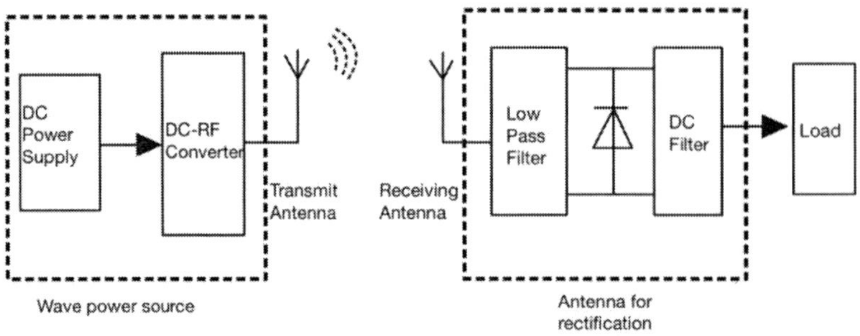

Fig4. Basic structure of WPT via microwave radiation
(Source: Research Status and Application of Wireless Power Transmission Technology)

2.2.5. Ultrasonic coupling wireless power transfer
This WPT is based on the electroacoustic conversion technique and the transmitter can convert electricity at certain frequency into acoustic energy, which will be used as electricity again in load with transducer.

With the fundamental systems of the wireless power transfer, several applications in this field have been used in life. Then, this paper will illustrate some development related to WPT technique.

2.3. The current development of WPT via magnetically-coupled resonance

2.3.1. two methods of analyzing
Magnetically induced coupled and magnetically coupled resonant WPT use the alternating magnetic field generated by the transmission coil to couple the energy to the receiving coil, thus realizing the radio power transmission to the load. These two transmission technologies are relatively mature. This paper mainly describes the magnetically-coupled resonance wireless energy transfer technology, which can be regarded as a special case of WPT via inductive coupled. It achieves high efficiency non-radiation energy transmission through the magnetic coupling resonance of the transmission receiving coil. The transmission distance is larger than the magnetic induction type, it belongs to the medium distance WPT technology and has a wide application[14].

So far, there are two main methods to quantitatively analyze the working principle of magnetic coupled resonance WPT, which can be divided into coupled mode theory and circuit theory. For the former one, we could established the equations of the coupled mode, which describes the transmission characteristics of the system. The equations are shown below[15].

$$\begin{cases} \frac{da_1}{dt} = -j\omega_1 a_1 - \tau_1 a_2 + jka_2 \\ \frac{da2}{dt} = -j\omega_2 a_2 - \tau_2 a_2 + jka_1 \end{cases}$$

a1 and a2 are the field amplitudes of resonant coils 1 and 2 respectively. ω 1, ω 2 is the natural frequency of resonance coils 1 and 2. τ 1、τ 2 is the inherent loss rate of resonant line, coil 1 and 2. k is the coupling coefficient between resonant coils 1 and 2.

And for circuit theory, it is easy to get the mathematical expression of the power and transmitting efficiency with KVL equations.

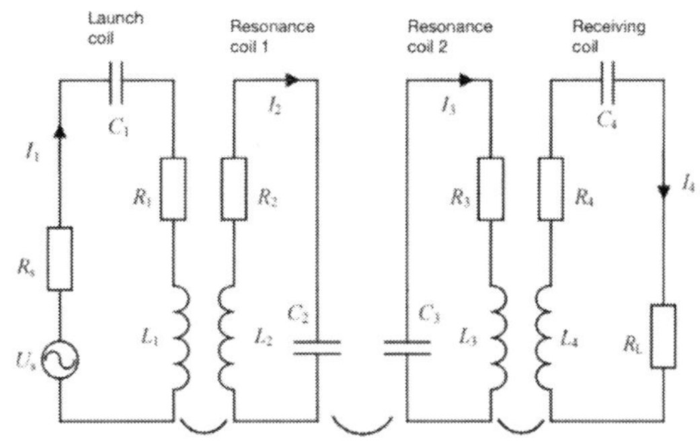

Fig5. Equivalent circuit of WPT via magnetically-coupled resonance
(Source: Research Status and Application of Wireless Power Transmission Technology)

2.3.2. *Optimization scheme of WPT via magnetically-coupled resonance*

As for of magnetically coupled resonance WPT, the transmitting and receiving antenna is composed of induction coil and resonance coil, so that update the structure of transmitter and receiver or the material of the coil can be the Optimization scheme of improving transmission power and transmission efficiency.

In the magnetically coupled resonant WPT system, the influence varies according to the location of the relay resonant coil and the material of the relay coil [16]. Rafif E. hamam of Massachusetts Institute of technology first proposed adding a relay resonant coil between transmitting and receiving antennas. The results show that the electromagnetic radiation outside the system is reduced and the transmission efficiency of the system is improved. The University of Pittsburgh in the United States simulated and analyzed the relay resonator network and put forward the optimization strategy of relay resonator placement .

2.3.3. *Problems of current development*

The main problems of magnetically coupled resonant WPT usually refer to the sensitivity of system parameters, the identification of different loads and the impedance matching of multiple loads, the safety of electromagnetic environment and electromagnetic compatibility. The frequency splitting phenomenon exists in the magnetic coupling resonant WPT. The change of some key parameter like the self inductance, equivalent capacitance and the distance between resonant coils change can always obviously change the transmission characteristics of the system.

2.4. *Two applications of the WPT with magnetic field*

With the recently development the WPT is used in public's daily life, here are two typical examples using wireless power transfer technology.

2.4.1. *WPT via magnetically-coupled resonance in medical use*

The applications of WPT in medical use is still in development in recent years . As for some certain tiny robots, the wireless power transfer technique provides convenience to the battery charging. For example, the intestinal robot used for testing, in the human gut fixed-point repair, the power supply has always been a tough problem. But the magnetic coupled resonance WPT make the technology become possible. The power transmit system is mainly consist of three parts: high frequency current transmitting part, wireless energy transmitting part and receiving coil power part. The system which is responsible for high frequency current is composed of high-frequency oscillation circuit and power amplification circuit. And before the current is eventually transmitted and consumed in the robot, the current need to be converted in the bridge rectifier circuit, filter capacitor and voltage stabilizing circuit[16].

2.4.2. WPT for mobile devices

At present, the WPT has some important develop directions. First is the way of transmitting. Electromagnetic induction and EMR wireless power transfer are quite normal, for the mobile phones are familiar to individuals. For a simple example, iPhone started to use wireless from iPhone X and iPhone 8/8 plus. A coil made of copper are installed behind the back. With the electromagnetic conversion and, the inner receiver use the coupling power to charge the battery with DC power. The back cover of the mobile phones turned the aluminum into a kind of material like tempered glass, which allows more electromagnetic wave to get through the back and reach the inner receiver with less energy loss, leading to relative higher transmitting efficiency.

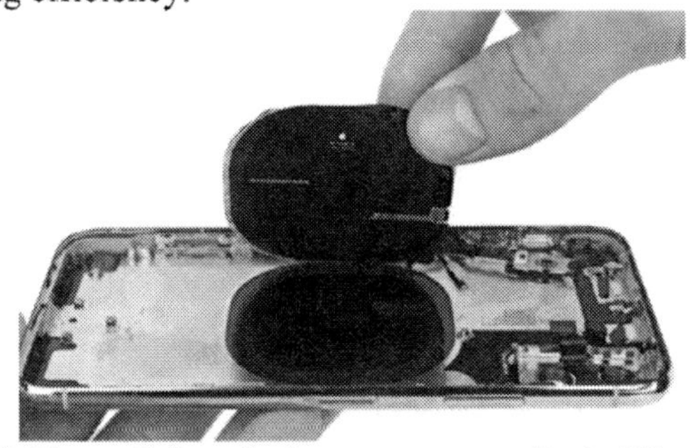

Fig6. The coil for wireless power transfer in iPhone X

iPhone 12 installed the magnet which support the technique of 'Magsafe'. So that with the application of electric vehicles and mobile phones battery charging, the technology is gradually matured and widely used. Take the smart home as an example, the wireless power transfer now can support multiple devices to charge at the same time, so that the electric home appliances could be better controlled with the system of WPT, and WPT technique will definitely have greater development in the future.

3. conclusion

The basic description and development of the wireless power transfer technique are shown in this paper, except the concept related to the development and types, the current application is discussed as well. The modern lifestyle makes mobile devices more portable, this trend requires a charging method with less limitation. But for WPT, the main problem at present stage is that the efficiency is relatively low when compared with cable transmitting. With the continuous development, the potential advantages and broad prospects of wireless energy transmission technology are gradually

reflected. WPT technology will be an important research trend of energy transmission in the future. Hope this paper can help readers to have a general understanding of WPT.

Acknowledgement

Firstly, I would like to express my appreciation to my supervisor, Prof. Lucyszyn, a respectable, responsible professor, who taught me the basic knowledge of the radio frequency subsystems. And I'm lucky to have the assistant teacher Yihao Wu, he helped me with valuable guidance in every stage of the writing of this thesis. Without his enlightening instruction, impressive kindness, I could not have completed my thesis. His enthusiasm and observation enlightens me not only in this thesis but also in my future study. I shall extend my thanks to Mrs. Han for all her kindness and help, she guided me with the thesis, and correct all my little mistakes patiently. Last but not least, I' d like to say thank you to the people who helped me.

References

[1] J. Shin, S. Shin, Y. Kim, S. Lee, B. Song and G. Jung, "Optimal current control of a wireless power transfer system for high power efficiency," 2012 Electrical Systems for Aircraft, Railway and Ship Propulsion, 2012, pp. 1-4, doi: 10.1109/ESARS.2012.6387426.

[2] H. Zheng, Z. Wang, Y. Li and P. Deng, "Data transmission through energy coil of wireless power transfer system," 2017 IEEE PELS Workshop on Emerging Technologies: Wireless Power Transfer (WoW), 2017, pp. 1-4, doi: 10.1109/WoW.2017.7959373.

[3] Marinicic A， Budimir D． Tesla's Contribution to Radiowave Propagation[C]//TELSIKS 5th International Conference on Telecommunications in Modern Satellite, Cable and Broadcasting Service． Nis， Serbia： IEEE Computer Society， 2001： 327-331．

[4] Brown W C． The history of wireless power transmission[J]． Solar Energy， 1996， 56(1)： 3-2．

[5] Glaser P E． Power from the sun： its future[J]． Science， 1968， 162(3856)： 857-861．

[6] Matsumoto H． Research on solar power satellites and microwave power transmission in Japan[J]． IEEE Microwave Magazine， 2002， 3(4)： 36-45．

[7] Nansen R H． Wireless power transmission： the key to solar power satellites[J]． IEEE Transactions on Aerospace and Electronic Systems Magazine， 1996， 11(1)： 33-39．

[8] Boys J T， Green A W． Inductively coupled power transmission： concept， design and application[C]//IPENZ Annual Conference． Nelson， New Zealand： The Institution of Professional Engineers New Zealand (IPENZ)， 1995： 1-9．

[9] Brown S, Pyke D, Steenhof P (2010) Electric vehicles: the role and importance of standards in an emerging market. Energy Policy 38: 3797–3806. https://doi.org/10.1016/j.enpol.2010.02.059

[10] Scheible G, Smailus B, Klaus M, Garrels K, Heinemann L (2003) System for wirelessly supplying a large number of actuators of a machine with electrical power. US Patent No. 6597076, USA

[11] NI Wang, DING Fei, ZONG Jun, JI Wei-wei, LIU Xing-jiang Research progress on electronic energy wireless transfer and energy interconnection National Key Laboratory of Science and Technology on Power Sources, Tianjin Institute of Power Sources, Tianjin 300384, China

[12] Soljacic M， Kurs A， Karalis A， et al． Wireless power transfer via strongly coupled magnetic resonances [J]． Science Express， 2007， 323(1)： 34-48．

[13] Zhao Zhengming， Zhang Yiming， Chen Kainan． New progress of magnetically-coupled resonant wireless power transfer technology[J]． Proceedings of the CSEE， 2013， 33(3)： 1-13(in Chinese)．

[14] Kurs A， Moffatt R， Soljacic M． Simu dltaneous mid-range power transfer to multiple devices[J]． Applied Physics Letters， 2010， 96(4)： 23-30．

[15] College of Electrical Engineering, Zhejiang University, Hangzhou 310007, Zhejiang Province, China; 2.Department of Electrical Engineering & Automation, Guilin University of Electronic and Technology, Guilin 541004, Guangxi Zhuang Autonomous Region, China

[16] S. Guo, P. Zhang, J. Guo, L. Wang and G. Sun, "Design of Wireless Power Transmission System based on magnetic coupling resonant for the capsule endoscopy," 2017 IEEE International Conference on Mechatronics and Automation (ICMA), 2017, pp. 23-28, doi: 10.1109/ICMA.2017.8015782.

Compressed Air Energy Storage and Future Development

Jingyue Guo[1, †], Ruiman Ma[2*, †], Huiyan Zou[3, †]

[1] Environmental engineering, Hubei University, Wuhan 430061, China

[2] Environmental and energy engineering, The University of Sheffield, Sheffield S10 2TN, the UK

[3] Changzhou No.1 High School International Division, Changzhou 213004, China

[*]Corresponding author's e-mail: rma21@sheffield.ac.uk

[†]These authors contributed equally.

Abstract. Power generation around the world is changing dramatically as a consequence of the demand to lower greenhouse gas releases and present a mix of power supplies. Energy storage technology is considered to be the fundamental technology to address these challenges and has great potential. This paper presents the current development and feasibilities of compressed air energy storage (CAES) and provides implications for upcoming technology advancement. The paper introduces various primary categories of CAES (Advanced Adiabatic-CAES, Liquid Air Energy Storage and Supercritical CAES). Compared with other energy storage technologies, CAES is considered a fresh and green energy storage with the distinctive superiorities of high capacity, high power rating, and long-term storage, and shortcomings of low power density, high transportation losses, and geological restriction. CAES is regarded as a promising technology that is able to be applied in renewable energy production, cogeneration, and distributed energy and microgrid systems. It's also considered to be integrated with other technologies, such as renewable energy, gas turbine, solid oxide fuel cells, and other systems in the future.

1. Introduction

The world relies heavily on traditional power supplies to generate electricity. As a result, two significant relevant problems are brought about, namely depleting fossil fuels, which are expected to be exhausted within ten decades[1], and the emissions of greenhouse gases (GHGs) and other contaminants with adverse effects on environment facilities and individual wellbeing[2]. Renewable energy is a prominent energy supply that can significantly reduce greenhouse gas emissions and meet global energy needs[3]. However, the inclusion of renewable energy into the grid presents upcoming issues[4]. Demand response (DR) and electric energy storage (EES) applications holds great promise[5]. EESs can absorb changes in renewable energy[6], and allow renewable energy to be allocated[7]. Energy storage technique is exerting vital impacts on the grid over the past few decades[8]. Another crucial facet associated with the adoption of storage technologies is represented by the air-quality concerns[9], minimizing the use of hydrocarbons for electricity transportation and circulation[10].

There are several suggested ways to classify the various EES technologies, for example, according to their functionality, response time, and appropriate storage time. A broadly used means is in the light of the accumulated in the device in the form of energy[11]. One thorough explanation and analysis of each class of EES technology will be offered later[12]. Among these energy storage technologies, CAES is considered a fresh and green energy storage with the distinctive superiorities of high capacity.

CAES represents the power stored as high-pressure compressed air and converted into diverse forms of energy consumption. This is a physical energy storage method with a large scale and can expand the utilization rate of sustainable energy[13]. When the demand is less than the output, the excess energy generated by renewable energy can be stored by compressed air energy storage technology[14]. The paper introduces three main types of CAES, including their operating principles, unique advantages, and developments. Then, the paper concludes the general advantages and disadvantages of CAES and its possible application, which has not been done systematically by others.

2. Overview and types of CAES

2.1. Overview

For maintaining the robustness and reliability of the energy devices, CAES has a significant comparative advantage, and has become the most promising large-capacity of power storage machineries. The gas turbine device uses air and gas as the working medium. The compressed air and fuel are mixed to form high-temperature and high-pressure gas in the burning space to drive the turbine, which then runs the generator. Since the CAES device is composed of two different operating stages, compression and expansion, and these two stages run at different times, the efficiency of air compression energy storage technology is higher than that of traditional gas turbine systems[15]. Hitherto, tremendous different concepts of compressed air energy storage technology are developed to adapt to the application requirements of different scenarios[16].

2.2. Types of CAES

2.2.1. Advanced Adiabatic-CAES (AA-CAES). In recent years, advanced AA-CAES has attracted people's attention because of its advantages: no fossil fuel employment, low cost, high start-up speed, and large partial load capacity[17]. In a traditional CAES system, high temperature produced by compression is discarded during charging and must be replaced by heat generated by combustion during power generation, which means that heat cannot be reused and limits storage efficiency[18]. However, in AA-CAES systems, as shown in Figure 1, the heat is released from the compression phase and can then be stored and reused in an adiabatic vessel during the expansion phase. Heat can be stored before the air enters the tank[19]. Therefore, through heat storage, the fossil fuel can be reduced by the heat of compression to reheat the air gas turbine increase and improve device efficiency. In conclusion, AA-CAES systems have the potential to improve efficiencies and emit fewer greenhouse gasses[20]. AA-CAES is regarded as a great choice to deliver extra services with reduced prices. Currently, several AA-CAES demonstration plants have been developed to some extent or in its infancy globally [17, 18].

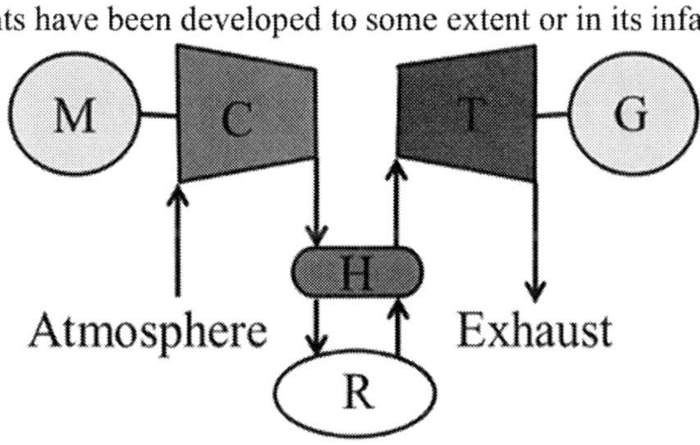

Figure 1. Scheme of AA-CAES system[19]

2.2.2. Liquid Air Energy Storage (LAES). LAES aims to increase the power storage density. Figure 2 reveals the typical scheme of AA-CAES system[21]. When operating LAES, both heated and cold thermal streams are generated, respectively, during charge-discharge process[22].

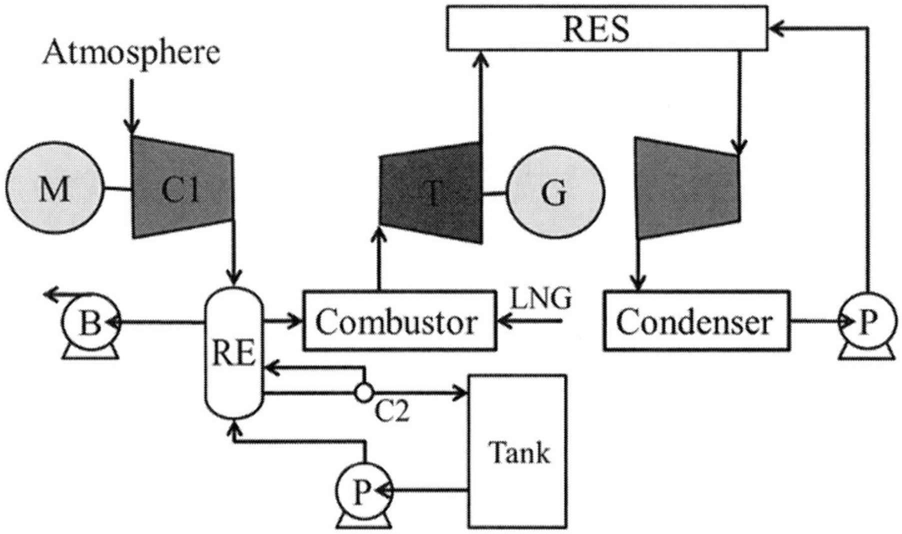

Figure 2. Schematic diagram of LAES system[19]

LAES is a promising class of sustainable power technique[23], whose components are strong with decent conservation constraints, and can be compatible with plant scale at scales ranging from dozens to hundreds of megawatts. The foremost components of a LAES device are likely to be individually sized and designed according to the specific presentation requirements[24].

2.2.3. Supercritical CAES (SC-CAES). The SC-CAES incorporates the benefits of AA-CAES and LAES. Figure 3 indicates the scheme of one typical SC-CAES system. According to Guo, Xu, Chen, Zhou's research[25], the energy densities are approximately 18 times higher than that of conventional CAES with round-trip efficiency of 67.41%.

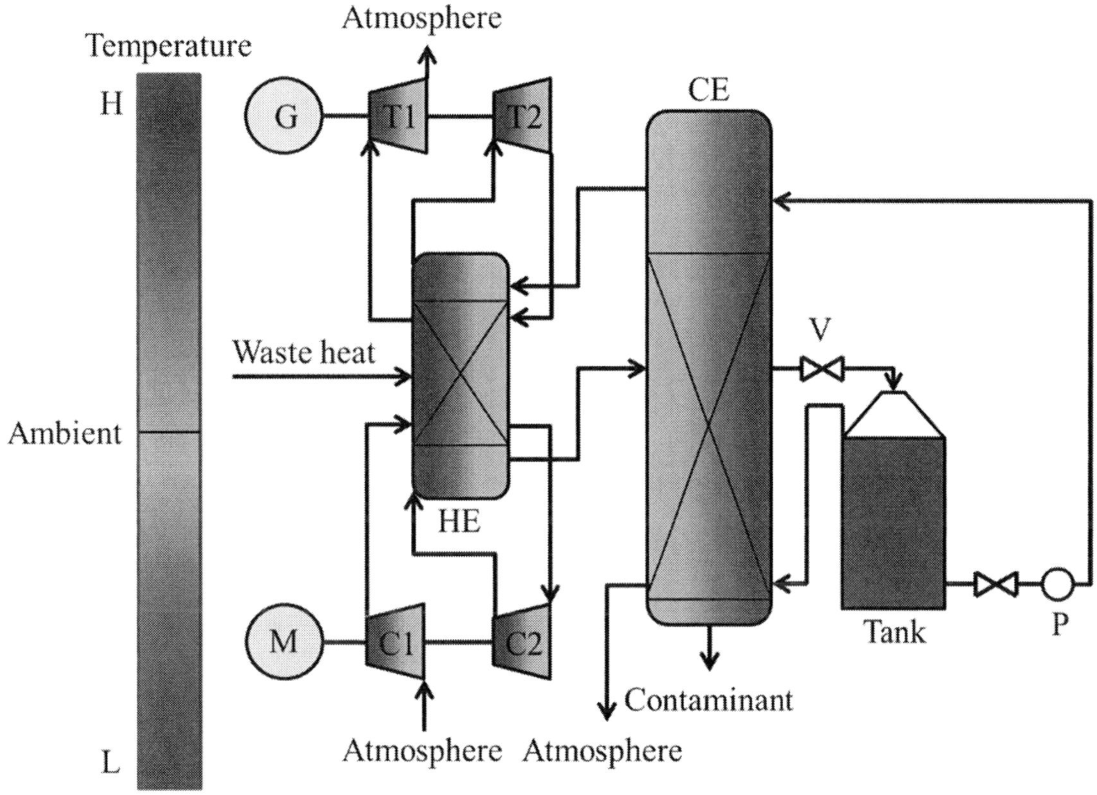

Figure 3. Scheme of SC-CAES system[19]

2.3. Merits and limitations

2.3.1. Merits. CAES possess the scales from small to large with wide storage stability and exceptional part-load properties[26]. Compared with all the technologies listed in table 1 below, the power rating of CAES is one of the highest technologies. Also, the table shows that its suitable storage duration is longer than most of the technologies such as the flywheel, capacitor, Superconductor Magnetic Energy Storage, Solar fuel, and so on. Instead of too long or too short, its lifetime is also relatively moderate. Besides, the table indicates that CAES is relatively mature, meaning that scientists have already had a lot of experience with it.

Table 1. Technological traits of electrical energy tools[26]

Technology	Energy Density (Wh/L)	Power Rating (MW)	Suitable Storage Duration	Lifetime (years)	Discharge Time	Cycling Times (cycles)	Maturity
PHS	0.5-2	30--5000	H-Mon	40-60	1--24 H+	10,000--30,000	Mature
Flywheel	20-80	0.1--20	Sec-Min	15--20	Sec-15 Min	20,000	Early Com
CAES	2--6	≧300	H-Mon	20--40	1--24 H+	8000--12,000	Early Com
Capacitor	2--6	0--0.05	Sec-H	1--10	Millis-1 H	50,000+	Com
SMES	0.2--6	0.1-10	Millis-H	20--30	≧30 Min	10,000+	Demo/Early Com

TES	80--500	0.1-300	Min-Days	5--30	1--24H+	-	Demo/Early Com
Solar fuel	500--10,000	0-10	H-Mon	-	1--24 H+	-	Developing
Hydrogen	500--3000	0-50	H-mon	5--20	Sec-24 H+	1000+	Developing /Demo
fuel cell Li-ion	150--500	0-100	Min-Days	5--20	Min-H	1000--10,000	Demo
Lead-acid	50--90	40	Min-Days	5--15	Sec-H	500--10,000	Mature

Abbreviations: SMES, Superconducting magnetic energy storage; TES, Thermal energy storage.

2.3.2. Limitations. Compared with gas, heat, and electricity, CAES's power density is lower and transportation losses are higher[19]. Table 2 also shows that its discharge time is comparatively long and its cycling time is shorter than some of the other technologies. Besides, compressed air energy storage is also restricted by the size and the availability of suitable geological locations[27, 28].

3. Applications

3.1. Cogeneration system

The combined heat and power system is also called the combined cooling, heat and power (CCHP) system. The main feature of this system is that it can reduce primary energy consumption and reduce the emission of pollutants. However, it still has difficulties integrating renewable energy into the grid, and as a result of the high volatility and instability of green power, the quality of the generated power cannot be guaranteed. In order to control the frequency, guarantee the grid robustness, and adjust the load, the system needs to be started and shut down frequently, which makes the cost higher. The air-storage compression use in the combined heat and power system can effectively improve the above-mentioned problems, storing excess energy during off-peak intervals to supply the requirement during peak intervals sooner or later. Thereby improving and stabilizing energy efficiency, reducing costs, etc.[29].

3.3. Distributed energy and microgrid systems

Distributed energy systems and microgrid systems are one of the main development trends of high-efficiency, low-carbon, and high-safety energy systems in the future. However, compared with large power grids, distributed energy systems have disadvantages such as large load fluctuations, poor system regulation capabilities, and high failure rates. air-storage compression can be used as a load-balancing device and a backup power source to effectively solve the above problems, improve the dependability and durability of the system's power supply, and realize black start and isolated grid operation. Due to the heat produced in compressed air energy storage technology, it could be merged with refrigeration and heating systems to realize the combined cooling, heating and power generation of distributed energy systems, which has a good application prospect[30].

4. Future development of CAES

From the perspective of technology itself, CAES is promising and is being considered to be integrated with other technologies, such as renewable energy, gas turbine, solid oxide fuel cells, and other systems[19]. For instance, the developing ADELE AA-CAES project in Germany[26] utilizes the CAES technology together with the wind farm. Also, CAES has a lot of space to develop and problems to overcome, including the urgent need for cooperation of engineers and scientists in various fields, technology innovation in air compressors and expanders, further substantial improvement in efficiency, necessity for lowering the cost of manufacturing air reservoirs and the study of the potential environmental impact[26]. From the perspective of economics, it is found that CAES investment can have good investment potential in the future since CAES operates both on the spot market and the controlling energy market, suggesting prospective practicability[31]. The better the technology develops, the less the operation will depend on higher price fluctuations[32]. From the perspective of the whole energy system and the whole society, the need for flexible technologies to maintain electricity balance is continuously increasing. In this way, it will have a well political and financial condition, which contributes to its better development.

5. Conclusion

The need for energy storage has been increasing rapidly recently and CAES plays a dominant role to address this. This paper has presented the current development of CAES according to its principles and advantages and the future development as one of the most encouraging energy storage technologies is also predicted. The advantages of CAES include high capacity, high power rating and long-duration storage. However, it has the disadvantages of low power density, high transportation losses, and restriction by the size and the availability of suitable geological locations. This study has identified three potential applications for CAES, including Renewable power generation, cogeneration system and distributed energy and microgrid systems. CAES is a mature technology that scientists have already had a lot of experience on and have more clear goals. It's also promising and is considered to be integrated with other technologies, such as renewable energy, gas turbine, solid oxide fuel cells, and other systems in the future. Finally, on the other hand, based on the summary of the status, the future research directions of CAES are prospected like further substantial improvement in efficiency, need for reduction in the cost of constructing air reservoirs, and the study of the potential environmental impact.

Acknowledgments

The authors strongly thank the advice provided by Professor Slav Hermanowicz in University of California, Berkeley. Also, the authors would love to give their thanks to the instructive and useful support from other thesis tutors in the project—Sustainable Development: Ethics, Physics and Technology.

References

[1] M.A.H. Baky, M.M. Rahman, A.S.J.R. Islam, S.E. Reviews, Development of renewable energy sector in Bangladesh: Current status and future potentials, 73 (2017) 1184-1197.
[2] F.J.I.j.o.e.r. Perera, p. health, Pollution from fossil-fuel combustion is the leading environmental threat to global pediatric health and equity: Solutions exist, 15(1) (2018) 16.
[3] M.M. Rahman, M.M.-U.-H. Khan, M.A. Ullah, X. Zhang, A.J.E. Kumar, A hybrid renewable energy system for a North American off-grid community, 97 (2016) 151-160.
[4] M.M. Rahman, A.O. Oni, E. Gemechu, A.J.E.C. Kumar, Management, Assessment of energy storage technologies: A review, 223 (2020) 113295.
[5] S. Tewari, N.J.I.T.o.P.S. Mohan, Value of NAS energy storage toward integrating wind: Results from the wind to battery project, 28(1) (2012) 532-541.
[6] A. Gabash, P.J.I.T.o.P.S. Li, Active-reactive optimal power flow in distribution networks with embedded generation and battery storage, 27(4) (2012) 2026-2035.
[7] D. Akinyele, R.J.S.E.T. Rayudu, Assessments, Review of energy storage technologies for sustainable power networks, 8 (2014) 74-91.
[8] P.F. Ribeiro, B.K. Johnson, M.L. Crow, A. Arsoy, Y.J.P.o.t.I. Liu, Energy storage systems for advanced power applications, 89(12) (2001) 1744-1756.
[9] A.-V. BOICEA, G. CHICCO, P.J.B.S.a.U.P.B. MANCARELLA, Seria C, Optimal operation of a 30kW natural gas microturbine cluster, 73(1) (2011) 211-222.
[10] V.A.J.P.o.t.I. Boicea, Energy storage technologies: The past and the present, 102(11) (2014) 1777-1794.
[11] H. Zhao, Q. Wu, S. Hu, H. Xu, C.N.J.A.e. Rasmussen, Review of energy storage system for wind power integration support, 137 (2015) 545-553.

[12] X. Luo, J. Wang, M. Dooner, J.J.A.e. Clarke, Overview of current development in electrical energy storage technologies and the application potential in power system operation, 137 (2015) 511-536.
[13] Q. Zhou, D. Du, C. Lu, Q. He, W.J.E. Liu, A review of thermal energy storage in compressed air energy storage system, 188 (2019) 115993.
[14] A. Olabi, T. Wilberforce, M. Ramadan, M.A. Abdelkareem, A.H.J.J.o.E.S. Alami, Compressed air energy storage systems: Components and operating parameters–A review, (2020) 102000.
[15] S. Succar, R.H.J.P.e.i.r. Williams, Compressed air energy storage: theory, resources, and applications for wind power, 8 (2008) 81.
[16] Z. Tong, Z. Cheng, S.J.R. Tong, S.E. Reviews, A review on the development of compressed air energy storage in China: Technical and economic challenges to commercialization, 135 (2021) 110178.
[17] Y. Li, S. Miao, S. Zhang, B. Yin, X. Luo, M. Dooner, J.J.I.J.o.E.P. Wang, E. Systems, A reserve capacity model of AA-CAES for power system optimal joint energy and reserve scheduling, 104 (2019) 279-290.
[18] X. Luo, J. Wang, C. Krupke, Y. Wang, Y. Sheng, J. Li, Y. Xu, D. Wang, S. Miao, H.J.A.e. Chen, Modelling study, efficiency analysis and optimisation of large-scale Adiabatic Compressed Air Energy Storage systems with low-temperature thermal storage, 162 (2016) 589-600.
[19] J. Wang, K. Lu, L. Ma, J. Wang, M. Dooner, S. Miao, J. Li, D.J.E. Wang, Overview of compressed air energy storage and technology development, 10(7) (2017) 991.
[20] H. Mozayeni, M. Negnevitsky, X. Wang, F. Cao, X.J.E.P. Peng, Performance study of an advanced adiabatic compressed air energy storage system, 110 (2017) 71-76.
[21] E. Borri, A. Tafone, G. Zsembinszki, G. Comodi, A. Romagnoli, L.F.J.A.S. Cabeza, Recent trends on liquid air energy storage: a bibliometric analysis, 10(8) (2020) 2773.
[22] A. Vecchi, Y. liang Li, Y. Ding, P. Mancarella, A.J.A.i.A.E. Sciacovelli, Liquid air energy storage (LAES): a review on technology state-of-the-art, integration pathways and future perspectives, (2021) 100047.
[23] H. Peng, X. Shan, Y. Yang, X.J.A.E. Ling, A study on performance of a liquid air energy storage system with packed bed units, 211 (2018) 126-135.
[24] R. Morgan, S. Nelmes, E. Gibson, G.J.P.o.t.I.o.C.E.-E. Brett, An analysis of a large-scale liquid air energy storage system, 168(2) (2015) 135-144.
[25] H. Guo, Y. Xu, H. Chen, X.J.E.c. Zhou, management, Thermodynamic characteristics of a novel supercritical compressed air energy storage system, 115 (2016) 167-177.
[26] X. Luo, J. Wang, M. Dooner, J. Clarke, C.J.E.P. Krupke, Overview of current development in compressed air energy storage technology, 62 (2014) 603-611.
[27] B. Cárdenas, A. Hoskin, J. Rouse, S.D.J.J.o.E.S. Garvey, Wire-wound pressure vessels for small scale CAES, 26 (2019) 100909.
[28] J. Chen, W. Liu, D. Jiang, J. Zhang, S. Ren, L. Li, X. Li, X.J.E. Shi, Preliminary investigation on the feasibility of a clean CAES system coupled with wind and solar energy in China, 127 (2017) 462-478.
[29] F.S. Vieira, J.A.P. Balestieri, J.A.J.E. Matelli, Applications of compressed air energy storage in cogeneration systems, 214 (2021) 118904.
[30] H. Ibrahim, K. Belmokhtar, M.J.E.P. Ghandour, Investigation of usage of compressed air energy storage for power generation system improving-application in a microgrid integrating wind energy, 73 (2015) 305-316.
[31] H. Lund, G.J.E.c. Salgi, management, The role of compressed air energy storage (CAES) in future sustainable energy systems, 50(5) (2009) 1172-1179.
[32] G. Salgi, CAES Future Scenarios; An Investment Analysis for Denmark, MSc. Thesis in Energy Planning in Aalborg University, 2006.

Characteristics and Applications of Superconducting Magnetic Energy Storage

Yuyao Huang [1,†], Yi Ru [2,*,†], Yilan Shen [3,†] and Zhirui Zeng [4,†]

[1]School of Civil and Architectural Engineering, Heilongjiang Institute of Technology, Harbin, Heilongjiang, 150050, China

[2]School of Civil and Environmental Engineering, Harbin Institute of Technology, Shenzhen, Shenzhen, Guangdong, 518055, China

[3]School of environment and natural resources, Ohio State University, Columbus, Ohio, 43210, United States

[4]School of civil engineering, Beijing Jiaotong University, Beijing, 100044, China

[*]Corresponding author's e-mail: 180420104@stu.hit.edu.cn

[†]These authors contributed equally.

Abstract. Energy storage is always a significant issue in multiple fields, such as resources, technology, and environmental conservation. Among various energy storage methods, one technology has extremely high energy efficiency, achieving up to 100%. Superconducting magnetic energy storage (SMES) is a device that utilizes magnets made of superconducting materials. Outstanding power efficiency made this technology attractive in society. This study evaluates the SMES from multiple aspects according to published articles and data. The article introduces the benefits of this technology, including short discharge time, large power density, and long service life. On the other hand, challenges are proposed for future study. The high energy requirement of the cooling system and carbon emissions are some of the drawbacks of SMES. It's found that SMES has been put in use in many fields, such as thermal power generation and power grid. SMES can reduce much waste of power in the energy system. The article analyses superconducting magnetic energy storage technology and gives directions for future study.

1. Introduction

Nowadays, resources use and storage have played important roles all over the world. Besides resources like water and fossil, electricity is more used widely in human society. In addition, with a growing population, the needs for energy rise quickly. Thus, high-effective energy storage technology would be so crucial to modern development. Superconducting magnetic energy storage (SMES) has good performance in transporting power with limited energy loss among many energy storage systems. Superconducting magnetic energy storage (SMES) is an energy storage technology that stores energy in the form of DC electricity that is the source of a DC magnetic field. The conductor for carrying the current operates at cryogenic temperatures where it is a superconductor and thus has virtually no resistive losses as it produces the magnetic field. The overall technology of cryogenics and superconductivity today is such that the components of a SMES device are defined and can be constructed. The integrated unit appears to be feasible for some utility applications at a competitive cost with other technologies. SMES is the only technology-based on superconductivity that applies to electric utilities and is commercially available today.

With such high efficiency, SMES requires particular circumstances to be installed. The device needs to function under low temperature, which made it cost a lot. In this situation, seeking cheap materials for cooling SMES becomes a popular topic. Various teams have worked on improving energy-saving technology and the wide application of SMES at a lower cost. It's reported that there are some existing examples that SMES has put in use in many fields. For instance, the U.S Department of Energy has sponsored the improvement of SMES at the grid level. They try to develop SMES technology at the commercial sale level, storing power at appropriate costs [1]. Their study aims to develop a 1-2 MWh SMES system, which can offer the price of lead-acid batteries. Additionally, Bonneville Power Administration in Washington installed a 30MJ SMES unit, absorbing and releasing 10 MJ of power at a frequency of 0.35 Hz. In 2003, a 5 MW 7 MJ SMES was installed in Japan to balance the voltage dips in a crystal factory. These cases proved that SMES could provide us high-quality power and practicability.

Considering the high energy demand, the advantages and limitations of superconducting magnetic energy storage are discussed in the article. The advantages, limitations, and sustainability are discussed with the help of tables and figures. Diverse applications of SMES are assessed in the study and are analyzed. The improvement of SMES can be evaluated better with these pros and cons. Also, sustainable development is regarded in the research, which specifies improving water quality. Different applications are discussed in our study. Therefore, the evaluation of superconducting magnetic energy storage may be useful in various fields and applications.

2. Advantages

As an emerging energy storage technology, SMES has the characteristics of high efficiency, fast response, large power, high power density, long life with almost no loss. These advantages make SMES a potential solution for addressing the challenges in the future.

2.1 Energy Storage Efficiency

Due to the AC losses in the superconducting coil and eddy current losses in the cooling system, some energy is lost in the SMES system. But these two contributions can be reduced to a very low level if there is a suitable design of the superconducting conductor and the cooling system. As a result, SMES exhibits a very high energy storage efficiency in the region of 90% to 99% (typically more than 97%) [2]. That means it has little energy loss during the discharge and the charging, which can also be interpreted that SMES shows excellent energy conversion efficiency. The value is really excellent compared to other energy storage systems (batteries 70 to 90%, pumped hydro up to 70%). This high efficiency can be attributed to the lack of energy conversion from and to another form.

2.2 Discharge Time

Besides its high efficiency, SMES has very fast discharge times, only for milliseconds to 8 seconds for a full discharge. Discharging is possible in milliseconds if it is economical to have a Power Conversion System (PCS). That is really a rapid response compared to other technologies. Therefore, it can be utilized for power quality, such as the instantaneous voltage drop, flicker mitigation, and short duration UPS.

2.3 Power Rating

The typical rating of SMES is about 1-10MW now, but many researches are conducted on larger SMES systems in the range of 10-100MW. A tiny minority can even reach more than 100MW. Figure 1 compares the ranges of power and discharges time for different storage technologies. It is obvious that SMES has a higher level of power compared to some traditional technologies.

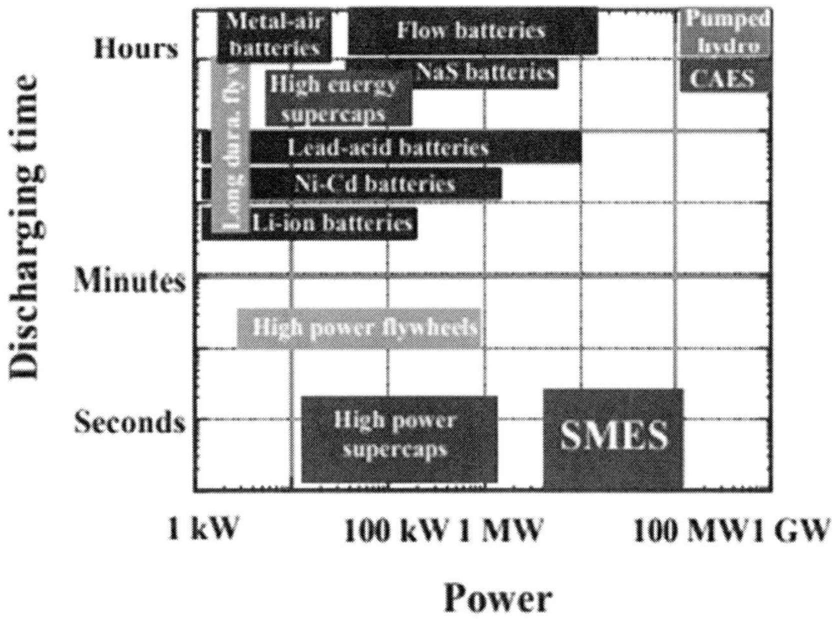

Figure 1. Discharging time versus power for various energy-storing devices [3]

2.4 Power Density

SMES shows a relatively low energy density of about 0.5-5Wh/kg currently, but it has a large power density. The power per unit mass does not have a theoretical limit and can be extremely high (100 MW/kg). While Batteries present higher values in energy density but lower values in power density. These two technologies are also combined to compensate for their drawbacks. Figure 2 shows the power and energy per unit mass for SMES and two more mature technologies: capacitors and batteries. For SMES, the grey zone indicates the presently attained values. The black zone covers theoretically possible ranges, which require more research and development. The high power density of SMES makes it a promising candidate for pulse power sources in the military and civil fields, such as the electromagnetic launcher [4], magnetic forming (use of electromagnetic forces to form a metal) [5].

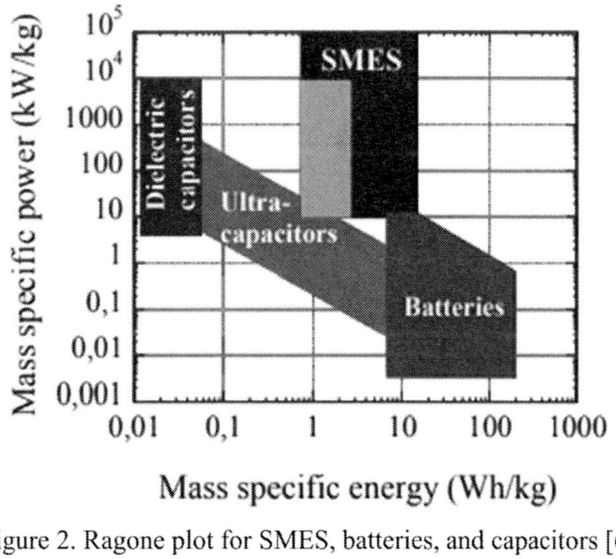

Figure 2. Ragone plot for SMES, batteries, and capacitors [6]

2.5 Life Span and Cycle Time

The lifetime of technology is mainly determined by the lifetime of the mechanical components. The life span of SMES is more than 20 years, which is also superior to most energy storage technologies because SMES has almost no loss characteristics through large numbers of discharge and charging cycles. That means SMES can afford more than 100000 cycles throughout its life.

3. Application

SMES can be combined with multiple fields and give play to its advantages under the combination of different fields to make up for the shortcomings in these fields. Many power generation systems can convert other energy into electric energy. Then the electric energy flows to the grid along the cable for users to use, while SMES stores users' remaining unused electric energy. The electric energy is released to make up for the shortage of power generation for users to use during the low power generation period. This reduces the waste of excess power generated while still keeping the grid well-supplied.

3.1 Application of power generation field

3.1.1 Photovoltaic power generation

Photovoltaic power generation is a technology that converts light energy directly into electric energy by using the photovoltaic effect of the semiconductor interface. It is mainly composed of three parts: solar panel (module), controller, and inverter. A system of combination of Photovoltaic power generation (PVPGS) and SMES in Fig.1 show how they are combined. SMES can not only store energy but also play a benign role in PVPGS. In the last few years, our power system has had a number of different challenges, such as compensation of induced harmonics, reactive power, unbalance current, neutral current, unbalance voltage, voltage sags & swells, power fluctuations, and load fluctuations [7, 8]. And these challenges stem from the rapid growth in the use of static power converters, Switched-Mode Power Supply (SMPS), nonlinear loads, ever-growing and congested power lines, and renewable energy sources (RESs) with high permeability (e.g., the solar, the wind, etc.) [9, 10]. Especially among all kinds of RESs, PVPGS with no pollution, no geographical restrictions, and abundant reserves account for an increasing proportion of microgrids [11, 12]. Nevertheless, unfortunately, energy solar energy is highly random and obviously intermittent, resulting in large fluctuations in photovoltaic output power [13-15]. SMES can smoothly suppress the power fluctuation caused by load mutation, pulse load, and time-varying solar irradiance [16]. Byung-Kwan Kang and Sun-Ho Bae have done simulation tests on the combined SMES and PV system based on PSCAD/EMTDC [17]. They analyzed the impact of SMES on system stability, several important factors such as the capability of SMES, its grid connection supervision, power reserve, output power fluctuation of a photovoltaic system, and seasonal load requirements, etc.

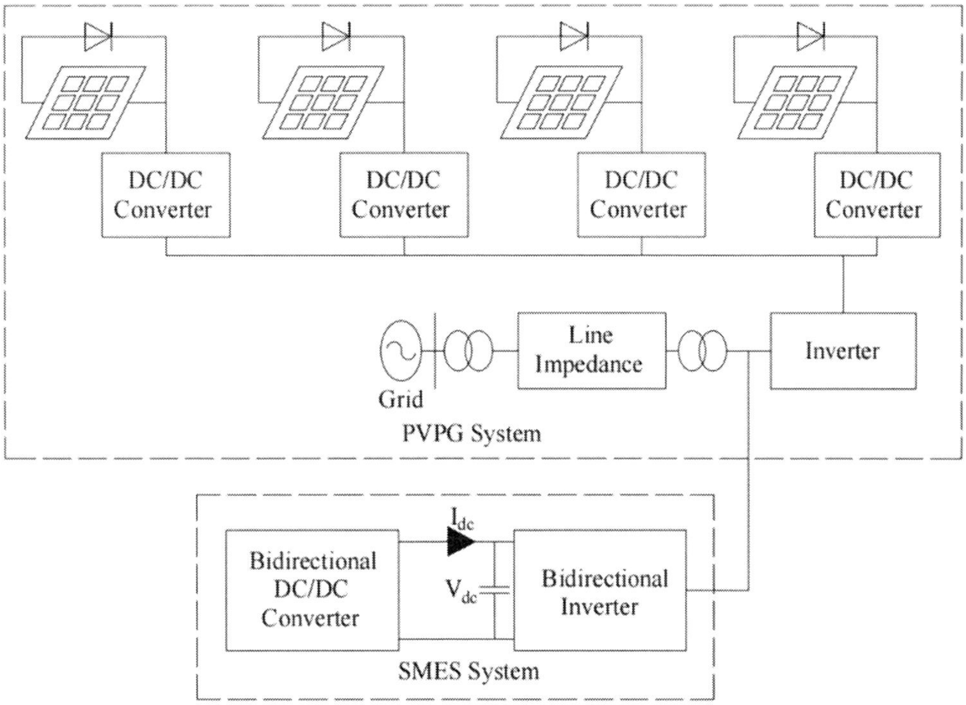

Figure 3. A system combining PVPGS and SMES

3.1.2 Wind power generation.

In essence, wind power generation converts wind energy into mechanical energy, which is then converted into electricity to generate electricity. The principle of wind power generation is to use the wind to drive the windmill's blades to rotate. Then, the machine will rotate the speed to promote the generator to generate electricity through the speed increase. Modern wind power generation system (WPGS) mainly faces two main problems: the output power of wind power generation is greatly affected by wind speed, and the output power fluctuates randomly; When the power grid voltage drops, if the wind power generation equipment is simply cut off, the stability of the power grid will be seriously affected. There's an experiment of SMES in controlling the voltage of power systems integrated with wind farms at variable wind speeds. SMES, WPGS, and the research system were simulated using MATLAB/Simulink program and Simpowersystem Package [18]. And through this experiment, they draw the following conclusions to prove the following advantages of SMES in the application combined with WPGS: (1) When WPGS is installed with SMES, the voltage value remains constant regardless of wind speed changes, indicating that SMES and WPGS combined have the ability to stabilize the voltage. (2) SMES can reduce reactive power and reduce electric energy loss in the line [18]. Therefore, SMES plays an important role in stabilizing the wind power generation system. Meanwhile, Similar to PVPGS, when SMES is combined with WPGS, voltage and frequency fluctuations caused by external factors can be suppressed. In this combined system, the external factor should be the wind speed. Thus, Wind power systems with SMES in them can operate more stably [19].

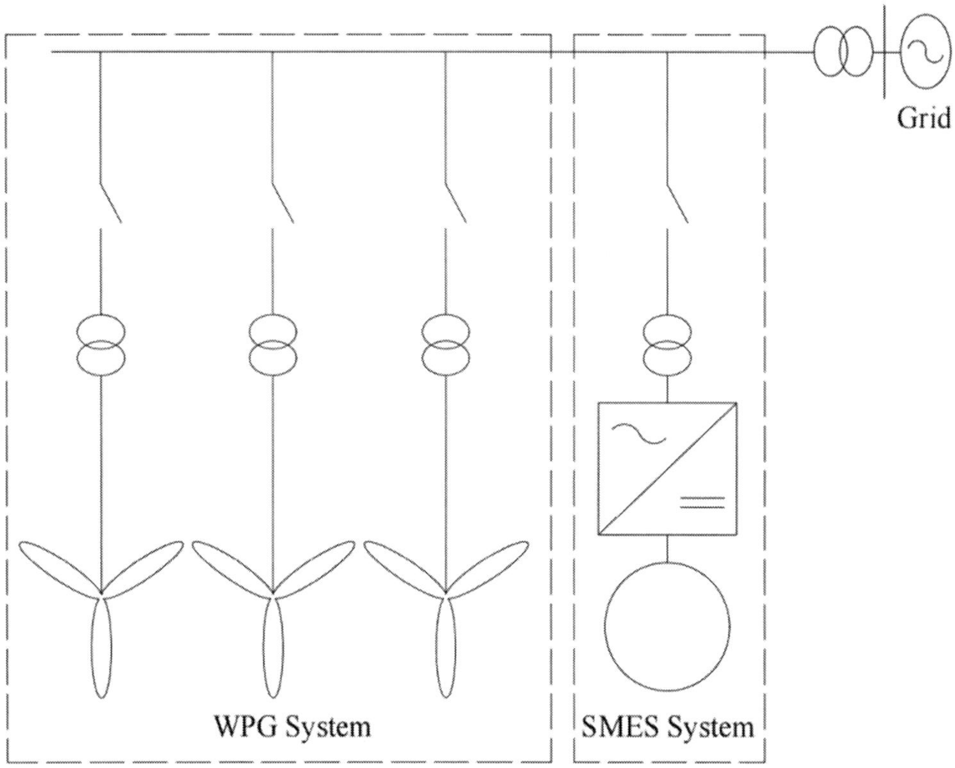

Figure 4. A system combining WPGS and SMES

3.1.3 Thermal power generation

Thermal power generation is actually the use of burning combustibles when the heat energy, through the power generation device into electricity generation. Thermal power generation systems are usually designed with Automatic Gain Control (AGC) that is an important function in the energy management system Energy Management System (EMS), it controls the output of the frequency modulation unit to meet the changing power demand of users and make the system in the economic operation state. The purpose of the AGC is to keep the system frequency and the actual power between different control areas at their respective specified nominal values when the system is subjected to load changes. If we connect the thermal power system with AGC and SMES, the role of AGC in the thermal power system will be enhanced [20]. The superiority of combining SMES with thermal energy generation was derived from an experiment by Sabita Chaine and M. Tripathy. The sensitivity and robustness of this experiment under different operating conditions demonstrate the effectiveness of tachyon energy storage devices such as SMES in suppressing power system oscillations with properly adjusted controllers [20].

3.2 Application of power grid

The growing demand for electricity around the world is putting enormous pressure on the environment and the economy. As a result, scientists in various countries have invented many renewable energy generation technologies to solve power supply pressure and environmental pollution problems. But economic challenges remain, such as the loss of grid lines and the dissipation of electricity before it is used during peak generation periods. These situations can lead to high costs of circuit repair and waste of precious resources such as electricity.

If there is an imbalance between the power load and the power supply or the frequency is unstable in an N-1 emergency, the power system may have Cascading outages. And these problems are exacerbated with the growth of RESs, which has some effects on microgrid (MG) performance, such as the reduction of system inertia. So, the voltage and frequency fluctuations increase. Moreover, the RESs exchange electrical power to MGs through electronic power inverters, which cause higher power fluctuations than the traditional synchronous generators. Higher power fluctuations mean more pressure on the grid and a more erratic supply. Therefore, if the RESs have greater penetration, it is difficult to stabilize the frequency and voltage of the system [21].

Superconducting Magnetic Energy Storage (SMES) is one way to solve these problems. First, it is a form of energy storage that can store electricity that customers do not use during peak generation times and solve the pressure on the power supply during the generation downturn and during peak consumption hours. Secondly, it can be used in combination with various power generation systems and power grids to adjust the power grid, such as reducing the losses of the power grid, improving the stability of the grid, etc.

The superiority of SMES was verified in the experiments done by J.B.X. Devotta and M.G. Rabbani, and S. Elangovan [22]. The energy transfer between the power system and the SMES unit is based on the simultaneous control of the active and reactive power modulation of the SMES unit. An artificial neural network-based controller controls the active power, and a conventional regulator controls the reactive power. They were tested using a synchronous oscillation benchmark model. The results of this experiment demonstrated the ability of the SMES unit to effectively restore the stability of the power system. Moreover, another experiment done by Emad A Mohamed and Yasunori Mitani, tested by load frequency controller and SMES together, yielded that SMES has good robustness, stability, coordination, and responsiveness to the grid [21].

4. Sustainable Development Goals

4.1 Reducing poverty

China's new countryside is trying to apply superconducting magnetic energy storage to rural electric energy storage. China's new rural commune includes many rural families. The power source of rural families is often solar power generation. Due to the small daily power consumption, there may be a lot of power waste. Therefore, superconducting magnetic energy storage can be used to store the remaining electricity and transfer it to the national power grid, which can reduce the burden of local power generation. The advantages of using superconducting magnetic energy storage are: solar power generation is characterized by high power generation efficiency when the sunlight intensity is maximum. In this case, if the excess electric energy is directly input into the power grid, the electric energy loss is also the highest. Due to superconducting magnetic energy storage characteristics, the system can compensate for the load fluctuation and reduce the loss. In this way, when the surplus electric energy is input into the national grid when the sun is sufficient, the government can allocate funds to drive the development of poor areas. After the scale is formed, it will be a virtuous circle. When the sunlight is insufficient, the stored electricity can also be used for self-sufficiency to achieve the purpose of green and sustainable development.

4.2 Improving the aquatic environment

Taking hydropower as an example, the principle is that the dam intercepts the flowing water. On this basis, the stored water passes through the turbine to convert potential energy into electric energy to generate electricity. When the construction of a dam destroys the mobility of water, the reduction of downstream flow may lead to drought. The upstream impoundment of the dam may inundate agricultural land and some plants or cause flooding, all of which may destroy animal habitat and agricultural land[23]. Hydropower also changes the river structure, slowing down the water flow, depositing sand and gravel, and raising the riverbed, which constitutes an adverse living environment for some aquatic organisms and affects the water storage efficiency of the dam[24, 25]. The higher the riverbed, the less the water storage capacity.

Therefore, if we want to slow down or even solve these problems, we must take measures. The superconducting magnetic energy storage system can be used. Every time the dam carries out hydropower generation, power is generated through the energy conversion effect of the turbine. When the power generation exceeds the required power, the remaining power can be collected by the superconducting energy storage system to store power. In this way, the utilization efficiency of power generation is improved. When the utilization efficiency is improved, the water storage capacity of the dam is reduced. That is, the power generation cost is reduced. Therefore, it can ensure a certain fluidity of the basin and protect aquatic organisms and the basin's ecosystem.

To sum up, superconducting magnetic energy storage has a wide range of applications, and its use has a positive impact not only on the ecological environment but also on society. If experts and scholars in various fields worldwide continue to work together, it can increase the possibility of realizing the goal of promoting the common sustainable development of mankind.

5. Challenges

Although the SMES system has considerable advantages, it still faces challenges on many fronts. The following text shows the main challenges of SMES at present.

(1) The short storage duration is the most obvious one in its drawbacks. SMES can only be implemented for short cyclic periods of several minutes to hours which is further shorter than the storage duration of some batteries. SMES has a short energy storage duration because it shows a very high self-discharge ratio of 10-15% per day.

(2) SMES has a low energy density which is only about 0.5-5Wh/kg. The energy content of current SMES systems is very small. Ways to increase the energy storage capacity of SMES are often to use large energy storage units.

(3) A robust mechanical structure is usually required to withstand the very high Lorentz forces generated by the magnet coils on the magnet coils without degrading the superconductivity.

(4) The substantial capital cost of SMES ranges from 1000 to 10000 dollars per kWh which is far more than most other technologies.

(5) There are also challenges in the materials and configurations of a superconducting coil. For instance, although the first generation of high-temperature superconductor, BSCCO (BiSrCaCuO), had a high critical current, it was difficult to use in power applications at high magnetic fields. This is because the critical current of magnetic tape decreases rapidly due to the magnetic field.

(6) The monitoring of the quench of the material needs to be more effective. In the event of the quench of superconducting material in the system, it can be costly in terms of time and money to restore the system to operate.

(7) The high energy requirements of cooling systems and the carbon emissions they generate mean that SMES is not a fully environmental-friendly energy storage technology.

(8) The strong magnetic field is likely to threaten people's health. Exposure to high magnetic fields often leads to abnormal heart rates, irregular changes in brain activity, damage to the immune system, the appearance of skin diseases, etc.

6. Conclusion

In general, the demand for energy increases faster than we can imagine. Better energy storage can offer a better supply of power. Superconducting magnetic energy storage did a great job in transferring power efficiently. The technology can achieve maximum efficiency due to its fast discharge speed, great power density, and long life span. Although SMES has low energy density currently, its great power density balances the storage condition. Besides these advantages, SMES is also uniquely suited for multi-domain applications because of its strong superiority. For example, it has good robustness, stability, responsiveness, and coordination in power systems. It can be combined with multiple power generation methods and can compensate for the shortcomings of multiple power generation methods. And the application of SMES at the commercial sale level would be a future topic to study. It meets the UN Sustainable Development Goals of improving the aquatic environment and reducing poverty. In addition, the choices of convenient materials to cooling down SMES systems are big issues in this field. The monitoring of quench needs to be improved for less loss. More research and development are needed to better integrate with other technologies to compensate for its weakness and make it more competitive. The evaluation of superconducting magnetic energy storage technology provides a vision of future development.

References

[1] R.V. Holla, Energy storage methods-Superconducting magnetic energy storage-A Review, Journal of Undergraduate Research 5(1) (2015) 49-54.
[2] H. Chen, T.N. Cong, W. Yang, C. Tan, Y. Li, Y. Ding, Progress in electrical energy storage system: A critical review, Progress in natural science 19(3) (2009) 291-312.
[3] www.electricitystorage.org.
[4] H.D. Fair, Electric launch science and technology in the United States, IEEE Transactions on magnetics 39(1) (2003) 11-17.
[5] T.E. Motoasca, H. Blok, M.D. Verweij, P.M. van den Berg, Electromagnetic forming by distributed forces in magnetic and nonmagnetic materials, IEEE transactions on magnetics 40(5) (2004) 3319-3330.
[6] P. Tixador, Superconducting magnetic energy storage: Status and perspective, IEEE/CSC&ESAS European superconductivity news forum, 2008.
[7] M.H. Bollen, R. Das, S. Djokic, P. Ciufo, J. Meyer, S.K. Rönnberg, F. Zavodam, Power quality concerns in implementing smart distribution-grid applications, IEEE Transactions on Smart Grid 8(1) (2016) 391-399.
[8] S. Rönnberg, M. Bollen, Power quality issues in the electric power system of the future, The electricity journal 29(10) (2016) 49-61.
[9] T. Penthia, A.K. Panda, S.K. Sarangi, M. Mangaraj, ANN controlled 4-leg VSC based DSTATCOM for power quality enhancement, 2015 Annual IEEE India Conference (INDICON), IEEE, 2015, pp. 1-6.
[10] W.U. Tareen, S. Mekhilef, M. Seyedmahmoudian, B. Horan, Active power filter (APF) for mitigation of power quality issues in grid integration of wind and photovoltaic energy conversion system, Renewable and Sustainable Energy Reviews 70 (2017) 635-655.
[11] J.T. Bialasiewicz, Renewable Energy Systems With Photovoltaic Power Generators: Operation and Modeling, IEEE Transactions on Industrial Electronics 55(7) (2008) 2752-2758.
[12] A. Gholami, M. Ameri, M. Zandi, R.G. Ghoachani, S. Eslami, S. Pierfederici, Photovoltaic potential assessment and dust impacts on photovoltaic systems in Iran, IEEE Journal of Photovoltaics 10(3) (2020) 824-837.
[13] J.J. Justo, F. Mwasilu, J. Lee, J.-W. Jung, AC-microgrids versus DC-microgrids with distributed energy resources: A review, Renewable and sustainable energy reviews 24 (2013) 387-405.
[14] K.W. Kow, Y.W. Wong, R.K. Rajkumar, R.K. Rajkumar, A review on performance of artificial intelligence and conventional method in mitigating PV grid-tied related power quality events, Renewable and Sustainable Energy Reviews 56 (2016) 334-346.
[15] W. Ma, W. Wang, X. Wu, R. Hu, F. Tang, W. Zhang, X. Han, L. Ding, Optimal allocation of hybrid energy storage systems for smoothing photovoltaic power fluctuations considering the active power curtailment of photovoltaic, IEEE Access 7 (2019) 74787-74799.
[16] J.X. Jin, J. Wang, R.H. Yang, T.L. Zhang, S. Mu, Y.J. Fan, Y.Q. Xing, A superconducting magnetic energy storage with dual functions of active filtering and power fluctuation suppression for photovoltaic microgrid, Journal of Energy Storage 38 (2021) 102508.
[17] B.-K. Kang, S.-T. Kim, S.-H. Bae, J.-W. Park, Effect of a SMES in power distribution network with PV system and PBEVs, IEEE transactions on applied superconductivity 23(3) (2012) 5700104-5700104.
[18] S.M. Said, M.M. Aly, B. Hartmann, Application of SMES for voltage control of power systems with high wind power penetration, 2018 International Conference on Innovative Trends in Computer Engineering (ITCE), IEEE, 2018, pp. 461-466.
[19] S. Said, M. Aly, B. Hartmann, A robust SMES control for enhancing stability of distribution systems

fed from intermittent wind power generation, Turkish Journal of Electrical Engineering & Computer Sciences 27(5) (2019) 3883-3898.

[20] S. Chaine, M. Tripathy, Design of an optimal SMES for automatic generation control of two-area thermal power system using Cuckoo search algorithm, Journal of Electrical Systems and Information Technology 2(1) (2015) 1-13.

[21] E.A. Mohamed, Y. Mitani, Load frequency control enhancement of islanded micro-grid considering high wind power penetration using superconducting magnetic energy storage and optimal controller, Wind Engineering 43(6) (2019) 609-624.

[22] J. Devotta, M. Rabbani, S. Elangovan, Application of superconducting magnetic energy storage unit for damping of subsynchronous oscillations in power systems, Energy conversion and management 40(1) (1999) 23-37.

[23] B.R. Deemer, J.A. Harrison, S. Li, J.J. Beaulieu, T. DelSontro, N. Barros, J.F. Bezerra-Neto, S.M. Powers, M.A. Dos Santos, J.A. Vonk, Greenhouse gas emissions from reservoir water surfaces: a new global synthesis, BioScience 66(11) (2016) 949-964.

[24] R.J. Schmitt, N. Kittner, G.M. Kondolf, D.M. Kammen, Deploy diverse renewables to save tropical rivers, Nature Publishing Group, 2019.

[25] M.L. Thieme, D. Khrystenko, S. Qin, R.E. Golden Kroner, B. Lehner, S. Pack, K. Tockner, C. Zarfl, N. Shahbol, M.B. Mascia, Dams and protected areas: Quantifying the spatial and temporal extent of global dam construction within protected areas, Conservation Letters 13(4) (2020) e12719.

Simulation Experiment on Energy Tower Coupled with Buried Pipe System of Ground-source Heat Pump for Cross-season Heat Storage

Jianhui Wang[1]*, Ziqiang Liu[1], Yingkai Liang[1], Guohui Peng[1], Weiyu Wang[1], Wei Liu[1]

[1]Institute of Energy Resources, Hebei Academy of Sciences, Shijiazhuang Hebei 050081, China

*Corresponding author's e-mail: 9985718@qq.com

Abstract. It is proposed to use energy tower instead of solar collector system to store heat for buried pipe system of ground-source heat pump in summer, so as to solve the problem that soil temperature is too low due to unbalanced cold and heat load of GSHP in cold and severe cold area. The simulation results show that the thermal storage power of energy tower is affected by soil temperature and environmental temperature. The lower the soil temperature, the better the thermal storage effect of energy tower system for soil. The thermal storage power and environmental temperature change in a positive relationship. It is proved that the application of cross-season heat storage is feasible for energy tower coupled with buried pipe system of ground-source heat pump in cold and severe cold area.

1. Introduction

The traditional ground source heat pump system uses underground rock and soil as a container for storing heat. The heat pump system stores the heat inside the building underground in summer, to reduce the temperature inside the building, and extracts the heat stored underground in winter to heat the building[1-2]. However, in specific practical engineering applications, it is difficult to balance the cooling load of the building in summer and the heat load in winter. Especially in cold and severe cold regions, the cooling load of the building in summer is much smaller than the heat load in winter, many years of operation will cause the heat pump to take more heat from the soil in winter than the system runs to the soil in summer. The heat balance of the soil will be destroyed, the soil temperature will gradually decrease, and the heat pump performance coefficient also will gradually decrease[3-4]. In areas where the ground source heat pump is used alone for heating in winter without cooling in summer, this problem will be more serious.

In order to overcome the problem of soil heat balance being damaged, many previous studies have proposed a combination of solar and ground source heat pumps to address this problem[5-7]. Solar energy is used as the auxiliary heat source of the soil-source heat pump, and the heat of the solar energy is to be stored into the soil seasonally, to make up the heat required for the imbalance of cold and heat. It will effectively increase the temperature of the soil around the buried pipe, thereby increasing the heating performance coefficient of the ground source heat pump during the heating period, makes the system more reliable. However, in practical applications, the use of solar energy is restricted by many factors, such as meteorological conditions, geographic location, collector price and collector installation conditions. For buildings with few solar resources or no space for collector installation, solar energy systems cannot be used.

In recent years, Zhang Shu and others have studied the ground coupled heat pump system with seasonal air source heat storage suitable for cross-season applications[8-9]. The system uses summer high-temperature air instead of solar energy as an auxiliary heat source, the outdoor air heat exchanger is used to store the heat in the air into the soil in summer, and then the heat pump is used to take heat from the soil into the room in winter. Compared with solar heat storage, air source heat storage equipment is simple, investment and maintenance costs are lower. When there is no central heating pipe network, and there is not enough roof or other space to install solar collectors, the air source heat storage system can also achieve cross-season utilization of natural energy, provides a new way for the application of ground source heat pump heating technology in severe cold areas. However, the outdoor air heat exchanger only relies on the outdoor fan coil for simple heat exchange with air, although it can store heat when the air temperature is high, the heat storage effect is lower than that of solar heat storage.

For this reason, we propose to use energy towers instead of solar heat collection system to store heat for ground-source heat pump underground pipe systems for cross-season heat storage in summer, to solve the problem that the soil temperature is too low due to the imbalance of the cold and heat load of the ground source heat pump in cold and severe cold regions.

2.Energy tower coupled with buried pipe system of ground-source heat pump

In recent years, in the hot summer and cold winter regions such as Central China and East China, a new type of heat pump form—energy tower heat pump system has been gradually developed. It uses an energy tower as a heat exchange device with the air, assists the heat pump system to dissipate heat in summer, and extracts low-temperature heat energy contained in the air in winter. Energy towers are also called heat source towers, derived from the theory of reverse heat absorption of cooling towers, it is a tower-shaped new device that uses air as a source of cold and heat. Through the heat exchange between the tower and the air, the energy tower realizes multiple functions such as cooling, heating, and domestic hot water[10-12].

Since the energy tower has the function of absorbing the low-temperature heat energy stored in the natural air, we consider using the Energy tower coupled with buried pipe system of ground-source heat pump. In summer, the energy tower is used to absorb the heat of natural air and transfer it to the soil through the buried pipe heat exchanger to realize the heat storage of the soil; in winter, the heat stored in the soil is taken out by the ground source heat pump through the buried pipe heat exchanger to heat the building, which solves the problem of low soil temperature caused by the imbalance of cold and heat load in cold and severe cold areas. The schematic diagram of the cross-season heat storage energy tower coupled with buried pipe system of ground-source heat pump is shown in Figure 1.

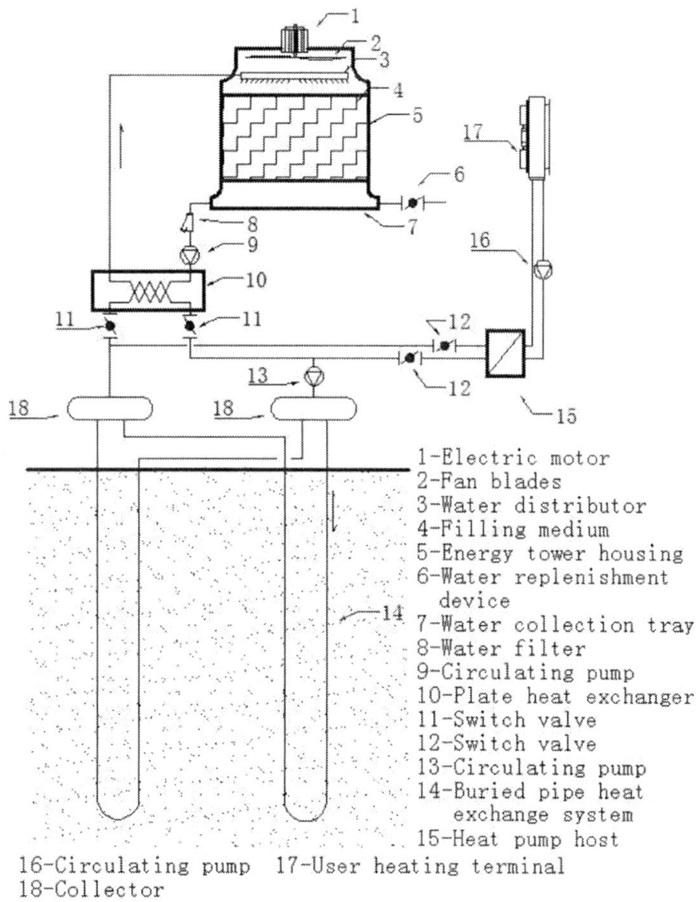

Figure 1. Schematic diagram of cross-season heat storage system for energy tower coupling buried pipe system

In the non-heating seasons when the temperature is high, the system closes the switching valve 12 and opens the switching valve 11. At this time, the energy tower and the buried pipe system form a soil cross-season heat storage system, the energy tower absorbs the heat of natural air and the latent heat of condensation of water vapor in the air through the working medium, and transfers the heat to the soil to achieve heat storage.

During the heating period in winter, the switching valve 11 is closed and the switching valve 12 is opened. Under the action of the heat pump, the system takes out heat from the underground soil through the buried pipe heat exchange system, and after the heat pump host heats up, the user terminal supplies heating to the building. The system uses the soil as a huge heat accumulator, storing heat in summer and taking heat in winter, realizing the cross-season utilization of energy, and also solving the problem of unbalanced cold and heat load of the ground source heat pump system in winter and summer.

3.Experiment system

In order to simulate and verify the heat storage effect of the energy tower coupled with buried pipe system of ground-source heat pump, we have built an experimental system. As shown in Figure 2, the experimental system consists of an energy tower, a plate heat exchanger, an integrated water source heat pump unit, a control system and a data acquisition system. We use a set of integral water source heat pump unit to replace the buried pipe system, the integral water source heat pump unit extracts the heat from the natural air absorbed by the energy tower through the plate heat exchanger, and then uses the fan coil to dissipate the heat to simulate an underground pipe system to store the heat in the underground soil. The control system can adjust the power of the heat pump unit through the frequency converter to control the outlet water temperature of the heat pump unit, simulate the different outlet water temperature of the buried pipe system under different soil temperature conditions, can better simulate the heat storage effect under different soil temperature conditions. In the experimental system, the energy tower is transformed by a cooling tower with a circulating water flow of 3t/h, the circulating pump power is 200W, and the energy tower fan power is 35W. In the system, temperature sensors and flow sensors with measurement accuracy of 0.2 are installed on the inlet and outlet pipes of the heat pump unit, and a data acquisition system is established. It can display and record inlet and outlet temperature and flow in real time, and calculate and record the heat storage power of the system in real time based on temperature and flow data.

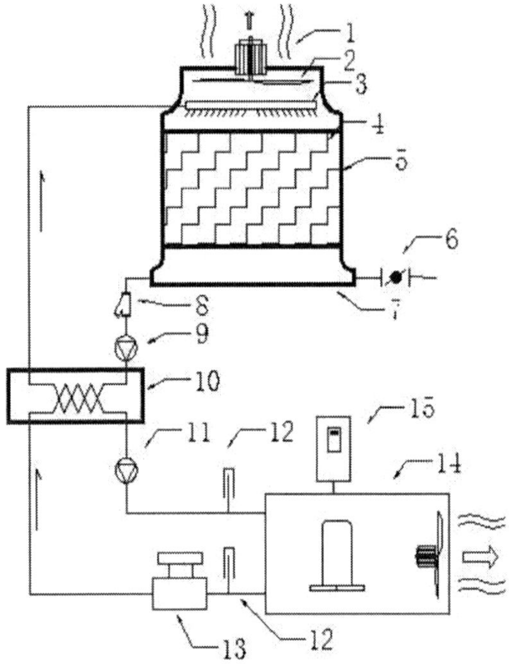

1-Electric motor 2-Fan blades 3-Water distributor
4-Filling medium 5-Energy tower housing
6-Water replenishment device 7-Water collection tray
8-Water filter 9-Circulating pump 10-Plate heat exchanger 11-Circulating pump 12-Humidity Sensor
13-Ultrasonic flowmeter 14-Water source heat pump unit 15-Frequency converter

Figure 2. Schematic diagram of energy tower coupled buried pipe heat storage experiment system

4.Experimental data and analysis

After the installation and commissioning of the experimental system was completed, in the summer of 2018, we began to conduct experiments on simulating energy towers for soil heat storage. The test experiments were carried out under the conditions of 9°C, 10°C, 12°C, 14°C, 16°C and 18°C at the outlet water temperature of the heat pump unit. The lower the outlet water temperature of the heat pump unit, the lower the soil temperature, to simulate the heat storage effect of the soil under different temperature conditions. Figure 3-Figure 8 are curves of ambient temperature and heat storage power under different outlet water temperature conditions of the heat pump unit, Table 1 is a summary table of the data analysis of the energy tower heat storage experiment.

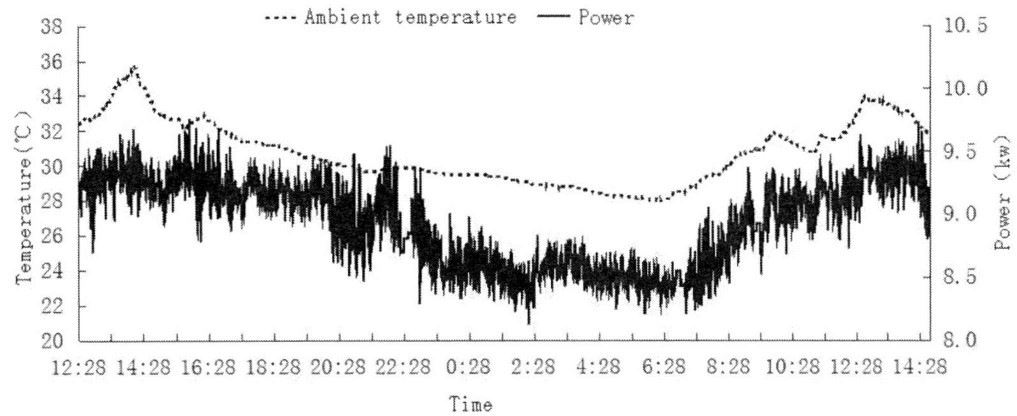

Figure 3. Curve of environmental temperature and thermal storage power at outlet temperature of 9°C

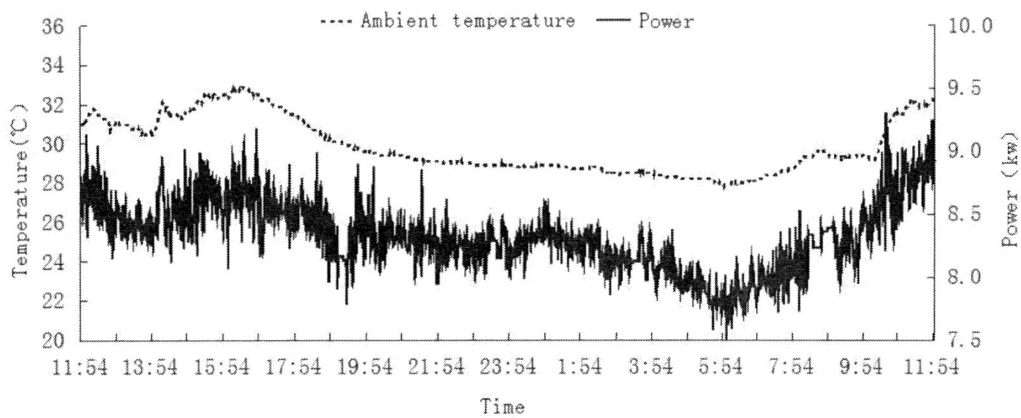

Figure 4. Curve of environmental temperature and thermal storage power at outlet temperature of 10°C

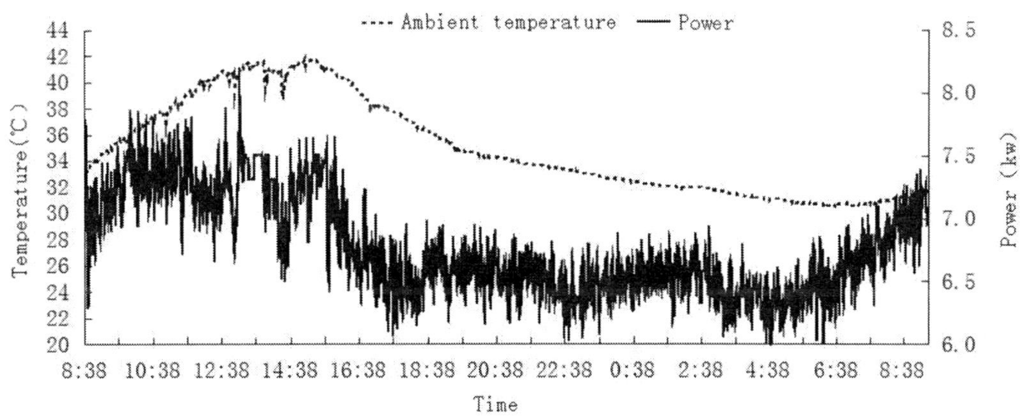

Figure 5. Curve of environmental temperature and thermal storage power at outlet temperature of 12°C

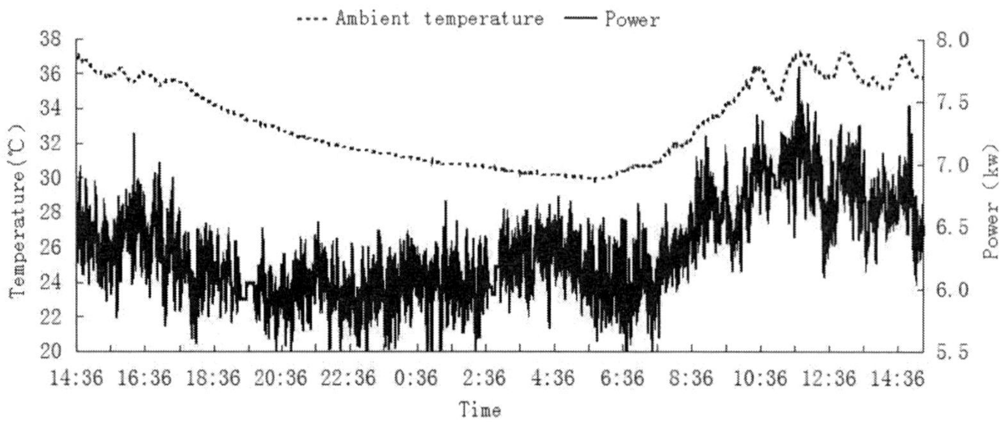

Figure 6. Curve of environmental temperature and thermal storage power at outlet temperature of 14°C

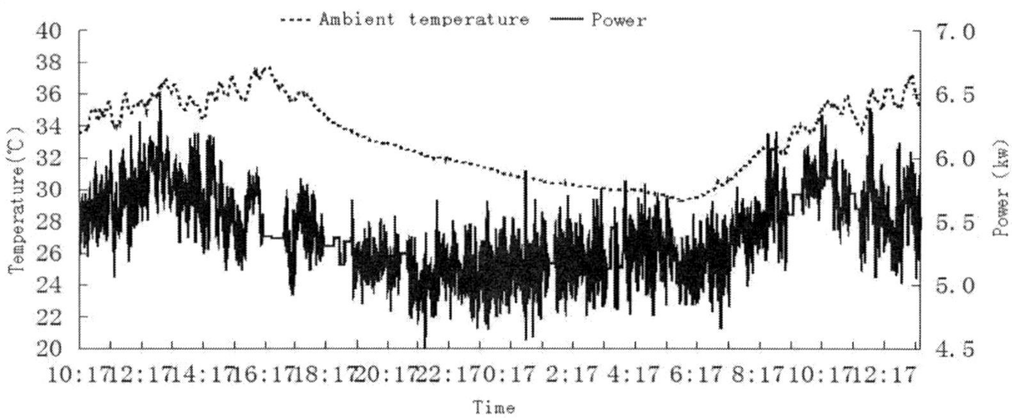

Figure 7. Curve of environmental temperature and thermal storage power at outlet temperature of 16°C

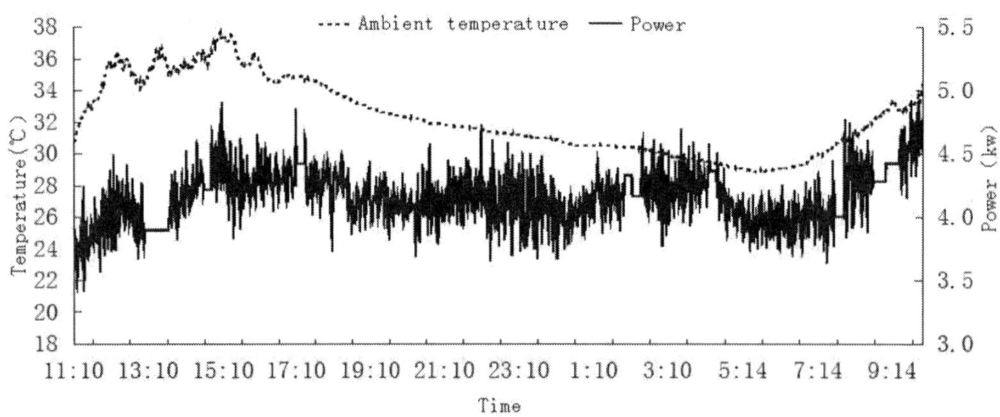

Figure 8. Curve of environmental temperature and thermal storage power at outlet temperature of 18°C

Table 1. Summary table of energy tower heat storage experiment

The outlet water temperature of the water source heat pump unit (°C)	Maximum ambient temperature (°C)	Thermal storage power (kW)	Minimum ambient temperature (°C)	Thermal storage power (kW)	Ambient temperature average (°C)	Thermal storage power (kW)
9	35.7	9.34	28.0	8.45	30.8	8.94
10	32.9	8.74	27.8	7.76	29.8	8.33
12	41.8	7.31	30.4	6.43	35.0	6.77
14	37.3	6.92	29.9	6.12	33.2	6.31
16	37.7	5.55	29.3	5.15	33.2	5.40
18	37.7	4.49	28.8	3.93	32.3	4.16

From Figure 3 to Figure 8 and Table 1, it can be seen that in the experiment of simulating the energy tower for soil heat storage, the power of the energy tower for soil heat storage is seriously affected by the underground soil temperature. The outlet water temperature of the water source heat pump unit represents the corresponding soil temperature, when the outlet water temperature of the water source heat pump unit is 9°C, 10°C, 12°C, 14°C, 16°C, 18°C, the average heat storage power is 8.94kW, 8.33kW, 6.77kW, 6.31kW, 5.40kW, 4.16kW. The heat storage power decreases with the increase of the water temperature at the outlet of the water source heat pump unit. When the water temperature at the outlet of the water source heat pump unit is 9°C, the heat storage power of the energy tower is more than twice the heat storage power when the outlet water temperature of the water source heat pump unit is 18°C. It shows that the lower the temperature of the soil, the higher the heat storage power of the energy tower, and the better the effect of the energy tower for storing heat in the soil. This is because when the soil temperature is low, the temperature difference with the ambient temperature is large, so the heat storage is large, the heat storage power of the energy tower is also increased.

In addition, the heat storage power of the energy tower is also affected by the ambient temperature, the change of the heat storage power and the change of the ambient temperature are positive changes. When the ambient temperature increases, the heat storage power increases; conversely, when the ambient temperature decreases, the heat storage power also decreases. This is because the higher the ambient temperature, the more heat stored in the natural air is absorbed by the energy tower, and the higher the heat storage power. And when the ambient temperature is low, the heat storage power decreases.

5.Conclusion

1)Proposed the cross-season heat storage of the buried pipe system of the energy tower coupled with the ground source heat pump. In summer, the energy tower is used to absorb the heat in the natural air and the latent heat of condensation of the water vapor in the atmosphere to achieve soil heat storage, which can avoid the restrictions of meteorological conditions, geographical location, installation conditions and investment conditions that are encountered by the use of solar thermal storage systems, and it also has the advantages of continuous operation 24 hours a day and convenient application. It can be used as another cross-season soil heat storage method in addition to solar-soil heat storage to solve the problem of low soil temperature caused by imbalanced heat and cold loads of ground-source heat pumps in cold and severe cold regions.

2)The power of the energy tower to store heat for the soil is severely affected by the temperature of the underground soil. The lower the temperature of the soil, the higher the heat storage power of the energy tower, and the better the effect of the energy tower system for storing heat in the soil.

3)The heat storage power of the energy tower is affected by the ambient temperature, and the change of the heat storage power and the change of the ambient temperature become a positive change, when the ambient temperature increases, the heat storage power increases; conversely, when the ambient temperature decreases, the heat storage power also decreases.

4)The experimental results prove that the cross-season heat storage of the energy tower coupled with buried pipe system of ground-source heat pump is feasible in cold and severe cold regions.

References

[1] Hu XW, Shi XZ, Zhang HD, et al.(2010) Research on heat recovery to reduce the imbalance between cold and heat of ground-source heat pumps[J]. Building Thermal Ventilation and Air Conditioning. 2010,29 (2): 49-52.

[2] Chen HW, Yang CZ.(2007) Study on the changes of underground soil temperature field in ground-source heat pump air-conditioning systems. Building Energy. 2007,35(4):51-54

[3] Zhang YJ, Liu D, Chen ZH, et al.(2018) Research on soil cold accumulation during long-term operation of ground source heat pump in cold regions[J]. Journal of Hebei Institute of Civil Engineering and Architecture, 2018, 36 (1); 79-84.

[4] Wang SQ, He SJ, Ma KY.(2016) Research on the time domain method operation strategy of ground source heat pump system in severe cold area[J].Fluid Machinery, 2016,44(6);76-80.

[5] Song ZW, Hou CQ, Xu LG.(2016) Research on Numerical Simulation of Solar-Ground Source Heat Pump Combined Heating System[J].Energy Conservation of Buildings, 2016,44(10);23-28.

[6] Wang EY, Li X, Wang M, et al.(2017) Research on operation strategy of solar-assisted ground-source heat pump system[J]. Renewable Energy, 2017,35(7);998-1006.

[7] Lu ZJ, Qu SL, Han JQ, et al.(2017) The applicability of the thermal storage solar ground source heat pump coupling system in severe cold areas[J]. Building Science, 2017, 33(10); 97-103.

[8] Zhang S, Zheng MY, Wang X, et al.(2012) Simulation and experimental verification of soil heat storage characteristics of cross-season air-U-shaped buried pipe in severe cold area[J].HVAC, 2012,42(3); 97-102.

[9] Zhang S, Zheng MY.(2012) Simulation research on air source regenerative ground source heat pump system[J]. Fluid Machinery, 2012,40(8); 69-74.

[10] Xu GQ, Huang DX, Zhang XD.(2016) Application advantages of heat source tower heat pump in hot summer and warm winter area[J].Refrigeration, 2016,35(2);52-57.

[11] Feng R, Meng X, Deng JP, et al.(2018) Current status of research on cooling tower reverse use endothermic as heat source tower technology[J]. Progress in Chemical Industry, 2018,37(11); 4135-4142.

[12] Song YQ, Ma HQ, Long WD.(2011) Application and analysis of energy tower heat pump technology in air conditioning engineering[J].HVAC, 2011,41(4); 20-23.

Application of Digital Twin Technology in Energy and Power Industry

Xueqing Liang [1], Shuming Du [1], Xiaofan Zhao[1] and Chao Liu[1*]

[1] Guangzhou Power Supply Bureau of Guangdong Power Grid Co., Ltd., Guangzhou City, Guangdong Province

*Corresponding author's e-mail: liuchao1@gzps.corp.csg

Abstract. Constructing digital twin of physical entity and analyzing twin data, and transfer learning method can be used to transfer the characteristics of twin to physical entity for analysis. This paper mainly studies the application of digital twinning technology in the energy and power industry. Aiming at the energy saving and control problem of building energy demand terminal, the fuzzy adaptive PID control algorithm is used to simulate the indoor HVAC system and lighting system, so as to solve the indoor temperature and illumination intensity are greatly affected by the weather, time-varying and unpredictable factors. By comparing with the conventional PID, we can see that the fuzzy adaptive PID controller adjusts faster, the output duty ratio is more accurate, and can achieve better control effect.

1. Introduction

CPS aims to integrate and collaborate organically with 3C (Communication, Control) technologies to achieve the interconnection and interoperation between the virtual and physical worlds and support more decentralized and enhanced production Control. Among them, as an important link or one of the key technologies to realize CPS, digital twinning realizes the digital model mapping from the physical system to the cyber space from the aspects of geometry, physics, behavior and rules to ensure the harmony between the digital world and the physical world. On the basis of constructing the digital model, it can also realize real-time monitoring, data collection, simulation and analytical reasoning of physical objects based on the digital model [1-2]. With the rapid development of the information technology of smart power grid, the connection between information and communication technology and modern power grid is getting closer and closer, which leads to the enhancement of the connectivity and automation degree of power grid, as well as the exponential increase of its scale and complexity. As a result, the modern power system gradually forms a complex system integrating perception, communication and control. In order to support the reliable operation of modern power system, 3C (Computation, Communication, Control) technology is introduced in traditional power system on a large scale. Through this technology, the coupling of Communication system and monitoring system with traditional power system is realized. This is characterized by the integration of Cyber Physical System (CPS) [3-4]. Therefore, from the perspective of CPS, the modern power system has developed into a power CPS composed of the interaction between the physical grid system and the information system, which contains frequent information flow and energy flow exchange [5-6].

With the development of informatization and digitalization, the concept of Digital Twin has been put forward. Digital twin first come from NASA's Apollo program, in the project, NASA made two identical aircraft, space one mission, another on the earth, accurate simulation of the flight on a mission in outer space conditions, another spacecraft mission to feedback, the situation of the aircraft on the earth will be fed back to the astronauts so as to make the best choice [7]. With the development of information technology, some scholars began to use computers to simulate real objects in digital space.

In this paper, based on the higher flexibility and other functions and characteristics of CPS enabled production system, and based on the unique attributes and characteristics of digital twin technology, the energy and power control system technology driven by digital twin is studied.

2. Energy Control Based on Digital Twin Technology

2.1. Improvement of Digital Twin Technology

2.1.1 Definition of digital twin

The definition of digital twinning has also changed over the years. In AFRL's definition, digital twin is defined as an integrated system composed of data models, which can deploy the entire aircraft formation as well as a single aircraft [8]. According to the definition of NASA, digital twin is a digital system integrated with multi-physics, multi-scale and probabilistic simulation, which can reflect the actual condition of the real aircraft through simulated physical model, real-time sensors and service history [9]. In GE's definition, digital twins focus on the software representation of assets, which are used to predict and optimize the business, and are composed of data, algorithms and knowledge instead of physical entities [10]. Therefore, there are still many differences in the definition of digital twinning. Different companies may have different understandings of digital twinning due to their different priorities.

2.1.2 Digital twin model

In the process of constructing the digital twin model, the traditional digital three-dimensional model cannot meet the current requirements. In order to promote the development of digital twin and the application of digital twin in many fields, a five-dimensional digital twin model is proposed. Different from the traditional manufacturing industry, which only needs to pay attention to production factors and products, the emerging manufacturing industry also needs to pay more attention to the data, information, knowledge and product services in the production process. Products in the physical world are no longer enough to meet the development of manufacturing industry. Only the combination of virtual and real can comply with the development of the market.

The digital twin five-dimensional model is composed of physical entity, virtual entity, link, twin data and service system. The physical object and virtual model are a pair of corresponding levels, in which the interaction is carried out by the information connection channel. The data generated by physical object, virtual object and service system are collectively referred to as twin data. By analyzing and calculating the twin data, valuable information can be obtained to drive the operation of multiple objects in turn. The whole digital twin model is mutual, and the decision optimization can also promote each other [11-12].

With the digital twin five-dimensional model, the advantages of digital twin are clear. The interaction between information and the physical world is the bridge for the development of digital twins, and virtual modeling is an indispensable part of the process of digital twins. In other words, virtual modeling is the foundation for digital twins to be truly different from traditional manufacturing. Twin data is generated in the process of digital twinning, and analysis of twin data is the driving force of the whole model. Connectivity is also essential for the digital twin model, which holds the data of the whole model and the interaction of information. The ultimate goal of digital twin is to target application service, and digital twin works for application service as a whole. Information collection of physical objects is essential as the primary source of information for digital twinning. By analyzing the digital twin model, the key points of digital twin can be clearly recognized, which provides theoretical support for the subsequent model construction.

When the digital twin model is built, the goal is to build a bridge between the virtual end and the physical end by comprehensively considering the various information of the model, so that the twin can fully reflect the information of the physical entity, and further realize the twin beyond the physical limit and predict the information of the physical entity.

2.2. Energy Demand Terminal Control Method

2.2.1 Fuzzy PID control principle

Fuzzy control is an intelligent nonlinear control system. Different from the traditional control system, it uses the fuzzy controller instead of the traditional controller.

One dimensional fuzzy controller is a single input single output fuzzy controller, the control logic is not complicated, the system requirements are easy to achieve, but the control effect is not ideal. Compared with the one-dimensional fuzzy controller, the three-dimensional fuzzy controller has good control effect and performance, but it also has some problems such as complex logic, difficulty in designing fuzzy rules and difficulty in realizing control. Since the two-dimensional fuzzy controller can well reflect the dynamic characteristics of the error signal, the control effect is better than that of the one-dimensional fuzzy controller, but it is simpler than that of the three-dimensional fuzzy controller, so the control system in this paper adopts the two-dimensional fuzzy controller.

2.2.2 Fuzzy PID controller design

The fundamental domain is the actual range of the precise input and output of a fuzzy controller. The fuzzy set theory domain after fuzzification is called fuzzy set theory domain. The theoretical domain of fuzzy sets is generally symmetric, that is, {-n, -n+1... , 0,... , n-1, n}. Theoretically, the more n choices there are, the more elements in the domain, which can both improve the control accuracy and increase the amount of calculated data. In general, fuzzy subsets have better domain coverage when there are about twice as many elements in the domain as fuzzy subsets.

After fuzzing input and output, it is also very important to choose appropriate membership function to describe fuzziness. Triangle membership function is widely used because of its fast calculation speed and strong reliability, so we choose trigonometric function, and we will not consider other functions here. Therefore, the trigonometric membership function is selected to fuzzize the input variables of the fuzzy controller from the basic domain to the corresponding fuzzy domain.

The system error and the rate of change of the error are taken as input variables, and the fuzzy conditional statements are obtained through the language variable values.

Completion of work, the on-line adjustment of PID controller parameters, in front of the control, according to a mathematical model of the case, first using optimization method to optimize a set of initial PID parameter, control according to the error e and error change rate ec again when find out the three parameters of PID controller correction, finally according to the following formula to calculate the actual parameters of PID.

$$\mathbf{K_P = K_{P0} + G_P * \Delta K_P} \quad (1)$$

$$K_I = K_{I0} + G_D * \Delta K_I \quad (2)$$

$$K_D = K_{D0} + G_D * \Delta K_D \quad (3)$$

3. Controller Simulation Test

3.1. HVAC System Control

Because the temperature of an air-conditioned room is affected by many factors, such as the opening and closing of shutters and doors and Windows and weather factors, it is difficult to describe the control process of room temperature with an accurate mathematical model. By analyzing the model of air-conditioned room and terminal air-conditioned box, and combining with relevant materials, the mathematical model of temperature control in air-conditioned room is simplified to a typical dynamic transfer function of first-order inertia plus pure lag.

The self-adaptive fuzzy PID controller completes the self-adjustment and self-correction of the PID control variables online through fuzzy reasoning and fuzzy operation on the real-time temperature of the room, and obtains the frequency of the air blower of the air conditioner box to control the air supply volume of the system, so that the room temperature is always maintained at the set value.

The deviation between the terminal room temperature and the set value and the deviation rate of change are taken as the input variables of the adaptive fuzzy PID controller, and the parameter adjustment values of the conventional PID controller Δ Kp, AmiKi and Ang Kd are taken as the output variables. The set value of the air conditioning room temperature in summer is 25°C. According to the summer temperature variation in the north, The room temperature deviation E is set as [-15,15], and the temperature deviation change rate is set as [-10,10].

3.2. Lighting System Control

Due to the time-varying and easy interference characteristics of natural light, it is difficult to complete the detection of actual data, so it is impossible to build an accurate mathematical model for control. In this study, the intelligent dimming system model in related studies was selected.

The self-adaptive fuzzy PID controller completes the self-adjustment and self-correction of the PID control variables through fuzzy reasoning and fuzzy operation on the room real-time illumination, and gets the illumination of the output LED light to control the indoor illumination, so that the indoor illumination is always maintained at the set value.

The deviation between the room illumination intensity and the set value and the deviation rate of change are taken as the input variables of the adaptive fuzzy PID controller, and the parameter adjustment values of the conventional PID controller ΔKP, aliki and lf Kd are taken as the output variables. The set value of the room illumination is 300lx. The basic domain of room illumination deviation E is set as [-100,100], and the basic domain of change rate of illumination deviation is set as [-50,50].

4. Simulation Test Results

4.1. HVAC Experimental Results

Table1. HVAC test results

	20	50	100	150	200
Traditional PID	4.9	22.8	24.7	24.5	24.6
Fuzzy PID	5.2	17.3	23.6	24.8	24.8

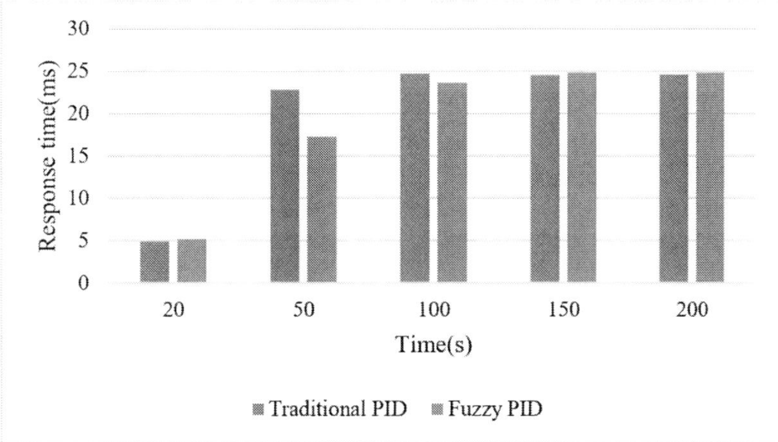

Figure1.HVAC test results

As shown in Table 1 and Figure 1, the effect of conventional PID control and adaptive fuzzy PID adjustment is very good. However, the traditional PID control system has slight oscillation, starting from the 50s to the 150s, and the adjustment time is a little longer than the adaptive fuzzy PID control system by about the 50s. Therefore, the simulated control effect is better, because the system adjusts quickly and steadily, without overshoot, and is stable without oscillation.

The traditional PID controller can basically meet the temperature control requirements of the air-conditioned room when the parameters are reasonable and the system control accuracy is not very high. However, the temperature regulation of HVAC system is complex, and it is difficult to use an accurate but fixed mathematical model to describe it. If there is disturbance, the traditional PID control method is not effective. Therefore, this experiment verifies the effectiveness of fuzzy PID control.

4.2. Experimental Results of Lighting System

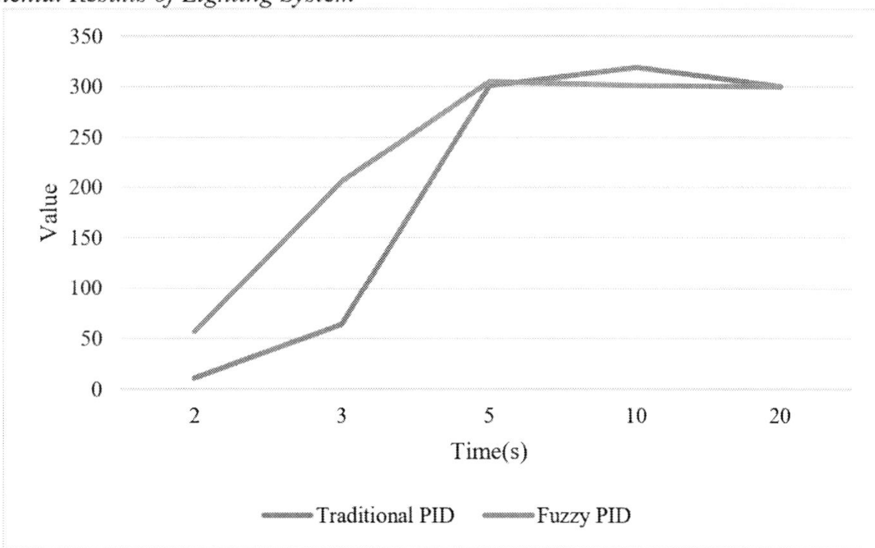

Figure2.Lighting system test results

Figure 2 shows the illumination intensity response curves of the traditional PID controller and the adaptive fuzzy PID controller in the control room under the condition that the transfer function parameters of the control object remain unchanged and there is no jitter factor. As can be seen from the figure, the effect of conventional PID control and adaptive fuzzy PID adjustment is also very good. However, the traditional PID control system has slight oscillation and overshoot phenomenon. Although the adaptive fuzzy PID control also has a slight overshoot phenomenon, the adjustment time is much smaller than the traditional PID control system, and the adjustment response is fast. Therefore, the fuzzy control effect is better, because the system adjustment is fast and stable, the overshoot is small, and the stability is no oscillation.

5. Conclusions

With the complexity of the physical network to improve electric power information, maintain the safe and stable operation of power system becomes increasingly important, according to various information security problem caused by the development of intelligent, we should give high attention, and the problems were analyzed, and the only combined with the actual situation, put forward effective solutions, can improve the level of information security management, promote the long-term and stable development of the power system. This paper applies the CPS system building energy system of energy supply and energy demand in the end, use the reliable awareness and precision control of the CPS system, the characteristics of real-time perception within the building on the physical environment, and according to the actual situation for the building of renewable energy production and building HVAC control system and lighting system, so as to achieve the "maximum energy supply, minimum energy consumption, the best comfort" purpose.

References

[1] Ross D . Digital twinning [Information Technology Virtual Reality][J]. Engineering & Technology, 2016, 11(4):44-45.
[2] Lu R , Brilakis I . Digital twinning of existing reinforced concrete bridges from labelled point clusters[J]. Automation in construction, 2019, 105(SEP.):102837.1-102837.16.
[3] Hiskey T . From Blockchain to Digital Twinning-Technologies That Can Revolutionise Business Models in Manufacturing[J]. Software world, 2019, 50(1):8-9.
[4] Kang M , Park J , Sohn S S , et al. Interpretation of quasi-static and dynamic tensile behavior by digital image correlation technique in TWinning Induced Plasticity (TWIP) and low-carbon steel sheets[J]. Materials Science & Engineering A, 2017, 693(May2):170-177.
[5] Keii, GI, Fuminori, et al. An Analysis on Global Energy-Related CO2 Emission Reduction and Energy Systems by Current Climate and Energy Policies and the Nationally Determined Contributions[J]. Journal of the Japan Institute of Energy, 2018, 97(6):135-146.
[6] Wang K , Zhou L , Deng Y , et al. Application range analysis and implementation of the logic-processed CPS-PWM scheme based mmc capacitor voltage balancing strategy[J]. CPSS Transactions on Power Electronics and Applications, 2019, 4(1):1-9.
[7] Mo C , Kontsos A . Twinning Contributions to Strain Localizations in Magnesium Alloys[J]. Materials Science and Engineering A, 2018, 722(apr.11):206-215.
[8] Lu L , Sun T , Fezz Aa K , et al. Simultaneous multiscale measurements on dynamic deformation of a magnesium alloy with synchrotron x-ray imaging and diffraction[J]. Materials Science & Engineering A, 2017, 701(jul.31):143-148.
[9] A T W M , B A V , A K B , et al. Power handling of a liquid-metal based CPS structure under high steady-state heat and particle fluxes[J]. Nuclear Materials and Energy, 2017, 12(C):210-215.
[10] Heqin, Tong, Ming, et al. Flexible hardware-in-the-loop testbed for cyber physical power system simulation[J]. IET Cyber-Physical Systems: Theory & Applications, 2019, 4(4):374-381.
[11] Luu P T , Lee J , Lee J , et al. Selection of Contactless-Power-Supply Housing Material Considering Eddy Current Loss, Thermal Load, and Structure Safety[J]. Journal of Magnetics, 2018, 23(1):86-92.
[12] Hanif M A , Marchisio A , Arif T , et al. X-DNNs: Systematic Cross-Layer Approximations for Energy-Efficient Deep Neural Networks[J]. Journal of Low Power Electronics, 2018, 14(4):520-534.

Hierarchical Control Strategy for Microgrid Based on Finite-time Consensus Algorithm

Binfeng Tang[1a], **Ying Huang**[1b*]

[1] School of Automatic Control, Liuzhou Railway Vocational Technical College, Liuzhou 545616, China;

[a] bingo361@foxmail.com, [b*] 1786278630@qq.com

Abstract—In order to overcome the shortcomings of centralized control strategy for microgrid, a hierarchical control strategy based on finite-time consensus algorithm is proposed in this paper. Based on the droop control, the control strategy uses the finite-time consensus theory of multi-agent system to solve the optimal power generation value of the system, and adjusts the droop coefficient to make the power generation power of the power supply the optimal value. Then the compensation of frequency and voltage is introduced by multi-agent technology to stabilize the system frequency and voltage at rated value.

1. Introduction

In recent years, due to the increasingly prominent environmental problems, the depletion of fossil energy and the increase of energy demand, distributed generation (DG) has developed rapidly [1].

Due to the difference of power generation cost characteristics and working characteristics of each DG, it is necessary to study the economic dispatch and operation control of the microgrid system[2]. The Economic Dispatch Problem (EDP) of microgrid is to minimize the total power generation cost of the system by coordinating the output between DGs under the condition of meeting the system and equipment level constraints. At present, many achievements have been made in the research on the optimization strategy to solve the micro grid EDP: Ref.[3] uses an improved particle swarm optimization algorithm to solve the EDP. This algorithms belong to centralized optimization algorithms, relying too much on the control center, and their failure will lead to system paralysis. In view of the shortcomings of centralized optimization algorithms, some scholars have proposed using distributed algorithm to solve the EDP of microgrid. Ref.[4] proposed a distributed control strategy of DC microgrid based on finite time consistency algorithm;Ref.[5] proposed a distributed microgrid algorithm based on multi agent system consistency theory, and used the incremental cost of each power generation unit as a consistent variable. The above documents do not consider the system response convergence time, which has some problems, such as slow convergence time, low control accuracy or no optimization of AC microgrid.

The MAS finite-time consensus algorithm has been successfully applied in fields such as robot formation and swarming problems[7] It is a distributed control method that can be achieved the purpose of global control by communicating with neighboring agents. Using the finite time theory of MAS, aiming at the problem of optimal operation control of micro-grid, this paper proposes a distributed hierarchical control strategy for micro-grid based on MAS finite-time consistency algorithm. The control strategy includes three contributions: First, the finite-time consistency algorithm is applied to make the micro-increase rate of each DG consistent, and the optimal value of each DG power generation is solved, so that the total power generation cost of the system is minimized; Secondly, introduce frequency and voltage compensation, and apply MAS technology to achieve frequency and voltage stability;Finally,based on the optimal value of each DG power generation, the actual power generation of each DG is consistent with the optimal value by adjusting the droop coefficient of each DG.

2. MAS

2.1. Graph theory

Generally, for a microgrid system containing n DG's, if the DG is regarded as a node in the agent communication network, its communication topology can be represented by an undirected graph $G=\{V,E,A\}$. Where, $V=\{v_1,v_2,v_3,\cdots,v_n\}$ represents the set of nodes, n is the number of nodes in the graph; $E\subseteq V\times V$ represents a set of edges, $e_{ij}=(v_i,v_j)$ represents the edges of the graph; In the graph G, the adjacency matrix $A=(a_{ij})_{n\times n}$ is usually used to describe the relationship between nodes, where a_{ij} is the connection weight from node i to node j. If the node i can receive the information of the node j, then $a_{ij}>0$, otherwise $a_{ij}=0$ [7]. For any node v_i, define its neighbor node set as $N_i(G,i)=\{v_j\in V:(v_i,v_j\in E)\}$. In agent communication, Laplacian matrix is generally used to describe the communication between agents, that is $L=(l_{ij})\in R^{n\times n}$, where $l_{ij}=-a_{ij}$, $l_{ii}=\sum_{j=1,j\neq i}^{n}a_{ij}$.

2.2. Finite-time consensus algorithm

For a MAS composed of n agents, the state of each agent can be expressed as:

$$\dot{x}_i(t)=u_i(t) \tag{1}$$

Where, x_i is the state variable of the i-th agent; $u_i(t)$ is the control variable;

Each agent can only communicate with its surrounding neighbors. The state of the agent depends on its current state and the current state of its neighboring agents. The following finite-time consistency protocol is designed.

$$u_i(t)=\sum_{j=1}^{n}a_{ij}(x_j-x_i)+\sum_{j=1}^{n}a_{ij}\,\mathrm{sign}(x_j-x_i)\left|x_j-x_i\right|^{\alpha} \tag{2}$$

Where, α is the control parameter, usually $0<\alpha<1$, $\mathrm{sign}(x)$ is a symbolic function.

Under the control of $u_i(t)$, each agent exchanges information with neighboring agents through its current state, and when the communication topology satisfies the condition of undirected connectivity, all agents finally reach the same state.

Theorem[9]:Assuming that the MAS is undirected connected and its corresponding Laplacian matrix must be a real symmetric matrix, then agreement (2) can achieve the finite time average consistency of the system, that is:

$$x_{1\infty}=x_{2\infty}=\cdots=x_{n\infty}\approx\sum_{i=1}^{n}\frac{x_{i0}}{n} \tag{3}$$

Where, $x_{i\infty}$ is the state quantity of the final convergence of the i-th agent; x_{i0} is the initial state quantity.

3. Hierarchical control Based on Finite-time Consistency Algorithm

The traditional hierarchical control mostly adopts centralized control, its failure is likely to lead to the paralysis of the whole system, so the reliability is not high The distributed control algorithm based on finite-time consistency can achieve the consistency of global state only by communicating with its adjacent agents[7], so as to realize global control and better adapt to the needs of microgrid in terms of rapidity and economy. So, a hierarchical optimization strategy based on finite-time consensus algorithm is proposed in this paper. This strategy overall control block diagram is shown in Fig.1.

The primary control layer is based on droop control strategy to perform *P-f* and *Q-V* control to realize DG power control and voltage and frequency adjustment. The secondary control layer is to eliminate the deviation of frequency and voltage generated by the primary control layer, and maintain the frequency and voltage of the system at the rated value. The three-level control layer is the economic optimization layer, which solves the optimal value of each DG power generation during the economic optimal operation of the system, so that the total power generation cost is the lowest.

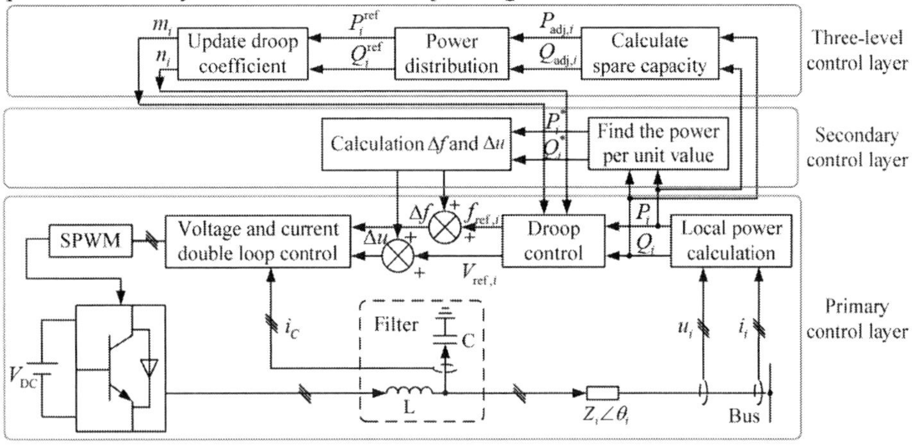

Fig.1 Overall block diagram of hierarchical control strategy for microgrid

3.1. Primary control layer

When the microgrid is in islanding mode, the droop control of P-f and Q-V is adopted for the parallel inverters, and the control equation is:

$$f_i = f_n - m_i(P_i - P_{in}), V_i = V_n - n_i(Q_i - Q_{in}) \tag{4}$$

Where, f_n, V_n, P_{in}, Q_{in} are the frequency, voltage, active and reactive power under rated conditions; m_i, n_i is the droop control coefficient; P_i, Q_i is the active and reactive power output;Generally, $f_n+m_iP_{in}$ the settings are the same. For a microgrid composed of multiple DGs in parallel, the frequency of each DG is equal. Then, the expression of output power of each DG can be obtained as follows:

$$P_i = \mathrm{K} / m_i \tag{5}$$

Where, K is a constant;It can be seen from Eq.(5) that the output power of each DG can be adjusted by adjusting the droop coefficient of the droop equation. In the primary control layer, the inverter droop control is adopted to realize the power control of each DG. However, since droop control is a kind of differential control, it will produce frequency and voltage deviations, so it is necessary to correct the frequency and voltage deviation in the secondary control layer.

3.2. Secondary control layer

In order to make the frequency and voltage of each DG reach the rated value of the system, the frequency and voltage compensation required for the introduction of the i-th DG are $\Delta f_i = m_i(P_i - P_{in})$ and $\Delta u_i = n_i(Q_i - Q_{in})$. In order to avoid the inconsistency between the frequency and voltage compensation of each DG during the operation of the microgrid, the average value of the compensation of each DG is taken as the frequency and voltage compensation of each DG, that is:

$$\Delta f = \frac{1}{n}\sum_{i=1}^{n}\Delta f_i = \frac{1}{n}\sum_{i=1}^{n}(m_i(P_i - P_{in})), \Delta u = \frac{1}{n}\sum_{i=1}^{n}\Delta u_i = \frac{1}{n}\sum_{i=1}^{n}(n_i(Q_i - Q_{in})) \tag{6}$$

Among them, Δf, Δu is the average value of the frequency compensation amount and the voltage compensation;Simplify Eq.(6), which can be rewritten as:

$$\Delta f = m_i P_{in}\left(\frac{1}{n}\sum_{i=1}^{n}\left(\frac{P_i}{P_{in}} - 1\right)\right) = m_1 P_{1n}\left(\frac{1}{n}\sum_{i=1}^{n}(P_i^* - 1)\right), \Delta u = n_i Q_{in}\left(\frac{1}{n}\sum_{i=1}^{n}\left(\frac{Q_i}{Q_{in}} - 1\right)\right) = n_1 Q_{1n}\left(\frac{1}{n}\sum_{i=1}^{n}(Q_i^* - 1)\right) \tag{7}$$

Where, P_i^*, Q_i^* is the unit value of the actual output active and reactive power of the i-th DG;Taking P_i^* and Q_i^* as state variables, substituting into Eq.(1-2), the consensus algorithm for P_i^* and Q_i^* is:

$$\begin{cases} \dot{P}_i^*(t) = \sum_{j=1}^{n} a_{ij}(P_j^*(t) - P_i^*(t)) + \sum_{j=1}^{n} a_{ij}\,\mathrm{sign}(P_j^*(t) - P_i^*(t))\left|P_j^*(t) - P_i^*(t)\right|^{\alpha} \\ \dot{Q}_i^*(t) = \sum_{j=1}^{n} a_{ij}(Q_j^*(t) - Q_i^*(t)) + \sum_{j=1}^{n} a_{ij}\,\mathrm{sign}(Q_j^*(t) - Q_i^*(t))\left|Q_j^*(t) - Q_i^*(t)\right|^{\alpha} \end{cases} \tag{8}$$

The final convergent value is substituted into Eq.(7) as the average value of the unit value of active and reactive power respectively to obtain the compensation amount of system frequency and voltage. After adding the compensation amount, the frequency and voltage are stabilized at the rated value.

3.3. Three-level control layer

The three-level control layer adjusts the power generation of each DG to make the total power generation cost of the system the lowest, and obtains the optimal value of the power generation of each DG at this time. The power generation cost of DG is composed of fuel cost and maintenance cost. Maintenance cost is a linear function of power generation, and fuel cost is a quadratic function of power generation. Assuming that the power generation P_i of the i-th DG, a_i, b_i, c_i are the power generation cost coefficient of the DG, the power generation cost function can be obtained as follows:

$$\mathrm{C}_i(P_i) = a_i P_i^2 + b_i P_i + c_i \tag{9}$$

Taking into account that the microgrid system should also meet a series of constraints such as the balance of supply and demand power and the limitation of DG power generation during the operation of the microgrid system, there are:

$$\sum_{i=1}^{n} P_i = P_{\mathrm{D}}, P_{i,\min} \le P_i \le P_{i,\max} \tag{10}$$

Where, $P_{i,\min}$ and $P_{i,\max}$ are the minimum and maximum power generation of the i-th DG respectively; P_{D} is the total active load of the system;For a microgrid system with n DGs, the overall goal of economic optimal operation is to minimize the total power generation cost of all DGs, that is $\min \mathrm{C}_{\mathrm{total}} = \sum_{i=1}^{n} \mathrm{C}_i(P_i)$.

In traditional power systems, the micro-increment rate is often used to solve the optimization problem of Eq.(12-14). By calculating the partial derivative of the active power from the power generation cost the micro increment rate. When the micro-increment rate of each DG is equal, the total power generation cost of the system is the lowest. At this time, the relationship between the system micro-increase rate and the optimal value of power generation is as follows:

$$\lambda_{\mathrm{typ}} = \lambda_i^{\mathrm{typ}} = 2a_i P_i^{\mathrm{typ}} + b_i = \frac{P_{\mathrm{D}} + \sum_{i=1}^{n} \frac{b_i}{2a_i}}{\sum_{i=1}^{n} \frac{1}{2a_i}} \tag{11}$$

Among them, $\lambda_{\mathrm{typ}}, \lambda_i^{\mathrm{typ}}, P_i^{\mathrm{typ}}$ are respectively the optimal value of the system micro-increase rate, the optimal value of the i-th DG micro-increase rate and the optimal value of the generation power when there is no power generation limit. Eq.(11) is the optimal value obtained by the centralized algorithm. Because the actual total active power load of the system is related to the operating point of the system, it is difficult to obtain it directly. This paper substitutes λ_i as a state variable into Eq.(1-2). Through local communication and calculation, the global optimal value of λ_i can be obtained.

$$\dot{\lambda}_i(t) = \sum_{j=1}^{n} a_{ij}(\lambda_j(t) - \lambda_i(t)) + \sum_{j=1}^{n} a_{ij}\,\mathrm{sign}(\lambda_j(t) - \lambda_i(t))\left|\lambda_j(t) - \lambda_i(t)\right|^{\alpha} \tag{12}$$

The final convergence value of Eq.(12) is taken as the optimal value λ_{typ} of the system's marginal increase rate under the condition of no generation power limitation, and it is substituted into Eq.(11) to obtain the optimal value of the i-th DG generation power:

$$P_i^{\mathrm{typ}} = \frac{\lambda_{\mathrm{typ}} - b_i}{2a_i} \tag{13}$$

If the limitation of DG generation power is considered, the DG exceeding or below the limited output power can only operate at the maximum or minimum power. It is assumed that Ω represents all DG sets where the optimal value of generation power does not reach the generation power limit under the condition of no generation power limit; λ_i^{opt} and P_i^{opt} are respectively the optimal value of the i-th DG micro increment rate and the optimal value of generation power under the condition of generation power limitation. According to the balance of total power supply and demand of the system:

$$\sum_{i\in\Omega} P_i^{\text{opt}} + \sum_{i\notin\Omega} P_i^{\text{opt}} = P_{\text{D}} \tag{14}$$

It can be seen from Eq.(13) that in order to make the system economically and optimally run, DG that does not reach the power limit should satisfy Eq.(15):

$$P_i^{\text{opt}} = \frac{\lambda_i^{\text{opt}} - b_i}{2a_i}, i \in \Omega; \lambda_i^{\text{opt}} = \lambda_j^{\text{opt}} = \lambda^{\text{opt}}, i, j \in \Omega, \tag{15}$$

Among them, λ^{opt} is the micro-increase rate when the system is economically optimal under the condition of limited power generation;Substituting Eq.(13) into Eq.(10), we know:

$$P_{\text{D}} = \sum_{i=1}^{n} P_i^{\text{typ}} = \sum_{i=1}^{n} \frac{\lambda_{\text{typ}} - b_i}{2a_i} = \sum_{i\in\Omega} \frac{\lambda_{\text{typ}} - b_i}{2a_i} + \sum_{i\notin\Omega} \frac{\lambda_{\text{typ}} - b_i}{2a_i} \tag{16}$$

As part of the power that exceeds or falls below the power limit is still provided by the DG that has not reached the power limit, there are:

$$P_{\text{D}} = \sum_{i\in\Omega} P_i^{\text{opt}} + \sum_{i\notin\Omega} P_i^{\text{opt}} = \sum_{i\in\Omega} \frac{\lambda_{\text{opt}} - b_i}{2a_i} + \sum_{i\notin\Omega} P_i^{\text{opt}} \tag{17}$$

Combining Eq.(16-17), it can be obtained that when the system economy is optimal under the condition of power limitation, the optimal value of the system micro-increase rate is shown in Eq.(18):

$$\text{if } i \in \Omega, \lambda_{\text{opt}} = \lambda_{\text{typ}} + \frac{\sum_{i\notin\Omega} \left(\frac{\lambda_{\text{typ}} - b_i}{2a_i} - P_i^{\text{opt}} \right)}{\sum_{i\in\Omega} \frac{1}{2a_i}}; \text{else } \lambda_{\text{opt}} = \left(2a_i P_{i,\min} + b_i\right) \text{ or } \left(2a_i P_{i,\max} + b_i\right) \tag{18}$$

From Eq.(10) and Eq.(15), the optimal power generation of each DG under the condition of power generation power limitation can be obtained as:

$$\text{if } i \in \Omega, P_i^{\text{opt}} = \frac{\lambda_{\text{opt}} - b_i}{2a_i}; \text{else } P_i^{\text{opt}} = P_{i,\min} \text{ or } P_{i,\max} \tag{19}$$

In the three-level control layer, the optimal value of each DG's power generation is obtained by Eq.(13) and Eq.(19), and it is output to the primary control layer.

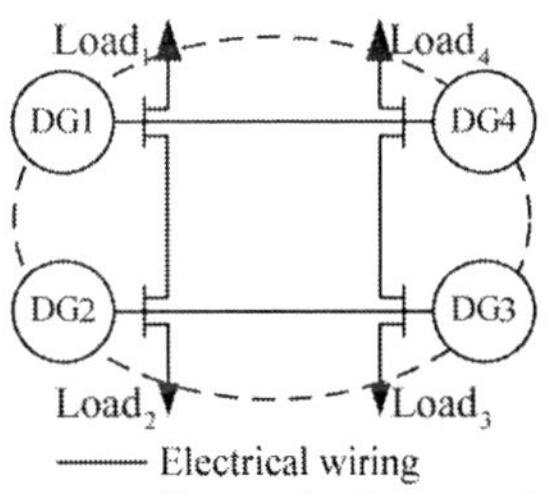

Fig.2 Simulation model

Tab.1 Simulation parameters and cost coefficients of DGs

	$P_{i,n}$ / kW	$Q_{i,n}$/ kVar	m_i ×10-5	n_i ×10-4	Load_i	a_i ×10-3	b_i ×10-2	c_i	$P_{i,min}$ /kW	$P_{i,max}$ / kW
DG1	80	50	1	1.2	60kW+j10kvar	1.6	2.03	0.85	50	120
DG2	70	40	1.14	1.5	70kW+j20kvar	1.2	5.66	0.52	50	100
DG3	60	30	1.33	2	80kW+j0kvar	1.74	4.2	1.295	40	80
DG4	50	20	1.6	3	90kW+j0kvar	1.36	6.7	0.7	30	60
Others parameters:L=1mH C=500μF $V_n = 311\text{V}$ f_n=50Hz k_{pc}=10 $k_{pv} = 5$										

4. Example and Simulation

4.1. Simulation model

In this section, the four-node microgrid system shown in Fig.2 will be simulated on the Matlab/Simulink platform to verify the layered control strategy proposed above.The microgrid system is composed of 4 DGs. The simulation parameters, cost coefficients and loads of each DG are shown in Tab.1. Next section analyzes and verifies the control strategy proposed in this paper through an example, and compares it with droop control. Assume that Load_1~Load_4 are put into the system.

4.2. Uses droop control

In droop control, the power distribution of each DG is distributed according to the ratio of the droop coefficient,which may result in the high power generation cost of the DG power generation, but the low power generation cost the low power generation of DG is not conducive to the economic operation of the power system. Fig.3 shows the simulation results using droop control. From the figure, we can that the micro-increase rate of each DG is inconsistent, which does not meet the "equivalent micro-increment rate criterion" in the economic operation of the power system. Meanwhile,we can see that the system frequency and node voltage are 49.87Hz and 310.3V respectively, which also shows that droop control is a differential control.

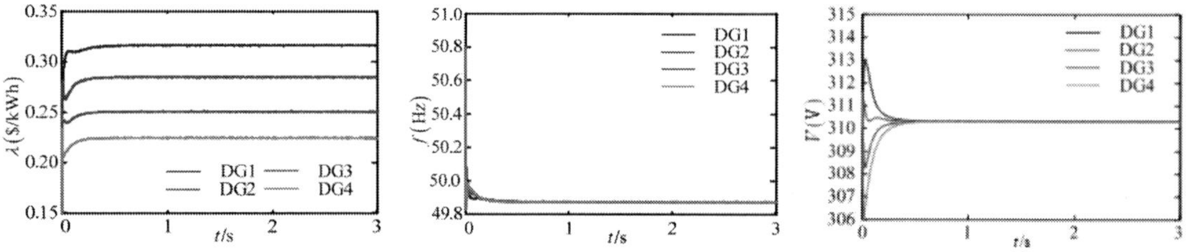

(a) Micro-increase rate of DG (b) Frequency change of DG (c) Voltage change of DG

Fig.3 Simulation results using droop control

4.3. Uses hierarchical control strategy for microgrid based on finite-time consensus algorithm

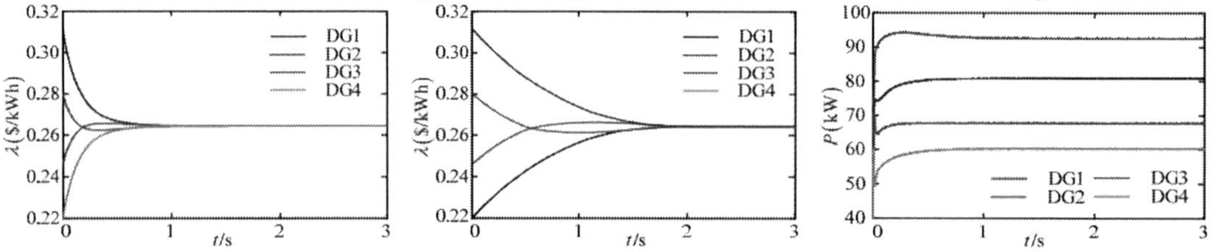

(a) Convergence of Eq.(2) (b) Convergence of Ref.[8] (c) Output active power of DG

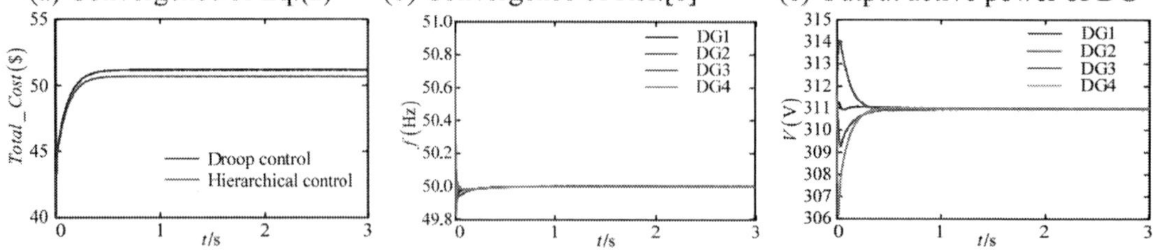

(d) Total cost comparison (e) Frequency change of DG (f) Voltage change of DG

Fig.4 Simulation results using hierarchical control

When the generation power limitation conditions are not considered, the economic optimal operation of the system is when the micro-increase rate of each DG are equal, as shown in Fig.4(a), the optimal value of the system increment rate is λ_{typ}=0.246$/kWh.Fig.4(b) is the simulation result obtained by using the consensus algorithm proposed in the Ref.[8]. Compare Fig.4(a) with the Fig.4(b) It can be seen that the convergence time of the finite-time consensus algorithm proposed in this paper is nearly 60% faster. From Eq.(13), the optimal value of each DG4 is 63.81kW already exceeded its maximum power generation 60kW, so it is necessary to consider cost optimization under the conditions of power generation limitations. When considering the generation power limit, Fig.4(c) is the curve of the actual power generation of each DG. It can be seen that each DG can operate near the optimal value of the generation power and power generation is within the power limit, at this time the system economy is optimal. Fig.4(d) shows the comparison between the hierarchical control strategy proposed and the total power generation cost of the droop control system. The figure shows that the optimization of the total power generation cost of the system is effective. Fig.4(e) and Fig.4(f) show the frequency and voltage changes of each DG. After adopting the layered optimization proposed, the frequency and voltage are stable at the rated value of the system.

5. Conclusion

This paper proposes a hierarchical control strategy for microgrid based on finite-time consensus algorithm, which realizes the economically optimal operation of the micro-grid system and ensures that the frequency and voltage of the system are stable at the rated value. This optimization algorithm only needs to communicate and exchange information between each distributed power source and its adjacent power source, which not only saves the investment of centralized control on communication equipment and lines, but also avoids that the failure of centralized control equipment will cause the entire system to be paralyzed. The control of the microgrid is more flexible and reliable. Finally, the simulation verification is carried out on the Matlab/Simulink platform, and the simulation results show the effectiveness of the optimization strategy proposed in this paper.

Acknowledgments

This work was financially supported by the 2018 Guangxi University High- level Innovation Teamand Excellence Scholars Program(Guijaioren[2018]No.35).

References

[1] AI Qian, LIU Siyuan, WU Renbo, FAN Songli,et al. Research and Prospect of Multi-agent System in Energy Internet System [J]. High Voltage Engineering, 2016, 42(9):2697-2706.

[2] Wang Chengshan,Wu Zhen,Li Peng. Research on Key Technologies of Microgrid[J]. Transactions of China Electrotechnical Society,2014,29(02):1-12.

[3] ZU Qiwu, NIU Yugang, CHEN Bei. Study on multi-objective economic operating strategy of microgrid based on improved particle swarm optimization algorithm[J]. Power System Protection and Control,2017,45(14):57-63.

[4] GU Wei,XUE Shuai,WANG Yong,et al. Finite-time Consensus Based Distributed Cooperative Control for DC Microgrids[J]. Automation of Electric Power Systems, 2016, 40(24):49-55.

[5] XIE Jun, CHEN Kaixuan, YUE Dong,et al. Distributed economic dispatch based on consensus algorithm of multi agent system for power system[J]. Electric Power Automation Equipment,2016,36(02):112-117.

[6] YU Zhiwen,AI Qian,XIONG Wen. Hierarchical Distributed Real-time Optimization Strategy for Microgrid Based on Consistency Protocol of Multi-agent System[J]. Automation of Electric Power Systems, 2017, 41(18): 25-31.

[7] Ji Lianghao. On Consensus Problem of Multi-agent Networks[D]. Chongqing:College of Computer Science,Chongqing University,2014.

[8] Wang L, Xiao F. Finite-time consensus problems for networks of dynamic agents[J]. arXiv preprint math/0701724, 2007.

[9] Xin H, Liu Y, Qu Z, et al. Distributed control and generation estimation method for integrating high-density photovoltaic systems[J]. IEEE Transactions on Energy Conversion, 2014, 29(4): 988-996.

Journal of Physics: Conference Series **2108** (2021) 012042 doi:10.1088/1742-6596/2108/1/012042

A Novel Control Method of Low-voltage-ride-through Technique for Asymmetrical Hybrid Energy Storage System

Xiaomeng LI[1*], Ze Gao[1], Zhen'an ZHANG[1] and Yufei RAO[1]

[1] The Electric Power Research Institute of the State Grid Henan Electric Power Company, Zhengzhou City, Henan, China

[*] Corresponding author's e-mail: lixiaomeng@ha.sgcc.com.cn

Abstract. Low-voltage-ride-through (LVRT) capability is an important criterion for the stability of cascaded multilevel energy storage system (ESS). Based on asymmetrical hybrid ESS, a coordinated operating method is proposed to maintain the phase angle of positive-sequence-voltage of battery cells and capacitor cells. To deal with the asymmetrical chains among three phases, its influence on the power distribution is eliminated by modifying the command of battery cells, then the traditional zero-sequence-voltage injection is feasible for asymmetrical situations. Proposed methods can stabilize the dc-side voltage of capacitor cells, maintaining the output current. Finally, the correctness and effectiveness of proposed methods are proved by experiment.

1. Introduction

Cascaded multilevel converters are widely used in energy storage systems due to their characteristics of modularization, low harmonic, and low switching stress [1-2]. Cascade multilevel energy storage system can be matched with different energy storage devices on the DC side, which can be mainly divided into battery energy storage system (BESS) with battery as the DC side, static synchronous compensator (STATCOM) with capacitor as the DC side and hybrid energy storage system [3-4]. The hybrid energy storage system based on batteries and capacitors has been applied to off-grid energy storage and motor drive [5-6]. This structure can maximize the use of each link, improve equipment efficiency and save costs.

Low voltage crossing is a key technical problem in cascined multilevel converters. The control methods of low voltage crossing for cascined multilevel converters connected with star can be divided into zero sequence voltage injection and negative sequence current injection. Literature [7] proposed a control method based on zero sequence current and voltage injection to realize the phase-to-phase average active power of h-bridge converter is zero, but the influence of unbalanced power network is not considered, so the balanced control of three-phase DC side voltage cannot be realized. Literature [8] proposes that the average active power control, DC voltage control and reference current tracking control can be realized under unbalanced power network by injecting appropriate phase voltage. However, this control method is based on ABC coordinate system, so the control system is more complex and requires a large amount of calculation. Literature [9] established a zero-sequence voltage compensation model based on coordinate system, and constructed a virtual vertical voltage vector participating in coordinate transformation to simplify the calculation process. Literature [10] further converted to DQ coordinate system for modeling, making the calculation process more simple. For hybrid cascaded multilevel circuits, especially when the three-phase system is asymmetric, its control targets are many, coupling terms are complex, and its low voltage crossing control is difficult.

In this paper, a low voltage crossing control method is proposed for the asymmetric hybrid energy storage system consisting of a three-phase asymmetric cascade multilevel hybrid energy storage system. By separating the zero-sequence components of the rated order of the battery, the three-phase asymmetric system can use zero-sequence voltage injection to cross the unbalanced power grid. The reliability of the method is verified by experiments to ensure the stable operation of the system under complex faults.

2. A Symmetric Cascaded Multilevel Hybrid Energy Storage System

The asymmetric cascade multilevel hybrid energy storage system studied in this paper is shown in Figure 1. Each phase is composed of battery chain and capacitor chain. The total number of each phase is the same, but the number of each phase battery and capacitor chain is different. L is the filter inductance, VS and IS are the grid voltage and grid-connected current respectively, m1,m2,m3 and n1,n2,n3 are the number of battery chains and capacitor chains respectively, and Vbat and Vcap are their DC side voltages respectively.

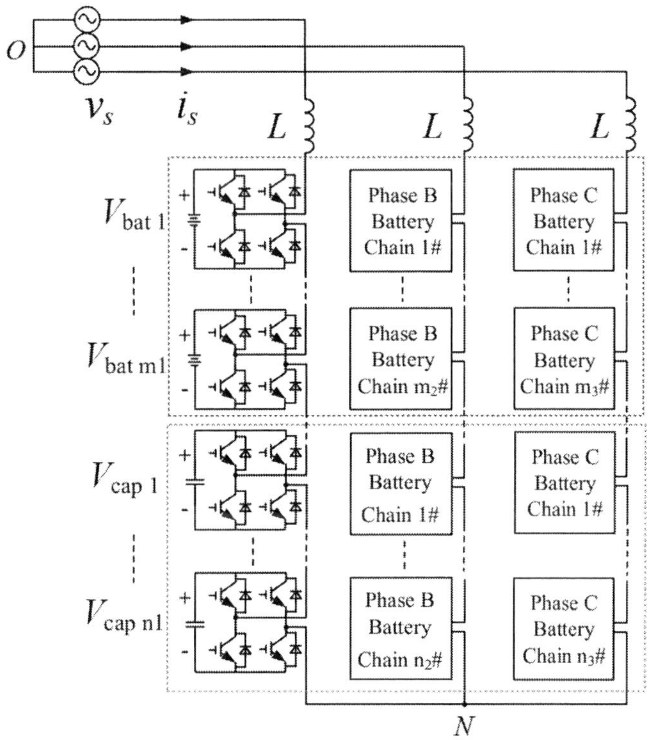

Figure 1. Hybrid energy storage system structure

The same type of link in each phase converter can be modeled by merging the DC side. The three-phase converter with cascaded topology as independent is controlled by split phase, which further simplifies the analysis of the three-phase hybrid energy storage system.

When the system is running normally, the vector relationship of each phase is shown in Figure 2. Vr1 and Vr2 are respectively the output voltage of the battery and capacitor chain, Vs is the grid voltage, and V_L is the inductance voltage. I_S is the grid-connected current, which has a close vertical relationship with V_{r2} to ensure the active power balance at the DC side of the capacitor chain. V_L and V_{r2} are combined into vector V_q. δ and δ_1 are the angles between V_{r1}/V_{r2} and VS.

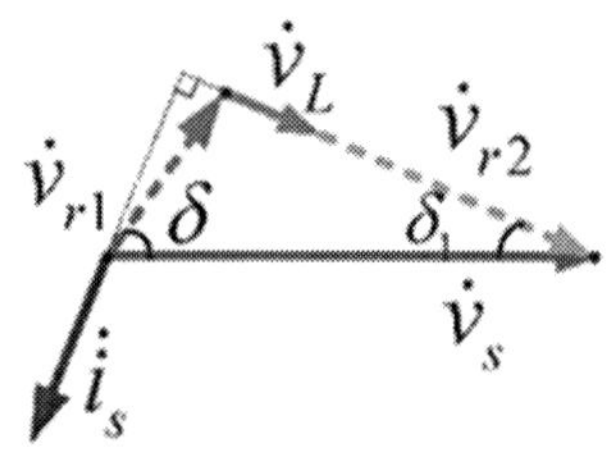

Figure 2. Hybrid energy storage system structure

3. Low Voltage Crossing Control Method for Asymmetric Hybrid System

To realize the low voltage crossing of three-phase asymmetric hybrid energy storage system, the control method for symmetric hybrid energy storage needs to be extended. In order to maintain the pre-fault output current of the symmetric hybrid energy storage system under the unbalanced grid, the battery chain output voltage Vr1 needs to track the grid voltage Vs. In the case of asymmetric equipment, the three-phase Vr1 tracking V_S is shown in Equation (1). K is the rated modulation ratio of the battery chain; Vp and Vn are the positive sequence and negative sequence components of Vs, respectively.

$$\begin{cases} v_{r1A} = m_A v_{bat} \cdot K'_A \left(v_{pA} + v_{nA}\right) e^{-j\delta_A} \\ v_{r1B} = m_B v_{bat} \cdot K'_B \left(v_{pB} + v_{nB}\right) e^{-j\delta_B} \\ v_{r1C} = m_C v_{bat} \cdot K'_C \left(v_{pC} + v_{nC}\right) e^{-j\delta_C} \\ K'_x = \dfrac{V_{sx}}{V_{rated}} \cdot K_x \quad x = A,B,C \end{cases} \tag{1}$$

The rated modulation ratio K of V_{r1} under the balanced grid is independently adjusted by the three phases according to the number of each battery chain. At this point, the sum of the three-phase V_{r1} modulated waves is no longer 0, and the amplitude and phase of the zero-sequence component also change with the change of the unbalanced degree of the power grid. Since Vs does not contain a zero sequence in the star circuit, the zero sequence component of V_{r1} will act on the capacitor chain and make V_{r2} contain an undesired zero sequence component, which will be interfered with by the zero sequence of V_{r1} regardless of the latter's low voltage crossing mode.

The positive sequence component of three-phase V_{r1} is shown in Equation (2). V_{r1pA} is obtained by the rotation and vector superposition of three-phase V_{r1}, and the relationship between V_{r1pA} and v_{pA} and v_{nA} is further decomposed. The positive sequence is affected by both the positive sequence and negative sequence of grid voltage.

$$\begin{cases} v_{r1pA} = (v_{r1A} + v_{r1B} \cdot e^{j\frac{2}{3}\pi} + v_{r1C} \cdot e^{-j\frac{2}{3}\pi})/3 \\ \quad = (k_1 v_{pA} \cdot e^{-j\delta_A} + k_2 v_{pA} \cdot e^{-j\delta_B} + k_3 v_{pA} \cdot e^{-j\delta_C})/3 \\ \quad + (k_1 v_{nA} \cdot e^{-j\delta_A} + k_2 v_{nA} \cdot e^{-j120} \cdot e^{-j\delta_B} + k_3 v_{nA} \cdot e^{j120} \cdot e^{-j\delta_C})/3 \\ v_{r1pB} = v_{r1pA} \cdot e^{-j\frac{2}{3}\pi} \\ v_{r1pC} = v_{r1pA} \cdot e^{j\frac{2}{3}\pi} \end{cases} \tag{2}$$

So that the vector of arbitrary tracking three-phase V_S in a balanced power grid satisfies the relation shown in Equation (3) for V_{SA}, it has the relation shown in Equation (4) for V_{nA} in an unbalanced power grid.

$$\begin{cases} k_1 v_{sA} \cdot e^{-j\delta_A} + k_2 v_{sB} \cdot e^{-j\delta_B} + k_3 v_{sB} \cdot e^{-j\delta_C} = 0 \\ k_1 v_{sA} \cdot e^{-j\delta_A} + k_2 v_{sA} \cdot e^{-j\frac{2}{3}\pi} \cdot e^{-j\delta_B} + k_3 v_{sA} \cdot e^{j\frac{2}{3}\pi} \cdot e^{-j\delta_C} = 0 \end{cases} \tag{3}$$

$$k_1 v_{nA} \cdot e^{-j\delta_A} + k_2 v_{nA} \cdot e^{-j\frac{2}{3}\pi} \cdot e^{-j\delta_B} + k_3 v_{nA} \cdot e^{j\frac{2}{3}\pi} \cdot e^{-j\delta_C} = 0 \tag{4}$$

Equation (4) is the second term of V_{r1pA} in Equation (2). Therefore, V_{r1P} is only determined by the positive sequence component V_P of the grid voltage, regardless of whether V_{r1} contains zero sequence or not, as long as the component of the tracking grid voltage does not contain zero sequence. If the phase of V_{r1p} is unchanged, then the corresponding V_{r2p} in Equation (5) also remains unchanged. In this case, the degree of imbalance of V_{r2} is the same as that of V_S.

$$v_{px} = v_{r1px} + v_{r2px}, \quad x = A, B, C \tag{5}$$

To sum up, when the battery chain tracks the unbalanced power network in the case of equipment asymmetry, the zero-sequence component in the rated state of V_{r1} can be separated to avoid the influence of the amplitude and phase of V_{r1p} on the positive sequence Vp and negative sequence Vn of the power network simultaneously, as shown in Equation (6). At this point, the positive sequence phase of the battery chain and the capacitor chain is a constant value no matter how the imbalance of the power grid changes.

$$\begin{cases} v_{r1z\,rated} = \frac{1}{3}(v_{r1A\,rated} + v_{r1B\,rated} + v_{r1C\,rated}) \\ v_{r1trc\,x} = v_{r1x\,rated} - v_{r1z\,rated} \\ \qquad = m_x v_{bat} \cdot K_x' \left(v_{px} + v_{nx}\right) e^{-j\delta_x} \\ v_{r1x} = v_{r1trc\,x} + v_{r1z\,rated} \cdot \dfrac{V_p}{V_{rated}} \\ x = A, B, C \end{cases} \tag{6}$$

When the battery chain is operated by this method, the working state of the capacitor chain is shown in Figure 3. N is the center of gravity of the voltage triangle, and the three-phase vector starting from this position does not have a zero sequence. F is fermat point, and the Angle between the three vectors starting from this position is 120° of each other. Vz is the zero-sequence voltage used by the symmetric system to cross the unbalanced power grid. As shown in Formula (7), V_{n2} is the effective value of the negative sequence component of V_{r2}, and ϕ_p and ϕ_n are the phases of the positive sequence and negative sequence components of the power grid respectively. This voltage transfers the star connection neutral point potential of the capacitance chain from N to F. In the unbalanced state, the capacitor chain node should also subtract the built-in zero-sequence $-V_{r1z}$ to transfer the neutral point potential from P point to N, and then combine with V_Z to make the output voltage of the three-phase capacitor chain node be perpendicular to Is is before each phase fault, so as to ensure the constant grid-connected current and active power balance of the capacitor chain node.

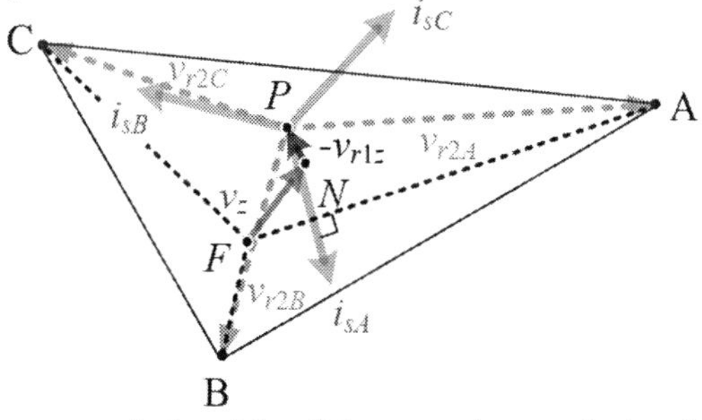

Figure 3. The voltage vector relationship of the capacitance chain of an asymmetric system

$$\begin{cases} V_z = V_{n2} \\ \varphi_z = \varphi_p - \varphi_n + \delta_1 = \Delta\varphi + \delta_1 \end{cases} \tag{7}$$

The battery chain is controlled by open-loop voltage. Vr1trc tracking imbalance vs is obtained after separating the rated zero-sequence component, and its output voltage is adjusted. The separated Vr1 zero-sequence component is modified according to the change of positive sequence amplitude during grid sag to obtain Vr1Z Add, so as to ensure that the system maintains a working state similar to the rated grid during symmetrical grid sag. The modulated battery chain wave is obtained by combining with Vr1TRC. Due to the different control targets on the DC side of the asymmetric three-phase capacitor, the capacitor chain adopts the split current control. According to the imbalance degree of the power network, the corresponding VZ balance power of the three phases is calculated to maintain the dc side voltage and output current unchanged. This builds up the overall control method for the three-phase asymmetric hybrid energy storage system.

4. Experimental Verification

The effectiveness of the proposed low voltage crossing method is verified by rT-LAB based hardware on the loop experimental platform. The experimental parameters are shown in Table 1. The system is simplified in the experiment.

The low voltage crossing process of asymmetric hybrid system is shown in Figure 4. In case of unbalanced power network, the positive sequence component drops from 20.2kV to 15kV at time T1, accompanied by additional negative sequence component. The grid-connected current is restored to the state before the power grid failure after 0.1s. Each phase voltage controller independently regulates the DC side voltage of the capacitor chain, and the voltage is stabilized at the reference value of 19.6kV, 21.1kV and 23.5kV during the fault.

Table 1. Experimental parameters of three-phase asymmetric hybrid energy storage system

	Phase A	Phase B	Phase C
Number of battery links m	20	18	15
DC side voltage of the battery $m \cdot V_{bat}$ (kV)	15.7	14.1	11.7
Number of battery links n	25	27	30
Rated DC side voltage $n \cdot V_{cap}$ (kV)	19.6	21.1	23.5
DC side link voltage V_{bat}, V_{cap} (V)		783	
Filter inductance L (mH)		60	
DC side capacitor C (mF)		5	
Number of links per phase $m+n$		45	
Rated RMS value of phase voltage Vs (kV)		20.2	
Magnitude of Current I_{sm} (A)		100	

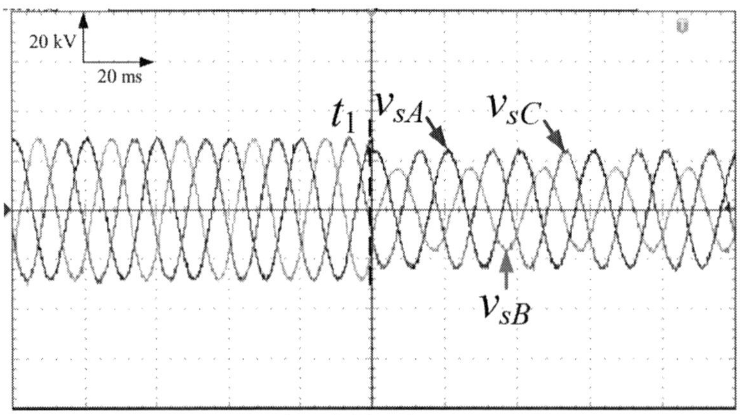

(a) Grid voltage

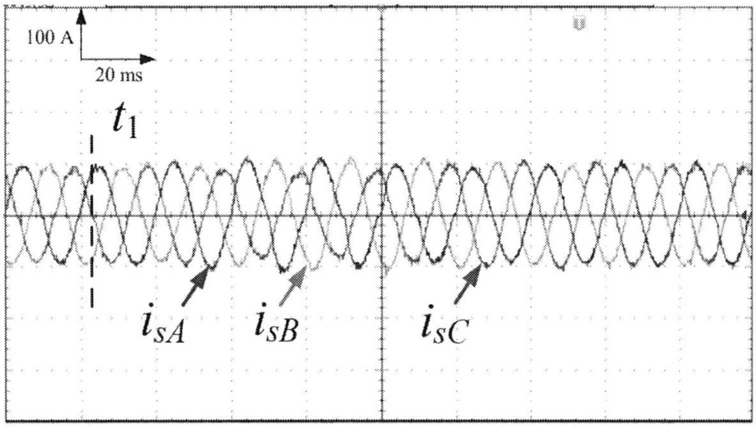

(b) Grid current

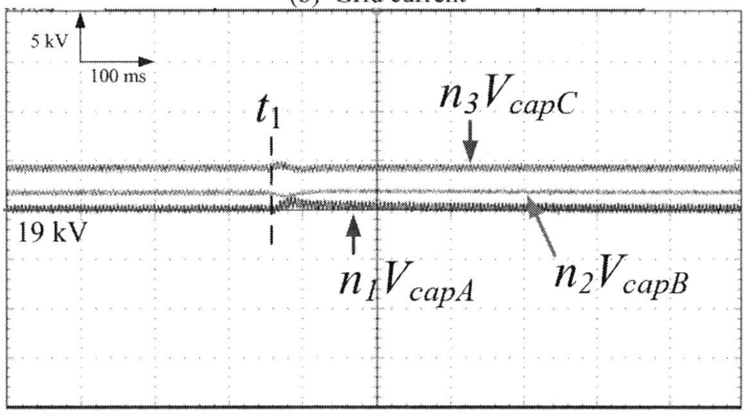

(c) Total DC side voltage of capacitor link

Figure 4. Asymmetric hybrid energy storage system low voltage crossing waveform

Instead of adopting the low voltage crossing mode of the asymmetric system presented in this paper, the current waveform of direct injection of zero sequence voltage is shown in Figure 5. At this point, the positive sequence of each physical quantity in the system is affected by Vs negative sequence, the closed-loop control and the calculation of zero sequence voltage are affected, and zero sequence voltage injection fails. The voltage controllers of each phase alone generate inaccurate active current instructions, so that the sum of the three phase current instructions is not zero, which causes the current controller instruction conflict in the star circuit. Even if the regulated current can achieve active power balance within a period and keep the voltage stable, the current waveform is no longer sine, and the harmonics contained in it will seriously damage the power quality. The experimental results verify the effectiveness of the proposed low voltage crossing method.

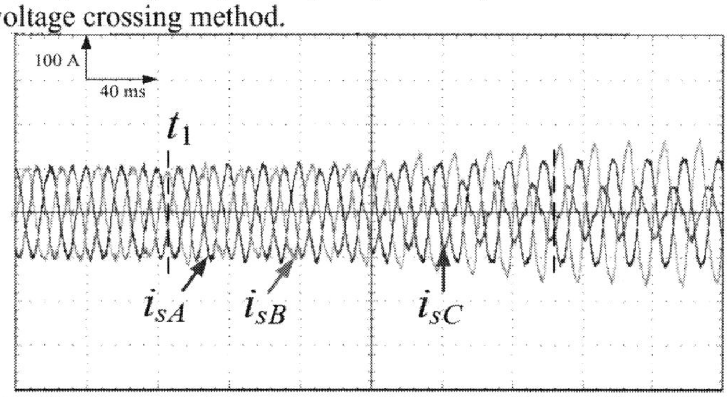

Figure 5. The asymmetric system low voltage crossing method is not used

5. Conclusions

This paper studies the low voltage crossing control of three-phase asymmetric hybrid energy storage system under unbalanced power grid. By making the battery link track the grid voltage, the battery and capacitor link can work together in case of grid failure and maintain the positive sequence phase invariance of each link. By separating the zero-sequence components in the battery control instructions, the system can operate stably in the case of three-phase asymmetry and cross the unbalanced power grid with zero-sequence voltage injection. The proposed method can maintain the dc side voltage and grid-connected current of the capacitor chain at the time of voltage sags and cross the unbalanced condition. Experimental results verify the correctness and effectiveness of the proposed method.

References

[1] Maharjan L, Inoue S, Akagi H, et al. A transformerless battery energy storage system based on a multilevel cascade PWM converter[C]. Power Electronics Specialists Conference, 2008.

[2] Novakovic B, Nasiri A. Modular Multilevel Converter for Wind Energy Storage Applications[J]. IEEE Transactions on Industrial Electronics, 2017, 64(11):8867-8876.

[3] Yifan Yu, Student Member. Delta-Connected Cascaded H-Bridge Multilevel Converters for Large-Scale Photovoltaic Grid Integration[J]. IEEE Transactions on Power Electronics, 2016, PP (99):1-1.

[4] Shibano Y, Ota J I Y, Akagi H. A Phase-Shifted PWM STATCOM Using the Modular Multilevel Cascade Converter (MMCC-SSBC)[J]. 2013, 133(9):936-942.

[5] Vahedi H, Sharifzadeh M, Al-Haddad K, et al. Single-DC-source 7-level CHB inverter with multicarrier level-shifted PWM[C]. Conference of the IEEE Industrial Electronics Society. IEEE, 2015.

[6] Chiasson J N, Ozpineci B, Tolbert L M. A Five-Level Three-Phase Hybrid Cascade Multilevel Inverter Using a Single DC Source for a PM Synchronous Motor Drive[C]. Applied Power Electronics Conference, APEC 2007. IEEE, 2007.

[7] Betz R E, Summers T, Furney T. Symmetry Compensation using a H-Bridge Multilevel STATCOM with Zero Sequence Injection[C]. Industry Applications Conference, 2006. 41st IAS Annual Meeting. Conference Record of the 2006 IEEE. IEEE, 2006.

[8] Qiang Song, Wenhua Liu. Control of a Cascade STATCOM With Star Configuration Under Unbalanced Conditions[J]. IEEE Transactions on Power Electronics, 24(1):45-48.

[9] Sano K, Takasaki M. A Transformerless D-STATCOM Based on a Multivoltage Cascade Converter Requiring No DC Sources[J]. IEEE Transactions on Power Electronics, 2012, 27(6):2783-2795.

[10] Chen H C, Wu P H, Lee C T, et al. Zero-Sequence Voltage Injection for DC Capacitor Voltage Balancing Control of the Star-Connected Cascaded H-Bridge PWM Converter Under Unbalanced Grid[C]. Energy Conversion Congress & Exposition. IEEE, 2014.

Journal of Physics: Conference Series **2108** (2021) 012043 doi:10.1088/1742-6596/2108/1/012043

Study on control strategy of output stability of wind-solar reservoir thermal system

Shengshan Li

Qinghai University for Nationalities, Xining 810007, Qinghai, China

Abstract—In view of the landscape reservoir heat output coordinated control demand, based on the topology of the hybrid energy storage system of the three ports heating model, using the sunlight with the electricity output of the complementary and heat accumulation can be regulatory, intends to research a kind of based on photo-thermal storage heat and scenery electricity heating output port control method, in order to achieve the goal of fast and smooth regulating heat output fluctuations, The coordinated output control of integrated wind-landscape storage and heat collection is realized.

1.INTRODUCTION

At present, scholars at home and abroad have put forward an idea: Various forms of three-port DC-DC converter topologies can achieve the optimal drive of the converter through the reasonable distribution of the energy of the hybrid energy storage, and also achieve the goal of coordinated control of the wind-solar DC microgrid. Based on this idea, this article starts with the operating principles and characteristics of the solar collector system. Finally, through simulation examples and energy flow analysis, based on the complete acceptance of the thermal energy output by the solar thermal collector, under the premise that the system's multi-energy combined output minimizes thermal energy fluctuations, and combined with wind power output, a daily output strategy for the combined operation of wind and solar heat storage is worked out.

2.Output principle of wind and solar heat storage and collection system

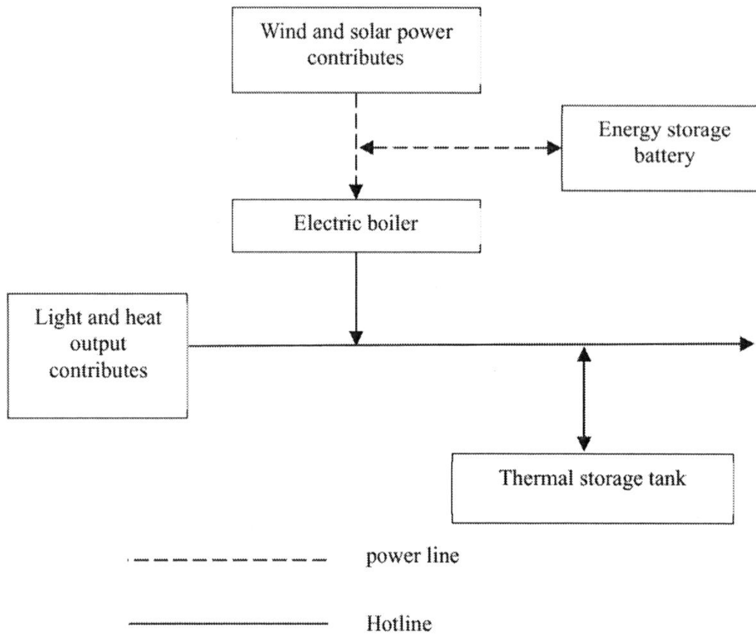

Figure 1 Output principle of wind-solar storage thermal

Solar heat collection is affected by solar radiation, resulting in discontinuous heat supply. Therefore, a large amount of supplementary heat is required from a spare heat source. The wind energy is huge, renewable, and pollution-free. The output of wind power also fluctuates due to changes in wind speed. Therefore, the application of these two clean energy technologies alone is not effective. This article proposes a solar and wind energy complementary heating system. The principle is shown in Figure 1. The solar thermal energy output by the solar collector is used as the basic output heat, and the wind-solar complementary power generation electric energy is converted into thermal energy supplement through the electric furnace, and the proportion of wind-solar complementary energy in the heating system is increased. Another part of the electricity generated by wind-solar hybrid power generation is allocated to energy storage batteries to improve wind-solar absorbing capacity. When the electric energy of the energy storage battery is insufficient, the electric energy is output to the electric heating furnace.

3. ANALYSIS AND STRATEGY OF OUTPUT STABILITY OF WIND AND SOLAR THERMAL STORAGE SYSTEM

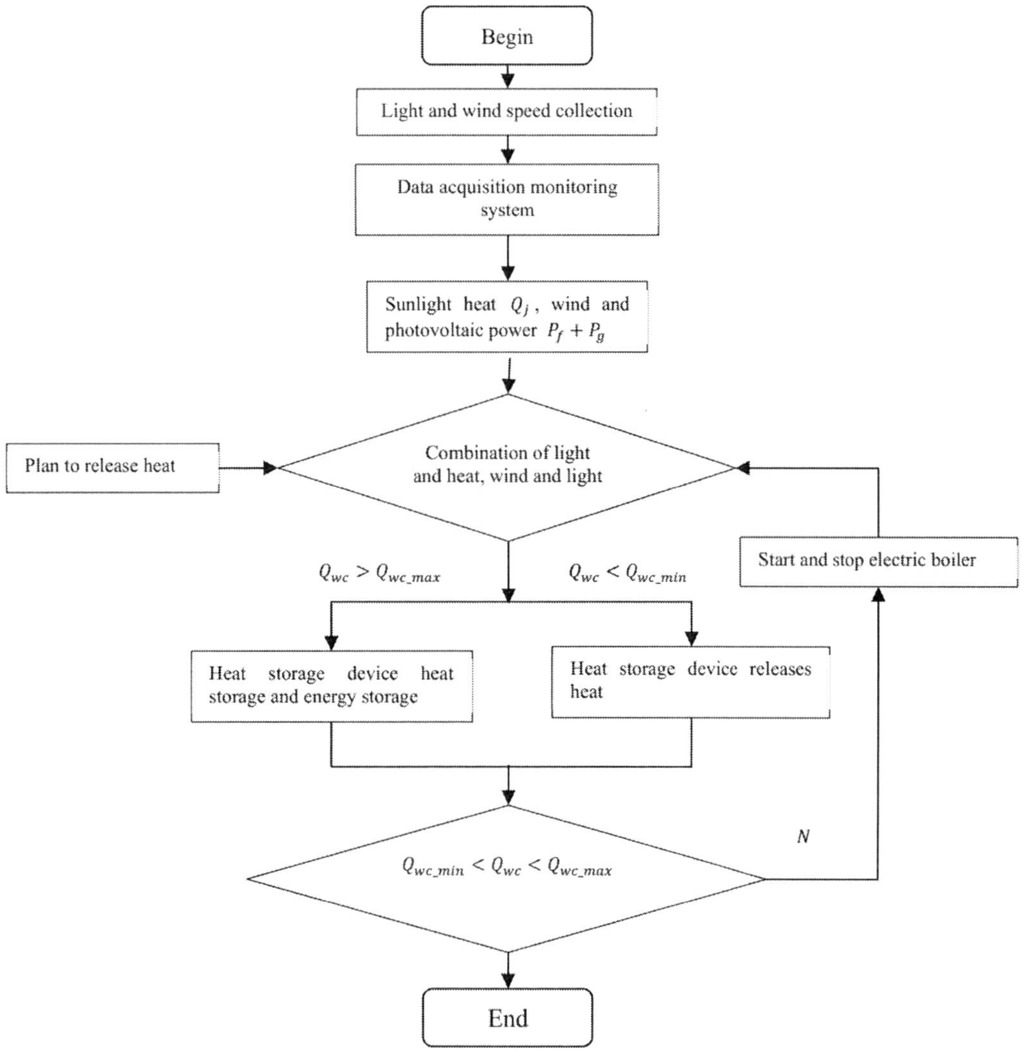

Figure 2 output joint scheduling strategy of wind-solar storage heat system

In order to alleviate the environmental fluctuation of the solar heat collection system, the heat collected by the solar heat collector is first collected into the hot water storage tank, and then the heat storage tank supplies heat to the user; According to the volatility of solar and wind energy, the output electric energy is transmitted and distributed to energy storage batteries and electric boilers; Use electric boilers to generate heat to supply heat to users. The energy storage battery can provide secondary power to the electric boiler according to the heating situation. According to the thermal load prediction curve determined by the thermal load variation range in the control period, and based on factors such as wind power and solar thermal forecast output and heat storage capacity, optimize the operating position of the solar thermal-wind power-storage combined system within the thermal load curve within the dispatch period. In order to ensure the sufficiency of the heat output of light and heat, when there is light in the daytime, the light and heat should be in a heat storage or heat release mode while outputting heat according to the light conditions. During this period, wind and wind power generation can be used for adjustment or fine adjustment. When the light is weak or the heat load is the most at night, the heat storage device is used to adjust the wind power according to the heat load and wind power output.

The joint scheduling strategy is shown in Figure 2

1) Predict the light irradiance and wind speed data in the next day. According to the limiting factors such as wind speed, sunlight, ambient temperature and heat load in a day, estimate the upper and lower ranges of the planned output and heat output of the solar thermal, wind and solar combined storage system.

2) Smoothly fill the solar thermal energy storage through the output of wind and solar complementary power generation, collect the relevant light, wind speed, and temperature data of the thermal system to obtain the real-time output of the solar thermal-wind photovoltaic energy storage system. Then compare the combined system's output with the user's calories. If the output heat energy of the combined system is greater than the upper limit, that is, when wind and light are sufficient, the heat storage device stores heat and energy, and the electric energy storage device stores energy. On the contrary, the heat storage device releases heat, and the insufficient part is output by the electric energy storage device through the electric furnace [1].

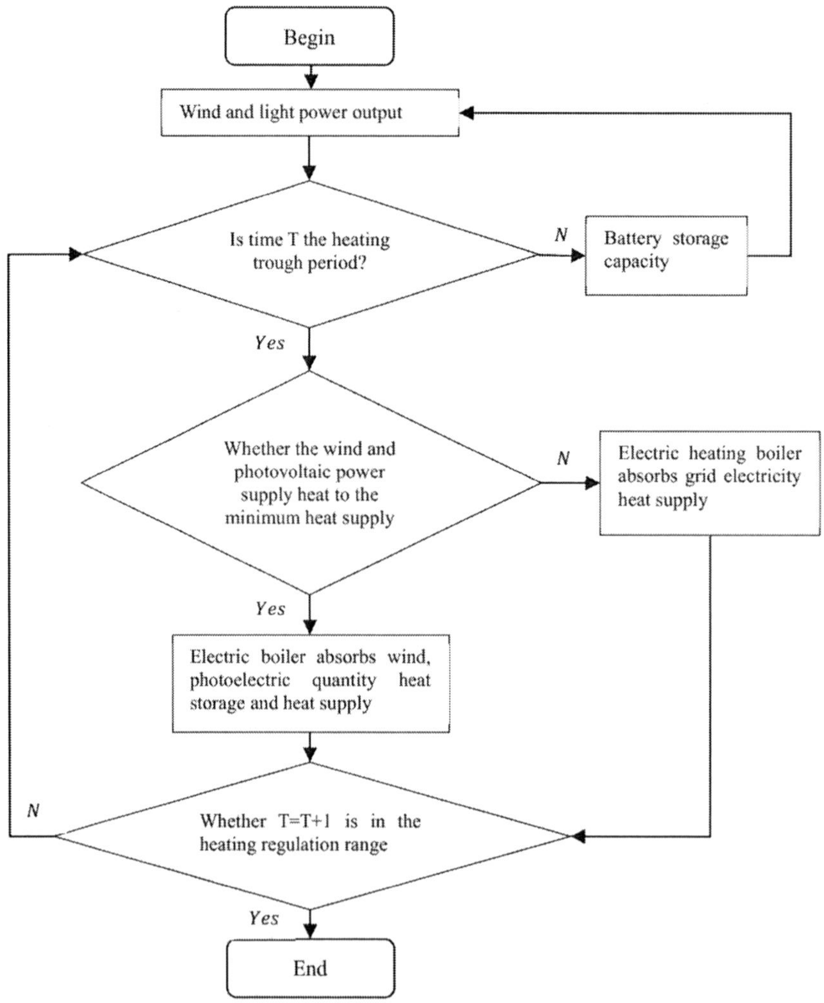

Figure 3 Start and stop control strategy of heat storage electric boiler

3) The start and stop of the thermal storage electric boiler is shown in Figure 3. The start and stop of the heat storage electric boiler is mainly used during the low heat supply period. In special weather conditions, wind-solar and electric energy storage or grid supplementary output plans can be arranged according to the power output needs of the solar thermal-wind power combined system. This article takes 24 hours as a scheduling period, and the operation control strategy is to judge the trough period. If during the trough period, the electric boiler will mainly supply heat by wind-solar complementary power generation. When wind-solar hybrid power generation does not meet the heating demand of the thermal load, conventional power is used; During non-trough periods, the electric energy storage device stores energy.

4.Heat dispatch model of wind and solar heat storage system

When the output of the solar thermal collection system is basically stable, the capacity adjustment is limited and the cost is high, wind photovoltaic energy storage is used to supplement it. This requires consideration of the wind and photovoltaic power storage depth peak shaving when the power of the solar thermal collection system is fully adopted, and the start-up and shutdown of the power thermal boiler. The adjustability of light and heat as the basic heat energy, and wind and photoelectricity as a complementary output of thermal energy storage should be fully utilized. Through the adjustment and scheduling of solar thermal and wind photovoltaic, the output of the combined system has the smallest fluctuation in the range. Based on the above analysis, a model is established.

4.1Solar collector model

Using solar collectors, flat-plate collectors and evacuated tube collectors are usually selected. Choose a flat plate collector here.

Useful heat of flat plate collector

$$Q_u = A_a I\eta_{Lc} = Gc_f\Delta T = Gc_f(T_{t,o} - T_{t,i}) \tag{1}$$

Among them, Q_u is the useful energy output of the collector per unit time. η_{Lc} is the heat collection efficiency of the flat plate collector. Take 0.855; G is the flow rate of the heat collecting medium flowing through the heat collector; c_f is the specific volume of the heat transfer working fluid; $T_{t,o}$ The outlet temperature of the heat collecting working medium in the heat collector; $T_{t,i}$ The inlet temperature of the collector working medium in the collector; A_a is the lighting area of the collector; I is the solar irradiance.

The heat output of the photothermal collection system at time t is Q_{c_t}. In the case of flat plate collectors, the heat output is Q_u [2-3].

4.2Model of hot water storage tank

The heat supply of the hot water storage tank per unit time is:

$$Q_r = \Delta m_c c_s T_{qs} \tag{2}$$

Among them, Δmc=mj-mc, is the difference between the hot-end water supply flow mj and the return water flow mc, kg/s; Cs is the specific heat capacity of water at constant pressure; Tqs is the initial temperature of water storage in the hot water storage tank in Δt, °C; Qr is the heat supply per unit time of the hot water storage tank, W[4-5].

4.3Wind photoelectric model

(1) The basic power equation of wind turbines is based on the wind power equation:

$$P_w = \frac{1}{2}\rho A V^3 \tag{3}$$

The output power of the wind turbine is

$$P_o = P_w C_p \eta_t \eta_m \tag{4}$$

Among them, ρ is the mass density of the air; V is the wind speed; A is the area swept by the wind turbine blades; η_t is the power conversion efficiency of the generator, taking 0.98; η_m is the mechanical transmission efficiency, taking 0.98; C_p is the wind energy utilization coefficient.

(2) Since the photoelectric output is greatly affected by factors such as temperature, light intensity, etc., the solar radiation intensity and the power under the ambient temperature are calculated here. Let U_m and I_m be the photovoltaic module parameters under the conditions of solar irradiance and ambient temperature at a certain time, and the relationship is:

$$\mathrm{I}_m = \mathrm{I}_m(\frac{S}{S_{ref}})[1 + a(t - t_{ref})] \tag{5}$$

$$\mathrm{U}_m = \mathrm{U}_m[1 - c(t - t_{ref})]\left[1 + b(\frac{S}{S_{ref}} - 1)\right] \tag{6}$$

Among them, S is the real-time light intensity; S_{ref} is the standard sunshine intensity, taking 1kW/m^2; t is the real-time temperature of the photovoltaic module; t_{ref} is the standard temperature for photovoltaic module testing, which is 25°C. The coefficients a, b, and c take typical values. : a=0.0025/°C, b=0.5/°C, c=0.00288/°C. The optimal output power of the photovoltaic array at any solar radiation intensity and ambient temperature is: $\mathrm{P} = \mathrm{I}_m \mathrm{U}_m$. The conversion efficiency of solar cells is generally about 15% [6].

4.4Objective function

It is required to control the output heat fluctuation of the combined solar-thermal-wind-photoelectric storage system to minimize, that is, the variance of the output heat of the combined solar-wind and solar storage system is the smallest.

$$\min f = \frac{1}{T}\sum_{t=1}^{T}[Q_{wc\_t} - Q_{wc\_av}]^2 \tag{7}$$

$$Q_{wc\_t} = (P_{f\_t} + P_{g\_t})t\eta_{fg} + Q_{c\_t} \tag{8}$$

$$Q_{wc\_av} = \frac{1}{T}\sum_{t=1}^{T} Q_{wc\_t} \tag{9}$$

In the formula: Q_{wc_t} is the actual output of the photothermal-wind photoelectric joint at time t; Q_{wc_av} is the average value of the combined output of light, heat and wind in the calculation period T; P_{f_t} is the actual wind power output at time t; P_{g_t} is the actual photovoltaic output at time t; η_{fg} is the conversion efficiency of the electric boiler.

4.5Constraints

1) Constraints on the output heat balance of the combined solar thermal-wind photovoltaic energy storage system

$$Q_{c\_t} + Q_{b\_t} + Q_{r\_t} = Q_{l\_t} \tag{10}$$

$$Q_{b\_t} = (P_{f\_t} + P_{g\_t})t\eta_{fg} = (P_{f\_t} + P_{g\_t})c_{dh} \tag{11}$$

Where: Q_{c_t} the output of the solar thermal collection system at time t; Q_{r_t} is the output of the hot water storage tank at time t; Q_{l_t} is the heat output at time t; Q_{b_t} is the output of the electric energy storage device at time t; c_{dh} is the electric heating conversion coefficient of the electric boiler, taking 0.98.

2) Output constraints of the CSP system

$$Q_{c\_min} + Q_{c\_t}^{Down} \leq Q_{c\_t} \leq Q_{c\_max} - Q_{c\_t}^{Up} \tag{12}$$

In the formula: Q_{c_max} and Q_{c_min} are the maximum and minimum output of the heat collection system respectively; $Q_{c\_t}^{Down}$ and $Q_{c\_t}^{Up}$ are the lower limit output standby and upper limit output standby of the CSP system at time t, respectively.

3) The capacity of the hot water storage tank is restricted as

$$Q_{r\_t}^{min} \ll Q_{r\_t} \ll Q_{r\_t}^{max} \tag{13}$$

Where: Q_{r_t} is the thermal energy capacity stored in the hot water storage tank at time t; $Q_{r\_t}^{max}$, $Q_{r\_t}^{min}$ are the maximum and minimum heat storage capacity of the hot water storage tank.

4) Wind farm output constraints

$$0 \le P_{f\_t} \le P_{f\_max} \tag{14}$$

Where P_{f_max} is the maximum output power of wind power generation.

5) Constraints on the output of photovoltaic power plants

$$0 \le P_{g\_t} \le P_{g\_max} \tag{15}$$

Where P_{g_max} is the maximum output power of photovoltaic power generation.

6) The constraints of the electric boiler are the constraints of the electric-heat conversion coefficient and the constraints of the power consumption [7].

$$Q_{b\_t} = c_{dh} P_{b\_t} \tag{16}$$

$$0 \le P_{b\_t} \le P_{b\_max} \tag{17}$$

Where: Q_{b_t} is the heat output of the electric boiler at time t; P_{b_t} is the power consumption of the electric boiler at time t.

5. EXAMPLE ANALYSIS AND RESULTS

In order to verify the feasibility of the scheduling model, two flat-plate collectors are selected. A 100L hot water storage tank, an electric boiler, an electric energy storage battery, an inverter controller for wind and photovoltaic power generation, and a user thermal load. According to the actual situation of Xining, the load is divided into 2 categories according to the season: Heating season, transition season. Here, select the heating season on January 3, 2021, and analyze the heat transmission situation of the system. The heating chamber has an area of 15m^2. Here we analyze the integrated system of solar thermal storage and wind photoelectric thermal output. The analysis period is set to 24h.

5.1 Wind speed, solar radiation, and outdoor temperature on a certain day in the winter heating season

The environmental data measured on January 3, 2021 are shown in Figure 4, Figure 5, and Figure 6.

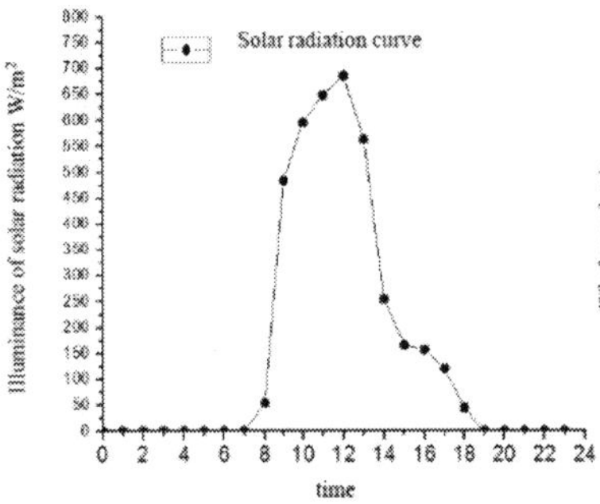

Fig. 4 Solar radiation on January 3

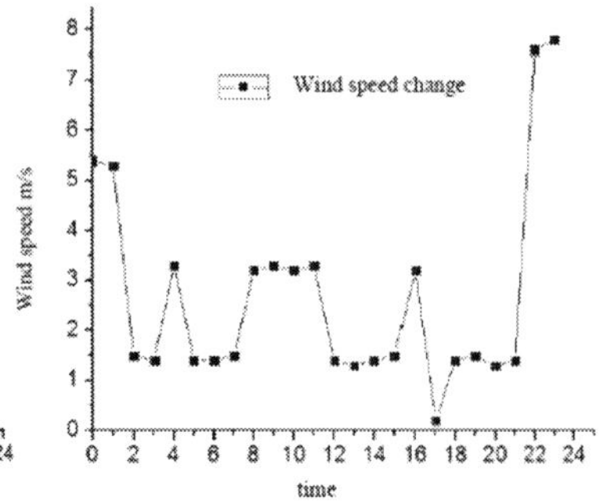

Fig. 5 Wind speed on January 3

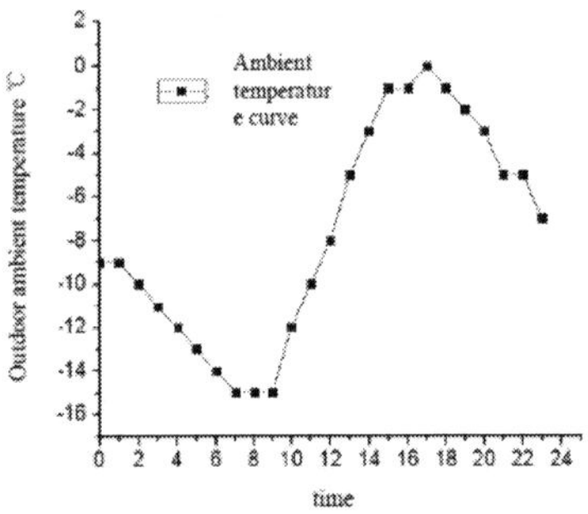

Fig. 6 Ambient temperature on January 3

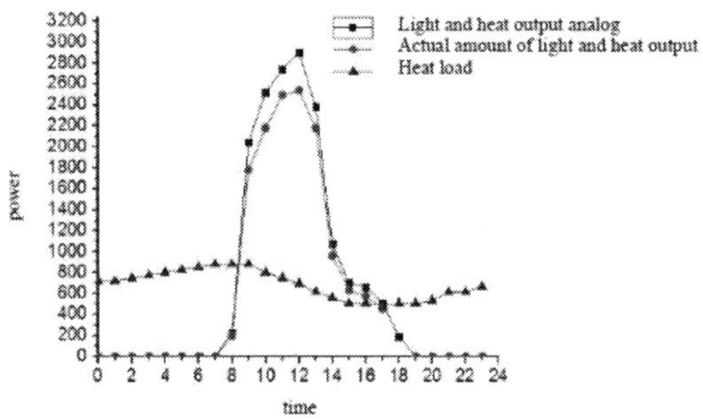

Fig. 7 Independent operation of collector

5.2. Independent operation of the collector system

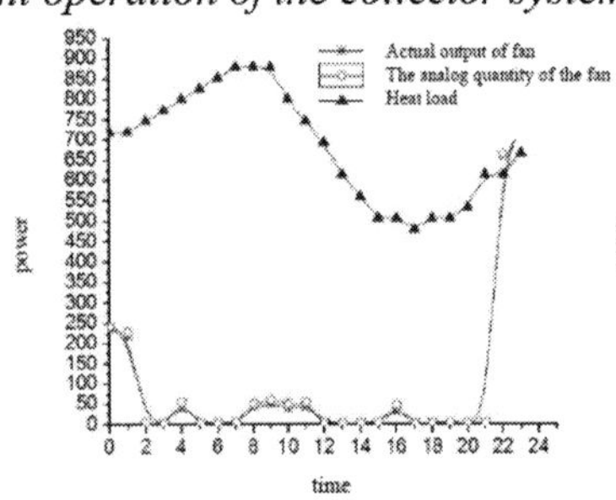

Figure 8 vIndependent operation of wind power generation

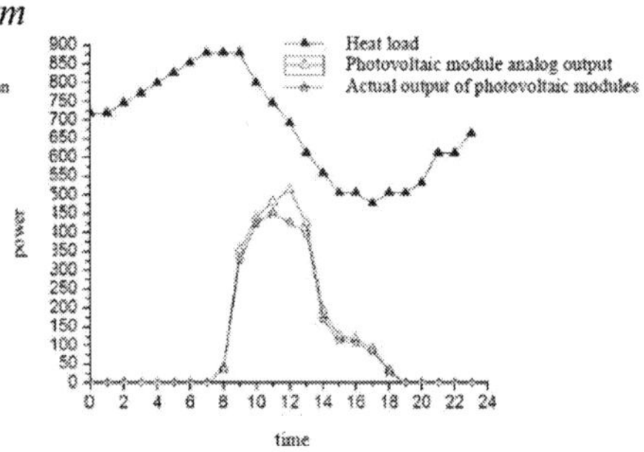

Fig.9 Independent operation of photovoltaic power generation

Figure 7 According to heating regulations, building heating generally takes the range of 45 ~ 70W/m^2. Here, Q_Y is taken as 45W/m^2, and the actual heat load index is expressed as $Q_{fz} = \frac{Q_Y(t_n - t_w)}{(t_n - t_{cp})}$. Among them, the calculated indoor temperature of t_n is 18°C. t_w is the outdoor ambient temperature. t_{cp} is the monthly average temperature, here is -7.3°C [8]. The surface temperature of the collector plate enters a stable state after being powered on for 10 minutes. Under different air volume, the surface temperature of the collector plate decreases with the increase of the system air volume [9]. Under different solar radiation, the surface temperature of the collector increases with the increase in irradiance. The lower the outdoor ambient temperature, the more heat is actually used.

5.3. Independent operation of wind power

Figure 8 selects a wind turbine with a power of 1KW. The diameter of the wind wheel is 2.6m. The atmospheric pressure in Xining in winter is 77.406Kp. The air density is calculated with ρ=1.29×(77.406×273.15)/(101.325×(273.15+t)), where t is the outdoor air temperature. The power of wind turbines increases with the increase of wind speed, and the temperature and humidity increase during the day. The actual power generation output has been reduced.

5.4. Independent operation of photovoltaic power generation

Figure 9 selects two monocrystalline silicon solar photovoltaic modules with a power of 90W. The maximum current is 5.45A. The maximum voltage is 16.6V. The short-circuit current is 6.11A. The open circuit voltage is 21V. During the day, the output power increases and decreases with the change of irradiance.

The Qinghai-Tibet Plateau has a relatively long heat output due to the winter weather, and the heat load is generally output within 24 hours a day. There is a high-load floating standard (including heating) in each of the three time periods of 06:00-8:30, 14:00-15:00, and 19:30-22:00. Prescribe a thermal float standard. There is a load trough in each of the three time periods of 9:00-13:30, 16:00-19:00 and 22:30-05:30. Provide a low-heat floating standard (not including heating). The solar and thermal energy storage system in the joint operation of the wind and solar energy storage system regulates the wind and solar energy through the heat storage box. When the output of wind and photoelectricity increases, the output of light and heat is reduced. On the contrary, increase the output of light and heat through the heat storage box to ensure that the heat energy output by the wind and solar energy storage combined system is stable within the range of system adjustment. Two issues need to be paid attention to during the adjustment process of the light and heat output by the collector: First, the heat output of the wind and solar storage system cannot exceed the upper and lower limits of output; Second, the capacity of the hot water storage tank when outputting heat cannot be lower than the limited capacity.

Reference

[1] Song Wenqin; Lu Jinli; Zhao Lingxia Research on Economic Dispatching Strategy of Power System for Combined Operation of Solar Thermal and Wind Power [J]."Power System Protection and Control". 2020/5

[2] Cao Lihua; Zhang Lai; Jiang Tieyi Numerical analysis of serpentine flow channel solar flat panel collector[J]. *Journal of Northeast Dianli University*. 2018-1

[3] Zhang Weiwei; Zhu Yuezhao; Jiang Jinzhu; Qian Qiang Design and simulation analysis of a new type of CPC heat pipe collector [J]. "Thermal Energy and Power Engineering". 2009/3

[4] Li Qiang; Chen Hongbing Model establishment and performance research of solar water heating system [J]. "Shanxi Architecture". 2017-27

[5] Chen Hongbing; Li Qiang; Chen Xilin Research on the performance of centralized solar water heating system [J]. "Renewable Energy". 2017/4

[6] Yang Yuexin, Che Yanbo, Li Shengshan Modeling and analysis of characteristics of photovoltaic modules on barren hills and slopes [J]. Electronic Design Engineering. 2017, 25(18)

[7] Fu Zhongzhou; Yuan Tiejiang; Pi Xia Optimal Scheduling Model of Electric and Heating Combined System with Heat Storage [J]. "Electrical Appliances and Energy Efficiency Management Technology". 2018-14

[8] 8. Liao Hancheng, Shen Yuan Discussion on solar heating technology in Xining area[J]. Journal of Qinghai University (Natural Science Edition). 2008 Vol. 26 October No. 5

A Smart Grid Planning Method with Coal-to-electricity Based on Multifactor Adaptability

Jiakun An[1], Shuqiang Yang[1a*], Tao Wang[1], Chungaung He[1], Hongjie Jia[2b*], Yunfei Mu[2c*], Xiaohong Dong[3], Yang Zhao[1], Xiaolin Tan[1], Wei Guo[1]

[1] State Grid Hebei Economic Research Institute, Shijiazhuang, Hebei 050000, China

[2] Key Laboratory of Smart Grid of Ministry of Education, Tianjin University, Tianjin 300072, China

[3] State Key Laboratory of Reliability and Intelligence of Electrical Equipment, Hebei University of Technology, Tianjin 300130, China

[a*]jyy_yangsq@163.com, [b*]hjjia@tju.edu.cn, [c*]yunfeimu@tju.edu.cn

Abstract—In order to reduce coal pollution and improve air quality in winter, many cities in northern China began to promote coal-to-electricity, and a series of policy subsidies were released accordingly. With the implementation of coal-to-electricity, the characteristics of power network has been changed significantly and the traditional planning method is no longer applicable. To meet the needs of coal-to-electricity engineering, a smart grid planning method with coal-to-electricity based on multifactor adaptability is proposed in this paper. First, the coal-to-electricity load characteristics is analysed, and typical parameters is proposed based on the actual operation data, which considers the impacts of different types of equipment and breaks through the modelling difficulties of coal-to-electricity characteristics. Then, a smart grid planning method with coal-to-electricity based on multifactor adaptability is set up, which combines the influence of transmission capacity for each equipment, the investment capacity for each participant, and the social acceptance of the electricity price. Based on this model, the technical, economic, and environmental benefits of coal-to-electricity modification are coordinated. The application shows that the proposed method is concise and effective and provides useful suggestions for determining the coal-to-electricity scale, network planning and compensation policy, which supports the implementation of coal-to-electricity.

1. Introduction

With the severe challenge of serious haze weather and increasing demand of energy reform, increasing the proportion of power consumption in terminal energy has become an important strategy in China. And thus, the shift from coal heating to electric heating, which is named coal-to-electricity project develops rapidly form year of 2016[1].

The types and development potential of coal-to-electricity are analyzed in [2]-[4] and the technical characteristics like air/ground source heating pumps, phase change heat storage are introduced in [5]-[7]. The load characteristics of coal-to-electricity are obtained by statistically analyzed in [8], which shows strong negative correlation between heating loads and temperature. And the forecasting method for short-term electric heating power load is proposed in [9] and the impacts of coal-to-electricity on power grid are analyzed in [10]-[12]. Furthermore, the economic and environmental benefits, evaluation index system and engineering harmonics of coal-to-electricity are mentioned in [13]-[15]. A method for evaluating the adjustable capacity of household heating load based on temperature forecast is presented in [16]. The planning method for enhancing the capacity of wind power consumption by coal-to-electricity equipment and the operation strategy are introduced in [17]-[18], which lays a foundation for demand side management of coal-to-electricity and is helpful for the development of the integrated energy system and enhancing of the power grid resilience.

In general, some research has been carried out in the field of coal-to-electricity. However, there is no load forecasting method of different types of coal-to-electricity for power grid planning. And most of the research focuses on the impact analysis, operation scheduling of the coal-to-electricity, and few studies are made on the power grid planning method with coal-to-electricity, aiming at promoting the implementation of coal-to-electricity and ensuring the interests of all participants like governments, enterprise, and household users. From the perspective of power balance, we have put forward a smart grid planning method with coal-to-electricity in [19], which is suitable for the regional overall scale planning but is difficult to determine the specific construction needs of substations and transformation lines. Further research is still needed today.

To scientifically plan the construction plan of power grid with coal-to-electricity, a smart grid planning method based on multifactor adaptability is proposed in this paper. Firstly, the typical parameters of different coal-to-electricity equipment for power prediction are extracted. Then, the technical constraints, economic constraints, and social constraints are considered comprehensively to maximize energy efficiency with minimum investment. Finally, taking the actual planning work of power network with coal-to-electricity in a certain region of Hebei province as an example, this paper makes an empirical analysis of the implementation scale, the construction scale of smart grid planning, investment analysis and the added electricity price of different types of coal-to-electricity.

2. Load prediction integrated with coal-to-electricity

2.1. Load prediction model

The coal-to-electricity power load in each region is the summary of electric heating load of public enterprises and residents.

$$l_c = \sum l_{c.m}^{unit} + \bar{L}_{user} \times N_{last.c} \times \alpha \times \beta \quad (1)$$

Where, l_c represents the coal-to-electricity load integrated into point c; $\bar{L}_{user}$ represents the increased average capacity for each household; α represents the load rate of electric heating equipment; β represents the simultaneity rate of electric heating equipment; $N_{last.c}$ represents the implementation scale of coal-to-electricity integrated into point c; and $l_{c.m}^{unit}$ represents the increased load of public enterprises, which is calculated by equipment power or heating area.

$$l_{c.m}^{unit} = \bar{L}_{unit} \times \alpha \times \beta \quad (2)$$

$$l_{c.m}^{unit} = S_{unit} \times p_{unit} \quad (3)$$

Where, $\bar{L}_{unit}$ is the increased power capacity of public enterprises; S_{unit} is the electric heating area; p_{unit} is the heating load per square meter.

2.2. Typical parameters for load prediction

The main coal-to-electricity equipment includes centralized heat storage electric heaters, air source heat pumps (centralized or distributed), distributed heat storage electric heaters, distributed heat storage electric boilers. It is different in electric heating area, equipment features and power for different types of coal-to-electricity equipment, which causes the typical parameters such as load rate and simultaneity rate differently. To make the load prediction model more general and universal, the typical parameters in the model for different types of coal-to-electricity equipment are analyzed based on the historical data.

$$\beta=\begin{cases}1 & \text{centralized heat storage electric heaters}\\ 1 & \text{air source heat pumps}\\ 0.8 & \text{distributed heat storage electric boilers}\\ 0.6 & \text{distributed heat storage electric heaters}\end{cases} \tag{4}$$

$$\alpha=\min\sum(P_{c.m}\text{-}\bar{P}_{user.m}\times N_{user}\times\alpha_i\times\beta) \tag{5}$$

Where, $P_{c.m}$ is the measured load power of coal-to-electricity; $\bar{P}_{user.m}$ is the equipment installation capacity; α_i represents the range of load rate recommended by technical guidelines for coal-to-electricity access. Taking the actual load of a certain area in Hebei province as an example, the typical parameters of various types of coal-to-electricity is reached and shown as follows.

Table 1 Typical parameters for load prediction of coal-to-electricity

Classification		Heating area (m^2)	Several equipment capacity(kW)	Typical parameters
Centralized	heat storage electric heaters	tens of thousands～Hundreds of thousands	50～600	α=0.9; β=1
	air source heat pumps	Several hundred～over ten thousand	20～70	α=0.75; β=1
Distributed	heat storage electric boilers	100～180	12、15、18	α=0.8; β=0.8
	air source heat pumps	100	6.4、4.8、3.7	α=0.6; β=1
	heat storage electric heaters	Less than 30	2.4、3.2	α=0.8; β=0.6
Public enterprises		-	-	p_{unit}=80

3. Proposed model formulation

3.1. Objective function

The objective function is minimizing the investment while maximizing the benefits of environmental improvement.

$$\min \alpha_1\frac{F_{grid}^{bas}}{F_{coal}}+\alpha_2\frac{CRQ}{CRQ^r} \tag{6}$$

Where, α_1 and α_2 are the weight coefficients respectively; F_{grid}^{bas} is the coal-to-electricity investment of matching power projects; $F_{coal}$is the heating cost of burning coal before coal-to-electricity ; CRQ^r and $CRQ$are the costs of controlling pollutants before and after the coal-to-electricity, as in

$$CRQ=(\sum N_c\text{-}N_{last})(\eta_s\chi_s+\eta_N\chi_N+\eta_D\chi_D+\eta_C\chi_C) \tag{7}$$

Where η_s, η_N, η_D, $\eta_C$describe the unit controlling costs of sulfur dioxide, nitrogen oxides, soot and carbon dioxide; χ_S, χ_N, χ_D, χ_C are the average household emissions of sulfur dioxide, nitrogen oxides, soot and carbon dioxide per heating season; N_{last} represents the implementation scale of coal-to-electricity. $N_c$describes the number of residents in node *c*.

3.2. Technical constraint

From the perspective of safe and stable operation of power grid, the technical constraint conditions are proposed, that is the power flow should meet the transmission capacity demands of transformers, lines, and other equipment, and operate normally in the N-1 mode.

(1) Normal operation mode

Under the normal operation mode, the power capacity needs to satisfy the power load needs of coal-to-electricity, while the output of each node and the transmission power of each operating line meet the relevant constraints.

$$\begin{cases} S_{ks}^T P + G = L_0 + L_n + L_c \\ p_{ij} - b_{ij}^{ks}(n_{ij}^0 + n_{ij} + n_{ij}^{ks})(\theta_i - \theta_j) = 0 \\ |p_{ij}| \le (n_{ij}^0 + n_{ij} + n_{ij}^{ks}) p_{ij.max} \\ g_{i.min} \le g_i \le g_{i.max} \\ 0 \le n_{ij} \le n_{ij.max} \\ 0 \le n_{ij}^{ks} \le n_{ij.max}^{ks} \\ i, j \in \Omega \\ ks \in V \end{cases} \tag{8}$$

Where S_{ks}^T is the node line incidence matrix; P, G, L_0, L_n, L_c are the vectors for power flow of lines, power output of generators, present power loads, increased natural power load, predicted power load of coal-to-electricity; p_{ij} is the line flow between node *i* and node *j*; b_{ij}^{ks} is the imaginary part of admittance matrix in DC power flow; n_{ij}^0 is the number of existing lines between node *i* and node *j*; n_{ij} is the number of original planned lines between node *i* and node *j* without considering newly constructed substations; n_{ij}^{ks} is the changes in the number of existing and original planned lines between node *i* and node *j* with newly constructed substations of mode *ks*; *Ω* is the set for power system nodes; *V* is the set for newly planned substations nodes.

When the capacity of each substation meets the load demand integrated with coal-to-electricity, there is no need to add new substations, and thus

$$\begin{cases} S_{ks}^T = S^T \\ n_{ij}^{ks} = 0 \end{cases} \quad 0 \le l_{0.i} + l_{n.i} + l_{c.i} \le l_{i.\max} \tag{9}$$

Where, S^T is the current node line incidence matrix; $l_{0.i}$, $l_{n.i}$ and $l_{c.i}$ are the present power load, increased natural power load and predicted power load of coal-to-electricity for note *i*; $l_{i.max}$ is the maximum capacity that can be transformed into lower voltage for the substation at node *i*.

(2) N-1 operation mode

It is similar in the N-1 operation mode that the power capacity needs to satisfy the power load needs of coal-to-electricity, while the output of each node and the transmission power of each operating line meet the relevant constraints.

$$\begin{cases} (S^{mn})_{ks}^T P^{mn} + G^{mn} = L_0 + L_n + L_c \\ p_{ij}^{mn} - b_{ij}^{ks}(n_{ij}^0 + n_{ij} + n_{ij}^{ks} - 1)(\theta_i^{mn} - \theta_j^{mn}) = 0, ij = mn \\ |p_{ij}^{mn}| \le (n_{ij}^0 + n_{ij} + n_{ij}^{ks} - 1) p_{ij.max}, ij = mn \\ p_{ij}^{mn} - b_{ij}^{ks}(n_{ij}^0 + n_{ij} + n_{ij}^{ks})(\theta_i^{mn} - \theta_j^{mn}) = 0, ij \ne mn \\ |p_{ij}^{mn}| \le (n_{ij}^0 + n_{ij} + n_{ij}^{ks}) p_{ij.max}, ij \ne mn \\ g_{i.min} \le g_i^{mn} \le g_{i.max} \\ 0 \le n_{ij} \le n_{ij.max} \\ 0 \le n_{ij}^{ks} \le n_{ij.max}^{ks} \\ (i, j) \in \Omega, (m, n) \in \Omega \\ ks \in V \end{cases} \tag{10}$$

where *mn* represents the constraint condition in case of line breaking between node *m* and node *n*.

Different from the traditional planning, this model considers the constraints of newly planned substations. The coal-to-electricity load is innovatively integrated into the mode calculated in (1).

(3) Implementation scale constraints

The regional total implementation scale of coal-to-electricity is the summary of implementation scale at each node, which is less than or equal to the number of residents in this node.

$$\begin{cases} N_{last} = \sum N_{last.c} \\ 0 \le N_{last.c} \le N_c \end{cases} \tag{11}$$

3.3. Economic constraint

(1) Investment capacity constraint

To guarantee the implementation of coal-to-electricity, the investments of governments, enterprises and users should be within their investment capacity.

$$\begin{cases} 0 \le F_G^{bas} \le F_{G.max} \\ 0 \le F_U^{bas} \le F_{U.max} \\ 0 \le F_{grid}^{bas} \le F_{grid.max} \end{cases} \tag{12}$$

where $F_{G.max}$, $F_{U.max}$ and $F_{grid.max}$ stand for the acceptable maximum investment for government, users and enterprises respectively; F_G^{bas}, F_U^{bas} and F_{grid}^{bas} are the coal-to-electricity investments for government, users and enterprises and can be calculated as follows[19].

$$\begin{cases} F_G^{bas} = N_{last} F_{user}^C + \sum w_{grid.u}^V F_{grid.u}^V + w_{grid}^{land} F_{grid}^{land} \\ F_U^{bas} = N_{last}(F_{user} - F_{user}^C) \\ F_{grid}^{bas} = \sum (1 - w_{grid.u}^V)\ F_{grid.u}^V + (1 - \omega_{grid}^{land})\ F_{grid}^{land} \end{cases} \tag{13}$$

where F_{user} and F_{user}^C stand for the average investment for heating equipment purchase and housing insulation repairmen, and average subsidy for users per household; $F_{grid.u}^V$ stands for the investment for enterprises at voltage u; F_{grid}^{land} represents the investment for land acquisition and demolition; $w_{grid.u}^V$ and ω_{grid}^{land} stand for the subsidy percentage for power grid investment at each voltage level and subsidy for land acquisition and demolition.

(2) Investment of matching projects constraint

The investment for enterprises at each voltage grade is relevant to the technical constraints and is the investment summary of planned lines and substations at this voltage grade.

$$F_{grid.u}^V = \sum n_{ij} LN_{ij.u}^V C_L + \sum n_{ij}^{ks} T_{ij.u}^V C_T \tag{14}$$

Where, $LN_{ij.u}^V$ is the planned line length between node i and node j; $T_{ij.u}^V$ is the power capacity of planned substations under mode ks; C_L is the average investment of line per kilometer; C_T is the average investment of substations per kVA.

3.4. Social constraint

According to the national policy in China, the investments of power enterprises will be split to the whole society via adding the sales electricity price. Therefore, the added electricity price should be within the range of social tolerance.

$$P_T \le P_{T.max} \tag{15}$$

where $P_{T.max}$ is the acceptable maximum electricity price; P_{NT} is the added electricity price (after-tax), which is a function of investment of enterprises, depreciation fee, operation and maintenance fee, loan interest and the allowable income and other factors, as detailed showed in [19].

$$P_T = f(F_{grid}^{bas}) \tag{16}$$

4. Solving algorithm

The planning method of coal-to-electricity access distribution network based on multi factor adaptation includes current situation mapping, initialization, environmental benefit analysis, technical adaptability analysis, economic benefit analysis, evaluation, and decision-making, etc.

(1) Find out the current situation

We should collect the relevant information of users, enterprises, and governments in the coal-to-electricity area, including the user's coal-to-electricity mode, equipment power, quantity, power consumption characteristics, power supply capacity of regional power grid and relevant compensation policies of the government.

(2) Initialization

The initial transformation scale shall be set according to the implementation of coal-to-electricity conversion for all bulk coal users and central heating users in the area. Considering the need to meet the natural growth load, the projects under implementation, planned and planned are included in the project library before the implementation of coal-to-electricity. Initialization of coal-to-electricity supporting power grid transformation project is empty.

(3) Environmental benefit analysis

According to the statistical information on the average household sulfur dioxide, nitrogen oxide, smoke, and carbon dioxide emissions in the heating season, and according to the current implementation scale of coal-to-electricity conversion, the current environmental benefits of coal-to-electricity conversion are calculated and converted into pollution control costs according to the unit treatment costs.

(4) Technical adaptability analysis

Step 1: calculate the new load of replacing coal with electricity according to formula (1) and obtain the total load of each node of each system in combination with the natural growth load and the current grid load.

Step 2: consider the commissioning of coal-to-electricity supporting projects and natural growth load supporting projects and carry out power flow calculation of power grid.

Step 3: judge whether the line power flow and node load meet the constraints in combination with the line transmission capacity limit and substation capacity limit of each voltage level. If all constraints are met, skip to (5) economic benefit analysis. If the constraints are not met, plan the new project of the line or substation with this voltage level in combination with the load distribution, and perform the following steps.

Step 4: to ensure the power demand of coal-to-electricity load, judge whether the power grid construction capacity can ensure the operation of the current project. If it cannot be met, it is necessary to reduce the implementation scale of coal-to-electricity and jump to (3) environmental benefit analysis. If the demand is met, the project will be included in the construction and transformation project library of coal-to-electricity supporting power grid and skip to step 2 to recalculate the power flow.

(5) Economic benefit analysis

Step 6: calculate the total investment cost of coal-to-electricity supporting project according to the construction length, substation capacity and typical project cost.

Step 7: calculate the coal-to-electricity investment of the government and users according to the subsidy policy and the type of coal-to-electricity equipment. The diversion electricity price is calculated according to the supporting power grid investment, enterprise equipment depreciation, interest rate and other financial data.

Step 8: judge whether the investment capacity of the government, enterprises and users meets the needs of coal-to-electricity construction according to the calculated diversion electricity price and investment; Judge whether the diversion price level can be accepted by the society. If not, it is necessary to reduce the implementation scale of coal-to-electricity conversion and jump to (3) environmental benefit analysis.

(6) Evaluation decision

Repeat the above steps until the optimal coal-to-electricity implementation scale and power grid project coordinated by environment, technology and economy are determined. Then, calculate the enterprise diversion electricity price, investment payback period and reduced pollutant emission, and evaluate the implementation benefits of coal-to-electricity from the three aspects of government, enterprises, and society as the decision-making basis for the implementation of coal-to-electricity.

5. Case study

The above model has been applied in Hebei province. The planning period is 2017~2020 and the proposed model are analyzed with different electric heating modes, shown as follows.

Case 1: Power grid without coal-to-electricity.

Case 2: Power grid with coal-to-electricity of distributed heat storage electric heaters.

Case 3: Power grid with coal-to-electricity of air source heat pumps.

Case 4: Power grid with coal-to-electricity of distributed heat storage electric boilers.

5.1. Basic information

There are 2404 public utilities and 1.9 million households with scattered coal users in the case study region. The substations of 500kV, 220kV and 110kV are 21, 177, 564; and the transformer capacity are 4.5, 6.35, 46.89 million kVA in this region. The substations of 220kV are divided into six sub-regions and the current maximum power load in the total region is 34.57 million kW in summer, and the maximum load reduces 2 million kW in winter. It is estimated that the traditional power load will increase 10 million kW during the power planning period.

In this region, the coal-to-electricity work is demanded to finish before November 15th of 2020. During this period, the original planned distribution of 500kV, 220kV and 110kV are 7, 49, 139; and the transformer capacity are 11.75, 16.38, 12.78 million kVA in this region. The planned investment of 10kV and below is 14.6 billion Yuan.

5.2. Load forecasting with coal-to-electricity

If coal-to-electricity is implemented by all residents in the region, the power load integrated with coal-to-electricity is expected to reach 48.01, 47.45 and 53.54 million kW respectively as shown in Fig.1. The newly added power load of coal-to-electricity and the maximum power load differ widely under different electric heating modes. Thereafter, the employed electric heating modes and the implemented scale of coal-to-electricity needs to be scientifically determined.

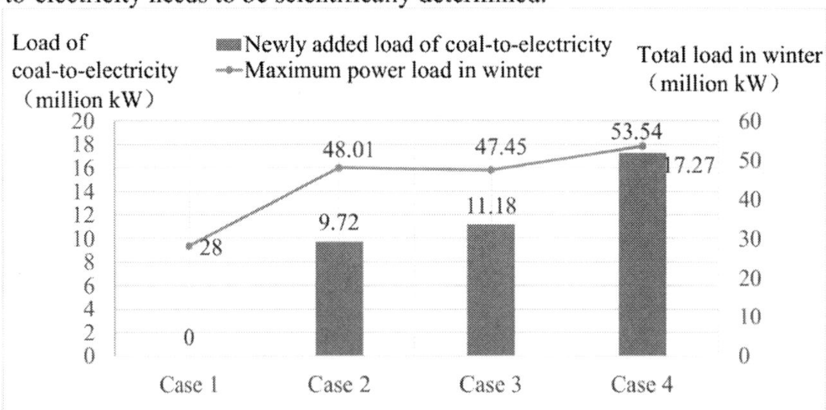

Fig.1 Load prediction of different coal-to-electricity methods

5.3. Smart grid planning with coal-to-electricity

(1) Taking environmental benefit as optimal

If $\alpha_1 \ll \alpha_2$, the objective function focuses on environmental benefits, and then power grid planning scheme can be obtained. In this model, coal-to-electricity is implemented by all residents of 1.9 million households. The specific construction scale is shown in Fig.2.

Only considering the environmental benefits, the number of substations needed to be built is larger than that of without coal-to-electricity in the same region. For example, there are 16, 77 and 338 substations for 500kV, 220kV and 110kV in Case 4, which are far more than 7, 49 and 139 substations of Case 1. The substations needed to be built are 2.2 times of the construction scale without coal-to-electricity and are far beyond the power grid construction capacity in a certain period. Therefore, the implementation scale of coal-to-electricity should consider the construction capacity of enterprises while considering the environmental benefits.

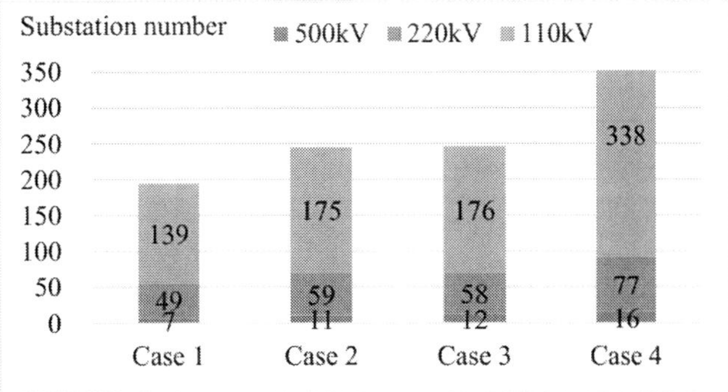

Fig.2 Construction scale of different types of coal-to-electricity

(2) Taking economic benefit as optimal

If $\alpha_1 \gg \alpha_2$, the objective function focuses on the investment cost. As the regional power load at daytime is higher than the nocturnal load, the acceptable capacity at daytime is lower than that at night. Meanwhile, the electric heating modes should also be considered. The heating equipment of distributed heat storage electric heaters and distributed heat storage electric boilers usually start operating when the power load is running low (20:00~8:00), and thus the newly added power load of coal-to-electricity is in the night. Thereafter, it is necessary to use the power supply ability at night to calculate the acceptable capacity. And the air source heat pump is the opposite. The specific construction scale is shown as follows.

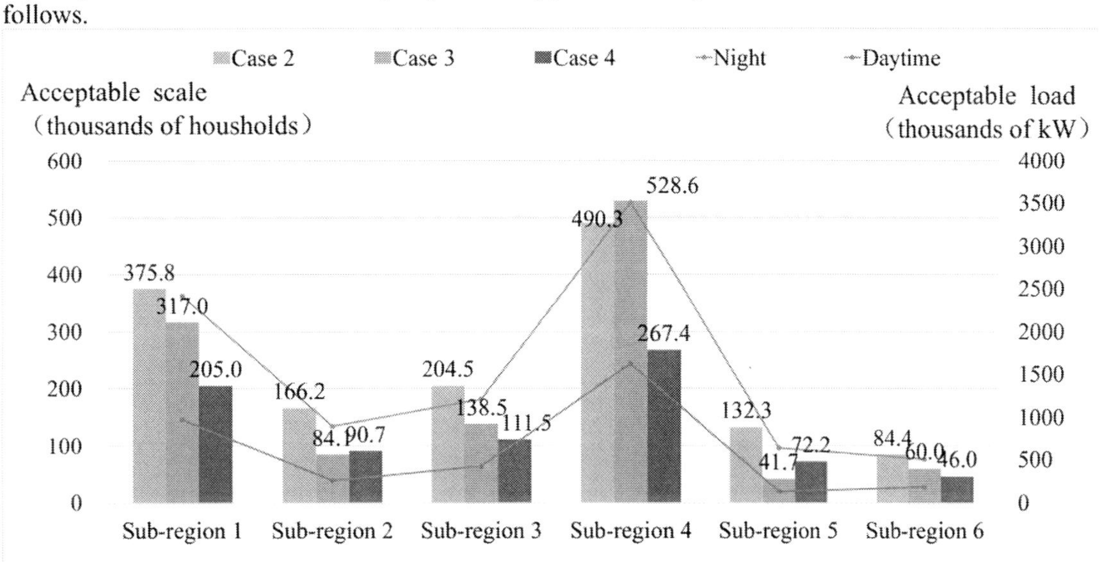

Fig.3 Acceptable scale of different types of coal-to-electricity

The acceptable capacity of coal-to-electricity is relevant to the power load characteristics and the acceptable load difference between day and night. When the acceptable load is relatively small, the implementation scale of coal-to-electricity by Case 2 (distributed heat storage electric heaters) is relatively large, followed by Case 3 (air source heat pumps), and the number of Case 4 (thermal storage electric boilers) is minimal, as shown in Sub-region 1. However, with the increase of acceptable load and the decrease of load difference in the daytime and at night, the difference between the implementation scale of Case 2 and Case 3 is reduced, and it is even possible that the implementation scale of Case 3 is higher than that of Case 2, such as Sub-region 4. The region can accommodate 1.45, 1.17 and 0.79 million households to implement coal-to-electricity, which are all lower than the regional total number of 1.9 million households.

Table 2 The acceptable scale of different types of coal-to-electricity

	Acceptable added load (thousands of kW)		Case 2 (thousands of households)	Case 3 (thousands of households)	Case 4 (thousands of households)
	Daytime	Night			
Sub-region 1	973	1443.1	375.8	317.0	205.0
Sub-region 2	258	638.2	166.2	84.1	90.7
Sub-region 3	425	785.1	204.5	138.5	111.5
Sub-region 4	1622	1882.8	490.3	528.6	267.4
Sub-region 5	128	508.1	132.3	41.7	72.2
Sub-region 6	184	324.1	84.4	60.0	46.0
Summary	3591	5581.4	1453.5	1169.8	792.8

(3) Benefit analysis after coal-to-electricity

Taking multifactor adaptability as optimal, the optimal implementation scale of coal-to-electricity is usually the largest acceptable capacity of coal-to-electricity. Under this implementation scale, the investment cost is minimal and within their investment capacity and construction capacity. The investment cost of all participants after coal-to-electricity is analyzed by taking the least amount of implementation scale of 0.79 million households in the region as an example. As shown in Fig.4, the increased investment cost of Case 3 is least reaching 36.2 billion Yuan for power grid companies, followed by Case 2 reaching 37.1 billion Yuan, and Case 4 is most reaching 40.9 billion Yuan. For users, the adoption of air source heat pumps (Case 3) will greatly increase the heating cost reaching about 17 billion Yuan, which is 8.3 times that of the distributed heat storage electric heaters (Case 2). For governments, under the current subsidy policy (7400 Yuan/household), different types of heating modes are subsidized by 5.8 billion Yuan. However, it is necessary to increase the sales electricity price to dredge the power grid investment cost, which is about 3.21~4.04 cents/kWh, which increases the social cost. Therefore, it is recommended that the heating mode of distributed heat storage electric heaters is preferred to be adopted to balance the investment cost of users, enterprises, government and social cost.

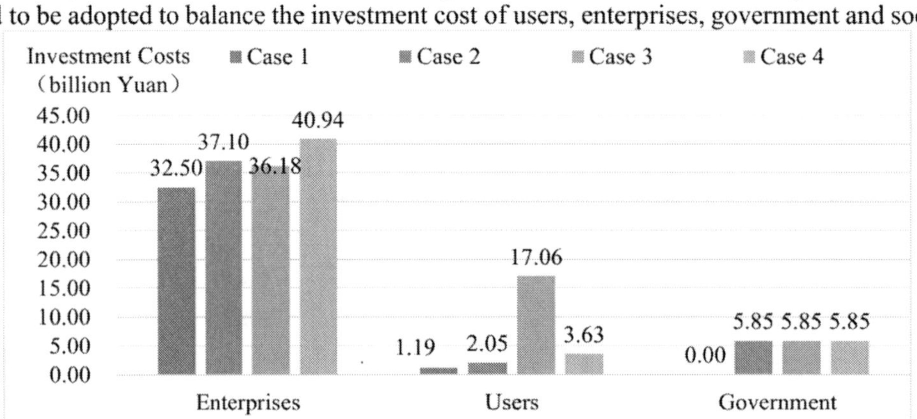

Fig.4 Investment of different types of coal-to-electricity

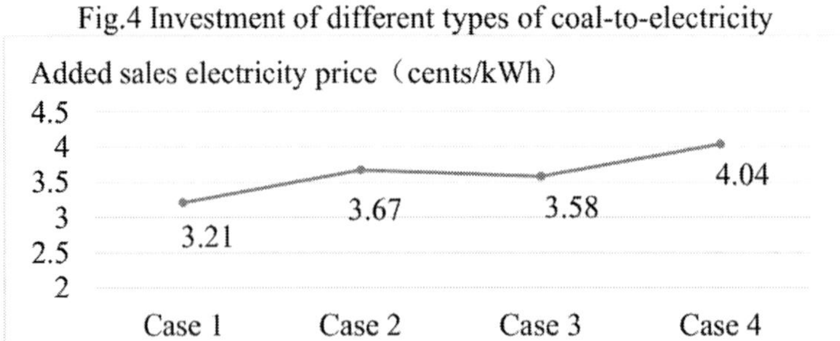

Fig.5 Added sales electricity price of different types of coal-to-electricity

From the perspective of environmental protection, the reduction amount of burning coal is 1.99 million tons in every heating season. As the emission reductions for each ton of coal are 2.493 tons of carbon dioxide, 0.68 tons of dust, 0.075 tons of sulfur dioxide and 0.0375 tons of nitrogen oxides, the pollutant emissions are reduced that of 4.95 million tons of carbon dioxide, 1.35 million tons of dust, 150,000 tons of sulfur dioxide and 70,000 tons of nitrogen oxides, which bring obviously environmental benefits.

6. Conclusions

In this paper, a smart grid planning method with coal-to-electricity based on multifactor adaptability is proposed to consider the influence of coal-to-electricity. A power prediction method with coal-to-electricity is proposed and typical parameters for different types of heating modes are extracted based on historical operational data, which lays a foundation for the study on coal-to-electricity integrated power grid. Furthermore, a smart grid planning method with coal-to-electricity is introduced, considering the investment cost of each participant, the construction capacity of power grid, and the social acceptance degree of electricity pricing comprehensively. The proposed model is universal and practical and meets the needs of smart grid planning and engineering practice under the new situation in China, which provides references for the subsidy policy formulation and is conducive for government to implement coal-to-electricity.

The proposed model and calculation method mainly aims for residents and public enterprises. But the above model does not consider industrial electrical boilers, electrical kilns, and other types of coal-to-electricity. The key point for further research is how to comprehensively study the power load characteristics and influences of industrial type coal-to-electricity.

References

[1] Liu ZY (2015) Global energy internet., China Electric Power Press, Beijing China.
[2] Tong ZQ (2016) Related issues of electric energy replacement implementation in Hebei Southern Power Grid, North China Electric Power University, Jul. 2016.
[3] Liang XL, Lu WB, Zhou HM (2015) Alternative energy in the energy network. Smart Grid, 2015, 3(12): 1192-1196.
[4] Yuan XR, Wu L, Zhang J, et al (2015) Situation of energy substitution and economic analysis of electric heating in Tianjin. Power demand side management, 2015, 2(5):24-29.
[5] Craciun VS, Trifa V, Bojesen C, et al (2012) Air source heat pump a key role in the development of smart buildings in future energy systems: Low cost and flexible experimental setup for air source heat pumps. In: Proceedings of the 2012 International Conference and Exposition on Electrical and Power Engineering, Iasi, Romania, 25-27 Oct. 2012, 984-989.
[6] Lin KP, Zhang YP, Di HF, et al (2004) Thermal performance of underfloor electric heating system with shape-stabilized PCM plates. Journal of Tsinghua Uni-versity(Science and Technology), 2004, 44(12):1618-1621.
[7] Lin KP, Zhang YP, Di HF, et al (2005) Experimental study of underfloor electric heating system with latent thermal storage air supply. Acta Energiae Solaris Sinica, 2005, 26(6):820-824.
[8] Han JL, Ma GZ, Hu SY, et al (2018) Analysis of electric heating load characteristics in south hebei power grid. In: Proceedings of the 2nd IEEE Conference on Energy Internet and Energy System Integration (EI2), Beijing, China, 22-23 Oct. 2018, 1094-1098.
[9] Chen H, Zhu J, Zhang B, et al (2017) A Novel combined forecasting method for short-term distributed electric heating power load. In: Proceedings of the 2017 IEEE Conference on Energy Internet and Energy System Integration (EI2), Beijing, China, 26-28 Nov. 2017, doi: 10.1109/EI2.2017.8245378.
[10] Ma LF, Long GB, Li XL, et al (2017) Research on influence of large-scale air-source heat pump start-up characteristics to power grid. In: Proceedings of the 2017 IEEE Conference on Energy Internet and Energy System Integration (EI2), Beijing, China, 26-28 Nov. 2017, doi: 10.1109/EI2.2017.8245377.

[11] Liu YR, Yang WH, Wang J, et al (2017) Load characteristic analysis of rural network before and after the coal-to-electricity project. Electrical Engineering, 2017, 18(4): 110-115.
[12] Shao H, ZHOU LM, Yuan JP, et al (2018) Analysis on impacts of domestic air source heat pumps to distribution networks in coal-to-electricity projects. In: Proceedings of the 6th Annual International Conference on Material Science and Engineering (IOP Conference Series: Materials Science and Engineering 397 (2018) 012001), 2018, 563-568.
[13] Weu JX, Ma ZB, Huang G S, et al (2016) Economic and environmental benefit from "electric power replacement" project in kilns production. North China Electric Power, 2016(3):66-70.
[14] Hou B, Zhang HY, Yu XJ, et al (2018) Research on comprehensive benefits evaluation of "Coal-to-Electricity" project based on combination weighting TOPSIS method. In: Proceedings of the 2nd IEEE Conference on Energy Internet and Energy System Integration (EI2), Beijing, China, 22-23 Oct. 2018, 1870-1875.
[15] Li ZW, Zhu WL, Sun PF, et al (2018) Study on power quality impacts of rural distribution network with large-scale heat pumps. In: Proceedings of the 2018 2nd International Conference on Power and Energy Engineering (ICPEE 2018), Xiamen, China, 3-5 Sept, 2018, 1-7.
[16] Huang YF, Zhu YJ, Mu G, et al (2018) Assessment of adjustable capacity of household heating load based on temperature forecast. doi: 10.13335/j.1000-3673.pst.2018.0547.
[17] Zhang N, Lu X, Mcelroy M B, et al (2016) Reducing curtailment of wind electricity in China by employing electric boilers for heat and pumped hydro for energy storage. Applied Energy, 2016, 184:987-994.
[18] Zhao J (2016) Study on the capacity of central heating and electric heating based on thermal scheduling to consume wind power. North China Electric Power University, 2016
[19] An JK, Sun PF, Han Jl, et al (2018) A power network planning method with coal-to-electricity considering technical and economical adaptability. In: Proceedings of the 2nd IEEE Conference on Energy Internet and Energy System Integration (EI2), Beijing, China, 22-23 Oct. 2018, 1110-1115.

Smart grid oriented Internet of Things technology and application analysis

Wenting Zhou[1], **Jingfeng Xiao**[1]*

[1]State Grid Xinjiang Electric Power Co. LTD, Urumqi 830002, Xinjiang, China

*email: 1158852782@qq.com

Abstract: In the development of modern society, the Internet itself with the characteristics of diversification and virtualization, has played a good application value. The Internet of Things can realize the interaction and construction of related network resources through the effective integration of physical and virtual bodies. It makes the overall information network communication and power industry structure to achieve the corresponding transformation and industrial upgrading, and ultimately promotes the transformation of social production structure. This provides good convenience for people and promotes the development of the whole society.

1. Introduction

The Internet of Things technology has been well applied in many fields. This modern technology is used in smart grid power generation, transmission and distribution, electricity consumption and other links. The Internet of Things technology enables the corresponding integration of various resources in the power grid, which is directly related to People's Daily life. The power grid industry will have higher requirements on the stability and security of the network. The application of Internet of Things technology can effectively guarantee the smooth completion of relevant functions and realize the application value of modern smart grid.

2. Scientific understanding of smart grids and the Internet of Things

First of all, smart grid is a kind of staggered management internetworking based on physical grid, using more advanced information communication and analysis, decision making, energy configuration, automatic control and other functions. It makes use of terminal sensors and power grid companies to form corresponding internetworking with customers. At the same time, it is necessary to analyze the high consumption of electricity generated due to uneven energy supply in the process of power resource transportation, as well as the high cost of power grid construction and long-distance transmission. Compared with the traditional high-voltage power transmission and transformation mode, smart power grid is characterized by observability, compatibility, real-time analysis, decision making and controllability. It can further optimize the modern power grid structure, power grid control and resource allocation, improve the overall service quality of the power grid, and finally realize the reasonable allocation of resources, so that different parts of assets can be optimized and used accordingly, and bring more stable power resources to users.

Secondly, the application of Internet of Things technology is generated with modern mobile communication technology and computer protection technology. Internet of Things is a network system connected with things. This technology itself is a technology that extracts and identifies information in time through sensors, GPS and other technical means, and carries out information transmission. In the application process of Internet of Things technology, it is necessary to collect sound, light, heat, chemistry and location information, and connect the information of articles to the network system according to specific protocols to complete the information exchange function, so that wireless and wired networks can fully integrate to transfer relevant information. Through cloud computing technology and related pattern recognition technology embodies the intelligent recognition and management of many aspects of the function.

3. Terminal acquisition layer

The most critical technology in the application of the Internet of Things itself is wireless sensor network technology, which can produce an overall effect on the development of the Internet of Things, but also enable the integration of other technologies. On the whole, the iot itself for smart grid reflects three levels of content (Figure 1).

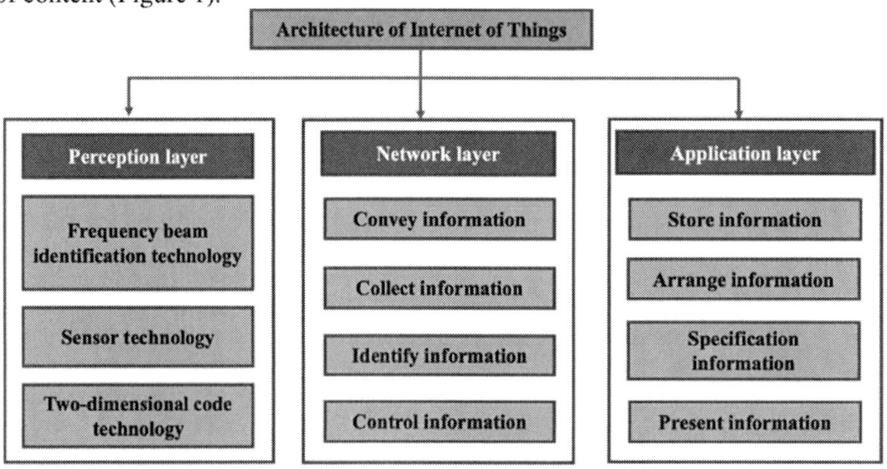

Figure 1. Iot for smart grid reflects three levels of content.

3.1 Perception layer

In this link, mainly through a variety of RFID technology and sensors and TWO-DIMENSIONAL code technology to collect relevant information content. After collecting information data at the perception layer, automatic identification technology is used to automatically identify the information, and effective transmission of data information content is carried out based on wired and wireless communication network to ensure that data information can be transmitted more safely and reliably, reflecting good transmission effect.

3.2 Network layer

The main task of the network layer is to transmit information, collect information, identify information and control information, so it is also called the information transmission layer. In the process of information transmission, it is necessary to make the coverage of various transmitted information more comprehensive and extensive. At the same time, the network layer itself needs to be based on the power fiber network to cooperate with the communication network, wireless network and broadband network, so as to realize the corresponding connection between the Internet of Things and the communication network dedicated to the smart grid. According to the specific situation to flexibly adopt communication methods, the corresponding data transmission. On this basis, the network layer formed by the specific data and information access network system and the core network system can realize high-speed transmission of the Internet of Things itself and the communication network dedicated to the power grid, so that data and information can be expanded [1].

3.3 Application layer

The existence of this level is mainly to comprehensively store, sort out and standardize the obtained information data content, so that the final data information can be presented on the correct basis. Through the effective application of data information, intelligent result analysis can also be carried out, and provide good basic decision for the relevant operators. In the service process of application layer, basic facilities can provide basic services and information processing for smart grid. Internet of things technology for smart grid itself can include smart grid power generation and transmission and the various links such as electricity, in order to realize the intelligent power grid control decision-making and service functions, the use of some modern intelligent computing technology and modern identification technology, the comprehensiveness of the content of different data in smart grid, arrangement and analysis [2].

4. Application analysis of Internet of Things technology for smart grid

The application of the Internet of Things is reflected in different links of the power system, especially in different links of power transmission. There are many links related to the Internet of Things, including the automation of power distribution and power acquisition, from the access of the initial power generation link to the subsequent detection and production management and supervision of the substation. Through the Internet of things technology in the smart grid power information content to realize intelligent control, and carries on the further optimization and configuration, can eventually realize tidal power, solar, and biomass power and wind power and other diversified power effect, make it play out in People's Daily life the good result, Ultimately, the overall power planning capability of state Grid will be improved to maximize the effect in the smart grid system in Figure 2 [3].

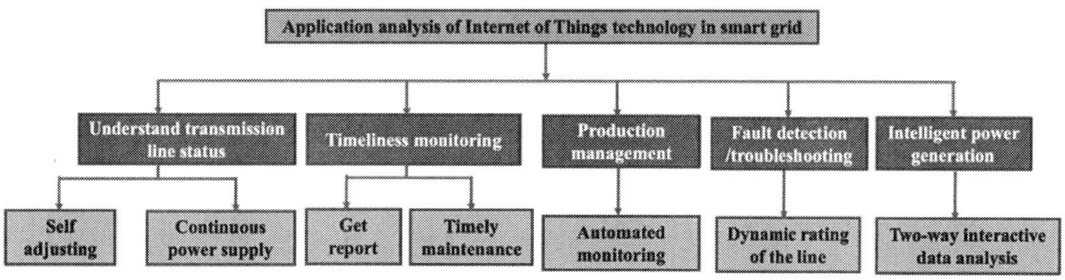

Figure 2. Application analysis of Internet of Things technology for smart grid

4.1 Knowledge of transmission line status

The understanding and monitoring of transmission line status is one of the most important application values of Internet of Things technology. Through the monitoring of different nodes in the whole transmission line by Internet of Things technology, various signs existing in the transmission line area, conductor temperature, circuit capacitance, tower tilt, etc., can be timely understood. The actual running state of transmission line is well sensed and analyzed. Smart grid can understand some abnormal operation conditions, self-regulate them, and provide users with the basis for continuous power supply, effectively preventing power outages due to delayed maintenance in the subsequent transmission line operation process, which affects People's Daily life [4].

4.2 Timely detection of equipment status

Through the Internet of things technology itself is equipped with some of the sensor and microelectronics technology, timely monitoring, on the whole line for understanding the operation parameters and running state of equipment, various technical indicators in the process of equipment operation and key parameters are obtained and analysis, and be able to real-time detection of power station and substation equipment data produced solutions, Obtain the specific data test report, the operating condition of the equipment for the corresponding prediction, and the related problems for real-time maintenance. In this case, not only can the actual level and link of each voltage in the traditional electrical equipment be detected correspondingly, but also the corresponding dynamic characteristics of the equipment in each link of wind power generation and solar power generation can be monitored and controlled at the same time. To be specific, sensors installed on the dam of hydropower station can monitor the actual situation of the dam itself, understand the problems existing in the dam itself, and prevent the dam break risk of the reservoir, making a significant contribution to water conservancy construction [5].

4.3 Management of power production

In practice, the management of electric power production originally reflects certain complexity characteristics, but in the management process of electric power production, full application of Internet of Things technology can reflect good monitoring effects. Some key steps and process and important equipment can through the Internet of things technology to realize the automation of a comprehensive and timely monitoring, and the corresponding coding way, every line taken identity recognition technology to prevent others in the process of electric power production to the operation of the power the entire production process execution error, or have some mistake. In addition, through the establishment of artificial intelligence database system, it is also possible to find some intelligent system operation problems existing in the equipment and timely report them so that the responsible person can understand them and effectively control these problems [6].

4.4 Monitor and troubleshoot power failure in time

In fact, with the increasing scale of modern power system, the role in People's Daily life is more and more obvious, the number of equipment has become more and more abundant, for the state detection of power equipment has generated higher requirements. Based on the climate conditions, in the condition of electric power equipment operate, make full use of the Internet of things technology to dynamically on the specific line, the application of this technology can be in time to face the problem of transformer oil temperature is too high or too negative charge, and start the emergency plan, make timely diagnosis for the fault of the problems associated with and response to, so that they can promote the troubleshooting, Solve the fault problem and provide good stability for the whole power system [7].

4.5 For intelligent power generation

People are increasingly demanding for the stability of the power grid, need attention in the process of the smart grid construction of the practical operation of electric power equipment and power distribution site operation situation, timely detection, to take effective safety protection measures, so that they can make intelligent electric network and realize the two-way interaction between users and data analysis, to achieve the effect of energy conservation and emissions reduction, At the same time, it can also timely collect users' electricity consumption data, realize remote monitoring and economic intelligent power dispatching, and provide stable power resources for schools, hospitals and other places with high electricity consumption [8].

5. Conclusion

To sum up, in the process of the development of modern society in China, the Internet of things technology itself reflects the good value of application of time in the Internet of things technology research is relatively short, compared with advanced countries there are some shortcomings, and technology of Internet of things has been applied directly to the smart grid construction has a great difficulty. This requires professionals in related fields to analyze better Internet of Things technology based on the objective situation of China, and apply it to smart power grid to play a good effect and provide people with more stable power resources.

About the author:

Wenting Zhou, female, Han nationality, born in March 1974 in Yichun, Jiangxi province, master candidate, senior engineer/deputy director, mainly engaged in digital new technology application research and management.

Jingfeng Xiao, male, Han nationality, xinyuan, Xinjiang, master's degree, deputy senior engineer/specialist, mainly engaged in digital new technology application research.

Reference

[1] Yang Yiyue. Research on Internet of Things technology application and security threat for smart grid [J]. China Informatization, 2019, 12:90-92.]
[2] Ma Jun. Smart grid oriented Internet of Things technology and its application [J]. Digital Communication World, 2020, 02:202.
[3] Leng Di. Technology and application of Internet of Things for smart grid [J]. Information System Engineering, 2020, 03:78-80.
[4] Zhou Chunlei. Smart grid oriented Internet of Things technology and its application [J]. Intelligent Building and Intelligent City, 2018, 9:69-70.
[5] Zeng Min. Application analysis of Internet of Things technology in smart power grid [J]. Information and Computer science, 2019, 13:150-151.
[6] Leng DI. Application analysis of Internet of Things technology in smart power grid [J]. Digital Communication World, 2019, 11:84.
[7] Xu Jing. Application analysis of Internet of Things technology in smart power grid monitoring [J]. Computer Products and Circulation, 2019, 12:81.
[8] Liu Youtang. Application analysis of Internet of Things technology in smart power grid [J]. Digital Communication World, 2020, 12:176-177.

Transient overvoltage characteristics in submarine cables of multi-terminal flexible HVDC transmission system

Man Feng[1a*], Tao Li[1b*], Jinjuan Liu[1c*]

[1] Wuhan NARI Limited Company, State Grid Electric Power Research Institute, Wuhan 430074, China

[a*]fengman@sgepri.sgcc.com.cn, [b*] litao3@sgepri.sgcc.com.cn, [c*]17363885@qq.com

Abstract—The overvoltage transient characteristics of submarine cables are one of the key technologies for submarine power transmission. This article takes the submarine cable of Zhoushan Multi-terminal Flexible HVDC Transmission Project as the research object. The MMC-HVDC simulation models at both ends of converter stations are built in PSCAD. The parameters and equivalent models of the flexible HVDC cables are determined, and the submarine DC cable operating overvoltage is calculated by using statistical methods. The simulation results show that the maximum overvoltage of the submarine cable, that is, the maximum overvoltage of the conductor to sheath, the sheath to the armour, and the armour to the ground are 400 kV, 7.27 kV and 0.45kV respectively. The maximum overvoltage when it occurs faults on grid-side or valve-side appears in the middle of the submarine cable. Maximum overvoltage of HVDC single pole grounding fault appears at the connection point of land cable and submarine cable. The simulation results and the discovery of features in this article provide reference in the design, selection, and testing of submarine power transmission equipment.

1. Introduction

Submarine cables are the critical part of trans-sea power transmission projects, which play an important role in realizing the regional and international power network interconnection. Once the submarine cable has an insulation breakdown due to overvoltage, it will cause power outages for maintenance over several days or even months, and cause significant economic losses. Therefore, the overvoltage transient characteristics of submarine cables are one of the key technologies for submarine power transmission, and its research has an important guiding role in the design, selection, manufacturing, and testing of submarine power transmission equipment.

For the multi-terminal flexible HVDC system, due to its short application history, the current research mainly focused on the basic principles such as Modular Multilevel Converter (MMC) modeling, power modulation strategies, control system designing, start-stop controlling, internal circulation suppression, fault characteristics, and protection configuration. Furthermore, some researchers started from the actual engineering. The Zhoushan multi-terminal flexible DC transmission project [1]-[3] and the Trans Bay Cable project [4],[5] is took as examples to research the insulation coordination of the converter station. Researchers found that the most serious cases of overvoltage in the converter station included single-phase grounding at transformer valve-side, short-circuit of the converter valve, grounding of the valve roof or DC lines. As the connection unit of the flexible HVDC system, the transient overvoltage characteristics of the submarine cable are related to the safe operation, yet there are few relevant literature reports.

This article takes the submarine cable from Dinghai Converter Station to Daishan Converter Station in the Zhoushan Multi-terminal Flexible HVDC Transmission Project as the research object. The statistical method is used to simulate the overvoltage distribution of the MMC flexible HVDC transmission cable, and the transient overvoltage characteristics of the submarine cable is obtained.

2. Simulation Model and Conditions

2.1. Simulation model selections

Zhoushan Multi-Terminal Flexible HVDC Project is a typical model of MMC. The project uses a flexible HVDC submarine cable for interconnecting, and the grid-side of Dinghai and Daishan converter stations is connected to a 220 kV AC system. The DC rated voltage is ±200 kV. The main connection of the DC field adopts a pseudo-bipolar DC connection.

A two-terminal MMC-HVDC simulation model is built in PSCAD as shown in Fig.1. The converter station control and protection include the station controlling and converter controlling. The station controlling determines the control strategy in the station, and the converter controlling generates the trigger pulse of the converter valve basic module. The simulation step is set to 20 μs, and the drawing step is set to 250 μs. The main wiring scheme of the converter station, the main circuit parameters of the flexible HVDC system, the technical parameters of the equipment, and the operating power range refers to the literature [3]. For arrester configuration, parameter selection and modelling methods refer to the literature [4] and [5].

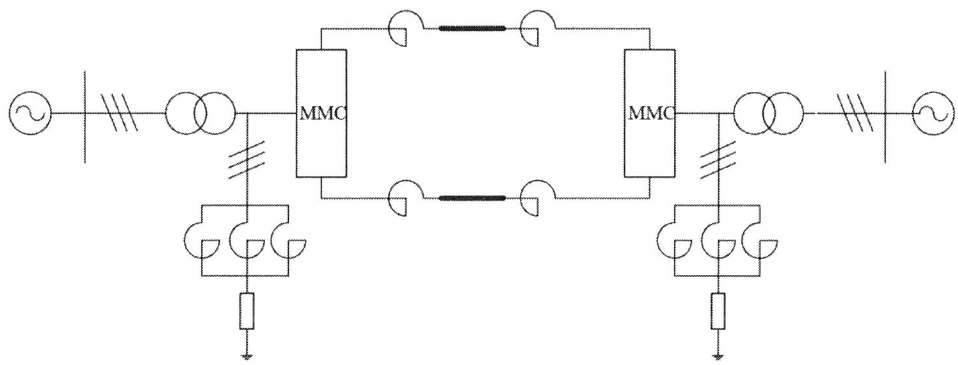

Fig.1 Two-terminal MMC-HVDC simulation model

The Dinghai-Daishan HVDC cable is 52 km long and is the longest cable segment in the five-terminal network. It includes a 2 km land cable and a 50 km submarine cable, both of which are cross-linked polyethylene DC cables. Among them, the submarine cable model is HVDC-YJQ41-200kV-1×1000mm^2. Its longitudinal section has 6 layers, including the conductor, insulation Ⅰ, lead sheath, insulation Ⅱ, armour, insulation Ⅲ. The parameters of each layer are shown in Table 1.

Table.1 Parameters of DC submarine cable

layers	thickness / mm	resistivity / (Ω·m)	relative permittivity
conductor	0.0191	1.75×10-8	
insulation Ⅰ	0.0160		2.30
lead sheath	0.0029	2.20×10-7	
insulation Ⅱ	0.0110		2.25
armour	0.0060	0.55×10-6	
insulation Ⅲ	0.0040		2.50

There are three equivalent models of cables in PSCAD: PI model, Bergeron model, and frequency-dependent characteristic model (FDC). Among them, the FDC model considers all frequency-related parameters and can accurately reflect the characteristics of the line in a larger frequency range. It can be further divided into Frequent Dependent (Mode) Model and Frequent Dependent (Phase) Model. For ideal transposed lines, both models can give sufficiently accurate results. For non-ideal transposed lines, the latter model is more applicable [6]. Because the DC-side line needless to consider the commutation, this article adopts the Frequent Dependent (Phase) Model.

This article takes the insulation level of the submarine cable from Dinghai Converter Station to Daishan Converter Station as the object, and makes the following simplifications:

(1) The equipment on the DC-side of the converter station is installed inside, so lightning overvoltage intruding from the DC-side is not considered.

(2) Due to the arresters are deployed in the incoming lines, AC busbar, and GIS separately, it can be considered that the lightning overvoltage on the AC-side can be limited, so the lightning overvoltage is not considered.

The statistical method is used to simulate the HVDC submarine cable operating overvoltage. The parameters are set through the Multiple Run component in the PSCAD. In this article, the time of failure is set to be randomly and uniformly distributed within 1.2~1.22s. The Multiple Run component can set the number of sampling calculations. This article is set to 120, and the program will automatically sample and calculate 120 times. After the calculation is completed, the Multiple Run component automatically finds the maximum operating overvoltage of conductor to sheath, sheath to armour, and armour to ground.

2.2. Computing conditions

The Zhoushan multi-terminal flexible DC transmission project is a 5-terminal flexible DC network, in which Daishan Converter Station and Dinghai Converter Station are connected by submarine cables. This article considers the following:

(1) The HVDC cable connecting Daishan Converter Station and Dinghai Converter Station is selected as an object due to it is the longest section of the 5-terminal system. Consider the most severe case, namely the outage of submarine cables between Daishan Converter Station and other converter stations.

(2) The electric field strength of the HVDC cable insulation layer is proportional to the insulation resistance coefficient and changes exponentially with temperature. When the cable is running at full load, the temperature of the insulation layer close to the conductor is higher, whereas the insulation resistance coefficient is lower and the electric field strength is smaller. In this condition, the fault overvoltage poses the greatest threat to the insulation layer. Hence, the maximum active load of 300 MW and the reactive power of 180 MVA are selected for calculation.

(3) 3 types of fault conditions: grounding fault on the grid-side of the converter transformer, grounding fault on the valve-side of the converter transformer, and grounding fault on the DC-side are mainly considered. Under such conditions, the maximum operating overvoltage of the conductor to sheath, sheath to armour, and armour to the ground is calculated. Because the greater the fault resistance, the less severe the fault caused. To strictly consider the grounding resistance, this article assumes that the grounding resistance is 0.01Ω, which is a metallic grounding fault.

The cable connecting Dinghai Converter Station to Daishan Converter Station includes two parts: submarine cable and land cable which are connected by flexible joints. Due to the terminals and connection positions of the cable have a higher failure risk, they are all set as over-voltage observation positions. To calculate operating overvoltage of conductor to sheath, sheath to armour, and armour to ground, a total of 7 calculation points is set up along submarine cable every 10km (as shown in Fig.2), where is, 0 km (Dinghai Station), 2km (land-submarine cable connection point), 12 km, 22 km, 32 km, 42 km, and 52 km (Daishan Station).

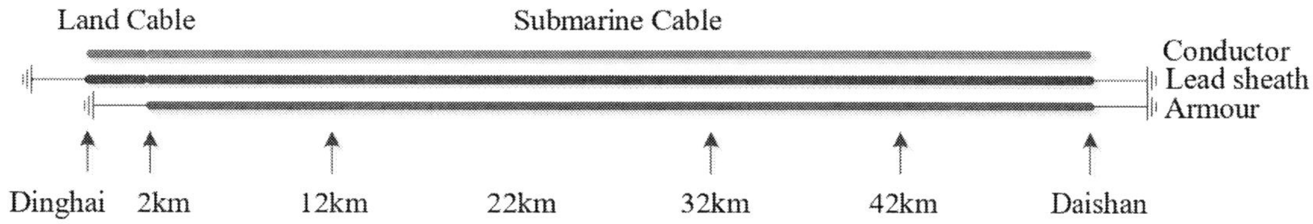

Fig.2 Positions of calculation points

3. Simulation Results and Analysis

3.1. Transient overvoltage characteristics of unipolar grounding

DC unipolar grounding faults are generally permanent faults. This type of fault is more serious, which will cause the converter station blocking after it occurs 20 ms. After blocking, the AC system can still form a fault loop through the fault point, the grounding point, and the freewheeling diodes of the three lower bridge arms. Thus, power on the AC circuit breaker after 100 ms due to the fault point can hardly extinguish itself [7]. Assuming that the negative pole of the DC cable is grounded at a single-pole, this situation changes the zero-potential reference point of the valve-side AC system, and the valve-side AC voltage has a DC bias. The fault current forms a loop through the grounding branch, the bridge arm reactor, the DC smoothing reactor, and the fault point. After the fault occurs, the negative pole voltage quickly drops to zero. Due to the existence of the module capacitance between the positive and negative poles of the converter station, the DC voltage between the poles will remain roughly unchanged, resulting in a large overvoltage to the ground and current on the DC pole line of the healthy pole. After the converter valve blocking, the pole line voltage current decreases rapidly. According to the cable continuous operating voltage 202kV, the charge rate is 0.7, the DC reference voltage is calculated as 289kV.

3.2. Transient characteristics of operating overvoltage

In this article, the operation overvoltage simulation of submarine cable is carried out for the asymmetric grounding fault of the outgoing line of AC field, grounding fault of the coupling transformer of the bridge arm reactor, short circuit fault of the bridge arm, AC phase grounding, AC phase-to-ground and phase-to-phase operating overvoltage, and grounding fault along the DC cable. The simulation results are shown as Fig.3.

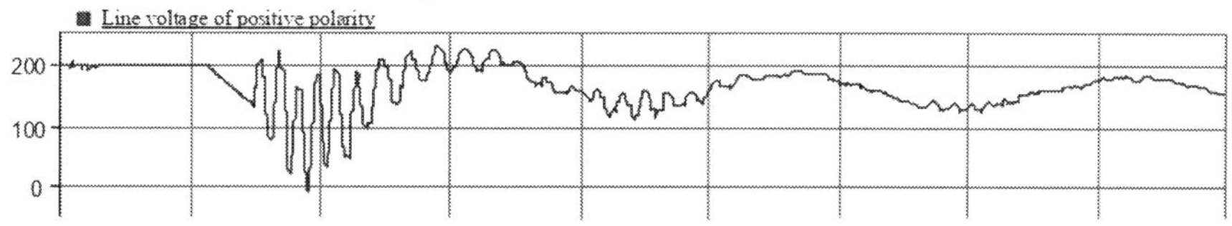

(a) Voltage in the cable when occurs the asymmetric grounding of the outgoing line in AC field (kV)

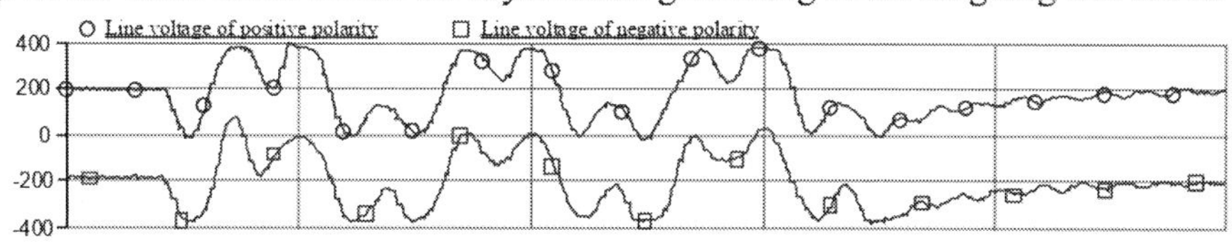

(b) Voltage in the cable when occurs grounding fault of the coupling transformer of the bridge arm reactor (kV)

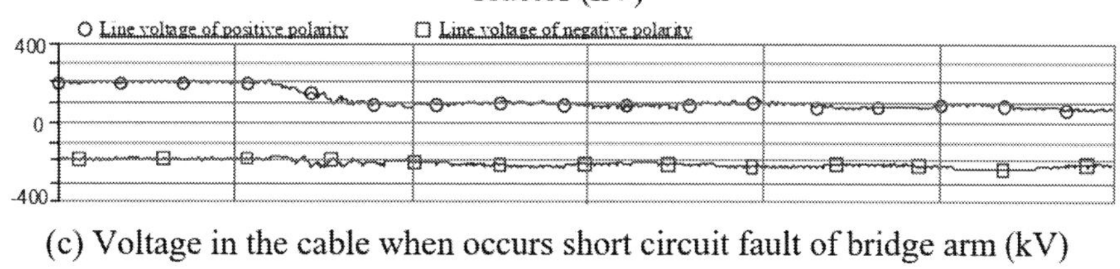

(c) Voltage in the cable when occurs short circuit fault of bridge arm (kV)

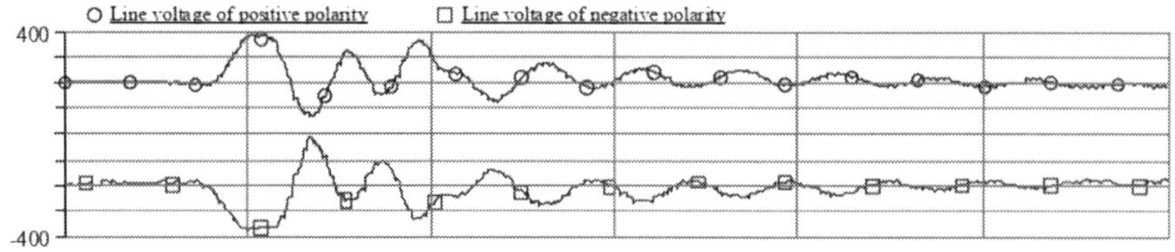

(d) AC phase-to-ground and phase-to-phase operating overvoltage (kV)

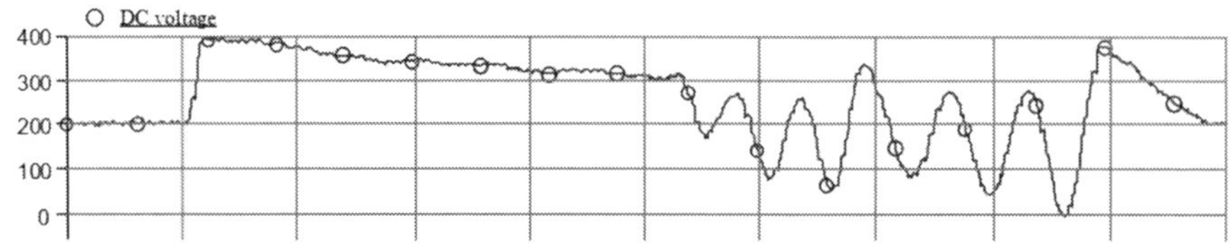

(e) Voltage in the cable when breakdown of faulty pole conductor to sheath and armour (kV)

Fig.3 Simulation results of transient characteristics of operating overvoltage

The operating overvoltage simulation is performed at the 7 observation points set up for the submarine cable. The calculation result shows that the highest overvoltage in the middle of the cable is 400kV, which is 1.98p.u (1p.u=202kV), whereas the lowest overvoltage is 398kV, and the standard deviation is 0.6kV. The statistical distribution of the central overvoltage is shown in Fig.4. The maximum overvoltage at the ends of the cable is 395p.u. The DC overvoltage decays with time and decays to the operating voltage within 40ms. The calculated maximum overvoltage of the sheath-armour is 6.69kV. The statistical conductor-to-sheath overvoltage distribution along the line is shown in Fig.5.

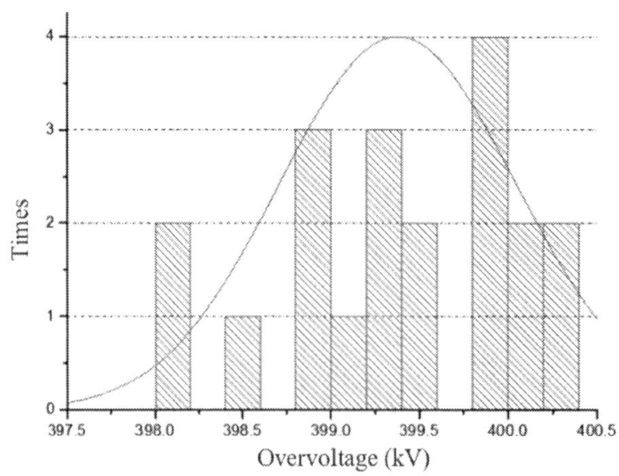

Fig.4 Statistical distribution of overvoltage in the middle of submarine cable

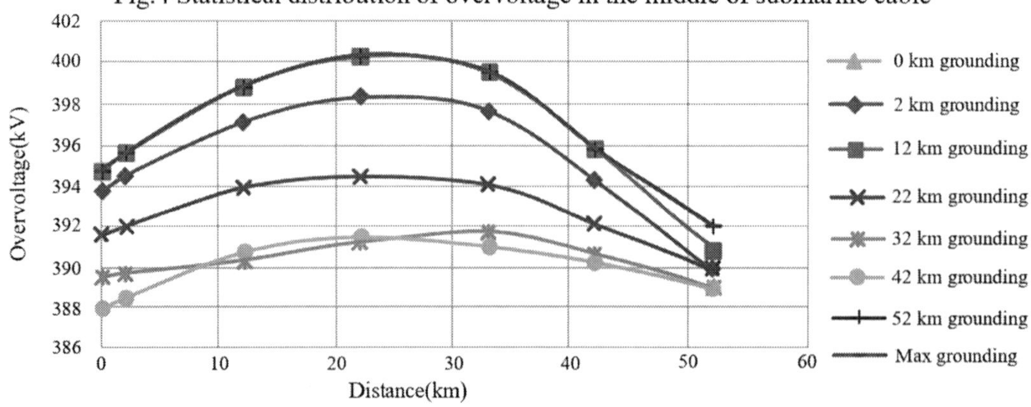

Fig.5 Overvoltage distribution of conductor to lead sheath when occurs monopolar grounding fault

The summary simulation result of the operating overvoltage is shown in Table 2. The maximum operating overvoltage of the same polarity as the operating voltage between the conductor to the sheath is 400kV (1.90p.u). The maximum operating overvoltage appears in the middle of the sound pole cable when it occurs the cable terminal head grounding failure.

Table.2 Parameters of DC submarine cable

Cause of fault	Operating overvoltage in parts of		
	conductor to sheath	sheath to armour	armour to ground
The asymmetric grounding of outgoing line in AC field	234 kV	5.9 kV	0.38 kV
Coupling transformer grounding of the bridge arm reactor	394 kV	7.12 kV	0.44 kV
Short circuit of the bridge arm	210 kV	-	-
AC phase-to-ground and phase-to-phase operating overvoltage	383 kV	7.27 kV	0.45 kV
Grounding fault along the DC cable	400 kV	6.69 kV	0.13 kV
Maximum value	400 kV	7.27 kV	0.45 kV

4. Conclusion

The MMC-HVDC simulation models at both ends of the Dinghai and Daishan converter stations are built in PSCAD software, the parameters and equivalent models of the flexible HVDC cables are determined, and the DC cable operating overvoltage is simulated using statistical methods. The main conclusions are as follows:

(1) In the case of three-phase grounding faults on the converter transformer grid-side, three-phase grounding faults on the valve-side, and HVDC single pole grounding faults, the overvoltage of the HVDC cable conductor to sheath fault is more serious, which causes the converter valve to be quickly blocked and the commutation. The maximum overvoltage when it occurs faults on grid-side or valve-side appears in the middle of the submarine cable. Maximum overvoltage of HVDC single pole grounding fault appears at the connection point of land cable and submarine cable

(2) The maximum operating overvoltage of the same polarity as the operating voltage between the conductor to the sheath is 400kV. The maximum operating overvoltage appears in the middle of the sound pole cable. The statistical maximum overvoltage along the cable is distributed in an umbrella shape, with the largest in the middle and decreasing to both sides. The overvoltage of the HVDC cable has the same polarity as the operating voltage The induction operating overvoltage of sheath to armor in the middle of the cable is the largest, and the maximum overvoltage is 7.27kV.

References

[1] Li, Y., Jiang, W., Yu, S. (2014) System design of Zhoushan multi-terminal VSC-HVDC transmission project. High Voltage Engineering, 40: 2490-2496.

[2] Zhou, H., Shen, Y., Li, M. (2013) Research on insulation coordination for converter stations of Zhoushan multi-terminal VSC-HVDC transmission project. Power System Technology, 37: 879-889.

[3] Deng, X., Wang, D. (2013) Research on transient overvoltage for converter station of Zhoushan multi-terminal VSC-HVDC project. Power System Protection and Control, 41: 111-119.

[4] Zhang, Z., Xu, Z., Xue, Y. (2013) Study of overvoltage protection and insulation coordination for MMC based HVDC. Power System Protection and Control, 41: 58-64.

[5] Huang, J., Zhao, C. (2013) Study on overvoltage and insulation coordination of MMC-HVDC converter station. Journal of North China Electric Power University, 40: 1-6.

[6] Liu, X., Wang, X. (2014) Simulating experiment and transient analysis on switching surges in cable collection grid of wind power plant. High Voltage Engineering, 40: 61-66.

[7] Wang, S., Zhou X., Tang, G. (2011) Modeling of modular multi-level voltage source converter. Proceedings of the CSEE, 31: 1-8.

Research on Data Acquisition Technology of Distribution Automation Based on Internet of Things Communication Protocol

HuiChao Wu[1a*], MingYi He[1b*], HaiZeng Wei[1c]

[1] Nari Group Co.Ltd, Nanjing 211106, China

[a*]wuhuichao@sgepri.sgcc.com.cn,[b*]hemingyi@sgepri.sgcc.com.cn, [c]weihaizeng@sgepri.sgcc.com.cn

Abstract: With the gradual application of the Internet of Things technology in the field of distribution automation, distribution terminals gradually adopt the Internet of Things protocol to communicate with the master station system of distribution automation. while the master station system of distribution automation supports the access of the traditional power dedicated protocol terminal, it faces how to solve the problem which how to access the distribution terminal of the network communication protocol. This paper proposes the MQTT protocol data acquisition architecture based on the existing master station system of distribution automation system. Through the research on the transmission characteristics of power distribution service data and MQTT protocol, the message structure of the MQTT protocol carrying power service data and the source maintenance process of the equipment model are designed. , It realizes the direct collection of the terminal data of the distribution Internet of Things in an object-oriented way, gives full play to the value of the existing distribution master station, reduces investment costs, and saves system maintenance workload.

1. Introduction

The low-voltage distribution network is at the end of the power system, directly undertaking the power supply tasks of grid users, and is a key link that affects the level of power supply services. For a long time, the low-voltage distribution network has lagging behind in automation, and at the same time it faces the challenges of rapid changes in management requirements, large-scale management equipment, and high service requirements [1]. In recent years, with the massive access of electric vehicles, distributed energy sources, micro grids, energy storage devices and other facilities, higher requirements have been placed on the automation level of low-voltage distribution networks [2].

To improve the automation level of the low-voltage power distribution network, communication protocols are the key link. traditional power communication protocols face a wide variety of terminal equipment problems, lack effective solutions, equipment access and business applications are subject to many restrictions, at the same time traditional The electric power communication protocol is not an object-oriented protocol, and a lot of work is added when terminal equipment is connected, which cannot meet the needs of fast access and plug-and-play of mass equipment [3-4]. Distribution Internet of Things is a new type of power distribution network form produced by the deep integration of power distribution technology and Internet of Things technology. Through the interconnection and intercommunication of low-voltage equipment in the distribution network, comprehensive perception, data fusion and intelligent applications of the distribution network are realized [5-6]. With the gradual application of Internet of Things technology in the field of power distribution, new power distribution terminals have gradually developed into Internet of Things terminals, using Internet of Things protocols to send data, and deploying an Internet of Things management platform at the cloud master station to be specifically responsible for accessing such terminal devices [7]. The master station system of distribution automation obtains the data of the IoT terminal through the IoT management platform. Due to the large investment and long construction period of the IoT management platform, the existing master station system of distribution automation does not support direct access to the power distribution through the IoT protocol Terminals, causing some power distribution Internet of Things terminals to fail to connect to the main station system in time, affecting the construction progress of the power distribution automation system.

This paper proposes to expand the communication protocol library in the master station system. Through the analysis and application research of the characteristics of the Internet of Things communication protocol MQTT, it realizes the direct collection of the data of the power distribution Internet of Things terminal, and gives full play to the value of the existing master station system of distribution automation. , Reduce investment costs and save system maintenance workload.

2. The communication mechanism of MQTT protocol

2.1. The features of MQTT protocol

MQTT released by IBM in 1999, is a lightweight communication protocol based on the Publish/Subscribe mode, and its message header has a minimum of 2 bytes , The message body is binary and compact, which is very suitable for the resource-constrained Internet of Things field. It can provide real-time and reliable message services for remote devices with very few codes and limited bandwidth. MQTT is used in Internet of Things, small devices, and mobile applications. It has a wide range of applications. The protocol runs on TCP/IP's orderly, reliable, and bidirectional network connection.

2.2. The communication model of MQTT protocol

The MQTT protocol supports point-to-point, group sending, group sending and other message interaction scenarios through topic-based subscription/publishing modes. The message routing method is flexible, the message body is in binary form, and the message format can be customized for each business scenario, which realizes message distribution and application. Decoupling between. Message distribution does not need to know the payload content. There are three identities in the MQTT protocol: Publisher, Broker, and Subscriber. Among them, the publisher and subscriber of the message are both clients, the message broker is the server, and the message publisher can be the subscriber at the same time. the communication process is shown in the following figure:

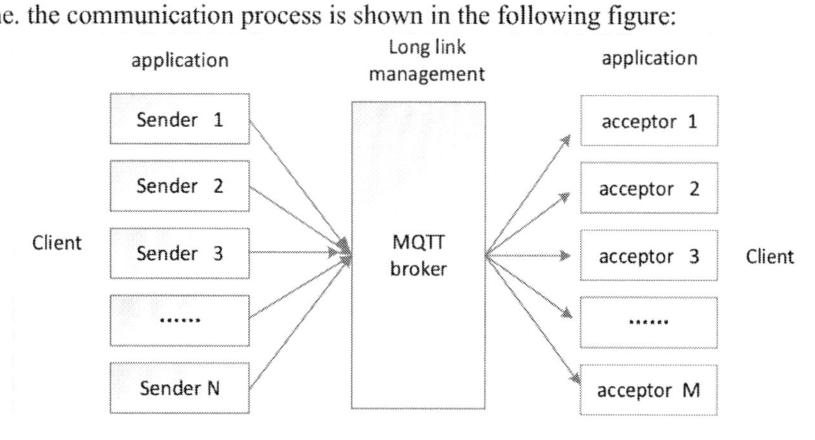

Fig. 1 The MQTT protocol communication process

When sending data, it sends data as a publisher, and when receiving data, it receives data as a subscriber. MQTT proxy is similar to the role of a post office or messenger. Its main function is to receive network connections from the client and receive messages published by the client. Application messages, processing client subscription and unsubscription requests, forwarding application messages to eligible client subscriptions, etc.

3. The MQTT data acquisition architecture of the master station system of distribution automation

The data acquisition function of the master station system of distribution automation is mainly responsible for the real-time data communication processing tasks between the power grid monitoring center and the power distribution terminal. The basic tasks include information exchange, command transmission, protocol organization and interpretation, channel encoding and decoding, and satellite Time synchronization, reasonable allocation of acquisition resources, etc.. the master station system of distribution automation converts the terminal data of different communication protocols and different data types into the internal data format of the master station system in advance to provide real-time data for the application functions of the master station.

3.1. The data acquisition software architecture of MQTT

The data acquisition function is composed of functional modules such as parameter management, collection task management, communication management, protocol processing, and data interaction. The parameter management module realizes the synchronization management of the parameters of data acquisition, and performs parameter synchronization in time. the collection task management module makes comprehensive judgments according to the collection server status, process status, and channel status, and manages the results of multiple machines, multiple channels, and multiple sources. data interaction module Realize the interaction of data and control commands between front and backend. data acquisition man-machine service provides manual operation interface for automation personnel to maintain and monitor the collection function. traditional protocol processing module converts data and control commands according to the configuration definition of different protocols. The above modules cooperate with each other to complete the data acquisition of the traditional power distribution terminal.

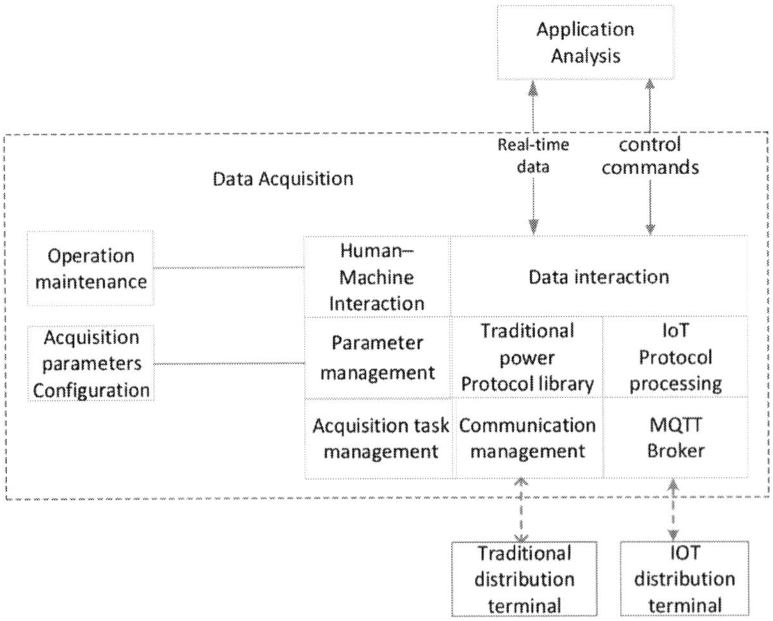

Fig. 2 The software architecture diagram

The Internet of Things MQTT protocol data acquisition is based on the original collection function, adding MQTT proxy module and Internet of Things protocol processing module to realize data acquisition on Internet of Things power distribution terminals. The MQTT proxy module uses Mosquitto middleware, which mainly implements TCP link management and the publishing/subscription of MQTT messages. the IoT protocol processing module parses out the power business data according to the definition of the MQTT message format and sends it to the data exchange module or exchanges data from it. The module obtains the master station's downlink control information and converts it into an MQTT protocol message and sends it to the MQTT agent. the detailed software architecture is shown in the figure below.

3.2. The MQTT data flow of master station system of distribution automation

The field Internet of Things power distribution terminal data is sent to the main station data acquisition software in the form of MQTT messages via the remote network. The data acquisition software converts the MQT messages into a unified data format inside the system and sends them to the main station background application. , The specific data flow is shown in the figure below:

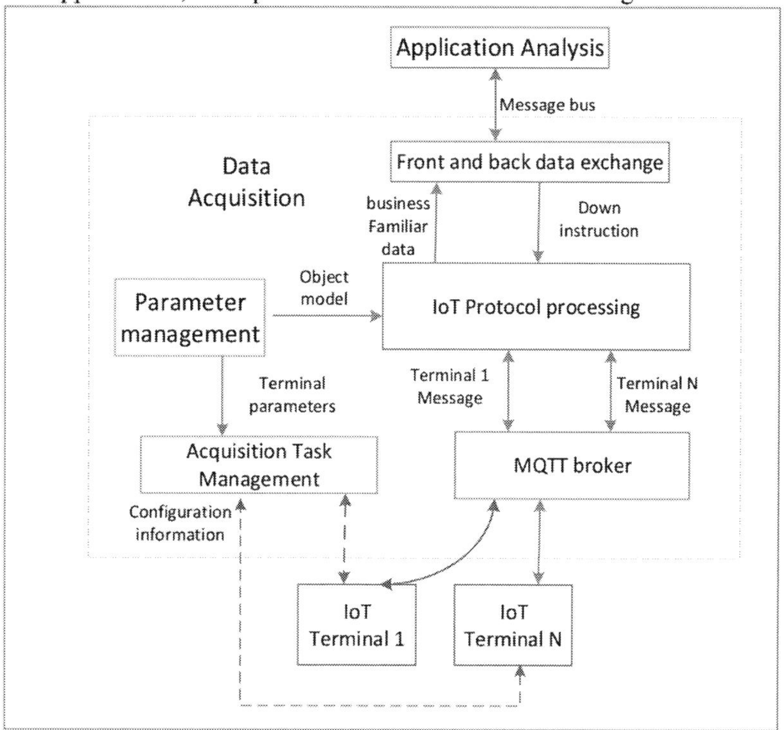

Fig. 3 MQTT protocol data flow diagram of distribution master station

(1) Before the Internet of Things terminal is connected to the main station system, it needs to provide the terminal's own ledger information and object model files, which are imported to the main station system by the parameter management module of the data acquisition software.

(2) Each time before the Internet of Things terminal goes online, first access the collection task management module to obtain the IP address and communication port of the front-end processor to be connected.

(3) After the IoT terminal obtains the communication parameters, it actively connects to the MQTT proxy module of the front-end processor through the MQTT connection message.

(4) The IoT terminal encapsulates the real-time data of the device into an MQTT message and publishes it to the MQTT proxy module in front of the master station.

(5) The IoT protocol processing module cyclically subscribes to the MQTT message sent from the terminal from the MQTT module, and parses the real-time data from the message according to the protocol structure, and simultaneously receives the real-time data and writes it into the data buffer queue.

(6) The front-end and back-end data interaction module reads the business data from the real-time data buffer queue, converts it into a unified data format inside the system, and sends it to the message bus. the back-end application obtains the real-time data of the entire network from the message bus to complete various business analysis Processing function to realize real-time monitoring of the entire power grid.

4. The key technologies for data acquisition of IoT MQTT

The MQTT protocol is not a transmission protocol designed for the power industry. The protocol does not define the format of business data messages and cannot be directly applied to the power distribution automation system. Compared with the traditional power communication protocol IEC104 protocol, the MQTT protocol does not have the concept of a point table. The master station system of distribution automation system cannot continue to be used by linking the primary and secondary equipment of the power system through the point table. the application of the Internet of Things MQTT protocol to realize the collection of power distribution automation data requires the following key technologies to be solved.

4.1. The message structure of distribution

The MQTT protocol classifies messages by topics. In order to avoid mutual interference between messages, the distribution automation master station assigns three topics to each terminal in advance. The first topic is defined as /device/data/{termId}, where termId is The ID number of the terminal itself is used for the power distribution terminal to send power business data and the platform to receive the data. the second topic is defined as /device/command/{termId}, which is used for the master station to send control commands in advance, and the power distribution terminal to receive the master Station command. the third topic is defined as /device/response/{termId}, which is used for the power distribution terminal to send the response master station command message, and the master station receives the command response message. If there are N power distribution terminals in a power distribution automation system, the total number of topics needs to be set to 3N. Each terminal needs to subscribe to one topic and publish two topics. and the master station system of distribution automation needs to subscribe to the topics published by 2N terminals and publish N topics for different power distribution terminals.

The distribution business data message structure adopts the object-oriented description method to directly transmit the object-oriented data. The distribution master station can directly identify the specific business information after receiving the MQTT message, without the need to analyze the data through the point table like the traditional power communication protocol. It saves a lot of work of checking information point by point between the master station and the terminal. The distribution service data message structure mainly includes three levels of equipment, measurement type, and data value to represent real-time data. For example, the A-phase voltage value of a switch is 10.56, the B-item voltage is 10.24, the switch state is closed, and the description of the message structure as follows:

```
{
    "currentTime":"2021-03-10  13:30:50"
    "break": /*device type */
    [
        {
            "rId": "1234567", /*device   rdf_id*/
            "Analog":
            [
                {
                    "measurementType":"Ua",
                    "value":"10.56",
                    "status":"1",
                }
                {
                    "measurementType":"Ub",
                    "value":"10.24",
                    "status":"1",
                }
            ]
            "Discrete":
            [
                {
                    "measurementType":"YX",
                    "value":"1",
                    "status":"1",
                }
            ]
        }
    ]
}
```


Fig. 4 MQTT message payload data structure

4.2. Device model source maintenance

The MQTT protocol of the Internet of Things uses an object-oriented way to transmit real-time data. How the master station system of distribution automation and the power distribution terminal use a common device model is the key to realize the plug-and-play access of the IoT terminal. The collection object of the low-voltage distribution network is mainly the distribution transformer station area. The inside of the station area includes various traditional measurement and control devices such as low-voltage branch switches, concentrators, smart meters, and low-voltage fault indicators, as well as new types of power stations such as charging piles and photovoltaics. Electrical equipment. The equipment model maintained at the source end includes the primary power equipment model, the secondary sensor equipment model, the primary equipment model and the secondary equipment model association relationship in the station area.

The sensor device ID is managed uniformly in the asset management system (PMS system), and the GIS system is obtained from the asset management system and is guaranteed to be unique. the source of source maintenance is in the GIS system, and the primary device topology in the GIS comes from the design and construction drawings. Any changes to the primary or secondary equipment in the district should be maintained in the GIS first, and the correctness of the topology should be ensured. in the GIS, the primary equipment is manually associated with the corresponding sensor device ID, and the information flow is transferred to the master station system of distribution automation system, and the master station system of distribution automation system Distribute the archives of the station area to the station area power distribution terminal, and finally realize the unification of the model between the main distribution station and the terminal. The source-side maintenance process of the equipment model is shown in the following figure:

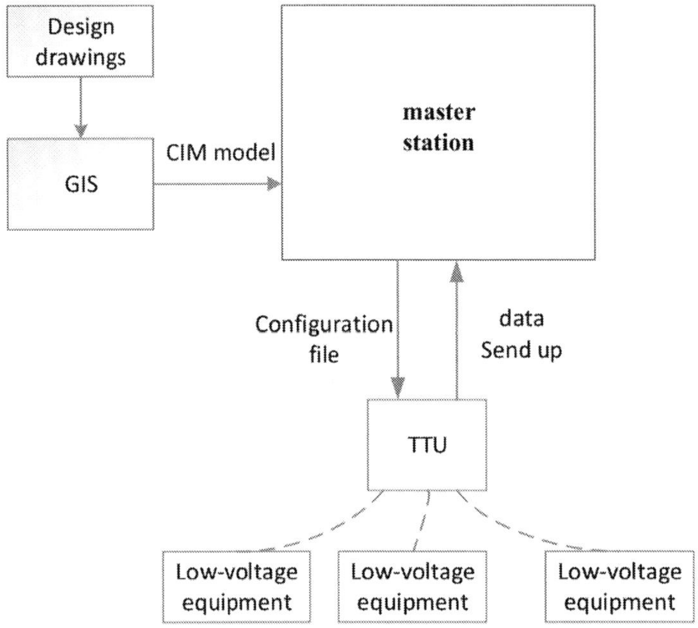

Fig. 5 Equipment model source maintenance flowchart

5. The MQTT protocol data acquisition test

In order to test the feasibility of the traditional master station system of distribution automation to collect data through the Internet of Things MQTT protocol, a test environment as shown in Figure 4 was built. The main equipment of the test environment includes 2 PC servers, 2 IoT terminal stations, and 1 switch.

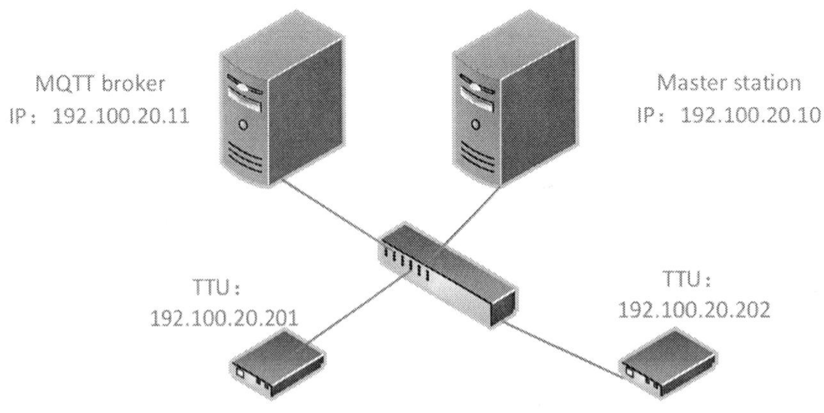

Fig. 6 test environment

The configuration of the master station system of distribution automation server is shown in the table below:

Table 1. Equipment list table

SN	Device Name	Model and configuration
1	MQTT broker	CPU: Intel Xeon CPU E7-4820 v2 RAM: 64GB operating system: Linux 3.10.0-862.el7.x86_64

2	FES	CPU: Intel Xeon CPU E5-2450 v2 RAM: 64GB operating system: Linux 2.6.32.41-Rocky4.2-x86_64

The master station system of distribution automation selects a PC server, deploys the OPEN5200 master station system of distribution automation development platform, and enables applications such as DB_SERVICE, PUBLIC, SCADA, FES. MQTT proxy server deploys Mosquitto message agent. power distribution terminal selects smart station terminals that support MQTT protocol.

In the test process, the process of importing the CIM model from the GIS system is simplified, and the primary equipment model and sensor equipment model of the grid area are manually entered directly at the main distribution station, and the association between the primary equipment and the sensor secondary equipment is completed. The station area terminal actively connects to the collection task management module in front of the master station, obtains the IP address and communication port of the MQTT proxy server, and downloads the sensor device model of the station area. the station area terminal organizes real-time data based on the sensor device model An object-oriented MQTT message structure is published to the MQTT proxy server. the front-end server cyclically subscribes MQTT messages to the MQTT proxy server to complete the data acquisition of the terminal in the station area. As the master station eliminates the need for meter maintenance work, the average access time of the two smart station terminals has been prompted from 2 hours to 15 minutes, and the access efficiency is increased by about 90%.

6. Conclusion

The State Grid Corporation of China is vigorously promoting the development and construction of the power Internet of Things. The power distribution field is directly related to users in the "generation, transmission, transformation and distribution". The construction of the Internet of Things in the field of power distribution is the key to the real realization of the Internet of Things in the entire power grid The application of the Internet of Things MQTT protocol to the power distribution automation system has significantly improved the flexibility and access efficiency of various power equipment data access, and solved the inconsistent access protocols of various terminal equipment such as video monitoring, temperature and humidity collectors, and capacitor equipment. At the same time, it reduces the workload of operation and maintenance personnel and improves the level of distribution automation operation management. However, the MQTT protocol is not a dedicated communication protocol for the power system. The business characteristics of the power system are not considered enough. The master station system and terminal devices must be adapted and modified. Further research is needed to increase the automatic conversion technology of the power equipment model to the MQTT message format. Gradually promote the promotion and application of the MQTT protocol in the power industry.

References

[1] Shi Jie, Ai Qian. Research on Several Key Technical Issues of Smart Grid Realization[J]. Power System Protection and Control, 2009, 37(19): 1-4.
[2] Li Peng, Dou Pengchong, Li Yuwei, et al. Application of microgrid technology in active distribution network[J]. Electric Power Automation Equipment, 2015, 35(4): 8-16.
[3] Gu Jianwei, Zhao Cuiran, Zhang Tiefeng. Research on Plug and Play Realization Technology of Power Distribution Terminal Based on IEC61850[J]. Power System Communication, 2012, 32(4): 31-35.
[4] Wu Huichao, Wu Yi, Zhu Haibing, et al. Research on plug-and-play system of distribution automation terminal based on IEC61850 standard [J]. Power Supply, 2015, 01: 60-63.
[5] Lu Jun, Luan Wenpeng, Liu Riliang, et al. Distribution Internet of Things Architecture Based on Comprehensive Perception and Software Definition[J]. Power System Technology, 2018, 42(10): 3108-3115.

[6] Hui Xiaolin, Sun Zhenquan. The integration of intelligent distribution network and electric Internet of Things [J]. Internet of Things Technology, 2011, 1(8): 31-35.
[7] Zhang Jichuan, Chen Lei, Zhang Mingyu, etc. The concept and application of intelligent terminals for power distribution Internet of Things [J]. High Voltage Technology, 2019, 45(6): 1729-1736.

Research on Multi-point Hopping Communication System in Power Network Based on Information Fusion Model

Lifeng Yu[1a*], Kaibo Hu[1b], Jiachen Lv[2c], Zhiling Xia[1d]

[1]ZHEJIANG ZHENENG LANXI ELECTRIC POWER GENERATION CO.,LTD., Jinhua, 321100, China

[2]ZHEJIANG ZHENENG ELECTRIC POWER CO., LTD., Hangzhou, 311200, China

[a*]13738993386@139.com

Abstract—Multi-source heterogeneous information will cause access burden to the power network, resulting in poor performance of multi-point hop communication. Therefore, a multi-point hop communication system based on information fusion model is designed. The IEC61970/61968CIM is selected as the integrated bus general data acquisition scheme of the information fusion model. The local-global distributed information fusion mechanism is used to realize the communication of the power multi-point hop communication system. In this paper, the artificial neural network algorithm is used to fuse local information features, and according to D-S evidence theory, the global decision-level fusion is carried out from both the spatial domain and the time domain. Through the information fusion model, the multi-point hop communication information in the power network realizes efficient transmission. The experimental results show that the application of the system in the power network to implement multi-point hop communication, the packet loss rate is less than 0.25%, the transmission delay is less than 30ms, and the communication performance of the power network is improved

1. Introduction

With the continuous improvement of human living standards, electricity has become an indispensable and important energy in human life and work. The security and effectiveness of power system determine the development of power network. How to run the power network efficiently, safely and economically is the most important part in power field [1]. At present, the information transmission and processing technologies of power networks lag behind the growth rate of the scale of power grids, and the phenomenon of repeated collection of the same information and isolation of part of the data of power grids is likely to occur, which leads to the long or missing transmission time of the edge data of power networks and affects the control decision of power systems [2-3].

Multi-point hop communication is the main communication technology in power network. Multi-point hop communication is easy to cause high packet loss rate. Information fusion technology can effectively integrate multi-source information in a comprehensive and accurate manner, and facilitate efficient transmission in power network [4]. An efficient multi-point hopping communication system is an important way to solve the problems existing in current power information system. Information fusion model can solve the problem of repeated collection of information and edge data isolation [5]. Based on the hierarchical, self-organizing and adaptive intelligence technology of information fusion technology, the model is applied to the multipoint hopping communication system of power grid.

2. Power Network Multipoint Hop Communication System

2.1. Overall System Structure

The overall structure diagram of the multipoint hop communication system based on the information fusion model is shown in Figure 1.

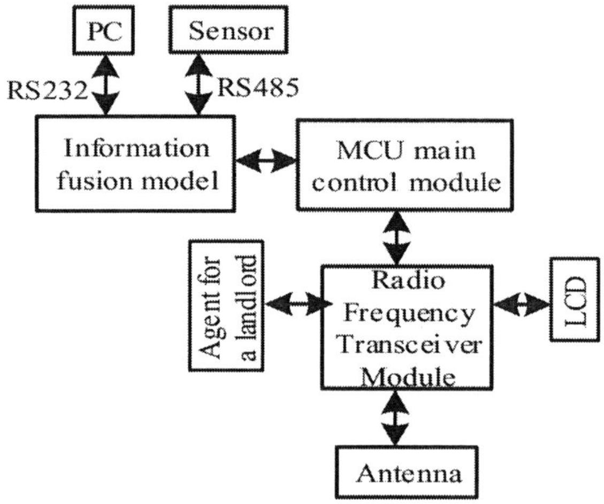

Fig. 1 Overall system structure

As can be seen from Fig. 1, the multi-point hopping communication system of power network adopts the information fusion model to send the multi-point information of the internal and external of power network to the MCU main control module after multi-level and multi-class processing, and then sends the main control module to the RF transceiver module, and realizes the multi-point hopping communication of data through antenna [6].

The MSP430 series MCU F1491 of TI Company is selected as the main control chip of the system, and the nRF905 of Nordic Company is selected as the RF transceiver module of the system to realize wireless data transceiver [7]. Using RS-232 interface to connect and communicate with PC, RS-485 communication interface can meet the connection requirements of different instruments in power system, choose high-gain external antenna or PCB antenna to realize long-distance multi-point hopping communication in power network.

2.2. Information Fusion Model

The information fusion model can combine the information of local space and temporal redundancy characteristics of different media, different modes and different sources, and make the information in the multi-hop communication system of the power grid consistent and comprehensive through effective information combination rules [8], so as to realize the successful communication of multi-hop communication of the power grid. Artificial Neutral Network (ANN) algorithm is applied to local information feature fusion, D-S evidence theory is applied to global decision level fusion, and information in different parts of the power network is efficiently transmitted through information fusion model. The overall structure of the information fusion model is illustrated in Figure 2.

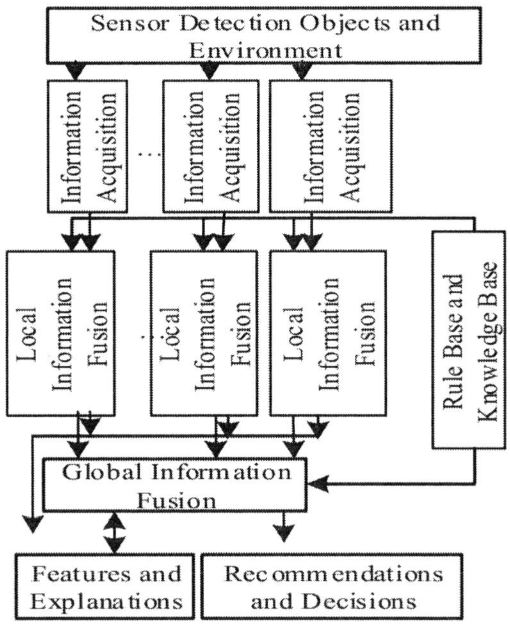

Fig. 2 General structure of the information fusion model

The local information fusion part is used to describe and explain the state features of power system [9]. The global information fusion part fuses the state features obtained from the local information fusion part and outputs them through effective decision making. Because different information fusion parts have different information types and fusion targets, different fusion methods are needed.

ANN fusion method can make use of the mapping relationship between identity and observation parameters of multi-hop communication system to mark the required information. D-S evidence theory is an efficient intelligent technology fusion method, which applies artificial intelligence method to explain and obtain state and assistant decision [10].

After the original information is collected by the information fusion model, the operation state of power network is described by preprocessing, and the complete and accurate feature state is judged by complementary fusion and redundant fusion. The model can effectively correlate the result of data fusion with the operation mode of power network [11], and realize multi-point hopping communication by using the rules of power network communication. The optimization strategy and communication state evaluation are realized by knowledge fusion, and the communication strategy and optimization strategy are effectively demonstrated by suggestion and phenomenon explanation.

2.3. General-purpose Information Integration Bus

The rapid collection and processing ability of power network information is an important part of the information fusion model in the multi-point hop communication system. IEC61970/61968 CIM is selected as the integrated bus general data acquisition scheme of the information fusion model. The IEC61970CIS interface standard is used as the data access interface standard of the information fusion model. Through the TAO/ACE platform with a language-based distributed environment, the end-to-end multi-hop communication configuration of the power network can be effectively realized [12]. The power general information integration bus frame diagram is shown in Figure 3.

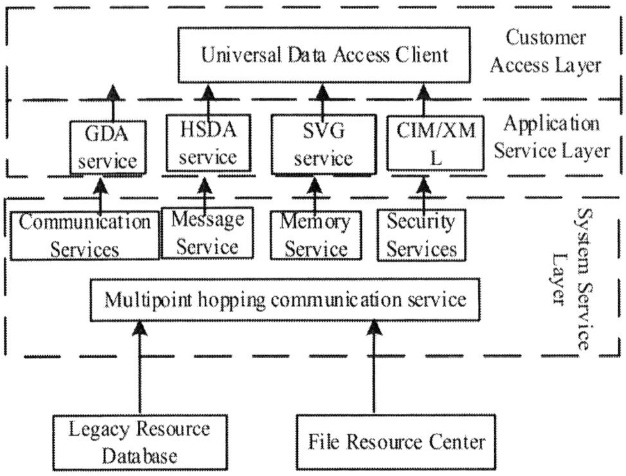

Fig. 3 General information integration bus framework

The services provided by the application service layer of the information integration bus include selecting GDA universal data access service, using HSDA high-speed data access service as power real-time database access service, using scalable SVG graphics to provide conversion service, and XML import and export service of data files.

General Standard for Data Access Forms of Heterogeneous Information Conversion as an Information Integration Bus [13], Using Pluggable Access as an Application for Legacy Resource Databases. Through the application service layer of the information integration bus, the access burden of the upper fusion center caused by multi-source heterogeneous information can be reduced and the edge data transmission time can be reduced.

2.4. ANN Algorithm and D-S Evidence Theory

2.4.1 ANN Algorithm

The ANN algorithm is selected as the feature fusion algorithm of the information fusion model, and the feature obtained by multiple sensors is set as the input of ANN. After neural network training, the initial pattern recognition and the inference knowledge stored in the weights and thresholds of pattern classification can be realized by sending the fusion results to the decision level.

A three-layer neural network is used to achieve feature-level fusion, and the error of feature fusion is used to modify the threshold $\boldsymbol{\theta}$ and the network weight $(\boldsymbol{\lambda_{ij}}, \mathbf{T_{li}})$ to make it fall along the gradient direction. The input node, hidden node and output node of the neural network are represented by $\mathbf{y_j}$, $\mathbf{x_j}$ and $\mathbf{o_l}$ respectively.

The sigmoid function is selected as the incentive function $\boldsymbol{\delta_{kj}}$ of the hidden layer of the network, and the linear function is selected as the function $\mathbf{x_{ki}}$ of the output layer of the network[14]. The iterative formula of the neural network weight coefficient is as follows:

$$\boldsymbol{\lambda_{ij}(k+1) = \lambda_{ij}(k) + \mu\delta_{kj}x_{ki}} (1)$$

The neural network output layer error formula is as follows:

$$\mathbf{E = (x'_{kj} - x_{kj})y_j(net_{kj})\left[\left(1 - y_j(net_{kj})\right)\right]} (2)$$

The error formula of hidden layer of neural network is as follows:

$$\mathbf{E_1 = y_j(net_{kj})\left[\left(1 - y_j(net_{kj})\right)\right]\sum_l \delta_{kj}\lambda_{ij}} (3)$$

The BP neural network can not converge when there are local minimum points. The weighted factor $\boldsymbol{\alpha}$ is introduced into the neural network to avoid the oscillation and accelerate the convergence of the neural network.

In the above formulas, α and μ represent the weighting factor and step size respectively, and $\mathbf{0} < \boldsymbol{\alpha}$, $\mu < \mathbf{1}$.

The mean square error of neural network can be obtained by using $\mathbf{E(w)}$ to express the average value of all training samples as follows:

$$\mathbf{E(w)} = \sum_j (\mathbf{x'}_{kj} - \mathbf{x}_{kj})^2 / 2 \quad (4)$$

2.4.2 D-S Evidence Theory

Decision-level recognition framework based on neural network output is represented by $\{(R_1), (R_2), (R_3)\}$, neural network training BP network sample output is not 1 or 0, usually between 1 and 0 rational number. D-S evidence theory is used to solve the uncertainty problem. D-S evidence theory can set the output of neural network as independent evidence [15], making it become multipoint hopping communication state reliability allocation.

U denotes the set of domains in which all possible values of X are included and all the elements of U are incompatible. In this case, the recognition framework of X is represented by U, and the 2^U power set of U is the set of all subsets of U.

Definition 1: Function n:

When $2^U \rightarrow [0,1]$ satisfies the $n(\emptyset) = 0$ and $\sum n(R) = 1$ conditions, the basic probability assignment of R is n(R), and the function n effectively shows that the given conditions support the conclusion R.

Definition 2: Identifying box U:

The likelihood function $1 - Bel(\overline{R})$ of R is expressed in terms of pl(R) and the degree of determination of non-$\overline{R}$ is expressed in terms of $Bel(\overline{R})$. The formula is as follows:

$$\mathbf{pl(R)} = \mathbf{1} - \mathbf{Bel}(\overline{\mathbf{R}}) \quad (5)$$

At this point, **R**'s trust interval is represented by $\mathbf{Bel(R)}$, $\mathbf{pl(R)}$.

$\mathbf{R_1}, \cdots, \mathbf{R_k}$ and $\mathbf{B_1}, \cdots, \mathbf{B_r}$ represents the set elements of $\mathbf{n_1}$ and $\mathbf{n_2}$ on $\mathbf{2^U}$, which are independent of each other in basic probability, respectively.

$$\mathbf{K_1} = \sum \mathbf{n_1}(\mathbf{R_i})\mathbf{n_2}(\mathbf{B_j}) < \mathbf{1} \quad (6)$$

Then:

$$\mathbf{n(c)} = \frac{\sum \mathbf{n_1}(\mathbf{R_i})\mathbf{n_2}(\mathbf{B_j})}{\mathbf{1} - \mathbf{K_1}}, \forall \mathbf{C} \subset \mathbf{U}\ \mathbf{C} \neq \emptyset \quad (7)$$

In equation (7), K_1 represents the conflict factor.

Based on formula (7), the output of feature set can be effectively fused in both space and time domains. By using the information fusion model, multi-hop communication information in power network can be transmitted efficiently, so as to improve the fusion performance of the information fusion model, reduce the rate of misjudgement, and solve the problem of repeated collection of information and isolation of edge data.

3. Experimental Results And Analysis

3.1. Experiment Platform

In order to validate the communication effectiveness of the multipoint hopping communication system based on information fusion model, a power network environment is simulated by Matlab, and programmed by Java language. Figure 4 shows the experimental platform:

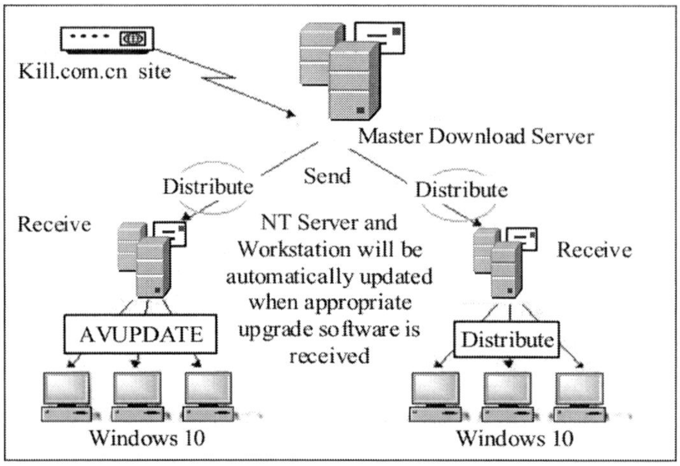

Fig. 4 Power grid communication system experiment platform

Through the experimental platform, the number of nodes in the power network is set to 100, each node contains 10 neighbor nodes, and the number of channels in the power network is 30. According to the number of different nodes and the size of different communication data packets, the throughput rate, packet loss rate, and transmission delay of multi-point hop communication are used as experimental indicators to test the communication performance of the system in this paper.

3.2. Throughput test under different number of nodes

Under different number of nodes, test the communication throughput rate of the system in this paper for 60 seconds. The test result is shown in Figure 5.

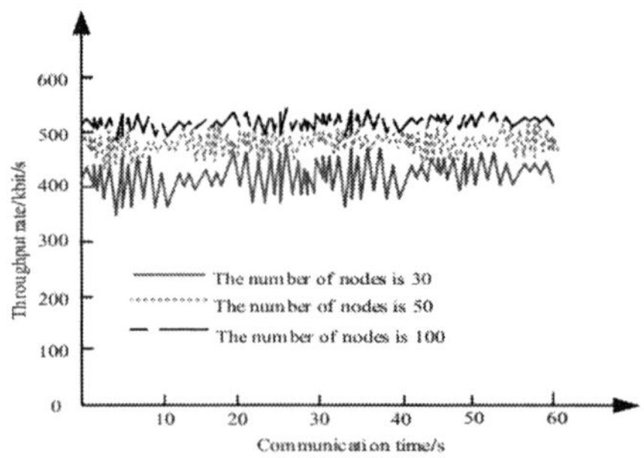

Fig. 5 Communication throughput rate under different number of nodes

It can be seen from the experimental results in Figure 5 that when the number of statistical nodes is 30, 50, 100, the communication throughput range of the multi-point hop communication tested by this system is 350kbit/s~550kbit/s. It shows that the system in this paper can still maintain high communication efficiency when the number of nodes is different.

3.3. Packet loss rate test under different number of nodes

Under different number of nodes, the communication packet loss rate of the communication system in this paper is tested for 60 seconds, and the test result is shown in Figure 6.

2108 (2021) 012048 doi:10.1088/1742-6596/2108/1/012048

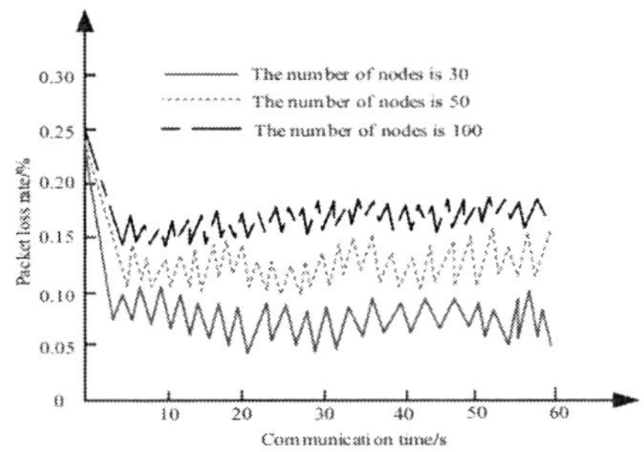

Fig. 6 Communication packet loss rate under different number of nodes

It can be seen from the experimental results in Figure 6 that when the number of statistical nodes is 30, 50, 100, the communication packet loss rate of the multi-point hop communication tested by the system in this paper is all lower than 0.25%. It shows that the system in this paper can still maintain a low communication packet loss rate when the number of nodes is different.

3.4. Transmission Delay Test Under Different Number Of Nodes

Under different number of nodes, test the transmission delay of 60 seconds of communication of the system in this paper, and the test result is shown in Figure 7.

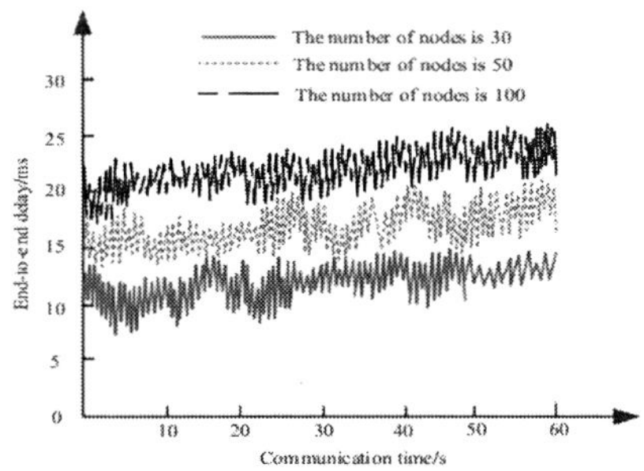

Fig. 7 Transmission delay with different number of nodes

It can be seen from the experimental results in Figure 7 that when the number of statistical nodes is 30, 50, and 100, the transmission delays of the multi-point hop communication tested by this system are all less than 30ms. It shows that the system in this paper can still maintain a low transmission delay when the number of nodes is different, which effectively verifies the effectiveness of the multi-point hop communication using the information fusion model in this paper.

3.5. Throughput Test with Different Packet Sizes

Set the number of nodes to 100 and test the communication throughput of the system for 60 seconds under different packet sizes, as shown in Figure 8.

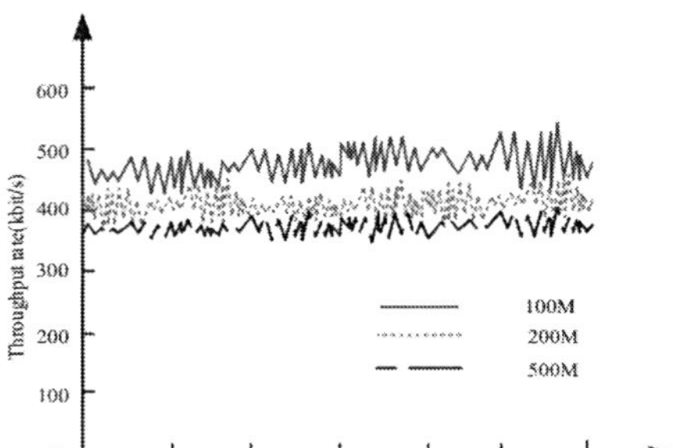

Fig. 8 Different packet large hour communication throughput rate

The experimental results in Fig. 8 show that when the packet size is 100M, 200M, 500M, the throughput of the system is still in the range of 350kbit/s ~ 550kbit/s. Once again, the system in this paper can maintain a high communication efficiency in different packet sizes.

3.6. Packet Drop Rate Test Under Different Packet Sizes

Set the number of nodes to 100, and test the 60 second packet loss rate of the system communication under different packet sizes. The test results are shown in Figure 9.

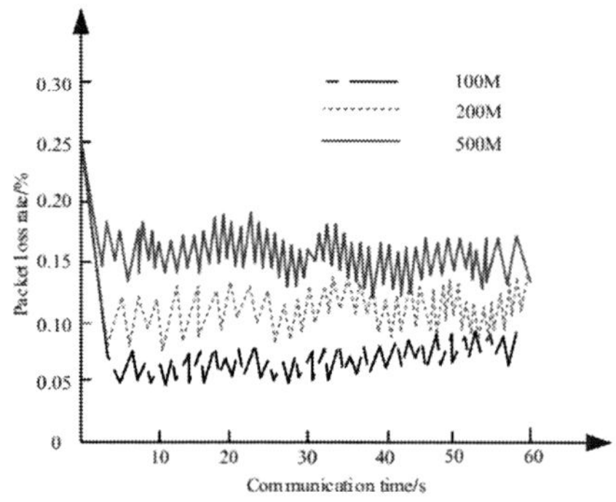

Fig. 9 Packet loss rate for different packet sizes

The experimental results in Figure 9 show that when the packet size is 100M, 200M and 500M, the packet loss rate is still less than 0.25%, which shows that the system can keep a low packet loss rate.

3.7. Transmission Delay Test Under Different Packet Sizes

Set the number of nodes to 100 and test the 60 second transmission latency of the system communication in this article with different packet sizes, as shown in Figure 10.

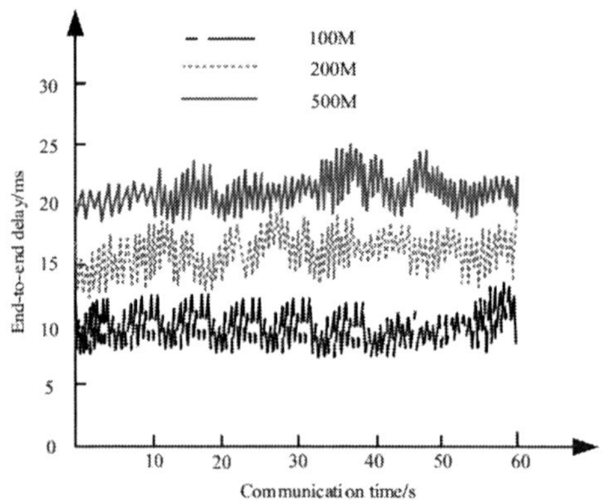

Fig. 10 Transmission delay for different packet sizes

The experimental results in Figure 10 show that when the size of the statistical packets is 100M, 200M, 500M, the transmission delay of the multi-hop communication is less than 25ms. It is shown that the system can maintain low transmission delay even when the packet sizes are different.

3.8. Communication Performance Test with White Noise

Set the number of nodes to 100, add 0-50dB white noise into the power communication network, test the performance of the system with white noise. The statistical results are shown in Table 1.

Tab. 1 Communication performance of the system with white noise

White noise /dB	Throughput rate /kbit/s	Packet loss rate /%	Transmission delay /ms
5	486.5	0.07	12
10	476.2	0.08	14
15	471.8	0.09	15

20	469.5	0.11	17
25	465.7	0.11	19
30	459.1	0.12	21
35	457.2	0.13	23
40	453.6	0.13	25
45	448.6	0.15	26
50	446.2	0.16	27

Table 1 Experimental results show that the proposed system can maintain high performance of multipoint hopping communication with different white noise levels. Moreover, the increase of white noise has little effect on the communication performance of the system, which shows that the system has high robustness.

4. Conclusion

The application of multi-point hopping communication system in smart grid communication can effectively improve the self-healing and self-organization of power network. The application of information fusion model in multi-point jump communication system of power grid can improve the reliability, integrity and accuracy of data acquisition, and solve the problem of large amount of data in emergency. The simulation experiment validates that the system can improve the timeliness of power network data communication, make power network data collection and data communication more efficient and reliable, and can effectively reduce packet loss rate and improve the performance of power network multi-hop communication.

Acknowledgments

This work was supported by the Zhejiang energy group cp ltd. "Research and development of power station network related data under power market environment" (Grant Number: ZNKJ-2020-032)

References

[1]Sun C B, Cheng S, Yuan K, et al. (2019) Real Time Simulation Platform of Power Cyber-physical System Based on Node Mapping Model[J]. Power System Technology, 43(07):2368-2377.

[2]Wang T, Zhou Y, Wang L F, et al. (2017) Design and implementation of electricity information remote meter reading system based on Beidou communication SRP mechanism[J]. Journal of Hefei University of Technology(Natural Science), 40(12):1630-1636.

[3]Wu J H, Liu X. (2017) Design of coal mine safety monitoring system based on ZigBee and information fusion[J]. Coal Engineering, 49(10):55-58.

[4]Liu S, Yu J, He Z Y, et al. (2018) Research on the Topology and Characteristic of Multi-terminal HVDC Based on VSC and LCC[J]. Proceedings of the CSEE, 38(10):2980-2988+3148.

[5]Zhang H, Zhao X W, Lu W B, et al. (2019) Research on Broadband MIMO Power Line Channel Characteristics Based on Graph Theory[J]. Proceedings of the CSEE, 39(07):2041-2049.

[6]Wang Z W, Li J Q, Huang B Y. (2019) Communication Bandwidth Prediction Based on Hybrid Queue Model for Condition Monitoring Service in Distribution Network[J]. Electric Power Construction, 40(11):1-7.

[7]Chen X, Huang S D. (2019) A Research on Via-tire Electric-Field Coupled Power Transfer System[J]. Automotive Engineering, 41(10):1145-1151.

[8]Chen S Z, Liang J L, Lu D Y, et al. (2019) Multi-source fault data comprehensive analysis method for power system based on COMTRADE model[J]. Journal of Electric Power Science and Technology 34(03):92-100.

[9]Long X F, Yang P, Guo H X, et al.(2019) Fault Diagnosis of Wind Turbine Gearbox Based on KELM and Multi-sensor Information Fusion[J]. Automation of Electric Power Systems, 43(17):132-139.

[10]Li Y G, Wang L, Li J Q, et al. (2019) Identification of Inter-turn Short-circuit Fault in Rotor Windings of Synchronous Generator Based on Multi-source Information Fusion[J]. Automation of Electric Power Systems, 43(16):162-170+191.

[11]Xue S M, Sun Y B, Liu B B, et al. (2019) Longitudinal Travelling Wave Differential Protection for Flexible HVDC System Based on Marti Model[J]. Proceedings of the CSEE, 39(21):6288-6300.

[12]Shi J C, Hu Z W, He D M, et al. (2019) Transmission Characteristic Recognition Method of Power Line Channel Based on Deep Learning[J]. Power System Technology, 43(12):4283-4290.

[13]Wang X G, Zhao F. (2019) Research on power line carrier communication routing algorithm considering communication resource optimization[J]. Electrical Measurement & Instrumentation, 56(22):79-83.

[14]Zhou W T, Wang T, Yuan M F, et al. (2017) Development of electricity information acquisition system based on Beidou short message communication[J]. Electric Power Automation Equipment, 37(12):211-217.

[15]Han Y Q, He Y Q, Lou F D, et al. (2018) Analysis and Application of SDN Based Dynamic Optimal Route Strategy for Cyber Layer in Cascading Failures of Cyber-Physical Power System[J]. Power System Technology, 42(08):2620-2629.

Design of distributed timing task scheduling system for smart grid

WenSheng Tang[1a*], JinZhi Wang[1b], Tao Fan[2c]

[1] Marketing Department of State Grid Corporation of China, Beijing, 100031, China

[2] Internet Department of State Grid Corporation of China, Beijing, 100031, China

[a*]mufengqin0603@163.com

Abstract—With the rapid development of information technology and the growing scale of enterprise development, more and more enterprise application systems appear due to business scenarios, requiring the system to execute the specified business operations at the planned time, which is the demand of "timed tasks". The rapid growth of business volume makes the number of timed tasks also increase massively, and the importance of timed task application scenarios requires more and more strict guarantee of reliable triggering of tasks. In this paper, to address these problems, we design and implement a distributed system providing timed task scheduling service based on Quartz, an open source lightweight job scheduling framework, and realize business modules such as timed task control service, trigger execution, lock grabbing trigger, unified configuration management, task load balancing and log management. The system has been tested and put into operation, and has completed the requirements well and achieved the design goals.

1. Introduction

Grid dispatching is an important element of power supply enterprises in the management of grid operation and maintenance, power supply, transmission, substation, distribution and power safety management, and the overall quality and efficiency of grid dispatching work directly determines the level of power service capability of power supply enterprises. Efficient and scientific grid dispatching specification and implementation mode is an important guarantee for power supply enterprises to achieve good economic and social benefits. With the rapid development and popularization of smart grid, the informationization of grid dispatching operation has made rapid development in power supply enterprises all over China. The traditional microcontroller-based timing scheduling system uses the establishment of scheduling pairs to schedule according to priority. Not only is it difficult for the system to handle a large number of dispatch requests, but it also tends to cause scheduling system anomalies [3-5]. The scheduling system from the perspective of economic value constraint only considers the influence of a single economic variable and ignores the carrying capacity of grid scheduling, which is prone to the fact that some scheduling tasks are always in the waiting state and it is difficult to meet the actual scheduling demand when applied independently [6]. The scheduling algorithm using particle swarm and fuzzy logic has improved the scheduling efficiency of the grid to a certain extent, but there are limitations when applied to large smart grids. Therefore, in this paper, we will use Quartz, a lightweight job scheduling framework, to design a distributed timed task scheduling system for smart grids for the actual needs of task scheduling for current smart grid operations.

2. Design of Distributed Timed Task Scheduling System for Smart Grid

The overall system architecture consists of five main components, which are a scheduling cluster based on Quartz implementation, a cluster of executors responsible for executing tasks, a message queue RabbitMQ for connecting the scheduling cluster to the executor cluster and distributing tasks[7], a Zookeeper cluster for executor cluster management and as a Dubbo registry, and a MySQL database .

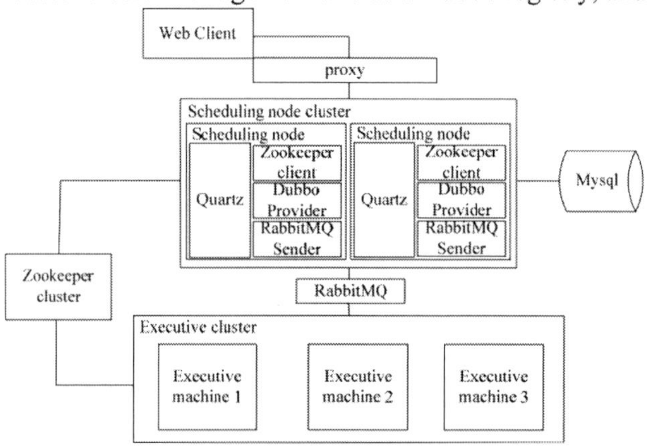

Fig. 1 Block diagram of the overall design of the distributed timed task scheduling system for smart grids

Scheduling node cluster: The task scheduling service in this system is based on Quartz, but with an improved working model. RabbitMQ: RabbitMQ is a bridge between task senders and task receivers, and the queues in RabbitMQ act as staging points for tasks, with each queue storing a different type of task. Executor clusters: Executor clusters are specific executors of tasks in the system, the number of nodes in the cluster can be added and removed dynamically according to specific needs, and the executor nodes can be deployed in a simple and fast way.

2.1. *Scheduling system hardware design*

The distributed timed task scheduling system for smart grid designed in this paper adopts a top-down design approach, adding FPGA chips in the message storage process to improve the stable reliability[8] . Figure 2 below shows the overall hardware framework schematic of the timed task scheduling system.

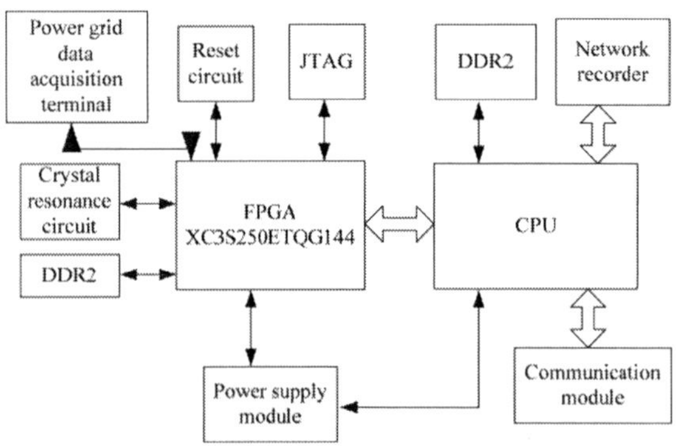

Fig. 2 System hardware architecture

In this paper, the system uses DDR2 SDRAM as the cache area for data, while DDR2 read/write control is implemented with an FPGA master controller chip. Considering leaving a certain redundancy, MT47H128M16RT-25E is selected as the data cache chip in the design, whose capacity size is 2GB and the maximum data operation rate of the chip can reach 800Mbps.The hardware part of the distributed timing scheduling system communicates through PCIE bus and adopts independent point-to-point communication links to transmit data so that it can work at a very high frequency, and each pair of differential lines has a data transfer rate of 2.5 Gbps to realize the transmission communication of scheduling commands.

2.2. *Task load balancing scheduling module design*

Assume that in the operation of smart grid, there are n different task servers to process m different task requests[9]. Let the task execution node set be $N=\{p_1,p_2,\ldots,p_n\}$, and the energy consumption set per task time of each task execution node is $C=\{c_1,c_2,\ldots,c_n\}$, where c_i represents the energy consumption of node p_i. When scheduling tasks in power grid operation, the task scheduling objective function is established with the minimum total operating energy consumption of the system scheduling server as follows:

$$\min_{\lambda_n^s, f_n^s, Y_n^s} E=\sum_{n=1}^{N}\sum_{s=1}^{S_n} C_n^s T_i \quad (1)$$

In the above equation, C_n^s denotes the energy consumption of the s th server as it performs its task. T_i is the task scheduling execution time for the s th server on the grid. λ_n^s is the total number of tasks assigned to the n th task execution node by grid scheduling. f_n^s is the frequency of operation of the s th task execution node of the nth server. The binary variable Y_n^s indicates the activation status of a server. If a server is activated, the binary variable takes the value of 1, otherwise the binary variable takes the value of 0. In order to achieve load balancing among server nodes during the grid task scheduling process, the execution nodes avoid congestion in the task scheduling queue due to uneven task load, which affects the system task scheduling efficiency. Therefore, the following scheduling constraints are set in this study. (1) Load balancing constraints:

$$\lambda_m=\sum_{n=1}^{N}\sum_{s=1}^{S_n}\lambda_n^s \quad (2)$$

In the above equation, λ_m is the task flow transmission speed submitted by the task demand side during grid scheduling. The above formula indicates that the total tasks to be performed by all server nodes should be equal to the total tasks reaching all front-end nodes. (2) Scheduling service delay constraint:

$$\frac{Y_n^s}{rf_n^s}+\frac{\lambda_n^s\overline{sz_n^{s2}}}{2\left(1-\frac{\lambda_n^s}{rf_n^s}\right)\left(krf_n^s\right)^2}\le T \quad (3)$$

Where , $\overline{sz_n^{s2}}$ is the average size of the task. r is a scaling factor. k is the scaling factor, which can be considered as a known parameter that indicates the expectation of the task size. T is the maximum delay allowed by the activated server.

(3) The operating frequency constraint of each server, i.e., the operating frequency of each server cannot exceed the maximum frequency $f_{\max}$.

$$0\le f_n^s\le f_{\max}, Y_n^s\in\{0,1\} \quad (4)$$

After setting up the power grid task scheduling objectives and constraints in the Quartz framework, the genetic algorithm is used to solve the above minimization problem, and a scheduling strategy for different mission requirements is obtained.

2.3. Task scheduling and trigger management module design

Task creation table TskTbl, task readiness table TrdTbl, event request table ErqTbl and timed waiting table TmrTbl are four independent tables. Each task list is a 64 bit one-dimensional table, and each task in the system occupies a binary bit in this one-dimensional table. The specific task list is shown in table 1.

Tab. 1 Description of the list of tasks

Symbolic	Name (of a thing)	Instructions
TskTbl[N]	Task creation form	Flag that a task has been created, as a flag for operations associated with the task. [n]=1: nth task was created; [n]=0: nth task was not created/deleted.
TrdTbl[N]	Mission Ready Form	Flag that a task is ready to be used by the task scheduler. [n]=1: the nth task is ready; [n]=0: the nth task is not ready.
ErqTbl[N]	Event Request Form	Flag that a task has an event request, used by the task scheduler. [n]=1: nth task has an event request; [n]=0: nth task has no event request.
TmrTbl[N]	Timed Waiting List	Flag that a task has timed waits, used by the task scheduler. [n]=1: nth task has timed waits; [n]=0: nth task has no timed waits.

When managing the task trigger records, cold execution data separation is used to separate the instance data from the historical data in the trigger run, which can improve the efficiency of accessing the instance data of the running tasks.

2.4. Task Timing Trigger Module Design

The main implementation classes are Executor, SimplePool, ParallelPool, etc., where SimplePool and ParallelPool are the task execution pool classes corresponding to simple tasks and parallel computing tasks respectively, and there are ConcurrentHashMap data structures in these execution pool classes. executorUnitTable is used to cache the Job instances running in the current Client, each instance will correspond to an ExecutorUnit class. In addition, the Executor class is responsible for connecting the clientSerivce (RPC service interface class) and the task execution pool class, and is responsible for forwarding the RPC requests from the Server side to the corresponding task execution pool class for business processing. The parallel computing task execution pool working timing diagram is shown in figure 3 below.

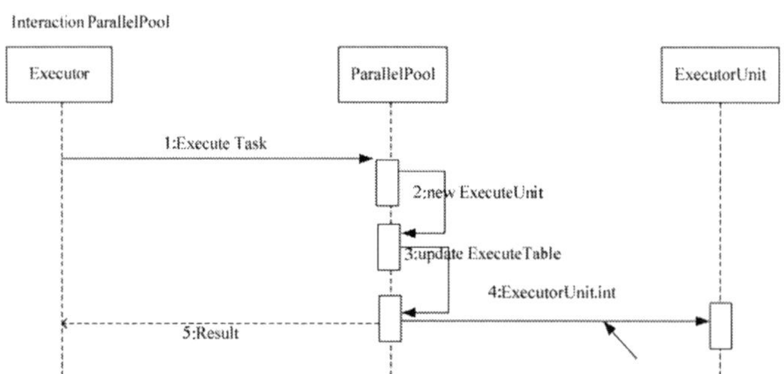

Fig. 3 Timing diagram of parallel computing task execution pool work

When the system performs timed task scheduling, it sets the task trigger timing and performs parallel task scheduling calculations according to the working timing diagram shown in figure 3 above.

2.5. Log management module design

The log management module is responsible for recording task execution logs, providing information update services for task execution records and task log query functions. The classes and descriptions related to the log management module are shown in table 2.

Tab. 2 Description of classes related to the log management module

Interface	Kind	Instructions
Public String updateJobRecord	LogRPCService	This interface defines the interface used to provide the remote call service
Public String uploadLog(String)	LogRPCServiceImpl	This class provides a concrete implementation of the remote call service

The task execution records are updated and logs are uploaded mainly through Dubbo, a distributed RPC framework. The scheduling node is a service provider and the execution machine is a service consumer. The LogRPCService interface defines the remote call service interface provided by the module. This completes the design of the distributed timed task scheduling system for smart grid.

3. System experiments

3.1. Experimental content

In this paper, we have designed a distributed timed task scheduling system for smart grid and after completing the system design, the online system is tested for the functionality of each module of the system.

3.2. Experimental results

Test the scheduling management module and task timing trigger module of the scheduling system using test data. Access to the simulated Server side was used as the scheduling request for each functional module to run. The execution of the test cases by each functional module of the system shows that all test cases pass and each functional module functions normally. In the integration test environment, during the integration test of each software functional module of the system, the business timing scheduling logic is normal and the test results are as expected. For high availability testing, the system is brought online into a small smart grid with different sizes and numbers of dispatch requests generated by the computer. The average response time of the scheduling system is tested for different scheduling task requests. Figure 4 below shows the response time test of the system.

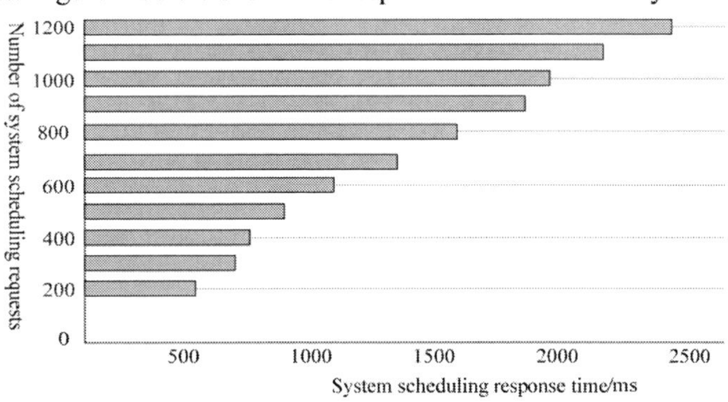

Fig.4 Scheduling system response time test

Analysis of the information in the above figure shows that as the number of scheduling task requests in the smart grid increases, the scheduling response time of the system shows a growing trend, but the overall growth is small, and the maximum response time is only 2463 ms. It shows that the system can respond to the scheduling requests in the process of grid operation in a timely manner and perform task scheduling processing quickly. That is, the distributed timed task scheduling system of smart grid designed in this paper has good operation performance, and the on-line test results show that the system works well.

4. Conclusion

Grid system automation is an important means to ensure the safety, stability, reliability and economic operation of the power system. The management of power grid's operation state is adjusted, changed and maintained, so that the power supply and distribution of power grid can be orderly implemented according to the requirements. On the basis of maintaining the overall stable and safe operation of the power grid, the quality and mode of power supply in the power supply area can be achieved. In this study, a smart grid distributed timing task scheduling system is designed. After the on-line test, the functions of each module are in good condition, which can meet the task scheduling requirements of smart grid. Since our current working environment is a local area network, the impact of network delay and communication is not considered in the scheduling algorithm. Therefore, further research will focus on the distributed task scheduling of the power automation system in the wide area network, and the determination of the node load after the network delay is added.

References

[1] Zhou Hailing. (2021). A multi-objective scheduling method for smart grid loads based on improved GSO algorithm. Journal of Physics: Conference Series, 1907(1), 012018.

[2] Ali Sajjad ; Khan Imran; Jan Sadaqat; Hafeez Ghulam. (2021). An Optimization Based Power Usage Scheduling Strategy Using Photovoltaic-Battery System for Demand-Side Management in Smart Grid. Energies, 14(8), 2201-2201.

[3] Hao Liang ; Sharad Sinha ; Wei Zhang. (2018). Parallelizing Hardware Tasks on Multicontext FPGA With Efficient Placement and Scheduling Algorithms.. IEEE Trans. on CAD of Integrated Circuits and Systems, 37(2), 350-363.

[4] Paul Suraj ; Chatterjee Navonil ; Ghosal Prasun. (2021). Dynamic task allocation and scheduling with contention-awareness for Network-on-Chip based multicore systems. Journal of Systems Architecture, 115(2),102020.

[5] Abbasi Sohaib Iftikhar ; Kamal Shaharyar; Gochoo Munkhjargal; Jalal Ahmad; Kim Kibum. (2021). Affinity-Based Task Scheduling on Heterogeneous Multicore Systems Using CBS and QBICTM. Applied Sciences, 11(12), 5740-5740.

[6] Duan Jiandong; Liu Fan; Yang Yao; Jin Zhuanting. (2021). Flexible Dispatch for Integrated Power and Gas Systems Considering Power-to-Gas and Demand Response. Energies, 14(17), 5554-5554.

[7] Jo Haesung ; Park Jaemin ; Kim Insu. (2021). Environmentally Constrained Optimal Dispatch Method for Combined Cooling, Heating, and Power Systems Using Two-Stage Optimization. Energies, 14(14), 4135-4135.

[8] Zare Mohsen; Narimani Mohammad Rasoul; Malekpour Mostafa; Azizipanah-Abarghooee Rasoul; Terzija Vladimir. (2021). Reserve constrained dynamic economic dispatch in multi-area power systems: An improved fireworks algorithm. International Journal of Electrical Power and Energy Systems, 126(PA), 106579-.

[9] Dey Bishwajit ; Raj Saurav; Mahapatra Sheila; MSheila Fausto Pedro Garcro. (2022). Optimal scheduling of distributed energy resources in microgrid systems based on electricity market pricing strategies by a novel hybrid optimization technique. International Journal of Electrical Power and Energy Systems, 1 34(10):107419.

Research on residual current detection system of substation alternating current power supply

You Li[1a*], Chuanfu Xia [1b], Yuan Zhuang[1c], Baoquan Liao[1d], Shuai Huang[1e]

[1] State Grid Sijishenwang Location Based Service(Beijing) Co.,Ltd., Beijing, 102209, China

[a*]liyou@sgitg.sgcc.com.cn, [b]xiachuanfu2897@163.com, [c]zhuangyu1@sgitg.sgcc.com.cn, [d]317685243@qq.com, [e]17710187156@163.com

Abstract—Residual current monitoring of substation is essential to prevent electrical fire and plays an important role in the safety of substation. This paper studied residual current detection technology of substation AC power supply, and proposed a residual current detection system composed of three devices: sensing terminal, management terminal and monitoring terminal. The design could solve some difficult problems such as the secondary line caused by equipment installation, fault wave recording, fault location analysis, and early warning based on analysis of big data. The study could improve the intelligent level and maintenance efficiency of residual current monitoring in substation.

1. Introduction

Residual current is also leakage current which is the electric current between the cable and the ground because the insulation of the cable or load device to the ground has broken [1]. Leakage current due to insulation damage of electrical lines or equipment probably cause electrical fire and electric shock accidents [2-3]. At present, residual current monitoring technology has been widely used in power system [4-9]. Residual current monitoring is essential for substation and other important facilities of power system, which can effectively prevent electrical fire, electric shock and other accidents, and plays an important role in the safety and stability of substation [10-11]

In May 2019, a fire accident occurred in 500KV Zhangjiaba Substation. After investigation and analysis, it was found that the incident was a typical electrical fire accident caused by cable fire, which ignited the control cable mixed in the same ditch and led to the trip of equipment. Through analysis, the cause of the fire was found that the insulation of power cable was weak which was from the AC central distribution panel to the relay protection cell in the station, and the relative armored layer was short-circuited and discharged, and the AC molded case circuit breaker in the circuit failed to jump off in time. There were some problems exposed in the accident that the lack of insulation monitoring means for low-voltage AC cables used in the station and the lack of sensitivity of low-voltage AC circuit breakers, make it impossible to find and remove metal grounding and high-resistance grounding faults in time. The monitoring technology of substation AC power system based on residual current has a good prevention effect on the cable initial fire caused by high resistance grounding.

The paper proposed residual current detection scheme for substation AC power system and developed the devices which were with the function of fault detection and protection for a substation low-voltage AC cable insulation. The system can on-line monitor the insulation status and running status of cable and load equipment online in real time. It could effectively realize the early warning of substation fire, timely find out fire hazards in substation, reduce the risk of fire, improve the efficiency of maintenance and reduce the cost of maintenance.

2. System architecture

The residual current monitoring system adopts three-layer architecture design, including data acquisition layer, collection management layer and edge calculation layer. The three-layer architecture is composed of three devices: sensing terminal, management terminal and monitoring terminal. The overall architecture of the system is shown in Figure 1.

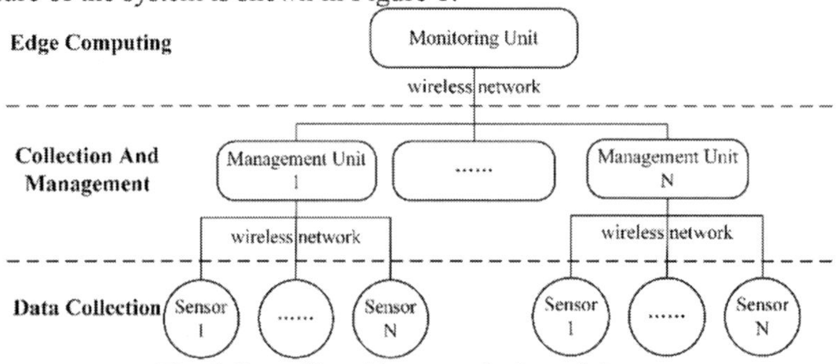

Fig. 1 Overall architecture design of the system

The design ideas of the system are introduced as follows:

(1) Sensing terminal can acquire residual current value, ambient temperature and humidity of each feeder branch of the AC power system.

(2) Management terminal can collect the data from multi-channel sensing terminal, be used for steady data record, leakage current alarm monitoring and fault alarm transient data record.

(3) Monitoring terminal can integrate various monitoring data of the sensing terminals in all systems and realize residual current monitoring, early warning, strong graphic display functions and human-machine interaction based on big data analysis.

(4) Communication between system devices, can support wireless communication such as LORA and RS-485. Wireless communication can effectively solve the problems such as secondary lines, the difficulty and low efficiency due to installation and maintenance of communication lines between the devices. RS-485 is compatible with existing equipment to meet the needs of wired communication.

3. Design of terminals

3.1. Sensing terminal

Based on the investigation and analysis of the existing residual current monitoring equipment in the market, this paper proposed the integrated design of the sensing terminal, which had functions residual current monitoring, inductive power-taking, wireless communication, temperature monitoring and humidity monitoring.

The design of sensing terminal is introduced as follows. The schematic diagram is shown in Figure 2.

(1) The power module composed of inductive power taking module, power conversion module and power function module can provide stable power for other functional hardware modules.

(2) Residual current sensor and signal processing modules, can support residual current values acquisition and A/D sampling and processing.

(3) The T/H sensor hardware directly transmits IO signals to the processor for data transmission.

(4) Processor unit can support local calculation of residual current, temperature and humidity data.

(5) Communication unit has two-way data exchange with the processor unit and it can realize wireless communication and data transmission.

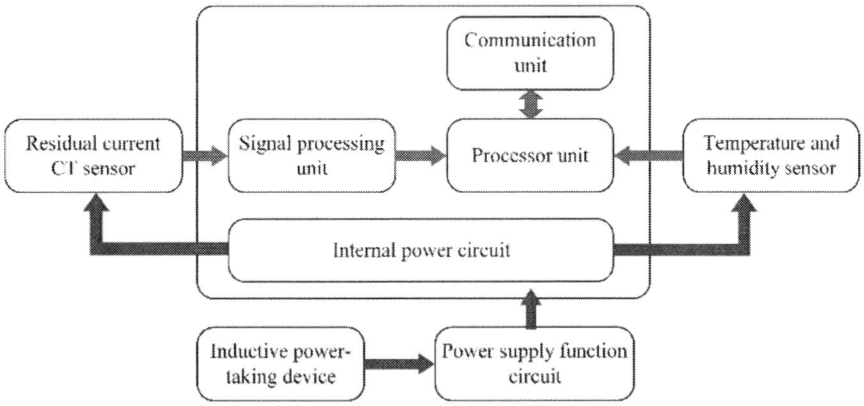

Fig. 2 Schematic diagram of sensing terminal

The main characteristics of sensing terminal are shown as follows:

(1) Inductive power-taking. It can ensure power supply for the device and reduce greatly secondary cables compared with other forms of power-up. It can improve the efficiency of the installation, operation and maintenance for the low-voltage AC power system, because there are too many feeder in AC power system.

(2) Wireless communication. Intelligent terminal with wireless communication is the development trend of power Internet of Things. And wireless communication can also effectively avoid using cable.

(3) Acquisition of temperature and humidity. Ambient temperature and humidity are one of the important parameters for residual current analysis. The integration of ambient temperature, humidity and residual current monitoring can provide effective data support for the early warning analysis of residual current.

The key technical indicators are shown in the Table 1.

Table 1. Key technical indicators of sensing terminal

Item	Residual current	Ambient temperature	Relative humidity
Measurement range	10mA~10000mA	-20°C ~ +85°C	0-100%
Measurement accuracy	0.5 level	±1°C	±1%

3.1.1. Power Module

The power module includes inductive power-taking module, power conversion module, and power function module. The induction power is obtained by current transformer (CT). According to the conventional load of substation AC power supply system, the starting current of CT is defined between 5A-500A. Considering the installation environment, the traditional high-voltage CT power-taking method is not appropriate, which will occupy a large space and cost too high. Therefore, use soft-power-taking method.

The design scheme is introduced. CT is used for power extraction, and super capacitor is equipped as backup power supply. In order to ensure that the capacity of super capacitor is fully utilized, DCDC is used as power conversion module. The key designed performance indexes are that maximum power is 1W and voltage output is 5V.

3.1.2. Residual current detection module

Residual current detection is based on the traditional electromagnetic induction principle. First, an ABCN line is used to pass through the magnetic loop coil for induction, then the acquisition circuit is used to signal transformation.

The key designed performance indexes are that residual current sampling accuracy should reach 0.2% and measurement range is 10mA-10A.

3.2. Management terminal

The management terminal can collect the residual current and other data from the sensing terminal, and to store and record steady-state recording and transient recording of fault alarm using wireless communication.

The fault wave-recording is for management terminal to record steady-recording of residual current and dynamic recording of the fault data in a specific time when the fault occurs. The function is used mainly to achieve data traceability and data backup, and to provide accurate and reliable data support for fault analysis. The leakage current, temperature and humidity. The leakage current, temperature and humidity are recorded for wave-recording.

The key technical indicators are as follows:

(1) Steady-state recording. The recording interval is 1s, and the recording time is over 6 months;

(2) Transient recording. The sampling frequency of recording is 4K Hz. It can record the transient waveform from 1s before leakage to 10s after leakage. A maximum of 500 transient recording waves can be recorded

The design idea of the management terminal is to collect the residual current and other data from the sensing terminal, and to store and record steady-state recording and transient recording of fault alarm. Therefore, the hardware architecture of management terminal mainly includes processor unit, memory unit, display unit and communication unit. The main functions of the management terminal include wireless communication, fault alarm, fault wave-recording and management interface.

3.2.1. Wireless communication

When communicating with the sensing terminal, the management terminal acts as the host to realize one-to-many information interaction and mainly collects the remaining current, temperature and humidity data from sensing terminals. When communicating with the monitoring terminal, the management terminal acts as a slave and realizes many-to-one information interaction, and upload the collected data to the monitoring terminal.

In order to avoid mutual interference of wireless communications with sensing terminal and monitoring terminal, it is needed that two independent wireless communication modules are used for the management terminal.

(1) Because ultra-low power wireless communication is necessary for the communication between sensor, the wireless communication module in the 433MHz band is selected. Monitoring terminal can be connected to a maximum of 24 sensing terminals.

(2) For the communication between the management terminal and the monitoring terminal, it is necessary to consider the large amount of communication data and data security, the expansibility and universality. Lora is chosen for wireless communication module of management terminal and the encryption module of State Grid is applied to ensure the communication security.

3.2.2. Fault Alarm

The management terminal provides the local alarm function. When the leakage current exceeds the threshold or the device status is abnormal, LED indicator or buzzer are used to notify user for the alarm status.

3.2.3. Fault wave-recording

Fault wave-recording is for management terminal to record steady-recording of residual current and dynamic recording of the fault data in a specific time when the fault occurs. The function is used mainly to achieve data traceability and data backup, and to provide accurate and reliable data support for fault analysis. The leakage current, temperature and humidity. The leakage current, temperature and humidity are recorded for wave-recording. The flow chart of wave recording function is shown in Figure 3.

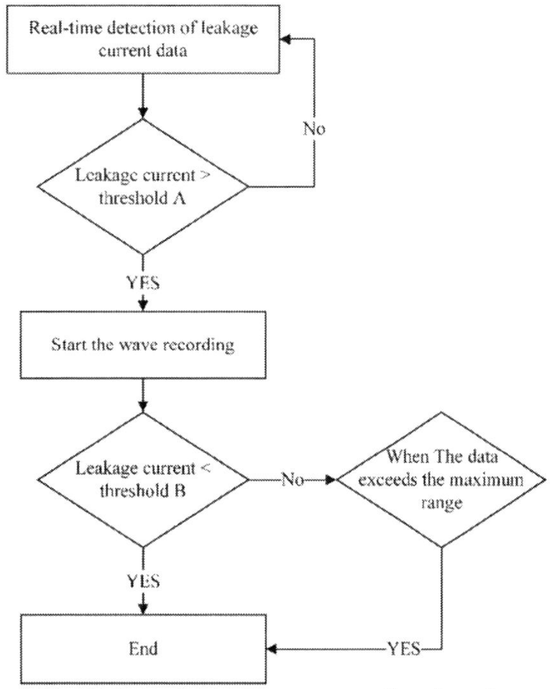

Fig. 3 Flow chart of wave recording function

Based on the effective value of leakage current, the management terminal can make real-time warning and alarm judgment. When the leakage current is greater than the threshold for a while, the wave-recording is started. The management terminal records the valid values of the current alarm branch within a fixed period and stores them until the alarm disappears or the recorded data exceeds the maximum range.

3.2.4. Management interface

(1) Real-time data display. The monitoring terminal obtains the leakage current and temperature and humidity values through wireless communication with sensing terminals of each branch and displays them on LCD.

(2) Configure parameter Settings. User can set the IP address, real-time time, and leakage current limit of the management terminal by pressing buttons.

(3) Data query. The management terminal can switch the interface menu by pressing buttons to query the data.

3.3. Monitoring terminal

The monitoring terminal is the core of data collection, calculation, and management for residual current monitoring system. There are the functions including data collection, data analysis and management interface.

3.3.1. Data collection

(1) Real-time data collection. The monitoring terminal can collect the leakage current, temperature and humidity and device status. The monitoring component can collect and display all the data and status information.

(2) History data collection. It includes historical alarm data and wave recording data. The history data is stored in the form of files in SD cards through the file system. User can view a maximum of 1000 pieces of historical data through the device interaction and copy data through to USB storage.

3.3.2. Data analysis

(1) Fault location. The monitoring terminal saves the sensor ID configured on the connected sensing terminal and identifies the specific branch ID. When the leakage current alarm exceeds the alarm threshold, the monitoring terminal determines the branch alarm to locate the specific branch.

(2) Fault warning. The thresholds are set for the monitoring terminal according to the real-time and history monitoring data values. Based the early-warning algorithm of big data analysis, it can analyze the trend change of leakage current, temperature, and humidity. When fault risks are found, the early warning is sent for user.

3.3.3. Management interface

(1) Operation monitoring. The monitoring terminal can monitor the operating status, communication status, semaphore, leakage current value, temperature and humidity value in real time. It can display the data value and alarm information on the interface.

(2) Operation and maintenance management. Device parameters can be set on the user interface, such as equipment number, alarm threshold through management interface.

(3) Remote control function. User can set the parameters of the sensing terminal through communication and manually control opening and closing through the remote-control function on the interface.

4. System testing

According to the system design scheme, three kinds of equipment products including sensing terminal, management terminal and monitoring terminal have been developed. The residual current monitoring system was tested in the laboratory simulation environment. The sensing terminals were installed on each feeder of the AC power supply. The management terminals were installed on the two AC distribution panels. And The monitoring terminal was installed in the transformer room.

When various leakage faults were simulated, the monitoring system was tested and the residual current values were measured electrical measuring instrument. Comparing the monitoring system with measurement values, The accuracy of residual current reached 1%. The monitoring system could trigger the alarm response time within 10 seconds and reflect timely and accurately the leakage situations. The test results were approved by experts from State Grid Corporation.

Fig. 4 Schematic diagram of management terminal installation

5. Conclusion

The research work has successfully designed the residual current monitoring scheme for substation AC power supply and developped three kinds of equipment including sensing terminal, management terminal and monitoring terminal. The system can solve the pain points and difficulties of secondary line due to the equipment installing, fault wave-recording, fault location analysis and fault warning based on big data analysis. It can realize the real-time on-line monitoring the insulation state and running state of each cable in the substation AC power supply system, improve the operation and maintenance efficiency, and reduce the cost of maintenance. In the future, the algorithm for residual current monitoring and early warning will be optimized to improve the monitoring accuracy.

Acknowledgments
This work is supported by State Grid Information & Telecommunication Group CO.,LTD (52680020002J).

References

[1] Hu K. (2017) Discussion on the Application of residual current Protection device in China. Scientific and Technological Innovation, 16:91-92.
[2] Zhang L. (2017) Application status and development trend of residual current protection technology. Industrial & Science Tribune, 16:61-62.
[3] Huang B, Han J, Zhou T L. (2019) Research on residual current electric fire monitoring system. China Computer & Communication, 31:86-87.
[4] Liu Y. Y. (2019) Analysis and application of residual current monitor. Building Electricity, 38:14-18.
[5] Tai Y. J., Chen C. Y. (2018) Design and implementation of rural grid residual current fault on-line monitoring platform. Industrial control Computer, 31:8-9.
[6] Yan Y. H., Li X. X., Zhou R. X., Li P., Zhao Y. (2017) Research on the design of residual current on-line monitoring system based on electricity information collection. Information & Communications, 4:131-134.
[7] Sun J. (2017) Pilot application of residual current online monitoring system in rural Area. Shanxi Agricultural Economy, 6 :74-76.
[8] Qian L. J., Lu H. X., Yin J. Y. (2015) Research on residual current monitoring function based on smart electricity meter. Electrotechnical Application, 34:204-207.
[9] Wang C. K., Yang Z. Q., Fu J. Y. (2012) Research and application of intelligent Remote Monitoring and Protection device for residual current in low-voltage power grid. Electrical & Energy Management Technology, 24:23-27.
[10] Sun J. P. (2021) Application of intelligent residual current detection technology in substation. Electric Power and Energy, 42:178-181.
[11] Zhang K. W., Chen X. P., Qu G. D. (2021) Measuring method of grounding and residual current of substation AC power supply. Rural Electricians, 29:45-46.

Intelligent detection method of abnormal state of power equipment based on infrared and visible image information

Tao Liu[1], Bo Qu[1], Guolong Li[1*]

[1]State Grid Tieling Electric Power Supply Company, Tieling, Liaoning 112000, China

[*]1142101056@ncepu.edu.cn

Abstract—Infrared thermal imaging cameras are usually used to detect the thermal state of the equipment. According to the heat radiation situation and temperature value emitted by the equipment, the thermal failure of the equipment is inferred, and whether the equipment is working is verified properly. This paper analyzes the thermal imaging principle of infrared technology and the characteristics of infrared images, and quantifies the surface temperature indicators of insulators. Through infrared and visible light image information, the pollution degree of insulators can be detected, and the abnormal condition of the equipment can be checked. The evaporation of moisture on the surface of the insulator mainly depends on the assumption that the surface heats up. A method for analyzing heat on the surface of wet-contaminated insulators is proposed, and the judgment conditions for arc generation in the drying zone and the drying zone and the heating models of the insulators in different operating conditions are established.The computer simulation results reveal the distribution of heat on the surface of insulators under different operating conditions, and the influence of the occurrence of drying zone and drying zone arc on leakage current and heating. The infrared thermal imaging test results of wet-contaminated insulators show that the model is reasonable and provides theoretical support for infrared thermal imaging detection of insulator contamination levels.

1. Introduction

The root cause of the transmission line pollution flashover accident is due to the deposit of pollutants on the surface of the insulator. After it absorbs the moisture in the humid air, the insulation strength drops sharply, and it cannot withstand the working voltage and flashover occurs. Pollution flashover is a regional problem. The probability of multi-point tripping is high, and it is easy to cause large-scale and long-term malignant power outages. If the power failure cannot be resolved in time after the power failure occurs, the scope of the failure will expand, and even cause a large-scale blackout, which will have a serious impact on people's lives and the national economy. Taking effective measures to prevent and reduce pollution flashover accidents is an important guarantee for the safe operation of the power grid.

In addition to strengthening the research on pollution flashover characteristics [1], optimizing the insulator structure and the design of the equalizing ring [2]. The main methods that currently exist to prevent pollution flashover are: Literature [3] proposes a simplified equivalent circuit of a porcelain insulator string, and obtains the heating power distribution of the insulator string and the insulation resistance value during maximum heating. Literature [4] established the equivalent circuit of semiconductor glaze insulator string. Solve the temperature distribution of the insulator string through the heat balance equation. Conduct heat generation research on the insulator as a whole.The detection of zero-value insulators has certain guiding significance. But it cannot provide information on the heat distribution along the surface of the insulator. It is also impossible to judge the formation and positioning of the drying belt. The numerical calculation results in literature [5] are all related to multiple physical parameters that need to be determined manually. The actual value is difficult to determine. And the model does not consider the drying zone and drying zone arc that may occur in the actual situation.

Leakage current method is a hot spot of current research, and a large number of researchers and scholars have done a lot of research.At present, there are many practical leakage current sensors, and the measurement accuracy is relatively high. There is a lot of interference in the measurement of leakage current.The adaptive noise cancellation technology and wavelet denoising method have also achieved good results. Therefore, there is no major problem in the measurement of leakage current, and it has been verified in practical applications.This paper proposes a method for analyzing the heat generation of wet-contaminated insulators. Experiment to obtain infrared and visible images of insulators with different pollution levels. Extract the surface of the insulator through image processing, and calculate the temperature and color characteristics of the surface respectively.It avoids manual setting of parameters, and the feature vector is concise. The results show that the information fusion method in this paper can improve the accuracy of the insulator contamination level identification. The field test results verify its practicability and effectiveness, and it is of positive significance to the field insulator contamination status detection.

2. Heating Model of Wet Contaminated Insulator

2.1. Modeling of insulator surface heating model

The pillar ceramic insulator is composed of iron caps, tempered glass parts and steel feet, and is glued together with cement glue. This article takes ZSW-12/4 type post ceramic insulator as an example. The basic parameters are listed in Tab.1.

Tab.1. Structural parameters of ZWS-12/4

Parameter name	Total height/mm	Diameter /mm	Creepage distance/mm	Mounting size of upper accessory/mm	Attachment installation size/mm
Parameter value	210	155	240	Center distance 36	Center distance 130

The insulator is contaminated and damp, but R_j and R_v are basically unaffected. Evenly wet and dirty surface resistance.

$$R_s = \int_0^L \frac{\rho_{ws}}{2\pi r(l)} dl = f\rho_{ws} \tag{1}$$

In the formula, L is the total creepage distance of the insulator, m; dl is the creepage distance of the outer insulating surface, m; $r(l)$ is the distance from the creepage distance to the insulator's rotation axis, m; f is the form factor, which is defined as

$$f = \int_0^L \frac{dl}{2\pi r(l)} \tag{2}$$

where R_s is proportional to ρ_{WS}. As the degree of contamination increases, ρ_{ws} drops sharply, so the contamination and dampness of the insulator can be simplified as a parallel circuit of C_0 and R_s. The heating power of the insulator is mainly changed from dielectric loss heating to surface leakage current heating.

It can be seen from the equivalent circuit of the damp and dirty insulator that the leakage current amplitude is large at this time, and the heat generated is consumed by convection, conduction, radiation and water evaporation. The author's experimental observations and the results of the literature all show that evaporation heat dissipation is the main heat dissipation method of wet-contaminated insulators. When the amount of water falling is less than the amount of water evaporation, the surface of the insulator is dried to form a dry zone; when the amount of water falling is greater than the amount of water evaporation, there will be no drying zone.If the electric field intensity on the drying belt exceeds a certain value, an arc will be generated in the drying belt. Therefore, the operating states of wet pollution insulators can be divided into three types, namely uniform wet pollution as a whole, wet pollution + dry zone and wet pollution + dry zone arc.

A uniform wet pollution insulator can be equivalent to a wet pollution layer resistance R_{SW}, as shown in Fig.1(a); after the dry belt is formed, the insulator is equivalent to the series connection of dry pollution layer resistance R_{sd} and wet pollution layer resistance R_{SW} (assuming the dry belt It appears on the first umbrella skirt), as shown in Fig.1(b); if a dry arc occurs, the insulator is equivalent to the series connection of the arc resistance R_{sa} and the wet dirt layer resistance R_{sw} as shown in Fig.1(c). Since the contamination state of insulators in actual operation is very complicated, it is assumed that:

1) The surface of the insulator is evenly polluted and damp, and the surface resistivity ρ_{ws} is constant;

2) The insulator is smooth and axisymmetric, and the drying belt is in a circular ring shape;

3) At most one drying zone and one arc can be formed at the same time;

4) The arc current waveform is a sine wave when the arc is generated in the drying zone;

5) When heat generation and heat dissipation reach a thermal balance, the surface temperature of the insulator does not change with time.

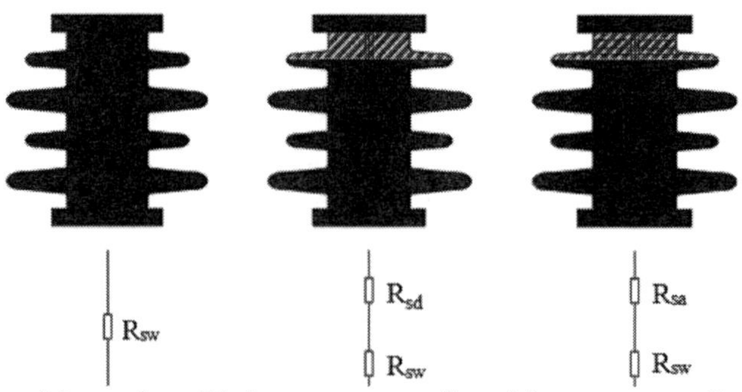

(a) evenly (b) dry zone + wet dirt (c) arc + wet pollution
Fig.1. Operation status description

Based on the above assumptions, the mathematical expressions of the three models of uniform wet pollution, dry zone + wet pollution, and arc + wet pollution can be obtained, as shown below:

$$U = If\rho_{sw} \tag{3}$$

$$U = I\left[\int_0^{l_d} \frac{\rho_{sd}}{2\pi r(l)} \mathrm{d}l + \int_{l_d}^{L-l_d} \frac{\rho_{sw}}{2\pi r(l)} \mathrm{d}l\right] \tag{4}$$

$$U = I\left[Ala\left(\sqrt{2}I\right)^{n-1} + \int_{l_a}^{L-l_a} \frac{\rho_{sw}}{2\pi r(l)} \mathrm{d}l\right] \tag{5}$$

In the formula: U is the working voltage of the insulator, V; I is the effective value of the surface leakage current, A; l_d is the width of the drying zone, m; I_a is the arc length, m; ρ_{sd} and ρ_{sw} are the surface resistivity of the drying zone and the wet contaminated layer, respectively. Ω; A and n are constants.

2.2. Model solution results and analysis

(1) Uniform wet soil model

The surface resistivity of the wet-contaminated insulator is about 10^6-10^8Ω. In this paper, the resistivity ρ_{sw} of the wet-soil layer on the surface of the insulator is taken as 5×10^6, 8×10^6, 2×10^7, 5×10^7. Solve the heating power ρ_s and temperature T along the disc diameter of the three heating models respectively.

The solution results of the uniform wet pollution model of insulators with different surface resistivities are shown in Fig.2(a) and Fig.2(b). The more serious the degree of wet contamination of the insulator, the lower the surface resistance, the greater the leakage current, and the higher the temperature rise at the same disk diameter. The change trend of T is slightly slower than that of ρ_s, which is due to the non-linear heat dissipation of thermal radiation, which causes faster heat dissipation in areas with increased temperature.

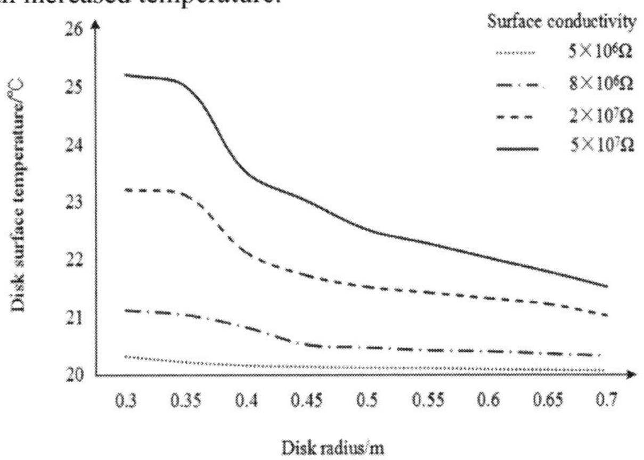

(a) Evenly wet and dirty heating power

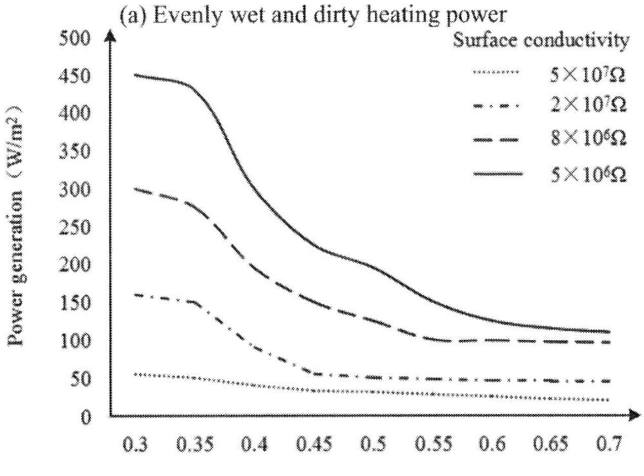

(b) Evenly wet and dirty pan surface temperature pan surface temperatureheating power

Fig.2. Heating power and temperature distribution on disk sueface of the three models

(2) Dry zone + wet pollution model

Suppose the resistivity of the drying zone is 10 times that in the wet and dirty state, that is, $\rho_{sd} = 10\rho_{sw}$. The model solution results show that only when $\rho_{sw} = 5\times10^{6}\,\Omega$, the leakage current is large enough to completely evaporate the moisture on the drying zone and maintain the stability of the drying zone exist. Due to the large surface resistance of the drying belt, the heating power caused by its leakage current exceeds 2×10^{3} W/m^2, which is much higher than that of the wet dirt layer; so the temperature rise of the dry zone and the temperature rise of the wet soil layer are maintained at the same order of magnitude.

(3) Arc + wet pollution model

The heating power and temperature distribution of the arc + wet pollution model when $\rho_{sw} = 5\times10^{6}\,\Omega$ are analysed. The comparison shows that the arc heating power is higher than the leakage current heating power, which causes a higher temperature rise in the drying zone, and the maximum temperature rise of the disk surface reaches nearly 20℃. It should be noted that the solution of this model is based on the assumption that the insulator can stably generate a dry zone arc and reach a thermal equilibrium state, which is difficult to appear in this operating state in practice. Combining the above three models, it can be seen that the higher the pollution degree of the insulator, the greater the gradient of the surface temperature along the radius, and it is more likely to produce drying zone and drying zone arc, thereby generating a greater temperature gradient. Therefore, theoretically, the temperature characteristics of the insulator disk surface, such as the maximum temperature variance, etc., can be used to identify its pollution level.

3. Experimental Results and Analysis

The post-ceramic insulator ZSW-12/4 is used for contamination test. Simulate the pollution status of insulators operating on-site in our country. Sample preparation, sample contamination and test methods are carried out in accordance with the provisions of GB/T 4585-2004. The salt density is respectively taken as 0.04, 0.8, 0.15, mg/cm^2; the pressurization experiment is carried out in a 3m×3m×3m artificial airtight fog chamber, and the laboratory environment temperature is 20℃. The electrical connection is shown in Fig.3.

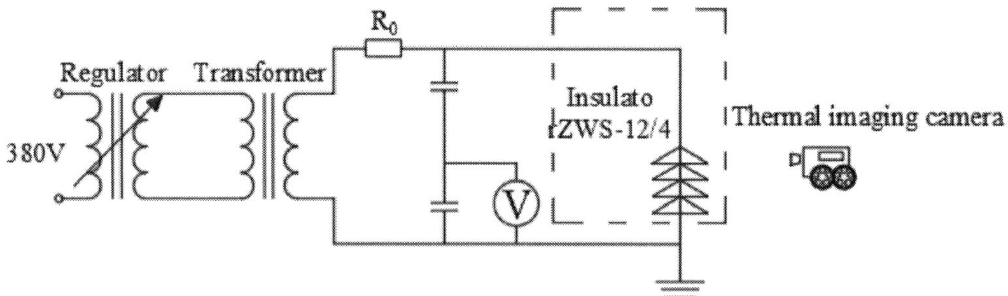

Fig.3. Experimental electrical wiring diagram

Infrared instrument Daipu X3 was used for shooting to shoot the insulators. Each insulator is subjected to a pressure test with a relative humidity H of 0%, 80%, and 90%:After humidifying the artificial airtight fog chamber to the required humidity and stabilizing it with a humidifier, add 10 kV voltage for 2 h to make the insulator reach a thermally stable state. After that, it will be taken every 10 minutes, the shooting distance is 2 m, and the shooting depression angle is 15°~30°. Fig.4 is an example of infrared samples under different pollution levels and relative humidity.

The gray value of the infrared thermal image is proportional to the temperature of the surface of the object. The larger the gray value, the higher the temperature value. The analysis shows that the temperature rise decreases with the increase of r, the upper surface of the insulator is smooth along the radial temperature distribution curve, and the lower surface has a jump change, which is basically consistent with the heat distribution. The test results are basically consistent with the simulation results. The infrared thermal image analysis of the artificial pollution test of insulators preliminarily verifies the validity of the model.

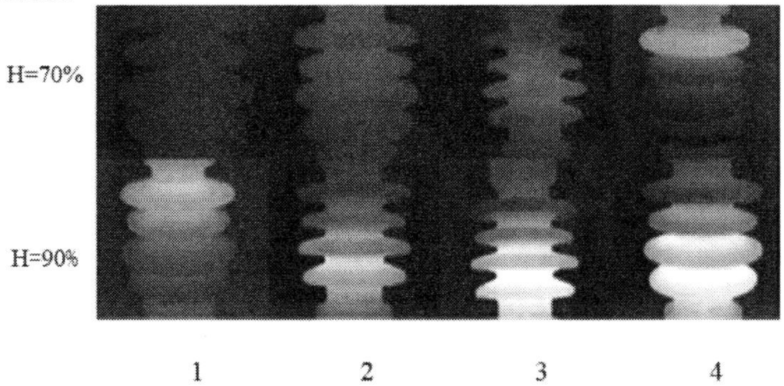

Fig.4. Infrared image

4. Conclusion

By using the two parameters J and ρ_{sw} to measure the moisture and contamination status of the insulator surface, a new method of heat analysis of wet and dirty insulators is proposed. The heat generation models are established for different operating conditions, and the numerical analysis method is applied to solve them, revealing the insulators The heat distribution law on the surface. This method realizes the discrimination of the pros and cons of filth characteristics, the method is simple, easy to implement, easy to understand, and strong in practicability. Using this method to select features prevents bad features from entering the classification feature set, ensures the optimal design and performance of the classifier, and improves the accuracy of classification. The drying zone and the arc generation conditions of the drying zone are deduced.

Acknowledgements

This work was supported by the Science and Technology Project of State Grid Corporation of China (Project No. 2021YF-33).

References

[1] Jin, L.J, Zhang, D. (2014) Research of Insulator Contamination Grades Recognition Methodology Based on Visual Image. Journal of System Simulation, 26(9): 2073-2078.

[2] Shi, J.J. (2017) Substation equipment identification and thermal fault diagnosis based on infrared image processing . Shanghai Dianji University.

[3] Huang, Q.D, Yang, Z.Y, Song, H.Y. (2016) Measurement and ion-pairing analysis of natural contamination constituents on insulators with different material. High Voltage Engineering, 42(12): 3917-3923.

[4] He, Z.H, Wang, Y.Q. (2016) Influence of salt deposit density and salinity on AC artificial pollution flashover performances of various types insulators . High Voltage Engineering, 42(12): 3810-3815.

[5] Zhang, Y, Tang X.Y, Zhang, Y. (2017) Determination and analysis of soluble anions in insulator contamination by capillary ion chromatography . High Voltage Engineering, 43(4): 1296-1301.

Research on the optimal strategy of pumped storage power station to provide multiple time-scale reserves

Xiaomeng LI[1a*], Xiaopeng Yu[1], Ze Gao[1]

[1]The Electric Power Research Institute of the State Grid Henan Electric Power Company, Zhengzhou City, Henan, China

[a*] lixiaomeng@ha.sgcc.com.cn

Abstract—In the background of peak carbon emissions and carbon neutrality, renewable energy generation (REG) will become the main generation form in the future power system. Simultaneously, the randomness and volatility increase the reserve requirement in the different time scale. Increasing importance has been attached to energy storage in the aspect of reserve, as energy storage has the advantages of power flexibility and relatively low reserve cost. Trading off the benefits of energy storage in the energy market and the reserve market to maximize its benefits is of great significance to the economic operation and investment of energy storage. In this regard, taking the pumped storage power station (PSPS) as an example, this paper establishes an optimal decision-making model for PSPS to participate in the energy market and to provide reserve services. In addition, an optimal decision model for PSPS to provide multiple reserve services is established. The analysis finds that the power reserve capacity provided by PSPS at different time scales have little impact on each other, but their storage capacity requirements are mutually restricted. Case studies show that the total revenue of the PSPS is significantly increased through providing reserve service. The PSPS may even bid all its capacity to provide reserve service when the compensation price reserve reaches a certain level. In addition, the total revenue of PSPS when providing multiple time-scale reserves is higher than that when providing reserve service at single time scale.

1. Introduction

Under the goals of peak carbon emissions and carbon neutrality, The generation structure and grid operation characteristics in China will undergo profound changes. The proportion of renewable energy generation (REG) in generation side will continue to increase and add difficulty to the power balance in power systems. It also increases the demand for ancillary services such as reserve at different time scales. Therefore, the literature [1] proposes a new structure of power system reserve at multiple time scales. The development of energy storage technology will provide reserve support to the power system at different time scales. However, the economy and market strategy of energy storage participating in various ancillary services and electric energy market need to be further studied.

Energy storage is an important support to the development of REG and the construction of new power system. From the perspective of power grid, an ultra high voltage (UHV) AC and DC interconnection network structure has been constructed between regional power grids in China. However, UHV DC block will cause serious power imbalance to the power system operation, posing a huge challenge to the rapid reserve demand at second to minute level in regional grids. Due to the high level of REG access, its intermittency and volatility calls for more and more reserve demands at minute level, hour level, and hours level. It indicates that under the construction of new power system, the reserve at various time scale need to be strengthened.

From the perspective of response time and duration, various national power grids defined and categorized various reserve related ancillary services. Europe continental power grid classifies power system reserve into primary reserve, secondary reserve and tertiary reserve corresponding to different time scales [2] and [3]. The reserve capacity in UK includes primary frequency response capacity, secondary frequency response capacity, rapid reserve and short-term operating reserve capacity [4]. California power grid in the United States divides reserve resources into frequency regulation reserve, spinning reserve, non-spinning reserve and alternative reserve [5]. In the early stage, reserve services in the power system are mainly provided by various power plants. However, with the increasing demand of reserve services, the reserve resources only from the generation side can hardly meet the reserve demand in future power systems. Therefore, more and more attentions have been paid to energy storages and demand-side reserve.

Energy storages have the characteristics of fast power regulation speed and low reserve cost. Their potential in the new power system will be gradually found. Relevant literatures have studied the operation and benefits of energy storages and flexible loads in ancillary services market, such as in frequency regulation (FR) and reserve market [6]-[9]. Literature [10] studied the peak regulation market model considering the participation of energy storage aggregators. Literature [11] studied the operation strategy of energy storage power stations and electric vehicles participating in the FR market in the background of California market rules. Literature [12] proposed a distributed autonomous control method to realize the application of battery storage in FR service. Literature [13] studied a cooperative control method of wind power and compressed air energy storage system. Compressed air energy storage is used for FR in charging and discharging mode, so as to alleviate the influence of wind power uncertainty on system frequency. In order to improve the AGC response performance of the power system, a dual-objective optimization control model with energy storage participating in FR was established in reference [14], aiming at minimizing the total power deviation and frequency modulation cost.

Literature [15] builds a market model in which multi-types of power supply i.e. wind, PV, water, coal-fired generation and energy storage participate in frequency modulation. The sequential bidding model marginal price clearing model is adopted. This paper analyzed the improvement effect of multi-types of power supply on the system FR performance. Literature [16] compares the FR performance of thermal power units and energy storages. It concludes that energy storage has the advantages of fast response speed and high stability.

In addition to the FR, the role of energy storage in reserve ancillary services and the corresponding price strategy also gain great attention. A lot of researches have been done regarding the problem of power grid reserve. Literature [2] puts forward systematic suggestions for power grid reserve services. Literature [17] puts forward the management idea of "fine emergency re" and the idea of "special time standby" to deal with new energy power generation. Based on robust optimization, Literature [18] proposed the bidding strategy of wind power and pumping storage joint participation in the day-ahead and real-time market.

With the deepening of research on energy storage to provide ancillary services, the advantages of flexibility and rapidity of energy storage are gradually highlighted. More and more attentions are paid to the potential of energy storage to provide ancillary services. Battery storage and pumped storage stations, as representative batteries, have the characteristic of rapid power regulation. Their main ways to make profits are arbitraging the price spread and providing ancillary services [19]. The cost of energy storage participating in the electricity market comes from the degradation cost and maintenance costs. When it participates in the reserve service market, it needs to reserve certain reserve capacity. Therefore, the reserve cost comes from the opportunity cost in the energy market.

This study takes pumped storage power station as an example to deal with the decision making problem for energy storage between energy market and reserve market, especially considering its providing reserve services at different time scales. The decision making model of energy storage participating in the electricity market and reserve market is constructed. The optimal strategy is also discussed considering the uncertain price of electricity and compensation prices of different reserve services.

2. The Profit Analysis of Pumped Storage in Electricity Market

2.1. Pumped storage operation mode

To satisfy the operation requirement of the pumped storage power station, it is necessary to consider its operation constraints in the actual dispatching stage.

Power generation constraint of pumped storage power stations:

$$0 \leq p_{\mathrm{dch},t} \leq b_{\mathrm{dch},t} P_{\mathrm{dch,max}} \tag{1}$$

where $p_{\mathrm{dch},t}$ is the generation power of pumped storage; $P_{\mathrm{dch,max}}$ is the maximum generating power of pumping storage; $b_{\mathrm{dch},t}$ is a 0-1 variable denoting the discharging state of the pumped storage.

Constraints on pumping power of pumped storage power station:

$$0 \leq p_{\mathrm{ch},t} \leq \ b_{\mathrm{ch},t} P_{\mathrm{ch,max}} \tag{2}$$

where $p_{\mathrm{ch},t}$ is the pumping power of the pumped storage station; $P_{\mathrm{ch,max}}$ is the maximum pumping power of the pumped storage; $b_{\mathrm{ch},t}$ is a 0-1 variable denoting the pumping state of the pumped storage station.

Mutual exclusion constraints of pumping and discharging states of the pumped storage station:

$$b_{\mathrm{ch},t} + b_{\mathrm{dch},t} \leq 1 \tag{3}$$

Constraint of reservoir capacity of pumped storage station:

$$0 \leq e_t \leq E_{\mathrm{max}} \tag{4}$$

where e_t is the current state of the reservoir; η_{dch} is generation efficiency, T is time length of a step; E_{max} is the maximum reservoir capacity of the pumped storage station; η_{ch} is the pumping efficiency;

Correlation constraints between the storage state and the pumping/generation power:

$$e_t = e_{t-1} + p_{\mathrm{ch},t}\eta_{\mathrm{ch}} - p_{\mathrm{dch},t}/\eta_{\mathrm{dch}} \tag{5}$$

The periodic constraint of pumped storage capacity: namely, the storage state at the end of each day is equal to the state at the beginning:

$$e_1 = e_{\mathrm{N}} \tag{6}$$

2.2. Benefit analysis of pumping storage in energy market

Pumped storage makes profits by arbitraging the peak-valley price difference in the energy market. In the day-ahead energy market, pumped storage participates in the market as an independent player. Its income is closely depends on the day-ahead clearing prices. In this study, a profit analysis model for energy storage was established considering the uncertainty of the energy prices. The modeling process is shown in Figure 1.

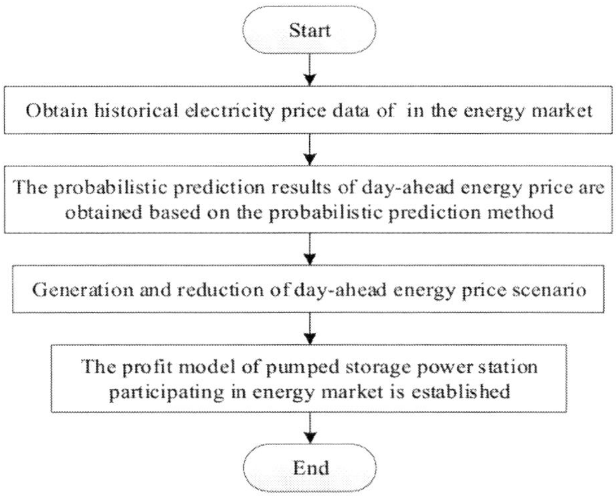

Fig.1 Steps of the revenue calculation for pumped storage power station under uncertainty of day-ahead energy price.

The modeling process includes:

1) Obtain the historical electricity price data in the energy market and use it for the energy price prediction of the next day;

2) Obtain system information of the next day, combined with the historical data and using the probabilistic prediction method; then, obtain the probabilistic prediction result of the day-ahead energy price;

3) Construct price samples of the day-ahead energy prices based on the probability prediction results; obtain certain number scenario curves acceptable for the actual solution through scenario reduction technology;

4) Construct the profit model of pumped storage stations participation in the energy market;

Since this study mainly analyzes the income situation and market strategy of the pumped storage power station, it is assumed that the scenario data of energy price is known. The profit model of pumped storage power station established in this paper is as follows.

The objective function is:

$$(\text{P1}) \max f_1 = \sum_{s=1}^{n_s} \pi_s \sum_{t=1}^{\mathrm{N}} \lambda_{s,t} (p_{\mathrm{dch},t,s} - p_{\mathrm{ch},t,s}) \, \mathrm{T} \tag{7}$$

Where n_s is the number of scenarios; $\lambda_{s,t}$ is the energy price at time t in scenario S. In each scenario, there are a set of variables of pumping power, generation power and storage state $\{p_{\mathrm{ch},t,s}, p_{\mathrm{dch},t,s}, e_t, t = 1,2,\ldots,\mathrm{N}\}$, which need to satisfy constraints (1-6).

3. Revenue Modeling of Pumped Storage Power Station under Multi-time Scale Reserve

3.1. Operational modeling of the pumped storage station when providing a single reserve

Literature [1] defines four types of reserve resources and defines their responding time and continuous supply duration $\tau_j, j = 1,2,3,4$. In this study, four alternative operational models and optimization strategies for pumped storage will be investigated.

When pumping storage provides reserve service, it is necessary to reserve certain power capacity and inventory capacity to be invoked. Therefore, its operating boundary in the energy market will change. In this study, the reserve resources are composed of upward and downward parts. It is assumed that the compensation price of reserve in each time scale is known.

Considering the reserve capacity from pumped storage, its declared upward and downward reserve capacity is denoted by $r_{j,\mathrm{up}}$ and $r_{j,\mathrm{dn}}$. Pumped storage power stations declare its reserve capacity after comprehensively considering the income in all scenarios. Therefore, in the multi-scenario model, there is only one set of reserve capacity variables. The sum of generation power and declared reserve capacity in each scenario should be less than its maximum generating power, and the sum of pumping power and declared reserve capacity should be less than its maximum pumping power, namely:

$$0 \le p_{\mathrm{dch},t,s} + r_{j,\mathrm{up}} \le P_{\mathrm{dch,max}} \tag{8}$$

$$0 \le p_{\mathrm{ch},t,s} + r_{j,\mathrm{dn}} \le P_{\mathrm{ch,max}} \tag{9}$$

Reserved capacity constraints:

$$0 \le r_{j,\mathrm{up}} \le R_{j,\mathrm{upmax}} \tag{10}$$

$$0 \le r_{j,\mathrm{dn}} \le R_{j,\mathrm{dnmax}} \tag{11}$$

Pumped storage has flexible adjustment characteristics and can change its pumping and generation status within a day. Therefore, different operating state variables are used for each scenario. However, in a specific scenario, the pumped storage power station still needs to meet the mutually exclusive constraint of pumping and generation states, namely:

$$b_{\mathrm{ch},t,s} + b_{\mathrm{dch},t,s} \le 1 \tag{12}$$

When pumped storage provides reserve capacity, a certain amount of water storage and storage capacity needs to be reserved for invoking. Therefore, the upper and lower limits of the storage capacity of the pumped storage station should be updated as follows:

$$r_{j,\mathrm{up}}\tau_j \le e_{t,s} \le E_{\mathrm{max}} - r_{j,\mathrm{dn}}\tau_j \tag{13}$$

Equation (13) indicates that a pumped storage station should reserve a certain amount of storage capacity for generation or pumping when playing a reserve role.

The correlation constraints between the storage state and the pumping/generation power is updated as (14). The periodic constraints of the storage capacity in each scenario is expressed as (15):

$$e_{t,s} = e_{t-1,s} + p_{\mathrm{ch},t,s}\eta_{\mathrm{ch}} - p_{\mathrm{dch},t,s}/\eta_{\mathrm{dch}} \tag{14}$$

$$e_{1,s} = e_{\mathrm{N},s} \tag{15}$$

The total revenue of the pumped storage station participating in the energy market and providing the jth type of reserve service, namely, the objective function of the optimization model is:

$$(\mathrm{P2})\ \max f_2 = \sum_{s=1}^{n_s} \pi_s \sum_{t=1}^{\mathrm{N}} \lambda_{s,t}(p_{\mathrm{dch},t,s} - p_{\mathrm{ch},t,s})\mathrm{T} + (\rho_{j,\mathrm{up}} r_{j,\mathrm{up}} + \rho_{j,\mathrm{dn}} r_{j,\mathrm{dn}})\mathrm{NT} \tag{16}$$

where, $\rho_{j,\mathrm{up}}$, $\rho_{j,\mathrm{dn}}$ is the compensation price of the jth type of reserve service, including upward and downward reserve.

The above objective function should satisfy constraints (1~2) and (8~15). By solving Model P2, the optimal declaration strategy for pumped storage to provide the jth type of reserve service can be obtained.

3.2. Operation modeling of pumped storage power station in multiple reserve situations

When the reserve services at multiple time scales is provided by energy storage, considering the correlation relationship of all types of reserve in the time horizon, the declared capacity of all types of reserve needs to satisfy constraint (1-2) and (8-11). Different from the power constraint, its water storage and storage capacity need to meet the reserve demand at each time scale at the same time. Therefore, the upper and lower limits of the storage state of the pumped storage station should be updated as:

$$\sum_{j=1}^{4} r_{j,\mathrm{up}}\tau_j \le e_{t,s} \le E_{\mathrm{max}} - \sum_{j=1}^{4} r_{j,\mathrm{dn}}\tau_j \tag{17}$$

The total revenue of the pumped storage station when it participates in the energy market and provides various reserve services, namely, the objective function of the optimization model is:

$$(\mathrm{P3})\ \max f_3 = \sum_{s=1}^{n_s} \pi_s \sum_{t=1}^{\mathrm{N}} \lambda_{s,t}\left(p_{\mathrm{dch},t,s} - p_{\mathrm{ch},t,s}\right)\mathrm{T} + + \sum_{j=1}^{4}(\rho_{j,\mathrm{up}} r_{j,\mathrm{up}} + \rho_{j,\mathrm{dn}} r_{j,\mathrm{dn}})\mathrm{NT} \quad (18)$$

The above objective function should satisfy constraint (1~2), (8~12), (14), (15) and (17).

By solving Model P3, the optimal declaration strategy for pumped storage station to provide multiple reserve services can be obtained.

4. The Example Analysis

4.1. Introduction of tested case

Parameters of a pumped storage power station are as follows: the maximum generation power and the maximum pumping power are 20MW, and the storage capacity is 100MWh. Its efficiency in a cycle is 75%. The multi-scenario energy price curve adopted in this study is shown in Figure 2. Energy prices at night are relatively low, while energy prices at noon and evening are relatively high.

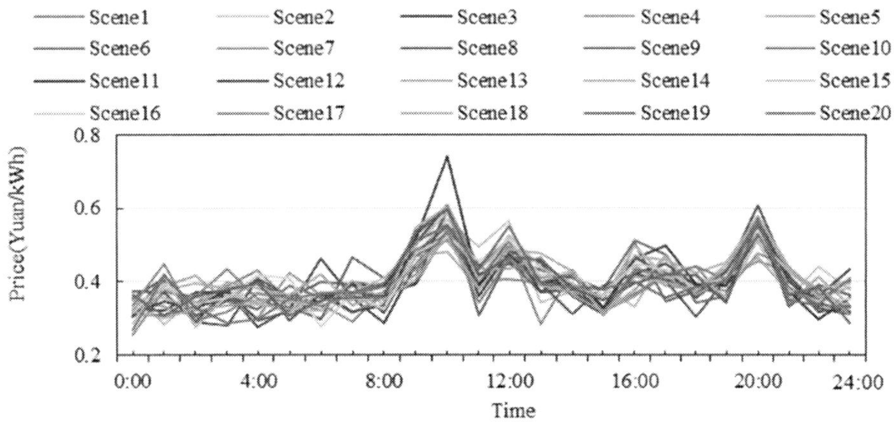

Fig.2 Electricity price curves in 20 scenarios.

4.2. Operation results participating in the of the energy market

When not providing reserve service, the pumping and generation power under different energy price scenarios are shown in figure 3. Positive values represent generation power and negative values represent pumping power. The pumped storage station pumps mainly at night, and generates at noon and evening when energy prices are high. It makes profit from the energy price spread. The total expected revenue is 8357 yuan in a day. The maximum pumping and generation power of are 20MW. The maximum and minimum storage state are 100 and 0 MWh, respectively, indicating that the capacity of the pumped storage station has been fully utilized.

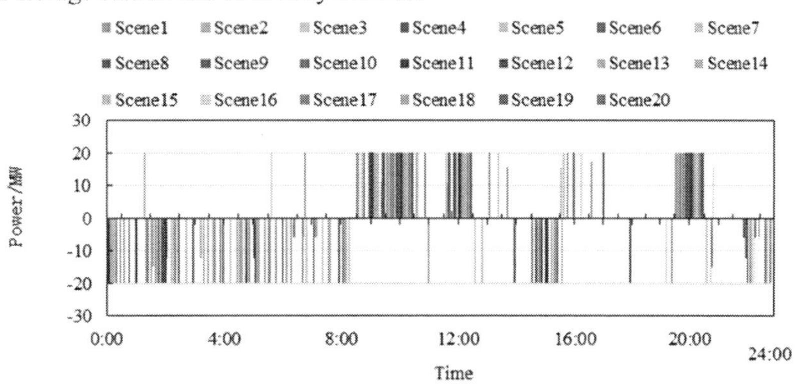

Fig.3 Operation results of pumped station s in energy market.

4.3. Operation strategy when providing reserve

When only considering one type of reserve, for example, primary frequency regulation reserve is provided. The duration is not less than one minute, and the upper limit of reserve capacity is set as 3 MW. The declared reserve capacity and income of the pumping and storage station are shown in Table 1 under different reserve compensation prices. When the reserve compensation price is low, the capacity of pumped storage power station is fully used to participate in the energy market. When the reserve compensation price is higher than 0.012 yuan/hour, the capacity of pumped storage station will be used to provide reserve service as far as possible, so as to obtain more profits. Under this price, the total revenue of pumped storage station is 8835 yuan, higher than the total expected revenue in the energy market, namely 8,357 yuan.

Table 1. The decision result of PSPS under different compensation prices of reserve type 1.

compensation price (Yuan/hour)	**declare spare capacity** (MW)		**total revenue** (Yuan)	**maximum pumping power** (MW)	**maximum discharge power** (MW)	**minimum storage capacity state** (MWh)	**maximum storage capacity state** (MWh)
0.002	0	0	8357	20	20	0.00	100
0.004	0	0	8357	20	20	0.00	100
0.006	0.8	0	8360	19.15	20	0.01	100
0.008	3	0	8479	17	20	0.05	100
0.01	3	0	8623	17	20	0.05	100
0.012	**3**	**3**	**8835**	17	17	0.05	99.95
0.014	3	3	9123	17	17	0.05	99.95
0.016	3	3	9411	17	17	0.05	99.95
0.018	3	3	9699	17	17	0.05	99.95
0.02	3	3	9987	17	17	0.05	99.95

When the minute-level reserve, the duration of 10 minutes, and the upper limit of reserve capacity of 10MW are considered, the declared reserve capacity and income of the pumping and storage station under different reserve compensation prices are shown in Table 2. When the reserve compensation price is higher than 0.0064 yuan/hour, the pumping and storage stations begin to gain more profits by providing reserve. When the reserve compensation price is 0.016 yuan/hour, the declared capacity of the pumping and storage station is 10MW, that is, the maximum declared reserve capacity. At this time, the maximum available capacity in the electricity energy market is 10MW. At this time, the operating state of the pumping and storage station in each scenario in the electricity energy market is shown in Figure 4. In addition, under the reserve compensation price, the total income of the pumping and storage station can reach 10,860 yuan, the minimum storage capacity is 1.67MWh, and the maximum storage capacity is 98.33MWh.

Table 2. The decision result of PSPS under different compensation prices of reserve type 2.

compensation price (Yuan/hour)	**declare spare capacity** (MW)		**total revenue** (Yuan)	**maximum pumping power** (MW)	**maximum discharge power** (MW)	**minimum storage capacity state** (MWh)	**maximum storage capacity state** (MWh)
0.0016	0	0	8357	20	20	0	100
0.0032	0	0	8357	20	20	0	100
0.0048	0	0	8357	20	20	0	100
0.0064	**2.4**	**0**	**8376**	18	20	0.40	100
0.008	4.9	0	8510	15	20	0.82	100
0.0096	10	6.8	8873	10	13	1.67	98.87
0.0112	10	8.6	9559	10	11	1.67	98.56
0.0128	10	10	10324	10	10	1.67	98.33
0.0144	10	10	11092	10	10	1.67	98.33
0.016	**10**	**10**	**11860**	10	10	1.67	98.33

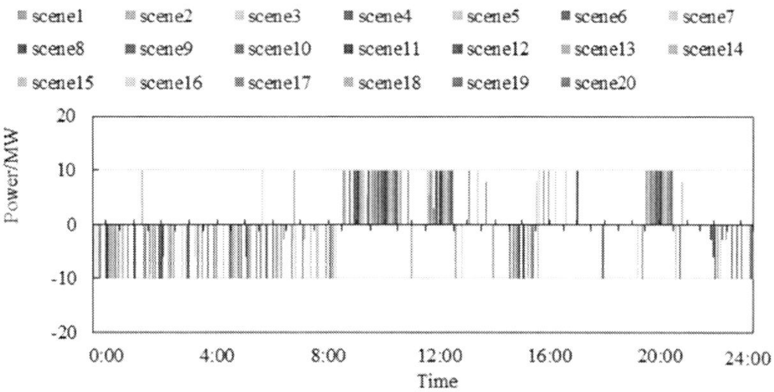

Fig.4 Operation results of PSPS in the electric energy market after providing reserve service.

When the pumped storage station provides 10 minutes of reserve, the duration is required to be 2 hours, and the limit of reserve capacity is set as 20 MW. The declared reserve capacity and income of the pumped storage station under different reserve compensation prices are shown in Table 3. When the reserve compensation price is lower than 0.0048 yuan/h, the pumped storage station will not participate in the reserve service market. When the reserve compensation price is higher than 0.096 yuan/h, all the capacity of the pumped storage station is used to provide reserve, namely 20MW, and more benefits can be obtained. In this condition, the minimum storage capacity is 40 MWh, and the maximum storage capacity is 60MWh.

Table 3. The decision result of PSPS under different compensation prices of the 3rd type of reserve.

compensation price (Yuan/hour)	**declared reserve capacity** (MW)		**total revenue** (Yuan)	**maximum pumping power** (MW)	**maximum generation power** (MW)	**minimum storage capacity state** (MWh)	**maximum storage capacity state** (MWh)
0.0012	0	0	8357	20	20	0	100
0.0024	0	0	8357	20	20	0	100
0.0036	0	0	8357	20	20	0	100
0.0048	0	0	8357	20	20	0	100
0.006	**0.9**	**0**	**8359**	19	20	1.71	100
0.0072	2.4	0	8420	18	20	4.77	100
0.0084	5.3	0	8558	15	20	10.66	100
0.0096	**20**	**20**	**9216**	0	0	40	60
0.0108	20	20	10368	0	0	40	60
0.012	20	20	11520	0	0	40	60

When the pumped storage station provides 30 minutes of reserve, the duration is required to be 2 hours. The limit of reserve capacity is set at 20MW. The declared reserve capacity and income of the pumped storage station under different reserve compensation prices are shown in Table 4. When the pumped storage station provides reserve, the factors that affect its profit in the energy market are the reserve capacity and duration of reserve. Therefore, the market strategy under this type of reserve is the same as that under the reserve of "10-minute response and duration of 2 hours" (see Table 3). When the reserve compensation price is 0.006 yuan/h, the pumped storage station starts to provide reserve service, thus increasing the total revenue.

Table 4. The decision result of PSPS under different compensation prices of the 4th type of reserve.

compensation price (Yuan/hour)	declare spare capacity (MW)		total revenue (Yuan)	maximum pumping power (MW)	maximum discharge power (MW)	minimum storage capacity state (MWh)	maximum storage capacity state (MWh)
0.001	0	0	8357	20	20	0	100
0.002	0	0	8357	20	20	0	100
0.003	0	0	8357	20	20	0	100
0.004	0	0	8357	20	20	0	100
0.005	0	0	8357	20	20	0	100
0.006	**0.9**	**0**	**8359**	19	20	1.71	100
0.007	2.4	0	8409	18	20	4.77	100
0.008	4.9	0	8508	15	20	9.80	100
0.009	6.8	0	8653	13	20	13.58	100
0.01	20	20	9600	0	0	40	60

When considering that pumped storage station can simultaneously provide four types of reserve services, the optimal market strategy is shown in Table 5. The compensation prices of four types of reserve are set as 0.01, 0.008, 0.006 and 0.005 yuan/h, respectively. At this time, the optimal declaration strategy of pumped storage station is that: primary frequency regulation is 3MW, reserve capacity of 1-minute is 10MW, and reserve capacity of 30-minute is 10MW. The operation results in the energy market is shown in Figure 5. In this case, the total revenue of pumped storage station is 12,340 yuan. At the corresponding prices, the revenues under conditions of providing single reserve are 8,632, 8,510, 8,359 and 8,357 yuan, respectively. It indicates that when the pumped storage station provides reserve services at different time scales its storage capacity and power capacity will be fully utilized, and the total revenue will be significantly increased.

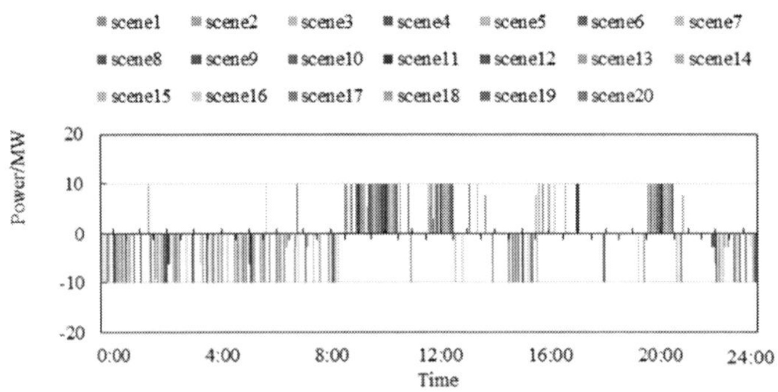

Fig.5 Operation results of PSPS in the electric energy market after providing multiple reserve services.

Table 5. Comparison of the operational decision and profit when providing different reserve services.

compensation price (Yuan/hour)	Reserve 1	Reserve 2	Reserve 3	Reserve 4	All Reserve
compensation price (Yuan/hour)	0.01	0.008	0.006	0.005	/
maximum pumping power (MW)	17	15	19	20	10
maximum discharge power (MW)	20	20	20	20	10
minimum storage capacity state (MWh)	0.005	0.82	1.71	0	21.72
maximum storage capacity state (MWh)	99.95	100	100	100	78.28
total revenue	8623	8510	8359	8357	12340

(Yuan)

5. Conclusions

This paper studies the optimization problem of pumped storage power station participating in energy market and reserve services market. Considering the uncertainty of energy prices, a scenario-based optimization model is established for the decision making. In the model, a variety of reserve services are considered. The power capacities provided by reserve services at different time scales does not affect each other. But the reserved storage capacity for various types of reserve are mutually restricted.

In the energy market, the pumped storage stations arbitrage the price spread to make profits. When providing reserve services, the pumped storage station needs to reserve certain power capacity and storage capacity. When the compensation price of reserve service is higher than a certain level, the pumped storage stations can get more profits through participating in the reserve service market, or even use all of its capacity to provide reserve services.

Simulation analysis is made about the optimal decisions and revenues considering pumped storage power station providing multiple types of reserve service. Even though the compensation price of each type of reserve is relatively low, the pumped storage power station can increase its revenue by providing multiple reserves. The total revenue of the pumped storage power station when providing multiple types of reserve services is obviously higher than that when providing single type of reserve service.

References

[1] Li X M, Zhang Z, Zhao H, et al. Reserve Services System Design in the Background of Energy Internet [J], Electric Power Construction, 2021, 42(5): 57-68.
[2] Zheng X, Xia C, Lin Y, et al. Suggestion on Operation Contigency Reserve Standards for Power Grids of China[J]. Automation of Electric Power Systems, 2018, 42(001):151-159.
[3] Zhao Y, Lv Q, Zhu Q S, et al. Overview on issues related to reserve capacity by wind power connected to the grid [J]. Renewable Energy Resources, 2015, 033(004):565-571.
[4] Ancillary services report 2017[R], Energy UK, 2017.
[5] Yu J, Liu Y, Yang J, et al. Analysis of Development of California Ancillary Service Market and Its Enlightenment to China's Power Market [J]. Power System Technology, 2019, 43(008): 2711-2717.
[6] Xue Y, Luo Y, Li B, et al. A Review of Interruptible Load Participating in System Reserve [J]. Automation of Electric Power Systems, 2007, 31(010): 1-6.
[7] Han B, Lu S, Xue F, et al. Day-Ahead Electric Vehicle Aggregator Bidding Strategy using Stochastic Programming in an Uncertain Reserve Market[J]. IET Generation, Transmission & Distribution, 2019, 13(12): 2517-2525.
[8] Konda S R, Panwar L K, Panigrahi B K, et al. Optimal Offering of Demand Response Aggregation Company in Price-Based Energy and Reserve Market Participation[J]. IEEE Transactions on Industrial Informatics, 2018, 14(2): 578-587.
[9] Tang Z, Liu Y, Wu L, et al. Reserve Model of Energy Storage in Day-ahead Joint Energy and Reserve Markets: A Stochastic UC Solution[J]. IEEE Transactions on Smart Grid, 2020, Early Access.
[10] Liu D N, Gao Y, Zhang T T, et al. Research on strategy of distributed energy storage aggregators participating in peak load regulation auxiliary service[J]. IOP Conference Series: Earth and Environmental Science, 2021, 687(1): 012140.
[11] Vatandoust B, Ahmadian A, Golka M, et al. Risk-averse optimal bidding of electric vehicles and energy storage aggregator in day-ahead frequency regulation market[J]. IEEE Transactions on Power Systems, 2019, 34(3): 2036-2046.
[12] HyungSeung K, Hong J H, Choi I S. Implementation of Distributed Autonomous Control Based Battery Energy Storage System for Frequency Regulation[J]. Energies, 2021, 14(9): 2672-2672.

[13] Abouzeid S I, Guo Y F, Zhang H C. Cooperative control framework of the wind turbine generators and the compressed air energy storage system for efficient frequency regulation support[J]. International Journal of Electrical Power and Energy Systems, 2021, 130. (Early Access)
[14] He T Y, Li S N, Wu S J, et al. Biobjective Optimization-Based Frequency Regulation of Power Grids with High-Participated Renewable Energy and Energy Storage Systems[J]. Mathematical Problems in Engineering, 2021. (Early Access)
[15] Sun Z, Zhao J Q, Yang Y H, et al. Clearing Model of Regulation Market with Participation of Multi-Type Source of Wind-PV-Hydro-Thermal-Storage [C]// Proceedings of the 2019 Annual Meeting of the Power Market Committee of the Chinese Society of Electrical Engineering and the Forum of the National Alliance of Power Trading Institutions. 2019.
[16] Wang K F, Xie L R, Qiao Y, et al. Analysis on Frequency Modulation Performance of Power System Improved by Battery Energy Storage [J]. Automation of Electric Power Systems,, 2021, 1-13.
[17] Ma J W, Li W D, Liu J T,et al. Design of Operation Reserve Classification System for China under New Circumstances [C]. Proceedings of the 2018 Annual Meeting of the Power Market Committee of the Chinese Society of Electrical Engineering and the Forum of the National Alliance of Power Trading Institutions. 2018.
[18] Jiang T, Wang X, Jiang C W, et al. Optimal Hybrid Stochastic Robust Bidding Strategy of Wind and Hydraulic Pumped Storage Jointly Participating in Day-ahead and Real-time Market Using Data-driven Method [J]. Power System Technology, 2021.
[19] Krishnamurthy, D; Uckun, C; Zhou, Z, et al. Energy Storage Arbitrage Under Day-Ahead and Real-Time Price Uncertainty[J]. EEE Transactions on Power Systems, 2018, 33(1): 84-93.

Online Monitoring of Transmission Line Operation Based on Power Internet of Things

Xiaohua Zhang[1a*], Wenxiang Xue[1], Shuyuan Wang[1], Yi Lu[2,3], Hui Wang[1]

[1] State Grid Jibei Electric Power Co., Ltd., Beijing, 100052, China

[2] State Grid Jibei Electric Power Research Institute, Beijing, 100045, China

[3] North China Electric Power Research Institute, Beijing, 100045, China

[a*] 11810003@zju.edu.cn

Abstract—Current monitoring methods of transmission line operation status can not obtain real-time data of distributed distribution transmission line, which leads to a large error in monitoring results. Therefore, a multi-state on- line monitoring method based on power Internet of Things is proposed. Using the gateway of power internet of things to set up network control access mode, build edge computing model, and using AD chip of ADS8365W5300 to obtain the real-time data of massive distributed distribution network, then make a decision on the fault after processing and analyzing the data. This paper constructs an edge computing model which can complete the data processing and analysis in the edge node, and designs the deployment of the edge computing model. By evaluating the faults in the dynamic incremental fault set, the risk state of transmission line in the danger control area is obtained, and a multi-state on- line monitoring method is designed. The experimental results show that the proposed method can monitor the transmission line running state accuratel

1. Introduction

With the rapid development of power grid construction and the gradual expansion of power grid scale, transmission lines will face a variety of risks in the process of operation, affecting the safety and reliability of the power grid [1]. Therefore, it is necessary to warn the risk of transmission line operation. The danger control area of transmission line refers to some common fault areas which are divided by some power supply companies according to their experiences, which is convenient for management and maintenance. The risk early warning of transmission line danger control area can effectively avoid the occurrence of transmission line faults, and provide a strong guarantee for the safe operation of power system [2]. Because the quantity of risk- related information of power transmission line is small, it can not reflect the risk of power network very well, but also cause certain economic losses [3]. However, the existing early warning technology is vulnerable to natural disasters and manual intervention, resulting in poor early warning effect. So the paper [4] studies the model of early warning and assessment of transmission line fire prevention based on AHP, constructs the hierarchical structure by using risk factors, constructs the judgment matrix by combining the actual operation and maintenance data of transmission lines, and obtains the weights to realize the early warning of transmission line risk. reference [5] Studies the intelligent monitoring method of transmission line hazard based on deep learning, and uses the intelligent identification algorithm of transmission line hazard to realize the risk early warning of transmission line. But the two methods have the problem of poor precision in the collection of risk information of transmission lines, which leads to the low precision of early warning results and the poor timeliness of risk early warning. In order to realize the risk early warning of transmission line danger control area, the risk early warning technology of transmission line danger control area based on ubiquitous IOT is proposed. Ubiquitous power IOT technology is real-time, accurate and comprehensive, which can be widely used in the monitoring of power facilities and communication infrastructure, and can get the main information in different cases. Based on the overall information perception and overall configuration information of ubiquitous power IOT, the dynamic failure rate of components and the incremental dynamic failure set are calculated, and the precise risk warning results are obtained after the failure, and the risk status of all power grids is known according to the results, so as to realize the risk warning of transmission line danger control area.

2. Multi state online monitoring of transmission line operation based on power Internet of things

2.1. Set network control access mode based on power Internet of things gateway

In order to control the access mode of the device, we need to start with the sense network and communication network. Perceived input layer and perceived network interactive task data, and the number of accessible perceived network nodes, and the network has a fast transmission speed, long transmission distance, strong confidentiality, anti-jamming and other characteristics, so based on the basic characteristics of the perceived network to establish the perceived network routing control protocol [6-7], as shown in Figure 1 below.

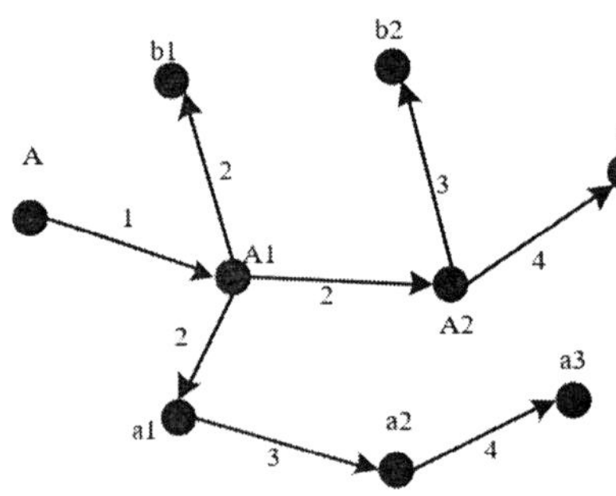

Fig. 1 Network-aware routing protocols

In Figure 1, A1-A3 represents the control center node; a1-a3, b1-b2 are the branch control nodes. According to the above figure, the source node in the protocol transmits data to the branch nodes on both sides, which includes segmented transmission mode and phased transmission mode. Through different routing protocols, the Internet of Things gateway can identify the types of power tasks and improve the processing speed of task data according to the different transmission methods [8].

Network control access mode based on power IOT gateway through optimization and improvement of IOT awareness layer and communication layer [9-10].

2.2. Construction of edge computing model

Edge computing refers to an open platform that integrates network, computing, storage and application core capabilities and provides edge intelligent services nearby at the edge of the network close to objects or data sources, so as to meet the key needs of industry digitalization in such aspects as agile connection, real-time business, data optimization, application intelligence, security and privacy protection. Based on ADS8365W5300 high-speed AD acquisition chip, a mass of real-time data of distributed distribution network is obtained and transmitted to local Moss agent service center through 4G wireless channel. According to the current requirements of real-time and accuracy of fault detection, the network transmission channel is unable to deal with such a large scale of real-time data.

Therefore, an edge computing model that can complete the data processing and analysis of distribution network at edge nodes is established to alleviate network load and unreliability. The edge computation model is shown in Figure 2.

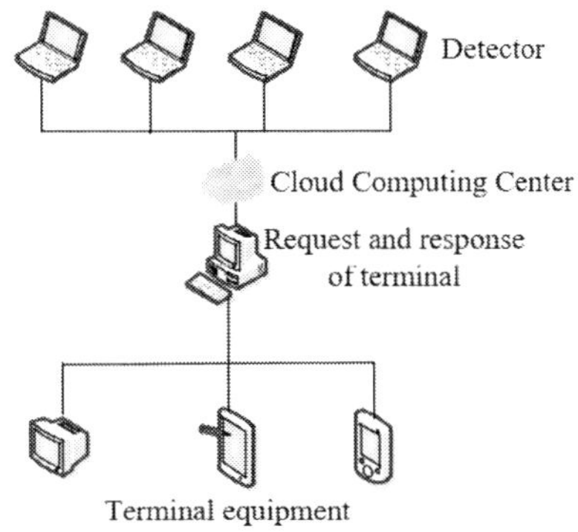

Fig. 2 Structural diagram of edge calculation model

As shown in Figure 2, the distribution network data in the model is generated by multiple fault detectors at the top and multiple terminal devices at the bottom, and transmitted to the cloud computing center through the network. In particular, the terminal equipment at the lower end (i.e. the edge node) needs to make a request to the cloud computing center, and the client may feed back relevant grid data to the edge node only after it receives a response.

2.3. Design of multi state on-line monitoring method for transmission line

Based on the transmission characteristics of transmission line fault current, the catenary formula is used to calculate the sag and stress data of fault current. Most of the catenary formulas are hyperbolic functions, and the calculation process is complicated. When the transmission line is abnormal, the maximum length of fault current in stall distance is about 5 ‰ larger than that in stall distance. Therefore, the calculation process can be simplified, and the distribution of fault current along the line direction of the hanging point can be regarded as uniform, so the parabolic formula is used to calculate the fault current data.

(1) Calculation of sag of fault current

As shown in Figure 1, in a transmission line with different height of suspension points, the shift l between points A and B is higher than that of point h. Based on the sag formula of oblique parabola, the maximum sag of fault current in the transmission line is:

$$f = \frac{l^2 \cdot G}{F \cos \varphi} \quad (1)$$

In the formula, F represents the horizontal tension at the lowest point of the fault current, and N, G represents the gravity force per unit length of the fault current.

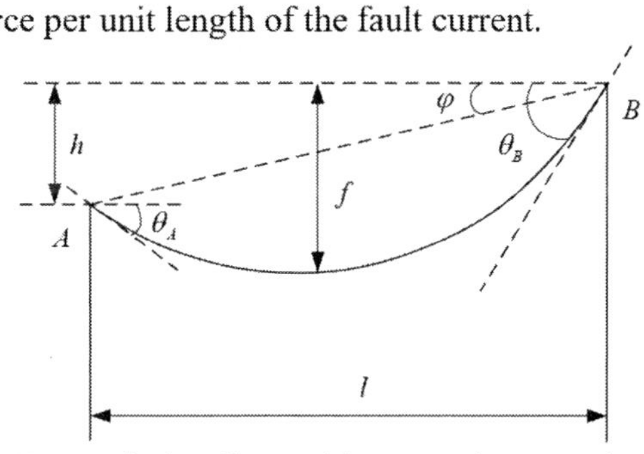

Fig. 3 Transmission lines with unequal suspension points

The formula for calculating the inclination angle of fault current at hanging points A and B is as follows:

$$\theta_A = \arctan\left(\frac{l \cdot G}{F\cos\varphi} - \frac{h}{l}\right) \quad (2)$$

$$\theta_B = \arctan\left(\frac{l \cdot G}{F\cos\varphi} + \frac{h}{l}\right) \quad (3)$$

By substituting the $\frac{G}{F}$ in the formulas (2) and (3) into the formula (1), the relationship between the fault current sag and dip angle can be obtained as follows:

$$f = \frac{\left(\tan\theta_A + \frac{h}{l}\right)}{8} \quad (4)$$

$$f = \frac{\left(\tan\theta_B - \frac{h}{l}\right)}{8} \quad (5)$$

By measuring the relationship between the fault current sag and the inclination angle, the maximum sag in the fault current stall is calculated.

(2) Load calculation of transmission lines

The change of temperature will cause the expansion or contraction of transmission line, and the change of load will make the value of fault current rebalance, and the result will change the value of fault current. Under different working conditions, the relationship between transmission line and current can be expressed by state equation, that is:

$$\begin{cases} H_n = \frac{l^2 w_n^2 ES}{12F_n^2} \\ H_m = \frac{l^2 w_m^2 ES}{12F_m^2} + \alpha ES\left(t_n - t_m\right) \end{cases} \quad (6)$$

In the formula, H_m and H_n respectively represent the magnitude of fault current under m and n working conditions, w_m and w_n respectively represent the magnitude of fault current under m and n and unit length of fault current under m and n working conditions, t_m and t_n respectively represent the conductor temperature of fault current section under m and n working conditions, E represents the elastic coefficient of conductor of fault current section, α represents the coefficient of temperature expansion, and S represents the cross-section area of conductor of fault current section. The energy consumption of abnormal data during the fusion process is:

$$E_{DA}(\cdot) = E_\eta \quad (7)$$

In formula (7), E_{DA} represents the energy consumed by fusing 1 bit abnormal data, and η represents the fusion coefficient. Next, the multi state on-line monitoring of transmission line is realized through the transmission line condition monitoring process.

3. Experimental comparative analysis

3.1. Setting experimental parameters

In order to analyze the performance of the proposed monitoring method based on fault current propagation, reference [4] method and reference [5] method, three monitoring methods are modeled in MATLAB environment. Three transmission line towers shall be installed within the monitoring area of 60 meters * 1100 meters. The communication master station is located on the first tower pole, and the distance between the three towers is 500 meters. 20 sensor collection nodes are deployed near each tower pole. Two of the nodes use induction to power the monitoring process with unlimited energy and all the remaining nodes use batteries to power the monitoring process. The initial energy is set at 1 joule. The main communication station is located at (30, 30), and all nodes cannot be moved once installed.

3.2. Comparative test of transmission line condition monitoring error

The multi-state on- line monitoring method, reference [4] method and reference [5] method based on the Internet of Things are adopted to monitor the state of transmission lines, and the comparison result of the monitoring error is shown in figure 4.

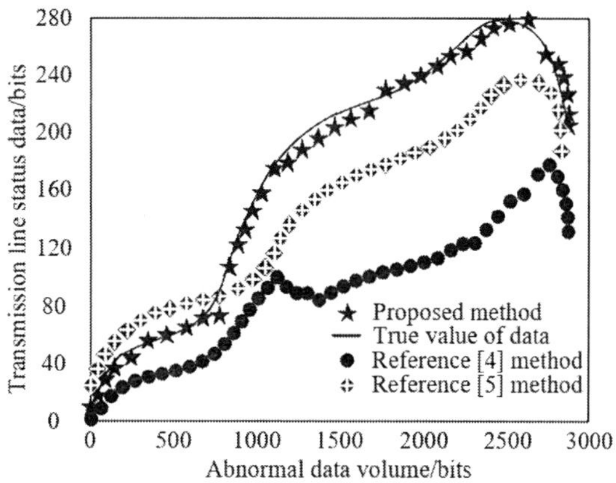

Fig. 4 Comparison results of transmission line condition monitoring errors

It can be seen from the experimental results in Fig. 4 that the monitoring method in reference [4] does not calculate the transmission line fault current data, so that the transmission line condition monitoring results affected by the sag of the fault current are lower than the real value of the transmission line condition data, and the monitoring error becomes larger; The monitoring method in reference [5] is affected by the transmission line load, which makes some values of the transmission line condition monitoring results close to the real value, but the final monitoring error is still very large; The transmission line condition monitoring method based on the propagation characteristics of fault current calculates the sag of fault current and the load of transmission line respectively, which makes the data value of transmission line fault current more accurate, so as to reduce the transmission line condition monitoring error.

4. Conclusion

In order to solve the problem of large errors in the monitoring results of current transmission line operation status monitoring methods, a multi-state online monitoring method for transmission line operation based on power Internet of Things is proposed. Build edge computing model, obtain massive real-time data of distributed distribution network, and design the deployment of edge computing model. The risk state of transmission line risk control area is obtained, and the multi-state online monitoring method of transmission line is designed. Experimental results show that the multi-state monitoring results of the proposed method are consistent with the real situation and meet the application requirements in this field. In this paper, the analysis and processing of the operation status of the distribution line is not in-depth enough. In the follow-up research process, we can consider the elimination and correction of erroneous data.

Acknowledgments
The study was supported by "Science and Technology Project of State Grid Corporation of China, Empirical research for construction and operation of substation transmission line and distribution grid based on internet of things technology (Grant No. 52010119006S)".

References
[1] Mueller C E . (2020) Why do residents participate in high-voltage transmission line planning procedures? Findings from two power grid expansion regions in Germany[J]. Energy Policy, 145:111779.
[2] Zhang Y, Bansal M, Escobedo A R . (2020) Risk-neutral and risk-averse transmission switching for load shed recovery with uncertain renewable generation and demand[J]. IET Generation, Transmission and Distribution, 14(21):4936-4945.
[3] Lee C K, Lee G S, Chang S J . (2021) Solution to Fault of Multi-Terminal DC Transmission Systems Based on High Temperature Superconducting DC Cables[J]. Energies, 14(5):1292.
[4] Zhu Q, Guo J, Zeng B, et al. (2018) An early-warning evaluation model against mountain fire of transmission line based on analytic hierarchy process[J]. Electrical Measurement & Instrumentation, 55(6):71-75,88.
[5] Liu Y P, Pei S T, Wu J H, et al. (2019) Deep Learning Based Target Detection Method for Abnormal Hot Spots Infrared Images of Transmission and Transformation Equipment[J]. Southern Power System Technology, 13(2): 27-33.
[6] Conlon T, Waite M, Modi V . (2019) Assessing new transmission and energy storage in achieving increasing renewable generation targets in a regional grid[J]. Applied Energy, 250(PT.1):1085-1098.
[7] Pei J, Yao J, Liu R, et al. (2020) Characteristic Analysis and Risk Assessment for Voltage–Frequency Coupled Transient Instability of Large-Scale Grid-Connected Renewable Energy Plants During LVRT[J]. IEEE Transactions on Industrial Electronics, 67(7):5515-5530.
[8]A Cortés, S Martínez. (2019) A Hierarchical Algorithm for Vehicle-to-Grid Integration under Line Capacity Constraints - ScienceDirect[J]. European Journal of Control, 47:53-63.
[9] Li B, Cui H, Li B, et al. (2020) A permanent fault identification method for single-pole grounding fault of overhead transmission lines in VSC-HVDC grid based on fault line voltage[J]. International Journal of Electrical Power & Energy Systems, 117:105603.
[10] Chung H M, Li W T, Yuen C, et al. (2019) Local Cyber-Physical Attack for Masking Line Outage and Topology Attack in Smart Grid[J]. IEEE Transactions on Smart Grid, 10(4):4577-4588.

Research on the Application of Renewable Energy in DC Data Center

Xingyu Bai

Chengdu University of Information Technology

No.24, Section 1 of Xuefu Road, Southwest Airport Economic Development Zone, Chengdu, Sichuan

2628569415@qq.com

Abstract-With the rapid development of the information technology industry, the demand for data centers is gradually expanding. As the foundation of carrying many information technology services, data centers have been built and put into use on a large scale. At the same time, the characteristics of huge energy consumption are more obvious. Due to the defects of low utilization rate, heavy pollution and limited reserves of traditional fossil fuels when supplying energy for data centers, the development of green renewable energy has become an inevitable demand. This paper studies the key technologies involved in supplying power to DC power supply data center with renewable energy, and makes a case to compare the DC power supply system and the AC power supply system. This paper also pays attention to the technology of energy storage in data center, and makes an analysis to prove the advantages of energy storage data center. In addition, this paper studies and summarizes the application of multi-station integration technology and other issues.

1.Introduction

Developing green data center is an important part of national strategy, technological development and social life. The energy consumption of data center is huge, and the annual economic input cost is very high. The data shows that the total power consumption of China's data centers in 2017 exceeded 120 billion kWh, exceeding the sum of the annual power generation of the Three Gorges Dam and the power generation of Gezhouba Power Plant in the same year. In 2020, the number of data centers in China has exceeded 80,000, with a total area of over 30 million square meters. Correspondingly, the energy consumption is also increasing year by year. Therefore, it is a major problem to be solved to study the appropriate energy supply scheme and promote the green development of data centers.

The traditional power supply mode uses UPS, but its efficiency is low, energy consumption is high and economy is poor relatively. Thus, a new technology represented by DC power supply mode is born. The DC control and protection system is mainly in the form of industrial computer [1], which has entered the stage of large-scale application in China.

In view of the high energy consumption, heavy energy pollution and existing problems in the original data center power supply mode, this paper analyzes the application and development of renewable energy, DC power supply technology and energy storage technology in the data center. Literature [2] introduces the application of renewable energy in data center. Literature [3] lists the huge energy consumption data of the data center, which shows the necessity and urgency of using renewable energy to power the data center. Literature [4] is a summary of the application of energy storage technology in renewable energy systems. Literature [5] pointed out the requirement of equipping data centers with energy storage, specifically pointed out the defects of traditional power standby mode, and put forward an energy storage data center. Literature [6] introduces the application of lithium battery for energy storage systems. Literature [7] puts forward the concept of '5G-source network-load storage' multi-station integration which mainly focusing on digital services and energy services. Literature [8] proposes a distributed coordination control strategy for load sharing and energy balancing between heterogeneous energy storages. Literature [9] evaluates the long-term impact of grid level energy storage on the penetration of solar and wind energies and on CO2 emissions reduction in Chile.

2. Renewable energy application in data center

2.1 Application of renewable energy

For a long time, fossil energy, as the main power generation source, has been applied in the power supply of data centers. Fossil energy such as oil, natural gas and coal account for about 86% of the total energy consumption in the world. However, fossil energy has caused serious pollution to the global environment. Long-term use of fossil energy will bring many irreversible harms to human society and face the risk of energy depletion. Therefore, this paper studies the application prospect and technical methods of renewable energy for data center power supply.

Renewable energy is widely used in power generation, especially in data centers [2]. The report shows that as early as 2015, the power consumption of big data centers nationwide has reached 100 billion kWh, which is equivalent to the annual power generation of the Three Gorges Power Station. In 2018, this figure reached 160.9 billion kWh, exceeding the annual social electricity consumption in Shanghai. It is estimated that by 2030, global big data centers will consume about 30% of the world's electricity [3]. The energy consumption of data centers is very high. In order to achieve the common goals of saving energy, reducing pollution and protecting the ecological environment, the world's leading technology companies have joined the RE100 organization and advocated the vigorous use of renewable energy.

2.2 Application of Energy Storage Technology in Data Center

2.2.1 Energy storage technology in data center

In recent years, the development of energy storage has been widely reported. Facing the problem of huge energy consumption in data centers, energy storage is an important means to effectively reduce energy consumption, save energy and reduce emissions, and is applied to the construction and development of data centers [4].

There are many types of energy storage technologies, which can be divided into gaseous energy storage (air compression, hydrogen energy), liquid energy storage (pumped energy storage, photo-thermal liquid oil), solid energy storage (battery, capacitor), etc. It can be divided into physical energy storage (flywheel energy storage), electrochemical energy storage (battery) and electromagnetic energy storage (superconducting energy storage). Among them, electrochemical energy storage technology is widely used in data centers. Energy storage technology plays a very important role in energy transformation. Its application degree not only determines the development level of renewable energy, but also determines the success or failure of data center construction and development.

Traditional data center power backup system consists of UPS host and lead-acid battery. When the mains supply is normal, UPS supplies power to the load after filtering the mains impurities, and charges the battery at the same time. When the mains is abnormal (power supply is interrupted), the battery is discharged and supplied to the load through UPS to ensure power supply [5].

Thanks to the development of energy storage technology in recent years, the construction of energy storage data center has become an important means to solve the above problems.

The traditional energy storage form of data center is lead-acid battery, which has stable voltage and low price. It has been used and developed for more than 100 years, but it is short in service life, easy to collapse instantaneously, pollutes the environment and has high demand for maintenance. In order to improve the security of the power supply system of the data center computer room, the operators of the data center computer room gradually use lithium batteries to replace the traditional storage batteries, so as to better realize the modernization process of the data center computer room industry [6]. At present, the purchase cost of lithium iron phosphate battery is higher than that of lead-acid battery, which is about 1.5 ~ 3 times that of lead-acid battery. However, due to its advantages of long service life, low maintenance cost and high working efficiency, it can save the total cost in the long run.

2.2.2 New application of energy storage in data center

The introduction of energy storage technology in the data center has the following advantages compared with the traditional backup power supply in the data center:

(1)The battery investment can be recovered by peak shaving and valley filling mode. Taking a UPS with a power of 500kW and a power factor of 0.8 as an example, for convenience of calculation, the discharge parameters of Siendi SHC12 250FT battery are adopted, the battery price is 1 CNY/Wh, the electricity price difference is 0.78 CNY as a reference, and the residual value of the battery is calculated by 25%. 120 SHC12 250FT batteries are required for standby power of 500kW UPS for 15min. Assuming that three more batteries are added, the discharge depth is 50% DOD, and the battery is used for 10 years. The preliminary income is calculated as shown in Table 1:

Table 1 10-year revenue estimation of data center with energy storage technology

	Traditional UPS	Energy storage UPS
Number of batteries connected in series (AH/group)	40	40
Number of parallel battery groups (groups)	3	6
Total battery capacity (kWh)	288	576
Battery price (CNY/AH)	2400	2400
Discharge depth of battery.		50%
Battery life (times)		3600
Total battery price (CNY)	288,000	576,000
Peak-valley electricity price difference (CNY/kWh)	0.78	0.78
Peak-valley arbitrage in life cycle (CNY)		808,704
Residual value of battery (CNY)		144,000
Total income in life cycle (CNY)		376,704

It can be seen from the table that transforming the traditional UPS power supply into the energy storage UPS voltage can not only recover the battery investment, but also obtain a large amount of benefits during the battery life cycle, thus greatly saving the cost of the data center.

(2)The battery is discharged every day, and the health state of the battery can be known through the cut-off voltage after discharge. It is helpful to eliminate bad batteries in time, and at the same time, it also saves the cost of doing false load test every year. However, batteries in traditional data centers are rarely discharged, and the state of batteries is unknown.

(3)The standby time of data center is longer. Because more batteries should be configured, the standby time of batteries under the same power is much longer than that of conventional configuration.

Besides, the energy storage data center should be built in a place with peak-valley price difference, and the larger the peak-valley price difference, the better the income will be. In addition, the factors affecting the installation area should also be considered.

2.3 System design based on application of renewable energy in DC data center

2.3.1 Case analysis

Taking an 8kW data equipment as an example, there has a concrete comparative analysis between the high voltage DC power supply system and the UPS power supply system. It can be seen that every 8kW of data equipment adopts high voltage DC power supply system, which saves 42,398 kWh of electricity every year, and the power saving rate is about 25.3%. In the case cited in this paper, the total power consumption of data equipment in a data center computer room is 16000kW, and the annual power saving is about 84.8 million kWh. According to the price of 1 CNY/kWh, the electricity cost can be saved by 848,000,000 CNY a year. If combined with the construction scheme of 'multi-station integration', self-built photovoltaic sub-station can lower the electricity cost than the ordinary case, and if the price of 0.7 CNY/kWh is adopted, the annual energy-saving electricity cost can be about 593,600,000 CNY. After the introduction of 'multi-station integration' mode, although the difference in electricity consumption between UPS system and high voltage DC system is reduced, the electricity consumption of both systems is lower than that of the original photovoltaic substation, and the electricity consumption of high voltage DC system is always lower than that of UPS system.

Table 2 Comparison of energy expenditure between UPS power supply system and high-voltage DC power supply system

Power consumption (kW)	UPS power supply system	high voltage dc power supply system	remarks
Data power consumption	8	8	
Loss of rectifier cabinet in data center	3.33	1.36	UPS power supply efficiency is 75%, and high voltage DC is 88%.
System rectification loss	2.35	0.86	UPS power supply efficiency is 85%, and high voltage DC is 93%.
System input power consumption	13.68	10.22	
Air conditioning power consumption	5.47	4.09	The air conditioning power consumption is 40% of the system input power consumption.
add up to	19.15	14.31	
Energy consumption for 1 year	167754	125356	

3. Key technologies for future development

3.1 Multi-station integration technology

'Multi-station integration' mode is an important development mode in the construction of data center, and also an important application of power Internet of Things.

'Multi-station integration' is based on the existing substation resources, and uses the data integration center to build various substations such as charging and replacing power plants, energy storage stations, 5G base stations, photovoltaic power stations, wind power stations, etc., which can expand the value of power Internet of Things at a deeper level from the existing value of substations. The energy storage station, as a backup energy station, can realize the business functions of peak shaving and frequency modulation and renewable energy consumption, and can also reduce the electricity cost of data center operation. Under the peak-valley electricity price policy, shifting the power consumption of the data center from peak hours to peak hours can save more electricity bills. Energy storage station can also reduce the peak load of charging and distribution capacity, and improve the utilization rate of equipment. Figure 1 is a schematic diagram of 'multi-station integration' mode.

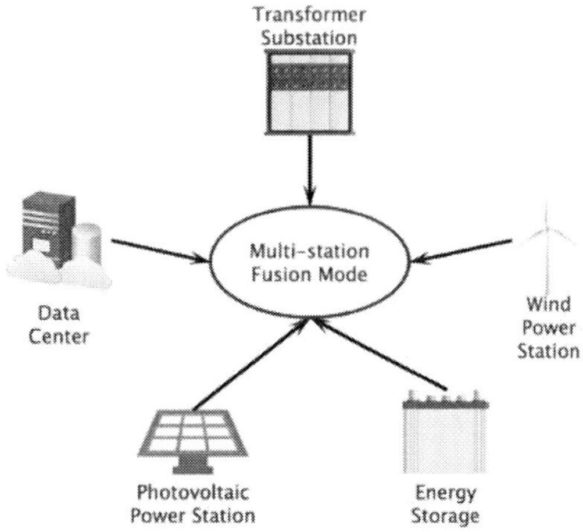

Fig.1 Schematic diagram of 'Multi-station Convergence'

The 'multi-station integration' architecture of full DC bus is a practical architecture. At present, the high integration of energy storage stations, new energy substations, substations and data centers can be realized through full DC. The system architecture is shown in Figure 2. On the power supply side, 10kV commercial power is rectified and connected to DC buses 1 and 2, with circuit breakers. The new energy substations are directly connected to DC buses 1 and 2 through inverters and circuit breakers, and a contact switch S1 is provided between the two buses. At the distribution end, the energy storage station and data center are connected to two buses through chopper, loop switch control equipment and loop switch. Compared with traditional discrete systems, this architecture can provide large-scale new energy consumption, reduce economic expenses and improve system reliability.

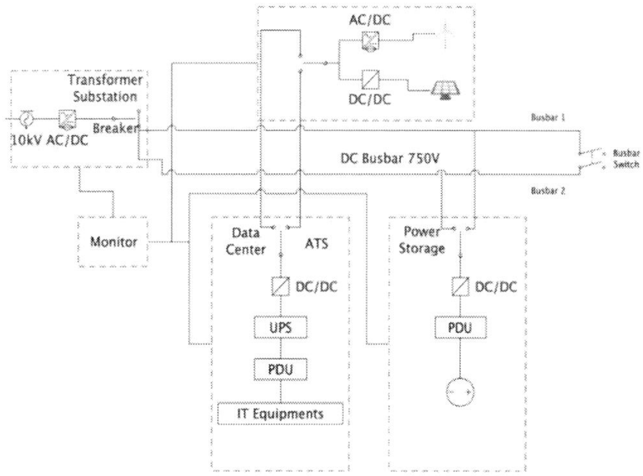

Fig.2 'Multi-station integration' architecture of full DC bus

At present, 'multi-station integration' mainly includes data center operation mode, energy storage station operation mode and charging station operation mode suitable for new energy vehicles. The energy storage station is applied to the power generation side and distribution network side, which improves the utilization rate of various energy sources, and new energy can be better accepted. It also effectively solves the intermittent and uncertain problems of wind energy, photovoltaic power generation, etc. when the energy storage station is configured on the power generation side or distribution network side. Data centers are for large Internet enterprises, such as Apple, Google, Facebook, Baidu, Ali and Tencent. At present, data centers built and operated by third-party enterprises can provide cloud data storage, cloud computing, data analysis and processing services to government, enterprises and individual users, and obtain corresponding benefits by providing rental services of cabinets and cloud services to various users.

4. Conclusion

This paper introduces the data center power supply mode, energy storage configuration and the development status of renewable energy application, with emphasis on the advantages of DC power supply mode, and the future prospect of gradually replacing AC UPS power supply as the main power supply mode. Thanks to the vigorous development of the Internet industry in recent years, the scale of the data center is expanding rapidly, which is followed by huge energy consumption and expensive expenses. The power supply mode based on renewable energy and advanced energy storage technology can alleviate this series of problems to a great extent, and will also become an important research direction of the data center in the future.

References

[1]Liang Xuming, Zhang Ping, Chang Yong. Current situation and development prospect of HVDC transmission technology [J]. Power Grid Technology, 2012,36(04):1-9.
[2]Soongeol Kwon. Ensuring renewable energy utilization with quality of service guarantee for energy-efficient data center operations. 2020.
[3]LONGi. 30% power consumption in global big data centers in 2030? How does the new infrastructure deal with the 'energy consumption behemoth' [EB/OL]. https://careers.longi.com/index.php? m=content&c=index&a=show&catid=155&id=665, 2020.
[4]D. Rekioua b, T. Rekioua b, S. Bacha. Overview of energy storage in renewable energy systems. Ould Amrouche a,b, 2016.
[5]Zhou Bijun. Data center consumes too much energy, so it is time to consider energy storage data center [EB\ol]. http://www.elecfans.com/tongxin/20180219637542.html, 2018-07-18.
[6]Ding Ningxiang, Huang Xuqing. Analysis of Application Trend of Lithium Battery for Energy Storage in Data Center Computer Room, 2018.
[7]Siwu, Li et al. "The Analysis of Business Scenarios and Implementation Path of '5G+Source-Network-Load-Storage 'Multi-Station Integration." E3S Web of Conferences. Vol. 248. EDP Sciences, 2021. 02031–.
[8]Zhang Runfan, Branislav Hredzak, Thomas Morstyn. "Distributed Control With Virtual Capacitance for the Voltage Restorations, State of Charge Balancing, and Load Allocations of Heterogeneous Energy Storages in a DC Datacenter Microgrid." IEEE transactions on energy conversion 34.3 (2019): 1296–1308.
[9]Maximov Gajardo, Serguey, Gareth Harrison, Daniel Friedrich. "Long Term Impact of Grid Level Energy Storage on Renewable Energy Penetration and Emissions in the Chilean Electric System." (2019)

Journal of Physics: Conference Series **2108** (2021) 012055 doi:10.1088/1742-6596/2108/1/012055

Research on the Formation and Solution of the Bullwhip Effect of Electricity Material Reserve

Wenlong Zeng[1a*], Fei Peng[1b*], Qiaozhang Hong[1c], Jinsheng Liu[2d]

[1]Energy Development Research Institute, CSG, Guangzhou 510000, China

[2]China Energy Engineering Group Guangdong Electric Power Design Institute Co.,Ltd., Guangzhou 510000, China

[a*]shzz2008@163.com, [b*]pengfei001001@163.com, [c]hongqz@csg.cn, [d]liujinsheng@gedi.com.cn

Abstract—With the deepening of the power system reform together with the transformation and development of power companies, the importance of supply chain management has become increasingly prominent. As an important part of the power supply chain management of power companies, the accuracy of electricity material reserves has an impact on the entire process of the company's supply chain. This article analyzes the reasons for the bullwhip effect in the process of electricity material storage, points out the harm caused by the bullwhip effect, and proposes ways to weaken the bullwhip effect such as improving the accuracy of demand reserves, strengthening inventory management.

1. Introduction

In recent years, with the continuous deepening of the power system reform together with the transformation and development of power companies, supply chain management as a new concept and driving force for the development of power companies, and an important part of strategic planning has been highly valued. As the source of supply chain management improvement, the material reserve demand plan will have a significant impact on its accuracy and comprehensiveness on subsequent procurement, warehousing and other links. While the existence of bullwhip effect and other phenomena will lead to a certain degree of demand information authenticity. Distorted variants make the quantity of material reserves inconsistent with the actual situation, and bring about corporate inventory risks. Therefore, combing through the various links of the electricity material reserve demand stage, analyzing the causes of the bullwhip effect, and proposing solutions to weaken the bullwhip effect, is significant to improve the accuracy of material demand and the level of electricity material management.

2. The Bullwhip Effect and Formation Analysis of Electricity Material Reserve

2.1. *Bullwhip effect*

The bullwhip effect is a phenomenon of demand variation and amplification. When demand information in the supply chain is transmitted from the customer to the supplier, the information cannot be effectively shared, and it is distorted and amplified step by step during the transmission process, resulting in a small amount of downstream demand. The change will be magnified in the upstream. This phenomenon acts like a swinging bullwhip on the graph, so it is called the bullwhip effect[1].

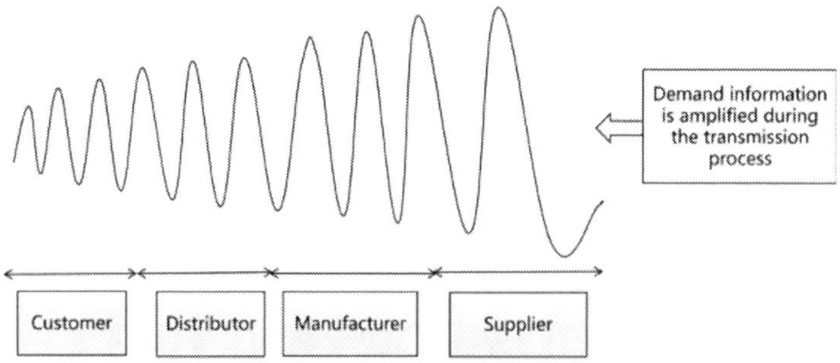

Fig.1 The bullwhip effect

2.2. Analysis on the Causes of the Bullwhip Effect in the Reserve Demand of Electricity Materials

Compared with other traditional enterprises, electricity materials have the characteristics, like large quantities, high value, and a wide range of departments. Electricity material reserves are stored at a fixed amount to meet the needs of production, operation, emergency rescue, and repair.

Electricity material reserves are organized by the professional departments of the prefectural and municipal bureaus to analyze the operation of the unit's reserves over the years, and put forward the required reserves of materials and materials. The supply chain management department summarizes the reserve quota requirements of each professional department and formulates the reserves at this level. The plan is reported to the provincial companies and professional departments of the Internet company for integration and optimization, and the supply chain department at the same level participates in the review of the list. Finally, the supply chain department will integrate the company's reserve materials list according to the list of professional reserve materials. Supply chain departments at all levels implement material reserves. According to literature and case analysis[2-5], in the process of material demand and storage, the following factors may cause the bullwhip effect risk:

(1) Demand expectation effect. In the entire process of demand reserve, there are many uncertain factors involved. When the departments make material reserve, various professional departments leave some margin for the reported reserve demand out of safety. Because of the company's professional departments at all levels have a margin when making the reserve demand, the reserve demand in head department will become very large.

(2) The lead time is too long. Due to the long period between the summary of material reserve requirements and actual purchases, the longer the lead time, the greater the deviation of the actual demand forecast in the later period. Besides the more inventory required, the purchase order is larger than the actual demand, and the bullwhip effect is also happened.

(3) Lack of information sharing. Due to the lack of real-time communication and sharing of material information with suppliers, companies cannot grasp the actual supply capabilities of upstream suppliers. In order to obtain sufficient and stable supply of materials, increase demand during storage to prevent supplier delivery problems. As a result, demand information is incorrectly amplified, which will also lead a bullwhip effect.

(4) The external environment changes. When facing the uncertainties caused by changes in natural disasters, external policies, and social environment, most of companies choose to deal with high inventories. The greater the uncertainty, the higher the corresponding inventory, which also leads to the demand is magnified.

3. The Harm of Bullwhip Effect in the Reserve Demand of Electricity Materials

The existence of the "bullwhip effect" makes the inventory of power enterprises remain high and occupy a lot of funds. At the same time, it also reduces the utilization and sales of materials, reduces the operation efficiency of the entire supply chain, and increases the difficulty and risk of material management, which may cause the following harms:

(1) Reserve demand is inconsistent with actual demand. The entire material reserve demand has been effected by subjective and objective factor. From the professional department reserve to the actual procurement process, the changing demand is significantly higher than the actual demand, which cannot objectively reflect the actual material demand of the enterprise.

(2) Increase enterprise costs. As an asset-intensive enterprise, power companies have high inventory requirements due to high reserve requirements. When the materials are actually used, the inventory cannot be quickly reduced. This makes the utilization of inventory materials low, increases the cost of corporate inventory, and occupies a large amount of fund, increases business costs.

(3) It is not conducive to the establishment of corporate trust. The existence of the bullwhip effect reduces the trust between power companies and suppliers, which is not conducive to the establishment of long-term cooperation between the upstream and downstream of the supply chain, resulting in an imbalance in the power supply chain and negatively affecting business operations.

4. Research on the Solution to the Bullwhip Effect in the Reserve of Electricity Materials

The existence of the bullwhip effect in the reserve of electricity materials has led to an increase in inventory, an imbalance between supply and demand, and a waste of resources, which endangers the operation of the supply chain and production operations of enterprises. Therefore, it is necessary to fully integrate the reasons for the bullwhip effect in the electricity material reserve, and propose effective methods to weaken the bullwhip effect, shorten the length and vibration amplitude of the "bullwhip", and reduce the risk of demand information distortion.

4.1. Improve the accuracy of demand reserves

Improve the accuracy of demand reserve by establishing a demand reserve forecasting model. On the basis of analyzing historical consumption, with combining factors such as demand characteristics, warehouse layout, investment plan, service level, replenishment timeliness and other factors, construct a demand reserve forecasting model. By comparing the gap between forecast consumption and actual consumption to analyze the accuracy of the forecast model and look for the cause of the obvious abnormality. Improve the accuracy of the forecast by improving the forecasting method or revising the forecasting model.

4.2. Strengthen reserve management

Agreement reserves and joint reserves are effective means to strengthen reserve management. Through the implementation of these reserve management, it can strengthen the cooperation between enterprises, short the lead time, reduce the gap between demands and supply, reduce the inventory, weaken the bullwhip effect.

(1) Agreement reserve. Agreement reserve refers to a model in which a framework purchase agreement is signed with a supplier based on the forecast of material demand for a certain period in the future, and the supplier manages the user agent to exercise decision-making of inventory. Different from the previous model of replenishment supply based on reserve demand, the agreement reserve directly allocates suppliers according to the actual or predicted material demand based on the agreement, which eliminates the amplification and distortion of traditional electricity material reserve demand information in the process of transmitting to suppliers, and reduces the impact of the bullwhip effect.

(2) Joint reserve. Joint reserve is a risk-sharing reserve management model that establishes a coordination and cooperation model with suppliers. The joint reserve emphasizes the simultaneous participation of both supply and demand parties to jointly formulate a demand reserve plan and integrate it into a coordination management center. Through the coordination management center, both the supply and demand parties share the demand information, so as to partially eliminate the demand fluctuation caused by the uncertainty between supply chain nodes and the distortion of demand information, ensure the consistency of demand expectations between the supply and demand parties, and improve the stability of supply.

4.3. Establish information sharing

The direct cause of the bullwhip effect is that demand information has been distorted and amplified during demand transmission. Upstream suppliers cannot recognize and obtain real demand due to insufficient information. Therefore, it is necessary to establish an information sharing mechanism with suppliers. It is possible to eliminate the asymmetry of demand information in the process of demand information transmission. Specifically, it can be shared from the following aspects:

(1) Demand forecast information. In the traditional supply-demand relationship, the supply parties only rely on orders to communicate demand information. However, order is the product of processing market demand information, there is a large deviation between the supply based on the order and the actual demand. Suppliers can share demand forecast information and actual consumption data, and they can reasonably amend the demand forecast to communicate and confirm, thereby avoiding the amplification of demand deviation caused by repeated forecasts.

(2) Inventory information. Since inventory information is not shared, both the supply and demand sides will set up a certain safety stock in order to avoid shortages, resulting in a high overall inventory level, and corresponding demand information has been further enlarged. Shared inventory information includes inventory data and inventory strategies. By sharing information, both the supply and demand parties can clarify each other's inventory levels, which can effectively reduce the safety stock level and reduce unnecessary inventory.

4.4. Build a partnership

Through the establishment of partnerships between supply chain members, mutual benefit and win-win results can be achieved. Different from simple short-term contractual relations, the establishment of cooperative partnerships within the supply chain requires supply chain members to realize information sharing, risk sharing, overall coordination, and common pursuit of maximum benefits. In the establishment of a partnership, incentive mechanism and trust mechanism are two indispensable elements. The incentive mechanism is based on the possible conflict of interest, looking for ways to maximize common interests, using the incentive mechanism to fully mobilize the enthusiasm of partners, and eliminate opportunistic behaviors caused by information asymmetry. The trust mechanism requires mutual trust between partners, coordination of collective and individual interests, and stable cooperative relations. The establishment of trust relations is a long-term process. The establishment of partnerships generally needs to be initiated and led by core members with an important position and influence in the supply chain. Power companies can play its leading role in upstream and downstream industries and build an ecosystem of risk sharing and benefit sharing with partners.

Fig.2 The partnerships between supply chain members

5. Conclusion

Based on the results and discussions presented above, the conclusions are obtained as below:

(1) This paper analyzes the causes of the "bullwhip effect" of electricity material reserve. Due to the expected effect of demand, too long lead time, lack of information sharing, change of external environment and other factors, there is a risk of bullwhip effect in the process of power material demand reserve.

(2) The bullwhip effect risk in the process of electricity material reserve will cause reserve demand is inconsistent with actual demand, increase enterprise costs, and it is not conducive to the establishment of enterprise trust.

(3) In order to weaken the harm of bullwhip effect, this paper puts forward countermeasures such as improving the accuracy of demand reserve, strengthening inventory management, establishing information sharing, and building a partnership, so as to further promote the supply chain of power enterprises to high-efficiency and high-quality development, promote development of power enterprises.

References

[1] Yinliang, Y. (2013) Research on the "Bullwhip Effect" in the Supply Chain of Power Enterprises and Its Solutions.Shenhua Technology, 11: 11-14.
[2] Jie, S., Jishun, N. (2009) Research on the Causes and Governance of Bullwhip Effect in Supply Chain.Market Modernization, 03: 112-113.
[3] Jia, S. (2014) Research on the Bullwhip Effect of Supply Chain Based on Complexity Theory.Beijing Jiaotong University.
[4] Huijuan, R., Shoufeng, J.,Yuxin, W. (2020) Causes and preventive measures of bullwhip effect in supply chain.Times Finance, 778: 124-125.
[5] Guang, F. (2015) Research on logistics system optimization of Jibei Electric Power Co., Ltd.North China Electric Power University.

Insulator filth monitoring and situation awareness based on ultraviolet pulse and least square method

Tao Liu, Bo Qu, Guolong Li*

State Grid Tieling Electric Power Supply Company, Tieling, Liaoning 112000, China

*1142101056@ncepu.edu.cn

Abstract—To monitor the filth state of insulators in time, a method based on ultraviolet pulse method is proposed to monitor the filth status of insulators, which can detect the early corona discharge of insulators by detecting UV signals and determine the degree of insulator filth. On this basis, the response characteristics of spot area to electrical pulse signal are studied and theoretical analysis is carried out, the relationship between spot area and electrical pulse signal amplitude is studied, and the correlation curve is obtained. By observing the change law of image size with the observation distance in ultraviolet image. the experiment found that the image size of UV channel and the image size of visible light channel increased or decreased in approximate proportion with the observation distance and proposed a new method to estimate the "discharge imaging area " of discharge on the surface of the insulator by using the image information of the visible light channel. According to the discharge "relative spot area mean", one minute time "number of large spot area image frames" two ultraviolet imaging parameters used to characterize the filth discharge characteristics, combined with the ambient humidity to establish a fuzzy logic reasoning model of the filth state, to achieve the assessment of the filth state of the insulator. By using the proposed UV image processing method, the accuracy of insulator filth detection is improved.

1. Introduction

Surface discharge of insulators is a common discharge phenomenon, and the analysis of their discharge strength can be used to assess the hazards of discharge, the stage of discharge development and the realization of flash warning. At present, the most common method for detecting and quantifying discharge strength is pulse current method, pulse current signal is also the most direct physical signal to characterize discharge, and the relevant parameters have become the parameters of quantitative discharge strength in fact [1-4].

Day-blind UV imaging method is a hot research topic of non-contact discharge detection in recent years, the imager uses dual-optical imaging technology, in which the detection of visible light signal to the device body imaging, and the other way to detect 240 to 280nm band of ultraviolet light signal, the discharge of the luminescent area imaging, the discharge of the luminescent area image superimposed on the visible light channel image is obtained ultraviolet image [5]. Compared with the pulse current method, UV imaging has a good discharge positioning ability, and can directly observe the discharge image invisible to the naked eye, very intuitive, but the optical signal detected by this method is an indirect physical parameter that reflects discharge, what kind of relationship exists between UV imaging signal and electrical signal, whether ultraviolet imaging can be used to quantify discharge is a matter of great concern in scientific research and engineering [6]. At present, there are the following difficulties in quantifying discharge by ultraviolet imaging: first, what parameters should be selected to quantify UV imaging, second, the establishment of the relationship curve between UV imaging quantitative parameters and electrical signals, and the observation distance and gain have obvious effects on UV imaging, so that the results of detection are not comparable, which brings some difficulties to quantitative discharge [7-8].

At present, the characterization and quantitative discharge intensity of ultraviolet imaging method is generally using the parameters of "photons", according to the characteristics of UV images with the intensity of discharge and change, the digital image processing algorithm is used to divide out the spot area, calculate the area of light spot and use it for the characterization of discharge strength and then test and study the change characteristics of electrical pulse signals and ultraviolet images under different discharge intensity and establish the relationship between the peak of the typical instrument gain and the light area at fixed observation distance. On this basis, the relationship between spot area and observation distance is studied, the method of estimating the area of discharge imaging area by using visible light channel information is proposed, the corresponding calculation formula is obtained, and the minimum second-square support vector regression model is established based on the data obtained from the experimental results, and the intensity of discharge signal based on ultraviolet image spot area is realized.

2. Insulator Discharge Detection Method Based on Ultraviolet Imaging

This paper analyzes the mechanism of dirty insulator discharge photoramasing, the attenuation of ultraviolet light signal in the air from the perspective of plasma, establishes the relationship between radiation power of ultraviolet light signal and discharge power, and puts forward the method of video analysis and digital image to extract the characteristics of relevant parameter quantitative discharge and characterization discharge, and studies the influence of gain on spot area. Through a quasi-synchronous acquisition and analysis of electrical pulse signal and UV image parameters, the response characteristics of ultraviolet image to electrical signal are analyzed, the relationship between the area frame of large spot and the number of discharge pulses is analyzed, and the relationship between the area of light spot and the amplitude of electrical pulse signal is studied. The relationship between spot area and visible light image and observation distance in ultraviolet image is studied, the method of "discharge imaging area area area" is proposed by using the image information of visible light channel, the method of area calculation is derived, the relationship between the area of discharge imaging area and the magnitude of electrical pulse is established, and the estimation of the magnitude of discharge pulse of composite insulator is realized by the least-square support vector regression method. Based on the artificial climate chamber, the ultraviolet image characteristics of the discharge of porcelain insulators at different filth and humidity levels are studied, and the "relative spot area mean" and "number of image frames of large spot area" are proposed, and two UV imaging parameters are used to characterize the filth discharge characteristics.

3. The Relationship Curve Between UV Pulse Parameters, Filth and Humidity

Using ultraviolet pulse method to evaluate the soiling state of insulators, the relevant characteristic parameters need to be extracted, in the current prediction study on the degree of insulator filth, the method of measuring leakage current has been widely used, and the most commonly used leakage current characteristics at home and abroad are: the maximum pulse amplitude of leakage current; Each discharge of the insulator corresponds to a set of ultraviolet pulse signals, but unlike the leakage current method, it dose not reflect the current frequency component, can only get the discharge of the pulse high frequency component, that is, can only reflect its discharge characteristics, so the extracted parameters and leakage current method there is a certain difference. It can be seen from the relevant experimental research that, in the case of pollution shift, with the increase of humidity, the amplitude of its optical pulse signal increases significantly, the number of pulses is more in the corona discharge stage, but the duration of a single pulse is very short, and in the stage of spark and arc discharge, not only the magnitude of the pulse increases, the duration of its individual pulse increases significantly, the pulse width is wider, and the number of pulses may be reduced. According to the above characteristics, two parameters are extracted here to characterize the discharge optical signal characteristics: the maximum discharge pulse in one minute, L_m., the average of discharge pulses in one minute, L_{av}. The two parameters are defined as follows:

$$L_m = \frac{\sum_{j=1}^{M} \max(l(i))}{M} \tag{1}$$

$$L_{av} = \frac{\sum_{j=1}^{N} l(i)}{N} \tag{2}$$

where $l(i)$ is the optical signal series collected by the data set card, M is the number of half the frequency cycles in one minute, N is the maximum number of points of data collected in one minute for the pulses in that half of the frequency period. In fact, L_m. averages the maximum pulse value over a half-hour period, while L_{av}. refers to the average of the light pulse signal over a one-minute period. According to the above parameters, the ultraviolet pulse parameters under different filth and humidity are calculated, and the corresponding parameters and humidity and soiling degree of the relationship curve, Fig.1 and Fig.2 are the insulator with the increase of humidity, different soiling conditions of the maximum light pulse, L_m. and pulse mean, L_{av} change curve.

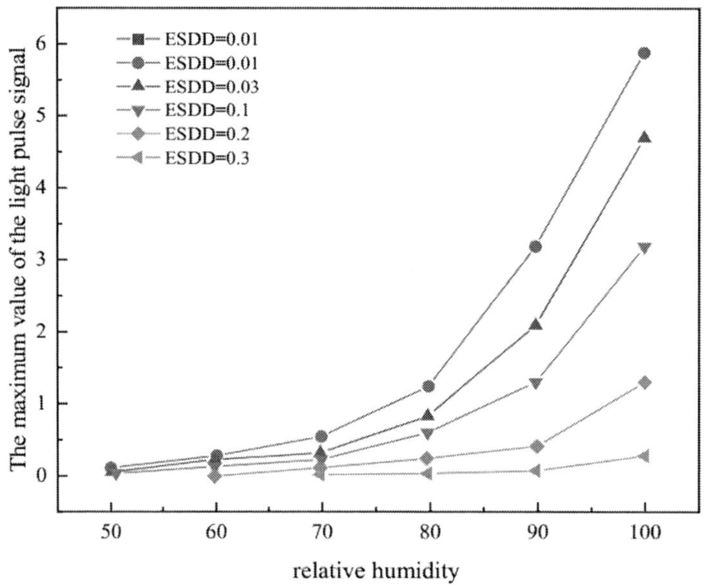

Fig.1 The change curve of the maximum pulse of the optical signal

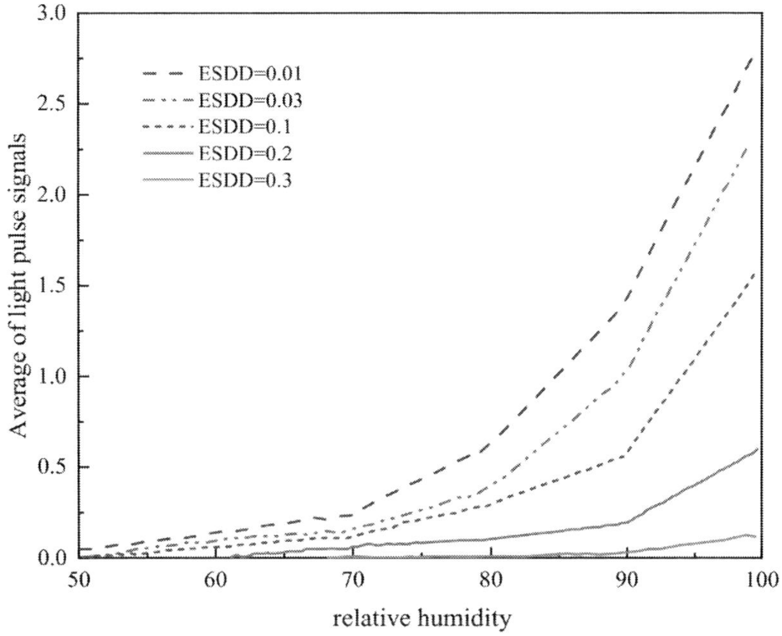

Fig.2 The change curve of the average pulse of the light signal

4. Estimation of the Discharge Strength of Insulators Based on Ultraviolet Imaging

From the above theoretical analysis and practical tests, it can be seen that, with the increase of the observation distance, the ratio of the diameter (or radius) of the spot area in the ultraviolet image to the height of the insulator is approximately a constant, while the type and type of insulator commonly used in power systems are relatively fixed, and the relevant geometric parameters are known, in this paper, the method of estimating the discharge "imaging area" of the insulator surface is proposed. At the same time, the area of the discharge luminescent area is determined by the ionization area, but the ionization area is unknown, the discharge "imaging area" described herein refers to the area of the light-emitting area detectable by the ultraviolet imager under a certain gain of the UV imager, obviously when the discharge intensity remains constant, the larger the area of the light-emitting area that can be detected, that is, the larger the area of the imaging area.

Suppose that the spot area in the ultraviolet image is circular, with a radius of R and a structural height of L for the insulator. It is assumed that at a gain of the UV imager, the radius of the discharge imaging area that can be detected by the UV imager is r, and the true height of the insulator corresponding to L in the UV image is l, which is approximately satisfied at different observation distances:

$$\frac{r}{l} = \frac{R}{L} \tag{3}$$

Therefore, at this gain, the radius of the discharge imaging luminous area is:

$$r = \frac{R}{L} \times l \tag{4}$$

The size of the area of the corresponding imaging area is:

$$S = \pi r^2 = \frac{\pi R^2}{L^2} \cdot l^2 \tag{5}$$

where S is the area of light spots in ultraviolet images, L is A height of the insulator in the image, l is the actual height (cm) of the insulator corresponding to L.

Therefore, as long as the light spot area value s in the UV image, a height L of the insulator and the corresponding actual height l, the size of the imaging luminous area that can be detected under a certain gain can be estimated by (5). In order to reflect the validity of the above formula, this paper makes the actual calculation of the area of the luminous area when the insulator corona discharges, and selects L as the height of the composite insulator, whose actual structure length is l-38cm, and the area of the luminous imaging area obtained by (5) is shown in Fig.3.

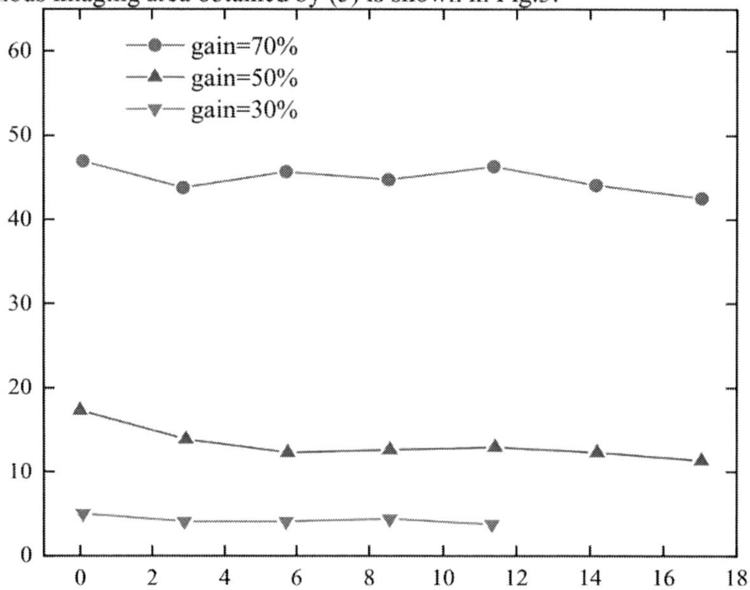

Fig.3 Calculated discharge imaging area under different distance and gain

As can be seen from Fig.3, under the same gain, the area of discharge imaging estimated by the UV image at different distances is basically the same, with small fluctuations; This shows that in ultraviolet imaging detection, it is feasible to estimate the size of the imaging area by combining the image information of the visible light channel within a certain detection distance, and the estimated area value of the imaging area is basically the same, with little to do with the detection distance. The above method avoids the problem that the detection results are incomparable when detecting discharge in engineering because it is not easy to determine the observation distance.

5. An Estimate of Discharge Strength Based on the Least Square Method

The estimation of pulse peak is used as an example to illustrate the estimation process of discharge intensity. The electrical signal peak of the surface of the insulator, i_{max}. is related to the area of the discharge imaging area, s_r. and the gain g of the instrument, expressed as a function

$$i_{\max} = f(s_r, g) \tag{6}$$

This problem belongs to the regression problem, this paper uses the least square regression to establish the corresponding regression model, the input of the model is g and the area of the imaging area, s_r. is the nuclear function selects Gauss radial base function. The input and output sample data are entered into the above model for training, and after the training is completed, the average error value of the training sample is 1.2% and the average error of the check sample is 1.8%, which shows that the model has good prediction accuracy and generalization ability. Using the trained support vector machine to predict the electrical signal of the surface of other actual insulators in the experiment, the results show that the error of discharge of the surface of the insulator estimated by the method proposed in this paper is generally within 15%, the cause of the error may come from two aspects, on the one hand, because the discharge itself has a certain randomness; That is, the corresponding current signal under the same area in practice is actually fluctuating up and down around the curve, with some randomness.

Although the use of ultraviolet imaging to quantify the analysis of discharge, the error is still somewhat large, but the actual engineering often does not require a very accurate current value, so the use of ultraviolet imaging to estimate the discharge of the surface of the insulator is feasible, with engineering application value.

6. Conclusion

This paper analyses the characteristics of ultraviolet pulses discharged by insulators at different filth and humidity levels, and on this basis extracts the parameters of "maximum light pulse" and "average light pulse" over a period of time and obtains the curve between the above parameters and the filth and humidity of insulators. According to the law of the reduction of the image size of the insulator onto body of the ultraviolet channel and the insulator body size of the visible light channel with the increase of the observation distance, the problem of ultraviolet imaging method detecting the effect of the discharge time spot area on the observation distance in practice is solved. At the same time, the regression relationship between the area of the imaging area, the gain of the instrument and the amplitude of the electrical pulse signal is established by using the least-square support vector regression algorithm, and the estimation of the amplitude of the electrical pulse signal is realized. The results of this paper are helpful to improve the intelligent monitoring level of power system.

Acknowledgements

This work was supported by the Science and Technology Project of State Grid Corporation of China (Project No. 2021YF-33).

References

[1] Mohammadnabi, S., Rahmani, K. (2021) Influence of humidity and contamination on the leakage current of 230-kv composite insulator. Electric Power Systems Research, 194: 107083.

[2] Wang, S.H. (2011) A Filthy suspended insulator discharge detection and evaluation based on ultraviolet imaging. North China Electric Power University.

[3] Ouchen, L., Bayadi, A., Boudissa, R. (2019) Dynamic model to predict the characteristics of the electric arc around polluted insulator. IET Science Measurement & Technology, 14(1).

[4] Fang, C.H, Wang, J.G, Fu, P.M. (2012) Analysis of the correspondence between the local arc of the surface of the filth insulator and the waveform characteristics of the leakage current. High voltage technology, 38: 609-615.

[5] Zhang, M.M. (2019) Study on the assessment method of insulator filth state based on UV pulse parameters. Southwest Jiaotong University.

[6] Bai, X.C, Yang, Y.K, Jing, G.R. (2021) The discharge detection system of power equipment based on ultraviolet pulse method. Science and Technology Innovation, 09: 61-63:

[7] Yuan, C.W. (2007) Experimental study of insulator discharge based on ultraviolet pulse detection. Chongqing University.

[8] Wang, C.L, Shen, Z.H, Liu, P. (2020) On-line detection of ultraviolet pulse method of electric corona discharge in power station switch cabinet. Electrician Technology,14: 67-70.

Dynamic data processing method based on generalized least square method

Zhaoxin Yang*, **Zhenghua Gu, Wenqing Zhang**

Facility Design and Instrumentation Institute China Aerodynamics Research and Development Center, Mianyang, Sichuan,621000, China

*Corresponding author's e-mail: yangzx@cardc.cn

Abstract. In this paper, a time domain-processing method is introduced for the dynamic calibration based on Hopkinson bar and shock tube. With a group of accelerometer and pressure sensor data processed in our work, the algorithm for establishing dynamic mathematical model is described in detail. "Generalized least square method for special whitening filter" and the simulation results are given. It shows that this method has the characteristics of simplicity and accuracy, and is especially suitable for the establishment of difference equation model in the data processing of dynamic calibration experiment.

1. Introduction

It is of great significance to study and improve the dynamic characteristics of the test system for the measurement effect of the dynamic signal with wide band and to ensure the accuracy of the dynamic test results. Currently, the commonly used time domain modeling methods include[1-3]: system identification method[4-5], Walsh transformation method[6-7], maximum entropy spectrum method, adaptive method and neuron method[8]. Among them, system identification modeling method is widely used. In contrast, the Walsh transformation method uses less data and the order of the model is low. It can directly obtain the coefficient of the differential equation and reduce the transformation error. However, it is not a recursive method and has strict requirements on the number of data[9]. The maximum entropy spectrum method is a time series analysis modeling method, which only needs the output data of the sensor dynamic calibration, but its modeling accuracy is not high. In long sequence modeling, adaptive method has obvious advantages over maximum entropy spectrum method. Although the neuron method has the advantages of low order and high accuracy, it is still far from practical application in the field[10].

In this paper, on the basis of studying various existing modeling methods, a sensor dynamic modeling method based on "generalized least square method of special whitened filter" is used. The method first uses "SIM algorithm which can simultaneously identify the order and parameters of linear difference equation model" to calculate the initial values of model parameters. The exact parameters of the model are calculated iteratively by the generalized least square method of a special whitening filter, and then the transient response is calculated recursively by the difference equation model. In this paper, the mathematical models of 4367 piezoelectric acceleration sensor and 6213 piezoelectric pressure sensor of Kistler company based on special whitened filter are established by means of Hopkinson bar and shock tube respectively. The established model has high accuracy and low order, and it is easy to realize the recursive algorithm, which lays a foundation for improving the dynamic characteristics of the sensor and realizing the dynamic compensation, to ensure the measurement accuracy within the allowable range.

2. System identification theory

The difference equation model of single input - single output linear system:

$$\begin{aligned} &A(d^{-1})y(k)=B(d^{-1})u(k) \\ &A(d^{-1})=1+a_1d^{-1}+a_2d^{-2}+...+a_nd^{-n} \\ &B(d^{-1})=b_0+b_1d^{-1}+b_2d^{-2}+...+b_nd^{-n} \end{aligned} \tag{1}$$

where $u(k)$ is the observation result of the system input, $d^{-1}$is the post-shift operator.
The observation equation with noise:

$$B(d^{-1})u(k)-A(d^{-1})z(k)=\varepsilon(k) \tag{2}$$

where $z(k)$ is the observation result of the system output, $\varepsilon(k)$ is the residual error.
According to a given sequence of observations $u(k)$, $z(k)$, the order of the model $\hat{n}$ and the model parameters of the least squares estimation can be acquired.

3. The procedure of dynamic calibration

3.1. Dynamic calibration of accelerometers

The impact device used in the experiment is shown in Figure 2, which is mainly composed of Hopkinson bar, digital oscilloscope and computer system.

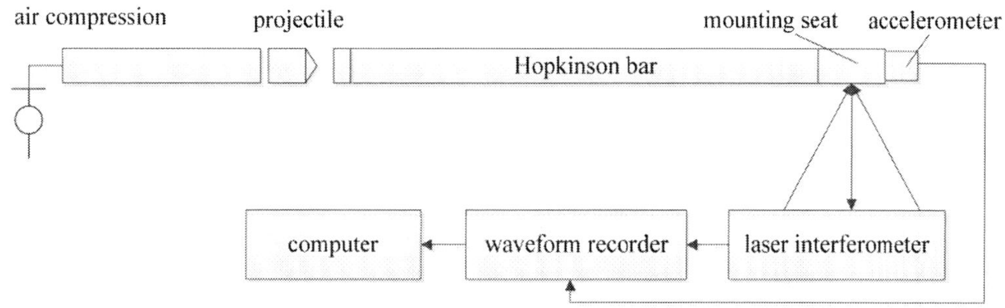

Figure 1. Hopkinson bar impact calibration system

The projectile body with a certain taper at the end is pushed by (0.05-0.8MPa) compressed air and vertically impinges on the aluminum pad. The impact causes a right half sine compressive stress pulse generated in Hopkinson bar. The accelerometer is fixed on the end face of the mounting seat through bolts, and the axis direction of the mounting seat surface is affixed with a reflection grating, and the other end face is tightly attached to the end of the Hopkinson bar by a vacuum fixture. When the compressive stress pulse is reflected as a tension pulse at the free end of the mounting seat and the net tension appears at the interface between Hopkinson bar and the mounting sea, the mounting seat and the calibrated accelerometer will fly away, and the accelerometer will acquire accelerated motion.

3.1.1. Establish dynamic model of accelerometer

The dynamic calibration experiment of 4367 piezoelectric acceleration sensor from B&K Company is carried out, and the sampling frequency was 1MHz, and the experimental data were obtained. Curve 1 in Fig. 2 is the recorded curve of the experimental impact response, and curve 2 is the calculated curve of the model. From this group of experimental data, the difference equation (third-order model) can be established:

$$G(s)=\frac{0.036s^3+2.124\times10^4s^2+9.1902\times10^{11}s+2.337\times10^{17}}{1.533s^3+1.228\times10^7s^2+1.174\times10^{12}+2.795\times10^{17}} \tag{3}$$

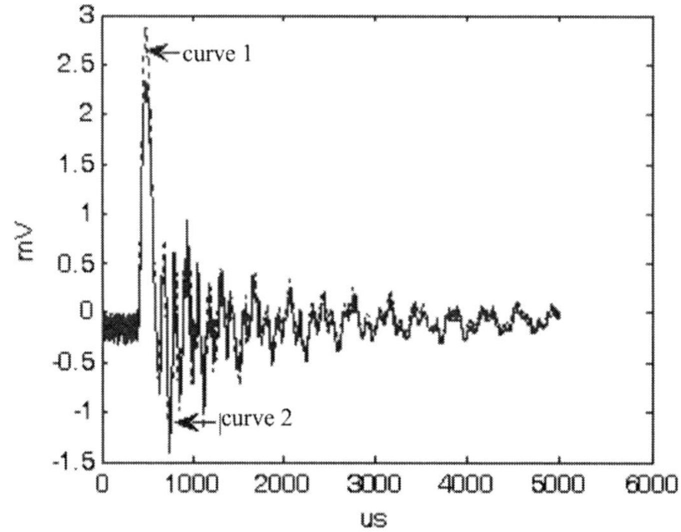

Figure 2. The shock response curve of the piezoelectric acceleration sensor

The experimental FFT frequency characteristic curve is shown in Fig. 3, and the characteristic curve calculated by the model is shown in Fig. 4.

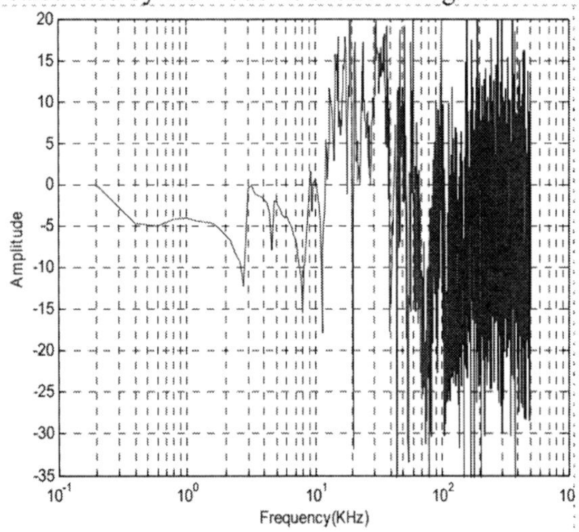

Figure 3. Logarithmic frequency characteristic curve of piezoelectric accelerometer with FFT results

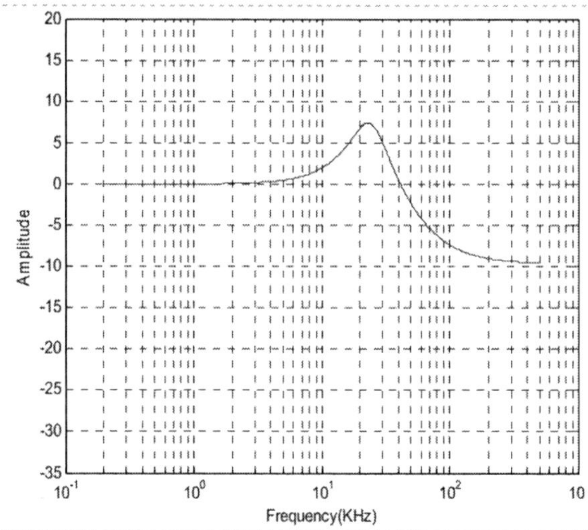

Figure 4. Logarithmic frequency characteristic curve of piezoelectric accelerometer with mdelling results

3.1.2. Dynamic performance index

According to the frequency characteristic curve in Fig. 4, the dynamic performance indexes of the piezoelectric pressure sensor in the frequency domain can be obtained as follows: the operating band of the ±10% amplitude error is ω_{g1}=6.387kHz, the operating band of the ±5% amplitude error is ω_{g2}=4.391kHz, the resonant frequency is ω_x=23.15kHz, and the resonant peak is A_m=7.457.

3.2. Dynamic calibration of pressure sensor

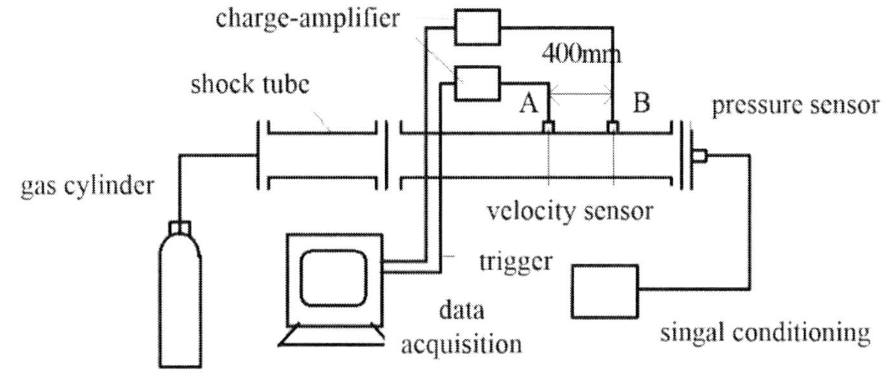

Figure 5. Shock tube dynamic pressure calibration system

At present, among all dynamic pressure calibration devices, shock tube is regarded as the best calibration device for pressure sensor. When shock tube is used for dynamic calibration of pressure sensor, the shock tube system generates a step pressure as a standard signal and adds it to the pressure sensor under test. The pressure sensor is calibrated by analyzing its output response, and the actual working performance of the pressure sensor is analyzed. The shock tube is here the electrical equivalent of a signal source, providing an ideal step signal. Figure. 6 is a shock tube dynamic pressure calibration device, which generates a typical step pressure signal. In general, the response of the first pulse on the pressure sensor is used to calibrate and analyze the characteristics of the pressure sensor.

3.2.1. Establish the dynamic model of pressure sensor

The 6213 piezoelectric pressure sensor of Kistler Company was selected to conduct the shock tube experiment to obtain the experimental data with a sampling frequency of 1MHz. Curve 1 in Figure. 6 is the recorded curve of the experimental impact response, and curve 2 is the model calculation curve. From this group of experimental data, the difference equation can be established:

$$G(s)=\frac{-3.211s^3+6.008\times10^6s^2+2.043\times10^{12}s+5.395\times10^{16}}{7.219s^3+6.690\times10^5s^2+1.761\times10^{12}+5.384\times10^{16}} \quad (4)$$

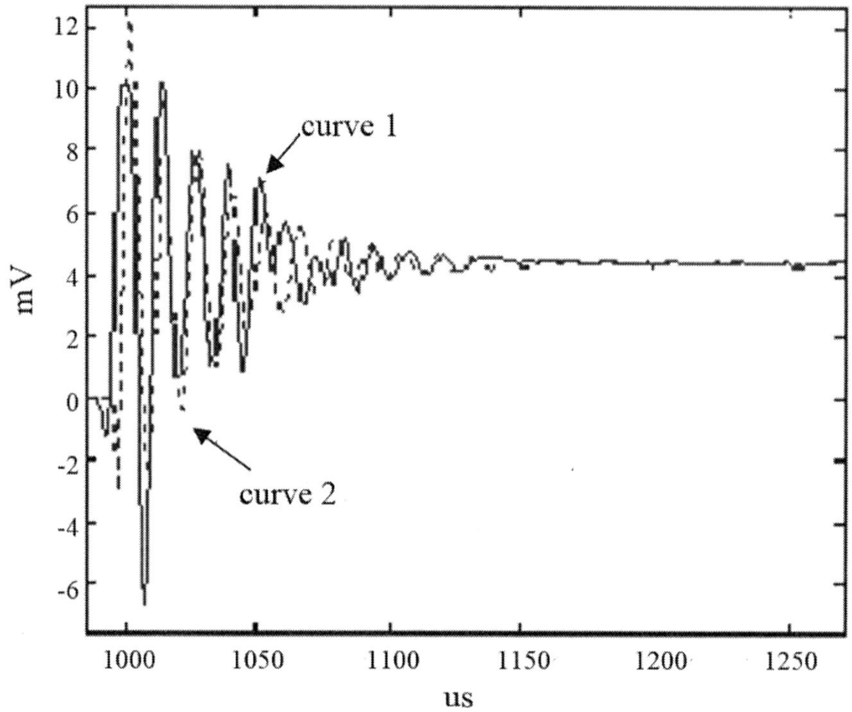

Figure 6. Step response of piezoelectric pressure sensor

The experimental FFT frequency characteristic curve is shown in figure. 7, and the characteristic curve calculated by the model is shown in figure. 8.

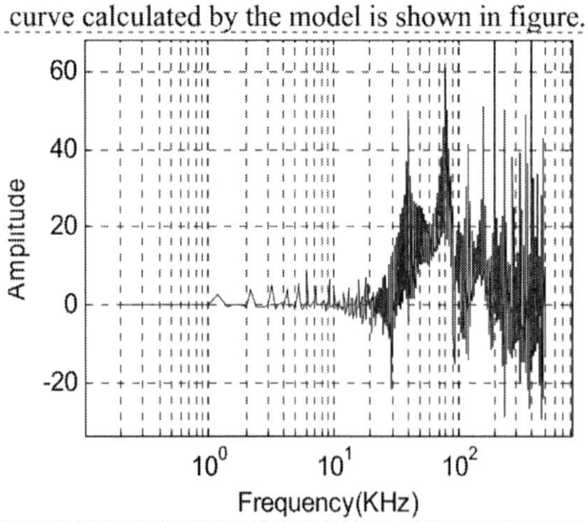

Figure 7. Logarithmic frequency characteristic curve of piezoelectric pressure sensor with FFT results

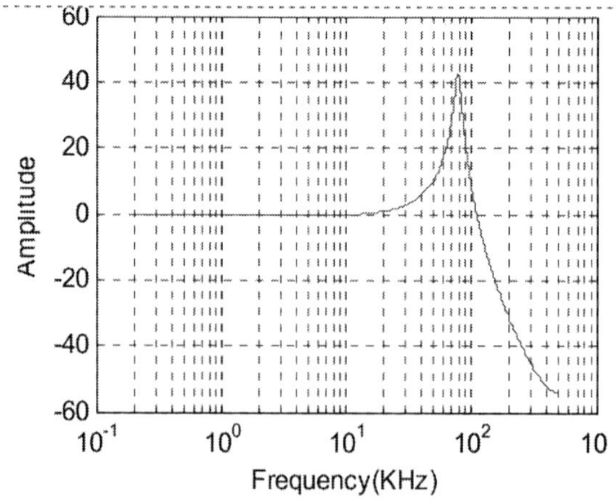

Figure 8. Logarithmic frequency characteristic curve of piezoelectric pressure sensor with mdelling results

3.2.2. Dynamic performance index

According to the frequency characteristic curve in Figure. 8, the dynamic performance indexes of the piezoelectric pressure sensor in the frequency domain can be obtained as follows: the operating band of the ±10% amplitude error is ω_{g1}=18.7kHz, the operating band of the ±5% amplitude error is ω_{g2}=16.3kHz, the resonant frequency is ω_x=77.4kHz, and the resonant peak is A_m=42.51.

4. Conclusion

Based on the dynamic response of Hopkinson bar and shock tube calibration system to acceleration and pressure sensor, a dynamic model of sensor based test system is established in this paper. Through many simulation experiments, various structural factors affecting identification accuracy are analysed and some rules are summarized. This method is suitable for the case where the input interference is negligible and the output measurement noise is almost white noise. The conditions of dynamic calibration experiment are close to this situation. Generally, the excitation signal of dynamic calibration is better, and its interference can be ignored. The output signal of the calibrated system can be approximated as white noise. Therefore, this method is especially suitable for the establishment of difference equation model in the data processing of dynamic calibration experiment.

References

[1] H. Y. Peng, H. J. Liu, J. H. Yang. (2021) A review on the wake aerodynamics of H-rotor vertical axis wind turbines, Energy, Volume 232.

[2] H. Su, H. R. Meng, T.M. Qu, L.P. Lei. (2021) Wind tunnel experiment on the influence of array conFigureuration on the power performance of vertical axis wind turbines, Energy Conversion and Management, Volume 241.

[3] H. Q. Jing, F. Y. Huang, X. H. He, C. Cai. (2021) "Wake-induced vibrations of tandem flexible cable models in a wind tunnel," Ocean Engineering, Volume 233.

[4] Y. F. ZHANG, Q. A. LI, X. Y. ZHU, X. Y. SONG, C. CAI, Z. GUO. (2021) Wind Tunnel Experiments and Numerical Study on Performance Characteristics of an H-type Vertical Axis Wind Turbine in the Spanwise Direction, Journal of Thermal Science, Volume 30, Issue 03.

[5] B. Birch, D. Buttsworth, F. Zander. (2020) Time-resolved stagnation temperature measurements in hypersonic flows using surface junction thermocouples," Experimental Thermal and Fluid Science, Volume 119, Issue C.

[6] Y. L. Tang, M. K. Yue, X. F. Fang, X. Feng. (2019) Overcoming high luminance gradient using serial exposure time method for synchronous full-field measurement of temperature and deformation.," Applied Optics, Volume 58, Issue 25, pp. 6966-6974.

[7] L. Měšťanová, O. Bartoš. (2019) A study of the coarse water droplets formation in the nozzle," EPJ Web of Conferences, Volume 213.

[8] Y. W. Zhu, F. J. Yi, S. H. Meng, L. J. Zhuo, W. Z. Pan. (2017) Long-duration heat load measurement approach by novel apparatus design and highly efficient algorithm," Measurement Science and Technology, Volume 28, Issue 11.

[9] M. H. Eisenbies, T. A. Volk, D. DeSouza, K. Hallen, B. Stanton, J. Espinoza, A. Himes, R. Shuren, R. Stonex, B. Summers, J. Zerpa. (2021) An assessment of the harvesting and fuel performance of a single-pass cut-and-chip harvester in commercial-scale short-rotation poplar crops as influenced by crop and weather conditions," Biomass and Bioenergy, Volume 149.

[10] A. Singh; I. A. Khan; M. Z. Khan; P. Mahto. (2021) The effect of low Reynolds number on coefficient of S-type pitot tube with the variation in port to port distance," Materials Today: Proceedings, Volume 45, Issue P9, pp. 7810-7815.

Journal of Physics: Conference Series **2108** (2021) 012058 doi:10.1088/1742-6596/2108/1/012058

Research on Intelligent Metrological Verification Technology Based on Internet of Things

Meng Yu[1*], Jianmin Liu[2], Xiaowen Li[3], Nana Wang[4], Fan Wang [5]

STATE GRID NANCHANG ELECTRIC POWER SUPPLY COMPANY. Nanchang, Jiangxi, 330200, China

STATE GRID ANQING ELECTRIC POWER SUPPLY COMPANY. Anqing, Anhui, 246000, China

*Email: tg778899@xzcstudio.com

Abstract. Intelligent metrological verification is of great significance to improve the accuracy of equipment. Especially in the current situation of rapid increase in residential electricity consumption, electric energy meters have received widespread attention from the society as an important device that affects electricity trade settlement. In this research, this article will be based on the technological environment of the Internet of Things. After the analysis of the intelligent metrological verification system, the key technical means will be highlighted, including automatic sealing technology, automatic connection and disconnection technology, and cycle simulation system design. This article provides support for the verification system construction of the metrological center.

1. Introduction

At present, with the further development of China's electric power industry, it has become an unstoppable development trend for electric power companies to realize the intensive management of electric energy measurement. In this context, how to improve the accuracy of metrological testing data has become a key concern of the academic community. As a representative technology in the modern information age, the Internet of Things can improve work efficiency and ensure the accuracy of verification results in intelligent measurement verification. It has a far-reaching impact and is worthy of attention.

2. Research on intelligent metrological verification of the Internet of Things

2.1. Basic design ideas

In accordance with the requirements of intelligent metrological verification, the ultimate goal of this design is to build an uninterrupted assembly line operation system without human intervention. The design not only can effectively control various risks in the process, but also control construction costs. It is possible to implement the national technical standards for metrological verification, and complete many testing links such as power-on inspection, identification, connection and disconnection, and tribute withstand voltage test, so as to ensure the realization of the verification goal. Therefore, in order to achieve this goal, with the support of the Internet of Things, technologies such as laser sensing and RFID should be combined through software and hardware design methods. Finally, a complete metrological verification structure is formed. The detailed information is shown in figure 1.

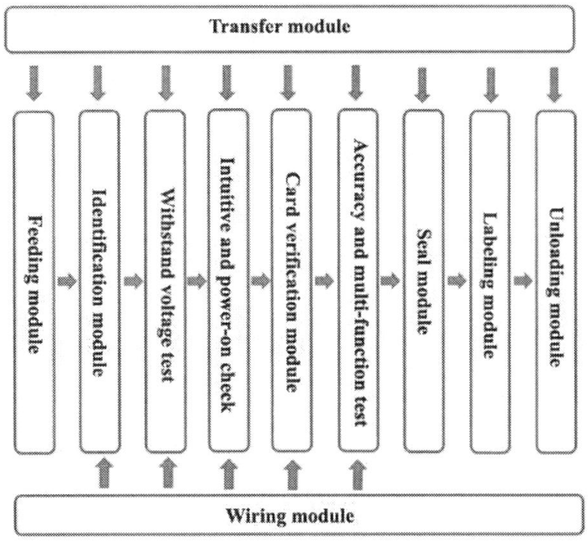

Figure 1. Structural diagram of intelligent metrological verification of the Internet of Things

2.2. The principle of design

In this study, the industry 4.0 IoT verification system is adopted. Compared with other systems, this system has significant advantages in intelligence and automation. It integrates modern technology such as computer network and photoelectric information. Among them, with the support of optoelectronic information technology, the system can detect the status of the equipment to be measured, and complete a series of functions such as positioning control and information collection. Through information technology, the binding and interconnection between the verification equipment and the production data can be realized. In addition, relevant personnel can further improve the content of the system according to the requirements of use, such as adding barcode information to realize anti-counterfeiting, or verifying the execution effect of the system through PLC servo technology.

2.3. Functional module design of the verification unit

2.3.1 Design of loading unit. The feeding unit is the initial function of the intelligent measurement verification, by taking out the equipment to be inspected from the turnover box and putting it into the electric energy meter conveying unit. The entire unit uses a constant speed motion mode. Its detection cycle meets the requirements of intelligent verification, and can be automatically adjusted after a deviation in the delivery cycle of the equipment to be inspected is found. In terms of function settings, the operating speed of the equipment can be automatically adjusted within the range of 0-120%. The positioning priority is controlled within ±2.0mm, and the success rate is higher than 99.0%.

2.3.2 Identification. This unit checks the barcode information and RFID tags of the equipment to be inspected by scanning the equipment. After the identity information is bound, the RFID module of the tooling board can be used to quickly read the information to be verified, helping the system to extract the identity information in a short time.

2.3.3 Connection and disconnection unit design. After the electric energy meter is positioned, the connection and disconnection unit can connect the terminals of the electric energy meter with communication ports and multi-function pulses. The terminal has the advantages of uniformity, reliability and stability, and can be effectively separated from the button box of the electric energy meter after the wire is disconnected. It can take protective action against the abnormal operation and avoid damage to the electric energy meter and other equipment. The connection success rate of each meter position of the connection and disconnection unit should be higher than 99%. It can meet the requirements of Q/GDW

1356, Q/GDW 1355 and other standards. The current terminal can be tested continuously for 10 minutes during the inspection, and the temperature rise detection range is within 35°C. The operating pressure of the voltage terminal is less than or equal to 60N, and the operating pressure of the auxiliary terminal of the electric energy meter is less than or equal to 10N, which ensures the adaptability of the system [1].

3. Key technical analysis

3.1. Automatic disconnection technology

In the intelligent verification of electric energy meters, it is necessary to realize the docking of several electronic terminals under a large current of 100A. There will be no problems such as displacement and heating during the detection process. This is the basis for effective verification.

3.1.1 The self-calibrated wiring support design. During the design period, the standard extension current terminal has an eccentric semi-circular arc surface structure. Effective contact is made with the electric energy meter column, wiring hole, etc., so the problem of insufficient contact area under the traditional mode can be changed. At the same time, because the resistance is inversely proportional to the area of the conductive material. Therefore, choosing the multi-surface design mode can ensure effective matching with the power meter terminals. According to the above-mentioned standard, the reliability connection test is performed under laboratory conditions, and the test is performed by an infrared imager by applying a current of 100A. The detailed test results are shown in table 1.

Table 1. Test results

Test time	Common table temperature（℃）	Common table temperature（℃）
0min	20	20
3min	49	32
6min	74	41
9min	79	52

It can be found from the data that after 100A of current is applied to the verification device, the temperature change of the multisurface watchholder is less, so the equipment safety can be better guaranteed.

3.1.2 Air cylinder. Air cylinder is an important power structure for automatic disconnection. It must have the advantages of fast response speed, stable and reliable action, etc. The double electronically controlled solenoid valve used in this design can make the disconnection action more reliable with the support of the Internet of Things. However, how to define the size of the cylinder thrust has become a key factor. In order to be able to determine the value, formula (1) can be used for verification.

$$F = \eta p S \quad \text{Formula (1)}$$

In formula (1), F is the thrust value of the cylinder. η is the load rate under static load and negative pressure. p is the pressure. S is the area under pressure.

3.1.3 Self-protection device. Intelligent metrological verification based on the Internet of Things needs to work in a continuous high-current environment. Therefore, security issues have always been the focus. In order to avoid damage to the equipment caused by temperature rise, a temperature control device needs to be added. The device can continuously detect and sample the temperature changes after connection. When an abnormal temperature is found during the verification process, an alarm can be issued in time and isolation can be performed to ensure the safety of the entire system.

In the design, the current terminal of the gauge holder recognizes the temperature of the gauge holder through the DS18B20 temperature sensor, monitors other abnormal data by paralleling the single-chip microcomputer, sets the temperature safety threshold on the upper computer software.

When the temperature is higher than the threshold, it can be cut off directly, ensuring the safety of the system.

3.2. Automatic sealing technology

When the test result of the electric energy meter shows that it is qualified, the system needs to automatically seal it. In the traditional mode, the electric energy meter and the meter body adopt the method of screw fastening, and the watch case and the screw are provided with through holes to facilitate the lead sealing line to pass through. Although this method is feasible, it has problems such as complicated operation methods. It is difficult to realize automatic processing. Therefore, it is necessary to design a new automatic sealing technology suitable for the automation mode of the Internet of Things.

3.2.1 Embedded design scheme. The embedded design scheme structure introduced in this paper is shown in figure 2.

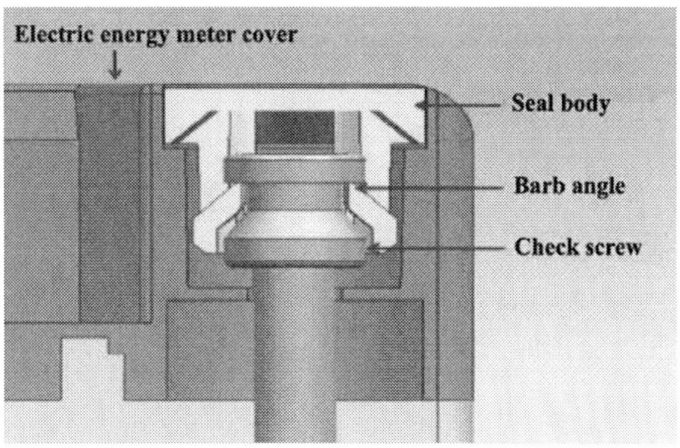

Figure 2. Embedded structure diagram

In the structure described in figure 2, the designed shape has a diameter of 8mm, a height of 7mm, and a thickness of 1mm. The length of the undercut button is 1.5mm, and the angle of the undercut angle is 45°C. The screw thread adopts a one-way feed method, so it has a non-reverse effect. The cap is composed of a circular cap and a plurality of arc-shaped plates arranged at the edge position. An undercut angle is set under the arc-shaped plate, which can be directly embedded in the screw hole by vertical pressing. The curved plate can be squeezed directly after being resisted by the watch case and produce an elastic bending effect, so the passing process is smoother. The arc-shaped plates are evenly distributed around, so that they can be directly pressed into the screw holes [2].

According to the relevant standards of the Electric Energy Metering Sealing Technical Specification (DL/T 1496-2016), the sealing materials used are all environmentally friendly materials that comply with the RoHS standard. The material has good toughness and good performance stability. No deformation was found during the continuous 6-hour 90°C temperature test, which ensures that the seal can remain stable under complex field conditions and is feasible.

3.2.2 The design scheme of the two-dimensional code of riding seam. On the basis of the above-mentioned sealing scheme, with laser technology, data such as the date and serial number of the verification can be recorded on the seal and the sewing position of the electric energy meter. After the seal is unsealed, the anti-counterfeiting mark is destroyed and not be restored. With the help of laser and pressing equipment, the sealing speed can reach up to 3 pieces/s, which has a higher sealing efficiency compared with the traditional mode.

3.3. Beat simulation design

The intelligent metrological verification system based on the Internet of Things involves multiple fields, and each verification function module is connected to each other. The unreasonable configuration of the number of devices in any one link will directly affect the final operating effect of the system, which is worthy of attention.

3.3.1 Beat measurement formula. Taking into account the complexity of themetrological verification, the entire operation unit needs to be equipped with many functional modules such as accuracy inspection, card verification, power-on inspection, multi-function experiment, and blanking. According to the content of the national metrology regulations, identification, card insertion verification, power-on inspection and multi-function testing can be completed, and each project has a detailed schedule. The time consumption of sealing, labeling and blanking are all affected by the operating efficiency of automated equipment [3]. In order to meet the requirements of flow operation, it is necessary to maximize the efficiency as much as possible on the basis of ensuring the functions of the entire module, and do a good job in the docking with the verification unit, and cooperate with the online data transmission method to transfer the blanking data, and finally realize precise feeding. Therefore, in response to this requirement, calculations can be made in accordance with the verification link and the time-consuming work of each module. According to the requirements of the empirical formula, the checking calculation can be carried out according to the content of formula (2) and formula (3).

$$\frac{m_1 n_1}{t_1} = \frac{m_2 n_2}{t_2} = \cdots \frac{m_x n_x}{t_x} \qquad \text{Formula (2)}$$

$$\frac{n_{tr}}{t_{tr}} = k\frac{n_{bu}}{t_{bu}} \qquad \text{Formula (3)}$$

In formula (2) and formula (3), m_1—m_x Represents the number of power meters per batch. n_1—n_x is the number of configurations for each module. t_1—t_x is the ime consumption generated for each batch test. tr is time consumption for storage lines. bu is the time-consuming verification. k is the consumption of the basic verification unit.

3.3.2 Develop a simulation test of beats. In the intelligent metrological verification, in order to realize the effective operation between each verification unit and realize the optimization of the equipment action, the simulation operation of the intelligent verification tempo can be developed. Through simulation technology, the operating data of each system can be checked and the technical requirements of each key process node can be identified. It is possible to simulate any failure event according to the technical parameters of the functional module, and find the most effective emergency treatment plan based on all discovered process bottlenecks. The function can identify operational risks, optimize the on-site debugging mode through on-site debugging and simulation methods, finally eliminate the debugging time, and deal with various possible risk events in advance.

3.4. Clock standard network set control

3.4.1 Design method of clock multiple-route frequency division. According to the error verification of traditional electric energy meters, the high-precision clock in each verification device is used as a standard metering clock, and the tested clock is compared and calculated. It is necessary to add a clock crystal oscillator to pass the large-scale centralized application of the verification device. This method will lead to an increase in equipment investment and a significant increase in maintenance workload. In response to this problem, this article uses a single crystal oscillator signal in the standard clock source multi-channel frequency division setting. The method of multi-channel frequency division output is completed through the network channel. The standard clock signal is transmitted to each individual verification device to ensure that the clock signal can complete effective transmission and synchronize the update time.

3.4.2 Clock network centralized control system. The network centralized control of the clock generally includes many functional modules such as frequency dividing circuit, shaping circuit, optical isolation circuit and clock error processing. The specific structure of this module is shown in figure 3.

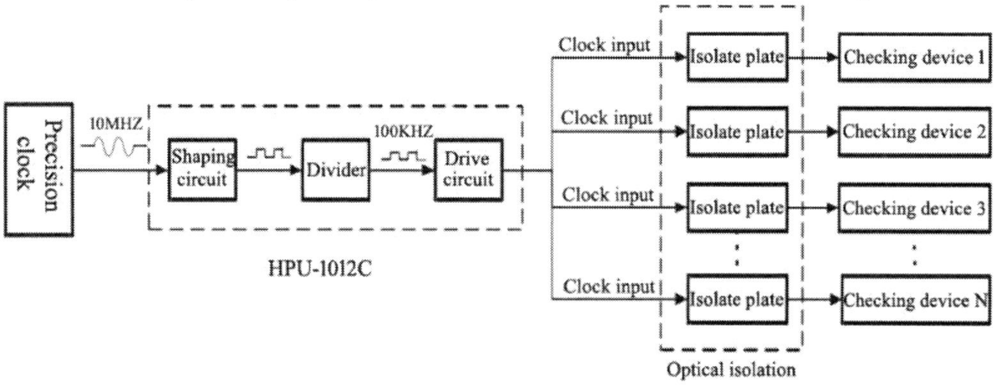

Figure 3. Network set control structure diagram of the clock standard

In the content introduced in figure 3, the precision clock reference can be converted into a square wave signal with the same frequency through a shaping circuit after sending out a sine wave high frequency clock signal. Through the frequency divider circuit, the high-frequency signal is converted into a signal with a lower frequency. After frequency division, the signal can be converted into multiple clock signals with the same frequency under the action of the drive circuit, and the signal can be uploaded to the clock error system to ensure that the time error can be recorded at any time and recognized by the system [4].

3.4.3 Shaping circuit. A sine wave output signal with a parameter of 10MHz is used in the standard clock signal setting. The square wave drive method verifies the constant error method in the device. With the support of the system, the square wave signal can be directly counted. Therefore, in order to achieve this effect, it is necessary to reshape the clock signal. The sine wave can be transformed into a rectangular wave or square wave with the same frequency through a voltage comparator. It needs to select MAX903 during signal conversion. This comparison circuit has the advantage of low power consumption, and the conversion time reaches 10ns.

3.4.4 Drive circuit. The clock signal after the frequency division of the drive circuit can realize long-distance data transmission. In the technical implementation, the 74HC14 trigger inverter structure is adopted. A gate drive mode is set on each channel. This method can complete the drive and shaping of the input signal， and set a signal source input mode port can output 10 signals through the same frequency. At the same time, by setting a photoelectric isolation receiving board on the verification device to receive the clock signal, this design method can avoid the problem of transient pulse interference and strengthen the stability of the system.

4.Conclusion

In the intelligent metrological verification technology of the Internet of Things, the functions of the system can be significantly enhanced through diversified functional modules such as the beat simulation system. The system introduced in this article has stronger system operation capability, satisfactory stability of the system, better identification of the function of the electric energy meter. The aim is to improve the verification efficiency of the equipment, and finally lay the foundation for improving the function of the electric energy meter, which is worthy of promotion.

References

[1] Wu Yan, Song Chaofan, Liu Zhe. Practice of intelligent quality control technology in the verification process of natural gas flow meter[J]. Oil and Gas Storage and Transportation, 2021, 40(06): 637-642.
[2] Guo Yuhua. Analysis of Intelligent Metrological Verification Techniques for the Internet of Things[J]. Instrument Standardization and Metrology, 2019(06): 29-30+42.
[3] Chen Lei, Lan Weiyong, Geng Shuqiu. Research on Intelligent Metrological Verification Technology Based on Internet of Things[J]. Science and Technology Innovation Herald, 2019, 16(25): 86+88.
[4] Qin Zhengyang, Qin Han. Analysis on the method of intelligent metrological verification technology for the Internet of Things[J]. Electroacoustic Technology, 2019, 43(05): 39-40+48.
[5] Wen Xiaohui, Yi Fanglei, Jiang Jiachen. Research and design of virtual concentrator system based on Internet of Things technology[J]. Northeast Electric Power Technology, 2021, 42(08): 25-29+38.
[6] Gu Haitong, Zhang Yuanliang, Lu Xiangzhi. Development and application of high-stability electric energy measurement terminal software based on the Internet of Things[J]. Journal of Southwest China Normal University (Natural Science Edition), 2021, 46(06): 133-138.

Energy Storage Planning of Park Energy System Based On Multi-Dimensional Digital Twin Technology

Tianlong Xiong[1*], Chao Yang[1], Qing Cheng[1], Xianlong Yang[1], Shu Lin[1]

[1]Sichuan Energy Internet Research Institute Tsinghua University, Chengdu, 610020, China

[*]xiongtianlong@tsinghua-eiri.org

Abstract—The existing energy storage planning methods have the problem of imperfect equipment mathematical model, resulting in small installed capacity of renewable energy. An energy storage planning method of Park energy system based on multi-dimensional digital twin technology is designed. This article explains the basic connotation of multi-dimensional digital twin technology and park energy system, and obtains feedback information. Combined with the energy consumption of industrial users, the park's electricity load is predicted. We used the multi-dimensional digital twin technology to construct the mathematical model of the equipment, extracted the energy conversion law, optimized the energy storage mode of the energy system, and achieved zero heat emission. Case analysis results: the average installed capacity of renewable energy after optimization is 2349.6 kW, 626.6 kW higher than that before optimization, indicating that the energy storage planning method integrating multi-dimensional digital twin technology has a broader application prospect.

1. Introduction

The energy system of the park refers to the integrated system composed of power supply, district heating, cooling and various energy conversion equipment. The energy system of the industrial park is composed of energy station and cold / heat / electricity / gas energy supply pipe network, in which the energy station is the core of the regional comprehensive energy system. According to the energy stations are distributed on the user side or distributed in the park by users, the park can be divided into distributed energy stations and centralized energy stations [1-3]. For the centralized energy station, it is generally required to meet the cooling, heating, power supply and other load requirements in the park at the same time. It is the hub of the energy supply system of the park. The centralized energy station in the regional integrated energy system is called the integrated energy station. From the perspective of energy supply, the energy station mainly carries out energy supply and energy storage. The energy system consists of many different types of energy, energy production equipment, energy conversion equipment and energy storage equipment [4]. Energy forms include cold, heat, electricity, gas, etc. terminal energy load demand includes electricity demand, heating demand, hot water demand and cooling demand. At the input end of the system, its energy can be wind energy, natural gas, diesel, solar energy, geothermal energy, etc. Through different energy conversion equipment, the energy of these energy sources is finally transformed into electricity, heat and cold for end users through utilization and conversion at all levels. Existing studies at home and abroad have proposed the physical model level of the integrated energy system, including all kinds of energy forms such as electricity, cold, heat and gas in all links from production, conversion, storage to consumption [5-6]. The development of energy system in foreign countries started earlier, its research on energy system planning considers more factors, and the construction of model is relatively complex. A stochastic programming method for comprehensive energy system is proposed. The uncertainty of different energy prices is described in the form of stochastic normal. Taking energy system investment planning as the first stage optimization problem and energy system operation as the second stage optimization problem, a two-stage model for energy system optimization is constructed. The development of domestic integrated energy system started later than that of foreign countries, but it has also achieved fruitful results. The factors considered in the research of integrated energy system planning are relatively simple. At present, the research data on the application of multi-dimensional digital twin technology to the park energy system is not very rich, which needs to be further discussed.

2. Energy storage planning of park energy system based on multi-dimensional digital twin technology

2.1. Forecast the power load of the park

When planning the equipment capacity of the comprehensive energy station in the industrial park, it is first necessary to analyze the power consumption characteristics of users and obtain the energy load curve of users[7]. Then, according to the user's economic development level and battery power storage capacity, the user's energy load in a certain period in the future is predicted, and the user's load prediction results are obtained[8]. When adopting the load forecasting method based on index method, for typical industrial enterprises, it should be composed of two parts: building envelope environmental heat exchange and equipment processing and production heat dissipation, which are usually dominated by building cooling / heating load. The calculation formula of total building cooling load is:

$$L=\sum_{n=1}^{m} H_n \times W_\gamma \quad (1)$$

In formula (1), H represents the sample building area, W represents the heat load, m represents the cold index of building heating, n represents the building type, and γ represents the comprehensive heat index. The thermal steam load and electric load of industrial users are highly related to product types, so they need to be analyzed and predicted according to the production plan of the enterprise. It is generally believed that the steam load / electric load is approximately proportional to the output. Therefore, the output value unit consumption method is used for analysis and prediction. The expression formula of industrial electric load is:

$$Fr=\sum_{g=1}^{t2} E \times r \quad (2)$$

In formula (2), E represents steam load, r represents steam consumed per unit product, t represents product output, and g represents power consumption. In general, in engineering practice, if there is a lack of sufficient analysis data, the proportional growth method is often used for simple load forecasting according to the existing historical data and combined with engineering experience. The load forecasting formula is as follows:

$$K=P' \times \left(1+d\right)^{j} \quad (3)$$

In formula (3), P represents historical load, d represents load forecasting results, and j represents annual load growth rate. Combined with formulas (1) ~ (3), and after obtaining the load curve of park users according to collection or analysis, further analyze the load change trend and obtain the load prediction results. Based on this, complete the steps of predicting the power load of the park.

2.2. Construction of equipment mathematical model by multi-dimensional digital twin Technology

The operation principle of multi-dimensional digital twin system is: when the production system is running, the service system controls the physical production line to carry out real production activities according to the production plan. In the future, the results of analysis and calculation can be fed back to the park energy system to realize the alarm, control optimization and energy storage planning of the production process. The power supply equipment in the energy system of the park mainly includes distributed renewable energy generation, energy storage battery, electric vehicle charging pile and distribution network. This kind of equipment can realize the production, transmission and storage of clean electric energy and is an important part of the comprehensive energy system of the industrial park. The power network power flow of the energy system in the park shall adopt the AC power flow of the distribution network to extract the energy conversion law[9]. The expression formula of its physical and mathematical model is:

$$\begin{cases} T = S_v \times \left(Q_v \cos\theta + l \right) \\ Y = L_v \times \left(Q_v \sin\theta + l \right) \end{cases} \quad (4)$$

In formula (4), S represents the active power on the distribution network line, L represents the amplitude of node voltage in the distribution network, v represents the power consumption of equipment, Q represents the number of equipment, θ represents the simultaneous working coefficient of equipment, and l represents the active power on the distribution network line. Ntural gas suppling includes natural gas source, pipeline, compressor and other equipment. Generally speaking, the relevant parameters will not change with time, such as pipeline pressure, fluid velocity, etc[10]. the parameters in the dynamic power flow model will change with time. The expression formula of physical and mathematical model of transmission capacity of natural gas pipeline network is:

$$Z = 67.86\sqrt{\left(\frac{K_0^2}{M_0^2} - \frac{K_1^2}{M_1^2} \right) \frac{\phi^3}{B}} \quad (5)$$

In formula (5), K_0 represents the mass flow of natural gas, K_1 represents the absolute pressure rating at the starting point of the natural gas pipeline, M_0 represents the compression factor of natural gas at the starting point of the pipeline, M_1 represents the natural gas potential energy factor function, ϕ represents the absolute pressure rating of natural gas at the end of the pipeline, and B represents the resistance coefficient along the pipeline. At the same time, electricity to gas is to obtain hydrogen by water electrolysis, and further synthesize hydrogen and carbon dioxide into methane and other gases, so as to convert electric energy into relatively convenient gas for storage. With the rapid development of renewable energy and supply and demand, the current in the power grid flows to the battery after rectification, and the battery discharge flows back to the power grid through inverter, so as to realize the two-way flow of electric energy between the battery and the power grid. The structure of battery energy storage system includes energy storage unit, battery energy management system, filter, etc. Battery is the core of energy storage unit, which is composed of multiple batteries for electric energy storage and release. The battery management system monitors and controls the output voltage, current, temperature and state of charge of the energy storage system in real time to ensure the safe, efficient and economic operation of the energy storage system. Based on this, the steps of building the mathematical model of the equipment are completed.

2.3. Optimizing energy storage mode of energy system

The energy storage mode of the energy system must meet the dislocation distribution of the same energy in time and can adjust the mismatch between supply and demand of energy supply in time. Therefore, it can be used as a secondary regulation means of the comprehensive energy system in the functional area to make up for the mismatch between energy supply and demand caused by multi energy complementarity. During the operation of the energy system, electric energy storage mainly plays the role of stabilizing the output uncertainty of renewable energy and improving the economy of system operation. It can store electric energy when the power grid load is low and release electric energy when the power grid load is high, which can be used to cut peak and fill Valley and reduce power grid fluctuation. At present, the electric energy storage equipment is mainly lithium-ion battery, lead-acid battery, sodium sulfur battery and liquid flow battery, and has realized megawatt demonstration applications in the fields of renewable energy grid connection, distributed generation and Microgrid. The electric boiler in the regenerative electric boiler is a coupling unit for energy form conversion. It generally uses new energy such as photovoltaic and wind power as the power supply, generates heat through electromagnetic induction or resistance, and outputs hot water or high-temperature steam. The installation of heat storage tank can break the conventional operation mode of "Determining electricity by heat", make the electric load of the electric boiler controllable, and increase the electric load during the low load period, Convert excess electricity into heat and store it. In the peak load period, the power load demand is reduced and the stored heat energy is used for heating. This will not only make more effective use of wind power in the trough period, but also play the role of peak shaving and valley filling. In extreme cases, it can even shut down the CHP unit and completely use the electric boiler for heating, so as to realize zero emission of heating. Among various cold energy storage methods, ice storage stores the cooling capacity by making water into ice. Compared with water storage, the volume of ice storage will be much smaller than that of water storage. Assuming that lead-acid battery, heat storage tank and cold storage tank are used as physical equipment for storing electric heat energy, the expression formula of the relationship between physical stored energy and charging / discharging power is as follows:

$$I=\left(1-p_{\varepsilon}\right)\times D_{t-\Delta\varepsilon} \quad (6)$$

In formula (6), p represents the self loss coefficient of physical energy storage, D represents the scheduling time interval, t represents the discharge efficiency, and ε represents the charging efficiency. Through the space-time coupling of cooling, heating and power and the complementarity of energy demand, the problem of users' comprehensive energy demand is solved, and the cold and hot electric energy is generated, transmitted, transformed and supplied in one system. At the same time, the cross complementarity of various energy sources and the flexible and reliable characteristics of thermoelectric units can efficiently absorb new energy sources such as photovoltaic and wind power with strong volatility. However, the energy storage planning of regional integrated energy system including multi-functional areas is no longer the capacity allocation of a single independent energy system. It is necessary to comprehensively consider the capacity requirements of energy supply, transmission, transformation and storage equipment of various energy forms. Considering that the functional area is equivalent to the whole as a decomposable state, the expression formula of electric quantity distribution characteristics in a certain area is obtained as follows:

$$U=\sum_{i=1}^{q} h^{i} \quad (7)$$

In formula (7), h represents the power supply period, q represents the initial power consumption of the functional area, and i represents the re decomposition period. How to plan according to different user needs during multi-functional interval collaborative planning. Based on the above analysis, considering the differences and space-time complementarity of "source network load storage" of cooling, heating and power in different types of functional areas, this chapter proposes to optimize EV charge and discharge, room temperature and cold storage temperature by using the space-time distribution characteristics of EV, the flexibility of temperature load and the thermal inertia of heating system, so as to build an entity considering the differences of "source network load storage" in multi-functional areas, Multi entity energy storage equipment planning model of regional integrated energy system with coordinated virtual energy storage. Based on this, the steps of optimizing the energy storage mode of the energy system are completed.

3. Case description

3.1. Selecting the basic parameters of equipment in the park
This paper selects an industrial park in southern China as the research object. There are 13 industrial users in the industrial park, which can be equivalent to 13 load nodes in the whole park. The four load nodes include four load types: electricity, heat, cooling and gas. The three load nodes include three load types: electricity, heat and cooling. The three load nodes include electrical and thermal load types. In this case, the heat load refers to the hot water for industrial production with water supply temperature of 140-170 ℃ and return water temperature of 30-45 ℃, and the heat for building space, and the cooling load refers to the cold for building space. The basic parameters of equipment in the park are shown in Table 1:

Table 1 Basic parameters and prices of equipment

Equipment	Fan	P2G	Water storage tank	CCHP
Cost（yuan/Kw）	19200	3700	35	9700
Maintenance Costs （yuan /Kw）	2200	1800	-	2400
Efficiency	-	0.5	0.9	0.42
Life（year）	22	8	18	14

Based on table 1, case analysis is carried out. In addition to the above conditions, the industrial park contains 24 road network nodes. According to the local resource conditions, geographical environment and policy factors, energy alternative stations are selected, including combined cooling and heating power supply, photovoltaic, electric heating boiler, gas boiler, absorption refrigerator, electric refrigerator, natural gas station and energy storage alternative station. For the cooling and heating load of buildings, this area is located in the subtropical climate zone. There is no heating demand all year round, which is cooling demand. Besides the Spring Festival, there is cooling load demand in all periods of the year. In summer, June, July, August and September are the peak period of cooling, and the cooling load demand in this period accounts for more than 55% of the total amount of the whole year.

3.2. Optimization results
Combined with the multi-dimensional digital twin technology, the energy storage planning method of the park energy system designed this time is used to optimize the energy storage scheme of the park, and the installed capacity of renewable energy before and after optimization is obtained, as shown in Table 2:

Table 2 Installed capacity of renewable energy（kW）

Scene	Before optimization	After optimization
Energy storage configuration is not considered	1268	1836
Only consider adding electric energy storage configuration	654	1022
Gees is considered	2648	3684
Only cooling load is considered	1680	2300
Only heat load is considered	2365	2906

It can be seen from table 2 that under different scenario conditions, the average installed capacity of renewable energy after optimization is 2349.6 kW and the value before optimization is 1723kW, indicating that the designed energy storage planning method has higher performance.

4. Conclusions
The result of electric / thermal energy storage planning is not only more economical, but also can ensure the safety and reliability of system operation. The electric / thermal energy storage planning

model is established to consider the balance of energy supply and demand, cold / thermal / electrical coupling and energy complementarity to meet the load demand. At the same time, the electric energy storage capacity is reduced and the waste of renewable energy is reduced, which brings both environmental and economic benefits. Due to the limited research conditions, this paper does not conduct in-depth research on electromagnetic energy storage, and will focus on relevant topics in the future.

Acknowledgment
The study was supported by the Ministry of industry and information with advanced manufacturing project in 2020 "innovative service platform of cloud-edge callaborative hardware in-the-loop simulation for energy internet based on digital twin" (TC200J023/001).

References
[1] Sperber E , Frey U , Bertsch V . (2020) Reduced-order models for assessing Demand Response with heat pumps – insights from the German energy system[J]. Energy and Buildings, 223(223):110144.
[2] Aldabesh A , MazmulHussain, Khan N , et al. (2020) Thermal variable conductivity features in Buongiorno nanofluid model between parallel stretching disks: Improving energy system efficiency[J]. Case Studies in Thermal Engineering, 23(2021):100820.
[3] Wang Z , Teng Y , Jin H , et al. (2021) Urban Waste Disposal Capacity and Its Energy Supply Performance Optimal Model Based on Multi-energy System Coordinated Operation[J]. IEEE Access, PP(99):1-1.
[4] X Yao, Xu F , Chen L , et al. (2020) Scheduling of electric-thermal integrated energy system considering environmental cost[J]. IOP Conference Series Earth and Environmental Science, 512:012001.
[5] Shao C , Ding Y , Siano P , et al. (2019) A Framework for Incorporating Demand Response of Smart Buildings Into the Integrated Heat and Electricity Energy System[J]. Industrial Electronics, IEEE Transactions on, 66(2):1465-1475.
[6] Tesema S L , Bekele G . (2019) High Wind Power Penetration Large-Scale Hybrid Renewable Energy System Design for Remote Off-Grid Application[J]. Journal of Power and Energy Engineering, 7(3):11-30.
[7] Li Q W , Jiang P , Li H . (2020) Prognostics and health management of FAST cable-net structure based on digital twin technology[J]. Research in Astronomy and Astrophysics, 20(5):067.
[8] Chen S , Wang T , Li X , et al. (2020) Research on the improvement of teachers' teaching ability based on machine learning and digital twin technology[J]. Journal of Intelligent and Fuzzy Systems, (1):1-12.
[9] Bianchi M , Branchini L , Pascale A D , et al. (2021) Performance modelling and greenhouse impact assessment of a micro-ORC energy system working with HFCs, low GWP fluids and mixtures[J]. E3S Web of Conferences, 238:10002.
[10] Zhang N , Jian X , Ding Y , et al. (2021) Day-ahead Low-Carbon Economical Dispatch of Integrated Energy System With CSP and ORC Based on Improved Imperialist Competitive Algorithm[J]. IOP Conference Series: Earth and Environmental Science, 814(1):012004 (5pp).

Lithium battery loss model and economic optimal control strategy for secondary frequency regulation in power system

XiaoChi Zhang[1a*], Xu Sha[1b*], ZhiTong Zheng [2a], Jie Gong[1c], Yuan Li[1d], Kai Sun[1e]

[1]Department of Operation and Maintenance, State Grid Jiangsu Electric Power Co., Ltd. Huaian Hongze District Power Supply Branch, Huaian 223000, China

[2] China Three Gorges University, Yichang, 443000, China

[a*]zhangxiaoch1@163.com, [b*]shaxusgcc@163.com, [c*]gongjiesgcc@163.com, [d*]liyuansgcc@163.com, [e*]sunkaihu@163.com,

Abstract—Lithium battery energy storage system (ESS) has fast response speed, but it is in service for a long time. Therefore, in the research of secondary frequency regulation, it is necessary to establish a battery model that can reflect the short-term polarization and estimate *SOC* accurately under long-term service. Therefore, we propose a method of variable parameter loss model of lithium battery suitable for secondary frequency modulation of power system and optimize its control strategy based on the relationship between the independent variable SOC and the internal electrical parameters of the battery. Firstly, we discretize the state equation of single cell to obtain the PNGV model with variable parameters. Then the SOC estimation method is improved by using the modified extended Kalman filter method. Finally, we did an analysis about the economy of the ESS participating in secondary frequency regulation by using the loss model, and the optimal control strategy is designed. The HPPC experiment of the ESS unit is designed by Simulink, and the accuracy of the simulation and experimental curves has been analyzed. Finally, the economic differences between the traditional control strategy and the optimal control strategy considering loss are compared. The results show that the error of the variable parameter model in the fitting experiment is no more than 0.41%, and the accuracy of SOC estimation is 12.96% higher than that of the traditional estimation. The economy of the optimized control strategy is 15.38% higher than that of the traditional control strategy.

1. Introduction

By the end of 2020, the total installed capacity of wind power in the world is 702GW [1]. When more and more renewable energy is connected to the power grid, the inertia will be greatly reduced [2]. Large scale ESS has become a necessary method to solve the problem of new energy interconnection [3]. At present, lithium battery ESS accounts for 81.4% of the total capacity [4]. Since the period of each secondary frequency regulation is in short cycle and the ESS unit must be in operation for a long time. It is necessary to analyze the battery characteristics under the two conditions accurately [5]. Before designing control strategy, most researchers must establish an accurate model of ESS. Therefore, we must establish a model that can not only reflect the transient polarization phenomenon of instantaneous operation of the ESS unit, but also accurately estimate *SOC* estimation under long-time working conditions, and use the model to optimize the economy of control strategy [6].

The most research on the participation of the ESS in auxiliary power grid frequency regulation uses the first-order inertia link, and it does not change the parameters with time. Secondly, the problem of capacity loss under long-time operation cannot be considered. Moreover, the *SOC* estimation of the ESS does not get rid of the sensitivity to the initial value. Reference [7-8] compares different modeling methods. PNGV model is suitable for power system simulation, but the parameters are fixed and not suitable for the long-term service in power system. Reference [9] uses PNGV to create the battery model. They research its parameter identification, and give the parameters of each element parameter of lithium battery through experiment. Reference [10] selects the model with variable parameters for research and designs the control strategy of the ESS, but it needs to further analyze the power flow of power system. Reference [11] proposes that the acquisition in the control system of the management system is periodic. And the model will be more accuracy in the discrete domain. It is necessary to discretize the continuous equation. References [12-13] give the *SOC* estimation method of Kalman filter under time-domain model, but it does not consider the attenuation of initial capacity of ESS battery in large power system, so the capacity need to be corrected on this basis. Reference [14] calculates the economy and capacity loss of ESS, and formulates the control strategy of wind power participating in the main frequency regulation of power system. However, it does not have the capacity loss of each ESS unit and effectively reduce the impact of large-scale wind power networking on frequency.

In this paper, we discretize the PNGV model to obtain a single model with variable parameters, which makes the model have better response. Then we use the EKF to correct the initial capacity by taking the capacity loss of single lithium battery as the noise error, so that the model can satisfy the application of power system simulation. Then the loss model is used to optimize the economy of the traditional control strategy. Finally, we use Simulink to simulate the single battery model and compare the control strategy.

2. Modeling of lithium battery with variable parameters

2.1. PNGV model of lithium battery in discrete domain

The traditional lithium battery model can not reflect the instantaneous polarization in the regulation simulation. To solve this problem, it is necessary to establish a dynamic model with variable parameters and good response. PNGV is selected as the basic model, because it can reflect the internal polarization process of the battery, and can also satisfy the simulation of ESS, which is composed of an ideal voltage source and multiple capacitors and resistors in series. Its electrical topology is shown in Fig.1.

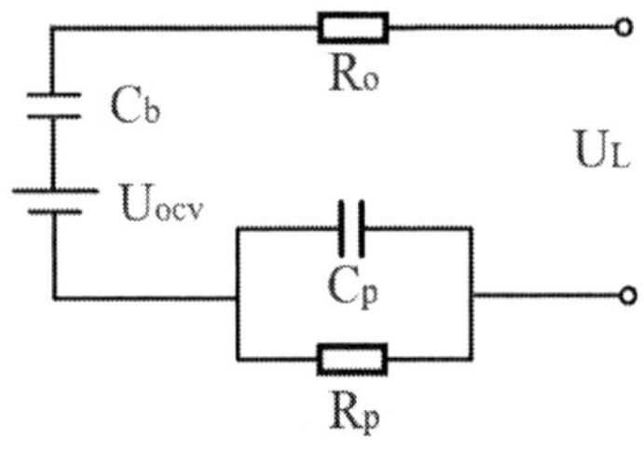

Fig.1: Equivalent circuit model topology of lithium battery

According to the topology and parameters given in Fig.1, the mathematical description of state equation and output equation in continuous system in PNGV equivalent circuit can be given as follows:

$$\begin{bmatrix} \dot{U}_b \\ \dot{U}_p \end{bmatrix} = \begin{bmatrix} 0 & 0 \\ 0 & -\frac{1}{\tau} \end{bmatrix} \begin{bmatrix} U_b \\ U_p \end{bmatrix} + \begin{bmatrix} C_b[SOC(t)] \\ C_p[SOC(t)] \end{bmatrix} I(t) \tag{1}$$

$$U_O = R_o[SOC(t)] \times I(t) \qquad U_L = U_{ocv}[SOC(t)] - (U_b + U_p + U_o) \tag{2}$$

$C_b[SOC(t)]$, $C_p[SOC(t)]$, $R_p[SOC(t)]$ and $R_o[SOC(t)]$ are the polarization capacitance, polarization resistance and ohmic internal resistance varying with *SOC*, and U_b is the cumulative voltage drop of terminal voltage with current in time domain. U_p is the polarization voltage of the battery and the time constant on the whole RC ring τ also changes with *SOC*:

$$\tau_p = R_p[SOC(t)] \times C_p[SOC(t)] \tag{3}$$

However, the model in the continuous time domain is not suitable for the simulation of power system, so the discretization of the time-varying state equation in the continuous domain is as follows:

$$M(s) = e^{Ms} = \begin{bmatrix} 1 & 0 \\ 0 & e^{-\frac{1}{R_p C_p}T} \end{bmatrix} \qquad N(s) = \left(\int_0^{s} e^{N s} ds\right) B = \begin{bmatrix} \frac{s}{C_b} \\ R_p(1 - e^{-\frac{1}{R_p C_p}}) \end{bmatrix} \tag{4}$$

We can derive the time-varying state equation of lithium battery in discrete domain:

$$\begin{bmatrix} U_b(k+1) \\ U_p(k+1) \end{bmatrix} = \begin{bmatrix} 1 & 0 \\ 0 & e^{\frac{-T}{\tau}} \end{bmatrix} \begin{bmatrix} U_b(k) \\ U_p(k) \end{bmatrix} + \begin{bmatrix} \frac{T}{C_b[SOC(k)]} \\ R_p[SOC(t)](1-e^{\frac{-T}{\tau}}) \end{bmatrix} I(k) \tag{5}$$

k is the sampling number in the control system; T is the sampling period. The SOC under each period can be obtained by using an accurate estimation method, then the parameters of the battery model can be updated each period, which can make the model more accuracy.

2.2. SOC estimation method of extended Kalman filter algorithm

Because the SOC and other electrical parameters is related to OCV in the batter, when SOC is determined, the values of OCV and other electrical parameters can be obtained according to the first experiment fitting curve. Therefore, we used the basic the EKF algorithm to estimate the SOC.

The EKF algorithm iterates the results of current integration method by using the observations of voltage. The state equation of ampere hour integration method is as follows:

$$Ah(k) = \int_{(k-1)T}^{kT} I_t(k)dT \tag{6}$$

Because of the long-term service of the ESS, its initial capacity is attenuated. The current integration method is a typical time-varying linear calculation method, so the EKF is used to predict the data of the next cycle, as shown in Fig.2.

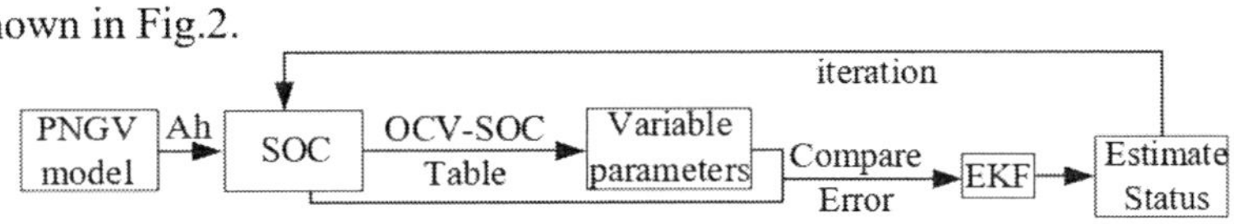

Fig.2: The EKF algorithm for SOC estimation

The state equation of SOC estimation is shown in equation (7):

$$SOC(kT) = \frac{Q_{initial} + \sum_{k=1}^{k} Ah(kT)}{Q_{useful}} \tag{7}$$

C_i is the capacity throughput of lithium battery in unit time; $I_t(T)$ is the charge-discharge current in each period. Because the curve of OCV and SOC is fitted in the first simulation, we can use the EKF to predict the future SOC of the system according to this curve and current data, but this estimation method does not select the appropriate noise, so we will select the appropriate error for correction.

2.3. SOC estimation method based on loss correction

The reason for the decline of SOC estimation accuracy about the traditional KFM is the attenuation of initial capacity C_{useful}. Therefore, we proposed the EKF based on loss correction in this paper.

First, we correct the nominal capacity and the error of nominal capacity loss as equation (8).

$$C'_{useful}(k+1) = C_{useful}(k) + \Delta C(k+1) \qquad \Delta C_i(k+1) = B \cdot \exp[\frac{-a_1 + a_2 I_t(k)}{RT_e}][Ah_t(k)]^z \tag{8}$$

The state equation and observation equation in the discrete domain are as follows:

$$X_{t+1} = P_t X_t + Q_t U_t + R_t \qquad Y_t = S_t X_t + W_t U_t + V_t \tag{9}$$

According to the polarization reaction, the equation of state with U_{sum} is as follows:

$$U_{sum}[(k+1)T] = [U_{sum}(kT) - i_t(kT)R_P]e^{-\frac{T}{\tau(kT)}} + I_t(kT)R_P \tag{10}$$

In this way, the capacity loss can be input into the error equation as an error. Combined with the discrete model and the state equation of equation (11). The input error Rt is the correction of the nominal capacity of the current and next cycle, as also shown in equation (11):

$$P_k = \frac{\partial f(U_{sum}, I_t)}{\partial U_{sum}} = \begin{bmatrix} 1 & 0 \\ 0 & e^{-\frac{T}{\tau(kT)}} \end{bmatrix} \qquad Q_k = [-\frac{T}{C_{useful}(kT)} \quad R_P(1-e^{-\frac{T}{\tau(kT)}})]^T \qquad R_t = \begin{bmatrix} C_{iXuseful}(kT) \\ C_{iXuseful}[(k+1)T] \end{bmatrix} \tag{11}$$

According to equation (10-11), the state equation of SOC estimation can be obtained as follows:

$$\begin{bmatrix} SOC[(k+1)]T \\ U_{sum}[(k+1)]T \end{bmatrix} = \begin{bmatrix} 1 & 0 \\ 0 & e^{-\frac{T}{\tau(kT)}} \end{bmatrix} \begin{bmatrix} SOC(kT) \\ U_{sum}(kT) \end{bmatrix} + \begin{bmatrix} -\frac{T}{C_{useful}(kT)} \\ R_P(1-e^{-\frac{T}{\tau(kT)}}) \end{bmatrix} I_t(kT) + \begin{bmatrix} C_{iXuseful}(kT) \\ C_{iXuseful}[(k+1)]T \end{bmatrix} \tag{12}$$

The improved EKF SOC estimation process based on loss correction is shown in Fig.3.

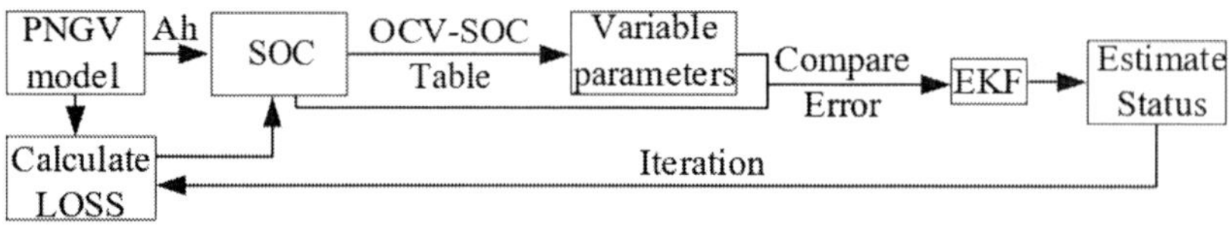

Fig.3: The flow chart of the improved EKF *SOC* estimation

Because the control period is generally about seconds, the charge-discharge current of the ESS unit can be linearized in each sampling period. Therefore, the modified EKF estimation method is still effective for complex nonlinear current.

3. Optimal economic control strategy of secondary frequency regulation

3.1. Analysis of the Grid connection with the ESS

During the operation of the ESS unit, the voltage is relatively stable when the *SOC* is between 10% - 90%. Therefore, the *SOC* of the ESS unit is generally partitioned, as shown in the Fig.4.

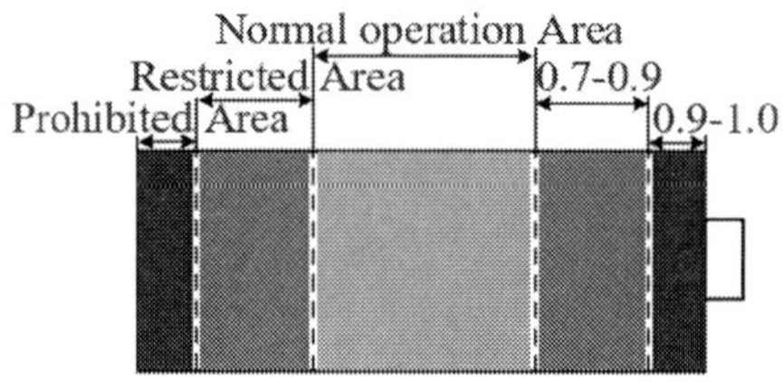

Fig.4: The diagram of *SOC* area division

Therefore, the constraints on *SOC* are as follows:

$$SOC_{i(\min)} \le SOC_i(kT) \le SOC_{i(\max)} \tag{13}$$

For the power limitation of the ESS PCS, the constraint on the output power of each ESS unit in each sampling period is given as equation (15). And the grid connected voltage of each ESS unit also needs to correspond with the public system voltage. At the same time, the ESS unit model has the conditions for operation, and the current of the ESS unit can be converted as equation (14).

$$|P_{Ei}| \le |P_{PCS}| \qquad U_{i(\min)} \le U_{i(sum)} \le U_{i(\max)} \qquad I_i(k) = \frac{P_{Ei}(k)}{U_{isum}(k-1)} \tag{14}$$

3.2. Analysis of loss economy and optimization of control strategy

When the ESS unit participates in the secondary frequency regulation, there will be capacity fading in each control cycle. In a single sampling cycle T, linearize the secondary frequency regulation power and take the average value. Combined with equation (15), the capacity loss of each cycle is as follows:

$$C_i(k) = B_{ci} \cdot \exp[\frac{-a_1 + a_2 C_{rate}(k)}{RT_e}][\int_{(k-1)T}^{kT} \frac{P_{Ei}(k)}{U_{isum}(k-1)} dT]^{\lambda} \tag{15}$$

Assuming that the current economic loss per unit capacity loss is E_K and the auxiliary regulation profit per unit power is E_y, we can derive instantaneous economic profit as follows:

$$E_{isum}(k) = \int_{(k-1)T}^{kT} E_k C_i(kt) dT + \int_{(k-1)T}^{kT} E_y P_{Ei}(kt) dT \tag{16}$$

Based on the above, each ESS unit shall satisfy the following equation in each sampling period participating in secondary frequency regulation. And during the calculation sampling period, in addition to the minimum capacity loss, it shall also be executed according to AGC command, minimum current combination and the requirements of equations (13) - (14), as shown in the equation (17).

$$\begin{cases} \max\{E_{isum}(k)\} \\ s.t: \min\{\sum_{i=1}^{n} C_i(k)\} \end{cases} \qquad \begin{cases} \left|\sum_{i=1}^{n} P_{Ei}(k)\right| \ge |P_{AGC}| \\ \min\{\sum_{i=1}^{n} I_i(k)\} \end{cases} \tag{17}$$

4. Example analysis

The example is divided into two aspects: lithium battery model simulation, *SOC* estimation accuracy and control strategy economy comparison. The lithium battery is the type of 18650. The experiments data are given by conventional HPPC experimental measurement. The parameters are shown in Tab 1.

Tab.1: Experimental parameters of lithium battery charge-discharge

SOC	Cb/F	Cp/F	Rp/mΩ	Ro/mΩ	Uocv/V	SOC	Cb/F	Cp/F	Rp/mΩ	Ro/mΩ	Uocv/V
0.1	551098	221944	0.001309	0.00119	1.976	0.6	490031	26198	0.001598	0.00047	2.071
0.2	521307	149783	0.001267	0.00098	1.988	0.7	489746	22516	0.001731	0.00047	2.082
0.3	503144	130006	0.001285	0.00089	2.016	0.8	481031	19098	0.001908	0.00051	2.101
0.4	485211	39926	0.001516	0.00049	2.034	0.9	470869	17014	0.002216	0.00055	2.124
0.5	491989	33453	0.001535	0.00048	2.056	1.0	612446	5096	0.003882	0.00031	2.159

4.1. Lithium battery model simulation verification

Firstly, the HPPC simulation is realized by Simulink and compared with the experimental fitting curve. The charge-discharge ratio is 0.1, as shown in Fig.5 and Fig.6.

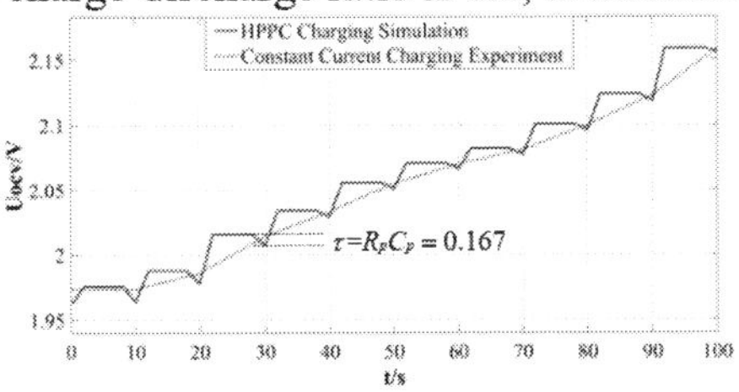

Fig.5: HPPC charging simulation curve

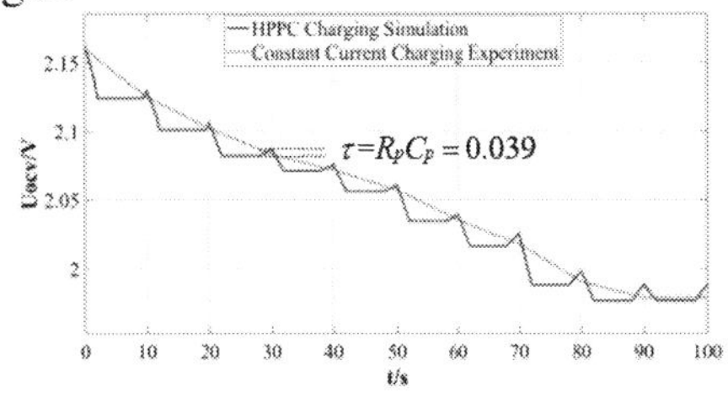

Fig.6: HPPC discharging simulation curve

The figure shows that in the charge-discharge experiment, at *SOC*=30% and 70%, the U_L of the single battery is 2.019V and 2.087V, and U_{ocv} is 2.082V and 2.016V, because the model in this paper takes into account the voltage generated by the current in C_B through accumulation. In the gap of HPPC experiment, the voltage drop of differences in the battery explains the model τ is time-varying. When *SOC*=10%, the simulated U_{ocv} is 1.985V, while the experimental data is 1.976V, and the error value is 0.41%. Therefore, the improved model not only fits well in the simulation accuracy, but also can accurately reflect the instantaneous polarization phenomenon of the battery.

Then, we use 8-branch single cells in 8 parallel and 6 series to obtain the ESS unit, and its reference voltage is 760V. We used the constant power to charge the Unit. The curve of the U_L is shown in Fig.9. Finally, we used the 0.6h actual condition, comparing the two methods. The results are shown in Fig.8.

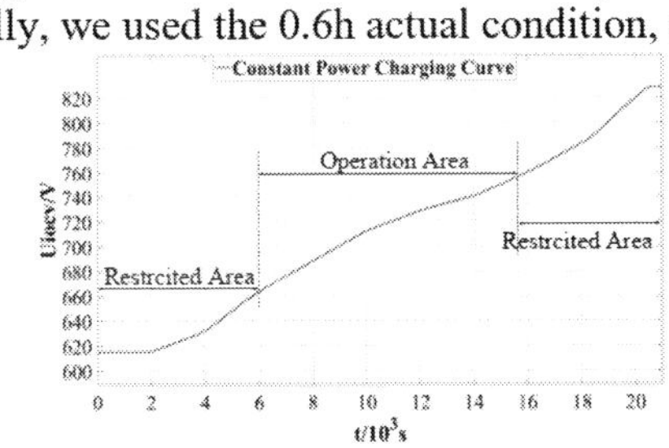

Fig.7: Constant power simulation curve

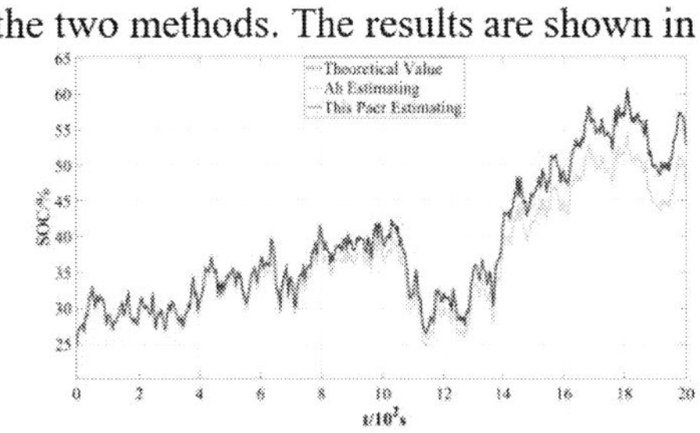

Fig.8: Comparison of the two SOC estimation

As shown in the figure, the U_L of the ESS unit is between 619V and 824V. When the ESS unit is in the working area, the working voltage fluctuates within ±11% of the reference voltage. The accuracy of the improved model in large-scale service condition simulation is also accurate.

It can be seen that the *SOC* estimation value of the EKF with capacity loss correction is close to the theoretical value, while the estimated value of the current integration method has deviated from the theoretical value by 12.96% due to the influence of capacity loss after 0.33h. This is because the method in this paper is not sensitive to the initial value and greatly improves the estimation accuracy.

4.2. Economic verification of optimized control strategy

We also use 0.6h AGC signal for comparative analysis. The instantaneous economy curve of traditional control strategy and economic optimal control strategy considering loss is shown in Fig.9. And the total economic benefits of five ESS units within 0.6h are shown in the Fig.10.

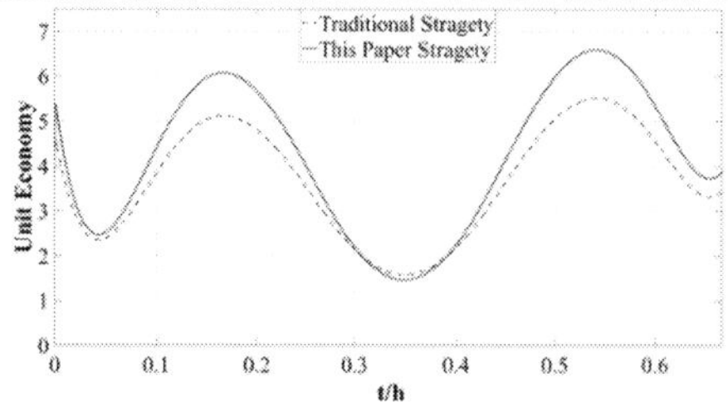

Fig.9: Instantaneous economy curve of the ESS

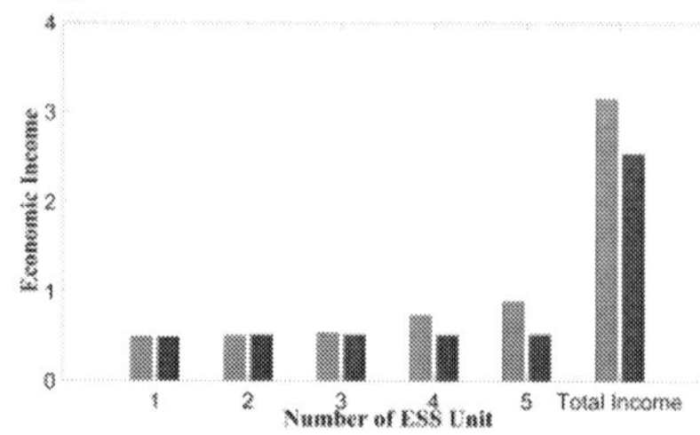

Fig.10: Total economic benefits of the ESS

It can be seen from Fig.9 that under the optimization strategy considering capacity loss, the instantaneous economy of auxiliary power system of the ESS is higher than that of traditional strategy in most periods, and the difference of maximum economic benefits is 15.38%.

Fig.10 shows that the economy of No. 4 and No. 5 ESS units is high because the initial value of the nominal capacity status of the two ESS units is high. It can be seen that the total income of this control strategy is increased by 10.21% compared with the traditional control strategy.

5. Conclusion

This paper proposes a lithium battery model for power system secondary frequency regulation, and uses the loss model to optimize the control strategy. We give the following conclusions:

(1) In HPPC and constant power charge discharge simulation experiments, the fitting error between the two simulation and experimental curve is less than 0.41%. The simulation accuracy and speed of the model can satisfy the use of secondary frequency regulation in power system.

(2) In the 0.6h simulation, the estimation accuracy of the extended Kalman method based on capacity loss correction is 12.96% higher than that of the traditional estimation method. This method can satisfy the long-term service of secondary frequency regulation.

(3) The economic optimal control strategy considering loss in this paper improves the instantaneous economy by 15.38%, and the total income by 10.21% compared with the traditional strategy.

References

[1] Hongbo Liu. (2021) Overview of wind power participating in power system frequency modulation control strategy. Electric Power Automation Equipment, pp.32.1318.
[2] J. Li, X. Feng, G. Yan et al. (2018) Survey on frequency regulation technology in high wind penetration power system, Power System Protection and Control, pp.163-170 Jan.
[3] C. Li, S. Zhang, and J. Zhang et al. (2017) Method for the Energy Storage Configuration of Wind Power Plants with Energy Storage Systems used for Black-Start. Energies, pp.3394, Dec. 2018.
[4] Li Cuiping, Cao Pujia, et al. (2018) Review on Reactive Voltage Control Methods for Large-Scale Distributed PV Integrated Grid. Journal of Northeast Dianli University, pp.37(02): 82-88.
[5] SUN Yushu, YANG Min, SHI Changli, et al. (2020) Analysis of application status and development trend of energy storage. High Voltage Engineering, pp.46(01): 80-89.
[6] Li Jianlin, Ma Huimeng, Hui Dong. (2016) Present development condition and trends of energy storage technology in the integration of distributed renewable energy. Transactions of China Electrotechnical Society, pp.31(14): 1-10.
[7] Liu X, Li W, Zhou A. (2018) PNGV equivalent circuit model and SOC estimation algorithm for lithium battery pack adopted in AGV vehicle, IEEE Access, pp.2018: 1-1.
[8] YANG Yang, TANG Taofeng, QIN Datong, et al. (2012) PNGV equivalent circuit model and SOC estimation algorithm of lithium batteries for electric vehicle, Journal of System Simulation, pp.24(04): 938-942.
[9] Gao W, Jiang M, Hou Y. (2011) Research on PNGV model parameter identification of LiFePO4

Li-ion battery based on FMRLS, Industrial Electronics & Applications, IEEE Publishing.
[10] Romaus C, et al. (2009) Optimal energy management for a hybrid energy storage system combining batteries and double layer capacitors, ECC & Exposition, IEEE Publishing.
[11] XIE Junwen, LU Jiming, MAO Chengxiong, et al. (2013) Design and implementation of digital battery simulator, Electric Power Automation Equipment, pp.33(05): 156-163.
[12] WANG Dangshu, WANG Xinxia. (2019) SOC estimation of lithium-ion battery based on extended Kalman filter[J]. Chinese Journal of Power Sources, pp.43(09): 1458-1460.
[13] ZHONG Yuan, MENG Tao, et al. (2017) SOC dynamic estimation of Lithium-iron battery based on improved Kalman filtering algorithm, Jilin Electric Power, pp.45(03): 24-27.
[14] CUI Hongfen, YANG Bo, JIANG Ye. et al. (2019) Strategy based on fuzzy control and self-adaptive modification of SOC involved in secondary frequency regulation with battery energy storage[J]. Power System Protection and Control, pp.47(22): 89-97.

Smart substation network quality monitoring and fault prediction

Chao Li[123a*], Qi Peng[123b*], Dehui Wang[123c*], Linglu Luo[123d*], Huanhuan Zuo[123e*]

[1]NARI Group Corporation (State Grid Electric Power Research Institute), Nanjing 211006, China

[2]NARI Technology Co., Ltd., Nanjing 211006, China

[3]State Key Laboratory of Smart Grid Protection and Control, Nanjing, 210061, China

[a*]lichao8@sgepri.sgcc.com.cn,[b*]pengqi@sgepri.sgcc.com.cn,[c*]wangdehui@sgepri.sgcc.com.cn,[d*]luolinglu@sgepri.sgcc.com.cn,[e*]zuohuanhuan@sgepri.sgcc.com.cn

Abstract—The smart substation communication network is the basis for information sharing of various devices in the substation. Its operation status has an important impact on the safe operation of the substation and even the power grid. Therefore, real-time status monitoring of the smart substation communication network is becoming more and more important. Aiming at the problems of single dimension, insufficient real-time performance and manual fault analysis in existing substation communication network state monitoring technology, this paper proposes a method of smart substation communication network state monitoring and fault prediction based on network communication quality. This paper uses switch ACL technology and coloring technology to obtain communication quality indexes such as bandwidth utilization, delay, and packet loss rate in real time; based on a multi-dimensional evaluation algorithm, a comprehensive evaluation model of network communication quality is constructed; the model of the relationship between abnormal network communication quality and failures is established. Finally, real-time monitoring of network communication quality and fault prediction are realized. The application analysis in a typical 110 kV substation shows that this method can effectively evaluate the network communication quality and accurately predict failures, and can guide operationer and maintenaner to quickly restore the normal operation of the communication network.

1. Introduction

The smart substation communication network provides an efficient, safe and reliable information exchange channel for all kinds of equipment in the substation[1-2]. It is the link of information exchange between equipment in the substation and between equipment and system. It is the basis for the realization of smart substation information digitization, information sharing standardization, and advanced application interaction. Its reliability and safety have an important influence on the safe operation of substations and even power grids. With the development of smart substation technology, the types and quantities of communication network access devices are increasing, the amount of service data carried is increasing, and the communication links are becoming more and more complex[3-4]. These changes have led to an increasing number of network failures, which seriously threaten the normal operation of substations[5-6]. Real-time status monitoring of the communication network within the station is becoming more and more important.

However, the existing technology is difficult to quickly and accurately locate and troubleshoot network faults. The network analyzer (hereinafter referred to as NA) is the only way to realize the monitoring and fault diagnosis of the communication network[7]. Its essence is the centralized real-time recording of network messages. It only has a simple alarm function and cannot provide real-time information for the nodes in the network. Troubleshooting and analyzing the cause of the failure still need to rely on manual labor. At the same time, because it can only give an alarm after a failure, the real-time performance is poor, and there is no failure early warning function, which will lead to the expansion of the scope of the failure. The smart substation network operation status monitoring and troubleshooting method proposed in the literature[8-10] is based on NA to realize message flow statistics and abnormal message analysis, and it has the same limitations as the NA. In the previous article[11-12], I and my collaborators proposed a process-level network dynamic monitoring method, which can quickly locate physical link failures and virtual link disconnections, and expand the fault diagnosis scope of NA. However, this method still alarms after the failure, with poor real-time performance, and can only locate physical topology-related failures.

The fundamental reason for the limitations of current network fault monitoring and diagnosis is that faults are only analyzed from the perspective of a single frame of messages, and there is a lack of analysis from the dimension of communication quality. The quality of network communication reflects the running status of network communication in real time. When network communication fails, network communication quality will also be abnormal. Abnormal communication quality is closely related to communication failure[13-14]. However, there are many difficulties in real-time monitoring of network communication quality. There are many evaluation indexes, and most indexes are difficult to obtain in real time[14]. A single index cannot comprehensively and accurately reflect the quality of network communication, and the degree of influence of abnormalities and failures of different indexes is different[14-15]. It is also difficult to evaluate the quality of network communication in a multi-dimensional manner and reveal the relevance of abnormal indicators and faults.

This paper conducts research from the dimension of the network communication quality in the substation, and provides a new perspective for the real-time monitoring of the communication network status of the smart substation. First, the article introduces the construction of the network quality monitoring and fault prediction system framework for smart substation business. Then, it introduces the realization of real-time monitoring of key indicators of network communication quality, and proposes a comprehensive evaluation model of network communication quality and a correlation model of network communication quality abnormalities and faults. Finally, the application of the smart substation network monitoring system based on this technology in the 110kV substation in Guanyinqiao, Sichuan is introduced. The results show that the smart substation network communication quality monitoring and fault prediction technology can realize the timely warning of abnormal conditions, and effectively improve the reliability of communication network operation and the efficiency of maintenanc.

2. Smart substation network quality monitoring and fault prediction system framework

The system framework of the smart substation communication network quality monitoring and fault prediction in this paper is shown in Fig.1, which is mainly composed of two key parts: real-time acquisition and predictive analysis.

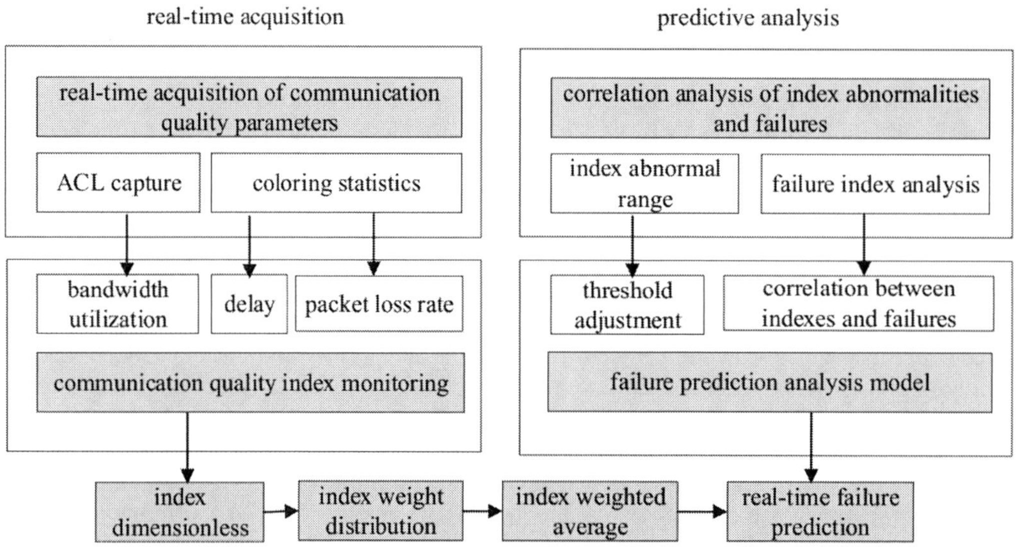

Fig.1 Framework of the smart substation communication network quality monitoring and fault prediction

In the smart substation network, the switch as the transit medium for the transmission of various types of messages, plays an important role in information transmission. Therefore, this paper proposes to use the ACL technology and coloring technology of the switch to obtain the bandwidth utilization, delay, and packet loss rate of virtual links in real time. So as to realize real-time monitoring of physical links and network communication quality index.

Since a single index cannot comprehensively reflect the network communication quality of smart substation network, this paper proposes a comprehensive evaluation method for multiple indexes. First, the three communication indexes of bandwidth utilization, time delay, and packet loss rate are processed in a dimensionless manner. Secondly, according to the importance of influencing communication and the characteristics of smart substation network, weight of each index is assigned. Finally, the comprehensive evaluation value of network communication quality is calculated after the weighted average algorithm.

The quality evaluation value of network communication reflects the operation status of network communication, and abnormal communication quality parameters are closely related to communication failures. This paper proposes a method to adjust the abnormal threshold of quality communication index, analyzes the correlation between abnormal communication indexes and failures, and builds a failure prediction model. When the comprehensive evaluation value of network communication quality exceeds the threshold, the fault prediction module can quickly and accurately predict the cause and scope of the fault.

3. Real-time monitoring of communication key index data

There are many indexes to evaluate the quality of network communication. Common ones are: delay, delay jitter, packet loss rate, device port throughput, bandwidth utilization and load rate, reliability, etc. Different indexes reflect the quality of network communication from different aspects. However, due to the many types of index , obtaining them in real time at the same time will reduce the operating efficiency of the switch. This paper selects three key indexes of bandwidth utilization, time delay, and packet loss rate to comprehensively evaluate the communication quality of the smart substation network. The index selection considers the following three aspects: 1) principles of comprehensiveness, ease of testability, and index-related smallness; 2) characteristics of zero packet loss required for smart substation network transmission; 3) characteristics of high requirements for SV packet transmission delay by protection devices .

In the article in [11-12], the ACL technology of the switch is used to realize the flow statistics of

GOOSE and SV packets. Based on the literature [11-12], the bandwidth utilization rate can be calculated approximately by dividing the traffic statistics of the virtual link and the physical link by the time slice.

This article uses the packet coloring technology to realize the information statistics of characteristic packets. The specific process is as follows: the switch at the packet sending end adds a flag to the packet for coloring, and the switch at the packet receiving end removes the flag to restore the packet, that is, decoloring. The transmitter and receiver switches respectively count the number of colored packets in the same measurement interval, and can obtain the number of packets lost and the packet loss rate. At the same time, each switch measures the entry time and exit time of the colored packet one by one, and writes the accumulated delay information into the reserved fields of the GOOSE, SV and MMS packets. The receiver switch calculates the delay value of the packet according to the reserved fields . When the network communication is normal, in order not to affect packet forwarding function of the switch, the packets are periodically colored to collect communication quality indexes. But when the network fails, the measurement period can be reduced to obtain more communication quality index values for locating the fault. Fig.2 shows the process of coloring reserved fields in GOOSE, SV packets.

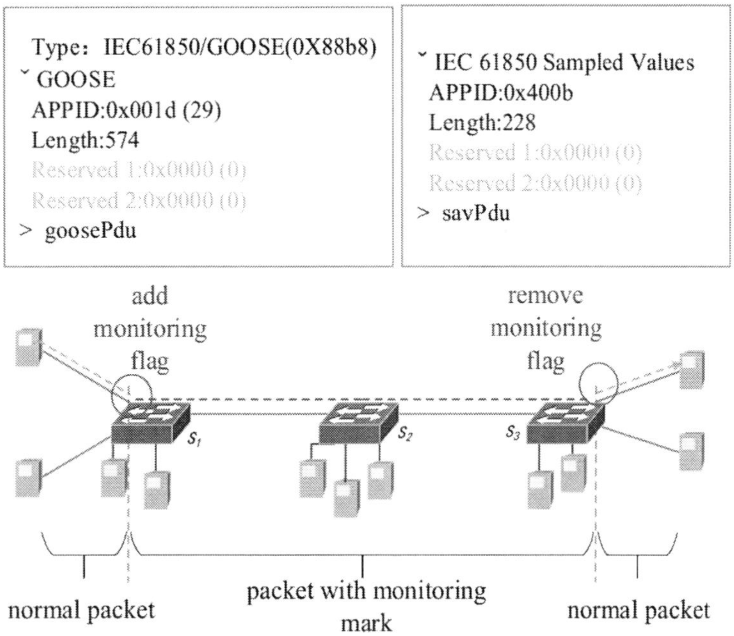

Fig.2 Process of coloring reserved fields in GOOSE, SV packets.

So far, this article uses the ACL technology and coloring technology of the switch to obtain key indexes such as the bandwidth utilization, delay, and packet loss rate of the virtual link in real time, and realizes the real-time monitoring of the communication quality of the physical link and the virtual link.

4. Establishment of network communication quality evaluation model

According to the indexes selected in the previous chapter, each index reflects the performance of the network from different aspects. Due to the complexity of the smart substation business, when a single index is abnormal, it is impossible to accurately determine whether there is a fault [13] and the cause and scope of the fault. In order to scientifically and comprehensively evaluate the communication quality of the smart substation network, this paper proposes an evaluation model for the communication quality of the smart substation network. The three indexes of delay, packet loss rate and bandwidth utilization are calculated through the evaluation model to obtain the comprehensive evaluation value of network communication quality.

Because the units of the indexes of delay, packet loss rate and bandwidth utilization are different, they are of different dimensions. The data values vary greatly and cannot be calculated directly without processing. Therefore, referring to the non-dimensional calculation method of the unit value of the power primary system, the measurement value of the evaluation index is processed in a non-dimensional manner. The size of the measured value is standardized in the interval of [0, 1], and the measured value of each index is dimensionless to form a parameter matrix. Then assign weights to each index according to the characteristics of the smart substation network. Finally, the comprehensive evaluation value of the smart substation network is calculated after weighted average.

4.1 Dimensionless communication indicators

First of all, combined with a large number of project site measured values and the requirements of standards and specifications, the evaluation standard of three indexes are formulated. The evaluation standard of bandwidth utilization are shown in Table.1 below. The evaluation standard for packet transmission delay are shown in Table.2 below. The evaluation standard of the packet loss rate, the score is 1 when there is no packet loss, and the score is 0 when there is packet loss.

Table.1 Bandwidth utilization index grade standard

Grade	1	0.9	0.8	0.7	0.6	0.5	0
Bandwidth utilization%	<10	10-20	20-30	30-50	50-70	70-100	>100

Table.2 Delay utilization index grade standard

Grade	1	0.9	0.8	0.7	0.6	0.5	0
SV/ms	<0.02	0.02-0.05	0.05-0.1	0.1-1	1~8	8~10	>10
Variable-bit GOOSE/ms	<0.02	0.02-0.05	0.05-0.1	0.1-0.2	0.2-0.6	0.6-1	>1
Heartbeat GOOSE/ms	<0.05	0.05-0.5	0.5-1	1~10	10-50	50-100	>100

4.2 Communication index empowerment

In this paper, the weighting method based on the principle of "function-driven" assigns weights to the three indexes of delay, packet loss rate and bandwidth utilization. The weight coefficient is set according to the importance of the indicator's influence on the network communication quality. The importance of each index is graded from 1 to 9, with 9 being the most important, decreasing step by step, and 1 being the least important.

Since the protection device in the smart substation network has strict requirements on the transmission delay of SV messages, and the delay fluctuation is very obvious when the network communication fails [13], the delay importance level is set to the highest. Because the smart substation network has high redundancy and good reliability, the possibility of packet loss is very small. Once a packet is lost, it can be quickly found through Network Analyzer. Therefore, the importance level of packet loss is set to the lowest. According to the importance of the three indexes of delay, bandwidth, and packet loss, the importance levels of the three indexes of delay, bandwidth utilization, and packet loss rate are divided into 8, 6 and 2. According to the formula (1) ,the weight coefficients of the three indexes of delay, bandwidth utilization, and packet loss rate are calculated to be 0.5, 0.375 and 0.125 respectively.

Suppose a system is composed of m "modules", which are F_1, F_2...F_m, where F_i(i=1,2...m) is the "important" of the i-th "module" ", then the F_i weight coefficient w_i can be defined as formula (1):

$$w_i = F_i / \sum_{k=1}^{m} F_k, i = 1,2, \cdots m \tag{1}$$

In the comprehensive evaluation of network message transmission delay indicators, the delay weight values of different types of messages are not the same. According to the priority of packet, the importance level is assigned. The shifted GOOSE message level is 9, the SV message level is 7, the heartbeat GOOSE message is level 4. The weight values are 0.45, 0.35, and 0.2, respectively.

In the comprehensive evaluation of the network bandwidth utilization index, the link between the switch and the switch is the communication trunk line, and the link between the IED device and the switch is the communication branch line. Since the influence of the disconnection of the communication trunk line is much larger than that of the communication branch line, the importance level of the communication trunk line is divided into 9, and the importance level of the communication branch line is 6. The weight values are 0.6 and 0.4 respectively.

4.3 Communication quality evaluation calculation

Suppose the measurement index has k measurement points (virtual links), and each measurement point is t time, totaling k*t samples, and the corresponding matrix is represented by V, which is shown in matrix (2).

$$V = \begin{bmatrix} v_{11} & v_{12} & \cdots & v_{1t} \\ v_{21} & v_{22} & \cdots & v_{2t} \\ \vdots & \vdots & \vdots & \vdots \\ v_{k1} & v_{k2} & \cdots & v_{kt} \end{bmatrix} \tag{2}$$

The measured value of each index is processed in a dimensionless manner according to the index evaluation standard, and the relative value of each index is obtained. Then this paper uses the weighted average algorithm to calculate the comprehensive evaluation value of the communication quality at a certain measurement point at a certain time. The calculation method is shown in formula (3).

$$P_j = (\sum_{i=1}^{n} w_i x_{ij} + \sum_{i=1}^{m-n} w_i x'_{ij}) / \sum_{i=1}^{m} w_i \tag{3}$$

In formula (3), P_j is the communication quality evaluation value of the j-th measurement point, m is the total number of evaluation index, n is the number of positive index in the evaluation index, and m-n is the number of inverse index in the evaluation index. w_i is the weight of the i-th index, v_{ij} is the normalized relative value of the i-th positive index at the j-th measuring point, and v_{ij}' is the normalized relative value of the i-th inverse index at the j-th measuring point. A positive index is an index that is better evaluated with a larger value, such as available bandwidth, and an inverse index is an index that is better evaluated with a smaller value, such as delay and packet loss rate.

$$x_{ij} = \frac{v_{ij} - min}{max - min}$$

$$x'_{ij} = \frac{max - v_{ij}}{max - min} \tag{4}$$

v_{ij} is the measurement index value of the i-th item of the j-th measuring point. min is the minimum value of the measurement index value of the i-th item of the j-th measuring point in t samples. max is the maximum value of the measurement index value of the i-th item of the j-th measurement point in t samples.

The comprehensive evaluation value of network communication quality at a certain moment is obtained by comprehensively weighting the comprehensive evaluation values of all measurement points in the network. The calculation formula is shown in equation (5).

$$P = (\sum_{i=1}^{m} P_i w_i) / \sum_{i=1}^{m} w_i \tag{5}$$

In formula (5), P is the network performance evaluation index, m is the number of measurement points, P_i is the comprehensive evaluation value of the communication quality of the i-th measurement point, and w_i is the weight value of the i-th measurement point.

According to the above research in this article, the evaluation results of the communication quality of each virtual link, physical link and the entire network can be calculated. When the smart substation communication network is operating in real time, the comprehensive evaluation value of network communication quality is also constantly changing. The operation and maintenance personnel can set the threshold value according to the experience value or the reference value of the specified range of the parameter. When the calculated comprehensive evaluation value of network communication quality changes beyond the threshold, the abnormal situation of the network is early warned.

5. Failure prediction analysis

Due to the burst of GOOSE packets in the smart substation network and a large number of abnormal index during failure, the accuracy and timeliness of failure prediction is reduced. At the same time, due to differences in the scale of substation business, and personnel experience, the thresholds set by operationer or maintenaner are also quite different, which further reduces the accuracy of fault prediction. This paper designs a fault prediction module based on communication quality indexes, proposes a general standard process for threshold setting, and analyzes the relationship between common communication faults and communication quality indexes.

5.1 Threshold setting adjustment

The fault prediction module first detects the communication quality parameters of the entire station, which is divided into passive detection and active detection. Passive detection refers to the periodic detection of the communication quality of all virtual links and physical links of the entire station. Active detection refers to the active detection of the communication quality of one or several specific virtual links and physical links when the network fails or according to engineering needs. The measured value of the detected indexes are calculated by the communication quality evaluation model.The calculated evaluation results are merged, and then filtered and analyzed. According to the merge result, judge whether there is an abnormal event. According to the correlation between abnormal communication quality indexes and communication failures, predict the cause of the failure. Finally, a fault alarm is generated.According to the accuracy of the fault alarm, the maintenaner can adjust the fault prediction threshold. After the threshold is adjusted, maintenaner need verify the overall status of the network to ensure the correctness of failure prediction. The general standard process of threshold setting is shown in Fig.3.

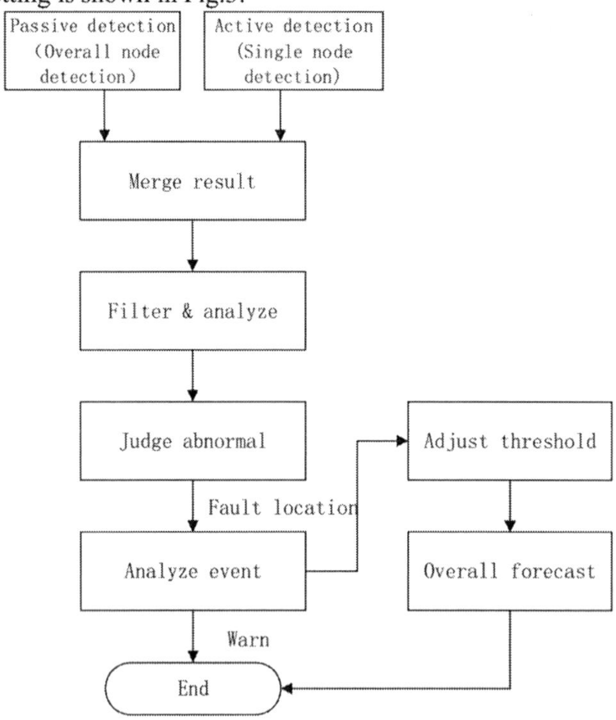

Fig.3 Process of threshold setting

5.2 Judgement of the cause of failure

Before the failure of the smart substation network (such as equipment damage, cable degradation, broadcast storm), the network communication quality will be abnormal. Since communication quality index abnormalities have a certain correlation with communication failures, the cause and scope of the failures can be predicted through the communication quality index abnormalities. The article [11-12] published earlier by myself and my collaborators also proposed a fault diagnosis model. The focus of the diagnosis is physical topology and virtual link connection faults, and it is impossible to diagnose complex communication network faults. This article will further study the relationship between abnormal communication quality indexes and common network communication failures.

Common smart substation communication failures include physical link failures (line disconnection, line aging), switch equipment failures (crash, power-off restart, high working temperature, etc.), network storm (ring network storm), network congestion (SYN Flood storm, ARP request storm) and so on.

When the comprehensive evaluation value of network communication quality is abnormal, check the packet loss rate, bandwidth utilization rate, and delay indexes of each physical link and virtual link. Define the abbreviation of packet loss rate as loss, bandwidth utilization abbreviation as band, delay abbreviation as delay, device abbreviation as IED, switch abbreviation as SW, single physical link abbreviated as SL, multiple physical links abbreviated as ML, virtual link abbreviated as VL.According to the following steps to predict and diagnose the fault, the specific process is shown in Fig.4.

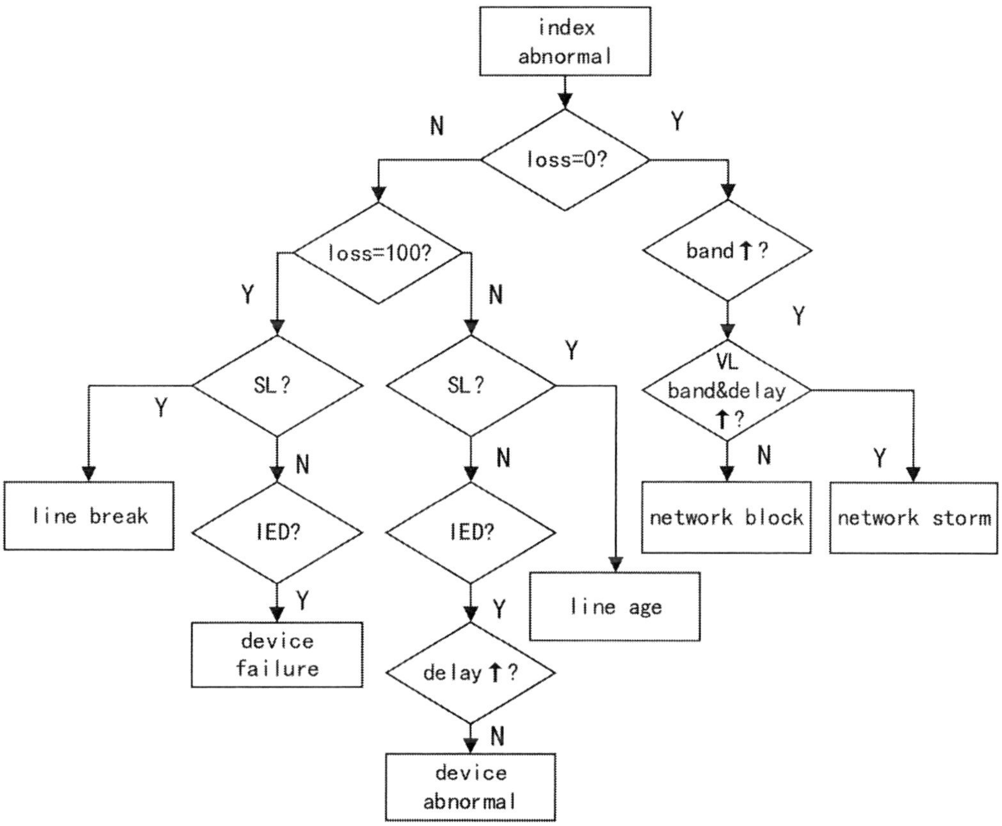

Fig.4 Process of failure prediction

(1) If the packet loss rate of a single physical link is 100%, the cause of the failure is judged to be a disconnection of the physical link (cable disconnection or interface disconnection);

(2) If the packet loss rate of a single physical link is not 100% and the delay change is small, the cause of the fault is judged to be the aging of the line or the loose interface;

(3) If the packet loss rate of multiple physical links is 100%, and these physical links are connected to the same device, the cause of the failure is judged to be an abnormal device (halt or communication interruption);

(4) If the packet loss rate of multiple physical links is not 100% and the delay change is small, and these physical links are connected to the same device, the cause of the failure is judged to be an abnormal device (restart the device or packet loss due to excessive temperature);

(5) If the bandwidth utilization of the physical link gradually increases, and the bandwidth utilization and delay of each virtual link do not increase significantly, the cause of the fault is judged to be network congestion (non-business packets);

(6) If the bandwidth utilization of the physical link gradually increases, and the bandwidth utilization and delay of each virtual link also increase significantly, the cause of the fault is judged to be ring network storm;

(7) If the bandwidth utilization of the physical link gradually increases, and the bandwidth utilization and delay of the goose virtual link increase significantly, and the bandwidth utilization and delay of the sv virtual link do not increase significantly, the cause of the phenomenon is the burst of GOOSE packets. It is not a fault .

The above process analyzes the relationship between common network faults in smart substation networks and abnormal communication parameters. As for unknown or unusual failures, abnormal values of communication quality indexes can be collected when the failure occurs. By these abnormal values, the communication quality change characteristics of the fault can be geted. In this way, the scope of the failure prediction model is continuously expanded.

6. Case verification

The research results of this paper have been successfully used in the pilot project of the 110kV substation in Guanyinqiao, Sichuan. This paper uses the data collected during the trial operation to verify the effectiveness of the method. The topological structure of the process layer network of the station is shown in Fig.5. The process layer network adopts star connection and consists of five subnets, including two subnets of 110KV line, two subnets of main transformer and a subnet of bus. In Figure 5, the twenty one ports connected to the IED and the switch are numbered in sequence, and the IED devices connected to the port numbers are shown in Table.3.

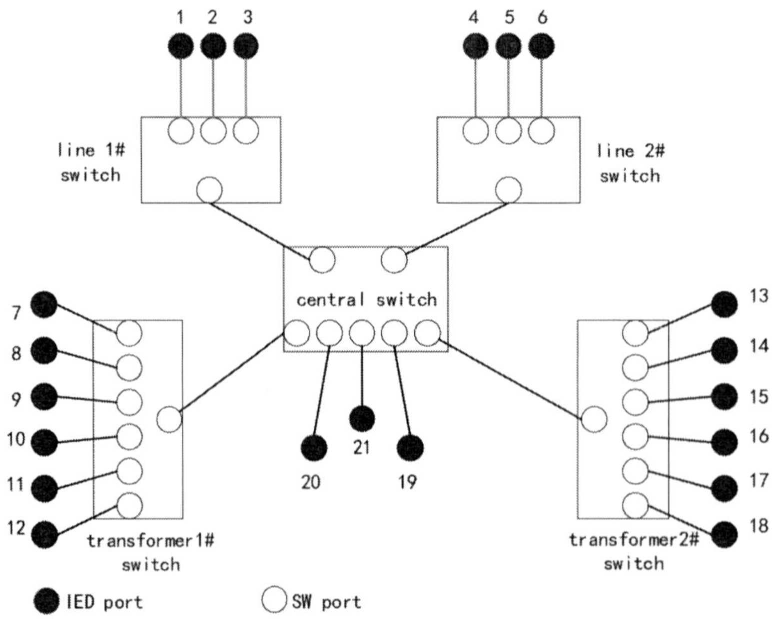

Fig.5 Topology structure of the process layer network

Table.3 IED name and port number

Port	IED name	Port	IED name
1、6	line protection & measure	10、16	transformer high side intelligent terminal
2、5	line intelligent terminal	11、17	transformer low side intelligent terminal
3、4	line merging unit	12、18	transformer protection & measure
7、13	transformer low side merging unit	19	bus protection & measure
8、14	transformer high side merging unit	20	bus intelligent terminal
9、15	transformer itself intelligent terminal	21	bus merging unit

As shown in Table.4. The bandwidth and delay values of the goose and sv packets sent by some IED devices collected by the switch, including the bandwidth and delay values of the service virtual links in steady-state operation and service failure.

Table.4 bandwidth and delay values of virtual links in steady-state and fault-state

Virtual link		Bandwidth of steady state (bit/s)	Delay of steady state (us)	Bandwidth of fault state (bit/s)	Delay of fault state (us)
line 1# MU	SV	7.751M	47.308	9.75M	61.286
line 1# PL	GOOSE	314	52.586	420	84.358
line 1# intelligent terminal	GOOSE1	604	58.240	726	89.295
	GOOSE2	1597	81.538	1916	152.586
	GOOSE3	291	30.460	397	164.556
transformer 1# MU	SV	7.751M	49.216	8.35M	204.947
transformer 1# PL	GOOSE	376	34.295	587	59.275
transformer 1# high side intelligent terminal	GOOSE1	601	59.275	926	82.725
	GOOSE2	1596	82.725	1816	91.521
	GOOSE3	292	31.521	501	48.214
bus MU	SV	7.751M	47.356	9.23M	101.736
bus PL	GOOSE1	274	99.748	486	186.318
	GOOSE2	269	96.992	677	197.256
	GOOSE3	271	92.482	583	193.573

Through the network communication quality evaluation model, the network evaluation index values under steady state and failure are calculated respectively, as shown in Table.5. Through comparison, it is found that the network communication quality index drops significantly when a fault occurs. According to the method of fault prediction in this paper, because the bandwidth and delay of each virtual link are increasing, the fault is judged as a ring network storm fault. According to the results of the failure prediction, the operation and maintenance personnel found that the line 1# switch and the main transformer 1# switch were interconnected, forming a ring network. This case proves the accuracy of failure prediction.

Table.5 Comprehensive evaluation value of index

Comprehensive evaluation value	Steady state	Fault state
Bandwidth utilization	84.15	63.78
Delay	83.93	59.31
Packet loss rate	100	100
Network communication quality	86.049	66.631

It can be seen, the method of smart substation network communication quality monitoring and fault prediction in this article can scientifically evaluate the network communication quality and accurately predict the cause of the fault. This method can help operationer or maintenaner to quickly find the fault location, quickly restore the normal operation of the process layer network, and improve the operational reliability of the secondary system of the smart substation.

7. Conclusion

This paper studies the method of smart substation network communication quality monitoring and fault prediction. This method can obtain the communication parameters of the virtual link and physical link of the packete in real time through the switch. This method constructs a network communication quality evaluation model. This method can predict the cause and scope of the failure based on the abnormality of the network communication quality evaluation index. The case results show that the method can effectively improve the operation and maintenance efficiency of smart substations, promptly warn of abnormal conditions, and improve the reliability of communication network .

The method proposed in this paper can provide a new perspective for smart substation network monitoring and fault diagnosis, and has a wide range of application prospects in engineering applications. In future research, we will further monitor more communication quality indexes and expand fault models to improve the comprehensiveness and accuracy of smart substation network communication quality detection and fault prediction technologies.

Acknowledgments

This work was financially supported by the science and technology project of State Grid Corporation (2021). Fund Project Name: "Research and Application of Substation Automation Equipment Technology Based on Localization Chips and Independent Controllable".

References

[1]. Hao Shaohua, Li Yong, Zhang Tiefeng. Scheme of communication network and network management system for new generation smart substation. (2017) Automation of Electric Power Systems, 41(17): 148-154.

[2]. Zhang Yanxu, Cai Zexiang, Li Xiaohua, et al. Analytical modeling of traffic flow in substation communication network. (2015) IEEE Transactions on Power Delivery, 30(5): 2119-2127.

[3]. Zhu Lin, Wang Pengyuan, Shi Dongyuan. Status monitoring information model and configuration description of communication network in smart substation. (2013) Automation of Electric Power Systems, 37(11): 87-92.

[4]. Wang Haizhu, Zhang Yanxu, Cai Zexiang, et al. Information flow calculation model and method for process bus network in smart substation. (2013) Power System Technology, 37(9): 2602-2607.

[5]. Zhang Yanxu, Cai Zexiang, Long Pianpian, et al. Real-time fault diagnosis models and method for communication network in smart substation. (2016) Power System Technology, 40(6): 1851-1857.

[6]. Gao Lei，Yang Yi, Liu Yu, et al. Proof table based fault location method for process level channel in smart substations. (2015) Automation of Electric Power Systems, 39(4)：147-151.

[7]. Yuan Hao, Qu Gang, Zhuang Weijin, et al. Discussion on condition monitoring contents of secondary equipment in power grid. (2014) Automation of Electric Power Systems, 38(12): 100-106.

[8]. Xu Weiguo, Zhang Liang. Design and application of on-line fault diagnosis system for network communication of digital substation. (2010) Electric Power Automation Equipment, 30(6): 121-124.

[9]. Zhang Jinsong, Yu Jianyu. Application of network analyzer in smart substations. (2011) East China Electric Power, 39(4): 665-668.

[10]. Ding Xiuling, Zhang Yanxu, Cai Zexiang, et al. A protection method of abnormal information flow in process layer network based on packet analysis. (2013) Power System Protection and Control, 41(13): 58-63.

[11]. Li Chao, Luo Linglu, Wang Dehui, et al. Design and implementation of process layer network monitoring and fault location system for smart substation. (2019) Power engineering technology, v.38; No.184(02):123-128+147.

[12]. Luo Linglu, Peng Qi, Wang Dehui, Li Chao, et al. Monitoring method of process level network in smart substation. (2018) Automation of Electric Power Systems, 42(11): 151-156.
[13]. Chen Qingtao, Yang Haitao, Ding Guocheng, et al. Qualitative analysis of data flow in smart substations and simulation of communication network performance. (2017) Electrical Application, 036(023): 22-25.112-115.
[14]. Ouyang Fan, Liu Haifeng, Zhao Yongsheng, et al. Analysis of the communication network congestion fault of smart substation and its preventive measures. (2011) Power System Technology, 11: 7-11.
[15]. Tan Xiliu. State evaluation and abnormal analysis of smart substation communication network . (2017) North China Electric Power University (Baoding).

Journal of Physics: Conference Series **2108** (2021) 012062 doi:10.1088/1742-6596/2108/1/012062

Topology identification method of distribution network based on branch active power

Biqi Liu[1*], Danni Wang[1], Yunpeng Li[1], Lin Qiao[1], Shuo Chen[1]

[1]State Grid Liaoning Information and Communication Company, Shenyang, Liaoning 110006, China

*1142101056@ncepu.edu.cn

Abstract: Because of low measurement redundancy and frequent switch changes, it is difficult to identify the correct topology structure. In this paper, a topology recognition method of distribution network based on branch active power is proposed. Firstly, branch active power residual algorithm is used to identify the topological structure. The topology obtained by this method has the highest matching degree with the real-time measured data. Then genetic algorithm is used to optimize the inverse recognition of power grid topology. The numerical example shows that the method is reasonable, effective, rapid and simple. It also has good adaptability with a large number of measurement errors.

1. Introduction

Topological structure analysis is the basis of all kinds of advanced analysis functions of the distribution system. Therefore, the accuracy of topology structure has an important influence on the rationality of power flow calculation and state estimation[1]. In order to achieve the goal of flexible operation and reliable power supply, power distribution network is usually installed a large number of switches, and switch opening and closing statement will change according to different needs. In addition, measurement redundancy of the distribution network is very low, which leads to the unreliable topology of the distribution network, thus affecting the application of other advanced analysis functions of the distribution network[2]. Therefore, it is necessary to find a topology identification method which can adapt to the dynamic change of the distribution network structure.

In this paper, a distribution network topology identification method based on branch active power is proposed. Compared with the existing analytical methods, this method has better adaptability to the dynamic changes of the distribution network structure, and makes reasonable use of the load data in the distribution network metering system and the real-time measurement data of each branch[3]. The numerical example shows that the method is reasonable, effective, quick and simple, and has good practicability.

2. Residual value of branch activity power

2.1. Concept of branch activity power residual value

Distribution networks mostly operate in a radial ring structure. If a branch in a certain ring network is arbitrarily broken, the power flow change caused by the change of topology structure can be simulated by introducing a specific circuit power into the ring network. The active power of the branch of the distribution network can reflect the operation state of the switch to some extent. For example, when the switch is in the disconnected state, the true value of the active power of the branch is 0. Therefore,

 2108 (2021) 012062 doi:10.1088/1742-6596/2108/1/012062

the branch with the greatest possibility of interruption of the distribution network can be found by superposition of the circuit power, the specific methods are as follows. Firstly, the whole network switch is closed to form several ring networks, and the initial power of each branch is obtained by power flow calculation. Then adjust the loop power of each ring network. Superposition the initial power of each branch and the loop power of each ring network, so that the active power of the branch equipped with measuring device is closest to the real-time measurement data of the active power of the branch. The active power of the branch is called the residual value of the active power of the branch. Finally, by comparing the residual value of active power of each branch to determine the branch interrupted by each ring network (take the minimum residual value), the topology structure obtained by this method has the highest matching degree with the current real-time measurement data.

2.2. Branch active power residual value algorithm

It is necessary to calculate the optimal solution of the active power of each circuit, so that the matching degree between the actual measured value of the branch active power and the residual value of the branch active power is the highest. Therefore, it should satisfy the following relation:

$$\min S = \sum_{j\in M} w_j \left(p_{m,j} - p_j\right)^2 \tag{1}$$

where S is the objective function; W_j is the weight of real-time measurement of branch power, $w_j = 1/\delta^2$; m is the branch set for setting the power measuring device; $P_{m,j}$ is the actual measured value of branch active power; P_j is the active power residual value of the branch. To minimize S, take the partial derivative of (1) :

$$\frac{\partial S}{\partial P_j} = 0 \tag{2}$$

Through mathematical derivation, the residual value of active power of each branch at this time can be written as:

$$P_{opt,j} = \sum_{j\in M} w_j P_{m,j} / \sum_{j\in M} w_j \tag{3}$$

where $P_{opt,j}$ is the residual value of the branch active power after the superposition of the active power of each circuit and the initial active power of each branch. Therefore, (4) is always satisfied:

$$P_{opt,j} = P_{0,j} + \sum_{i=1}^{n} \Delta P_{opt,Li} b_{ij} \tag{4}$$

where P_{0j} is the initial active power of the branch, that is, the power obtained by power flow calculation after closing the whole network switch; $\Delta P_{opt,\,Li}$ is the optimal solution of active power of each ring network; b_{ij} is an element in the relationship matrix between branch and independent circuit in the distribution network (1 when branch j belongs to independent circuit i and both directions are the same; -1 when branch j belongs to independent circuit i and both directions are opposite; 0 when branch j does not belong to independent circuit i).

In the field of electrical engineering, the relationship between variables can be fitted according to the known data, and the method with better fitting effect is the least square fitting. The fitting method is as follows.

Assume that there is a linear relationship between n variables x and variable y. To make the linear fitting of them, as showed in (5), then the key point of fitting is to find out a_j ($j = 0,1...n$).

$$y = a_0 + \sum_{j=1}^{n} a_j x_j y = a_0 + \sum_{j=1}^{n} a_j x_j \tag{5}$$

Let x_{ij} be the ith measured value of variable x_j, and its corresponding function value is y_i ($i = 0,1...m$), then the sum of squares of residuals is:

$$s\left(a_0, a_1, \cdots a_n\right) = \sum_{i=1}^{m} \left(y_i - y\right)^2 = \sum_{i=1}^{m} \left(y_i - a_0 - \sum_{j=1}^{n} a_j x_{ij}\right)^2 \tag{6}$$

In order to minimize the square sum of residual difference between the variable y and the actual value y_i, the partial derivative of (6) is obtained, and the measurement data (x_{ij}, y_i) is substituted into it to obtain the unknown parameter a_j (j = 0,1... , n).

According to (4), $P_{opt,j}$ and n+1 variables (P_{0j}, b_{1j}... ,b_{nj}), the objective function of the branch active power residual value based on the least square method can be established by linear fitting:

$$\min J = \sum_{j\in M} w_j \left[p_{opt,j} - \left(p_{o,j} + \sum_{i=1}^{m} \Delta p_{opt,Li} b_{ij} \right) \right]^2 \quad (7)$$

In order to minimize J, the partial derivative of (7) is taken and the known data is substituted into it. The optimal solution of active power of each circuit can be obtained as $\Delta P_{opt,\,L1}$,..., $\Delta P_{opt,\,Ln}$, and after fitting, the active power residual value equation of each branch can be obtained as:

$$P = P_0 + \sum_{i=1}^{m} b_i \Delta p_{opt,Li} \quad (8)$$

3. Genetic algorithm for inverse topology identification of power network

3.1. Overview of genetic algorithm

Due to the large scale of the power network and the high requirement of real-time performance, all possible solutions should not be exhaustive in the process of solving branch correlation matrix[4]. The initial solution can only be evolved continuously under the constraint conditions, and the global optimization search can be carried out to gradually and quickly find the optimal solution to meet the online identification requirements[5]. Therefore, the algorithm's evolution ability and global optimization search ability are required to solve the grid topology reverse identification.

Genetic Algorithm (GA) is a parallel, efficient and global search method that simulates the genetic and evolutionary process of biology. Inheritance can make the good traits of the population continuously passed to the next generation, maintain the characteristics of the population. Variation can change the characteristics of the population and make the population develop towards the direction of adapting to the new environment. Under certain constraints, a series of genetic operations such as selection, crossover and mutation of the current population can generate a new generation of the population, and the fitness calculation can be carried out to judge the adaptability of the population to the environment. Solving process of this algorithm is consistent with the solving requirements of the reverse identification problem of grid topology structure, and GA can be chosen to better realize the solution of the reverse identification problem of grid topology structure.

3.2. Initial population generation

The initial solution of grid topology reverse identification is the branch correlation matrix C=(C_{ij}). When incoming line i and outgoing line j are combined, C_{ij}=1, otherwise C_{ij}= 0.

However, the solution of the initial branch correlation matrix formed according to the above ideas is not necessarily the real topological correlation matrix of the power network. That is, the loss $d_{i,ki}$ corresponding to the combination of incoming line i and outgoing line k_i is not necessarily the line loss corresponding to incoming line i of the real power network. At some point, $d_{i,ki}$ may be less than the actual line loss of grid incoming line i. That is, the combination of line i and line k_i outbound is not a real branch of the grid, and may lead to interlocking combination errors in the loss calculation of line i+1 and later in the incoming line matrix. In other words, there is no guarantee that the global optimal solution can be obtained.

3.3. Fitness calculation

After the initial branch correlation matrix population is generated, the fitness fc^r of each branch correlation matrix C^r in the population is calculated to evaluate the fitness degree between the correlation matrix and the real grid topology structure. The correlation matrix with the highest fitness

 2108 (2021) 012062 doi:10.1088/1742-6596/2108/1/012062

in the population R is selected as the optimal solution at the current moment. In order to facilitate comparison, the fitness of the association matrix in the population is normalized, as showed below:

$$f_{C^r} = 1 - \frac{F'\left(C^r\right) - F_{min}'(R)}{F_{max}'(R) - F_{min}'(R)} \tag{9}$$

where $F'(C^r)$ is the objective function value of the branch correlation matrix C^r; $F'_{min}(R)$ is the minimum fitness of the branch correlation matrix in the population; $F'_{max}(R)$ is the maximum fitness of the branch correlation matrix in the population.

3.4. Algorithm process

Solving process of power network branch correlation matrix based on the genetic algorithm is shown in Fig.1.

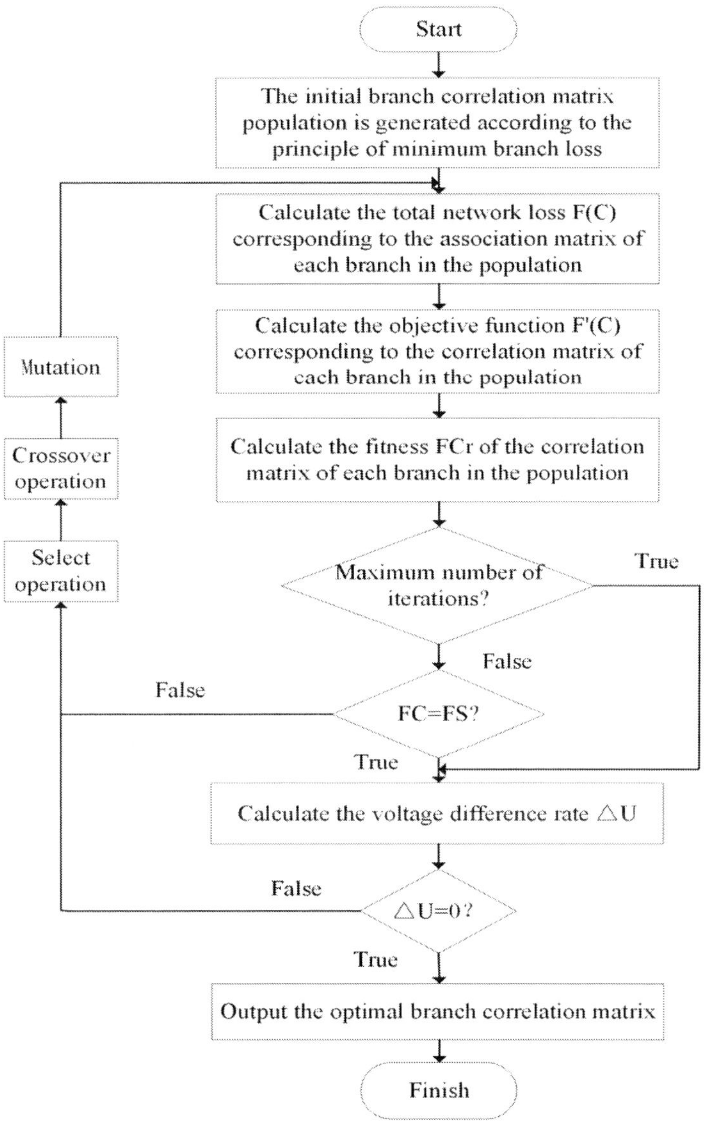

Fig.1 Flowchart of genetic algorithm

The specific implementation steps are as follows:

1) According to the idea of Section 2.2, an initial branch correlation matrix population R of N_P size is generated according to the principle of minimum branch loss;

2) Calculate the total network loss $F(C)$ corresponding to each branch correlation matrix C^r in population R, the improved objective function $F'(C)$, and the fitness f_C^r;

3) If the objective function $F(C)$ of the correlation matrix in the initial population is equal to the total network loss F_S, and the solution is unique, then the optimal correlation matrix is output;

4) Judge whether the maximum number of iterations has been reached, if so, calculate the voltage difference rate ΔU, if ΔU=0, output the optimal correlation matrix, otherwise go to Step 5);

5) Determine whether the loss $F(C)$ corresponding to the branch correlation matrix is equal to the grid measurement calculation loss F_S. If not, select, cross, immunize and other operations are carried out, and then go to Step 2). If equality into step 6);

6) Calculate the voltage difference rate ΔU, if ΔU=0, it means that the correct grid topology solution has been obtained, otherwise go to Step 5) until the solution is unique;

7) Output the correlation matrix and end it.

4. Analysis of the example

4.1. Example of actual distribution network

In order to verify the practicality of the proposed topology identification method to the actual distribution network, three feeders R1, R2 and R3 at the voltage level of 10 kV in a certain area were selected for analysis. Their network structure can be referred to [6].

In the calculation example, the measurement data of phase A at 8:00 a.m. on a certain day is selected as real-time measurement and pseudo-load measurement, the load data collected by the electricity charging system is taken as load measurement, and the power measurement updated every 15 minutes by the medium voltage load collected by the metering system is taken as real-time measurement. There are six possible topologies, as showing in Table 1. The objective function values of the six topologies are calculated respectively, and the calculation results are shown in Fig. 2.

Table 1 Possible topology

Topological number	R1 and R2	R1 and R3
1	LN R1-21-R2-2	LN R1-13-R3-12
2	LN R1-21-R2-2	LN R1-2-R1-13
3	LN R1-21-R2-2	LN R3-2-R3-12
4	LN R1-16-R1-21	LN R1-13-R3-12
5	LN R1-16-R1-21	LN R1-2-R1-13
6	LN R1-16-R1-21	LN R3-2-R3-12

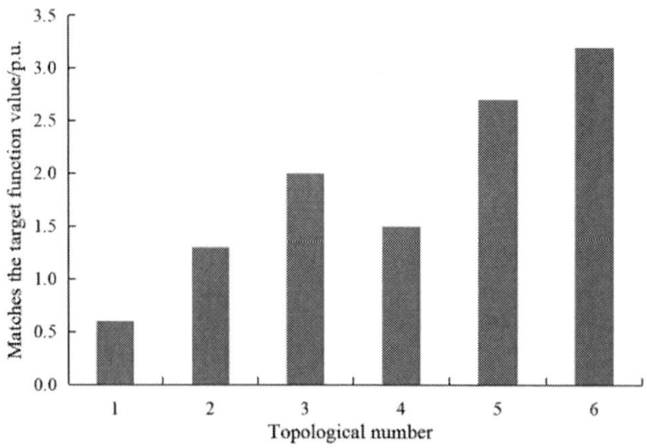

Fig.2 Objective function values of different topologies

It can be seen from Fig. 2 that the objective function value of No. 1 topology is the smallest, which is the same as the assumed benchmark network topology, proving that the method can accurately identify topology in the actual distribution network.

4.2. Comparison of global optimization efficiency of intelligent algorithms

In order to verify the superiority of the algorithm presented in this paper in global optimization, the immune algorithm, a commonly used optimization algorithm, was selected to invert and identify the topological structures of the IEEE 39-node system and the IEEE 118-node system by using the combinatorial optimization mathematical model proposed in the text, and the time needed to obtain the final inversion topological structures was recorded, as shown in Table 2.

Table 2 Comparisons of search efficiency of each algorithm

The system category	Time/s	
	Immune algorithm	GA
IEEE 39	20	0.30
IEEE 118	35	0.33

It can be seen from Table 2 that the global optimization time of the genetic algorithm used in this paper on the IEEE39-node system and the IEEE118-node system is 0.3 and 0.33s respectively, which is far lower than the optimization time required by the immune algorithm. The superiority of genetic algorithms in solving the global optimal combination problem in this paper is verified. The method presented in this paper can realize the online identification of grid topology structure and has a wide range of engineering application value.

5. Conclusion

Aiming at the current situation of topology error detection difficulty of distribution network, this paper proposes a distribution network topology identification method based on branch active power. A topology recognition method of distribution network based on combined branch active power and GA is proposed. The branch active power residual algorithm is used to identify the topological structure. The topology obtained by this method has the highest matching degree with the real-time measured data. Numerical examples show that this method can effectively identify the accurate topology structure of the distribution network, and have good adaptability in the case of a large number of measurement errors. It is simple and fast to realize, especially suitable for a small scale distribution network.

Acknowledgements

This work was supported by the Science and Technology Project of State Grid Liaoning Electric Power Co., Ltd. (Project No. 2021YF-76).

References

[1] Ahmad F and Rasool A 2017 *Successive-over-relaxation based recursive Bayesian approach for power system configuration identification* Vol 36 (The international journal for computation and mathematics in electrical and electronic engineering) pp 1043-1058

[2] Zhu H and Giannakis G B 2011 *Estimating the state of ac power systems using semidefinite programming* 43rd North America Power Symp (NAPS) pp 1–7

[3] Lavaei J and Low S H 2012 *Zero duality gap in optimal power flow problem* vol 1 (*IEEE Trans Power Syst*) pp 92–107

[4] Butler P K L and Sarma N D R 2004 *Self-healing reconfiguration for restoration of naval shipboard power systems* vol 2 (IEEE Transactions on Power Systems) pp 754-762

[5] Malini Ghosal 2018 *European Control Conference* (ECC) pp 1-6

[6] Liu Di, Zhang Qiang, Lu Ganyun, Wu Jialu, Xu Huimin 2021 *Power Engineering Technology* pp 92-98

An optimization method of clock synchronization for large-scale regional power network based on IEEE 1588

Yue Zuo[1], Xingcai Wang[1*], Bo Zhang[1]

[1]State Grid Huludao Electric Power Supply Company, Huludao, Liaoning 125000, China

*1142101056@ncepu.edu.cn

Abstract—At present, mobile devices generally use GPS, Beidou and other satellite time service methods to obtain time, but the clock synchronization based on IEEE 1588 protocol still has deviation. To solve this problem, a clock synchronization method is proposed to improve IEEE 1588 protocol. Based on the analysis of IEEE 1588 protocol, the clock deviation and frequency deviation which affect the synchronization accuracy are modeled. The second-order Kalman filtering algorithm is used to recursively deduce the clock deviation and frequency deviation, and the Allan variance is used to verify the noise characteristics and constantly correct the clock deviation. Finally, the improved effect is verified by relevant experiments. The results show that the improved system can improve the synchronization accuracy.

1. Introduction.

As the research on Beidou satellite navigation technology started late in China, and the relevant technologies are not mature enough at present, the navigation chips in China are purchased through domestic agents of foreign manufacturers, and the core technology has not been formed for a long time, which also limits the development of China's navigation system[1]. At this stage, China's power enterprises mainly rely on GPS time synchronization technology, and satellite time service products are mainly single products, with only a small number of dual system products, Beidou three system products and single Beidou products[2]. However, limited by the United States and free of charge, it has low reliability and poor autonomy[3]. With the development and improvement of Beidou satellite navigation system in China, and in order to avoid the incalculable serious consequences to the power industry, many domestic enterprises and scholars have carried out the research on Beidou satellite navigation technology and its related products.

The accuracy of clock synchronization largely determines the performance of distributed network measurement and control system, so clock synchronization is very important for distributed network measurement and control system. IEEE 1588 clock synchronization protocol is based on network clock synchronization[4]. Assuming that the network line between master and slave clock modules has equal delay, the synchronization deviation correction of slave clock relative to master clock is completed by calculating clock deviation. In many network environments, the assumption of peer delay of master-slave clock lines can not be established, which leads to large deviation of synchronization accuracy[5]. The root cause of clock deviation is frequency deviation. The protocol only compensates the clock deviation of master-slave clock, but not the frequency deviation.

The time service function of Beidou system is to broadcast the time service information regularly by the ground central control system to provide the time service user with the time delay correction value. According to the time delay correction value, the time service user continuously adjusts the local clock to align it with Beidou standard time, or control the time difference within a certain range. Beidou user machine provides two time service modes: vehicle direction time service and two-way time service.

Based on the deep analysis of IEEE 1588 protocol, this paper models the clock deviation and frequency deviation which affect the synchronization accuracy. Second order Kalman filtering algorithm is used to recursively deduce the clock deviation and frequency deviation, and Allan variance is used to verify the noise characteristics and constantly correct the clock deviation. The clock synchronization accuracy of the improved method is tested through experiments to verify the effectiveness of the improved method.

2. IEEE 1588 protocol.

IEEE1588 protocol is also called precise synchronization clock protocol, that is, PTP (Precision Time Protocol). PTP system is a distributed network structure, and the logical range of protocol operation is called a domain. PTP system consists of one or more PTP subdomains. In a PTP system, the grandmaster clock provides the time source of the whole system. The implementation protocol only needs to add time synchronization messages to the original network, which only takes up a small amount of network resources. By calculating the network transmission delay accurately, the time deviation from the clock to the master clock is calculated, and then the local time is adjusted to match the time of the clock source.

IEEE1588 synchronization system is a system composed of publisher and receiver. During the operation of the system, the master clock acts as the time publisher, publishes the local time to the network every other period of time, and the slave clock receives the time according to its own domain and priority. At the same time, the line delay is calculated from time to time to ensure accurate synchronization according to the network situation.

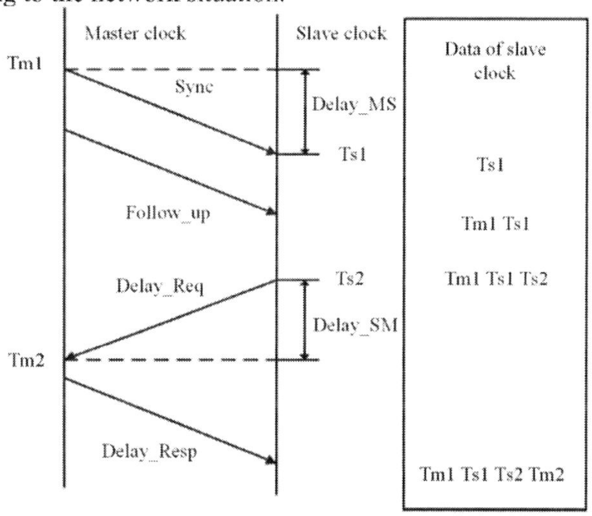

Fig.1. Schematic diagram of IEEE1588 synchronization

The process of IEEE 1588 clock synchronization is as follows: (1) the master clock sends a sync message to the slave clock every certain time (sync interval). The master node records the sending time stamp TM1, but TM1 is not sent with sync. The slave clock records the time stamp TS1 when it receives sync. (2) The master sends a follow to the slave node_ Up message containing the TM1 timestamp mark recorded the previous time sync was sent (3) Send delay from node to master_ The request message and the timestamp TS2 at the time of the message are recorded. The time stamp TMS2 is recorded after the master node receives the message. (4) The master node returns a delay-response to the slave node. This message contains the TM2 mark of the previous record. After the slave node receives this message, it uses the TM1, TS1, TS2, TM2 to calculate the time delay between it and the master node. The specific synchronization process is shown in Fig.1.

The synchronization calculation process includes delay value calculation and offset value calculation. The delay value is the average value of message transmission delay between slave nodes, and the offset value is the time deviation between slave nodes. The calculation formula is as follows:

$$Delay = \frac{delayMS + delaySM}{2} = \frac{Ts1 - Tm1 + Tm2 - Ts2}{2} \tag{1}$$

$$offset = \theta(k) = Ts1 - Tm1 - Delay = \frac{Ts1 - Tm1 + Tm2 - Ts2}{2} \tag{2}$$

When the clock offset value between master clock and slave clock is obtained, the time or frequency can be further adjusted by servo control system, and finally the adjusted slave clock and master clock are nearly synchronized.

3. Improved clock synchronization method.

Aiming at the problem that IEEE 1588 only compensates the clock deviation of master-slave clock, this paper establishes the state equation and measurement equation of clock deviation and frequency deviation, and uses Kalman filtering method to recursively correct the clock deviation, so as to reduce the deviation statistics and improve the synchronization accuracy. Kalman filtering algorithm is suitable for the situation that process noise and measurement noise are white noise. Therefore, Allan variance is used to test the noise characteristics according to the experimental data.

Kalman filtering algorithm was proposed by Kalman in 1960, and its discrete filtering and prediction process is shown in Fig.2.

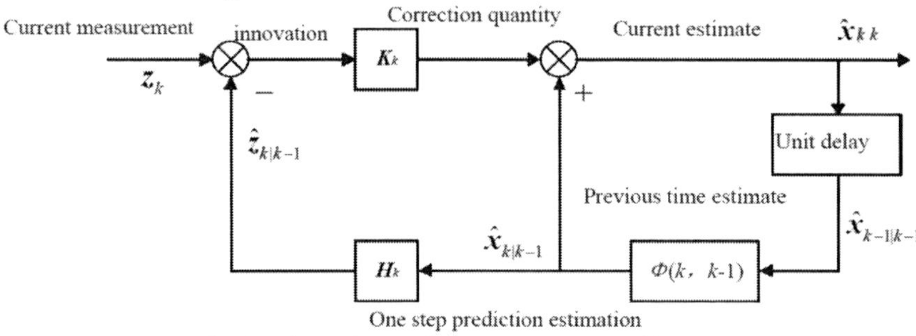

Fig. 2. Flow chart of discrete Kalman filter

3.1 State transition equation model of crystal oscillator clock

Rewrite the ground cabinet expression of the crystal oscillator clock mathematical model in the second-order state into the following form:

$$\theta(k+1) = \theta(k) + \gamma(k)\tau_k + \omega_\theta(k) \tag{3}$$

$$\gamma(k+1) = p \cdot \gamma(k) + \omega_\gamma(k) \tag{4}$$

where K represents the k-th update of the oscillator clock, which $\theta(k)$ is actually to correspond to the continuous update of the offset, that is, to correspond to the clock offset rate. When the physical clock is continuously updated from the k-th time to the K+1st time, the time interval of the clock τ_k is updated to the corresponding number of times. It can be seen from the above formula that in an asymmetric sensor network environment, the time value of the clock, that is, the state of the crystal oscillator clock, can be described by the clock skew and offset. In other words, the parameter variables in the state model of the crystal oscillator clock can be regarded as the clock offset or offset rate.

Equations (3) and (4) can be changed into vector form to express the clock of the oscillator. Therefore, the basic state transition equation in the process of clock change can be obtained, which is described by the above vector equation.

3.2 Observation equation model of PTP time synchronization protocol

The true value of clock offset at the nth time synchronization is:

$$\theta(n)=\frac{[t_2(n)-t_1(n)]-[t_4(n)-t_3(n)]}{2}+\frac{d_{sm}(n)-d_{ms}(n)}{2} \tag{5}$$

In the asymmetric wireless network environment, there are serious differences between the actual wireless network and the theory, so the propagation delay obtained according to PTP is different from that in reality. The clock calculation value of PTP is only an observation value of the clock offset in the real environment, not a synchronization value. In order to more conveniently describe the difference between the calculated value of the clock offset and the real value 0, the calculated value of PTP can be measured in the research and regarded as an observation value of clock synchronization.

Therefore, in the nth update of time synchronization, the calculated value of clock offset is as follows:

$$\theta_M(n)=\frac{[t_2(n)-t_1(n)]-[t_4(n)-t_3(n)]}{2} \tag{6}$$

Comparing the above two formulas, it is not difficult to find that θ is equal to 0 only when d_{sm} is equal to d_{ms}. Because a variety of factors affect the clock synchronization, these factors mainly include the asymmetric delay of the clock, the processing factors to obtain the clock delay parameters and so on. Then the real value of time offset 0 and the estimated value of clock offset are different in the real environment. There is a difference between them.

$$\Delta \mathrm{d}(n)=\frac{d_{\mathrm{sm}}(n)-d_{\mathrm{ms}}(n)}{2} \tag{7}$$

Therefore, from the essence of clock synchronization, the calculated value of clock offset actually has an observation error of θ_M, which is an observation error corresponding to the deviation of the real value.

(3) Optimal modeling of time synchronization based on Kalman filter

Vector $\mu(n)=\begin{bmatrix}\mu_\theta(n) & \mu_\gamma(n)\end{bmatrix}^{\mathrm{T}}$ is introduced to correct the clock. The clock correction process can be described by mathematical formula as follows:

$$\begin{cases}\theta(n)=\theta(n-1)-\mu_\theta(n-1)\\ \gamma(n)=\gamma(n-1)-\mu_\gamma(n-1)\end{cases} \tag{8}$$

Therefore, the new clock state transition equation with clock correction output is:

$$x(n)=Ax(n-1)+B\mu(n-1)+\omega(n) \tag{9}$$

Therefore, the Kalman filter iterative algorithm is described as:

$$\hat{x}(n|\ n-1)=A\hat{x}(n-1)+B\mu(n-1) \tag{10}$$

4. Analysis of clock synchronization test results

Through the back-to-back test, the synchronization accuracy is analyzed, and the advantages of IEEE 1588 protocol clock synchronization method after the improved algorithm are verified.

The clock synchronization test process starts to record data 2 minutes after the system is powered on. At this time, the clock synchronization system enters a basically stable state, and its waveform data has reference value. The clock synchronization error is updated once every second, and the waveform data of 2 h is selected for recording time. The clock deviation recorder is used to record the synchronization deviation waveform and generate the deviation curve and deviation distribution histogram

The waveform data of back-to-back test of the improved algorithm is shown in Fig.3 and Fig.4. The clock deviation at any time is about 79 ns. The deviation curve is shown in Fig.3, and the deviation distribution histogram is shown in Fig.4. From the deviation distribution histogram, it can be seen that the main deviation distribution is in the range of (-20 ns, 20 ns).

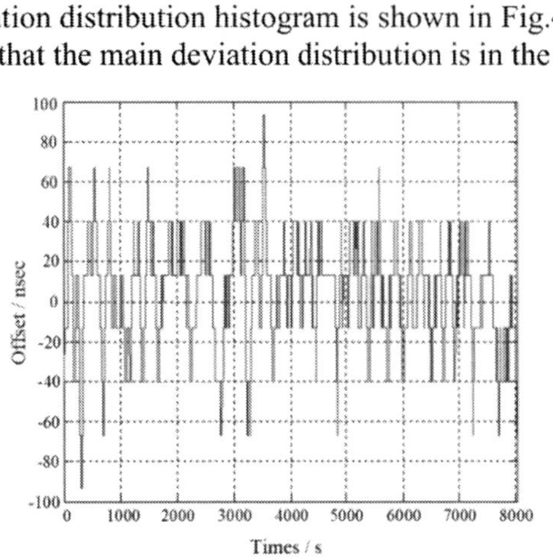

Fig.3. Deviation curve

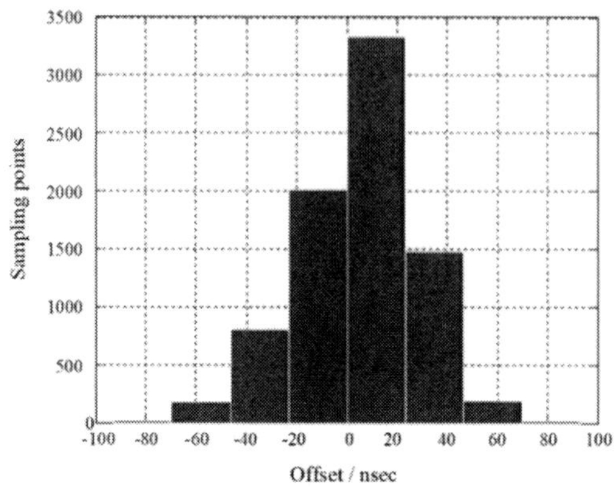

Fig.4. Deviation distribution

5. Conclusion.

Aiming at the problem that IEEE 1588 protocol's equal delay assumption is easy to cause large synchronization error in complex network environment, this paper analyzes the factors that affect the synchronization accuracy, establishes the second-order Kalman state estimation model of relative frequency deviation and frequency drift, and estimates the noise characteristics of the state estimator by using Allan variance. The hardware design of the clock synchronization module is implemented. The synchronization precision of the improved synchronization protocol meets the requirements of high precision clock synchronization.

Acknowledgement
This work was supported by the Science and Technology Project of State Grid Corporation of China (Project No. 2021YF-37).

References

[1] Qian, W., Chao, G ., Mei J . (2013) Research of the Precision Clock Synchronization Based on IEEE 1588. Journal of Software, 8(2):451-458.
[2] Dong, Q., Shao, P. (2018) Error analysis and improvement measures of single-phase power supply method to measure the impedance parameters of single-circuit asymmetric line. Electrical Measurement & Instrumentation.
[3] Cho, J. H., Kim, H., Wang, S., Lee, J., Lee, H. (2009). A novel method for providing precise time synchronization in a distributed control system using boundary clock. IEEE Transactions on Instrumentation & Measurement, 58(8): 2824-2829.
[4] Giorgi, G., Narduzzi, C. (2011). Performance analysis of kalman-filter-based clock synchronization in ieee 1588 networks. Instrumentation & Measurement IEEE Transactions on, 60(8): 2902-2909.
[5] Ruiqing, P., Peng, H., Wenxue, Y., Min, G., Bin, Z. (2010). The optimization techniques for time synchronization based on ntp. IEEE, V2-296-V2-299.

Capacity configuration optimization and analysis for multi-energy complementary microgrid in hydropower station considering the renewable energy accommodation

Yang Gao[1a], Bin Liu[1b], Jinyu He[1c*], Kai Zhang[1d], Li Liu[1e], Li Jin[2f*], Juguang Ren[2g]

[1]Nationality Skill Training Centre of State Grid Sichuan Electric Power Company, Chengdu 611133, China

[2]Key Laboratory of Fluid and Power Machinery, Ministry of Education (Xihua University), Chengdu 610039, China

[a]chengdudaxueaa@163.com, [b]1187138354@qq.com, [c*]15882101586@163.com, [d]1084756540@qq.com, [e]292275785@qq.com, [f*]jinli@stu.xhu.edu.cn, [g]renjuguang@stu.xhu.edu.cn

Abstract—Aiming at the problems of large short-term fluctuation, strong randomness and difficulty to consume of renewable energy generation, it is an important way to integrate abundant wind and photovoltaic (PV) power into hydropower stations to promote the consumption of renewable energy sources (RES). A hydro-wind-PV and energy storage multi-energy complementary microgrid (MECM) model is proposed to meet the demand of load supply and RES consumption. Firstly, according to the characteristics of load and resource endowment, the MECM is established in a hydropower station. Secondly, according to the two different objectives of load self-sufficiency and RES consumption, the mathematical model for the capacity configuration optimization of MECM was proposed. Finally, combined with the practical case study, the effectiveness and practicability of this scheme are verified, and different configuration schemes are discussed based on different objectives.

1. Introduction

The traditional hydropower stations only have hydropower as a single form of energy supply [1]. Energy supply varies with weather conditions and seasons [2]. Therefore, WT and PV can be introduced into traditional hydropower stations. The regulation ability of hydropower can overcome the intermittent and unstable disadvantages of WT and PV power output [3-4].

At present, many scholars combine hydropower, WT and PV power to establish a MECM model with hydropower as the main part on the power generation side [5-8]. However, they did not consider the linkage with the load side. With the increasing demand of local load and the emergence of new-type loads, provided a new solution for the consumption of RES [9], considering the linkage with the load side can effectively solve the problem of RES consumption.

In this paper, with a study on the output characteristic and complementary relationship of the hydro, wind and PV, a capacity configuration optimization method of MECM is proposed, which takes the load self-sufficiency and RES consumption as the objective function. Using hydro-wind-PV coordinated operations to achieve RES complement each other, the introduction of pumped storage increases power supply reliability, access to the EV, data center and other new-type loads increase the local consumption ability of RES.

2. System structure and analysis of energy supply mode of MECM

2.1. System structure of MECM

The MECM system designed is shown in Fig. 1, which takes hydropower as the main part and is composed of WT, PV, pumped storage, EV charging station, data centre and other parts.

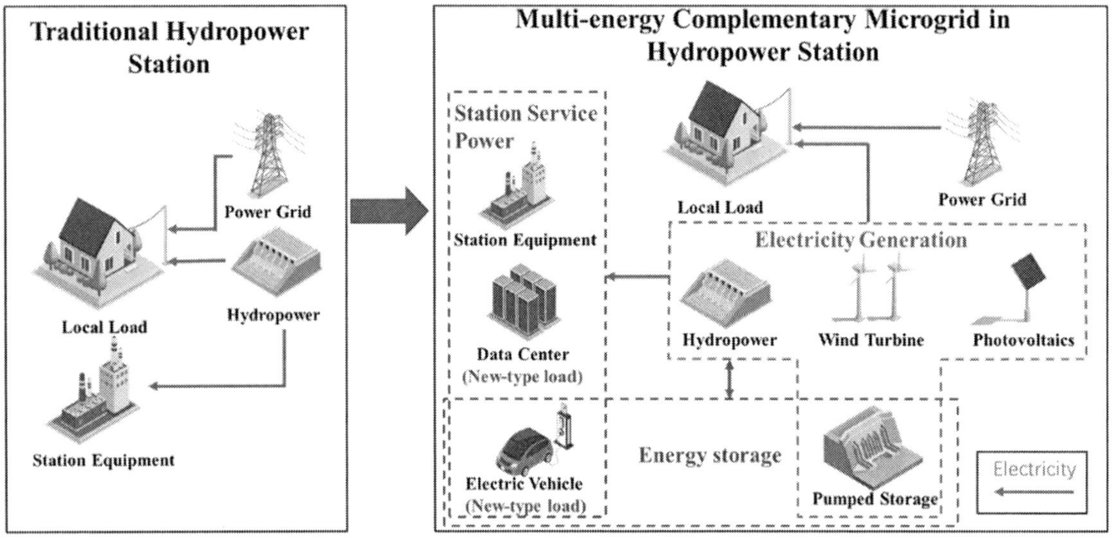

Fig.1 System structure of traditional hydropower station and MECM

2.2. Operation characteristics and operation mode analysis of MECM

The energy supply mode of the MECM is mainly hydro, WT and PV complementary operation, supplemented by peak-cutting and valley-filling of pumped storage [10]. According to the load characteristics and the output characteristics of RES, the whole day can be divided into three periods.

(1) 1:00-12:00 time period: the output of PV and hydropower units increases continuously, while the output of WT increases first and then decreases. The total output power of RES is high, and the load demand is low. Prioritize WT and PV power supply, and then hydropower is used to supply the remaining load, and the power is stored by pumped storage.

(2) 12:00-19:00 time period: the output of PV and hydropower is decreasing, and the output of WT is low. The total RES output is low, but the load demand is high. Prioritize WT and PV power supply, and then hydropower is used to supply load, the pumped storage generates and EVs discharge to meet the remaining load. If the load demand is still unable to meet, electricity can be purchased from grid.

(3) 19:00-24:00 time period: the energy supply mode of this period is the same as that of the 1:00-12:00 time period, but due to the higher load during this period, RES abandonment is relatively small.

3. Methodology

According to the system structure of MECM and the analysis of system operation characteristics, the equipment model is constructed, and the objective function and constraints are set for the two objectives of load self-sufficiency and RES consumption, in which evaluation index are listed to evaluate the economy and effectiveness of the two schemes finally.

3.1. Equipment model

(1) Power generation equipment model

The main generating equipment of MECM is hydraulic generator, WT and PV. The model of three generating equipment is showed as follows:

$$P_{hy} = 9.81\eta_{hy} Q_d H_d \tag{1}$$

$$P_w = \begin{cases} 0 & v < v_{ci}, v \ge v_{co} \\ \frac{P_R}{v_R^3 - v_{ci}^3} v^3 - \frac{P_R}{v_R^3 - v_{ci}^3} v_{ci}^3 & v_{ci} \le v < v_R \\ P_R & v_R \le v < v_{co} \end{cases} \tag{2}$$

$$P_{pv} = \alpha I_T \tag{3}$$

Where, P_{hy} is the output power of the turbine; H_d is the design head; Q_d is power generation discharge; η_{hy} is the efficiency of hydraulic turbine, which is 0.9; P_{pv} is output power, I_T is irradiation intensity, and α is the conversion coefficient of PV power; P_w is the output power of the WT; v is the actual wind speed of the WT; v_{ci}, v_{co} and v_R are respectively the cut to the wind speed, cut out of the wind speed and rated wind speed of the WT. P_R is the rated power of the WT.

(2) Storage model

The water volume of the upper reservoir after storage and discharge of the pumped storage power station is expressed as follows:

$$W(t+1) = W(t) + \frac{3600 \times 1000 \eta_p \eta_{wp} P_p(t)}{\rho g H} - \frac{3600 \times 1000 P_h(t)}{\eta_h \eta_{wp} \rho g H} \tag{4}$$

Where, $W(t)$ is the remaining water of the reservoir at the time t; η_p and η_h are respectively the pumping efficiency and power generation efficiency of the hydro-generator set, taking 80% and 90% respectively; η_{wp} is pipeline efficiency, which is 95%; $P_p(t)$ and $P_h(t)$ are respectively the pumping power and power generation power of the hydro-generator set; H is head height;

3.2. Configuring the objective functions of optimization

Objective 1 is that the load is mainly supplied by MECMs, and the purchase of electricity from large power grids is minimized. Objective 2 is that the MECM system can fully consume its own renewable energy generation, and the power abandonment is minimized.

$$\min F_{zj} = \min \sum_{t=1}^{24} P_b(t) \tag{5}$$

$$\min F_{loss} = \min \sum_{t=1}^{24} P_{loss}(t) \tag{6}$$

Where, $P_b(t)$ is the power purchased from the power grid; $P_{loss}(t)$ is power abandonment.

3.3. Configuring the constraints of the optimization

The constraint conditions are as follows: Eq. 7 and Eq. 8 are power constraints, Eq. 9 and Eq. 10 are water flow constraints, and Eq. 11 and Eq. 12 are EV charge and discharge constraints.

$$P_{hy}(t) + P_w(t) + P_{pv}(t) + P_h(t) + P_{evf}(t) + P_b(t) = P_L(t) + P_{evc}(t) + P_p(t) + P_{loss}(t) \tag{7}$$

$$P_{i,\min} \le P_i(t) \le P_{i,\max} \tag{8}$$

$$V_{cap,\min} \le W(t) \le V_{cap,\max} \tag{9}$$

$$W(\text{start}) = W(\text{end}) \tag{10}$$

$$E_{c,\min} \le P_{evc}(t) \le E_{c,\max} \tag{11}$$

$$E_{f,\min} \le P_{evf}(t) \le E_{f,\max} \tag{12}$$

Where, $P_L(t)$, $P_p(t)$ and $P_{evc}(t)$ are respectively the power that the wind power supplies to the local load, pumped storage pumping and EV at time t ; $P_{i,\min}$ indicates the lower limit of output power.; $P_{i,\max}$ indicates the upper limit of output power; $V_{cap,\min}$ is the lower limit of water storage capacity; $V_{cap,\max}$ is the upper limit of water storage capacity; $E_{c,\max}$ and $E_{v,\min}$ are the maximum and minimum value of EV charge. $E_{f,\max}$ and $E_{f,\min}$ are the maximum and minimum value of EV discharge.

3.4. Evaluation index

Self-sufficiency rate refers to the percentage of station power load and local load supplied by hydropower stations in total load. The power abandonment rate is equal to the power abandonment divided by the total generation of RES The cost of MECM configuration for hydropower station can be divided into investment cost (C_{ic}) and operation and maintenance cost (C_{om}).

$$L_{zj} = \frac{\sum_{t=1}^{24} P_{hy}(t) + P_w(t) + P_{pv}(t)}{\sum_{t=1}^{24} P_L(t) + P_c(t)} \tag{13}$$

$$L_{loss} = \frac{\sum_{t=1}^{24} P_{loss}(t)}{\sum_{t=1}^{24} P_w(t) + P_{pv}(t) + P_{hy}(t)} \tag{14}$$

$$C_{ic} = C_w N_w + C_{pv} N_{pv} + C_{cap} N_{cap} \tag{15}$$

$$C_{om} = C_{om,w} N_w + C_{om,pv} N_{pv} + C_{om,cap} N_{cap} + C_{om,hy} N_{hy} + P_b C_{grid} \tag{16}$$

Where, $P_c(t)$ is the station power load; C_w , C_{pv} , C_{cap} are the unit prices of each unit of WT, PV, pumped storage respectively; $C_{om,hy}$, $C_{om,w}$, $C_{om,pv}$, $C_{om,cap}$ are the operation and maintenance costs of hydraulic turbine, WT, PV, pumped storage unit; N_{hy} , N_w , N_{pv} , N_{cap} are the unit capacity of hydraulic turbine, WT, PV, pumped storage unit; C_{grid} is the electric price purchased from the grid.

4. Case study

This study takes a hydropower station in China as a case, which contains two 1250kW hydroelectric generating units and a reservoir of 4000 m^3. The load demand and the typical daily hydro, WT and PV output per kW capacity as shown in Fig. 2. The power purchase cost of the MECM system is 0.54 CNY/kWh. Related parameters of other equipment are shown in the following table1.

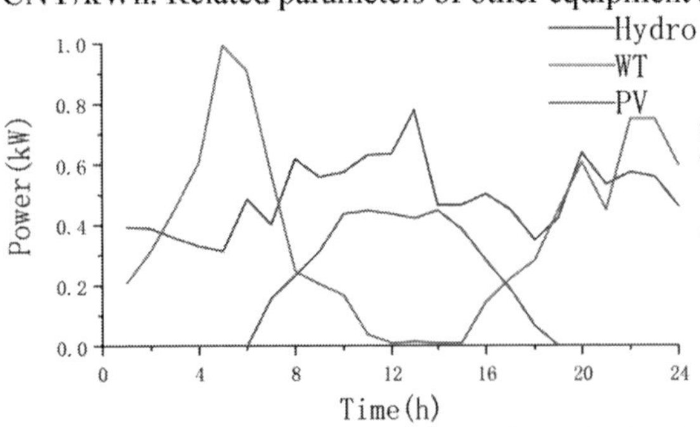

(a) Output per kW capacity of Hydro, wind and PV

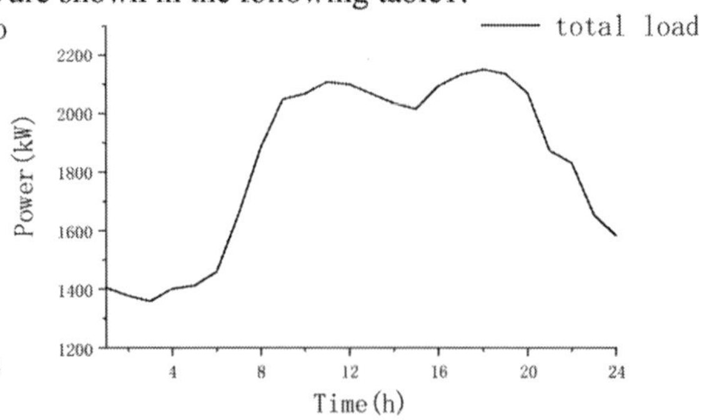

(b) Load demand

Fig. 2 Hydro-wind-PV power system output and load demand

Table 1: Equipment price parameters

Equipment	Construction cost (CNY/kW)	Operation and maintenance cost (CNY/kW)	Life cycle (Years)
WT	1695	51	20
PV	1040	9	25
Reversible pump turbine	2100	21	15

5. Result Analysis and Discussion

The configuration capacity calculated under the two conditions of energy self-sufficiency and complete local consumption is shown in Tab. 2: Scheme 1 is that loads are supplied as far as possible by MECM; Scheme 2 is to reduce more renewable energy power abandonment of MECM.

Table 2: Configuration optimization results

Equipment	Scheme 1: Capacity (%)	Scheme 2: Capacity (%)
Hydropower Unit	39.16	49.54
WT	24.05	18.90
PV	33.92	25.77
Pumped storage	2.87	5.78

As shown in Tab. 2, in order to reduce power purchase from the power grid, improve the self-sufficiency rate, newly configured WT and PV capacities account for 59.68% of total generating units' capacity. In order to reduce RES power abandonment, the configured WT and PV capacities are relatively small, accounting for 47.42% of the total generating unit capacity. In addition, in order to further reduce power abandonment of RES, more capacity of pumped storage is configured.

Table 3: Performance evaluation analysis of configuration optimization

Scheme	Investment cost (CNY/D)	Operational cost (CNY/D)	Total cost (CNY/D)	Self-sufficiency rate	Power abandonment rate
1	1111.4	890.88	2002.3	98.48%	14.95%
2	780.11	1816.5	2596.6	94.29%	1.83%

As shown in Tab. 3, increasing self-sufficiency rate requires configuration of more RES, the investment cost is relatively high and power abandonment account for 14.95% of the total generated energy. To reduce power abandonment of RES, MECM only supplies 94.29% of the load. Compared to scheme 1, more electricity needs to be purchased from the grid, resulting in higher operating costs.

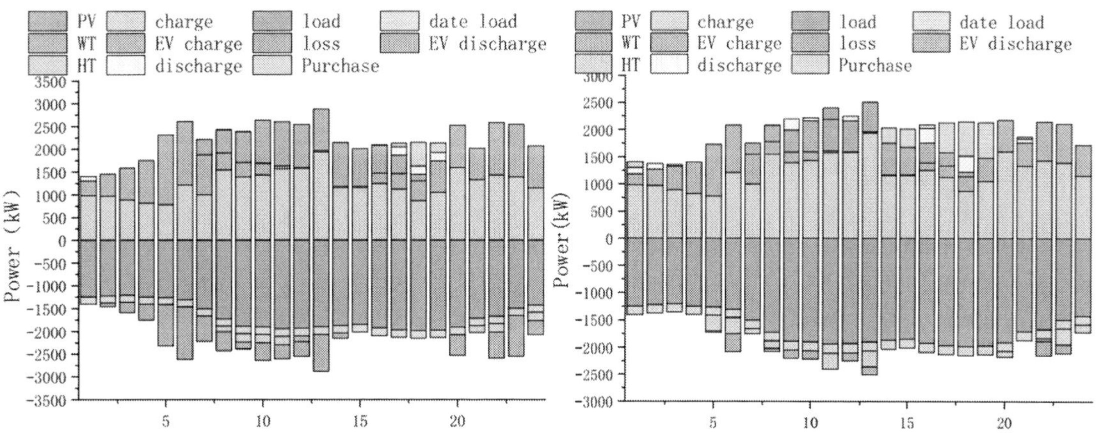

(a) Scheme 1: self-sufficient (b) Scheme 2: complete consumption

Fig.3 Power output of equipment under two configuration objectives

As shown in Fig. 3, scheme 1 has massive power abandonment at 1:00-12:00. Due to the small renewable energy generation and large load at 17:00-19:00, it is necessary to purchase electricity to maintain the balance between supply and demand. In scheme 2, only a small amount of electricity was abandoned at 6:00, 22:00 and 23:00. However, due to the capacity of RES configuration is small, more electricity power would be necessary to be purchased from the grid at 14:00-19:00.

6. Conclusion

In order to improve the RES utilization and power supply reliability, this paper establishes a mathematical model of MECM. According to different optimization objectives of MECM, two different capacity configuration schemes are proposed respectively. Based on the results and discussions presented above, the conclusions are obtained as below:

1) Under the two configurations, the MECM has effectively played the complementary advantages of the three RES and improved the reliability of energy supply while maintaining stable operation.

2) Access to new-type loads can effectively increase the utilization rate of RES and improve the reliability of the energy supply.

In the future work, the influence of multiple climate conditions on the optimal configuration of MECM will be considered.

Acknowledgments

This work was financially supported by the Science and Technology Project of State Grid Sichuan Electric Power Company (2021kjc07051515).

References

[1] Yi Min, Mo Li, Shen Qin. (2020) Study on Generation Scheduling of Cascade Hydropower Stations Based on SAPSO. Journal of Coastal Research, 2020, 104 (sp1).

[2] S. Shivashankar, Saad Mekhilef, Hazlie Mokhlis, M. Karimi. (2016) Mitigating methods of power fluctuation of photovoltaic (PV) sources-A review. Renewable and Sustainable Energy Reviews, 59(Jun.): 1170-1184.

[3] Zifa Liu, Zhe Zhang, Ranqun Zhuo, Xuyang Wang. (2019) Optimal operation of independent regional power grid with multiple wind-solar-hydro-battery power. Applied Energy, 235(FEB.1): 1541-1550.

[4] Yajun Tang, Guohua Fang, Qiaofeng Tan, and etc. (2020) Optimizing the sizes of wind and photovoltaic power plants integrated into a hydropower station based on power output complementarity. Energy Conversion and Management, 206 (2020) 112465: 1-11.

[5] Xuebin Wang, Jianxia Chang, Xuejiao Meng, Yimin Wang. (2018) Short-term hydro-thermal-wind-photovoltaic complementary operation of interconnected power systems. Applied Energy, 229: 945-962.

[6] Xianxun Wang, Yadong Mei, Yanjun Kong, Yuru Lin, Hao Wang. (2017) Improved multi-objective model and analysis of the coordinated operation of a hydro-wind-photovoltaic system. Energy, 134(Sep.1): 813-839.

[7] Shafiqur Rehman, Luai M. Al-Hadhrami, Md. Mahbub Alam. (2015) Pumped hydro energy storage system: A technological review. Renewable and Sustainable Energy Reviews, 44(Apr.): 586-598.

[8] Xiao Xu, Weihao Hu, Di Cao, Qi Huang, Cong Chen, Zhe Chen. (2020) Optimized sizing of a standalone PV-wind-hydropower station with pumped-storage installation hybrid energy system. Renewable Energy, 147: 1418-1431.

[9] Xinzhou Li, Yitong Tan, Xinxin Liu, and etc. (2020) A cost-benefit analysis of V2G electric vehicles supporting peak shaving in Shanghai. Electric Power Systems Research, 179 (2020) 106058: 1-9.

[10] Aeidapu Mahesh, Kanwarjit Singh Sandhu. (2015) Hybrid wind/photovoltaic energy system developments: Critical review and findings. Renewable and Sustainable Energy Reviews, 52: 1135-1147.

Application of Trusted Computing Technology in Active Defense of Smart Substation

Liandong Chen[1], Tao Yang[2*], Gang Li[2], Xin Liu[3], Ning Lu[3], Kai Cheng[2], and Xiaopeng Xin[2]

[1]State Grid HeBei Information and Telecommunication Branch, Shijiazhuang, Hebei, China

[2]North China Electric Power University/School of Control and Computer Engineering, Baoding, Hebei, China

[3]State Grid HeBei Electric Power Co. Ltd, Shijiazhuang, Hebei, China

[*]Corresponding author's e-mail: lg_seek@ncepu.edu.cn

Abstract—Nowadays, most security systems mainly use firewall, intrusion detection, virus killing and other technologies to defend against external attacks. This passive defense mode has low efficiency and high false positive rate, and active defense technology can solve these problems to the greatest extent. At present, the intelligent substation is in the development stage with weak security. After a large number of terminals are connected, although the operation efficiency of the whole station is improved, it also brings more risks. This paper takes the access security of terminal equipment in intelligent substation as the research object. By mining the security vulnerabilities of substation, an active defense method based on trusted computing is proposed, the applicability and efficiency of trusted computing technology in the field of substation active defense are proved and the effectiveness of the method is verified to realize the trust guarantee of the whole process of terminal access to intelligent substation.

1. Introduction

In recent years, cyber attacks have occurred frequently. In the "Earthquake Network" virus incident in 2010, a large number of nuclear power plants were paralyzed; The power outage in Ukraine in 2015 reflected the actual power of hackers attacking the power grid. It was a power control system paralyzed by taking the terminal as the attack springboard[1], which became the world's first publicly reported large-scale power outage. As an important national energy conversion facility, intelligent substation has weak protection measures for a large number of terminal equipment connected, so it can become the object of attack. Terminal equipment plays an important role in substation and is full of crisis. Terminal equipment has already become the key target and main springboard to attack substation.

Intelligent substation has the characteristics of Cyber physical system(CPS). The close coupling between information system and physical system makes the asset information and real-time data of substation face the risk of theft. Network attacks can cause data leakage or data tampering, resulting in the risk of invading the substation with the terminal.

The research on trusted computing has become an international research hotspot. The National Science Foundation (NSF) focuses on how to build and manage the security and privacy of the network; Microsoft has also proposed a 10-year "trusted computing" strategy, with the goal of making computers more secure and stable[2]; China also attaches great importance to the research of system credibility.

There are relevant topics of credibility in the National Natural Science Foundation and national key basic research projects. In 1995, on FTCS-15, IEEE fellow, Professor A.Avizienis and others proposed the concept of trusted computing[3], and implanted a security chip on the PC hardware platform to improve the security of terminal equipment when accessing the system.

In recent years, trusted computing technology has been widely used in the field of power security. Literature[4]uses trusted computing to build a multi-agent active defense framework to better protect substation communication in Wan; Literature[5]combines trusted computing with immune system theory to build a trust chain transmission mechanism suitable for power platform, which greatly improves the security of power grid dispatching control system; According to the idea of trusted network connection, document[6]analyzes the behavior of power grid users and determines their continuous credibility, so as to realize the whole process security control of terminal access; In reference[7], considering the Intranet environment of terminal operation, a terminal security inspection and access control system is designed. The trusted access control of terminal is realized by checking and repairing the operating environment of terminal equipment.

In this paper, the hierarchical structure and data flow of intelligent substation are studied by using the information physical system theory, and the vulnerability of intelligent substation under the extensive use of terminal equipment is analyzed; Then, based on trusted computing technology, the terminal trusted access process is constructed; Finally, the performance of trusted computing platform is tested by simulating the corresponding experimental environment, and the effectiveness of this technology for active defense of intelligent substation is verified.

2. Safety problems of intelligent substation under CPS

2.1. Structure and data flow of Intelligent Substation

Smart substation is an important pillar in the construction of smart grid, which can realize power conversion and transmission and distribution control. However, the cost of high intelligence is to face many security threats, Because the asset information and real-time data of substation can bring great benefits to attackers, once leaked, the consequences are difficult to estimate, so it is necessary to strictly ensure the information security of substation.

The data transmission of intelligent substation mainly depends on the network between layers. In order to ensure the transmission efficiency and security, different networks use different transmission protocols. The data transmitted between layers and the transmission mode used are shown in Tab. 1. It can be seen that the division of work at each layer of substation is clear, which is also the basis for the stable operation of substation.

Tab.1 Data flow of Intelligent Substation

Data flow direction	Data content	Transmission mode
Process layer to interval layer	Merge unit voltage and current sampling values, circuit breaker switching status, diagnostic information, etc	SV
Interval layer to process layer	Trip command, control command, status information, etc	GOOSE
Bay layer to station control layer	Equipment protection data, etc	MMS

2.2. Communication mode of intelligent substation under CPS

Since the primary equipment and secondary equipment of intelligent substation are perfectly corresponding to CPS, the communication mode of substation can be analyzed by CPS theory. CPS realizes the close combination and deep cooperation between information system and physical system, and can realize the real-time perception of large-scale system operation state and the dynamic control of interactive process. Intelligent substation has CPS characteristics to a great extent, such as the combination of virtual and real energy flow and information flow, and the dynamic interaction of data generated by primary equipment and secondary equipment. Although CPS has not been fully integrated into the intelligent substation, it still improves the working efficiency of the equipment to a certain extent and ensures its stable operation.

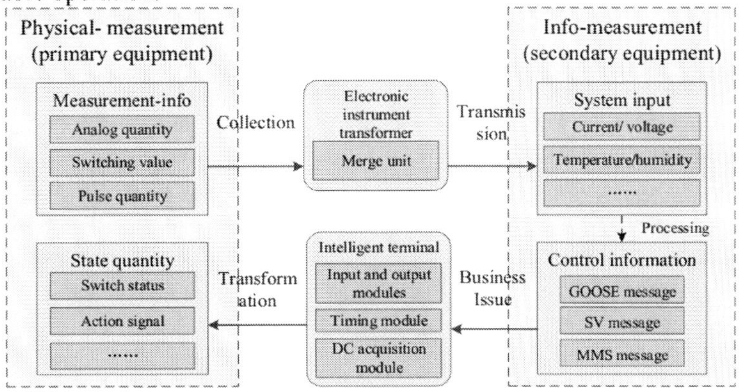

Fig.1 Communication mode of intelligent substation under CPS

As shown in Fig.1, CPS communication mode of intelligent substation is divided into physical measurement and information measurement. Physical measurement corresponds to primary equipment and information measurement corresponds to secondary equipment. The measurement information generated by the physical measurement is collected and integrated by the electronic transformer, and then transmitted to the information side as its system input. These data form control information through processing and conversion. The control information is in the form of message, and the control information is sent to the intelligent terminal. The intelligent terminal converts several types of messages to form the state quantity of the equipment on the physical side.

The operation state of the primary equipment determines the system input of the secondary equipment, and the instructions of the secondary equipment determine the operation state of the primary equipment[8], showing a strong coupling relationship. Although CPS improves the operation stability and communication efficiency of the substation, the frequent interaction between the information side and the physical measuring equipment will also have some impact, mainly as follows: the normal operation of the information side equipment requires the physical measuring equipment to supply power for it, and the safe and reliable operation of the physical measuring equipment depends on the normal operation of the information measuring equipment[9].

2.3. Vulnerability analysis of Intelligent Substation

A large number of network attacks will seriously damage the stability of CPS in intelligent substation. Common attack means include eavesdropping messages, interrupting processes, tampering with data, forging records, etc. the real-time data of substation becomes almost completely transparent to attackers, and its vulnerability is mainly reflected in the following aspects.

(1) Imperfect physical isolation: the level of firewall, anti-virus software and intrusion detection system is too low to resist external attacks;

(2) The security protocol is not involved: the IEC 61850 standard of intelligent substation improves the interoperability between different terminal equipment, but there are few security protocols in terms of terminal equipment access, resulting in many security vulnerabilities in the control message transmitted on the information side[10];

(3) Insecure communication protocol: goose, SV and manufacturing message protocol have low protocol confidentiality in the transmission process, and lack of encryption measures in access control and identity authentication, so the message is easy to be intercepted by attackers.

3. Trusted access model of intelligent substation terminal

3.1. Trusted computing platform for intelligent substation terminal

As the terminal equipment works in the communication module in the substation CPS model, security protection is extremely important, but the current terminal lacks effective isolation protection. After being attacked, it will bring hidden dangers into the substation, so as to steal confidential data or control equipment action. Therefore, it is necessary to ensure that the process of connecting the terminal to the substation is credible. For terminal equipment, if all data interaction and action signals are in normal state in the process of accessing the substation, the terminal is called trusted, and the process can be called trusted access. Trusted computing is one of the core technologies of trusted access.

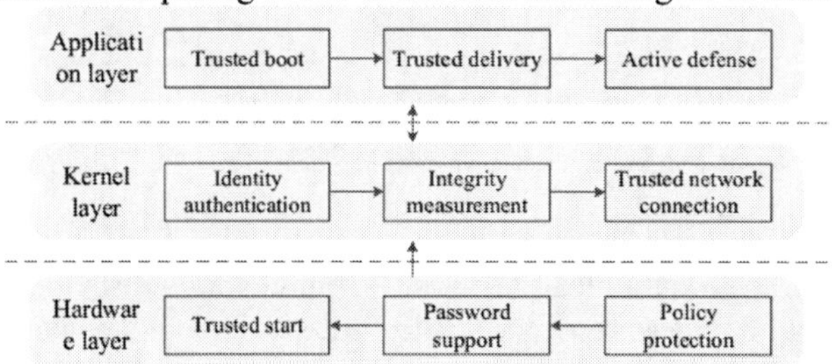

Fig.2 Trusted computing platform structure of Intelligent Substation

As shown in Fig. 2, the trusted computing platform structure of intelligent substation is composed of hardware layer, kernel layer and application layer. Among them, the hardware layer is mainly responsible for the power on of the trusted cipher module, self-test[11]and the generation of measurement root, that is, the trusted start process; The kernel layer has the largest workload. First, the terminal is authenticated to detect its basic information, then the integrity measurement is carried out to ensure the credibility of its running state, and finally the trusted network connection is used to realize the secure communication between the terminal and the network nodes; The work of the application layer is almost carried out at the same time as that of the kernel layer. In this layer, the trusted password module completes the trusted guidance and trust chain transmission to the terminal operating system.

3.2. Trusted access process of terminal equipment

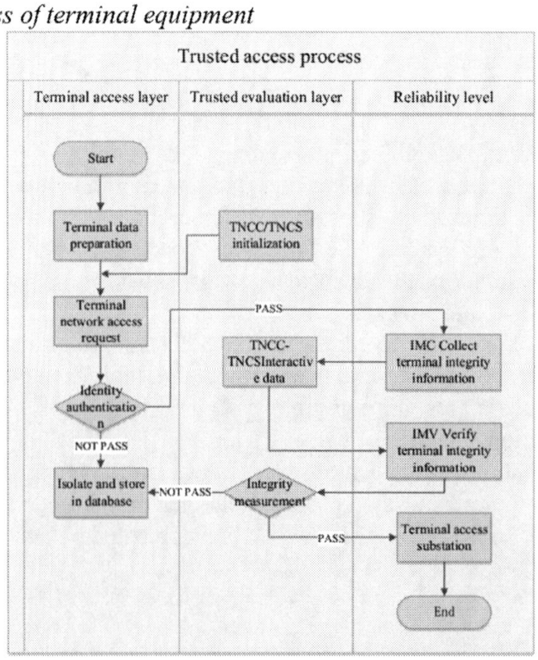

Fig.3 Trusted access process of terminal

As shown in Fig.3, the process of terminal access to substation is divided into the following steps:

(1) The terminal shall prepare the data and sort out the data interacting with the substation;

(2) Trusted Network Connection Client (TNCC) and Trusted Network Connection Server (TNCS) are initialized to detect the connection status of trusted network. If it is in safe status, the terminal can apply for network access;

(3) The terminal accesses the substation access gateway and carries out identity authentication at the same time. If it passes the authentication, it will carry out integrity measurement. If it fails, it will be isolated and marked as a risk terminal and stored in the database;

(4) IMC collects the terminal integrity information and transmits it to TNCC. TNCC interacts with TNCS and processes the integrity information. If the integrity requirements are met, it is transmitted to IMV;

(5) IMV verifies the integrity information of the terminal, and measures the integrity after passing the verification. If the measurement passes, it agrees to the terminal network access request, and the terminal can access the substation. If it passes, it can also isolate it, mark it as a risk terminal, and store it in the database.

4. Experimental analysis

Smart substation is an important part of smart grid and carries most of the power transportation work. Therefore, the CPS system of smart substation has extremely high requirements for the real-time performance of various businesses. Trusted computing technology introduces an integrity measurement mechanism[12]in the operating system, startup and execution links and the loading of executable programs, which is suitable for security inspection during the access of different types of terminals to large-scale systems and active defense against network attacks. By simulating the operation environment of the intelligent substation control system, this paper deploys the terminal server, database, primary equipment, secondary equipment and terminal equipment, and implants the trusted computing platform into the terminal equipment to provide it with trusted password module and trusted software base. The test environment adopts the current mainstream server, and the terminal equipment is installed with a virus-free operating system in advance, Ensure the safety of the initial state.

After the simulation environment is built, the performance test of the intelligent substation control system is carried out. The test items include the CPU utilization of the trusted chip and the data transmission time between the devices in different processes when the terminal equipment is connected to the secondary equipment (measurement and control equipment, metering equipment and protection equipment). After the test, the influence of the trusted computing platform on the two test items in different processing time is analyzed.

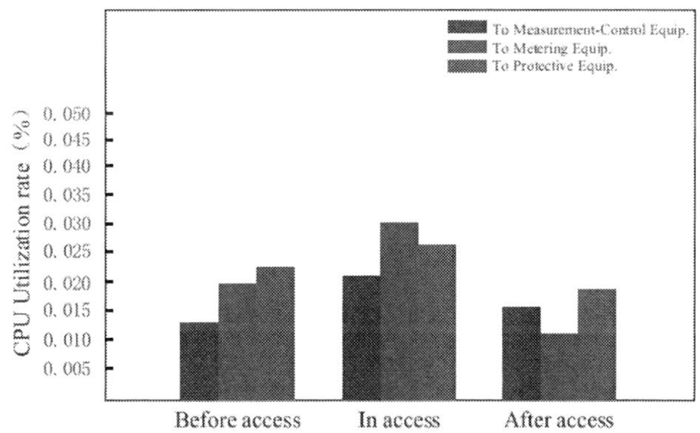

Fig. 4(a) CPU utilization of trusted chip

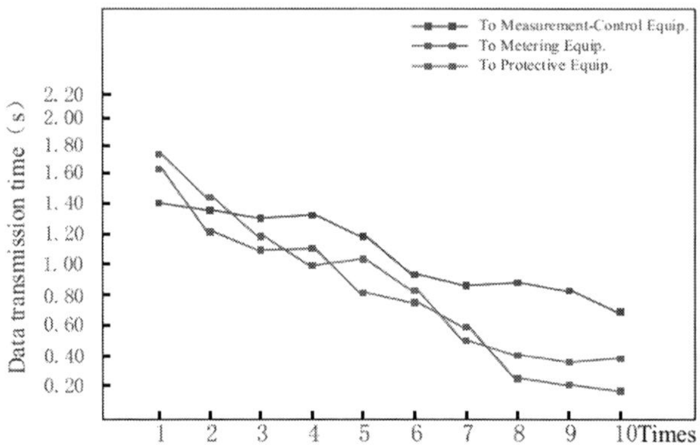

Fig. 4(b) Impact on data transmission time

As shown in Fig. 4(a), when the terminal equipment accesses the secondary equipment, the CPU utilization of the trusted chip is the highest during access, followed by before access and the lowest after access, which indicates that the terminal equipment is most vulnerable to network attacks during access. In order not to damage the experimental equipment, the malicious program is implanted into the terminal device to simulate the network attack.

As shown in Fig. 4(b), with the increase of test times, the data transmission time between the terminal and secondary equipment becos shorter and shorter, because the longer the data transmission time, the more time left for the attacker, and the more vulnerable the terminal is to attack[13]. When the same malicious program is implanted into the terminal device every time, the defense autonomy of the trusted computing platform is improved every time it is tested. Under this effect, the data transmission time is shortened.

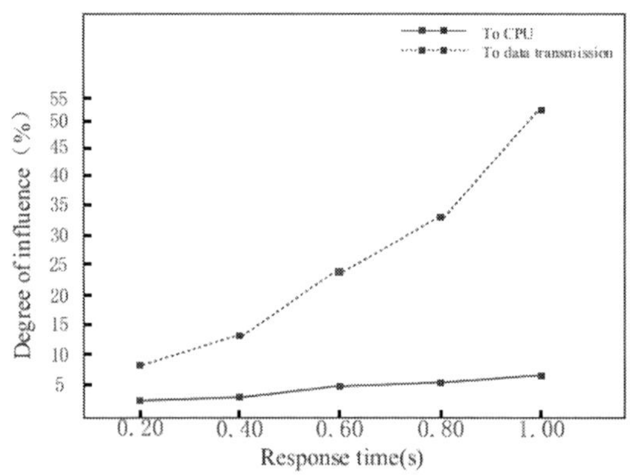

Fig.5 Performance of trusted computing platform

As shown in Fig. 5, the longer the response time, the greater the impact of the trusted computing platform on the trusted chip CPU or data transmission between devices. The curve change rate of CPU impact is small, one is because the simulated network attack power is limited, and the other is because the trusted chip has strong computing power. The curve change rate of data transmission between devices is large because the terminal equipment used in the experiment is low in intelligence. The longer the response time, the greater the probability that the terminal will be attacked, and the more the trusted computing platform will play its role.

5. Conclusion

Trusted computing technology has absolute advantages and broad application prospects in the field of active defense. Especially for large-scale industrial control systems, only active can resist external intrusion to the greatest extent. Due to the shallow research on network security technology, the network attacks simulated in the experimental test in this paper are not strong enough to verify the ability of trusted computing technology to a greater extent, but this can not deny its position in active defense. In the future research, we will continue to improve and strengthen the encryption algorithm to improve its efficiency.

There is a saying on the Internet that "Active defense is essentially equivalent to network attack", because active defense can not only block the attack behavior, but also counterattack the intruder according to the collected effective information, so as to repel the intruder. However, whether these defense methods are legal and whether they have reached a social consensus is a problem.

In short, although active defense is considered to be an effective security strategy, there are still technical problems. In my opinion, in order to make effective use of active defense technology in the future, we not only need to strengthen technology, but also severely crack down on cyber crime and increase the punishment of cyber attackers.

Acknowledgments

This paper is supported by the Science and Technology Project of State Grid HeBei Information and Telecommunication Branch (No.SGHEXT00WLJS2000144). We would like to thank anonymous reviewers for their helpful comments and suggestions.

References

[1] Zhao J H, Liang G Q, Wen F S and Dong Z Y 2016 Automation of Electric Power Systems vol. 40, no. 7, pp 149-151
[2] Long D, Sheng W X, Wang Y, Lu Y M and Sun C 2015 Chinese Society for Electrical Engineering vol. 35, no. 14, pp 3522-3531
[3] Ji H, He Z, He Z, Jiang H, Cai S and Yang T 2019 IEEE 25th International Conference on Parallel and Distributed Systems PP 917-920
[4] Tong X Y 2011 Automation of Electric Power Systems vol. 35, no. 20, pp 53-58
[5] Gao K L, Wang Z H, An N Y and Zhao B H Advanced Engineering Sciences vol. 49, no. 2, pp 28-35
[6] Chen L and Qiao W D 2016 Electric Power Information and Communication Technology vol. 14, no. 4, pp 70-74
[7] Yang G L, Dai X and Mao H D 2013 Computer Engineering and Applications vol. 49, no. 6, pp 109-113
[8] Chen Y R, Li J E, Xu A, Yuan K, Liu K P and Ni P 2021 Automation of Electric Power Systems vol. 45, no. 3, pp 102-110
[9] Chen L J, Mei S W and Chen Y 2012 Control Theory and Applications vol. 29, no. 2, pp 240-244
[10] Zhang Y H, Ni M, Sun Y H and Li M L 2019 Automation of Electric Power Systems vol. 43, no. 21, pp 12-22
[11] Gao K L, Xin Y Z, Li Z, Sun W, Nan G L, Tao H T and Zhao B H 2015 Automation of Electric Power Systems vol. 39, no. 1, pp 48-52
[12] Chen Y Q, Wu X P and Fu Y 2013 International Workshop on Cloud Computing & Information Security vol. 52, no. 1391, pp 336-340
[13] Zhang Z, Zhang W and Wang J 2014 Information and Communication Technology-EurAsia Conference pp 59-69

Journal of Physics: Conference Series **2108** (2021) 012066 doi:10.1088/1742-6596/2108/1/012066

Research on key technology of 3D digitization for secondary system of power grid based on model application analysis

Qingquan LIU [1,*], Tiecheng LI [1], Qian ZANG [1], Jiagnbo REN [2], Yuhao ZHAO [1], Xiaoguang HAO [1]

[1]State Grid Hebei Electric Power Co., Ltd. Electric Power Research Institute, Shijiazhuang, 050000 China

[2]State Grid Hebei Electric Power Co., Ltd, Shijiazhuang, 050000 China

* Corresponding author: jiangzhuoyan@kemov.com

Abstract. In order to solve the problems of long development time, poor expansibility and small application scope in the digital design of secondary system of intelligent substation at present, this paper puts forward the 3D assembly techniques based on the expanded model by analyzing the integration of secondary model files of intelligent substation. In the design process of this method, the relationship between primary equipment and secondary equipment, the mapping between logical circuit and real circuit are fully considered. By using standard family library techniques, file analysis techniques and automatic mapping techniques, the coupling between physical model and logical relationship is realized. Taking the actual 220kV intelligent substation as the pilot, the three-dimensional model in secondary system simulation analysis, online monitoring and the application of fault diagnosis shows that the techniques can effectively improve the efficiency of three-dimensional design and maintenance of secondary system.

1. Introduction

Three dimensional digital design of power grid is an integrated innovation of modeling technology, information technology and network technology, which is conducive to the optimization of design, construction and installation, production and operation, and is conducive to improving the intrinsic safety level of power grid engineering in the whole life cycle. It is an important technical support for the construction of smart grid. Three dimensional technology can realize the full information simulation of power grid, with the characteristics of digitization and visualization, which can effectively improve the management level of engineering construction and operation inspection, and lay the foundation for the construction of digital power grid [1-3].

In China, digital design of power transformation projects based on three-dimensional information technology has entered the promotion stage, and relevant departments of State Grid Corporation of China have successively issued the specification for digital design transfer of power grid projects [4, 5] However, the three-dimensional design specification for substation secondary system has not yet been established. At present, some scholars have carried out relevant work on the three-dimensional design of substation secondary system. Reference [6] proposed a three-dimensional visualization information interaction technology of substation based on common information model (CIM). This technology can solve the problems of non intuitive display of equipment information and difficult interaction in the traditional way, but the traditional collision method is still used in the construction of three-dimensional scene, which is not adaptable to the secondary system. Reference [7] proposed a three-dimensional design method based on substation terminal block diagram, cable inventory and other information. This method needs to collect a large number of original data, and the modeling process is time-consuming, labor-consuming and inefficient. Reference [8] proposed a design scheme of secondary specialty cooperating with other specialties. Although the scheme can fully consider the three-dimensional design of secondary system, it has high requirements for the compatibility of design software and the coordination of various departments, which is not conducive to the actual implementation. Literature [9] adopts the 3D design method of virtual reality technology, but this method is not conducive to the expansion and maintenance of the built model. Once the substation structure changes, all the related designs need to be modified, which brings difficulties to the maintenance and update of the system. Literature [10-11] focuses on the application of 3D model in the intelligent operation and inspection of the secondary system, but it is not conducive to the rapid and effective establishment of 3D model Without specific elaboration, it can not provide effective modeling methods for designers.

In view of the shortcomings of the above research content, this paper takes several key model files of the secondary system of intelligent substation as the starting point. Through the application integration analysis of the secondary model files, the relationship between the model file and the 3D design is studied. The 3D assembly technology based on the extended model, including the standard library technology, the file analysis technology and the automatic drawing technology, is proposed. The technology can realize the high-efficiency integration of the three-dimensional model of the secondary system in substation, the coupling between the physical model and the logic, which lays the foundation for the advanced application of 3D model in the secondary system, such as simulation analysis, online monitoring and fault diagnosis.

2. Overall thinking of 3D digital design for secondary system

The establishment of three-dimensional digital design model of intelligent substation aims at the practical application scene of secondary specialty, so as to provide more direct and efficient means for operation and maintenance personnel in the aspects of simulation analysis, online monitoring and fault diagnosis. At present, the establishment of information interaction model of intelligent substation is based on IEC61850 standard, which is applicable to intelligent electronic equipment (IEE) of intelligent substation The interaction language and information interaction mode of equipment (IED) equipment are specified in detail [12]. The normal operation and maintenance of the secondary system of intelligent substation are based on the secondary system model file. The key models include the substation configuration description (SCD) file and the system specification file Description (SSD) and grid information model (GIM).

In this paper, three-dimensional digital design is constructed and applied based on the above three core model files, as shown in Figure 1: (1) firstly, the application analysis of three types of model files of intelligent substation is carried out, and the internal relationship mechanism of three-dimensional digital design of secondary equipment is constructed. Among them, SCD model file provides logical connection information for the communication between secondary equipment 3D models; SSD model file provides hierarchical attribute information for the association between primary and secondary equipment 3D models; GIM model file provides document information for the handover and application of secondary system 3D models. (2) Thirdly, based on the above model files, through the study of the secondary equipment 3D model standard family library technology, file analysis technology and automatic mapping technology, so as to complete the secondary equipment 3D external visualization model. (3) After completing the three-dimensional digital design of the secondary system, the advanced application of the three-dimensional model of the secondary system in the actual operation and maintenance is completed by studying the interface technology between the three-dimensional model of the secondary system and the operation information, online monitoring and diagnosis technology.

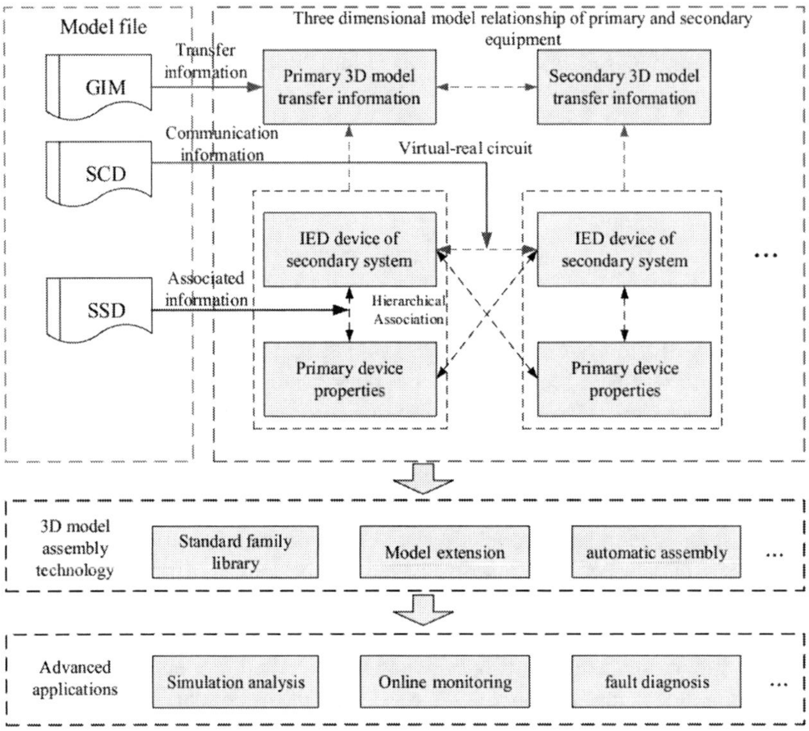

Fig. 1. Three dimensional digital design of secondary system

3. Application analysis of key model files in secondary system

3.1 Application analysis of SCD file

The SCD model file of Smart Substation is based on IEC61850 specification and is written in XML language. It specifies the information sharing and interoperability between IEDs in Smart Substation. Based on the method introduced in reference, the SCD file is analyzed to obtain the communication relationship among IEDs in substation, including ied model information, subnet allocation information and virtual circuit information The file contains five elements: header, substation, IED, communication and datatype templates [13]. At the same time, according to IEC61850 standard, SCD file establishes three kinds of information service models: MMS (Manufacturing Message Specification), goose (general substation oriented time object) and SV (sampling value) The goose information contains the interactive information between IED and intelligent terminal, and the SV information contains the sampling value between IED and merging unit.

By analyzing the node information of goose control block, inputs and SV control block in ln0 under different ied nodes, the virtual circuit information about goose / SV of the IED device can be obtained, as shown in Figure 2, including the failure trip information between protection devices, the voltage and current sampling information of protection devices, the trip information of protection devices, etc.

SCD model file can provide logical relationship description for information transmission between 3D models of secondary system. IEC 61850 engineering relay protection application model provides optical fiber physical connection relationship between ied device ports. By obtaining logical loop relationship between ied devices and matching with actual optical fiber, virtual real association mapping of secondary system in 3D model can be realized.

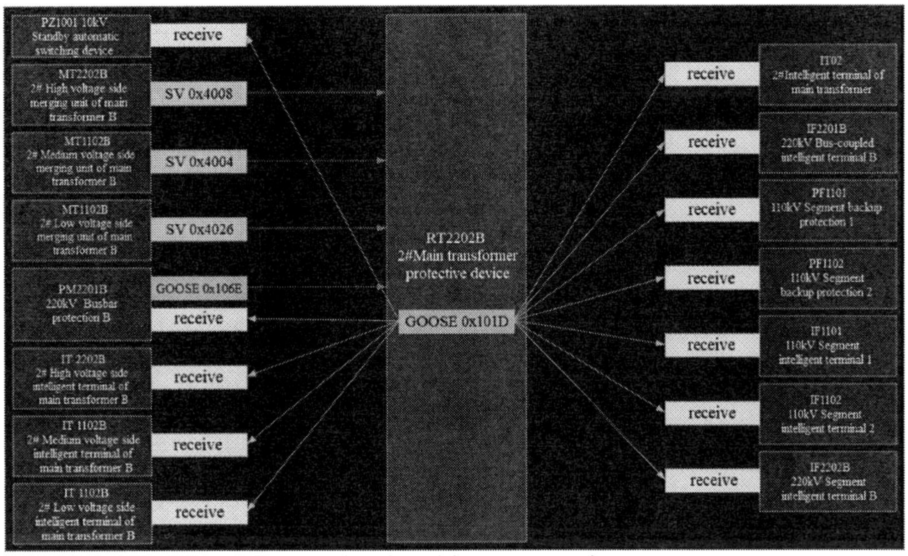

Fig. 2. Virtual loop connection

3.2 Analysis of SSD file application

The SSD file of Smart Substation describes the structure of the primary system and the associated logical nodes, which provides an effective basis for the device attributes and connection relationship. At the same time, the primary and secondary device relationship of the system is described in detail, and the system can be graphically realized through the configuration tool. Reference [14] describes the SSD file structure, which mainly includes the following object models: substation, voltage level, transformer, bay, equipment, Subequipment, function, connectivity node, terminal, Logical node (lnode), etc.

SSD file describes the model of primary equipment and secondary device in substation, establishes the hierarchy structure of "power grid, station / voltage level, interval, equipment / component, attribute", and sets parameters for primary equipment of power grid. SSD files follow the requirements of gb/t4728 to establish primary equipment elements of power grid. In engineering, we should adopt the modeling method based on Bay (interval), that is, organize equipment or function into a series of intervals according to their relationship or function relationship.

After the primary device interval modeling is completed, the primary and secondary association is carried out, and the lnclass, iedname and other attributes of lnode are assigned. The primary and secondary device association is established according to the following rules:

(1) The functions and information in the secondary equipment are associated with the primary equipment through the logical node, and the lphd configuration of the protection equipment logical node is associated with the corresponding primary equipment;

(2) All the four remote data associated with the primary equipment are associated with the corresponding primary equipment object through the logical node;

(3) All the protection action information and setting information associated with the primary equipment are associated with the corresponding primary equipment object through the logical node;

(4) All the process layer information associated with the primary device is associated with the object of the corresponding primary device through the logical node;

Through the above process, the association of two-dimensional graphics between primary equipment and secondary equipment is realized, as shown in Figure 3. The graphic display of the primary and secondary equipment is realized by using the integration of graphics and modeling technology, as shown in Figure 4.

Therefore, in the construction of 3D model of secondary equipment, in order to make the assembly process as simple as possible, this paper uses the integration of graphics and modeling technology to associate the 3D model of secondary equipment in the construction process of visual components of primary equipment, so as to ensure the consistency of 2D module and 3D module state of equipment.

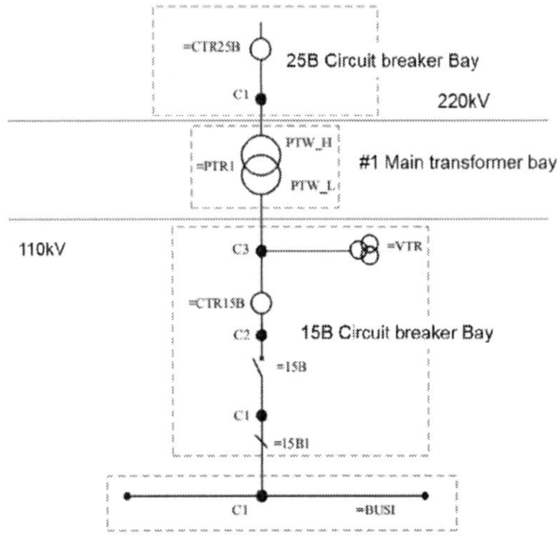

Fig. 3. Primary device associated with secondary node

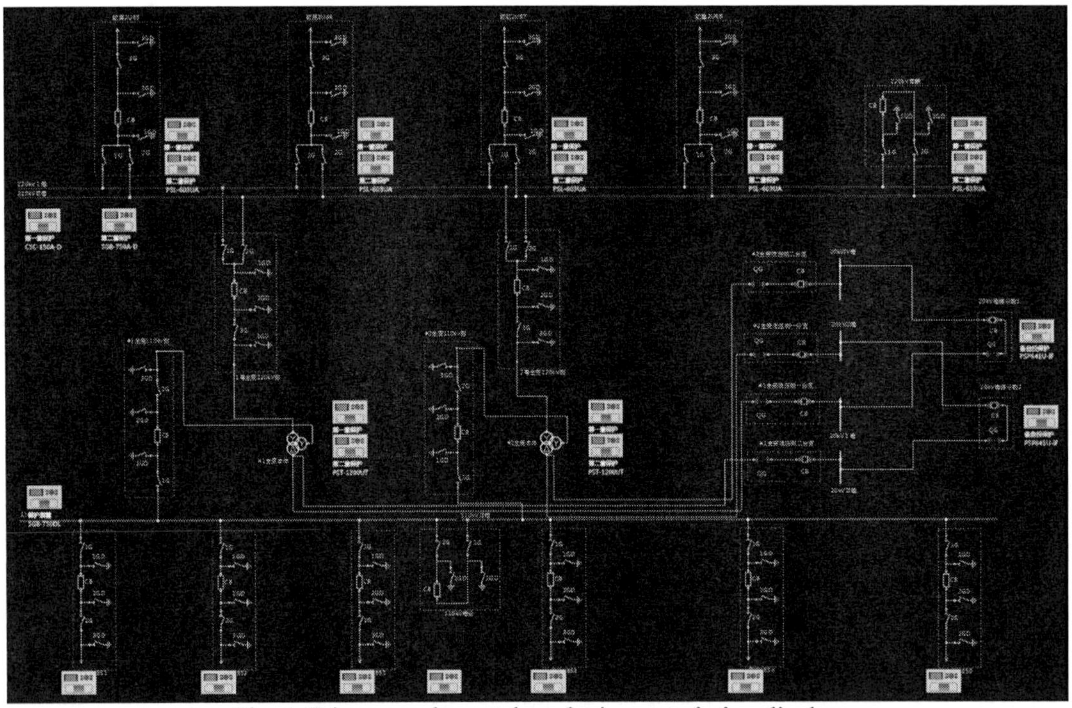

Fig. 4. Primary and secondary device association display

3.3 Application analysis of GIM file

Power grid GIM technology is mainly used in the design and construction of power transmission and transformation engineering [15]. This paper expands the content of secondary system according to the GIM specification of State Grid, including protection devices, intelligent terminals, merging units, measurement and control devices, switches, stability devices, recorders and other secondary equipment of various manufacturers, as well as related secondary auxiliary components. For example, "geometric fineness of protection device model", "engineering information of protection device" and "attribute fineness of protection device" are specified for secondary protection device, as shown in Table 1.

Table 1. Engineering information of protection device

Attribute	**Data Type**	**unit**	**Example**	**General model**	**Produtct model**
Project name	Charater type	-		√	√
Power grid project identificaion system code	Charater type	-	Software antomatic genration	√	√
Dispatching code	Charater type	-			
Real ID	Charater type	-			
Attachment No.	Charater type	-		√	√
Project number	Charater type	-	Software automatic generation	√	√

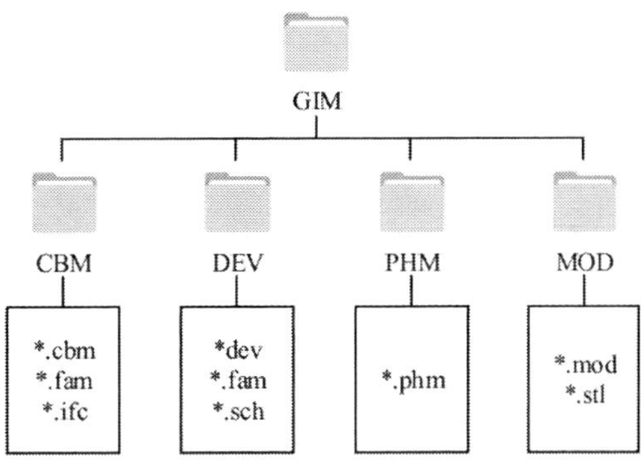

Fig. 5. Handover model composition of GIM

The established 3D model of substation secondary system is handed over according to the GIM standard format of State Grid, as shown in Figure 5. The models include digital elevation model (MOD), engineering model (CBM), physical model (DEV) and composite model (PHM).

Through the transfer of standard GIM model, we can achieve: (1) data normalization check. Including 3D model integrity check, GIM format normative check, 3D model and design drawings consistency check. (2) Model visual editing. The established three-dimensional model is visualized to display the three-dimensional model of the secondary system, and the wrong part of the model can be modified. (3) Model change process control. Through the normative review and consistency verification of the model, the model is archived and managed. When there are changes on site, the warehousing model is modified in time and marked to ensure the consistency between the warehousing model and the site.

4. Key technology of 3D model assembly for secondary system

In order to improve the efficiency and quality of the 3D design of the secondary system and ensure the integrity of the 3D model design, as shown in Figure 6, this paper develops the standard family library and material library of the secondary system in an object-oriented way, expands the 3D information of the secondary model file, and studies the model file identification and analysis technology, automatic screen assembly and editing technology, so as to realize the automatic installation of the 3D model of the secondary system match.

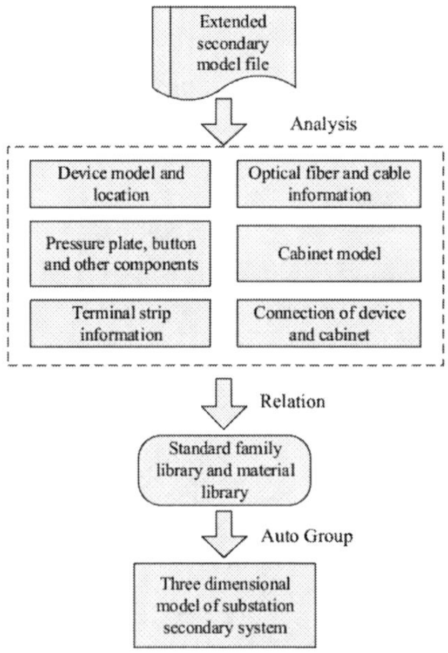

Fig. 6. 3D modeling process of secondary system

4.1 3D component family library and material library technology

In this paper, the three-dimensional modeling of the secondary system is carried out by using Autodesk Revit software [16]. The general model family library is established, including the secondary panel cabinet and secondary accessory component family library. The secondary accessory components include: pressure plate, button, indicator light, air switch, terminal strip, handle, optical fiber and distribution frame. The basic attribute information of all secondary Bay cabinets and secondary equipment accessory components includes name, type, basic shape and size scaled according to standard scale. The family library also includes different states of components.

In order to create the 3D model quickly and accurately, this paper establishes the material library of packaged secondary equipment according to different manufacturers and different types of secondary equipment. The material library contains three-dimensional models of different manufacturers and models of secondary equipment. The three-dimensional model contains the manufacturer information, model information, protection type information, device plug-in information and three-dimensional space information of secondary equipment, as shown in Table 2. The 3D model of the secondary equipment in the material library is shown in Figure 7.

In order to ensure that the three-dimensional information of the protective device in the material library has good reusability, the three-dimensional model of the secondary equipment in the material library is updated regularly according to the verified equipment list issued by the State Grid, so as to ensure that the equipment in the material library is consistent with the actual application scene.

Table 2. Equipment and material library information

N0	Name	Type	Manufactor
1	220kVline protection	RCS-931ADMM	Nanrui
2	Main transformer measurement and control device	CSC-282	Beijing Sifang
3	Ethernet switch	2D-PCS-9882BD	Beijing Sifang
...	...	...	...

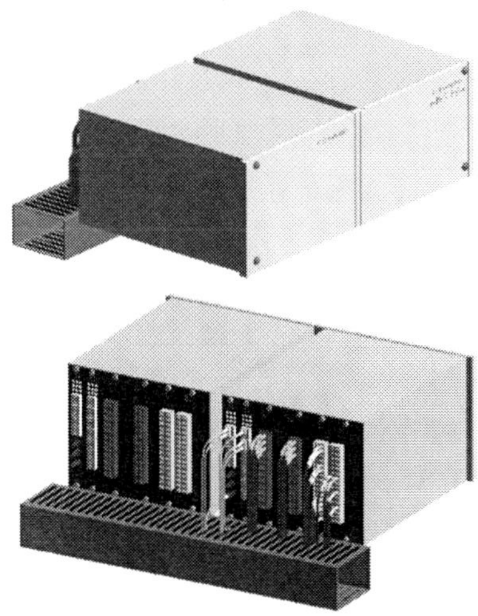

Fig. 7. Merge unit model in material library

4.2 Three dimensional information expansion of secondary equipment model

Because the secondary model file of Smart Substation does not contain relevant three-dimensional information at this stage, based on the SSD file of substation, this paper expands it, improves the three-dimensional attribute information of secondary equipment, provides model support for the construction of three-dimensional model, and ensures the consistency of two-dimensional information and three-dimensional information of secondary equipment of substation.

According to the hierarchical structure of primary and secondary equipment described in SSD file, this paper establishes the attribute information of secondary room, cabinet and unit, as shown in Table 3, forming a complete hierarchical structure of "power grid. station / voltage level. Room. Bay. Cabinet. device / element and attribute".

In terms of three-dimensional spatial information, the specific location of the secondary cabinet in the small room adopts the spatial coordinate system (x, y, z) The secondary device and auxiliary components are described in terms of coordinates (F level, u level, L level), in which F is the level arranged from front to back, u is the level arranged from top to bottom, and l is the level arranged from left to right, so as to complete the attribute addition of the spatial location information of cabinets, devices and components, as shown in Figure 8.

Table 3. Secondary system object information

No	Object	Attribute information extension
1	room	name, describe (des) , Voltage level (vol)
2	cabinet	name, describe (des) , Voltage level (vol) , bay, position (pos)
3	device	name, describe (des) , type, cabinet, position (pos)
4	Unit	name, describe (des) , type, class, cabinet, position (pos)

```
<?xml version="1.0" encoding="UTF-8"?>
<SLCL>
<Substation name= "LS220" desc= "Lingshou 220kV substation" >
   <Region name= "R2201" desc= "220KV line room" type= "Indoor" >
      <Cubicledesc="220 Line intelligent control cabinet" name="R220#01_1P">
         <Device desc="220KV Line intelligent terminalA" name="IL2203A" type="" pos=
 "F1-U3-L2" >
            <Borad name="X16" desc=" Trip plug in" class="E" type= " " >
             <Terminal name="c2" desc=" Trip outlet+" class = " " />
```

Fig. 8. Extended SSD model file

4.3 Model analysis and automatic screen assembly technology

In this paper, the parsing method based on document object model (DOM) is used to parse the extended SSD file [13]. In this method, the entire SCL file is read into the memory at one time, and the configuration file information is transformed into the object node tree, so as to extract the attribute and position information of the secondary cell, secondary cabinet, secondary device and components. By analyzing the SCD model file of intelligent substation, the logical connection relationship between secondary equipment and the connection address between optical fiber and cable are obtained.

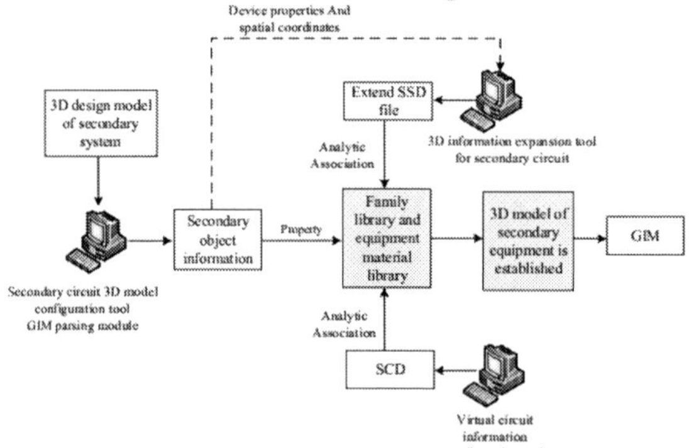

Fig. 9. 3D modeling process based on extended SSD file

According to the basic rules of electrical secondary design, the number in the system logic data has the unique principle in the same level directory. Through the model analysis results, according to the module family in the mapped standard modular family library, the secondary equipment family model in the system design is extracted. By using the logical relationship of cabinet box, cable and optical fiber in the model file, as well as the numbered attribute information of cabinet, component, equipment and wiring, it is constructed from the outer cabinet to the internal device and terminal strip equipment, calls the matching module family, analyzes its location layout, automatically generates the three-dimensional model of the secondary system, and finally generates the GIM file of the three-dimensional model of the substation secondary system The overall process is shown in Figure 9.

5. Application cases

The three-dimensional model assembly technology of secondary system proposed in this paper has been applied in the 220 kV Qiushan substation of Hebei Province. As a key construction project of three types and two networks of Hebei electric power company, it has achieved good results, which proves the feasibility and effectiveness of the method.

(1) In this paper, the 3D modeling of intelligent substation secondary system is carried out by using Autodesk Revit software, and the device attributes are assigned by std-r to form a universal modular family library of standard cabinet box, device equipment, terminal strip, components, cable and optical fiber, and the effect rendering is carried out, which lays the foundation for the establishment of 3D model of pilot station. The general model family library of substation secondary system can be loaded repeatedly in any model project. As the resource accumulation of three-dimensional model of substation secondary system, it realizes the reuse of three-dimensional model resources, reduces the workload of modeling and improves the efficiency of modeling.

(2) The SSD model file of 220kV Qiushan station is expanded with three-dimensional information. Through the actual measurement of the position of each voltage level chamber and panel cabinet in Qiushan station, the position information of each panel cabinet is determined, the equipment parameterization setting is carried out, and the relative position of each secondary device and accessory component is determined to complete the addition of three-dimensional information. The SCD model file of 220kV Qiushan station is analyzed to determine the position information of each panel cabinet The logical information of each ied device in process layer and bay layer is connected to construct the mapping between virtual circuit and real circuit of the whole station.

(3) According to the analysis of the three-dimensional attributes of the secondary equipment in the established extended SSD file and the equipment connection relationship in the SCD file, combined with the associated information of the standard equipment material library and the standard modular family library, the model integration design is carried out to realize the instantiation of the secondary three-dimensional model cabinet in the engineering design stage of Qiushan station, as shown in Figure 10.

(4) The 3D model of Qiushan station is applied to 3D panoramic visualization. The visualization data support includes extended SSD model file, standard SVG graphics file, standard GIM file, online monitoring real-time monitoring data. The 3D engine display technology can visually display the 3D panorama of the substation by parsing the standard GIM file; it can integrate with the online monitoring system platform through TCP protocol and interface access technology to interact and display with the 3D engine when the fault information is detected, so as to realize the panoramic perception and fault location in the 3D scene, as shown in Figure 11-13.

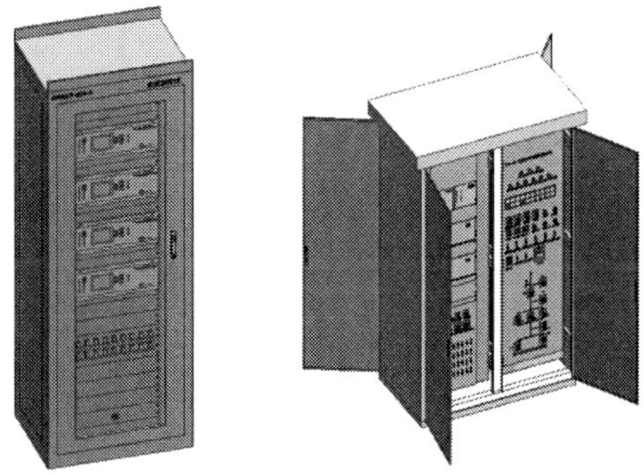

Fig. 10. 3D model of secondary panel cabinet and central control cabinet in qiushan station

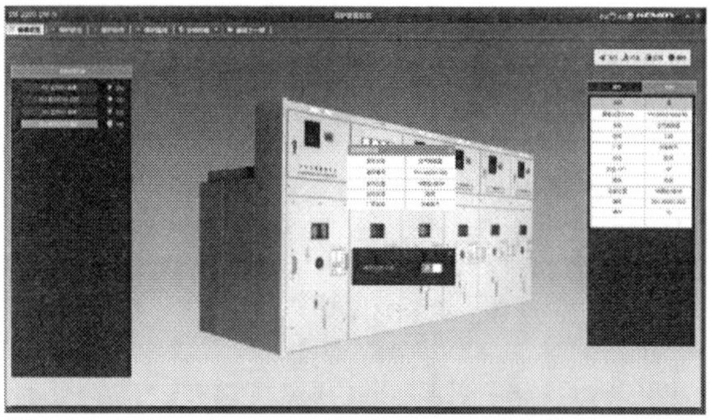

Fig. 11. 3D online monitoring of secondary system

Fig. 12. fault location of 3D model

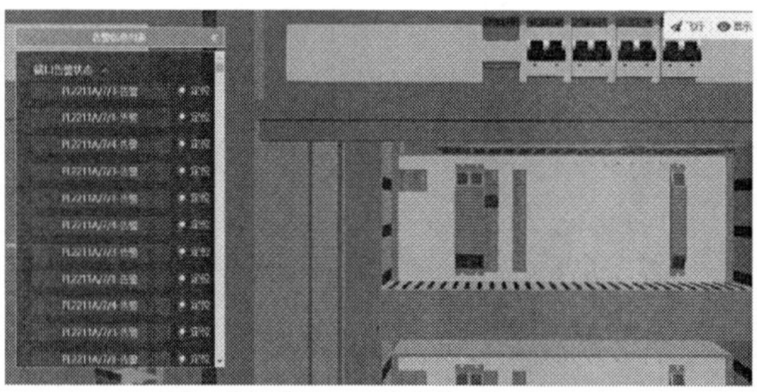

Fig. 13. 3D model fault alarm

The 220kV Qiushan substation adopts the three-dimensional modeling method proposed in this paper to realize the rapid establishment of the three-dimensional model of the secondary system. Because the establishment of the three-dimensional model is based on the model file of the secondary system, the consistency of the three-dimensional model and the two-dimensional model is ensured, and the consistency of the three-dimensional online monitoring information and the actual operation information of the substation is ensured, and the secondary system simulation of the intelligent substation is realized The effective application of analysis, online monitoring and fault diagnosis in the three-dimensional scene can provide more friendly and intuitive experience for the staff and greatly improve the operation and maintenance efficiency of the secondary system.

6. Conclusion

This paper introduces a three-dimensional assembly technology for secondary system of intelligent substation. Firstly, the current situation of 3D modeling of secondary system is analyzed, and the shortcomings of existing assembly technology are pointed out. In view of these shortcomings, a 3D model assembly technology based on substation secondary system model file is proposed. In the design process, the relationship between primary equipment and secondary equipment, and the mapping between logical circuit and real circuit are fully considered. Through the use of standard family library technology, file analysis technology and automatic mapping technology, the coupling between 3D model object and logical relationship is realized to ensure the reliability. Finally, the effectiveness of the technology is verified by a pilot case. Through this technology, it can not only improve the development efficiency of the current three-dimensional model of substation, but also the key technology has strong popularization, which can greatly reduce the use cost, and lay the foundation for the future application to other new generation of intelligent substation. At the same time, under the background of energy Internet construction, it plays a significant demonstration role in improving the 3D digital design of substation secondary system and promoting the application of 3D visualization technology.

Acknowledgments
This work is supported by State Grid science and technology projects (Research on Key Technologies of intelligent operation and maintenance of secondary system based on physical loop general model, No.5100-202013020A-0-0-00).

References
[1] Opinions of State Grid Co., Ltd. on the Collection of Framework for Energy Internet Technology Research (keji [2019] No. 22).
[2] Wu Zhili, HAN Wenjun, "Necessity and urgency of power grid project digital transfer work," Electric Power Construction, vol. 35(2), pp. 66–69, 2014.
[3] Lin Zhihe, "Research on advanced application function of substation based on 3D visualization," Energy and Environment, vol. 38(6), pp. 38-40, 2018.
[4] State Grid Corporation of China. Code for 3D design and modeling of substation (converter station) in power transmission and transformation engineering. Beijing: China Standards Press, 2019.
[5] State Grid Corporation of China. Code for 3D design and modeling of overhead transmission lines. Beijing: China Standards Press, 2019.
[6] Tao Songmei, Zhang Wei, "Application of Substation 3D Visual Interaction Technology Based on CIM model," Guangxi Electric Power, vol. 37(6), pp. 22-24, 2014.
[7] Qi Wei, Chen Keji, "Research on Digitization and 3D Interface Design of Secondary System in Substation," Zhejiang Electric Power, vol. 20(37), pp. 53-56, 2018.
[8] Tu Furong, Chen Xiaojie, Wang Xiaowen, "Discussion on 3D Digital Collaborative Design of Electrical Secondary System of Substation," Electrician Technology, vol. 24(6), pp. 56-58, 2018.
[9] Gong Qingwu, Jiang Fangfang, Chen Yifei, "Virtual reality technique based substation training simulator," Power System Technology, vol. 29(24), pp. 74-77, 2005.
[10] Pang Sirui, Xu Hongfei, Du Jianwen, et al, "Research and Design of Intelligent Substation of Assist Control System Based on Internet of Things and Three-Dimensional Visualization Technology," Telecommunications Science, vol. 13(8), pp. 1-4, 2015.
[11] Zhou Yuanqiang, Chen Gang, Fei Yijun, et al, "Development and Application of 3D Real-Time Monitoring and Display System for Substation Intelligent Operation and Maintenance," Power System and Clean Energy, vol. 34(12), pp. 41-47, 2018.
[12] Zheng Guangxu, "Research on ied information modeling and communication based on IEC61850," Beijing: North China Electric Power University, 2014.
[13] Guo Jiansheng, Wu Wenxuan, Wang Yunmao, et al, "Parse for SCD file of smart substations based on TinyXML," Electric Power and Electrical Engineering, vol. 36(3), pp. 7-10, 2011.
[14] State Grid Corporation of China, technical specification for Smart Substation System Configuration description file. Beijing: China Standards Press, 2018.
[15] HU Junhui, SHENG Dakai, QIE Xin, et al, "Establishment of digital design system of power Transmission &transformation project for power grid construction and development," Electric Power Construction, vol. 33(12), pp. 1-5, 2012.

[16] WANG Yindi, LIU Hongbin, "Application Exploration of the analysis software of importing revit into ecotect," Intelligentilize and Informatization, vol. 27(5), pp. 275-277, 2019.

Online cleaning method of power grid energy anomaly data based on improved random forest

Ke Chen[1*], Hongkai Wang[1], Zhangchi Ying[1], Chengxin Zhang[1], Jiaqi Wang[1]

[1]State Grid Zhejiang Electric Power Corporation Information & Telecommunication Branch, Hangzhou, Zhejiang, 310016, China

[*]ricemice1227@163.com

Abstract—Aiming at the problem of high root mean square error of traditional power grid energy anomaly data online cleaning, a power grid energy anomaly data online cleaning method based on improved random forest is designed. Firstly, an outlier data recognition model of isolated forest is designed to identify outliers in the data. Secondly, an improved random forest regression model is established to improve the adaptability of random forest to mixed abnormal data, and the data trend is fitted and predicted. Finally, the improved random forest data cleaning method is used to compensate the missing data after removing the mixed abnormal data, so as to clean the abnormal energy data of the power grid. The experimental results show that when the amount of power grid energy anomaly data increases, the cleaning root mean square error of the experimental group is significantly lower than that of the control group. The method in this paper solves the problem of high root-mean-square error in the online cleaning of abnormal data of traditional grid energy.

1. Introduction

The power grid contains huge real-time data, which is closely related to the safe operation of the power system and plays a decisive role in the safety and reliability of the power grid. The power grid needs to generate and consume a lot of electric energy in the process of working. Under the condition of normal work, it will still produce bad data, affect the decision-making of the system and threaten the safety of the whole power grid. Data cleaning is not only applied in power grid, but also in the fields of data mining and total data quality management. The definitions of data cleaning are different in different fields, and there is no unified definition of data cleaning. Therefore, using professional technology to detect and correct bad data will help to ensure the normal operation of power grid and the normal supply of power energy. So far, scholars at home and abroad have done a lot of theoretical and practical research on the detection and correction of power grid energy big data, and summarized two widely used cleaning methods. One is the identification and cleaning method of abnormal data of power grid operation based on Spark framework. Abnormal data are extracted from big data through data mining algorithms such as clustering or neural network, and cleaned under Spark framework. The other is the missing value cleaning method of power grid data based on improved low rank matrix completion. This method mainly sets the threshold according to a certain confidence level, carries out hypothesis test combined with the relevant knowledge of probability theory, and then completes the missing value of data by improving the low rank matrix to realize anomaly cleaning. In the actual operation of the power grid, it is very difficult to ensure the accuracy and completeness of user electricity data collection. Affected by sensor failures, transmission line failures, natural weather and other complex factors, users' electricity consumption data will have varying degrees of data omissions and errors. This article refers to these user electricity consumption data as abnormal data. In the above two methods, the data recognition is abnormal due to too many repeated searches, or the normal data is judged as bad data, resulting in a high root mean square error in practical applications.

The data set does not need standardization, and the training speed is fast. It can be applied to large-scale data sets. Default values can be processed (as a separate class) without additional processing. Due to the out of bag data (OOB), the unbiased estimation of the real error can be obtained in the process of model generation without losing the amount of training data. In the training process, the interaction between features can be detected, and the importance of features can be obtained, which has a certain reference significance. Because each tree can be generated independently and simultaneously, it is easy to make a parallelization method. Because of its simple implementation, high precision and strong anti overfitting ability, it is suitable to be used as a benchmark model in the face of nonlinear data. Based on this, this paper designs an online cleaning method of power grid energy anomaly data based on improved random forest to solve the problems existing in the previous cleaning methods.

2. Improved random forest

In order to adapt to the compensation of mixed missing data, the random forest algorithm is improved to obtain an improved random forest algorithm with strong adaptability [1-2]. The X excluding abnormal data is linearly interpolated, and then the filling matrix containing target compensation variables is obtained through matrix transformation. Then, the random forest regression model is used to predict the filling matrix, and the integration idea is used to output t_b times[3-4]. The specific methods for improving random forest are as follows:

In step 1, the missing value in X is interpolated by linear interpolation method to obtain matrix R, which is expressed as:

$$R=\left[x_1, x_2, x_3, \dots, x_n\right] \quad (1)$$

In formula (1): the sample size of R and X is the same, which is m × n dimensional matrix, n is the number of variables, and m is the number of data samples contained in a single variable.

In step 2, take column i in X as the target filling column, and the remaining n-1 column in R except column i as the relevant variable column to form the filling matrix R_{fill}, I = 1, 2, 3,..., n, expressed as:

$$C(u)=2(In(u-1)+\xi)-\frac{2(u-1)}{u} \quad (2)$$

In step 3, a training matrix D_{zv} is randomly selected from D to complete the establishment of random forest regression model.

In step 4, take the relevant variable column in R_{fill} as the input and the target filling column as the output to predict the missing value of column i to obtain the t-th prediction value y(t), t = 1, 2, 3,..., t_b.

Step 5 if $t < t_b$, then t = t + 1, return to step 4; otherwise, take the average value of t_b predictions as the final compensation value y_z, and replace the missing value in x_i with y_z to complete the update of matrix X.

$$y_z=\frac{1}{t_b}\sum_{t=1}^{tb} y(t) \quad (3)$$

Step 6 if i < n, i = i + 1, repeat the above steps; otherwise, stop prediction and complete data compensation for X [5-6]. The above is to improve the overall step flow of random forest algorithm.

3. Online cleaning method of power grid energy anomaly data based on improved random forest

3.1. Classification of grid energy anomaly data

According to the analysis of the original data of power grid energy, the abnormal data of power grid energy are mainly divided into the following three categories: outlier data, duplicate data and missing data. Outliers are data that deviate significantly from the rest of the values[7]. Duplicate data is the data with the same value at multiple times. According to the dynamic characteristics of power grid operation process, this kind of data is determined as abnormal data. Missing data refers to data not collected at a certain time or data vacancy[8]. However, there are many energy data variables and parameters in the power grid, and the abnormal data do not appear in a single type, but often appear in a variety of abnormal characteristics at the same time. At present, the research on online data cleaning of power grid energy anomalies is limited to cleaning methods for single type of abnormal data, and there is no mixed abnormal data cleaning method for different types. When the traditional probability distribution, clustering algorithm and intelligent algorithm are used to clean different types of mixed abnormal data, the effect is still difficult to meet the requirements of data quality during power grid operation. Therefore, this paper proposes a data cleaning method based on improved random forest to clean different types of mixed abnormal data.

3.2. *Identification of grid energy anomaly data based on improved random forest*

In order to quickly and accurately identify abnormal data, an abnormal data identification model based on isolated forest is constructed. IF is a decision tree ensemble learning method. IF algorithm first divides the known and continuous time data set randomly, and then realizes fast and accurate identification by using the difference between abnormal data and normal data. The IF model consists of a isolation tree, and the algorithm steps are as follows:

Step 1 randomly select a sub sample matrix X_z from the training data matrix X as the set of root nodes of the b-th tree, where b = 1, 2, 3,..., a and a are the number of root nodes, and the training data matrix X and sub sample matrix X_z are expressed by the formula:

$$X=\left[x_1,x_2,x_3,...,x_n\right] \quad (4)$$

$$X_z=\left[x_1,x_2,x_3,...,x_{nz}\right] \quad (5)$$

In formula (4, 5), the matrix is m × n dimensional matrix, n is the number of variables, and m is the number of data samples contained in a single variable; u is the number of samples, u = mn; X_z is $m_z \times n_z$ dimensional matrix, n_z is the number of variables, m_z is the number of data samples contained in a single variable, $0 < m_z \leqslant m$, $0 < n_z \leqslant n$.

Step 2 perform binary segmentation on Xz: randomly select a column vector xj, j∈{1, 2, 3,..., mz} from Xz, and randomly select a cutting point T from the set of column vectors xj, as shown in formula (6).

$$T = \min(x_j) + (\max(x_j) - \min(x_j))r \quad (6)$$

In formula (6), r is a random number between 0 and 1. If $X_z(i, j) < T$, all variables in row I of matrix X_z are divided into left subtree nodes. If $X_z(i, j) \geqslant T$, all variables in row i of matrix X_z are divided into right subtree nodes, i = 1, 2, 3,..., m_z. When all the data in X_z is divided, the left subtree node set forms matrix X_{left} and the right subtree node set forms matrix X_{right}.

Step 3 record the path length h_n of the node where X_{left} and X_{right} are located. h_n is the number of edges from the root node to the current node. If h_n is greater than or equal to the tree height h_{max} or the number of sets in nodes is less than or equal to m_z, stop training and complete the construction of a single isolation tree; otherwise, perform binary division for X_{left} and X_{right}, and repeat step 3. The tree height h_{max} is:

$$hmax = Ibu \quad (7)$$

Step 4 if b < a, b = b + 1, repeat step 2 and step 3, otherwise stop training and complete the establishment of if model. IF divides and isolates the data in X. the normal data needs to be divided and isolated for many times and is in a high-density area; abnormal data needs to be isolated by a few divisions and located in low-density areas.

3.3. *Eliminate abnormal data of power grid energy*

After the data is calculated by the IF model, different high and low density areas are formed. The density area where the data is located is reflected by calculating the abnormal value score of the data, and the data with high score is eliminated. The path length of x_{ij} output from IF model is h_{ij}, and x_{ij} is the element in matrix X. calculate the outlier score of x_{ij} through h_{ij}, and the formula is:

$$S(h_{ij},u)=2^{\frac{-E(hij)}{C(u)}} \quad (8)$$

$$C(u)=2(In(u-1)+\xi)-\frac{2(u-1)}{u} \quad (9)$$

In formulas (8 and 9), c(u) is the average path length of all data in X; ξ Is Euler constant; $E(h_{ij})$ is the average path length of data x_{ij} in a isolation tree. When the value of $S(h_{ij}, u)$ approaches 0.5, it indicates that the data has no obvious abnormal state; When the value of $S(h_{ij}, u)$ approaches 1, it indicates that the data is an abnormal value. After obtaining the outlier score of each data, combined with the equivalence and missing characteristics of the data, the following three kinds of data are eliminated from X: For the first data, xij with outlier score S(hij, u) greater than 0.75 in X is eliminated and replaced by 0. For the second data, replace the missing data in X with 0. For the third data, c1 or more consecutive xij with the same value in X are eliminated and replaced with 0.

3.4. Complete the online cleaning of abnormal energy data of power grid

From the above analysis, it can be seen that the identification and cleaning of non clean data is difficult to ensure the complete elimination of non clean data, especially the multi-dimensional nature of power grid energy abnormal data, which brings difficulties to cleaning. With the gradual improvement of the fineness of the cleaning, the operation performance of the knowledge base system will gradually decrease, and the information contained in the abnormal energy data of the power grid will be gradually lost, which can not quickly and effectively solve the problems caused by the non clean data. Data cleaning across data sources is mainly to ensure the correlation between metadata by matching the semantic constraints between multiple data sources. Assuming that the metadata feature set is sdfrewa, the following two constraints need to be met to complete metadata cleaning. First, if the two vertices of the edge representing the metadata relationship must be metadata objects, they are expressed by a formula, as shown in formula (10).

$$\forall e(f,t)\in E, f,t\in V \quad (10)$$

In formula (10), e refers to the directed edge of the metadata feature; (f,t) refers to the coordinate point corresponding to the metadata feature; E refers to the directed edge combination of metadata features; V refers to the vertex set of metadata features. Second, if the relationship between two metadata objects can only be one-way, it can be expressed by formula, as shown in formula (11).

$$\forall e(f,t)\in E, e^{-1}(f,t)\notin E \quad (11)$$

On the basis of satisfying formula (11), the syntax and semantics of energy metadata characteristics in the power grid operation process can be analyzed, and then the energy effective information can be extracted and tracked, so as to complete the online cleaning of power grid energy abnormal data. Therefore, through data cleaning, the quality of power grid energy data can be improved.

4. Experiment

4.1. Experimental preparation

The grid energy data information of a cross-border power enterprise in recent 5 years is selected and introduced into the simulation experiment software to build a virtual power database. Two computers with Windows 10 and 16GB memory capacity are selected to simulate the sending and receiving of power grid energy data. Another computer with Windows 10 and 32GB memory capacity is selected to simulate the transmission environment of power grid energy data, and the experiment is carried out with the abnormal power grid energy data as the sample data. This paper uses the online cleaning method of power grid energy abnormal data based on improved random forest design to clean the power grid energy abnormal data online, and then uses the traditional method to clean the power grid energy abnormal data online. 1000 grid energy anomaly data are randomly selected, of which 500 grid energy anomaly data are cleaned online by using the design method in this paper and set as the experimental group, and the other 500 grid energy anomaly data are cleaned online by using the traditional method and set as the control group. The root mean square error of the experimental group and the control group in the online cleaning of power grid energy anomaly data was compared. The lower the root mean square error, the higher the accuracy of the online cleaning method of power grid energy abnormal data, so as to obtain a higher accuracy online cleaning method of power grid energy abnormal data.

4.2. Experimental results and analysis

According to the above experiment preparation, complete the comparative experiment, record the data information generated during the experiment, and display the experimental results in the form of tables. The specific experimental results are shown in Table 1.

Table 1 Comparison of experimental results between experimental group and control group

Number	Data volume	Root mean square error of cleaning in the experimental group	Root mean square error of cleaning in the control group
1	50	0.1058	0.8574
2	100	0.1414	0.9978
3	150	0.1238	0.7569
4	200	0.1824	0.7581
5	250	0.1460	0.8534
6	300	0.2052	0.9580
7	350	0.1437	0.9573
8	400	0.1841	0.8581
9	450	0.1609	0.9560
10	500	0.1513	0.8577

According to Table 1, when the amount of grid energy abnormal data increases, the cleaning root mean square error of the experimental group is significantly lower than that of the control group, which can realize high-precision online cleaning of grid energy abnormal data. The main reason is that the experimental group can realize intermittent deletion mixed continuous deletion data cleaning based on the improved random forest. Therefore, the comparative experiments show that the online cleaning method proposed in this paper can effectively reduce the root mean square error of online cleaning of power grid energy anomaly data and improve the online cleaning accuracy. The mean square error of the cleaning data using this method is less than 0.21, while the mean square error of the cleaning data of the control group is higher than 0.75. It further ensures the quality of power grid energy data, meets the needs of power grid construction at the present stage, and has higher application value.

5. Conclusion

Aiming at the problem of energy abnormal data in the process of power grid operation, an online cleaning method of power grid energy abnormal data based on improved random forest is proposed in this paper. Firstly, the random forest algorithm is used to identify the outlier data. Then, outlier data, duplicate data and missing data are eliminated to obtain the mixed type missing data set. Finally, the improved random forest algorithm is used to compensate the missing data sets of mixed types. The effectiveness of IRF data cleaning algorithm is verified by taking the actual sewage data, and the following conclusions are obtained: the random forest algorithm can eliminate the outlier data in the power grid energy data. In the mixed type missing data set, the improved random forest algorithm has better cleaning effect than other algorithms, and is suitable for cleaning mixed type abnormal data. For data cleaning algorithms, some research institutions have proposed data preprocessing, sorting neighbor method, multiple traversal data cleaning method, cleaning with domain knowledge, integrated data cleaning with database management system and so on. Based on the differences between Chinese data and Western data, Chinese data cleaning not only transplants the cleaning methods of Western data, but also has its own unique cleaning methods. According to the deficiency of current data cleaning research, the main research directions of data cleaning in the future are: the research and development of Chinese data cleaning tools; do in-depth research on the application of data mining methods in the field of data cleaning; the efficiency of duplicate record identification needs to be further improved; cleaning of unstructured data; interoperability between data cleaning tools; universality of data cleaning scheme. The improvement and optimization of Chinese data cleaning technology will be further studied in the follow-up.

References

[1]Govindarajan S , Ardila-Rey J A , (2021) Krithivasan K , et al. Development of Hypergraph Based Improved Random Forest Algorithm for Partial Discharge Pattern Classification[J]. IEEE Access, (99):1.

[2] Sihag P , Angelaki A , Jain B C . (2020) Estimation of the recharging rate of ground water using Random Forest Technique[J]. Applied Water Science, 10(7).

[3] Yao D , X Zhan, Kwoh C K . (2019) An improved random forest-based computational model for predicting novel miRNA-disease associations[J]. BMC Bioinformatics, 20.

[4] Asadi S , Roshan S E , Kattan M W . (2021) Random forest swarm optimization-based for heart diseases diagnosis[J]. Journal of Biomedical Informatics, 115(24):103-690.

[5] Su Y , K Weng, Lin C , et al. (2021) An Improved Random Forest Model for the Prediction of Dam Displacement[J]. IEEE Access, (99):1.

[6] Canion A , Mccloud L , Dobberfuhl D. (2019) Predictive modeling of elevated groundwater nitrate in a karstic spring-contributing area using random forests and regression-kriging[J]. Environmental Earth Sciences, 78(9).

[7] Xianjun, Liu, Xixiang, et al. (2019) Robust data cleaning methodology using online support vector regression for ultra-short baseline positioning system.[J]. The Review of scientific instruments, 90(12):124901-124901.

[8] Martinez-Luengo M , Shafiee M , Kolios A . (2019) Data management for structural integrity assessment of offshore wind turbine support structures: data cleansing and missing data imputation[J]. Ocean Engineering, 173(1):867-883.

[9] Ridzuan F , Wan M . (2019) A Review on Data Cleansing Methods for Big Data[J]. Procedia Computer Science, 161(3):731-738.

[10] Neira-Rodado D , Nugent C , Cleland I , et al. (2020) Evaluating the Impact of a Two-Stage Multivariate Data Cleansing Approach to Improve to the Performance of Machine Learning Classifiers: A Case Study in Human Activity Recognition[J]. Sensors (Basel, Switzerland), 20(7).

A FPGA-based parallel method of bridge data compression

Jianlong Zhang[1*]

[1] Henan Provincial Expressway Network Management Center, Zhengzhou 450016, China

*Corresponding author's e-mail: zhengxdhw@gmail.com

Abstract. With the rapid development of computer technology, sensor technology, and communication technology, people have gradually designed and continuously improved the health monitoring system of bridge structure. Due to the uninterrupted long-term work of the bridge monitoring system, the long-term operation of a large number of sensors will generate massive amounts of data, and a lot of data to be transmitted from the data collector at the bridge site to the data monitoring center requires the use of a wireless network with limited speed, which will consume a lot of time, thereby bring such negative effects as reducing transmission efficiency, reducing throughput, increasing transmission costs, and affecting the entire bridge inspection system. This paper proposes a parallel compression method of LZSS data based on FPGA, which can compress the incoming data from the sensor system in parallel in the data acquisition system to improve the transmission efficiency, increase the throughput, and reduce the transmission cost.

1. Introduction

In China, many bridges built in the 1950s and 1960s are still in operation. Due to the impact of the environment and the aging of the bridges, various safety hazards in these "old bridges" that have been in operation for more than half a century are gradually exposed [1]. In order to ensure the normal operation of roads and railways in our country, it is very necessary to grasp the health status of bridges in time.

In recent years, with the rapid development of computer technology, sensor technology and communication technology, people have gradually designed and continuously improved the health monitoring system of bridge structure [2]. Figure 1 is a schematic diagram of the current mainstream health monitoring system of bridge structure. The system deploys various sensors that monitor different locations, types, and monitoring items of the bridge to the corresponding locations of the bridge, collects various real-time field data of the bridge through the data collector located at the bridge site, and then passes the collected data transmitted to the back-end data monitoring center through the public network of the data transmission subsystem. Finally, through the analysis of these data, the health, reliability, bearing capacity and durability of the bridge are evaluated. The current advanced detection system can also be based on these data to predict the situation that will happen, and promptly report potential safety hazards to prevent them before they happen.

However, due to the uninterrupted and long-term work of the bridge monitoring system of 7*24 hours, the long-term operation of a large number of sensors will generate massive amounts of data, because the data acquisition subsystem is usually far away from the monitoring center, and many bridges are located in complex terrain and cannot be reached by wired networks. Therefore, a large amount of data to be transmitted from the data collector at the bridge site to the data monitoring center requires the use of a wireless network with limited speed, which will consume a lot of time, thereby reducing transmission efficiency, reducing throughput, and increasing transmission costs. The entire bridge inspection system will be effected negatively.

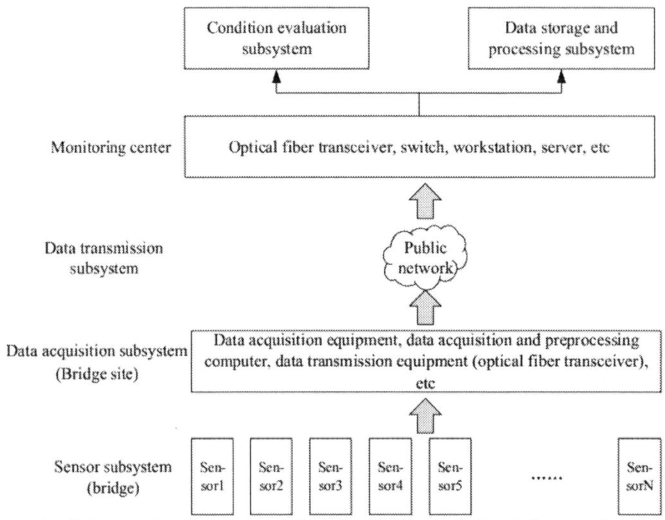

Figure 1. Schematic diagram of bridge structure health monitoring system

Through observing the data collected by the data collection center, it can be shown in Figure 2, which is a part of the temperature data of Sutong Bridge on January 12, 2018. It is not difficult to find that these data structures are monotonous and have a very high repetition rate. Thus, Compression necessitates most, while good and effective compression method can reduce or even eliminate the above-mentioned negative effects and greatly reduce the amount of data. However, due to the large amount of raw data, it takes time to compress these data. The time cost of data compression is included in the total transmission cost from the sensor system to the monitoring center, so the realization of data compression has also become very important.

0.587800	0.587800	0.551769
0.587800	0.587800	0.551769
0.587800	0.587800	0.551769
0.587800	0.587800	0.587800
0.587800	0.551769	0.587800
0.587800	0.551769	0.587800
0.587800	0.551769	0.551769
0.587800	0.587800	0.551769
0.587800	0.587800	0.551769
0.587800	0.587800	0.587800
0.587800	0.587800	0.587800
0.587800	0.587800	0.587800
0.587800	0.587800	0.587800
0.587800	0.587800	0.587800
0.587800	0.587800	0.587800
0.587800	0.587800	0.551769
0.587800	0.587800	0.551769

Figure 2. Bridge temperature data

This paper proposes a parallel compression method of LZSS bridge data based on FPGA [3,4], and formulates a computing architecture using this method for parallel computing. By burning the LZSS algorithm [5] into the FPGA method, N pieces of data are compressed in N channels in parallel, which speeds up the compression, thereby improving the transmission efficiency of the entire system, increasing the throughput, and reducing the time cost of transmission.

2. METHODS

2.1. LZSS Algorithm

The LZSS algorithm is a dictionary-based compression algorithm, which is an enhancement and improvement to the LZ77 algorithm. The LZ77 algorithm maintains a forward buffer and a sliding window. For example, if the forward buffer contains character sequences (A, B, C), then phrase in the buffer has {(A),(A,B),(A,B,C)}, and if the sliding window contains the phrase (A,B,C), then the phrase in the dictionary has {(A), (A,B),(A,B,C),(B,C),(B),(C)}. The data in the forward buffer is compared with the phrases in the dictionary, and if there is a match, the offset B and the matching length L are output. The LZSS algorithm improves the phenomenon that LZ77 may output null pointers and characters that may match in the next iteration although they do not match in this iteration. The LZSS algorithm sets the minimum matching length M. If there are matches greater than or equal to M characters in the iteration process, then (B, L) is output, otherwise the first character of the forward buffer is output and moved forward by one character.

Table 1. LZSS algorithm flow

Enter the data to be encoded:
POS: 1 2 3 4 5 6 7 8 9 10 11
A A A A B B B A A A A
Assuming that the initial window is 4, the sliding window is:
AAAA, and the forward buffer is: BBBAAAA;
Assume that the minimum matching length M is 2.

STEP	POS	MATCH	OUTPUT
1	0	--	B
2	0	--	B
3	2	BAA	(2,3)
4	1	AA	(1,2)

Taking the second column of the bridge temperature data shown in Figure 2 as an example (this column is the most difficult column of the temperature data shown in the figure, because this column contains different data), the specific process of the LZSS algorithm can be expressed as Table 1, where A represents the data 0.587800, and B represents the data 0.551769. In the first step, a character string that matches the dictionary is not found in the forward buffer, so B is output, the sliding window becomes AAAB, and the forward buffer becomes BBAAAA. In the second step, because only B that matches the dictionary is found in the buffer, its length is less than the minimum matching length, so B is output directly, the sliding window becomes AABB, and the forward buffer becomes BAAAA. In the third step, the character string BAA that matches the dictionary is found in the buffer, so (2,3) is output, which means that two characters are reverted forward, three characters are copied, the sliding window becomes BBAA, and the forward buffer becomes AA . In the fourth step, the character string AA that matches the dictionary is found in the buffer, so (1,2) is output, which means that one character is reverted forward, and two characters are copied. At this point, the buffer becomes empty, and the final compression result is AAAABB (2,3)(1,2).

2.2. The advantages of FPGA compression

Programmable gate array FPGA has been widely used in various fields where technology has spread due to its advantages such as flexible design, short cycle, low cost, rich logic resources, high parallelism, and strong reconstruction. Traditional bridge data acquisition subsystems mostly use CPU as computing power to compress sensor data. Its basic structure is shown in Figure 3 (Left). FPGA contains a large number of configurable logic units (CLB), and its basic structure is shown in Figure 3 (Right). It shows that because the CPU processor adopts the Von Neumann structure, the operation process is serial, and the parallelism is poor.

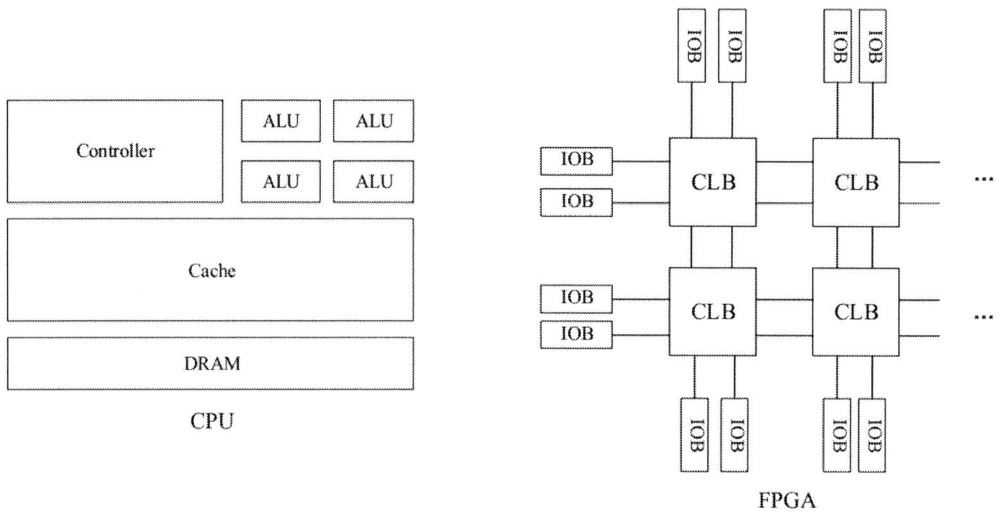

Figure 3. Basic structure of CPU and FPGA

The characteristics of the bridge data acquisition subsystem receiving data from the sensor subsystem are analyzed, as shown in Figure 4, which is a part of the temperature data generated by a certain bridge in a certain day, where WD stands for temperature, the number after WD denotes sensor number, and the following underlined number is referred to the time when the temperature data was generated. It can be seen that there will be a lot of data from the sensor subsystem at the same time.

The computational efficiency of using CPU for compression will be reduced, and FPGA contains a large number of logic operation units, which can process a large amount of data at the same time, analyze the LZSS data compression process, and perform a large number of logic and operations to compare data whether it matches during dictionary matching. FPGA is adept at logic operations, therefore, the use of FPGA for data compression is very feasible and efficient. The parallel compression method of LZSS data based on FPGA can improve the computational efficiency of the data acquisition subsystem and reduce the bandwidth used for data transmission.

WD080101_090000.txt	WD080102_090000.txt	WD080103_090000.txt
WD080101_100000.txt	WD080102_100000.txt	WD080103_100000.txt
WD080101_110000.txt	WD080102_110000.txt	WD080103_110000.txt
WD080101_120000.txt	WD080102_120000.txt	WD080103_120000.txt
WD080101_130000.txt	WD080102_130000.txt	WD080103_130000.txt
WD080101_140000.txt	WD080102_140000.txt	WD080103_140000.txt
WD080101_150000.txt	WD080102_150000.txt	WD080103_150000.txt
WD080101_160000.txt	WD080102_160000.txt	WD080103_160000.txt
WD080101_170002.txt	WD080102_170002.txt	WD080103_170002.txt
WD080101_180000.txt	WD080102_180000.txt	WD080103_180000.txt
WD080101_190000.txt	WD080102_190000.txt	WD080103_190000.txt
WD080101_200000.txt	WD080102_200000.txt	WD080103_200000.txt
WD080101_210000.txt	WD080102_210000.txt	WD080103_210000.txt
WD080101_220000.txt	WD080102_220000.txt	WD080103_220000.txt
WD080101_230000.txt	WD080102_230000.txt	WD080103_230000.txt

Figure 4. Partial temperature data of the bridge on a certain day

2.3. A parallel compression method of LZSS based on FPGA

Firstly, the computing architecture for the parallel compression method of FPGA-based LZSS bridge data is defined as shown in Figure 5. The above LZSS algorithm is designed using Vivado HLS, and the FPGA chip is burned to the FPGA chip according to the FPGA chip model. Secondly, it is assumed that N pieces of data are received from the bridge at the same time and input them to the off-chip DRAM of the FPGA, then the controller transmits the input data from the off-chip DRAM to the on-chip buffer of the FPGA. Finally, after the transmission is completed, the on-chip processing unit (Processing unit) is divided into N small on-chip processing units according to the number N of input data to realize N-way parallel execution.

The specific steps for using the parallel compression method are as follows:

1) HLSC/C++ realizes LZSS compression program.

2) Vivado HLS is used to burn the program in step 1 into the data acquisition subsystem and monitoring center.

3) At a certain moment, N pieces of sensor data passed in from the sensor subsystem are read into the off-chip DRAM of the FPGA.

4) The controller controls the on-chip cache to read the data in the DRAM.

5) At this time, DRAM can read N pieces of sensor data from the next moment.

6) N pieces of data in the cache are handed over to N on-chip processing units for compression processing.

7) The compressed data first passes through the on-chip buffer area, and finally is output to the off-chip DRAM by the controller.

8) Data is transmitted from the data acquisition subsystem to the monitoring center through the public network of the data transmission subsystem.

9) The data is read into the off-chip DRAM of the FPGA.

10) Repeat Steps 4) ~ 5)

11) N pieces of data in the cache are handed over to N on-chip processing units for decompression processing.

12) The decompressed data first passes through the on-chip buffer area, and finally is output to the off-chip DRAM by the controller.

13) These data are finally analyzed and processed by other modules of the monitoring center.

Among them, Step 1 burns the LZSS compression algorithm into the FPGA, and Steps 2-12 complete the data reading, data compression, and data writing operations in the data acquisition subsystem. Due to the parallelism of the FPGA, the compression process can be executed in parallel to shorten the compression time. Step 13 transmits the compressed data from the data acquisition subsystem to the monitoring center. Steps 14-24 complete the data read-in, data decompression, and data write-out operations of the monitoring center. The same decompression process can be executed in parallel. Step 25 passes the data to other subsystems for processing.

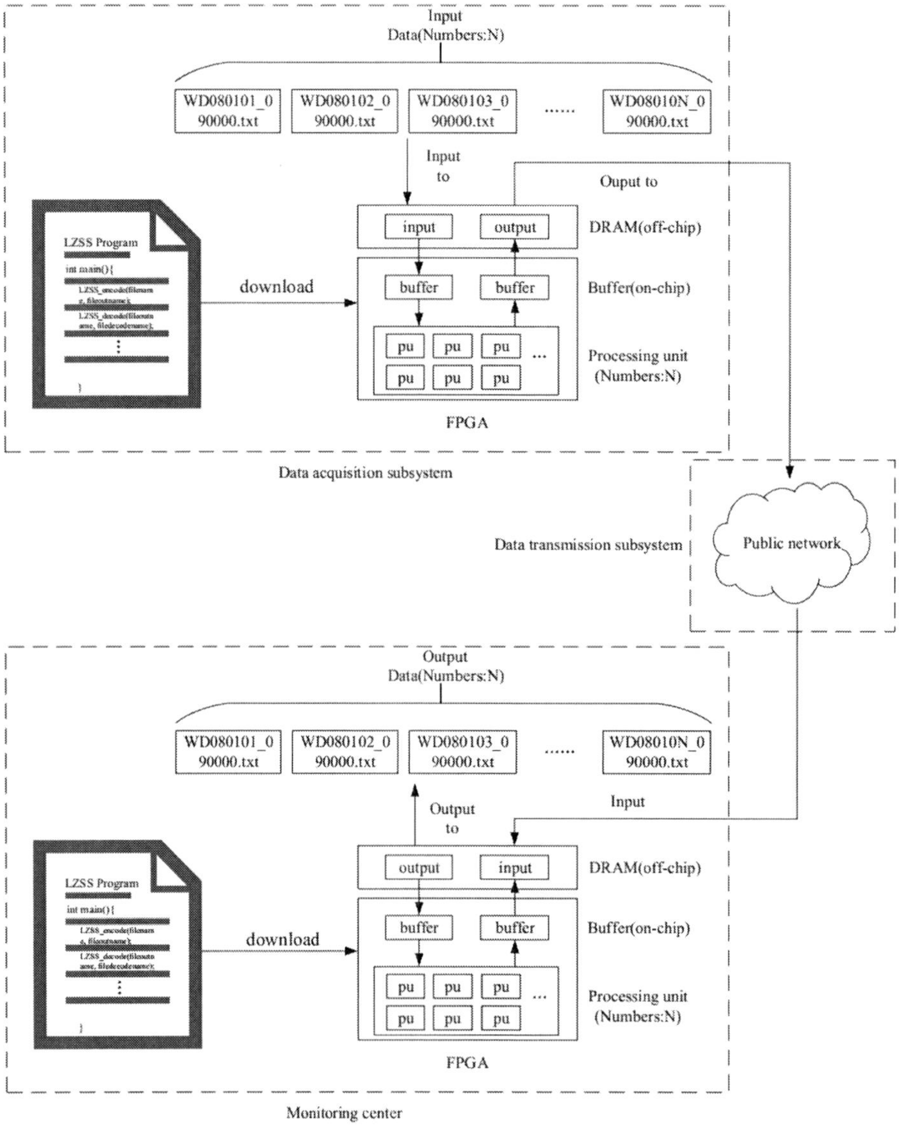

Figure 5. FPGA-based LZSS bridge data compression parallel computing architecture

The pseudo code for the above steps is as follows:

FPGA.burn(LZSS);
FOR all ni∈N
DRAM.read(ni);
END FOR
FOR all ni∈N
Buffer.read(ni);
LZSS.encode(ni, ni_encode);
Buffer.write(ni_encode);
END FOR

```
FOR all ni_encode∈N_encode
  DRAM.write(ni_encode);
END FOR
Do transform(N_encode) FORM Acquisition subsystem TO Monitoring center
FOR all ni_encode∈N_encode
  DRAM.read(ni_encode);
END FOR
FOR all ni_encode∈N_encode
  Buffer.read(ni_encode);
  LZSS.dencode(ni_encode, ni);
  Buffer.write(ni);
END FOR
FOR all ni∈N
  DRAM.write(ni);
END FOR
Processing(N) By other subsystems;
```

3. Results

The DEV-C++ tool is used in the Windows system to implement the LZSS algorithm, and the data of the Sutong Bridge at 9 am every day from January 12 to January 31, 2018 is compressed and decompressed. The data is transmitted from the sensor system at the same time. There are 40 temperature data files, that is, 40 temperature sensors transmit temperature data files to the data acquisition subsystem at the same time. Compare the LZSS algorithm using the traditional system with the LZSS algorithm implemented using ten-way parallel FPGA (ie N=10). The time comparison diagram of compression at the data acquisition system side is shown in Figure 6, and the abscissa axis represents the date. The vertical axis represents the time of compression at nine o'clock in the morning of the day. It can be seen that the compression time of FPGA is about 8 times shorter than that of traditional CPU under the same number and amount of data files. Similarly, the time comparison chart for decompression in the monitoring center is shown in Figure 6.

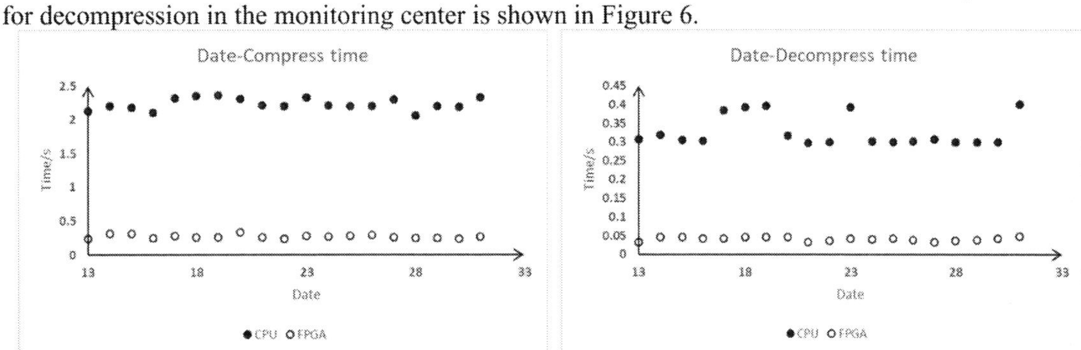

Figure 6. Comparison of compression and decompression time between traditional CPU and FPGA on January 13-31

Based on the specific analysis of the data at 9:00 am on January 12, 2018, Figure 7 shows the relationship between the number of input files and the time when the traditional system is used for compression and decompression and the system embedded with FPGA. It can be seen that the compression using FPGA The time is about 8 times shorter than that of a traditional CPU.

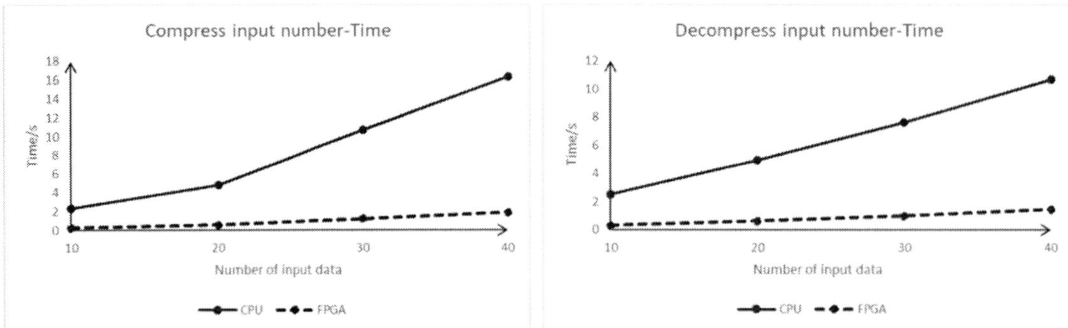

Figure 7. The relationship between the number of input files and the time of compression and decompression on January 12

4. Conclusion

To deal with the problems of low transmission efficiency, low throughput and high transmission cost of the large amount of data to be transmitted from the sensor subsystem of the bridge health monitoring system to the data monitoring center, this paper proposes a parallel compression method LZSS bridge data based on FPGA, formulates a computing architecture using this method for parallel computing, and implements a compression parallel method based on this architecture, which solves the problems of long data compression and low transmission efficiency caused by high time costs, thereby improving bridge health monitoring. Thus, the system transmission efficiency is increased, the throughput is boosted, and the transmission cost is saved.

Acknowledgments

This paper is supported by Henan Provincial Department of Transportation Science and Technology Planning Project (2019J-2-5).

References

[1] Booth, N. and Smith, A. S., [Infrared Detectors], Goodwin House Publishers, New York & Boston, 241-248 (1997).

[2] Davis, A. R., Bush, C., Harvey, J. C. and Foley, M. F., "Fresnel lenses in rear projection displays," SID Int. Symp. Digest Tech. Papers 32(1), 934-937 (2001).

[3] Van Derlofske, J. F., "Computer modeling of LED light pipe systems for uniform display illumination," Proc. SPIE 4445, 119-129 (2001).

[4] Myhrvold, N., "Confessions of a cybershaman," Slate, 12 June 1997, <http://www.slate.com/CriticalMass/97-06-12/CriticalMass.asp> (19 October 1997). www.optics4yurresearch.com/7752.html

[5] Jones, C. J., Director, Miscellaneous Optics Corporation, interview, Sept. 23 2011

[6] FamilyName, GivenName Initial., "Title," Source, pg# (year).

Application of millimeter wave radar in safety obstacle avoidance of crane in electric power construction

Wei He[1] , Shenglei Hu[1] , Wang Pan[2] and Lijuan Zhang[2*]

[1] Shanghai High Voltage Industry Co. Ltd, Shanghai ,200331, China

[2]School of Electrical and Electronic Engineering, Shanghai Institute of Technology, Shanghai ,201418, China

[*]Corresponding author's e-mail: zljsmx@sit.edu.cn

Abstract. When the traditional electric construction crane safely avoids obstacles, obstacles are only detected according to the parameter estimation results, which is greatly affected by the external environment and takes a long time to avoid obstacles. Aiming at the above problems, the safety obstacle avoidance method of electric construction crane based on millimeter wave radar is studied. The millimeter wave radar is used to obtain the millimeter wave radar detection signal for crane safety obstacle avoidance. According to the impulse response of millimeter wave radar detection echo, the obstacle location mark and distance estimation are carried out, and the distance and azimuth information between the crane and the obstacle are calculated. By optimizing the safe obstacle avoidance path of electric construction cranes, the safe obstacle avoidance of electric construction cranes can be realized. Through comparative simulation experiments, it is verified that the design method takes less time to avoid obstacles, has higher efficiency, and does not collide with obstacles in the workspace in the process of movement, which verifies the effectiveness of the method.

1. Introduction

Crane is the main construction machinery widely used in electric power construction, and it is indispensable and important for electric power construction. With the wide use of the crane, its safety problem cannot be ignored[1]. The density of obstacles around electric power construction cranes is also increasing, which makes the electric power construction environment complex. There are often cross operations of multiple cranes, even close cross operations. In this case, it is easy to cause collision between cranes due to human operation errors, or collision accidents between cranes and surrounding buildings, and even fatal accidents. The traditional safety obstacle avoidance method of electric construction crane is to set ultrasonic sensor or infrared detector on the boom and hook of tower crane, and use ultrasonic or infrared to monitor the distance of surrounding obstacles. Although it can play a certain obstacle avoidance effect within a limited range, it has the problem of low obstacle avoidance rate in actual detection. Millimeter wave radar has the advantages of high resolution, high precision, high efficiency, and small antenna aperture. So millimeter wave radar is applied to the safety obstacle avoidance of electric power construction crane, and a safety obstacle avoidance method of electric power construction crane based on millimeter wave radar is designed.

2. Safety obstacle avoidance method of electric power construction crane based on millimeter wave radar

The flow chart of safety obstacle avoidance method of power construction crane based on millimeter wave radar designed in this paper is shown in figure 1. The five-step process shown in the figure will be described in detail below.

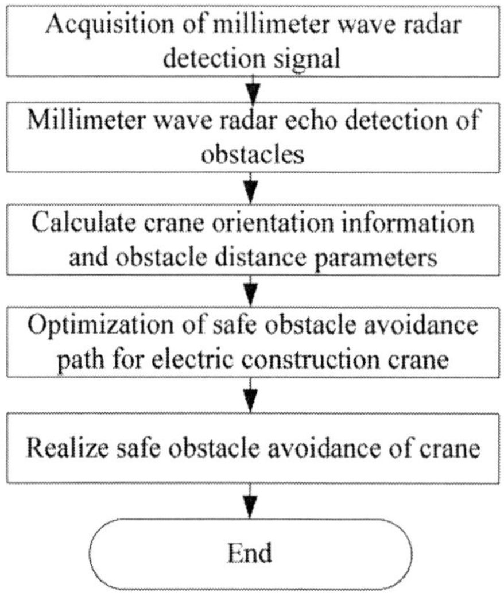

Figure 1. Safety obstacle avoidance method flow of electric construction crane based on millimeter wave radar

2.1. Obtain millimeter wave radar detection signal for crane safety obstacle avoidance

In order to realize the safety obstacle avoidance method of electric power construction crane based on millimeter wave radar, firstly, the millimeter wave radar pulse detection signal is obtained. The wireless sensor device is adopted at the bottom of the safety obstacle avoidance method of electric construction crane for signal acquisition. The spacing of millimeter wave radar signal acquisition array elements for ranging in the safety obstacle avoidance of electric construction crane in the x-axis direction is d_x, and the spacing of millimeter wave radar ranging pulse signal array elements for safety obstacle avoidance of electric construction crane in the y-axis direction is d_y. The array element spacing of safety obstacle avoidance of electric construction crane meets the requirements, as shown in equation (1).

$$d_x=d_y \tag{1}$$

The correlation spectrum detection method is used to separate the electromagnetic interference factors of millimeter wave radar pulse signal, and $n(t)$ is separated. In the safety obstacle avoidance of electric power construction crane, the scattering function R_z of crane millimeter wave radar ranging and echo detection output is:

$$R_z=R_x(\varphi)+n(t) \tag{2}$$

In equation (2), $R_x(\varphi)$ is the angle information representing the obstacle position, and the angle information of the obstacle position after introducing the cross-correlation core detected by radar millimeter wave can be expressed as equation (3):

$$R_x(\varphi)=d_y d_x \sum_{i=1}^{p}\sum_{j=1}^{p}\alpha(\theta) \tag{3}$$

Where p represents the total number of signal fluctuations; i represents the fluctuation times of x-axis direction signal; j represents the fluctuation times of y-axis direction signal; α represents the initial cross-correlation kernel function of radar millimeter wave detection; θ represents the termination cross-correlation kernel function of radar millimeter wave detection. The matched filter detection method is used to process the millimeter wave radar pulse signal of crane obstacle avoidance[2]. Multi-scale feature of millimeter wave radar detection echo pulse signal is decomposed, and then crane safety obstacle avoidance is carried out according to the feature decomposition results. Assuming that the running speed of crane is v, the relationship between target azimuth and expansion angle of crane safety obstacle avoidance millimeter wave radar detection signal is $g(t)$:

$$g(t) = R_s(\varphi)\sqrt{s}v(s(t-\tau)) \tag{4}$$

In equation (4), s is a two-dimensional spectral peak, abbreviated as scale; t is the termination coefficient of characteristic decomposition; τ is the initial coefficient of characteristic decomposition. The problem of target parameter estimation for crane safety obstacle avoidance is transformed into the problem of distributed source azimuth estimation. The linear feedback equalizer is used to design the channel model of millimeter wave radar transmission data for crane safety obstacle avoidance, modulate and demodulate the transmission signal, and the scale expansion change method is used for obstacle control.

2.2. Millimeter wave radar echo detection of obstacles

Assuming that there are N obstacle distributed nodes, the detection signal of crane safety obstacle avoidance millimeter wave radar is detected under the background of interference. By estimating the scale and time delay of the millimeter wave radar detection signal for crane safety obstacle avoidance, the time scale W of the millimeter wave radar detection signal is obtained by using the signal covariance estimation method, that is:

$$W = \sqrt{s} \cdot r(t)^*(s(t-\tau)) \tag{5}$$

Where $*$ represents complex conjugate. Combined with the minimum mean square error estimation method, the mean value of crane safety obstacle avoidance navigation control is recorded as D:

$$D = W \cdot x(t+\frac{\tau}{2}) \tag{6}$$

The time domain and frequency domain of millimeter wave radar detection signal for crane safety obstacle avoidance are combined through equation (6). According to the maximum peak detection results, windowing operation is carried out. The windowing function is N_f. The transmission channel pulse broadening T_s of millimeter wave radar for crane safety obstacle avoidance is as follows:

$$T_s = N_f D \tag{7}$$

According to the pulse broadening results of the transmission channel of the millimeter wave radar for crane safety obstacle avoidance, the discrete characteristic component of the millimeter wave radar detection signal for crane safety obstacle avoidance is detected. Using the spectrum separation detection method, the steep gradient function expression k of millimeter wave radar pulse signal for crane safety obstacle avoidance is obtained as follows:

$$k = T_s(x^4) - 2T_s(x^2) \tag{8}$$

According to equation (8), the obstacle location mark and distance estimation are carried out according to the impulse response of millimeter wave radar detection echo, so as to accurately estimate the distance and azimuth information between the crane and the obstacle[3-4]. In the millimeter wave radar system for crane safety obstacle avoidance, the frequency component feature decomposition method is used for signal feature decomposition, and the transmission signal of millimeter wave radar for crane safety obstacle avoidance is sampled. The time-frequency decomposition method is used to realize the signal conversion from time domain to frequency domain, complete the millimeter wave radar echo detection of obstacles, and improve the ability of detection and obstacle avoidance.

2.3. Calculate the orientation information of electric construction crane and obstacle distance parameters

Based on millimeter wave radar echo detection of obstacles, the azimuth information of electric power construction crane and obstacle distance parameters are calculated. The time domain characteristics of millimeter wave radar signal for crane safety obstacle avoidance are obtained by setting the array element spacing of millimeter wave radar signal sampling for crane safety obstacle avoidance. Under the quasi-stationary random detection model, the discrete characteristic component e of the detection control output of millimeter wave radar is calculated by using equation (9).

$$e = \frac{\vartheta}{\vartheta u}[\vartheta D] \tag{9}$$

Where ϑ represents the probability of electric construction crane avoiding obstacles; u is a discrete function. According to the relevant characteristics of obstacle avoidance millimeter wave radar detection signal, in the millimeter wave radar detection of crane safety obstacle avoidance, the obstacle location mark and distance estimation are carried out according to the impulse response of millimeter wave radar detection echo. If the objective function is A, there is equation (10).

$$A = \frac{f^2}{\vartheta u^2}[t_0] + \frac{t}{\gamma} \tag{10}$$

Where f is the frequency domain decomposition feature; t_0 is the time of obstacle location mark and distance estimation; γ is the angle information component of the obstacle[5-6]. Through equation (10), the orientation information of electric construction crane and obstacle distance parameters are calculated as the key parameters for safe obstacle avoidance of electric construction crane.

2.4. Optimization of safe obstacle avoidance path for electric construction crane

According to the calculated orientation information of electric construction crane and obstacle distance parameters, plan the next travel direction of crane. Support vector machine algorithm is used to train the optional travel direction data of crane. If the training set is U, there is $U = \{(x_1, y_1), (x_2, y_2), \ldots, (x_m, y_m)\}$, and $U \in (x \times y)^m$, x, y are kernel functions; m refers to the traveling direction of the crane. After obtaining the training set, the training set is defined as C in the moving space of the crane. The midpoint at both ends of the C space is taken as the moving control point of the crane. The potential field value of the crane obstacle avoidance space is calculated through the support vector machine algorithm, and the potential field value of the crane obstacle avoidance space is set as $j_1(v)$:

$$j_1(v) = \begin{Bmatrix} {}^{tj_1} \\ {}^{tj_1}x + {}^{tj_1}z \\ {}^{tj_1}y - {}^{tj_1}z \end{Bmatrix} \tag{11}$$

Where ${}_tj_1$ refers to the number of crane moving joints; ${}^{tj_1}x$ refers to the torsional linear speed of the crane moving joint; ${}^{tj_1}y$ refers to the torsional linear velocity vector of the crane moving joint; ${}^{tj_1}z$ refers to the torsional angle velocity vector of the crane moving joint[7-8]. Obtain the C space potential field value of the crane configuration corresponding to each node through equation (11), and select the node with the smallest potential field value as the successor node of the current node. In this way, the greater the potential field value after recursive iteration, the greater the possibility of safe obstacle avoidance of the crane. Through path optimization, select the current node and subsequent nodes without collision[9]. The support vector machine algorithm is used for the path optimization of the nearest priority, that is, the RBF kernel function is obtained. If the expression of the RBF kernel function is $K(x, y)$, the equation is shown in the following.

$$K(x, y) = \exp(-r\|x_i - x_j\|^2) \tag{12}$$

Where r refers to the optimal training times. The RBF kernel function can not only reduce many unnecessary calculations in the traditional crane safety obstacle avoidance method, but also obtain the optimal solution, but also effectively solve the discontinuity of crane safety obstacle avoidance path planning.

2.5. Realize safe obstacle avoidance of crane

Through the optimization of the above electric construction crane safety obstacle avoidance path, signal analysis and obstacle avoidance, it is obtained that the decision function H of crane safety obstacle avoidance.

$$H = K(x, y)(t) \tag{13}$$

On this basis, the Internet of things for crane safety obstacle avoidance is developed and designed with Revit software, and the bus development and hardware integration design for crane safety obstacle avoidance are carried out with ADSP-BF537 as the core processor[10]. ISA/EISA/Micro Channel expansion bus is used to load the instructions of crane safety obstacle avoidance, and the program loading and cross compilation control of crane safety obstacle avoidance are carried out in the embedded environment. The JTAG debugging interface is used to read and write the real-time program of the crane safety obstacle avoidance code, ADSP-BF537BBC-5A is used to collect the original information of the crane safety obstacle avoidance system, and the collected automatic obstacle avoidance control commands are imaged and loaded. To sum up, the design method of safety obstacle avoidance of electric construction crane based on millimeter wave radar is realized.

3. Simulation experiment

3.1. Experimental preparation

In order to test the application performance of this method in realizing safe obstacle avoidance of electric construction crane, simulation experiments are carried out. The simulation experiments are carried out in INTEL (R) core (TM) i7-6820 CPU@3.60 GHz, 16 GB memory, on a computer equipped with 64-bit windows7 operating system. To obtain the accurate speed performance parameters of the obstacle avoidance method proposed in this paper, the method proposed in this paper is implemented in C++ language by using Visual Studio 2013 software platform. At the same time, considering the strong matrix operation and data visualization ability of MATLAB platform, the joint angle value sequence obtained after obstacle avoidance operation by C++ program is input into MATLAB platform and visualized to obtain reliable and intuitive results. This experiment selects an electric construction crane as the experimental object, and the top view of plane 2R crane drawn by MATLAB software. It takes about 0.15s to establish the external collision database OCQD and self-collision database SCQD of this 2R planar manipulator in Visual Studio software platform. The configuration space obstacle model can be obtained through the unit serial number index OCQD and SCQD of square obstacles in the workspace, and the color of the configuration space obstacle model corresponds to the workspace obstacle model one by one. The obstacle avoidance path planning problem of plane 2R crane in workspace is transformed into the problem of finding a path connecting the starting point and target point in the free region of configuration space.

Firstly, the traditional obstacle avoidance method is used for the safe obstacle avoidance of electric construction crane, and then the obstacle avoidance method based on millimeter wave radar is used to implement the same operation steps. The response time of millimeter wave radar ranging and echo detection is set to 0.25ms, the length of millimeter wave radar signal acquisition is 1024, the bandwidth is 24dB, and the interference signal-to-noise ratio is 10dB. According to the above simulation environment and parameter settings, millimeter wave radar ranging and echo detection are carried out to realize crane safety obstacle avoidance and obtain millimeter wave radar echo signal. The obstacle avoidance method designed in this paper is set as the experimental group, and the traditional obstacle avoidance method is the control group. At the same time, test the total obstacle avoidance time under the two obstacle avoidance methods. The shorter the total time, the higher the obstacle avoidance efficiency of the method. According to the simulation experiment steps designed above, the experimental data under the two obstacle avoidance methods are collected, and the status of crane obstacle avoidance path planning is displayed by embedding IMU in SRTYE rendering window through human-computer interaction interface.

3.2. Analysis and conclusion of experimental results

To show the difference more intuitively in the total time of obstacle avoidance between the two methods, the experimental results under the two obstacle avoidance methods are compared together. The experimental results are shown in table 1:

Table 1. Comparison results of total obstacle avoidance time

Number of experiments	Experience group			Control group		
	Obstacle avoidance path planning takes time (s)	Obstacle avoidance path search time-consuming (s)	Total time (s)	Obstacle avoidance path planning takes time (s)	Obstacle avoidance path search time-consuming (s)	Total time (s)
1	0.036	0.015	0.051	0.075	0.035	0.110
2	0.035	0.014	0.049	0.078	0.036	0.114
3	0.033	0.015	0.048	0.080	0.029	0.109

It can be seen from table 1 that the obstacle avoidance path planning time and obstacle avoidance path search time of the design method in this paper are faster than those of the control group. Therefore, it is proved that the total obstacle avoidance time of the design method is shorter, the efficiency is higher, and there is no collision with the obstacles in the workspace during the movement, which verifies the effectiveness of the method proposed in this paper and has good application value.

4. Conclusion

In this paper, millimeter wave radar is applied to the safety obstacle avoidance design of electric construction crane, and a new safety obstacle avoidance method of electric construction crane is proposed, which improves the obstacle avoidance ability of electric construction crane. Through the above research, obstacle location and distance estimation based on millimeter wave radar detection echo impulse response is feasible and accurate. The method provides theoretical support for improving the working efficiency of electric construction crane. However, due to the limited research time, the proposed method still has some shortcomings. In the future, the method needs to be continuously optimized and improved through practical application. In addition, other equipment obstacle avoidance related contents will be further studied in the future, including the detection and optimization of crane obstacle avoidance technology and the research on crane obstacle avoidance method in three-dimensional state, to promote the scientific development of power construction.

References

[1] Zhang W , Chen H , Chen H , et al. A Time Optimal Trajectory Planning Method for Double-Pendulum Crane Systems With Obstacle Avoidance[J]. IEEE Access, 2021, (99):1.

[2] Cao C , J Gao, Liu Y C . Research on Space Fusion Method of Millimeter Wave Radar and Vision Sensor[J]. Procedia Computer Science, 2020, 166:68-72.

[3] Lingard H , Cooke T , Zelic G , et al. A qualitative analysis of crane safety incident causation in the Australian construction industry[J]. Safety Science, 2021, 133:105028.

[4] Bagheri S M , Taghaddos H , Mousaei A , et al. An A-Star algorithm for semi-optimization of crane location and configuration in modular construction[J]. Automation in Construction, 2021, 121:103-447.

[5] Zhang Z , Pan W . Virtual reality supported interactive tower crane layout planning for high-rise modular integrated construction[J]. Automation in Construction, 2021, 130:103-854.
[6] Shentu S , Xie F , Liu X J , et al. Motion Control and Trajectory Planning for Obstacle Avoidance of the Mobile Parallel Robot Driven by Three Tracked Vehicles[J]. Robotica, 2020:1-14.
[7] He Y , Wu M , Liu S . A cooperative optimization strategy for distributed multi-robot manipulation with obstacle avoidance and internal performance maximization[J]. Mechatronics, 2021, 76:102-560.
[8] Li W , Xiong R . Dynamical Obstacle Avoidance of Task- Constrained Mobile Manipulation Using Model Predictive Control[J]. IEEE Access, 2019, (99):1.
[9] Mu Z , Liu T , Xu W , et al. A Hybrid Obstacle-Avoidance Method of Spatial Hyper-Redundant Manipulators for Servicing in Confined Space[J]. Robotica, 2019, 37(6):998-1019.
[10] Xi X , Shi Y , Jin Y , et al. Obstacle avoidance path control method for agricultural machinery automatic driving based on optimized Bezier[J]. Nongye Gongcheng Xuebao/Transactions of the Chinese Society of Agricultural Engineering, 2019, 36(19):82-88.

Ethernet transmission method based on modified format for distribution domain protection system

Haijiang Wu[1], Jinrong Chen[1], Feng Liao [1],Yonghao Li [1], Shaoheng Zhong[1], Shuaibing Wang[2*]

[1]Guangdong power Grid Corporation, Guangdong, Foshan, 52800

[2]Beijing NENET technology development Co., LTD, Beijing, 100192

*Corresponding author's e-mail: wangshuaibing@nenetsystem.com

Fund Item: Science and Technology Project of China Southern Power Grid Corporation 030600KK52200019[GDKJXM20200486]

Abstract: This paper proposes an Ethernet transmission method based on modified format, which is applied to the distribution domain protection system. By tagging Ethernet data new identifications, the data has unique identity information in transmission. The switch replicates data and transmit it in two directions. When the ring network is destroyed, there is no data loss caused by switching the transmission direction. The distribution domain protection system realizes ring network with seamless redundancy.

1. Introduction

At present, distribution network protection is in an important period of reform. In the past, the local protection that detects short-circuit current cannot accurately locate faults according to the overall information of the region in the scenario of multi-power supply access. However, the station-domain protection system can solve this problem well. The station-domain protection system requires peer to peer data interaction between devices, and Ethernet is the mainstream communication mode at present. In the distribution network, network equipment generally uses RSTP, STP, ring network protection protocol and so on to form ring network extension[1]. When the ring network fails, the topology self-healing time ranges from 10 ms to 500 ms. During this period, the local system cannot transmit instructions properly. In this paper, an Ethernet transmission method based on format modification is proposed to locate, copy and eliminate Ethernet data. The two copies of data are transmitted in the two directions of the ring network at the same time, so that there is no self-healing process and data is not lost when the ring network fails [2].

2. Principle and Design

The paper reports the operation principle of Ethernet ring network, and forming a ring network on the physical connection of the switch. The data transmission logic is single-direction transmission. Therefore, when a link fails, data must be switched to another direction for transmission, which results in data loss.

In this paper, two copies of the same data were formed through the data replication of the switch, and the two copies of data were transmitted in two directions of the ring network at the same time. In the case of a link failure in one direction, data in the other direction can still reach the destination, thus eliminating the process of data transmission direction reversal in the case of a switch ring network failure to achieve seamless redundancy [3].

2.1. Ethernet data format modification
In the method discussed in this paper, the two-way transmission of the switch ring network depends on the replication of data by the starting switch and the elimination of duplicate data by the ending switch. Whether to copy data or eliminate redundant data, accurate identification of Ethernet data must be done first. The Ethernet format identifies only the source device (source MAC) and destination device (destination MAC) of the data (Figure 1). On the switch, the characteristics of the data sent from the same source MAC device to the same destination MAC are the same. Switches cannot data format distinguish between each data through Ethernet, the switch cannot to accurate identification of data, the starting switch to the format of data replication, end switches cannot distinguish between the same source MAC and destination MAC data replication data and original data of one-to-one correspondence relationship, also can't do the correct copy data to eliminate. Therefore, this paper first defines a modified Ethernet data format, so that the switch can accurately identify Ethernet data [4].

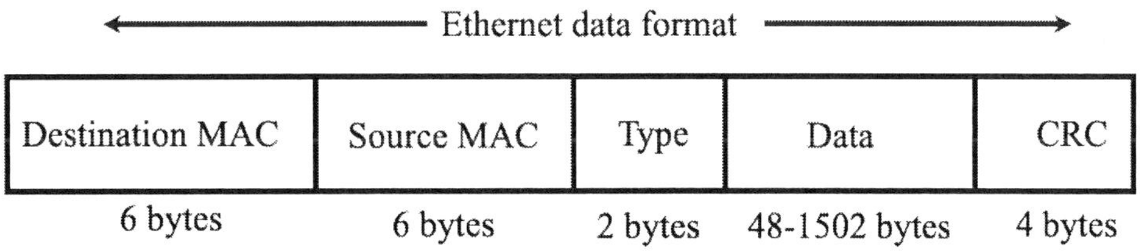

Figure 1. Ethernet data format.

2.1.1 Modified Ethernet data format
The modified *Ethernet* format adds three additional identifiers: source switching device ID, target switching device ID, and data frame ID. The switching equipment of the fast protection system of distribution network website can identify the unique identity of data by the source switching equipment ID, target switching equipment ID, data frame ID, source MAC and destination MAC. The added Ethernet data format is shown in Figure 2.

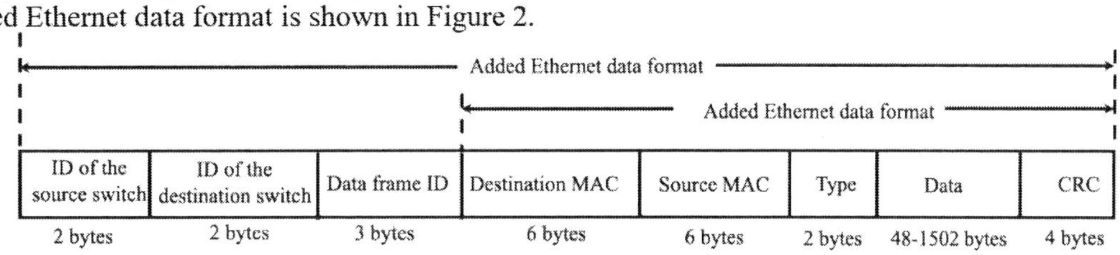

Figure 2. Added Ethernet data format.

In the new data format, the source switch ID refers to the ID of the switch connected to the source network device. Destination switch ID Indicates the ID of the switch to which the destination network device connects. The data frame ID refers to the sequence ID of data frames sent by the switching device. The sequence starts from 1 to a certain value and then repeats from 1.

2.2 Process the modified data on the switch
The modified data format workflow is shown in Figure 3. Power distribution Site Domain protection devices send data in the original Ethernet format and the connected switches add new data in the format. Source switch ID Indicates the ID of the switch. Destination switch ID indicates the ID of the last switch during data transmission (the ID of the destination switch can be all F if the data is broadcast). Frame ids are incremented from 1 to 1 based on the sequence of data entering the switch. The data format is added by the source switch and the data is copied into two copies for transmission.

The switch records data identification during the intermediate process. The destination switch records the identifier of the incoming data, reformatting the first incoming data, and sends the source-to-source-to-source-to-start Ethernet data to the domain-type protection device on the power distribution website to complete data transmission. The destination switch discards the data with the same id that reaches the destination switch for the second time.

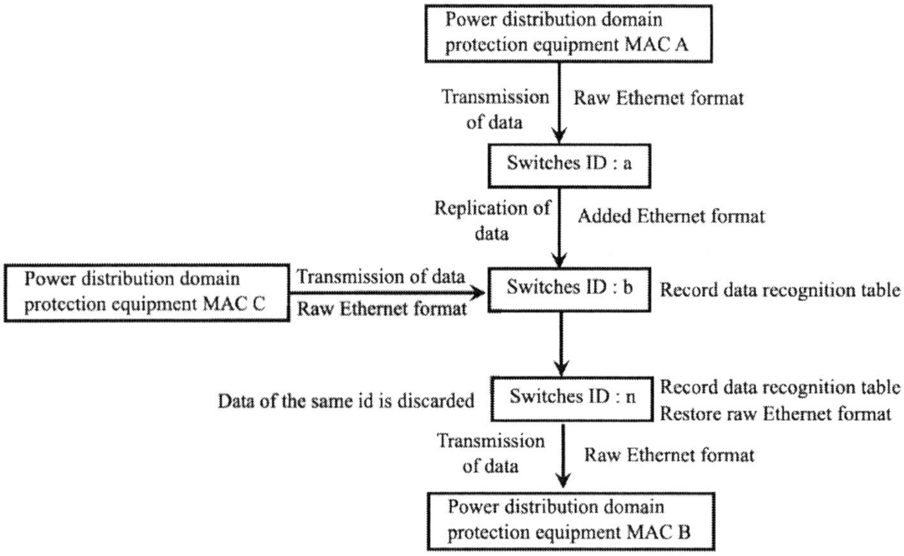

Figure 3. Process after the format is changed.

Table 1. Data information received by each device

Project	Equipment	Source Switches ID	Destination switch ID	Frame ID	Destination MAC	Source MAC
Protection device A>B first data	Protection A				B	A
	Switches a	a	n	1	B	A
	Switches b	a	n	1	B	A
	Switches n	a	n	1	B	A
	Protection B				B	A
Protection device A>B second data	Protection A				B	A
	Switches a	a	n	2	B	A
	Switches b	a	n	2	B	A
	Switches n	a	n	2	B	A
	Protection B				B	A
Protection device C>B first data	Protection A				B	A
	Switches a	b	n	1	B	C
	Switches b	b	n	1	B	C
	Switches n	b	n	1	B	C

	Protection B				B	C
Protection device C>B second data	Protection A				B	A
	Switches a	b	n	2	B	C
	Switches b	b	n	2	B	C
	Switches n	b	n	2	B	C
	Protection B				B	C

Table 1 shows the format of data received on each device (station domain protection device, switch) in the system.

2.3 Redundant data transfer based on format modification of switch ring network

In a switch ring network, after receiving data A sent by the device, the original switch modifies and replicates the data format to form data A1 and A2 with the same content. A1 and A2 transmit in different directions of the ring network respectively, and finally, A1 and A2 collide at the destination switch (Figure 4). The seamless redundancy of ring network can be realized by copying and transferring data and eliminating redundant data [5]. Eliminate loop network switchover time caused by link failure in the loop network structure of switches, increase the reliability of Ethernet data transmission in the domain protection system of power distribution sites, reduce communication time increase and control command failure caused by data loss and retransmission of data by protection devices.

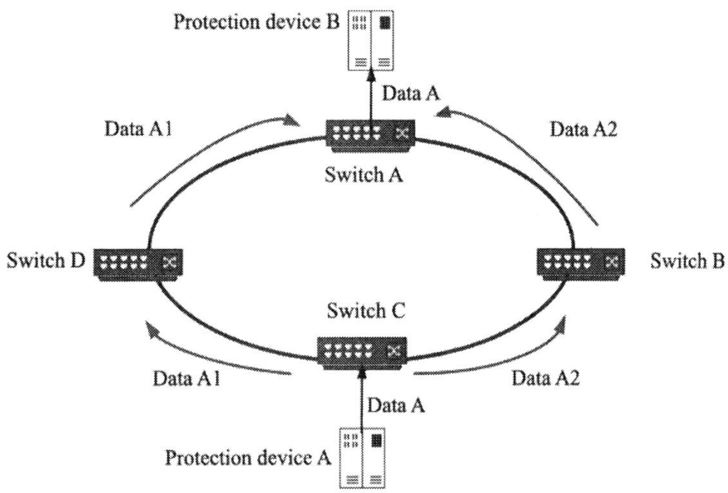

Figure 4. Ring network transmission method.

3. Conclusion

In this paper, the Ethernet transmission method based on data format modification can be used to identify the unique identity of the data transmitted in the domain fast protection system of distribution website, so that the switch can identify the characteristics of each data. The data can be copied and transmitted, and redundant data can be eliminated to achieve seamless redundancy of ring network. Eliminate loop network switchover time caused by link failure in the loop network structure of switches, increase the reliability of Ethernet data transmission in the domain protection system of power distribution sites, reduce communication time increase and control command failure caused by data loss and retransmission of data by protection devices.

Reference

[1] Liu Xingyong, Wang Tao, Li Kuo, et al. Science and Wealth, 2018,(23).284.
[2] Xu Aidong, Wang Hong, Xing Zhihao. Information and control, 2005,34 (1) : 60-64.
[3] Meng Fei. Implementation of Loop Network Redundancy for Industrial Ethernet Switch [J]. Electronic Technology and Software Engineering, 2019(13):11-11.
[4] Wang JIANjun, Li Yongfan, Xiao Jie, et al. Design and Implementation of Dynamic Simulation for Ethernet Protocol Data Encapsulation [J]. Computer Technology and Development, 2010(06):133-136.
[5] Mi Zhengkun, Yang Guomin. Switching technology [M]. Beijing: Tsinghua University Press, 2006. 15-18.

Short term wind power interval prediction based on VMD and improved BLS

Yang Zhao, Chuanbo Wen

School of Electrical Engineering,Shanghai Dianji University.Shanghai201306, China

Zhao Yang's e-mail: 196001010121@st.sdju.edu.cn

*Wen Chuanbo's e-mail: chuanbowen@163.com

Abstract. Aiming at the problem of wind power interval prediction, a short-term wind power interval prediction model based on VMD and improved BLS is proposed. Firstly, the complex wind power time series are decomposed by variational mode decomposition to reduce the non stationarity of wind power. Then an improved broad learning system (BLS) is established to predict the power and error of each component, and a weight is given to the prediction error of each component. The sparrow search algorithm (SSA) is used to optimize the weight, and the width of the prediction interval is obtained by adding the power and error prediction values. The experimental data show that the proposed model improves the accuracy of prediction interval.

1. Introduction

Due to the intermittence, randomness and volatility of wind power itself, point prediction cannot be accurately carried out. Therefore, it is also very important to obtain the predicted wind power fluctuation range for power system planning and operation decision-making.

At present, many literatures have studied the interval prediction of wind power. In reference [1] Deterministic and probabilistic interval prediction for wind farm based on VMD and weighted LS-SVM can effectively improve the prediction accuracy and reduce the prediction width.In reference [2] A broad learning system is proposed. The model has the advantages of simple structure, fast training time and high prediction accuracy. In reference [3] proposed Wind Power Interval Prediction Based on Improved PSO and BP Neural Network, but particle swarm optimization has poor search ability and is easy to fall into local optimal solution. In reference [4] proposes the wind speed interval prediction based on the error prediction method. The point prediction is carried out through the combination of vmd-gru. On this basis, the prediction error interval is constructed, and the value parameters are optimized by PSO algorithm. This method does not depend on the distribution characteristics of prediction error.

Based on the above research, this paper proposes a short term wind power interval prediction based on VMD and improved BLS, Improved BLS has the characteristics of low complexity and strong generalization ability. Vmd and improved BLS combined model can obtain higher precision point prediction value and prediction error. On this basis, the wind power interval prediction model is constructed, and the sparrow algorithm optimizes the prediction interval. The simulation results show that the proposed model has low complexity, strong robustness and can obtain better interval width.

2. Variational mode decomposition

VMD is an adaptive and non recursive decomposition method [5]. The main idea is to construct the variational problem and solve it iteratively. The complex signal $f(t)$ is decomposed into modal components $u(k)$ with different bandwidths and center frequencies by setting the modal number K in advance, and each modal component and its center frequency are updated. When the sum of the estimated bandwidths of each modal is the smallest, the update is completed to obtain the optimal solution. The process is as follows:

- The unilateral spectrum of each modal component is calculated by Hilbert transform, and the corresponding center frequency ω_k is estimated
- The exponential term $e^{-j\omega_k t}$ of the center frequency of each modal analysis signal is aliased, and the spectrum of each modal is converted to the corresponding fundamental frequency band.
- The bandwidth of each mode is estimated by Gaussian smoothing method and transformed into a variational solution problem with constraints:

$$\begin{cases} \min\limits_{\{u_k\},\{\omega_k\}} \{\sum_{k=1}^{K} \left\| \partial_t [(\delta(t)+\frac{j}{\pi t} * u_k(t)] e^{-j\omega_k t} \right\|_2^2 \\ \sum_{k=1}^{K} u_k(t)] e^{-j\omega_k t} = f(t) \end{cases} \tag{1}$$

where: $\delta(t)$ is the unit pulse function; $*$ represents convolution; $\partial(t)$ represents partial derivative; $u_k(t)$ represents K components; ω_k represents the center frequency of K components; $f(t)$ represents the time data of wind power.

Lagrange multipliers λ and penalty factors α are introduced into the constrained problem in the above equation (1) above to change it into an unconstrained variational problem, as shown in equation (2):

$$\zeta(\lambda,\{u_k\},\{w_k\}) = a\sum_{k=1}^{K} \left\| \partial_t [(\delta(t)+\frac{j}{\pi t}) * u_k(t)] e^{-j\omega_k t} \right\|_2^2 + \left\| f - \sum_{k=1}^{K} u_k \right\|_2^2 + \langle \lambda, f - \sum_{k=1}^{K} u_k \rangle \tag{2}$$

The saddle point in equation (2) is solved by alternating direction multiplier method (ADMM), and the updated formulas of corresponding variables u_k, ω_k and λ are obtained, as shown in equation (3-4):

$$\hat{u}_k^{n+1}(\omega) = \frac{\hat{f}(\omega) - \sum_{i<k} \hat{u}_k^n(\omega) - \sum_{i>k} \hat{u}_k^n(\omega) + \frac{\hat{\lambda}^n(\omega)}{2}}{1 + 2\alpha(\omega - \omega_n^k)^2} \tag{3}$$

$$\hat{\omega}_k^{n+1} = \frac{\int_0^\infty \omega \left| \hat{u}_k^{n+1}(\omega) \right|^2 d\omega}{\int_0^\infty \left| \hat{u}_k^{n+1}(\omega) \right|^2 d\omega} \tag{4}$$

where: $\wedge$ is Fourier transform; n is the number of iterations.

3. Improved broad learning system

In this paper, based on the structure of improved BLS,as shown in Figure1. In view of the flexibility of BLS structure, the feature layer of BLS is connected in a cascade way, which can improve the recognition ability of the network to data features. Generate m group feature mapping as $[Z_1, Z_2, \cdots, Z_m]$. The feature mapping node Z_1 is defined as:

$$Z_1 = \phi(XW_{f1} + \beta_{f1}) \tag{5}$$

where: X is the input data, $\phi(\cdot)$ is the linear activation function, and W_{f1} and β_{f1} are the weights and offsets of group 1 random initialization

Group 2 feature node Z_2 takes Z_1 as the input through cascading. The generated Z_2 is:

$$Z_2 = \phi(Z_1 \cdot W_{f2} + \beta_{f2}) = \phi(\phi(X \cdot W_{f1} + \beta_{f1})W_{f2} + \beta_{f2}) = \phi^2(X; \{W_{fi}, \beta_{fi}\}_{i=1,2}) \quad (6)$$

Therefore, the m-th mapping feature is represented as:

$$Z_i = \phi(F_{i-1} \cdot W_{fi} + \beta_{fi}) = \phi^i(X; \{W_{fi}, \beta_{fi}\}_{i=1}^m), i = 1, 2, ..., m \quad (7)$$

At the same time, in order to enhance the dynamic capture characteristics of the system, this paper introduces the reserve pool layer in ESN network[6] into the enhancement layer of BLS, so that the network has the function of memory. The state matrix of the output h of the enhancement layer is shown in equation (8):

$$H(t+1) = \xi(W_h Z^i(t+1) + W_c H(t) \quad (8)$$

where, $t = 1, 2, \cdots l$ (l is the number of data samples), $\xi(\cdot)$ Represents the nonlinear activation function, W_h is the weight matrix from the mapping layer to the reserve pool layer, and W_c is the neuron connection matrix of the reserve pool

The output y of improved BLS is shown in equation (9), and the output weight is also calculated by ridge regression, as shown in equation (10):

$$Y = [Z_1, Z_2, ..., Z_m, H]W = AW \quad (9)$$

$$W = [(A)^T A + \alpha I]^{-1}(A)^T Y \quad (10)$$

where, W is the output weight matrix, α is the regularization parameter

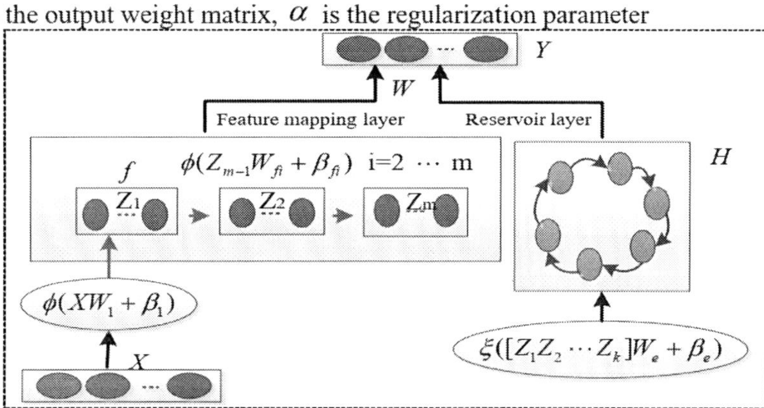

Figure1. Improved BLS network

4. Construction and optimization criteria of short-term wind power range

4.1. Interval construction

The wind power and its error are predicted by VMD and improved BLS. Then the wind power is included in the (U_{pre}, L_{pre}) interval composed of power prediction value and prediction error as much as possible. In order to further improve the level of prediction interval. The prediction error E_{pre} is given a weight to adjust the interval width. The interval mathematical model is constructed as follows:

$$\begin{cases} U_{pre} = \sum_{i=1}^{K} W_{pre} + \sum_{i=1}^{K} w_i |E_{pre}| \\ L_{pre} = \sum_{i=1}^{K} W_{pre} - \sum_{i=1}^{K} w_i |E_{pre}| \end{cases} \quad (11)$$

where $W_{pre}i$ and $E_{pre}i$ represent the wind power prediction and error prediction of the ith mode respectively. w_i is the corresponding weight of $E_{pre}i$.

4.2. Evaluation interval prediction criteria

The prediction interval coverage (PICP) and normalized root mean square width (PINRW) are used to test the interval prediction effect of the proposed model.

(1) Forecast interval coverage(PICP)

It is used to evaluate the probability of the test set falling within the prediction interval, that is, the probability between interval U_{pre} and L_{pre}. The larger the value of PICP, the more the test set falls into the prediction interval, and the better the prediction effect.

$$\mathrm{PICP} = \frac{1}{n}\sum_{t}^{n} c_t \tag{12}$$

Where: n represents the number of test sets, and c_t represents Boolean value

(2) Normalized root mean square width of prediction interval (PINRW)

In order to prevent the prediction interval from being too wide and meaningless by only seeking the coverage of the prediction interval, the PINRW value is used to evaluate the prediction interval. The smaller the PINRW value, the narrower the prediction width and the better the prediction.

$$PINRW = \frac{1}{R}\sqrt{\frac{1}{n}\sum_{i=1}^{n}(U_{pre}i - L_{pre}i)^2} \tag{13}$$

where: $U_{pre}i$ 、 $L_{pre}i$ is the upper limit and lower limit of the ith prediction interval respectively

4.3. Short term wind power interval prediction model

In order to obtain higher quality wind power interval width, SSA is used to optimize parameter w of interval prediction model .The fitness of sparrow optimization algorithm is selected as follows:

$$\begin{cases} \min \quad \mathrm{PINRW}(w) \\ \mathrm{s.t} \quad \mu \le PICP(w) \le 100\% \end{cases} \tag{14}$$

where μ is the nominal confidence level and $w = (w_1, w_2, \cdots, w_K)$.

The basic flow of VMD and improved BLS interval prediction model is shown in Figure2. The specific steps are as follows:

- VMD is used to decompose the original wind power data into a series of subsequences. Each sub serialization is divided into training set and test set.
- The improved BLS model is used to predict the wind power and calculate the training error of each sub sequence.
- The improved BLS model is established for the error sequence to predict the error.
- The prediction interval is composed of wind power prediction, error prediction and corresponding weight of each subsequence. The sparrow algorithm[7] is used to search the optimal value of the weight w_i
- The performance of the proposed model is evaluated by PICP and PINRW.

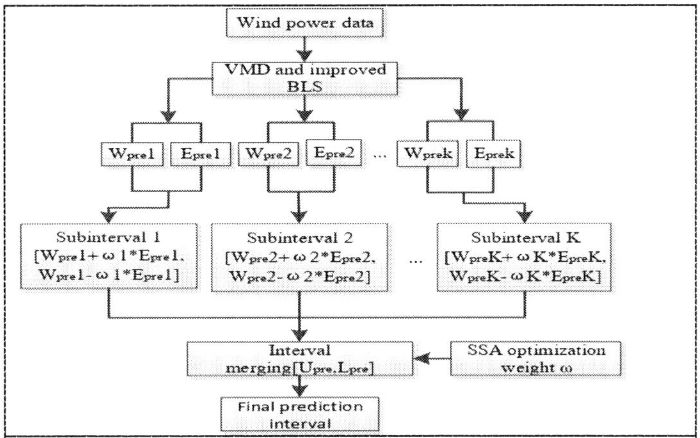

Figure2. Short term wind power interval prediction model

5. Example analysis

A total of 3072 samples were selected from the actual data collected in July 2012 with a wind farm capacity of 16mw and a sampling interval of 15min in Northeast China to verify the effectiveness of the proposed model. Firstly, the samples are processed by hourly averaging. For the 803 processed samples, the data of the first four times are selected as the input, the prediction point is the fifth time, and so on. 803 data samples were processed into 799 groups of data, of which the training data set was the first 703 groups of data and the test data set was the last 96 groups.

VMD decomposition is performed on the processed original wind power time series. During VMD decomposition, the value of modal number K needs to be determined. Excessive K value will over decompose the signal modal component, and too small K value will under decompose the signal modal component. The penalty factor α can effectively avoid the phenomenon of mode aliasing of each component. In this paper, the values of K and α are determined by instantaneous frequency and overall orthogonality index. K is 11 and α is 644. The grid search method is used to optimize the parameters of improved BLS. Finally, the prediction model established in this paper is used to predict the test data set, and the prediction results are shown in Figure 3.

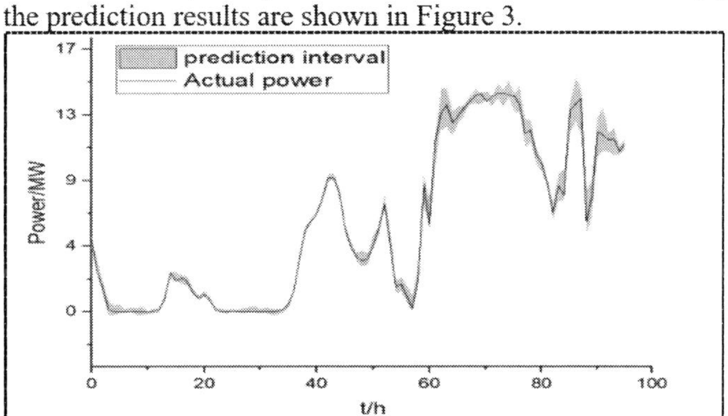

Figure3. VMD and improved BLS interval prediction

In order to verify that the model in this paper has good interval prediction effect.Compared with VMD and ELM model and VMD and BLS model, The prediction results of each model are shown in Figure4- Figure5.

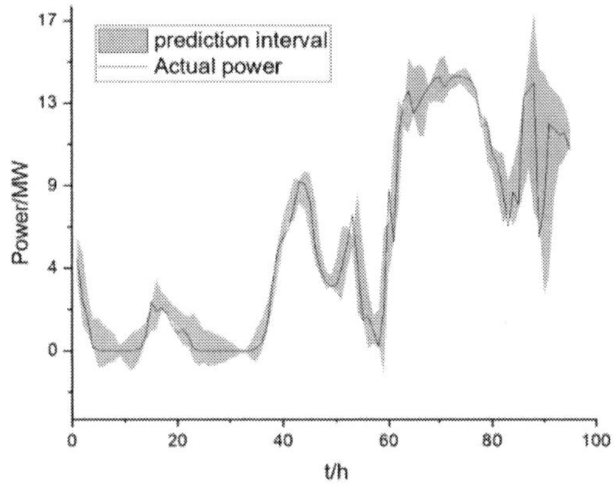

Figure4. VMD and ELM interval prediction

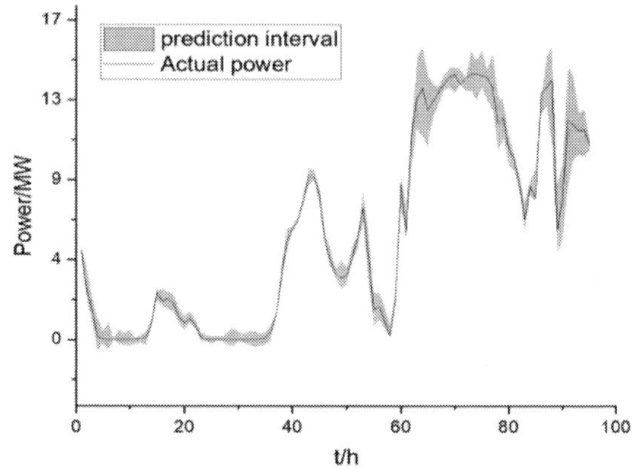

Figure5. VMD and BLS interval prediction

According to the interval prediction results in Fig. 4 and Fig. 5, the prediction interval of the proposed model has the smallest width and contains the most actual power, and its fluctuation trend is around the actual output value. The effect is most obvious at the inflection point of wind power series change. Therefore, the proposed model further improves the prediction accuracy of wind power interval. The interval prediction performance index results of the three models are shown in Table1.

Table1. Interval prediction and evaluation indexes of each model

Interval prediction model	PICP	PINRW
VMD and ELM	0.9375	0.4454
VMD and BLS	0.9479	0.3002
VMD and improved BLS	0.9587	0.2590

As can be seen from table1, compared with VMD and ELM interval prediction model and VMD and BLS interval prediction model, PICP of the model in this paper is improved, PINRW is reduced, and the interval width is narrowed. Therefore, the proposed model can well reflect the change of wind power uncertainty in wind power interval prediction.

6. Conclusion

The proposed model is compared with VMD and ELM and VMD and BLS. It is verified that the proposed model has better prediction effect on wind power range. VMD can adaptively decompose and reduce the non stationarity of data, and improved BLS model can improve the prediction ability. SSA algorithm has strong optimization ability for interval model parameters.

Acknowledgments

Fund projects: National Natural Science Fund (no.61973209),
Shanghai Natural Science Fund (No. 20zr1421200)

References

[1] Gendeel Mohammed et al. Deterministic and probabilistic interval prediction for wind farm based on VMD and weighted LS-SVM [J]. Energy Sources, Part A: Recovery, Utilization, and Environmental Effects, 2021, 43(7): 800-814.

[2] Jidong Wang et al. Wind Power Interval Prediction Based on Improved PSO and BP Neural Network [J]. Journal of Electrical Engineering & Technology, 2017, 12(3) : 989-995.

[3] Chen C L Philip and Liu Zhulin. Broad Learning System: An Effective and Efficient Incremental Learning System Without the Need for Deep Architecture.[J]. IEEE transactions on neural networks and learning systems, 2018, 29(1): 10-24.

[4] G. Tang, Y. Wu, C. Li, P. K. Wong, Z. Xiao and X. An, "A Novel Wind Speed Interval Prediction Based on Error Prediction Method," in IEEE Transactions on Industrial Informatics, vol. 16, no. 11, pp. 6806-6815, Nov. 2020, doi: 10.1109/TII.2020.2973413.

[5] Dragomiretskiy K，Zosso D. Variational modedecomposition［J］. IEEE Transactions on Signal Processing，2013，62（3）：531—544
[6] Mohammad Amin Chitsazan and M. Sami Fadali and Andrzej M. Trzynadlowski. Wind speed and wind direction forecasting using echo state network with nonlinear functions [J]. Renewable Energy, 2019, 131: 879-889.
[7] Xue Jiankai and Shen Bo. A novel swarm intelligence optimization approach: sparrow search algorithm [J]. Systems Science & Control Engineering, 2020, 8(1): 22-34.

Intelligent Driving Detection System Based on unmanned Driving Background

Hongqi Wang[1*], Shouman Chen[1], Yahong Wen and Chaofan Zhou[1]

[1] School of Electronics and Information Engineering, Ankang University, Ankang 725000, Shananxi Province, China

[*]Corresponding author's e-mail: 2020990043@aku.edu.cn

Abstract. Unmanned driving is the future development direction of the automobile industry, and intelligent driving detection is one of the important research topics. In order to improve driverless technology and traffic safety, this design is based on FPGA technology to design a driving detection system, combined with off-chip SDRAM's high-speed, large-capacity cache and image processing related algorithms, and it will achieve the function of acquiring target image, cache, detection and display. Finally, the FPGA development board is used to test the driving detection system, which demonstrates that this driving detection system can better improve the safety monitoring problems in unmanned driving. At the same time, the system's frame rate is close to 55pfs, which has practical significance and application value.

1. Introduction

With the rapid development of artificial intelligence and unmanned driving, unmanned driving systems and intelligent traffic control systems have been widely used in transportation, medical and other fields[1]. In daily traffic, the intelligent driving detection system is an important part of the driverless field[2,3]. Intelligent vehicle detection that uses image processing, pattern recognition and visual technology, increased in the monitoring and control system for for intelligent detection analysis module[4]. With the aid of computer powerful data processing ability of filtering useless video screen, automatic identification of moving object, judging monitoring images of the abnormal situation, so as to advance warning effectively, make handling of the affair[5]. At the same time, the driverless car must be monitored and be able to be switched to manual driving mode at any time, which can effectively control the speed and direction of the vehicle in emergency situations and avoid traffic accidents[6,7].

In this context, this article develops an intelligent driving detection system for unmanned driving based on FPGA chips[8]. It uses the characteristics of FPGA parallel computing efficiency and low hardware cost to perceive the surrounding vehicles, so as to avoid and deal with obstacles quickly and accurately, reduce the probability of accidents greatly, and ensure the safety of people's lives and property. In the current era, it has great development significance and research value[9,10].

2. Implementation of Intelligent Driving Detection System

The FPGA-based intelligent driving detection system adopts a top-down design idea, and the system is roughly divided into four modules, which includes CMOS video image acquisition module, SDRAM memory module, FPGA image processing module and VGA display module[11,12]. Firstly, the OV5640 CMOS camera is controlled by the FPGA core to capture the video data stream in real time, and the grayscale image is obtained after image preprocessing, which is subjected to median filtering and Sobel edge detection, and then cached in the SDRAM memory. Secondly, the SDRAM is read through the control of the SDRAM, the two adjacent frames of images in the memory are sent to the target detection module to perform inter-frame difference processing on the adjacent two frames of images, and one frame of image data is corroded and expanded in a parallel manner. Finally, VGA display of the video images and moving target detection are realized. The overall system frame diagram is shown in Figure 1.

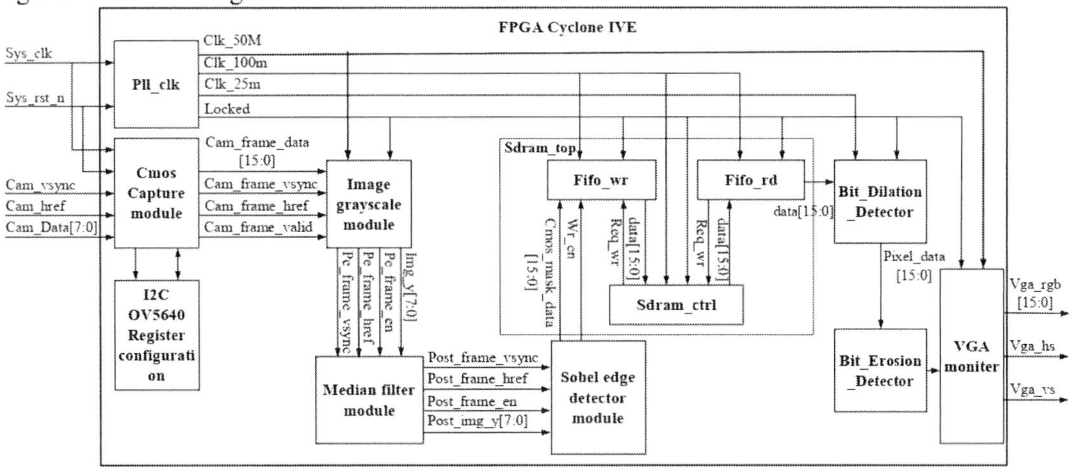

Figure 1. Block diagram of the driving detection system

3. Simulation Analysis and Realization of Intelligent Driving Detection System

3.1. Image acquisition module simulation

For the image acquisition module, the Testbench simulation program is written to simulate and verify it on the modelsim software. In order to better observe the simulation results and analysis, effective pixels were set as 12 rows and 8 columns in the simulation of image acquisition. Figure 2 shows a partial enlarged view of the CMOS image acquisition module.

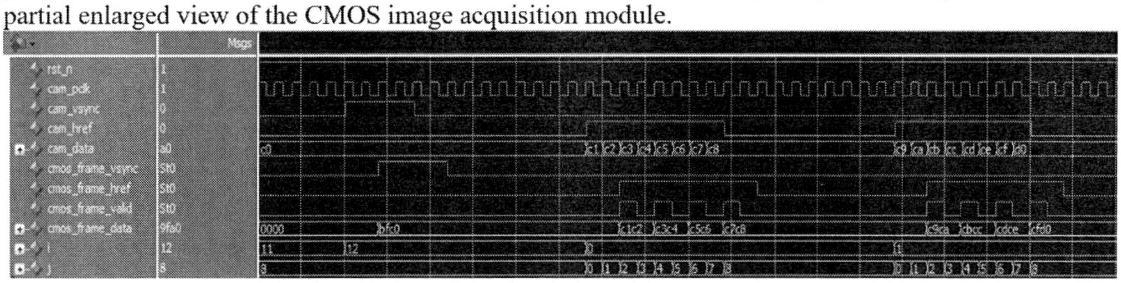

Figure 2. A partial enlarged view of the simulation of the CMOS image acquisition module

It can be seen from the simulation waveform Figure 2, when cam_href is low, the data of cam_data remains unchanged, and coms_frame_data does not output valid data. when cam_href is high, the coms_frame_vaild signal starts to change periodically. Cam_data starts to input valid values. after two clocks cycle, coms_frame_data starts to output valid data. The 8-bit data of two cmos_data in each row were converted into 16-bit data in cmos_frame_data, and the data mosaic in RGB565 format was completed. The image acquisition module was successfully implemented.

3.2. Sobel edge detection module simulation

For the Sobel edge detection module, the Testbench simulation program is written to simulate and verify it on the Modelsim software. In order to better observe the simulation results and analysis, effective pixels were set as 12 rows and 8 columns in the simulation of Sobel edge detection. Figure 3 shows a partial enlarged view of Sobel edge detection.

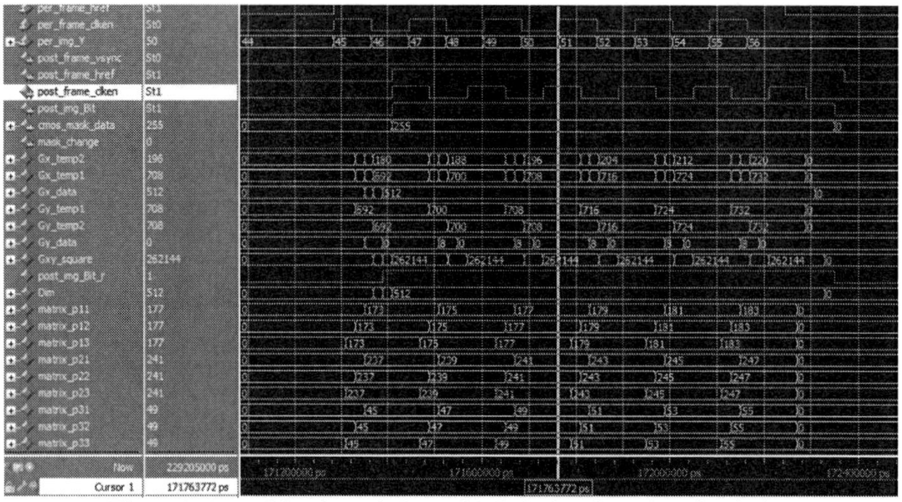

Figure 3. A partial enlarged view of Sobel edge detection simulation

It can be seen from the simulation waveform Figure 3, the data in the first row, second row and third row were set to 173, 237, 45; 175, 239, 47; 177, 241, 49. this design sets the threshold Sobel_Threshold of 250, which obtains the values of Gx_tmp1, Gx_tmp2, Gy_tmp1, Gy_tmp2 after one clock cycle, and then obtains the values of Gx_data and Gy_data after one clock cycle. When the third rising edge of the clock arrives, the value of Dim (that is, Gxy) is 512. Compared with the preset threshold of 250, the result post_img_bit is 1, which is consistent with the actual value. This results show that Sobel edge detection algorithm is correct.

4. Hardware test of driving inspection system

4.1. System simulation test

The simulation verification of this design will play the driving video through the computer to achieve the simulation effect. The monitored video results and processed driving video images are shown in Figures 4 and 5.

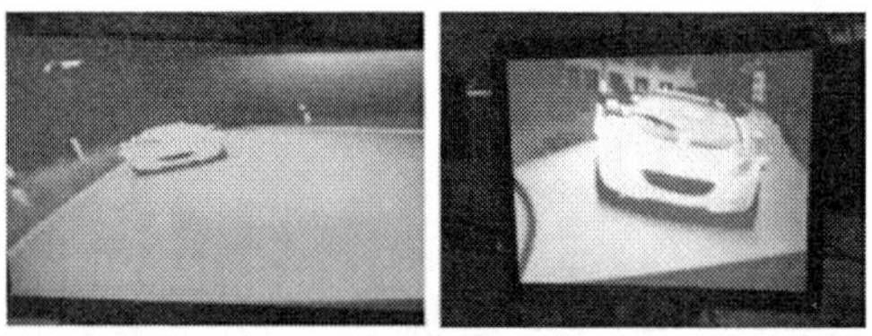

Figure 4. Driving video image before processing

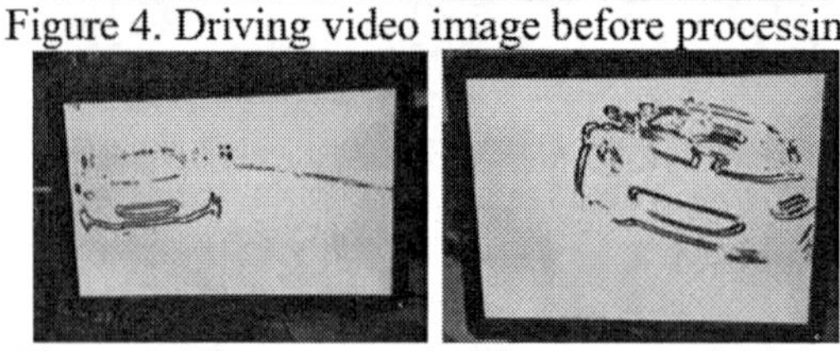

Figure 5. The processed video image of driving

It can be seen from Figure 5 that the results processed by the driving detection system can better verify the realization of the driving detection function, and have better practicability for driverless driving detection.

4.2. System hardware test

After comprehensive compilation, the project is downloaded to the development board, and the camera is used to monitor the traffic on the road in real time during the day and night. The monitored and processed traffic images are shown in Figures 6, 7, 8 and 9.

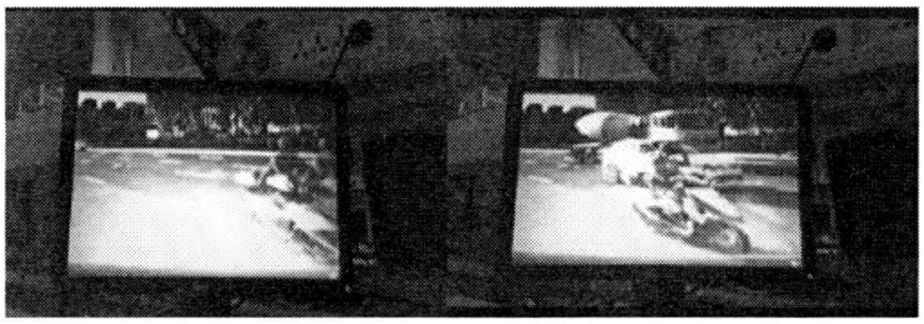

Figure 6. Snapshot images of driving during the day

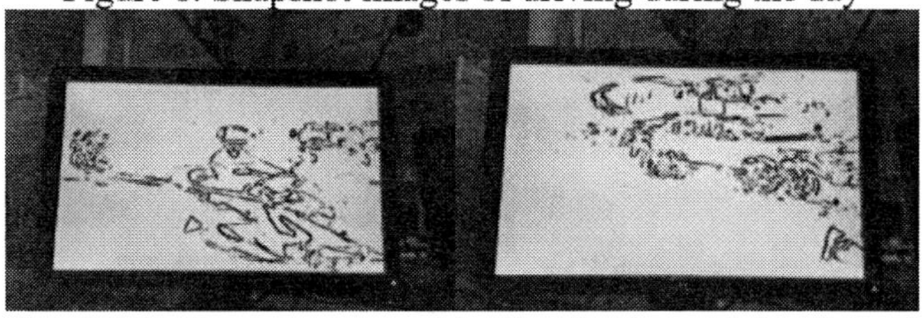

Figure 7. The driving image processed by the daytime driving detection system

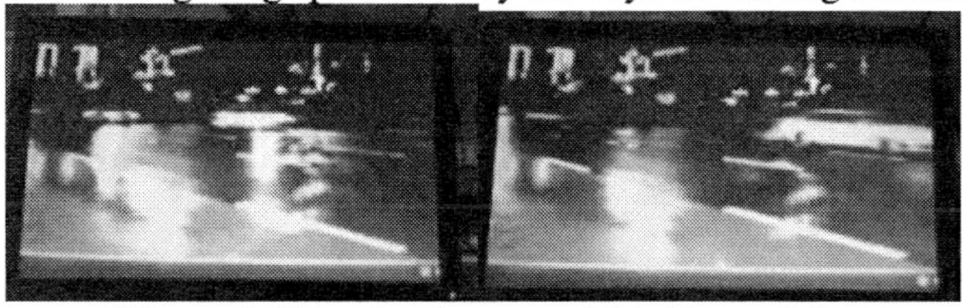

Figure 8. Snapshot images of driving at night

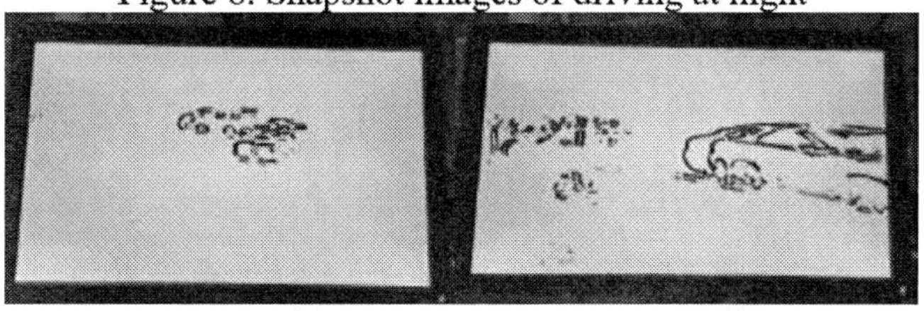

(a) (b)

Figure 9. The driving image processed by the night driving detection system

It can be seen from Figures 7 and 9, that this driving detection system can realize the driverless driving detection function during the day and night. However, due to the low pixels of the camera and the influence of the weather, the detection image is blurred, but the detection function of unmanned driving can be realized overall. As it can be seen from the test, a mirror image will appear after the image is processed by the system since there is water on the ground in rainy days(Figure 9(b)). In the daytime, the low camera pixel of the camera reduces the definition of the original image, resulting in the double image of the detected image (Figure 7). At night, due to the influence of the camera pixels and the lights, part of the information of the detected vehicle is lost, and the driving test result is not very clear (Figure 9(a)).

5. Conclusions

This paper mainly designed a kind of driving detection system with FPGA as the core, which improved the calculation speed by using the characteristics of FPGA parallel structure. Combined with edge detection and frame difference image processing, realizing the driving detection on VGA. The system has good practicability, it can improve the safety index of unmanned driving in life, improve the problems which exists in intelligent transportation.

Acknowledgments

This work was supported in part by the National Natural Science Foundation of China under Grant 61801005, in part by the Ankang University School Foundation under Grant 2021AYQN013.

References

[1] Yang, G., Chen, W., Zhaojian L. (2021) Method for detecting lane lines of driverless cars based on deep learning. Science Technology and Engineering, 21(24): 10401-10406.
[2] Zhang, Z., Jiang, Q., Wang, R., et al. (2019) Research on Management System of Automatic Driver Decision-Making Knowledge Base for Unmanned Vehicle. International Journal of Pattern Recognition and Artificial Intelligence, 33(4).
[3] Xie, J., Liu, R. (2019) The study progress of object detection algorithms based on deep learning. Journal of Shaanxi Normal University(Natural Science Edition), 47(5): 1-9.
[4] Kim, D. H., Choi, C. H., Kim, Y. J. (2018) Analysis of driving performance evaluation for an unmanned tractor - ScienceDirect. IFAC-PapersOnLine, 51(17):227-231.
[5] Zhou, X., Zhou, J. (2020) Data-Driven Driving State Control for Unmanned Agricultural Logistics Vehicle. IEEE Access, (99): 1-1.
[6] Harada, K., Kanazawa, K., Yasunaga, M. (2019) FPGA-Based Object Detection for Autonomous Driving System. 2019 International Conference on Field-Programmable Technology (ICFPT), IEEE, Tianjin. pp. 465-468.
[7] Simmonds, H., Carlisle, N., Li, X. (2019) Autonomous Vehicle Development using FPGA for Image Processing. 2019 International Conference on Field-Programmable Technology (ICFPT). IEEE, Tianjin. pp. 449-452.
[8] Kai, X., Binhua, L. (2020) Real-time detection and tracking algorithm of moving target based on FPGA and its implementation technology. Optical Technology, 46(2): 1002-1582.
[9] Chen, Y., Hu, N., Yu Q.Y. (2019) Moving target detection combining background difference method and inter-frame difference method. Modern Computer, (34):50-53.
[10] Guo, Z., Zhang, Q.G. (2021) Design of high-definition target rapid detection system based on FPGA. Electronic Design Engineering, 29(04): 150-154.
[11] Wei, K.J., Honda, K. (2018) FPGA Design for Autonomous Vehicle Driving Using Binarized Neural Networks. 2018 Internation Conference on Field-Programmable technology(FPT 2018). Japan. pp. 428-431.
[12] Lei X. (2020) Implementation of FPGA-based tunnel driving detection system. Guizhou University.

Analysis and Design Based on the Operation Mode of Power Electronic Transformer in Smart Grid

Runzhuo Ma

Electronic and Electrical Engineering, School of Engineering, University of Glasgow, Glasgow ,G12 8QQ,UK,Email: Joker679@163.com

Abstract. Power electronic transformers(PET) are the key energy conversion equipment in the operation of modern smart grids, the main function of PET is to achieve the conversion of AC voltage to AC voltage, while taking into account the DC ports. This article mainly studies three-stage power electronic transformers based on three-phase uncontrolled rectifier, full-bridge isolated DC-DC converter and three-phase inverter. The operation mechanism and actual working process of the three parts of the PET are analyzed respectively, and the transformer is simulated and analyzed based on the Matlab/Simulink simulation platform. The rectifier converts the AC voltage on the grid side into a rippled DC voltage; the DC-DC converter transforms the obtained DC voltage, taking into account the access of the DC ports; the inverter converts the obtained DC voltage into AC voltage through unipolar modulation and connects to the grid. The experimental results show that the PET constructed in this way can operate safely and stably, which has good voltage conversion and electrical isolation functions, and can be connected to DC loads.

1.Introduction

In the modern society, electricity is essential to people's lives. The transformer is an important part of the power distribution system, the main function of the transformer is electrical isolation and voltage conversion. With the continuous development of technologies such as active distribution networks, smart grids, and new energy power generation, traditional power frequency transformers can no longer meet the demand of smart power generation[1].

With the in-depth study of power electronic technology, PET have emerged as a new type of smart transformer. Power electronic transformer, also known as solid-state transformer, is composed of two parts, namely power electronic converter and high-frequency transformer, which realize the function of voltage conversion through electronic components. Compared with traditional transformers, which has some disadvantages such as large size, heavy weight, and single transformation function, PET have the following advantages[2-4]:

Small size, light weight, and will not use insulating oil, will not pollute the environment; 2.PET have DC and AC links and can be connected to AC power grids and DC loads, new energy equipment that can be connected to DC power supplies power to the grid; 4. Higher reliability, more accurate and flexible adjustment of the power output terminal.

The concept of PET was first proposed by American W. McMurray, and the AC-AC converter he proposed is the embryonic development of today's PET[5]. When the concept of "smart grid" was proposed, power electronic technology has made great development and progress, and PET have also been widely studied and paid attention by more countries and enterprises. Literature [6] proposed a PET that can be used in medium and high voltage distribution networks. The structure of this PET is three-phase structure, which is divided into three parts: high voltage, medium voltage, and low voltage. This method effectively verifies the accuracy and validity of the PET topology and has a wide range of application prospects in smart grids.

In addition to the functions of traditional transformers, they can also control current and voltage and improve power quality[7]. In today's era, new energy power generation has become a new trend. The world is advocating the concept of energy saving, environmental protection, and green power generation. PET can also connect the new energy power generation system with the power system and input the generated electrical energy into the power system[8]. Literature [9] proposed a three-level PET, in the topology of this PET, a three-phase and three-level topology is used. In this type of transformer, the voltage that the switch components bear is only one-half of the voltage that the two-level circuit bears, which increases the range of input voltage levels and is better suited for high-voltage power distribution. At present, the application of PET in power systems includes two aspects: distributed power grid connection and improvement of power quality[10].

2.Rectifier

PET are mainly divided into three parts: AC-DC rectifier, DC-DC converter and DC-AC inverter, in order to realize AC-DC-AC voltage conversion[11]. The rectifier is the input part of the entire pet. Since the input voltage is three-phase voltage during the operation of the power grid, a three-phase three-wire structure is adopted, and the inside of the rectifier adopt an H-bridge full bridge structure. The bridge arm connected by each phase voltage is composed of two diodes in series, and a capacitor and a resistor are connected in series with the three bridge arms[12]. The topological structure of the rectifier is shown as in Figure 1.

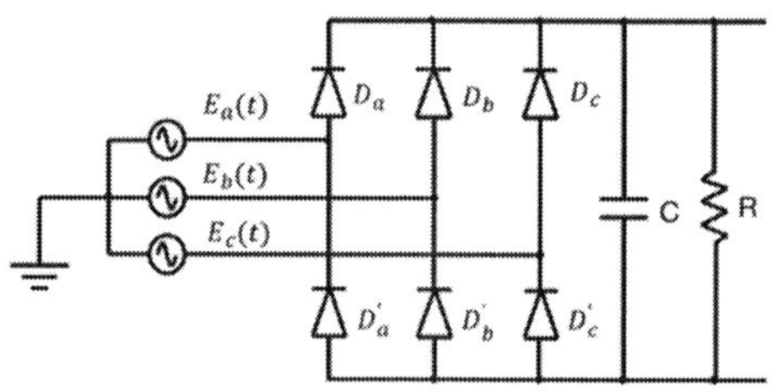

Figure 1. Topological structure diagram of rectifier

In Figure 1: $E_a(t)$ 、 $E_b(t)$ 、 $E_c(t)$ are the three-phase voltage of the three-phase power supply, D_a 、 D_b 、 D_c are the upper diode of each corresponding bridge arm in the three-phase power supply, D_a' 、 D_b' 、 D_c' are the lower diode of each corresponding bridge arm in the three-phase power supply, C is the filter capacitor for output DC voltage, R is the equivalent load. The goal is to output 1000V direct current, the input three-phase power is 730 volts, C is 0.0033 farads, and R is 10 ohms.

The rectifier is controlled by the conduction characteristics of the diode. The upper diode of the bridge arm connected to the phase with the higher phase voltage is turned on, and the diode below the bridge arm of the phase with the lower phase voltage is turned on, thereby converting the input three-phase sinusoidal AC voltage into DC voltage.

The voltage ripple obtained in this way is relatively large, and an inductor connected in parallel with the bridge arm needs to be added for filtering. The existence of capacitor can eliminate the pulsation of DC voltage, thereby ensuring that the conversion can obtain a stable DC voltage. In order to get an output voltage of 1000V, after experiments, the input voltage needs to be adjusted to 730V. The voltage waveform after filtering is shown as in Figure 2.

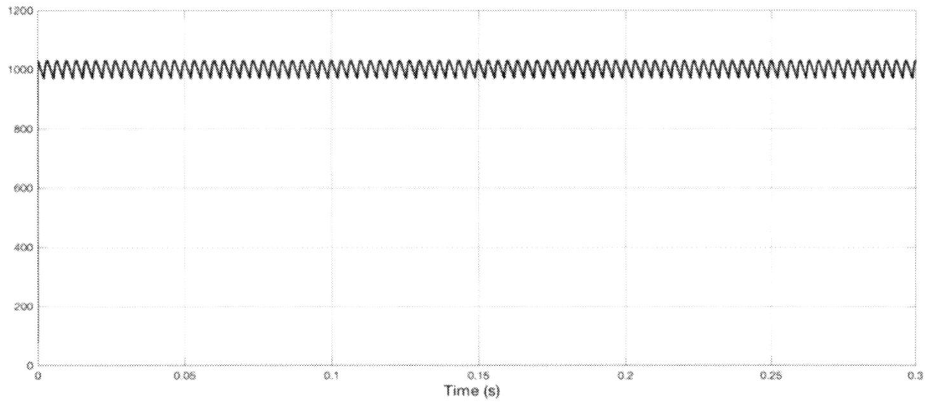

Figure 2. Inverter output voltage waveform after filtering

3.DC-DC Converter

As the secondary transmission path of the PET, The DC-DC converter adopts a full-bridge isolated DC-DC converter, and each part on both sides has an H-bridge structure[13]. Its topological structure is shown as in Figure 3.

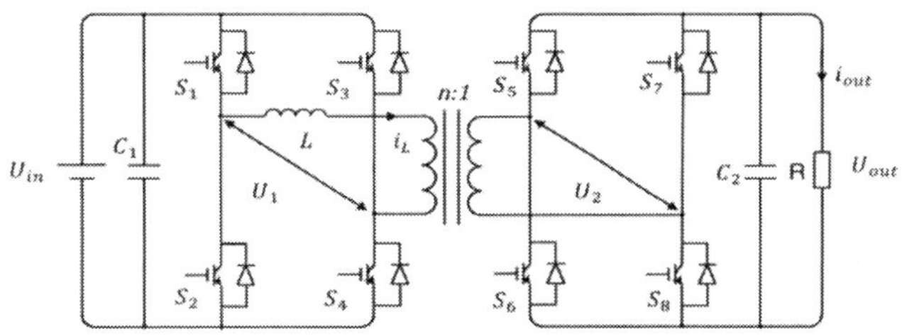

Figure 3. Topological structure diagram of full-bridge isolated DC-DC converter

In Figure 3: U_{in} is the input power supply voltage, i_{out} and U_{out} are load current and output side voltage respectively, $S_1 - S_8$ are mosfet switch tubes, L is auxiliary inductance, i_L is the current on the inductor side, U_1 and U_2 are the output voltages of the H bridge at both ends of the DC-DC converter, C_1 and C_2 are the buffer capacitors on the power supply side, R is equivalent load. U_{in} is 1000 volts, C_1 and C_2 both are 0.0033 farads, L is 0.001 Henry, and R is 10 ohms.

The equivalent circuit diagram of the full-bridge isolated DC-DC converter is shown in Figure 4.

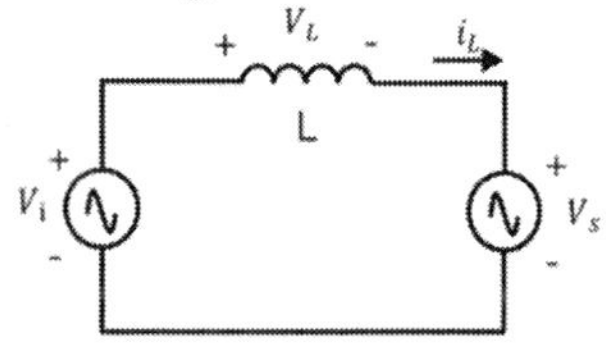

Figure 4. The equivalent circuit diagram of a full-bridge isolated DC-DC converter

In Figure 4: V_i and V_s are respectively represent two AC power sources, L is the inductor, V_L and i_L are respectively represent the voltage and current of the inductor.

Assuming that the phase of AC power V_i leading V_s is θ, calculate the current $\overrightarrow{I_L}$. By $\overrightarrow{V_i} = V_i e^{j\theta}$, $\overrightarrow{V_s} = V_s e^{j0}$, then introduce the formula $e^{ix} = \cos x + i \sin x$, then can get: $\overrightarrow{I_L} = \frac{\overrightarrow{V_i} - \overrightarrow{V_s}}{jX} = \frac{V_i e^{j\theta} - V_s}{jX} = \frac{V_i \sin\theta - j(V_i \cos\theta - V_s)}{X}$, Where X is the inductive reactance of the inductor, X = ωL, ω = 2πf

From the formula: $\overrightarrow{I_L} = I_P + jI_Q$, where I_P is the active power current and I_Q is the reactive power current, then can get: $I_P = \frac{V_i \sin\theta}{X}$, $I_Q = \frac{V_s - V_i \cos\theta}{X}$, calculated active power: $P_S = V_s I_P = \frac{V_s V_i \sin\theta}{X}$

Due to V_i and V_s are both positive values, so it can be seen from the calculation result that the positive or negative of active power P_S depends on $\sin\theta$. When the value of θ is 0-180 degrees, the power is positive, and it reaches the maximum when θ is equal to 90 degrees. From the characteristic curve of the sine function, it can be concluded that when two power supplies are connected, the power supply with the leading phase is charged and the power supply with the lagging phase is discharged.

The input signal of the converter is a square wave signal, and the phase difference is controlled by the square wave inverter. On the output side of the DC-DC converter, it is also necessary to add a filter capacitor. The voltage output waveform of the DC side is shown in Figure 5.

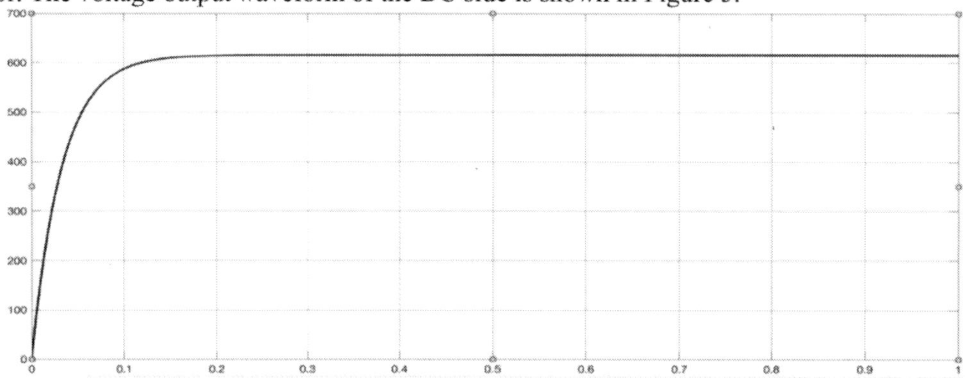

Figure 5. DC-DC converter DC side voltage output waveform

4.Inverter

The inverter is the last stage of the PET. The function of this part is to convert the DC voltage into the AC voltage that can be used in the grid system.

The input end of the inverter is a DC voltage, and it needs to be output as an AC voltage, which requires adding a suitable input signal. Input a sine wave (modulation wave) and a triangle wave (carrier wave), and compare the amplitudes of the two waveforms, When the amplitude of the sine wave is large, the output is high level 1, and when the amplitude of the triangular wave is large, the output is low level 0. That is, Pulse Width Modulation (PWM modulation).

PWM modulation is to adjust the switches in the inverter circuit to make the multiple pulses obtained at the output of the circuit have the same amplitude, so that the equivalent voltage of each pulse is the waveform of a sine wave, which is used instead of generating a sine wave. Due to the PET is to be connected to the power grid for work, the three-phase voltage at the output must conform to the law of phase difference, which is "360 /N"degrees. As it is three-phase, so N is 3, then the phase difference is 120 degrees, and three input signals are obtained after offsetting 120 degrees and 240 degrees.

The modulation mode of the inverter is divided into unipolar modulation and bipolar modulation. The inverter is connected through an H bridge, and there are two switch tubes on each bridge arm. Bipolar modulation means that two bridge arms share the same input signal. When the upper switch tube of one bridge arm is turned on, the lower switch tube of the other bridge arm is turned on, so that the waveform obtained in this way has both positive and negative polarities within half a cycle. For unipolar modulation, each bridge arm connect an input signal, and the conduction of the switch tube on each bridge arm is independent of each other, which is not affected by other bridge arms. The waveform obtained in this way has only a single polarity in each half cycle[14].

Assuming that the frequency of the fundamental wave is 50Hz and the frequency of the carrier wave is 1000Hz, put them into the inverter of unipolar modulation and bipolar modulation respectively. Using the "Powergui" module in the Matlab/Simulink simulation platform to perform Fast Fourier Transform Analysis(FFTA) on the two methods. The results are shown in Figure 6.

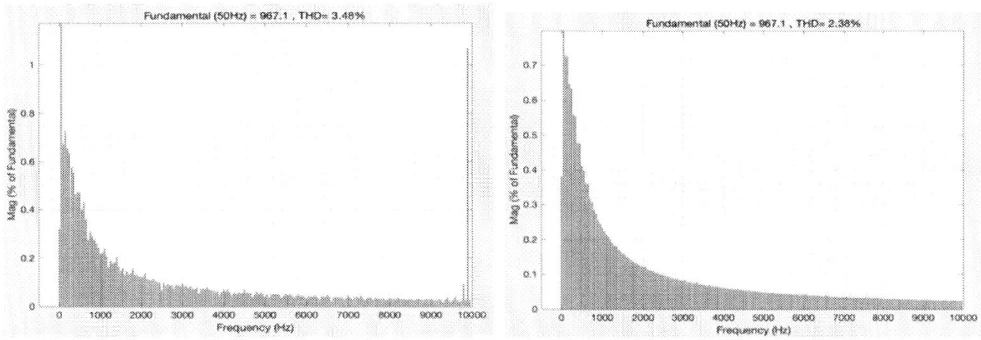

Figure 6. FFTA diagram of bipolar and unipolar modulation.

THD represents the proportion of harmonic distortion, the large value means the great harmonics. The comparison shows that the unipolar modulation method is better than the bipolar modulation.

Since the output voltage needs to match the grid, a three-phase inverter is used in the PET to convert the constant DC voltage into AC voltage after filtering[15]. The topology of the inverter is shown in Figure 7.

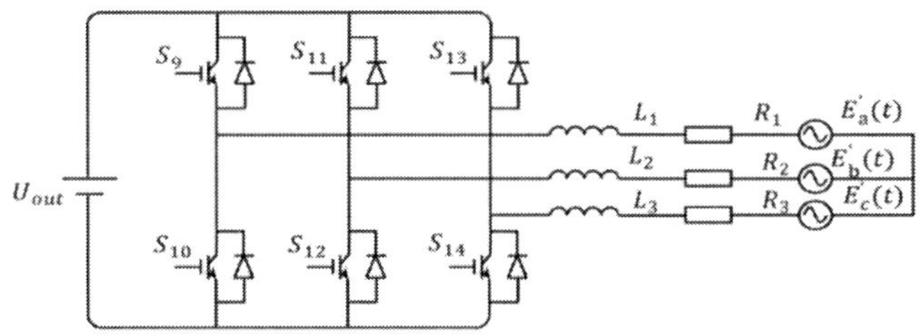

Figure 7. Inverter topology structure diagram

In Figure 7:U_{out} is the output voltage of the DC-DC converter, $S_9 - S_{14}$ are 6 mosfet switch tubes, $L_1 - L_3$ is the three filter inductors, $R_1 - R_3$ are the equivalent load resistances, $E_a'(t)$、$E_b'(t)$、$E_c'(t)$ are the output three-phase power supply. The inductance $L_1 - L_3$ are 0.005 farads, the resistance values are all 10 ohms, and U_{out} is 1000 volts.

In a three-phase inverter, each phase is connected by an H bridge arm, which is composed of two switch tubes in series. Each bridge arm is connected to an equivalent load resistor and an filter inductance, the inductance and resistance are connected in parallel, which will play a role in voltage division.

From the formula $X = \omega L = 2\pi fL$ we can get the resistance of the inductor is proportional to the inductance. If L increases, the inductor's partial voltage will increase, the generated harmonics will be smaller, and the output voltage curve will be smoother. However, an increase in the inductance voltage divider will cause the output voltage to decrease, and the voltage amplitude on the equivalent resistance will be reduced. The output voltage should also meet the voltage requirements for grid operation, so the inductance should be adjusted appropriately.

At high frequencies, the inductive reactance is bigger than the impedance, then the carrier frequency plays a major role. At low frequencies, the result is just the opposite. Therefore, the choice of inductance must meet two conditions: $2\pi f_C L >> R$ and $2\pi f_M L << R$, simplify and get $\frac{R}{2\pi f_C} << L << \frac{R}{2\pi f_M}$. A suitable inductance can be obtained by taking an intermediate value within this range.

After being processed by the inverter, the output AC voltage waveform is shown in Figure 8.

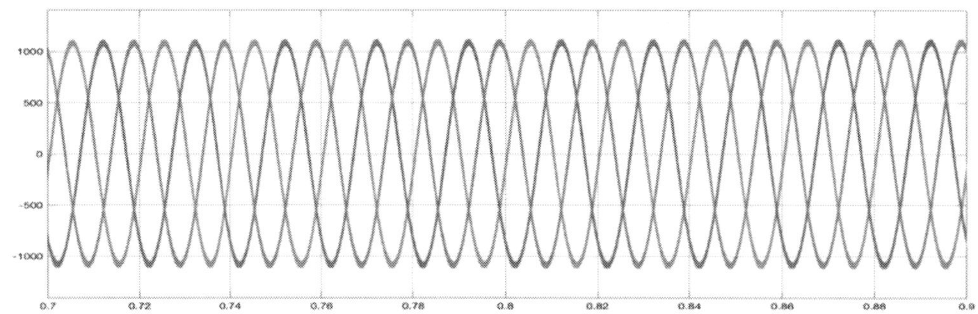

Figure 8. Waveform diagram of inverter output AC voltage

The overall simulation system diagram of the PET is shown in Figure 9.

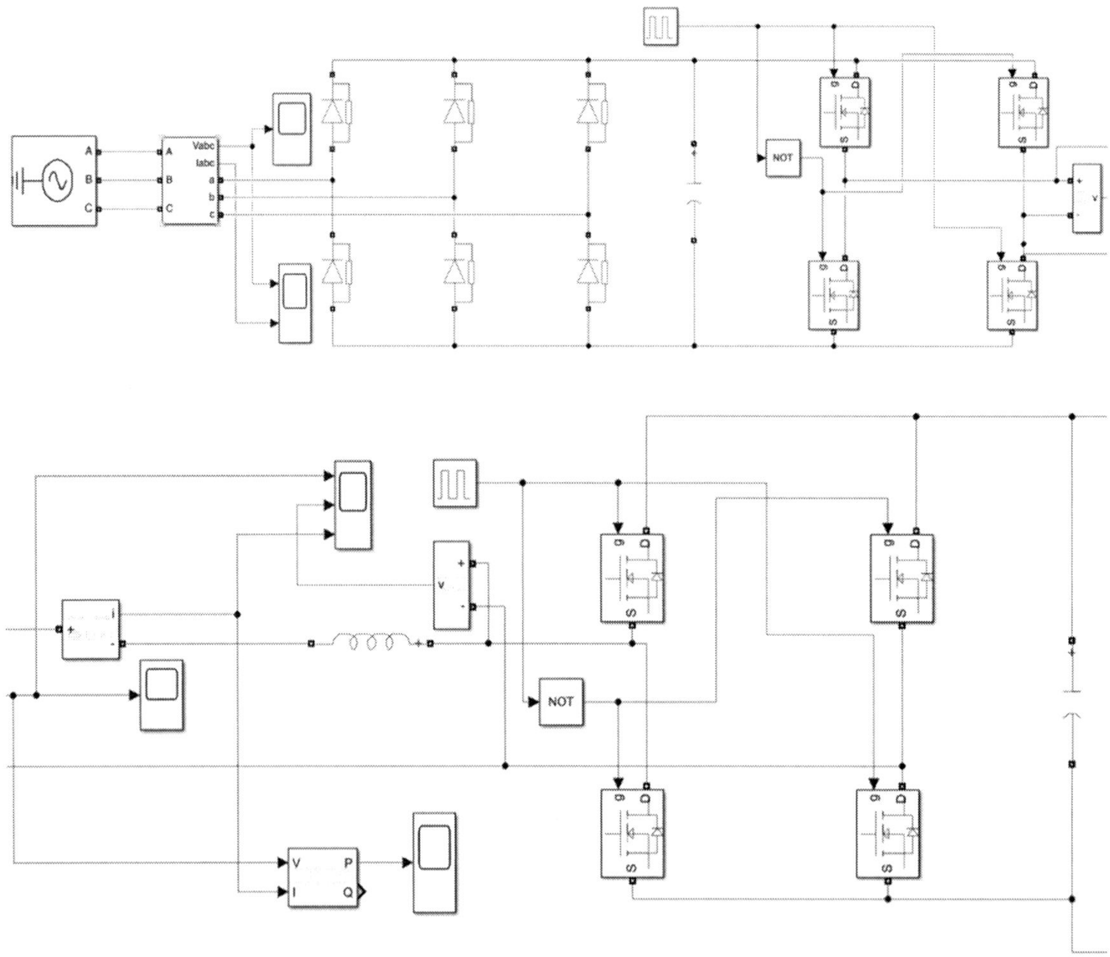

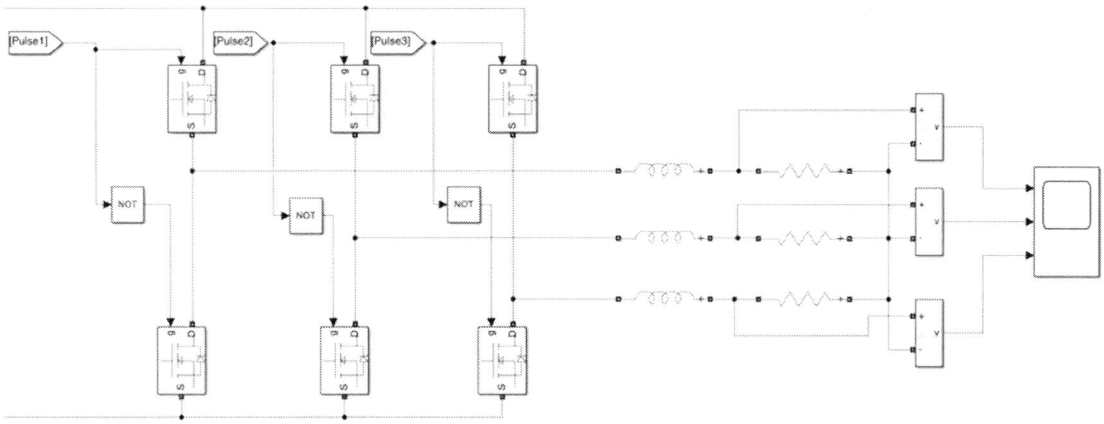

Figure 9. Simulation system diagram of PET

5.Conclusion

In order to build a PET with DC ports, this paper builds a three-stage PET based on a three-phase uncontrolled rectifier, a full-bridge isolated DC-DC converter, and a unipolar modulation three-phase inverter. And build the simulation system of this transformer through Matlab/Simulink simulation platform, carry on simulation analysis and verification to it. Theoretical analysis and experimental results show that:

1. Three-phase uncontrolled rectifier can convert the AC voltage on the grid side into DC voltage, and the resulting DC voltage has a small ripple, which is convenient to connect the full-bridge isolated DC-DC converter.

2. The full-bridge isolated DC-DC converter is mainly used for voltage conversion and electrical isolation. During operation, the power conversion, the charging and discharging functions are realized by controlling the phase difference of the voltage across the DC-DC converter.

3. The output voltage of the DC-DC converter is used as the input voltage of the three-phase inverter, the unipolar modulation method is adopted to adjust the inverter, which can make the AC voltage at the output end have smaller ripple, and the obtained voltage is suitable for connecting to the power grid.

4. When designing the inverter, it is necessary to add a filter inductance. The selection of the inductance is related to the frequency of the fundamental wave and the carrier wave. The value of the inductance is usually within the range of $\frac{R}{2\pi f_C} \ll L \ll \frac{R}{2\pi f_M}$.

References

[1] Zixin Li, Fanqiang Gao, Cong Zhao, Zhe Wang, Hang Zhang, Ping Wang, Yaohua Li. Summary of Research on Power Electronic Transformer Technology[J]. Proceedings of the Chinese Society of Electrical Engineering, 2018, 38(05): 1274-1289.

[2] Weihua Deng, Bo Zhang, Zongbo Hu. Research on Circuit Topology and Control Strategy of Power Electronic Transformer[J]. Automation of Electric Power Systems, 2003, 27(20): 40-45.

[3] Yang Yan. A Summary of Theoretical Research on Power Electronic Transformers[J]. Electrician Electrical, 2012, (3): 5-8.

[4] Qichao Chen, Yanchao Ji, Yanlin Pan, Jianze Wang. Summary of Topological Structure of Power Electronic Transformer in Distribution System[J]. New Electrician and Energy Technology, 2015, 3.

[5] Mcmurray W. Power converter circuits having a high frequency link[P]. U.S. Patent, 3517300, June 23, 1970.

[6] Zixin Li, Wang Ping, Zunfang Chu, Haibin Zhu, Yaohua Li. Research on power electronic transformer for medium and high voltage intelligent distribution network[J]. Power Grid Technology, 2013, 37(09): 2592-2601.
[7] Jianpeng Cheng, Chengxiong Mao, Shu Fan, Dan Wang. Research on Principle and Simulation of Electronic Power Transformer[J]. Power Automation Equipment, 2004, 24(12): 23-26.
[8] Lei Chen, Jiaxiang Ou, Qiuyan Zhang, Xufeng Yuan, Sheng Hu, Fei You, Guo Shi. Summary of Research on Power Electronic Transformers[J].Power Grid and Clean Energy, 2015, 31(12): 36-42.
[9] Chen Ling. Research on the Control Strategy of Electronic Power Transformer[D]. Beijing: Beijing Jiaotong University, 2011.
[10] Lihua Ji. Research on Modular Multi-level Control Strategy Applied to Power Electronic Transformers[D]. Jinan: Shandong University, 2017.
[11] Guohu Liao, Xufeng Yuan, Guoyue Qiu. Simulation Research on Power Electronic Transformers in Distribution Networks[J]. Electrical Measurement and Instrumentation, 2014, 51(17): 35-41.
[12] Chunming Tu, Zheng Lan, Fan Xiao, Jun Ge, Yang Meng, Yi Yang. Design and Implementation of Modular Power Electronic Transformer[J]. Advanced Technology of Electrical Engineering and Energy, 2017, 36(05): 42-50.
[13] Biao Zhao, Qiang Song, Wenhua Liu, Guowei Liu, Yuming Zhao, Senjing Yao. High-frequency link DC solid-state transformer for flexible DC distribution[J]. Proceedings of the Chinese Society for Electrical Engineering, 2014, 34 (25): 4295-4303.
[14] Zhe Li. Application Research of Power Electronic Transformer[J]. Coal Mine Electromechanical. 2019, 40(3): 56-59.
[15] Kaichang Xue, Zongxin Luo. Analysis of Output Distortion and Efficiency of Unipolar PWM Inverter. 2019, 53(7).
[16] Ronan E R, Sudhoff S D, Glover S F, et al. A Power Electronic-Based Distribution Transformer[J]. IEEE Power Engineering Review, 2002, 22(3): 61-61.

Intelligent alarm data noise reduction for power systems based on singular value decomposition

Chen Chen [1a*], **Hongren Man** [1b], **Xiu Liu** [1c]

[1] Information center of Yunnan Power Grid Co., Ltd

[a*]654689833@qq.com , [b]349205635@qq.com , [c]154071502@qq.com

Abstract: The noise types of power system intelligent alarm data are complex. When reducing the intelligent alarm data, the profile noise statistics of the noise data are large, resulting in the actual noise reduction value is too small. To solve this problem, a power system intelligent alarm data noise reduction method based on singular value decomposition is designed. The selected normalized decomposition matrix iteratively processes the original matrix, the singular value decomposes the power system alarm data, sets an estimation quantity within the paradigm of the alarm data, controls the noise profile noise statistics, characterizes the noise alarm data structure, uses the SC algorithm to process the cluster basis vectors in the noise data structure, and constructs a repeated iterative convergence process to realize intelligent data noise reduction processing. The original alarm data within a known power system is used as test data, the power system alarm window is set, and the power system alarm data singular values are circled. The data mining-based alarm data noise reduction method, the regularized filter-based alarm data noise reduction method and the designed data noise reduction method are applied to the noise reduction process, and the results show that the designed data noise reduction method has the largest noise value and the best noise reduction effect.

1. Introduction

Singular value decomposition belongs to the category of linear algebra. In the existing matrix decomposition structure, the singular value obtained by decomposition is determined in the conjugate transpose formed by diagonalization similar processing method [1]. As a tool of power system security supervision, alarm data is an important index to judge the normal operation of power system. There are many operating environments in the power system in the substation [2]. In the process of monitoring the substation system, the analog quantity or switching value intelligent parameters representing the operating state will produce the influence of external noise. How to reduce the noise and deal with the intelligent alarm data of the power system has become a research hotspot at present. Therefore, the numerical optimization process in the singular value decomposition process is used as support [3], design the noise reduction method of intelligent alarm data.

Foreign research on alarm technology started early. According to the potential relationship between power systems, researchers mine the alarm data in the system by means of sequential processing, set a fixed alarm condition [4], and use post-processing to determine the transfer parameters of noise data to form a noise reduction process. The domestic research on alarm data started late. Combined with the site parameters of power system, set the alarm threshold of different numerical levels [5] according to the parameters, set the threshold priority, and use the Gaussian accounting method to convert the noise data binary into the alarm data with small noise value, forming a noise processing process. The noise reduction method in reference [6] extracts the trend item in the data by empirical mode decomposition, and controls the superposition of the trend item with the data containing noise, so as to eliminate the noise in the data. In the data denoising method in reference [7], the S-transform is used to obtain the time-frequency matrix in the noise data, and a noise threshold is set in combination with the singular value difference spectrum theory, so as to eliminate the noise in the data. Through phased application exploration, it can be seen that the noise reduction value obtained by the existing data noise reduction processing methods is too small. Therefore, it is of research and development significance to design the power system intelligent alarm data noise reduction method based on singular value decomposition.

2. Noise reduction of power system intelligent alarm data based on singular value decomposition

2.1. Singular value decomposition power system alarm data

The actually obtained power system intelligent alarm data is processed into a data set that can be processed directly, and the data set is processed into an original matrix. The standardized decomposition matrix is selected to process the original matrix iteratively [8]. The iterative processing process can be expressed as:

$$\begin{cases} H_n \to \dfrac{V^T}{\sum\limits_{r\in T} H_r} \\ K_n \to \dfrac{K^T}{\sum\limits_{r\in T} H_r} \end{cases} \qquad (1)$$

Wherein, H_n and K_n respectively represent the iteration formed in the positive and negative directions, V^T represents the original matrix of alarm data, K^T represents the standardized matrix, and H_r represents the constructed decomposition matrix. Combined with the decomposition of intelligent alarm data [9], for the alarm matrix obtained by the above decomposition, decompose the singular value in the numerical relationship, and the singular value decomposition process can be expressed as follows:

$$A = U\begin{bmatrix} 0 & V^T \\ 0 & 0 \end{bmatrix} H_r \qquad (2)$$

Where A represents singular value decomposition process, U represents left singular value vector, and the meaning of other parameters remains unchanged. According to the iteration period generated in the above positive and negative directions, the column vector [10] in the singular value is extracted, and the extraction process can be expressed as:

$$\begin{cases} Av^T = \gamma^T u^T \\ Au^T = \gamma^T v^T \end{cases} \qquad (3)$$

Where v^T represents the column vector generated by the forward iteration process, u^T represents the column vector generated by the reverse iteration process, γ^T represents the singular value extraction disturbance parameters, and the meaning of other parameters remains unchanged. Take the column vector extracted above as the rotation direction [11], define the norm of singular value of alarm data, and the numerical relationship can be expressed as:

$$\kappa = \frac{\|A\|_2}{\sqrt{A^T}} \qquad (4)$$

Where κ represents the singular value norm of alarm data, $\|A\|$ represents the rotation direction of column vector, and the meaning of other parameters remains unchanged. Sort out the singular value norm in the alarm data, and characterize the noise alarm data structure according to the norm change in the time period.

2.2. *Characterization of noise alarm data structure*

Corresponding to the singular value norm calculated above, set an estimator in the norm according to the order of generating the norm corresponding to the rotation direction [12], and the value relationship of the estimator can be expressed as:

$$\tau^2 = \frac{\left\|\widehat{A} - r(z)\right\|^T}{F} \qquad (5)$$

Where τ represents the calculated noise estimator, $\widehat{A}$ represents the singular value estimation parameter, $r(z)$ represents the low order threshold, F represents the rank value parameter, and T represents the estimator period. Combined with the calculated estimator period value, the noise generated by the corresponding alarm data is obtained. The acquisition process is shown in Figure 1 below:

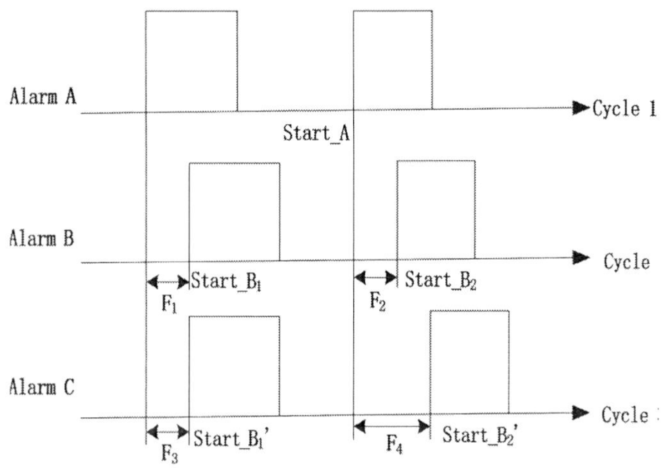

Fig.1 noise generated by periodic alarm data of adjacent estimators

When characterizing the alarm data structure, set a soft threshold operator into the estimator. At this time, the numerical relationship can be expressed as:

$$g^* = \sqrt{m}\left(\frac{Y}{\tau}\right) \qquad (6)$$

Where g^* represents the characterization processing function, m represents the set soft threshold operator, Y represents the actually generated noise sequence, and the meaning of other parameters remains unchanged. Taking the noise alarm data structure of the above characterization processing as the processing object, when realizing the noise reduction of alarm data, the sparse coding processing process is adopted to sort out the fusion generated by vector basis [13], so as to realize the noise reduction of alarm data.

2.3. *Noise reduction of alarm data*

Under the alarm data structure after the above characterization processing, the SC algorithm is used to determine the cluster base vector in the data structure, and the numerical relationship can be expressed as:

$$\|x\| = \min_{x} \left\| \frac{y}{D_x} \right\|^2 \quad (7)$$

Where $\|x\|$ represents the calculated cluster base vector, y represents the alarm data structure parameters, and D_x represents the over complete data node. The dimension value in the numerical tensor is defined, and the dimension value is treated as a cyclic matrix by using independent Gaussian distribution. The numerical relationship can be expressed as:

$$p(y \mid x) = \prod_{k=1}^{K} (\Gamma_k) \quad (8)$$

Where Γ_k represents the cyclization function, K represents a posteriori parameter, and the meaning of other parameters remains unchanged. The multiplier algorithm is set in the circularized matrix, and the structured values in the matrix are treated as equivalent numerical differences by variational method [14], and the numerical relationship can be expressed as:

$$\ln P = \frac{S_i}{w_i} \quad (9)$$

In the above numerical relationship, P represents the equivalent parameter, S_i represents the regularization function, and w_i represents the singular numerical parameter. Through the equivalent numerical difference obtained above and the defined tensor, the noise reduction update process is set, and the penalty coefficient is quoted to form an alternating convergence preset process. The numerical relationship can be expressed as:

$$k_i = \frac{\ln P + \rho_h}{Z_k} \quad (10)$$

Where k_i represents the constructed convergence preset function, ρ_h represents the augmented tensor, and Z_k represents the set update penalty coefficient. Control the above convergence preset process, continuously update the above set penalty coefficient, and correspond to the convergence conditions obtained by the penalty coefficient, so as to finally achieve the noise reduction processing of alarm data [15]. Based on the above processing process, the noise reduction of power system intelligent alarm data based on singular value decomposition is finally completed.

3. Noise reduction effect test

3.1. Setting power system alarm window

Randomly select the original alarm data of one month in the power system as the test data, set five alarm sequences with unit time in the operation and maintenance alarm scenario in the power system according to the time cycle of the original alarm data, and form rules between the given alarm time and alarm sequence according to the alarm time actually generated in the sequence alarm, Set the confidence actually generated in the capture rule, process the confidence parameter as a mining parameter, and the value relationship of confidence can be expressed as:

$$c(X \to Y) = \frac{\sigma(X \cup Y)}{\sigma(X)} \quad (11)$$

Among them, X represents the given alarm time, Y represents the set alarm sequence, $\sigma(X)$ represents the numerical variance of the given alarm time, $\sigma(X \cup Y)$ represents the variance value

 2108 (2021) 012074 doi:10.1088/1742-6596/2108/1/012074

between parallel data sets, the alarm sequence corresponds to a specific confidence value, which is taken as the threshold value of the alarm sequence, and sorted into five alarm monitoring rules with different capture contents. Therefore, the capture alarm sequence with time window is set in the actual capture process. The time window and corresponding alarm monitoring items are shown in the table below:

Table 1 set time window alarm sequence

Grab sequence number	time window	Alarm item capture rules
Z-01	(60,0,60)	Rule4, Rule5
Z-02	(60,120,180)	Rule2
Z-03	(60,180,240)	Rule4, Rule5
Z-04	(60,240,120)	Rule1
Z-05	(60,300,180)	Rule3, Rule1
Z-06	(60,360,240)	Rule4, Rule5
Z-07	(60,400,300)	Rule4
Z-08	(60,120,360)	Rule3, Rule4
Z-09	(60,180,420)	Rule1
Z-10	(60,60,120)	Rule1 , Rule2 , Rule3

According to the time window alarm sequence set in the above table, for the data with timing attribute, the numerical interval between the start time of the alarm sequence and the time window is agreed and defined as one time unit. Sort out the data obtained by the alarm item capture rules as the actual processing object, and delineate the singular value of the power system alarm data.

3.2. Delineate singular values of power system alarm data

Using the alarm data collected from the above window, the alarm data under different capture rules are weighted and fused. The processing process can be expressed as follows:

$$w_{ij} = \exp\left(\frac{\left\|p_{ij} - p_{i-k}\right\|}{2T}\right) \quad (12)$$

In the above numerical relationship, w_{ij} represents the weighted fusion function, p_{ij} represents the acquired data information block, p_{i-k} represents the alarm data information block acquired in the next time cycle, and T represents the time interval between alarm capture Windows. For the alarm data after data fusion, when delineating the singular value of power system alarm data, calculate the mean value of the data in the unified capture rule, take the value as the singular value, and calibrate the data noise in the alarm singular value. The noise value relationship can be expressed as:

$$f(x, y, z) = \frac{w_{ij}}{\lambda(\phi_1 - \phi_2)} \quad (13)$$

Where λ represents the introduced noise contrast parameter, ϕ_1 represents the obtained noise ratio, ϕ_2 represents the confidence measurement parameter, and the meaning of other parameters remains unchanged. Kriging difference method is used to convert the structure in the mean value of noise data. Sparse indicates that the alarm data is a three-dimensional structure, as shown in the following figure:

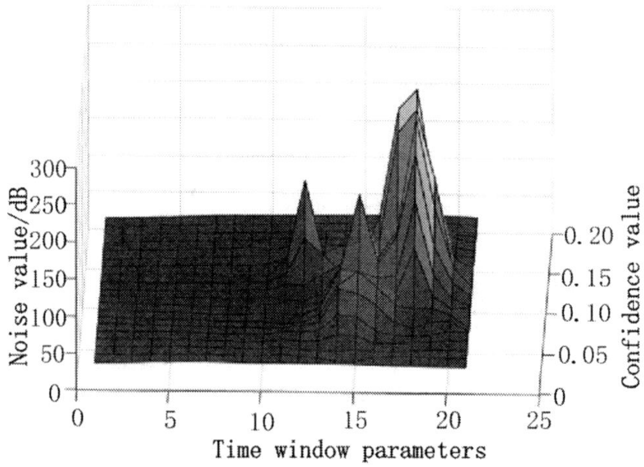

Fig.2 three dimensional structure of sparse alarm data

According to the three-dimensional structure of alarm data shown in the above figure, the data containing noise is highlighted in the acquisition diagram. Corresponding to the time window parameters, noise values and confidence values in the three-dimensional structure, a noise data contour processing process is constructed. The numerical processing process can be expressed as:

$$\nabla n = \frac{\partial \phi(x, y, z)}{\partial l} \qquad (14)$$

Where ∇n represents the contour gradient, $\phi(x, y, z)$ represents the amplitude of noise value, and l represents the noise control point. Keep the axis of time window parameters and the axis of confidence parameters unchanged. The equal noise numerical structure of alarm data in the plane is shown in the following figure:

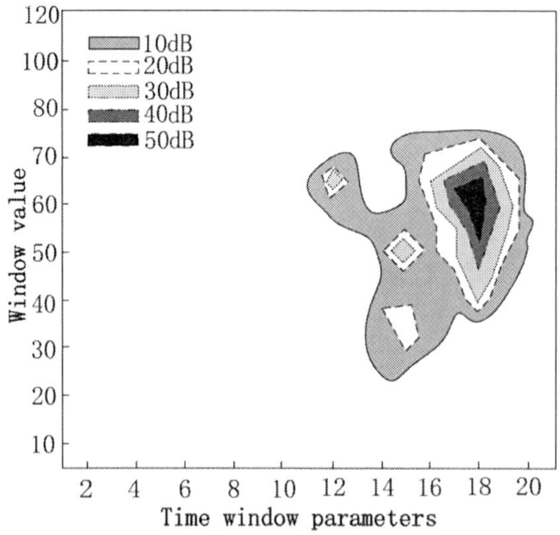

Fig.3 equal noise numerical structure of alarm data

Corresponding to the equal noise numerical structure in the figure above, the alarm data noise reduction method based on data mining, the alarm data noise reduction method based on regularization filtering and the designed data noise reduction method are applied for noise reduction, and the noise reduction effects of the three data noise reduction methods are compared.

3.3. Noise reduction results and analysis

Based on the above processing process, sort out the noise value in the above figure as the noise change of alarm data. Corresponding to the above set capture window, sort out three alarm data noise reduction methods to reduce the size of noise value structure. The results are shown in the following figure:

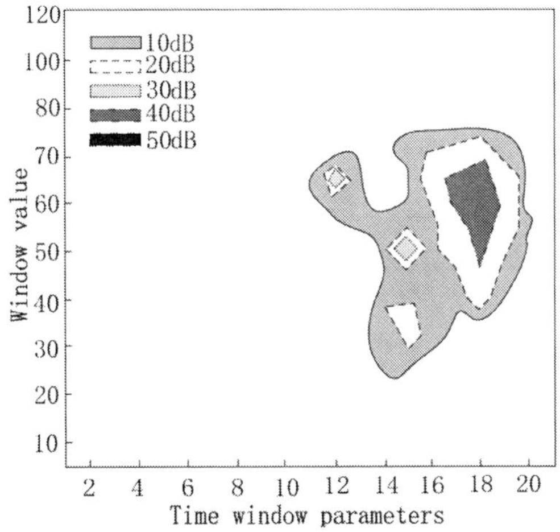

(a) Noise reduction method of alarm data based on Data Mining

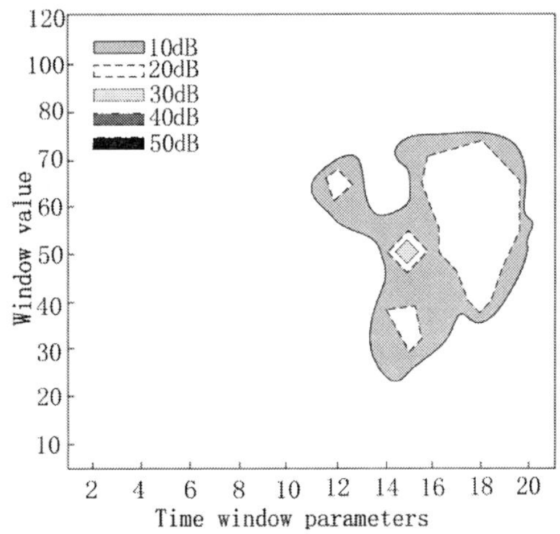

(b) Noise reduction method of alarm data based on regularization filtering

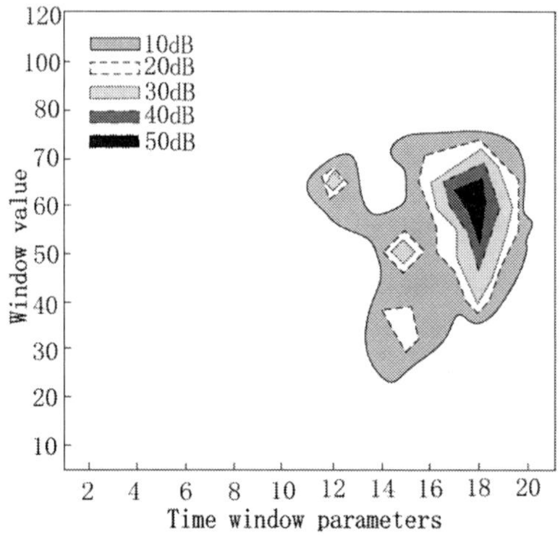

(c) The designed alarm data noise reduction method can deal with the noise structure

Fig.4 noise reduction structure results of three alarm data

According to the numerical noise reduction results shown in the above figure, according to the power system alarm data processed, the numerical range of noise is calculated according to the calculated numerical relationship, and the noise figure corresponding to the corresponding area is calibrated. The three alarm data denoising methods are used to identify the structure. According to the test results shown above, The alarm data denoising method based on data mining can successfully identify the area with noise between 10dB ~ 40dB, and the noise data area that can be denoised is large. The alarm data denoising method based on regularization filtering can successfully identify the alarm data structure with noise value between 10dB and 30dB. The designed alarm data noise reduction method can identify the noise value range of 10dB ~ 50dB. Compared with the two selected noise reduction methods, this alarm data noise reduction method can be applied to alarm data sets with a wider range of noise values.

Sort out the noise value in the equal noise numerical structure of the alarm data and take it as the standard noise value change in the alarm data. Apply the noise reduction algorithm of the three alarm data noise reduction methods to calculate the noise value of the actual noise reduction captured alarm data by the three noise reduction methods according to the data processing area covered by the noise structure. The numerical relationship can be expressed as:

$$R_{SSN} = S \times n - \frac{e_{s_i}}{W^T} \quad (15)$$

Among them, R_{SSN} represents the actual noise reduction value of the noise reduction method, e_{s_i} represents the noise value generated by the alarm data in the structural area, W^T represents the kernel function of the three noise reduction methods in the same time window, S represents the area of noise generated by the window, and BB represents the noise number value corresponding to the area. Corresponding to different grab numbers, sort out the noise values of the noise reduction treatment of the three noise reduction methods. The noise reduction results are shown in the figure below:

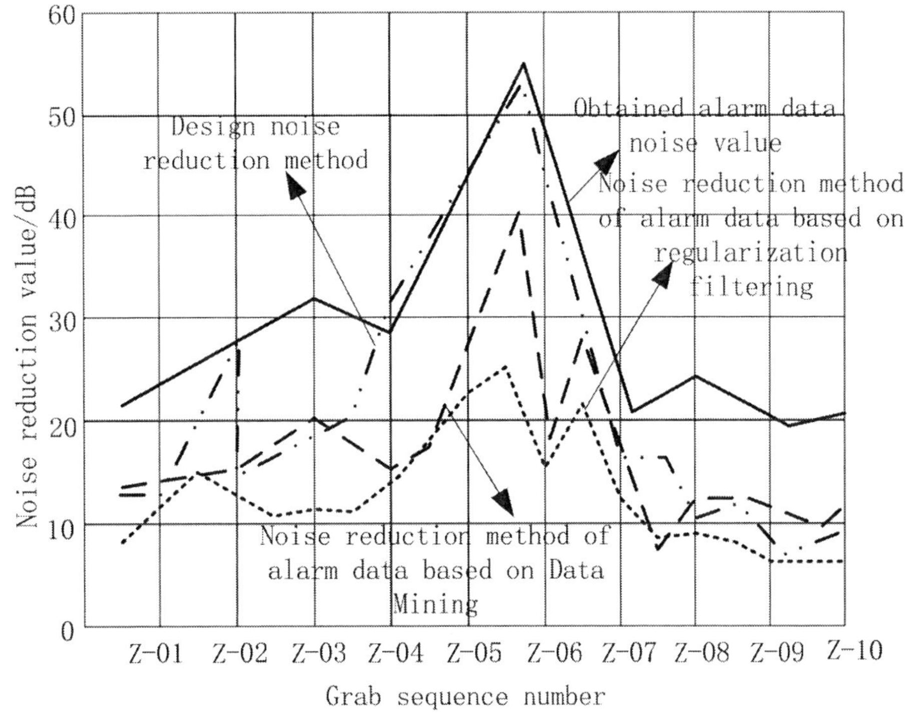

Fig. 5 noise reduction results of three noise reduction results

According to the defined noise reduction numerical relationship, the alarm capture data sets processed by the three noise reduction processing methods are transformed, and the singular values of power system alarm data are delineated correspondingly to obtain the noise value change in the capture data set, and this value is used as the starting point of noise value processing. Sort out the noise reduction values obtained by the algorithms of the three noise reduction methods, and take the numerical difference between the value and the actually obtained noise value processing points as the comparison index. When the numerical difference is larger, it means that the noise treatment effect of the noise reduction method is better. According to the numerical difference results shown in the figure above, the noise reduction value of the alarm data noise reduction method based on regularization filtering is about 25dB, and the noise reduction effect is poor. The noise reduction value obtained by the noise reduction processing method based on data mining is about 40dB, which has poor noise reduction effect on alarm data. The noise reduction value obtained by the designed noise reduction method is about 52dB. Compared with the two applied noise reduction methods, the noise reduction value of the data of this noise reduction me

4. Conclusion

Under extreme weather conditions, the alarm data of power system is an important medium to evaluate the operating conditions of the system. With singular value decomposition as the numerical processing support, the design of power system intelligent alarm data noise reduction method can improve the existing noise reduction process, which has the disadvantage of small noise processing value. In the future work, it is hoped that the designed alarm data noise reduction method can provide support for data noise reduction.

Reference

[1] Yang Zhiwei, Liu Hao, Bi Tianshu, et al. PMU data recovery method based on singular value decomposition [J]. Chinese Journal of electrical engineering, 2020,40 (03): 812-821
[2] Fan Xianguang, Wu Tengda, Zhi Yuliang, et al. Raman imaging data denoising method based on

singular value decomposition and median absolute deviation [J]. Spectroscopy and spectral analysis, 2020,40 (02): 436-440
[3] Yang Juhua, Zhang Linjing, Chen Guangwu, et al. Error analysis and noise reduction of MEMS gyroscope based on SVD / wavelet [J]. Journal of Chongqing University of Posts and Telecommunications (NATURAL SCIENCE EDITION), 2020,32 (02): 322-328
[4] Gu Tingyun, Gao Yunpeng, Wu Cong, et al. Power quality disturbance detection based on improved wavelet threshold function and singular value decomposition [J]. Electrical measurement and instrumentation, 2020,57 (21): 111-118
[5] Wang Hongwei, Xie Lirong. Fuzzy model identification of non uniformly sampled nonlinear systems based on singular value decomposition [J]. Control and decision making, 2020,35 (03): 757-762
[6] Liu Shuzhen, Chen Zhixing. Noise reduction method of nonlinear vibration signal based on singular value decomposition and EEMD [J]. Journal of detection and control, 2019,41 (03): 37-42
[7] Niu Haiqing, song Tinghan, Luo Xin, et al. Partial discharge signal denoising method based on S-transform and singular value decomposition [J]. Journal of South China University of Technology (NATURAL SCIENCE EDITION), 2020,48 (02): 9-15
[8] Shen Zhongting, Ding Renjie. Power system low frequency oscillation modal identification based on Improved EMD denoising and matrix beam [J]. Journal of Applied Science, 2019,37 (06): 761-774
[9] Zhang Qi, Liu Ling, Wen Junhao. A singular value decomposition recommendation algorithm based on domain trust and distrust [J]. Computer science, 2019,46 (10): 27-31
[10] Wei xiulei, Liu Shuyong. Signal denoising method of communication radar based on multilevel singular value decomposition and SG [J]. Journal of Wuhan University of Technology (traffic science and Engineering Edition), 2020,44 (04): 658-662
[11] Hu Chaofan, Wang Yanxue. Research on multi-channel signal noise reduction method for composite fault of rolling bearing based on tensor decomposition [J]. Journal of mechanical engineering, 2019,55 (12): 50-57
[12] Xing Tingting, Guan Yang, Liu Zihan, et al. Frequency close signal separation method based on variational modal decomposition and singular value decomposition [J]. Journal of metrology, 2020,41 (11): 1404-1409
[13] Li Weijiang, Luo Panhu. Diversified recommendation algorithm integrating kernel density estimation and singular value decomposition [J]. Small microcomputer system, 2020,41 (01): 56-60
[14] Yin Fang, song Yao, Li Ao. Collaborative filtering algorithm based on locally optimized singular value decomposition and K-means clustering [J]. Journal of Nanjing University of technology, 2019,43 (06): 720-726
[15] Shao meiyang, Wu Junyong, Li Baoqin, et al. Power system transient stability assessment based on two-stage integrated deep confidence network [J]. Power grid technology, 2020,44 (05): 1776-1787

Remote acquisition of insulator icing flashover signal based on frequency optimization of ring-shaped intelligent current sensor

Qiancheng Lv[1a*], Bing Tian[1], Zhong Liu[1], Zhiming Wang[1], Xu Yin[1], Hongdi Sun[1], Bofeng Luo[1]

[1]Digital Grid Research Institute, China Southern Power Grid, Guangzhou, 510663, China

[a*] lvqc@csg.cn

Abstract—In order to improve the safety of power grid operation and avoid the hidden danger in the process of power supply, a remote acquisition method of insulator ice-covered flashover signal based on the frequency optimization of ring intelligent current sensor was proposed. According to principle of take electricity by capacitance, transmission line, current sensor, current sensor B and double rectifier circuit composed of parallel resonant magnetic circuit for electric model. By monitoring the current sensor and current sensor B the secondary side of induction electromotive force, magnetic core KaiQi gap equivalent permeability values, such as through the calculation of current sensor parameters of iron core and coil number of turns. In this paper, the current sensor is optimized to improve the power monitoring effect, and the remote acquisition of the flashover signal of insulator icing is realized. The experimental results show that. This monitoring method can effectively monitor the maximum load power, load resistance and load voltage of the model when the current value is different. In the face of impedance branch resonant, resonant and open air gap and other circumstances, can effectively monitor the power taking situation, a wide range of applications.

1. Introduction

The development of the power grid system has become increasingly mature and complete, and the power consumption in various fields has greatly increased [1]. With the continuous improvement of the voltage level of the transmission line, the demand for the safety of the power grid has also increased rapidly [2]. At present, the more commonly used methods of taking power mainly include battery power supply, laser power supply, capacitor voltage division power supply, current sensor induction power supply, etc. [3]. There are certain technical shortcomings and difficulties in the above-mentioned methods of obtaining electricity, in order to improve the reliability and safety of the transmission line to obtain electricity, and to ensure the safe operation of the transmission line [4,5]. Liu and others proposed a non-intrusive load monitoring method based on improved chicken flock algorithm. This method uses an improved chicken flock algorithm to monitor power grid load. Due to the influence of steady-state harmonic current and power, this method has low load identification and the monitoring data is not accurate enough[6]. Zhao and others proposed a four-way four-state modulation-based fiber optic gyroscope optical path power monitoring method. This method uses four-way four-state square wave modulation and software independent monitoring to achieve power monitoring, but the method has poor stability and performance. Broadband has a high degree of impact and cannot meet most power demand [7].

In order to solve the problems of the above methods, a method for remote acquisition of ice-covered flashover signal of insulator based on frequency optimization of annular smart current sensor is presented. The method achieves remote acquisition of ice-covered flashover signal of insulator by means of frequency optimization of shunt resonance dual magnetic circuit and monitoring principle.

2. Remote acquisition method for insulator icing flashover signal under frequency optimization of ring-shaped intelligent current sensor

2.1. Remote signal extraction for insulator icing flashover

The peak value, effective value and frequency are extracted from the remote signal of the insulator icing flashover. The characteristic quantity information such as the accumulated charge amount, the maximum pulse amplitude, the average value of the peak value in a certain time, the average value of the effective value and the number of pulses are extracted from the current signal. The software can display and store the voltage and current waveforms and characteristic quantities, and can call out the observation data for analysis afterwards. The calculation formula of signal extraction induced electromotive force is as follows:

$$K_2 = \frac{2\pi f V \mu_0 \mu S \lambda i_1}{l} \quad (1)$$

In equation (1), the induced electromotive force on the secondary side of the current sensor A and the relative permeability of the magnetic core material are represented by K_2 and μ, respectively. Power frequency 50Hz is represented by f. The cross-sectional area of the core, the core lamination coefficient, the number of turns on the secondary side of the core, the average magnetic path length, and the vacuum permeability are represented by S, λ, V, l, and μ_0 respectively [10]. In addition, due to the addition of the climbing skirt, the ice coating state is effectively improved, so that most of the ice is not bridged, and more air gaps are formed. At the same time, due to the installation of the climbing skirt, the overall climbing distance of the insulator is increased. In the case of the umbrella skirt without bridge, it also has a certain effect on the improvement of ice flash voltage.

The relative permeability μ and the transmission line current i_1 are the main parameters of the induced electromotive force on the secondary side of the current sensor A. And the induced electromotive force of the secondary side of the current sensor A, the relative permeability and the current of the transmission line are all proportional. The current sensor A uses an open air gap, and the equivalent permeability calculation equation for the open air gap of the magnetic core is as follows:

$$\mu_{eq} = \frac{\mu}{\mu \frac{\delta}{l} + 1} = \frac{1}{\frac{\delta}{l} + \frac{1}{\mu}} \approx \frac{l}{\delta} \quad (2)$$

It can be seen from equation (2) that the equivalent permeability of the magnetic core drops rapidly after the magnetic core has an air gap, which reduces the induced electromotive force E_2 of the secondary side of the current sensor A. Therefore, the conventional current sensor takes less power [11].

In order to avoid the problem of small electric power of conventional current sensors, the new electric circuit topology is introduced into the electric circuit [12]. Using the principle of circuit shunting, suppose μ' represents the relative permeability in parallel. Connect a non-open air-gap magnetic core with a relative permeability of μ' to the branch of current sensor A, and the current expression equation input to current sensor B is as follows:

$$i_2 = I - i_1 \quad (3)$$

In equation (3), I represents the current of the high-voltage transmission line.

Assume that the magnetic core magnetizing impedance is represented by $\omega L_{m1} + R_{m1}$. The capacitance is represented by C. Parallel the capacitor on the secondary side of the current sensor A. When the magnetic core excitation impedance and the capacitor generate parallel resonance, it means that the total impedance of the branch circuit of the current sensor A is higher than the total impedance of the current sensor B. Affected by the current shunt, most of the high-voltage transmission current I flows into the power-taking branch i_2, so the calculation equation of the induced electromotive force $K_2^{'}$ on the secondary side of the current sensor B is as follows:

$$K_2^{'} = \frac{2\pi f V \mu_0 \mu' S \lambda i_2}{l} \quad (4)$$

It can be seen from equation (4) that the induced electromotive force $K_2^{'}$ of current sensor B is higher than the magnetic induced electromotive force K_2 of current sensor A, and the power obtained by the load is relatively high [13]. The reason is that the relative permeability of the core without air gap is higher, and the current μ' input to the current sensor B is higher than the current μ input to the current sensor A.

2.2. Current sensor parameter optimization

In order to obtain accurate monitoring results of integrated power supply, the current sensor core parameters and coil turns are calculated and optimized [14]. At the moment of current start, the load voltage has a certain limit, and the maximum power obtained by the load is the most reasonable power. Assuming U_d is the lowest operating voltage of the device, the load resistance can be obtained by the ratio of the lowest operating voltage of the device to the operating load current of the device [15].

Let the primary side current of the current sensor be I_q, according to the principle of the transformer, there are:

$$I_s = \frac{I_q}{V} \quad (5)$$

In equation (5), the number of turns on the secondary side of the electric current sensor is represented by V. When the value I_s is not extremely small, then:

$$Y_m = R_L \quad (6)$$

In equation (6), the excitation reactance is represented by Y_m. It can be seen from equation (6) that the power can be maximized when most of the currents have different values. When the difference between the electric power and the load power when the maximum power condition is met is low, there are:

$$Y_{m,\min} = R_L \quad (7)$$

When the circuit satisfies the condition of equation (7), the value of the secondary turns V of the current sensor can be obtained through equation (5).

In order to make the starting current of the circuit within the set range, the core material must meet the following conditions:

$$\frac{\mu S}{l} \geq \frac{Y_{m,\min}}{2\pi f V} \quad (8)$$

Affected by the iron consumption and heating factor of the current sensor, there are:

$$U_d + 1.4 = 4 f N B_{d,m} S \quad (8)$$

In equation (8), when the voltage is U_d, the peak value of magnetic induction is represented by $B_{d,m}$. Therefore, the peak value of the magnetic induction intensity at U when the voltage is U_d can be calculated from the cross-sectional area of the iron core as $B_{d,m}$. After determining the core material, the no-load experiment can be performed to obtain the magnetic permeability μ value under the corresponding magnetic induction intensity. According to the iron core shape, calculate the average magnetic path length l value, and finally obtain the best iron core.

The selection of the wire diameter of the secondary winding of the current sensor can be calculated by the following equation:

$$d = \sqrt{\frac{I_{1\max}}{V} / \pi} \quad (9)$$

In equation (9), the maximum allowable current effective value of the line and the copper wire radius are represented by $I_{1\max}$ and d respectively.

3. Experiment and analysis

In order to verify the effectiveness of the proposed remote acquisition method of insulator icing flashover signal based on ring-shaped intelligent current sensor frequency optimization, a simulation experiment is designed. The experiment uses PSIM software to establish the power model and circuit topology, and select 0.02A and 0.04A respectively from the primary side conversion to the secondary side current value. Set two sets of current values to adjust the load size to 350 Ω.

Use the monitoring method in this paper to monitor the calculation results of the load power curve corresponding to the power model at different current values, as shown in Figure 1.

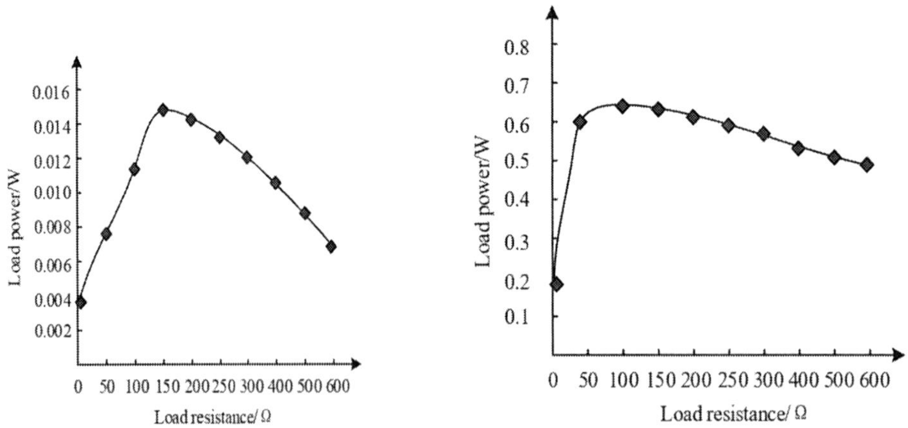

(a) The current value is 0.02A (b) The current value is 0.04A

Fig. 1 Load power curve of electric model under different current values

Analyzing Figure 1(a), we can see that when the current value is 0.02A and the load resistance is 150 Ω, the load power of the electrical model reaches 0.015W. It can be seen from Figure 1(b) that when the current value is 0.04A and the load resistance is 70 Ω, the load power of the electrical model reaches 0.65W. It can be seen that the method in this paper can monitor the maximum load power of the power-taking model under different current values, and has a high monitoring capability.

Use this method to monitor the load resistance and load voltage of the electric model under different current values, and the results are shown in Figure 2.

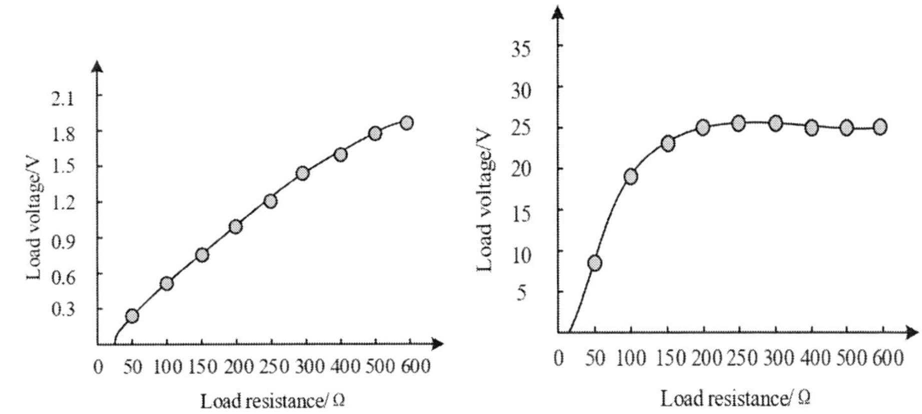

(a) The current value is 0.02A (b) The current value is 0.04A

Fig. 2 Load voltage curve of the electrification model under different current values

Analysis of Figure 2 shows that when the current value is 0.02A, the load voltage curve shows a gentle upward trend. When the load resistance is 600 Ω, the load voltage reaches the maximum value of 1.99V. When the current value is 0.04A, the load voltage curve rises first and then remains flat. When the load resistance is 250 Ω, the load voltage reaches the maximum value of 27V. It can be seen that the method in this paper can effectively monitor the load resistance and load voltage of the power model under different current values.

Test the method in this paper when there is no resonance in the impedance branch of the current sensor A, resonance occurs, and the air gap is open. The result is shown in Figure 3.

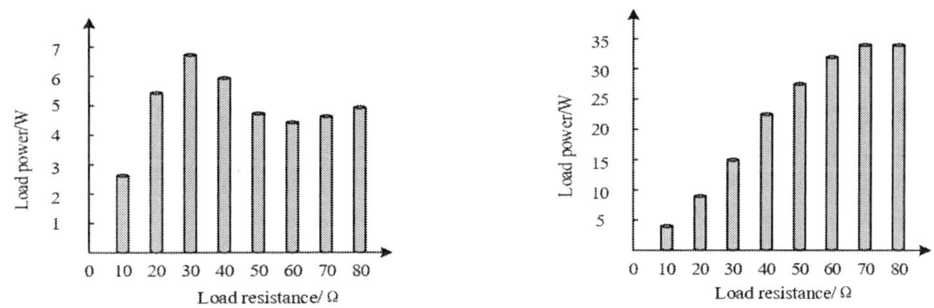

(a) Current sensor A does not resonate (b) Current sensor A resonates

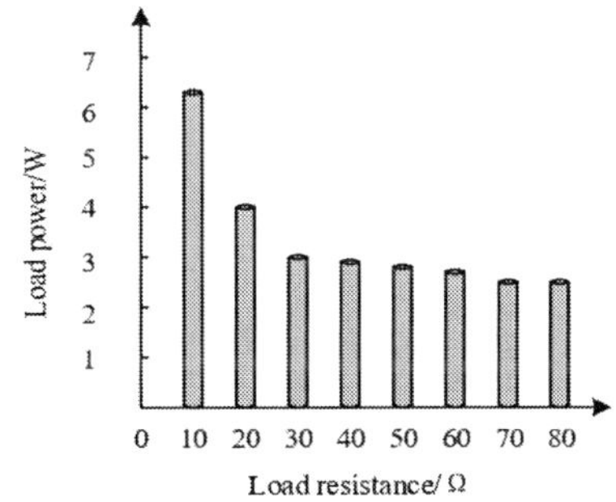

(b) When the air gap is open, the sensor takes the power of the electric model

Fig. 3 This article monitors the circuit load power results under three conditions

Analyzing Figure 3(a), it can be seen that when the current sensor A does not resonate, the power of the electrification model rises rapidly from 2.8W to 7W. After a slow decline, it showed an upward trend, and the overall load power curve appeared undulating and wavy. In Figure 3(b), after the resonance of the current sensor A, the minimum load power of the electrification model is 4.8W. As the load resistance increases, the load power also increases. When the load resistance is between 70V and 80V, the load power remains stable. In Figure 3(c), when the air gap is opened, the power of the sensor's electrical model is inversely proportional to the load resistance. When the load resistance is between 10Ω~30Ω, the load power drops rapidly. When the load voltage exceeds 30Ω, the load power of the electrification model will decrease gradually. Comprehensive analysis of Figure 5 can be obtained, the monitoring method in this paper can monitor the occurrence of impedance branch and the sensor's electrical power situation when resonance and open air gap are not occurred, and its application has a wide range of applications.

Test the input voltage and output voltage of the current sensor in the case of different coil turns, and the results are shown in Table 1.

Tab. 1 Current sensor voltage monitoring results under different coil turns

Coil turns	Current sensor input voltage/V	Current sensor output voltage/V
3	18.92	5.11
6	22.3	5.13
9	24.2	5.13
12	27.1	5.13
15	29.9	5.13
18	31.3	5.13

Analysis of Table 1 shows that when the number of turns of the coil increases, the input voltage of the current sensor also increases. When the number of turns of the coil is 3, the output voltage of the current sensor is 5.11V. As the number of coil turns continues to increase, the output voltage of the current sensor remains at 5.13V. It can be seen that the method in this paper can monitor the input and output of the current sensor when the number of turns of the coil is different.

Take communication distance as the index of test monitoring method, set up different communication distance in PSIM software, and record the results of different communication distance monitored by three monitoring methods respectively by using method of this paper, the method of reference [6] and the method of reference [7] method, as shown in Table 2. The method of reference [6] represents a noninvasive load monitoring method based on the improved flock algorithm, and the method of reference [7] represents a fiber optic gyroscope power monitoring method based on quadripartite four-state modulation.

Tab. 2 Communication distance test results of three monitoring methods

Communication distance/m	The method of this paper monitors the distance/m	Reference [6] Method of monitoring distance/m	Reference [7] Method of monitoring distance/m
200	200	200	200
400	400	394	395
600	600	586	593
800	800	781	766
1000	999	957	980
1200	1198	1157	1182
1400	1396	1309	1350
1600	1592	1503	1499
1800	1792	1730	1785
2000	1981	1824	1909

Analysis of Table 2 shows that with the increase of communication distance, the distance monitored by the three methods gradually appears error, and the difference gradually increases. When the monitoring distance is ≤800 m, the monitored distance is the same as the communication distance and there is no error. When the monitoring distance is more than 1000m, the error of this monitoring method appears gradually, and the maximum error is 19m. Reference [6] methods and Reference [7] methods both have monitoring errors at communication distances of 400m. The maximum error of the monitoring distance was 157m and 101m, respectively, and the monitoring error was higher and lower.

The results show that the monitoring stability of the method [6] and the method [7] is poor, and the method has higher monitoring stability than the method.

4. Conclusion

This paper proposes a remote acquisition method of insulator icing flashover signal based on frequency optimization of ring-shaped intelligent current sensor, and designs an integrated monitoring model of parallel resonant dual magnetic circuit for power extraction. Through the optimization of the frequency of the ring-shaped intelligent current sensor that meets the monitoring requirements, the remote acquisition of the flashover signal of insulator ice coating is realized. In order to improve the power monitoring effect, the current mutual inductance parameters of the current sensor are optimized during the monitoring process. After experimental verification, the conclusions are as follows:

(1) When the current value is 0.02A and 0.04A respectively, the maximum load power is 0.015W and 0.64W, and the load resistance is 150Ω and 70Ω, respectively.

(2) When the current value is 0.02A, the load voltage curve of the model rises gently, and the maximum load voltage value is 1.99V. When the current value is 0.04A, the load voltage curve first rises and then maintains a smooth state, and the maximum load voltage is 27V.

(3) When resonance, resonance and open air gap occur in current sensor A, this method can effectively monitor the current load power of the model, and the monitoring ability is high.

(4) In the case of different turns of current sensor coil, the input and output of current sensor can be effectively monitored.

(5) When the communication distance is different, the monitoring distance is the closest to the actual communication distance, and the error value is the lowest, the monitoring stability is strong.

References

[1]Zhou L W, Li S Y, Tang W S, et al. (2020) An evaluation scheme for the performance of anti-ice composite insulators[J]. Electrotechnical Application, 39(02):59-64.

[2]Yu H, Chang W Z, Du F, et al. (2020) Research on Anti-Interference Principle of Current Sensor Based on Circular Array of TMR[J]. Electric Power Information and Communication Technology, 18(10):11-18.

[3]Mao X Y, Ma X H, Peng C, et al. (2019) Research on extracting power from sectionalized insulation ground wire[J]. Electric Power Automation Equipment, 39(03):219-223.

[4]Wang Y, He Y F, Mo C F. (2018) Design of power supply based on LC resonance and zero sequence voltage measurement system[J]. Electrical Measurement & Instrumentation, 55(11):78-82+93.

[5]Huang X, Liao J Q, Wang S, et al. (2019) The Harmonic Suppression Method of High Voltage Power Extraction Device Based on Inductive Filtering[J]. Science Technology and Engineering, 19(25):191-196.

[6]Liu X J, Cao M H, Xu Y J. (2018) Non-intrusive load monitoring based on improved chicken swarm optimization algorithm[J]. Electric Power Automation Equipment, 38(05):235-240.

[7]Zhao Y F, Li Y, Zhang Y F, et al. (2020) Monitoring method of optical part power of FOG with one quarter of frequency and four-state modulation[J]. Journal of Chinese Inertial Technology, 28(03):391-396.

[8]Tian Y, Yang H T, Xie J, et al. (2020) Study and Design on Current Coupled Power Supply with Positive and Negative Output[J]. Power Capacitor & Reactive Power Compensation, 41(03):141-146+155.

[9]Huang H B, Li W, Hu W, et al. (2019) Design and Simulation on Power Supply Circuit of Getting Electricity from Grounding Wire System Based on Single Port Forward Converter[J]. Chinese Journal of Electron Devices, 42(03):657-661.

[10]Li M L, Hu G Z, Zhang H, et al. (2020) Research on Power Transfer Technology of IPT System Based on U-type Power Collection Structure[J]. ower Electronics, 54(02):24-27.

[11]Deng Y, Sha Q K, Yang Y, et al. (2019) Research on High Voltage Busbar Operation Monitoring

Journal of Physics: Conference Series **2108** (2021) 012075 doi:10.1088/1742-6596/2108/1/012075

System Based on STM8[J]. Industrial Safety and Environmental Protection, 45(05):51-54.

[12]Liu X L, Song T B, Liu W X, et al. (2019) Development of a Energy Storage Mobile Test Platform for Electric Vehicle Charging Piles[J]. Industrial Safety and Environmental Protection, 45(09):88-93.

[13]Wu S H, Meng Y F. (2019) Intelligent Optimization Design for Test Stimulus of Nonlinear Analog Circuits[J]. Electronics Optics & Control, 26(01):17-20.

[14]Li L J, Wen Q F, Zhang C, et al. (2020) Research on Ice-snow Covering Characteristics and Flashover Prediction of Coastal-region Transmission Lines[J]. Proceedings of the CSU-EPSA, 32(05):98-103.

[15]Fang Z Q, Tang G D, Lv L M, et al. (2018) Design of a Novel S-Band Bandpass Filter Based on Silicon Integrated Passive Device Technology[J]. Semiconductor Technology, 43(05):347-352.

Research on Synchronization Mechanism of Artificial Intelligence Model Based on Belief Measure and Its Application in Pumped Storage Power Station

Jinliang Dong[1a], Xu Zhang[1b], Haijiang Li[1c], Wenzhi Song[2d], Jinglin Guo[1e*]

[1]Xinjiang Hami pumped storage Co., Ltd., Hami 839099, China

[2]Beijing Guodiantong Network Technology Co., Ltd., Beijing 100000, China

[a]137265241@qq.com, [b]94698198@qq.com, [c]2206496336@qq.com, [d]songwenzhi@sgitg.sgcc.com.cn, [e*]478817379@qq.com

Abstract—For the security monitoring of pumped storage power station, a model synchronization mechanism for cloud edge cooperation framework is proposed. The method uses the belief function to describe the threshold and uses the ping-pong operation strategy to update the model alternately, which solves the problem of artificial intelligence model synchronization and update of edge equipment. The cloud is based on Baidu BML platform, the edge uses customized servers, and the average model update cycle is about three months.

1. Introduction

With the implementation of the double carbon policy, the construction of energy Internet has accelerated. Pumped storage power station(PSPS) is an important part of energy Internet. In China, about more than 60 PSPSs have been built or are under construction. "Few people on duty" is a feature of PSPS. Therefore, artificial intelligence technology has a wide application prospect in PSPS. This work proposes a model synchronization mechanism for cloud edge cooperation framework, which solves the problem of artificial intelligence model synchronization and update of edge devices.

2. System overview

2.1 Basic ideas

The artificial intelligence models used in the security monitoring of PSPS can be divided into two categories: (1) applications less related to the environmental background, such as not wearing safety helmets and smoking, etc.; (2) Applications related to environmental background, such as plant scaffold construction, etc. For (1), all PSPSs adopt the general model; for(2), personalized model is adopted for different PSPS. In the cloud, the basic sample warehouse and model warehouse are realized. At the same time, personalized samples are saved for different PSPS, and the corresponding models are trained. At the Edge, the customized equipment is used to realize the real-time image detection, and the model synchronization and update mechanism and sample return mechanism are designed and implemented.

Fig.1 PSPS environment

2.2 Introduction to system architecture

Edge AI has the potential to benefit both the data and the network infrastructure itself. At a network level, it could be used to analyze the flow of data for network prediction and network function management, while enabling edge AI to make decisions over the data itself offers significantly reduced backhaul to the cloud, negligible latency and improved security, reliability and efficiency across the board. Another key function of edge AI is sensor fusion: combining the data from multiple sensors to create complex pictures of a process, environment or situation. Consider an edge AI device in an industrial application, tasked with combining data from multiple sensors within a factory to predict when mechanical failure might occur. This edge AI device must learn the interplay between each sensor and how one might affect the other and apply this learning in real-time.

In this paper, Baidu BML platform is used as the bottom support[1], and customized server is used to realize the edge computing support.

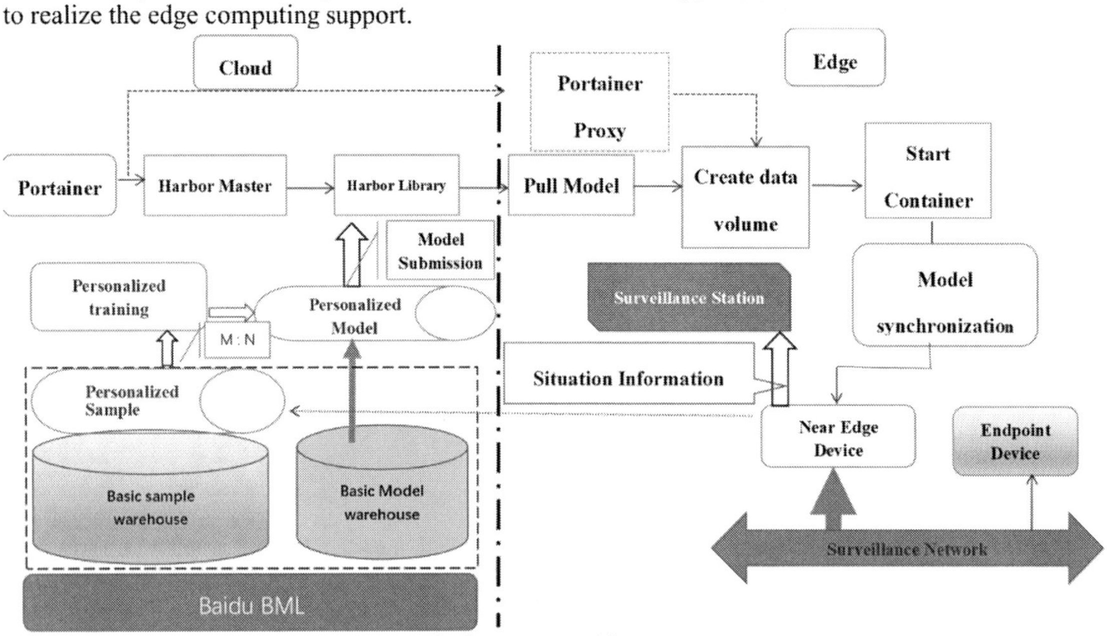

Fig.2 system architecture

In Figure 2, M: N means the comparison of basic samples and personalized samples in the training set.

2.3 Near edge device

In the implementation of Near Edge device, the heterogeneous computing[2] mode is adopted, the main FPGA accelerates the common model, and the h.265 decoding card is used to improve the speed of video coding and decoding, and GPU is used to realize the image recognition based on deep learning. Furthermore, in the airborne equipment of the project, the Yolo V4[3] framework is adopted. A large number of calculations are performed at the edge, and the cloud is mainly responsible for training the model.

2.3.1 PCI-E communication module

PCI-E is a high-speed serial point-to-point dual channel high bandwidth transmission. In many applications, PCI-E switching is the main implementation way in heterogeneous systems. Its main advantage is high data transmission rate. The highest 16X2 version now achieves 10GB/s. In fact, in general applications, high speed Ethernet technology can be used in data communication, and the advantage of using PCI-E is that the real time is guaranteed.

2.3.2 FPGA module

FPGA is more efficient than CPU even GPU, mainly due to its architecture without instruction and no memory sharing. There are two functions of memory in the von Neumann structure. One is the save state, and the two is the inter cell communication. Because memory is shared, we need to do access arbitration. In order to make use of the locality of access, every executive cell has a private cache, which means we need to maintain consistency between components. For the needs of saving state, registers and on chip memory (BRAM) in FPGA belong to their own control logic, without unnecessary arbitration and caching. For communication needs, the connection between each logical unit of FPGA and surrounding logic units has been determined during reprogramming, and no communication is required through shared memory. In this work, Xilinx Virtex-6 is used.

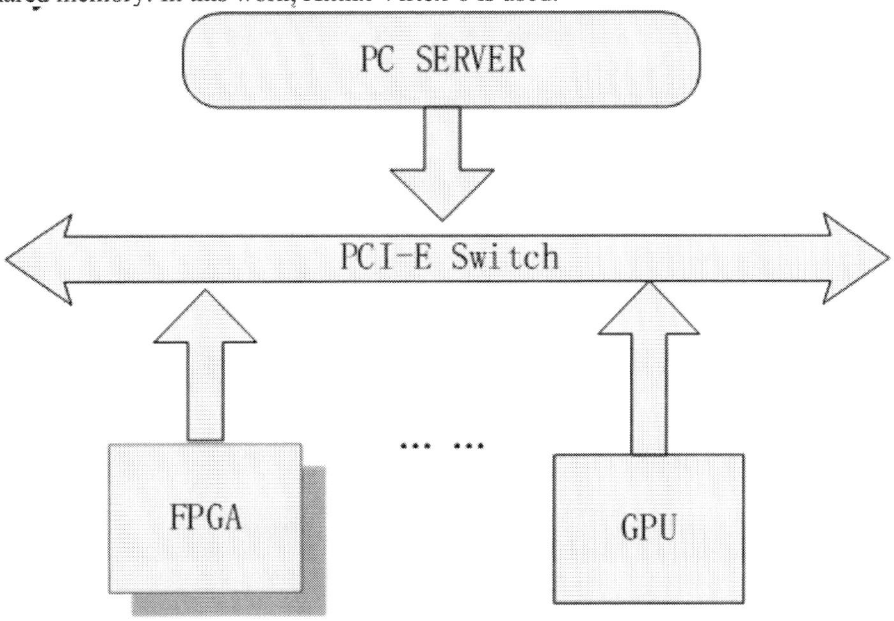

Fig. 3. Realization of Near Edge device adopting FPGA and GPU

2.3.3 Performance optimization

In order to improve the performance, in the case of engineering application precision, this paper made three changes during the implementation process.

The decode image is completed by using the H.265 decoding card.

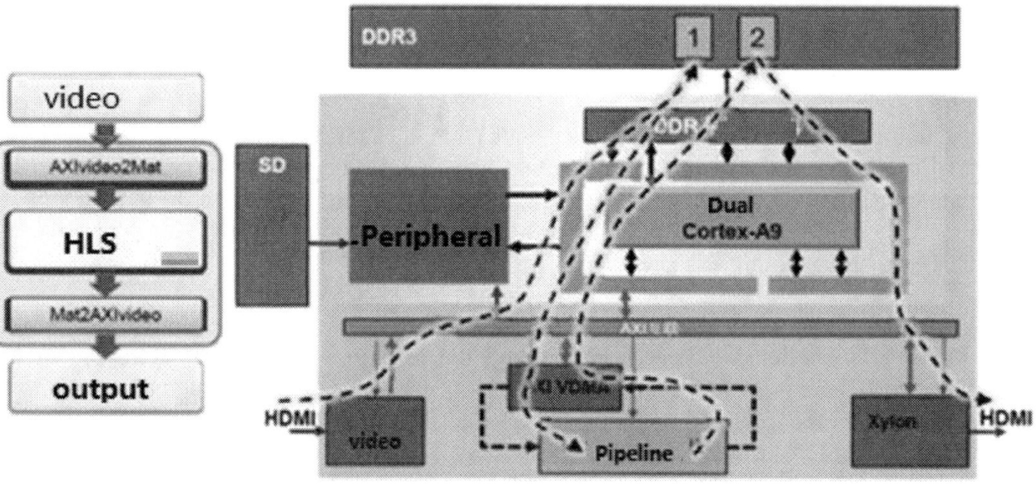

Fig.4 Image decoding process

2.3.4 Communication in the microservice

In the near edge server, it is a better choice to support endpoint devices with microservice architecture. in the microservice architecture, the patterns of communication between clients and the application, as well as between application components, are different than in a monolithic application. There are two main approaches to inter-process communication in a microservice architecture. One option is to a synchronous HTTP-based mechanism such as REST(Representational State Transfer) or SOAP (Simple Object Access Protocol). This is a simple and familiar IPC(Inter-Process Communication) mechanism. It's firewall friendly so it works across the Internet and implementing the request/reply style of communication is easy. The downside of HTTP is that it doesn't support other patterns of communication such as publish-subscribe.

In this work, we use one FPGA and six GPUs (each about 20Tflops), which can support about 128 channels of 1080p video monitoring at the same time.

3.Ping-pong operation

Ping Pong operation is a data buffer optimization design technology in FPGA development[4], which can be regarded as another form of pipeline technology. This paper uses its design idea for reference, combines it with reliability function and applies it to model updating.

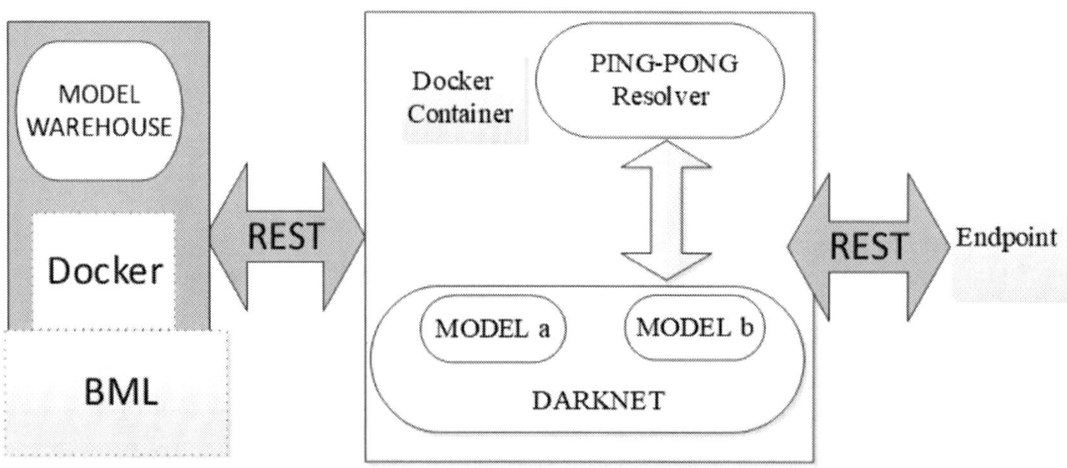

Fig.5 PING-PONG resolver architecture

Strategy of Ping Pong operation resolver

(1)Define the belief function BF (which can be replaced by Bayesian formula). In case of simplification, F1 score can be used instead of this function[5].

(2)For the same target problem, two yolov4 models are trained with different sample sets without losing generality, model a and model b

(3)According to the performance of model a and model b in the test set, the priority is defined

(4)Model a and model b are deployed on edge devices

(5)In the process of edge target recognition, when the results of model a and model b are different, the difference image is uploaded to the cloud for manual confirmation, and the confirmed results are used as the evaluation of model a and model b and accumulated.

(6)When the difference reliability function is greater than the threshold, the model with low evaluation score is updated

4.Conclusion

Based on the results and discussions presented above, the conclusions are obtained as below:

(1)Ping Pong operation strategy can effectively improve the accuracy of artificial intelligence model

(2)The performance of customized server is better than general server in cloud side collaborative application.

References

[1]Zhong Z , Zheng L , Luo Z , et al. Invariance Matters: Exemplar Memory for Domain Adaptive Person Re-identification[J]. 2019 IEEE/CVF Conference on Computer Vision and Pattern Recognition (CVPR), 2019.

[2]Shuai C , Boyer M , Meng J , et al. Rodinia: A benchmark suite for heterogeneous computing[C]// Proceedings of the 2009 IEEE International Symposium on Workload Characterization, IISWC 2009, October 4-6, 2009, Austin, TX, USA. IEEE, 2009.

[3]Bochkovskiy A , Wang C Y , Liao H . YOLOv4: Optimal Speed and Accuracy of Object Detection[J]. 2020.

[4]Zhang Z H , Chu W , Huang S Y . Ping-Pong Methodology for Boosting the Resilience of Cell-Based Delay-Locked Loop[J]. IEEE Access, 2019, PP(99):1-1.

[5]Huang H , Xu H , Wang X , et al. Maximum F1-Score Discriminative Training Criterion for Automatic Mispronunciation Detection[J]. IEEE/ACM Transactions on Audio, Speech, and Language Processing, 2015, 23(4):787-797.

Journal of Physics: Conference Series **2108** (2021) 012077 doi:10.1088/1742-6596/2108/1/012077

Research on output power quality control technology of three-phase inverter

Yinfeng Zhang[1 2], Ying Shao[1], Kang Lei[1*] and Kongxian Wang[1]

[1]College of Electrical Engineering, Naval University of Engineering, Wuhan City, Hubei Province, 430033, China

[2]Air Force Radar Sergeant School, Air Force Early Warning Academy, Yichang City, Hubei Province, 443111, China

[*]Corresponding author's e-mail:1549623735lk@gmail.com

Abstract—With the wide application of new energy and distributed generation system, high-performance inverter plays an important role as the key link of power system access. Its performance determines the quality of output power. Therefore, the output power quality control technology of three-phase inverter has become one of the important research directions in this field. This paper introduces the current main inverter power quality control methods, summarizes and compares the advantages and disadvantages of a single control method, and points out that the composite control method is the development trend of inverter output power quality control technology.

1. Introduction

The formulation of the goal of "carbon peaking and carbon neutralization" and the implementation of relevant policies have greatly promoted China's new energy revolution and the diversification of energy structure. With the continuous improvement of the penetration rate of new energy and distributed generation system, high-performance inverter plays an important role as the key link of power system access. Its performance directly or indirectly determines the quality of output power and is the key to the normal operation of new energy generation system[1]. With the continuous popularization of various complex, precise and power quality sensitive electrical equipment, load mutation, load nonlinearity and load imbalance put forward higher requirements for the design of high-performance inverter power supply. Therefore, the output power quality control technology of three-phase inverter has become one of the important research directions in this field.

Generally speaking, power quality refers to high-quality power supply, including voltage quality, current quality, power supply quality and power consumption quality. Power quality problem can be defined as the deviation of voltage, current or frequency that leads to the failure or failure of electrical equipment, including frequency deviation, voltage deviation, voltage fluctuation and flicker, three-phase imbalance, instantaneous or transient overvoltage, waveform distortion, voltage sag, interruption, rise and power supply continuity.

An ideal power system should supply power to users at a specified voltage level with constant frequency and sinusoidal waveform. In three-phase AC power system, because the parameters of each element of the system are not ideal linear or symmetrical, the load properties are different and change randomly, coupled with the imperfection of regulation means, operation, external interference and various faults, this ideal state does not exist in reality, resulting in power grid operation various problems in electrical equipment and power consumption also produce the concept of power quality.

2. National standard for power quality

Among the national standards for power quality formulated by the National Technical Committee for voltage, current level and frequency Standardization, the following five are mainly related to inverter output, as shown below is the abstract of national standards for power quality[2].

1.GB/T14549-1993 Power quality-Harmonics in public Power Grid: The total harmonic distortion rate limit of harmonic voltage of 0.38kv public power grid is 5%, the odd harmonic voltage content is 4%, and the even harmonic voltage content is 2%.

2. GB/T12325-2008 Power quality-Supply voltage deviation: The three-phase power supply of 10kV and below shall be less than ± 7%. The single-phase power supply of 220V shall be 7% ~ -10%.

3. GB/T15945-2008 Power quality-Power system frequency deviation: The normal allowable deviation is ± 0.2Hz. If the system capacity is small, it can be relaxed to ± 0.5Hz. The frequency change caused by user impact shall generally not exceed 0.2Hz.

4. GB/T12326-2008 Power quality-Voltage fluctuation and flicker: Allowable voltage fluctuations are 2.5% for 10kV and below; 2% for 35~110kV; 1.6% flicker for 220kV and above; Flicker: 0.4% in case of high requirements and 0.6% in general.

5. GB/T15543-2008 Power quality-Three phase voltage unbalance: The negative sequence voltage imbalance shall not exceed 2%, and shall not exceed 4% in a short time. The allowable value of negative sequence voltage imbalance of each user is generally 1.3%.

3. Inverter output power quality control method

Focusing on the control goal of inverter output power quality, a variety of control technologies have been applied to the design of inverter power supply. As shown in Figure 1, the working principle framework of the inverter system is shown. Common inverter output power quality control methods mainly include the following:

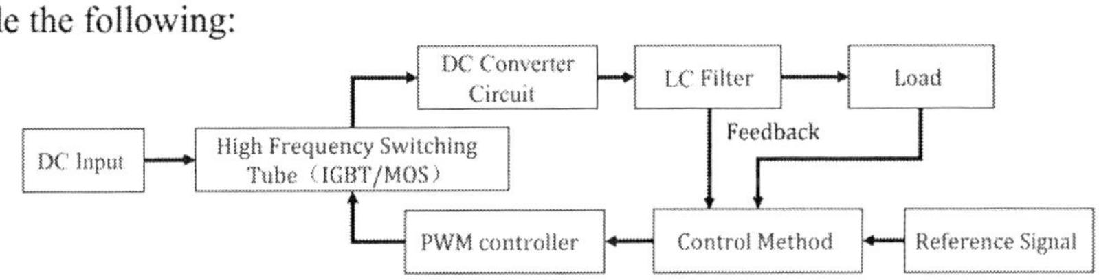

Figure 1. Inverter system working principle frame diagram

3.1. PID control

PID control has the advantages of simple structure, simple algorithm, high reliability, high robustness and convenient debugging process. It has become the most classic and widely used control method in modern control theory. Figure 2 shows the PID control principle block diagram of the inverter. In order to improve the effect of inverter waveform control, the feedforward control of reference sinusoidal signal can be combined with the PID control of output voltage error to improve the dynamic response performance of the inverter and the quality of steady-state output waveform.

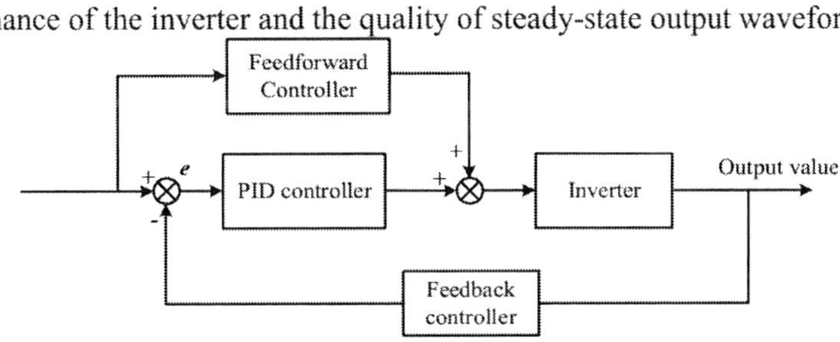

Figure 2. Inverter PID control principle block diagram

According to the principle of internal mode, PID control can not realize the non-static error tracking of sinusoidal signal, so it is usually necessary to add external loop mean feedback to improve the steady-state accuracy of the system[3]. With the appearance of DSP and ARM controller, the instantaneous value feedback digital PID control of inverter becomes possible. Because the inverter system has strong oscillation when no load, the proportion of PID control must be limited, but this will obviously reduce the dynamic performance of the control system; As a result, the controller cannot eliminate errors in real time, and the output performance and waveform quality are affected, especially the THD value of the output voltage is large under nonlinear load conditions.

3.2. PR (proportional resonance) control

PR controller is a controller based on resonance effect. Its control loop has an infinite gain mode at a special frequency. In this way, any nonlinear influence or interference influence will be completely eliminated at this frequency. Therefore, zero error steady-state tracking of sinusoidal signal can be realized at this resonance frequency. For a complex inverter, the output voltage contains multiple harmonics in addition to the fundamental wave. In order to effectively suppress harmonics of different frequencies, PR controllers with different resonant frequencies can be added[4]. As shown in Figure 3, the principle diagram of multiple PR control of the inverter is shown. However, this control method needs to connect the resonant controllers under multiple resonant frequencies in parallel, which will increase the complexity of control, and the design of proportional link in PR control is basically the same as that of PI control, so the dynamic performance of the system is still general.

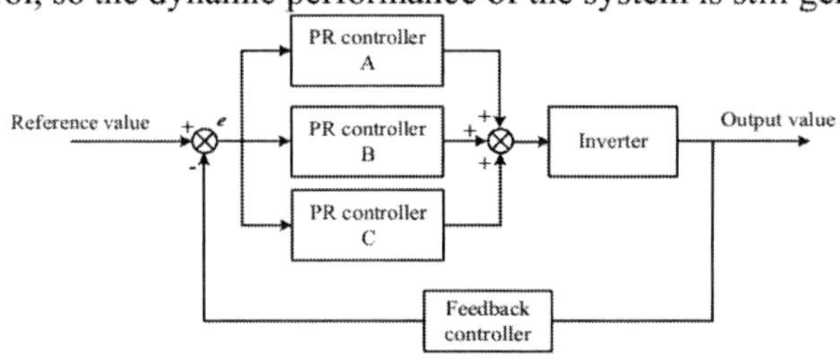

Figure 3. Inverter multiple PR control principle block diagram

3.3. Hysteresis control

At present, the most widely used control method in tracking harmonic current is hysteresis comparison control. The principle of hysteresis comparison control is to compare the controlled value with its given value within a given range to determine the switching timing of inverter switching elements. Hysteresis comparison control has the advantages of fast response speed, high control accuracy, easy implementation and no need to understand the load characteristics; The main disadvantage is that the switching frequency is not fixed, there is serious phase to phase interference when used in three-phase three wire system, and the controlled quantity can not be effectively controlled when the load is changed[5]. Combining with vector control can effectively overcome the above shortcomings.

3.4. State feedback control

In the linear system state space theory of automatic control theory, feedback quantity includes output quantity and state quantity. State feedback can get more feedback information and system change rule from the system through system pole allocation, and the controller will become more accurate and better, and can compensate disturbance better. Pole assignment can improve the damping of the system, and thus improve the dynamic performance of the system, and reduce the transition response time. However, the disadvantage of state feedback control is that it has poor improvement effect on the steady-state performance of the system, and it has poor ability to suppress the output waveform distortion caused by various disturbances. In the actual inverter control, state feedback is often combined with other control methods to improve the dynamic performance of the system, and compound control to suppress the output waveform distortion caused by interference to improve the steady-state performance of the system[6].

3.5. Double loop control

Figure 4 shows the principle block diagram of inverter double loop control. Because the single loop control of inverter is not ideal for the regulation of AC voltage, an inductor / capacitor current inner loop can be added on the basis of PWM inverter voltage single loop to improve the dynamic and static performance of the system.

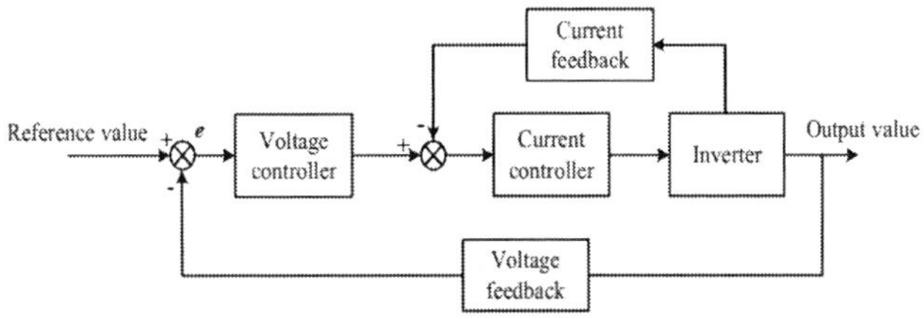

Figure 4. Inverter double loop control principle block diagram

Reference[7] realizes the control of inverter based on the predictive voltage and current regulator with the same structure, which shows good dynamic and static performance. However, the dual loop control requires a comprehensive analysis of the open-loop and closed-loop frequency characteristics of the inverter under different controller parameters. In addition, in order to suppress the disturbance of output voltage and nonlinear load, the current inner loop must have a sufficiently high bandwidth to obtain satisfactory performance, which increases the difficulty of realizing the digital controller.

3.6. Adaptive control

The inverter system is bound to be affected by load disturbance and other environmental factors in the actual operation process. Using a conventional controller to adapt to various changes with a set of constant controller parameters is obviously difficult to achieve satisfactory results. Figure 5 shows the principle block diagram of inverter adaptive control.

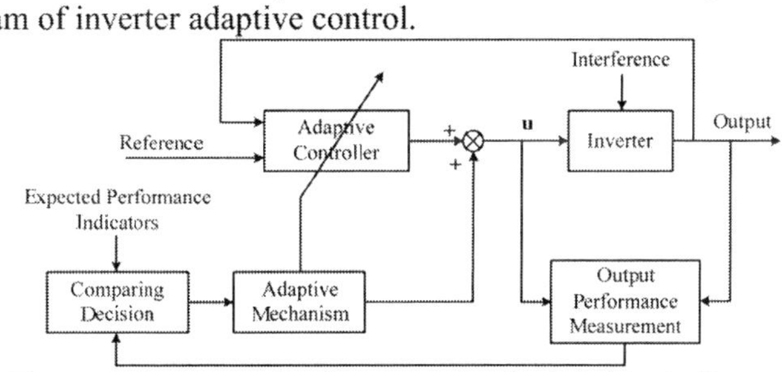

Figure 5. Inverter adaptive control principle block diagram

The adaptive control method can identify the system model on line[8], and then adjust the controller parameters according to the system model and control index to achieve high precision control. The disadvantages mainly exist in the following aspects: the present parameter estimation methods are asymptotically convergent with time approaching infinity under ideal conditions, but the practical engineering application needs the fast convergent parameter estimation methods in finite time; The dynamic performance of some adaptive controllers during startup or transition process can not meet the actual requirements. Low order controller has high frequency unmodeled; Because of the contradiction between the control precision and parameter estimation, the measurement precision directly affects the controller parameters, and then affects the system performance.

3.7. Sliding mode variable structure control

The principle of sliding mode variable structure control is to design the switching phase plane of the system according to the expected dynamic characteristics of the system, and force the state variables of the system to move from outside the phase plane to a pre designed "sliding mode" trajectory in the switching phase plane through the sliding mode controller, so as to achieve the expected performance. Sliding mode variable structure control has the advantages of fast response, insensitive to parameter changes and disturbances, no system on-line identification, simple implementation and good dynamic performance. The main difficulty of applying sliding mode control in inverter system is that it is difficult to select the ideal sliding mode switching phase plane of the system; If sliding mode control is realized by analog circuit, the hardware circuit is too complex. Therefore, only digital circuit can realize its application value, and higher system sampling frequency is required to achieve better control effect, so as to eliminate the system instability caused by discrete sampling[9].

3.8. Deadbeat control

K. P. Gokhale et al. First proposed the deadbeat control method of inverter in 1987. Its main idea is to calculate the switching control value of the next cycle according to the state equation of the system and the current state information, and finally achieve the purpose of making the output value track the input value. Its principle block diagram is shown in Figure 6.

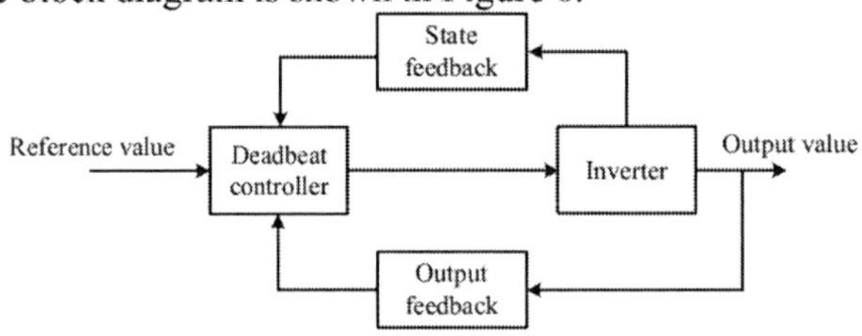

Figure 6. Block diagram of inverter beat free control principle

Deadbeat control[10] has the advantages of good dynamic response performance, good output voltage tracking effect, low harmonics and high waveform quality. The disadvantages of the controller are that the accuracy of the system mathematical model is high, the robustness is poor, the transient response overshoot is large, and the real-time calculation is strong, so the hardware requirements are very high. The performance of deadbeat control can be greatly improved by adopting disturbance state observer or optimal predictive control.

3.9. Repetitive control

Repetitive control is mainly used to eliminate the influence of periodic disturbances such as inverter dead time effect and nonlinear load, and improve the waveform quality of output voltage. Its basic idea is to assume that the distortion in the previous fundamental cycle will repeat in the next cycle, and the controller determines the control signal according to the error between the given signal and the feedback signal, This signal is then superimposed on the original control signal of the next cycle to eliminate repetitive distortion. Figure 7 shows principle block diagram of inverter repetitive control.

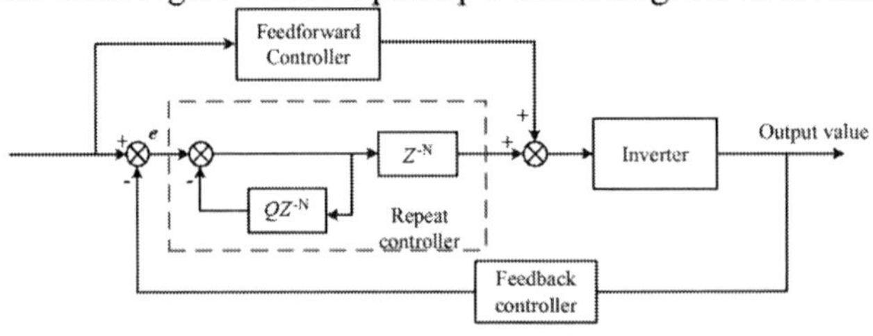

Figure 7. Inverter repetitive control principle block diagram

The repetitive control algorithm is simple. Only by sampling the output voltage, the inverter can obtain the steady-state output waveform with low THD under periodic disturbance. However, due to the fundamental cycle delay in the repetitive controller, the repetitive controller does not produce any regulation effect in the first fundamental cycle of load step change, and the dynamic response is poor. If the disturbance is aperiodic, repetitive control will increase the error of inverter output voltage. In order to solve this problem, reference[11] proposes a new frequency adaptive repetitive control (FARC) method applied to grid connected inverters, which uses all-pass filter to replace the delay unit in the repetitive controller. When the grid frequency changes, it can effectively suppress voltage harmonics by adjusting the coefficient of all-pass filter online.

3.10. Fuzzy control

The inverter system is a nonlinear time-varying system. The traditional control method usually needs to give a very accurate system model in order to accurately suppress the disturbance signal, which greatly improves the control difficulty. The fuzzy controller has no requirements for model accuracy. It can accurately suppress the disturbance signal and optimize the inverter voltage waveform without being affected by the modeling accuracy of the controlled object. Fuzzy control look-up table occupies a short time of digital chip, so a higher sampling frequency can be used to compensate the control deviation caused by limited fuzzy rules. Literature[12] combines fuzzy control with repetitive control. Fuzzy control has the characteristics of parameter adaptation. When sudden disturbance or distortion occurs, it can quickly adjust parameters according to fuzzy rules to adjust the waveform in time, so as to overcome the cycle delay caused by repetitive control. However, due to its use of fuzzy principle and fuzzy rule base, Therefore, the adjustment of waveform is not very accurate.

Although it has been proved theoretically that fuzzy control can simulate any nonlinear function with arbitrary accuracy, due to the lack of perfect theoretical basis, the formulation of fuzzy controller rules still depends on human experience. Therefore, there are many irrationalities, which seriously affect the performance of the controller. The application of fuzzy control technology mainly lies in further research Improve fuzzy variable segmentation and fuzzy rules.

3.11. Neural network control

Neural network control belongs to the category of intelligent control. Its control idea is to obtain the optimal control law of the system through off-line analysis of inverter experimental and simulation data, and apply it to the inverter system to realize on-line control[13]. The control has low requirements for the calculation accuracy of the mathematical model of the system control object, good robustness and strong ability to adapt to the environment. However, neural network control is difficult to be realized in hardware, so it is rarely used in the control of power electronic equipment such as inverter power supply.

4. Development trend of inverter output power quality control technology

Each power quality control scheme has its advantages, at the same time, various methods also have shortcomings. Therefore, a single control method is difficult to meet the performance requirements of high-performance inverter power supply. In order to make the inverter output waveform have satisfactory dynamic and static characteristics, compound control can be selected to make different control methods interact and learn from each other. Compound control is the inevitable trend of the future development of inverter output power quality control.

Reference[14] proposes a new resonance control method with frequency adaptive ability for three-phase grid connected inverter. In the two-phase static coordinate system, the whole control system can be equivalent to a first-order frequency servo system by introducing frequency tracking link and pre current feedback link, The problem that the traditional proportional resonance controller can't realize the zero steady-state error tracking control of AC flow when the power grid frequency is offset is solved. Reference[15] proposed an optimal control method combining state feedback accurate linearization and quadratic type. According to the passivity control idea, the analysis and design method of quadratic optimal controller is given, and the values of Q and R are optimized by using the passivity energy function, so as to reduce the output voltage harmonic and improve the harmonic performance of inverter system. Reference[16] proposes an optimal control strategy based on linear quadratic theory, which not only eliminates the inherent resonance phenomenon of LCL filter, but also realizes the compensation of harmonic, reactive and unbalanced current at the parallel node. The feedforward control method suitable for linear optimal control is used, the stability of the system and the power quality at the parallel node are improved.

5. Conclusions

In order to improve the quality control method of the inverter, this paper summarizes and compares the advantages and disadvantages of multiple single control methods, and points out that the composite control method is the development trend of the inverter output power quality control technology, and the selection of the composite control method is the result of the selection of the performance indicator requirements of specific goals, and the advantages and advantages are complementary.

Acknowledgments

Project Supported by National Natural Science Foundation of China Youth Program(52007195).

References

[1] Zhou Yuanxiang, Chen Jianning, Zhang Ling, et al. Development opportunities of UHV transmission technology under the background of "double carbon" and "new infrastructure construction" [J]. High voltage technology, 2021,47 (07): 2396-2408.
[2] Cheng haozhong. Power quality [M]. Beijing: Tsinghua University Press, 2006.
[3] Wang Jibiao, Chen Qihong, Zhang Liyan, et al. Dual mode PI control of grid connected inverter based on internal model principle [J]. Journal of electrotechnics, 2018,33 (23): 5484-5495.
[4] Zhao qiangsong. Research on new proportional integral multi resonance control and its application to grid connected inverter [D]. Nanjing University of Aeronautics and Astronautics, 2016.
[5] Wang Jiali. Research on constant frequency ZVS inverter technology based on hysteresis current control [D]. Zhejiang University, 2018.
[6] Lei pengjuan, Zhao Qinglin, Han Yanlong, Zhang haiduo. Research on state feedback active damping control of grid connected inverter using LCL filter [J]. Journal of solar energy, 2019,40 (08): 2354-2359.
[7] Li Jiajian. Research on dual loop predictive control strategy of single-phase voltage source inverter [D]. South China University of technology, 2019.
[8] Nhan T. Nguyen. Model-Reference Adaptive Control[M]. Springer, Cham, 2018.
[9] Shihua Li et al. Advances in Variable Structure Systems and Sliding Mode Control-Theory and Applications[M]. Springer, Cham, 2018.
[10] Li Donghui, Kong Xiangjie, Liu Lingling. Deadbeat fast repetitive control of single-phase double buck inverter [J]. Power grid technology, 2019,43 (10): 3671-3677.
[11] Chen Lei, Ji Liang, Yang Xingwu, Ou Yangwu, Niu Mengjiao. New frequency adaptive repetitive control method for LCL grid connected inverter [J]. Power system protection and control, 2017,45 (23): 57-64.
[12] Wang Yang. Research on Inverter Based on repetitive and fuzzy control [D]. Tianjin University of technology, 2020.
[13] He Zongyuan. Research on grid connection control of micro inverter based on FPGA and neural network inverse control [D]. North China Electric Power University, 2017.
[14] Ma Wenjie, ou YANGSEN, Yang Lin. new resonance control method for three-phase grid connected inverter [J]. Power grid technology, 2019,43 (08): 2884-2891.
[15] Chen Jianghui, Xie Yunxiang, Wang Jianmin. State feedback linearization optimal control of single buck inverter [J]. Journal of motor and control, 2014,18 (08): 75-80.
[16] Cheng Chong, Zeng Zheng, Tang Shengqing, Yang Huan, Zhao Rongxiang. Linear optimal control of compound function grid connected inverter [J]. Power automation equipment, 2016,36 (01): 135-142.

Influence of antenna half-power beam width on equipment under test volume in radiation disturbance measurement

Yi Zhou [1, a], **Dong Zhang** [1, b], **Xiaoguang Xu** [1, c], **Jun Li** [2, d*]

[1]Testing and Technology Center for Industrial Products of Shenzhen Customs, Shenzhen, Guangdong, China

[2]Shenzhen Academy of Inspection and Quarantine, Shenzhen, Guangdong, China

[a]email: zhouyith2004@163.com, [b]email: dzhang1@qq.com, [c]email: xu.xg@163.com

*Li Jun: [d]email: szxxzx@customs.gov.cn

Abstract: Through the study of the latest radiation disturbance measurement standards, the difference between test volume and equipment under test volume was compared and analyzed. Different types of antennas will be used in the radiated disturbance measurement in terms of the different frequency range. Various antennas have kinds of directional characteristics. This article analyzes the influence of antenna half-power beam width on the equipment under test volume, especially on the height, from three frequency ranges. Furthermore, taking the typical horn antenna as the starting point, as well as the formula of volume with antenna half-power beamwidth, the relationship between equipment volume and frequency is calculated. Finally, the influence of antenna half-power beam width on the equipment under test volume is obtained.

1. Introduction

The latest version of CISPR 16-2-3 standard adds the specification of influencing factors about the equipment under test (EUT) volume. According to the standard, the EUT volume depends on the test distance and frequency range in the radiation disturbance measurement. That means the EUT volume suffers the restrictions of field strength underestimation effect, near-field effect, test antenna half-power beam width (HPBW) and site validation results [1]. For radiation disturbance measurement, different types of test antennas are used for different frequency bands. Thus test antennas has the different directivity. The half-power beam width or 3dB beam width is an important technical index of antenna directivity. The narrower the half-power beam width, the better the directivity and the stronger the anti-interference ability [2-3], and the smaller the volume of EUT is required. Matching degree between EUT volume and antenna HPBW determines the accuracy of the measurement, which has the impact on the evaluation of uncertainty. Therefore, it is necessary to study the influence of half -power beam width of receiving antenna on EUT volume.

2. Methods and Materials

2.1 Comparison of test volume and EUT volume

In radiation disturbance test, test volume and EUT volume are two different concepts. Test volume refers to the effective volume to place EUT in a set of test system. The test volume is the volume containing EUT volume. Before the radiation measurement, the test site needs to be validated. The results of site validation determine the size of test volume, which is maximum range meeting the site requirements in the anechoic chamber [4].

EUT volume is a cylindrical space defined by EUT diameter and height. The EUT volume consists of a cylinder formed by the rotation of the EUT around its geometric center. The cylinder includes the every EUT part, even the cables and cable racks. The length of included cable is related to test frequency. When the frequency range is from 30MHz to 1GHz, the EUT volume includes 1.6m long cable. When the frequency range is above 1GHz, the EUT volume includes 0.3m long cable. The diagram of test volume and EUT volume is shown in Figure 1.

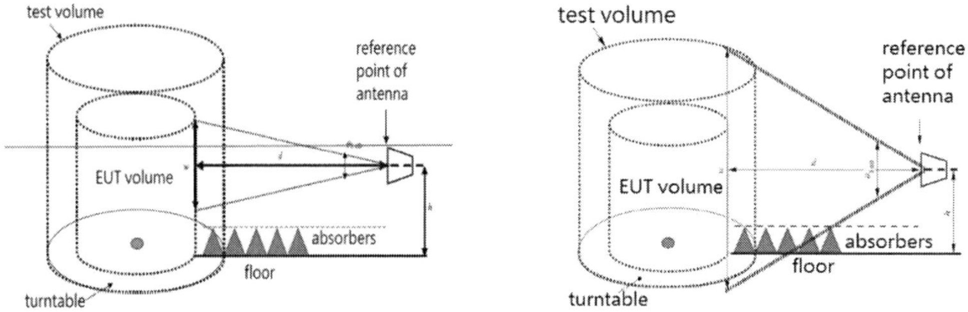

Figure 1. Diagram of test volume and EUT volume

Generally, to find the maximum field strength generated by EUT, the receiving antenna needs height scanning during the measurement [5]. If the half-power beam width of the antenna cannot completely cover the EUT at each scanning height, the test data might be inaccurate, which will increase the uncertainty. Therefore, testers should know whether the HPBW of the antenna can fully cover the EUT before testing. When the HPBW cannot completely cover the EUT, step scanning is required. The height scanning is like a field strength probe, which sniff the radiation disturbance by approaching the EUT. This method can effectively judge whether there is radiation or not, but it can not precisely measure the field strength of radiation.

2.2 9kHz - 30MHz measurement method

In the range of 9 kHz – 30 MHz, radiation disturbance measurement can be carried out with large loop antenna system (LLAS) or in semi anechoic chamber. The LLAS system uses three loop antennas to measure the magnetic field components in three directions perpendicular to each other. Due to the weak directivity of the loop antenna, the HPBW of the antenna is not the main limiting factor of the EUT volume in this frequency range. As the EUT shall be placed in the center of the LLAS system during measurement, the EUT size is mainly limited by the diameter of the LLAS system. When the test is conducted in a semi anechoic chamber or open site, the separating distance will be a major limitation on the EUT volume.

2.3 30MHz - 1GHz measurement method

EUT volume mainly includes two parameters: diameter and height. In the frequency range of 30 MHz – 1 GHz, log periodic antenna is generally used. The directivity of this kind of antenna has an important impact on the uncertainty of radiation disturbance test. Take the log periodic dipole array antenna (LPDA) as an example [6], the antenna has strong directivity above the 200MHz. The typical value of antenna HPBW is 60 °, according to the trigonometric function, the diameter of EUT volume can reach 6m when the separating distance is 3m as shown in Figure 2. The EUT volume is large enough. It can also be said that the diameter limit of the antenna HPBW on the EUT volume is relatively loose.

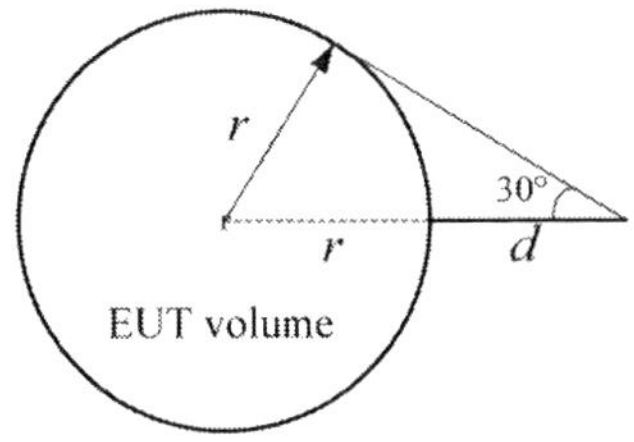

Figure 2. Calculation of typical EUT volume diameter（30MHz - 1GHz）

However, the parameters of EUT volume include not only the diameter of the cylinder, but also the height. When the HPBW θ_{3dB} is 60 ° and the test distance d is 3m, it can be calculated that the volume height w of EUT is 1.73m according to the trigonometric function formula $w = d\,tan\,0.5\theta$ as shown in Figure 3.

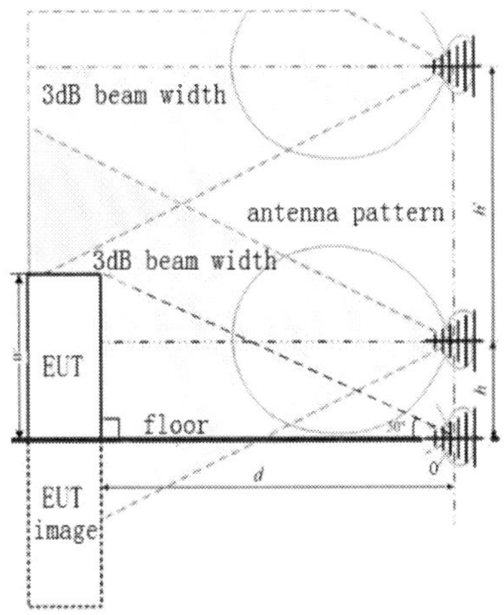

Figure 3. Calculation of typical EUT volume height（30MHz - 1GHz）

When measured in a semi anechoic chamber, the HPBW of the receiving antenna shall completely cover the EUT and the EUT image in the floor. If the HPBW is 60 °, the antenna can achieve full coverage of EUT and its image only at the position of O point on the ground. Generally, the radiation disturbance is measured by non-pitch antenna, which means the receiving antenna always remains parallel to the reference plane at each scanning height. Thus there is a problem. It is impossible for the antenna to reach point O in Figure 3, because the receiving antenna has a certain size. So, the HPBW of the antenna cannot completely cover the EUT and its image. The bore-sight test method may be adopted to ensure that the axis of the test antenna always points to the center of EUT and mirror image with the change of antenna height. Therefore, it can be seen that the height of EUT volume is a more important parameter than the diameter in the radiation disturbance measurement. In the frequency range of 30MHz - 1GHz, the HPBW of the antenna limits the height of EUT more than the diameter.

2.4 1GHz - 18GHz measurement method

The radiation measurement above 1GHz is generally carried out in the full anechoic chamber with horn antenna. The directivity of horn antenna is very strong, and the beam width is much narrower than that below 1GHz [7]. Taking one horn antenna as an example, under different distances, Table 1 shows the limit of antenna half-power beam width on EUT volume.

Table 1. Requirements of HPBW for EUT volume（1GHz - 18GHz）

distance /m	3	5	10
EUT volume D×h /m	1.5×1.5	2×2	5×3
HPBW θ_{3dB} /°	28	22.6	22.6

D is the diameter of EUT volume, h is the height of EUT volume, θ_{3dB} is the half-power beam width of antenna.

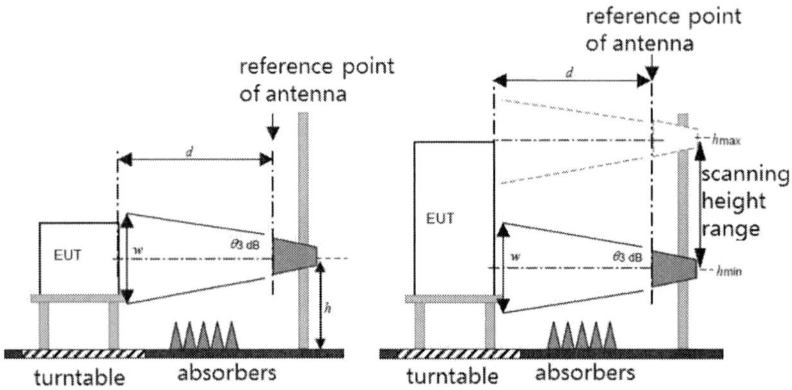

Figure 4. Height scanning requirements for EUT

As shown in the Figure 4, if the test distance d and the antenna HPBW θ_{3dB} are known, w is set the width of the antenna beam on the EUT, then the corresponding EUT size can be calculated by Formula 1 below.

$$w = 2d\tan(0.5\theta_{3\text{dB}})\cdots\cdots\cdots\cdots\cdots\cdots\cdots \quad (1)$$

It can be seen from Table 1 that the smaller the separating distance, the smaller the half-power beam width of the antenna, and the smaller the EUT volume. The larger the measurement distance, the larger the HPBW of the antenna, which means the larger EUT volume. In the radiation disturbance test, the distance and antenna type must be reasonably selected. That means the antenna beam width irradiated on EUT can completely cover the height of EUT volume, or to increase the separating distance or replace the antenna with larger beam width.

3. Results & Discussion

In the radiation disturbance measurement, the separating distance and antenna type must be selected in advance to ensure that EUT is within the coverage of antenna pattern. When the EUT size is larger than the HPBW, the receiving antenna needs to height scan to obtain the accurate field strength.

According to the formula (1), the size w of EUT is decided by the separating distance d and the half-power beam width $\theta_{3\text{dB}}$. In order to study the relation between EUT size and antenna HPBW, we select the measuring distance of 3m and the frequency range of 1 GHz – 18 GHz by means of the double ridged horn antenna.

According to the antenna pattern provided by the manufacturer, we can obtain the relation between HPBW, EUT volume and frequency as shown in the Table 2

Table 2. Relation between HPBW, EUT volume and frequency

Frequency f/GHz	HPBW θ_{3dB} (E plane) /°	Width w_E /m	HPBW θ_{3dB} (H plane) /°	Width w_H /m
1	96.4	6.71	57.9	3.32
2	52.9	2.98	41.3	2.26
3	37.6	2.04	32.1	1.73
4	32.4	1.74	37.3	2.03
5	51.7	2.91	35.8	1.94
6	46.2	2.56	37.3	2.03
7	51.7	2.91	37.5	2.04
8	49.2	2.75	35.3	1.91
9	46.5	2.58	35.3	1.91
10	48.7	2.72	36.9	2.00
11	49.5	2.77	33.3	1.79
12	45.7	2.53	35.9	1.94
13	45.8	2.53	36.1	1.96
14	46.9	2.60	34.1	1.84
15	15	0.79	9.0	0.47
16	13	0.68	9.0	0.47
17	43	2.36	16.1	0.85
18	12	0.63	11.1	0.58

Antenna pattern refers to the pattern in which the field strength of the radiation changes with the direction at a certain distance from the antenna. Due to the spatial complexity of the pattern, the antenna pattern is usually represented by two mutually perpendicular plane patterns in the maximum radiation direction of the antenna. Therefore, w_E and w_H in Table 2 represent the corresponding widths on the E plane and H plane perpendicular to each other, respectively.

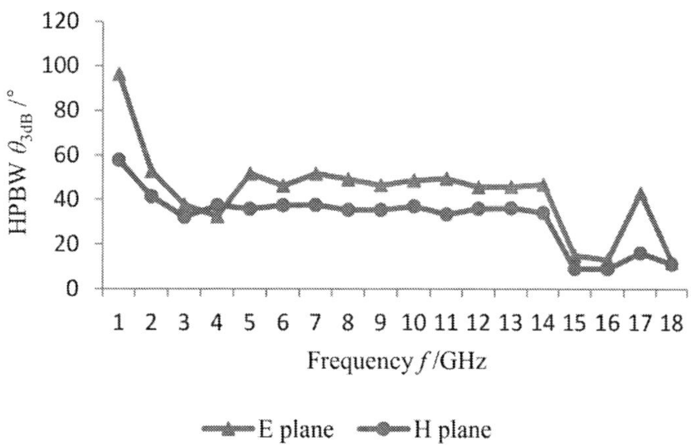

Figure 5. Relation between HPBW and frequency

As illustrated in the Figure 5, for E-plane and H-plane, at 1 GHz - 4GHz, the HPBW of the antenna is large and varies sharply with the frequency, and the antenna pattern angle is large. With the increase of frequency, the HPBW decreases rapidly. In the frequency range of 4GHz – 14GHz, the beam width is relatively stable with no significant change. When the frequency reaches above 14GHz, the HPBW in the H plane decreases gradually, following the decline in the E plane.

Using the data in Table 2 and Formula 1, the relation between EUT volume and antenna HPBW can be calculated. It is shown in the Figure 6 and Figure 7.

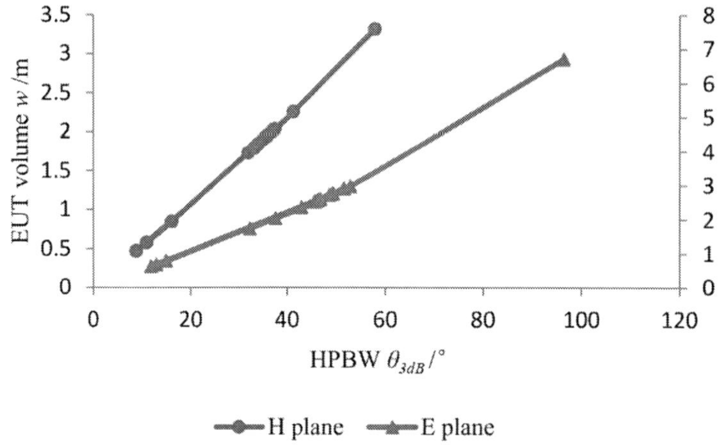

Figure 6. Relation between EUT volume and antenna HPBW

As shown in the Figure 6, whether in E or H plane, the higher the half-power beam width of the antenna, the larger the EUT volume, which is consistent with the actual situation. The difference is about the range and rate of change. For E plane, the variation range of HPBW is large. The difference between the maximum and the minimum is 84.4 °. The difference between the maximum and minimum of EUT volume is up to 6.08m. In the whole frequency range, the situation in E plane changes more largely. The situation in the H plane is much more relaxed. The antenna HPBW varies from 0 to 48.9 °, while the maximum and minimum of EUT volume is only 2.85m.

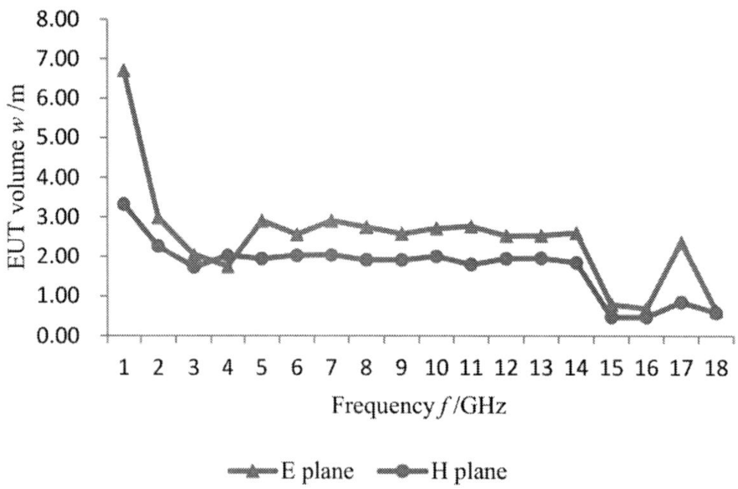

Figure 7. Relation between EUT volume and frequency

As shown in the Figure 7, the relation between EUT volume and frequency is similar to that between HPBW and the frequency. The situation in the high-frequency band and low-frequency band all changes sharply and it remains relatively stable in the middle part. This is because, according to Formula 1, the EUT volume is positively correlated with the antenna half-power beam width. When the antenna HPBW is large, the EUT volume should be large and vice versa.

4. Conclusions

The differences between test volume and EUT volume are analyzed and compared. In different frequency bands, antennas with certain patterns will be used in radiation disturbance measurement. So the influence of antenna HPBW on EUT volume is deeply analyzed from frequency ranges. Finally, taking the typical horn antenna as the starting point, in terms of the relation between antenna HPBW and frequency, and the formula between EUT volume and HPBW, the relation between EUT volume and frequency is calculated, and the influence of antenna HPBW on EUT volume is obtained eventually.

It is concluded that the antenna half-power beam width has little effect on the volume of EUT in the range of 9 kHz – 30 MHz. The EUT volume is mainly limited by the diameter of LLAS antenna system in this band. In the frequency range of 30MHz-1GHz, especially above 200MHz, the antenna HBPW has a great impact on the height of EUT volume. If the HPBW cannot completely cover the EUT, the bore-sight method may be used. Above 1GHz, the horn antenna with stronger directivity is usually used. Basically, as the antenna HPBW increases, the applicable EUT volume also increases. In the 1 GHz - 4 GHz band and the 14 GHz - 18 GHz band, due to the large fluctuation of HPBW, the EUT volume also changes sharply. At middle part of 4 GHz - 14 GHz, the volume changes relatively stable. In conclusion, the half-power beam width of antenna is an important factor for the volume size of EUT in radiation disturbance measurement. You can use the bore-sight method or increase the separating distance to overcome the limitation of the HPBW of the antenna on the EUT volume.

Acknowledgments
This work was supported by National Key Project of Science and Technology (China, no.2018YFC080910005) and Shenzhen Science and Technology Project (China, no.KJYY20160229141621130).

References
[1] CISPR 16-2-3, Specification for radio disturbance and immunity measuring apparatus and methods – Part 2-3: Methods of measurement of disturbances and immunity – Radiated disturbance measurements[S].2019.
[2] Zhong, S.S.(2015) Directional coefficient and gain of antenna .In: Lan, M.(Eds.), Antenna Theory and Techniques. Publishing House of Electronics Industry, Beijing. 47 - 51.
[3] Song, Z., Zhang, J.H., Huang, Y. (2011) Electrical parameters of transmitting antenna. In: Daping, Xia. (Eds.), Antenna and Propagation. Xidian University Press, Xi'an. 7 - 10.
[4] CISPR 16-1-4, Specification for radio disturbance and immunity measuring apparatus and methods – Part 1-4: Radio disturbance and immunity measuring apparatus – Antennas and test sites for radiated disturbance measurements[S].2019.
[5] GB/T 6113.203-2008, Specification for radio disturbance and immunity measuring apparatus and methods Part 2-3 : Radio disturbance and immunity measuring apparatus-Ancillary equipment- Radiated disturbances[S].2008.
[6]Xu, X., Ju, H. (2019) Analysis about the influence of log periodic antenna on radiation disturbance measurement. Automobile Technology, 20: 53-54.
[7]Zhang, F., Zhang, X.C., Zhang, X.H.(2020) An ultra wideband double ridge horn antenna for electromagnetic compatibility test. Information communication, 12: 96-99.

Thermal Analysis of Main Components in The Design of High-power Tetrode

Haoran ZHONG[1], Wenbo CHEN[1*], Xueyu GONG[1], Honghu WU[2], Puqiong YANG[1], Yongjie LUO[1]

[1] School of Electrical Engineering, University of South China, Hengyang, Hunan ,421001, China

[2]Chengdu Xuguang Electronic Co. Ltd, Chengdu, Sichuan, 610500, China

[*]Corresponding author's e-mail: afwind3@gmail.com

Abstract. High-power tetrode is a key component in ion cyclotron resonance heating system, as the final amplifier component to increase the RF power to the megawatt level. Thermal analysis is of important significance to the development and use of high-power tetrodes. In this paper, the trajectory of particles during the operation of the tetrode is obtained by CST particles studio, and its energy conversion is analyzed through the simulation results. Then the grid, anode, and support structure of the tetrode are analyzed thermally in Ansys Workbench, and the results are helpful to the design of the heat dissipation structure of the tetrode.

1. Introduction

1.1. Research background

Ion cyclotron resonance heating is a commonly used auxiliary heating method in controlled nuclear fusion. As the final amplification pole, the high-power tetrode a is an important component in the ion cyclotron resonance heating system, to amplify the radio frequency signal transmitted from the front pole to the megawatt level, and finally inject microwave energy into the plasma of EAST(Experimental and Advanced Superconducting Tokamak)[1].Thermal analysis is of great significance to the design and improvement of high-power tetrodes. While the tetrode is working, the hot cathode will continue to radiate heat to the internal structure of the tube, the heat accumulation in the tube will cause thermal deformation of the tetrode structure, affect the stability and service life of the tetrode, and even cause irreversible physical damage. It is very important to optimize the design of a good heat dissipation structure for making a tetrode with higher operating frequency and higher output power, Therefore, it is necessary to conduct thermal analysis on the high-power tetrode.

At present, there are relatively few researches on thermal analysis of tetrodes. Yaqi Zhao's research simplifies the tetrodes into four simple plates whose grid is composed of tungsten and copper. The temperature distribution of the four plates is analyzed [2]. The influence of the tetrode support structure and anode on temperature is not considered. In another study, they analyzed the relationship between particle motion and radiation in the tube, and verified that the temperature radiation trend is consistent with the direction of particle motion [3], The effect of particle heat loss on the temperature in the tube is not analyzed.

This article will analyze a simplified model of a typical tetrode provided by Chengdu Xuguang Electronics Co. Ltd. First introduce its structure and analyze the significance of each part in the design of its heat dissipation. introduce its working principle and the corresponding heat transfer principle. Analyze the effect of the support structure and anode on the heat dissipation of the tetrode in the simulation. Considering that the electron flow of the tetrode has a certain effect on the temperature of the tetrode when it is working, the movement state of the electrons in the tetrode will be further analyzed, and then simulated by CST particle studio and Ansys Workbench.

1.2 The tetrode structure
The main components of the tetrode are cathode, anode, control grid, screen grid, and the support structure. Figure 1 shows the structure of the tetrode model.

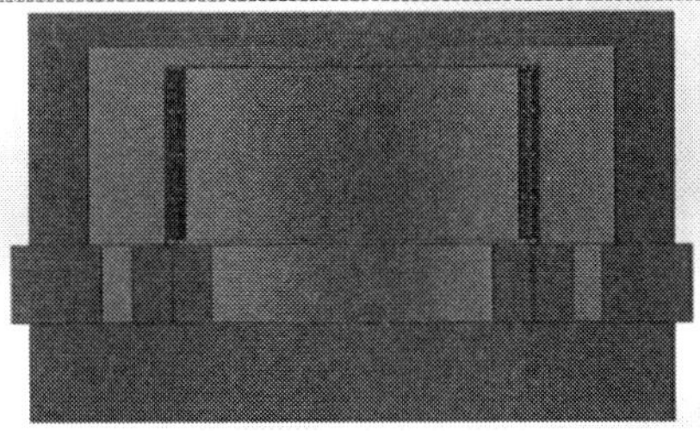

Figure 1. The tetrode simplify model

The material of the cathode is tungsten. When the cathode is heated to 2200K, the cathode will generate a very large emission current. Its emission current density is 0.7~1.5A/cm^2. The cathode is the component that generates the most heat in the entire tetrode. Due to the vacuum environment in the tube, the heat generated by the cathode will only be transferred to other components through heat radiation and heat conduction.

Both the control grid and screen grid use pyrolytic graphite as the material and are made after deposition and photolithography. The grid made of pyrolytic graphite not only has good electrical and thermal properties, but the thermal conductivity parallel to the deposition surface is similar to that of copper. And it has good radiation characteristics (80% of the ideal black body), so its heat can be effectively dissipated through radiation and heat conduction. Since the grid determines the performance of the tetrode to a large extent, it is necessary to focus on its heat dissipation capacity and thermal deformation.

The anode is not only a collector of electrons, but also a part of the tetrode shell, Heat dissipation is another important task of the anode. The temperature of the anode will affect the temperature of the grid in the form of thermal radiation, and it will also affect the stability of the tetrode due to thermal deformation. In several cases, it will directly affect the working life of the tetrode. A part of the focus of thermal analysis is to analyze the anode temperature to further design more optimized heat dissipation methods. The anode cooling method usually has four methods: natural cooling, forced air cooling, forced water cooling, and evaporative cooling.

2. Particle Simulation

2.1. The Theory of Particle Simulation
The movement of electrons will affect the temperature distribution of the tetrode. The tetrode emits electrons by heating the cathode. The electrons emitted from the cathode will accelerate under the action of the electric field, and finally hit the anode according to a certain trajectory. The electrons are in the electric field. A certain amount of kinetic energy is obtained under the action of the force. When it hits the inner surface of the anode, the kinetic energy is converted to the anode in the form of heat energy, which increases the temperature of the anode [4]. The energy conversion of the tetrode can be expressed:

$$P_{in} = P_T + P_{out}。$$

In static working conditions, the kinetic energy of electrons under the acceleration of the anode voltage is equal to the work done when the electric field force pulls the electrons from the cathode to the anode:

$$E_k = \frac{1}{2}mv^2 = eE_a$$

Where m is the electron mass, v is the speed at which the electron reaches the anode, e is the electron charge, and E_a is the anode voltage.

Suppose the electron beam reaching the anode is n. I_a is anode current.

$$I_a = ne$$

then I_a is the total energy W converted into heat.

$$W = neE_a$$

2.2. Result of Particle Trajectory

The outer surface of the cathode is set as the particle emission surface, and the electron emission density is $1.2A/cm^2$. In order to simulate the real thermionic emission, set the thermionic emission with an initial temperature of 2200K, set the potential distribution of each pole, and perform particle trajectory simulation after completing the preceding steps, The simulation results are shown in Figure 2.

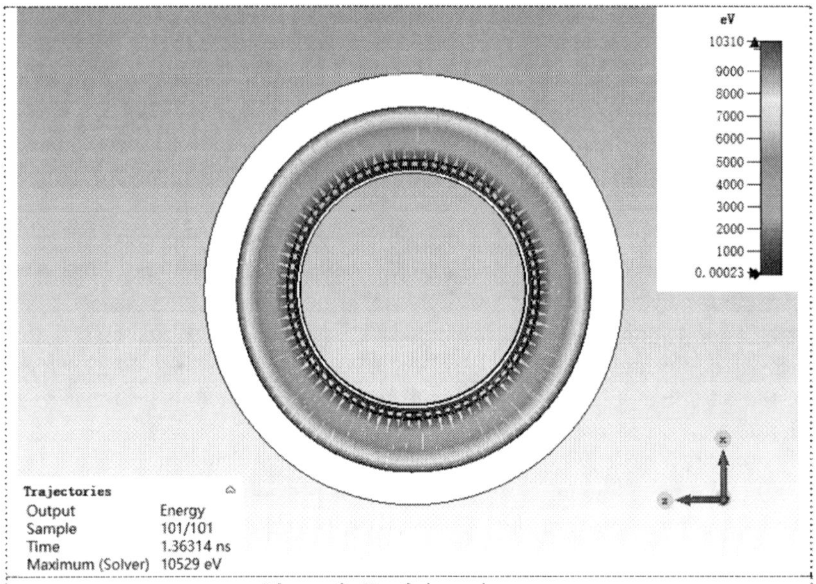

Figure 2. Particle trajectory

It can be seen from the particle trajectory that in an ideal state, when control grid's voltage is negative, its intercepted current is zero. Changing the control grid voltage can control the density of particles passing through. In the design of the tetrode, in order to allow more electrons to pass through the grid, the grid design adopts the Correspond-grid structure, which the grid lines of the control grid and the screen grid are aligned. The electric field of the control grid will converge the electrons emitted from the cathode to form an electron beam. Therefore, in theory, the screen will not intercept the electrons. The electron beam will pass between the grid wires of the screen and move toward the anode.

In general, the current of the tetrode is controlled by control grid voltage, and screen grid voltage and the anode voltage determine the energy of the electrons to the anode. The post-processing of CST particle studio can obtain the total heat loss of electrons, which is similar to the heat loss Q on the inner surface of the anode under ideal conditions. In the thermal analysis, Q will be converted into heat flux, which is applied to the inner surface of the anode as a thermal boundary condition.

3. Thermal Analysis

3.1. The Theory of Thermal Transfer

There are three ways of heat transfer: heat conduction, heat convection, and heat radiation. The inside of the tetrode is a vacuum environment, and the heat transfer is mainly divided into two ways: heat conduction and heat radiation. Thermal conduction means that due to the temperature difference between two objects in contact, the temperature will be transferred from a high temperature to a low temperature. Expressed by the Fourier formula:

$$\frac{KA(T_{hot}-T_{cold})}{d}=\frac{Q}{t}$$

where K is thermal Conductivity, A is object contact surface area, Q is the heat transferred in time t. The heat Q radiated from plane i to plane j be expressed as:

$$Q=\sigma\varepsilon F_{ij}A_i(T_i^4-T_j^4)$$

where Q is Heat flux, σ is Boltzmann constant, ε is Emissivity, A_i is the area of radiating surface i, F_{ij} is angle coefficient of surface i and j.

For heat conduction, it follows the governing equations for thermal analysis:

$$q+\frac{\partial}{\partial x}\left(k_{xx}\frac{\partial T}{\partial x}\right)+\frac{\partial}{\partial y}\left(k_{yy}\frac{\partial T}{\partial x}\right)+\frac{\partial}{\partial z}\left(k_{zz}\frac{\partial T}{\partial x}\right)=\rho c\frac{dT}{dt}$$

and

$$\frac{dT}{dt}=V_x\frac{\partial T}{\partial x}+V_y\frac{\partial T}{\partial y}+V_z\frac{\partial T}{\partial z}+\frac{\partial T}{\partial t}$$

where V_x, V_y, V_z represents the medium conduction rate [5].

3.2. The Result of Thermal Analysis

Table 1. are main materials' parameter. Since pyrolytic graphite is an anisotropic material.

Table 1. Main Materials' parameters

Materials' parameters	Main Materials			
	Tungsten	Copper	Pyrolytic Graphite	
			ab	c
Coefficient of thermal expansion (10^{-6}/°C)	4.6	5.2	0.7	22
coefficient of thermal conductivity($W/m\cdot k$)	174	340	382	2.8
Poisson's Ratio	0.28	0.3	0.25	0.45
Young's modulus (GPa)	411	200	28	13

Simulate two structure models separately. the ambient temperature is normal, the cathode is used as the heat source. The outer surface of the cathode and the inner surface of control grid, the outer surface of control grid and the inner surface of the screen, the outer surface of the screen and the inner surface of anode, are four sets of radiation pairs [6]. The heat loss caused by particle bombardment is applied to the inner surface of the anode in the form of heat flux with a size of 0.015W/mm².

Figure 5. are temperature distribution without supporting structure after hiding the maximum temperature of cathode:1926°C. In the figure, the control grid has a maximum temperature of 1253.2℃, the temperature of the screen grid is evenly distributed at 555℃, and the overall temperature of the anode is approximately 554℃.

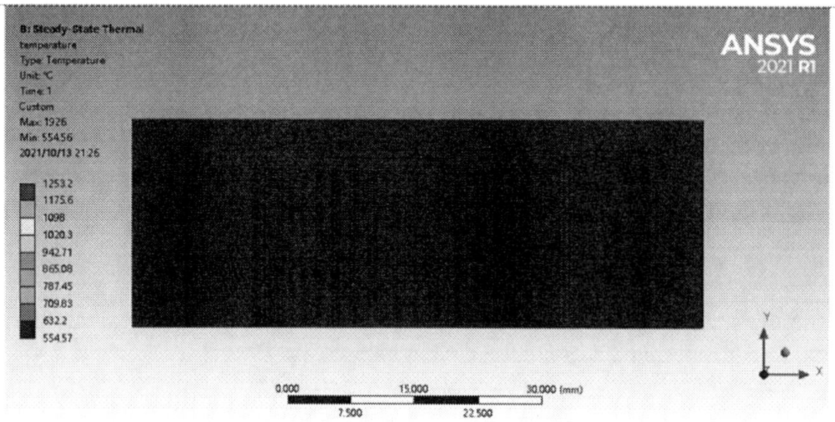

(a)

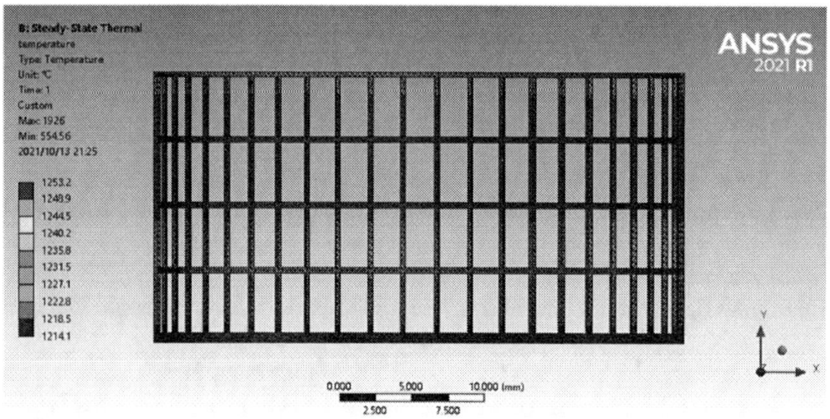

(b)

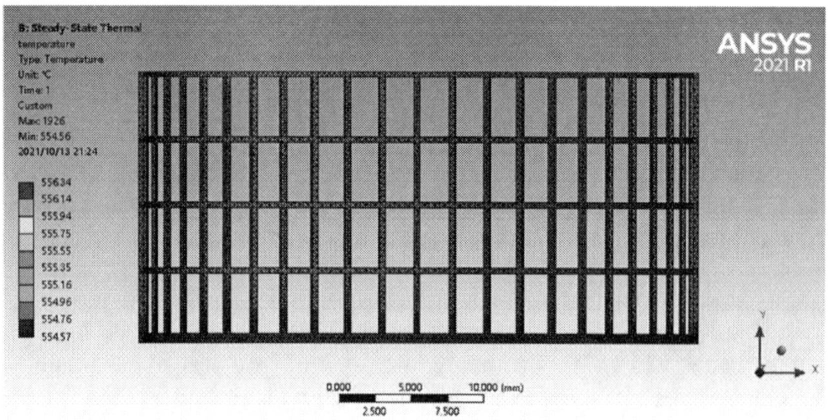

(c)

Figure 3. (a) Temperature distribution without supporting structure (b) Control grid (c)Screen grid

Figure 6. are temperature distribution with supporting structure, The highest temperature is also 1158.7℃ in the control grid. However, the lowest temperature of the control grid is much lower than unsupported one. because the temperature is transferred to the supporting structure through heat conduction and diffuses out. The anode temperature is much higher than the former, Therefore, the temperature of the screen is also more balanced. As a part of the heat dissipation structure, anode can bear as much temperature as possible, so that the temperature in the tube can be released through heat dissipation measures. It can be seen that adding a support structure to the thermal analysis can not only make the results more accurate, but also provide more ideas for heat dissipation design.

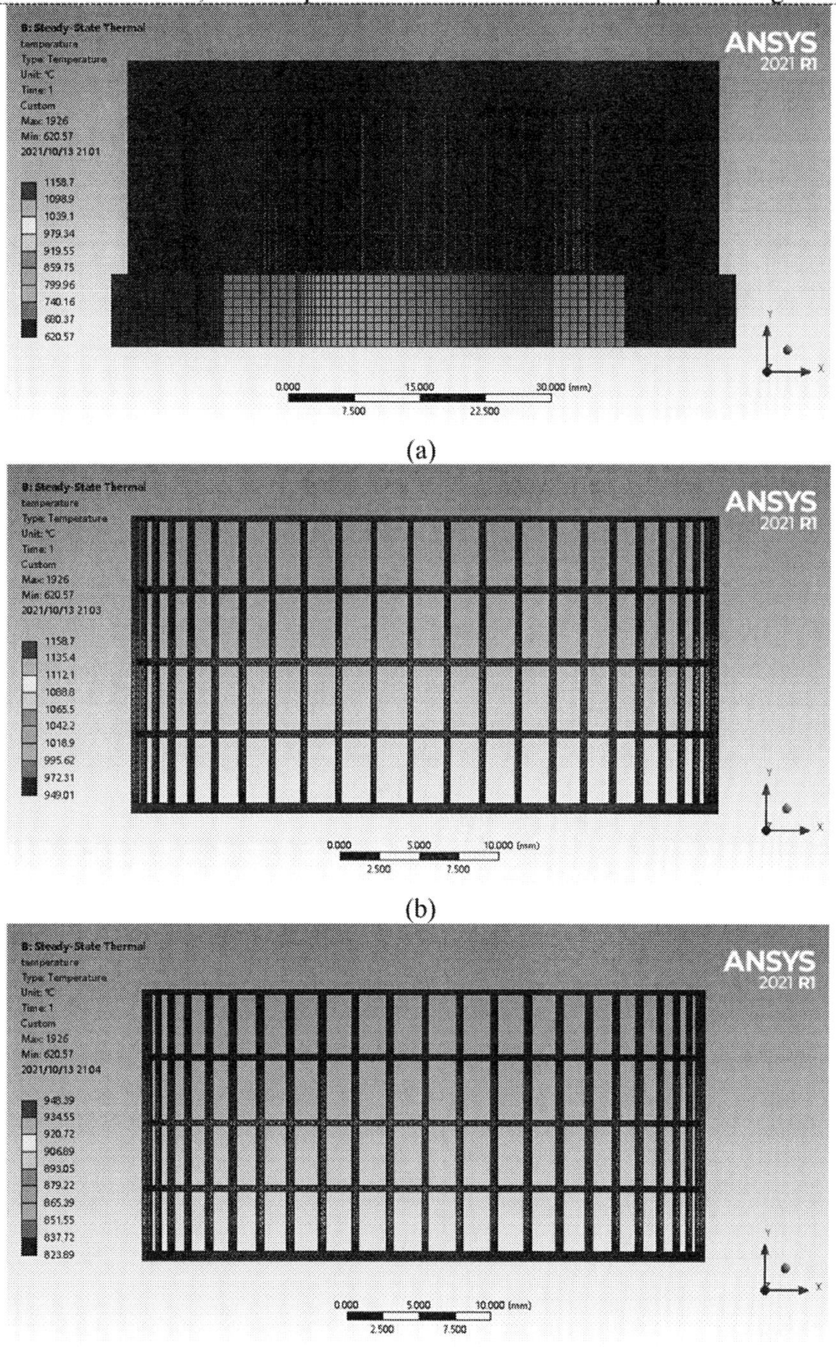

Figure 4. (a) Temperature distribution with supporting structure (b) Control grid (c) screen grid

4. Conclusion

The main purpose of this article is to provide thermal analysis simulation data for high-power tetrode design. Therefore, the main components of the high-power tetrode are analyzed. Using the CST particle studio to simulate the particle trajectory to obtain the heat loss of the particle during the movement.Then set the boundary conditions in Ansys Workbench to get the temperature distribution of tetrode, which conforms to the actual situation. The importance of the support structure and anode to the heat dissipation of the tetrode is analyzed, which provides a basis for the heat dissipation design in the design process of the high-power tetrode.

Acknowledgments

This work was supported by the National Magnetic Confinement Fusion Energy Development Research Project (Grant No.2019YFE03070000), the National Natural Science Foundation of China (Grant No.11675074), and the Science and Technology Planning Project of Hengyang City (Grant No.202002042159).

References

[1] Chen G., Mao Y. Z., Zhao Y.P. (2010) Study on 1.5 MW RF amplifier for EAST ICRF heating. Nuclear Fusion and Plasma Physics,30(1):67-70(in Chinese).
[2] Y. Zhao, R. Zhang, Y. Wang and X. Yang, "Thermal Analysis of Cathode and Grid Components of 500kW Tetrode," 2020 Cross Strait Radio Science & Wireless Technology Conference (CSRSWTC), 2020, pp. 1-3.
[3] Z. Yaqi, Z. Rui, W. Yong and Y. Xiudong, "Study on Electron Trajectory of Cathode of 500kW Tetrode and Thermal Analysis Verification," 2020 IEEE 6th International Conference on Computer and Communications (ICCC), 2020, pp. 2161-2165.
[4] Tian D.M. (1994) Practical Maintenance Manual for Large Launch Tube[M]. China Radio Television Publishing, Beijing (in Chinese).
[5] Jingyuan Che. Research on Multiphysics Simulation Technology of Traveling Wave Tube[D]. University of Electronic Science and Technology of China, 2020.
[6] Wenjun Dai. Thermal analysis of traveling wave tube electron gun and collector[D]. University of Electronic Science and Technology of China, 2016.

A New Hybrid Magnetic Coupler for Large Misalignment Tolerance in Inductive Power Transfer System

Zhongxiu Hao, Pengcheng Fu and Pengfei Cao*

Information Science and Engineering, Lanzhou University, Lanzhou, Gansu, 730000, China

*Corresponding author's e-mail: caopf@lzu.edu.cn

Abstract. Wireless charging has been widely used in a variety of applications. However, pad misalignment can influence system parameters that can lead to increased losses as well as reduction in power throughput. In this article, a new coupler based on solenoid-quadrature coils (SQ) is proposed. By combining with hybrid compensation networks, this coupler can stabilize the system mutual inductance within a certain offset range in order to keep power throughput at a certain level. As a result, this coupler can greatly improve the anti-misalignment capability of the wireless charging system. A 20W prototype is developed and tested in the lab. The experimental results validate that the proposed hybrid SQ-SQ coupler system has an improved misalignment performance.

1.Introduction

Recently, inductive power transfer (IPT) technology has been well adopted in various applications [1]–[3]. However, the inevitable misalignment between the transmitter and the receiver can cause parameter variations. Especially with the change of mutual inductance, output voltage / current varies widely leading to high power losses. Therefore, the technique to improve misalignment tolerance is a critical problem to solve.

Researchers have demonstrated that a load-independent output and a stable output power under the condition of pad misalignment can be achieved by combining LCL-LCL and S-S compensations [4]–[7]. Furthermore, some novel couplers, such as double-D (DD), double-D quadrature (DDQ), and bipolar (BP), have been designed to improve the magnetic coupling and tolerance to pad misalignment [8, 9]. For example, the coupling structure proposed in Ref. [10] can improve the misalignment tolerance in the winding direction. However, there is not much improvement in the vertical direction of the winding, and this method costs more magnetic core. In order to increase the misalignment tolerance in the vertical direction of the winding, some methods have been proposed [11-12]. The coupling structure proposed in [12] adds a Q coil outside the solenoid to improve the misalignment tolerance in the direction of the vertical solenoid winding. In addition, the IPT system's three-dimensional (3-D) misalignments tolerance can be increased by combining magnetic pads design with compensation circuits [13]. Unfortunately, tolerance to the direction of magnetic core (longitudinal) misalignment of polarized charging pads is still unsatisfactory due to lack compensation in the direction of magnetic field coupling.

This paper presents a novel hybrid magnetic coupler combined with LCL-S circuit for inductive power transfer system with high misalignment tolerance in vertical /horizontal, including diagonal. The proposed coupler based on solenoid-quadrature coils (SQ) can compensate for the system mutual inductance within a certain offset range, which can keep power throughput at a certain level.

Laboratory measurements show excellent agreement with the simulated values. Overall, the proposed hybrid SQ-SQ coupler provides a practical method to improve the longitudinal misalignment performance of most inductive power transfer systems.

2. Magnetic Coupler Design and Misalignment Analysis

In order to improve the anti-misalignment ability of the system, the coupler SQ-SQ is proposed in this paper. A solenoid coil (S coil) is added to the Q coil to form the solenoid-quadrature pad (SQ) as shown in Fig. 1 (a). The 3-D model is shown in Fig. 1 (b).

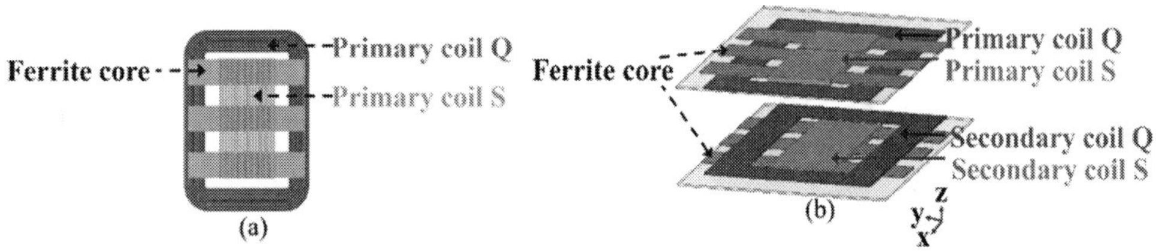

Fig. 1. Proposed coil structures. (a) SQ primary pad model. (b) 3-D model.

The Q coil sits below the S coil, its position ensures that it is magnetically decoupled from the S coil and it can be treated independently. The reason is that the S coil and Q coil in the same pad are decoupled because the solenoid structure is polarized producing parallel flux, while Q coil is non-polarized producing perpendicular flux. Therefore, there is no energy transfer between the transmitter to ensure more power being transferred to the receiver. In addition, when the pad is accurately aligned, the S/Q coil in transmitter and the Q/S coil in receiver are decouple because the net flux is zero. In this case, there are only main couplings (S/S and Q/Q, namely k12 and k34) without cross couplings(S/Q and Q/S, namely k14 and k32). When the receiver is moved along the z directions, the position of S and Q in the coil is always placed symmetrically, so the two coils decouple and cross couplings are always zero. Viewed from the magnetic field direction, however, when the pad is moving along the center line in the x direction, the S coil is always aligned below the Q coil, so the net magnetic flux is zero. The S and Q coil are still decoupled. Consequently, there are only main couplings. Conversely, as the receiver is moved along the y-axis (longitudinal), the S coil in primary or second pad and Q coil in second or primary pad are coupled. In this case, main couplings decreases while the cross couplings increases, which can compensate for magnetic field coupling. Ferrite cores are placed close to the SQ pad in order to increase the coupling of the magnetic field. To eliminate leakage fluxes, the aluminum plate is placed under coupling structures.

To verify that, MAXWELL was used for magnetic field simulation and experiment to verify the change of coupling coefficient. The finished coil pads are shown in Fig. 2(a). The plane is referenced to the axes shown in Fig. 2(b). The coupling coefficient of the coupler structure simulated by MAXWELL with an air gap of 70 mm and experimental results are shown in Fig. 3. The main coupled coefficients (k12 and k34) and the cross coupled coefficients (k23 and k14) vary greatly when pads are misaligned along y direction as shown in Fig. 3(a) and (b). The main coupled coefficients are dropping because that single transmitter S or Q coil and single receiver S or Q coil pass through the process from aligning and misaligning. The cross coupled coefficients progress from uprising to subsiding because that single transmitter S or Q coil and single receiver Q or S coil pass through the process from aligning to misaligning. The reason is that when the S and Q coil aligned, no coupling occurs because the net flux is zero. The k12 and k34 increase after zero coupling point due to the continuous strip of ferrites [8]. The results will suffice to indicate that when the misalignment occurs in y-axis, the coupling coefficient of coils 14 and 23 supplement the coupling coefficient variation of coils 12 and 34. This is so, in order to stabilize the system total coupling within a certain offset range. As shown in Fig. 3(c) and (d), when misalignment occurs along the x direction, the main coupling coefficients (k12 and k34) are slowed down since the magnetic flux reduces as the misalignment between transmitter and receiver increases. The cross coupling (k14 and k23) is always zero, which also confirms the previous theoretical analysis. It can be seen from Fig. 3 that the simulation results are consistent with the experimental results, where the value is different due to the simulation coil design used a simplified model. Fig. 3(e) and (f) shows the coupling coefficient changes in diagonal directions. Like the misalignment processes along the *y* direction, the coil coupling variation is also complementary.

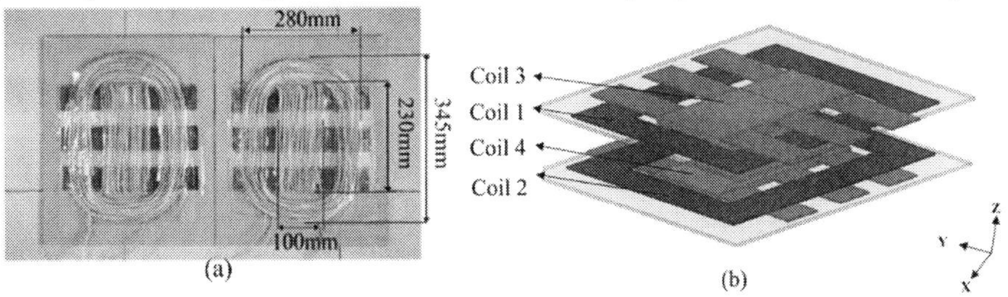

Fig. 2. (a)The structure of coil pads. (b) Transmitter and receiver with reference axes.

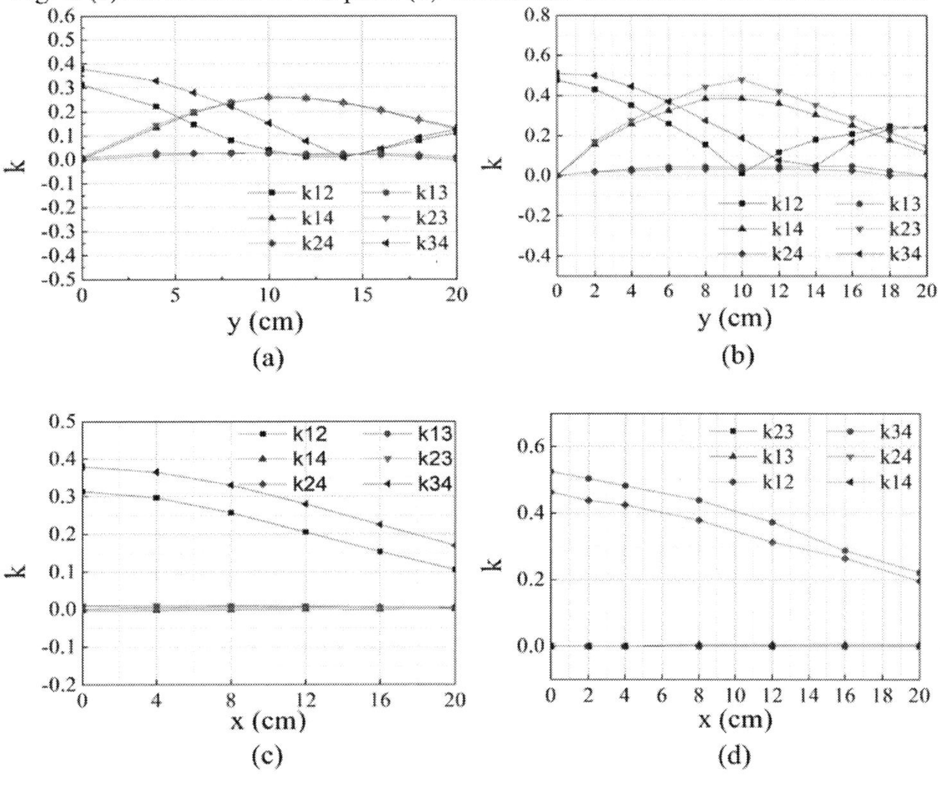

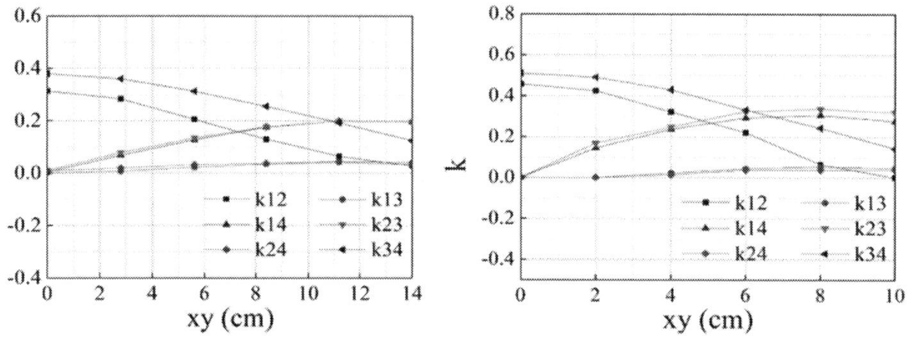

Fig. 3. Coupling coefficient varying with three directional misalignments. (a) longitudinal misalignment(simulation), (b) longitudinal misalignment (experiment), (c) lateral misalignment(simulation), (d) lateral misalignment(experiment).
(e) diagonal direction misalignment(simulation), (f) diagonal direction misalignment(experiment).

3. Experiment and Analysis

The LCL-S hybrid IPT topology [4] has certain anti-vertical-misalignment capability because the combination of converters with opposite changing trend can offset the change of total output voltage when the mutual inductance changes. Therefore, it can improve the anti-misalignment ability in the x and z directions. The circuit is shown in Fig. 4. As space is limited, the misalignment in the z and diagonal direction won't be described here. The parameters of LCL-S topology circuit are presented in Table 1.

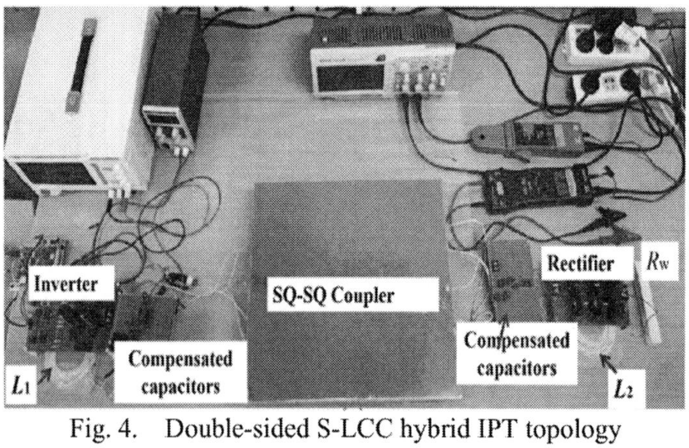

Fig. 4. Double-sided S-LCC hybrid IPT topology

Table 1. System specification and parameter values.

Parameter	Value	Parameter	Value
Vin	30.38 V	Lpt2	204.25μH
Rw	10 Ω	Lst1	97.6 μH
f	100 kHz	Lst2	196.21 μH
Lpt1	98.31 μH	C1	115 nF
Cp1	332 nF	Cs1	335 nF
Cpt2	12.4 nF	C2	115 nF

*C*st2	12.91 nF	*L*1	22.12 μH
*L*2	21.899 μH		

A laboratory setup is built to verify the proposed hybrid magnetic coupler. In order to confirm the above results, Fig. 5 show the output voltage diagram to demonstrate its high tolerance to pad misalignments. In the *y* direction, some coils have coupling coefficient decrease while others have coupling coefficient increase, so that output of overall system can be maintained at a high level within a certain misalignment range, as shown in Fig. 5(a). With the *x*-misalignment, the output voltage is drawn in Fig. 5(b). It can be seen that the output voltage remains relatively stable within the misalignment range from -10 cm to +10 cm. Due to the symmetry of the coil, the result is the same in the opposite direction misalignment.

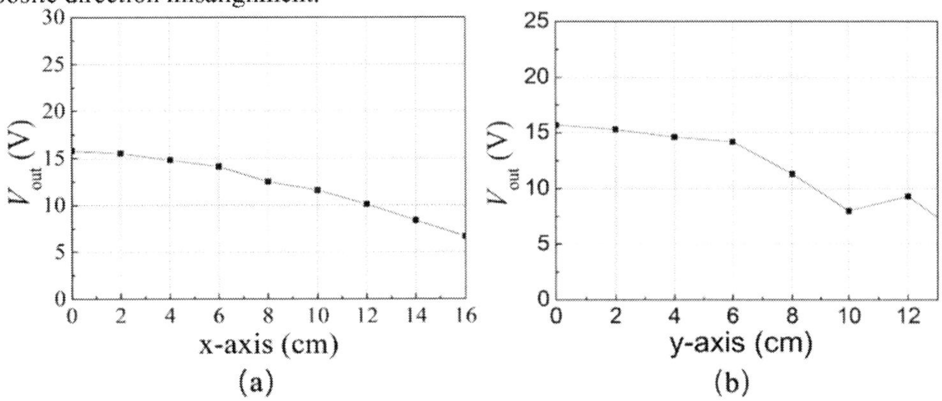

Fig. 5. (a) Load voltage along the *y* direction. (b) Load voltage along the *x* direction.

4. Conclusions

This paper introduced a new hybrid coupler called SQ-SQ which is added to the IPT system to improve the misalignment tolerance by compensating for the system mutual inductance within a certain offset range. The theoretical and experimental results validate that the proposed hybrid SQ-SQ coupler combined the double-sided LCL-S system has an improved misalignment performance along the *y*, *x* and diagonal direction. The results prove that hybrid SQ-SQ coupler is a good candidate for all IPT applications.

Acknowledgments

This work was supported in part by the National Natural Science Foundation of China under Grant 61804071, the Natural Science Foundation of Gansu Province (No. 20JR10RA614), the Open Fund of Key Laboratory of Time and Frequency Primary Standards, CAS, Fundamental Research Funds for the Central Universities (lzuxxxy-2019-tm04).

References

[1] S. Y. Hui, "Planar wireless charging technology for portable electronic products and Qi," Proc. IEEE, vol. 101, no. 6, pp. 1290–1301, 2013.
[2] P. Si, A. P. Hu, S. Malpas, and D. Budgett, "A frequency control method for regulating wireless power to implantable devices," IEEE Trans. Biomed. Circuits Syst., vol. 2, no. 1, pp. 22–29, 2008.
[3] H. H. Wu, A. Gilchrist, K. D. Sealy, and D. Bronson, "A high efficiency 5 kW inductive charger for EVs using dual side control," IEEE Trans. Ind. Informatics, vol. 8, no. 3, pp. 585–595, 2012.
[4] L. Zhao, D. J. Thrimawithana, and U. K. Madawala, "Hybrid Bidirectional Wireless EV Charging System Tolerant to Pad Misalignment," IEEE Trans. Ind. Electron., vol. 64, no. 9, pp. 7079–7086, 2017.

[5] L. Zhao, D. J. Thrimawithana, U. K. Madawala, A. P. Hu, and C. C. Mi, "A Misalignment-Tolerant Series-Hybrid Wireless EV Charging System with Integrated Magnetics," IEEE Trans. Power Electron., vol. 34, no. 2, pp. 1276–1285, 2019.
[6] Y. Chen, B. Yang, Z. Kou, Z. He, G. Cao, and R. Mai, "Hybrid and Reconfigurable IPT Systems with High-Misalignment Tolerance for Constant-Current and Constant-Voltage Battery Charging," IEEE Trans. Power Electron., vol. 33, no. 10, pp. 8259–8269, 2018.
[7] X. Qu, Y. Yao, D. Wang, S. C. Wong, and C. K. Tse, "A family of hybrid IPT topologies with near load-independent output and high tolerance to pad misalignment," IEEE Trans. Power Electron., vol. 35, no. 7, pp. 6867–6877, 2020.
[8] M. Budhia, J. T. Boys, G. A. Covic, and C. Y. Huang, "Development of a single-sided flux magnetic coupler for electric vehicle IPT charging systems," IEEE Trans. Ind. Electron., vol. 60, no. 1, pp. 318–328, 2013.
[9] A. Zaheer, H. Hao, G. A. Covic, and D. Kacprzak, "Investigation of multiple decoupled coil primary pad topologies in lumped IPT systems for interoperable electric vehicle charging," IEEE Trans. Power Electron., vol. 30, no. 4, pp. 1937–1955, 2015.
[10] Y. Yao, S. Gao, J. Mai, X. Liu, X. Zhang, and D. Xu, "A Novel Misalignment Tolerant Magnetic Coupler for Electric Vehicle Wireless Charging," IEEE J. Emerg. Sel. Top. Ind. Electron., vol. 9735, no. c, pp. 1–1, 2021.
[11] M. Budhia, G. Covic, and J. Boys, "A new IPT magnetic coupler for electric vehicle charging systems," IECON Proc. (Industrial Electron. Conf., pp. 2487–2492, 2010.
[12] G. Yang, C. Zhu, K. Song, K. Liu, R. Lu, and G. Wei, "Power stability optimization method of wireless power transfer system against wide misalignment," 2017 IEEE Transp. Electrif. Conf. Expo, Asia-Pacific, ITEC Asia-Pacific 2017, pp. 0–5, 2017.
[13] L. Zhao, S. Ruddell, D. J. Thrimawithana, U. K. Madawala, and P. A. Hu, "A hybrid wireless charging system with DDQ pads for dynamic charging of EVs," 2017 IEEE PELS Work. Emerg. Technol. Wirel. Power Transf. WoW 2017, pp. 3–8, 2017.

Muti-level voltage synthetic voltage direct power flow controller

Xinwei Cao[1](Cao), Youjun Zhang[1*](Zhang) and Zhengzheng Sun[3](Sun)

[1、2、3]Department of Mechanical and Electrical Engineering,Soochow University, Suzhou 215000, China

[*]Corresponding author's e-mail: zhangyoujun@suda.edu.cn

Abstract: Aiming at the defects of DC energy storage elements, complex structure and large volume of unified power flow controller (UPFC), the paper combines flying capacitor multilevel technology and voltage vector synthesis technology with unified direct power flow controller. A three-level direct power flow controller (3L-DPFC) and a new control strategy are proposed. By obtaining the power parameters of the receiving end and substituting them into the derived calculation formula, the compensation voltage can be deduced to adjust the phase and amplitude, and the change relationship between the compensation voltage and the power parameters of the receiving end can be calculated through multiple groups of data, so as to adjust the active power flow and reactive power flow in the power grid. Compared with the unified direct power flow controller, it has the advantages of no DC energy storage element, small volume, simple structure, and the voltage stress of each switch is reduced by half, which reduces the failure rate and economic cost of the controller. In the paper, the topology and control strategy of three-level synthetic DPFC are described in detail, and its buck AC conversion theoretical analysis is verified by simulation.

1.Introduction

Because the electric energy plays an increasingly important role in industrial production and daily life in modern society, the power generation and power supply industry must make profound changes to meet the needs of society. Therefore, compared with the previous power system, the power transmission capacity of the power grid must be improved. However, the previous power system facilities are difficult to meet the requirements of the current power grid due to the lack of infrastructure and long service time. Therefore, how to improve the power transmission capacity of the power grid and the stability of the system is of great significance to the economy, energy conservation and social life[1-3].

The appearance of flexible alternating current transmission system (FACTS)[4]and its development and application in recent decades have correspondingly solved the problems of insufficient power transmission capacity and system stability.

Unified power flow controller UPFC[5]is the most perfect FACTS controller at present. a single-stage phase and amplitude controllable AC/AC converter ACCPA without DC energy storage link is proposed in literature [6]. this converter is an excellent and simple structure of two-stage circuit π-type ACCPA[7]and cascaded ACCPA[8], with only a single-stage power conversion and no DC energy storage element, which not only has the functions of boosting and reducing voltage, and can lead and lag the phase of the regulated voltage.

On the basis of single-stage ACCPA, the orthogonal Direct power flow controller, DPFC) was proposed and studied in literature [9]. However, the orthogonal direct power flow controller has the problem of the third harmonic. Literature [10] puts forward a two-level composite direct power flow controller, but it will be limited by the voltage stress of the switch tube in high-voltage applications such as actual power grid. Based on the basic topology and control strategy of synthetic DPFC, this paper proposes a multi-level voltage synthetic direct power flow controller and a new power flow control strategy with three levels as an example. The three-level synthesized DPFC has only one-level power conversion and does not contain DC energy storage elements, the voltage stress of the switch tube is reduced by half, and its input voltage is connected in parallel with the power grid. The output voltage is connected in series with the power grid to form a compensation voltage for adjusting the power grid voltage. Three-level synthesis DPFC can effectively adjust the active power flow and reactive power flow in the power grid by controlling the compensation voltage, and the phase of the compensation voltage can be adjusted within 360.

2 Two-level synthesis direct power flow controller

2.1 Topological structure

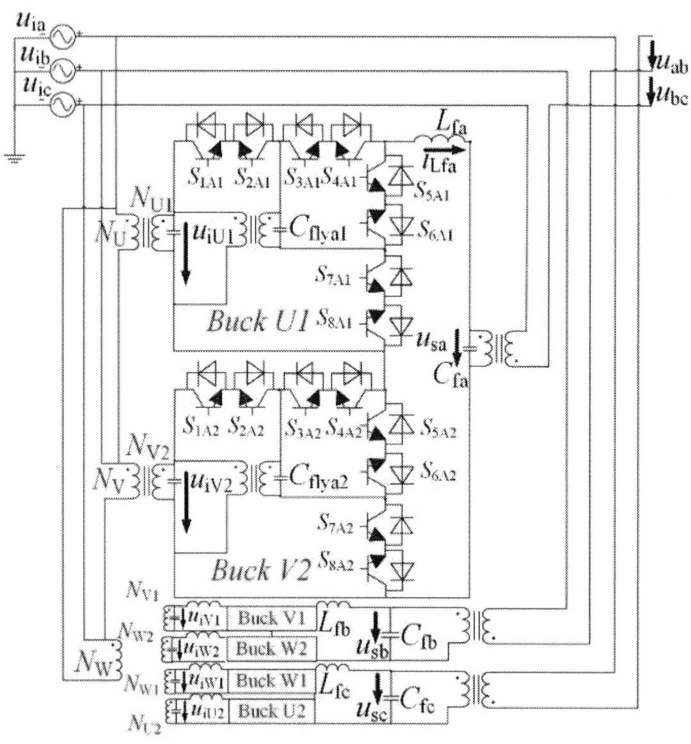

Fig. 1 Topology structure of three-level synthesis direct power flow controller.

Fig. 1 shows the main circuit topology of the three-level synthetic direct power flow controller. Considering the negative influence of the third harmonic, the primary and secondary sides of the input transformer are connected with Δ|Y, and the output transformer is connected with the same connection to avoid the third harmonic pollution of the power grid. The input transformer of the power flow controller is connected in parallel with the power grid, and the output transformer is connected in series with the power grid. In addition, it contains no DC energy storage element and an auxiliary transformer with the ratio of primary to secondary turns of 2:1 to ensure that the voltage of the self-following flying capacitor C_{y1}can automatically and stably follow $U_i/2$ ($U_i/2$ is the input voltage). A three-level synthesis direct power flow controller has six Buck AC conversion units U1, U2, V1, V2, W1 and W2 and three output LC filters. Each phase conversion unit has four pairs of switch tubes, and each switch tube is connected with a diode in reverse, so that the AC conversion unit can work in four quadrants, and the power factor is improved.

In practical application, the voltage withstand value of switching components has been paid close attention to due to the working requirements in high-voltage and high-power occasions. Therefore, it is of great significance to use the self-following flying capacitor multilevel synthetic direct power flow controller to reduce the voltage stress requirements of switching tubes by half, reduce the failure rate of the system and reduce the economic cost.

2.2 switch mode.

Here, for convenience of explanation, the U1Buck AC converter in Phase A is taken as an example, and the other five Buck AC converters are analogized. In a voltage input cycle, because the flow direction of input voltage U_i and filter inductor current i_{Lf} changes, when input voltage U_i>0 and U_i<0 and filter inductor current i_{Lf}>0 and i_{Lf}<0,U1 Buck AC converter has four different working modes and four switching modes. When the input voltage U_i>0, the switching tubes $S_{1A1}$and S_{7A1}, S_{3A1} and S_{5A1} are complementarily switched on and off at high frequency, and the switching tubes S_{2A1}, S_{4A1}, S_{6A1} and S_{8A1} are kept in synchronization. When the input voltage u_i is less than 0, the switching tubes $S_{4A1}$and S_{6A1}, S_{2A1} and S_{8A1} are complementarily switched on and off at high frequency, and the switching tubes S_{1A1}, S_{3A1}, S_{5A1} and S_{7A1} keep in constant synchronization. In which the duty ratio signals of switching tubes S_{1A1} and S_{3A1} and S_{5A1} and S_{7A1} are different by 1/2 of a single cycle of input AC voltage. Let's take U_i>0 and i_{Lf}>0 as examples. The four switching modes of U1Buck AC converter are shown in the following figure.

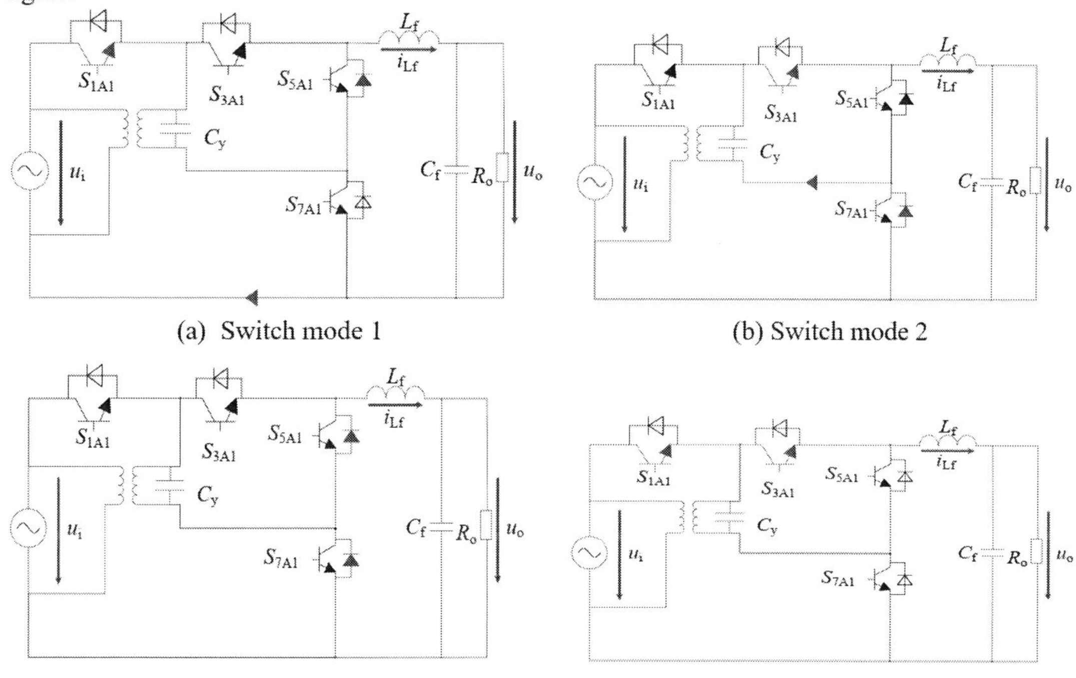

(a) Switch mode 1 (b) Switch mode 2

(c) Switch mode 3 (d) Switch mode 4

Fig. 2 Equivalent circuit diagrams of four switching modes.

As shown in fig. 2, in a switching cycle, the switching mode sequences involved are also different according to the duty cycle D<0.5 or D>0.5.

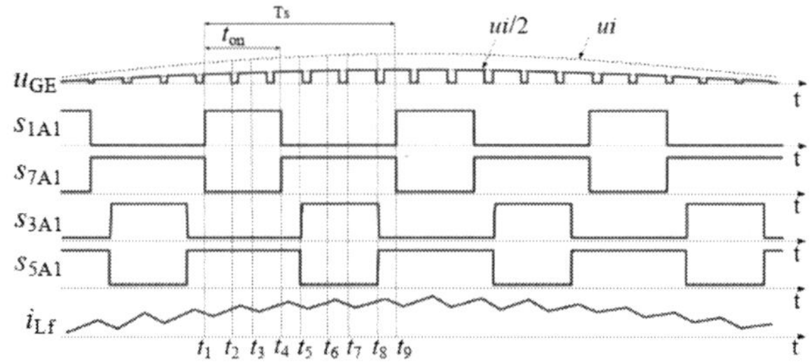

Fig. 3 Schematic waveforms of switch tube control signal, u_{GE} and i_{Lf} when d < 0.5

When the duty ratio D<0.5, the control signal waveform of each switch tube is as shown in the figure. The working modes of the converter are switched between switching modes 1, 2 and 3, respectively. At this time, the output voltage u_{GE} of the unfiltered inductor fluctuates between 0 and $1/2u_i$.

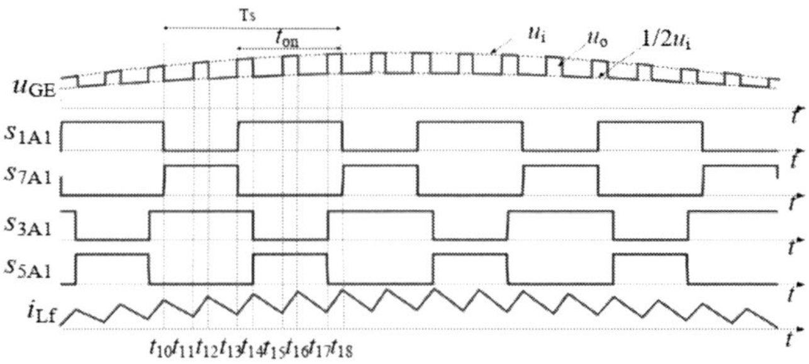

Fig. 4 Schematic waveforms of switch tube control signal, u_{GE} and i_{Lf} when d > 0.5

When the duty cycle D>0.5, the control signal waveform of each switch tube is as shown in the figure. The working modes of the converter are switched between switching modes 1, 2 and 3, respectively. At this time, the output voltage u_{GE} of the unfiltered inductor fluctuates between $1/2u_i$ and u_i.

2.3 Adjustment area of compensation voltage.

As shown in fig. 5, taking phase a as an example, the output compensation line voltage u_{ibc} synthesized by the U1 Buck AC converter and the V2 Buck AC converter is vector synthesized with the grid primary line voltage u_{iab} to obtain a new modulation line voltage u_{ab}, where The region I (O-F-H-J) represents the basic adjustment region of the compensation voltage when the input voltage of Buck AC conversion unit U1 is u_{iab} and the input voltage of V2 is u_{ibc}.

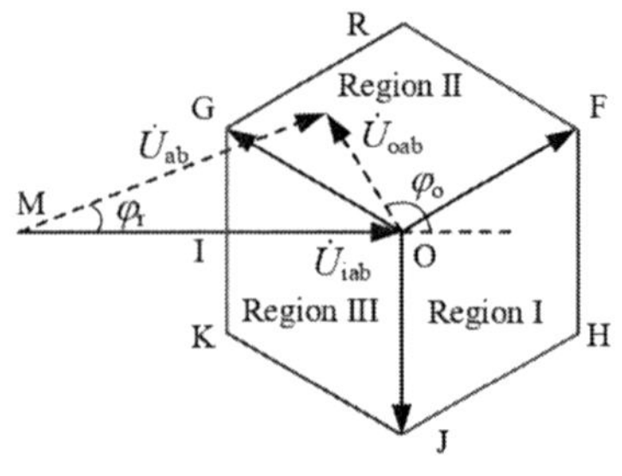

Fig. 5 Three regulation regions of compensation voltage u_{oab}

According to the analysis of fig. 5, the amplitude and phase of the new modulation line voltage u_{ab} of the power grid are related to those of the compensation voltage u_{oab}, and the amplitude and phase of the new modulation line voltage of the power grid can only be increased and the phase can only be adjusted in the interval of [-90, 30] compared with the original line voltage of the power grid in region I.In order to realize buck and other phase modulation, other adjustment areas must be obtained. To realize the adjustment range of the compensation voltage of 360, we have connected two selector switches SW_0 and SW_1 in the topology shown in fig. 1, and the two selector switches SW_0 and SW_1 are connected to the primary sides of input transformers T_{i1} and T_{i2} respectively. The selector switch changes the input voltage of the Buck AC conversion unit by changing the connection sequence between the primary side of the input transformer and the power grid, thus obtaining different output voltages and obtaining three different regulation regions.

3.Control strategy

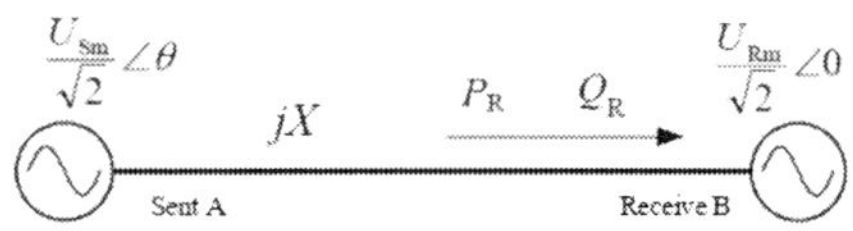

Fig. 6 Electric energy transmission system

The control strategy we usually use in power flow control analysis is shown in Figure 6. In the daily energy transmission process of the power grid, we can control the active power and reactive power of the power grid by adjusting one or several parameters. The specific relational expression is as follows:

$$\begin{cases} P = \dfrac{U_S U_R \sin\theta}{2X} \\ Q = \dfrac{U_S U_R \cos\theta - U_R^{\ 2}}{2X} \end{cases} \tag{1}$$

The usual control strategy is to determine the receiving terminal voltage U_{RM} and the sending terminal voltage U_{SM}, and then deduce the amplitude and phase angle of the compensation voltage U_B that the AC converter should provide. In actual power grid operation, the voltage at the receiving end is often affected by the power consumption of the load connected to the receiving end. For example, due to the heating demand in winter, the cooling demand in summer will inevitably lead to a significant increase in electricity consumption and a decrease in terminal voltage U_{RM}.

The new control strategy proposed here is based on the derivation of the basic relational expression, taking U_{SM}, x, p and q as known quantities, and finding U_{RM} and θ. The specific relational expression is as follows:

$$\begin{cases} U_{RM} = \sqrt{\dfrac{\left(U_S^{\ 2} - 4XQ\right) + \sqrt{-8XQU_S^{\ 2} + U_S^{\ 4} - 16X^2P^2}}{2}} \\ \theta = \tan^{-1}\dfrac{2XP}{U_S^{\ 2} + \sqrt{-8XQU_S^{\ 2} + U_S^{\ 4} - 16X^2P^2}} \end{cases} \tag{2}$$

Set the specific U_{SM} and obtain U_{RM} and θ by monitoring the receiving terminals p and q. According to U_{SM}, the relationship between U_B, p and q can be deduced by measuring multiple sets of data on the premise of monitoring U_{SM}. Adjust θ to adjust active power, adjust U_{SM} to adjust reactive power, and combine with multilevel synthetic direct power flow controller, θ and U_{SM} can be adjusted independently, so as to achieve the purpose of quickly adjusting power flow.

4.Simulation experimental results

We use the simulation software Saber Sketch to prove the correctness of the proposed theory, and always set the amplitude of the primary line voltage of the power grid as 538V during simulation. According to the three-phase symmetry of the input voltage, the new modulated phase voltages u_a, u_b and u_c of the power grid synthesized by them are naturally sinusoidal symmetrical and have a positive sequence difference of 120 between phases.

Fig. 7 shows that the input voltage of the A-phase Buck AC converter unit and the voltage stress borne by the switching tubes S_{1a} and S_{3a} are always half of the input voltage of the A-phase Buck AC converter when the actual duty cycle D<0.5 of the main circuit of the three-level synthetic direct power flow controller, which obviously reduces the voltage stress borne by the switching tubes during operation.

When the duty cycle D1 of Buck AC conversion unit is 0.2 and D2 is 0.2, and the regulation area of compensation voltage is in area I, the waveforms of voltage u_{ia}, u_{oa} and u_a are shown in Fig. 8. After measurement, the amplitude of voltage u_a is 406.6v, and the phase lag u_{ia} is 7.07 °, which is close to the amplitude of voltage u_a calculated by the formula is 406.7580v, and the phase lag is 7.0605 °.

Only change the connection form of the input transformer so that the compensation voltage is in area II. When the duty cycle of the buck AC conversion unit remains D1 = 0.2 and D2 = 0.2, FIG. 9 shows the waveforms of the voltages u_{ia}, u_{oa} and u_a at this time. At this time, the measured amplitude of the voltage u_a is 328.12v and the phase advance u_{ia} is 19.386 °, which is very close to the calculated value 328.14v and the phase advance 19.1417 °.

Fig. 10 shows the waveforms of voltages u_{ia}, u_{oa} and u_a when the compensation voltage u_{oa} is located in region III. when the duty cycle of Buck AC conversion unit remains D1 = 0.2 and D2 = 0.2, the measured amplitude of voltage u_a is 223.96v and phase lag u_{ia} is 13.7268 °, which is very close to the calculated value 223.3909v and phase lag 13.9358°.

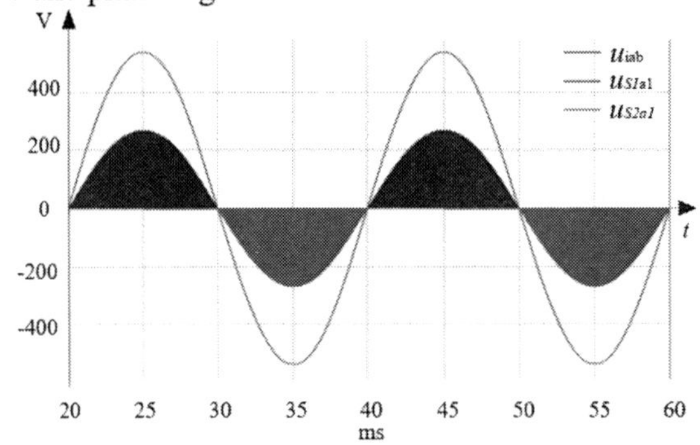

Fig. 7 Input voltage of phase A buck AC converter and voltage on both sides of switch tube

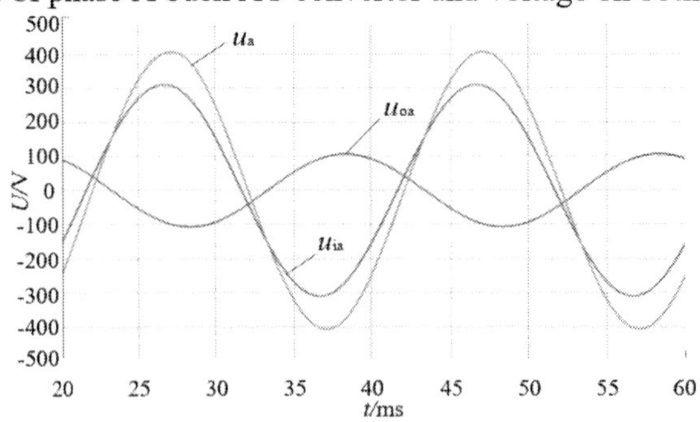

Fig. 8 The waveforms of modulated voltages u_a, u_{ia}, u_{oa} in area I when d_1= 0.2 and d_2=0.2.

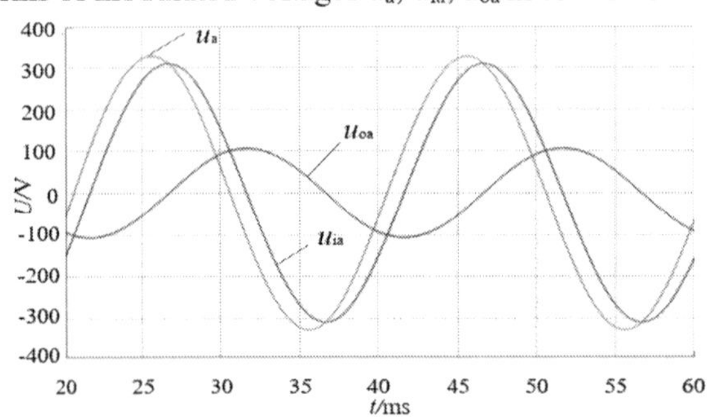

Fig. 9 The waveform of modulated voltages u_a, u_{ia}, u_{oa} in area II when d_1= 0.2 and d_2=0.2.

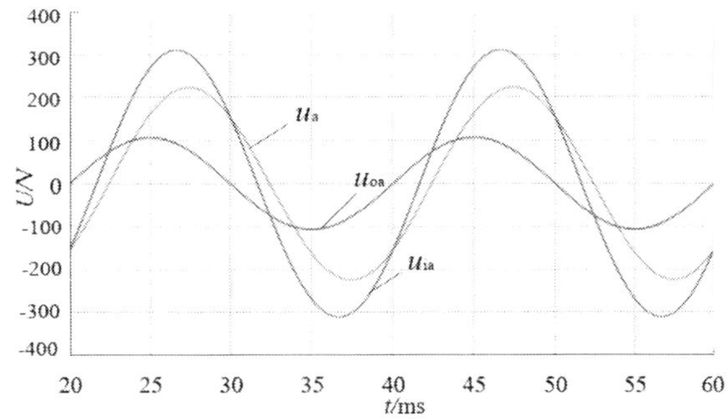

Fig. 10 d_1=0.2, d_2=0.2, and the waveforms of modulated voltages u_a, u_{ia}, u_{oa} in region III.

5.Conclusions

In order to effectively control the power flow in the power grid, this paper proposes a multi-level synthetic direct power flow controller and a new control strategy, taking three-level as an example. Similar to UPFC structure, they all contain input parallel transformers and output series transformers. However, the three-level voltage synthesis direct power flow controller does not contain DC energy storage components (DC energy storage components often lead to high failure rate of equipment). Set the specific U_{SM} and obtain U_{RM} and θ by monitoring the receiving terminals p and q. According to U_{SM}, the relationship between U_B, p and q can be deduced by measuring multiple sets of data on the premise of monitoring U_{SM}. Adjust θ to adjust active power and U_{SM} to adjust reactive power. Combined with the multilevel synthesis direct power flow controller, θ and U_{SM} can be adjusted independently, so as to achieve the purpose of quickly adjusting the power flow of the power grid. The topology theory of the three-level voltage synthesis direct power flow controller is demonstrated by simulation.

References

[1] Y. Zhang, H. Jin and Y. Zhang, "Full-duty-cycle regulated three-level AC-AC converter with self-following flying capacitor," IEEE Access, vol. 6, no. 9, pp. 48428–48437, Aug. 2018.

[2] Huang Fangfang, Huang Chengjun, Chen chen et al. Research and application of UPFC stability controller.J. Electric Power Automation Equipment, 2009, 29(3): 101-105.

[3] D. M. Divan and J. Sastry, "Voltage synthesis using dual virtual quadrature sources—A new concept in AC power conversion," IEEE Trans. Power Electron., vol. 23, no. 6, pp. 3004–3013, Nov. 2008.

[4] E. Ghahremani and I. Kamwa, "Optimal placement of multiple-type FACTS devices to maximize power system loadability using a generic graphical user interface," in *IEEE Transactions on Power Systems*, vol. 28, no. 2, pp. 764-778, May.2013.

[5] E. Gholipour and S. Saadate, "Improving of transient stability of power systems using UPFC," in *IEEE Transactions on Power Delivery*, vol. 20, no. 2, pp. 1677-1682, April.2005.

[6] Y. Zhang, S. Guan and Y. Zhang, "Single-Stage AC–AC Converter With Controllable Phase and Amplitude," in IEEE Transactions on Power Electronics, vol. 34, no. 7, pp. 6991-7000, July 2019, doi: 10.1109/TPEL.2018.2875093.

[7] Y. Zhang and X. Ruan, "π Model AC–AC Converter With Controllable Phase and Amplitude," *IEEE Trans. Ind. Electron.*, vol. 64, no. 8, pp. 6422–6431, Aug.2017.

[8] Y. Zhang and X. Ruan, "Three-Phase AC–AC Converter With Controllable Phase and Amplitude," *IEEE Trans. Ind. Electron.*, vol. 62, no. 9, pp. 5689–5699, Sep.2015

[9] Y. Zhang, G. Lu, W. A. Khan, Y. Zhang and Q. Zhu, "Direct Power Flow Controller—A New Concept in Power Transmission," in IEEE Transactions on Power Electronics, vol. 35, no. 2, pp. 2067-2076, Feb.2020.

[10]Yuan Qin. Two-phase voltage synthesis direct power flow controller.D. Soochow University, 2020.

Effect of F Functional Modification on the Stability and Electronic Structure of Borene

Cuicui Sun* and Mengmeng Liu

School of Civil Engineering, Shandong Jiaotong University, Jinan 250300, China.

*Corresponding author's e-mail: 204150@sdjtu.edu.cn

Abstract. Since the discovery of graphene, two-dimensional materials have quickly won widespread attention in the academic community. Borene is a two-dimensional isomer of boron and the lightest element Dirac material. It becomes the latest and promising two-dimensional material due to its unique structure and electronic properties. In the periodic table, B is a close neighbor of C and has a certain similarity with C. It can also form a hexagonal honeycomb structure. An additional B atom is added to the center of the ring to form a triangular lattice borene. The triangular borene has surplus electrons and belongs to a multi-electron state, which is equivalent to a metastable structure. In this paper, the first principles are used to study the F functionalized modification of the triangular borene. The aim is to transfer the surplus electrons in the system, and probe its structural stability and electronic structure characteristics. The study found that functional modification significantly improved the stability of borene. This can provide feasible ideas and practical guidance for the experimental synthesis of stable boronene.

1. Introduction

With the successful preparation of single-layer graphene in 2004, scientists have become very interested in two-dimensional materials. The physical and chemical properties of two-dimensional materials have also become a hot research project by scientists at home and abroad[1]. Among a series of unique two-dimensional materials, boronene has attracted widespread attention due to its unique chemical structure and impressive electronic conductivity and surface properties. In addition to the usual covalent bonds, boron can also form three-center double-electron bonds, which allows the formation of a large number of clusters of atoms with different sizes and atoms. The electronic band structure of boronene is different from that of graphene. Graphene is semi-metallic, phosphorene is semiconductor, and boronene is metallic. The ground state of electronic structure of boron is $[He]2s^2 2p^1$, the number of valence electrons is 3, but the number of atomic orbitals is 4. So the number of valence electrons of boron is one less than the number of atomic orbitals, and the valence electron layer cannot be filled when the bond is formed[2]. This ultimately leads to protruding and defective layers and complex bonds. Borene is the lightest element Dirac metal material[3], with negligible spin-orbit coupling. A recent simulation-based theory has shown that borene has partial ions of double Dirac cones near the Fermi level. Recently, scientists have proposed that the phase of borene sheet on the Ag(111) surface can be decomposed into two triangular sublattices. In a similar way, the honeycomb borene structure can be observed to be split in Dirac cones[4]. Scientists report that the electron bands of the borene monolayer sheet overlap each other at the Fermi level, resulting in a metallic structure. However, the boron sheet is easily degraded by the environment (due to defects, functionalization, hydrogen bonding, structural phase change and temperature change). And if the boron sheet is doped or adsorbed by other atoms or groups, the energy of the boron structure will be lower, the structure is more stable. Therefore, we expect a system with lower energy with our design materials.

Based on the first principles and density functional theory, this paper studies the effect of F functional modification on the stability and electronic structure of triangular borene by calculating energy bands, density of states, and differential electron density.

2. Calculation method

This article uses the atomic-scale simulation simulator Atomistix ToolKit (ATK) to perform first-principles calculations. In the calculation process, the Brillouin K point is selected as (10×10×1). In order to eliminate the periodicity of the nanoribbons, the vacuum distance between the nanolayers between adjacent unit cells is more than 10 Å. And the exchange correlation function between electrons is Generalized gradient approximation (GGA) and PBE (Perdew-Burke-Ernzerhof) functional for processing. All atoms use SZP (sigle zata polarization) to calculate the basis set, and the Brillouin zone cutoff energy used to describe the two consecutive steps of the periodic wave function is 150 Ry.

3. Results and discussion

3.1 Model and stability

By analyzing the structure of the triangular borene, we found that there are three different positions on the triangular lattice, namely, bridge, top and hollow, as shown in Figure 1.

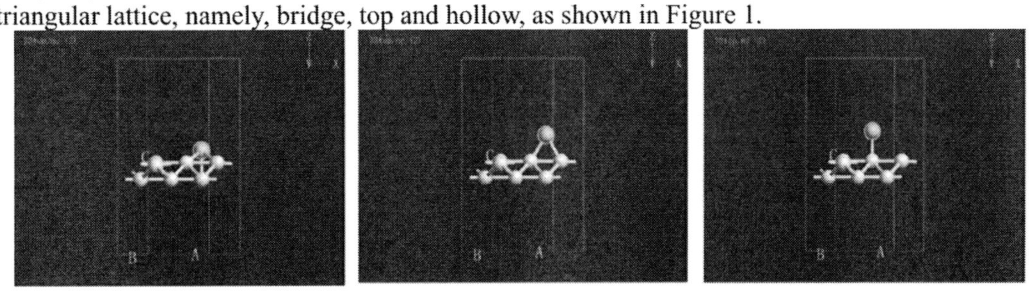

a b c

Figure 1. The schematic diagram of F adsorbed on triangular borene. (a) hollow site, (b) bridge site, and (c) top site. The pink represents the B atom, and the green represents the F atom.

After optimization, it is found that no matter which position, when F is adsorbed on the triangular borene, the energy is significantly reduced, which shows that our previous design ideas is correct and feasible. Among them, the total energy of the bridge position structure is the lowest and the structure is the most stable. Therefore, the energy E and the adsorption energy ΔE of the three positions and the triangular borene structure are listed here, as shown in Table 1. It can be clearly seen from Table 1 that the bridge system has the lowest energy and the most stable system.

Table 1. The total energy E and adsorption energy ΔE of the F adsorb at different positions on borene.

Adsorbed atoms	Adsorbed site	E/eV	ΔE/eV
F	hollow	-1268.44147	-0.20538
F	top	-1272.64347	-4.40701
F	bridge	-1274.21450	-5.97804
None	triangular borene	-630.38537	/
None	F	-637.85109	/

3.2 Electronic properties

The optimized composite structure has some changes to the structure before optimization, which is caused by the interaction between fluorine atoms and boron atoms. On this basis, the energy band and density of states are calculated for the composite structure, and the results are shown in Figure 2(b). In order to more clearly highlight the changes in the structure after F atoms are adsorbed, the energy band of the triangular borene is carried out, as shown in Figure 2(a). From the comparison of Figure 2, it is obvious that after F atoms are adsorbed, the electronic structure of the system changes significantly, and the dispersion of the conduction band minimum and valence band maximum becomes stronger, indicating that F atoms have a significant impact on the energy band structure. The Bloch states of the conduction band minimum and valence band maximum can further illustrate this problem. The triangular structure and F atoms have made great contributions to the electronic state of the system. Therefore, using the strong highly electronegative F atoms to absorb the surplus electrons in the borene is theoretically feasible.

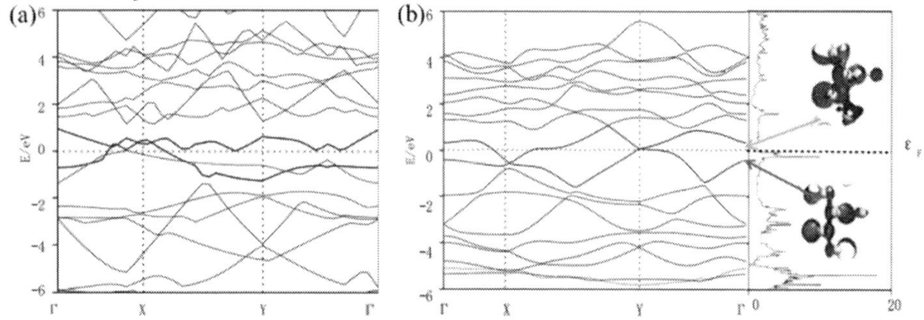

Figure 2. (a) Energy band of triangular borene, (b) The energy band, density of states and Bloch state of borene adsorbed by F atom.

Figure 3 analyzes the electron difference density of the composite system. The electron difference density refers to the difference between the self-consistent valence charge density and the superposition of atomic valence density, which indicates the coupling between the atoms in a certain system. The red part represents obtaining electrons, while blue region indicates losing electrons. Figure 3. clearly shows that boron atoms lose electrons and F atoms gain electrons. It also proves that there is electron transfer in the system to make the system more stable. Confirm the correctness of the previous conjecture.

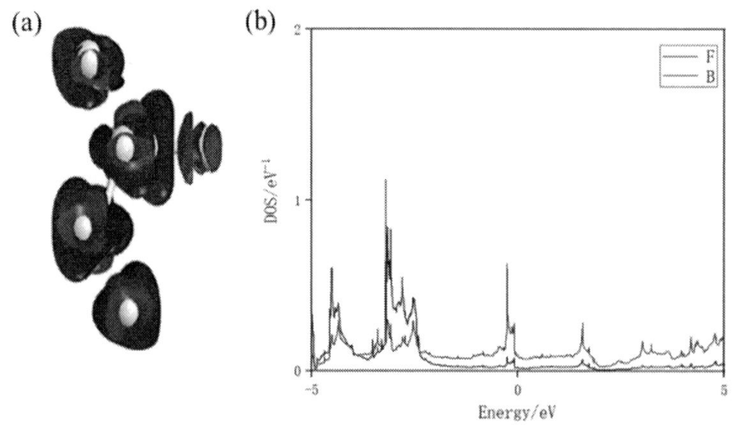

Figure 3. (a) Computed electron difference densities of borene adsorbed by F atom. (b) The projected density of states of borene adsorbed by F atom.

4. Conclusion

The method proposed in this paper to improve the stability of borene is to use the electronegative fluorine atoms to transfer the surplus electrons in the triangular borene structure, thereby achieving the purpose of stabilizing the triangular structure. Under this idea, three adsorption positions are proposed, hollow, bridge and top site. From calculations, it can be seen that all three are more stable than the original structure and can improve the stability of borene. The bridge position is more stable than the other two positions. A charge transfer occurs between the F atom and the triangular borene, which produces a hybrid effect. Therefore, the F changes the electronic structure of borene and improves its stability. That is to say, it is possible to conduct experimental synthesis based on the conjectures in this chapter, which can provide theoretical basis and design ideas for the experimental synthesis of more stable triangular borene.

Acknowledgments

This work was supported by the Doctoral Scientific Research Foundation of Shandong Jiaotong University (BS50004943), Scientific Research Fund Project of Shandong Jiaotong University (Z201916).

References

[1] Zhao, J., Ma, D., Wang, C., et al. (2021) Recent advances in anisotropic two-dimensional materials and device applications. Nano Research, 14(4): 897-919.
[2] Mathew, M., Radhakrishnan, S., Vaidyanathan, A., et al. (2021) Flexible and wearable electrochemical biosensors based on two-dimensional materials: Recent developments. Anal. Bioanal. Chem., 413(3): 727-762.
[3] Ranjan, P., Lee, J. M., Kumar, P., Vinu, A., (2020) Borophene: New Sensation in Flatland. Adv. Mater., 32(34): 2000531.
[4] Yao, Y., Liu, G., (2021) Density functional theory study on the electronic structure and optical properties of Li absorbed borophene. Mol. Phys., DOI:10.1080/00268976.2021.1966114.
[5]Wang, H., Zhao, Y., Xie, Y., et al. (2017) Recent progress in synthesis of two-dimensional hexagonal boron nitride. J. Semicond., 38(3): 031003.
[6]Beniwal, S., Hooper, J., Miller, D. P., et al. (2017) Graphene-like Boron–Carbon–Nitrogen monolayers. Acs Nano, 11(3): 2486-2493.
[7]Arnold, F. M., Seifert, G., Kunstmann, J., (2020) Thermodynamic stability of borophene, B_2O_3 and other $B_{1-x}O_x$ sheets. J Phys. Commun., 4(3): 031001.
[8]Chowdhury, S., Majumdar, A., Jana, D., (2019) Electronic and optical properties of the supercell of 8-Pmmn borophene modified on doping by H, Li, Be, and C: a DFT approach. Appl. Phys. A, 125(5): 360.
[9] ATK, Version 13.8, atomistix a/s, 2013, www.quantumwise.com.
[10] Perdew, J. P., Burke K., Ernzerhof, M., (1996) Generalized gradient approximation made simple. Phys. Rev. Lett., 77: 3865–3868.

Design and Analysis of CFETR CS Coil and Pretension Structure

Zitong Liang[1, a, †], **and Langzhou Jin**[2, b, †]

[1]College of Mechanical and Electrical Engineering, Central South University, Changsha, Hunan Province, 410083, China.

[2]Walter Scott, Jr. College of Engineering, Colorado State University, Fort Collins, Colorado State, 80521, United States.

[a]lzt6017@163.com, [b]davidjin@colostate.edu

[†]These authors contributed equally.

Abstract. The coil and the pretension rod play important roles in the central solenoidal coil system of Tokamak. Therefore, the analysis of the coil and the pretension rod in the normal operating environment is essential for the facility's construction. In this paper, the 3D model of the two parts mentioned above is built based on SolidWorks. Subsequently, the simulations are carried out to revert the real operation environments following the fields, loads, and boundary conditions. Then, we compare the official analysis results of the corresponding parts in the central solenoidal system of the China Fusion Engineering Test Reactor (CFETR) based on the results from COMSOL. Compared with the official analysis, the error is about 1.3% for the maximum magnetic field and about 0.53% error for the outer part, respectively. As for the pretension rod, simulations are carried out based on the finite element analysis software ABAQUS. The Von mises and the maximum principal stress are 140.1 MPa and 168.4 MPa, which is much lower than the material's yield stress (205 MPa). The safety factor is 1.78, larger than the critical failure value 1. According to our results, the rationality of the CFETR central solenoidal coil system design is re-examined and verified.

1. Introduction

The Tokamak is a toroidal system with an axially symmetrical shape, in which the high temperature plasma is contained by the magnetic field of an inner current circulating [1]. Besides, a strong longitudinal magnetic field parallel to the current is used to suppress the main magnetohydrodynamic instabilities [1]. The central solenoidal (CS) coil system is an important component in the Tokamak to provide the poloidal magnetic field. The system has several main parts to support the normal operation, including the coil, the pretension mechanism and other important parts. In the CS coil system, the coil will generate the poloidal magnetic field to support the plasma in the reaction chamber. On the other hand, the pretension mechanism will ensure stability with the accompanied specific loads and boundary conditions during the operation.

Due to the important roles of the CS coil system, the analysis is necessary to investigate various loads during the working process to ensure normal and safe operation. The specific analysis aims at the chosen parts in the system, e.g., the coil and the pretension rod will generate the key ideas about the precautions. The integral cognition of the properties and the conditions for the entire system could be generated after researching every component in the system. Thus, the generated and refined cognition will be meaningful insight to later researchers who study designs of the CS coil system.

The coil and the pretension rod in the China Fusion Engineering Test Reactor (CFETR) CS coil system are chosen as the analysis objects to fulfil this goal. In this analysis, the basic 3D models of the two parts are built based on SolidWorks. The different kinds of fields and loads are generated to create the simulations for a real operation environment to check whether the design is reasonable. Finally, with the comparison between the simulations and the official analysis results of CFETR, it is confident to claim the CFETR CS coil system design is valid. In this paper, section 2 makes a brief introduction to the research method, including the software platforms, guiding research ideology. In section 3, results, and discussions of the two objects in the paper are displayed in detailed descriptions, respectively. Specifically, we demonstrate the pre-setting, concept concretization, data generating methods, and visible results. Section 4 gives a brief summary based on the results eventually.

2. Methods

In this paper, several software platforms, including Solidworks2017, COMSOL5.6, and ABAQUS, are utilized to investigate different analysis parts, respectively. Specifically, SolidWorks, a solid modelling computer-aided design (CAD) and computer-aided engineering (CAE) computer program, is utilized to construct the 3D model of CS coil, the preload system, and the specific pretension rod [2]. COMSOL Multiphysics is a cross-platform finite element analysis, solver, and multi-physics simulation software used to create the magnetic field simulation around the coil and outer coil conductor [3]. For the pretension rod, ABAQUS, a Finite Element Analysis (FEA) software for finite element analysis and computer-aided engineering, is utilized to generate the potential loads and boundary conditions under the real operation situation [4].

2.1. Formatting the title Solidworks2017 setting/parameters

Solidworks2017 creates a coil part drawing with the parameters listed in Table 1 (collected from Ref. [5]). Based on the coil size, three-dimensional parts drawings of the tension rod, pressure plates, buffer zone, and load beam are established, respectively. Finally, the parts are assembled to obtain the general assembly drawing of the CFETR CS coil and its pre-tensioning mechanism, prepared for subsequent structural, magnetic field, and force analysis.

Table 1. CS coil parameters **(parameters are collected from Ref. [5])**

Parameter	Nb3Sn Coil		NbTi Coil
	Inner	Outer	
Conductor Size/mm	$49^2\times\Phi32.6$	$49^2\times\Phi32.6$	$51.9^2\times\Phi35.3$
Conductor Insulation/mm	2.6	2.6	2.6
Ground Insulation/mm	4.9	4.9	4.9
Radial Turns	4	4	10
Axial Turns	28	29	22
Total Turns	112	116	230
Inner Radius/mm	750	963.8	1217.6
Outer Radius/mm	953.8	1167.6	1760
Height/mm	1442.2	1493.8	1196.4
Conductor Length/m	600	777	2060

2.2. Magnetic field analysis by COMSOL5.6

The magnetic field is analysed by COMSOL5.6. Firstly, we construct a geometric model of the coil where the two-dimensional cross-sectional view of the coil is illustrated in Figure 1. Then, a rotational symmetry operation is performed to obtain a three-dimensional CS coil diagram, as shown in Figure 2.

Besides, it is feasible to build three-dimensional models of single-turn coils with different diameters layer by layer. Nevertheless, the COMSOL5.6 system will freeze if too many three-dimensional models are added, i.e., the former way is chosen in our case. Besides, the air domain is added with software default attributes. Materials and the magnetic field properties are entered following Table 1. Afterward, we construct the finite element mesh according to the parameters in Table 2.

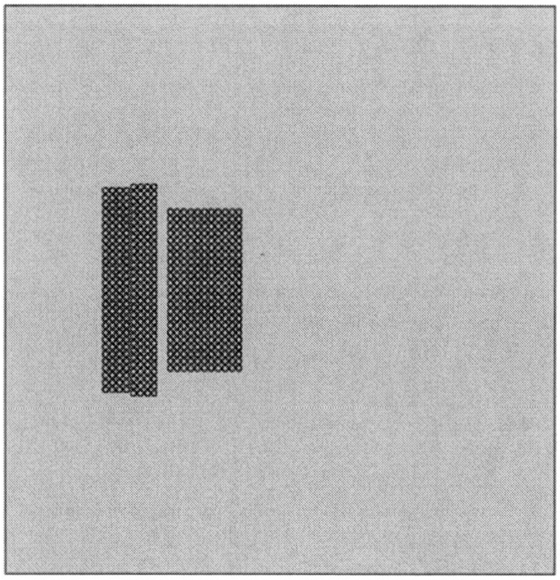

Figure 1. Coil two-position cross section and air domain diagram

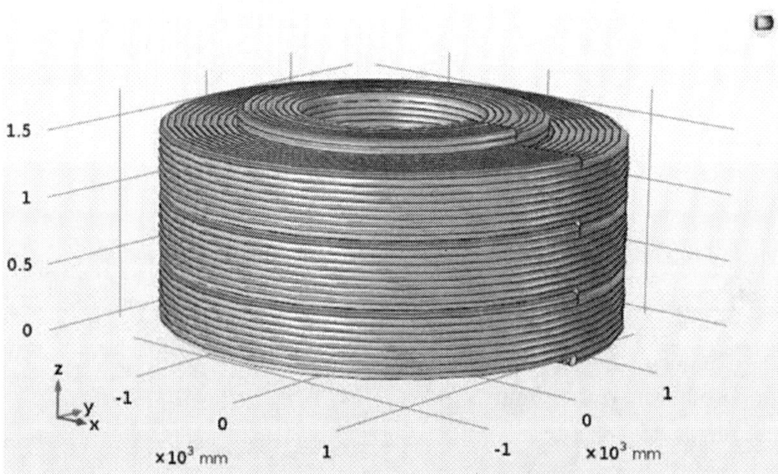

Figure 2. 3D model of CS coil in COMSOL5.6

Table 2. Finite element parameter

Parameter type	Numerical value
Maximum cell size	400 mm
Minimum cell size	0.08 mm
Maximum unit growth rate	1.1
Curvature factor	0.2
Narrow area resolution	1

2.3. The strength analysis for the pretension rod based on ABAQUS FEA Method

2.3.1. Background for ABAQUS.

ABAQUS is an FEA program with various functions designed specifically for high level simulation analysis, including structure, heat transfer, etc. ABAQUS is also designed for both nonlinear and linear stress analysis of different kinds of structures with versatility [4].

For the pretension rod strength analysis, the finite element analysis (FEA) is used based on the ABAQUS platform. With the guideline of FEA, the research project will be divided into numerous target elements to make a simulation. The user defines the properties of the element (e.g., shape, quantity, etc.). The different kinds of load and boundary conditions will be added with demands to the research project. With the reaction from loads and boundary conditions, the simulation based on the meshed project will be created as the simulation.

2.3.2. The settings of the simulation

For the material of the pretension rod, the stainless steel Grade 316LN (UNSS31653) is chosen because of the versatility. The existed design has a similar dimension to the CFETR. Other corresponding properties (e.g., the density, Young's modulus, etc.) are based on the properties of the chosen material. The adopted properties are all based on the ABAQUS unit system to fit the calculation and analysis based on the ABAQUS. The default length unit is in millimeters. The detailed data settings are shown in Table 3.

Table 3. The basic properties for the pretension rod

Properties	Values	Unit
Density	7.99×10^{-9}	Tonne/mm^3
Young's Modulus	200000	MPa
Poisson's Ratio	0.28	N/A
Thermal Expansion Coefficient	17.2×10^{-16}	1/ °C

2.4 The reaction settings on the pretension rod simulation

2.4.1 Loads and boundary conditions

Three kinds of load are considered in the stimulation: gravity, temperature, and pretension. For the pretension load, each pretension rod will be added to 3.125 MN for the load. The temperature load is assumed to decrease from room temperature (293 K) to 4 K. The gravity load is generated autonomously with the initial density input. For the boundary condition, since any kinds of displacement or rotation are not allowed for the ends of the pretension rod, the boundary condition "encastre" is chosen. The limitation of the present studies naturally includes the restricted licence of the ABAQUS, i.e., the electromagnetic loads will not be presented here.

2.4.2 Strength analysis category choice

The simulation could display different kinds of results with the demands of properties. For the pretension rod, the Von mises and the maximum principal stress will be chosen as the reference to be compared with the original stress properties of the material. Then, the deformation on the three axes (U1, U2, U3, corresponding to x, y, z axis) will be considered at the same time to check if the unacceptable deformation occurs.

3. Results and Discussion

3.1 3D Modelling results of CS coil and pretension system

3.1.1 3D model of preload system and draw assembly diagram

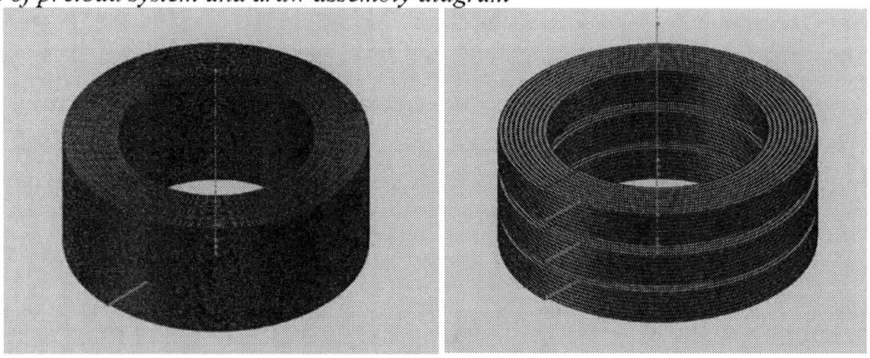

(a) (b)
Figure 3. 3D models of (a) inner and (b) outer coils from SolidWorks 2017

In order to achieve the design goals of the CS coil and reduce the manufacturing cost simultaneously, the coil is designed as a hybrid magnet coil with an Nb3Sn conductor in the inner high-field area and a NbTi conductor in the outer low-field area. Therefore, the coil winding design adopts the embedded package structure, which is divided into Nb3Sn coil module and NbTi coil module, and then packaged together. The two-dimensional cross-section design of the coil winding is shown in Figure 3. Generally, a total of 5 pie coils are constructed. Considering that the longest conductor length needs to be less than 1000m, the Nb3Sn coil module is wound and designed into two pie coils (namely the Nb3Sn inner coil and the Nb3Sn outer coil). Besides, NbTi, the coil module, is designed into three pie coils: NbTi upper coil, NbTi middle coil, and NbTi lower coil.

3.1.2 3D model of preload system and draw assembly diagram.

According to the preload structure provided by reference [5], we build the 3D model of preload system and assemble them with the coil models. Figure 4 illustrates the composition of the pretension system, including Upper and Lower Load Beam, Buffer Zone, Tension Rod, Pressure Plate. And it can be divided into 12 symmetrical structures.

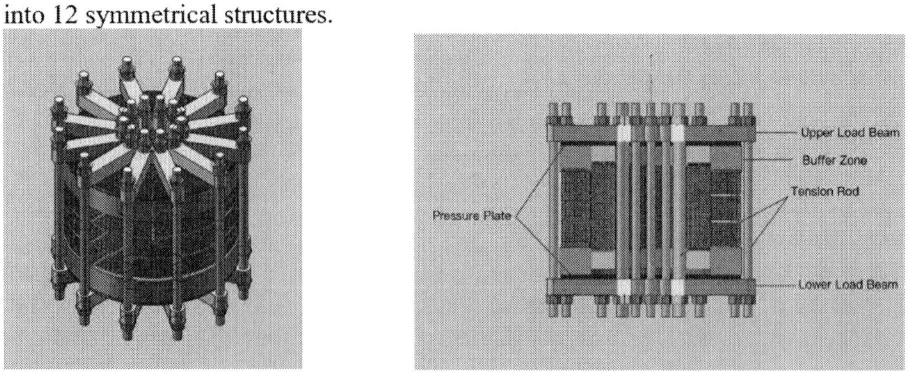

(a) (b)
Figure 4. The assembly diagram of CS coil and its preload system: (a) three-dimensional assembly diagram and (b) assembly section structure drawing.

3.2 Magnetic field simulation result

3.2.1. Finite element network & Magnetic field distribution around CS coil
The finite element network and the corresponding distribution of the magnetic field are shown in Fig. 5 and 6. The magnetic field distribution on the coil section under 49kA working current. The maximum magnetic field of the coil is 12.25T, which is near the inner center conductor of the inner coil. The maximum magnetic field on the outer coil conductor is 5.958T, located on the conductors at the upper and lower ends of the innermost layer.

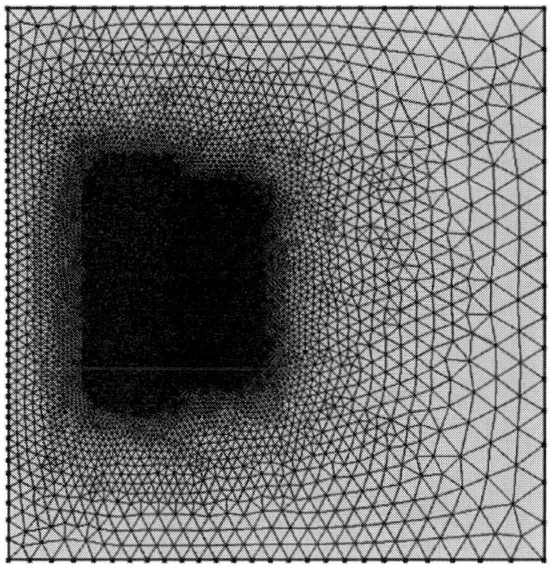

Figure 5. Finite element network

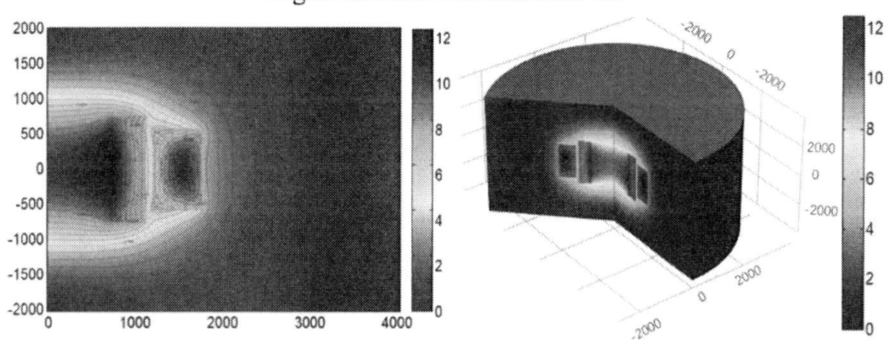

(a) (b)

Figure 6. Magnetic field distribution: (a) Three-dimensional profile of magnetic field distribution and (b) is magnetic diagrams of the magnetic field.

3.2.2. Comparison with the results of reference [5]

Table 4. Magnetic field comparison with ANSYS [5] and COMSOL results

	COMSOL	ANSYS	Program
Maximum magnetic field of CS coil/T	12.25	12.05	12.09
Maximum magnetic field on outer coil conductor/T	5.958	6.0	5.99

It can be seen from Table 4, the results of modeling and simulation using COMSOL are in agreement with the results of analyzing and programming using ANSYS. In detail, simulation results of ANSYS show that the maximum magnetic field of the coil is 12.05 with 1.66% error while it is 12.09T with 1.32% error in our results. The maximum magnetic field of the outer coil conductor obtained by ANSYS simulation is 6.0T with 0.70% error, while it is 5.99T with 0.53% error in our results.

3.3 The results of pretension rod strength analysis

Fig. 7-11 shows the simulation results for the stress (Von Mises and Maximum Principal) and deformation (U1, U2, U3), respectively. From Fig. 8 and 9, the max stress of the Von mises stress simulation and the maximum principal stress simulation is 140.1 MPa and 168.4 MPa correspondingly.

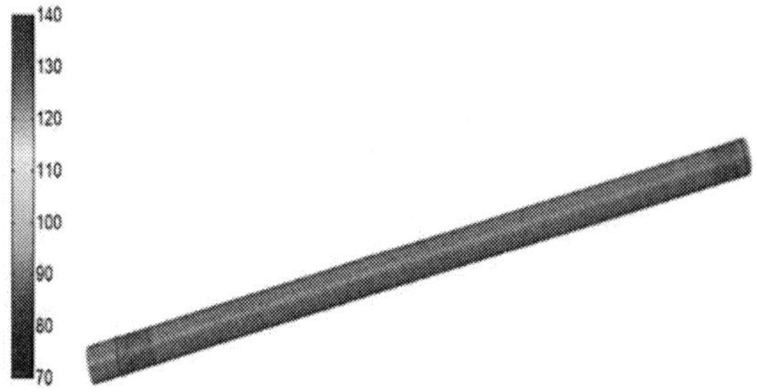

Figure 7. Simulation for Von Mises stress of pretension rods from ABAQUS.

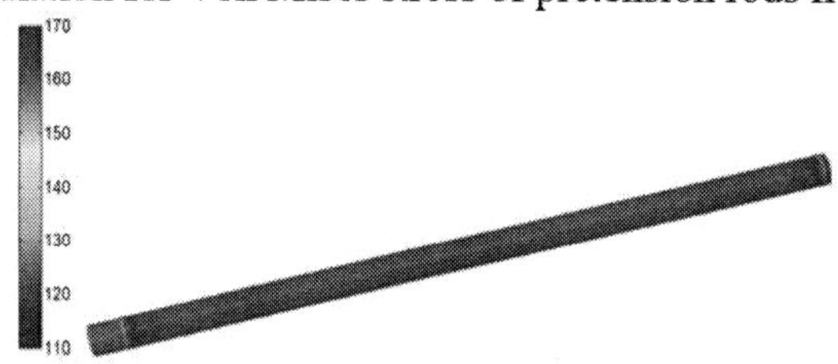

Figure 8. Simulation for Maximum Principal Stress of pretension rods from ABAQUS.

Figure 9. The deformation of pretension rods in the U1 direction (collected from ABAQUS)

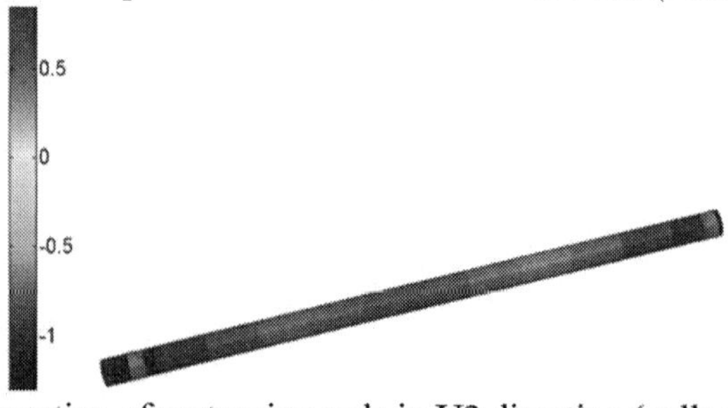

Figure 10. The deformation of pretension rods in U2 direction (collected from ABAQUS)

Figure 11. The deformation of pretension rods in the U3 direction (collected from ABAQUS)

Table 5. The simulation stress results compare with the UNSS31653

	Von mises stress	Maximum principal stress	Material yield
Maximum stress occurred (MPa)	140.1	168.4	205

For better verification of the stability of the pretension rods, we compare the results of the proposed method with those of the UNSS31653 yield stress. According to Table 5, both the simulation results of stress are lower than the UNSS31653 yield stress, i.e., reveal the safety and reliability of the current pretension rod design. On the other hand, based on the fact for the ductile material, the safety factor calculation is applied for the pretension rod with the maximum distortion energy theory:

$$SF = \frac{\sigma_{yield}}{\sigma_{Von\ mises}} = 1.78 > 1 \tag{1}$$

Since the safety factor is higher than the critical failure value 1, the design of the pretension rod is again proved to be safe and reliable enough.

4. Conclusion

In summary, the simulations are carried out for the coil and the pretension rod with the determinate field, loads, and boundary conditions to show the situation under a real operating environment. We identified the magnetic field around the pre-setting coil 3D model and working current, which align with prior literature. Meanwhile, the feasibility of the coil design has been further verified. We investigated three kinds of loads for the pretension rod and one boundary condition on the pretension rod. With the generated simulations, we found that the simulation result of the maximum magnetic field of the coil and the outer coil conductor has a high similarity with the official analysis results. For the pretension rod, the Von mises stress, and the maximum principal stress are all in the acceptable range of the material yield stress. The safety factor is also larger than the critical failure value. The specific analysis on the coil and the pretension rod of CFETR provides the basic cognition about the parts design and testing. In the future, the CS coil is looked forward to be an advanced design based on improving the innovation of the wiring, material, etc. With the help of the optimized FEA method, the more accurate and reasonable simulation closer to the real operation environment is possible to create, i.e., paves a path to improve the CS coil system.

References

[1] Artsimovich, L. A. (1972), Tokamak devices, Nuclear Fusion 12.2: 215.
[2] Schmitz, Barbara M. (1999), SolidWorks Animator, Computer-Aided Engineering.
[3] Multiphysics, C. O. M. S. O. L. (1998), Introduction to COMSOL Multiphysics®, COMSOL Multiphysics, Burlington, MA, accessed Feb 9: 2018.
[4] Abaqus, G. (2011), Abaqus 6.11, Dassault Systemes Simulia Corporation, Providence, RI, USA.
[5] Liu, Xiaogang, et al. (2016), Electromagnetic optimization and preliminary mechanical analysis of the CFETR CS model coil."" IEEE Transactions on Plasma Science 44.9:1559-1563.

Journal of Physics: Conference Series **2108** (2021) 012083 doi:10.1088/1742-6596/2108/1/012083

[6] Wei, Cai, et al. (2019), Magnetic field analysis of solenoid driven by alternating current. Journal of Physics: Conference Series. 1237:3.
[7] Börgesson, L. (1996), Abaqus, Developments in geotechnical engineering, 79: 565-570.
[8] JI, Feng, et al. (2015), Magnetic field and electromagnetic load calculation for CFETR central solenoid model coil, Journal of Hefei University of Technology (Natural Science) 2015: 06 (Chinese).

Study on Environmental Adaptability of Metal Plate Fuel Cell

Hailiang Wang[1a,*], Junxiao Feng[1b*], Lanyun Li[2c*], Zhengkai Tu[3], Bo Zhou[4], Guanqiong Wang[2], Qinshu Xu[2], Jianping Xu[5], Jinghui Gao[2]

[1]School of Energy and Environmental Engineering, University of Science and Technology Beijing, Beijing 100083, China

[2]Xinxing Jihua Science and Technology Development Co;Ltd., Beijing 100070,China

[3]School of Energy and Power Engineering, Huazhong University of Science and Technology, Wuhan 430074, China

[4]Xinxing Heavy Industry Group Co., Ltd., Beijing 100166, China

[5]Tellhow Jungong Group Research Institute, Beijing 100166, China

[a*]80063177@xxcig.com, [b*]jxfeng@ustb.edu.cn, [c*]173183384@qq.com

Abstract—As a main task of national key project (2018YFC081000)"emergency power under the special environment systems research and application demonstration",Study on Environmental Adaptability of Metal Plate Fuel Cell.the paper focus on the key technologies needed to be developed of researching PEMFC system to adapt to the special environment of high temperature, high humidity, high salt, high cold and high altitude;this topic reviews the impact of plasma exchange membrane fuel cell and climate environment on such emergency power supply equipment, and uses Au@TiN as a metal bipolar plate of proton exchange membrane fuel cell. After a series of tests, it proves the excellent operation characteristics of the fuel cell reactor in "4-high" environment.the study showed that after constant 1.0 V (vs SHE) potential polarization, the size order of different coating corrosion currents is: gold plated> stainless steel> carbonization;In a high altitude environment, the optimal air intake parameters are obtained by numerical simulation and optimization of air pressure, air volume, wind speed, air excess coefficient and other parameters;In a high-temperature environment, the problem of the metal bipolar fuel cell supply system can be solved by improving the flow structure of the air side to increase the heat exchange area and the air capacity;In the high-cold environment, the shutdown pulse sweep scheme is adopted,and the rapid start of hydrogen battery supply system is realized by using hydrogen catalytic reaction assisted heating.

1. Introduction

Fuel cell is known as one of the cleanest energy sources in the 21st century, water is the only product of power generation process, and it is quiet power generation and has high efficiency. The new hydrogen fuel cell power supply system has the advantages that high power generation efficiency (about 50%), clean, silent, low infrared characteristics, modular power generation, and strong environmental adaptability. In some technically developed countries, the hydrogen emergency power supply developed is being applied and verified. In recent years, with the strong support of the government and the capital market, the research and development of fuel cells, especially proton exchange membrane fuel cells (PEMFC) in China has made great progress, which has a series of characteristics of high efficiency, low noise, environmental protection, is a promising energy conversion equipment[1-3].

To satisfy the public security needs of border posts, remote reef scene practical environment in China, and far away from the grid area, "under the special environment of emergency power systems research and application demonstration" especially in the high altitude, high and low temperature, high humidity and high salt is one of the key projects from the establishment of the "national public security risk prevention and control and emergency technical equipment research". Most of the bipolar plates of fuel cells are made of graphite, but the graphite flow filed plate is brittle, which is easy to be damaged under vibration condition and has poor adaptability for field transportation. Fuel cell metal plate has characteristics of good mechanical properties, light weight and easy to mass production which is the key development direction of fuel cell membrane electrode. However, it is one of the key technologies needed to be developed of researching PEMFC system to adapt to the special environment of high temperature, high humidity, high salt, high cold and high altitude. Environmental conditions are divided into five environmental types, including climate environment, chemical environment, electromagnetic environment, biological environment and mechanical environment. Emergency power system is a complex equipment in long-term outdoor service, and these five environmental conditions will affect its operation.

2. Summary

We mainly considers the influence of climate environment on emergency power equipment in this paper. Environmental conditions of different environmental types can be divided into many different environmental factors, such as in the climate environment, temperature, humidity, salt, low pressure, storm, lightning, snow and other severe weather and other environmental elements.In view of the above environmental factors, environmental adaptability of emergency power supply can be improved by taking appropriate protective methods, mainly including the following ways.

(1) For high and low temperatures, taking heat prevention, anti-freezing methods. In high temperature weather, considering the impact of temperature rise, for emergency power supply equipment, taking "heat prevention" methods including strengthening ventilation management, to ensure heat dissipation of the equipment area and the equipment itself. Strengthen equipment inspection, timely reflect equipment abnormal temperature rise, sound, vibration, etc. For the power supply equipment that works in an environment with high temperature, the cooling fan can be used for forced cooling. In low temperature weather, it should be noted that the reforming hydrogen production equipment is easy to be unable to start and run normally due to the increase of diesel viscosity. It is necessary to put in the electric heater in advance and take "anti-freezing" methods.

(2) Take dehumidification methods for high humidity.The high relative humidity poses a serious threat to the normal operation of power supply equipment.Byungchan et al.[4] prove through experiments that humidity affects the performance and durability of PEMFC. Proper humidity can enhance the performance of PEMFC, but excessive humidity may cause overflow.This paper authors tested I-V performance, linear sweep voltammetry, cyclic voltammetry, and impedance diagrams of microchannel cells with changes in relative humidity (RH).The test data shows that the overflow phenomenon begin at RH=70%, and the ionic conductivity of the membrane reached its maximum value at RH=80%.When the relative humidity is 80%, the effect of ionic conductivity increasing is higher than its decreasing mass transfer,therefore, it is necessary to take reasonable methods to reduce the impact of ambient humidity on the equipment operation and carry out moisture-proof and dehumidification.Effective moisture-proof and dehumidification methods include: ventilation, local isolation, etc, take waterproof cover if neccessary,Or install industrial dehumidifier, quickly and automatically solve the moisture problem;To prevent condensation, a heating or drainage dehumidifier can be installed at a proper position in the cabinet.

(3) For high altitude and low-pressure environments, an air compressor is needed to increase the oxygen content in the air. When the altitude exceeds a certain value of the area of use, due to the corresponding decrease in atmospheric pressure, air density, and humidity, the air gap and external

insulation characteristics continue to decline. Therefore, the maximum allowable operating voltage of the device should be modified.

(4)Pay attention to prevent corrosion in high salt environment. On the one hand, it is better to use sealed and moisture-proof electrical component areas, or add cooling facilities to slow down the corrosion rate. On the other hand, it is better to treat the metal surface of the equipment to form a protective film, such as adding an anti-corrosion coating to slow down the corrosion rate of salt.

(5) The high cold from "four high" serious environments has the biggest influence on PEMFC start-up.Low temperature start-up at -10°C is difficult due to MEA stratification.Many studies on start-up operation under high cold conditions have been done at China and abroad.Cappadonia et al.[5] believe that the conductivity temperature of Nafion117 membrane is related to moisture content, and -13°C is the turning point from point to start-up because water in the membrane froze, and the conductivity of protons in the membrane is closely related to water form in the membrane. Therefore, -10°C is a critical point for membrane materials. Freezing/thawing process will reduce the toughness of the membrane, and the membrane molecules will be rearranged horizontally, leading to the deceasing of the membrane performance under freezing point.Berning, Mao, and Meng[6-9] have done a lot of work on cryogenic initiation mechanisms and three-dimensional multi-model.Aiming to explore the dynamic change of multi-dimensional temperature field inside the battery during low-temperature starting. However, since the water management of the battery has not been explored in the above studies, there is no mature operation method for low-temperature starting internationally.

The fuel cell which is researched in this study is made of metal plate flow field, the bipolar plate is processed into hydrogen and oxygen monopole plate by extrusion molding technology such as pressing or extrusion of metal sheet, and then the bipolar plate is formed by welding two monopole plates, after welding, the two sides of the bipolar plate are hydrogen and air circulation channels. Looking from the experimental results, the performance indicators are excellent.

3. Research on operation characteristics of fuel cell stack in "four high environment"

3.1. Research on operation charateristics of fuel cell stack in high humidity and high salt environments

In order to enhance the proton conduction ability and improve the performance of hydrogen fuel cell, hydrogen fuel cell needs water to ensure the wettability of membrane electrodes during operation. Therefore, high humidity environment is conducive to the improvement of battery performance[4] .In addition, according to the above mentioned, this fuel cell metal bipolar plate adopts Au@TiN metal surface anti-electrochemical corrosion coating process, which has excellent anticorrosion performance.

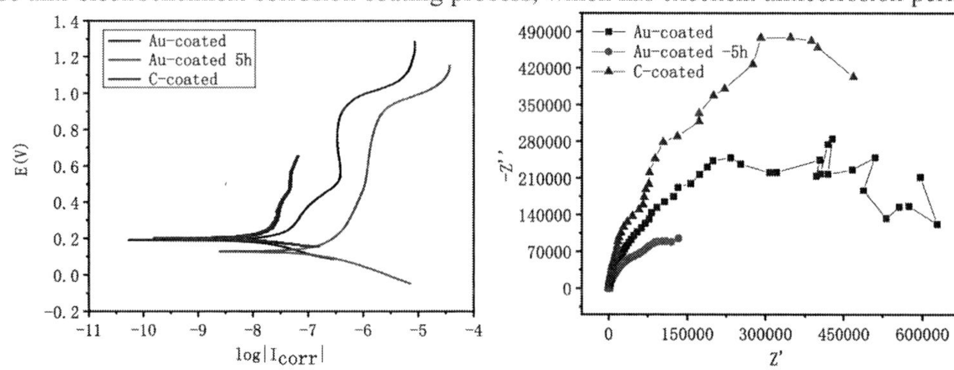

Figure 1 metal bipolar plate anticorrosion performance

Test the anticorrosion performance of gold-plated and carbon-plated bipolar plates was tested at pH=2 and voltage 0.8V (vs SHE) and pH=2 and voltage 1.0V (vs SHE). The test result is shown in Figure1 Tafel test result shows that the corrosion potential of the carbon plated plate is the highest and the corrosion current density is the lowest.After 5 hours of testing, the corrosion potential of the gold-plated plate decreased. It is speculated that pitting corrosion has occurred, leading to an increase in the corrosion current density and the corrosion rate.EIS test also shows that the corrosion resistance of the gold-plated plate decreased after 5 hours test.The three samples have no corrosion after 1 hour and 5 hours of constant 0.8Vvs SHE potential. In addition, it can be seen from Figure 2 that after 1 hour of constant 1.0V (vs SHE) potential polarization, the corrosion current of different coatings is in the order of gold plating > stainless steel > carbon plating.

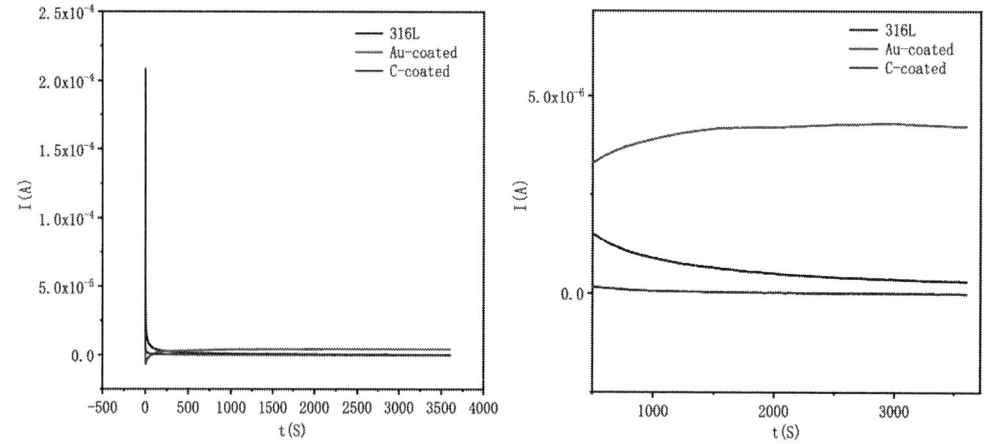

Figure 2 compare and analysis of corrosion current of different coatings

After test finished, we can see that, The gloss of plated plate decreases, and the gloss of plated carbon plate increases.With the help of metallographic microscope, we have observed obvious signs of pitting on the gold-plated plate, and the surface morphology of the carbon-plated plate has changed.The test result is shown in Figure 3.

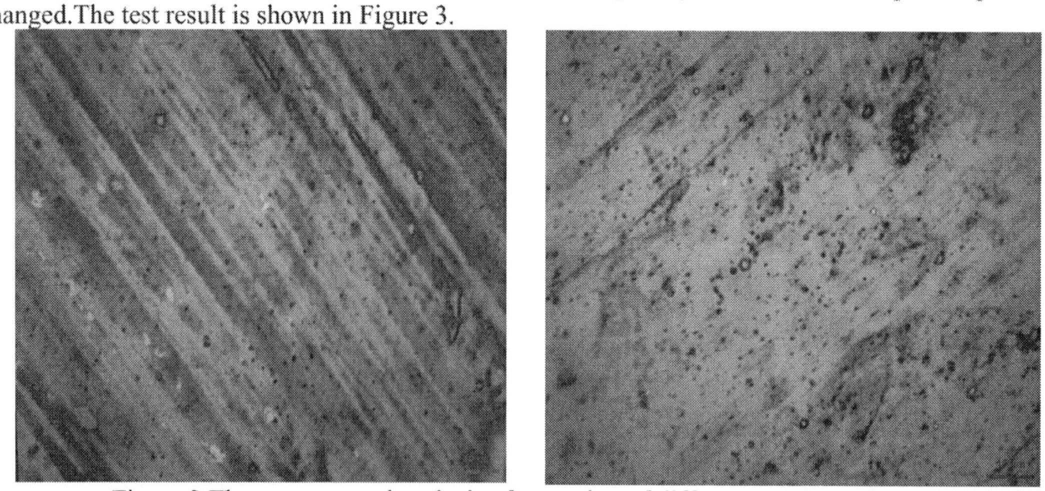

Figure 3 The compare and analysis of corrosion of different coatings after test

3.2. Operation characteristics of fuel cell stack in plateau environment

The two-phase flow in the membrane electrode will be affected to some extent when the air becomes thinner in the plateau environment， it is necessary to improve its pressure to ensure the performance of cell.To prove that, this topic adopts the method of installing a fan inside the hydrogen cell stack, which can improve the speed of the fan and the air pressure at high altitude.However, the battery needs to consume additional auxiliary power. In order to achieve high power generation of the cell, the intake parameters of the cell need to be optimized.At the same time, with the increase of altitude, oxygen content in the air also decreases significantly (shown in Figure 4), which is must to affect the supply of cathode reaction gas for hydrogen-air fuel cell.

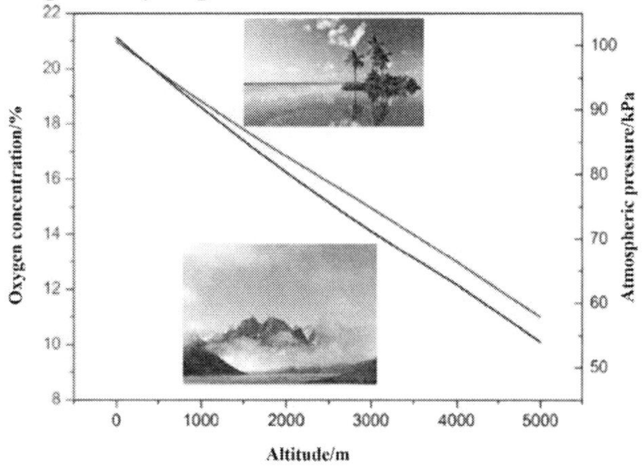

Figure 4 Atmospheric pressure and oxygen content in plateau environment

In order to ensure stable power output and high utilization rate of hydrogen, it is necessary to increase air volume to improve air excess coefficient. With the increase of air volume and air excess coefficient, heat and mass transfer processes and water and gas distribution in the cell at high altitude will be significantly different from normal conditions. Therefore, the influence of low atmospheric pressure, oxygen content and high air excess coefficient on heat and mass transfer and water and gas transfer in the battery should be analyzed deeply, and the change rule of electrochemical reaction mechanism of the membrane electrode module should be studied more.By studying the parameters of air pressure, air volume, wind speed and air excess coefficient, the optimal inlet parameter conditions are finally obtained.

The determination of optimal intake parameters adopts numerical simulation and optimization algorithm, which is used to effectively seek the optimal operating conditions of fuel cell performance from the global range.First, the fuel cell model is established, select the variables to be optimized and evaluate the fuel cell performance and treat the optimized variables to the initial value, import the calculated target function value into MATLAB by the genetic algorithm, inherit the good performance to the next generation, eliminate the poor performance, then produce new variable values, and then enter the COMSOL Multiphysics for continuous calculation, and finally get the best fuel cell parameters.

3.3 Study on high-temperature environmental operation characteristics of fuel cell reactor

In high-temperature environment, metal bipolar fuel cell is difficult to heat dissipation. About 50% of the chemical energy of the fuel cell is directly converted into thermal energy discharge. Fuel cell membrane electrode is sensitive to temperature, local temperature may lead to membrane electrode perforation, cause the cathode and anode poles gas string, and then cause fuel cell performance parasitic crab, seriously may cause hydrogen and oxygen should be too intense and explosion, bring safety problems, so must be the fuel cell operating temperature strictly controlled.

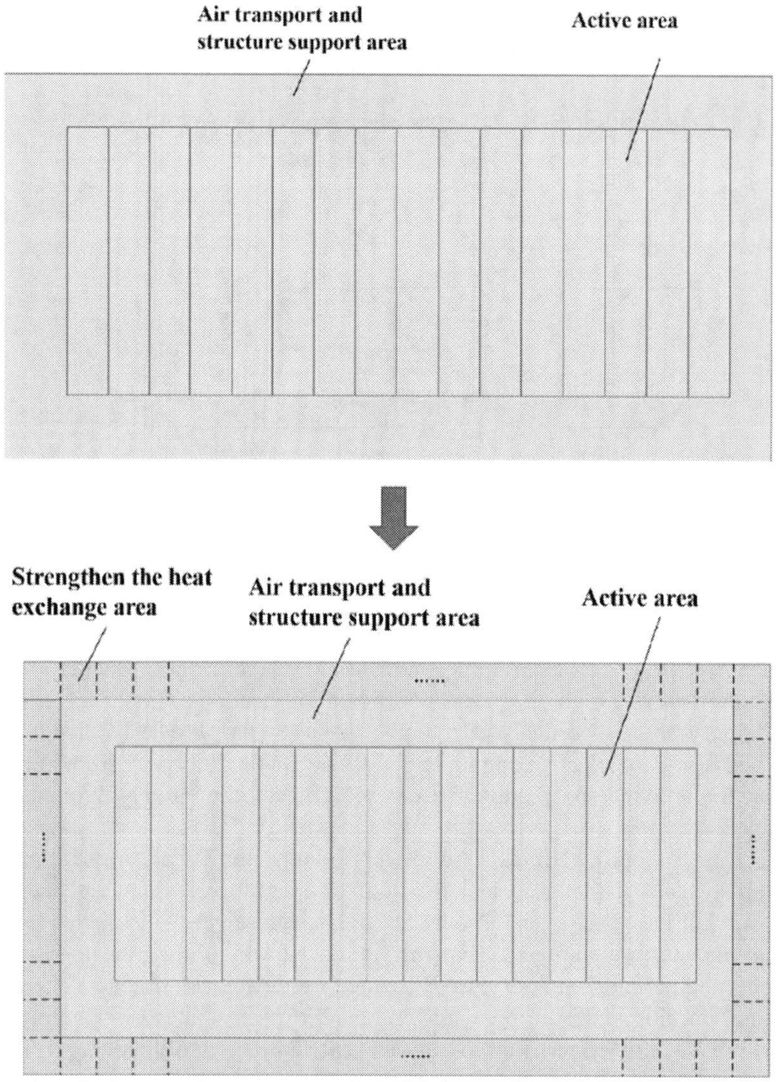

Figure 5 Increasing the heat exchange area under a high-temperature environment to strengthen the heat exchange pathway

In view of the problem of heat dissipation difficulties of metal bipolar fuel cell supply system in heat temperature environment, the following two solutions: on the one hand, prepare bipolar metal material with high thermal conductivity to enhance heat exchange, while improving the air side, increase the heat exchange area and increase the heat exchange area, as shown in Figure 5;On the other hand, similar to the plateau environment, the air volume is increased by the built-in fan of the hydrogen battery stack, thus achieving the purpose of strengthening the heat exchange.Enhance the heat exchange through the above scheme to realize the long-term and stable operation of the hydrogen fuel cell in a high heat environment.

Figure 6 shows the cathode cycle voltaman contrast before and after the fuel cell accumulates after 60h operation. We can see that both the cathode hydrogen adsorption peak and detachment peaks were decreased, and the cathode electrochemical activity area was reduced.

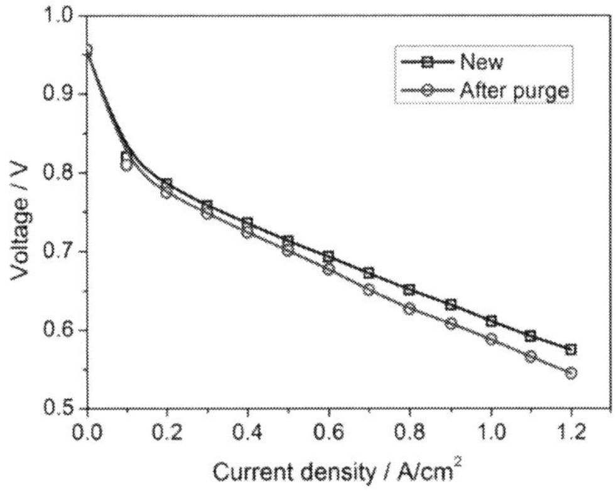

Figure 6 Voltaman contrast of the membrane electrode cathode cycle before and after 60h operation

Figure 7 shows the comparison of the polarization curve before and after 60h operation. After 60 hours of experimental accumulation, the performance of the battery first decreased, then gradually increased, and the current density was 2.3% when the current density was 0.6A/cm^2.

Battery performance decay was 1.2% at a current density of 5.2A/cm^2

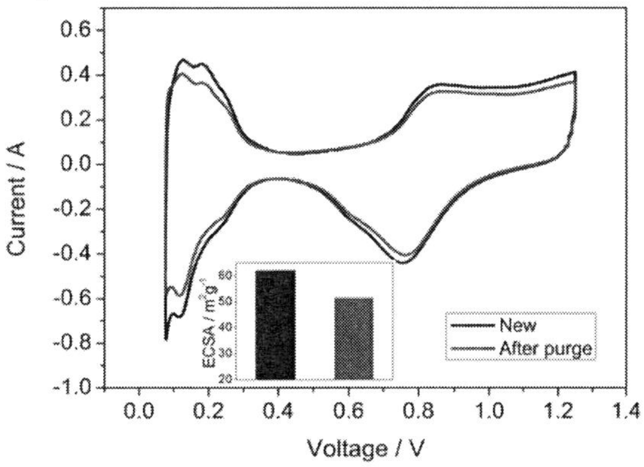

Figure 7 Comparison of the polarization curves before and after running 60h

3.4 Study on cold environment of fuel cell reactor

On cold environment, hydrogen fuel cells are facing the problem of storage and start difficulties, water in the process of battery cathode and external transmission, if the environmental temperature is less than 0°C, there may be freezing phenomenon, and water ice will produce 9% volume expansion, when the battery heat after the start will melt ice into water, volume will reduce, repeated phase change will have a great impact on the battery material structure, battery performance and life.Therefore, it is necessary to establish the reactor preheating and control technology, combined with the shutdown purging strategy, to realize the reactor low temperature storage and start.After shutdown, the residual water vapor inside the reactor to reduce the membrane electrode components for the next start. The schematic diagram of the shutdown blow scheme is shown in Figure 8.

Although removing the liquid water in the battery component structure during shutdown, reduces the free water content in MEA and avoiding the damage caused by ice to MEA, not the lower the water content inside the battery, the stronger the low temperature start performance. If excessive drainage may seriously dewater the proton exchange film and also reduce the battery performance. This topic determines the best purge parameters by monitoring the changes in the battery resistance online, as shown in Figure 9.It can be seen that the purge flow, purge time and purge times will have a great impact on the internal resistance of the fuel cell, and the change of internal resistance is the internal water content change and redistribution of the battery.

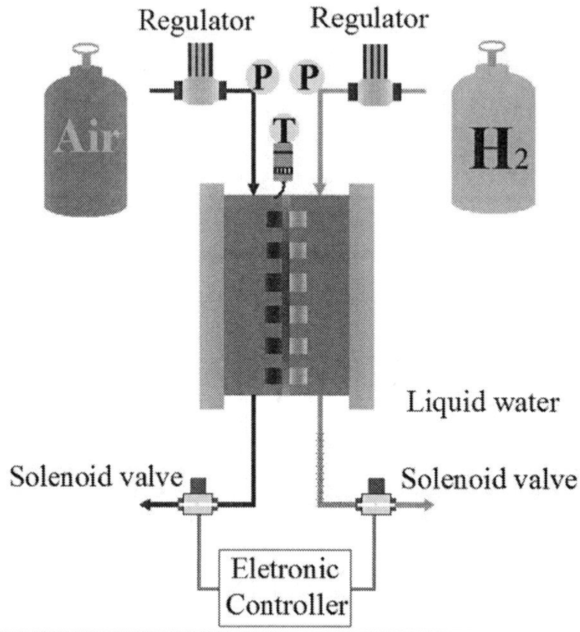

Figure 8. Schematic diagram of the shutdown pulse purge

Therefore, in the process of low temperature storage, the microstructure and performance of the key materials of dry air puraging and the reactor startup failure mechanism, and then analyze the impact of dry air purge flow and time parameters on the residual water vapor inside the battery, and finally obtain the optimal purging strategy.

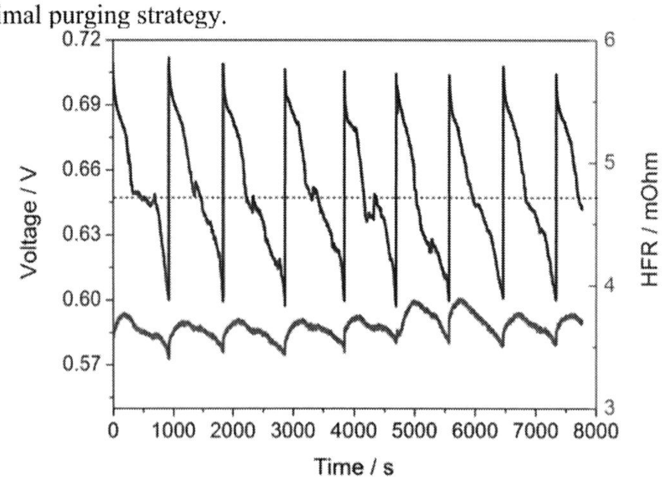

Figure 9. Pulse sweep battery performance and internal resistance changes

Due to the less heat generated by the electrochemical reaction in the initial stage of the fuel cell start, the water generated by the electrochemical reaction in the high cold environment is easy to freeze, block the flow channel and membrane electrodes, and prevent the electrochemical reaction, leading to the failure of the fuel cell cold start.Fuel cell in the cold environment, its core work is to avoid ice formation in the process of ice and gas channel blocked, so this project adopts a main and auxiliary two schemes to start fuel cell, the auxiliary scheme for the reforming hydrogen heat direct heating reactor, to ensure that the system can still safely and quickly start operation, not detailed here.

The main scheme adopts the rapid start of the metal bipolar hydrogen battery power system by the hydrogen catalytic reaction auxiliary heating. The basic principle is in the initial stage mainly by inputting a certain proportion of hydrogen and oxygen mixed gas in the anode, the volume ratio of hydrogen is usually about 95% (far higher than the explosion limit of hydrogen), the hydrogen electrochemical gas reaction under the catalyst, because the energy is not output, the electrochemical reaction is converted into heat, and then heating the reactor. The reaction gas input and control method during the cold start process are shown in Figure 10.

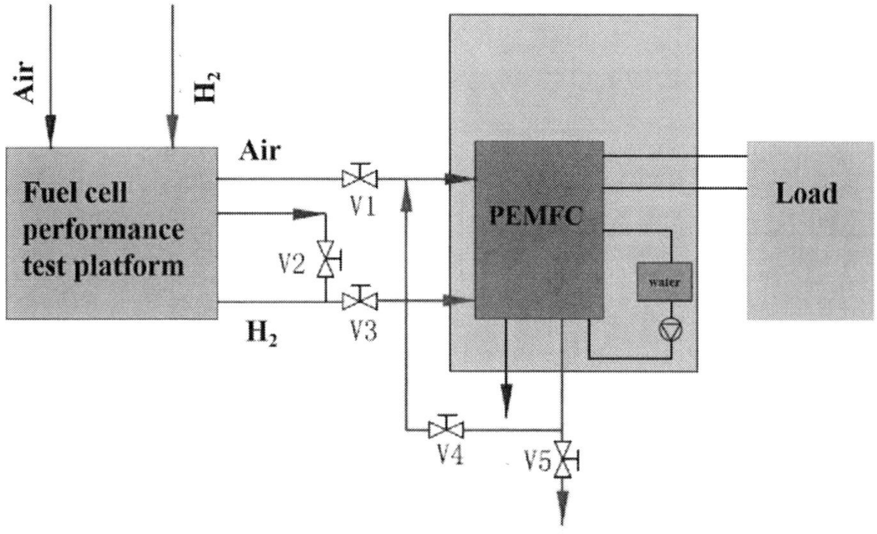

Figure. 10 A cold start scheme based on the hydrogen catalytic reaction-assisted heating

Specifically, the temperature inside the fuel cell before cold start is less than 0°C, when the system is in the battery stack self-heating-preheating stage, the valves 2,3,4 open, the anode directly into a certain proportion of hydrogen and oxygen mixed gas, and electrochemical reaction under the catalyst, the heat generated for heating the reactor; meanwhile, the anode exhaust gas through the pipeline and the valve, the anode hydrogen and oxygen mixed gas is not fully reacted in the cathode, on the other hand, the high temperature exhaust gas can heat the reactor again.When the reactor temperature is higher than the set value, the valves 1,3 and 5 are open, the valves 2 and 4 are closed, the mixed gas supply of the anode is stopped, and the fuel cell is started and operated normally.

At different temperatures, the cold start process of the metal bipolar hydrogen battery power system by hydrogen catalytic reaction assisted heating is shown in Figure 11 and 12, which show that the fuel cell can complete the battery start process of about 82s under-20°C and the battery temperature of about 71s can be increased to 0°C under-25°C environment.

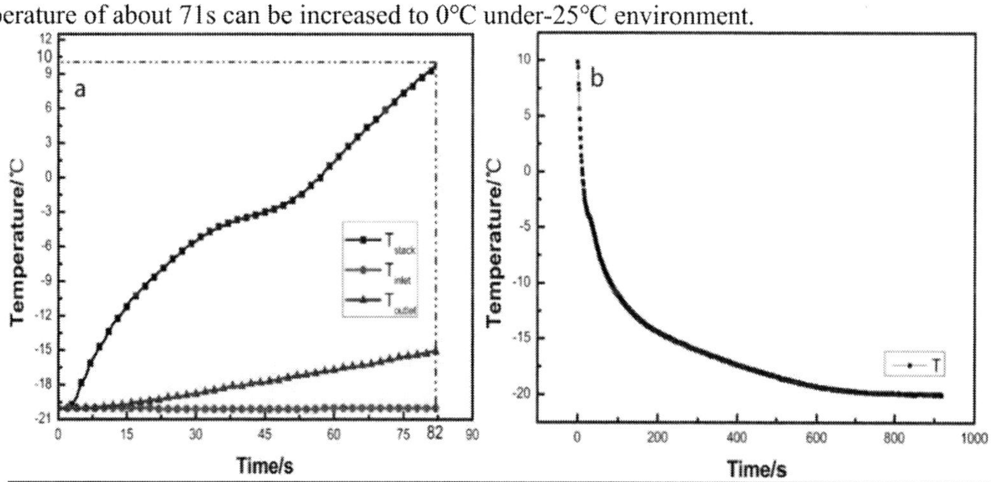

Figure . 11 Temperature change of the fuel cell during the 20℃ cold start process under the main scheme

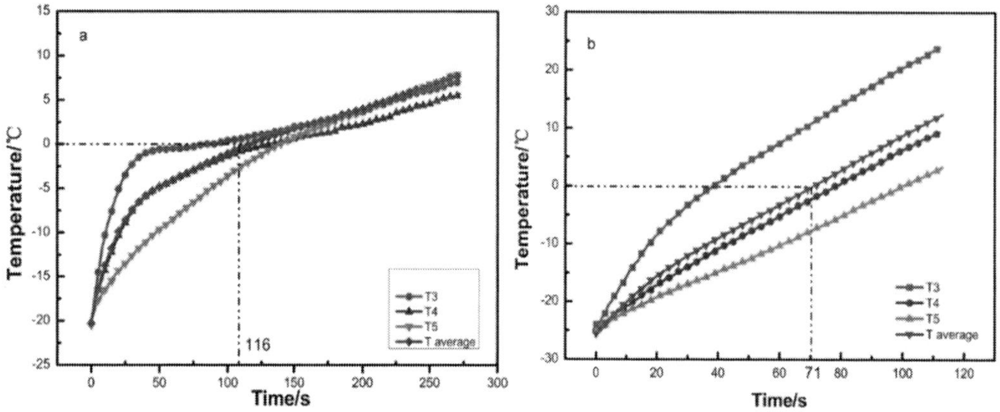

Figure 12 Cold start process at different temperatures under the main scheme

4.Conclusion

The research and development of hydrogen fuel cells, especially proton exchange membrane fuel cells (PEMFC) in China has made great progress In recent years,,and Fuel cell is the most key component in hydrogen power generation, and strong environmental adaptability of Fuel cell is very importent, basic on the need of project "emergency power under the special environment systems research and application demonstration",we studied on Environmental Adaptability of Metal Plate Fuel Cell,came to the Conclusion. uses Au@TiN as a metal bipolar plate of essential subexchange membrane fuel cell, after a series of tests, proved the excellent operating characteristics of the material under high humidity, high salt, high altitude, high temperature and high temperature, from which the following conclusions can be obtained:

(1).Studies in high wet and high salt environment showed that after constant 1.0 V (vsSHE) potential polarization, the size order of different coating corrosion currents is: gold plated> stainless steel> carbonization.

(2).In a high altitude environment, the optimal air intake parameters are obtained by numerical simulation and optimization of air pressure, air volume, wind speed, air excess coefficient and other parameters, and the fuel cell performance can achieve the optimal operating conditions.

(3).In a high-temperature environment, the problem of the metal bipolar fuel cell supply system can be solved by improving the flow structure of the air side to increase the heat exchange area and the air capacity.

(4).In the high-cold environment, the shutdown pulse sweep scheme is adopted. The best purge parameters were determined by monitoring the hydrogen battery supply system by using the method of hydrogen catalytic reaction assisted heating.

Above study found the way to solve the basic problem Environmental Adaptability of Metal Plate Fuel Cell, but we still continue research and improvement, and to do the reliability testing, through the application demonstration to form forming products.

Acknowledgments

This work was financially supported by the 2018 National Key Project(2018YFC081003).

References:

[1] Shen J, Tu Z, Chan S H. Performance enhancement in a proton exchange membrane fuel cell with a novel 3D flow field[J]. Applied Thermal Engineering, 2020,164.

[2] Chang H, Duan C, Xu X, et al. Technical performance analysis of a micro-combined cooling, heating and power system based on solar energy and high temperature PEMFC[J]. International Journal of Hydrogen Energy, 2019,44(38).

[3] Shen J, Tu Z, Chan S H. Enhancement of mass transfer in a proton exchange membrane fuel cell with blockage in the flow channel[J]. Applied Thermal Engineering, 2018,149.
[4] Byungchan H, Bum C H, Hyun S M, et al. Effect of Humidity and Flooding on the Performance of Proton Exchange Membrane Fuel Cell[J]. Korean Chemical Engineering Research, 2017,55(3).
[5] Cappadonia M, Erning J W, Niaki S M S, et al. Conductance of Nafion 117 membranes as a function of temperature and water content[J]. Solid State Ionics, 1995,77.
[6] Berning T, Djilali N. Three-dimensional computational analysis of transport phenomena in a PEM fuel cell—a parametric study[J]. Journal of Power Sources, 2003,124(2).
[7] Mao L, Wang C Y, Tabuchi Y. A Multiphase Model for Cold Start of Polymer Electrolyte Fuel Cells[J]. Journal of The Electrochemical Society, 2007,154(3).
[8] Mao L, Wang C Y. Analysis of Cold Start in Polymer Electrolyte Fuel Cells[J]. Journal of The Electrochemical Society, 2006,154(2).
[9] Meng H. Numerical studies of cold-start phenomenon in PEM fuel cells[J]. Electrochimica Acta, 2008,53(22).

Application of molecular dynamics simulation in self-assembly

Mingjia Fang[1,a]

[1]Department of Materials science and Engineering, Jiangsu University of Science and Technology, Zhenjiang, Jiangsu, China, 212121

[a]mjfang@live.com.

Abstract: During the past few years, numerous studies have been done in self-assembly. Among most of these studies, Molecular Dynamic Simulation is widely used to construct the experiment model. This work firstly introduced three practical applications of MD simulation in self-assembly. Then, two main kinds of simulation are discussed including all-atom simulation and coarse-grained simulation, together with the way of thoughts before the simulation start. It is found that researchers always start with the whole analysis of the substances that need to be studied. It helps to confirm the appropriate model that can apply in the simulation naturally. Besides, depended on the principles that need to be studied, the way of establishing the simulation system varies, ranging from separation experiment in both types of simulation to the change of essential parameters. Furthermore, the adoption of L-J potential in MD simulation proves to be a wise option on account of its convenient and simple model. It is remarkable that, considering some small details like the differences between implicit and explicit solution, classical Martini force field is replaced by Dry Martini force field.

1. Introduction

Molecular dynamics simulation is an excellent tool for prognosticating the way that atoms in different molecular systems move with the time coursing depended on a generic model of the interactions between atoms governed by physics. The basic concept of MD simulation is evident. It is feasible that, with the positions of all the atoms given, the forces exerted by all other atoms on each atom can be calculated. Newton's laws of motion can be used here to predict the position of every atom in space as a function of time. The forces are calculated repeatedly over a period of time, so that the position and velocity of each atom can be updated. Generally, Trajectory obtained from the simulation clearly shows the distribution of atoms at each stage of the simulation.

Self-assembly, also called ordering, is a well-known phenomenon where individual particles in a system spontaneously organize to form ordered structures. [1] Weak and non-covalent interactions often contribute to such process. Self-assembly tends to be controlled by the functionalization of the particle surface or particles' physical characteristics, or via the exertion of magnetic fields. For example, an external shear force can cause unidirectional ordering of lamellar bilayers formed by a polymeric surfactant, resulting in the formation of anisotropic hydrogels, which can be used as artificial muscles. [2] Besides, the behaviors of self-assembly usually happen in the cases like protein folding and the formation of cytomembrane. Thus, during the biophysical processes, one can capture the self-assembled structures.

It is no uncommon that researchers usually observe the colloidal self-assembly via the MD simulations. However, although numerous studies have been done in this field, the understanding of the recapitulative train of thought when it comes to how to combine the MD simulation with the self-assembly, still needs to be consummated. In this work, three applications of MD simulation in colloidal self-assembly are discussed including their individual research methods and some necessary preconditions and theories.

2. Application

2.1 MD simulation study of polyoxyethylated alcohols self-assembly in emulsion systems [3]

According to the paper, although numerous researches have been conducted to illustrate the effect on properties of the emulsion, the process that AEO self-assembled in water or oil systems seems vague to perceive. Besides, the understanding of the internal interactions between the branching structure made up of hydrophilic groups and self-assembly remains incomplete. Considering that MD simulation is an efficient method in the research of the structure of interface, especially oil/water. CGMD simulation, which are derived from the MARTINI force field, is adopted in view of its capability that it can carry large-scale calculation. However, CGMD turns out to be lacking accuracy. Therefore AAMD, also called all-atom molecular dynamic simulation, is used which is of great precision. With the aid of the multiscale combination of two kinds of MD simulation, it is possible that two simulation approaches complement each other, so that both can give full play to individual advantages.

GROMACS is chosen as the platform to do all the simulations. $C_{12}E_X$ (x=4,6,8,10,15, and 20), which is CmEn conventionally, is the oil-soluble part of Alcohol ethoxylate (AEO). It serves as the nonionic surfactant in the AAMD simulation with the oil component octane which is both described via the AA model. Moreover, the simple point charge extended model is applied in water molecules. As for the CGMD simulation, MARTINI force field plays the role of explaining the single particle combined by an atomic cluster. Besides, the simulation naturally integrates other components into the CG models, depending on the principles of single chemical groups like volumes and types.

Furthermore, in order to do the research, simulation system is a must. Thus, two kinds of simulation systems are designed to analyze two cases, including the comparison of AA model and CG model in the environment of biphasic interfaces without co-surfactants and components that distribute stochastically in advance under the CG model. Both cases are demonstrated by the diagram in Fig.1. It is worth mentioning that, considering the process of the simulation in mixed system is 10 times longer compared with the biphasic interfaces, the data are collected in the second half of the simulation for the sake of the equilibrium state of the whole system. Even more to the point, by applying the VMD, also called Visual Molecular Dynamics software, the visualization of all simulations can be achieved.

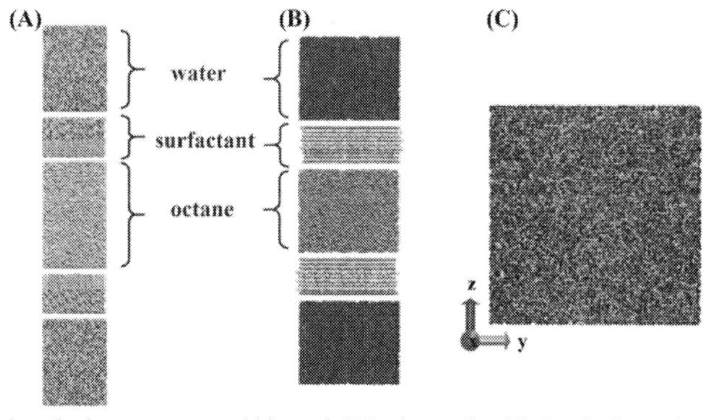

Fig.1. Diagrams of three simulation systems:(A) and (B) show the biphasic interfaces under AA model and CG model respectively, (C) shows the stochastically distributed system under the CG model

2.2 Role of repulsive forces on self-assembly behavior of amyloid β-peptide (1-40): MD simulation approach [4]

Aβ40 protein has the tendency to aggregate, which is the focus of a certain number of scientific studies in view of its significance in demonstrating the reason behind Alzheimer's disease. However, due to the restrictions like temperature the heterogeneity occurs in the process of aggregation and the failure of applying the NMR and X-rays in the experiment that analyze the monomer structure and oligomers of β amyloid, the researches in this area seems hard to be productive. Therefore, the necessary data of Aβ is obtained indirectly from other methods, like scanning TEM and FTIR method, which is used to capture the information about the mass of molecules and the secondary structures, respectively.

Among those methods, it is the transformation in the expression of potential that matters a lot. Numerous studies have been carried out in order to control the hydrophobic forces, which is generally the contributing factor during the procedure of Aβ protein folding. It is remarkable that the Lennard-Jones potential is widely used in various research models, due to its universality. Even more to the point, the participation of L-J potential naturally introduces the MD simulations which construct the whole experiment structure. It is not uncommon that L-J potential has been adopted in many molecular dynamic simulations due to its simple analytic form. However, collisions in repulsion are arisen from the adoption of L-J potential part which is unavoidable. Studies have been done to develop new methods, so that the collisions can be weakened or eliminated. For example, in a new potential conducted by a research, the repulsion part is separated from the whole to demonstrate the flexibility of protein. Besides, Gaussian model works as the tool for expressing the attraction part. With the help of the combination of two aspects, the researches are simplified.

As for Aβ40, the transition is made on the method from the idea of separation to the approach of doing adjustment of the repulsion term exponent, which is an essential parameter in Lennard-Jones potential. Reasons are as follows. Parameters like bond length and angles is impossible to migrate away from other restrictions. That explains the narrow applications in the method of separating the repulsion part and attraction part completely, including the practice in simple fluids. Therefore, in the complex system, such method turns out to be less effective. Hanley demonstrated some differences between the complex system and the simple system in fluids via such method. [5] The specific expression of L-J potential is as follows:

$$u_{LJ}(r) = 4\,\varepsilon\left[\left(\frac{\sigma}{r}\right)^{m} - \left(\frac{\sigma}{r}\right)^{6}\right]$$

The m in the expression ranges from 8 to 12. It represents the changes of the repulsion term. Therefore, OPLA-AA/L and GROMOS96 turn out to be the suitable and basic force fields through the simulation. Particularly, the adjustment of the repulsion term usually cast an influence on different force fields. Thus, the comparison between the effect on two different force fields is also taken into consideration.

2.3 Coarse-grained molecular dynamics simulation of the interface behavior and self-assembly of CTAB cationic surfactants [6]

Surfactant is a substance that significantly reduces the surface tension of the target solution. It has a fixed hydrophilic and oleophilic group, which can be arranged directionally on the surface of the solution. The molecular structure of the surfactant is amphiphilic. One end is a hydrophilic group, the other end is a hydrophobic group. Hydrophilic groups are often polar groups, such as carboxylic acid. Particularly, under certain thermodynamic conditions, the surfactant in water solution has to tendency to self-assemble. Such processes produce various structures. However, many studies although have been done on this theme, it remains incomplete in the understanding of surfactant self-assembly. Therefore, researches seem necessary to refine the theory. Since the self-assembly of surfactant takes place at the molecular level, MD simulation turns out to be a suitable option to carry out the research. Nevertheless, the commonly used AA model proves to be unqualified for such a mission. Mainly because the appropriate time and length scales which are used to study the principles of surfactants and micelles, are widely divergent with the AA model. Thus, sights are turned on the CG models. It is also called the molecular dynamic simulations of coarse-grained models. By reducing the degrees of freedom, the simulation time can be studied much longer than using the classical atomic model (all-atomic model). Besides, with the help of the increasingly advanced computational power, CG models are widely used for molecular modeling of biomolecules of various granularity levels.

Here, CTAB (Hexadecyl trimethyl ammonium Bromide) is chosen as the surfactant for the

Journal of Physics: Conference Series **2108** (2021) 012085 doi:10.1088/1742-6596/2108/1/012085

experiment. CTAB is a cationic surfactant. It has good surfactivity and stability. It is usually used to produce gargles, disinfectants and deodorants. Like other surfactants, CTAB also has the characteristics that one of its ends has a hydrophilic group and the other has a hydrophobic group. The specific structure is demonstrated in Fig.2. To make the simulation more accurate, both AA model and CG model are applied in the research including the behaviors of CTAB in explicit aqueous solvents via two models and the CTAB in implicit aqueous solvents via CG model.

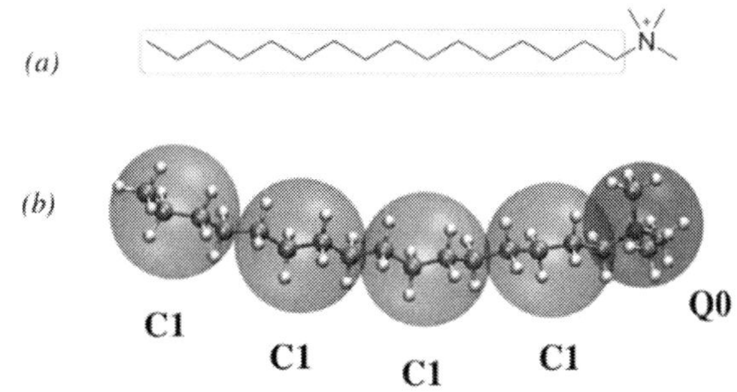

Fig.2. (A) shows the structure of CTA^+ surfactants. (B) explains the difference between the AA model and CG model on CTAB. CG model is characterized by the brown and blue spheres, while all the small single atoms in the spheres make up the AA model. C_1 is the hydrophobic headgroup and the Q_0 is the hydrophilic tail.

Worth reminding that MARTINI force field is also applied in the CG model of CTAB in implicit aqueous solvents. However, instead of the classical MARTINI force field, the Dry Martini with a CG force field is adopted, which is illustrated by Clément Arnarez et al. [7] Specifically, considering the implicity of aqueous solvents, changes should be made on the interaction strength while the expression of force fields remains unchanged.

3. Conclusion

In this review, we investigated three practical applications of molecular dynamic simulation in self-assembly. We also summarized their train of thoughts about how to reach a natural integration between the MD simulation and the self-assembly. We described the all-atom MD simulation and its characteristics like more accuracy and suitable for small time scales model. Meanwhile, we also introduced the coarse-grained MD simulation which mainly focuses on the basic features of particles and neutralize the less important part in order to simplify the experiment procedures. Also, due to such advantages, CGMD simulations are more suitable for experiments with large time and length scales. It is found that, among the various CG models, the one that combined with the MARTINI force fields proves to be more efficient. Besides, we introduced the Lennard-Jones potential. Because of its convenience, it is often used to describe the interaction of overlapping wave functions in gas properties, scattering model and molecular models. Description of inert gas atoms made by L-J potential is particularly accurate and the interactions between neutral atoms or molecules are also a good approximation. Therefore, it is always applied in various molecular dynamic simulations.

Two main analytical approaches are described. The first is the idea of separation. Since the substances to be studied are on the molecular level so that we can use the MD simulation. Then, because the AA model and CG model focus on different essentials, both models are carried out at the same time, respectively. Finally, a comparison and summary are made between the results of two ways of simulations. Another way of thought is employed when L-J potential is adopted in MD simulation. In case of the model combined with the L-J potential, collisions caused by repulsive forces are inevitable. In order to counteract or mitigate the collisions, we have to change the repulsion term exponent. Thus, the results of simulation tend to be more precise.

Acknowledgement
We acknowledge Erik Luijten, Ting Wang, Chaya Cheng and Mandy Han. This work is supported by World Class Education Program.

Reference

[1]. Boles, M.A.; Engel, M.; Talapin, D.V. Self-assembly of colloidal nanocrystals: From intricate structures to functional materials. Chem. Rev. 2016, 116, 11220–11289.
[2]. Haque, M.A.; Kamita, G.; Kurokawa, T.; Tsujii, K.; Gong, J.P. Unidirectional alignment of lamellar bilayer in hydrogel: One-dimensional swelling, anisotropic modulus, and stress/strain tunable structural color. Adv. Mater 2010, 22, 5110–5114.
[3]. Jule Ma, Xianyu Song, Baoliang Peng, Teng Zhao, Jianhui Luo,Ruifang Shi, Shuanglian Zhao, Honglai Liu; Multiscale molecular dynamics simulation study of polyoxyethylated alcohols self-assembly in emulsion systems.
[4]. J. Ma, X. Song, B. Peng et al., Multiscale molecular dynamics simulation study of polyoxyethylated alcohols self-assembly in emulsion systems, Chemical Engineering Science.
[5]. H.J. Hanley, M. Klein, On the Utility of the m − 6 − 8 Potential Function, US National Bureau of Standards, 1972.
[6]. Sı́lvia Illa-Tuset, David C. Malaspina, Jordi Faraudo; Coarse-grained molecular dynamics simulation of the interface behaviour and self-assembly of CTAB cationic surfactants. Phys. Chem. Chem. Phys.2018, 20, 26422—26430.
[7]. Clément Arnarez,Jaakko J. Uusitalo,Marcelo F. Masman,Helgi I. Ingólfsson,Djurre H. de Jong,Manuel N. Melo,Xavier Periole,Alex H. de Vries,Siewert J. Marrink;_Dry Martini, a Coarse-Grained Force Field for Lipid Membrane Simulations with Implicit Solvent. J. Chem. Theory Comput. 2015, 11, 1, 260–275

Research on the Control Mode of Homogeneous Charge Compression Ignition Combustion Working Process and Its Technical Prospect

Sirui Chen[1a*†], Yichen Deng[2b*†], Zhuojun Ma[3c*†] and Yujing Zhang[4d*†]

[1]School of Engineering and Computer Science, Oakland University, Rochester, Michigan, MI 48309, America

[2]Nansha College Preparatory Academy Guangzhou, Guangdong, 511458, China

[3]College of Engineering and Mathematical Science, University of Vermont, Burlington, Vermont, 05405, America

[4]Jilin University, Changchun, Jilin, 130022, China

[a*]siruichen@oakland.edu, [b*]2220023@ncpachina.org, [c*]zma7@uvm.edu, [d*]zhangyj1717@mails.jlu.edu.cn

[†]These authors contributed equally.

Abstract. The homogeneous charge compression ignition (HCCI) engine is considered an advanced technique, a form of internal combustion in which well-mixed fuel and oxidizer (typically air) are compressed to the point of auto-ignition. HCCI engines have higher thermal efficiency and lower emissions than Spark Ignition (SI) and Compression Ignition (CI) engines. The emissions of NOx can be neglected compared to the CI engine. In addition, a wide variety of fuels, combinations of fuels and alternative fuels can be used in this type of internal combustion engine. Moreover, when investigating the heat release rate of a HCCI engine for both single- and two-stage ignition fuels, the results show that for both fuel types, the cycle changes in the ignition and combustion phases increase with the delay of the combustion phase. Also, the cycle change of iso-octane (the single-stage ignition fuel) is higher than that of PRF80 (the two-stage ignition fuel). This paper will first introduce the control mode of the HCCI engine and then review its current status from the perspective of combustion, emissions, and consumption. After presenting the current status, the authors present suggestions about the prospect of further development with respect to the timing of ignition, the expansion of the engine operating range, and the choice of fuel mixture in this new mode of technology.

1. Introduction

Long since the first Industrial Revolution, vehicles driven by heat engines have gradually become indispensable to people [1]. Due to highly developed traffic and severely polluted environment, the pursuit of higher efficiency and lower emissions for engines has only become more urgent, regarding compression-ignition (CI) engines and spark-ignition (SI) engines as two popular options. On the one hand, CI engines vents polluting by-products due to incomplete combustion, in exchange for the highest thermal efficiency among all internal combustions [2]. Operating with premixed fuels such as gasoline and autogas (LPG), on the other hand, SI engines diminish potential burning residue but have a limited compression ratio and, therefore, lower efficiency. In the hope of an applicable alternative of both SI and CI, Homogenous Charge Compression Ignition (HCCI) later came into being. Combining features from SI and CI, HCCI can produce satisfactory efficiency and limit its emission at a low level [3]. It has not gone through a long enough period of time. However, HCCI is still limited by several technical difficulties such as cold start, autoignition timing, operation range, etc.

Homogeneous charge compression ignition (HCCI, homogeneous charge compression ignition) is the combustion method adopted by the next generation of piston internal combustion engines [4]. The thermal efficiency of this type of internal combustion engine is comparable to that of a compression-ignition direct-injection diesel engine, and the NO x and PM emissions are very low [5]. The HCCI internal combustion engine uses diluted pre-mixed gas [6]. When the mixed gas is compressed and ignited by the piston, the mixed gas in the cylinder is combusted in a volumetric manner. There is no flame propagation process, allowing the combustion to occur at a lower temperature. The HCCI internal combustion engine is innovatively designed based on homogeneous charge spark ignition gasoline engine and stratified charge compression ignition diesel engine, combining the advantages of the two. After the homogeneous charge is compressed, its pressure and temperature increase [7]. When it reaches the spontaneous ignition point, it ignites at several points almost at the same time. The combustion process is rapid, close to the ideal Otto Cycle (Otto Cycle), with high thermal efficiency and no throttling. Loss, full combustion, the low maximum temperature is conducive to control emissions [8].

This article analyzes the control mode of the HCCI internal combustion engine working process and summarizes its current status from the aspects of combustion, emission and fuel consumption. Finally, the future development of the HCCI internal combustion engine has been prospected.

2. Control mode

The HCCI engine's control mode can be highly efficient because it emits a limited amount of carbon dioxide, NOx, and particulate matter [9]. Thus, the usage of the HCCI engine is critical in cutting down on pollution. The limited amount of greenhouse gases emitted by the HCCI engine implies that the engine will potentially lower the rate of climate change. The reduction of the rate of climate change suggests that the engine will uphold the requirements of society concerning the mechanisms that can be adapted to maintain the quality of life. The greenhouse gases display to cause global warming that causes climate change. Climate change, in turn, triggers hazards that lead to the loss of biodiversity. Therefore, using HCCI engines will enhance the protection of biodiversity through the limitation of greenhouse gases emitted.

The HCCI control mode engine faces operations difficulties, especially when running the engine at high loads. Running the HCCI engine at high loads triggers the cold start aspect that makes the engine running idle. Besides, the attempts made to facilitate the control of the engine in the entire period of its operation limit the efficacy of its practical application. The reduction of the efficacy associated with applying the HCCI engine induces the limitation of usage quality. The solution to the problem of cold start in the HCCI engine can be realized by integrating the hybrid model. In the hybrid model, the engine operates at cruising, medium, and low loads, and while operating at high loads, idle, and cold start, the engine switches to a spark ignition mechanism [10]. Using spark ignition eliminates the need for advancement of monitoring of the engine throughout the cause of its operation.

In the process of advancing the operation of the HCCI engine in the hybrid model, effective transitions in its operations SI and HCCI and SI modes are required [11]. The shifts are mandatory due to the changes in the speed of the engine and the load. The effectiveness correlated with the changes displayed to promote the engine's operations' quality and thus uphold its performance. The changes also require the HCCI engine to be fitted with appropriate combustion parameters in excellent ranges. The proper integration of the combustion parameters manifests the transition strategy and thus integrates the fueling rate's efficiency. The switch in the control approaches of the HCCI engine is mandatory because it depends on pressure, composition, and charge temperature. Thus, using the direct control approach like the SI combustion is not applicable for the HCCI engine.

Moreover, to uphold the efficacy of the performance of the HCCI engine, adequate capacities of transition between the SI to HCCI to SI are required [12]. The integration of exhaust gas circulation

(EGC) portrays a smooth and faster transition between the SI – HCCI – SI. Enhancing the smooth and more rapid transition provides the engine with relevant aspects of operation, thus enhancing HCCI combustion quality. The increased efficacy of combustion implies that the objectives of operation of the HCCI engine will be attained. Additionally, the effective control of the HCCI engine portrays to limit the greenhouse emission and thus curbing the plight of global warming.

3. Current status

3.1.Combustion
HCCI uses a fuel-air mixture in a cylinder through the compression process, reaching the spontaneous combustion temperature to ignite everything. There is no spark plug and not allowed to inject fuel to control ignition timing. The combustion simply bases on the temperature inside the cylinder. The fuel injection and combustion conditions of HCCI will be described in the following.

HCCI usually uses port fuel injection (PFI) or early direct injection (DI) to make a relatively uniform fuel-air mixture [1]. Partial fuel stratification(PFS), an example of using a combination of PFI and DI, has been shown to prevent excessive ringing from allowing higher load from HCCI engines. This allows most of the fuels to mix well, with local regions of higher equivalence ratio exist.

When it comes to how to make combustion happen, although it depends on internal temperature, it can be implemented by high compression ratios, high intake temperatures, high intake pressure and highly reactive fuels through Samveg Saxena's article [13].

Here is an example of the heat release rate from an HCCI of different fuels. For different fuels of HCCI engine, the occurrence of single - or two-stage ignition is a way of characterizing the specific fuel. In general, low-octane fuels such as n-heptane, diesel, PRF80, and dimethyl ether (DME) are shown for two-stage ignition, while high-octane fuels such as ethanol or isooctane are shown for single-stage ignition [14]. Figure 1 illustrates the difference in the performance of single-stage and two-stage ignition fuels in HCCI engine combustion.

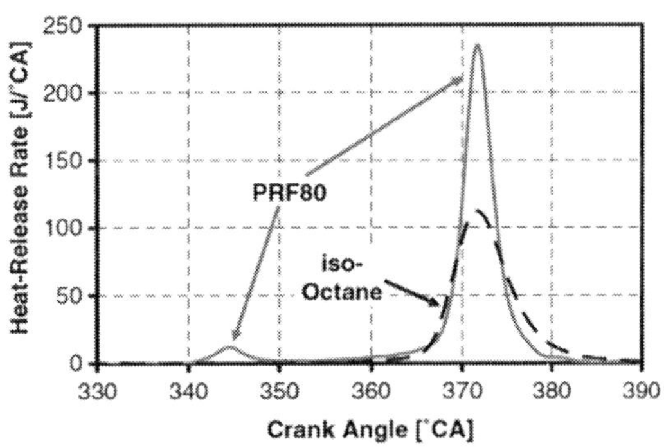

Figure 1. Heat release rate from an HCCI engine using two different fuels, PRF80 which exhibits two-stage ignition, and iso-octane which exhibits single-stage ignition [15]

3.2.Emissions and fuel consumption
HCCI typically has low nitrogen oxides, soot, and particulate matter emissions and higher emissions of unburned hydrocarbons and carbon monoxide [16]. The two aspects, unburned hydrocarbons and carbon monoxide and nitrogen oxides, and soot, will be described in the following.

The in-cylinder temperature is a key factor in determining unburned HC and CO [17]. Lower temperatures usually result in higher emissions of unburned HCs and CO. However, there is a temperature range for HC emissions when CO is reduced. Figure 2 and Figure 3 show a series of equivalent ratios and combustion timing for unburned HC and CO emissions. From Figure 4., the trend in both numbers shows that HC and CO emissions generally increase with lower equivalence ratios and delayed combustion times because these factors lead to lower cylinder temperatures. The hot spot fire and auxiliary co-oxidation threshold temperature are important considerations for HC and CO emission characteristics.

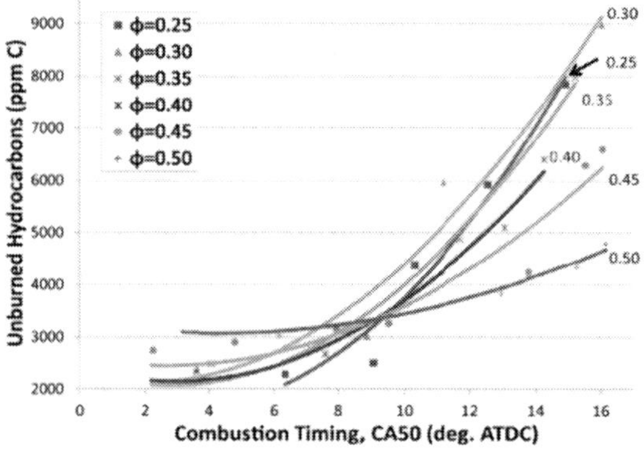

Figure 2. Experimental results of unburned hydrocarbon emissions for emissions different equivalence ratios and combustion timings at Pin¼ 1.8 bar combustion timings and 1800 RPM [18]

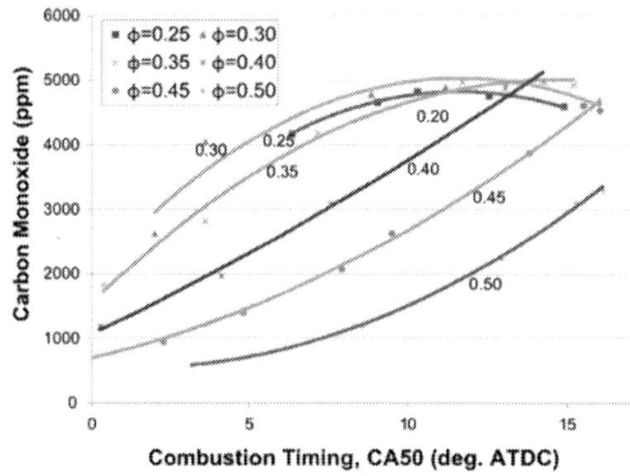

Figure 3. Experimental results of carbon monoxide for different equivalence ratios and at Pin¼ 1.8 bar and 1800 RPM [18]

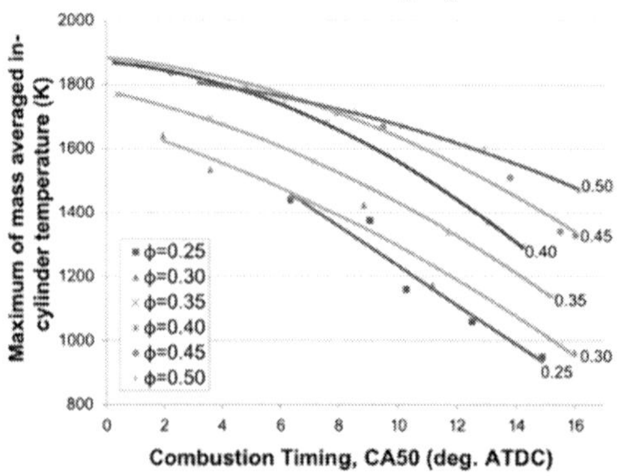

Figure 4. Experimental results of mass averaged peak in-cylinder temperature for different equivalence ratios and combustion timings at Pin¼ 1.8 bar and 1800 RPM[18]

Then NOx and soot are put together because they are both low in the HCCI engine. Nitrogen oxide emissions, NO and NO_2, are formed mainly through the Zeldovich chemical pathway by breaking strong triple bonds between nitrogen atoms in N_2 [19]. Given that the peak cylinder temperature of HCCI is still too low compared to diesel or spark-ignited engines to reach the activation temperature needed for the chemical reaction of NO and NO_2 to occur [20]. As a result, HCCI engines can achieve high power output with very low NO_x emissions.

Journal of Physics: Conference Series **2108** (2021) 012086 doi:10.1088/1742-6596/2108/1/012086

Soot is a major concern for diesel engines. When the fuel breakdown begins, it forms in the concentrated fuel spray area of the fuel spray, and there is insufficient oxygen. Given that the fuel is fully mixed before ignition of the HCCI engine, negligible levels of soot emissions occur.

In diesel engines, the high radiation of NO_x and soot requires the use of an expensive post-treatment system [21], such as particulate filters and oxidation catalysts [22], which are as expensive as the diesel engine itself. Thus, an obvious advantage of HCCI is that it can achieve the same level of efficiency with low CO_2 comparable to diesel engines [23] without the need for expensive NO_x and soot removal post-treatment systems.

4.Technology prospects

4.1.Timing for ignition

HCCI engines operate with autoignition, in which a clear control mechanism is absent. Several factors of this autoignition are fuel chemistry, burning duration, surrounding and intake temperature, compression ratio, and other parameters related to the engine. Earlier ignition can be caused by high intake temperature and pressure due to rapid chemical reactions. It is worth noting that either prematurity or delay on combustion phases may cause higher CO emissions, residues of incomplete combustion.

4.2.The expansion of the engine operating range

Certain reasons prevent the enlargement of the HCCI operation range, such as knocking, cold start, ignition timing, high pressure rate, and burning noise.

While the increasing frequency of knocking bounds the higher load limit, the lower limit is defined by the difficulty of autoignition at excessively low loads. Cold start, though it reduces emission, contradicts stable HCCI operation that prefers higher intake temperature. While ignition timing is arduous to control, the absence of preheating process may even forbid the firing due to unexpected heat loss. Furthermore, since autoignition inevitably produces high internal pressure, rapid heat release and pressure oscillation that follows may produce a high level of noise and damage engine components.

5.Conclusion

Review of present performance of HCCI engines remains crucial to anticipate and accomplish their future progress. That is, only after existing issues are settled or bypassed can engines be thoroughly utilized. Major conclusions are as follows:

(1) Though the ignition timing of HCCI seems to remain a critical obstacle, various proposed solutions have been put into use, which focuses on the timing of valves, compression ratio, injection timing, or fuel mixture. In the future, researchers may want to test the combinations of more factors in certain values, seeking better control of timing.

(2) Difficulty in cold start might be solved by adding more features from traditional engines. Existing methods are starting the engine in conventional modes, retrofitting a glow plug, or utilizing different fuels. At best, HCCI may someday achieve a stable cold start all by itself, without the assistance of the methods above. To realize it, fuels that burn at a relatively lower temperature may be the breaking point, requiring more advanced research.

(3) Efforts regarding restraining noise concentrate on its production or propagation. Although noise itself could not break the engine, it implies acute transmissions of heat and pressure and may cause accumulated damage to human hearing. Assuming that huge internal pressure is unavoidable during combustion, noise propagation must be intensively suppressed, which may be done by exploiting sound insulation materials in engine manufacturing.

(4) That draft animal was replaced by engine and steam engine was gradually substituted with internal-combustion one [24] indicates even HCCI might not be the ultimate resolution of transportation. Considering HCCI to be a promising choice presently, though, the use of technologies in it from wider various fields can be valued in the future. Such mixture may identity components that can either be added or removed from HCCI, promoting its perfection and even evolution.

References

[1] Sato, S., Yamasaki, Y., Kawamura, H. (2006) Research on the Influence of Hydrogen and Carbon Monoxide on Methane HCCI Combustion. Jsme International Journal.ser.b Fluids & Thermal Engineering., 48: 725–734.

[2] Yao, L. (2007) Research on the Technical Control Mode of Network Information Communication. Information Studies:Theory & Application., 34: 346–352.

[3] Jinsheng, S., Yingrui, B., Rongchao, C. (2020) Research progress and prospect of plugging technologies for fractured formation with severe lost circulation.Petrol Explor Dev+., 48: 12.

[4] Yang, X., Zhu, G. (2013) SI and HCCI Combustion Mode Transition Control of an HCCI Capable SI Engine. Control Systems Technology, IEEE T Pattern Anal., 21: 1558–1569.

[5] Tsutsumi, Y., Iijima, A., Yoshida, K. (2009) HCCI combustion characteristics during operation on DME and methane fuels. Int J Auto Tech-Kor., 10: 645.

[6] Komai, Y., Nakashima, F., Hashimoto, H. (2012) Studies on the direct control of the start of HCCI combustion with rapid compression expansion machine. Energe., 8: 446–451..

[7] Omair, A. (2016) Current Status and Prospects of Supercomputing Used for Gas Turbine Engines Design. J Energ Eng., 2:82–92.

[8] Kobayashi, Y., Gengo, T., Ando, Y. (2009) Current status and future prospects for large-scale SOFC combined cycle system. J of H Tem Society., 35: 256–262.

[9] Polat, S., Solmaz, H., Yılmaz, E., Calam, A., Uyumaz, A., and Yücesu, H. S. (2020) Mapping of an HCCI engine using negative valve overlap strategy. Energ Source Part A., 42: 1140–1154.

[10] Noh, H.K., and No, S.Y. (2017) Effect of bioethanol on combustion and emissions in advanced CI engines: HCCI, PPC and GCI mode–A review. Appl Energ., 208: 782–802.

[11] Duan, X., Lai, M. C., Jansons, M., Guo, G., and Liu, J. (2021) A review of controlling strategies of the ignition timing and combustion phase in homogeneous charge compression ignition (HCCI) engine. Fuel., 285: 119142.

[12] Rather, M.A., and Wani, M.M. (2018) A numerical study on the effects of exhaust gas recirculation temperature on controlling combustion and emissions of a diesel engine running on HCCI combustion mode. International Journal of Automotive Science and Technology., 2: 17–27.

[13] Saxena, S. and Bedoya, I.D. (2013) Fundamental phenomena affecting low temperature combustion and HCCI engines, high load limits and strategies for extending these limits. Prog Energ Combust., 39: 457–488.

[14] Curran H, Gaffuri P, Pitz W, Westbrook C, et al. Autoignition chemistry of the hexane isomers: an experimental and kinetic modeling study. SAE 952406; 1995.

[15] Sjöberg M, Dec JE. Comparing late cycle autoignition stability for single- and two-stage ignition fuels in HCCI engines. Proceedings of the Combustion Institute 2007;31(2):2895e902.

[16] Flowers, D.L., Aceves, S.M., Martinez-Frias, J., Dibble, R.W. (2002) Prediction of carbon monoxide and hydrocarbon emissions in iso-octane HCCI engine combustion using multizone simulations. P Combust Inst., 29: 687e94.

[17] Saxena, S., Chen, J.Y., Dibble, R. (2011) Maximizing power output in an automotive scale multi-cylinder homogeneous charge compression ignition (HCCI) engine. SAE., 54:1232–1241.

[18] Dec, J.E., Ji, C., Dernotte, J. (2015) Energy Distribution Analysis in Boosted HCCI-like/LTGC Engines - Understanding the Trade-Offs to Maximize the Thermal Efficiency[J]. SAE International Journal of Engines., 20: 3321–3325.

[19] Goward, G.W. (1998) Progress in coatings for gas turbine airfoils. Surf Coat Tech., 108–109: 73–79.

[20] Bose, S., Demasi-Marcin, J. (1997) Thermal barrier coating experience in gas turbine engines at Pratt & Whitney. J Therm Spray Techn., 6: 99–104.

[21] Demasi-Marcin, J.T., Gupta, D.K. (1994) Protective coatings in the gas turbine engine. Surf Coat Tech., 68–69: 1–9.

[22] Bahr, D.W. (2010) GAS TURBINE COMBUSTION lternative Fuels and Emissions. J Eng Gas Turb Power., 132: 116501.

[23] Bang-Møller, C., Rokni, M., Elmegaard, B. (2011) Exergy analysis and optimization of a biomass gasification, solid oxide fuel cell and micro gas turbine hybrid system. Energy., 36: 4740–4752.
[24] Buchanan, R. A. (n.d.). History of technology. *Britannica School.* Retrieved October 11, 2021, from https://www.britannica.com/technology/history-of-technology

Influence of Floating Ring Seal Working Condition Parameters on the Dynamic Characteristics of Transient Gas Film

Lishan Xu[1,2a*], Weizheng Zhang[1], Junjie Lu[2 b*], Zhu Liu[1]

[1] School of Petrochemical Engineering, Lanzhou University of Technology, Lanzhou 730050, Gansu

[2]Zhejiang University Ningbo Institute of Technology, Ningbo 315100, Zhejiang

[a*]xulishan229@163.com, [b*]loveljj4566@163.com,

Abstract—The high requirements for sealing performance in high-speed rotating machinery has led to the design of floating seal with annular spiral groove that offer the advantages of low leakage and extended stability. However, efforts to model the dynamic performance of these floating seal have suffered from the great complexity of the flow field. The present work addresses this issue by establishing a transient Reynolds formulation of a floating seal with annular spiral groove in a rotating coordinate system based on the small perturbation method. In addition, the influence of radial eccentricity and film thickness on the solution divergence and calculation accuracy is calculated. The dynamic stiffness and dynamic damping matrixes are built. Then the variation rules of the dynamic stiffness and damping coefficient of the gas film with structure and working conditions are investigated in detail.The results show that the floating ring seal is more suitable for the service conditions of small film thickness, low pressure, high speed and large eccentricity. Accordingly, the results obtained lay a theoretical foundation for evaluating real-world applications of floating ring seal.

1. Introduction

High-speed rotating machinery require ultra-high sealing performance[1], which led to the development of floating seal with annular spiral groove for applications such as aircraft engines and rocket turbine pumps[2-4]. As illustrated in Fig. 1, stable separation of the floating ring is supported by a gas film between the rotating and floating rings, and the floating capacity of this film is based on the hydrodynamic pressure effect generated by the spiral groove and the gas film wedge effect resulting from eccentricity.

Several studies have evaluated the steady-state sealing performances of floating seal from the perspective of seal structure based on numerical solutions of gas flow equations . For example, the National Aeronautics and Space Agency (NASA) developed the GCYLT computer program to analyze the gas film pressure and thickness of floating seal with different structures and dimensions[5]. The China Gas Turbine Research Institute optimized the design of floating seal [6]. Other studies have employed numerical analyses conducted at the microscale to investigate the lubricity of a groove floating seal[3], and the effects of the spiral groove angle and rotation speed on seal leakage[4]. These studies have demonstrated that the micro-gap between the rotating and floating rings and the characteristics of spiral groove profoundly influence gas flow through the seal. Other researchers have pursued the steady-state analysis of floating seal using computational methods. For example, Chochua et al.[7] developed a computational model for a honeycomb-stator gas floating seal. Mel'nik et al.[8] proposed a modified computational algorithm, and thereby analyzed the effects of changes in gas flow and inlet loss in the spiral groove of floating seal. More recently, the dynamic characteristics of floating seal have been investigated numerically. For example, Tokunaga et al.[9] solved the three-dimensional (3D) Reynolds equation for a floating seal using the finite difference method (FDM), and employed the solution to evaluate the hydrodynamic pressure distributions and leakage flow rate of the seal, as well as evaluate the Lomakin effect. Peng et al.[10] also applied the FDM to solve a coupled model of the clearance flow and elastic structure of floating seal, and analyzed the obtained principle stiffness and cross damping elements of the dynamic stiffness and damping matrices of the system. Banakh and Barmina[11] investigated the influence of bearing capacity and contact force in a dry friction micro-gap on oscillation in a rotor floating ring system. While these past studies have contributed greatly to modeling the dynamic characteristics of floating seal, the shear flow of the discontinuous spiral groove structure caused by variations in pressure and rotation further increased the complexity of the flow field. Therefore, numerical models for analyzing the dynamic performance of floating seal require further development.

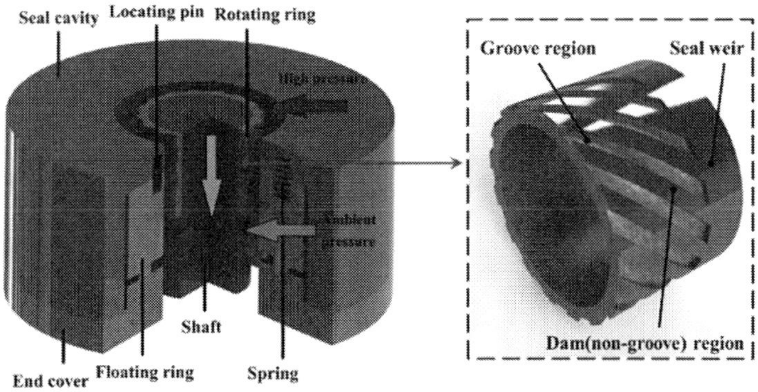

Fig.1 Illustration of an annular spiral groove floating seal

The present work addresses this issue by establishing a transient Reynolds formulation and film thickness equation of an annular spiral groove floating seal in a rotating coordinate system based on the small perturbation method, high-precision eight-point difference method and Newton-Raphson iterative method. The proposed formulation accounts for variations in the gas film thickness arising from eccentricity between the rotating and floating rings, and the effect of the Rayleigh step on the gas film shape. The dynamic stiffness and damping coefficients of the gas film in The floating ring seal are evaluated in detail based on the principles of eddy energy.

2. Calculation model

2.1. Dynamic Reynolds differential equation

The floating ring seal model employed in the present study is illustrated in Fig. 2. Here, the model regards the sealed gas film as several springs and dampers.

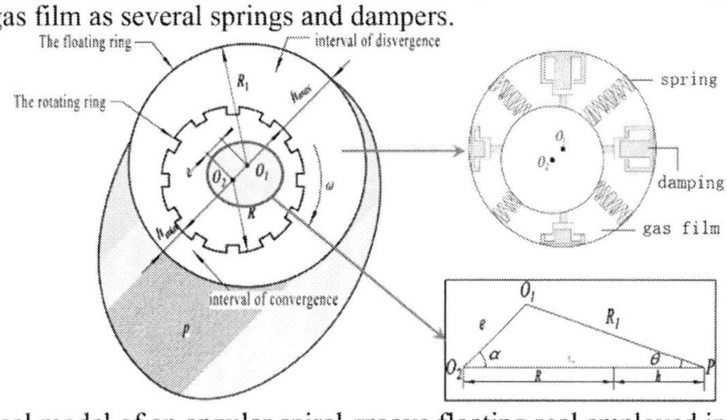

Fig.2 Physical model of an annular spiral groove floating seal employed in the analysis

The dimensionless dynamic Reynolds equation of the floating ring seal model is given as

$$\frac{\partial}{\partial\theta}\left(\overline{ph}^{3}\frac{\partial\overline{p}}{\partial\theta}\right)+\left(\frac{R}{L}\right)^{2}\frac{\partial}{\partial\overline{z}}\left(\overline{ph}^{3}\frac{\partial\overline{p}}{\partial\overline{z}}\right)=\Lambda_{x}\frac{\partial}{\partial\theta}\left(\overline{ph}\right)+2\gamma\Lambda_{x}\frac{\partial}{\partial\overline{t}}\left(\overline{ph}\right) \quad (1)$$

where is circumferential coordinate (rad), is the outer radius of the rotating ring (m), is the axial length of the floating ring (m), Λx is the compressibility coefficient, and γ is the perturbation ratio. The dimensionless parameters employed in Equation (1) above are defined as

$$\overline{z}=\frac{z}{L};\overline{h}=\frac{h}{C};\overline{p}=\frac{p}{p_{a}};\overline{\mu}=\frac{\mu}{\mu_{o}};\overline{t}=vt \quad (2)$$

where z is the axial coordinate (m), h is the gas film thickness (m), C is the average gas film thickness (m), which is the difference between the floating ring radius R_1 and the rotating ring radius R (i.e., $C = R_1 - R$), p is the gas film pressure (Pa), p_a is the atmospheric pressure (Pa), μ is the gas viscosity ($N\cdot s/m^2$), μ_0 is the initial gas viscosity ($N\cdot s/m^2$), v is the eddy velocity (rad/s), and t is the time (s).

In addition, we define $\Lambda_{x}=\frac{6\omega\mu_{o}R^{2}}{p_{a}C^{2}}$ and $\gamma=\frac{v}{\omega}$, where ω is the angular velocity of the shaft (rad/s).

Dimensionless pressure p and gas film thickness h under microscale perturbation can be given as follows.

$$\begin{cases}\overline{p}=\overline{p}_{0}+\overline{p}_{x}\Delta X+\overline{p}_{x}{}'\Delta\dot{X}+\overline{p}_{y}\Delta Y+\overline{p}_{y}{}'\Delta\dot{Y}\\ \overline{h}=\overline{h}_{0}+\Delta x\sin\varphi-\Delta y\cos\varphi\end{cases} \quad (3)$$

2.2. *Dynamic characteristic coefficients*

The model system of springs and dampers illustrated in Fig. 2 includes four stiffness coefficients k_{xx}, k_{xy}, k_{yx}, and k_{yy}, and four damping coefficients c_{xx}, c_{xy}, c_{yx}, and c_{yy}, which are defined as follows.

$$\begin{bmatrix}k_{xx} & k_{xy}\\ k_{yx} & k_{yy}\end{bmatrix}=\frac{RLp_{l}}{C}\int_{0}^{1}\int_{0}^{2\pi}\begin{bmatrix}\overline{p}_{x}\cos\theta & \overline{p}_{y}\cos\theta\\ \overline{p}_{x}\sin\theta & \overline{p}_{y}\sin\theta\end{bmatrix}d\theta d\overline{z} \quad (4)$$

$$\begin{bmatrix}c_{xx} & c_{xy}\\ c_{yx} & c_{yy}\end{bmatrix}=\frac{RLp_{l}}{Cv}\int_{0}^{1}\int_{0}^{2\pi}\begin{bmatrix}\overline{p}_{x}{}'\cos\theta & \overline{p}_{y}{}'\cos\theta\\ \overline{p}_{x}{}'\sin\theta & \overline{p}_{y}{}'\sin\theta\end{bmatrix}d\theta d\overline{z} \quad (5)$$

In addition, the pressure boundary condition is given as follows.

$$\begin{cases}z=0, p_{x}=p_{x}'=p_{y}=p_{y}'=p_{h}\\ z=L, p_{x}=p_{x}'=p_{y}=p_{y}'=p_{l}\end{cases} \quad (6)$$

Here, p_l is the medium pressure on the low pressure side, p_h is the medium pressure on the high pressure side. Finally, the cyclic circumferential boundary condition is given as follows.

$$\begin{cases}p_{x(\theta,z)}=p_{x(\theta+2\pi,z)} \quad p_{x(\theta,z)}'=p_{x(\theta+2\pi,z)}'\\ p_{y(\theta,z)}=p_{y(\theta+2\pi,z)} \quad p_{y(\theta,z)}'=p_{y(\theta+2\pi,z)}'\end{cases} \quad (7)$$

3. Results and Discussions

3.1. *Seal parameters*

The gas property parameters of the analyzed the floating ring seal are listed in Table 1.

Table 1 Gas property parameters of the seal

Symbol	Parameter	Value	Symbol	Parameter	Value
R	Outer diameter of the rotating ring	0.025 m	T_0	Environmental temperature	298 K
L	Axial length of the floating ring	0.052 m	p_1	Environmental pressure	101325 Pa
C	Average thickness of gas film	5×10^{-6} m	μ_0	Viscosity	1.8×10^{-5} Pa·s
ε	Eccentricity ratio	0.6	ρ_0	Density	1.1452 kg/m^3

3.2. Model validation
We first validate the accuracy of the proposed calculation method using the results obtained by Cheng et al.[12]The perturbation ratio γ is plotted with respect to the compressibility coefficient Λx in Fig. 3 for an eccentricity ratio of 0.2 and a width-to-diameter ratio of 0.5. The reference value of the perturbation ratio γ is close to 0.5 for all Λx, and the results obtained by the proposed method are consistent with those in reference Cheng et al. [12].

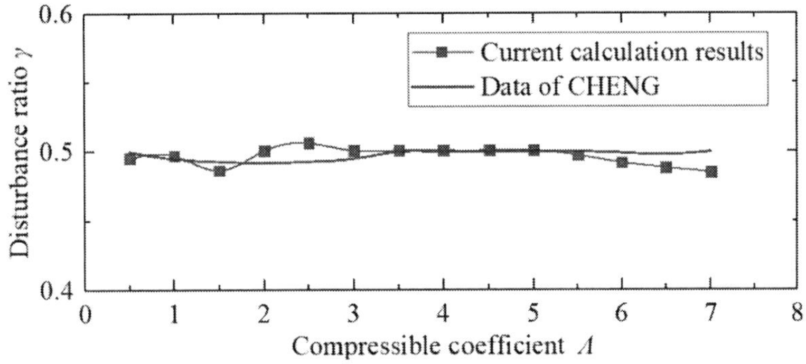

Fig. 3 The comparison between the calculational result and literature

3.3. Effect of rotation speed on the dynamic characteristic coefficients
From Fig. 4(a) the principal stiffness coefficients of the spiral spiral groove initially increased with increasing nr and then decreased until becoming negative for nr values greater than 38000 r/min, indicating that sealing system instability might occur from that point onward. As shown in Fig. 4(b), the cross stiffness coefficients exhibited substantial symmetrical variation, and the differences between the values were particularly large for nr values ranging between 10000 r/min and 20000 r/min and between 40000 r/min and 50000 r/min. Thus, positive work was applied to eddy motion in these intervals, resulting in a negative effect on sealing system stability. The results shown in Fig. 4(c) indicate that the principal damping coefficients underwent complex variation with increasing nr. In contrast, the difference between cross damping values initially increases with nr, and then decreases gradually until the nr value is greater than 38000 r/min and finally tends to zero, as shown in the shaded area in Fig. 4(d).

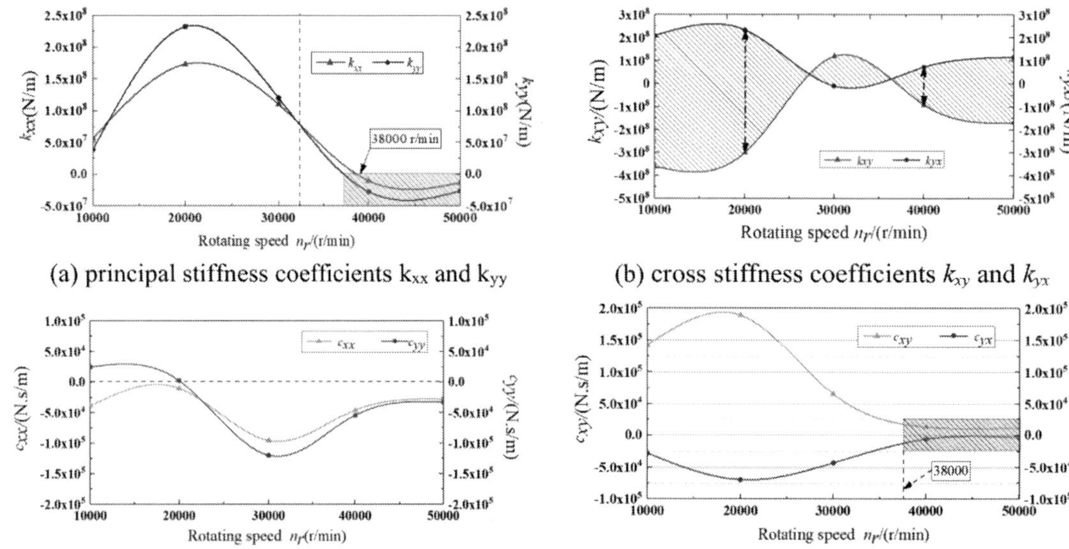

(a) principal stiffness coefficients k_{xx} and k_{yy} (b) cross stiffness coefficients k_{xy} and k_{yx}

(c) principal damping coefficients c_{xx} and c_{yy} (d) cross damping coefficients c_{xy} and c_{yx}

Fig.4 Dynamic characteristic coefficients with the rotation speed

3.4. Effect of pressure on the dynamic characteristic coefficients

As shown in Fig. 5(a), both principal stiffness coefficients initially increased with increasing ph to 0.42 MPa, and then decreased until finally obtaining a constant value. For the subsequently decreasing principal stiffness coefficients, according to Fig. 5(b), the difference between the two cross stiffness values began increasing with increasing ph above about 0.34 MPa, and this increasing difference for ph values greater than 0.42 MPa resulted in positive work being applied to eddy motion, which increased system instability. It can be seen from Fig. 5(c) that cxx decreases with the increase of ph, but increases gradually when ph > 0.3 MPa, and tends to be stable when ph > 0.5 MPa. On the contrary, cyy decreases with increasing ph, but increases rapidly when ph exceeds 0.28 MPa, and cyy shows a parabolic distribution until ph reaches 0.42 MPa. However, the cross damping coefficients shown in Fig. 5(d) varied more simply and symmetrically, where the values of cxy and cyx increasingly diverged with increasing ph above 0.42 MPa, and eventually maintained a fairly stable difference.

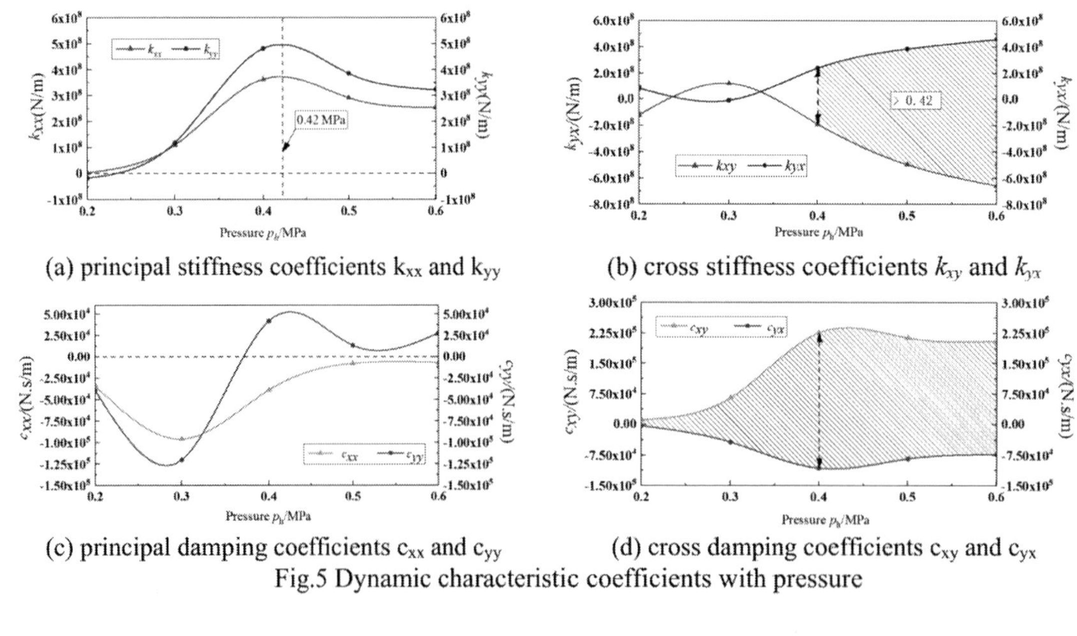

(a) principal stiffness coefficients k_{xx} and k_{yy} (b) cross stiffness coefficients k_{xy} and k_{yx}

(c) principal damping coefficients c_{xx} and c_{yy} (d) cross damping coefficients c_{xy} and c_{yx}

Fig.5 Dynamic characteristic coefficients with pressure

3.5. Effect of eccentricity ratio on the dynamic characteristic coefficients

It can be seen from Fig. 6(a) that with the increase of ε, the main stiffness coefficient increased at first and then decreased. When the value of ε was from 0.64 to 0.78, the main stiffness coefficient become negative. Therefore, the ring spiral groove floating seal held stable when ε was between 0.3 and 0.64 and remains stable when ε > 0.78. The two cross stiffness values were given in Fig. 6(b). Both of them were initially symmetrical, but then began to conform to the same increasing trend with increasing ε greater than 0.72, and the difference between them decreased progressively. As shown by the shaded region in Fig. 6(c), the sum of the principal damping values gradually increased for ε values between 0.64 and 0.72. As shown in Fig. 6(d), the damping coefficients fluctuated around zero, but the difference between the two cross damping values sharply increased for ε = 0.7. According to the observed influences of ε on the principal stiffness, cross stiffness, and principal damping coefficients that the floating ring seal would be more stable with a large eccentricity ratio (ε > 0.7).

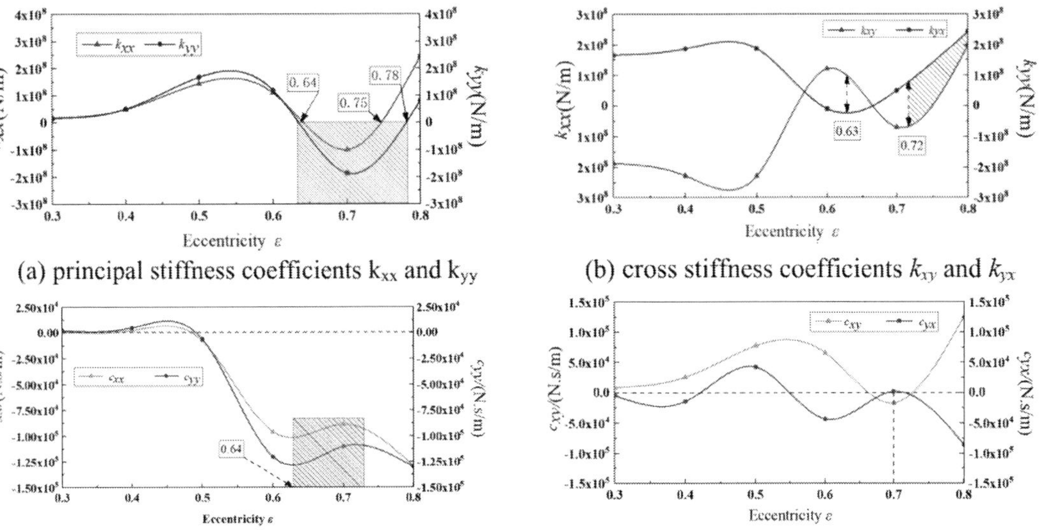

(a) principal stiffness coefficients k_{xx} and k_{yy} (b) cross stiffness coefficients k_{xy} and k_{yx}

(c) principal damping coefficients c_{xx} and c_{yy} (d) cross damping coefficients c_{xy} and c_{yx}

Fig.6 Dynamic characteristic coefficients with the eccentricity ratio

4. Conclusion

This study established a transient Reynolds formulation and film thickness equation for The floating ring seal in a rotating coordinate system based on the small perturbation method. The dynamic stiffness and damping matrices are thereby obtained. The dynamic characteristic coefficients were obtained under working conditions, and the results and analyses provided the following main conclusions.

(1) The circular and symmetrical geometry of the floating ring seal ensured that the cross stiffness coefficients (k_{xy} and k_{yx}) and cross damping coefficients (c_{xy} and c_{yx}) of the gas film were symmetrical and conformed to the requirements for sealing structures. Moreover, the xx and yy components of the principal stiffness coefficients (k_{xx} and k_{yy}) and principal damping coefficients (c_{xx} and c_{yy}) conformed to equivalent trends.

(2) The optimal sealing system became unstable for rotation speeds greater than 38000 rev/min. With respect to pressure, the divergent trends observed for the cross stiffness coefficients and the decreasing trends observed for the principal stiffness coefficients led to a decreasing stability of the sealing system with increasing pressure greater than 0.42 MPa. Accordingly, the increase in pressure had a noticeable influence on the pressure flow and shear flow of the seal, resulting in eddy divergence. Large eccentricity was observed to enhance the wedge effect, which resisted non-uniformities in the gas film pressure and thickness distribution, and was beneficial to eddy convergence.

Acknowledgments
This work was financially supported by the National Natural Science Foundation of China (51905480) and the Science and Technology Innovation 2025 Major Projects, China (2020Z112 and 2018B10006)

References

[1] Mayhew E, Bill R, Voorhees W. (1994),Military engine seal development: potential for dual use[C]. American Institute of Aeronautics and Astronautics 30th Joint Propulsion Conference. Indianapolis, USA, pp.2694-2699.
[2] Ding X X, He Z H, Zhang W Z, et al. (2018),Parameters analysis of steady micro-scale flow of cylindrical spiral groove dry gas seal[J]. CIESC Journal, Vol.69,pp. 1537-1546.
[3] Nagai K, Kaneko S, Taura H, et al. (2017), Numerical and experimental analyses of static characteristics for liquid annular seals with spiral grooves in seal stator[J]. Journal of Tribology, Vol. 140/032201-1.
[4] Shapiro W, Artiles A, Aggarwal B, et al.(2004) Numerical, analytical, experimental study of fluid dynamic forces in seals[C]. In: Final Interim Report No. 1 Mechanical Technology, Latham, NY, NASA/CR, Vol.2,pp.2004-213199.
[5] Wang W, Zhao Z J, ZHANG Z S. (2009), Design and application of single circumferential seal[J]. Aero-Engine Journal, Vol.35 No.4.
[6] Xu Q C, Jiang J B, Peng X D, et al. (2019),Unified model and geometrical optimization of bi-directional groove of dry gas seal based on genetic algorithm[J]. CIESC Journal, Vol. 70,pp. 995-1005.
[7] Chochua G, SHYY W, MOORE J. (2002), Computational modeling for honeycomb-stator gas annular seal[J]. International Journal of Heat and Mass Transfer, Vol.45,pp. 1849-1863.
[8] Mel'nik V A. (2013), Calculation of the characteristics of seals with floating rings[J]. Chemical & Petroleum Engineering, Vol.49,pp. 542-548.
[9] Tokunaga Y, Inoue H, Hiromatsu J, et al. (2016),Rotordynamic characteristics of floating ring seals in rocket turbopumps[J]. International Journal of Fluid Machinery & Systems, Vol.9,pp.194-204.
[10] Xia P, Zhang G, Zhao J, et al.(2017) Investigations on rotordynamic characteristics of a floating ring seal considering structural elasticity[C]. In: ASME Turbomachinery Technical Conference and Exposition, Charlotte, USA, V07AT34A021.
[11] Banakh L Y, Barmina O V. (2018), Rotor System Oscillations under the Rotor Rotation Inside Floating Elements[J]. Journal of Machinery Manufacture and Reliability, Vol.47,pp.6-12.
[12] Cheng H S, Pan C H T. (1965),Stability analysis of gas-lubricated, self-acting, plain, cylindrical, journal, bearings, of, finite, length, using Galerkin's method[J]. J. Basic Eng. Trans. ASME, Vol.87,pp.185-192.

Stress Analysis of HCPB BB under Ex-Vessel LOCA Condition

Mengdi Dai[1] [a, †], and Xiaomo Wang[2, b, †]

[1]School of Mechanical Engineering, North China University of Science and Technology, Tangshan, Hebei, 063000, China

[2] School of Joint Institute, Shanghai Jiao Tong University, Shanghai, 200240, China

[a]Daimengdi@stu.ncst.edu.cn

[b]xiaomo.wang@sjtu.edu.cn

[†]These authors contributed equally

Abstract. Helium Cooled Pebble Bed Breeding Blanket (HCPB BB) is a kind of concept for the European demonstration fusion reactor (DEMO). The blanket attachment system plays an important role in the mechanical connection of the BB and vacuum vessel. Typically, the mechanical and thermal loads should meet the requirement to avoid collapse of the system with off-normal conditions, e.g., under ex-vessel Loss of Coolant Accident (LOCA. This paper investigates the loading requirement corresponding to the maximum stress that can sustain to avoid the LOCA condition. Firstly, a model of the BB is constructed using SolidWorks. Then, stress analysis is carried out based on the cross section of the blanket. Through simulation, the critical condition for the LOCA case and the maximum stress value for the model are obtained. According to the relevant size dimension from the reference, the blanket's cross section is drawn, and one can get the stress field under the ex-vessel LOCA through stress analysis. The stress distribution under the ex-vessel LOCA condition is simulated to find out the maximum stress field that the blanket can sustain through this paper. The significance is to predict the possible conditions leading to an accident and find possible methods to avoid them.

1. Introduction

Contemporarily, there is an increasing demand for energy sources. However, the supply of traditional energy (e.g., oil and natural gas) is limited, which may only sustain for only hundreds of years [1]. Thus, new-type energy sources (e.g., wind and solar energy) have been developed to solve the crisis. Thereinto, nuclear fusion energy is a kind of new-type energy source, which produces a tremendous amount of energy during the process of compressing hydrogen atoms and heating them into a dense, hot plasma state [2]. The fusion reactor is often applied to control the process of fusion reaction and transfer the energy. One part of it is helium cooled pebble bed (HCPB), a European DEMO reference concept for nearly one decade [3]. Thereinto, Loss of Coolant Accident (LOCA) is one of the important components in the HCPB Breeding Blanket (HCPB BB), which converts heat into useful electrical power [4].

Generally, nuclear reactors generate heat internally, which requires a coolant system removes the heat and turns it into useful electrical power [5]. If the coolant flow is reduced, the nuclear reactor's emergency shutdown system will immediately stop the fission chain reaction. The breakdown of the system may lead to thermal expansion of the whole structure. The unsteady state of the structure is highly likely to cause dangerous accidents [4]. Thus, the stress analysis of the system is essential to ensure the safety of the nuclear reactor.

In this paper, we present the stress analysis of LOCA in the HCPB BB with ambient temperature. The motivation is to test the structural stability of the whole system before it starts to run, i.e., evaluate the reaction loads to the structure during the accident. Since the normal conditions have been studied a lot, it is more meaningful to make stress analysis under ex-vessel LOCA condition. It is also common in real situations and deserves consideration for further analysis. The rest parts of the paper are organized as follows: Sec. 2 introduces the simulation methods; Sec. 3 demonstrates the results and gives discussions; a brief summary is presented in Sec. 4 eventually.

2. Method

In this paper, SolidWorks [6] is applied to model the stress analysis, which constructs a cross section of the blanket about the global model (Figure 1) and part model (Figure 2). Each module is connected to the so-called back supporting structure (BSS). Each group of modules supported by a BSS forms a segment. The European DEMO segmentation is arranged in 7 outboards (OB) and 7 inboards (IB) blanket modules in each OB/IB blanket segment [7]. The BSS provides the coolant for the FW and BZ of each blanket module by BSS manifold pipes. The FW cooling channels have a cross-section of 12.5 mm × 12.5 mm. The FW front part (plasma side) thickness is 3 mm, resulting in a total FW thickness of 25 mm, which is thin to contribute to a good tritium breeding performance (TBR ~ 1.15) while still satisfying the structural requirement. The CP cooling channels have a rectangular cross-section of 3 mm × 5 mm. The constructed global model is shown in Figure 1.

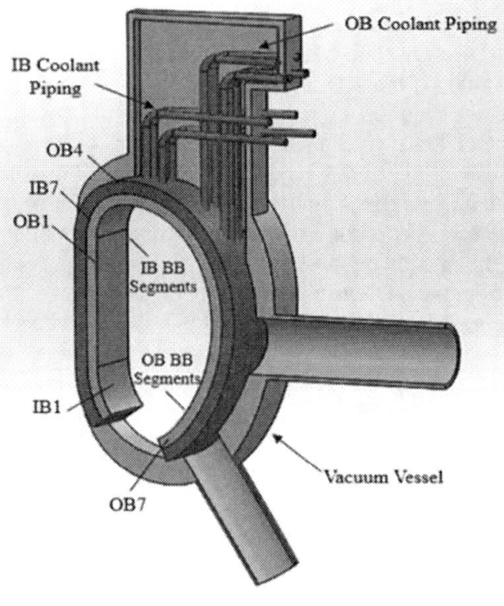

Figure 1. Global model of the blanket

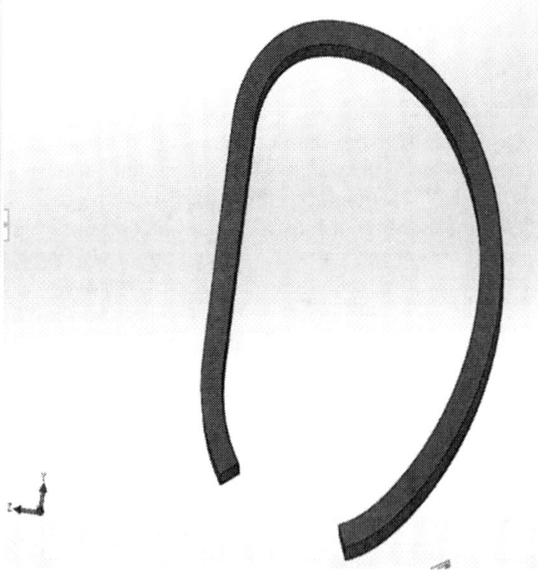

Figure 2. Part model of the blanket

Subsequently, stress analysis is carried out for the global model and part model about OB7 in the room temperature. The part model consists of 3 OB blanket segments with their attachments, 2 IB blanket segments with their attachments, the upper port shield plug, and the corresponding VV counterparts, as shown in Figure 2. During modelling, we first select materials (synthetic steel) and fix geometry and actual force surface following Ref. [3]. Then, the simulation is run to obtain the force analysis, displacement change, and strain. Based on the room temperature and LOCA, we pick up the data about the pressure and moment, where one obtains the maximum stress and moment that the blanket can sustain to avoid the LOCA. Figure 1 shows the global model of the blanket.

 2108 (2021) 012088 doi:10.1088/1742-6596/2108/1/012088

3. Results and Discussion

Table 1. Reaction forces and moments at attachments acting upon VV (F-force, M-moment)

	Fx (MN)	Fy (MN)	Fz (MN)	Mx (MNm)	My (MNm)	Mz (MNm)
OBR Key Bottom	3.2	-6.4×10^{-2}	5.7	-5.9×10^{-2}	0.13	-7.5×10^{-3}
OBR Pad	3.7×10^{-4}	-2.5×10^{-8}	5.2×10^{-3}	-5.6×10^{-8}	6.3×10^{-6}	-4.7×10^{-8}
OBC Key Bottom	3.1	-2.7×10^{-2}	5.7	0.18	0.13	3.4×10^{-2}
OBC Pad	1.7×10^{-13}	3.8×10^{-15}	5.4×10^{-3}	-2.1×10^{-8}	-2.5×10^{-6}	2.3×10^{-10}
OBL Key Bottom	3.2	5.0×10^{-2}	5.7	-9.1×10^{-2}	0.13	-4.2×10^{-2}
OBL Pad	5.1×10^{-4}	-1.6×10^{-8}	5.2×10^{-3}	-5.5×10^{-8}	9.9×10^{-6}	1.0×10^{-7}
OBR Key Extens	0.62	-9.4	-2.6×10^{-2}	0.77	4.7×10^{-2}	0.57
OBR Top	2.7	-0.22	-4.2	0.26	-0.96	0.23
OBL Key Extens	0.64	9.6	-3.1×10^{-2}	-1.4	8.3×10^{-2}	-1.3
OBI Top	2.7	1.5×10^{-2}	-4.5	-0.33	-0.93	-0.2
IBR Key Bottom	0.87	2.8×10^{-2}	3.7	-1.4×10^{-2}	7.4×10^{-3}	1.5×10^{-2}
IBR Key Mid-1	2.3	1.5×10^{-2}	3.4×10^{-2}	-2.5×10^{-3}	-0.19	2.8×10^{-3}
IBR Key Mid-2	5.7	-0.17	-0.25	2.0×10^{-2}	0.57	-7.7×10^{-3}
IBR Key Top	-0.25	0.14	-2.7	3.7×10^{-2}	-0.22	6.2×10^{-2}
IBL Key Bottom	0.78	5.4×10^{-2}	4.7	1.9×10^{-2}	4.2×10^{-3}	-2.1×10^{-2}
IBL Key Mid-1	2.4	-0.15	3.6×10^{-2}	-2.6×10^{-3}	-0.20	-1.4×10^{-2}
IBL Key Mid-2	6.1	0.32	-0.27	-5.7×10^{-2}	0.70	1.7×10^{-2}
IBL Key Top	-0.3	-0.21	-3.7	-5.9×10^{-2}	-0.36	-0.17

Note: OBR /IBR is the right of the out board/ in board, OBL/IBL is the left of the out board/in board.

Table 1 summarizes the three-dimensional forces and moment parameters acting on the blanket attachment. Based on this table, the maximum pressure can be acquired. According to the part model, one can find that the stress is lower in the outer plate while higher in the inner plate. In the global model, gravity acts on all objects. The maximum value of the IBLIBL Middle-2 is 6.4 MN. By comparing with the existing literature on stress analysis of HCPB BB in the unloaded state, it is found that the maximum stress is located in the region where the two cladding modules interact. It was found that the HCPB BB and attachment system exhibited sufficient structural strength under selected forward ship LOCA. For better illustration, the pressure and displacement distributions are presented in Figure 3, which further proves the structure is stable at room temperature since the effects are minor. As for the figure3, we first, add the Simulation plug-in. Click Simulation to perform the force analysis. Next, click on the new example in the example consultant and select the static stress analysis. After that, click Apply Materials and select the material. Next click Fixture Advisor, fix the geometry and select the fixed surface. Then, the external load consultant selects the force and surface, that is, the actual force surface, and then inputs the value. Finally, this example is run to obtain the force analysis, displacement change and strain.At last, we got the result of Figure 3.

Even though our estimation is relatively accurate, there are still some drawbacks for improvement. First, we only consider the stress analysis of the HCPB BB under ambient temperature, while high temperature environments might occur during operation. Thus, one ought to estimate the distribution of forces and moments under different temperatures and give a safe temperature range where the HCPB BB can still sustain. Second, we only use SolidWorks as the software to build up the model and stress analysis, i.e., it will be one of the limits about the result. Third, the model is slightly different from Ref. [3, 7], which might result in a little bit of error.

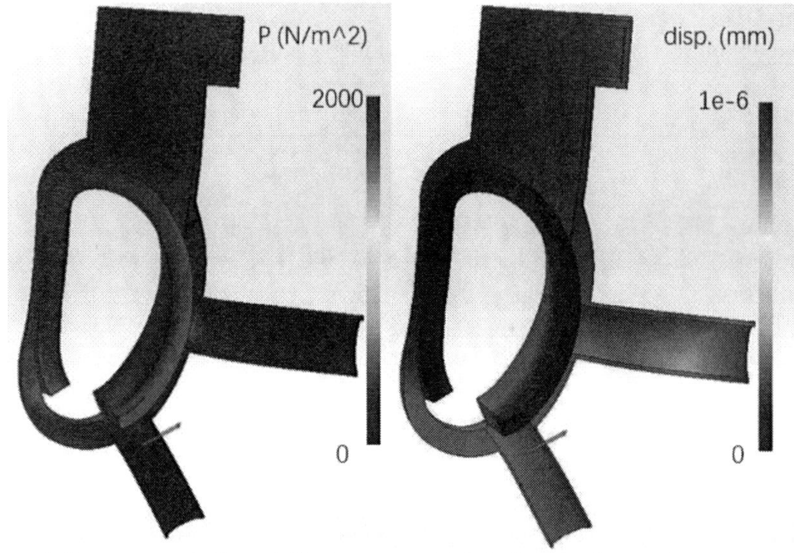

Figure 3. Part model's pressure(left) and displacement(right) distribution under ex-vessel LOCA in the room temperature.

4. Conclusion

In summary, stress analysis has been presented to re-exam the current design of EU DEMO baseline 2015 under an ex-vessel LOCA condition. Base on SolidWorks modelling, the reaction forces and moments at attachments acting on VV under selection accident conditions are obtained. According to our results, the designed structure is stable at room temperature, verifying the structure's stability under certain forces for the state-of-art design. These results will be useful for further improving the design of the HCPB BB and its attachment system.

References

[1] George E. Halkos, Eleni-Christina Gkampoura, (2021) Evaluating the effect of economic crisis on energy poverty in Europe, Renewable and Sustainable Energy Reviews, 144: 110981.

[2] Catrinel Turcanu, Ana Prades, Roser Sala, Tanja Perko, Christian Oltra, (2020) Fusion energy: A deeper look into attitudes among the general public, Fusion Engineering and Design, 161:111891.

[3] G. Zhou, F. A. Hern·ndez, C. Zeile, and I. A. Maione, (2018) Transient thermal analysis and structural assessment of an ex-vessel loca event on the eu demo hcpb breeding blanket and the attachment system, Fusion Engineering and Design, 136: 34–41.

[4] Suman S, Khan M K, Pathak M, et al, (2016) Rupture behaviour of nuclear fuel cladding during loss-of-coolant accident[J]. Nuclear Engineering and Design, 307: 319-327.

[5] A.A. Avramenko, N.P. Dmitrenko, I.V. Shevchuk, A.I. Tyrinov, M.M. Kovetskaya, (2021) Heat transfer and fluid flow of helium coolant in a model of the core zone of a pebble-bed nuclear reactor, Nuclear Engineering and Design, 377: 111148.

[6] SolidWorks- design and simulation tools, 2002. http://www.solidworks.com.cn

[7] M Dalle Donne, H Albrecht, L.V Boccaccini, et, al. European helium cooled pebble bed blanket: Design of blanket: module to be tested in ITER. Forschungszentrum Karlsruhe, Posffach 3640, D-76021 Karlsruhe, Germany.

Performance and Optimization Study of R290 as Alternative Refrigerant for R22 in Low Temperature Heat Pump System

Yun Zhang[1,2], Cichong Liu[1,2], Wanyong Li[1,2], Junye Shi[1,2], Jiangping Chen[1,2,*]

[1]Institude of Refrigerant and Cryogenics, School of Mechanical Engineering, Shanghai Jiao Tong University, Shanghai 200240, China;

[2]Shanghai High Efficient Cooling System Research Center, Shanghai 200240, China;

*Corresponding author's e-mail: Jpchen@sjtu.edu.cn

Abstract This paper mainly studies the replacement performance of R290 in R22 low temperature heat pump system from the experimental point of view. By comparing the performance differences under different working conditions, it is found that when R22 is directly extracted from the original system and filled with R290, the heat capacity and COP of the system are attenuated, and the compressor discharge temperature and pressure of the R290 system are higher than those of the original R22 system in low temperature environment. Through the analysis of the system components, it can be considered that the main reason for the above phenomenon is that the compressor displacement of the R22 system is too large and does not match the R290 system. Therefore, in order to meet the safety requirements of the system and improve the overall performance of R290 in the low temperature heat pump system at the same time, it is considered to replace the compressor with a smaller displacement which is more matched with R290 in the system. The experimental results show that the compressor displacement optimization of the R290 low temperature heat pump system can effectively reduce the exhaust temperature and pressure of the system and improve the overall performance of the system. The COP of the optimized R290 low temperature heat pump system is 6.5% higher than that of the original R22 system, and the exhaust temperature in the low temperature environment is reduced by 36% to below 80 °C.

1. Introduction

HCFC refrigerant R22 is widely used in household and commercial areas. However, due to its ability to destroy the atmospheric ozone layer and certain greenhouse effect potential, R22 has been banned by the Kigali amendment of the Montreal Protocol. At present, the replacement refrigerant of R22 at home and abroad is mainly divided into two directions: One is the HFCS replacement scheme dominated by the United States and Japan[0], and the other is the natural refrigerants such as R290 (propane) dominated by China and Europe[[2]].

The attitude of the industry to refrigerant replacement is to try not to make too many changes to the original system, in order to achieve optimal performance at a lower cost[[3]]. For the selection of R22 alternative refrigerants, many scholars have made the corresponding research [[4]-[6]]: Li[[7]] and others have used the new refrigerant mixture R417A to replace the R22 heat pump hot water system; Zhang[[8]] also used R417A to replace R22 in the field of household air conditioning; By comparing the thermophysical properties of R22 and R32, Shi[[9]] and others studied the alternative research of R22 air conditioning in household and commercial fields; Yao[[10]] and others simulated the thermophysical properties of R22 and R290 by establishing the corresponding parameter relationship equations. Among the alternatives in the field of household air conditioning, R290 stands out among many alternative refrigerants because of its excellent environmental performance and similar thermophysical properties to R22 due to its ODP and GWP values of 0[[11]]. Generally speaking, the existing R22 system does not need to be improved too much, and R290, as a by-product of the oil and gas industry, is easy to produce and low cost. Therefore, for low temperature heat pump system, R290 is an ideal substitute for R22[[12]].

The existing research on R290 replacing R22 refrigerant mainly focuses on the performance of R290 as an alternative refrigerant in the refrigeration system and the related problems and solutions. There is a lack of research on the alternative performance of R290 in heat pump systems, especially in low temperature environment. Therefore, this paper mainly explores the performance effect of R290 in the low temperature heat pump system from the experimental point of view, and gives the relevant optimization suggestions and direction for the alternative research of R22 in the low temperature heat pump system.

2. Theoretical analysis

It can be seen from Tab.1 that the main physical properties of R290 are very similar to those of R22. And the ODP value of R290 is 0, GWP value is very small, so it has the basic conditions to replace R22[[13]].

Tab.1 Physical properties of R290 and R22

Refrigerant	R22	R290
Molecular formula	CHCLF2	C3H8
Molecular weight	86.47	44.09
Boiling point (°C)	-40.78	-42.07
Freezing point (°C)	-160	-187.7
Critical pressure (Mpa)	4.974	4.254
Critical temperature (°C)	96.2	96.8
Critical specific volume (m^3/kg)	0.001904	0.004545
Latent heat of boiling point (kJ/kg)	234.5	430
Saturated liquid density (kg/m^3)	1412	582
ODP value	0.055	0
GWP value	1600	<20

It can be seen from Fig.1 that the saturated gas pressures of R22 and R290 are very close, and because of their similar pressure temperature curves, R22 can be directly extracted from the original R22 system to fill R290 without major changes[[14]].

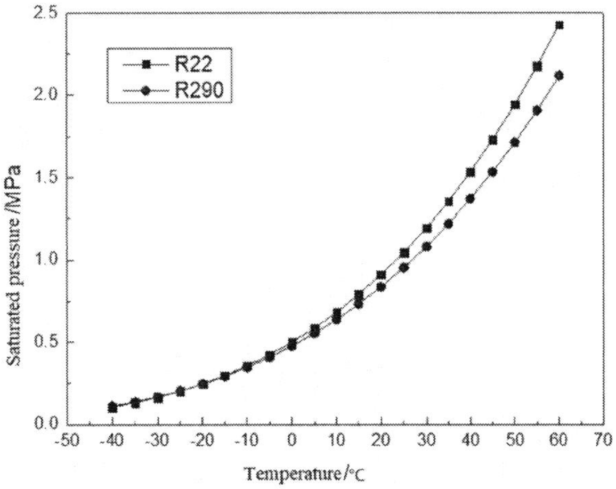

Fig.1 The diagram of saturated steam pressure and temperature of R22 and R290

Through the above comparison of R22 and R290, it can be concluded that the characteristics of R290 mainly focus on the following aspects[[15]~[16]]:

(1) The latent heat of vaporization of R290 is 1.84 times that of R22, which indicates that the circulation of working fluid can be reduced by using R290 under the same heating capacity.

(2) The adiabatic index of R290 is lower than that of R22. At the same compression ratio, the power consumption can be reduced, the exhaust temperature can be reduced, the gas transmission coefficient can be increased, the heat exchange between gas and cylinder can be reduced, the irreversible loss can be reduced, the exhaust temperature can be reduced, the working condition of condenser can be improved, the irreversible heat transfer loss can be reduced, and the energy consumption can be reduced.

(3) The thermal conductivity of R290 is higher than that of R22, which can improve the heat dissipation condition of the compressor, increase the heat transfer coefficient of the condenser and evaporator, and further reduce the energy consumption of the system.

3. Experimental study

In order to study the performance of the R290 low temperature heat pump system under different working conditions, especially in a low temperature environment, the original R22 heat pump system was replaced. The charging capacity of the original R22 system is 5kg, and that of the R290 system is 3kg, which meets the relevant standards of GB 4706.32-2012 '*safety of household and similar electrical appliances - special requirements for heat pumps, air conditioners and dehumidifiers*' about the refrigerant charging capacity M < (130m ^ 3) * LFL of heat pump system.

The heat pump system is mainly composed of a scroll compressor, tube fin heat exchanger, shell and tube heat exchanger, a thermal expansion valve, gas-liquid separator and a liquid storage tank. The experiment was carried out in the enthalpy chamber, which is mainly divided into an evaporation chamber and a condensation chamber. A wind tunnel is respectively arranged in the evaporation chamber and the condensation chamber, which is used to adjust the inlet air volume of the evaporator and the condenser and measure the dry and wet bulb temperature of those two chambers. The temperature and humidity of the evaporation chamber and condensation chamber are controlled by a refrigerating unit, an electric heater and humidifier respectively. In the pipeline of the original machine, holes are drilled in the front and back of each part, and sensors are arranged to measure the pressure and temperature of the system. The experimental platform is shown in Fig. 2 and Fig. 3, and the accuracy analysis of the experimental platform can be referred to Tab.2.

Tab.2 Experiments Measurement Accuracy

Equipments	Accuracy
Power meter	±0.2%
Electronic scale	±10g
Pressure sensor	±10kPa
Temperature sensor	±0.5K
Electronic nozzle pressure differences	±2Pa

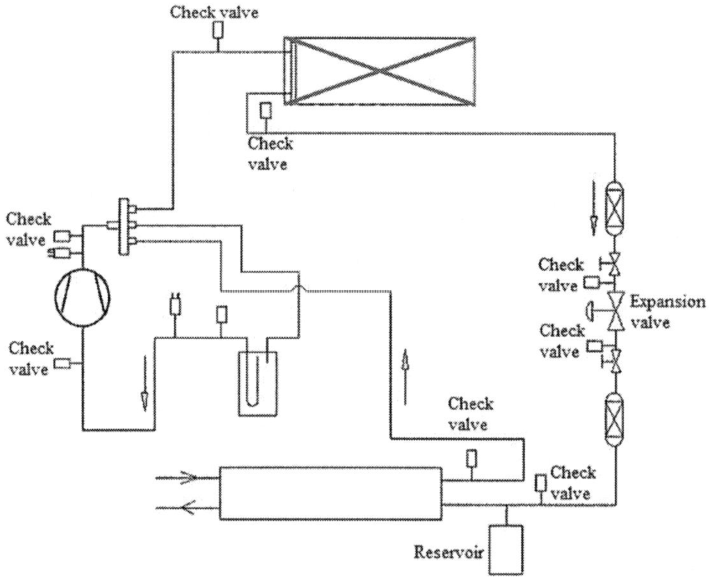

Fig.2 Experiment Test Rig

Fig.3 Experiment Test Rig

In order to analyze the performance of the heat pump system under different environmental conditions, a total of 20 groups of experimental conditions as shown in Tab.3 were designed to analyze and compare the performance differences of the two systems.

Tab.3 Experiments Test Conditions

Condition	Water-flow rates m^3/(h*kW)	Ambient temperature (°C)	Outlet water temperature (°C)
1	0.172	21	40
2	0.172	21	45
3	0.172	21	60
4	0.172	7	40
5	0.172	7	50
6	0.172	7	60
7	0.172	2	40
8	0.172	2	50
9	0.172	2	55
10	0.172	-6	40
11	0.172	-6	45
12	0.172	-6	50
13	0.172	-12	40
14	0.172	-12	45
15	0.172	-12	50

4. Experimental results and discussion

In order to analyze the specific performance differences between R290 and R22, the heating capacity, COP, exhaust pressure, exhaust temperature and other parameters under different ambient temperatures and the same outlet temperature were compared.

4.1 Comparison of system d and COP

The results show that the heat transfer of the system decreases with the decrease of the ambient temperature, and there is little difference between the two refrigerants. The COP of the R290 system is higher than that of the R22 system at 21 °C, but the performance degradation of R290 system is greater than that of R22 system. The COP of the R22 system is higher than that of the R290 system under 21 °C.

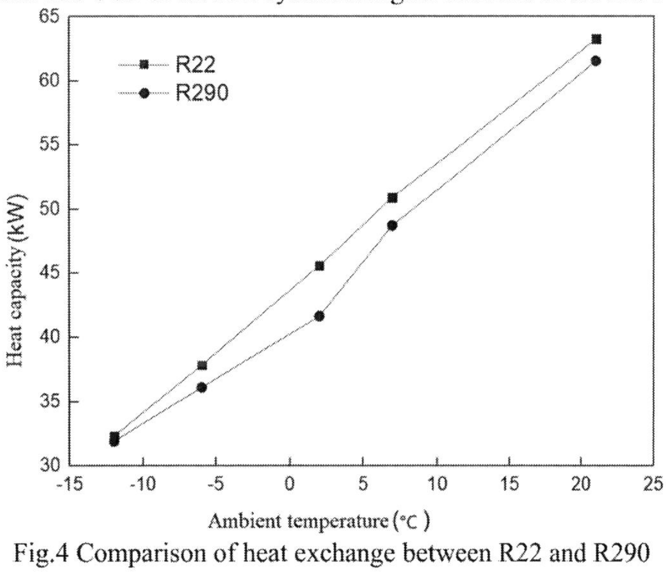

Fig.4 Comparison of heat exchange between R22 and R290

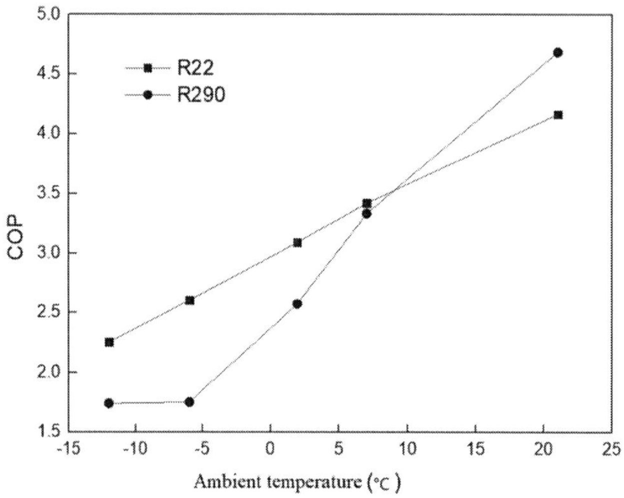

Fig.5 Comparison of COP between R22 and R290

Because the latent heat of vaporization of R290 is higher than that of R22, the displacement of the compressor used in the original R22 system is too large for R290. Larger refrigerant flow means that the heat exchange temperature difference required by the system is larger under the same heat exchange conditions. By comparing the relationship between the temperature difference and heat transfer between the condenser and outlet water temperature of the R22 and R290 systems, it can be seen that the overall heat transfer temperature difference of the R290 system is greater than that of the R22, and the overall heat transfer of R22 system is higher than that of R290 system, so the overall heat transfer efficiency of R22 system is higher than that of R290 system. It is worth noting that the heating capacity of R22 in a low temperature environment has obvious attenuation (three points in the lower part of Fig.6), which indicates that the thermal efficiency of R22 is low in a low temperature environment, while the efficiency of R290 is relatively stable at low temperature. This is also reflected in the changes of COP of R290 and R22: the change of COP tends to be stable when R290 is under - 6 °C, and the COP attenuation of R22 is more severe. In addition, the overall heat transfer demand and refrigerant flow of the system are small at indoor temperature of 21 °C, so the compressor displacement has little effect on the R290 system. With the larger latent heat of vaporization, the heat transfer efficiency of the R290 system is higher (as shown in the top three points on the left of Fig.7), which also explains why the COP of the R290 system is higher than that of the R22 system at indoor temperature of 21 °C.

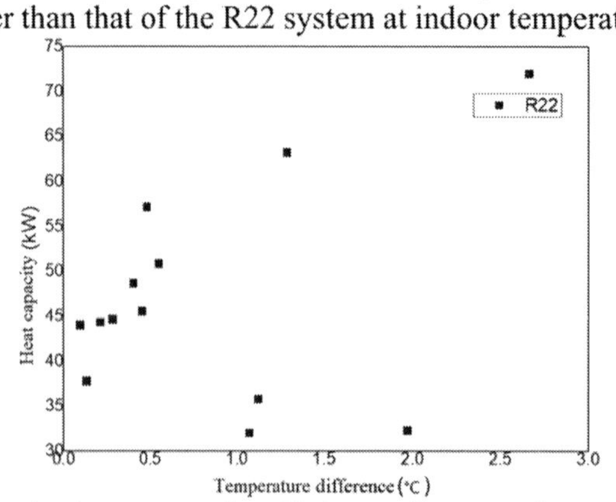

Fig.6 Heat transfer efficiency diagram of R22

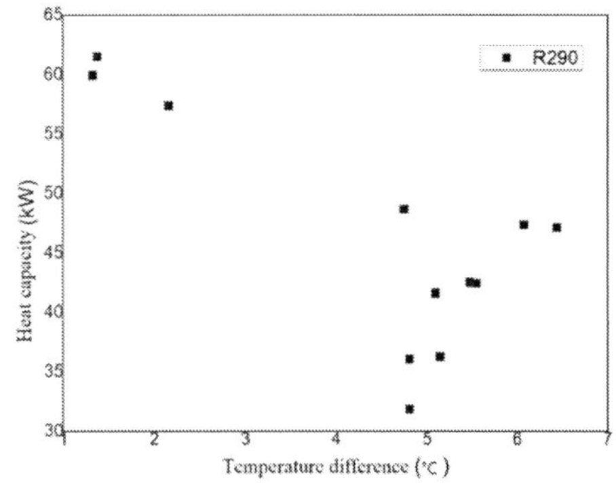

Fig.7 Heat transfer efficiency diagram of R290

4.2 Comparison of exhaust temperature and pressure

In order to analyze the influence of compressor displacement on system performance, the exhaust temperature and pressure of the two systems are compared. It can be seen that the overall discharge pressure of the R290 system is higher than that of the R22 system. The exhaust pressure of the R22 system decreases slightly with the decrease in ambient temperature. The exhaust pressure of the R290 system first increases and then decreases with the decrease of ambient temperature, and reaches the highest exhaust pressure at - 6 °C. The exhaust temperature of the R290 system is lower than that of the R22 system when it is above 2 °C, and higher than that of the R22 system when it is below 2 °C. The exhaust temperature of the R22 system increases with the decrease in ambient temperature. The exhaust temperature of the R290 system first increases and then decreases with the decrease in ambient temperature, and reaches the highest exhaust temperature at - 6 °C.

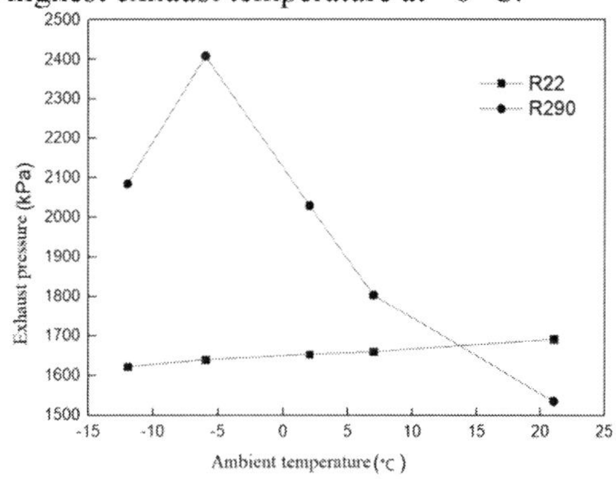

Fig.8 Comparison of system exhaust pressure between R22 and R290

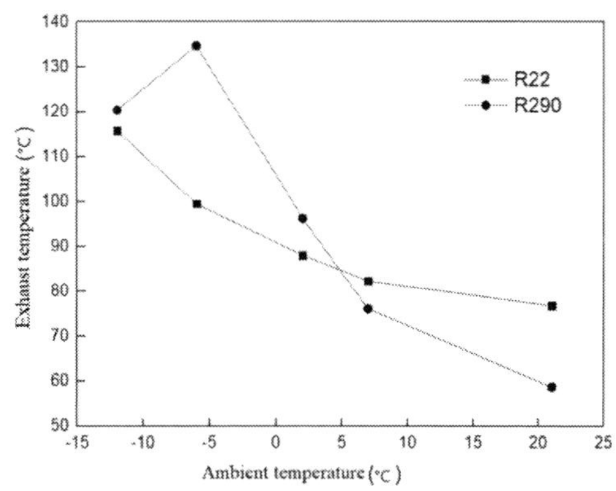

Fig.9 Comparison of system exhaust temperature between R22 and R290

This change in trend is mainly caused by the matching problem between the refrigerant and the compressor displacement. When the temperature is 21 °C, the overall heat exchange demand of the system is small and the refrigerant flow is small, so the compressor displacement has little effect on the R290 system. At this time, the discharge pressure and temperature of the R290 system are lower than those of the R22 system. With the decrease in ambient temperature, the heat load of the system becomes larger. At this time, the mismatched compressor makes the refrigerant flow of the R290 system much higher than that of the R22 system, which means that the compressor load of the R290 system is higher than that of the R22 system, and the discharge pressure and temperature of the R290 system are higher than that of R22 system. At - 6 °C, the exhaust temperature and pressure of the R290 system suddenly increase, which is mainly due to the overheating of the system caused by the too small opening of the expansion valve.

4.3 Data analysis
From the above data results, it can be concluded that when R290 is directly used in the R22 system, the overall performance of the system decreases, and the heat exchange and COP of the system decrease. In a low temperature environment, the compressor discharge temperature and pressure of the R290 system are higher than those of the R22 system. Because too high exhaust pressure and temperature will affect the safety of the compressor, it is more unfavorable for the operation of the system. Through the analysis of the system components, it can be considered that the main reason for the above phenomenon is that the compressor displacement of the R22 system is too large and does not match the R290. Too large displacement of the compressor means that the refrigerant flow of R290 is too large. Therefore, in order to achieve the same level of heating, the heat exchange temperature difference required by the system is larger, which results in higher discharge temperature and pressure of the R290 compressor [[17]]. Therefore, in order to meet the safety requirements of the system and improve the overall performance of R290 in the low temperature heat pump system at the same time, it is considered to replace the compressor with a smaller displacement which is more match with R290 in the system.

5. Optimization and improvement
In order to ensure the safe and efficient operation of the R290 low-temperature heat pump system, the original compressor with 120cc displacement in the R22 system is replaced by the compressor with 85cc displacement, which is more suitable for the R290. At this time, the optimal charge of the R290 system is reduced to 2.6kg. The test was completed under the same environmental conditions, and the performance of the improved R290 system is compared with that of the R22 system.

Due to the use of a smaller displacement compressor, the overall heat transfer of the improved R290 system is slightly less than that of the original R22 system, but the overall COP of the R290 system is about 6.5% higher than that of the original R22 system.

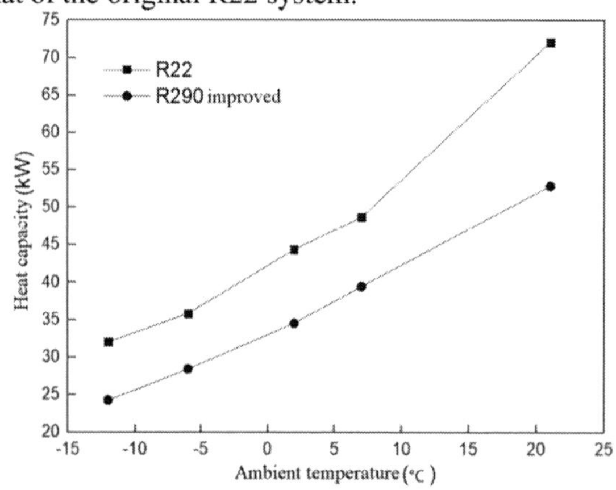

Fig.10 Comparison of heat exchange between R22 system and improved R290 system

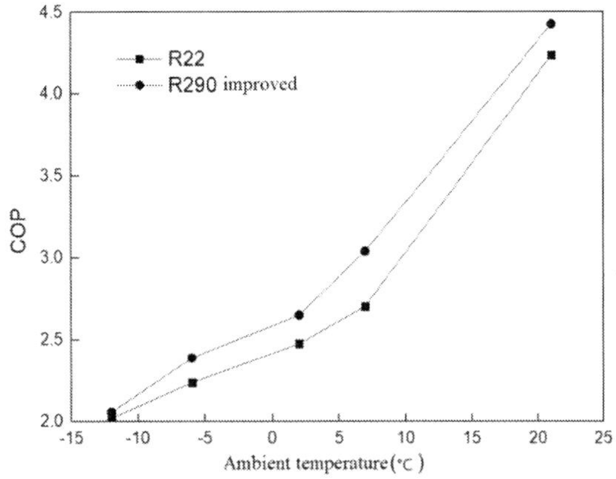

Fig.11 Comparison of COP between R22 system and improved R290 system

In the aspect of compressor operation, it can be seen that the discharge pressure and temperature of the improved R290 system have been greatly improved: When the temperature is above 2 °C, the improved R290 system can effectively reduce the discharge pressure of the compressor, and when the temperature is below 2 °C, the R290 system can also reach the same level as the R22 system. At the same time, the exhaust temperature of the improved R290 system can be greatly reduced, especially in the low temperature environment, the exhaust temperature of the original R22 system can be reduced by more than 36%.

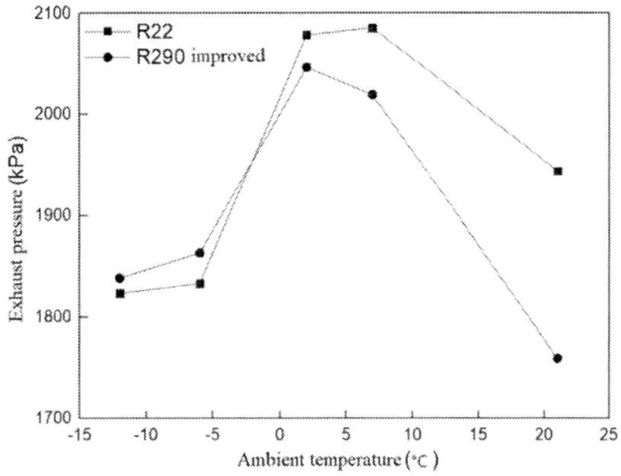

Fig.12 Comparison of system exhaust pressure between R22 system and improved R290 system

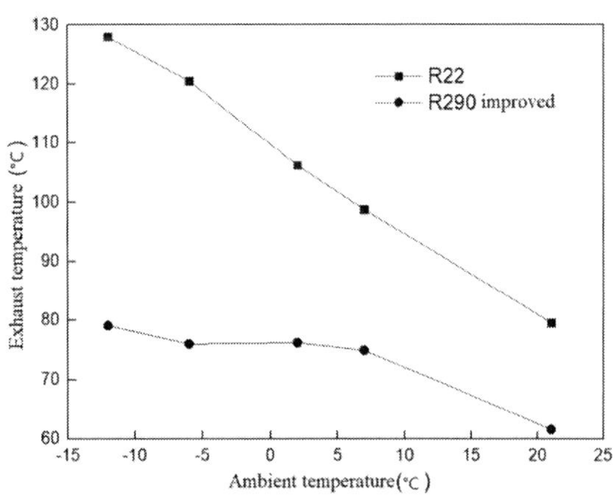

Fig.13 Comparison of system exhaust temperature between R22 system and improved R290 system

From the above data results, it can be seen that after the optimization of the compressor, the overall performance of the R290 low temperature heat pump system has been improved to a certain extent, especially greatly reducing the exhaust temperature of the compressor, effectively improving the overall safety performance of the system.

6. Conclusions

In this paper, the system performance tests of R290 and R22 were completed under different conditions. The heating capacity, COP, exhaust pressure and exhaust temperature were compared at different ambient temperatures and the same outlet temperature. The feasibility of R290 replacing R22 in low temperature heat pump system is analyzed from the experimental point of view:

(1) The thermophysical properties of R290 are similar to those of R22, and R290 is superior to R22 in terms of latent heat of vaporization, ODP, GWP, etc;

(2) The optimization of the compressor displacement in the R290 low temperature heat pump system can effectively reduce the exhaust temperature of the compressor and improve the overall performance of the system;

(3) Compared with the original R22 system, the performance of R290 low-temperature heat pump system after compressor optimization has been greatly improved: COP increased by 6.5%, and the exhaust temperature decreased by 36% under -12°C low-temperature environment to below 80 °C;

(4) In the low temperature heat pump system, R290 instead of R22 has the characteristics of high efficiency, good safety and low cost, so it has a good application prospect.

References

[1] Kim S G, Kim M S, Ro S T. Experimental investigation of the performance of R22, R407C and R410A in several capillary tubes for air-conditioners[J]. International Journal of Refrigeration, 2002, 25(5):521-531.

[2] Liu Z M. Study on alternative refrigerant of R22[J]. Refrigeration & Air-Conditioning, 2001, 1(3):47-53.

[3] Mohanraj M, Jayaraj S, Muraleedharan C. Environment friendly alternatives to halogenated refrigerants—A review[J]. International Journal of Greenhouse Gas Control, 2009, 3(1):108-119.

[4] Rocca V L, Panno G. Experimental performance evaluation of a vapour compression refrigerating plant when replacing R22 with alternative refrigerants[J]. Applied Energy, 2011, 88(8):2809-2815.

[5] Bolaji B O. Performance investigation of ozone-friendly R404A and R507 refrigerants as alternatives to R22 in a window air-conditioner[J]. Energy & Buildings, 2011, 43(11):3139-3143.

[6] Zhang J, Li D Y. A Study on Replacing Refrigerant R410a With R22[C]// Annual Meeting of Chinese Association of Refrigeration in 2011. 2011.

[7] Li X Y, Yan Z S. Experimental Study on Replacement of R22 with R417a in Heat Pump Heating Water System[J]. Refrigeration Journal, 2003, 24(4):1-4.

[8] Zhang M. Experimental Study on Room Air Conditioner Using R417A Instead of R22[J]. Refrigeration and Air-Conditioning, 2006, 6(5):80-82.

[9] Lin S, Zhu M S. Re-analysis on Using R32 to Substitute for R22 in Household/commercial Air-conditioning[J]. Journal of Refrigeration, 2010. 31(1):1-5.)

[10] Yao S G, Wang J. Analysis of Performance for R22 and the Alternative Refrigerant R290 in Unit Condition System[J]. Journal of East China Shipbuilding Institute, 2002, 16(5):70-73.

[11] Rahadiyan L, Gim Y G, You S C, et al. Experimental study of Propane (R290) as an alternative refrigerant of R22[J]. Korean mechanical Society 2006. Proceedings of the annual spring Academic Conference, 2006.

[12] Castro J B, Urchueguía J F, Corberán J M, et al. Optimized design of a heat exchanger for an air-to-water reversible heat pump working with propane (R290) as refrigerant: Modelling analysis

 2108 (2021) 012089 doi:10.1088/1742-6596/2108/1/012089

and experimental observations[J]. Applied Thermal Engineering, 2005, 25(14–15):2450-2462.

[13] Yang R, Yang Z, Sun R, et al. Investigation on the Miscibility of Lubricants with Natural Refrigerant R290[J]. Journal of Refrigeration, 2013, 34(5):23-27.

[14] Li T, Yang. Experiment on R290 Substituting for R22 in A Room Air-conditioner[J]. Journal of Refrigeration, 2010, 31(4):31-34.

[15] Zhang Y F, Tang R N. The Theoretical and Experimental Research for Natural Refrigerant Propane(R290) and Its Mixtures as Substitute for R22[J]. Refrigeration Journal, 1999(2):18-24.

[16] Qian W. The Feasibility Study of Propane(R290) as Substitute for R22 in Room Air Conditioner[J]. Journal of Guandong College Petrochemical Technology, 2005.

[17] He G G, Liu Q. Optimization of Rotary Compressor for Air-Conditioner Using R290 to Replace R22[C]//Annual Meeting of Chinese Association of Refrigeration in 2009. 2009.

Stability analysis of microgrid inverter off-grid mode considering nonlinear load

Xiaoyan Shi[1a*], **Bin Wang**[1b], **Shuijuan Yu**[1c]

[1]Department of Automation, College of Electrical and Information Engineering, Anhui University of Science and Technology, Huainan, Anhui, China

[a*]shixiaoyan1980@163.com, [b]285277712@qq.com, [c]7402805@qq.com

Abstract—With the large-scale utilization of distributed generation in microgrid, inverter as the connection hub of new energy grid connection, directly affects the operation performance of microgrid. In order to improve the output voltage quality and load capacity of the inverter in the off-grid mode of distributed energy, the stability region of the inverter with load is analyzed by using the impedance analysis method of cascade converter and control theory. The quasi proportional resonant (QPR) double loop control is adopted to realize no static error tracking voltage while increasing bandwidth, and the influence of control parameters on performance is analyzed. At the same time, in order to improve the capacity of inverter with nonlinear load, odd harmonics are introduced into the controller to suppress the influence of low harmonics of load current on output voltage. Finally, the influence of inverter output impedance change, load level, controller parameters and filter parameters on system stability is analyzed through impedance ratio Nyquist curve, which provides corresponding theoretical support and parameter optimization reference for the design of actual system.

1. Introduction

Nowadays, facing the dual crisis of fossil energy depletion and environmental degradation, all countries are targeting renewable energy to seek to establish a safer, efficient and green energy system[1]. As a new and flexible energy architecture, microgrid cooperates with photovoltaic, wind power energy, various new energy, energy storage equipment and various loads to conduct two-way energy flow with the power grid through the grid connection interface[2]. The operation mode of microgrid are divided into grid connection mode and island mode. It not only needs to give full play to the advantages of resource integration and friendly interaction with large power grid, but also promotes the local consumption of new energy and independent power supply to local loads. As a necessary energy buffer link of microgrid system, energy storage system is the key to the safe and reliable operation of microgrid[3]. Especially in off-grid mode, the output characteristics of microgrid inverter directly determine the load power supply quality. Therefore, the stability analysis of inverter is of great significance to the selection of microgrid control strategy, the design of control parameters and the switching and optimal control of microgrid operation mode[4].

For the stability analysis of microgrid inverter, most studies are based on the establishment of small signal model, using root locus analysis method, frequency domain analysis and other methods, but these methods have high order, many parameters and cumbersome calculation[5]. In microgrid operation, the existence of a large number of power electronic devices have an impact on the equivalent impedance of the power grid that can not be ignored. In off-grid mode, the loaded operation of energy storage inverter can be regarded as a cascade system included source module composed of

 2108 (2021) 012090 doi:10.1088/1742-6596/2108/1/012090

energy storage inverter and load module composed of load and line impedance. The impedance analysis method proposed by Middlebrook et al has become an effective tool to analyze the performance of inverter. This method can improve the performance of the system by establishing the input and output impedance of the inverter to match and improving the stability margin of the system without modeling and knowing the specific parameters. Therefore, it has attracted the attention of many scholars and put forward some improved impedance analysis criteria[6-7]. In reference[8] the stability criterion of grid connected inverter for impedance analysis have been presented, and pointed out that the grid connected system can be stable only when the ratio of grid impedance to inverter output impedance met the Nyquist stability criterion. In reference[9], the effects of PLL, current controller parameters and filter parameters on the output impedance characteristics of grid connected inverter were analyzed in detail. In reference[10] the small signal model, influencing factors and optimization design method of buck DC-DC converter output impedance were studied, which provided a reference basis for formulating the impedance standard of converter in distributed power system. In reference[11] from the perspective of impedance matching, the instability of distributed generation grid connected inverter in large power grid with high impedance characteristics was studied. At the same time, the detailed analysis method and the stability prohibition region of equivalent loop gain corresponding to the traditional impedance stability analysis method were also given.

Based on the above research, aiming at the operation condition of energy storage inverter in off grid mode, the system output impedance is derived according to the system structure, and the impedance ratio forbidden zone is used to analyze the stability of the system. Double closed-loop control strategy is adopted, and odd harmonics are introduced into the controller to ensure high gain, increase system bandwidth, reduce the influence of low harmonic current on voltage, and realize zero error tracking of voltage. Finally, through impedance ratio Nyquist curves, the influence of inverter output impedance change, load level, controller parameters and filter parameters on system stability is analyzed, and the correctness and effectiveness of the proposed stability criterion are verified, which provides corresponding theoretical support and parameter optimization reference for the design of practical system.

2. Main circuit of microgrid inverter level connection system

A large number of distributed generators are connected to the microgrid through power electronic equipment, which makes the system have the problems of low inertia, easy interference and instability. The energy storage unit adopts droop control, which can provide voltage and frequency support for the microgrid system, realize peak shaving and suppress the power fluctuation of new energy. The off-grid mode structure of the distributed energy storage system is shown in Fig.1. The energy storage inverter adopts voltage current double loop control, introduces virtual impedance and QPR control, realizes voltage zero static error tracking in a wide frequency band, and reduces the circulating current caused by voltage deviation between inverters. Where u_{abc} is the output voltage of the inverter and i_{abc} is the three-phase current on the filter inductor.

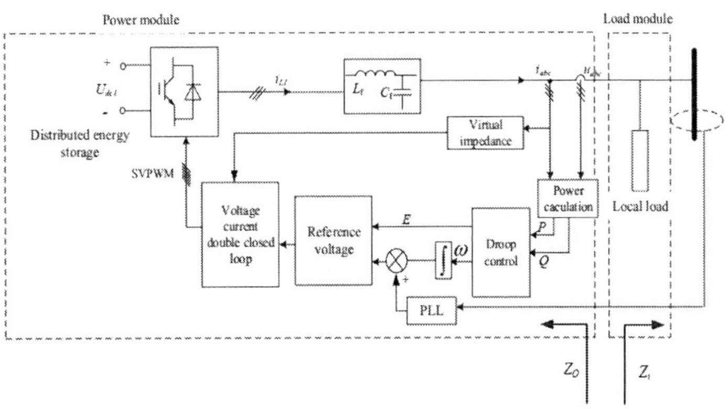

Fig.1 The structure diagram and control block diagram of the microgrids

The structure diagram and control block diagram of the microgrids composed of the inverter controlled by droop control. L_f is the filter inductance, C_f is the filter capacitor. The control system mainly includes *PQ* power outer loop and voltage and current double loop tracking link. According to the collected output voltage and current signals at the grid connected terminal, the voltage reference command *E* and angular frequency ω are generated through *PQ* calculation and control module. The virtual impedance link is introduced to compensate the voltage to achieve accurate power distribution, and then the voltage and current double loop tracking *E* modulates the *SPWM* drive signal required by the grid connected inverter components. Finally, the power is controlled and adjusted by the inverter to ensure the stability of voltage and frequency.

In the actual distributed power system, the front end of load converter usually needs to be equipped with input filter, and their overall input impedance characteristics are undoubtedly more complex, which makes it difficult to establish and analyze the input impedance model. Since the concept of impedance ratio forbidden region was put forward, the stability analysis process of power supply system has been simplified, and the stability of the system can be analyzed and determined[12]. The distributed energy storage power supply module, inverter and *LC* filter form the source module in the cascade system, and the line impedance and load form the load module in the system. The equivalent impedances of the original module and load module are Z_o and Z_i respectively.

3. Control strategy of inverter system

3.1. Double loop control structure

The inverter system controller adopts double loop control structure, which can increase the anti-interference of the system and improve the dynamic performance of the system. In the selection of control strategy, although the parameter setting of PI controller is convenient and easy to realize, there is steady-state error in tracking AC signal. The proportional resonance (PR) control can eliminate the steady-state error at the specified frequency, and its narrow bandwidth makes it impossible to track the signal when it deviates from the given frequency of the system.

The voltage outer loop adopts QPR controller $G_{PR}(s)$ to improve the voltage tracking performance of the inverter, while the current inner loop adopts inductive current proportional controller $G_i(s)$ to improve the response speed and anti-interference ability of the system. In the case of nonlinear load, the output voltage waveform is seriously distorted, and the load current is introduced into the inner loop, which transfer function is H(s). Thus, the block diagram of double closed-loop control strategy based on QPR control can be obtained, as shown in Fig.2.

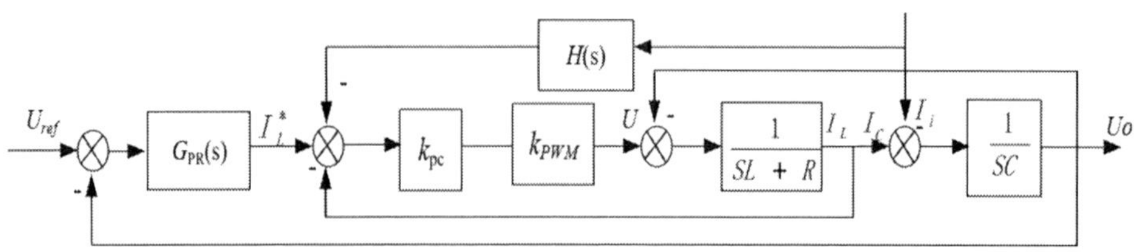

Fig.2 Double loop control structure of inverter

In Fig.2, *U* is the output voltage of inverter; I_i is the output current of the inverter, I_c is filter capacitor current, k_{pc} is the proportional coefficient of current loop, k_{pwm} is the inverter gain, *C* is the filter capacitance. U_{ref} and U_o are the voltage reference load side voltage .

3.2. Principle of QPR controller

In order to reasonably configure the parameters of the QPR controller, it is necessary to understand the influence of each parameter on the performance of the controller. A zero point is added to the PR controller to transform the proportional resonance controller into a QPR controller. The transfer function of the QPR controller is as following:

$$G_{PR} = k_{pr} + \frac{2k_r\omega_r s}{s^2 + 2\omega_r s + \omega^2} \tag{1}$$

Where: k_{pr} is the proportional coefficient and k_r is the integral coefficient, ω_r is the cut-off angle frequency; ω is the fundamental angular frequency.

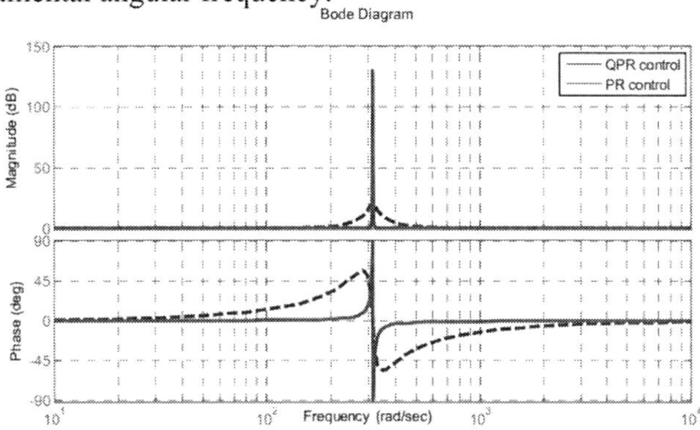

Fig.3 Bode diagram of PR and quasi PR control

Comparing PR control and QPR control, it can be found that PR is at the resonance point the gain at ω_0 is almost infinite, which can realize closed-loop static error free tracking of ac signal. However, it has large gain at the resonance point, narrow bandwidth and poor anti disturbance ability, which can easily lead to system instability. It is difficult to accurately adjust the resonance point in practical control. While the QPR has high gain at the resonance point, it increases the bandwidth of the controller and reduces the sensitivity of the system, but it still maintains high gain, has good tracking performance, increases the steady-state margin and improves the dynamic performance.

3.3. Parameter analysis of QPR controller

In order to reasonably configure the parameters of the QPR controller, then the influence of each parameter on the performance of the controller is analyzed.

3.3.1. Influence of ω_r on controller characteristics

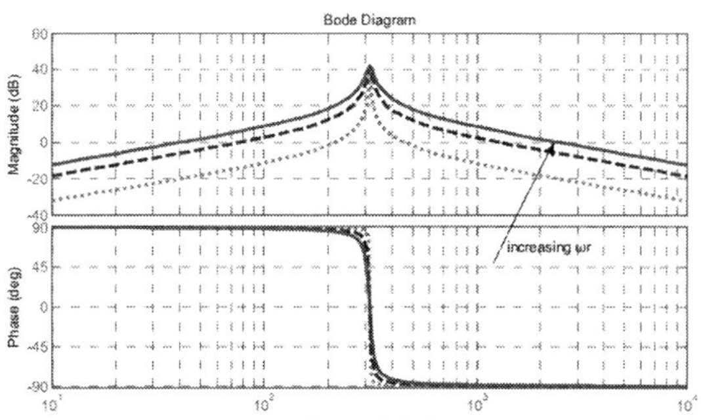

Fig.4 Impedance Bode diagram of ω_r variation

The bandwidth of QPR controller is mainly determined by ω_c, which is independent of the resonant point. The selection of frequency ω_r is generally between 5-20rad/s.

3.3.2. Influence of k_r on controller characteristics

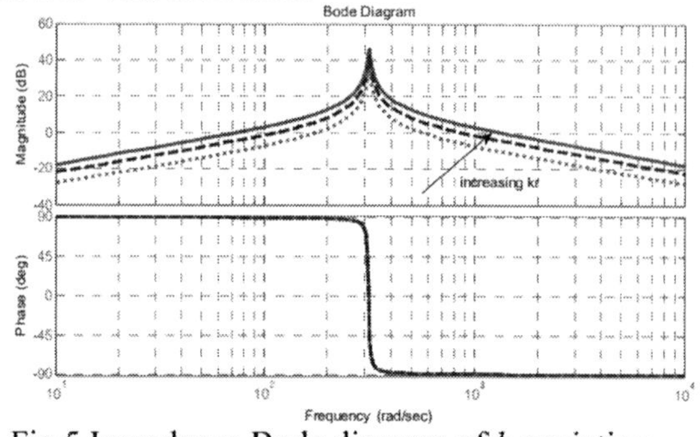

Fig.5 Impedance Bode diagram of k_r variation

As can be seen from Fig.5, k_r does not affect the bandwidth of the controller, but only the gain. When k_r increases, the gain of QPR controller increases too.

3.3.3. Influence of k_{pr} on controller characteristics

As can be seen from Fig.6, when k_{pr} is small, the bandwidth of the QPR controller and the gain of the full frequency band are affected. When k_{pr} is large, the QPR controller is approximately equal to the proportional controller.

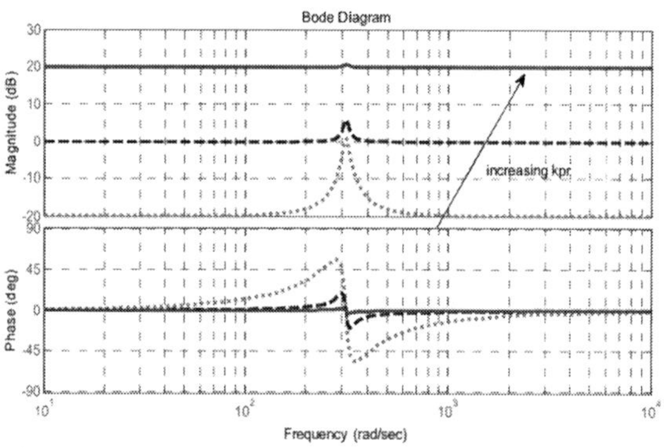

Fig.6 Impedance Bode diagram of k_{pr} variation

3.4. QPR controller with harmonic resonator

When the inverter is loaded with nonlinear load, the load current contains a large number of harmonics. It is necessary to implant the internal model of fundamental wave and all harmonics into the controller to make the voltage ring obtain large enough gain, so as to improve the output voltage accuracy of the inverter and the capacity with nonlinear load. After the 1,3,5,7 harmonics are implanted into the control, there are:

$$G_{PR}(s) = k_{pr} + \sum_{h=1,3,5,7} \frac{2k_{rh}\omega_{ch}s}{s^2 + 2\omega_{ch}s + \omega_h^2} \qquad (2)$$

Due to k_r, ω_r affects the gain and bandwidth of the resonance point respectively, so the design of QPR needs to determine k_{pr} and fundamental resonance parameters first. The harmonic resonance part increases the gain at the harmonic of the controller, reduces the output impedance at the harmonic, and realizes no static error tracking of the output current harmonic. The gain of the resonant point can be designed from the point of view of reducing the output impedance and eliminating harmonics. Fig.7 shows the voltage outer loop open-loop Bode diagram after adding 1,3,5,7 harmonic resonators. After adding harmonic resonance, the open-loop cut-off frequency increases and the phase angle margin decreases, but the system stability can still be guaranteed. The open-loop Bode diagram has large gain at the fundamental and harmonic frequency points, and can well track the signal at the specified frequency points.

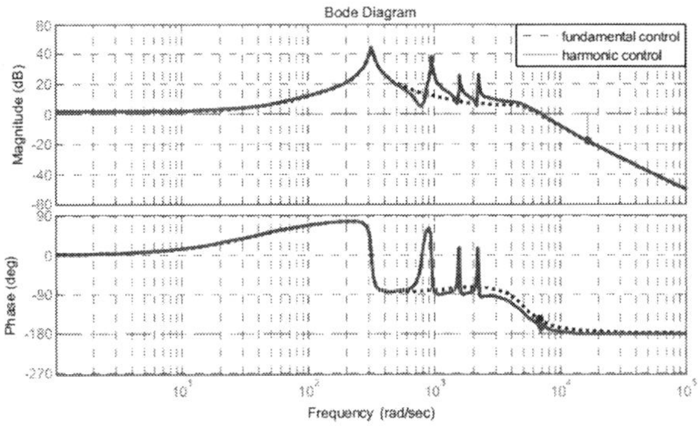

Fig.7 Open loop Bode diagram of voltage outer loop

3.5. Inverter output impedance analysis

Based on the parameter design of QPR controller, the outer loop voltage transfer function is decomposed into impedance form according to the structure diagram through impedance stability analysis:

$$U_o = G(s)U_{ref} - Z(s)I_i \tag{3}$$

If H(s)=1,then the output impedance of the inverter is:

$$Z(s) = \frac{Ls^2 + (R+2L\omega_r)s^2 + (2R\omega_r + L\omega^2)s + R\omega^2}{as^4 + bs^3 + cs^2 + ds + e} \tag{4}$$

Where the coefficients are:

$$\begin{cases} a = LC \\ b = (k_{pc}k_{PWM} + R + 2L\omega_r)C \\ c = 1 + 2(k_{pc}k_{PWM} + R)\omega_r C + L\omega^2 C + k_1 k_p \\ d = (k_{pc}k_{PWM} + R)\omega^2 C + 2(k_1 k_p + k_1 k_r)\omega_r + 2\omega_r \\ e = (k_1 k_p + 1)\omega^2 \end{cases}$$

The characteristic equation of the system is:

$$DG(s) = as^4 + bs^3 + cs^2 + ds + e \tag{5}$$

The parameter conditions required for system stability can be determined according to Rouse criterion, so as to solve the value range of control parameters k_p and k_r in QPR controller.

3.6. Impedance ratio forbidden zone

Since the concept of impedance ratio forbidden region has put forward, the stability of the system can be analyzed and determined only by the impedance ratio of source and load converter.The stability criterion proposed by Middlebrook considered the relative stability of the system on the basis of absolute stability, that is, the dynamic performance must meet a certain stability margin, including amplitude margin GM and phase marginγ. Among the many proposed impedance ratio prohibition areas, GM(dB)≥6dB and $\gamma \geq 60°$ are selected.

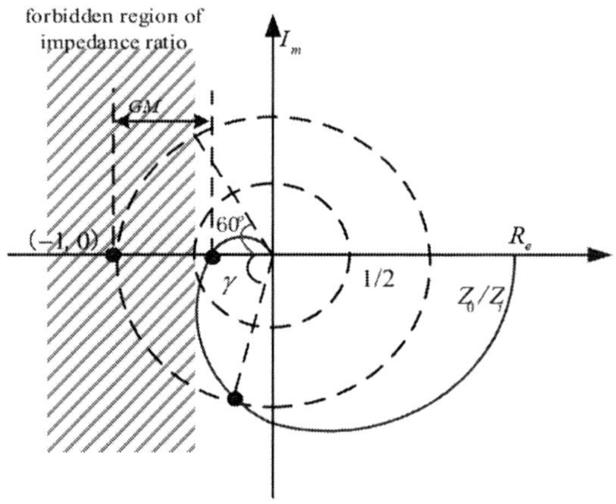

Fig.8 Forbidden domain of system stability

Through conversion, it is obtained that the impedance Nyquist curve meets re (Z_0/Z_i)≥-1/2, that is, it is unstable when it crosses the red region in Fig.8. In practical engineering application, the criteria for judging system stability can be appropriately relaxed according to the actual situation, so as to avoid unnecessary trouble to system design and research due to over conservatism on the premise of ensuring system stability.

4. Analysis of the system stability

Table 1 System parameters

parameters	value
DC voltage/V	800
L/mH	1
R/Ω	0.01
C/F	40×10^{-6}
k_{pc}	0.02
ω	100*π
k_p	0.4
k_r	120
ω_r	3.2
k_{PWM}	200

4.1. Influence of changing control parameters on system stability

4.1.1. Influence of ω_r

The influence of control parameters and filter parameters on system stability is verified. Only change ω_r value and other parameters remain unchanged. When ω_r Nyquist curve of impedance Z_o/Z_i when ω_r =13.

The Nyquist curve of impedance Z_o/Z_i is shown in the Fig.9. It can be seen that the curve has crossed the stable prohibition region. When the load impedance is certain, with ω_r increased, the Nyquist curve will gradually approach the rejection region, making the system stability worse.

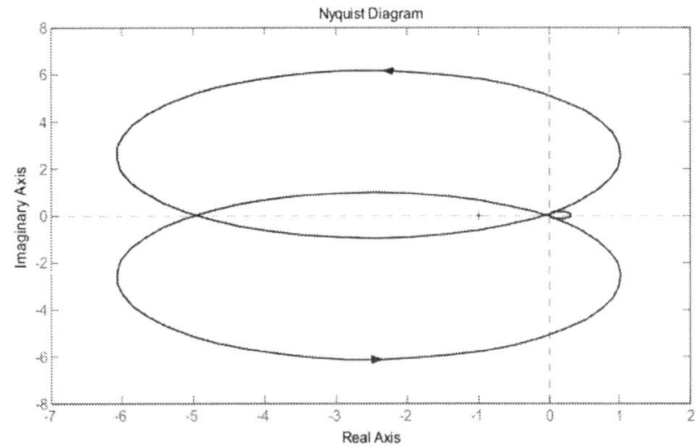

Fig.9 The Nyquist curve of impedance Z_o/Z_i (ω_r=13)

4.1.2. *Influence of k_r*

It can be seen that with the increase of $\boldsymbol{k}_r$, the area surrounded by the Nyquist curve of system impedance becomes larger and closer to the prohibition domain. Therefore, too large $\boldsymbol{k}_r$ will make the Nyquist curve of system impedance enter the prohibition domain, resulting in system instability. However, it should be noted that the larger the $\boldsymbol{k}_r$, the higher the steady-state accuracy of the system and the faster the response speed. Therefore, a larger $\boldsymbol{k}_r$ value can be selected on the premise of ensuring that the system has sufficient stability margin.

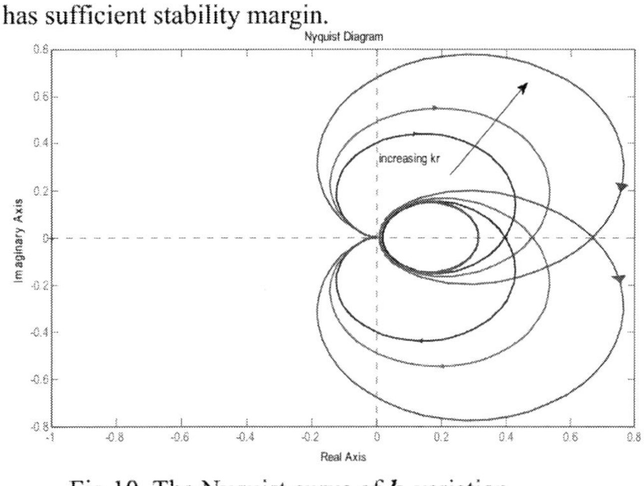

Fig.10 The Nyquist curve of $\boldsymbol{k}_r$ variation

4.1.3. *Influence of C*

In the process of increasing C, the small signal stability of the main energy storage power supply becomes better. When the filter capacitance increases, the small signal stability of the main energy storage power supply becomes better. The larger the C, the smaller the impact on the small signal stability of the main energy storage power supply.

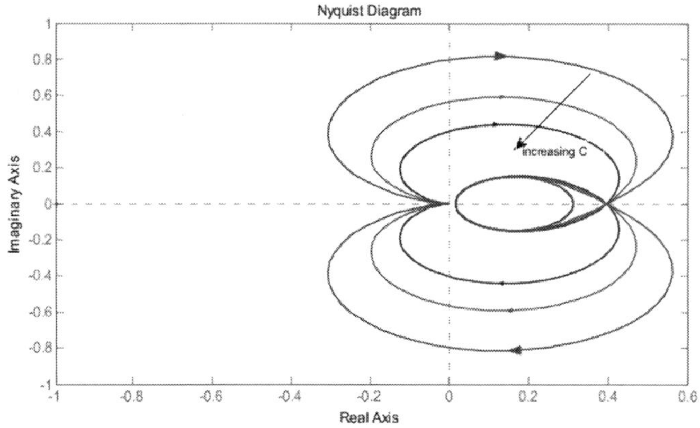

Fig.11 The Nyquist curve of C variation

4.2. Influence of load variation on system stability

The impedance ratio analysis method is used to analyze the impact on system stability by changing the load. The load parameters are:Z_L=1+0.005s

By changing the load parameters, the Nyquist curve of impedance Z_o/Z_i is obtained, as shown in Fig.12.

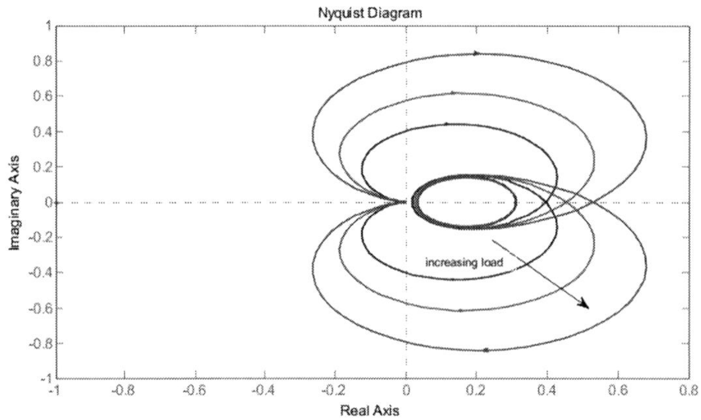

Fig.12 Impedance Nyquist curve varying with load level

As can be seen from Fig.12, with the increase of load level, the impedance Nyquist curve keeps approaching the rejection region. Therefore, when the energy storage system operates with load in off grid mode, the change of load level will have a great impact on the control stability. When the load level exceeds the system power level limit, the system will not operate stably.

5. Conclusion

The stability of microgrid inverter in off-grid mode is analyzed by impedance analysis method. The double closed-loop structure with QPR control can increase the bandwidth and realize no static error tracking voltage. The change of control parameters and load will affect the output impedance of the inverter. When the impedance Nyquist curve crosses the prohibition domain, it will cause the instability of the system. Based on the results and discussions presented above, the conclusions are obtained as below:

(1) In the off-grid mode, the microgrid energy storage inverter does not need to know the internal specific parameters of the subsystem, and the stability of the system can still be analyzed by the impedance ratio method of the cascade system.

(2) The double closed-loop QPR control can increase the bandwidth and gain of the controller, improve the tracking performance of the system, increase the steady-state margin and improve the dynamic performance

(3) The controller parameters and filter parameters will affect the output impedance of the inverter. When the impedance Nyquist curve crosses the prohibition domain, it will cause the instability of the system.

Acknowledgments

This work was financially supported by the Key Projects of Natural Science Research in Colleges and Universities of Anhui Province（KJ2018A0085）.

References

[1] ZENG Zheng, ZHAO Rongxiang, TANG Shengqing, et al. An overview on advanced grid-connected inverters used for decentralized renewable energy resources[J]. Proceedings of the CSEE, 2013, 33(24):1-12.
[2] H.J. Cheng, Z.K. Shuai, C. Shen, et al. "Transient angle stability of paralleled synchronous and virtual synchronous generators in islanded microgrids"[J]. IEEE Transactions on Power Electronics, 2020, 35(8): 8751-8765.
[3] WAGNER M, BARTH T, ALVAREZ R, et al. Discrete-time active damping of LCL-resonance by proportional capacitor currentfeedback[J]. IEEE Transactions on Industry Applications, 2014, 50(6): 3911–3920.
[4] LEI Yaxiong, LI Jianwen, LI Yonggang. Control strategy of three-phase LCL grid-connected inverter based on quasi-PR adjuster[J]. Power System Protection and Control, 2014, 42(12): 44-50.
[5] YANG Ping, ZHENG Yuanhui, XU Zhirong, et al. Study on photovoltaic grid - connected inverter based on quasi - PRI and repetitive control[J]. Renewable Energy Resources, 2015, 33(7): 993-998.
[6] Chen X, Wang Y C, Gong C Y, et al. Overview of Stability Research for Grid-Connected Inverters Based on Impedance Analysis Method[J]. Proceedings of the CSEE, 2018, 38(7): 2082-2094+2223.
[7] Sun J. Impedance-Based Stability Criterion for Grid-Connected Inverters[J]. IEEE Transactions on Power Electronics, 2011, 26(11): 3075-3078.
[8] Xu Shaohua, Li Jianlin, Hui Dong. Stability Analysis of Energy Storage Inverter Based on Quasi PR Controller Under Off-grid Mode[J]. Automation of Electric Power systems, 2015, 39(19):107-112.
[9] Li Yi Xin, Zhao Shuqiang, Ma Yanfeng, et al. Impedance modeling and characteristic analysis of three-phase LCL-type grid-connected inverters[J]. Electric Power Automation Equipment, 2019, 39(7): 107-113.
[10] JIANG Shigong, WANG Wei, LIU Guihua, et al. Admittance specification stability of islanded microgrid[J]. Electric Power Automation Equipment, 2013, 33(5):19-25.
[11] TONG Qiang, ZHANG Donglai, XU Dianguo. Output Impedance and Stability Analysis of Converters in Distributed Power Systems[J]. Proceedings of the CSEE, 2011, 31(12):57-64.
[12] CHEN Xin, WANG Yuncheng, GONG Chunying et al. Overview of Stability Research for Grid-Connected Inverters Based on Impedance Analysis Method[J]. Proceedings of the CSEE, 2018, 38(7):2082-2092.

,

A control strategy for battery/ supercapacitor hybrid energy storage system

Congzhen Xie[1a*], Jigang Wang[1b], Bing Luo[2c], Xiaolin Li[2d], Lei Ja[2e]

[1]School of Electric Power, South China University of Technology, Guangzhou 510000, China

[2]China Southern Power Grid Research Institute Co., LTD., Guangzhou 510000, China

[a*] xiecongzhen@scut.edu.cn, [b] 202111082884@mail.scut.edu.cn, [c] luobing@csg.cn,

[d] lixl@csg.cn, [e] jialei@csg.cn

Abstract—In DC microgrid (MG), the hybrid energy storage system (HESS) of battery and supercapacitor (SC) has the important function of buffering power impact, which comes from renewable energy sources (RES) and loads. This paper proposes a HESS control strategy with DC bus voltage self-recovery function. High and low frequency power decomposition based on virtual droop control, and DC bus drop voltage is compensated by added proportional integral regulation. The relationship between DC bus voltage recovery and super-capacitor (SC) state of charge (SoC) recovery is analyzed. The system can realize stable energy storage, supply under frequent load power impact. The effectiveness of the proposed control strategy is verified by simulation in MATLAB/Simulink.

1. Introduction

In a distributed DC microgrid (MG), super-capacitors (SCs) can effectively reduce the adverse impact of renewable energy sources (RESs) and loads on batteries. They are often combined with batteries to form a MG energy storage system (ESS), making the ESS has high power density as well as high energy density [1-2].

In the hybrid energy storage system (HESS) control strategies, the droop control has gradually become a popular solution due to its obvious advantages such as no additional communication, simple structure and strong practicability. However, under the action of this strategy, there will be a static error in the DC bus voltage and may cause the SC to cross the line under multiple power shocks, which is not conducive to system stability.

In recent years, the state of charge (SoC) recovery of SC and bus voltage stability control have become the research hotspots of scholars, and several improved droop control methods had been proposed [3-5]. Method 1: Added a compensation voltage source to the SC branch, so that the SC had the SoC self-recovery function [6-7], but this strategy does not have an effect on the bus voltage deviation. Method 2: According to the accurate line impedance information corrected the droop control resistance value, and performed bus voltage translation compensation to restore the DC bus voltage [8]. This method requires high accuracy of impedance measurement.

In order to solve the above problems, this paper proposes a simple and practical improved droop control strategy that does not require communication, which overcomes the problem of bus voltage steady-state deviation, and the SoC of SC has self-recovery capability under this strategy.

2. The traditional virtual droop control strategy for HESS.

2.1 The system structure

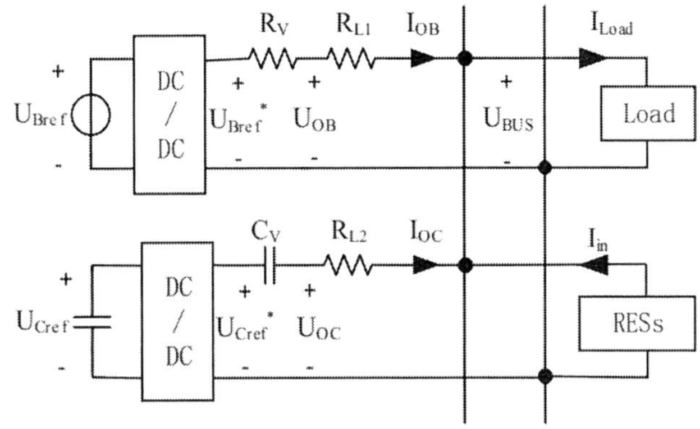

Fig.1 Structure diagram of MGs using virtual droop control strategy

Fig. 1 shows the system structure. The storage battery (energy type) and SC (power type) form the energy storage unit, they are connected to the DC bus in parallel through DC-DC converters. I_{OB}, I_{OC}, I_{load} and I_{in} are output current of battery and SC, load output current and RESs input current respectively. R_V is virtual droop resistance (VDR) and C_V is virtual droop capacitor (VDC), R_{L1} and R_{L2} are line resistance, U_{BUS} is reference voltage of DC bus. U_{bref} and U_{cref} are reference terminal voltage of battery and SC, they are equivalent to the DC converter bus side being U_{Bref}* and U_{Cref}*, The RESs input current and load output current are equivalent to the output current I_O.

$$I_O = I_{Load} - I_{in} \tag{1}$$

2.2 The virtual droop control principle.

Energy storage branches are very close to the DC bus, and the line impedance is far less than the VDR. In order to simplify the derivation process, the line impedances (R_{L1} and R_{L2}) are ignored [9-10].

For the circuit model writing KVL and KCL equations, they are shown in Formula (2).

$$\begin{cases} U_{BUS} = U_{Bref}^{*} - R_V I_{OB} \\ U_{BUS} = U_{Cref}^{*} - \dfrac{1}{sC_V} I_{OC} \\ I_O = I_{OB} + I_{OC} \end{cases} \tag{2}$$

The VDR satisfies Formula (3).

$$\frac{\Delta U_{BUS,\min}}{I_{OB,\min}} \le R_V \le \frac{\Delta U_{BUS,\max}}{I_{OB,\max}} \tag{3}$$

Formula (4) can be obtained by solving the output current gain of battery and SC. ω_c is the cutoff Angle frequency. When $\omega > \omega_c$, the high frequency current component required by the load will be provided by the SC. When $\omega < \omega_c$, the low frequency current component required by the load will be provided by the battery. The HESS has the dual characteristics of high-power density and high-energy density.

$$
\begin{cases}
|A_B(j\omega)| = \left|\dfrac{I_{OB}(j\omega)}{I_{OB}(j\omega)+I_{OC}(j\omega)}\right| = \dfrac{1}{\sqrt{1+(\dfrac{\omega}{\omega_c})^2}} \\
|A_C(j\omega)| = \left|\dfrac{I_{OC}(j\omega)}{I_{OC}(j\omega)+I_{OB}(j\omega)}\right| = \dfrac{1}{\sqrt{1+(\dfrac{\omega_c}{\omega})^2}} \\
\omega_c = \dfrac{1}{R_V C_V}
\end{cases}
\tag{4}
$$

2.3 SoC self-recovery conditions of SC.

According to formula (4), in case of sudden change of load or renewable energy output power, the SC needs to share high-frequency current to alleviate power impact. Therefore, there should be sufficient power to meet the requirements of frequent power shocks, so the SC needs to have the recovery ability of state of charge (SoC).

The SoC of SC is solved by ampere method, as shown in formula (5). Where SoC (t_0) is the initial charge state of the SC, Q is the rated total capacity, u_{CV} is the terminal voltage of the VDC, the start time is recorded as t_0, and the time to reach the steady-state again under the power impact is recorded as t_1

$$
\begin{cases}
SOC(t_1)=SOC(t_0)-\dfrac{1}{Q}\displaystyle\int_{t_0}^{t_1} i_{OC}dt \\
i_{OC} = C_V \dfrac{du_{C_V}}{dt}
\end{cases}
\tag{5}
$$

$$
SOC(t_1)=SOC(t_0)-\frac{1}{C_V Q}[U_{C_V}(t_1)-U_{C_V}(t_0)] \tag{6}
$$

The SoC calculation formula can be obtained by solving equations. Among them, Ucv(t_0) is the initial terminal voltage of the VDC, and Ucv(t_1) is the terminal voltage when the power shock reaches a new steady state.

According to formula 6, when Ucv(t_1) = Ucv(t_0), SoC(t_1) = SoC(t_0), the SoC of the super capacitor reaches the self-recovery condition. Setting the battery reference voltage, SC reference voltage and DC bus reference voltage to be equal, meeting formula (7), that is, the VDC voltage Ucv(t_1) = Ucv(t_0).

$$
U_{BUS} = U_{Bref}^{*} = U_{Cref}^{*} \tag{7}
$$

When the system reaches steady state again, if $U_{BUS} = U_{Cref}$*, the SC has SoC self-recovery ability.

2.4 The limitation of virtual droop control.

Assuming that the initial reference voltage of the DC bus is U_{BUSref}, which is equal to the output voltage of battery and SC. Under the action of virtual droop control, the relationship between DC bus voltage and battery output voltage is shown in formula (8). $U_{BUS} \neq U_{Cref}$*, therefore, the SC does not have the SoC recovery capability.

$$
\begin{cases}
U_{BUSref} = U_{BUS} + \Delta U_B \\
\Delta U_B = I_{OB}(R_V + R_{L1})
\end{cases}
\tag{8}
$$

3. Bus voltage compensation control strategy.

Due to the existence of VDR, the DC bus voltage is lower than the rated value, so that the SC does not meet the SoC recovery condition. Therefore, it is easy to cause the SoC to cross the line under multiple power shocks. Then it will affect the stability of power supply and even cause accidents, so it is necessary to compensate for the reduced DC bus voltage.

3.1 The improved droop control with bus voltage recovery.

In this paper, based on the traditional droop control, the DC bus voltage recovery is realized by PI control, the compensation voltage is equal to the DC bus drop voltage ΔU_B.

$$\Delta U_B = (k_p + \frac{k_i}{s})(U_{BUSref} - U_{BUS}) \tag{9}$$

KVL and KCL equations of the system obtained by PI control are listed as formula (10).

$$\begin{cases} U_{BUS} = U^*_{Bref} - \dfrac{s}{s + sk_p + k_i} R_V I_{OB} \\ U_{BUS} = U^*_{Cref} - \dfrac{1}{sC_V} I_{OC} \\ I_O = I_{OB} + I_{OC} \end{cases} \tag{10}$$

Then, the transfer function of battery output current can be obtained as shown in Formula (11).

$$A_B = \frac{s\dfrac{k_p+1}{R_V C_V} + \dfrac{k_i}{R_V C_V}}{s^2 + s\dfrac{k_p+1}{R_V C_V} + \dfrac{k_i}{R_V C_V}} \tag{11}$$

If $\omega_n = \sqrt{\dfrac{k_i}{R_V C_V}}$ and $2\alpha\omega_n s = \dfrac{k_p+1}{R_V C_V} s$, the above formula can be simplified as follows:

$$A_B = \frac{2\alpha\omega_n s + \omega_n^2}{s^2 + 2\alpha\omega_n s + \omega_n^2} \tag{12}$$

$$k_i = R_V C_V \omega_n^2 \tag{13}$$

$$C_V = \frac{k_p + 1}{2R_V \alpha\omega_n} \tag{14}$$

When $\omega = \omega_c$, the gain is approximately $\dfrac{1}{\sqrt{2}}$, therefore, the expression for ω_n can be calculated.

$$\left|A_B(j\omega)\right|_{\omega=\omega_c} = \frac{\sqrt{1 + 4\alpha^2(\frac{\omega_c}{\omega_n})^2}}{\sqrt{[1-(\frac{\omega_c}{\omega_n})^2]^2 + 4\alpha^2(\frac{\omega_c}{\omega_n})^2}} = 10^{-0.15} \approx \frac{1}{\sqrt{2}}$$

$$\omega_n = \frac{\omega_c}{\sqrt{1 + 2\alpha^2 + \sqrt{1 + (1+2\alpha^2)^2}}} \tag{15}$$

The transfer function of output current of SC is as follows:

$$A_C = 1 - A_B = \frac{s^2}{s^2 + s\frac{k_p + 1}{R_V C_V} + \frac{k_i}{R_V C_V}} \tag{16}$$

By substituting the parameters obtained in formulas (13) to (15) into formulas (11) and (16), the output current transfers function of battery and SC can be solved.

3.2 The system control strategy.

The topological structure of the DC-DC converter used in this paper is shown in Fig. 2, where U_{ref} is the voltage at the low voltage side and U_{ref}^* is the voltage at the DC bus side. The converter adopts voltage outer loop and current inner loop control. Control block diagrams are shown in Fig. 3.

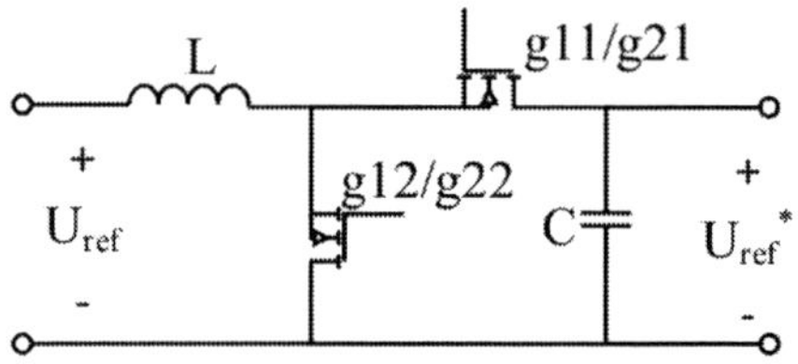

Fig. 2 The DC-DC converter topology

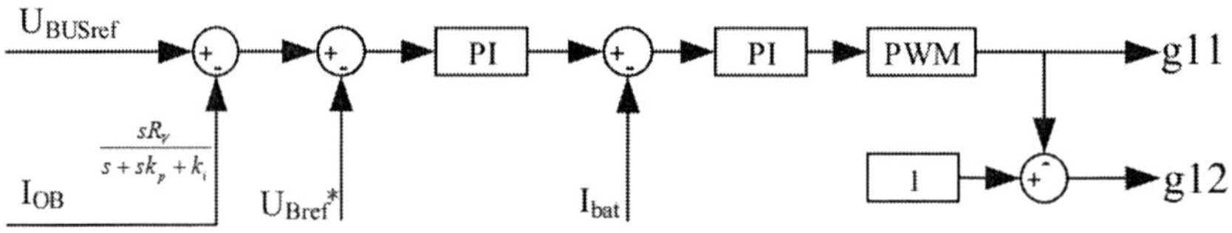

(a) Battery output control block diagram

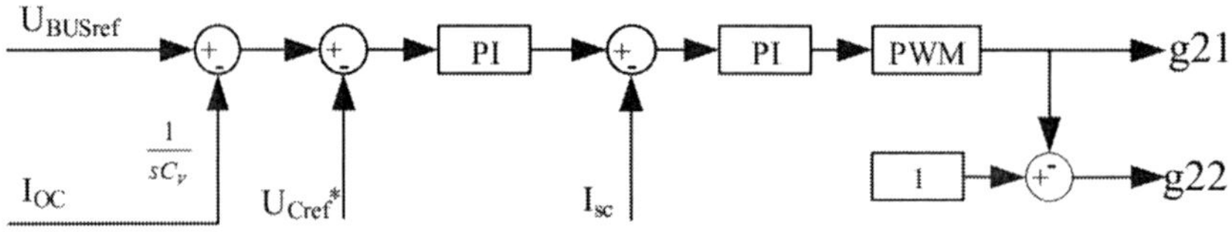

(b) SC output control block diagram

Fig. 3 The system control block diagram

4. Simulation

In order to verify the rationality of the proposed control strategy, MATLAB/Simulink was used for simulation. To study the influence of different parameters on the system, four cases were set, as shown in Table 1. All cases used the same cut-off frequency and virtual droop resistance. The parameter settings of system simulation circuit model are shown in Table 2.

Table 1 cases parameters

Cases	ω_c	α	VRD(R_V)	VCD(C_V)	*kp*	*ki*
1	0.4×2π	/	1	0.3979	0	0
2	0.4×2π	0.8	1	1.0862	1	1.4384
3	0.4×2π	1	1	0.9877	1	1.0124
4	0.4×2π	1.2	1	0.9312	1	0.7457

Case 1 represents the use of traditional virtual droop control, while cases 2 to 4 represent the use of improved virtual droop control. Cases 2 to 4 are used for comparative analysis of the impact of damping α on the system. The system output power change from 0 to 20W at 1s, increases the load power to 30W at 10s, and reduces the load power to 10W at 20s.

Table 2 Circuit model parameters of the system

Parameters	Instructions	Value
U_{Bref}、U_{Cref}、U_{BUSref}	Reference voltage of battery, ultracapacitor and DC bus	12V、12V、24V
$SoC_B(t_0)$	Battery Initial SoC	50%
$SoC_{SC}(t_0)$	SC initial SoC	50%
L	DC-DC converter filter inductance	2mH
C	DC-DC converter capacitance	1100uH
P_O	Output power of hybrid energy storage system	0W(0~1s) 20W(1~10s) 30W(10~20s) 10W(20s~30s)

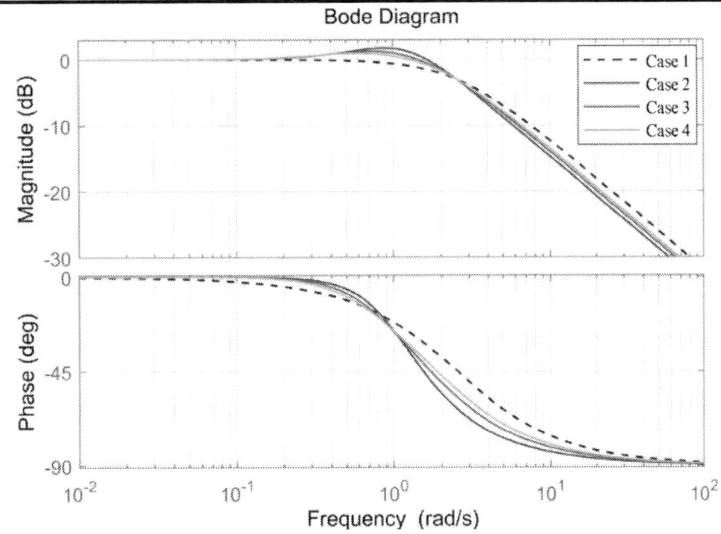

(a) Frequency response of battery branch current

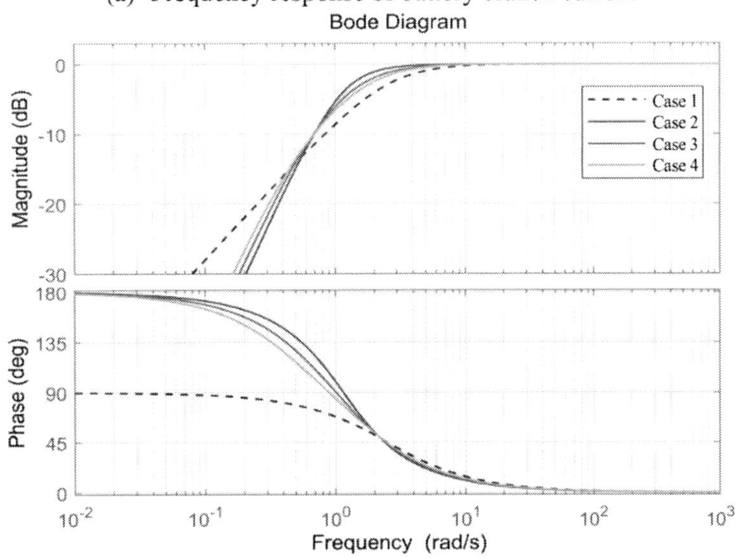

(b) Frequency response of SC branch

Fig. 4 frequency response characteristics

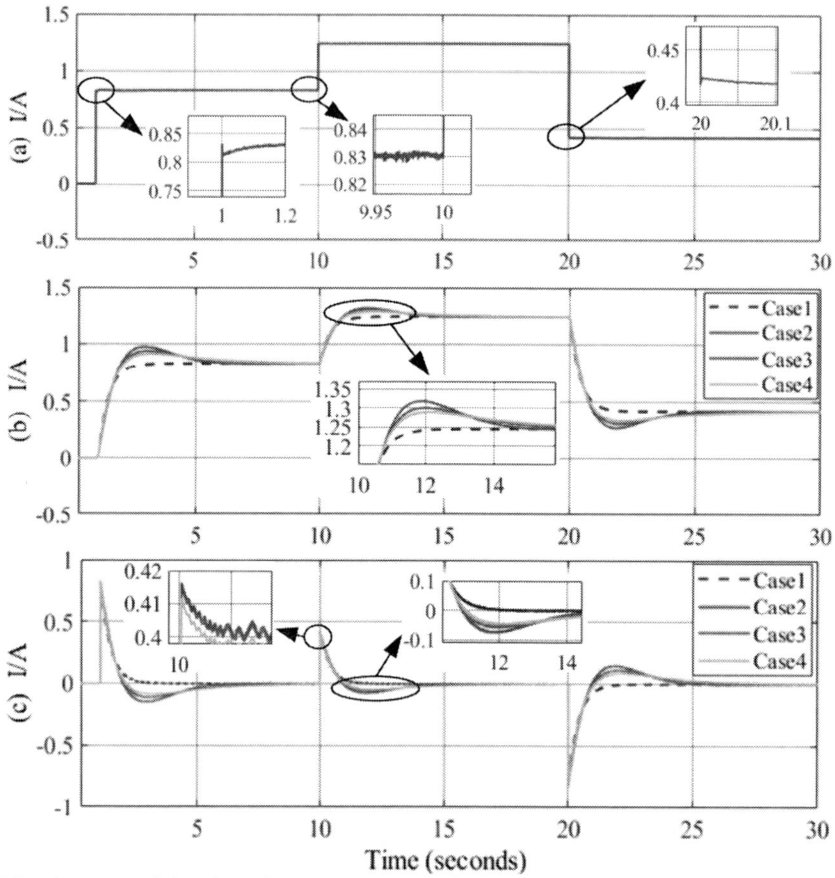

Fig. 5 load current (Fig. a) and output current of storage battery (Fig. b) and SC (Fig. c)

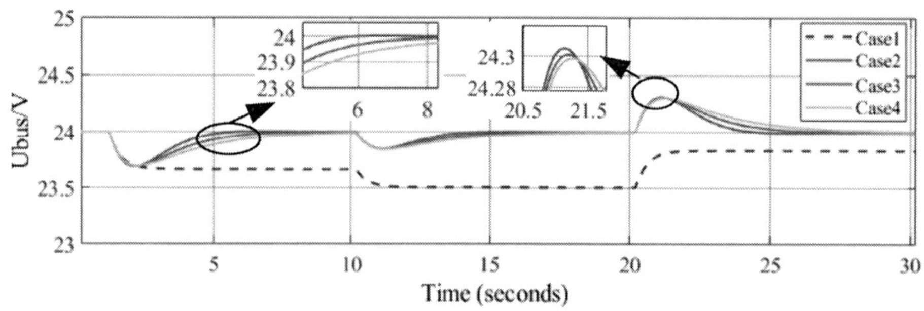

Fig. 6 DC bus voltage

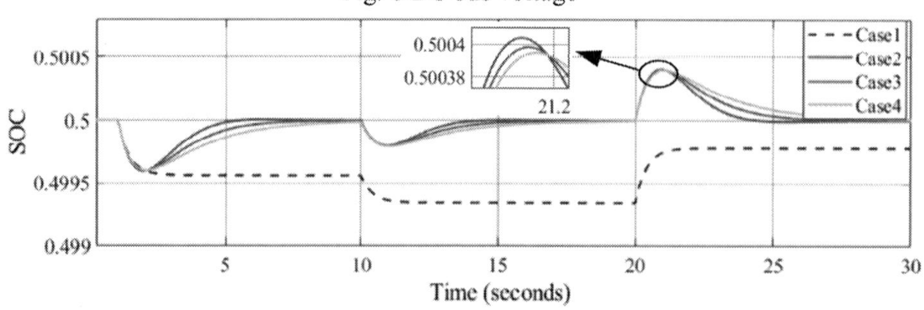

Fig. 7 The SoC of SC

Fig. 4 shows the output current frequency characteristic curves of the battery branch and the SC branch under the control of schemes 1 to 4. The dashed line is the simulation result of case 1 (traditional droop control). Through observation, it can be seen that all cases can achieve effective separation of high and low frequency power. In cases 2 to 4, the increase of α will reduce the peak value of the frequency response curve. The α has a certain blocking effect on the system regulation response.

Fig. 5 shows the load current waveform and the output current waveform of the battery and SC under the control of schemes 1 to 4. The time axis starts at 0.2s. The load current is 0 from 0.2 to 1s, 0.83A from 1 to 10s, 1.25A from 10 to 20s, and 0.42A from 20 to 30s. In all cases, the peak value of current change in case 1 is the smallest, and in cases 2-4, the peak value of current change gradually decreases with the increase of damping coefficient α.

Fig. 6 shows the changes in the DC bus voltage under the control of schemes 1 to 4. It can be seen that under the traditional droop control (dotted line), when the load power changes suddenly, the bus voltage will have an obvious static deviation when the system is stabilized again. The larger the load current is, the larger the bus voltage static deviation will be. In cases 2 to 4, the DC bus voltage has a self-recovery function. With the increase of α, although the peak value of the voltage change will decrease, the recovery time is significantly prolonged, which is not conducive to the rapid recovery of the system.

Fig. 7 shows the change curve of SC SoC. It can be seen that the SC SoC has the same change trend as the bus voltage, and the improved droop control can realize the self-recovery of SC SoC, and the peak value of SoC change and recovery time are affected by α.

Through the analysis of the above simulation results, it can be seen that the proposed control strategy has the functions of bus voltage self-recovery and SC SoC self-recovery while effectively separating high and low frequency power. Simulation results verify the effectiveness of the proposed strategy.

5. Conclusion

Based on the results and discussions presented above, the conclusions are obtained as below:

(1) The battery/ultracapacitor hybrid energy storage system can realize the high and low frequency power shunt under the traditional droop control, and can protect the energy storage system under the impact power. However, the traditional droop control will cause steady-state error of DC bus voltage, and it is easy for super power to go over line without intervention, which may cause serious harm to the system.

(2) Based on the traditional virtual droop control, this paper uses the method of PI control to compensate the dc bus drop voltage, which is caused by the existence of virtual resistance. The effectiveness of this method is verified by theoretical derivation and simulation. The system has the functions of dc bus voltage self-recovery and SC SoC self-recovery. It is of great significance to the stable and reliable operation of battery/SC hybrid energy storage system.

(3) When buffering impulse power, there will be a mutual charge and discharge process between battery and supercapacitor, which is redundant for the system. And it is possible to shorten the battery life due to this phenomenon. The current decoupling between supercapacitor and battery will become a research hotspot in the future.

Acknowledgments

This work was financially supported by the Open fund Project of National Engineering Laboratory for Ultra High Voltage Engineering Technology (Kunming、Guangzhou).

References

[1] T. S. Babu, K. R. Vasudevan, V. K. Ramachandaramurthy, S. B. Sani, S. Chemud, and R. M. Lajim. "A Comprehensive Review of Hybrid Energy Storage Systems: Converter Topologies,

Control Strategies and Future Prospects". *IEEE Access*, vol 8, pp. 148702-148710, Aug, 2020.
[2] M. Shi, X. Chen, J. Zhou, Y. Chen, J. Wen, and H. He, ''Advanced secondary voltage recovery control for multiple HESSs in a droop-controlled DC microgrid,'' *IEEE Trans. Smart Grid*, vol. 10, no. 4, pp. 3828–3839, Jul. 2019.
[3] S. Augustine, M. K. Mishra, and N. Lakshminarasamma. "A Unified Control Scheme for a Standalone Solar-PV Low Voltage DC Microgrid System With HESS". *IEEE Emerg. Electron*, vol. 8, no. 2, pp. 1351-1359, Jun. 2020.
[4] D. N. Huu, ''An adaptive control of hybrid battery-supercapacitor storage for integration of wind and solar,'' in Proc. IEEE Int. Conf. Sustain. Energy Technol. (ICSET), pp. 157–162, Nov. 2016.
[5] X. Chen, J. Zhou, M. Shi, L. Yan, W. Zuo, and J. Wen. "A Novel Virtual Resistor and Capacitor Droop Control for HESS in Medium-Voltage DC System," *IEEE Trans. Power System.* vol. 34, no. 4, pp. 2518-2527, Jul. 2019.
[6] Q. Xu, X. Hu, P. Wang, J. Xiao, P. Tu, and C. Wen, "A Decentralized Control Strategy for Autonomous Transient Power Sharing and State-of-Charge Recovery in Hybrid Energy Storage Systems," *IEEE trans. Sustain. energy*, vol. 8, no. 4, pp. 1443-1452, Oct. 2017.
[7] Y. Zhang, and Y. Li. "Energy Management Strategy for Supercapacitor in Droop-Controlled DC Microgrid Using Virtual Impedance," *IEEE trans. electron*, vol. 32, no. 4, pp. 2704-2716, Apr. 2017.
[8] W. Jiang, J. Zhao, M. Gao, and L. Mao, "Control Strategy With Voltage Self-recovery Characteristic for Autonomous DC Microgrid," *Power System Technology*, vol. 44, no. 9, pp. 3547-3555, Sep. 2020.
[9] Q. Xu, X. Hu, P. Wang, J. Xiao, P. Tu, C. Wen, and M. Y. Lee, ''A decentralized dynamic power sharing strategy for hybrid energy storage system in autonomous DC microgrid,'' *IEEE Trans. Ind. Electron.*, vol. 64, no. 7, pp. 5930–5941, Jul. 2017.
[10] Q. Xu, J. Xiao, X. Hu, P. Wang, and M. Y. Lee, ''A decentralized power management strategy for hybrid energy storage system with autonomous bus voltage restoration and state-of-charge recovery,'' *IEEE Trans. Ind. Electron*, vol. 64, no. 9, pp. 7098–7108, Sep. 2017.

A clock synchronization method in power dispatching based on BeiDou navigation system

Hongbo Lian[1], **Xingcai Wang**[1*], **Shuo Zhang**[1]

[1]State Grid Huludao Electric Power Supply Company, Huludao, Liaoning 125000, China

[*]1142101056@ncepu.edu.cn

Abstract—This paper introduces the problems of poor signal condition of a single time source in time synchronization or error in time source switching in the power system. It adopts the two-clock mode timing scheme of BeiDou 2 synchronous satellite and GPS synchronous satellite to ensure the stable output of the main clock within a certain accuracy. The system takes full advantage of the field programmable gate array (FPGA) hardware method, aligns the internal clock lock signal with the input second pulse phase, and proposes an improved method for the problem of large clock synchronization deviation based on the network, which proves the effectiveness and superiority of the system to accurately output the pulse and provide reliable time synchronization for the power dispatching system.

1. Introduction

The BeiDou Satellite Navigation System (BDS) refers to a global satellite navigation system that is being independently developed and operated by China. Together with the American Russian GPS and Galileo of the EU system, they are known as the four major satellite navigation systems in the world. The BeiDou Navigation provides continuous navigation, positioning and timing services since December 27, 2011. The BeiDou satellite navigation system consists of three parts: space end, ground end and user end. The space end includes 5 geostationary orbit satellites and 30 non-geostationary orbit satellites, and the ground end includes major control stations, injection stations, and monitoring stations. The user terminal consists of the BeiDou user terminal [1-3]. The BeiDou satellite system has the following advantages in timing over the US GPS system: it can provide better timing accuracy than the GPS system. The BeiDou navigation system has two one-way and two-way timing functions. Sync with BeiDou navigation according to different accuracy requirements.

2. Timiming Principle and Method of BeiDou Navigation System

2.1. Operating principle of the BeiDou Satellite Navigation System

The timing of the BeiDou satellite system is realized by sending standard satellite time information and satellite location information from the ground central station. The specific process is: First, The ground center station first sends time marking information and DUT1 information to the user at the start frame time of each superframe, The time benchmark per frame of these time information maintains an accurate time synchronization relationship with the atoms; Secondly, When the user needs the rights, For different time information codes are demodulated by different users, In the response to the query signal simultaneously, Enables the user terminal to measure the false clock difference between the user clock and the center station clock through the time counter, The central station may have the path delay of the standard clock reference signal sent to the user through the response signal information from different users, Digital digital sent to the user to correct false clock difference, Get the actual clock difference. After a series of actions, users can get accurate coordination of world time, adopt 220um BeiDou satellite/GPS dual-mode reception module to receive standby signals of BeiDou satellite and GPS satellite to ensure the stability of satellite signal and continuous and stable output of timing information. The synchronization time signal is obtained by receiving the GPRMC and BDRMC serial time information and 1pps-second pulse information sent by the satellite [4].

2.2. Timiming method of BeiDou satellite navigation system
The BeiDou navigation system adopts the synchronous time matching mode in most cases:

(1) Serial port pair timing. It can also be called software timing, which refers to the serial message of the autonomous clock cycle, accepts the clock and decoding work, and analyzes the year, month, day, hour, minutes, seconds and other information.

(2) Pulse-pair timing. It can also be called a difficult data processing module for high-precision positioning and common time, respectively. The covisual data processing module calculates the covisual format data using the time difference measurement data and the original observation data, and it is sent to the data processing center through the communication module [5]. The data processing center combines the co-visual reference source data to calculate the local time deviation and return the time difference result to the reference station receiver. Controlling the internal frequency source and output from this difference produces synchronous, stable, accurate, and continuous standard time signals.

2.3. System function
The main functions of the power system timing equipment include the output of various time information, a stable second pulse output, a timing function, and a delay compensation function. This design mainly performs the following functions:

(1) Sufficient to receive the time information of the BeiDou 2 and GPS satellite systems;
(2) The system can output an accurate second pulse signal;
(3) The system can complete the coding and output of the unified target time code IRIG-B code;
(4) The system can output various forms of time codes, such as a serial port time message;
(5) On punctual effect of time signal when satellite fails;
(6) With the transmission delay compensation function;
(7) The timing accuracy can reach the nanosecond level, about 100ns

3. Hardware Design
The transmission delay system device is organically connected together by each module via the Avalon bus technology. The BeiDou receiving module receives the BeiDou / GPS information first. The output information includes time information and a second pulse signal, read and encoded via the UART serial port via the FPGA, and a second pulse alignment via the PLL phase-locked loop inside the FPGA [6]. The information output of the time information reading module is encoded to the IRIG-B encoding module, encoded by the modulus 100 counter and 10 counter generated by the FPGA itself, constantly generating an IRIG-B code composed of 10010ms elements. After each function completes its responsibilities, various time information is converted into the format required for the power system via the TTL level and RS232, RS485, fiber. This design adopts ALTERA company FPGA development board, EP2C8Q208C type, hardware system contains 420,000 equivalent doors, 64-bit SDRAM, 25MHz provides the system main clock, support download design program.

4. Decoding Process Design
Due to the superior actual performance of one frame, the completion time information conversion to B code becomes the preferred standard code for timing system devices (see Fig.1). The B code contains

three codes, the P high level pulse width 8ms,1 high level pulse width 5ms,0 high level pulse width 2ms. 0,9,19,99 are index marker bits (P elements), the rising edge of the second element of two continuous index markers is the starting point of each frame, the second information occupies 1 to 8 bits, the segmentation information occupies 20 to 23,25 to 27 bits when the segmentation information occupies 10 to 17 bits, the time information is expressed in the form of BCD code, front and back. The decoding of the B code is to read out the BCD code information in each frame into the time.

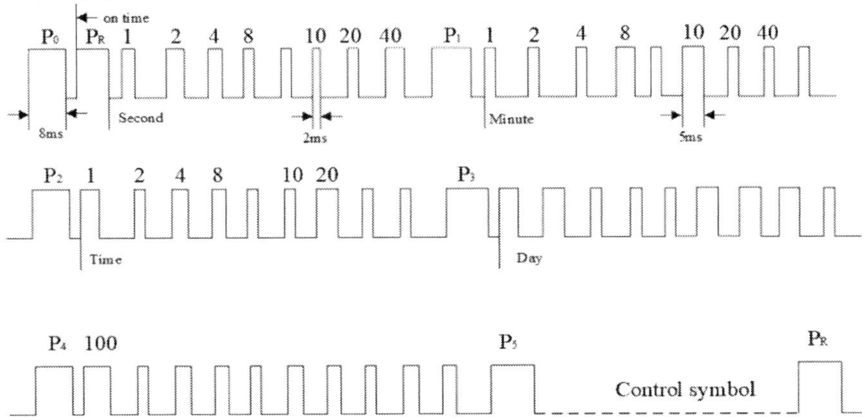

Fig.1 IRIG-B diagram

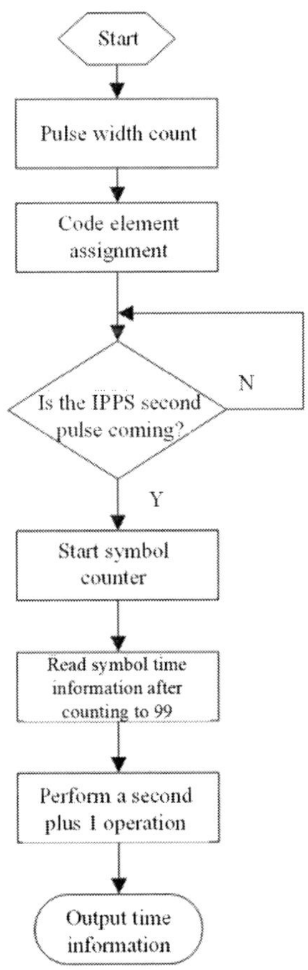

Fig.2 Decodes the flowchart

Firstly, the B code signal is filtered through 3 D triggers, To remove the possible spike signals, The pulse width within 10ms is then counted, 1 (01), 0 (00) and P(10) according to pulse width meter value, Then two P judgments, Create a high-level signal, Start the code element counter, Start the counting, The B code signal transmitted during this time is the time signal that we need to decode the output, When the counter counts to 99, The B code signal transmission time is exactly 1s(frame), Jump jump high level to low level, And reset the counter zero, For a second-plus-1 operation, And output the solved time information (can be seen Fig.2 for the specific process)

5. Improved Clock Synchronization Method

5.1. Specific methods:IEEE

The 1588 protocol only provides compensation for the clock deviation of master and slave clock. Establish the equation of state and measurement equation of clock deviation and frequency deviation, and adopt kalman filtering method to correct the clock deviation, thus reduce the deviation statistics and improve the synchronization accuracy. The kalman filtering algorithm is suitable when the process noise and measurement noise are white noise, so the noise characteristic is tested using Allan variance according to the experimental data.

Let the equation of state and the measurement equation be

$$\begin{cases} x_k = \Phi(k,k-1)x_{k-1} + G(k-1)w_{k-1} \\ y_k = H(k)x_k + v_k \end{cases} \tag{1}$$

The formula is the system state vector; the measurement vector; the acquisition noise and measurement noise are independent Gaussian white noise and the state matrix; G is the input matrix and H is the measurement matrix.The estimated variance is:

$$\hat{x}_{k|k-1} = \Phi(k,k-1)\hat{x}_{k-1,k-1} \tag{2}$$

The Kalman filter gain is:

$$K_k = P_{\hat{k}k-1}H^{\mathrm{T}}(k)\left[H(k)P_{\hat{k}k-1}H^{\mathrm{T}}(k)+R_k\right]^{-1} \tag{3}$$

The status update equation is

$$\hat{x}_{k|k} = \hat{x}_{k|k-1} + K_k\left[y_k - H(k)\hat{x}_{k|k-1}\right] \tag{4}$$

The updated covariance moment is

$$P_{kk} = \left[I - K_k H(k)\right]P_{kk-1} \tag{5}$$

The Kalman filtering algorithm is suitable for the case of process noise and measuring noise Gaussian white noise. In practice, in order to verify the applicability of the model algorithm, the Allan variance is used to test the noise characteristics, and the Allan variance model is defined as:

$$ADEV(\tau) = \sqrt{\frac{1}{2\tau^2}\left\langle [x(t+2\tau) - 2x(t+\tau) + x(t)]^2 \right\rangle} \tag{6}$$

Combining the kalman state estimation model and the Allan variance inspection model, the obtained deviation value is measured in the realization of synchronization. To=2s,, one cycle per 10 sampling data, n = 10, with the timestamp obtained as each observation, calculates the information to the model and the acute compensation correction of the clock deviation.

5.2. Improved synchronization deviation accuracy test after the algorithm.

The waveform data of the back-to-back test of the improved algorithm is shown in Fig.3, and the clock deviation of any one time is about 79 ns, the deviation curve is shown in Fig.4, and the deviation distribution histogram is shown in Fig.4. From the deviation distribution histogram, the main deviation distribution is within the range of (-20ns~20ns).

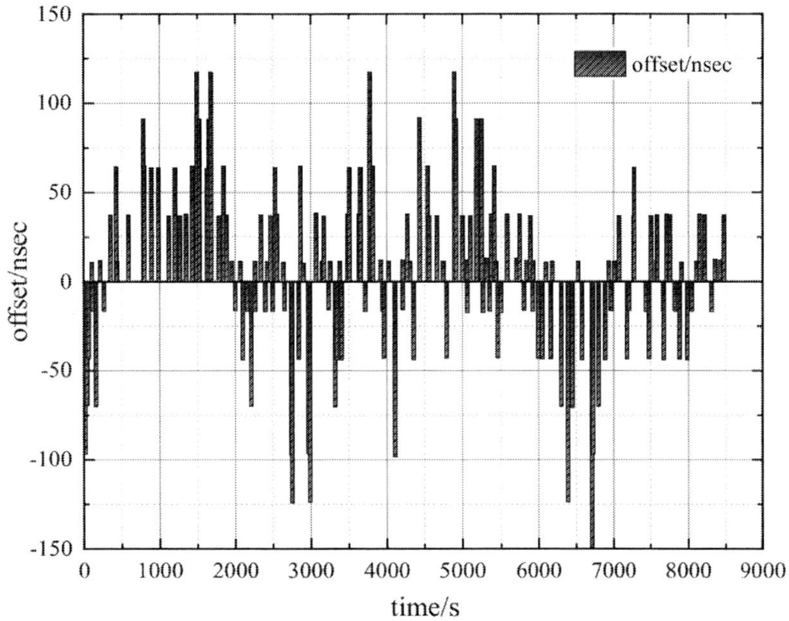

Fig.3 Deviation curve

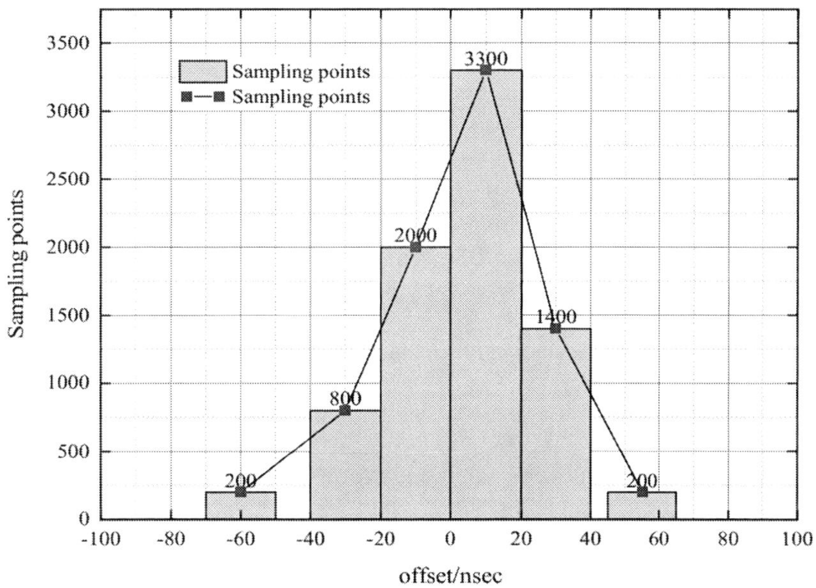

Fig.4. Deviation distribution

6. Operation Effect

The BeiDou Power Grid time synchronous management system can provide high-precision time synchronous timing equipment and the whole network time synchronization system in large areas, realizing the remote centralized state management of the clock equipment. The system is a high precision, high stability, and high reliable time synchronization management system that provides an overall solution of set management, monitoring and unified time timing. After commissioning, discovery and traditional time.

The Synchronization system has the following advantages:

(1) The BeiDou satellite adopts bidirectional timing to solve the inherent time and accumulated errors, such as the traditional one-way timing (one-way timing and GPS timing and one-way BeiDou), to ensure the unified time of the whole network and provide a reliable, safe and accurate time source for the development of the smart grid.

(2) By monitoring the state of the timing device and the entrusted equipment, compare the accuracy of the timing device and the timing equipment in real time, save and analyze the difference between the abnormal alarm, and send it to the central station monitoring system of the entrusted equipment through the BeiDou satellite to ensure the connection of the entrusted equipment. Then, the correct time was received.

(3) Taking the standard configuration of the BeiDou satellite navigation system as the main clock of satellite synchronization and a single GPS as the main time benchmark, it solves the huge security risks brought by the current time synchronization system and realizes the autonomy of the satellite timing of the power grid.

(4) Relying on the communication function of the BeiDou system, the equipment information such as its operation status and accuracy will be sent to the monitoring system of the central main station, monitoring and remote control the operation status and installation time of each station, so as to realize the centralized monitoring and management of the operation status of each clock equipment.

(5) The whole network time synchronization management system of BeiDou Power can integrate SDH, PTP and other local chain network methods, independently set the main preparation mode, intelligent switch, build a comprehensive time management platform, and realize the real "mutual preparation between heaven and earth".

7. Conclusion

This design is based on the BeiDou/GPS FPGA field programmable gate array dual-mode synchronous timing system. The stability of the system is improved through the BeiDou and GPS,. At the same time, the fast speed of the hardware codec ensures real-time, and the PLL phase-locked loop achieves the continuous temperature synchronous output of the second pulse. The system outputs time information such as serial information, IRIG-B coding, 1PPS second pulse and other formats, to meet the requirements of time information format and accuracy. At the same time, the BeiDou 2 satellite timing reduces the over-dependence on the GPS system.

Acknowledgements

This work was supported by the Science and Technology Project of State Grid Corporation of China (Project No. 2021YF-37).

References

[1] Gao, L. (2011) Design of multi-time source multi-time timing device of power system. Beijing: School of Electrical Engineering, Beijing Jiao Tong University.

[2] Wang, W.Y. (2008) Development of the timing awarding system based on the BeiDou Satellite . Beijing: School of Electronics and Communication Engineering, Beijing University of Posts and Telecommunications.

[3] Ping, B.Q. (2012) Design and implementation of IRIG-B Code System based on FPGA. Suzhou: School of Electronics and Communication Engineering, Suzhou University.

[4] Zeng, Y.F, Wu, S.Q. (2014) Design of conservation systems based on FPGA and finite-state machines . Computer measurement and control, 22: 1565-1567.

[5] Feng, Y, Guo, Y. (2010) GPS continuously operating reference station (CORS) principles and applications. Journal of Inner Mongolia University of Science and Technology, 29(4): 298-301.

[6] Mu, Y. (2014) Research and application review ofground-based augmentation system. Modern Navigation, 8(4): 306-308.

Research status of thermal management of hydrogen energy proton exchange membrane fuel cell

Jibao Liu[1a*], Huanyong Cui[1b*], Lei Gao[2c], Jiangtao Wang[3d], Meiyan Wang[2e], Fajia Li[1f]

[1]University of Jinan, Jinan 250022, China;

[2]Shanghai REFIRE Technology Co., Ltd, Shanghai 200000, China;

[3]Beijing jingtun Science & Technology Co.,Ltd, Beijing 101300, China;

[a*]1934458740@qq.com, [b*]cuihuanyong@163.com, [c]508712513@qq.com, [d]3450946913@qq.com, [e]2159017180@qq.com, [f]me_lifj@ujn.edu.cn

Abstract—The working principle and advantages of proton exchange membrane fuel cell using renewable energy hydrogen as fuel are introduced, the importance of fuel cell thermal management is emphasized, and the performance influencing factors, heat source and design requirements of proton exchange membrane fuel cell are summarized. According to the current research status of fuel cell related thermal management, the methods of simulation analysis of fuel cell with the help of mathematical model are summarized, and the existing problems and solutions of fuel cell heat dissipation and cold start are summarized.

1. Introduction

Facing the worsening global environmental pollution and energy shortage, new energy and renewable resources have become the focus of the world. In the national strategy of "made in China 2025", China mentioned to vigorously promote the breakthrough development in key fields such as energy saving and new energy vehicles [1], because the fuel cell reaction process does not involve combustion, breaking the restriction of Carnot cycle, The ideal energy conversion efficiency is more than 80%, and there is no noise and pollution. It is becoming an ideal way of energy utilization [2]. Among them, proton exchange membrane fuel cell (PEMFC) has the advantages of low operating temperature, no pollution, high specific power and rapid start at room temperature. It has great market potential and broad application prospects in the fields of aerospace power system, motor vehicle power energy conversion, small and medium-sized power generation system and portable mobile power supply [3-6].

Proton exchange membrane fuel cell belongs to low temperature (60 °C ~ 80 °C) fuel cell. Reasonable temperature conditions during operation have a great impact on the performance of fuel cell. Therefore, it is very necessary to study the heat generation and transfer, temperature field distribution and cooling mode in the fuel cell, including how to discharge the heat generated in the cell to the outside, ensure the uniform temperature distribution in time and space, avoid the occurrence of hot spots, and realize effective heat management.

2. Importance of PEMFC thermal management

PEMFC cell is mainly composed of proton exchange membrane, cathode/anode catalytic layer, cathode/anode gas diffusion layer and cathode/anode collector plate. The flow field is compounded on the collector plate, and cooling flow channel can also be added. The specific reaction process in the stack is shown in Figure 1 [7] below:

Hydrogen and oxygen are introduced into the anode and cathode respectively, and contact with the catalytic layer and proton exchange membrane through the gas diffusion layer. The following reactions occur under the action of the catalyst:

$$\text{Anode: } H_2 \rightarrow 2H^+ + 2e^- \quad (1)$$

$$\text{Cathode: } 2H^+ + 1/2O_2 + 2e^- \rightarrow H_2O \quad (2)$$

$$\text{Total reaction: } H_2 + 1/2O_2 \rightarrow H_2O + \text{thermal energy} + \text{electric energy} \quad (3)$$

Hydrogen protons and electrons generated by oxidation reaction of anode H_2 reach the cathode through proton exchange membrane and external circuit respectively, and water, thermal energy and electric energy are generated in this process [8-10]. Therefore, effective water heat management is very important to improve the performance of PEMFC.

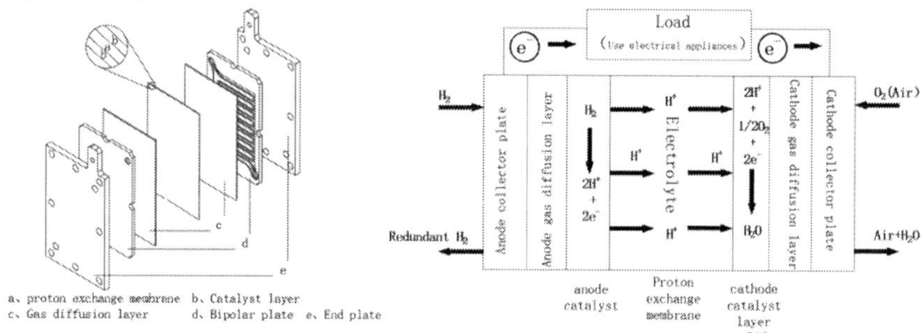

Fig. 1 Structure and working principle of PEMFC [7]

The importance of fuel cell thermal management comes from the evaluation of the performance of the fuel cell itself. The regulation and control of heat directly affects the temperature change of the fuel cell itself. To ensure the normal and efficient operation of the fuel cell, its temperature must be kept within the optimal operating temperature range. Therefore, effective and reliable thermal management directly affects the performance of the fuel cell, The importance of fuel cells is self-evident.

3. Factors affecting PEMFC performance

The performance of PEMFC depends on the influence of various factors on proton exchange membrane. Among the many factors, temperature has the most obvious impact on the performance of PEMFC. Proton exchange membrane has become the key to restrict its temperature characteristics. Temperature can cause the transportation of internal water components, affect the permeability of membrane, the activity of catalyst, the diffusion of gas in the battery, and the occurrence of "water flooding". It can be seen that temperature has a relatively large impact on the performance of PEMFC [11-12].

Mattinoponen et al. [13] studied the influence of PEMFC temperature on current distribution by using a current distribution measurement system, and learned that the main factors affecting its performance will also change with its temperature. Zhang Riqing et al. [14] used the self-made three in one membrane electrode and assembled into a single cell to analyze several factors affecting the performance of the cell, such as Nafion dosage, catalyst dosage, Nafion membrane thickness, anode hydrogen pressure and cathode oxygen pressure. This study provides a reference for the optimal value of each parameter of the single cell. Ma Haipeng et al. [15] used straight single channel to visualize proton exchange membrane single cell, studied the transfer behavior of humidification temperature and gas flow rate on liquid water in cathode channel, and analyzed the phenomenon of "water flooding" with the help of visualization technology.

To judge the performance of PEMFC, the influence of temperature is the top priority. The best temperature can not only ensure the normal operation of its own work, but also maintain the internal moisture content. The factors affecting the performance of PEMFC are directly or indirectly affected by temperature.

4. PEMFC heat source and heat dissipation mode

For the heat source of PEMFC, irreversible chemical reaction heat and entropy change of electrochemical reaction, Joule heat (heat generation of ohmic polarization), heat brought in by humidified gas, external radiant heat and condensation heat of water vapor generated by reaction. The first two chemical reaction heat and Joule heat are the main heat sources, and other heat is negligible [16].

In the actual working condition of PEMFC, the efficiency of converting internal fuel chemical energy into electrical energy is 40% ~ 60%. Most of the energy in the remaining is still converted into thermal energy, and about 50% of the total capacity will be dissipated through the cooling system [17]. It is necessary to select appropriate heat dissipation methods for PEMFC with different power, mainly as follows:

(1) Air cooling: take air as the cooling medium and share it with the reaction air or set a separate channel for transmission to realize heat dissipation. Low convective heat transfer coefficient of air cooling, large heat exchange area required, inconsistent air flow rate and other problems lead to uneven distribution of internal temperature and performance of the battery, which is mostly used in PEMFC power supply equipment with low power (≤ 5kW) [18-19].

(2) Liquid cooling: using deionized water or mixed solution of ethylene glycol and water as the cooling medium, relying on the forced convection heat transfer of the coolant and high heat transfer coefficient, the heat dissipation of high-power (> 5kW) PEMFC stack is realized, and the volume of its cooling system is also increased.

The above two cooling methods can be called active cooling or single-phase cooling. They are characterized by using the sensible heat of the cooling medium to take away the waste heat. Usually, a cooling channel is set between the two bipolar plates of the adjacent battery of the stack or a cooling plate with a cooling channel is added to realize forced convective heat transfer.

(3) Evaporative cooling: the latent heat of phase change during liquid evaporation is used to take away the heat inside the battery for temperature control. Generally, this method enters the battery from the cathode side together with the coolant and air to participate in the heat absorption and reaction process, which can not only realize the gas humidification but also realize the heat dissipation process, and simplify the complex structure of the system.

(4) Heat pipe cooling: with the help of heat pipe (such as copper or aluminum alloy) with good thermal conductivity to maintain uniform temperature on the heat source surface, heat transfer is realized by phase change of its internal working liquid without external equipment power input, and a large amount of heat is transmitted to the outside through its small cross-sectional area [20].

(5) Flow boiling cooling: the coolant needs to be set with a separate flow channel and the temperature is constant at the boiling point, changing from liquid to gaseous to realize heat dissipation. Due to the particularity of its cooling mode, deionized water or ethylene glycol aqueous solution cannot be used as the coolant, and the liquid with the characteristics of low boiling point / high latent heat of vaporization and low freezing point shall be selected [21].

The above three types are passive cooling or phase change cooling, which is completed by using the characteristics of absorbing a large amount of heat during phase change. Mramezanizadeh et al. [22] provided a reference matching with PEMFC power for various cooling modes of fuel cells, as shown in Figure 2.

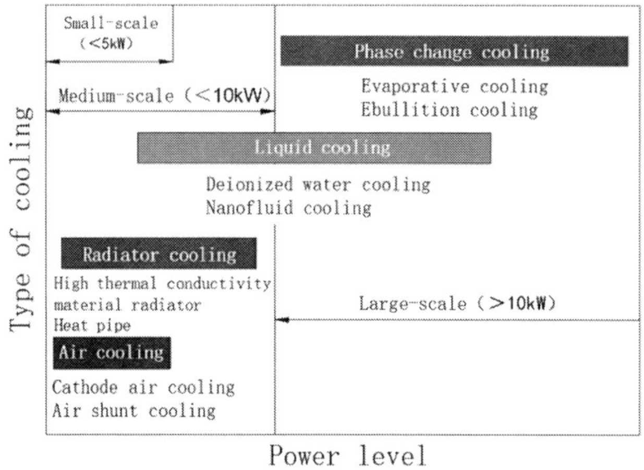

Fig. 2 Relationship between cooling mode and power level of PEMFC [22]

Through the analysis of the heat source and required heat emission of PEMFC, it can be seen that the overall cooling workload is large, and the heat power is basically the same as the power generation power. Therefore, reasonable recycling of PEMFC heat can improve its energy utilization efficiency to a greater extent, significantly reduce the burden of heat dissipation system and effectively ensure the performance of fuel cell. In order to ensure that PEMFC is in heat balance, it is necessary to adapt cooling mode, coolant type, flow and other factors to battery operating conditions.

5. Design requirements and modeling and Simulation of thermal management

(1) Ensure the overall temperature change. Perfluorosulfonic acid membrane is widely used in PEMFC. If the battery temperature is too low, it will aggravate the phenomenon of electrode internal polarization and reduce the output voltage, resulting in the deterioration of battery performance; If the temperature is too high, it will lead to dehydration, shrinkage and even rupture of the proton exchange membrane, which will seriously cause the water loss of the proton exchange membrane and accelerate the attenuation of the proton exchange membrane and catalyst.

(2) Keep component temperature balanced. The temperature uniformity of each internal part is conducive to the evaporation, condensation and distribution of water, so as to avoid dehydration and final rupture on one side of the proton exchange membrane and water flooding on the other side.

(3) Control the overall temperature bearing limit. Most components of PEMFC system have temperature tolerance limits, so it is necessary to ensure that the whole stack operates within its temperature tolerance. If the local temperature distribution of the stack is uneven and higher than 100 °C, micropores will appear in the proton exchange membrane, and hydrogen will penetrate into the air system, causing serious safety accidents [23].

At present, the research on the models to describe the output performance of fuel cells according to the changes of internal parameters and external characterization at home and abroad is complex. There are three common types: mechanism model, empirical model and equivalent circuit model [24]. The mechanism model is based on a reasonable assumption, and the influence of internal temperature, material distribution and other parameters on the battery output characteristics is analyzed by using the electrochemical reaction law. The establishment of simultaneous equations increases its complexity and accuracy. The empirical model does not need to analyze the internal reaction parameters of fuel cell, and summarizes the output characteristic law through its external characterization and big data statistical analysis. The external reference value is mainly obtained from the volt ampere characteristic curve. The model is easy to understand, but it also lacks the investigation of the internal reaction parameters of the battery. The equivalent circuit model is to construct an equivalent circuit through circuit elements to simulate the internal polarization characteristics of fuel cells and express the output characteristics of the system.

The importance of fuel cell thermal management is clarified in the previous article, so it is necessary to optimize the design. Through the constraints of design requirements and the analysis of modeling and Simulation of thermal management, the thermal management problem of PEMFC is improved by combining theory with practice, so that the whole design process is more perfect, which not only ensures the performance of fuel cell, but also shortens the design and research and development cycle, It reduces the consumption of research and development funds.

6. PEMFC cold start at low temperature

Low temperature cold start of PEMFC refers to the process of successful start-up and operation to normal working temperature (about 60 °C ~ 80 °C) at ambient temperature below 0 °C. In the below zero environment, the reaction product water of fuel cell will freeze due to low temperature, resulting in the decline of fuel cell performance and failure to start. Whether it is used as a portable mobile power supply or as a power supply equipment for new energy vehicles, PEMFC will inevitably encounter starting and operation in a low temperature environment below 0 °C. In this environment, the following problems mainly occur:

(1) The water generated by the reaction at the cathode catalytic layer will freeze, hinder or stop the reaction of the reaction gas, and the electrode active material is completely covered;

(2) The humidification of the reaction gas and the freezing of the water directly cause that the reaction gas cannot enter the reaction area through the flow channel;

(3) The densities of water and ice at 0 °C are 0.9998g/cm^3 and 0.9168g/cm^3 respectively. When water freezes, it will produce 9% volume expansion. If unreasonable repeated freezing/melting cycles are carried out inside the stack, it will cause the decline of internal single cell performance, and even lead to the decline of stack performance and irreversible damage such as physical deformation or breakdown of component structure [25].

At present, researchers have taken some measures to solve the cold start problem of fuel cell in low temperature environment:

(1) Safe shutdown: the residual water in the reactor is treated in advance by purging before and after shutdown to ensure that there is no freezing problem during startup. Tajirie et al. [26] used humidified gas to purge the internal gas of the stack for a long time in order to ensure that the membrane is not excessively dry. The experimental effect is good but time-consuming. Secondly, by using antifreeze materials or structures inside the stack, the internal damage caused by freezing can be reduced, resulting in performance degradation. Ko et al. [27] [28] proposed to alleviate the problem of internal expansion damage by adding a bifunctional microporous layer between the gas diffusion layer and the catalyst coating membrane, and verified that this method can guarantee the performance of the battery.

(2) Rapid cold start: the external heating device is installed to realize rapid temperature rise start, but it will increase the overall volume, cost and energy consumption of the battery. The internal heating method is not the same. Toyota Mirai uses short circuit or near short circuit to realize short-time heating and cold start, but the risk coefficient of short-circuit behavior is high. Some researchers also control the supply of reactants during the working process of the stack to cause the occurrence of internal electrode overpotential to realize the heating of internal resistance, or reduce the pressing force of the stack and increase the internal resistance to realize the heating [29].

The problem of fuel cell cold start cannot be avoided. Therefore, researchers have adopted different strategies to achieve the normal working state of fuel cell and ensure the overall safety and reliability. Obviously, the solutions to this cold start problem need to be realized in two aspects: safe shutdown and rapid cold start, This problem will also continue to be studied as a hot spot and technical difficulty.

7. Conclusion

Based on the results and discussions presented above, the conclusions are obtained as below:

(1) Thermal management is of far-reaching significance to the performance of proton exchange membrane fuel cells. It is the analysis and heat treatment of the fuel cell heat source, which directly

affects the overall temperature of the fuel cell. At the same time, the temperature has an impact on the internal water distribution and water content of the fuel cell. Therefore, water management and thermal management are closely related and coupled, Effective hydrothermal management ensures the stability and safety of fuel cell performance.

(2) Working temperature is one of the most important control parameters of fuel cell. From the internal water dispersion, membrane permeability, catalyst activity, gas diffusion, "water flooding" and other phenomena to the performance of fuel cell performance, it shows the importance of working temperature. At the same time, it is also the focus and difficulty of cell stack heat dissipation and cold start research.

(3) There are many sources of heat in fuel cells, among which the main heat sources are chemical reaction heat and Joule heat. In the research process, the heat brought by other heat sources accounts for too little and can be ignored.

(4) For PEMFC, the heat dissipation treatment method will be selected according to the power required for the application of fuel cell. Among them, the heat dissipation method of phase change material has obvious advantages and has become a research hotspot.

(5) With the help of mathematical modeling, select an appropriate modeling method to simulate and analyze the actual working conditions of fuel cell, significantly shorten the design optimization and operation test cycle of fuel cell, and reduce the investment of test funds in the process of research and development.

The hydrothermal management of PEMFC has always been a research direction of continuous innovation and optimization, but the investment in its research and development process has also increased. By constantly imposing design requirements on research and development products, with the help of mathematical model simulation analysis and theoretical guidance, it has obviously promoted the progress of this research and development direction.

Acknowledgments

This work was financially supported by the National Key R&D Program of China (2019YFB1504605).

References

[1] http://www. gov. cn/zhengce/zhengceku/2015-05/19/content_9784.htm

[2] Wang Ju, Zhu Xinyi. Overview of demonstration and application of fuel cell vehicles at home and abroad [J]. Solar energy, 2017 (8): 31-34.

[3] Huang Zhuo, Tu hailing, Zhang Jiqiang, et al. Research development and application of proton exchange membrane fuel cells [M]. Beijing: Metallurgical Industry Press, 2000:5-29.

[4] Yi Baolian. Fuel cells – Principle technology and application [M] Beijing: Chemical Industry Press, 2003.

[5] Ren Xueyou. Research progress of proton exchange membrane fuel cells [J]. China Engineering Science, 2005. 7 (01): 86-94.

[6] Fu Tiantian. Overview of the development of hydrogen fuel cells for electric vehicles [J]. Power supply technology, 2017,41 (4): 651-653.

[7] Wang Jihua, Ju Yusheng, Yi zhenggen, et al. Overview of fuel cell technology development and application status (I) [J]. Modern vehicle power, 2018 (5): 7-11.

[8] Appleby A J, Foukes F R. Fuel cell handbook[M]. New York:Van Reinhold, 1989.

[9] Yi Baolian. Principle, technical status and Prospect of fuel cell [J]. Battery industry, 2003,8 (1): 16-22.

[10] Hu Qingsong. Research progress of hydrogen fuel cell [J]. Automotive practical technology, 2017 (21): 114 - 116.

[11] OWEJAN J P, OWEJAN J E, GU W B, et al. Water transport mechanisms in PEMFC gas diffusion layers [J]. J. Electrochem.Soc,2010,157(10):1 456-1 464.

[12] Wang Xu, Zhang Rui, Wang xumin. Overview of research on performance thermal management of proton exchange membrane fuel cells [J]. Special purpose vehicles, 2020 (09): 76-80.

[13] Matti Noponen, Tuomas Mennola, Mikko Mikkola, et al. Measurement of current distribution in a free-breathing PEMFC [J]. J Power Sources, 2002,106(1).
[14] Zhang Riqing, He Yan, Zhou Zhentao. Research on the electrical performance of proton exchange membrane fuel cells [A]. Ministry of science and technology of the people's Republic of China, China Association for science and technology, hydrogen energy professional committee of China solar energy society International Hydrogen Energy Association. Proceedings of the second international hydrogen energy forum youth hydrogen energy forum [C]. China solar energy society, 2003:5.
[15] Ma Haipeng, Zhang Huamin, Hu Jun, et al. Effects of humidification temperature and gas flow rate on liquid water distribution and drainage in PEMFC cathode channel [J]. Journal of chemical industry, 2007 (09): 2357-2362.
[16] Kandlikar S G, Lu Z J. Thermal management issues in a PEMFC stack-A brief review of current status [J]. Applied Thermal Engineering,2009,29(7):1 276-1280.
[17] Li Zhonghua, Du Chuanjin, Hou Xianjun. Research on thermal management of proton exchange membrane fuel cells [J]. East China power, 2007 (02): 19-22.
[18] Meter Q, Ronaszegi K, Robinson J B, et al. Combined current and temperature mapping in an air-cooled, open cathode polymer electrolyte fuel cell under steady-state and dynamic conditions[J]. J Power Sources,2015,297(31): 315-322.
[19] Bu Qingyuan, Li Qi, Chen Weirong, et al. Experiment and Simulation of cathode fan system of air-cooled PEMFC reactor [J]. Journal of chemical engineering, 2015,66 (10): 4 211-4 217. Power sources, 2006156 (1): 114-118.
[20] Lei Dongqiang, Wang Xiuchun, Zhu Weili, et al. Research on the application of heat pipe technology in transformers [J]. Transformers, 2007,44 (1): 37-40.
[21] Hou Jian, Yang Zheng, He Ting, et al. Research progress on thermal management of proton exchange membrane fuel cells [J]. Journal of Central South University (Natural Science Edition), 2021,52 (01): 19-30.
[22] Ramezanizadeh M, Nazari M A, Ahmadi M H, et al. A review on the approaches applied for cooling fuel cells[J]. International Journal of Heat and Mass Transfer, 2019, 139(AUG.):517-525.
[23] Puji, Qin Xiaojin, Lu Yan, et al. Design and research of fuel cell thermal management system [J]. Automotive abstracts, 2019 (04): 24-27.
[24] Yang Duo, Pan Rui, Wang Yujie, et al. Dynamic modeling and simulation of proton exchange membrane fuel cell [A]. Professional Committee of system simulation of Chinese society of automation, professional committee of simulation technology application of Chinese society of system simulation. Proceedings of the 18th Annual Academic Conference of China system simulation technology and its application (18th CCSSTA 2017) [C]. Professional Committee of system simulation of Chinese society of automation, 2017:5.
[25] Ma Jianxin, Gao Xin, Zhang Cunman, et al. Cold starting Mechanism and Cold Starting Strategy of proton exchange membrane fuel cell [J]. Power technology, 2009,33(07): 533-540.
[26] Kazuya Tajiri, Yuichiro Tabuchi, Fumio Kagami, et al. Effects of operating and design parameters on PEFC cold start[J]. J Power Sources, DOI: 10.1016/j.jpowsour.2006.12.017.
[27] Johan Ko, Whan-Gi Kim, Young-Don Lim, et al. Improving the cold-start capability of polymer electrolyte fuel cells (PEFCs) by using a dual-function micro-porous layer (MPL): Numerical simulations[J]. Int J Hydrogen Energy, DOI: 10.1016/j.ijhydene.2012.06.026.
[28] Ko, Johan,Ju, Hyunchul. Numerical evaluation of a dual-function micro-porous layer under subzero and normal operating temperatures for use in automotive fuel cells[J]. Int J Hydrogen Energy, DOI: 10.1016/j.ijhydene.2013.07.060. membrane fuel cell [J]. Battery, 2018,48 (03): 202-205.
[29] Li Jie, Sun Tiesheng, Zhang Guangmeng, et al. Research progress on thermal management and cold start of PEMFC engine [J]. Battery, 2020,50 (04): 383-387.

Experimental study on pore forming characteristics of simulated hydrate broken by high pressure water jet

Zijian Dai[1a*], Hualin Liao[1b], Ke Yang[1c], Feng Sun[1d]

[1]School of Petroleum Engineering, China University of Petroleum (East China), Qingdao, 266580, China

[a*]s19020034@s.upc.edu.cn,[b]liaohualin2003@126.com,[c]2322589549@qq.com,[d]14622 42923@qq.com

Abstract—Natural gas hydrate is a research hotspot at present. However, the current exploitation technology can't meet the demand of commercial exploitation of natural gas hydrate. In order to improve the efficiency of hydrate production, this paper believes that the idea of using high-pressure water jets for sandblasting perforation is expected to constitute an effective way to extract natural gas hydrates. The experimental study on sandblasting perforation and hydraulic slitting of simulated reservoirs was carried out by using large-scale ground fracturing equipment and full-scale hydraulic blasting perforating equipment. The driving pressure is analysed under the action of high-pressure water jet. The influence of diameter on the effect of simulated reservoir fracture. The results show that the diameter of the perforation increases with the increase of pressure; This experimental study can provide an experimental basis for the use of abrasive jet blasting perforating technology to improve the efficiency of natural gas hydrate production.

1. Introduction

Natural gas hydrate, also known as combustible ice, is scattered in seabed sediments and land frozen soil areas. It has a series of advantages such as low pollution, high energy, shallow reservoir depth and large reserves. It has huge resource potential and high efficiency. It is considered to be the most ideal energy in the 21st century. In May 2017, China's first trial production of natural gas hydrate in Shenhu Haijiang, the South China Sea proved that China has made a breakthrough in the field of hydrate production and provided technical support for China to occupy the commanding height of global energy development strategy in the future [1]. Therefore, this paper holds that the idea of sandblasting perforation with high-pressure water jet is expected to constitute an effective way to exploit natural gas hydrate. Through this experimental study, the feasibility of natural gas hydrate exploitation by high-pressure water jet is verified, which can provide an experimental basis for using sand blasting perforation technology to improve the exploitation efficiency of natural gas hydrate[2].

2. Experimental scheme and device

The working principle of the simulated reservoir injection experimental device is as follows: firstly, the target is installed with a crane, and the jet tool is assembled and the experimental device is built according to the diameter of the nozzle required for each experiment [3]. The jet pump pressure and sand ratio are set through the console. The high-pressure jet output by the jet pump flows through the high-pressure pipeline to the nozzle and then shoots out to break the simulated reservoir. With the progress of the experiment, the water volume in the cavity increases, and the excess water carries the broken simulated reservoir and sediment particles and is discharged through the drainage hole[4]. After the completion of fracturing, after the jet pump is turned off through the console, the experimental device is disassembled manually, and the target is disassembled by the crane. Then, the simulated reservoir crushing effect is observed and recorded, and the experimental data are measured [5].

Fig.1 Experimental device

3. Analysis of experimental results of sand blasting perforation

3.1. Analysis of the influence of pump pressure on the fracture effect of simulated reservoir

In order to study the influence of pressure on the simulated reservoir crushing effect under the action of high-pressure water jet, four groups of experiments were carried out, the pressure was 15, 20, 25 and 30 MPa, the sand ratio was 6%, and the nozzle diameter was 6mm. Table 3-1 shows the jet operating parameters and test results.

Table.1 Fracture test results of simulated reservoir under different pressure conditions

No.	pressure /MPa	ratio /	Nozzle diameter/ mm	depth /cm	max/cm	min/cm
1	15	6%	6	40.5	-	-
2	20	6%	6	77.7	20.3	12.2
3	25	6%	6	96	22	20
4	30	6%	6	148	23.5	15

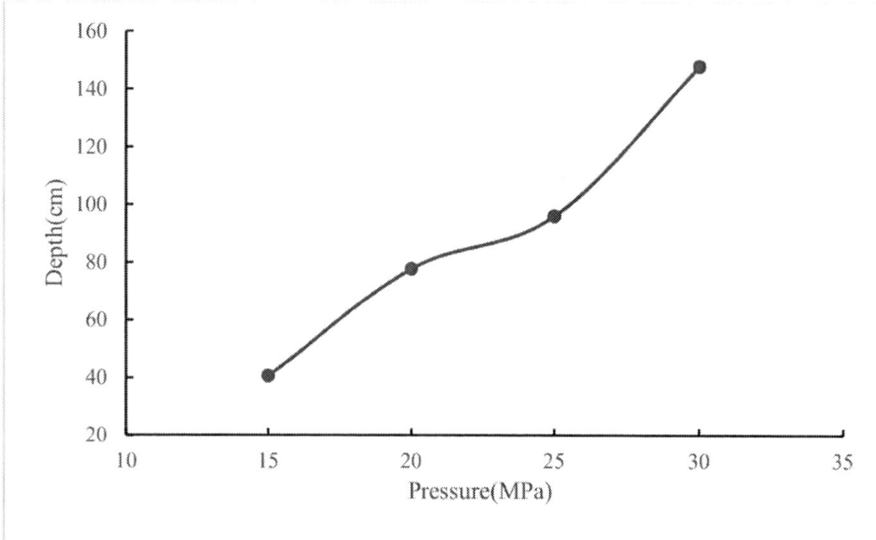

Fig.2 Effect of pressure on crushing depth

According to Fig.2, by analysing the crushing effect under different jet pressures, it can be seen that the jet pressure has a significant impact on the crushing effect when the high-pressure water jet breaks the simulated reservoir. As the pressure increases, the fracture depth of the simulated reservoir becomes larger and the size of the broken hole becomes larger.

3.2. Analysis of the influence of nozzle diameter on the fracture effect of simulated reservoir

In order to study the influence of nozzle diameter on the simulated reservoir crushing effect under the action of high-pressure water jet, the sand ratio is 6%, the pump pressure is 25MPa, and the nozzle diameter is 4mm, 5mm, 6mm and 7mm. Table 3-2 shows the jet operating parameters and experimental results.

Table.2 Fracture test results of simulated reservoir under different nozzle diameter conditions

No.	pressure /MPa	ratio /	Nozzle diameter/ mm	depth /cm	max/cm	min/cm
7	25	6%	4	115.8	11	9.5
3	25	6%	5	148	20.4	16.5
3	25	6%	6	96	22	20
7	25	6%	7	54	30	22

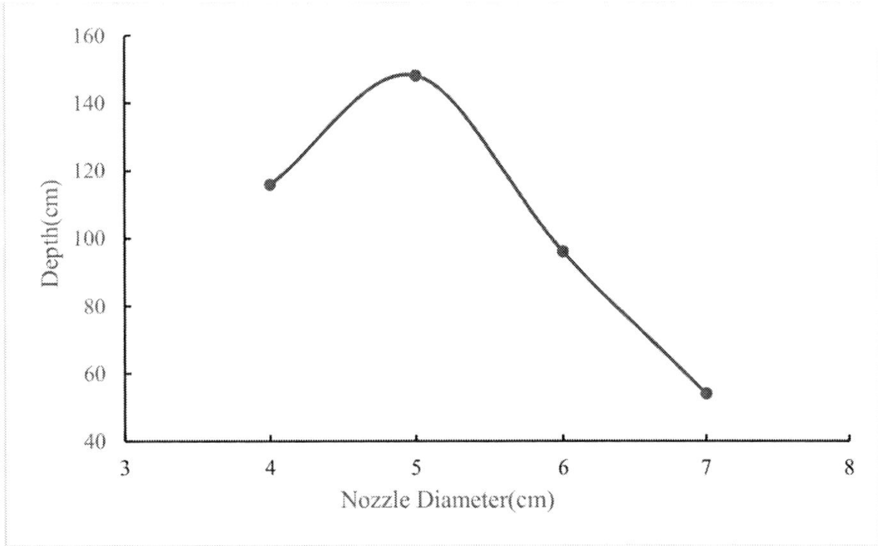

Fig.3 Effect of nozzle diameter on crushing depth

It can be seen from Fig.3 that when the high-pressure water jet breaks the simulated reservoir, the breaking depth first increases and then decreases with the increase of nozzle diameter, while the breaking size always increases. By analysing the crushing effect under the condition of different nozzle diameter, it can be seen that the nozzle diameter has an important influence on the crushing depth and size when the high-pressure water jet is used to simulate reservoir crushing. When the nozzle diameter is 5mm, it has a better crushing depth after perforation. The crushing size increases with the increase of nozzle diameter, because when the nozzle diameter increases, the jet flow will increase, and then the jet scouring effect will be stronger, and the flow rate of upward fluid will be faster, which will cause greater tensile stress on the contact surface of simulated reservoir, and finally lead to the increase of crushing size.[2]

4. Analysis of simulated reservoir fracture process under the action of high pressure water jet

When the high-pressure water jet breaks the simulated reservoir, the breaking process is roughly divided into three stages:

Step1: The upper part is broken.

Under the action of water jet, the contact surface of simulated reservoir is affected by tensile shear. When the tensile shear strength is greater than the tensile shear strength of sediment, broken holes appear on the contact surface[6]. When the jet continuously breaks the simulated reservoir, the fracture near the broken hole expands rapidly, and the high-pressure jet enters the fracture, reducing the tensile strength of the reservoir. At the same time, the upper part of the reservoir will also be subjected to the tensile shear action caused by jet[7]. When the tensile shear action reaches the maximum, the upper area of the reservoir will expand downward.

Step2: The middle part is broken.

Under the high-speed crushing of water jet, the crushing hole expands rapidly, and the crushing depth increases[8]. At the same time, the action distance of jet will also increase. Because of the resistance, the jet is dispersed and the section expands, resulting in water flow in the direction other than the vertical action surface, shearing and scouring the inner wall of the reservoir, resulting in the increase of fracture size[9].

Step3: The lower part is broken.

When the crushing depth increases to a certain extent, the strength of the jet decreases and the flow velocity in the tangential direction decreases, resulting in the reduction of the crushing size, the expansion speed of the hole slows down, and the upward return fluid carries more debris, the resistance increases and the energy decreases [10].

5. Conclusion

Based on the results and discussions presented above, the conclusions are obtained as below:

(1) Through the experiments conducted by the large-scale surface fracturing equipment and full-scale hydraulic sand blasting perforation experimental device, it is found that when the pressure is 30MPa, the sand ratio is 6% and the nozzle diameter is 6mm, the simulated hydrate reservoir is expected to be perforated 150cm deep and the maximum diameter of the hole is 25cm. The feasibility of natural gas hydrate production by high-pressure water jet is verified, which provides an experimental basis for using sand blasting perforation technology to improve the production efficiency of natural gas hydrate.

(2) Characteristics of simulated reservoir pores: there are abrasive particles and simulated reservoir debris in the pores; The inside of the hole is smooth, and there are obvious erosion characteristics around the hole under the action of abrasive jet. The surface shape of the eyelet is mostly oval or regular circle; The middle and upper parts of the eyelets are mostly near circular or cylindrical cylinders; The end of the eyelet is a conical cylinder.

(3) In the next step, it is necessary to carry out experimental research on the crushing of simulated hydrate sediments with different reservoir characteristics by high-pressure water jet under different working conditions, and deeply study the crushing mechanism of seabed hydrate sediments.

References

[1] Van der Geer, J., Hanraads, J.A.J., Lupton, R.A. (2010) The art of writing a scientific article. J. Sci. Commun., 163: 51–59.

[2] Chermenski F P, Davidyants G F. Pulsed water jet pressures in rock breaking[C]//Proceeding of the Fifth International Symposium on Jet Cutting Technology. Cranfield, BHRA, 1980: 155-163.

[3] Moridis G J, Collett T S, Dallimore S R, et al. Analysis and interpretation of the thermal test of gas hydrate dissociation in the JAPEX/JNOC/GSC et al. Mallik 5L-38 gas hydrate production research well[J]. Bulletin-Geological Survey of Canada, 2005, 585: 140.

[4] Osadetz K G, Chen Z. A re-evaluation of Beaufort Sea-Mackenzie Delta basin gas hydrate resource potential: petroleum system approaches to non-conventional gas resource appraisal and geologically-sourced methane flux[J]. Bulletin of Canadian Petroleum Geology, 2010, 58(1): 56-71.

[5] Yamamoto K, Terao Y, Fujii T, et al. Operational overview of the first offshore production test of methane hydrates in the Eastern Nankai Trough[C]//Offshore Technology Conference. Offshore Technology Conference, 2014..

[6] Annoni, M., Cristaldi, L., Lazzaroni, M., Ferrari, S.. Nozzles Classification in a High Pressure Water Jet Systems[P]. Instrumentation and Measurement Technology Conference Proceedings, 2007. IMTC 2007. IEEE,2007.

[7] V. V. Koryakina,I. K. Ivanova,M. E. Semenov,I. I. Rozhin,A. F. Fedorova,E. Yu. Shits. Specific features of the growth, composition, and content of natural gas hydrates synthesized in inverted oil emulsions[J]. Russian Journal of Applied Chemistry,2017,90(8).

[8] Wang Sheng,Zhang Chuan,Yuan Chaopeng,Chen Liyi. Rheological Properties of Polymer Drilling Fluid Developed for Permafrost Natural Gas Hydrate Drilling[J]. Chemistry and Technology of Fuels and Oils,2017,53(2).

[9] Azad Jarrahian,Ehsan Heidaryan. Natural gas hydrate promotion capabilities of toluene sulfonic acid isomers[J]. Polish Journal of Chemical Technology,2014,16(1).

[10] Boswell R. Japan completes first offshore methane hydrate production test—Methane successfully produced from deepwater hydrate layers[J]. Center for Natural Gas and Oil, 2013, 412: 386-7614

Recent progress for different inertial confinement fusion schemes: a systematical review

Yihong Qian[1, a, †] **and Botao Huang**[2, b, †]

[1] College of Power and Energy Engineering, Harbin Engineering University, Harbin, Heilongjiang Province, 150001 code, China

[2]Queen Mary Engineering School, Northwestern Polytechnical University, Xi'an, Shanxi Province, 710129, China

[a]yihong-qian@hrbeu.edu.cn, [b]botao_huang@mail.nwpu.edu.cn

[†]These authors contributed equally.

Abstract. The pursuing of controlled fusion energy has been continuously developed for more than half a century. Inertial confinement fusion (ICF) is one of two major approaches to actualize controlled fusion. Here, we systematically reviewed several typical forms of ICF on the part of their physical principles and encountering technical barriers currently. Besides, some great simulation results of the implosion for each ICF scheme are shown, and the simulation algorithm of Vlasov-Fokker-Planck (VFP) is introduced. In addition, several instabilities in the fusion process are analyzed. These results offer a guideline for future ICF research.

1. Introduction

Inertial confinement fusion (ICF) is an approach of fusion energy attempting to initiate nuclear fusion reactions conducted by way of heat treatment and compression of one fuel target [1]. It is mainly in the form of a pellet with a radius of around 400 μ m, which contains an admixture of deuterium and tritium [2]. To this end, the out layer of the fuel target gains energy from extreme intensity laser (in most cases) [3] or heavy ion beam [4]. In this case, the inner part is accelerated inwards, which results in compressing of the fuel. This process is well known as implosion [5], which can generate shock waves, leading to the occurrence of fusion reactions [6].

During the history of ICF development, researchers were devoted to the improvement of devices' efficiency and methods to meet Lason's criteria (so-called ignition) for the first 10 years since the first proposal of ICF in the 1970s [7]. The research continued in the late 1970s make use of infrared lasers for compression experiments that yielded thermonuclear neutrons [8]. Due to super thermal electrons and preheating of the target related to infrared laser wavelength, lasers with shorter wavelengths were required to be built after 1980 [9]. After the evolution of relevant research methods and improved fusion devices, substantial achievements in implosion performance have been accomplished in ICF [10].

With the great development of ICF, it is necessary to summarize the recent situation and issues. In this paper, several typical schemes are introduced to achieve ICF in Sec. 2, including ICF with extreme intensity laser (in the form of direct driven, indirect driven, and fast ignition), heavy ion beam, and Z-pinch. Specifically, part of their principles, several technical barriers encountering currently and some achieved progress are discussed. Apart from this, some details about insights into the current stage simulations related to the asymmetry of the radiation and implosion would be spread out in Sec. 3 and 4, respectively. Moreover, the instabilities of ICF are discussed in Sec. 5. Finally, a concise and critical summary is given in Sec. 6.

2. ICF applied with typical schemes

2.1. ICF with extreme intensity laser

2.1.1. Direct driven

The concept of compression of a small target with high-power laser to turn the thermonuclear fuel into a condition possible for ignition was first proposed by Nuckolls in 1972 [7]. As shown in Figure 1 [11], the pressure of the plasma is confined by the inertia for a relatively long time (a few ns). It is enough for the burning process of thermonuclear to result in enough fusion reactions producing abundant energy to overcome the Coulomb barrier. Among the approaches in fusion reactions, the deuterium-tritium (D-T) reaction [12] has the largest reaction cross section. It can be achieved under the circumstances with relatively low requirements compared to other reaction channels.

Generally, the laser drive ICF is divided into two methods. First is direct drive, where a spherical target containing fusion fuel is directly irradiated by laser beams [13]. Another one is the indirect drive. The laser beams heat the inside of a typically cylindrical enclosure, generating x rays that irradiate a spherical fuel-containing capsule [14, 15]. The physical processes of direct-drive implosion are illustrated in Figure 2 [16].

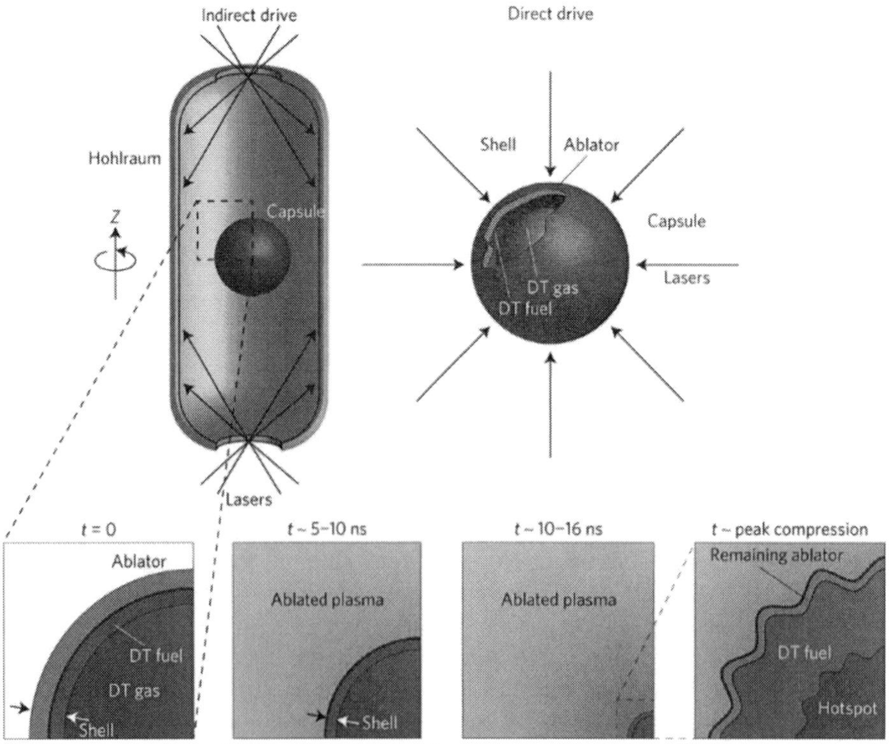

Figure 1. Schematics of indirect- and direct-drive ICF [11].

In the early stage, the target's energy is absorbed by the target, leading to laser ablation over the target material, which forms a hot plasma. A set of low-intensity pulses quantifying from one to three occurs. Then, the laser pulse intensity grows sharply, releasing a stronger shock that combines with the early ones breaking the inner surface of the ice layer. During the transition, target modulation transforms due to a Richtmyer–Meshkov (RM)-like uncertainty [17,18]. Afterward, the ice layer starts to accelerate inward to the target centre, laser-plasma interactions produce energetic electrons leading to fuel preheat.

Exponential growth on ablation-surface modulations occurs resulting from the Rayleigh–Taylor (RT) instability [19, 20], which will be harmful for the ignition (discussed in Sec. 5). The deceleration phase begins after its reflection from the target center and reaches the converging shell. The DT fuel is squeezed and heated with kinetic energy, converting into thermal energy. Eventually, the peak compression session begins, which converts the energy in the DT fuel to fusion neutrons. As a consequence, it further compresses fuel to fusion temperatures, i.e., triggers the catenated burn (known as ignition) [21].

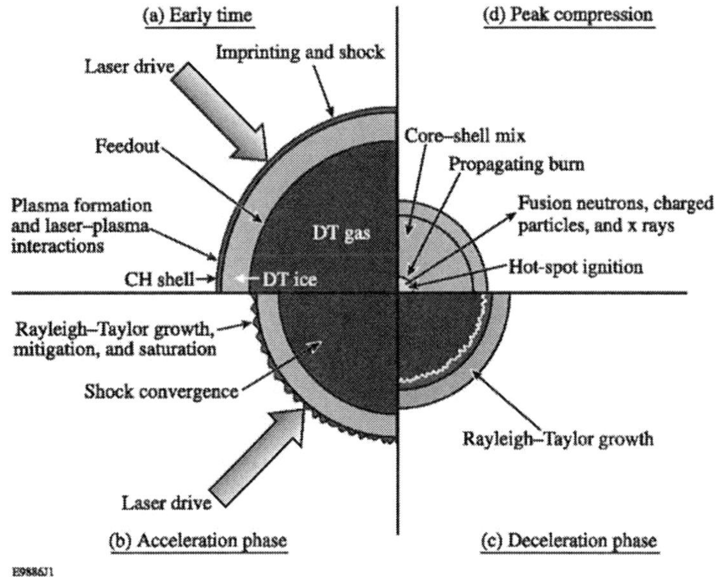

Figure 2. Illustration of 4 various stages of a direct-drive target implosion [16].

The main restrictions on the development of direct driven ICF consist of reduced uniformity of the illumination, low-mode asymmetries, and laser-plasma instabilities (mainly occurring during the interaction between the laser light and the ablated plasma) [22]. Another current research on low-mode asymmetries proposed an innovative asymmetry alleviation method, in which the assessed hot-spot velocity to ascertain the ideal target position to diminish mode-one asymmetries [23]. In regard to the progress in laser-plasma instabilities, recent research based on the SGIII-P facility played a key role [24].

2.1.2. Indirect drive

Indirect drive inertial confinement fusion applies small cylindrically-shaped, high-Z radiation cavities called hohlraums to transform extreme intensity laser into X-rays. At the center of the hohlraum locates a spherical capsule containing DT fuel. The physical principle of the hot spot ignition process for the indirect drive is similar to the direct drive, which has been covered in the 2.1.1 part (Figure 1.). The material for capsules was traditionally made of plastic or beryllium [25-29]. Fusion is initiated by compressing DT gas, creating a crucial hot spot. Under the right conditions, the self-heating could be achieved through capturing alpha particles from fusion reactions, resulting in the hot spot igniting. After the formation of the robust burning plasma, ablation, and heating of the inner surface of the ice layer, the target sets up a propagating burn producing multi-megajoule neutron yields [30].

The technical points which still need optimization mainly lie on low-mode asymmetries in the X-ray drive (implosion symmetry), X-ray transformation, and capsule hydrodynamic durability. Consequently, it may cause critical impacts on some recent experiments leading to ρ R asymmetries [31]. In X-ray conversion, literature proposed a unique means for adjusting the X-ray flux inside hohlraums applying burn-through obstacles, which utilizes multiple chambers separated by burn-through barriers [32]. Some current stage information related to capsule hydrodynamic stability can be found in Sec. 5[33].

2.1.3. Fast-ignition

Fast-ignition opens a new route to directly heat up the high-density fuel after compression, i.e., it can separate the heating phase and the compressing phase [34]. By this means, the target is first compressed as the process in other ignition schemes. Thereby, the target is dramatically heating with a high possibility to start fusion ignition [35].

There are 2 types of fast ignition: 1) the plasma bore-through method and 2) the cone-in-shell method. In method 1, the petawatt laser is designed to penetrate through the outer plasma of an imploding capsule and impact and heat the dense core [36]. For the latter, the capsule is installed with a small projecting high-z cone (e.g., Au shown in Figure 3) in the core. However, the interaction between the cone and the implosion process remains unclear, too complicated to clarify.

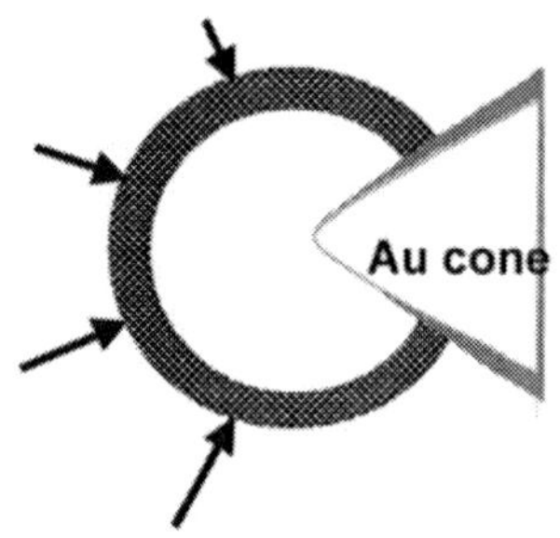

Figure 3. Cone in shell geometry for fast ignition [36].

2.2 ICF with heavy ion beam

To meet the criteria of ignition (producing high fuel densities and temperatures), About one milligram of DT fuel needs to be compressed to a very high density in several ns. In ICF with a heavy ion beam (seen from Figure 4), this compression is achieved by intense beams striking an external region of the fuel-containing pellet. The generated heat leads to compression of the fuel core and ignites the burning process.

HIB has several characteristics which own advantages on the construction of nuclear fusion power plant [37, 38]. It's flexible for accelerators to regulate the HIB constraints [38]. Depositing HIB energy inside material is another significant favourable feature. The collaboration of HIB ions with samples is nearly traditional, the Coulomb interaction [39-42]. In this case, HIB ion interaction is relatively understandable and could be specified inside materials [43,44].

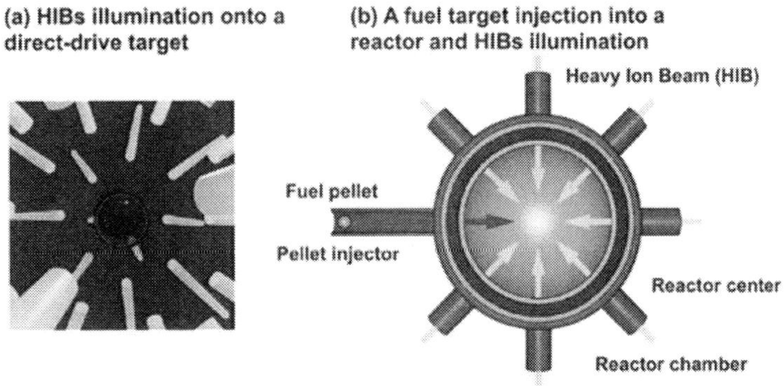

Figure 4. Schemes of (a) HIBs illumination onto a direct-drive fuel target and (b) a fuel target injection into a reactor [37].

2.3 Z-pinch
Note that Z-pinch is one application of Lorentz force [45]. Hopefully, as shown in Figure 5, the Z-pinch configuration is to become a compact fusion device due to the corresponding regular geometry, unity beta, and non-existence of external magnetic field coils. A rapid increase in the fusion reaction rate could be achieved via raising the axial current to compress the plasma. The Z-pinch has been restricted by rapid rising instabilities, limiting plasma lifetime. Latest development aiming at providing stability is supported by a method exploiting sheared-flow equilibrium to create an equilibrium Z-pinch [46,47].

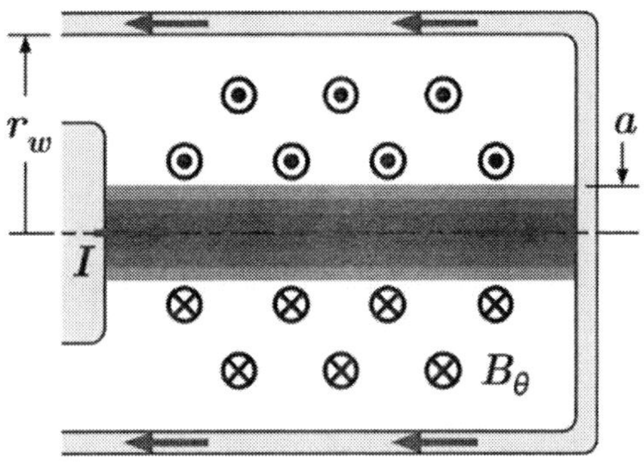

Figure 5. Scheme of the Z-pinch equilibrium [46].

3. The simulation of implosion in ICF

3.1. The simulation of laser-driven ignition devices
NIF is the world's largest laser-driven laser fusion device. To achieve the ignition, the driving of the fusion needs to be fairly symmetric and uniform. The hydrodynamic instabilities and mix are very important for the degradation of implosion in ICF. To study them further and more comprehensive, some focused platforms were developed for measurement in some planned experiments [33]. For example, the measurements of the velocity non-uniformities of shock fronts were conducted by the two-dimensional imaging velocimetry technique [48]. For the materials used, the beryllium is usually employed as the materials for the capsules in the implosion for its high mass ablation ratio. The initial comprehensive experimental investigation about beryllium capsules with a cryogenic deuterium-tritium fuel layer in the indirectly driven fusion was done in 2019[49]. Then, a study applied the same dimension hohlraums on tactics to increase the capsule diameter and found that using a 'picketless' pulse can improve performance [50]. Recently in 2016, a thin-glass-shell, D3He-filled exploding-

pusher implosion of ICF at NIF was developed and demonstrated as a proton resource, which is the first promising attempt to develop a monoenergetic proton, α, and triton backlighting platform at the NIF [51]. This is hopeful to develop further and create a new platform at the NIF for some related research. Another famous laser-driven device is LMJ (Laser Megajoule) of the French Atomic Energy Commission (CEA). To study the ion dynamic structure of the front of planar collisional shock for the fully ionized plasma, the Vlasov-Fokker-Planck code was first employed in 1991 [52]. In 1993, the code was used to simulate the plasma collision in aircraft geometry [53] and the ion dynamics simulation of formation and propagation of plasma shock [54]. Besides, the code for the calculation of thermonuclear fuel fluid dynamics in the ICF was added to solve the Vlasov-Fokker-Planck equations [55]. The simulation results show that the ion transmission between the target core and outer side is enhanced compared with standard hydrodynamics methods. Subsequently, a locally split-step explicit (LSSE) algorithm was developed to solve the highly non-uniform and highly non-isotopic diffusion tension problems in the multi-dimensional advection-diffusion type equation in 2007 [56]. The optimization made the application of the Vlasov-Fokker-Planck code in the simulation of the slowing-down process of fast charged-particle.

3.2. The simulation of heavy ion inertial fusion (HIF)

For the heavy ion inertial fusion, the code OK1 came up to simulate the irradiation of the heavy ion beam on the directly driven spherical fuel pellets [57]. From 2004-2010, the code was developed to fit the target with any shape and structure [58]. As shown in Figure 6, the significant wobbling beam illumination capability was added [59] to increase the stability. Finally, the codes are combined as a program system named O-SUKI [60, 61]. Then in 2020, the O-SUKI was applied in practice to research how the beam input pulse parameters affect the fuel implosion and the implosion non-uniformity [62], which proved that the codes are very meaningful and can be a tool to research the HIF target implosion with high efficiency.

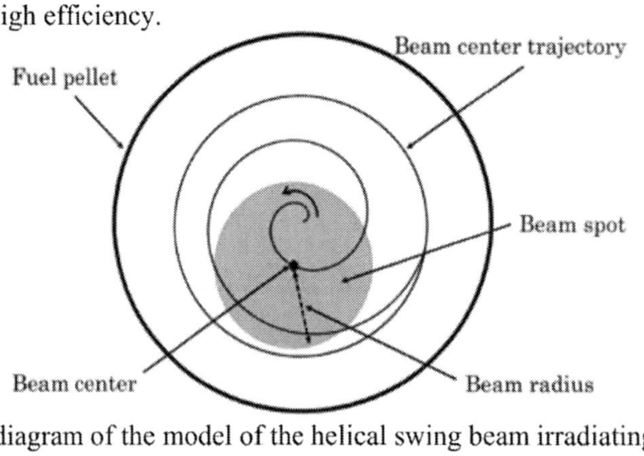

Figure 6. Schematic diagram of the model of the helical swing beam irradiating the fuel target [60].

3.3. The simulation of fast ignition devices

Simulations for fast ignition were developed by ILE group. The one-dimensional hydrodynamic implosion code and Fokker-Planck equation were used to study the kinetic energy effect of electron heat transfer in plasma [63]. The Fokker-Planck equation is solved by simplifying it into multiple sets of transmission equations in the implosion simulation. The reproduction findings are extremely coherent with the experiment. The results in the simulation experiment showed that the electron heat transfer ought to be explained using the Fokker-Planck (FP) equation in the fluid implosion code. In contrast, Spitzer-Helm (SH) heat conduction model is not appropriate [64]. Afterward, a simulation code was developed to describe the dynamics of spherically symmetric implosion water. The Fokker Planck equation of electron transmission was resolved to define non-local effects in 2008 [65].

4. Vlasov-Fokker-Planck (VFP) algorithm

VFP kinetic energy equation, together with Maxwell's equation, has wide applicability in the fusion experiments. One successful example is a high-fidelity Vlasov-Fokker-Planck code, iFP code, developed by the Los Alamos National Laboratory. In 2014, a new model for the analysis of fluid dynamics and burn of thermonuclear fuels was conducted in line with a two-velocity-scale Vlasov-Fokker-Planck kinetic model, which is specially for the supra-thermal α particles in the fusion [66]. Later, the model was applied to describe the transmission of super-thermal α particles in the ICF target [67], which means the Fokker-Planck codes can simulate a complete DT target implosion at the ion dynamics level, including ignition and burn processes. At the end of 2014, the ion Fokker-Planck model was applied to analyze the different behaviors of deuterium and tritium at different stages in the implosion of marginally-igniting cryogenic ICF [68]. Additionally, in 2015 and 2016, a fully implicit algorithm [69] and an adaptive velocity-space separation scheme [70] were proposed for the multi-species, multi-dimensional Rosenbluth-Fokker-Planck (RFP) equation. Besides, a collision particle intracellular simulation is used to explore the kinetic energy effect generated during the mixing process of unmagnetized plasma media to correct the diffusion speed conditions [71]. Moreover, a discretization approach was employed to maintain discrete in arbitrary multidimensional geometry [72]. In order to study the basic structure and the dependence of the parameters of the plasma, the state-of-the-art VFP code, iFP was applied to analyze the steady-state planar shocks in D-3He plasmas. The simulation results was compared with multi-ion hydrodynamic simulations and semi-analytic results [73]. Besides, in order to verify that the iFP can produce correct results in weak shocks with multi-ion hydrodynamics conditions. The electron and ion temperature of iFP and the multi-ion hydro results were compared and shown in Figure 7, where the results tally well with each other.

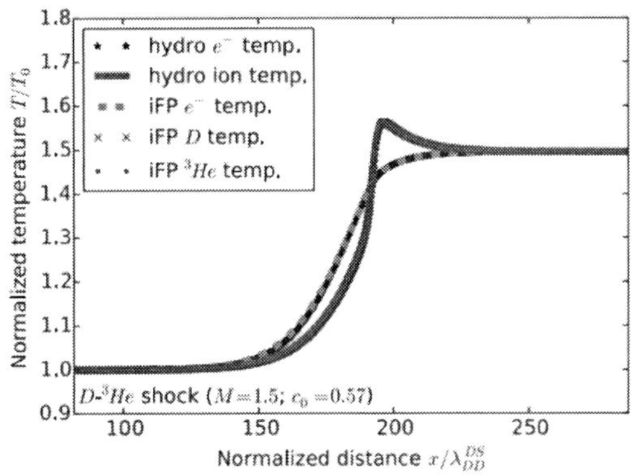

Figure 7. Electron and ion temperature profiles for weak shock [73].

Besides, some similar simulations based on the iFP codes were done to analyze the features of the plasmas shock in different environments [74] and the early transients of the interface between two species [75]. Besides, in 2018, the temporal stiffness was solved with implicit time stepping and the velocity-space scheme in 2016 was developed to an adaptive, implicit, conservative, spatially inhomogeneous system to build a 1D-2V multi-species VFP multi-scale solver in planar geometry [76]. Considering the non-thermodynamic equilibrium of ions composed of a mixture of fuel and strong thrust fuel, a mixed hybrid ion kinetic-fluid mathematical model was developed to simulate the kinetic behaviors of the implosion [77]. In 2020, the hybrid iFP codes were then employed [78]. Subsequently, the iFP code was used to confirm the pictures of two dynamical fuel implosion stages to reach ignition: a shock phase and an adiabatic compression phase [79]. Recently, a conservative configuration and moving-grid strategy of velocity-space for the planar geometrical VFP equation was developed. In this case, some more complex simulation cases were performed to show the beneficial features of the configuration and strategy with good integration capability of the algorithm [80]. Finally, the moving-mesh strategy was developed to spherical geometry to make a phase-space adaptive Eulerian VFP simulation of inertial confinement fusion capsule implosions available [81]. Some cases include the test problems were implemented to verify the strategy and finished the first ever reported verification of Guderley and Van-Dyke problem with VFP. The integrated algorithm can be applied to simulate the implosion of the Omega capsule in real cases.

5. Instabilities in the fusion

5.1. R-T instabilities

In practical there will be some fluid instability problems in the interface of the interface of thermal nuclear fuel medium because the shape of the pellet is not a uniformly distributed sphere, and the energy of the laser is hard to distribute evenly. One of the instabilities is called Rayleigh-Taylor instability which is a hot topic focused on by researchers. In 1993, the radiation effects on pellet implosion were studied by statistical method [82]. However, the classical method does not work in the reduction of the R-T instability [83], which shows that the perpendicular vibration of an R-T unstable interface will not benefit, and some other methods are needed. Then, a new mitigation method for the instability came out by a rotating or oscillating HIB illumination onto a pellet [84]. The principal is to add a dynamically adaptive perturbation to increase the total instability. Typically, a sketch of R-T instability is presented in Figure 8. For a perturbed plasma system, taking the time at Figure 8 (a) as the initial time point, and, if the perturbation added has the inverse phase as shown in Figure 8 (b), the total amplitude growth can be mitigated when time pass to $t = \Delta t$ (seen from figure 8 (c)). The effect of the method had been proved by the R-T instability analysis and numerical simulations. Another smoothing method by a high-intensity charger-particle beam to accomplish consistent illumination over a fitting region of the target in the HIF was proposed in a research in 2010 [85]. The wobbler's mitigation method was applied by controlling and decreasing the perturbation growth though defining the imposed perturbation phase and amplitude at every wobbling cycle in 2012 [86]. Furthermore, studies found that the original imprint of the wobbling HIBs will bring a great deplorable energy gathering nun-uniformity [87]. On this basis, the robustness of the method was discussed further in 2015 [88]. Later, the robustness and the alleviation of the non-uniform implosion process for the ICF target by the HIB wobbler are discussed together to prove that in a further step [89]. Moreover, the following filamentation instability of them was studied by the dynamic stabilization when an electron beam was introduced into plasma [90]. The results denote that the application is successful, and the filamentation instability can also be stabilized similarly. In 2019, the method was developed to a dynamic smoothing method in the plasma systems with perturbation. The results demonstrated that the HIBs wobbler would help improve the stability and uniformity of the fuel target implosion [91].

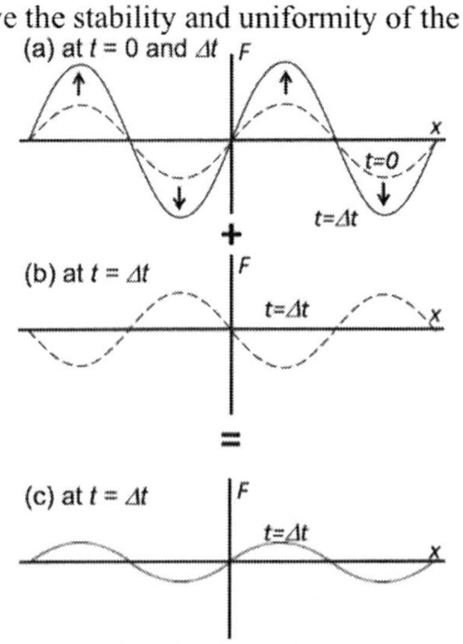

Figure 8. A sketch of R-T instabilities [91].

5.2. Laser plasma instabilities (LPI)

Due to the inevitable initial density disturbance in the plasma, LPI is very easy to be excited. Typical LPI includes Stimulated Brillouin Scattering (SBS) and Stimulated Raman Scattering (SRS) [92]. A theoretical research method based on Coupled Mode Equations is established for the three-wave coupled LPI process, starting from Maxwell's equations and fluid mechanics equations. The growth characteristics of typical processes such as SRS and SBS are investigated [93]. For SRS, it is reported that applying particle-in-cell simulations to study SRS excited by incoherent light showed that the growth of SRS and electron heating greatly could be decreased by large incoherent light bandwidth in the linear stage [94]. Subsequently, there might be some effects from the electron beam for the stimulated Raman side-scattering [95]. Some useful conclusions are obtained for the relationship between beam radius/plasma density and SBS [96]. The materials own a great impact on the SBS [97]. This can be employed for the material selection for SBS reduction.

6. Conclusion

In summary, we introduce the ICF ignition schemes and simulation of implosion and discuss the instabilities during the ignition. Specifically, three kinds of ICF driven by different methods (laser/particle beam/Z-pinch) are demonstrated. Subsequently, simulations of implosion in each type of ICF are mentioned, indicating some valuable methods or results in the field. Afterward, the development and example of the useful VFP algorithm are introduced. Finally, the vital problems instabilities that affect the ignition are reviewed, including R-T instabilities and some parameter instabilities. This paper provides some milestones of ICF, which offers a guideline for further studies.

References

[1] Johnson, T. H. (1984). Inertial confinement fusion: Review and perspective. Proceedings of the IEEE, 72(5), 548-594.

[2] Lindl, J. (1995). Development of the indirect - drive approach to inertial confinement fusion and the target physics basis for ignition and gain. Physics of plasmas, 2(11), 3933-4024.

[3] Betti, R., & Hurricane, O. A. (2016). Inertial-confinement fusion with lasers. Nature Physics, 12(5), 435-448.

[4] Deutsch, C. (1986). Inertial confinement fusion driven by intense ion beams. In Annales de Physique (Vol. 11, No. 1, pp. 1-111). EDP Sciences.

[5] Nakai, S., & Takabe, H. (1996). Principles of inertial confinement fusion-physics of implosion and the concept of inertial fusion energy. Reports on progress in physics, 59(9), 1071.

[6] Pfalzner, S. (2006). An introduction to inertial confinement fusion. CRC Press.

[7] Nuckolls, J., Wood, L., Thiessen, A., & Zimmerman, G. (1972). Laser compression of matter to super-high densities: Thermonuclear (CTR) applications. Nature, 239(5368), 139-142.

[8] Fraley, G. S., Linnebur, E. J., Mason, R. J., & Morse, R. L. (1974). Thermonuclear burn characteristics of compressed deuterium - tritium microspheres. The Physics of Fluids, 17(2), 474-489.

[9] Krupke, W. F. (1989). Solid state laser driver for an ICF reactor. Fusion Technology, 15(2P2A), 377-382.

[10] Bose, A., Woo, K. M., Betti, R. et al. (2016). Core conditions for alpha heating attained in direct-drive inertial confinement fusion. Physical Review E, 94(1), 011201.

[11] Betti, R., & Hurricane, O. A. (2016). Inertial-confinement fusion with lasers. Nature Physics, 12(5), 435-448.

[12] Fraley, G. S., Linnebur, E. J., Mason, R. J. et al. (1974). Thermonuclear burn characteristics of compressed deuterium - tritium microspheres. The Physics of Fluids, 17(2), 474-489.

[13] McCrory, R. L., Meyerhofer, D. D., Betti, R. et al. (2008). Progress in direct-drive inertial confinement fusion. Physics of Plasmas, 15(5), 055503.

[14] Lindl, J. (1995). Development of the indirect - drive approach to inertial confinement fusion and the target physics basis for ignition and gain. Physics of plasmas, 2(11), 3933-4024.

[15] Yamanaka, C. (1999). Inertial confinement fusion: The quest for ignition and energy gain using indirect drive.
[16] Craxton, R. S., Anderson, K. S., Boehly, T. R., et al. (2015). Direct-drive inertial confinement fusion: A review. Physics of Plasmas, 22(11), 110501.
[17] Richtmyer, R. D. (1954). Taylor instability in shock acceleration of compressible fluids (No. LA-1914 (del.)). Los Alamos Scientific Lab., N. Mex.
[18] Meshkov, E. E. (1969). Instability of the interface of two gases accelerated by a shock wave. Fluid Dynamics, 4(5), 101-104.
[19] Lord, R. (1900). Investigation of the character of the equilibrium of an incompressible heavy fluid of variable density. Scientific papers, 200-207.
[20] Middlebrooks, J. B. (2019). Shock tube experimentation utilizing advance diagnostics for the study of an impulsively accelerated multiphase cylinder (Doctoral dissertation, University of Missouri--Columbia).
[21] Lindl, J. (1995). Development of the indirect - drive approach to inertial confinement fusion and the target physics basis for ignition and gain. Physics of plasmas, 2(11), 3933-4024.
[22] Temporal, M., Canaud, B., Garbett, W. J. et al. (2014). Numerical analysis of the direct drive illumination uniformity for the Laser MegaJoule facility. Physics of Plasmas, 21(1), 012710.
[23] Mannion, O. M., Igumenshchev, I. V., Anderson, K. S. et al. (2021). Mitigation of mode-one asymmetry in laser-direct-drive inertial confinement fusion implosions. Physics of Plasmas, 28(4), 042701.
[24] Tikhonchuk, V. T., Gong, T., Jourdain, N. et al. (2021). Studies of laser-plasma interaction physics with low-density targets for direct-drive inertial confinement fusion on the Shenguang III prototype. Matter and Radiation at Extremes, 6(2), 025902.
[25] Wilson, D. C., Bradley, P. A., Hoffman, N. M. et al. (1998). The development and advantages of beryllium capsules for the National Ignition Facility. Physics of Plasmas, 5(5), 1953-1959.
[26] Haan, S. W., Amendt, P. A., Dittrich, T. R. et al. (2004). Design and simulations of indirect drive ignition targets for NIF. Nuclear fusion, 44(12), S171.
[27] Haan, S. W., Atherton, J., Clark, D. S. et al. (2013). NIF ignition campaign target performance and requirements: status May 2012. Fusion Science and Technology, 63(2), 67-75.
[28] Haan, S. W., Clark, D. S., Baxamusa, S. H. et al. (2016). Update 2015 on target fabrication requirements for NIF layered implosions, with emphasis on capsule support and oxygen modulations in GDP. Fusion Science and Technology, 70(2), 121-126.
[29] Stephens, R. B., Haan, S. W., & Wilson, D. C. (2002). Characterization specifications for baseline indirect drive NIF targets. Fusion science and technology, 41(3P1), 226-233.
[30] Kline, J. L., Batha, S. H., Benedetti, L. R. et al. (2019). Progress of indirect drive inertial confinement fusion in the United States. Nuclear Fusion, 59(11), 112018.
[31] Casey, D. T., MacGowan, B. J., Sater, J. D. et al. (2021). Evidence of Three-Dimensional Asymmetries Seeded by High-Density Carbon-Ablator Nonuniformity in Experiments at the National Ignition Facility. Physical Review Letters, 126(2), 025002.
[32] Trickey, W. (2021). Novel Approaches to Indirect Drive Inertial Confinement Fusion (Doctoral dissertation, University of York).
[33] Smalyuk, V. A., Weber, C. R., Landen, O. L. et al. (2020). Recent and planned hydrodynamic instability experiments on indirect-drive implosions on the National Ignition Facility. High Energy Density Physics, 36, 100820.
[34] Tabak, M., Clark, D. S., Hatchett, S. P. et al. (2005). Review of progress in fast ignition. Physics of Plasmas, 12(5), 057305.
[35] Campbell, R. B., Kodama, R., Mehlhorn, T. A. et al. (2005). Simulation of heating-compressed fast-ignition cores by petawatt laser-generated electrons. Physical review letters, 94(5), 055001.
[36] Tabak, M., Hinkel, D., Atzeni, S. et al. (2006). Fast ignition: Overview and background. Fusion science and technology, 49(3), 254-277.

[37] Kawata, S. (2021). Direct-drive heavy ion beam inertial confinement fusion: a review, toward our future energy source. Advances in Physics: X, 6(1), 1873860.
[38] Horioka, K. (2018). Progress in particle-beam-driven inertial fusion research: Activities in Japan. Matter and Radiation at Extremes, 3(1), 12-27.
[39] Jackson, J. D. (1999). Classical electrodynamics.
[40] Ziegler, J. F., & Biersack, J. P. (1985). The stopping and range of ions in matter. In Treatise on heavy-ion science (pp. 93-129). Springer, Boston, MA.
[41] Ichimaru, S. (2018). Statistical plasma physics, volume I: basic principles. CRC Press.
[42] Mehlhorn, T. A. (1981). A finite material temperature model for ion energy deposition in ion - driven inertial confinement fusion targets. Journal of Applied Physics, 52(11), 6522-6532.
[43] Doi, H., Fukaya, K., Oka, S. I. et al. (2019). Evaluation of detection probabilities at the water-filtering and initial PCR steps in environmental DNA metabarcoding using a multispecies site occupancy model. Scientific reports, 9(1), 1-8.
[44] Kawata, S., Karino, T., & Ogoyski, A. I. (2016). Review of heavy-ion inertial fusion physics. Matter and Radiation at Extremes, 1(2), 89-113.
[45] Haines, M. G. (2011). A review of the dense Z-pinch. Plasma Physics and Controlled Fusion, 53(9), 093001.
[46] Shumlak, U. (2020). Z-pinch fusion. Journal of Applied Physics, 127(20), 200901.
[47] Shumlak, U., Nelson, B. A., Claveau, E. L. et al. (2017). Increasing plasma parameters using sheared flow stabilization of a Z-pinch. Physics of Plasmas, 24(5), 055702.
[48] Ali, S., Celliers, P., Haan, S., Boehly, T., Whiting, N., & Baxamusa, S. et al. (2018). Probing the seeding of hydrodynamic instabilities from nonuniformities in ablator materials using 2D velocimetry. Physics Of Plasmas, 25(9), 092708.
[49] Zylstra, A., MacLaren, S., Yi, S., Kline, J., Callahan, D., & Hurricane, O. et al. (2019). Implosion performance of subscale beryllium capsules on the NIF. Physics Of Plasmas, 26(5), 052707.
[50] Zylstra, A., Ralph, J., MacLaren, S., Yi, S., Kyrala, G., & Bachmann, B. et al. (2020). Beryllium implosions at smaller case-to-capsule ratio on NIF. High Energy Density Physics, 34, 100747.
[51] Rosenberg, M., Zylstra, A., Séguin, F., Rinderknecht, H., Frenje, J., & Gatu Johnson, M. et al. (2016). A direct-drive exploding-pusher implosion as the first step in development of a monoenergetic charged-particle backlighting platform at the National Ignition Facility. High Energy Density Physics, 18, 38-44.
[52] Casanova, M., Larroche, O., & Matte, J. (1991). Kinetic simulation of a collisional shock wave in a plasma. Physical Review Letters, 67(16), 2143-2146.
[53] Larroche, O. (1993). Kinetic simulation of a plasma collision experiment. Physics Of Fluids B: Plasma Physics, 5(8), 2816-2840.
[54] Vidal, F., Matte, J., Casanova, M., & Larroche, O. (1993). Ion kinetic simulations of the formation and propagation of a planar collisional shock wave in a plasma. Physics Of Fluids B: Plasma Physics, 5(9), 3182-3190.
[55] Larroche, O. (2003). Kinetic simulations of fuel ion transport in ICF target implosions. The European Physical Journal D - Atomic, Molecular And Optical Physics, 27(2), 131-146.
[56] Larroche, O. (2007). An efficient explicit numerical scheme for diffusion-type equations with a highly inhomogeneous and highly anisotropic diffusion tensor. Journal Of Computational Physics, 223(1), 436-450.
[57] Ogoyski, A., Someya, T., & Kawata, S. (2004). Code OK1—Simulation of multi-beam irradiation on a spherical target in heavy ion fusion. Computer Physics Communications, 157(2), 160-172.
[58] Ogoyski, A., Kawata, S., & Someya, T. (2004). Code OK2—A simulation code of ion-beam illumination on an arbitrary shape and structure target. Computer Physics Communications, 161(3), 143-150.

[59] Ogoyski, A., Kawata, S., & Popov, P. (2010). Code OK3 – An upgraded version of OK2 with beam wobbling function. Computer Physics Communications, 181(7), 1332-1333.
[60] Stević, S. (2019). Solvability of a general class of two-dimensional hyperbolic-cotangent-type systems of difference equations. Advances In Difference Equations, 2019(1).
[61] Sato, R., Kawata, S., Karino, T., Uchibori, K., Iinuma, T., Katoh, H., & Ogoyski, A. (2019). Code O-SUKI: Simulation of direct-drive fuel target implosion in heavy ion inertial fusion. Computer Physics Communications, 240, 83-100.
[62] Uchibori, K., Sato, R., Karino, T., Iinuma, T., Kato, H., Kawata, S., & Ogoyski, A. (2020). Development of fuel target implosion simulation system in heavy ion inertial confinement fusion. High Energy Density Physics, 34, 100748.
[63] Nishiguchi, A., Mima, K., Azechi, H., Miyanaga, N., & Nakai, S. (1992). Kinetic effects of electron thermal conduction on implosion hydrodynamics. Physics Of Fluids B: Plasma Physics, 4(2), 417-422.
[64] Honda, M., Nishiguchi, A., Takabe, H., Azechi, H., & Mima, K. (1996). Kinetic effects on the electron thermal transport in ignition target design. Physics Of Plasmas, 3(9), 3420-3424.
[65] Mora, P., & Luciani, J. (1994). Nonlocal electron transport in laser created plasmas. Laser And Particle Beams, 12(3), 387-400.
[66] Oparin, A., Anisimov, S., & Meyer-Ter-Vehn, J. (1996). Kinetic simulation of DT ignition and burn in ICF targets. Nuclear Fusion, 36(4), 443-452.
[67] Peigney, B., Larroche, O., & Tikhonchuk, V. (2014). Fokker - Planck kinetic modeling of suprathermal α -particles in a fusion plasma. Journal Of Computational Physics, 278, 416-444.
[68] Inglebert, A., Canaud, B., & Larroche, O. (2014). Species separation and modification of neutron diagnostics in inertial-confinement fusion. EPL (Europhysics Letters), 107(6), 65003.
[69] Taitano, W., Chacón, L., Simakov, A., & Molvig, K. (2015). A mass, momentum, and energy conserving, fully implicit, scalable algorithm for the multi-dimensional, multi-species Rosenbluth–Fokker–Planck equation. Journal Of Computational Physics, 297, 357-380.
[70] Taitano, W., Chacón, L., & Simakov, A. (2016). An adaptive, conservative 0D-2V multispecies Rosenbluth–Fokker–Planck solver for arbitrarily disparate mass and temperature regimes. Journal Of Computational Physics, 318, 391-420.
[71] Yin, L., Albright, B., Taitano, W., Vold, E., Chacon, L., & Simakov, A. (2016). Plasma kinetic effects on interfacial mix. Physics Of Plasmas, 23(11), 112302.
[72] Taitano, W., Chacón, L., & Simakov, A. (2017). An equilibrium-preserving discretization for the nonlinear Rosenbluth–Fokker–Planck operator in arbitrary multi-dimensional geometry. Journal Of Computational Physics, 339, 453-460.
[73] Keenan, B., Simakov, A., Chacón, L., & Taitano, W. (2017). Deciphering the kinetic structure of multi-ion plasma shocks. Physical Review E, 96(5).
[74] Keenan, B., Simakov, A., Taitano, W., & Chacón, L. (2018). Ion species stratification within strong shocks in two-ion plasmas. Physics Of Plasmas, 25(3), 032103.
[75] Vold, E., Yin, L., Taitano, W., Molvig, K., & Albright, B. (2018). Diffusion-driven fluid dynamics in ideal gases and plasmas. Physics Of Plasmas, 25(6), 062102.
[76] Taitano, W., Chacón, L., & Simakov, A. (2018). An adaptive, implicit, conservative, 1D-2V multi-species Vlasov–Fokker–Planck multi-scale solver in planar geometry. Journal Of Computational Physics, 365, 173-205.
[77] Larroche, O., Rinderknecht, H., & Rosenberg, M. (2018). Nuclear yield reduction in inertial confinement fusion exploding-pusher targets explained by fuel-pusher mixing through hybrid kinetic-fluid modeling. Physical Review E, 98(3).
[78] Keenan, B., Taitano, W., Simakov, A., Chacón, L., & Albright, B. (2020). Shock-driven kinetic and diffusive mix in high-Z pusher ICF designs. Physics Of Plasmas, 27(2), 022704.
[79] Keenan, B., Taitano, W., & Molvig, K. (2020). Physics of the implosion up until the time of ignition in a revolver (triple-shell) capsule. Physics Of Plasmas, 27(4), 042704.

[80] Taitano, W., Chacón, L., Simakov, A., & Anderson, S. (2021). A conservative phase-space moving-grid strategy for a 1D-2V Vlasov–Fokker–Planck Solver. Computer Physics Communications, 258, 107547.
[81] Taitano, W., Keenan, B., Chacón, L., Anderson, S., Hammer, H., & Simakov, A. (2021). An Eulerian Vlasov-Fokker–Planck algorithm for spherical implosion simulations of inertial confinement fusion capsules. Computer Physics Communications, 263, 107861.
[82] Kawata, S., Sato, T., Teramoto, T., Bandoh, E., Masubichi, Y., & Takahashi, I. (1993). Radiation effect on pellet implosion and Rayleigh-Taylor instability in light-ion beam inertial confinement fusion. Laser And Particle Beams, 11(4), 757-768.
[83] Piriz, A., Piriz, S., & Tahir, N. (2011). Dynamic stabilization of classical Rayleigh-Taylor instability. Physics Of Plasmas, 18(9), 092705.
[84] Kawata, S., Iizuka, Y., Kodera, Y., Ogoyski, A., & Kikuchi, T. (2009). Robust fuel target in heavy ion inertial fusion. Nuclear Instruments And Methods In Physics Research Section A: Accelerators, Spectrometers, Detectors And Associated Equipment, 606(1-2), 152-156.
[85] Qin, H., Davidson, R., & Logan, B. (2010). Centroid and Envelope Dynamics of High-Intensity Charged-Particle Beams in an External Focusing Lattice and Oscillating Wobbler. Physical Review Letters, 104(25).
[86] Kawata, S. (2012). Dynamic mitigation of instabilities. Physics Of Plasmas, 19(2), 024503.
[87] Kawata, S., kurosaki, T., Koseki, S., Noguchi, K., Barada, D., & Ogoyski, A. et al. (2013). Wobbling Heavy Ion Beam Illumination in Heavy Ion Inertial Fusion. Plasma And Fusion Research, 8(0), 3404048-3404048.
[88] Kawata, S. (2012). Dynamic mitigation of instabilities. Physics Of Plasmas, 19(2), 024503.
[89] Karino, T., Kawata, S., Kondo, S., Iinuma, T., Kubo, T., Kato, H., & Ogoyski, A. (2016). Target implosion uniformity in heavy-ion fusion. Laser And Particle Beams, 34(4), 735-741.
[90] Kawata, S., Gu, Y., Li, X., Karino, T., Katoh, H., & Limpouch, J. et al. (2018). Dynamic stabilization of filamentation instability. Physics Of Plasmas, 25(1), 011601.
[91] Sato, R., Kawata, S., Karino, T., Uchibori, K., & Ogoyski, A. (2019). Non-uniformity smoothing of direct-driven fuel target implosion by phase control in heavy ion inertial fusion. Scientific Reports, 9(1).
[92] Kruer, W. (1991). Intense laser plasma interactions: From Janus to Nova. Physics Of Fluids B: Plasma Physics, 3(8), 2356-2366.
[93] Rosenbluth, M., White, R., & Liu, C. (1973). Temporal Evolution of a Three-Wave Parametric Instability. Physical Review Letters, 31(19), 1190-1193.
[94] Zhao, Y., Weng, S., Chen, M., Zheng, J., Zhuo, H., & Sheng, Z. (2017). Stimulated Raman scattering excited by incoherent light in plasma. Matter And Radiation At Extremes, 2(4), 190-196.
[95] Baliyan, S., Rafat, M., Panwar, A., Sharma, N., & Sajal, V. (2019). Suppression of stimulated Raman side-scattering of a laser in a plasma in presence of co-propagating electron beam. Optik, 179, 330-335.
[96] Walia, K. (2020). Stimulated Brillouin Scattering of high power beam in unmagnetized plasma: Effect of relativistic and ponderomotive nonlinearities. Optik, 221, 165365.
[97] Bai, Z., Yuan, H., Liu, Z., Xu, P., Gao, Q., & Williams, R. (2018). Stimulated Brillouin scattering materials, experimental design and applications: A review. Optical Materials, 75, 626-645.

Study on impulse and ablation of aluminum irradiated by millimeter spot of short pulse laser

Mingyu Li[1a], Jifei Ye[*1b], Chentao Mao[1c], Sibo Wang[1d], Chenghao Yu[1e]

[1]State Key Laboratory of Laser Propulsion & Application, Department of Aerospace Science and Technology, Aerospace Engineering University, Beijing 101416, China

[a]merle_lee@outlook.com, [b*]yjf1981@163.com , [c]794449157@qq.com,

[d]bosiwang1@163.com,[e]yuchenghao0536@163.com

Abstract—Aluminum is a high performance working medium for laser ablation micro propulsion. In order to study the propulsion performance and ablation of aluminum under millimeter light spot irradiation, a short pulse Nd: YAG laser with wavelength of 1064nm and pulse width of 8NS was used to irradiate aluminum target in atmosphere. The impulse, the impulse coupling coefficient and the ablation morphology of the aluminum target produced by 6 kinds of millimeter-level light spots are measured.The experimental results show that when the spot diameter reaches 6-7mm, the increasing trend of impulse and impulse coupling coefficient of aluminum target with the increase of laser energy slows down; A large number of ablation products began to accumulate on the surface of the target pit.

1. Introduction

The rapid development of micro-nano satellite has put forward the technical requirements of light weight and micro-element for space micro-propulsion technology. Laser ablation micropropulsion, as a space micropropulsion technology, uses solar energy as the main energy source and has the advantages of small impulse element, high specific impulse and high control precision, which can provide more accurate attitude and orbit control for micro/nano satellites[1][2]. In addition to laser, the main optimization direction of laser ablative micro-thruster is the iteration of more efficient working medium and the improvement of working medium supply mode. At home and abroad in recent years have been a variety of laser ablation thruster principle prototype test engineering prototype and some have space, Phipps[3][4] has been developed, such as ms - mu LPT I and ms - mu LPT II laser ablation principle prototype micro thruster, domestic Yanji Hong[5] disc type is developed, such as laser ablation principle prototype micro thruster, above the operation mode of the prototype are transmission type; Akira Kakami et al.[6] designed a reflective laser micropropulsion prototype, and the LDU-7 laser ablative microthruster engineering machine launched by the Russian Federal Space Agency (ROSCOsmos)[7] also works in reflective mode[8]. All of the above laser ablation microthruster types contain complex working medium supply systems, and all of them generate microthrust for laser focused ablation working medium. Although the micron spot after focusing can bring larger single-pulse impulse, it also brings some difficulties for accurate supply of working medium. In this paper, surface ablation is realized by increasing the diameter of laser spot, forming a one-dimensional working medium supply, increasing the space of laser system and working medium in the prototype, so as to improve the performance of thruster. Using nanosecond pulse width laser, the aluminum target is irradiated with different spot diameters. The impulse of aluminum target under different laser energy, impulse coupling coefficient, target pit morphology and target pit depth are measured. Combined with the physical process of laser irradiating metal aluminum, the propulsion performance and ablation morphology of aluminum target irradiated by millimeter spot are analyzed. The obtained propulsive performance and ablation morphology provide reference for laser surface ablation of solid targets.

2. Experimental device and measurement method

2.1. Experimental device

The experimental device for laser ablation of aluminum metal is shown in Figure 1, which is mainly composed of laser, impulse measurement torsion pendulum, isolation cover and laser rangefinder.

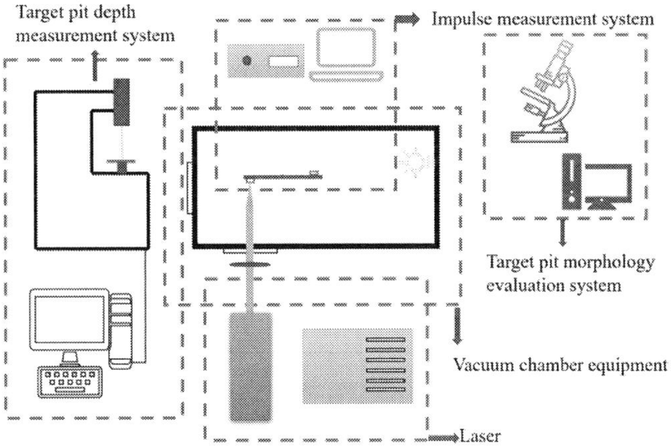

Fig.1 Schematic diagram of experimental equipment

As the increase of spot area will lead to the decrease of laser power density and energy density, in order to ensure the experimental effect, the Nd:YAG laser with the maximum single pulse energy of 900 mJ, wavelength of 1064nm and pulse width of 8 ns is used as the energy source in the experiment. The experimental material was a cylindrical 1060 aluminum alloy target with a diameter of 15mm and a thickness of 5mm, and its purity was greater than 99.6%.

2.2. Measurement method

2.2.1 Impulse measurement

Microthrust of the order $\mu N \cdot s$ was usually measured using a torsional pendulum. Figure 2 shows the impulse measurement torsion pendulum customized according to the size of the experimental material, mainly composed of precision flexible pivot, measuring beam, high-precision displacement sensor and stop rod (used to measure the fast balance of beam). The dynamic model of impulse measurement torsion pendulum is a second-order mass-spring-damping system. The maximum linear displacement of torsion pendulum can be measured by displacement sensor after the pulse laser beam is applied to the target, and the formula of impulse I can be obtained[9]:

$$I = \frac{k}{l_1 l_2 \omega_n} s \tag{1}$$

Type:

k : Torsional stiffness coefficient of torsional pendulum; ω_n :Angular frequency of torsion beam;

l_1 : The distance between displacement sensor and flexible pivot (measuring arm);

l_2 : The distance between pulse laser ablation point and flexible pivot (moment arm);

s: The maximum linear displacement of the first swing after pulsed laser loading;

The relationship between impulse coupling coefficient C_m and impulse I and incident laser energy E:

$$C_m = \frac{\Delta m v}{E} = \frac{I}{E} \tag{2}$$

Type: Δm : Ablative mass; v : Plume jet velocity; E : Incident laser energy

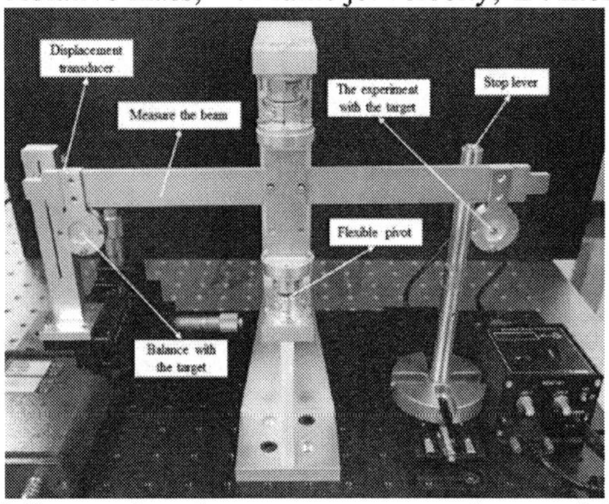

Fig.2 Custom impulse measurement torsion pendulum

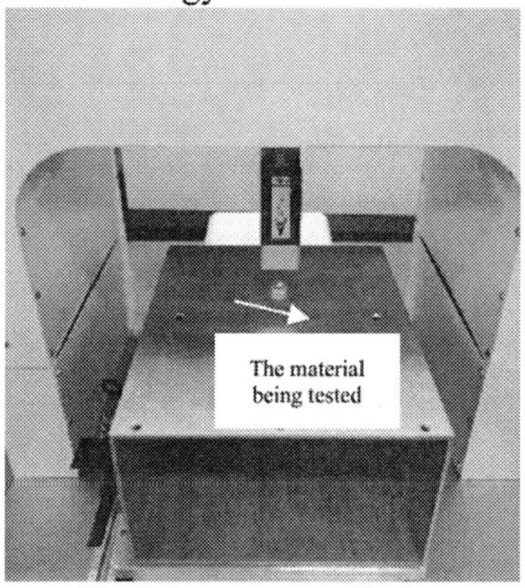

Fig.3 Laser rangefinder

2.2.2 *Target pit depth measurement*

Laser rangefinder (as shown in Figure 3) was used to measure the ablation depth of the target pit, and different sections of the ablation plane of the same target pit were measured at the distance of 100 μm.

3. Experimental process and results

3.1. *The experimental process*

1) Experimental study on the propulsion performance of aluminum ablated by millimeter light spot ablation by short pulse laser. The aluminum target fixed on the torsional pendulum measuring beam was irradiated independently for several times by 18 kinds of Nd:YAG laser with the output energy range of 100-1000MJ and six kinds of millimeter light spots with diameters of 9.6mm, 8.6mm, 7.7mm, 6.6mm, 5.5mm and 3.5mm. The experimental data were recorded by displacement sensor.

2) Experimental study on the morphology of aluminum target pits ablated by millimeter light spot ablation by repeated pulsed laser. A Nd:YAG laser with the output energy of 1000mJ laser beam, repeated frequency of 4Hz, and six millimeter light spots with diameters of 9.6mm, 8.6mm, 7.7mm, 6.6mm, 5.5mm and 3.5mm were used to irradiate the aluminum target on the fixed beam 7200 times, and the ablation depth was measured by laser rangefinders.

3.2. *The experimental results*

1) Experimental study on the propulsion performance of aluminum ablated by millimeter light spot ablation by short pulse laser.

Figure 4(a) shows the impulse of pulsed laser ablation of aluminum target with different energy and spot diameter. It can be seen from the figure that the impulse of aluminum target increases with the change of monopulse laser energy in an inverse and non-linear relationship with spot diameter. When spot diameter increases to 7.7mm, the impulse increases with the increase of monopulse energy gradually decreases.

Figure 4(b) shows the impulse coupling coefficient of aluminum target ablated by pulsed laser with different energy and spot diameter. It can be seen from the figure that the impulse coupling coefficient of aluminum target ablated by laser with millimeter light spot also increases with the change of laser energy and is inversely proportional to spot diameter and non-linear. When spot diameter increases to 6.6-7.7mm, The increase of single pulse energy decreases the increase of impulse coupling coefficient.

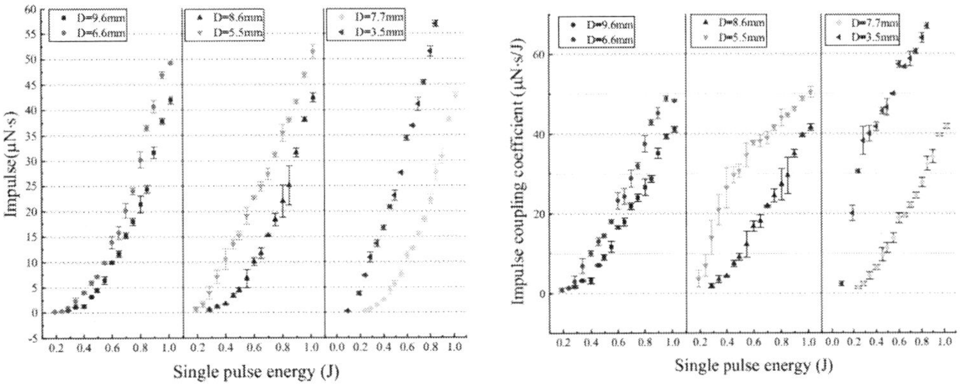

(a) (b)

Fig.4 The relationship of impulse and impulse coupling coefficient with single pulse energy and spot diameter

By the above experimental results it can be seen that the laser energy within the 1000 mj, as large as possible while assuring ablation area, under the condition of the laser spot diameter is 6.6-7.7 mm, mm level spot of laser ablation of aluminum impulse and impulse coupling coefficient can be considered to be the optimal solution, for subsequent repeat impulse millimeter level spot ablation experiments, The selection of spot size provides reference.

2) Experimental study on the morphology of aluminum target pits ablated by millimeter light spot ablation by repeated pulsed laser.

A laser with a single pulse energy of 1.02J and a repetition frequency of 4Hz is used to ablate the aluminum target loaded on a fixed beam with a number of 7200 pulses. The ablations of different spot sizes are shown in Figure. 5. Spot diameter D=9.6mm, low energy density $1.41 J/cm^2$,incomplete ablation, ablation product residues, target pit surface smooth and black; The spot diameter D= 8.6mm, the energy density was $1.76 J/cm^2$, the ablation was incomplete, there were ablative product residues, and the target pit surface was smooth and black, compared with the target pit surface of D=9.6mm, the black residue was reduced. The spot diameter D=7.7mm, the energy density was $2.19 J/cm^2$, the ablation is complete, the more clear Gaussian energy distribution ablation ring can be seen, the ablation plane is flat. The spot diameter D=6.6mm, the energy density was $2.98 J/cm^2$, the ablation is relatively complete, also has a clear Gaussian energy distribution ablation ring, the ablation plane is flat. The spot diameter D= 5.5mm, the energy density was $4.29 J/cm^2$ the ablation was complete, and the ablation plane began to concave. Spot diameter D=3.5mm, energy density was $10.6 J/cm^2$, ablation complete, ablation plane obviously concave.

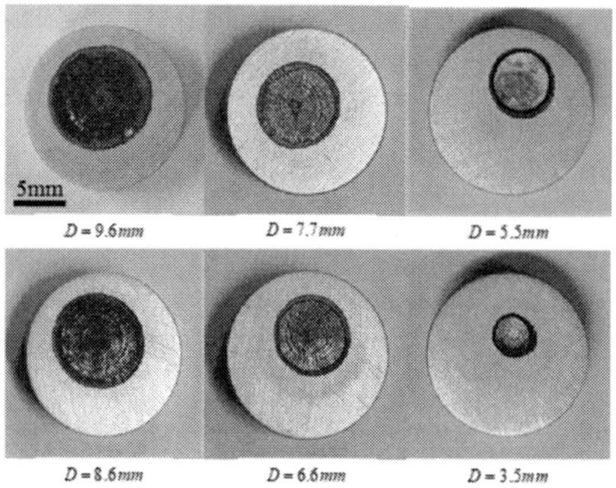

Fig.5 Target pit morphology of aluminum ablated by millimeter - grade laser spot ablation by repeated pulse laser with fixed energy

Figure 6 shows the measurement results of target pit depth after repeated pulse ablation experiments when spot diameters are 9.6 and 8.6mm. As can be seen from the figure, due to the low energy density and incomplete ablation, the cross section height is higher than the target surface due to the serious accumulation of ablation products. "O" refers to the area where the black ablative products are more densely packed, so that the laser can not be ranging, and the bad point appears. In the figure, there are more bad spots in the 9.6mm spot than in the 8.6mm spot, which indicates that the larger the spot area, the more incomplete ablation and more bad spots.

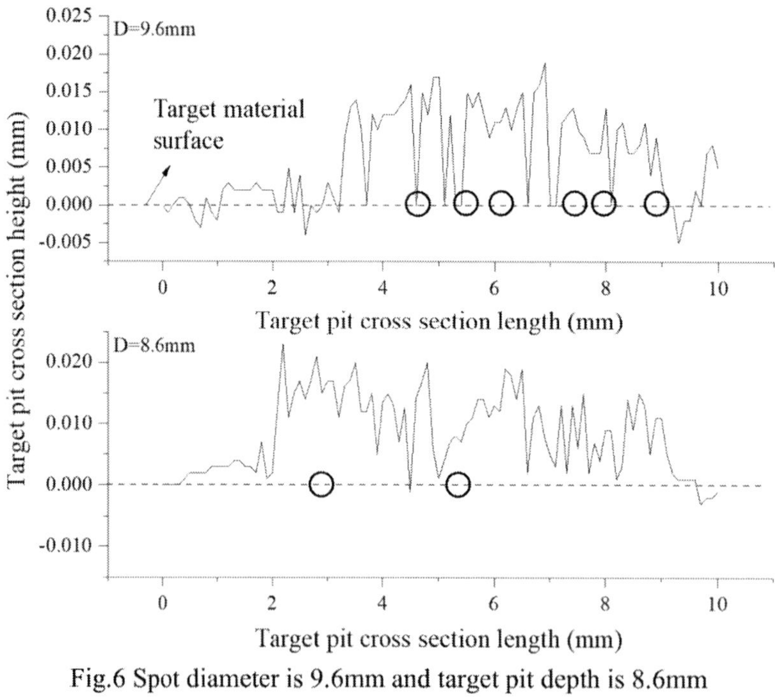

Fig.6 Spot diameter is 9.6mm and target pit depth is 8.6mm

Figure 7 shows the measurement results of target pit depth after repeated pulse ablation experiment when spot diameters are 7.7mm, 6.6mm, 5.5mm and 3.5mm. As can be seen from the figure, when the spot diameter decreases gradually, a clear target pit boundary formed by ablation melting begins to appear, and the depth of the target pit is also significantly lower than the target surface, indicating that the laser irradiation of aluminum target at this time produces effective ablation. In addition, when the spot diameter is 3.5mm, the inner surface of the target pit is smooth and concave obviously. The inner surface of target pits with spot diameters of 6.6mm and 5.5mm is smooth, but the aluminum target with spot diameters of 5.5mm is ablated more fully, so the inner surface of target pits is smoother, which is roughly consistent with the results obtained by direct observation.

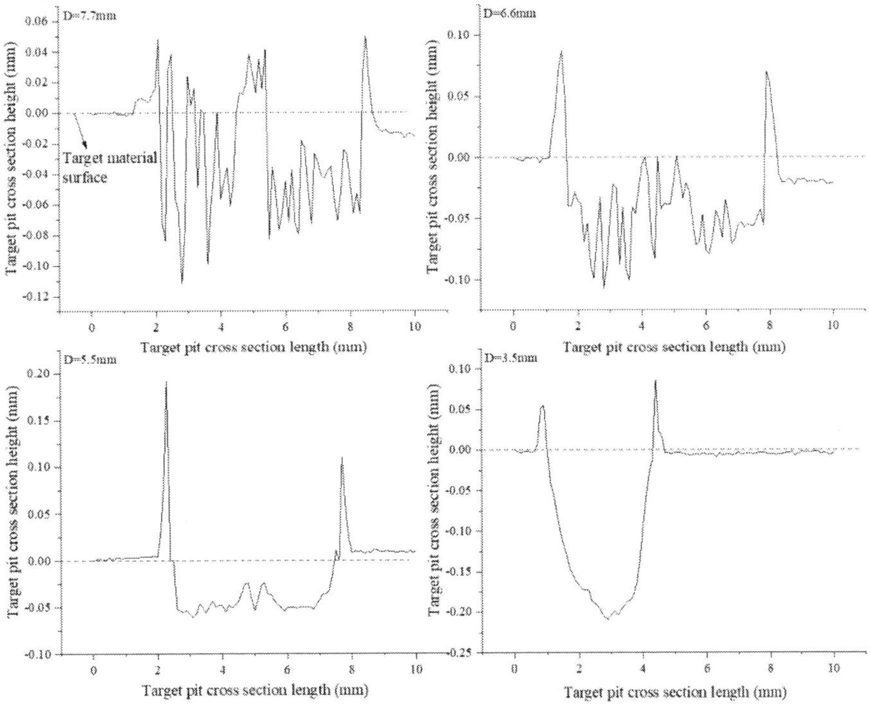

Fig.7 Spot diameter is 7.7mm, 6.6mm, 5.5mm, 3.5mm target pit depth

4. Analysis and discussion

The laser irradiates the target material to make the surface structure material absorb the laser energy and convert it into the internal heat energy of the target material, and then through the internal diffusion, it affects the physical and mechanical properties of the target material.

Generally, there is no body heat source when the infrared wavelength laser irradiates the metal target. The laser irradiation is a surface heating process, so the body heat source can be changed to surface heat source on the surface boundary conditions according to certain boundary conditions.

Therefore, the temperature field distribution of millimeter spot ablation of aluminum by fixed energy repeated pulse laser can be regarded as surface absorption, and the temperature field distribution of incident laser power density -- $q_{inc}(r,t)$ is gaussian distribution. Then its axisymmetric temperature distribution[10] is:

$$T(r,z,t)=\frac{Aq_{inc,0\max}\omega^2}{\kappa}\sqrt{\frac{\kappa}{\pi}}\int_0^t\frac{B(t-t_1)}{\sqrt{t_1}(4\kappa t_1+\omega^2)}dt_1\cdot\exp(\frac{z^2}{4\kappa t}-\frac{r^2}{4\kappa t_1+\omega^2}) \tag{3}$$

Type:

$q_{inc,0\max}$:Maximum power density of spot center; $B(t)$: The time distribution function;

κ :Thermal conductivity of materials; ω :Gaussian beam radius.

As the target continuously absorbs laser energy and converts it into its own internal energy, the surface temperature of the target continues to rise to the boiling point and gasification occurs. The gasification products will continue to absorb laser energy and further ionize to form plasma. The expansion movement of plasma generates impulse.

Clausius-clapeyron equation is used to calculate the saturated vapor pressure and density of the target surface through the target surface temperature[11]:

$$N_s(T_s) = N_a \exp[\frac{\Delta H_{lv}}{Q}(\frac{1}{T_B} - \frac{1}{T_S})]$$
$$\rho_s = \frac{N_s}{k_B T_S} \times m_v \quad (4)$$

Type:

N_s :Saturated vapor pressure on the surface of the target; T_s :Surface temperature of target;

T_B :Boiling point temperature of target material when $N_a = 1atm$; ΔH_{lv} :Gasification enthalpy;

Q :Gas constant; k_B :Boltzmann constant; m_v :Atomic mass of vapor.

Because the pulse width of laser is nanosecond, the duration of ablation pressure on target surface is very short, so impulse can be used to characterize its macroscopic mechanical effect. For a pulsed laser, the impulse coupling coefficient can be expressed as the ratio of the cumulative ablation pressure on the target surface to the power density of the incident laser:

$$C_m = \frac{N}{q_{inc}} = \frac{4Nt}{E\pi D^2} \quad (5)$$

It can be known from equations (3), (4) and (5):

$$T_s \propto q_{inc}, N \propto T_s$$
$$I \propto N, C_m \propto N \quad (6)$$
$$I \propto \frac{1}{D^2}, C_m \propto \frac{1}{D^2}$$

It can be known from Equation (6), the coupling coefficient of impulse and impulse increases with the increase of single pulse energy, while the increase of spot diameter accelerates the attenuation of the coupling coefficient.

The spatial distribution of laser energy is Gaussian distribution, so under the same laser energy, the larger the spot diameter, the more uniform the spatial distribution of laser energy. And the decrease of energy density will cause the ablation effect to decrease. Therefore, the plane of the target pits with spot diameters of 9.6mm and 8.6mm is relatively flat, and at the same time, black ablation products with a relatively uniform distribution appear on the surface of the target pit. However, target pits with smaller spots of 3.5mm have smooth surface and obvious concave phenomenon.

5. Conclusion

Based on the results and discussions presented above, the conclusions are obtained as below:

In the experiment of nanosecond pulse width laser irradiating metal aluminum, the 5 ~ 7mm diameter spot can take into account the propulsion performance, spot size and target pit plane flatness; Due to the low energy density of 7 ~ 10mm light spot, the target ablation is incomplete and the propulsion performance is poor; 3 ~ 5mm light spot has high energy density. Under repeated pulse ablation, the target pit plane will be obviously concave.

References

[1] Hong Y J.Laser Ablation Micro-thruster for Attitude and Orbit Control of Microsatellite [J]. Journal of Equipment Academy,2013,24(06):57-61.

[2] Hua Z H,Guo N,Mao C L et al. Development situation of laser solid micro-thrust technique [J]. Chemical Propellants & Polymeric Materials,2018,16(01):13-18.

[3] Phipps C R.Micro laser plasma thrusters for small satel-lites[R].AFOSR STTR Phase II2002, 2002.

[4] Hong Y J,Wang G Y,Dou Z G. Research progress of laser ablation microthrusters[J]. Acta Aeronautica et Astronautica Sinica, 2009,30(9):1555-1565.
[5] Ye J F, Hong Y J,Li N L et al. Laser ablative microthruster and its performance test[C], China Joint Conference on Aerospace Power,2017.
[6] Akira Kakami, Shinichiro Masaki, Hideyuki Horisawa and Takeshi Tachibana.Solid Propellant Microthruster Using Laser-Assisted Combustion[R].AIAA 2004-3797.
[7] A.V. Gurin,K.Yu. Kuvaev,E.Yu. Loktionov,Yu.S. Protasov,K.N. Sirenko,V.I. Zakharov. First attempt of a laser thruster space flight test: Lost at launch[J]. Optics and Laser Technology, 2019,120.
[8] Ye J F, Li N L,Chang H. Discussion on working modes of two laser ablative micro-thrusters[J], Journal of Rocket Propulsion, 2017,43(05):7-13.
[9] Ye J F,Hong Y J,Li N Let al. Laser ablation microthruster and its performance test [C]. Proceedings of the 38th Technical Exchange Conference and the 2nd Joint Conference on Space and Space Power of China Aerospace Third Professional Information Network, 2017:37-43.
[10] Sun C W, Lu Q S, Fan Z X et al. Laser irradiation effect [M].BeiJing: National Defense Industry Press, 2002:32-37.
[11] Chang H. Coupling characteristics of nanosecond laser ablation impulse and its application in space debris removal[D]. Academy of Equipment,2014.

Numerical simulation of the effect of different NaCl concent on boiler flame combustion process

Zhihai Cheng[1a*], Jiahao Wang[1b*], Xinhai Han[1c*]

[1]School of Energy and Mechanical Engineering, Shanghai Electric Power University, Shanghai 200120, China

[a*]czh.cn@163.com, [b*]wangjh231008@163.com, [c*]xinhaihan@126.com

Abstract—Zhundong coal has been widely concerned because of its high alkali metal content, which brings great danger to the combustion of boiler. Therefore, it is extremely necessary to study the laws and characteristics of alkali metal influencing combustion in the burning process of zhundong coal. A gas-solid two-phase flow combustion model of pulverized coal containing NaCl was established by using Fluent software and FactSage software in a hot experimental combustion furnace. The influence of different NaCl content in pulverized coal on pulverized coal combustion process was discussed. The results show that with the increase of NaCl content in pulverized coal from 0 to 1% and 2%, the flame center temperature in the furnace increases about 80°C and 120°C under the same coal content, so it can be concluded that the increase of NaCl content can promote the combustion process of pulverized coal in the furnace. At the same time, it can be calculated that, with the increase of NaCl content, the flame range of the combustion region inside the furnace increases by 1/3. Because NaCl is decomposed by heat during combustion to help combustion, and the radiation heat transfer increases, the flame radiation range inside the furnace will increase.

1. Introduction

As an important pillar of China's national economic development, coal power is still the main power supply in China for a period of time to come. "Zhundong Coal" is currently the largest integrated coal field in China, with an estimated coal reserve of 390 billion tons, which is of great significance to the modern industrialization process of the energy industry. Zhundong coal, however, high volatile coal, the characteristics of high moisture content, high alkali content, Na in zhundong coal and its compound content is higher, zhundong coal ash content usually contains about 5%, the highest can reach 15%, the content of alkali metal (ash) overall 5% or more, some of the alkali metal content as high as 10%[1]. Wei Xiaofang et al.[2] studied Australian high-sodium coal and found that most of the existing forms of sodium in coal are water-soluble NaCl salts. In the study of zhundong coal, Chen Chuan et al.[3] found that sodium in these high-sodium coals is mainly water-soluble sodium, with less organic sodium and insoluble sodium. Song Jinling[4] used mixed coal to control the change of alkali metal content of experimental coals, and found that the coal sample with higher alkali metal content had stronger catalytic effect on pulverized coal combustion, which was faster and more intense, and had stronger reaction ability in the early stage of combustion.

In this paper, by calculating the content of alkali metals in ash after conversion, 1% and 2% NaCl were added to raw coal to simulate different concentrations of alkali metals in pulverized coal[5]. Fluent software is used for numerical simulation of a hot state experimental combustion table to realize numerical simulation calculation of combustion state in the furnace, and obtain the combustion state, flame distribution and temperature distribution in the furnace. The influence of NaCl on the combustion process in furnace was analyzed by referring to the simulation results.

2. Introduction to experimental combustion furnace

The research object is a hot experimental combustion furnace. The diameter of the combustion furnace shell is 380mm, the external insulation layer is about 120mm, and the height of the furnace is 1200mm. The top area is the burner nozzle, the burner is the center of the primary air, the outer layer is adjacent to the secondary air structure layout, the primary air nozzle diameter is 20mm, the secondary air nozzle diameter is 10mm ring, the secondary air pipe in the insulation layer close to the outer wall of the furnace, depending on the temperature of the furnace heating secondary air temperature. The bottom of the combustion furnace is the outlet of water-cooled flue. Along the inner wall of the furnace, 8 auxiliary heating tubes with length of 800mm and diameter of about 80mm are arranged at equal spacing. Natural gas combustion is used for heating, which is used to adjust the temperature of the inner wall of the furnace and keep the furnace temperature constant.

3.Mathematical model and simulation calculation

3.1 Geometric models and grids

The structure of the test furnace was simplified and the auxiliary heating pipe was ignored. Gambit software was used to build a 1:1 3D geometric model and grid the furnace. In order to make the calculation result more accurate and reliable, the hexahedral structured grid is used to divide the grid, and the grid quality is good and meets the requirements of calculation. The total number of grids is 980,000, and the burner area is the starting point for the calculation of furnace combustion. The flow, heat transfer and other conditions are complicated, which have a great impact on the combustion, temperature field and velocity field in the furnace. Therefore, in order to improve the accuracy of calculation, the encryption method is adopted in the division of the regional network. The grid layout of primary and secondary air nozzles and the overall model is shown in Figure 1.

3.2 mathematical model

3.2.1 Realizable K-ε model

The k-ε two-way equation model[7] is selected to simulate the gas-solid two-phase flow of pulverized

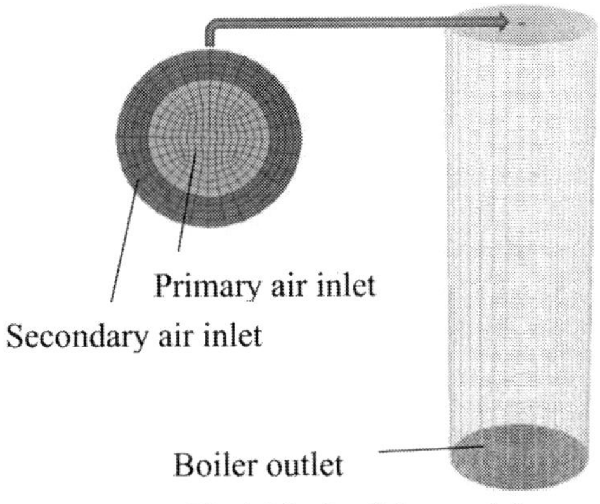

Fig.1 Mesh of the model

coal in the furnace. For the three-dimensional steady convection and diffusion problem, the governing equation is as follows:

$$\frac{\partial}{\partial x}(U\phi)+\frac{\partial}{\partial y}(\rho V\phi)+\frac{\partial}{\partial z}(\rho W\phi)=\frac{\partial}{\partial x}\left(\Gamma_\phi\frac{\partial\phi}{\partial x}\right)+\frac{\partial}{\partial y}\left(\Gamma_\phi\frac{\partial\phi}{\partial y}\right)+\frac{\partial}{\partial z}\left(\Gamma_\phi\frac{\partial\phi}{\partial z}\right)+S_\phi \quad (1)$$

Where: ϕ is the general dependent variable; Γ_ϕ is the transport coefficient; S_ϕ is the source term of general dependent variable; ρ is the gas phase density.

Equation (1) gives the transport equations of continuity and momentum and energy and turbulent kinetic energy and turbulent dissipation rate.The transport equation of turbulent kinetic energy K and turbulent kinetic energy dissipation rate ε for k-ε model can be written as follows:

k equation:

$$\frac{\partial}{\partial t}(\rho k)+\frac{\partial}{\partial x_i}(\rho k u_j)=\frac{\partial}{\partial x_i}\left[\left(\mu+\frac{\mu_t}{\sigma_k}\right)\frac{\partial k}{\partial x_j}\right]+G_k+G_b-\rho\varepsilon-Y_m+S_k \qquad (2)$$

ε equation:

$$\frac{\partial}{\partial t}(\rho\varepsilon)+\frac{\partial}{\partial x_j}(\rho\varepsilon u_j)=\frac{\partial}{\partial x_j}\left[\left(\mu+\frac{\mu_t}{\sigma_\varepsilon}\right)\frac{\partial\varepsilon}{\partial x_j}\right]+\rho C_1 S\varepsilon-\rho C_2\frac{\varepsilon^2}{k+\sqrt{\upsilon\varepsilon}}+C_{1\varepsilon}\frac{\varepsilon}{k}C_{3\varepsilon}G_b+S_\varepsilon \qquad (3)$$

Where:μ_t Is turbulence viscosity coefficient; μ Is the laminar viscosity number; G_b Is turbulent kinetic energy due to buoyancy; Y_m Is turbulence dissipation caused by pulsating expansion of compressible fluid,$Y_m=2\rho\varepsilon\frac{k}{a^2}$,a is sound velocity;$S_k$、$S_\varepsilon$ Is the source term; $C_1=\max\left[0.43, \frac{\eta}{\eta+5}\right]$, $\eta=Sk/\varepsilon$;$C_{1\varepsilon}$、C_2 are constants,σ_k、σ_ε are the turbulent Prandtl numbers of turbulent kinetic energy and its dissipation rate respectively.In Fluent,as the default constant value,$C_{1\varepsilon}=1.44$、$C_2=1.9$、$\sigma_k=1.0$、$\sigma_\varepsilon=1.2$.

3.2.2 Component transport and chemical reaction models

Component transport and chemical reaction were used to simulate the combustion process of pulverized coal with different concentrations of NaCl. The component transport model was selected as the finite rate/vortex dissipation model[6], indicating that chemical reaction was of finite reaction rate, and the reaction rate was controlled by both chemical reaction itself and turbulent mixing.

The reaction equation of pulverized coal combustion is calculated based on the coal quality characteristics used in the simulation in Fluent software,as shown in Formula (4) and Table 1.

$$C_{1.53}H_{1.69}O_{0.56}N_{0.0275}S_{0.0137}+1.68O2\longrightarrow 1.53CO_2+0.84H_2O+0.013N_2+0.012SO_2 \qquad (4)$$

Combined with the pure matter thermodynamics database in FactSage software,the thermal physical property parameters of eight optimized compounds containing Na were queried, as shown in Table 2.The 7-step elemental reactions of NaCl combustion reaction process were established, as shown in Table 3. The elemental reaction rate depends on Arrehenius' law[7]:

$$k=A\cdot T\beta\exp[-E/(RT)] \qquad (5)$$

Where:k is the rate constant; A is the pre-exponential factor; T is temperature; β is the index; E is the activation energy, J/mol;R is the gas constant,taking 8.314J/(mol·k).

Table 1 Industrial analysis and elemental analysis of coal used in simulation

Industrial analysis/%				Elemental analysis/%					Heat value
$F_{C,ar}$	V_{ar}	A_{ar}	M_{ar}	C_{ar}	H_{ar}	O_{ar}	N_{ar}	S_{ar}	(MJ·kg^{-1})
31.2	39. 6	5.2	24. 0	52. 51	2. 13	11. 36	0. 48	0.52	18. 57

Table 2 Reaction mechanism of optimized sodium-containing substances

No.	elementary reaction		A	n	E/R
1	$NaCl+H_2O=NaOH+HCl$[8]	(6)	1.7E14	0	0
2	$NaCl+NaCl=(NaCl)_2$ [8]	(7)	8.0E13	0	0
3	$NaO_2+HCl=NaCl+HO_2$ [8]	(8)	1.4E14	0	0
4	Na+HCl=NaCl+H [8]	(9)	1.3E15	-1	5030
5	Na+Cl+M=NaCl+M [8]	(10)	1.1E20	-1	0
6	$NaCl+SO_3(+M)=NaSO_3Cl(+M)$[8]	(11)	1.0E14	0	0
7	$NaHSO_4+NaCl=Na_2SO_4+HCl$ [9]	(12)	1.0E14	0	0

4. Simulation results

4.1 Effect of NaCl concentration on flame temperature

It can be seen from observation that with the increase of NaCl content in coal,the temperature of the flame center,zone and zone in the furnace also rises. When NaCl content increases from 0 to 1%, the flame center temperature rises by about 80°C; when NaCl content increases from 0 to 2%, the flame center temperature rises by about 120°C. When NaCl content increases from 0 to 1% and 2%,the overall furnace temperature increases by about 50°C. The temperature distribution of furnace flame is shown in Figure 2. Researchers think that electron transfer theory is the main reason why alkali metal acts on carbon atoms in pulverized coal and promotes the catalytic combustion of pulverized coal. Electron transfer refers to the heating of alkali metal thermal decomposition and activated enough, leave the electrons in the metal ions,metal a cationic hole,cause the change of the carbon surface electronic structure, so the metal ions change is incentive structure, coal molecular changes structure is the result,so that volatile release intensifies, and accelerate damage the structure of the carbon, accelerates a pulverized coal combustion. At the same time, in the flame burnout region, the flame temperature also increased with the increase of NaCl content, but the increase was less, because most of NaCl had already reacted and only a few remained.

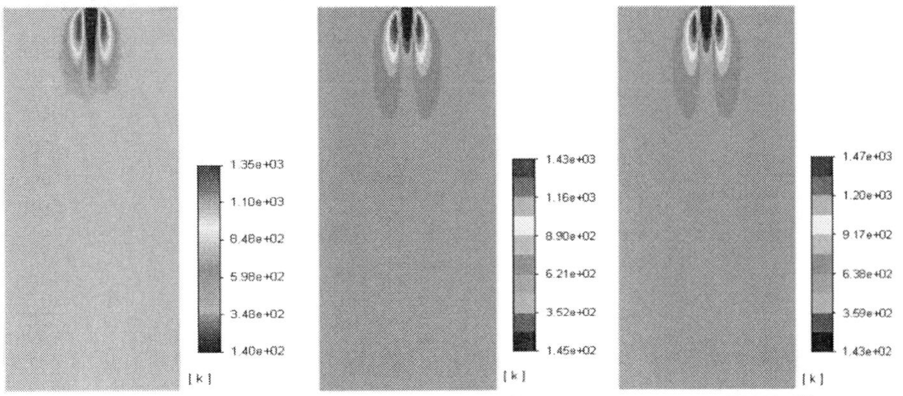

0%NaCl content 1%NaCl content 2%NaCl content

Fig.2 Flame combustion temperature at different NaCl content

4.2 Effect of NaCl concentration on flame combustion range

Through calculation, it can be found that with the increase of NaCl content in coal, the combustion range of the flame in the furnace also expands. When NaCl content increases from 0 to 1% and 2%, the combustion radius of furnace flame increases from 4.5cm to 6.8cm, and it can be calculated that the flame combustion range on the cross section is about S=81.6cm^2.The combustion length of the furnace flame increases from 26cm to 38cm, and the combustion range of the flame can be simplified into a cylinder with the same radius in terms of the combustion length. The radius is obtained by measuring respectively, and it can be known that the overall combustion range is expanded by about V=897.6cm^3. The flame temperature distribution in the cross-section of the furnace is shown in Figure 3. It can be seen from Figure 4 that NaCl enters the furnace mainly at the position 0-35cm at the nozzle of the burner, where the coal powder carrying NaCl decomposes into Na^+ ions and Cl^- ions at high temperature and releases a large number of free electrons to promote combustion.

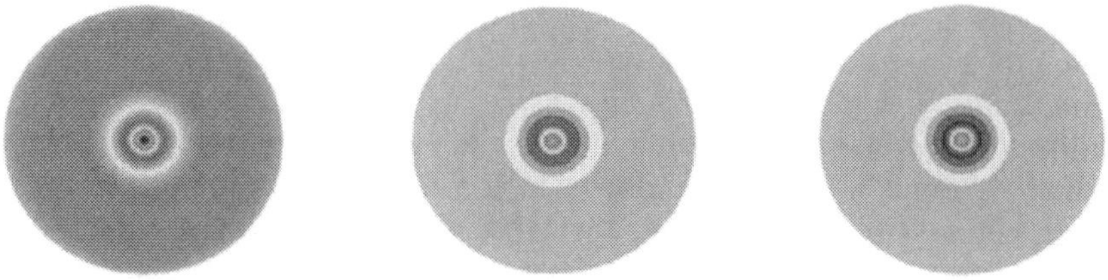

0%NaCl content 1%NaCl content 2%NaCl content

Fig.3 Flame combustion cross sections at different NaCl content (z=1050mm)

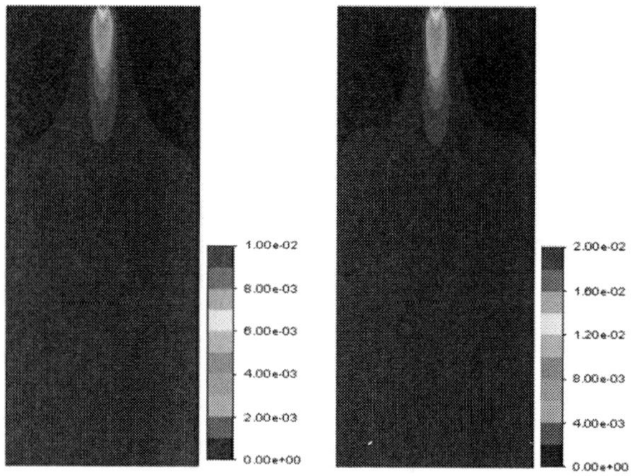

1%NaCl content 2%NaCl content

Fig.4 The distribution of NaCl in the furnace

5. Conclusion

Based on the results and discussions presented above, the conclusions are obtained as below:

(1) As the content of NaCl added to pulverized coal increases from 0 to 1% and 2%, the flame center temperature in the furnace also increases to about 80°C and 120°C, so it can be concluded that NaCl content can promote the combustion process of pulverized coal in the furnace.The increase of NaCl on the temperature of the furnace is mainly reflected in the main combustion region, because in the burnout zone, most NaCl has been completely decomposed, and there is less decomposition through the burnout zone, so the increase of the ambient temperature in the furnace is small.

(2) With the increase of NaCl content in pulverized coal from 0 to 1% and 2%, the overall combustion range of the flame in the furnace is about $897.6cm^3$, and the flame combustion range on the cross section is about $81.6cm^2$.

(3) Through the research of this paper, we can more deeply understand and explore the combustion characteristics and promoting combustion mechanism of alkali metals in the combustion process of zhundong coal, which can provide experimental and theoretical basis for the large-scale development and utilization of zhundong coal in the next step.

References

[1] Chen Xinwei, Zhuang Xinguo,Zhou Jibin,et al. Coal quality characteristics and distribution law of Zhundong Coalfield[J].XINJIANG DIZHI, 2013, 31(01): 89-93.

[2] Wei Xiaofang,Liu Tiefeng,Huang Jiejie,et al. Changes in the form of sodium in high salt Australian coal during pyrolysis[J].Journal of Fuel Chemistry and Technology, 2010, 38(02): 144-148.

[3] Chen Chuan, Zhang Shouyu, Liu Dahai, et al. The form of sodium in high sodium coal in xinjiang and its influence on combustion process[J]. Journal of Fuel Chemistry and Technology, 2013, 41(07): 832-838.
[4] Song Jinling. Experimental study on the effect of alkali metals in high alkali coal on coal combustion characteristics[D]. Shanghai Jiao Tong University, 2015.
[5] Ling Zhongqian, Zeng Xianyang, Hu Shantao, et al. Numerical simulation of optimization of SCR flue gas denitration system for power plant boiler.Journal of Chinese Society of Power Engineering, 2014(1): 50-55.
[6] Zhizhong Kang, Xian Ding. Numerical analysis on combustion process and sodium transformation behavior in a 660 MW supercritical face-fired boiler purely burning high sodium content Zhundong coal[J]. Journal of the Energy Institute, 2020, 93(2).
[7] Ji Jie Qiang, Cheng Leming, Liu Yanquan, et al. Chemical kinetic simulation of high temperature form of sodium in zhundong coal[J]. Journal of Chinese Society of Power Engineering, 2017, 37(10): 780-787.
[8] Peter Glarborg, Paul Marshall. Mechanism and modeling of the formation of gaseous alkali sulfates[J]. Combustion and Flame, 2004, 141(1).
[9] Lusi Hindiyarti, Flemming Frandsen, Peter Glarborg, Paul Marshall. An exploratory study of alkali sulfate aerosol formation during biomass combustion[J]. Fuel, 2007, 87(8).

Application of Step Motor Servo Control Technology in Integrated Thermal Management of Advanced Fighter System

Mingliang Hu[1*], Erjun Yu[1] and Wen Yan[1]

[1] AVIC Computing Research Institute, Xi'an, Shaanxi,710065,China

[*]Corresponding author's e-mail:minglhu@126.com

Abstract. The integrated thermal management of advanced fighter jets directly affects the combat performance of the aircraft. This paper develops a set of multi-cycle integrated thermal management control system, based on the stepper motor drive servo valve. The stepper motor control system is integrated in the aircraft Remote Execution Unit (REU). The controlled information, such as flow, pressure, temperature, is collected and combined through the Remote Interface Unit (RIU). Through the IEEE-1394 bus, feedback achieves deterministic transmission between control-action-response, and through inverse time protection,the high reliable power drive can be achieved. The open-loop vector micro-stepping driving strategy of the stepper motor is designed, simplifying software and hardware designs. The experimental results show that the strategy developed in this paper can better realize the characteristics of two-phase current sine wave, have better acceleration performance, improve the controllability of the opening and closing angle of the stepping motor servo valve, and satisfy the comprehensive thermal management of advanced fighters.

1. Introduction

As the continuous improvement of the performance of advanced fighter jets, a large number of highly integrated electronic equipment has been comprehensively applied, and the heat load generated by the power system, power supply system, and hydraulic system has continuously increased its heat dissipation requirements [1]. At the same time, with the integration and compactness of fighter aircraft design, a large number of heat sources are encapsulated in a small space. The surface of the aircraft fuselage is often made of composite materials with poor thermal conductivity to reduce weight [2]. The number of openings on the outer surface of the aircraft should be reduced as much as possible. The area is required to achieve stealth and drag reduction. The above factors limit the ability of advanced fighter jets to dissipate heat through convection between the surface and the outside air. Meanwhile, during the flight mission cycle of advanced fighters, there are some uncertainties in the timing and magnitude of heat sources, generated by the multi-mission system, scattered heat sources, and limited heat sink resources, posing great challenges to the comprehensive thermal management of advanced fighters [3].

Integrated thermal management includes rational utilization, distribution and schedules of heat and heat sinks. With the development of comprehensive and effective control technology, comprehensive utilization and management of energy and heat sinks can be realized by efficient heat collection, transmission and dissipation methods. The advanced fighter, integrated the thermal management technology, is mainly applied to the cooling subsystem of the aircraft ring control and to the thermal load subsystem of the fuel system to realize heat management of the entire aircraft. Air forced cooling is usually applied to key electronic equipment, liquid cooling is usually used in high heat flux and high-power task electronic equipment, and liquid cooling heat transfer methods are applied to power systems and hydraulic systems[4]. The traditional liquid cooling control valve generally takes a brushed DC motor as the drive. However, the life-cycle of this is short and its control accuracy is low, which limits the high-precision closed-loop application of the valve. The new generation of electric valve utilizes brushless motor as drive, but the control is complicated and the control accuracy is not high at low speed. According to the advanced fighter's servo valve reliability, opening and closing angle controllability and environmental requirements, the system uses a stepper motor as drive source, and adopts an incremental control method based on rotation speed and superheat to ensure that the refrigerant flow can meet requirements [5]. In this paper, the inherent low-frequency oscillation and overshoot of the stepper motor are studied, and an open-loop vector micro-stepping driving strategy for a two-phase hybrid stepping motor is designed. The stator magnetic field position information is obtained by open-loop cumulative rotation steps. Form a single current closed loop the control accuracy of the stepping motor is improved and control requirements of the advanced fighter integrated thermal management of the servo valve is fulfilled through the synthesis of current vector and discrete positioning angle value, which simplifies software and hardware designs.

2. System overall design

Figure 1 shows the architecture diagram of the multi-cycle heat integrated control system, which realizes real-time control of the system's heat dissipation flow by temperature-flow coupling control method. After pressurized by the fuel pump, the high-pressure fuel flows through the heat exchangers, after which it passes through the fuel supply pipe and enters the fuel pump again, forming an internal circulation. Through the internal circulation, the fuel supply temperature is continuously increased, so that the engine reaches the highest temperature. Fuel returns to the front fuel supply tank all the way through the system to ensure that the fuel supply temperature of the engine does not exceed the limit of the engine. When the system temperature exceeds the high temperature threshold, the stepper motor is controlled to increase the opening of the flow valve, and consequently to increase the fuel flow, and the heat dissipation capacity, thereby reducing the system temperature. When the system temperature is lower than the low temperature threshold, the stepping motor is controlled to achieve the flow valve. The opening degree is reduced to decrease the fuel flow, and as a result the engine consumes high-temperature hot oil as much as possible, and therefore the heat taken away by the engine's fuel consumption is increased. The stepping motor servo control system is integrated in the aircraft Remote Execution Unit (REU). The whole machine is equipped with sixteen REUs. Each REU is equipped with two stepping motor interfaces, which are distributed on the front fuselage, middle fuselage and Rear fuselage. By the IEEE-1394B bus interface, REU receives control instructions issued by the Vehicle Management Computer (VMC), to generate a fixed pulse sequence, drive the motor to rotate to a predetermined position, realize the control of the valve opening angle, and adjust the flow, pressure, as well as pressure of network nodes. The flow, pressure, temperature and other information of the aircraft's regional network nodes are collected through the Remote Interface Unit (RIU) and uploaded to the flight control computer by the IEEE-1394B bus to be included in the closed-loop control. The entire aircraft is equipped with eight RIUs to participate in the integrated heat and management of the closed-loop control system.

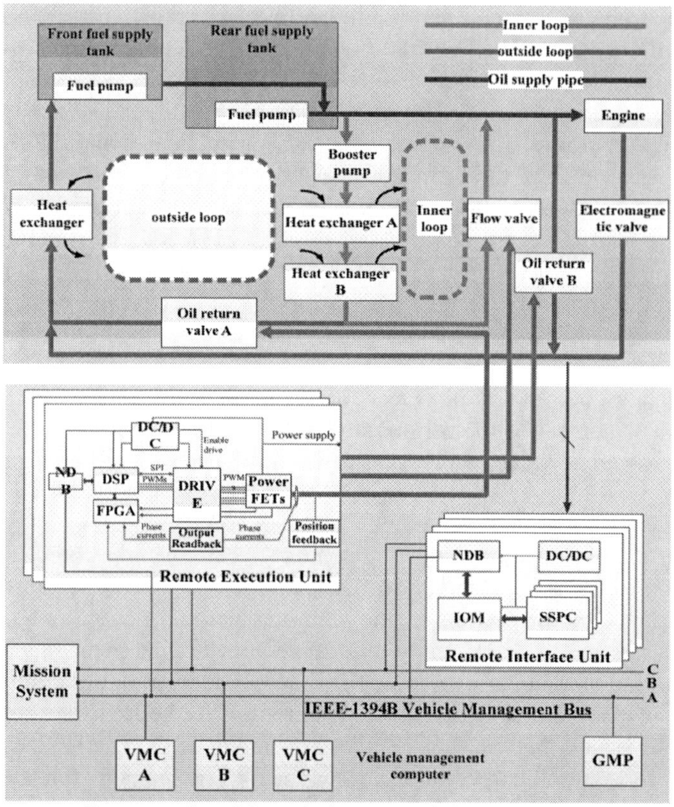

Figure 1. Architecture diagram of multi-cycle integrated thermal management control system.

3. Highly certain and reliable hardware architecture design

3.1. High-reliability network transmission delay design

The system uses IEEE-1394 bus network to realize the information exchange among sixteen REUs, ten RIUs and three VMCs. The IEEE-1394 bus adopts the basic topological structure with loops, and different REU, RIU, as well as VMC to form multiple loops, so that the system is able to deal with one-time disconnection. The IEEE-1394 bus uses pre-allocated bandwidth technology to improve the certainty of data transmission. Each node is allocated with a fixed transmission offset, reception offset and data offset in advance, and each node is planned in advance to arrive at the time of the transmission offset[6]. The number and size of asynchronous stream packets should be sent at the time, while each remote node changes its own predefined time offset according to the asynchronous stream packet tail information sent by the bus controller node in the first bus cycle. The communication sequence is reasonably arranged to ensure that the key data has a deterministic time delay that can be designed during the transmission process. Take the stepping motor control process in REU as an example. Assuming that the time RIU takes time from collecting flow pressure information to complete asynchronous flow packet grouping is $\Delta T1$, and VCM takes time from receiving RIU data to complete instructions for steering, speed, and rotation steps. $\Delta T2$ means the time REU takes from receiving the VMC instruction to the start of drive the actuator $\Delta T3$. By reasonably designing the time offset of RIU, CC and REU, the corresponding time delay of the motor control instruction in the electromechanical system is controlled within ($\Delta T1+\Delta T2+\Delta T3+2*\Delta T4$). Figure 2 shows the architecture diagram of the multi-cycle integrated thermal management control system.

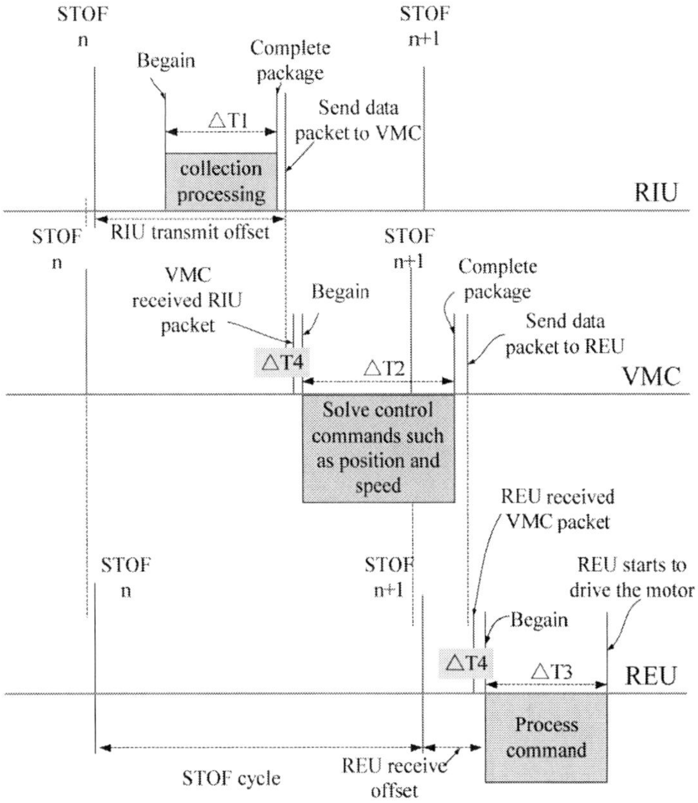

Figure 2. Description of system deterministic delay.

3.2. Highly reliable power drive design

When the hardware circuit is abnormal or the motor load is short-circuited, stepper motor interface overcurrent and short circuit protection can prevent damage from motor and controller. In the double H-bridge power drive of the stepping motor, the output of the upper bridge arm performs the over-current point anti-delay shutdown protection according to the load characteristics [7]. Figure 3 illustrates the block diagram of the inverse time protection of the upper arm power circuit. The inverse time protection controller obtains the load current by a sampling resistor, and simulates the high-order inverse time protection curve in a segmented manner. The inverse time limit energy accumulation calculation is realized by looking up the table to determine the off time. In the energy accumulation calculation, a fixed step is used as the energy accumulation value. When the energy accumulation value reaches the specified threshold, the output is turned off, a trip signal is issued, and the hardware is placed in a safe state until the REU main controller releases the trip signal.

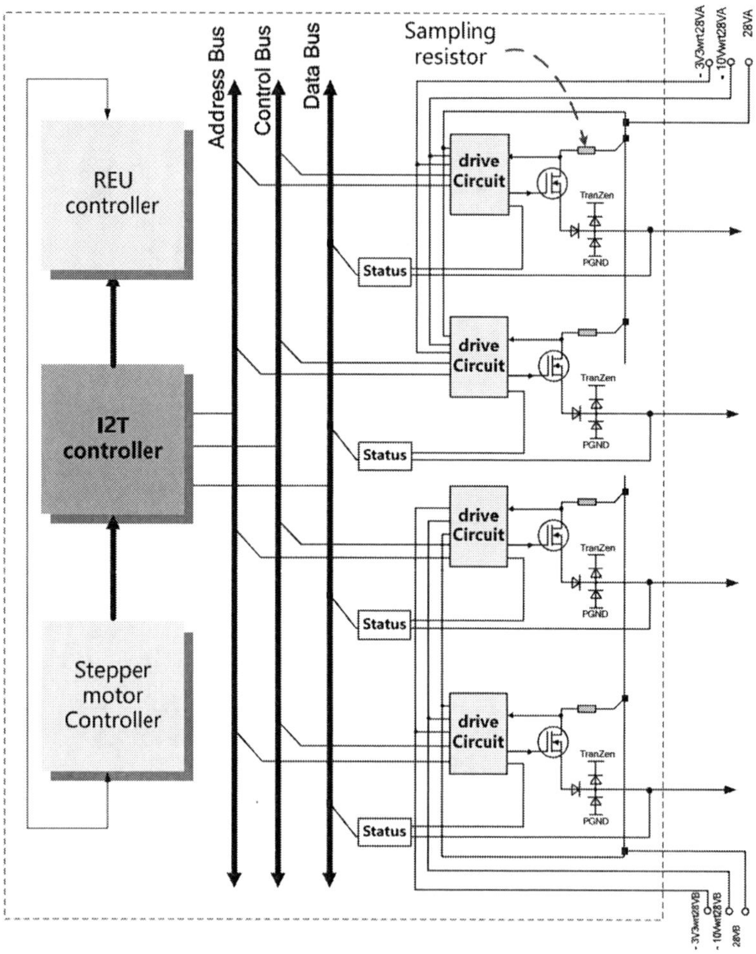

Figure 3. Block diagram of high-side bridge with inverse time protection.

The general mathematical model form of inverse time limit protection is:

$$t = \frac{k}{\left(\frac{I}{I_p} \right)^r - 1} \qquad (1)$$

Where, I and I_p represent the sampling current and the protection starting current respectively, r and k are constants, in which r takes a value between 0 and 2, and the dimension of the k value is time. Inverse time protection data processing methods mainly include direct data storage method and curve fitting method [8]. In this paper, direct data storage method is used. According to the sampling current value, the corresponding time can be obtained by looking up the table. The realization method is shown in figure 4. The action relationship between the inverse time protection time t and the current is:

- $I/I_p = 1$, means that the protection does not operate;
- $I/I_p < 1$, means that the protection does not operate;
- $I/I_p > 1$, indicates that the protection will operate. The larger the protection is, the shorter the protection operation time is.

 2108 (2021) 012098 doi:10.1088/1742-6596/2108/1/012098

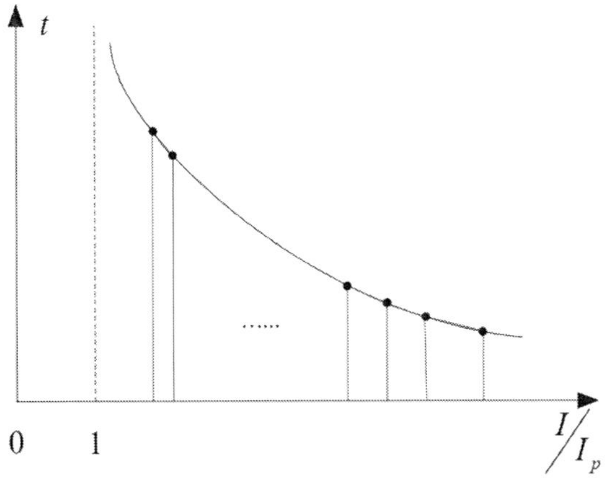

Figure 4. Direct data storage method.

4. The design of open-loop vector micro-stepping control strategy

4.1. Micro-stepping control principle

When the two-phase hybrid stepping motor A and B are connected with a sine wave current, the electromagnetic torque is calculated, and when the iron core saturation effect are ignored and the high-order harmonic effect in the main permeance is ignored, the moment angle characteristic is a sine wave[9]:

$$T = k_T \cdot (-I_a \sin\theta + I_b \cos\theta) \tag{2}$$

Where, k_T represents the constant of proportionality, and θ represents the electrical angle position of the rotor.

If the A and B phase windings are supplied with the following currents:

$$\begin{cases} I_a = I_m \cdot \cos\beta \\ I_b = I_m \cdot \sin\beta \end{cases} \tag{3}$$

Where, β means the electrical angle that the motor hopes to position.

After substituting the torque expression, the torque is expressed as:

$$T = k_T \cdot I_m \cdot \sin(\beta - \theta) \tag{4}$$

It can be seen from the torque equation that the micro-stepping driving of the two-phase hybrid stepping motor is to control the current in the two-phase windings, so that the synthesized magnetic field inside the motor is a circular space rotating magnetic field[10]. The ideal A and B phase currents are sine waves and the sine waves are fitted by multi-step waves to achieve the control effect of constant torque amplitude. The vector size of the synthesized magnetic field determines the torque of the motor. The angle between two adjacent synthesized magnetic field vectors is the step angle after micro-stepping. One tooth pitch of the two-phase hybrid stepping motor is 2π electrical angles. In the full-step operation mode, the motor needs to go through 4 state switching for every 2π electrical angles. If the micro-stepping number is n, the motor needs to go through 4n states for every 2π electrical angle, and the step angle after micro-stepping is $\pi/2n$. Figure 5 demonstrates the eight micro-step states current vector diagrams and two-phase current values.

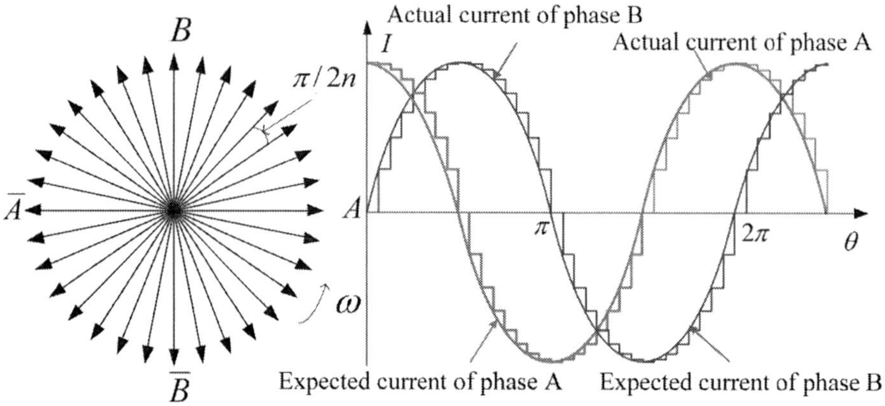

Figure 5. Eight micro-step states current vector diagrams and two-phase current values.

4.2. Closed-loop micro-stepping control strategy design

The traditional micro-stepping drive of a two-phase hybrid stepping motor is to store the subdivided two-phase current value in EPROM, obtain motor rotation steps, steer and speed control signals through the microprocessor, and take out the reference current value from the EPROM[11]. The magnitude is subtracted from the actual two-phase current value obtained by the A/DC sampling of the microprocessor respectively, and the difference is modulated by the PID link The modulation value is compared with the triangular carrier, and a PWM wave is generated to control the on and off of each power tube. This control strategy adopts dual current closed-loop control A and B phases are completely symmetrical, and each unit module requires two. Figure 6 shows a block diagram of traditional software to achieve micro-stepping control. The realization of the entire system is relatively complicated, and PID parameters need to be set for different motors[12]. The software portability and consistency are poor, which triggers greater challenges to the design of the control system.

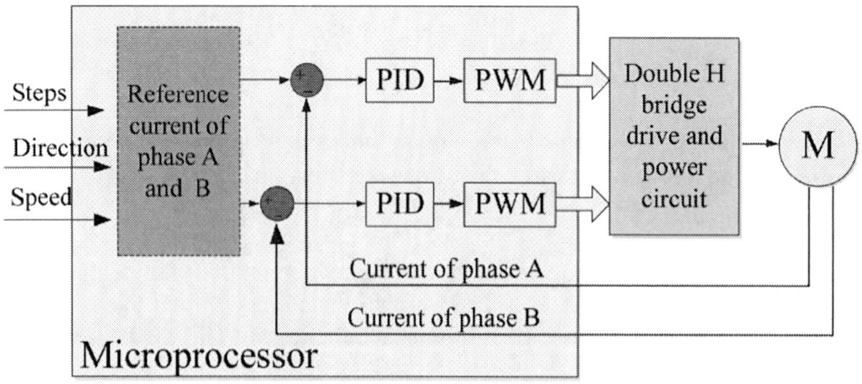

Figure 6. Traditional software to achieve subdivision control block diagram.

4.3. Open-loop vector micro-stepping control strategy design

The open-loop vector means that the electrical angle value, representing the position of the stator magnetic field, is not obtained in a closed-loop way, but is obtained in an open-loop way of accumulating the number of rotation steps [13]. Figure 7 illustrates the principle block diagram of the open-loop vector micro-stepping control system. After the two-phase winding current is sampled and modulated, it is sent to the A/D sampling port of the microprocessor, and the microprocessor performs software filtering and calculation on the sampled current. The amplitude of the synthesized current space vector is obtained and compared with the given current. The difference is modulated by the PI link to obtain the modulated voltage signal Vp. According to the motor positioning requirements, discrete electrical angle values of the sine and cosine waves are obtained, and sine and cosine values are multiple with Vp representatively to obtain A and B phase modulated signals of Va and Vb. In order to prevent the same bridge arm from being through, the dead time is added to complement two PWMs. To ensure the accuracy of the control signal, the dead time is compensated at the same time to obtain modulation signals of Va' and Vb' after the dead time compensation. Through micro-processing, eight unipolar PWM modulation waves are generated to control the on and off of the MOS tube of the H bridge.

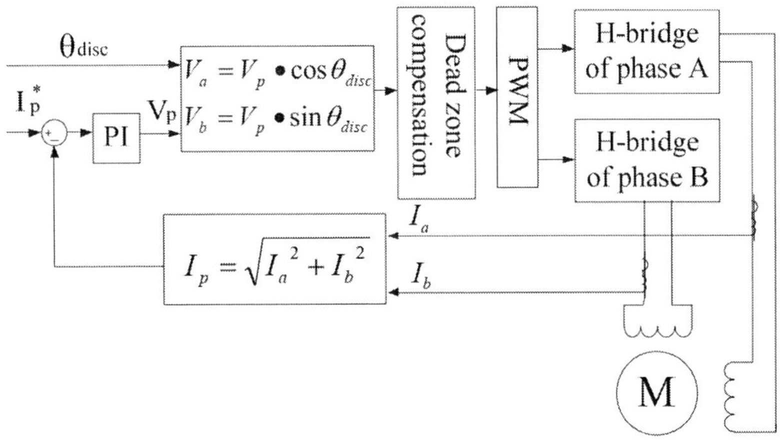

Figure 7. The principle block diagram of the open-loop vector subdivision control system.

5. Design of Open-loop Vector micro-stepping Control Software

The stepper motor open-loop vector micro-stepping control software is divided into the main program and the A/D interrupt service program. The interrupt service program includes sine cosine table generation subroutine, PI subroutine, square root subroutine, dead zone compensation subroutine and PWM generation. Firstly, the main program initializes the system configuration register, enables the first-level interrupt INT1, and uses the clock 1 underflow interrupt to start A/D sampling. When the underflow interrupt flag appears on the clock, the microprocessor samples A and B phase currents. The A/D interrupt service program is entered to calculate the sum of squares of A and B phase current values, solving the square root, and getting the combined current Ip. In the angle discretization program, look up the table of the current step number to get corresponding sine and cosine values. The two-phase modulation signal is calculated by using the given voltage signal and the discrete angle output in the PI link The dead zone compensation is performed on the modulation signal to obtain a unipolar PWM modulation wave. The PWM modulation wave adopts a constant frequency pulse width modulation chopping method. to avoid the direct connection between the upper and lower pipes of the same bridge arm, the dead zone time and dead zone compensation time are set at the same time. Figure 8 shows the flow chart of the open-loop vector micro-stepping control software.

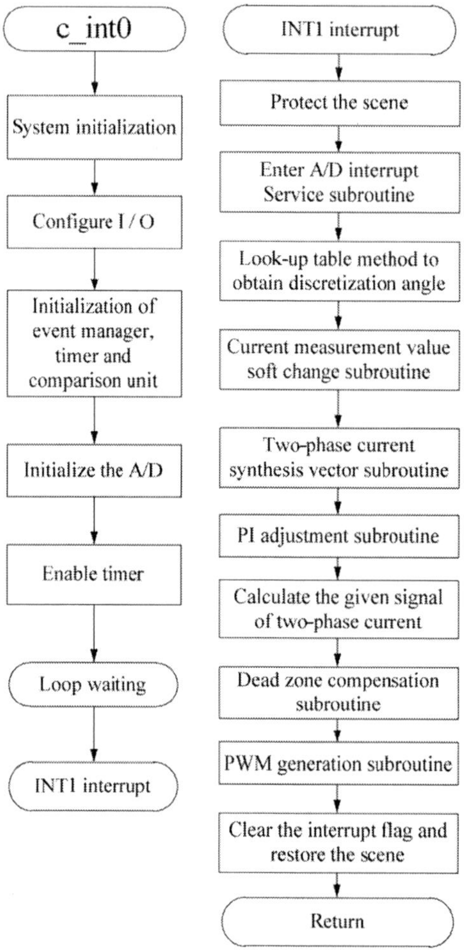

Figure 8. Open loop vector subdivision control software flow chart.

6. Experiment

The experiment adopts a two-phase hybrid stepping motor with a rated phase current of 3A, a power supply voltage of 36V, a phase resistance of 0.35Ω, a phase inductance of 1.6mH, and a step angle of 1.8°. In the experiment, the step angle is subdivided into various steps to verify the effect of micro-stepping. Figure 9 demonstrates the two-phase winding current waveform in the four-division state, Figure 10 shows the two-phase winding current waveform in the eight-division state, and figure 11 illustrates the two-phase winding current waveform in the sixteen-division state. As the number of micro-stepping increases, phase current fluctuations become smaller. Figure 12 shows that during the acceleration of a two-phase hybrid stepping motor, the two-phase current waveform of the motor presents a better sine wave and has a better acceleration performance.

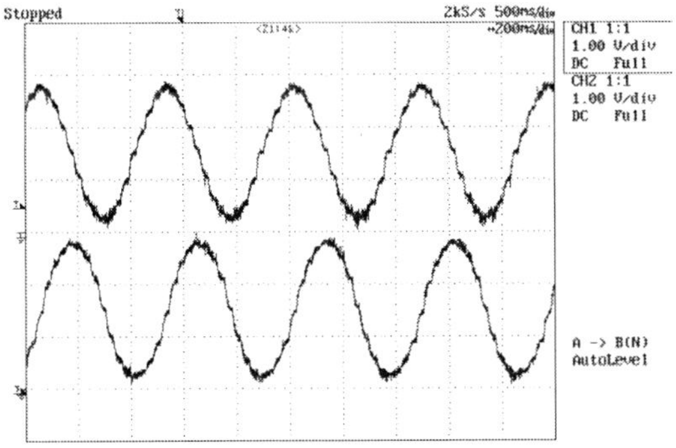

Figure 9.Two-phase winding current waveform in four micro-step states (2A/div)

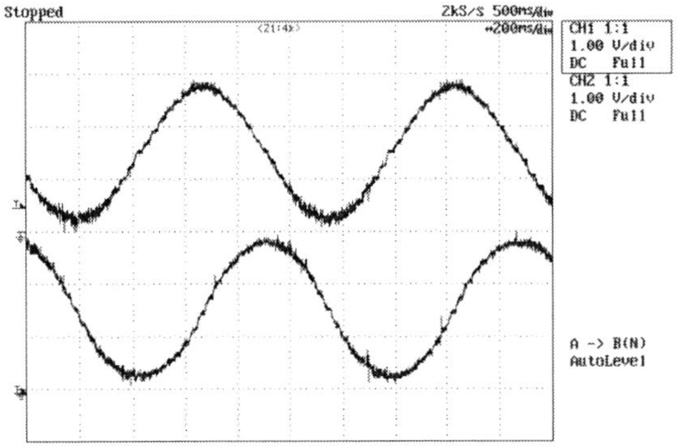

Figure 10.Two-phase winding current waveform in eight micro-step states (2A/div)

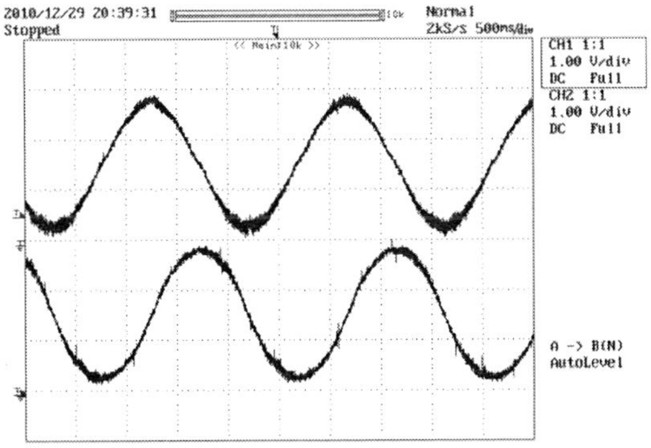

Figure 11.Two-phase winding current waveform in micro-step states state (2A/div)

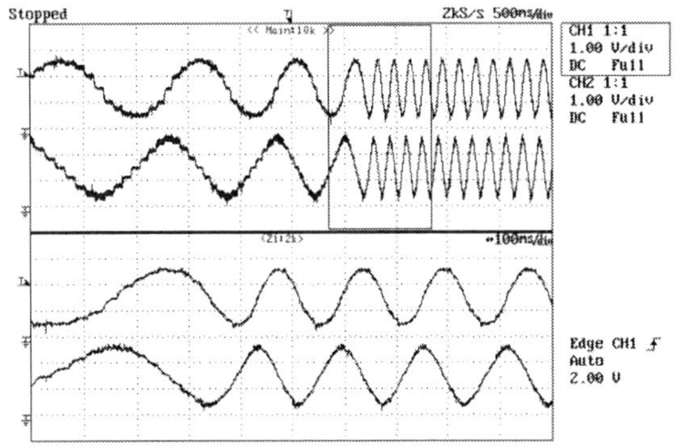

Figure 12.Current waveform under the condition of motor starting and accelerating (2A/div)

7. Conclusion

In this paper, based on stepper motor driven servo valve, a set of multi cycle integrated thermal management control system is developed. The stepper motor control system is integrated into the aircraft remote actuator. The controlled information, such as flow, pressure and temperature, is collected and feedback through the remote interface unit and the deterministic transmission between control actions driving of power is realized by inverse time limit protection. The open-loop vector micro-stepping driving strategy of stepping motor is designed, which simplifies the design of software and hardware. The experimental results show that the strategy developed in this paper can better realize the sine wave characteristics of two-phase current, with better acceleration performance It improves the controllability of opening and closing angle of stepper motor servo valve, and meets control requirements of advanced fighter integrated thermal management.

References

[1] Li, D., Dong, S., Wang, J., et al. (2020) Thermal dynamics and thermal management strategy for a civil aircraft hydraulic system. Thermal Science, 24(4): 2311-2318.

[2] Doman, D. (2018) Fuel Flow Topology and Control for Extending Aircraft Thermal Endurance. Journal Of Thermophysics And Heat Transfer,32(1): 35–50.

[3] Wang, J., Liu, Y., Dong, S. (2010) Advanced technologies of the integrated aircraft utility system and ECS. Journal of Beijing University of Aeronautics and Astronautics, 11: 959-963.

[4] Tu, M., Yuan, G.M, Xue, F., etal. (2020) The use of integrated thermal management in system development of advanced fighter aircraft. Acta Aeronautica et Astronautica Sinica, 41 (06): 136-146.
[5] Chen, N., Zheng, J., Jiang, X., et al. (2020) Analysis and control of micro-stepping characteristics of ultrasonic motor. Frontiers of Mechanical Engineering, 15(4): 585-599.
[6] Doman, D. (2018) Fuel Flow Topology and Control for Extending Aircraft Thermal Endurance. Journal Of Thermophysics And Heat Transfer, 32(1): 35–50.
[7] Pang, L, Zou, L. (2019) Optimization of fuel heat management system for high-speed aircraft. Journal of Beijing University of Aeronautics and Astronautics, 45(2): 252 (in Chinese).
[8] Saraya, S.F., Elksasy, M., Gadi, H.H. (2020) A New Technique for Controlling Hybrid Stepper Motor Through Modified Pid Controller (Dept. E). MEJ. Mansoura Engineering Journal, 35(1): 10-20.
[9] Rashmi, R. (2010) Design and Development of control strategy for stepper motor head lamp leveling device used in automobile. International Journal of Science, Technology, Engineering and Management-A VTU Publication, 2(3): 33-41.
[10] Qin, J., Zhang, S., Bao, W., et al. (2013) Thermal management method of fuel in advanced aeroengines. Energy, 49: 459-468.
[11] Hu, M., Che, J., Yu, F., et al. (2021)Design and Implementation of A Configurable Remote Analog Data Acquisition System Using the IEEE-1394B Bus. In: 2021 3rd International Conference on Intelligent Control, Measurement and Signal Processing and Intelligent Oil Field (ICMSP). Xi'an. pp. 141-146.
[12] Wang, X., Dang, X., Zhang, G. (2020)Design and simulation of thermal management system with liquid as the cooling media. In: CSAA/IET International Conference on Aircraft Utility Systems (AUS 2020). Beijing. pp.136-141.
[13] Erjun, Y., Mingliang, H.(2020)Design of high-precision and high-reliability aircraft stepper motor drive servo control system. In: CSAA/IET International Conference on Aircraft Utility Systems (AUS 2020). Beijing. pp.22-27.

Journal of Physics: Conference Series **2108** (2021) 012099 doi:10.1088/1742-6596/2108/1/012099

Experimental Analysis on Mechanical Properties of Aluminum Conductor Steel-reinforced for Transmission Line

Xing Li[1*], Jian Zhang[2], Haochen Wang[2], Chuanbin Liu[2], Yuhui Liu[2] and Chenrui Guo[2]

[1]Shandong Electric Power Research Institute, Jinan 250003, China

[2]Shandong Zhongshi Yitong Group Co., Ltd., Jinan 250003, China

[*]316242157@qq.com

Abstract—Under the background of "carbon neutralization", the power transmission quality of power grid transmission lines becomes more and more important. This paper discusses the mechanical properties of aluminum conductor steel reinforced(ACSR), the main medium of electric energy transmission. This test simulates the real environment, takes the stress-strain curve and elastic modulus as the starting point, divides the test process into four stages, and compares the mechanical properties of the same conductor under different loads in detail. On this basis, the data difference and performance comparison of three different sections of ACSR are obtained. Finally, the commonness and difference of different cross-section conductors are analyzed, and the precautions in the test process are given. The conclusion of this paper can promote the quality improvement of power transmission conductors.

1. Introduction

With the proposal of the "carbon peak and carbon neutralization" strategic index policy, the vigorous era of clean energy has quietly opened. During the 14th Five Year Plan period, the substitution of electric energy should be driven by clean energy. Electric energy has the advantages of high efficiency, safety and convenience. Electric energy is used to replace the energy consumption mode of bulk burning coal and oil, such as electric heating, industrial electric boilers (kilns), agricultural electric drainage and irrigation, electric vehicles, shore power of berthing ships, airport bridge equipment, electric energy storage and peak shaving. It is an important measure to promote the goal of carbon neutrality.

In modern power grid, power transmission mainly depends on overhead lines, and the most important transmission medium in overhead lines is conductors. Wires are components used to conduct current and transmit electric energy. The conductor is often tested by various natural conditions in operation. It must have the characteristics of good conductivity, high mechanical strength, light weight, low price and strong corrosion resistance. As China is richer in aluminum resources than copper, and the price difference between aluminum and copper is large, ACSR is almost used [1].

Overhead conductors are connected with poles, towers and insulators through connecting fittings, which mainly bear the tensile load in the axial direction. The conductor is affected by the natural environment in the line, and the natural environment is bad, so it must bear external loads such as conductor self weight, wind load, icing and so on [2]. Due to the importance of transmission lines, the State Grid Corporation of China lists the quality sampling inspection of conductors as the focus of its work every year. In order to obtain the complete mechanical properties of the conductor, the stress-strain curve and elastic modulus test need to be carried out [3]. The test is listed as a class a project by the State Grid. It simulates various environmental conditions that the conductor may withstand under real operation conditions. The most complete performance data of the conductor under real load conditions can be obtained through the test.

At present, the research on the stress and strain of conductor mainly focuses on the mechanical properties of single material [1] [4], the stress distribution between conductor strands [5-11] through finite element simulation, the stress between conductor layers [2] [12] through mathematical modeling, and the influence of Aeolian fatigue vibration on conductor characteristics [13-15]. The above research mainly depends on simulation software or mathematical modeling. Limited by test conditions, there is a general lack of real test data under long-time and large load.

In this paper, the products sampled from power grid materials are taken as physical samples. The test method is in strict accordance with the national standard GB / T 1179 and the national grid enterprise standard Q / GDW 13236.2. The real stress-strain curve and elastic modulus test are carried out in the detection center, the mechanical properties of ACSR with different sections are obtained, and the elastic modulus under different conditions is quantitatively analyzed.

2. Performance characteristics of ACSR

2.1. Structure of ACSR

ACSR is a composite conductor of aluminum and steel, which belongs to concentric lay stranded conductor. It is composed of multi-layer single wires spirally twisted around a central core, and the adjacent layers are twisted in the opposite direction. Taking the common ACSR model JL / g1a-240 / 30-24 / 7 in the material sampling inspection of the State Grid as an example, G1a represents ordinary strength steel wire, 240 represents the cross-sectional area of aluminum strand is approximately 240mm2, 30 represents the cross-sectional area of steel core is approximately 30mm2, and 24 represents the number of aluminum strands. In fact, it is composed of two layers, 9 in the inner layer, 15 in the outer layer, and 7 represents the number of steel cores. A real section is shown in Figure 1.

Figure 1. ACSR with 240 / 30 section

Figure 2. ACSR with 400 / 35 section

In power transmission lines, the cross-sections of conductors are different with different voltage levels. For example, JL / g1a-400 / 35-48 / 7 for higher voltage levels will increase its cross-sectional area, the number of aluminum wire layers and the number of wires per layer, as shown in Figure 2.

2.2. ACSR requirements for material procurement of State Grid Corporation

In order to ensure the quality of products connected to the grid, the State Grid Corporation of China has formulated special technical requirements for each type of materials and formed a series of enterprise standards. Among these technical requirements, several key items affecting the mechanical properties of the conductor are as follows.

- The aluminum wire shall be cold drawn aluminum wire for electrician. The cold drawn aluminum wire before stranding shall meet the bidding technical conditions and meet the requirements of GB / T 17048.

Journal of Physics: Conference Series **2108** (2021) 012099 doi:10.1088/1742-6596/2108/1/012099

- The surface of the aluminum wire shall be smooth and free from all defects that may affect the performance of the product, such as cracks, roughness, scratches and impurities.
- The pitch diameter ratio of conductor shall be within the range specified in GB / T 1179. For conductors and aluminum stranded wires with only one layer of aluminum wire structure, the pitch diameter ratio of the outer layer shall not be greater than 14; For conductors with two or more layers of aluminum wire structure, the pitch diameter ratio of the outer layer shall not be greater than 12. The twist direction of adjacent layers shall be opposite, and the twist direction of the outer layer shall be right. For stranded wires with multiple layers, the pitch diameter ratio of any layer shall not be greater than that of the immediately adjacent inner layer.
- During stranding, the steel core of the conductor and the strand to be strangled shall be placed in the factory for a long enough time to ensure that the steel core and the strand to be strangled should be at the same temperature, and the same temperature should be maintained during the whole stranding process.

2.3. Test items of conductor
Material sampling inspection mainly includes sampling test and type test items.

Sampling test is to ensure that the quality of conductor and ground wire meets the requirements of procurement standards. The test samples shall be randomly selected from the batch. The sampling test items include single wire performance, cross-sectional area, outer diameter, linear density, surface condition, pitch diameter ratio and twist direction, DC resistance at 20 °C and strand breaking force, etc.

The type test is to check whether the performance of conductor and ground wire meets the design requirements. The type test shall be carried out in a quality inspection institution with corresponding qualification. The conductor type test items include single wire performance, rated tensile strength of stranded wire, elastic modulus, DC resistance, pitch diameter ratio, mass per unit length, stress-strain curve, creep curve, linear expansion coefficient, ampacity, fatigue performance, tightness and flatness, etc.

3. Test setup of stress-strain curve

3.1. Selection of test samples
Three kinds of cross-section ACSR commonly used in material sampling inspection(Figure 3) are selected, which are hereinafter referred to as sample 1, sample 2 and sample 3. Their models are JL / G1a 240 / 30-24 / 7, JL / g1a-300 / 40-24 / 7 and JL / g1a-400 / 35-48 / 7 respectively.

Figure 3. ACSR for test

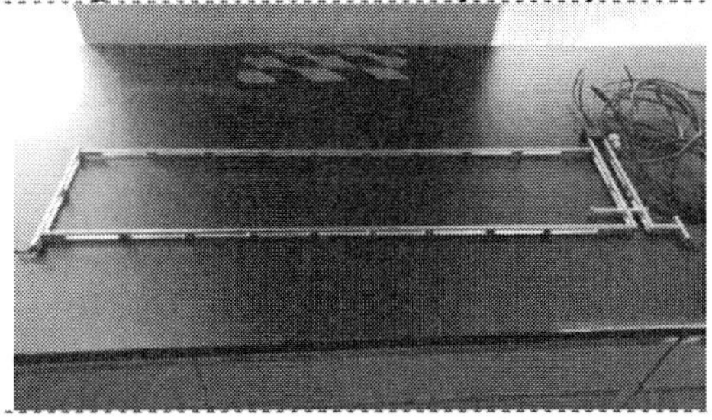

Figure 4. Strain gauge with large deformation

The mechanical property parameters of the three samples are shown in Table 1.

Table 1. Comparison of key parameters of three different specifications of ACSR

	Sample 1	Sample 2	Sample 3
Number of galvanized steel wires / diameter (PCs. / mm)	7/2.40	7/2.66	7/2.50

Number of aluminum single wire strands / diameter (PCs. / mm)	24/3.60	24/3.99	48/3.22
Calculated sectional area (mm^2)	276	339	425
Aluminum (mm^2)	244	300	391
Steel (mm^2)	31.7	38.9	34.4
Outer diameter (mm)	21.6	23.9	26.8
Rated tensile force (kN)	≥75.19	≥92.36	≥103.7
Nominal modulus of elasticity (GPa)	70.5	70.5	65.9

3.2. Configuration of software and hardware

According to GB / T 1179-2008, the sample length shall not be less than 400 times of the outer diameter of the conductor and not less than 10m. Here, the maximum diameter of the sample is 26.8mm, 26.8mm * 400 = 10720mm, so 11m long sample is selected. Use a large deformation strain gauge (Fig. 4) with a length of 1m and place it in the middle of the sample. The strain gauge is 5m away from both ends. The strain gauge shall be fixed when the sample is in a natural state, that is, after the prestress is removed. The horizontal tensile machine with the maximum bearing capacity of 2000kN (Fig. 5) is used, and the nondestructive fixture is configured. The automatic test software is customized, which can automatically record the test process data (Fig. 6).

The formula of engineering stress and strain is:

Stress

$$\sigma = \frac{P}{A_0}$$

Strain

$$\varepsilon = \frac{L - L_0}{L_0}$$

In the above formula, P is the load, A_0 is the original cross-sectional area of the sample, L is the length of the sample after deformation, and L_0 is the original gauge length of the sample. The stress-strain curve is abbreviated as σ-ε Curve. When to strain ε Is independent variable, stress σ When you draw a graph for a function, you get σ-ε Curve. Here, because the cross-sectional area is fixed and the gauge distance is 1m, it is also constant. In order to eliminate irrelevant interference and facilitate the comparison between different models, the load deformation curve is used to replace the stress-strain curve in this paper.

Figure 5. Horizontal tensile machine for test

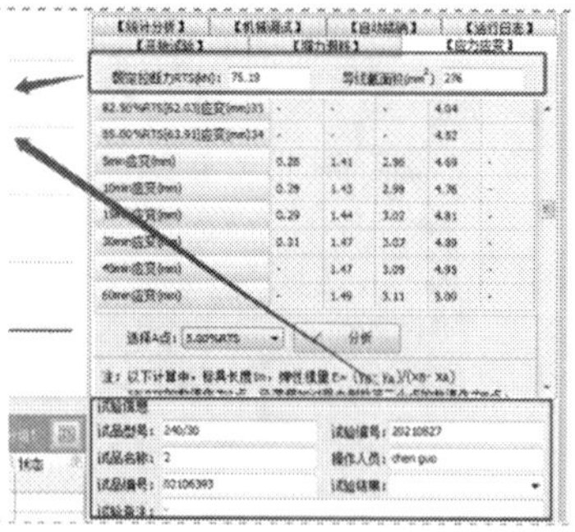

Figure 6. Customized software interface for test

Slope Tan of elastic stage (deformation is reversible) tan α = E is the elastic modulus or young's modulus of the material. The elastic modulus reflects the deformation resistance of the conductor in the elastic stage.

3.3. Method and process

The initial load is 2% RTS (rated breaking force), which is used to straighten the conductor, remove the load after straightening, and then install the strain gauge without tension. For discontinuous stress-strain data records, take a strain reading every 2.5% RTS, in kN.

The experiment is divided into four stages. ①The load applied to 30% RTS shall be maintained for 0.5h. During the test, the readings shall be taken after 5, 10, 15 and 30min, and unloaded to the initial load. ②Reapply the load to 50% RTS for 1 h, take readings after 5, 10, 15, 30, 45 and 60min, and unload to the initial load. ③Re apply the load to 70% RTS for 1 h, take readings after 5, 10, 15, 30, 45 and 60min, and unload to the initial load. ④Reapply the load to 85% RTS for 1 h, take readings after 5, 10, 15, 30, 45 and 60 min, and unload to the initial load.

During the test, the load growth rate shall be uniform, and the time to reach 30% RTS shall not be less than 1 min nor more than 2 min. the same load growth rate shall be maintained throughout the test.

Applying a certain prestress can make the conductor obtain a straight initial state and prevent abnormal points at the beginning of the test. The significance of keeping it for a period of time is that it can simulate the mechanical properties of the conductor under continuous stress in the overhead line. When the wedge clamp is used for the test, the removal of load may loosen the grip of the wedge clamp. Therefore, in this case, the initial load of 2% RTS shall be maintained when the strain gauge is reset to zero.

4. Test Results and Discussions

4.1. Acquisition of load deformation curve

The load deformation test data of three samples are shown in Table 2.

Table 2. Load deformation test data of three samples

	Sample 1				Sample 2				Sample 3			
	30%RTS	50%RTS	70%RTS	85%RTS	30%RTS	50%RTS	70%RTS	85%RTS	30%RTS	50%RTS	70%RTS	85%RTS
2.50%	0.09	0.51	1.05	1.59	0.48	0.71	1.16	2.02	0.05	0.2	0.75	1.35
5.00%	0.2	0.58	1.12	1.83	0.46	0.72	1.19	2.15	0.06	0.26	0.81	1.67
7.50%	0.3	0.67	1.21	2.08	0.5	0.76	1.24	2.25	0.1	0.33	0.87	1.88
10.00%	0.41	0.75	1.29	2.18	0.57	0.81	1.29	2.32	0.12	0.4	0.93	1.95
12.50%	0.51	0.84	1.38	2.27	0.65	0.87	1.36	2.39	0.15	0.48	1.01	2.03
15.00%	0.62	0.93	1.46	2.36	0.74	0.95	1.43	2.46	0.26	0.56	1.1	2.11
17.50%	0.74	1.02	1.55	2.44	0.83	1.02	1.51	2.54	0.38	0.64	1.18	2.2
20.00%	0.86	1.11	1.64	2.53	0.93	1.1	1.59	2.62	0.5	0.73	1.27	2.28
22.50%	0.96	1.2	1.73	2.63	1.03	1.19	1.68	2.71	0.61	0.81	1.36	2.37
25.00%	1.07	1.29	1.83	2.73	1.14	1.28	1.76	2.79	0.72	0.9	1.44	2.46
27.50%	1.2	1.39	1.92	2.82	1.26	1.37	1.85	2.88	0.84	0.98	1.53	2.55
30.00%	1.34	1.48	2.01	2.92	1.39	1.45	1.94	2.96	0.97	1.07	1.62	2.65

32.50%		1.58	2.11	3.01		1.54	2.02	3.05		1.17	1.7	2.73
35.00%		1.68	2.2	3.1		1.65	2.12	3.14		1.28	1.79	2.82
37.50%		1.8	2.3	3.2		1.77	2.21	3.23		1.39	1.88	2.91
40.00%		1.92	2.39	3.29		1.89	2.3	3.32		1.52	1.97	3.01
42.50%		2.06	2.48	3.39		2.01	2.4	3.41		1.65	2.06	3.09
45.00%		2.2	2.58	3.49		2.15	2.5	3.5		1.79	2.16	3.19
47.50%		2.37	2.67	3.59		2.31	2.59	3.59		1.96	2.26	3.28
50.00%		2.55	2.77	3.68		2.51	2.68	3.69		2.14	2.35	3.38
52.50%	-		2.87	3.79	-		2.79	3.78	-		2.46	3.48
55.00%	-		2.99	3.88	-		2.97	3.88	-		2.55	3.57
57.50%	-		3.11	3.98	-		3.11	3.97	-		2.69	3.6
60.00%	-		3.22	4.09	-		3.25	4.07	-		2.85	3.7
62.50%	-		3.37	4.18	-		3.42	4.17	-		3.03	3.8
65.00%	-		3.56	4.29	-		3.62	4.27	-		3.24	3.9
67.50%	-		3.8	4.39	-		3.82	4.36	-		3.45	4
70.00%	-		4.12	4.5	-		4.13	4.47	-		3.72	4.12
72.50%	-	-		4.62	-	-		4.58	-	-		4.23
75.00%	-	-		4.76	-	-		4.65	-	-		4.37
77.50%	-	-		4.9	-	-		4.79	-	-		4.55
80.00%	-	-		5.1	-	-		4.99	-	-		4.8
82.50%	-	-		5.36	-	-		5.22	-	-		5.12
85.00%	-	-		5.9	-	-		5.62	-	-		5.83
5min	1.41	2.64	4.27	6.04	1.44	2.59	4.26	5.83	1.03	2.25	3.88	6.08
10min	1.43	2.66	4.32	6.11	1.45	2.62	4.3	5.91	1.04	2.27	3.95	6.2
15min	1.44	2.68	4.35	6.15	1.45	2.63	4.33	5.95	1.05	2.29	3.99	6.28
30min	1.46	2.71	4.41	6.26	1.46	2.66	4.38	6.04	1.07	2.32	4.05	6.42

45min	2.73	4.45	6.33	2.68	4.41	6.09	2.34	4.09	6.52
60min	2.74	4.47	6.39	2.69	4.47	6.13	2.35	4.12	6.6

According to Section 3.3, the test is divided into four stages, 30% RTS, 50% RTS, 70% RTs and 85% RTs. In each stage, several points will be obtained to obtain an independent load deformation curve. Taking sample 1 as an example, the load deformation curves of its four stages are shown in Fig. 7-10.

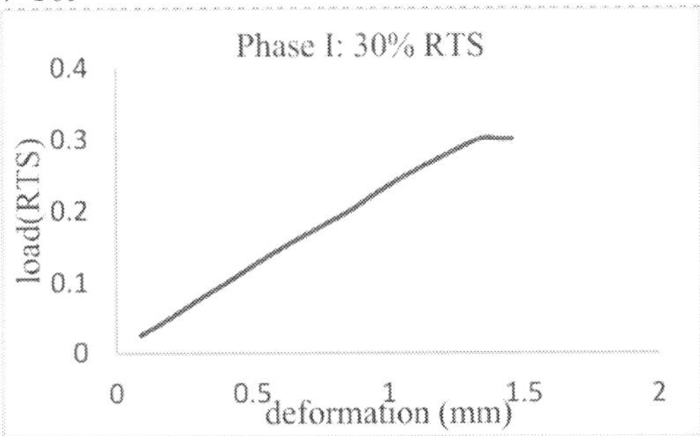

Figure 7. Phase I: 30% RTS

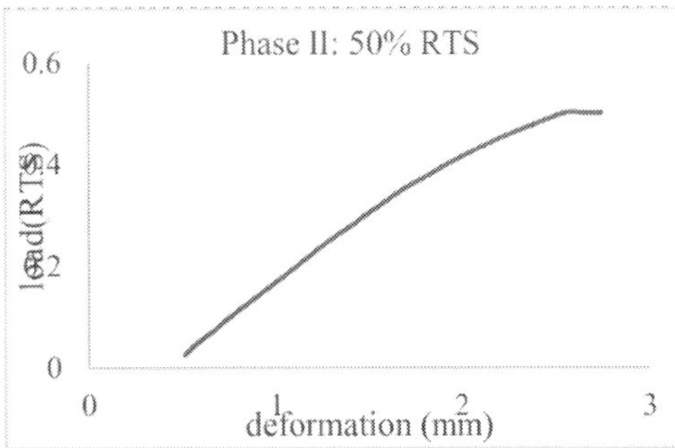

Figure 8. Phase II: 50% RTS

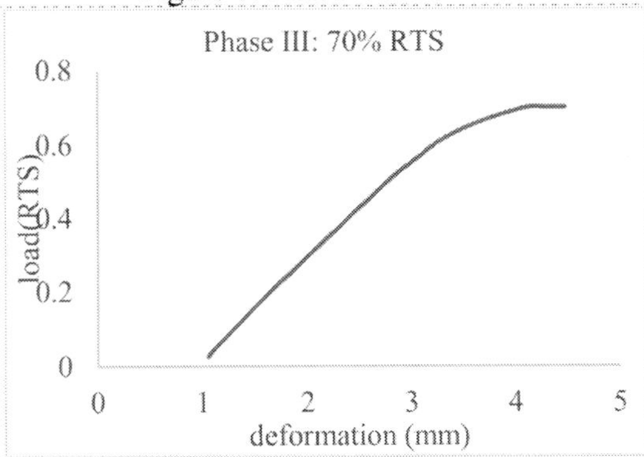

Figure 9. Phase III: 70% RTS

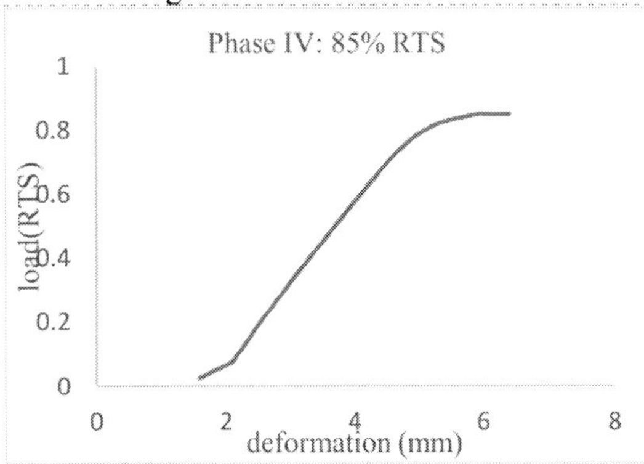

Figure 10. Phase IV: 85% RTS

All points of the above four stages are depicted on a graph, which can represent the complete test process of nearly 4 hours, so as to obtain our target load deformation curve. The target load deformation curves of the three samples are shown in Figure 11-13 respectively.

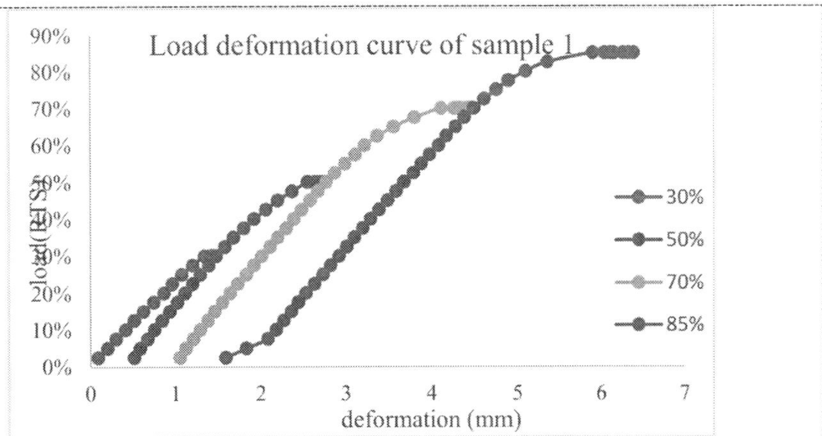

Figure 11. Load deformation curve of sample 1

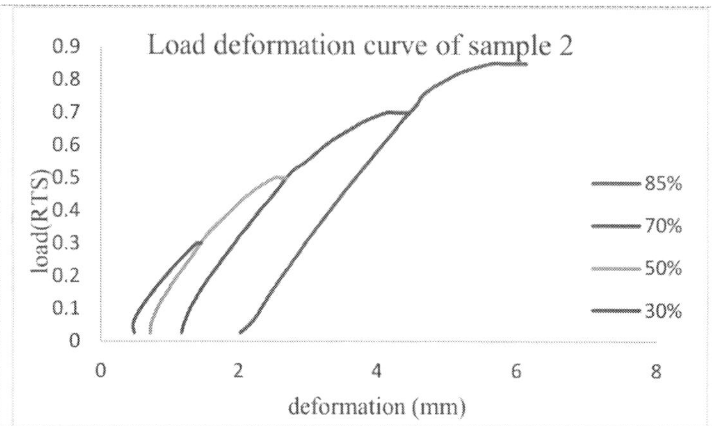

Figure 12. Load deformation curve of sample 2

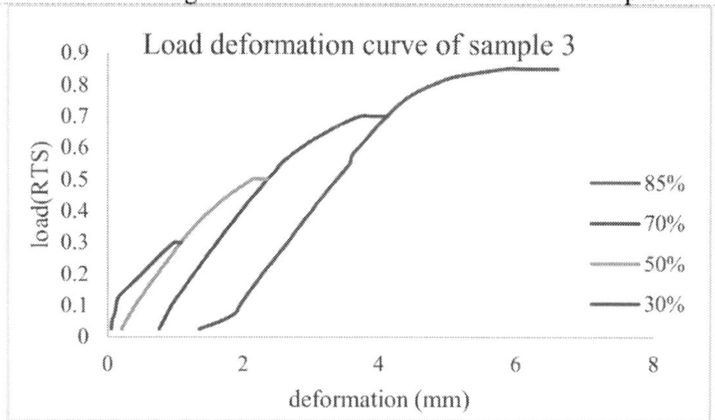

Figure 13. Load deformation curve of sample 3

4.2. Results and analysis

According to the test data and curves, the following conclusions are drawn:

(1)It can be seen from figure 7-10 that the deformation of the conductor will change after it is maintained for a period of time under a certain load. The change rate is different under different loads. The greater the load, the greater the change rate. The maximum change rate of deformation after holding for 1 hour at fourth stage(85% RTS) is close to 5%.

(2)It can be seen from figure 7-10 that the performance of the combined conductor is different from that of a single material. It is basically linear in the first and second stages, plastic deformation begins in the third stage, and yield begins in the fourth stage. After the third stage of force relief, the conductor is still elastic when it is stretched again.

(3)It can be seen from Fig. 11-13 that, different from the short-term tensile properties of a single material, after the ACSR is unloaded to the initial load (2% RTs), the sample does not return to the initial state, but has a certain amount of deformation, resulting in the right deviation of the starting point of the next stage in the graph. The maximum deformation of the previous stage is very close to the instantaneous deformation of the same load in the next stage. On the graph, these two points almost coincide. It seems that the curve of the latter stage intersects the curve of the previous stage. This intersection point indicates that the wire cannot remain elastic all the time, and local permanent deformation will occur.

(4)According to figure 11-13, the elastic modulus of the conductor at different stages can be calculated. When calculating the elastic modulus, select the appropriate starting point and end point. Due to the influence of compression joint, the curve is unstable in the initial stage, so the linear region of the curve is anchored here, and the starting point and end point of the linear region are selected to calculate the elastic modulus.For example, for the first stage of sample 1, the elastic modulus E = (0.275-0.075) / 276 / (1.2-0.3) * 10^5 = 80.5GPa. According to this method, the true elastic modulus of each sample at each stage is calculated successively, and the results are shown in Table 3.

Table 3. Calculated elastic modulus of each sample at each phase

Elastic modulus	Sample 1	Sample 2	Sample 3
Phase I (30.00% RTS)	80.5	**95.3**	97.9
Phase II (50.00% RTS)	85.3	93.5	**102.5**
Phase III (70.00% RTS)	**92.9**	90.5	95.9
Phase IV (85.00% RTS)	90.0	94.7	96.6

It can be seen that the distribution range of the maximum elastic modulus of conductors with different sections is different. This is related to the cross-sectional area of the conductor, stranding process and raw material properties. The real deformation resistance needs to be determined according to the actual situation.

(5)By observing the graph of each sample in the fourth stage in Fig. 11-13, it can be found that there is an obvious fault in the graph of sample 2 and sample 3 in the fourth stage (85% RTS), that is, the load suddenly increases and the deformation remains unchanged, indicating that this stage has separated from yield, began to appear hardening and entered the strengthening stage. The relative change of sample 1 is not obvious.

(6)The deformation of sample 1 and sample 2 before the third stage is larger than that of sample 3, but it is exceeded by sample 3 in the fourth stage. It shows that the large section sample 3 has relatively strong deformation resistance in the elastic stage, but once its elasticity is destroyed, its performance is even weaker than that of the small section sample.

(7)The load deformation curve in the image does not pass through 0 point in the first stage, which is related to the crimping end. Here, several factors affecting the test are analyzed. First, the end shall be crimped with wedge-shaped wire clamp to ensure that the end is not loose and the conductor does not slip in the tensile and unloading stages, and prevent the fracture of the conductor at the end due to uneven stress. Second, different rated tensile breaking force and elastic modulus of aluminum and galvanized steel lead to relative displacement between steel core and aluminum strand, which will affect the overall mechanical properties.

5. Conclusion

This paper discusses the mechanical properties of ACSR, the main medium of electric energy transmission, in the test. A series of comparative tests were set up to simulate the real environment. The mechanical properties of ACSR under different loads and the performance differences of ACSR with different sections were analyzed in detail, and the complete curve of ACSR load deformation with time was given, The precautions in the test process are suggested. This paper can not only provide test method guidance for power grid conductor quality sampling inspection, but also provide data reference for transmission line fault analysis. At the same time, it can be used as the evaluation basis for bad suppliers and provide support for conductor bidding and selection.

Acknowledgments

This work is supported by the independent research and development project fund of Shandong Electric Power Research Institute.

References

[1] Dong G, Wei T. (2020) Analysis and Research on Mechanical Properties of ACCC With Aluminum Coating. SHANDONG ELECTRIC POWER,47(04): 10–14.

[2] Cui Y.W. (2020) Study on Stress Characteristics between Transmission Conductors under Tension Load. North China Electric Power University, BeiJing.

[3] Q/GDW 13236.1—2019. Purchasing standard of conductor and earth wire—Part 1:General technical specification.
[4] Zhao J. X., Zhang Y. Z., Lv X. J. (2019)Analysis of Stress-Strain Curve of Metal Materials. Technology Wind, 2019,(09):171,182.
[5] Zhou X. K., Qu C, Liu S. S, Yao J. N., Wang J. J. (2020)Simulation Analysis on Mechanical Properties of Aluminium Cable Steel Reinforced under Tensile Load. JOURNAL OF GRAPHICS, 2020,41(02):288-294.
[6] Huang Y. C., Chen C.,Wang F. (2016)Stress Distribution Characteristics of Steel Core and Aluminum Strand of Large Span Overhead Transmission Line. J of China Three Gorges Univ. (Natural Sciences), 2016,38(04):76-81.
[7] Lin J. H., Zeng W. (2015)Research on Aluminum Conductor Steel Reinforced Tension Layered Characteristics of Overhead Conductors. Wire & Cable,2015,(04):30-33.
[8] Lao H. J. (2009)The Research on Surface Contact and Wear of Overhead Transmission Wires. China Three Gorges University, YiChang.
[9] Qin L., Li J. K., Fu Q. (2013)Research on Aluminum Conductor Steel-reinforced Mechanical Characteristic of Transmission Lines. Water Resources and Power,2013,31(05):194-197
[10] Zhu H., Liu Y. F., Zhang J. (2020)Modeling and Simulation Analysis of Layered Mechanical Characteristics of Bending Wires in Transmission Lines. Water Resources and Power, 2020,38(04):162-166.
[11] Zhang Q. Y. (2017)The Study on Stress and Stiffness of Conductors under Tensile Load. North China Electric Power University, BeiJing.
[12] Cai S. T., Rui X. M., Ni H. Y. (2009)Application and Study on Stratified Mechanical Model of Overhead Conductors. Electric Power Science and Engineering,2009,25(10):35-39+46.
[13] Li P. S., Cheng W. F., Liu G. P. (2018)Simulation analysis of stress-strain and fatigue of shaped wires under gentle wind vibration. Journal of Nanchang University(Engineering & Technology,2018,40(01):68-72.
[14] Zhang J. G., Liu J. J., Li C. G. (2017)Aeolian vibration property of large cross-sectional conduct for UHV transmission line. Journal of Hohai University(Natural Sciences), 2017,45(03):271-276.
[15] Ding L. L. (2019)Thermal Stress and Vibration Fatigue Life of Overhead Transmission Line.North China Electric Power University, BeiJing.

Effect of rapid thermal shock cycle on the thermomechanical reliability of 20Sn-80Pb solder bumps

Shiqi Chen[1], Guisheng Gan[12*], Qianzhu Xu[1], Zhaoqi Jiang[1], Tian Huang[1], Cong Liu[1], Peng Ma[1]

[1]Chongqing Municipal Engineering Research Center of Institutions of Higher Education for Special Welding Materials and Technology (Chongqing University of technology), Chongqing 400054, China;

[2]College of Materials Science and Engineering, Huazhong University of Science and Technology, Wuhan 430074, China)

*Corresponding author's e-mail: ggs@cqut.edu.cn

Abstract. The influence of rapid thermal shock(RTS) cycles on 20Sn-80Pb solder bumps was studied. In the study, 20Sn-80Pb solder bumps were prepared by desktop nitrogen lead-free reflow soldering machine. The prepared 20Sn-80Pb solder bumps were used for RTS test in the temperature rang of 0°C ~ 150°C. One cycle of RTS is 24 seconds, and the temperature rise and fall rate of RTS is 12.5 °C/s. The result indicated that when the cycle of RTS reached 1500T (here T is cycle, the same below), the shear strength of Sn-80Pb solder bump dropped by drastically 48.6%. Whereas, when the cycle of RTS reached 5500T, 20Sn-80Pb solder bumps' shear strength decreased to 18.35 MPa, which increased by 7.5% compared with that of 16.97 MPa at 4500T. With the increase of RTS cycles, 20Sn-80Pb solder bumps' shear strength was a decreasing trend and the fracture mechanism changed from ductile fracture to ductile-brittle mixed fracture, which could be subject to the thickening of the interfaical IMCs and the stress concentration caused by the growth of interfacial IMCs. To understand the changes of the mechanical properties of 20Sn-80Pb solder bumps, the influences of RTS on the crack and interfacial IMC of 20Sn-80Pb solder bumps were studied in details.

1. Introduction

Although people are trying to seek the right lead-free solders to replace traditional lead-based solder, there has not been found to be able to completely replace the lead-based solder[1]. Additionally, Jiang et al[2]. emphasized the dependability problems of Pb-free solder alloys in electronic applications, and pointed out some elements that lead-free solders still had challenges in many application fields. Although considerable researched has been done on lead-free solder alloys, there was currently no alternative to achieve characteristics of high-lead solder[3-5]. Hang et al.[6] made it clear that Pb-based solder alloys still existed in many fields, especially in the aerospace field without considering its related environmental concerns. In addition, high Sn-Pb solder is used in many high temperature applications[7,8] because of its resistance to formation of intermetallic compounds, high melting temperature, low cost, excellent long-term reliability and high thermal fatigue temperature[9,10]. During deep space exploration, electronic components must withstand extreme temperature changes[11]. Solder joints' integrity play a crucial role in the overall function of electronic components owing to solder joints providing mechanical, thermal and electrical continuity of electronic components[12]. Hence, it is a matter of urgency to research the thermomechanical dependability of high Sn-Pb solder bumps under extreme temperature variations. In this study, the influence of RTS cycle on 20Sn-80Pb solder bumps was studied.

2. Experimental methods

A ball grid array with dimensions of 6 mmx6 mmx1 mm containing 25 20Sn-80Pb solder bumps is the material used in this experiment, which had a diameter of 600 μm. 20Sn-80Pb solder bumps were prepared by desktop nitrogen lead-free reflow soldering machine. The peak reflow temperature for 20Sn-80Pb solder bumps was 357°C, and the duration above liquids (273°C) was about 120s. The prepared 20Sn-80Pb solder bumps were used for RTS test in the temperature range of 0°C ~ 150°C. RTS equipment mainly included heating system and cooling system, and heating system consisted of induction coil and sample placement table, and cooling system was comprised of sample placement table cooling fluid as shown in Figure 1. One cycle of RTS is 24 seconds, and the temperature rise and fall rate of RTS is 12.5 °C/s in Figure 2. The RTS cycles included 1500T, 2500T, 3500T, 4500T and 5500T, respectively (here T is cycle, the same below). Optical metalloscope(OM) and scanning electron microscope(SEM) were used to analyze the microstructures of interface. The thickness of the interfacial IMCs layer was measured using 'Image-J' software. The average shear strength of several joints (the shear velocity was 0.05 mm/sec) were identified by the PTR-1102 bonding tester.

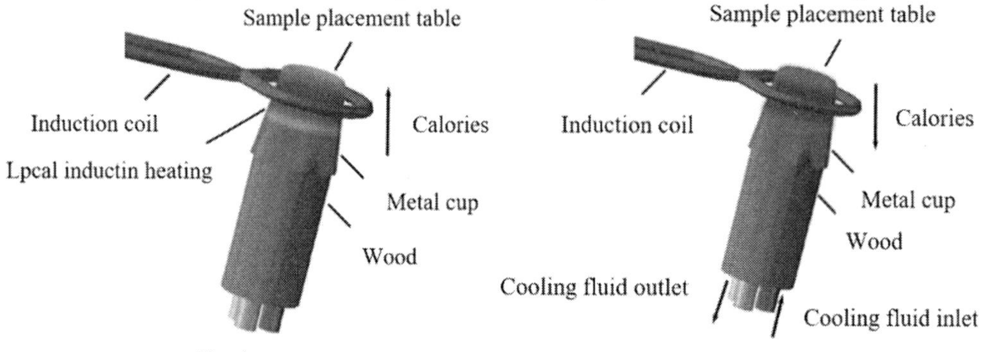

Figure 1. Schematic diagram of electromagnetic induction heating and cooling system.

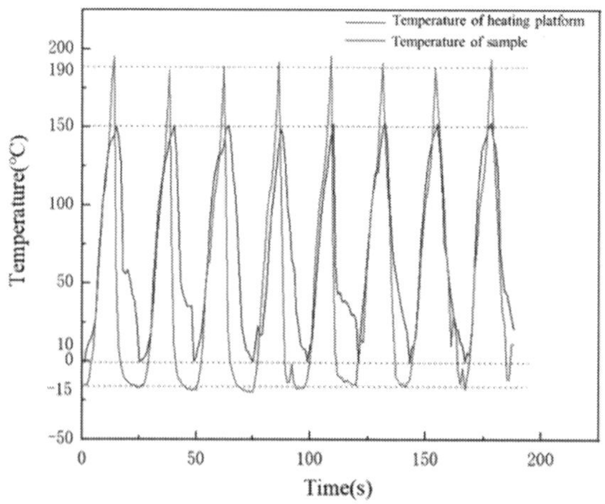

Figure 2. Degree curve of RTS.

3. Results and discussion

3.1 Microstructure of 20Sn-80Pb solder bumps after RTS

Microstructure morphology of overall of 20Sn-80Pb solder bumps after different RTS cycles could be seen from Figure 3, which showed that the Sn-rich appeared in the 20Sn-80Pb solders matrix. When the period of RTS reached 1500T, the Sn-rich phases in the 20Sn-80Pb solders matrix increased and distributed uniformly in the solder matrix. Whereas, as the period of RTS continued to increase, the number of Sn-rich phases in the solder matrix did not change significantly. In general, the microstructures of the high-Pb solder were stable and there was no significant change in the microstructure during thermal shock[13]. This stability, combined with the lack of brittle IMCs in the matrix, enabled a wide range of applications for high-lead solder.

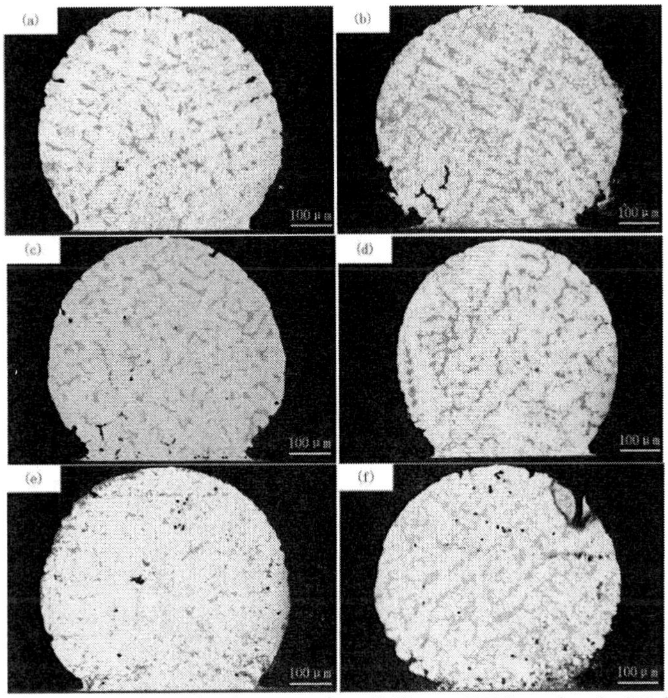

Figure 3. Overall of 20Sn-80Pb solder bumps after different RTS cycles.(a.0T, b.1500T, c.2500T, d.3500T, e.4500T, f.5500T).

3.2 Internal cracks of 20Sn-80Pb solder bumps after RTS

Figure 4 showed the corners of solder bumps after different RTS. It was obvious that no cracks appeared at the interface of solder bumps at 0T. When the cycle of RTS reached 1500T, a crack was found at the lower left of solder bumps. Cracks were formed gradually at the lower corners of solder bumps as the increase of RTS cycle, and the width and depth of the cracks kept increasing. The phenomenon could be explain by the fact that the stress concentration between the solder and IMCs continued to occur, which made cracks form here and eventually continued to expand with the continuous increase of RTS cycles. In the RTS process, solder bumps were hastily heated and cooled in a wide temperature range (0°C to 150°C) in this study, and the coefficient of thermal expansion (CTE) values of the lead, Cu_3Sn, copper were 29.3 x 10^{-6}, 0.067, 0.059 ppm/K, respectively, so there was a difference in CTE between lead and Cu_3Sn, which was significantly greater than the difference in CTE between lead and copper. Hence, the mismatch of CTE led to the concentration of shear stress at the interface of the solder/IMCs layer. Additionally, Cu_3Sn IMCs would shrink in volume during the growth process[14,15], which would also enhance the stress concentration between the solder and IMCs layer interface.

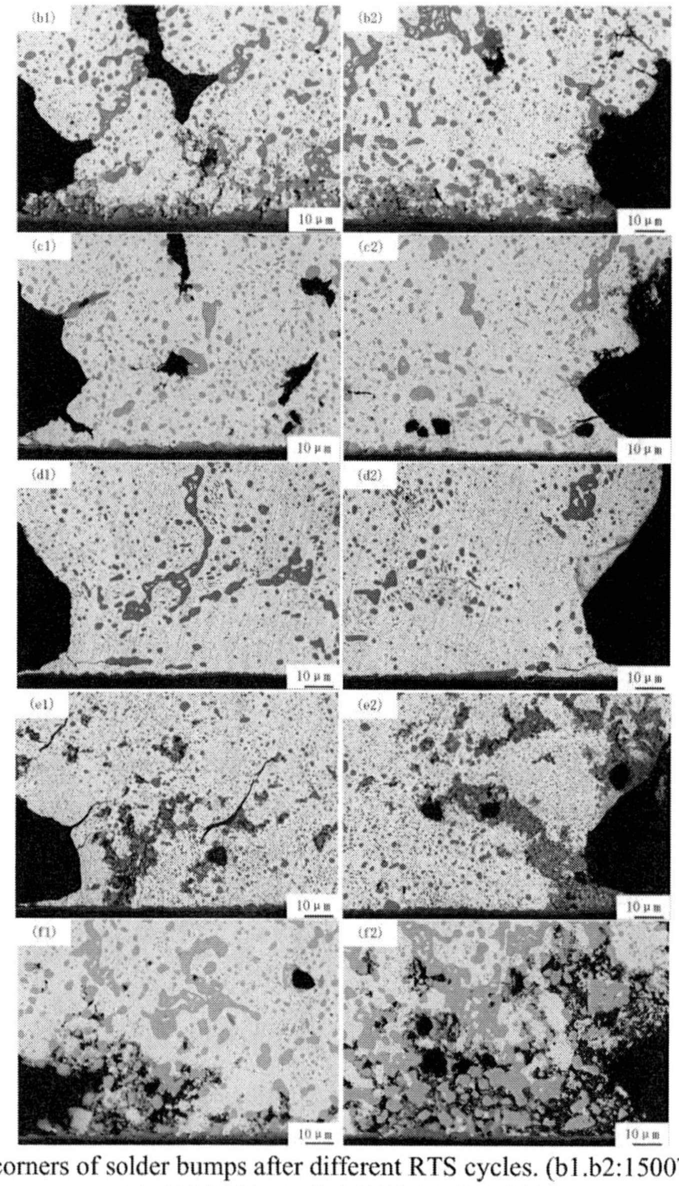

Figure 4. The corners of solder bumps after different RTS cycles. (b1.b2:1500T, c1.c2:2500T, d1.d2:3500T, e1.e2:4500T, f1.f2:5500T).

3.3 Interfacial IMCs of 20Sn-80Pb solder bumps after RTS

Figure 5 presented interfacial IMCs of solder bumps after different RTS cycles. It was noticed that the original morphology of IMCs at the solder bumps interface was plane-type, but after rapid thermal shock, the morphology of IMCs changed from plane-type to scallop-type. The empirical power-law equation was used to elucidate the mechanism of IMCs layer growth, and the formula was used as follows[16].

$$W=W_0+At^n \tag{1}$$

Where W is the IMCs layer thickness at time t, W_0 is the IMCs layer thickness in the as-soldered condition, A is growth constant and n is the time index. The value of the time index n is considered to be an indicator of the IMCs growth mechanism[17,18]. If n is closer to 0.33, the control mechanism for IMCs growth is considered to be grain-boundary diffusion[19,20]. The research determined the value of n by appropriate nonlinear regression analysis fitting to experimental results. The values of Cu-Sn IMCs layer time index n and related factor R^2 were shown in Table 1. The results indicated that the control mechanism of Cu-Sn IMCs growth was grain-boundary diffusion.

Table 1. Time index (n) of the IMC layers at solder bumps/Cu interface

IMC layer	n	R^2	IMCs Growth Mechanism
Cu-Sn	0.28	0.41	Grain-Boundary Diffusion

The calculation formula for the thickness of IMCs was as follows:

$$LIMC = (NIMC / NSEM) \times LSEM \quad (2)$$

Figure 6 showed that the thickness of interfacial IMCs of bumps after different RTS cycles, revealing that the thickness of interfacial IMCs of solder bumps generally went up as the increase of RTS cycles. Whereas, when RTS cycle reached 3500T, the thickness of interfacial IMCs of solder bumps dropped to 2.68 μm compared to the thickness of interfacial IMCs of solder bumps at 2500T. Additionally, the thickness of interfacial IMCs of solder bumps dropped to 3.59 μm at 5500T compared to the thickness of interfacial IMCs of solder bumps at 4500T. This was due to the formation of interfacial cracks in the solder/IMCs layer led to the further diffusion of Cu and Sn atoms to form the gaps, which hindered the growth of the interface IMCs layer [21]. The crack length of solder bumps interface at 3500T was obviously longer than that at 2500T as showed in the Figure 5, explaining that the thickness of interfacial IMCs of bumps at 3500T was lower than that at 2500T. Similarly, the thickness of interfacial IMCs of bumps at 5500T was lower than that at 4500T.

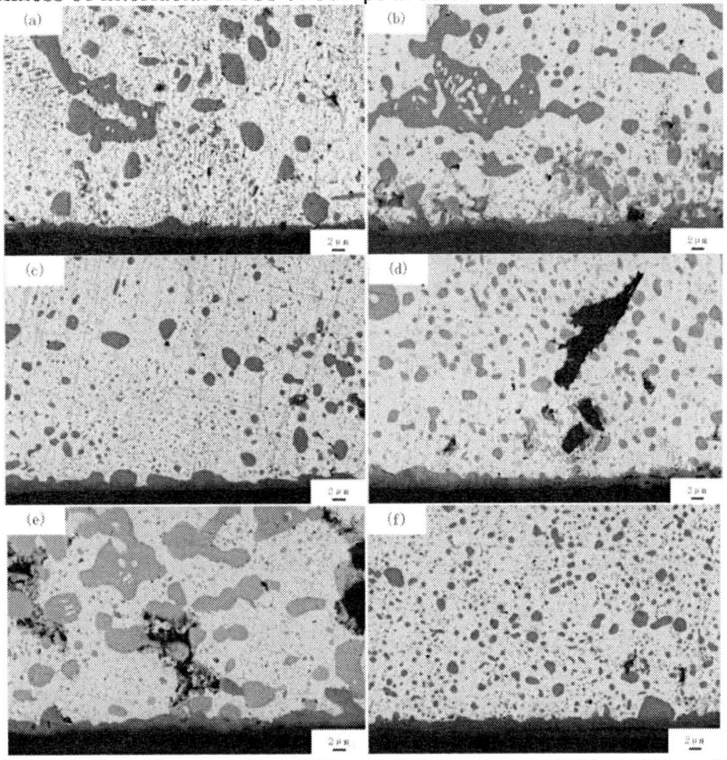

Figure 5. Interfacial IMCs of bumps after different RTS cycles.(a.0T, b.1500T, c.2500T, d.3500T, e.4500T, f.5500T).

 2108 (2021) 012100 doi:10.1088/1742-6596/2108/1/012100

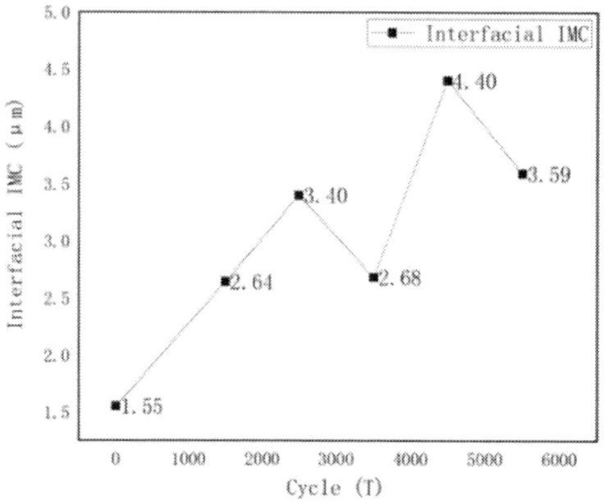

Figure 6. The thickness of interfacial IMCs of solder bumps after different RTS cycles.

3.4 The shear strength and fracture morphology of 20Sn-80Pb solder bumps after RTS

The effect of RTS on the shear strength of 20Sn-80Pb solder joints was studied through shear tests. The results were showed Figure 7, which presented solder bumps' shear strength after different RTS cycles. Before RTS, the initial solder bumps' shear strength was as high as 50.18MPa. Solder bumps' shear strength generally showed a downward trend as the increase of cycles of RTS. However, whilst the cycle of RTS reached only 1500T, solder bumps' shear strength was extremely reduced to 25.81MPa, which was 48.57% lower than that at 0T. As the cycle of RTS continued to increase, solder bumps' shear strength decreased slowly. Solder bumps' shear strength even reduced by approximately 63.43% (18.35MPa) at 5500T. Fracture morphology of solder bumps after RTS could be seen from Figure 8, the fracture mainly occurred in the solder. The fracture was a parabolic dimple, and its orientation was the same as the shear orientation. As the growth of interfacial IMCs and the thickening of Cu_3Sn IMCs layer led to stress concentration near the interface, cracks propagated more easily through Cu_3Sn IMCs layer. Therefore, the fracture mode of solder bumps changed from the inside of solder matrix to the mixed fracture mode of solder matrix and Cu_3Sn IMC layer as the increase of RTS.

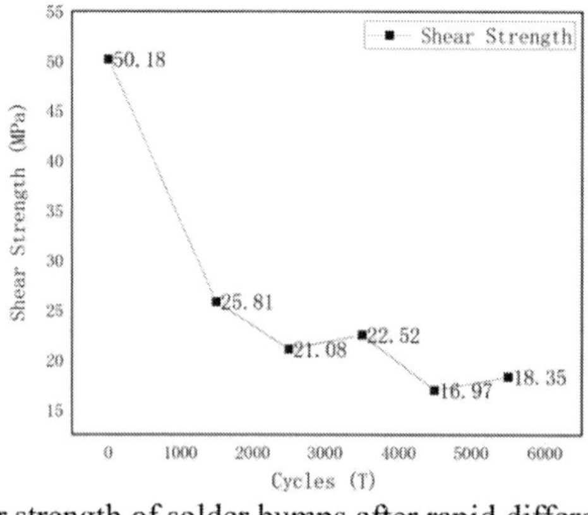

Figure 7. Shear strength of solder bumps after rapid different RTS cycles.

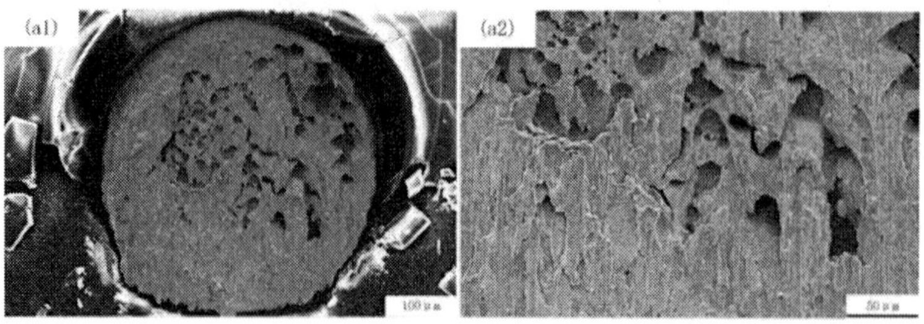

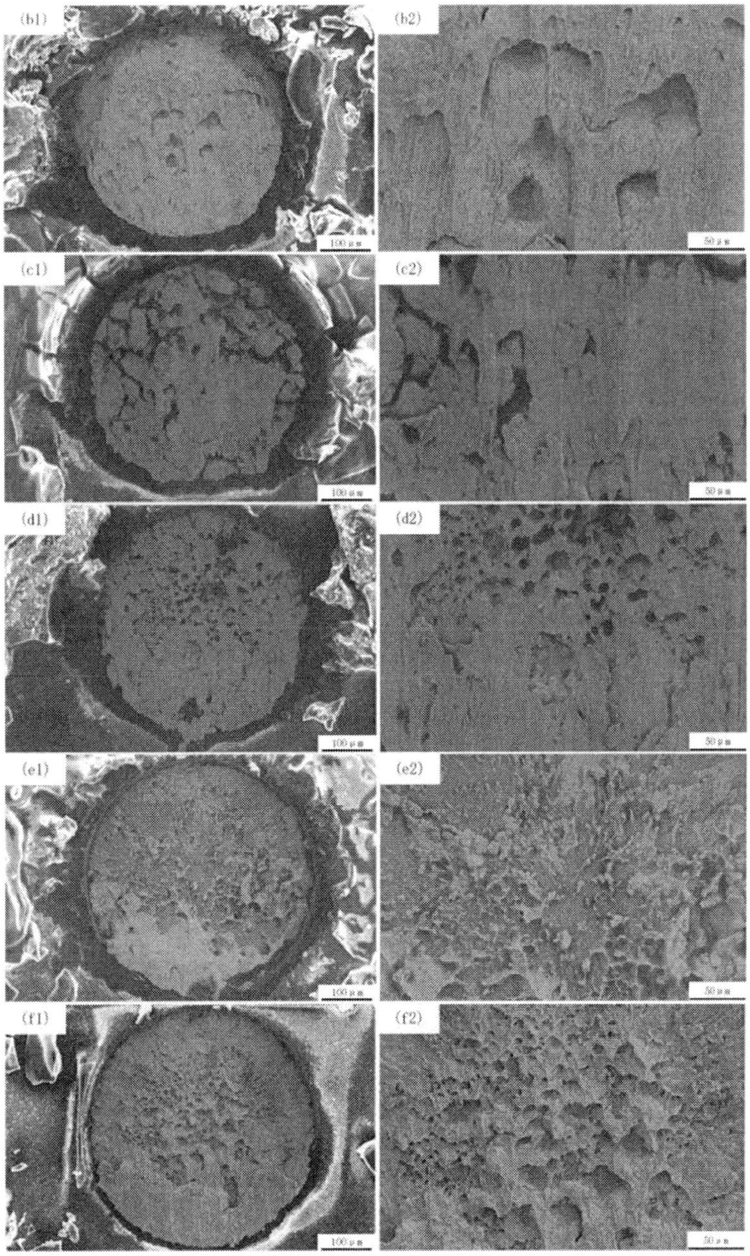

Figure 8. Fracture morphology of bumps after different RTS cycles. (a1.a2:0T, b1.b2:1500T, c1.c2:2500T, d1.d2:3500T, e1.e2:4500T, f1.f2:5500T).

4. Conclusions

(1) The microstructures of 20Sn-80Pb solder bumps were stable and there was no significant change in the microstructure during RTS due to the stability of high-lead content solders. When the cycle of RTS reached 1500T, a crack was found at the lower left of solder bumps. Cracks were formed gradually at the lower corners of solder bumps, and the width and depth of the cracks kept increasing as the increase of RTS cycle.

(2) The thickness of interfacial IMCs of solder bumps generally went up as the increase of RTS cycles. Whereas, when the RTS cycle reached 3500T, the thickness of interfacial IMCs of solder bumps dropped to 2.68 μm compared to the thickness of interfacial IMCs of solder bumps at 2500T.

Additionally, the thickness of interfacial IMCs of solder bumps dropped to 3.59 μm at 5500T compared to the thickness of interfacial IMCs of solder bumps at 4500T.

(3) Solder bumps' shear strength generally showed a downward trend as the increase of cycles of RTS. However, whilst the cycle of RTS reached only 1500T, solder bumps' shear strength was extremely reduced to 25.81MPa, which was 48.57% lower than that at 0T. As the cycle of RTS continued to increase, solder bumps' shear strength decreased slowly. Solder bumps' shear strength even reduced by approximately 63.43% (18.35MPa) at 5500T.

Acknowledgments

This work was supported by the National Natural Science Foundation of China (Grant No.61974013 and 61774066), the Chongqing Research Program Basic Research Frontier Technology (cstc2020jcyj-msxmX0819), the Scientific and Technological Research Program of Chongqing Municipal Education Commission(NO.KJQN201803701), Graduate Innovation Project of Chongqing University of Technology (clgycx20201001, clgycx20201002), respectively.

References

[1] Menon, S., George, E., Osterman, M. (2015) High lead solder (over 85%) solder in the electronics industry: RoHS exemptions and alternatives[J]. Journal of Materials Science Materials in Electronics, 26(6):4021-4030.

[2] Jiang, N., Zhang, L., Liu, Z.Q. (2019) Reliability issues of lead-free solder joints in electronic devices[J]. Science and Technology of Advanced Materials, 20(1):876-901.

[3] Kamal, M., Gouda, E.S. (2006) New lead-containing solder alloy with improved properties[J]. Radiation Effects and Defects in Solids, 161(8):461–466.

[4] Sidorov, V., Drápala, J., Uporov, S. (2011) Some physical properties of Al-Sn-Zn melts[J]. The European Physical Journal Conferences, 15 (2011) 01022.

[5] Galib, R.H., Sharif, A., Gafur, M.A. (2016) Mechanical and Thermal Properties of Zn-xMg Solder Alloys[J]. Applied Mechanics & Materials, 860:173-178.

[6] Hang, C., Tian, R., Zhao, L. (2018) Influence of Interfacial Intermetallic Growth on the Mechanical Properties of Sn-37Pb Solder Joint under Extreme Temperature Thermal Shock[J]. Applied Sciences, 8(11):2056.

[7] Aamir, M., Muhammad,R., Hanif, M. (2016) Relationships between Microstructure and Mechanical Properties in High Sn Content Pb-based and Pb-Free Solder Alloy after Thermal Aging[J]. Adv. Mater. Manuf, 1(1):33-41.

[8] Suganuma, K. (2018) Interconnection technologies[J]. Wide Bandgap Power Semiconductor Packaging.

[9] Galib, R.H., Hasan, R.A.,Sharif, A. (2016) Study of off-eutectic Zn–xMg high temperature solder alloys. J Mater Sci: Mater Electron, 27(8):8734–8744.

[10] Zeng, G., Mcdonald, S., Nogita, K. (2012) Development of high-temperature solders: Review[J]. Microelectronics Reliability, 52(7):1306–1322.

[11] Rahim, M.K., Suhling, J.C., Jaeger, R.C. (2006) Reliability of flip chip assemblies subjected to extreme low temperatures[C]. Thermal and Thermomechanical Phenomena in Electronics Systems, 1379–1389.

[12] Tian, R., Hang, C., Tian, Y., (2019) Brittle fracture of Sn-37Pb solder joints induced by enhanced intermetallic compound growth under extreme temperature changes[J]. Journal of Materials Processing Technology, 268:1-9.

[13] Schoeller, H., Bansal, S., Knobloch, A. (2009) Microstructure Evolution and the Constitutive Relations of High-Temperature Solders[J]. Journal of Electronic Materials, 38(6):802-809.

[14] Tang, Y., Luo, S.M., Wang, K.Q. (2016) Effect of Nano-TiO_2 particles on growth of interfacial Cu_6Sn_5 and Cu_3Sn layers in Sn-3.0Ag-0.5Cu-xTiO_2 solder joints[J]. Journal of Alloys and Compounds, 684:299-309.

[15] Yang, S.C., Chang, C.C., Tsai, M.H. (2010) Effect of Cu concentration, solder volume, and

temperature on the reaction between SnAgCu solders and Ni[J]. Journal of Alloys & Compounds, 499(2):149-153.
[16] Tu, P.L., Chan, Y.C., Hung, K.C. (2001) Growth kinetics of intermetallic compounds in chip scale package solder joint[J]. Scripta Materialia, 44(2):317-323.
[17] Deng, X., Piotrowski, G., Williams, J.J. (2003) Influence of initial morphology and thickness of Cu_6Sn_5 and Cu_3Sn intermetallics on growth and evolution during thermal aging of Sn-Ag solder/Cu joints[J]. Journal of Electronic Materials, 32(12):1403-1413.
[18] Xu, L.H., Pang, J.H.L., Prakash, K.H., Low, T.H. (2005) Isothermal and Thermal Cycling Aging on IMC Growth Rate in Lead-Free and Lead-Based Solder Interface.[J]. IEEE Transactions on Components & Packaging Technologies, 28(3):408-414.
[19] Schaefer, M., Laub, W., Sabee, J.M. (1996) A numerical method for predicting intermetallic layer thickness developed during the formation of solder joints[J]. Journal of Electronic Materials, 25(6):992-1003.
[20] Schaefer, M., Fournelle, R.A., Liang, J. (1998) Theory for intermetallic phase growth between cu and liquid Sn-Pb solder based on grain boundary diffusion control[J]. Journal of Electronic Materials, 27(11):1167-1176.
[21] Shen, J., Zhao, M., He, P. (2013) Growth behaviors of intermetallic compounds at Sn-3Ag-0.5Cu/Cu interface during isothermal and non-isothermal aging[J]. Journal of Alloys & Compounds, 574(Complete):451-458.

Analysis and Prevention of Boiler fire-extinguishing Caused b y the Thermal DCS Fault

Mingliang Zhao *[1a], Shuai Guo[1b], Zhiyu Song[1c], Xiu Liu[1d], Ping Guo[2e]

[1]Jilin Electric Power Research Institute Co., Ltd, Changchun， China

[2]Jilin University, Changchun， China

[a]4321105965@qq.com, [b]1508395553@qq.com, [c]18629983850@163.com, [d]13944866340@126.com, [e]1508395553@126.com

Abstract: The number 4 boiler of a certain power plant happened the boiler fire extinction events. Because of main and auxiliary DPU switched failure, more piece of card took off the network and malfunctioned, both of PAF(Primary Air Fan) stop, which triggered MFT,that lead to the boiler fire-extinguishing. Aimed at the course of the boiler fire-extinguishing and the measure that had been adopted, this paper expounds in detail , analyzes the cause of the incident and other aspects of the problem that the incident exposed to , finally ,in view of the existing problems and potential sa[1]fety hazard ,this paper gives specific recommendations .

1.Introduction

With the adjustment of the national energy structure and the increase of the proportion of new energy, higher requirements are put forward for the peak shaving flexibility of the power grid. The State Grid puts forward higher requirements for the reliability and stability of the operation of thermal power generating units, and the non-stop assessment of thermal power generating units is increasing day by day. Due to the poor reliability and stability of large thermal power generating units, unit tripping has seriously threatened the safe operation of power grid and main equipment of the system, increased the maintenance cost within the service life of the equipment, and seriously affected the profitability of thermal power generating units. However, unit tripping often occurs for some units due to defects in the system and equipment.

Aiming at a boiler fire extinguishing event in an ultra-high pressure unit, this paper analyzes the causes of the event and other problems exposed in this event, and gives specific preventive measures for the existing problems and potential safety hazards.

2.System Overview

The boiler of the unit is an ultra-high pressure primary intermediate reheat, natural circulation solid-state slag discharge pulverized coal boiler Π The pulverizing system is of the hot primary air positive pressure direct blowing type of the medium speed coal mill, the pulverized coal burner is of the swirling type, the front and rear walls are arranged, the opposed combustion mode is adopted, the balanced ventilation is adopted, the whole steel frame single drum suspension structure, and the design fuel is bituminous coal.

DCS system is a tcs3000 instrument electric integrated distributed control system of a company. Its main functions include: data acquisition system (DAS), analog control system (MCS),

sequence control system (SCS) and furnace safety monitoring system (FSSS) [1]. The controller adopts redundant configuration, that is, one controller chassis is equipped with two sets of main and auxiliary controllers. The whole controller consists of controller power supply, DPU controller and gateway. The hardware of DPU controller is provided by MOX company, and the software and controller are imported OEM products.

At 01:48 on a certain day, No. 2 main DPU failed, and the DPU automatically switched between main and auxiliary DPUs. During the switching process, the main and auxiliary DPUs operated alternately. In this process, 01:51:33 forced draft fan a tripped, 01:52:22 induced draft fan B tripped, 01:52:32 primary air fan B tripped, 01:52:42 primary air fan a tripped, MFT occurred in the boiler and the boiler was extinguished. Then, the load of unit 4 decreased from 201mw to the target value of 1.5MW, the steam turbine operated and the generator was not disconnected. 02:03 the thermal maintenance personnel manually cut off the power supply of the main DPU, the auxiliary DPU operated normally, then turned on the power supply of the main DPU, and the main DPU was in standby state. The main and auxiliary DPUs are switched. At 02:16, the operator started the primary air fan a, started the plasma ignition at 02:18, started the No. 1 coal mill at 02:20, gradually increased the load, increased the load to 100MW at 02:57, and the unit returned to normal operation. During the whole recovery process, the minimum parameters of the unit are: load 1.187 MW, main steam pressure 8.781mpa, main steam temperature 449.9 ℃ and reheat steam temperature 424.3 ℃.

Table.1 parameter status during fire extinguishing

Back wall of flue			
54	41	71	38
45	25	34	21
23	37	48	16
42	17	41	27
51	27	38	57
37	11	26	12
28	9	16	35
21	30	26	19
24	11	30	26
26	41	18	31

3.cause analysis

According to the time sequence of SOE records, it is found that at 01:48:08, the main and auxiliary DPUs begin to switch, and after this time, the cards begin to fail one after another. Through the analysis of the event process and SOE records and the test of cards, it is found that the load rate of No. 2 DPU exceeds 40% (standard value 40%) and the maximum exceeds 60%, which affects the system safety.

By referring to the function introduction in 2.1 DCS operation monitoring function and 2.3.3 DPU status [2] of tcs3000 operation and maintenance manual, the analysis shows that the reason why DPU can not complete automatic switching for a long time is that the data flow increases and the network is blocked, resulting in excessive load of main DPU and sharp decline in processing speed, thus triggering the redundant switching function of main and auxiliary DPU, At the same time, the "redundancy" between DPUs fails, the "redundancy" is lost, and the controller cannot be switched normally.

At the same time, the DPU hardware of unit 4 was upgraded to a higher version DPU, and the card was not replaced with a matching new card. The incompatibility between the old card and the new DPU is also one of the reasons for the card failure. The following figure is the schematic diagram 3 of the cause of the event.

In addition, the arrangement of important auxiliary control signal cards in DCS system does not consider the risk dispersion in case of failure. The control signals of forced draft fan, induced draft fan, air preheater and primary air fan are arranged in the same DPU (No. 2 DPU); The control signals of No. 1 and No. 2 feed pumps are arranged on the same control card (No. 1 card on the first floor of No. 9 DPU); The control signals of AC and DC lubricating oil pumps and high-pressure starting oil pumps are arranged on the same control card (No. 5 card on the third floor of No. 8 DPU); The control signals of No. 1 and No. 2 condensate pumps are on the same control card (No. 1 card on the first floor of No. 9 DPU); Once the control station or control card fails, it will cause the important auxiliary equipment of the unit to lose control.

It can be seen from the analysis process that the causal relationship leading to this event is: firstly, the data flow is large, the network is blocked, the DPU load rate increases, and the DP module fails, resulting in the failure of the main DPU. After the failure of the main DPU, the main and auxiliary DPUs operate alternately, and the old card is incompatible with the new DPU, resulting in the failure of multiple cards in this process. After the card fails, The wrong signal was input into the DPU, resulting in confusion in the internal program of the DPU, instantaneous misoperation of some commands, random full stop of two primary fans, triggering MFT, and the boiler fire extinguishing event occurred.

4. analysis conclusion

4.1 direct cause
The data flow increases, the network is blocked, the DPU load rate increases, the DPU fails, the DPU switching is unsuccessful, the DP module fails, and the card is offline.

4.2 indirect causes
Unreasonable arrangement of important auxiliary control signal cards of DCS system; The upgrading and transformation of DCS system in the whole plant is not complete; The DCS operation and maintenance is not in place.

5.preventive measures
Accident prevention should take measures from the aspects of technology, equipment and organization management to improve the ability of accident prevention on the whole, so as to effectively control events and ensure production safety.

5.1 according to DL / T 5175-2019 technical regulations for design of thermal control system of thermal power plant, important auxiliary equipment shall be redistributed on DCS [3]:

① The control signals of forced draft fan, induced draft fan, air preheater and primary air fan are arranged in different control stations respectively;

② The control signals of No. 1 and No. 2 feed pumps are arranged on different control cards respectively;

③ The control signals of AC and DC lubricating oil pumps and high-pressure starting oil pumps are arranged on different control cards respectively;

④ Arrange the control signals of No. 1 and No. 2 Condensate Pumps on different control cards respectively;

⑤ Test the changed important auxiliary control circuit to ensure that the auxiliary control function is normal.

5.2 due to the unstable operation of DCS system, high card failure rate, failure of automatic switching of DPU and long switching time, the following work shall be carried out:

① Contact the DCS system manufacturer to check and repair the controller and communication network;

② Strengthen the patrol inspection of DCS network operation and deal with problems in time;

③ The DCS system shall be completely transformed and upgraded at the right time. Strictly review and control the feasibility study, design, implementation scheme and other links of the project, eliminate hidden dangers, improve reliability and ensure safe operation;

④ After the upgrading, the DCS system shall be strictly accepted according to DL / T 659-2016 code for acceptance and test of distributed control system in thermal power plant, and the performance of DCS system shall be comprehensively tested [4].

5.3 network congestion and DPU failure. The following two aspects should be strengthened:
① Strengthen the monitoring of DPU load rate in DCS system, especially DPU with a large number of control equipment and large data flow. When it is found that DPU load rate is high, contact the operator, take safety measures and manually switch DPU. At the same time, it is necessary to strictly implement Article 9.3.1 of the 25 key requirements for the prevention of power production accidents (GNQ [2014] No. 161). For power plants equipped with distributed control system, an emergency treatment mechanism in case of distributed control system failure shall be established according to the specific conditions of units, and a practical and operable emergency treatment plan for distributed control system failure shall be formulated under various conditions, And conduct regular anti accident drills [5];

② Test the network load in DCS system. If it does not meet the performance requirements, it shall be reformed in time;

③ It is necessary to deeply study the tcs3000 instrument electric distributed control system - software manual and tcs3000 operation and maintenance manual.

④ Add bypass to the output command in the logic of important auxiliary equipment, and input the bypass into RS trigger for monitoring.

6.Conclusion

In view of the events occurred in the plant, the thermal control discipline should strengthen the management of DCS system, carry out thorough transformation, redistribute the distribution of equipment in the transformation process, and improve the reliability of DCS system.

Acknowledgments

Thanks for the test data provided by the technicians of Jilin Electric Power Research Institute, which has made a great contribution to my article.Supported by the science and technology project of State Grid Jilin Electric Power Co., Ltd KY-GS-21-01-03

References

[1] Tcs3000 instrument electric distributed control system - Software Manual
[2] Tcs3000 operation and maintenance manual
[3] DL / T 5175-2019 technical specification for design of thermal control system of thermal power plant
[4] DL / T 659-2016 code for acceptance test of distributed control system in thermal power plant
[5] Twenty five key requirements for preventing power production accidents (GNQ [2014] No. 161)

Research and Design of X-band Low Noise Amplifier Based on CMOS Process

Chao Ma[1], Hongjiang Wu[1*], Xudong Lu[1], Haitao Sun[1]

[1] The 13th Institute of CETC, Shijiazhuang, Hebei, 050051, China

*Corresponding author's e-mail: ludx@mail.cetc13.cn

Abstract. Based on CMOS process, a low noise amplifier(LNA) operating at 7.4GHz ~11.4GHz was designed. The two-stage differential cascode structure is adopted. Transformer was used to achieve inter-stage matching. Balun was used to achieve input and output matching, which reduces the number of inductors used, effectively reduces the chip size while ensuring good gain and noise figure. The actual measurement results show that the power gain at the center frequency of 9.4GHz is 27dB, the maximum noise figure is less than 3.82dB, the output power 1dB compression point is greater than 8dBm, the chip area is only 0.41mm×0.83mm(excluding PAD).

1. Introduction

With the rapid development of silicon-based semiconductor technology, the characteristic size of transistor is constantly decreasing. At present, the most advanced process node has reached less than 5nm. Silicon-based process has the advantages of the low power consumption, high integration and easy integration with digital circuit[1], which can make up for the deficiency of compound semiconductor. In recent years, the research on silicon-based microwave amplifier has been gradually enriched in China, but there is still a widening gap between domestic and foreign research results. As the first active circuit in the RF receiver, the noise performance of LNA directly determines the noise characteristics of the whole RF receiver system. The silicon-based process has the disadvantages of large substrate noise and low gain, and the parasitic effect of the transistor itself will further deteriorate its gain and noise performance, which makes the design of LNA more difficult.

In this paper, based on CMOS process, a two-stage differential structure LNA working in 7.4GHz~11.4GHz is designed and manufactured. The transformer is used for inter-stage matching, and the input and output matching is realized by balun.

2. Circuit Design

2.1. LNA Design

Aiming at the problems of poor noise figure(NF) and low gain of silicon-based process amplifier, the LNA designed in this paper adopts differential cascade structure. Through the source cascading inductor, the amplifier stability is improved while the input matching is improved[2]. The resistive negative feedback improves the gain flatness and facilitates impedance matching. Transformer is used for inter-stage matching and reducing the circuit size.

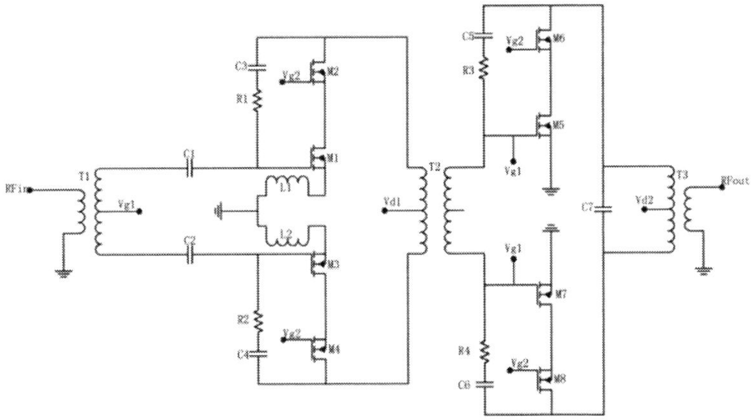

Figure1. Two-stage LNA schematic.

For LNA, the bias voltage will affect the gain, noise, power consumption and other aspects of the system. Gain, linearity, power consumption and other factors should be considered when selecting bias voltage. Through simulation of drain current changing curve with bias voltage, the bias voltage is selected from 0.5V to 0.9V, which ensures that the transistor operated in the saturated region, with good gain and linearity, while not consuming too much power. The change of maximum available gain(MAG) and NF_{min} of common source(CS) stage and common gate(CG) stage with bias voltage was simulated respectively. And the bias voltage of CS was 0.7V and the bias voltage of CG was 1.8V. This ensures that the amplifier has a low noise figure and sufficient gain.

In order to provide sufficient gain, the LNA designed in this paper adopts two-stage amplifier. The noise figure of the multi-level cascade circuit is expressed as:

$$\mathrm{NF} = \mathrm{NF}_1 + \frac{\mathrm{NF}_1 - 1}{\mathrm{G}_1} + \frac{NF_3 - 1}{G_1 G_2} + \ldots + \frac{NF_n - 1}{G_1 G_2 \ldots G_{n-1}}$$

In this equation, $\mathrm{NF_n}$ and $\mathrm{G_n}$ represent the NF and gain of the nth stage circuit, respectively.

It can be seen that the NF of the first stage amplifier directly determines the noise performance of the whole amplifier. Therefore, NF needs to be taken into special consideration when designing the first stage amplifier. The second stage amplifier provides high gain as far as possible while satisfying the NF.

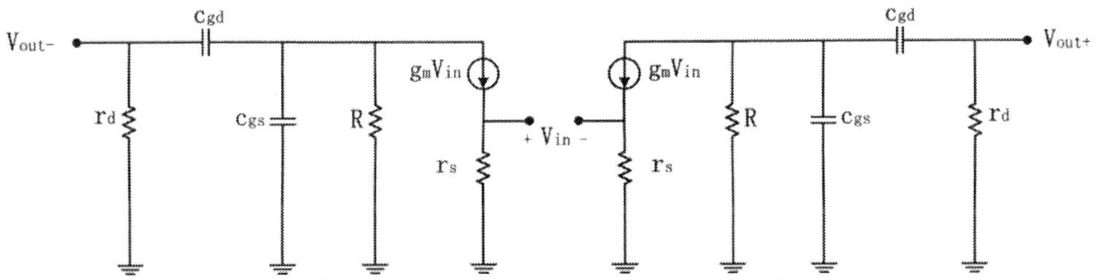

Figure2. Small-signal equivalent circuit.

According to the small signal equivalent circuit of the differential cascade amplifier(Fig2), the maximum voltage gain of the differential circuit is

$$\frac{V_{\mathrm{out}}}{V_{\mathrm{in}}}(\mathrm{j}\omega_0) = \frac{g_m R - j\omega_0 R C_{gd}}{1 + Z_1 + jZ_2}$$

$$Z_1 = R(r_s + r_d)(\omega_0 C_{gd})^2$$

$$Z_2 = \omega_0(r_s + r_d)[C_{gs} + C_{gd}(1 + g_m R)]$$

According to the equation, the parasitic capacitance affects both the numerator and the denominator and significantly reduces the voltage gain of the transistor. Moreover, the negative component introduced by the parasitic capacitance in the input admittance will affect the circuit stability. Cascode is a CS and CG cascade. On the basis of the CS stage with negative source feedback, the CG stage can not only improve gain, but also suppress the parasitic capacitance of the CS stage, which improves the noise performance and stability of the system.

2.2. Transformer Matching

The traditional matching method is LC matching. Impedance selection is carried out according to the equal noise circle and equal gain circle obtained by Smith chart.

In order to achieve better noise matching and input impedance matching, this design uses balun to replace the traditional LC network for input matching. As a passive device, balun can realize DC isolation, single-ended signal conversion to differential signal, impedance matching and other functions[3]. The matching network with LNA designed by balun can effectively reduce the number of inductors used, thus avoiding the loss caused by additional inductors, and greatly reducing the chip area. In addition, balun can replace the capacitor to achieve DC isolation, reducing the difficulty of layout design. This balun uses two layers of metal and is implemented in a planar structure. The main geometric parameters are microstrip line width, line spacing and external diameter. It is pretty difficult to design a balun with good noise matching and input impedance matching just by adjusting these parameters. Therefore, balun geometric parameters can be adjusted to a relatively reasonable value first, and then the capacitance or inductance can be connected outside the input end according to Smith chart to achieve noise matching and input impedance matching. So we choose a compromise method, in which the inner inductor of balun is connected, and the matching is further adjusted by adjusting the inner inductor to achieve good standing wave performance and noise performance.

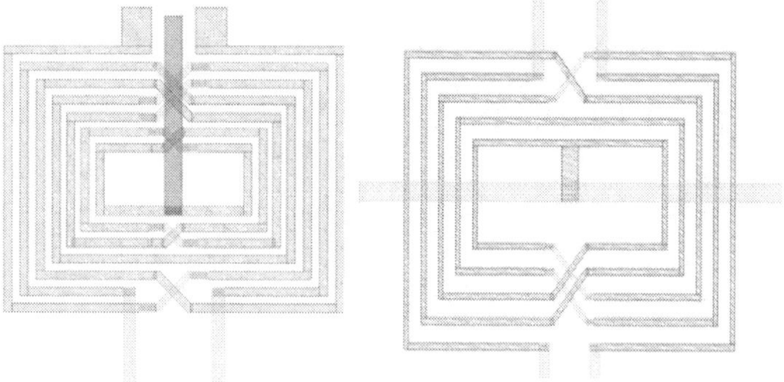

(a) (b)

Figure3. Input Balun layout (a) and Inter-stage Balun layout (b).

2.3. Stability problem

As the first stage of the receiver, LNA needs to be stable in any frequency band, otherwise the gain peak will appear, and the poor linearity will cause the circuit system to fail to work normally.

A common way to improve the stability of a LNA is cascade the inductor between the source and the ground. But the negative feedback due to series inductance will lead to gain loss. Therefore, the value of the source feedback inductance is generally small and can be replaced by the microstrip line. Another approach is to generate feedback by a parallel resistor between the gate and drain. The resistor determines the depth of the feedback. A zero point and two poles far apart are introduced to improve the gain flatness and the stability of the LNA. In addition, there is a method of series attenuator at the end of the multistage LNA, but it will bring great loss to the gain. This circuit design adopts the method of combining the source cascade inductance and the parallel resistance between gate and drain. The small value of cascade inductance can improve stability without bringing too much gain loss. Feedback is introduced through the shunt resistor to improve gain flatness and facilitate impedance matching.

The necessary and sufficient conditions for the circuit to be unconditionally stable are as follows.

$$|S_{11}| < 1$$

$$|S_{22}| < 1$$
$$|\Delta| = |S_{11}S_{22} - S_{21}S_{12}| < 1$$
$$K_f = \frac{1 - |S_{11}|^2 - |S_{22}|^2 + |\Delta|^2}{2|S_{12}||S_{21}|} > 1$$

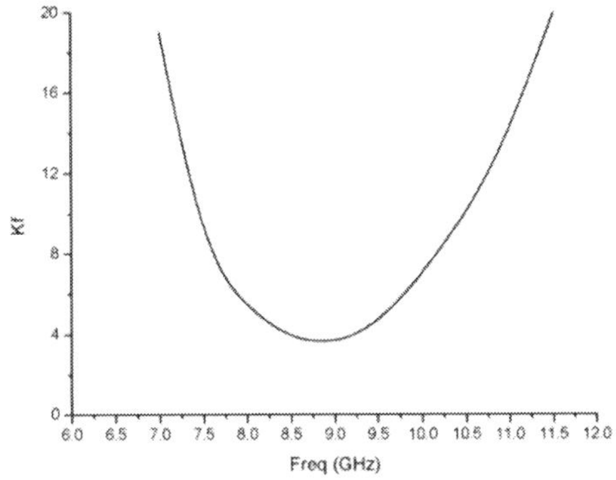

Figure4. Stability factor K_f characteristic curve.

As can be seen from the Fig4, K_f is greater than 3.5 in the full frequency band, and the circuit is unconditionally stable.

2.4. Layout Design

The MOSFET with different the number of fingers and gate width is simulated by Cadence. After comparison, the MOSFET with single finger gate width of 3 um and the number of fingers of 24 is selected for circuit design to achieve the expected effect.

It is necessary to design the layout symmetrically when designing the layout of the LNA. The top metal layer is thicker than other metal layers, with smaller resistivity and smaller parasitic capacitance. Therefore, the top metal is selected for signal wiring, the bottom metal for bias wiring, and the middle layer metal for connecting the lower layer, which can effectively reduce the influence of crosstalk between wires. A certain distance should be maintained between the active device and the transformer to avoid magnetic interaction between them. Dummy resistor can be drawn when drawing the feedback resistor to improve the accuracy of physical graphics in subsequent circuit fabrication.

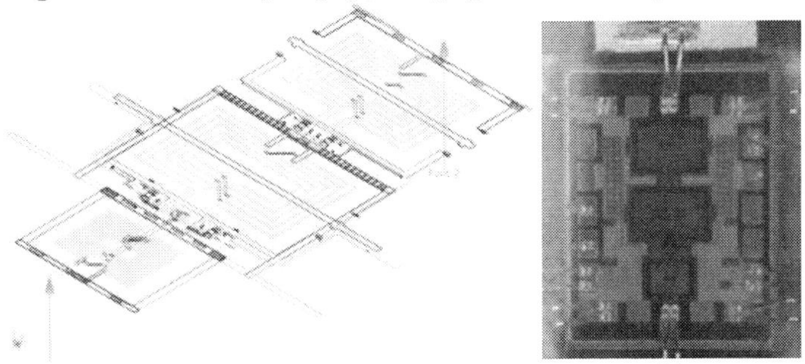

(a) (b)

Figure5. Overall layout picture (a) and Chip physical picture (b).

After the layout design is completed, the passive part of the layout is imported into ADS for EM simulation, and S parameters are extracted and imported into schematic through the nport component in Cadence for co-simulation with the transistor. After the simulation results meet the expected requirements, it is necessary to check the design rules and compare the layout with the schematic to ensure that the electrical properties are consistent. These are based on the DRC and LVS functions in the Calibre column in Cadence. The final design layout and the physical picture of the chip are shown in the Fig5, and the chip size is 0.41mm×0.83mm(excluding PAD).

3. Test Result

The S parameters, input-output standing wave ratio and noise figure of LNA were tested by microwave network analyzer. The comparison between the test results and the simulation results is shown in the Fig6.

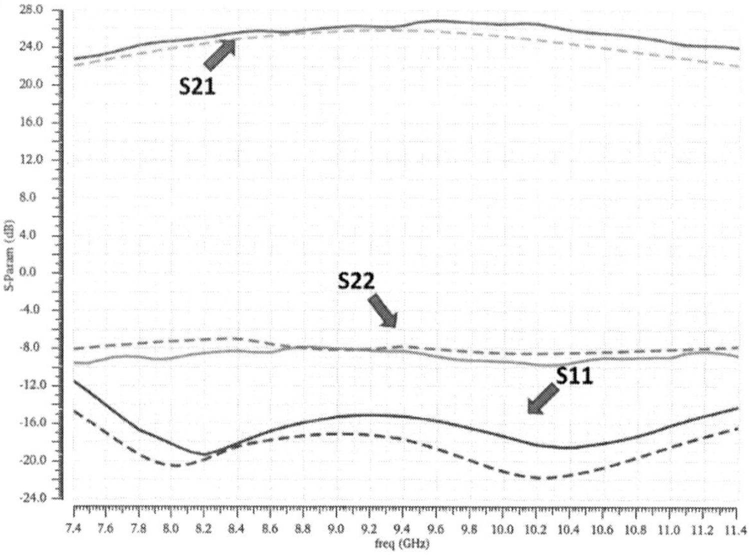

Figure6. Comparison of S parameter test results and simulation results of LNA.

According to the comparison curve between the test results and the simulation results, the measured results show that the gain of the LNA from 7.4GHz to 11.4GHz is (24.5±1.5dB), which basically coincidences with the simulation curve. According to the input-output standing wave comparison curve, the typical input standing wave ratio is 1.4:1, the typical output standing wave ratio is 2:1. The measured value of the input standing wave is basically consistent with the trend of the simulation curve with a slight deviation, while the measured value of the output standing wave is slightly worse than the simulated value, indicating that the measured data of the active device model is slightly different from the model data.

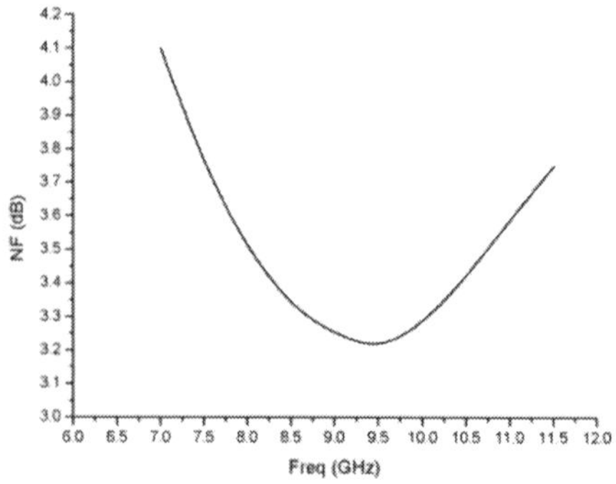

Figure7. NF test curve.

According to the noise figure test curve(Fig7), the NF is less than 3.82dB from 7.4GHz to 11.4GHz. Because it is difficult to achieve ideal performance in all aspects when designing LNA. It is necessary to make a trade-off among NF, gain, frequency band, input and output standing wave. Therefore, the output standing wave is relatively poor. The simulation result of DC power consumption is 41.08mA, and the test result is 41mA. At 9.4GHz, the output 1dB gain compression power is 8.16dBm.

Table 1. The performance comparison between the LNA in this paper and the LNA based the same process in other recent literatures..

	[4]	[5]	[6]	[7]	This design
Process	180nm CMOS	65nm CMOS	90nm CMOS	180nm CMOS	
Frequency（GHz）	6.4 ~7.4	9.65~10.65	8~12	15.1~16.6	7.4~11.4
Gain（dB）	18	28	24	15.2	27
NF（dB）	2.85(min)	1.8(min)	3.3~3.8	5(min)	3.2~3.82
S11（dB）	-13	-28.2	<-10	<-10	<-12
Area（mm^2）	0.55	0.2	0.96	0.144	0.34

As shown in Table 1, by comparing the LNA performance based the same process in recent years, it can be seen that the LNA designed in this paper ensures good gain, noise and standing wave ratio while the chip size is relatively small.

4. Conclusion

In this paper, we use CMOS process to design a two-stage differential structure LNA working in 7.4GHz~11.4GHz. Transformer is used for the inter-stage matching. Input and output matching is realized by balun. The test results show that the LNA has a working frequency range of 7.4GHz~11.4GHz. Gain is (24.5±1.5dB), noise figure is less than 3.82dB, output 1dB compression point is greater than 8dBm. This verifies that the implementation of matching using balun and transformer can ensure good performance while effectively reducing the chip size.

References

[1] Su Guodong. Design of Silicon-based CMOS Millimeter-Wave Front-End Key Circuits[D]. Hangzhou: Zhejiang University, 2018Podani, J. (1994) Multivariate Data Analysis in Ecology and Systematics. SPB Publishing, The Hague.

[2] Yang Jing. Design of Wideband Low Noise Amplifier for Globle Navigation Satelite Systems[D]. Tianjin: Tianjin University, 2018.

[3] Li C, et al. A high-efficiency 5G K/Ka-band stacked power amplifier in 45nm CMOS SOI process supporting 9Gb/s 64-QAM modulation with 22.4% average PAE[A].2017 Texas Symposium on Wireless and Microwave Circuits and Systems (WMCS)[C], Waco, TX, 2017.1-4.

[4] Alican Caglar. A 180-nm X-Band Cryogenic CMOS LNA[J]. IEEE Microwave and Wireless Components Letters, VOL30, NO.4, April 2020.

[5] Mohammad Reza Nikabakhsh. Two-stage current-reused variable-gain low-noise amplifier for X-band receivers in 65nm complementary metal oxide semiconductor technology[J]. IET Circuits, Devices & Systems, 15th June 2018.

[6] Ping-Yi wang. A low-power and low-noise X-band receiver MMIC in 90nm CMOS[A]. Proceeding of the 45th European Microwave Conference. Sept.2015.

[7] Zhengdong Jiang. Research on Silicon-Based Integrated Microwave and Millimeter-wave Amplifier Circuits[D]. Chengdu: University of Electronic Science and Technology of China, 2019.

Numerical Simulation Research on Inter-electrode Capacitance of High-power Electronic Tetrode

Yongjie LUO [1], Puqiong YANG [1*], Honghu WU [2], Wenbo CHEN [1] and Haoran ZHONG [1]

[1] School of Electrical Engineering, University of South China, Hengyang, Hunan ,421001, China

[2]Chengdu Xuguang Electronic Co. Ltd, Chengdu, Sichuan, 610500, China

[*]Corresponding author's e-mail: yangpuqiong@usc.edu.cn

Abstract. High-power electronic tetrode is the RF wave transmitting component in the ion cyclotron range of frequency (ICRF) heating system in EAST. In high-frequency use, its inter-electrode capacitance is an important factor that cannot be ignored, which will have a significant impact on the operating frequency and output power. Based on the theory of electrostatics and considering the influence of each electrode, this paper established a three-dimensional solid model of the electronic tetrode, and the numerical simulation about the inter-electrode capacitance is calculated by the finite element software COMSOL. The obtained inter-electrode capacitance simulation calculation values are all within the actual test value range. On this basis, the relationship between the inter-electrode capacitance and the inter-electrode distance is studied, and the adjustable range of the electrode diameter is calculated. The finite element numerical simulation method calculated the inter-electrode capacitance of high-power tetrodes and the quantitative relationship between the inter-electrode capacitance and inter-electrode distance more accurately, which can provide a reference for the simulation and design of megawatt-level electronic tetrodes in the future.

1. Introduction

High-power electronic tetrode is the core component of the ion cyclotron range of frequency (ICRF) heating system in EAST. It is often employed as the final RF power amplifier to provide megawatt-level RF output power [1-3]. In actual high-frequency applications, the inter-electrode capacitance of the electronic tetrode will become a non-negligible part of the oscillation circuit. It will have an impact on the operating frequency, which limits the operating frequency of the electronic tetrode. At the same time, the parasitic coupling generated by it will cause the instability of the amplification, which will have an adverse effect on the normal operation of the electronic tetrode. Therefore, in order to reduce some empirical experiments in the development process, it is of practical significance to use computer software to numerical simulate the inter-electrode capacitance of high-power electronic tetrodes for high-frequency applications.

This paper will discuss the high-frequency response of the electronic tetrode, analyze the relationship between the inter-electrode capacitance, the operating frequency and output power. According to the structure and parameters of a certain type of electronic tetrode produced by Chengdu Xuguang Electronics Co. Ltd, based on the concept of mutual capacitance in classical electrostatics, the electronic tetrode is considered as a multi-conductor system, and the finite element analysis software COMSOL is used to simulate and calculate the inter-electrode capacitance of the electronic tetrode.

2. High frequency response of electronic tetrode

When the electronic tube is used as an oscillator or amplifier, its output power will decrease continuously as the operating frequency increase, and even eventually it will unable to obtain effective oscillation or amplified output power [4], that is, the output power of the electronic tube tends to decrease with the increase of frequency. As shown in Figure 1, f_{lim} is the limit frequency of the electron tube, at which the tube has stopped working. One of the main reasons for this phenomenon is the oscillation system of the electron tube [4-5].

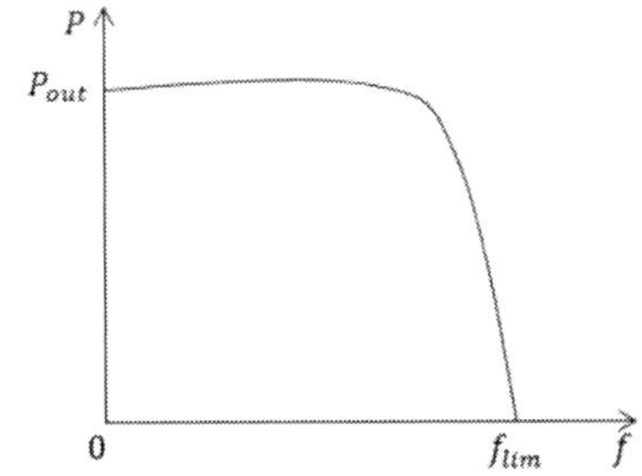

Figure 1. Relationship between output power and frequency at high frequency

The electrodes of electronic tetrode and its leads have certain capacitance and inductance, which can be represented by the equivalent circuit shown in Figure 2. The capacitances and inductances of C_{g2a}, C_{g1a}, C_{g1g2}, C_{g1k}, C_{g2k}, C_{ak}, L_a, L_{g1}, L_{g2}, L_k are all small, and their influence can be ignored during low-frequency operation. However, in high-frequency applications, the inter-electrode capacitance and lead inductance become the main factors that determine the frequency.

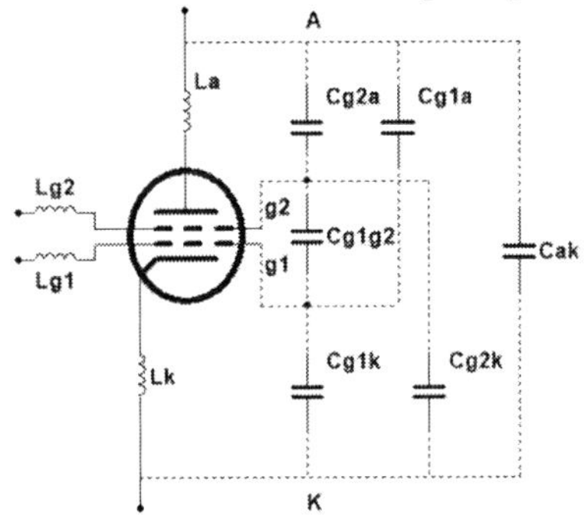

Figure 2. Equivalent circuit of tetrode at high frequency

If the electrodes work in case of a short circuit at the tube base, the highest operating frequency is given by

$$f = \frac{1}{2\pi\sqrt{LC}} \tag{1}$$

where L is the equivalent inductance in the tube, and C is the equivalent capacitance in the tube. When working at the highest operating frequency, the output power and efficiency of the tube is actually zero. It can be seen that the inter-electrode capacitance and lead inductance of the tube have certain restrictions on the high-frequency use of the tube. The existence of the inter-electrode capacitance will have an impact on the fundamental equivalent load impedance R_{oe}, which mainly depends on the characteristic impedance ρ and the quality factor Q of the loop [4], as expressed in equation (2).

$$R_{oe} = \rho \cdot Q \tag{2}$$

where $\rho = \sqrt{\frac{L}{C}}$, $Q = \frac{\rho}{\gamma}$. For the amplifier, if the R_{oe} is too low, the output will drop, causing the amplifier to fail.

At the same time, the existence of the inter-electrode capacitance will also affect the gain [4]. When the common cathode circuit is used, the gain expression is given by

$$G = S^2 / 4\pi C_{g1k} \cdot C_{g2a} \cdot \Delta f \quad (3)$$

In the formula, G is the gain (dB), S is the mutual conductance (mA/V), C_{g1k} is the capacitance between the cathode and the control grid (F), C_{g2a} is the capacitance between the anode and the screen grid (F) and Δf is the frequency bandwidth (Hz). Under the premise of ensuring bandwidth, according to Eq.（3）, we know the effective way to increase the gain is to increase the mutual conductance and reduce the inter-electrode capacitance of the tube [4-5].

Therefore, the inter-electrode capacitance is an important factor that affects the normal operation of the electronic tetrode. Improving the structure of the electronic tube to reduce the influence of distributed capacitance is an important way to improve the working performance of tetrode [5].

3. Simulation analysis of inter-electrode capacitance of tetrode

3.1 Introduction and testing of tetrode

The electronic tetrode studied in this paper is a coaxial electrode structure, ceramic-metal tetrode with a directly heated thoriated tungsten cathode, a pyrolytic graphite grid and a hypervapotron cooling anode. The actual product is shown in Figure 3.

Figure 3. Electronic tetrode

The electronic tetrode is a mature product of Chengdu Xuguang Electronics Co. Ltd. The company has a complete electronic tube static parameter test room. The inter-electrode capacitance test is carried out in accordance with the national standard electronic tube inter-electrode capacitance test method. Through the actual test and inspection of a large number of this type tetrode, we have obtained the reference range of the inter-electrode capacitance, as shown in Table 1.

Table 1. Reference range of inter-electrode capacitance

Inter-electrode capacitance	**Reference range（pF）**
C_{g1k}	73-85
C_{g1g2}	110-130
C_{g2a}	20-23

3.2 Calculation method of inter-electrode capacitance

In the simulation, the electronic tetrode is analyzed as a multi-conductor system [6], and the system satisfies the conservation of charge equation (4).

$$\frac{\partial \rho}{\partial t} + \nabla \cdot J = 0 \quad (4)$$

The potential of each electrode is not only related to the charge of the electrode itself, but also related to the charges of other electrodes. The relationship between charge, capacitance and potential [6-7] is given by

$$\begin{cases} Q_1 = C_{10}\varphi_1 + C_{12}(\varphi_1 - \varphi_2) + \cdots + C_{1n}(\varphi_1 - \varphi_n) \\ Q_2 = C_{21}(\varphi_2 - \varphi_1) + C_{21}\varphi_2 + \cdots + C_{2n}(\varphi_2 - \varphi_n) \\ \vdots \\ Q_n = C_{n1}(\varphi_n - \varphi_1) + C_{n2}(\varphi_n - \varphi_2) + \cdots + C_{n0}\varphi_n \end{cases} \tag{5}$$

where C_{ii} is the self-capacitance of the ith conductor, and C_{ij} is the mutual capacitance between the ith conductor and the jth conductor. The simulation in this paper will lead to the concept of mutual capacitance to solve the inter-electrode capacitance of the electronic tetrode. When calculating C_{ij}, the potential $\varphi_{i \neq 0}$, and the potential of all other electrodes is equal to zero, as expressed in equation (6) [6-8]. That is, during the solution process, the ith electrode is applied 1V and the other electrodes are grounded.

$$C_{ij} = \frac{Q_j}{\varphi_i}\bigg|_{(\varphi_i,\ \cdots,\ \varphi_{i-1},\ \varphi_{i+1},\ \cdots,\ \varphi_n)=0} \tag{6}$$

3.3 Simulation calculation of inter-electrode capacitance

According to the design drawings of the electronic tetrode provided by Chengdu Xuguang Company, the electrode components are drawn by computer 3D drawing software, and combined to establish a simulation model. The simulation model of tetrode diagram is shown in Figure 4. The simulation model consists of cathode, control grid, screen grid and anode. The simulation uses the electrostatic interface of the AC/DC module in the finite element analysis software COMSOL to calculate the inter-electrode capacitance of the electronic tetrode, import the built model into COMSOL, set the materials for each component, and then add four terminals of cathode, control grid, screen grid and anode in the electrostatic interface to set the boundary conditions and various physical parameters. The meshing is then completed, the meshing diagram is shown in Figure 5, and finally the steady-state solution is performed.

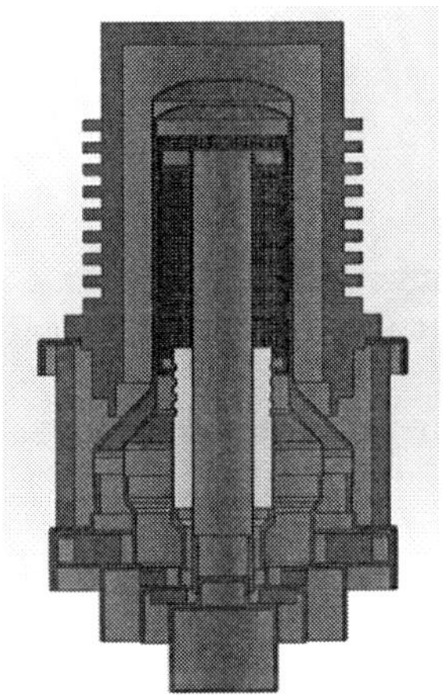

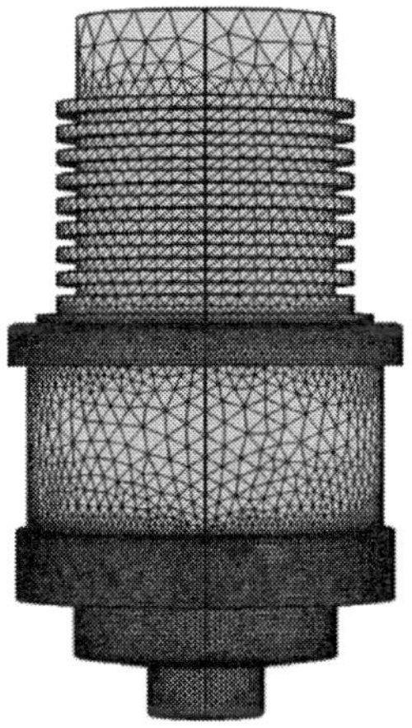

Figure 4. Cross section of simulation model Figure 5. Meshing diagram

Taking the calculation of the capacitance C_{g1k} between the control grid and the cathode as an example, according to the foregoing analysis, we set the control grid terminal voltage to 1V and other electrodes to 0V. After dividing mesh, steady-state solution is carried out to obtain the electric field distribution and the cathode terminal charge [8], as shown in Figure 6. According to Eq.(6), the value of C_{g1k} can be obtained. The simulation calculation methods of C_{g1g2} and C_{g2a} are the same. The results are shown in Table 2, and compared with Table 1, it can be seen that the simulation calculation results are all within the reference range.

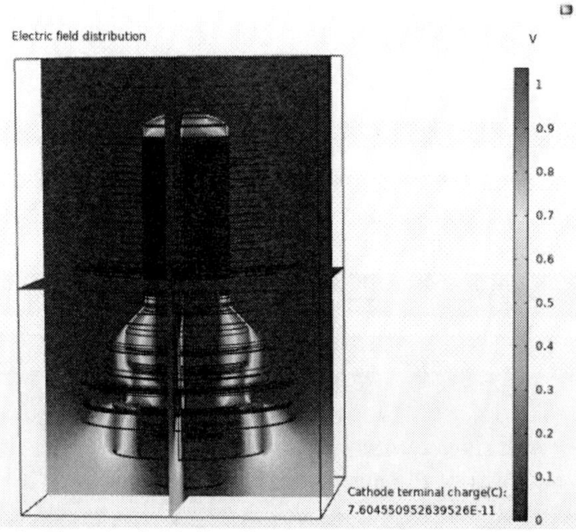

Figure 6. Electric field distribution and cathode terminal charge

Table 2. Inter-electrode capacitance simulation calculation value

Inter-electrode capacitance	Simulation results（pF）
C_{g1k}	76.04
C_{g1g2}	125.50
C_{g2a}	20.94

4. Analysis of the inter-electrode distance of tetrode

The cathode of the electron tube is responsible for emitting electrons, whose structure is the most complicated. The anode of the electron tube involves collecting electrons, and a reasonable anode cooling system has been designed, so the positions of the cathode and anode are not changed [5]. Firstly, adjust the distance between the cathode and the control electrode to obtain the distance range corresponding to the lower limit value 73pF and the upper limit value 85pF of C_{g1k}. Secondly, adjust the distance between the screen grid and the anode on the basis of the lower limit interval and the upper limit interval of C_{g1k} respectively. Finally, the distance between the screen grid and the anode corresponding to the lower limit of 110pF and the upper limit of 130pF of C_{g1g2} can be obtained. Through the simulation calculation, the simulation results are shown in Figure 7 and Figure 8. It can be seen that the inter-electrode capacitance shows a decreasing trend with the increase of the inter-electrode distance. Figure 7 shows that the adjustable range of the distance between the control grid and the cathode is 0.77-0.92mm, that is, the diameter of the control grid has an adjustable range of about 0.28mm. When the distance between the control grid and the cathode is 0.77mm and 0.92mm, the adjustable distance between the screen grid and the anode is 8.02-8.20mm and 7.85-8.03mm respectively, and the diameter of the screen grid has an adjustable range of about 0.36mm(Figure 8).

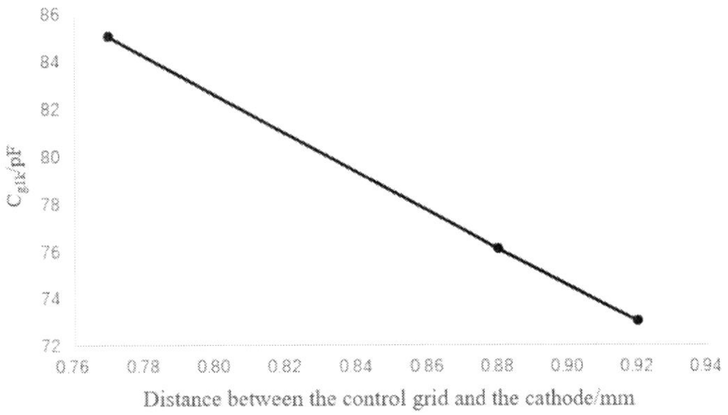

Figure 7. Relationship between the distance between the control grid and the cathode and the inter-electrode capacitance

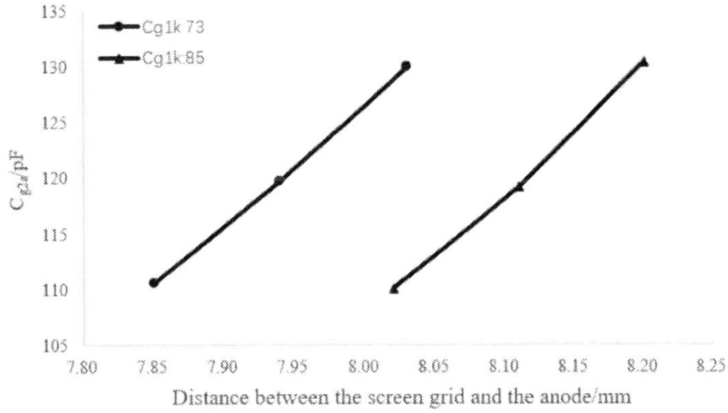

Figure 8. Relationship between the distance between the screen grid and the anode and the capacitance between the control grid and the screen grid

5. Conclusion

This paper has carried out a numerical simulation study on the inter-electrode capacitance of high-power electronic tetrode, and the simulation results are good. These studies provide a good reference for the next step of the nationalization of megawatt tetrode. The paper discusses that the operating frequency and output power of the tube are limited by the influence of the distribution parameters, and other factors such as the height of the cathode and the emission capability are not considered. Meanwhile, the electronic tetrode has complex structure with many parts. There is a slight deviation in the establishment of the three-dimensional solid model in the simulation experiments. In the following research, the three-dimensional solid model will be optimized, and considering the influence of multiple factors, the correlation between inter-electrode capacitance, operating frequency and output power will be further studied.

Acknowledgments

This work was supported by the National Magnetic Confinement Fusion Energy Development Research Project (Grant No.2019YFE03070000), the National Natural Science Foundation of China (Grant No.11675074), and the Science and Technology Planning Project of Hengyang City (Grant No.202002042159).

References

[1] Chen G., Mao Y. Z., Zhao Y.P. (2010) Study on 1.5 MW RF amplifier for EAST ICRF heating. Nuclear Fusion and Plasma Physics,30(1):67-70(in Chinese).
[2] Verma, S., Rajnish, K., Soni, D., Patel, H., Singh, R., Trivedi, R., et al. (2017) Real-time controller for research and development on ITER ion cyclotron heating and current drive source. IEEE Transactions on Nuclear Science.
[3] Powers, P.. (2010) Ultra-high-frequency tetrodes. Journal of the Institution of Electrical Engineers, 6(72), 722-723.
[4] Tian D.M.. (1994) Practical Maintenance Manual for Large Launch Tube[M]. China Radio&Television Publishing, Beijing(in Chinese).
[5] Carter, Richard, G.. (2018) Microwave and RF vacuum electronic power sources || triodes and tetrodes., 10.1017/9780511979231(6), 224-249.
[6] Shi N., Pi D.F.. (2002) Simulation of Electrostatic Field Capacitance for Vacuum Micro-triode[J]. Vacuum Electronics,01:13-18(in Chinese).
[7] Lekner, J. (2011) Capacitance coefficients of two spheres. Journal of Electrostatics, 69(1), 11-14.
[8] Cao, H. Y. , Duan, X. M. , Wang, H. X. . (2013). Study of sensor of electrical capacitance tomography (ECT) system based on comsol. Advanced Materials Research, 734-737, 3016-3021.

AUTHOR INDEX